10	11 IB	12 IIB	13 IIIA	14 IVA	15 VA	16 VIA	17 VIIA	18 O
							1 **H** 1.00794	2 **He** 4.00260
			5 **B** 10.811	6 **C** 12.011	7 **N** 14.0067	8 **O** 15.9994	9 **F** 18.9984	10 **Ne** 20.1797
			13 **Al** 26.9815	14 **Si** 28.0855	15 **P** 30.9738	16 **S** 32.066	17 **Cl** 35.4527	18 **Ar** 39.948
28 **Ni** 58.69	29 **Cu** 63.546	30 **Zn** 65.39	31 **Ga** 69.723	32 **Ge** 72.61	33 **As** 74.9216	34 **Se** 78.96	35 **Br** 79.904	36 **Kr** 83.80
46 **Pd** 106.42	47 **Ag** 107.868	48 **Cd** 112.411	49 **In** 114.82	50 **Sn** 118.710	51 **Sb** 121.75	52 **Te** 127.60	53 **I** 126.904	54 **Xe** 131.29
78 **Pt** 195.08	79 **Au** 196.967	80 **Hg** 200.59	81 **Tl** 204.383	82 **Pb** 207.2	83 **Bi** 208.980	84 **Po** (209)	85 **At** (210)	86 **Rn** (222)
110 **Uun** (269)	111 **Uuu** (272)	112 **Uub** (277)						

Metals ← → Nonmetals

63 **Eu** 151.965	64 **Gd** 157.25	65 **Tb** 158.925	66 **Dy** 162.50	67 **Ho** 164.930	68 **Er** 167.26	69 **Tm** 168.934	70 **Yb** 173.04	71 **Lu** 174.967

95 **Am** (243)	96 **Cm** (247)	97 **Bk** (247)	98 **Cf** (251)	99 **Es** (252)	100 **Fm** (257)	101 **Md** (258)	102 **No** (259)	103 **Lr** (262)

INTRODUCTION TO

General, Organic & Biochemistry

INTRODUCTION TO

General, Organic & Biochemistry

Fifth Edition

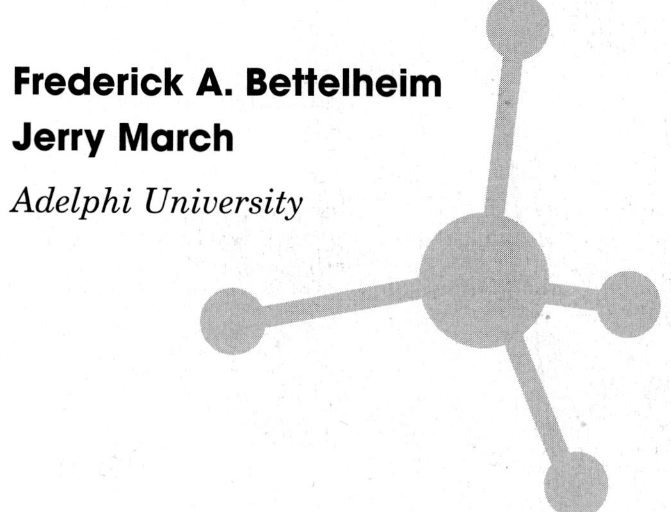

Frederick A. Bettelheim
Jerry March
Adelphi University

Saunders
College
Publishing

Harcourt Brace College Publishers

Fort Worth • Philadelphia • San Diego • New York
Orlando • Austin • San Antonio • Toronto • Montreal
London • Sydney • Tokyo

Publisher: John Vondeling
Publisher: Emily Barrosse
Product Manager: Angus McDonald
Developmental Editor: Beth Rosato
Project Editor: Anne Gibby
Production Manager: Charlene Catlett Squibb
Art Director: Joan Wendt/Lisa Caro
Text Designer: Ruth Hoover
Cover Design: Joan Wendt/Lisa Caro

Cover Credit: Angel Falls, Caraima National Park, Venezuela (Courtney Milne/Masterfile)

Printed in the United States of America

Introduction to General, Organic, and Biochemistry, Fifth Edition

0-03-020217-5

Library of Congress Catalog Card Number: 97-66892

7890123456 032 10 98765432

To our wives:

Vera S. Bettelheim and Beverly March

Contents Overview

Photograph by Charles D. Winters.

Photograph by Beverly March.

Photograph by Charles D. Winters.

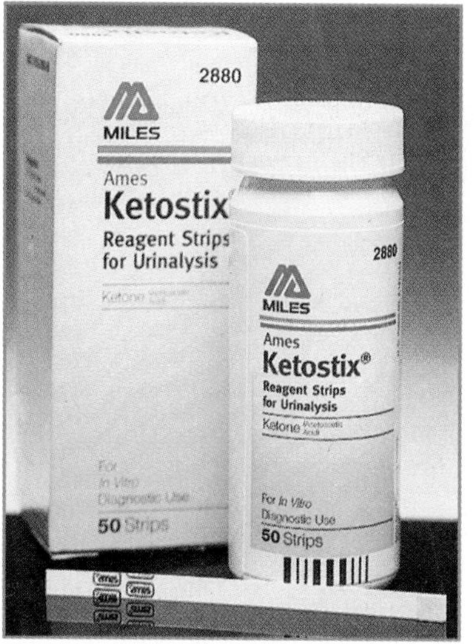

Photograph by Charles D. Winters.

Preface

Photograph by Leon Lewandowski.

> "Aurea prima sata est aetas, quae vindice nullo sponte sua, sine lege, fidem rectumque colebat."
>
> *Ovid: Quattor Aetatis.*

Ovid, with many ancient poets and thinkers, identifies a golden age when rights and trust were cultivated without the constraints of laws . . . an age of innocence. Perceiving order in the nature of the world as well as that of man is an ontological, not just a pedagogical, necessity. It is our primary aim to convey the relationship among facts and thereby to present the totality of the magnificent edifice science has built over the centuries.

". . . .matter is matter, neither noble nor vile, infinitely transformable, and its proximate origin is of no importance whatsoever. Nitrogen is nitrogen, it passes miraculously from the air into plants, from these into animals and from animals to us; when its function in the body is exhausted, we eliminate it, but it still remains nitrogen, aseptic, innocent." So says Primo Levi in his book *The Periodic Table.* In transforming a lifelong love affair toward his chosen profession, chemistry, into literature, he succeeded in expressing his enthusiasm for all to see.

In writing the Preface for this, the fifth edition of our textbook, we hope that somewhat similarly we manage to convey our delight in observing the chemical processes in the core of life sciences. The increasing use of our textbook has made this new edition possible, and we wish to thank our colleagues who adopted previous editions for their courses. Testimony from colleagues and students indicates that we manage to convey our enthusiasm to students, who find this book a great help in studying difficult concepts.

Thus, this fifth edition intends to be even more readable and understandable than earlier editions. While maintaining the overall organization of the textbook, we strive to produce more integration of the three domains of the text: general, organic and biochemistry. Chemistry, especially biochemistry, is a fast-developing discipline, and we include new relevant material in the text. This is done not just by upgrading information, but also by enlarging the scope of the book both in the text and in the boxes containing medical and other applications of chemical principles. At the same time, we are aware of the need to keep the book to a manageable size and proportion. Twenty percent of the problems are new; we have increased the number of more challenging, thought-provoking problems (marked by asterisks).

AUDIENCE

As were the previous editions, this book is intended for nonchemistry majors, mainly those entering health science and related fields (such as nursing, medical technology, physical therapy and nutrition). It also can be used by students in environmental studies. In its entirety it can be used for a one-year (two-semester or three-quarter) course in chemistry, or parts of the book can be used in a one-term chemistry course.

We assume that the students using the book have little or no chemistry background. Therefore, we introduce the basic concepts slowly at the beginning, increasing the tempo and the level of sophistication as we go on. We progress from the basic tenets of general chemistry to organic and finally to biochemistry. We consider this progression an ascent in terms of both practical importance and sophistication. However, the three parts are integrated by keeping a unified view of chemistry. We do not consider the general chemistry sections to be the exclusive domain of inorganic compounds, so we frequently use organic and biological substances to illustrate general principles.

While it is our aim to teach the chemistry of the human body as the ultimate goal, we try to show that each subsection of chemistry is important in its own right, besides being required for future understanding.

BOXES (Medical and Other Applications of Chemical Principles)

The boxes contain applications of the principles discussed in the text. Comments from users of the earlier editions indicated that these have been especially well-received, providing a much requested relevance to the text. The large number of boxes deal mainly with health-related applications, including many also related to the environment. A table following the Contents lists these medically relevant applications. Nineteen boxes from the fourth edition have been dropped. Five others have been incorporated into the text or other boxes. Seventeen new boxes have been added, dealing with diverse topics such as the human genome project, entropy and disorder, combinatorial synthesis, advanced glycation endproducts in aging, ubiquitin in protein targeting, promoter in genetic engineering, protection against oxidative damages, toxicity and lethal dose, among others. Many boxes have been enlarged and updated. For example, boxes on Alzheimer's disease, antacids, pain relievers, laser surgery, and AIDS now contain recent material, including the newly approved HIV protease inhibitors.

The presence of boxes allows a considerable degree of flexibility. If an instructor wants to assign only the main text, the boxes will not interrupt continuity, and the essential material will be covered. However, most instructors will probably wish to assign at least some of the boxes, since they enhance the core material. In our experience, students are eager to read the relevant boxes even without assignments and they do so with discrimination. From such a large number of boxes, the instructor can select those that best fit the particular needs of the course and of the students. Problems are provided for nearly all of the boxes.

ORGANIZATION

We have maintained the organization of the previous edition. Nine chapters deal with general chemistry, six with organic chemistry, and eleven with biochemistry. A few new organizational features are present in the fifth edition. We moved the presentation of exponential notation from the Appendix into Chapter 1 where measurements are discussed and its application is constantly used. We left the discussion of significant figures in the Appendix. We also introduce in Chapter 1 the use of estimation. We emphasize that the use of estimation provides not only a check on the calculated exact solution, but also a logical exercise in solving problems. Although we restrict the use of estimation to the problems presented in Chapter 1 only, we recommend its use throughout the book. Nuclear chemistry is presented in Chapter 9 at the end of the general chemistry sections. Some instructors feel that this topic is better presented immediately after Chapter 2 on Atoms. There are no difficulties in moving up nuclear chemistry, if the instructor so desires. We feel, however, that the complexity of this topic is better appreciated after the exploration of equilibrium and kinetics.

In organic chemistry we concentrate on the structure and properties and only the most important reactions of each class of compounds. As for the mechanisms of the reactions, we provide only one example: carbocation intermediates in addition reactions. We do this deliberately because we feel that in the relatively brief portion of the course devoted to organic chemistry, students do not have time to learn a large number of reactions or anything substantial about mechanisms. As stated before, we consider the progression from general to organic and to biochemistry an ascent. Therefore, we selected mainly organic compounds and reactions which have physiological activity of one sort or another and biological importance. In order to help students to learn the reactions, we include summaries of reactions at the ends of the chapters.

Within the biochemistry chapters we also maintain the traditional order. We find this a pedagogical imperative. Even though most of the important new developments in biochemistry occur in molecular biology (Chapter 23 on Nucleic Acids and Protein Synthesis), neurochemistry (Chapter 24), and immunology (Chapter 25), these chapters come late in the book. We feel that the appreciation of these topics requires a previous acquaintance with carbohydrate, lipid and protein chemistry and metabolism. We hope that each instructor, to his or her taste, will judiciously appropriate time to discuss the exciting developments presented in the late chapters. As nutritional concerns are gaining new prominence, we decided to expand Chapter 26 and discuss the role of vitamins, minerals, etc., in greater detail, not just as the tabulated compilation provided in previous editions.

NEW MATERIAL

In addition to several new boxes, we include new material in the text. Some examples are:

- We provide the isoelectric points of amino acids.
- We discuss non-coding DNA satellites and transcription factors and their role in gene regulation.
- We include a few examples of exceptions to the octet rule.
- We enlarge the discussion on immunochemistry and detail the role of B and T cells.
- We add new material on Zaitsev's rule.
- We extend the discussion of calcium ions in neurotransmission.
- We describe the truth-in-labeling requirements for packaged food.

METABOLISM; COLOR CODE

The biological functions of chemical compounds are explained in each of the biochemistry chapters and in many of the organic chapters. The emphasis is on chemistry rather than physiology. We have received much positive feedback regarding the way in which we have organized the topic of metabolism (Chapters 20, 21, and 22). We maintained this organization.

First we introduce the common metabolic pathway through which all food will be utilized (citric acid cycle; oxidative phosphorylation), and only after that do we discuss the specific pathways leading to the common pathway. We find this a useful pedagogic device, and it enables us to sum up the caloric values of each type of food because their utilization through the common pathway has already been learned. Finally, we separate the catabolic pathways from the anabolic pathways by treating them in different chapters, emphasizing the different ways the body breaks down and builds up different molecules.

The topic of metabolism is a difficult one for most students. We have tried to explain it as clearly as possible. As in the previous edition, we enhance the clarity of presentation by the use of a color code for the most important biological compounds discussed in Chapters 20, 21, and 22. Each type of compound is shown in a specific color, which remains the same throughout the three chapters. These colors are as follows:

ATP and other nucleoside triphosphates

ADP and other nucleoside diphosphates

The oxidized coenzymes NAD^+ and FAD

The reduced coenzymes NADH and $FADH_2$

Acetyl coenzyme A

The circled numbers in the figures showing the steps involved in the various metabolism pathway are always in yellow.

In addition to this main use of a color code, other figures in various parts of the book are color-coded, so that the same color is used for the same entity throughout. For example, in Chapter 19, enzymes are always shown in blue and substrates in orange in all of the figures in the chapter that show enzyme-substrate interactions.

INTERVIEWS

Each of the three sections—general, organic and biochemistry—opens with an interview with an individual who has made significant contributions in that particular field. Roald Hoffmann, Nobel Laureate in 1981, not only enhanced our understanding of chemical bonding through his theoretical work but also regularly teaches general chemistry at Cornell University. Carl Djerassi of Stanford University is a master of steroid chemistry, both in synthesis and analyses of organic compounds, and he is often cited as the father of the birth control pill. Jacqueline K. Barton of the California Institute of Technology has made important discoveries relating to the structure and conformation of the DNA double helix. These interviews are intended to give the student a human face of science and an insight into how science affects our lives.

FEATURES

One of the main features of this book, as in earlier editions, is the number of applications of chemical concepts presented in the boxes. Another important feature is the Glossary-Index. The definition of each term is given along with the index entry and the page number. Another feature is the list of key terms at the end of each chapter, with notation of the section number in which the term is introduced. Many students find these lists to be helpful study guides.

Other features are the summaries at the end of each chapter (including summaries of organic reactions in Chapters 11 to 15) and the substantial number of margin notes. In this, the fifth edition, we added subtitles to the problems, guiding the students to a section of the chapter where they can find the relevant material.

STYLE

Feedback from colleagues and students alike indicates that the style of the book, which addresses the students directly in simple and clear phrasing, is one of its major assets. We continue to make special efforts to provide clear and concise writing. Our hope is that this eases the understanding and the absorption of the difficult concepts.

PROBLEMS

About one fifth of the problems are new in this, the fifth edition. The number of starred problems, which represent the more challenging, thought-provoking questions, has increased. The end-of-chapter problems are grouped and given subheads in order of topic coverage. The last group, headed "Additional Problems," is not arranged in any specific order. The answers to all the in-text problems and to the odd-numbered end-of-chapter problems are given at the end of the book. Answers to the even-numbered problems are included in the Instructor's Manual and the Study Guide.

ANCILLARIES

This text is accompanied by a number of ancillary publications to help support your teaching and your students' learning:

- Study Guide by W. Scovell (Bowling Green State University). Includes review of chapter objectives, important terms and comparisons, focused review of concepts, self-tests, and answers to the even-numbered problems in the text.
- Flash Cards by H. Akers (Lamar University) Two hundred bi-directional flash cards offer drills on important reactions, terms, structures, and classifications.
- Lecture Outline by D. Neff (University of Louisville). Organized to follow class lectures, freeing students from extensive note-taking during lectures.
- Instructor's Manual and Test Bank by F.A. Bettelheim and J. March. Contains suggested course outlines, newly revised exam questions or-

ganized chapter by chapter, answers to the exam questions, and answers to the even-numbered problems.

- ExaMaster™ Computerized Test Bank is the software version of the printed test bank. Instructors can create thousands of questions in a multiple-choice format. A command reformats the multiple-choice question into a short-answer question. Adding or modifying existing problems, as well as incorporating graphics, can be done. ExaMaster™ has grade-book capabilities for recording and graphing students' grades.
- Approximately 150 overhead transparencies in full color are available. Figures and tables are taken from the text.
- Laboratory Experiments for *Introduction to General, Organic & Biochemistry,* 3e, by F.A. Bettelheim, and J. Landesberg (Adelphi University). Fifty experiments illustrate important concepts and principles in general, organic, and biochemistry. Simple equipment and inexpensive, common, and environmentally safe chemicals are used. The large number of experiments allows sufficient flexibility for the instructor to select the usual 24 experiments the students perform in a two-semester course.
- Instructor's Manual to accompany Laboratory Experiments for *Introduction to General, Organic & Biochemistry,* 3e, is available. This manual will help instructors in grading the answers to the questions as well as in assessing the range of the experimental results obtained by the students.
- Chemistry 1998 MediaActive™ CD-ROM provides still imagery from *Introduction to General, Organic & Biochemistry* and other quality Saunders College Publishing textbooks. Available as a presentation tool, this CD-ROM can be used in conjunction with commercial presentation packages, such as Powerpoint™, Persuasion™, and Podium™, as well as the Saunders LectureActive™ presentation software. Available in both Macintosh and Windows platforms.

Saunders College Publishing may provide complimentary instructional aids and supplements or supplement packages to those adopters qualified under our adoption policy. Please contact your sales representative for more information. If as an adopter or potential user you receive supplements you do not need, please return them to your sales representative or send them to:

Attn: Returns Department
Troy Warehouse
465 South Lincoln Drive
Troy, MO 63379

Acknowledgments

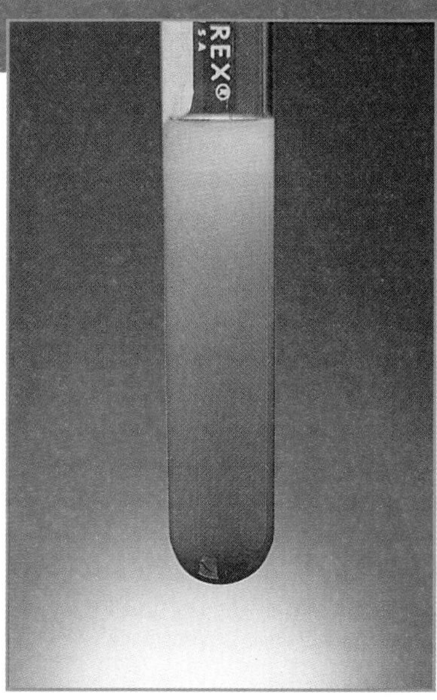

Photograph by Beverly March.

The publication of a book such as this requires the efforts of many more people than merely the authors. A number of reviewers have read all or significant portions of the manuscript in various stages. We thank the following people for their constructive criticism and helpful and valuable suggestions in revising the text for this fifth edition:

Lamar Anderson	Utah State University
Kathleen Ashworth	Yakima Valley Community College
Sharon Cruse	Northeast Louisiana University
Wendy Gloffke	Cedar Crest College
Eric G. Holmberg	University of Alaska–Anchorage
Judith Iriarte-Gross	Middle Tennessee State University
Peter J. Krieger	Palm Beach Community College
Kent S. Marshall	Quinnipiac College
Jane Martin	San Francisco State University
Elva Mae Nicholson	Eastern Michigan University
Howard K. Ono	California State University–Fresno
Thomas Pynadath	Kent State University
Edith Rand	East Carolina University

Many of our reviewers pointed out inadvertent errors or certain weaknesses in the previous editions. We have attempted to correct these, and thus we hope that the fifth edition will prove even more useful than the earlier editions.

We also wish to thank several of our colleagues at Adelphi University for their useful advice. These include Stephen Goldberg, Joseph Landesberg, Sung Moon, Donald Opalecky, Reuben Rudman, Charles Shopsis, Kevin Terrance and Stanley Windwer. We are grateful for the support of John Vondeling, Vice President/Publisher, Saunders College Publishing. We thank Beth Rosato, Senior Developmental Editor, and Anne Gibby, Senior Project Editor, for their congenial steady assistance. We would like to express our appreciation to Charlene Squibb for supervising the production of this edition and to Sue Kinney for supervising the art. We also thank Angus McDonald, Executive Product Manager, for the successful marketing of our textbook. J/B Woolsey Associates transformed our crude drawings into pieces of art. Last but not least we want to thank Beverly March and Charles D. Winters for their many excellent photographs.

Frederick A. Bettelheim and Jerry March
Adelphi University
June 1997

Contents

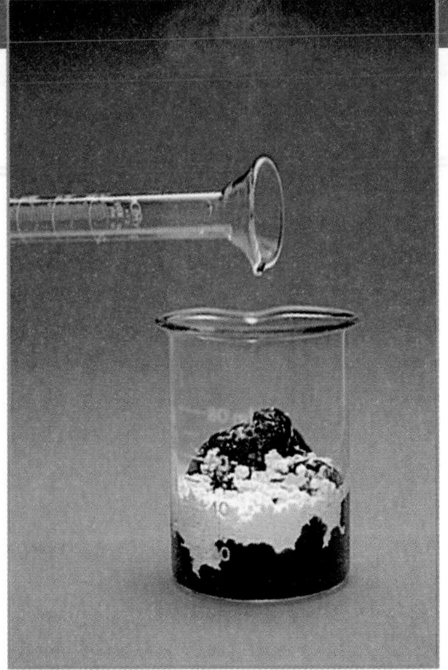

Photograph by Charles D. Winters.

Part One General Chemistry 1

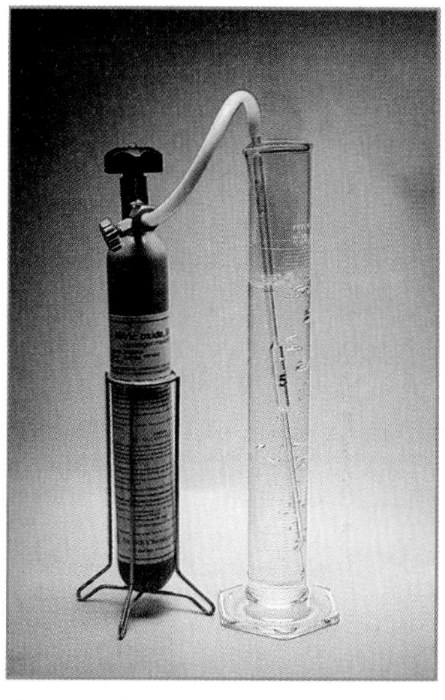

Photograph by Charles D. Winters.

Photograph by Beverly March.

CHAPTER 5 GASES, LIQUIDS, AND SOLIDS 141

CHAPTER 6 SOLUTIONS AND COLLOIDS 177

Photograph by Leon Lewandowski.

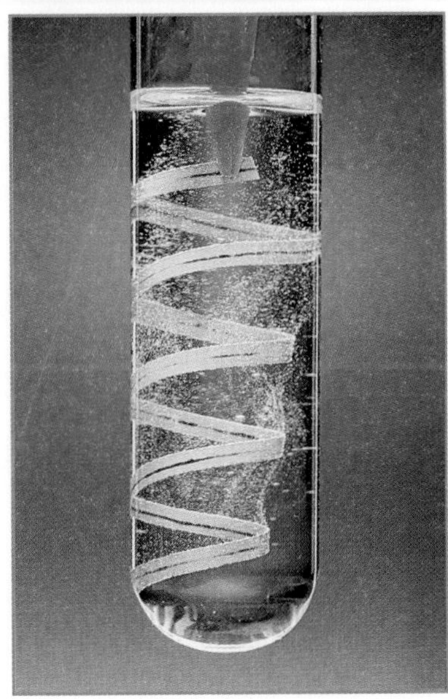

Photograph by Charles D. Winters.

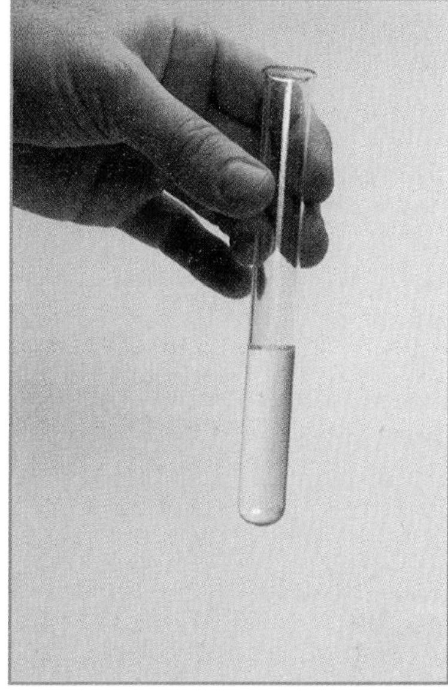

Photograph by Beverly March.

Photograph courtesy of U.S. Department of Agriculture.

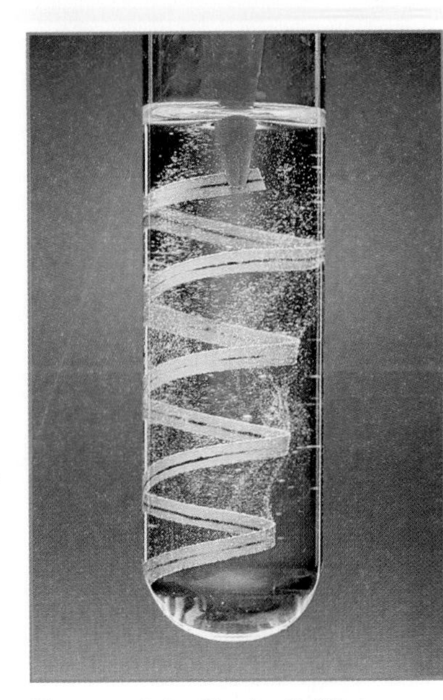

Photograph by Beverly March.

Photograph by Charles D. Winters.

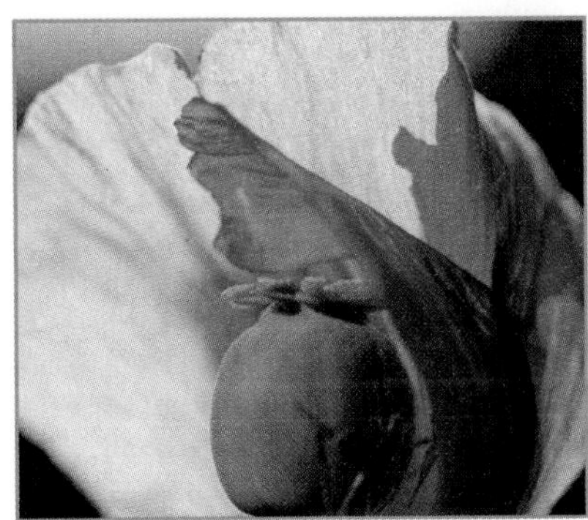

Photograph by FourByFive.

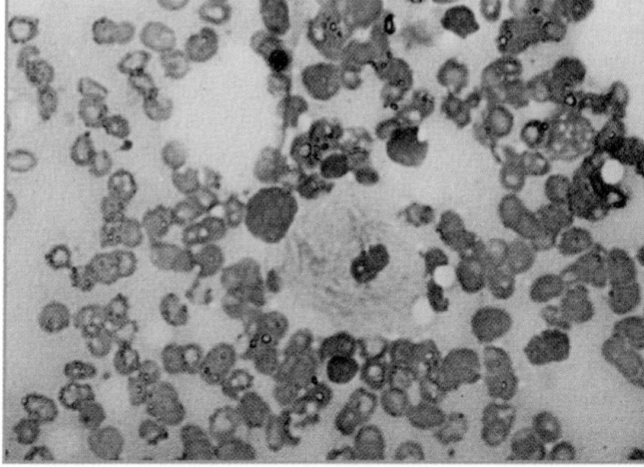

Photograph courtesy of Drs. B.G. Bullogh and V.J. Vigorita, and the Gower Medical Publishing Co., New York.

Photograph by Charles D. Winters.

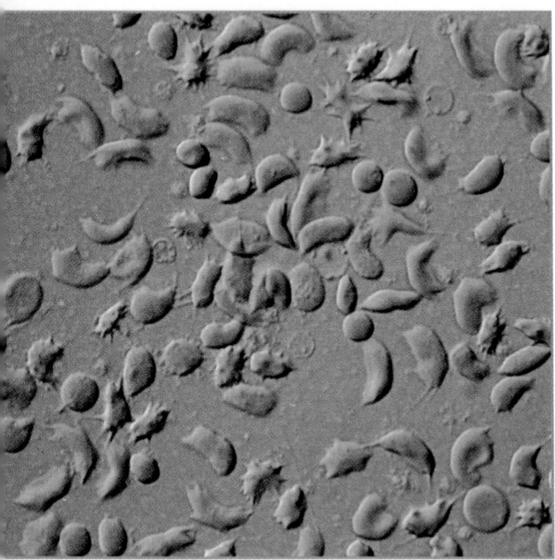

Photograph by G.W. Willis, M.D./Biological Photo Service.

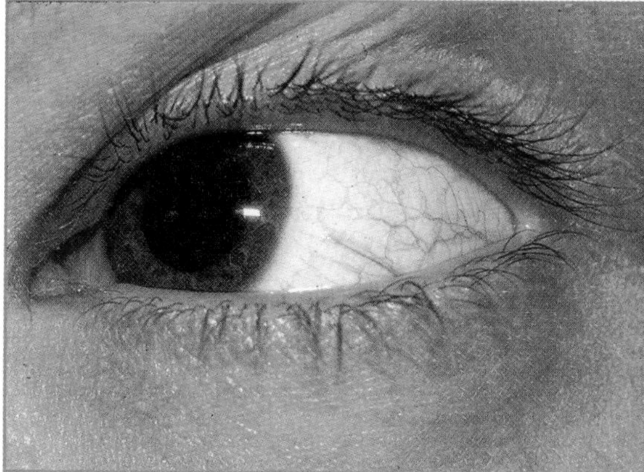

Photograph courtesy of Service Photo Library/Photo Researchers, Inc.

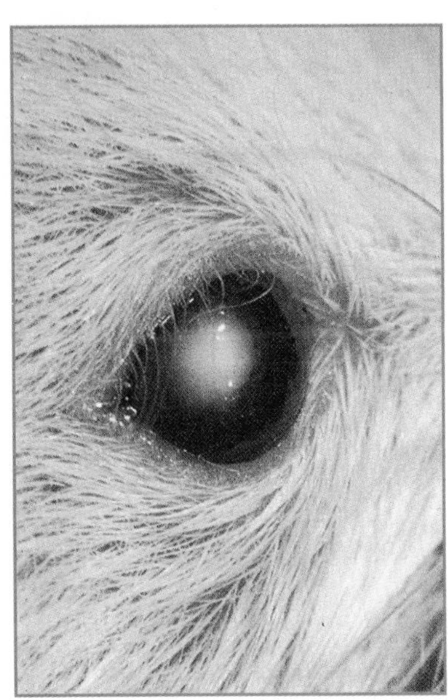

Photograph courtesy of Dr. Paul Russell, National Eye Institute, NIH.

CHAPTER 22 BIOSYNTHETIC PATHWAYS 691

CHAPTER 23 NUCLEIC ACIDS AND PROTEIN SYNTHESIS 705

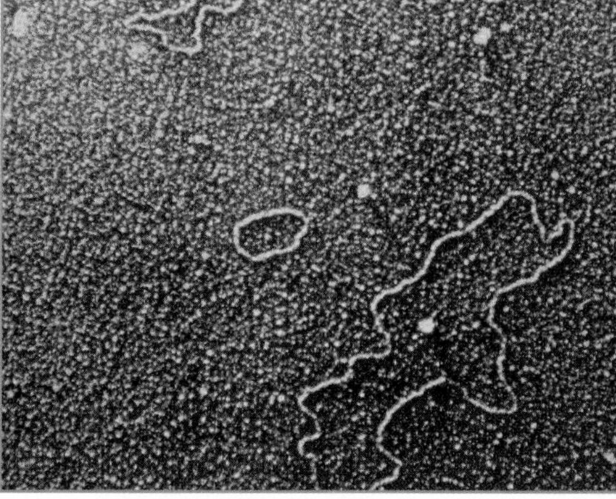

Photograph courtesy of Thomas Broker/Cold Spring Harbor Laboratory.

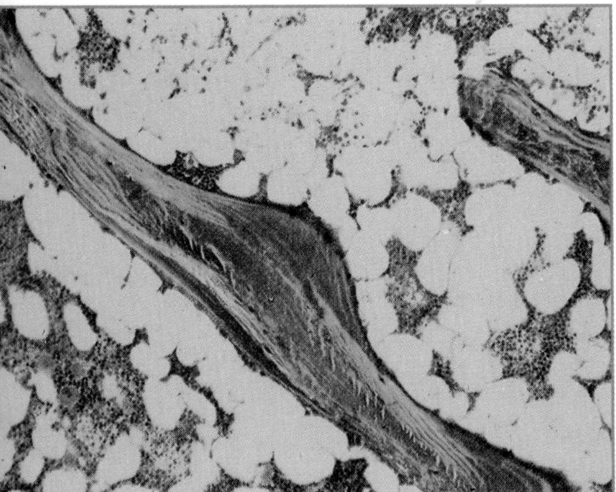

Photograph courtesy of Drs. P.A. Dieppi, P.A. Bacon, A.N. Bamji, and I. Wat and Gower Medical Publishing, Ltd., London, England.

Health-Related Topics

Part one

Photograph by Charles D. Winters.

General chemistry

Roald Hoffmann

Roald Hoffmann is a remarkable individual. When he was only 44 years old, he shared the 1981 Nobel Prize in Chemistry with Kenichi Fukui of Japan for work in applied theoretical chemistry. In addition, he has received awards from the American Chemical Society in both organic chemistry and inorganic chemistry, the only person to have achieved this honor. And, in 1990, he was awarded the Priestley Medal, the highest award given by the American Chemical Society.

The numerous honors celebrating his achievements in chemistry tell only part of the story of his life. He was born to a Polish Jewish family in Zloczow, Poland, in 1937, and was named Roald after the famous Norwegian explorer Roald Amundsen. Shortly after World War II began in 1939, the Nazis first forced him and his parents into a ghetto and then into a labor camp. However, his father smuggled Hoffmann and his mother out of the camp, and they were hidden for more than a year in the attic of a schoolhouse in the Ukraine. His father was later killed by the Nazis after trying to organize an attempt to break out of the labor camp. After the war, Hoffmann, his mother, and his stepfather made their way west to Czechoslovakia, Austria, and then Germany. They finally emigrated to the United States, arriving in New York on Washington's birthday in 1949. That Hoffmann and his mother survived these years is our good fortune. Of the 12,000 Jews living in Zloczow in 1941 when the Nazis took over, only 80 people, three of them children, survived the Holocaust. One of those three children was Roald Hoffmann.

Roald Hoffmann

On arriving in New York, Hoffmann learned his sixth language, English. He went to public schools in New York City and then finally to Stuyvesant High School, one of the city's select science schools. From there he went to Columbia University and then to Harvard University, where he earned his Ph.D. in 1962. Shortly thereafter, he began the work with Professor R. B. Woodward that eventually led to the Nobel Prize. Since 1965 he has been a professor at Cornell University, where he regularly teaches first-year general chemistry.

In addition to his work in chemistry, Professor Hoffmann also writes popular articles on science for the American Scientist *and other magazines, and he has published two volumes of his poetry. Finally, he appears in a series of 26 half-hour television programs for a chemistry course called "World of Chemistry," which began airing on public television and cable channels in 1990.*

From medicine to cement to theoretical chemistry

We visited Professor Hoffmann in his office at Cornell University, an office full of mineral samples, molecular models, and Japanese art. When asked what brought him into chemistry he said that "I came rather late to chemistry; I was not interested in it from childhood." However, he clearly feels that one can come late to chemistry, and that it can be a very positive thing. "I am always worried about fields in which people exhibit precocity, like music and mathematics. Precocity is some sort of evidence that you have to have talent. I don't like that. I like the idea that human beings can do anything they want to. They need to be trained sometimes. They need a teacher to awaken the intelligences within them. But to be a chemist requires no special talent, I'm glad to say. Anyone can do it, with hard work."

He took a standard chemistry course in high school. He recalls that it was a fine course, but apparently he found biology more enjoyable because, in his high school year book, "under the picture of me with a crew cut, it says 'medical research' under my name." Indeed, he says that "medical research was a compromise between my interest in science and the typical Jewish middle class family pressures to become a medical doctor. The same kind of pressures apply to Asian Americans today."

When he went to Columbia University, Hoffmann enrolled as a pre-med student, but says that there were several factors that shifted him away from a career in medicine. One of these was his work at the

National Bureau of Standards in Washington, D.C., for two summers and then at Brookhaven National Laboratory for a third summer. He says that these experiences gave him a feeling for the excitement of chemical research. Nonetheless, during his first summer at the Bureau of Standards, he "did some not very exciting work on the thermochemistry of cement." During his second summer there he visited the National Institutes of Health to find out what medical research was about. "To my amazement," he says, "most of the people had Ph.D.'s and not M.D.'s. I just didn't know. Young people do not often know what is required for a given profession. Once I found that out, and found that I did well in chemistry, it made me feel that I didn't really have to do medicine, that I could do some research in chemistry or biology. Later, what influenced me to decide on theoretical chemistry was an excellent instructor. Had I had some really good instruction in organic chemistry, I'm pretty sure I would have become an organic chemist."

"At the very same time I was being exposed to the humanities, in part because of Columbia's core curriculum—which I think is a great idea—that had so-called contemporary civilization and humanities courses. I took advantage of the liberal arts education to the hilt, and that has remained with me all my life. The humanities teachers have remained permanently fixed in my mind and have changed my ways of thinking. These were the people who really had the intellectual impact on me and helped to shape my life."

"To trace the path, I was a latecomer to chemistry and was inspired by research. I think *research* is the way in. It just gives you a different perception."

A love for complexity

Having discussed what brought him into chemistry, we were interested in his view of the qualities that a student should possess to pursue a career in the field. He said that "one thing one needs to be a chemist is a love for complexity and richness. To some extent that is true of biology and natural history, too. I think one of the things that is beautiful about chemistry is that there are 10,000,000 compounds, each with different properties. What's beautiful when you make a molecule is that you can make derivatives in which you can vary substituents, the pieces of a molecule, and we know that those substituents give a molecule function, give it complexity and richness. That's why a protein or nucleic acid with all its variety is essential for life. That's why to me, intellectually, isomerism and stereochemistry in organic chemistry are at the heart of chemistry. I think we should teach that much earlier. It requires no mathematics, only a little model building; you can do this without theory. I think it is no accident that organic chemistry drew to it the intellects of its time."

Experiment and theory

Professor Hoffmann has spent his career immersed in the theories of chemistry. However, he believes that fundamentally "chemistry is an experimental science, in spite of some of my colleagues saying otherwise. However, the educational process certainly favors theory. It's in the nature of things for teacher and student both to want to understand and then give primacy to the soluble and the understood at the expense of other things. We also have this reductionist philosophy of science, the idea that the social sciences derive from biology, that biology follows from chemistry, chemistry from physics, and so on. This notion gives an inordinate amount of importance to theoretical thinking, the more mathematical the better. Of course this is not true in reality, but it's an ideology; it is a religion of science."

There is of course a role for theory. "You can't report just the facts and nothing but the facts; by themselves they are dull. They have to be woven into a framework so that there is understanding. That's accomplished usually by a theory. It may not be mathematical, but a qualitative network of relationships." Indeed, Hoffmann believes that the incorporation of theory into chemistry "is what made American science better than that in many other countries. The emphasis in chemistry on theory and theoretical understanding is very important, but not nearly as important as the syntheses and reactions of molecules."

"Although I think chemists need to like to do experiments, that doesn't mean there is no role for people like me. It turns out that I am really an experimental chemist hiding as a theoretician. I think that is the key to my success. That is, I think I can empathize with what bothers the experimentalists. In another day I could have become an experimentalist."

Major issues in chemistry and science today

Professor Hoffmann has worked, and is presently working, at the forefront of several major areas of chemistry. He is currently quite intrigued by surface science. "For instance, there is the Fischer-Tropsch process, a pretty incredible thing in detail. Carbon monoxide and hydrogen gas come onto a metal catalyst, a surface of some sort, and off come long-chain hydrocarbons and alcohols. The richness of all these things happening is intriguing, and we are on the verge of understanding. We now have structural information on surfaces that's reliable, and we are just beginning to get kinetic information. Surface science is at a crossing of chemistry, physics, and engineering. The field is in some danger of being spun off on its own, but I would like to keep it in chemistry."

"Bioinorganic chemistry is another such field. In my group we are doing some work trying to understand the mechanism of oxygen production in photosynthesis, the last steps. What is known is very little. There is an enzyme in photosystem II that involves 3 to 4 manganese atoms. And they somehow take oxide or hydroxide to peroxide and eventually to molecular oxygen. That's all we know. Experimentally, not theoretically, I think bioinorganic chemistry is a very interesting field."

Finally, Hoffmann remarked that "there are going to be finer and finer ways of controlling the synthesis of molecules, the most essential activity of chemists. If I were to point to a single thing that chemists do, it would be that they make molecules. Chemistry is the science of molecules and their transformations. The transformations are the essential part. I think there are exciting possibilities for chemical intervention into biological systems with an ever finer degree of control. We need not be afraid of nature. We can mimic it, and even surpass its synthetic capabilities. And find a way to cooperate with it."

Scientific literacy and democracy

Roald Hoffmann is very concerned not only about science in general and chemistry in particular but also about our society. One of his concerns is scientific literacy because "some degree of scientific literacy is absolutely necessary today for the population at large as part of a democratic system of government. People have to make intelligent decisions about all kinds of technological issues." He recently offered his comments on this important issue in the *New York Times.* He wrote that "What concerns me about scientific, or humanistic, illiteracy is the barrier it poses to rational democratic governance. Democracy occasionally gives in to *technocracy*—a reliance on experts on matters such as genetic engineering, nuclear waste disposal, or the cost of medical care. That is fine, but the people must be able to vote intelligently on these issues. The less we know as a nation, the more we must rely on experts, and the more likely we are to be misled by demagogues. We must know more."

The responsibility of scientists

Our discussion of the importance of scientific literacy led to a conversation about the broader responsibilities of scientists. "Scientists have a great obligation to speak to the public," Hoffmann says. "We have an obligation as educators to train the next generation of people. We should pay as much attention to those people who are *not* going to be chemists, and sometimes need to make compromises about what is to be taught and what is the nature of our courses. I think scientists have an obligation to speak to the public broadly, and here I think they have been negligent. I think society is paying scientists money to do research, and can demand an accounting in plain language. That's why I put in a lot of time on that television show [*The World of Chemistry*]."

A teacher of chemistry—and proud of it

In the Nobel Yearbook, Professor Hoffmann wrote that the technical description of his work "does not communicate what I think is my major contribution. I am a teacher, and I am proud of it. At Cornell University I have taught primarily undergraduates—I have also taught chemistry courses to nonscientists and graduate courses in bonding theory and quantum mechanics. To the chemistry community at large, and to my fellow scientists, I have tried to teach "applied theoretical chemistry": a special blend of computations stimulated by experiment and coupled to the construction of general models—frameworks for understanding." His success in this is unquestioned.

Interview conducted by John C. Kotz, SUNY Distinguished Teaching Professor, State University of New York, College of Oneonta.

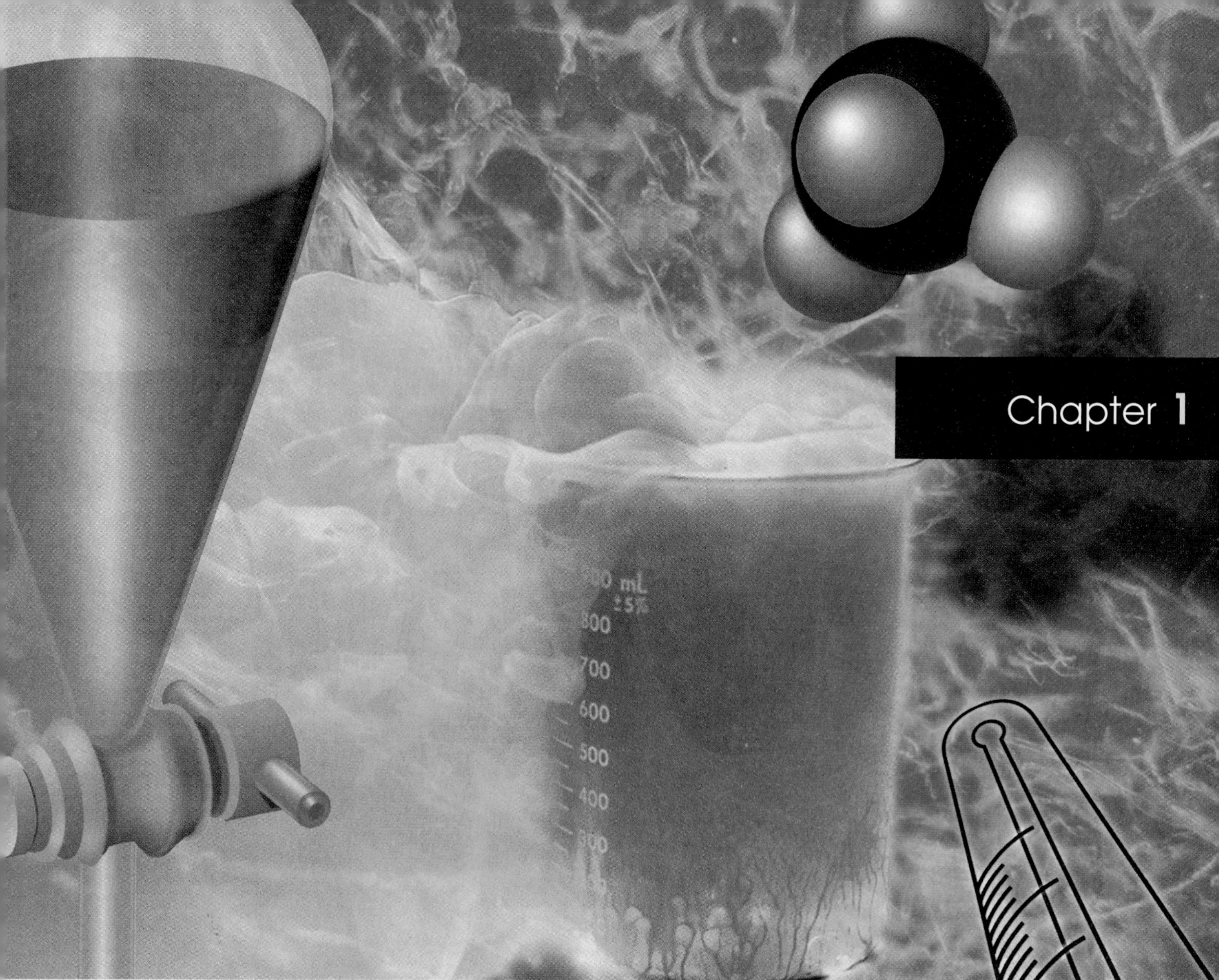

Chapter 1

Matter, energy, and measurement

1.1 Introduction

There was a time—only a few hundred years ago—when physicians were powerless to treat many diseases. Cancer, tuberculosis, smallpox, typhus, plague, and many other sicknesses struck people seemingly at random, and doctors, who had no idea how any of these diseases were caused, could do little or nothing about them. Between 1348 and 1350, the disease known as the Black Death (bubonic plague) wiped out about one third of the population of Europe (Fig. 1.1). Doctors treated it with magic as well as by such measures as bleeding (Fig. 1.2), laxatives, hot plasters, and pills made from powdered stag horn, saffron, or gold. None of these was of any use, and the doctors, because they came into direct contact with a highly contagious disease, died at a much higher rate than the general public.

Doctors of those times did have remedies for a few conditions, such as the quinine bark that was used to treat malaria, but all such remedies were discovered by trial and error. For most illnesses, human beings, including the doctors themselves, were entirely at the mercy of body processes and microorganisms they knew nothing about.

Medicine has made great progress since those times. Although people still get sick, they do on the average live much longer, and many once-feared diseases have been essentially eliminated or made easily curable. Smallpox has been completely eradicated, and polio, typhus, bubonic plague, diphtheria, and other diseases that once killed millions no longer pose a serious problem, at least not in the developed countries.

How has this medical progress come about? The answer is that diseases could not be cured until they were understood, and this under-

"Bring out your Dead"

Figure 1.1 • This old engraving shows a town crier asking people to bring out the bodies of plague victims. (Courtesy of the National Library of Medicine.)

Figure **1.2** • A woman being bled by a leech on her left forearm; a bottle containing more leeches is on the table. From a 1639 woodcut. (Courtesy of the National Library of Medicine.)

standing has come about through a knowledge of how the body functions. The study of bodily functions is an important part of the science of biology, but biologists know that because most body processes are chemical reactions it is also necessary to study chemistry. It is progress in our understanding of the principles of biology, chemistry, and physics that has led to these advances in medicine. When further advances are made—for example, the curing of cancer—they too will come from an increased understanding of biology and chemistry.

Because so much of modern medicine depends on chemistry, it is essential that students who intend to enter the health professions have some understanding of basic chemistry. This book was written to help you achieve that goal. Even if you choose a different profession, you will find that the chemistry you learn in this course will enrich your life.

Chemistry as the Study of Matter

The universe consists of matter, energy, and empty space. **Matter** is the stuff that everything is made of. It can be defined as **anything that has mass and takes up space. Chemistry is the science that deals with matter:** the structure and properties of matter and the transformations from one form of matter to another. We will discuss energy in Section 1.8.

It has long been known that matter can change, or be made to change, from one form to another. In a **chemical change,** substances are used up (disappear) and others are formed to take their places. An example is the burning of propane ("bottled gas"). When this chemical change takes place, propane and oxygen from the air are converted to carbon dioxide and water. Another chemical change is shown in Figure 1.3. Chemical changes are more often called *chemical reactions.* Many thousands of them are known, and the study of these reactions is the chief business of most chemists.

Digestion of food also involves chemical changes.

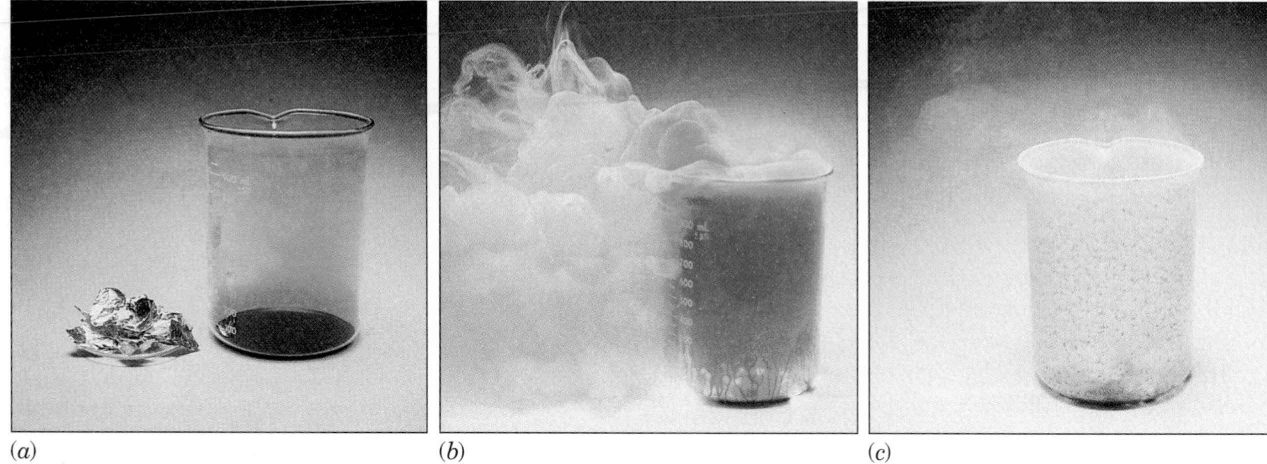

(*a*) (*b*) (*c*)

Figure 1.3
Bromine Br_2, an orange-brown liquid, and aluminum metal (a) react so vigorously that the aluminum becomes molten and glows white hot (b). The vapor in (b) consists of vaporized Br_2 and some of the product, white aluminum bromide. At the end of the reaction (c), the beaker is coated with aluminum bromide and the products of its reaction with atmospheric moisture. (*Note:* This reaction is dangerous! Under no circumstances should it be done except under properly supervised conditions.) (Photographs by C. D. Winters.)

Matter also undergoes other kinds of changes, called **physical changes.** These changes differ from chemical reactions in that substances do not change their identity. Most physical changes are changes of state—for example, the melting of solids and the boiling of liquids. Water remains water whether in the liquid state or in the form of ice or steam. Conversion from one state to another is a physical, not a chemical, change. Another important type of physical change involves making or separating mixtures. Dissolving sugar in water is a physical change.

The mixing of vinegar, salad oil, and eggs to make mayonnaise is a physical change.

When we talk of the *chemical properties* of a substance, we mean the chemical reactions it undergoes. *Physical properties* are all properties that do not involve chemical reactions; for example, density, color, melting point, and physical state (liquid, solid, gas) are all physical properties.

1.2 The Scientific Method

Scientists learn what they know by using a tool called the **scientific method.** The heart of the scientific method is the *testing* of theories. It was not always so. Before about 1600, philosophers believed statements because they sounded right. For example, the great philosopher Aristotle (384–322 BC) believed that if you took the gold out of a mine it would grow back. He believed this because it fitted in with a more general picture that he had about the workings of nature. In ancient times, this was the way most thinkers behaved. If a statement sounded right, they believed it without testing it.

Before about 1600, science as such did not exist except as a branch of philosophy.

About 1600 the scientific method came into use. Let us look at an example to see how the scientific method operates. The Greek physician Galen (AD 130–200) recognized that the blood on the left side of the heart somehow gets to the right side. This is a fact. A **fact** is a statement that is obvious to anyone who cares to look, that is, a statement based on direct experience. Having observed this fact, Galen then proposed a hypothesis to explain it. A **hypothesis** is a statement that is not obvious

For example, it is a fact that humans must breathe in order to live.

but is offered to explain the facts. Since Galen could not actually see how the blood got from the left side to the right side of the heart, he came up with the hypothesis that there must be tiny holes present in the muscular wall that separates the two halves.

Galen. (Courtesy of the National Library of Medicine.)

Up to this point, modern scientists and ancient philosophers would behave the same way. Each offers a hypothesis to explain the facts. From this point on, however, their methods differ. To Galen, his explanation sounded right and that was enough to make him believe it, even though he couldn't see any holes. His hypothesis was in fact believed by virtually all physicians for more than 1000 years. When we use the scientific method, we do not believe a hypothesis just because it sounds right. **We test it,** using the most rigorous testing we can think of.

Around 1600 William Harvey (1578–1657) tested Galen's hypothesis by dissecting human and animal hearts and blood vessels. He discovered that there are one-way valves separating the upper chambers of the heart from the lower chambers. He also discovered that the heart is a pump that, by contracting and expanding, pushes the blood out. Harvey's teacher, Fabricius (1537–1619), had previously observed that there are one-way valves in the veins, so that blood in the veins can travel only toward the heart and not the other way.

Harvey put all these facts together to come up with a better hypothesis: blood circulates throughout the body, pumped by the heart. This was a better hypothesis than Galen's because it fitted the facts more closely. Even so, it was still a hypothesis and, according to the scientific method, had to be tested further. One important test took place in 1661, four years after Harvey died. According to Harvey's hypothesis, there had to be a way for the blood to get from the arteries to the veins, and Harvey predicted that there must be tiny blood vessels connecting them. In 1661 the Italian anatomist Malpighi (1628–1694), using the newly discovered microscope, found these tiny vessels, now called **capillaries.**

Malpighi's discovery supported the blood circulation hypothesis by fulfilling Harvey's prediction. When a hypothesis passes the tests, we have more confidence in it and call it a **theory.** A theory, in this sense, is the same as a hypothesis except that we have a stronger belief in it because it is supported by more evidence. No matter how much confidence we have in a theory, however, if new facts are discovered that conflict with it or if new tests are devised that it does not pass, the theory must be altered or rejected. In the history of science, many firmly established theories have eventually been thrown out because they could not pass new tests.

The word "theory" has another meaning as well: a model that explains many interrelated facts and can be used to make predictions about natural phenomena. Examples are Newton's theory of gravitation and the kinetic-molecular theory of gases, which we will meet in Section 5.2. This type of theory is also subject to testing and will be discarded or modified if it is contradicted by new facts.

The scientific method is thus very simple. We don't accept a hypothesis or a theory just because it sounds right. We devise tests, and only if the hypothesis or theory passes the tests do we accept it. The enormous progress made since 1600 in chemistry, biology, and the other sciences is a testimony to the value of the scientific method.

1.3 Exponential Notation

Scientists often have to deal with numbers that are very large or very small. Many years ago, an easy way to handle such numbers was devised.

Box 1A

The Scientific Method in Medicine

The use of the scientific method is not confined to scientific laboratories. Good detectives use the scientific method in investigating a crime, and doctors use it in diagnosing illnesses. The following is an example of the use of the scientific method in a hospital.

A two-year-old child was brought to the hospital. The child had been vomiting for several weeks. No apparent abnormalities were visible. Blood tests showed low hemoglobin levels. The red blood cell count was also low. These tests indicated anemia of unspecified origin. X-rays showed some opaque material in the intestinal lining and along the growth areas of the bones. With this information the doctors were ready to advance a hypothesis: that the opaque material accumulated in the intestines and in the growth areas of the bones was a heavy metal and that the child was suffering from heavy metal poisoning.

It was now necessary to test the hypothesis. If there was a heavy metal, the child must somehow have ingested it. In this case, the first test consisted of asking the parents. The doctors learned that the family lived in a rundown house that had peeling paint and that the child had a habit of chewing on window sills. Since house paint can contain significant amounts of lead, a heavy metal, this lent support to the hypothesis, but it still needed further testing. The next test was for lead in the blood, and it was found that almost two and a half times the normal lead concentration was present.

The hypothesis could now be regarded as confirmed. As a final test, treatment for lead poisoning was provided. This was done by administering calcium EDTA, a chemical that extracts heavy metals, binds them tightly, and carries them through the kidney membranes into the urine. One day after the injection a 25-fold increase in lead concentration was found in the child's urine. A few days later the poisoning symptoms disappeared, and after a week the patient was discharged, cured.

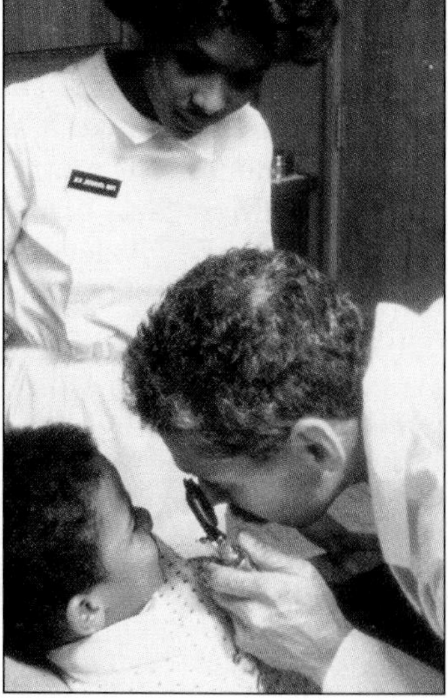

A physician examining a child.
(© Superstock, Inc.)

For example, an ordinary copper penny (before 1982, when pennies in the United States were still made of copper) contains approximately

29 500 000 000 000 000 000 000 atoms of copper

This is a very large number, but chemists and other scientists must often deal with very large and also with very small numbers. For example, a single copper atom weighs

0.00000000000000000000000023 pound

The **exponential notation** system is based on powers of 10 (Table 1.1). For example, if we multiply $10 \times 10 \times 10 = 1000$, we express this as 10^3. The 3 in this expression is called the **exponent** or the **power,** and it indicates how many times we multiplied 10 by itself and how many zeros follow the 1.

Exponential notation is also called scientific notation.

For example, 10^6 means a one followed by six zeros, or 1 000 000, and 10^2 means 100.

There are also negative powers of 10. For example, 10^{-3} means 1 divided by 10^3:

$$10^{-3} = \frac{1}{10^3} = \frac{1}{1000} = 0.001$$

Table 1.1 Examples of Exponential Notation

$10\,000 = 10^4$
$1000 = 10^3$
$100 = 10^2$
$10 = 10^1$
$1 = 10^0$
$0.1 = 10^{-1}$
$0.01 = 10^{-2}$
$0.001 = 10^{-3}$

Numbers are frequently expressed like this: 6.4×10^3. In a number of this type, 6.4 is the **coefficient** and 3 is the exponent, or power of 10. This number means exactly what it says:

$$6.4 \times 10^3 = 6.4 \times 1000 = 6400$$

Similarly, we can have coefficients with negative exponents:

$$2.7 \times 10^{-5} = 2.7 \times \frac{1}{10^5} = 2.7 \times 0.00001 = 0.000027$$

For numbers greater than 10 in exponential notation, we proceed as follows: *Move the decimal point to the left,* to just after the first digit. The (positive) exponent is equal to the number of places we moved the decimal point.

EXAMPLE

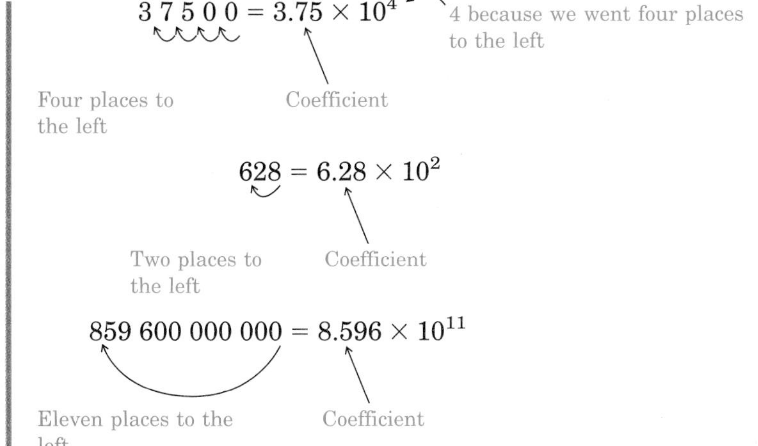

We don't really have to place the decimal point after the first digit, but by doing so we get a coefficient between 1 and 10, and that is the custom.

Using exponential notation, we can say that there are 2.95×10^{22} copper atoms in a copper penny. For large numbers the exponent is always *positive.* Note that we do not usually write out the zeros at the end of the number.

For small numbers (less than 1) we move the decimal point *to the right,* to just after the first nonzero digit, and use a *negative exponent.*

EXAMPLE

$$0.00346 = 3.46 \times 10^{-3}$$

Three places to the right

$$0.000004213 = 4.213 \times 10^{-6}$$

Six places to the right

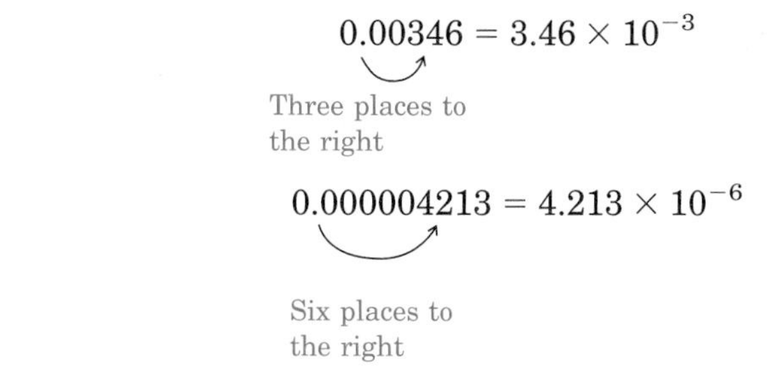

In exponential notation a copper atom weighs 2.3×10^{-25} pounds.

To convert exponential notation into fully written-out numbers, we do the same thing backward.

EXAMPLE

Write out in full: (a) 8.16×10^7 (b) 3.44×10^{-4}.

Answer

(a) $8.16 \times 10^7 = 81\ 600\ 000$

Seven places to the right (add enough zeros)

(b) $3.44 \times 10^{-4} = 0.000344$

Four places to the left

Problem **1.1**

Write out in full: (a) 4.71×10^4 (b) 7.93×10^{-6}.

When scientists add, subtract, multiply, and divide, they are always careful to express their answers with the proper number of digits, called significant figures. This method is described in the appendix.

Adding and Subtracting Numbers in Exponential Notation

We are allowed to add or subtract numbers expressed in exponential notation *only if they have the same exponent.* All we do is add or subtract the coefficients and leave the exponent as it is.

EXAMPLE

Add 3.6×10^{-3} and 9.1×10^{-3}.

Answer

$$\begin{array}{r} 3.6 \times 10^{-3} \\ + \quad 9.1 \times 10^{-3} \\ \hline 12.7 \times 10^{-3} \end{array}$$

The answer could also be written in other, equally valid ways:

$$12.7 \times 10^{-3} = 0.0127 = 1.27 \times 10^{-2}$$

Problem **1.2**

Add 6.1×10^5 and 5.6×10^5.

When it is necessary to add or subtract two numbers that have different exponents, we first must change them so that the exponents are the same.

A calculator with exponential notation changes the exponent automatically.

EXAMPLE

Add 1.95×10^{-2} and 2.8×10^{-3}.

Answer • In order to add these two numbers, we make both exponents -2. Thus, $2.8 \times 10^{-3} = 0.28 \times 10^{-2}$. Now we can add:

$$\begin{array}{r} 1.95 \times 10^{-2} \\ + \; 0.28 \times 10^{-2} \\ \hline 2.23 \times 10^{-2} \end{array}$$

Problem **1.3**

Subtract 3.21×10^4 from 1.77×10^5.

Multiplying and Dividing Numbers in Exponential Notation

To multiply numbers in exponential notation, we first multiply the coefficients in the usual way and then algebraically *add* the exponents.

EXAMPLE

Multiple 7.40×10^5 by 3.12×10^9.

Answer
$7.40 \times 3.12 = 23.1$
Add exponents:
$10^5 \times 10^9 = 10^{5+9} = 10^{14}$
Answer: $23.1 \times 10^{14} = 2.31 \times 10^{15}$

EXAMPLE

Multiply 4.6×10^{-7} by 9.2×10^4.

Answer

$4.6 \times 9.2 = 42$

Add exponents:

$10^{-7} \times 10^4 = 10^{-7+4} = 10^{-3}$

Answer: $42 \times 10^{-3} = 4.2 \times 10^{-2}$

Problem 1.4

Multiply: (a) 3.7×10^2 by 7.8×10^5 (b) 6.9×10^5 by 3.4×10^{-7}.

To divide numbers expressed in exponential notation, the process is reversed. We first divide the coefficients and then algebraically *subtract* the exponents.

EXAMPLE

Divide: $\dfrac{6.4 \times 10^8}{2.57 \times 10^{10}}$

Answer

$6.4 \div 2.57 = 2.5$

Subtract exponents:

$10^8 \div 10^{10} = 10^{8-10} = 10^{-2}$

Answer: 2.5×10^{-2}

EXAMPLE

Divide: $\dfrac{1.62 \times 10^{-4}}{7.94 \times 10^7}$

Answer

$1.62 \div 7.94 = 0.204$

Subtract exponents:

$10^{-4} \div 10^7 = 10^{-4-7} = 10^{-11}$

Answer: $0.204 \times 10^{-11} = 2.04 \times 10^{-12}$

Problem 1.5

Divide: (a) $\dfrac{3.75 \times 10^4}{1.93 \times 10^{-2}}$ (b) $\dfrac{9.24 \times 10^{-3}}{6.14 \times 10^3}$.

Scientific calculators do these calculations automatically. All that is necessary is to enter the first number, press $+$, $-$, $\times$, or $\div$, enter the second number, and press $=$. (The method for entering numbers of this form varies; consult the instructions that come with the calculator.) Many scientific calculators also have a key that will automatically convert a number such as 0.00047 to its scientific notation form (4.7×10^{-4}), and vice versa.

1.4 Measurements

In our daily lives we are constantly making measurements. We measure ingredients for recipes, driving distances, gallons of gasoline, weights of fruits and vegetables, the timing of TV programs. Doctors and nurses measure pulse rates, blood pressures, temperatures, and drug dosages. Chemistry, like other sciences, is based on measurements.

A measurement consists of two parts: a number and a unit. A number without a unit is usually meaningless. If you were told that a person's weight is 57, the information would be of very little use. Is it 57 pounds, which would indicate that the person is very likely a child or a midget, or 57 kilograms, which is the weight of an average woman or a small man? Or is it perhaps some other unit? Because there are so many units, a number by itself is not enough; the unit must also be stated.

In the United States most measurements are made with the English system of units: pounds, miles, gallons, and so on. In most of the rest of the world, however, this is not the case. In Germany, France, Russia, Japan, and most other countries, almost nobody could tell you what a pound or an inch is. These countries use the **metric system,** a system that was begun in France about 1800 and has since spread throughout the world. Even in the United States, metric measurements are slowly being introduced (Fig. 1.4). For example, many soft drinks and most alcoholic beverages now come in metric sizes (Fig. 1.5). *Scientists* in the United States have been using metric units all along.

Around 1960, international scientific organizations adopted another system, called the **International System of Units** (abbreviated **SI**). The SI is based on the metric system and uses some of the metric units. The main difference is that the SI is more restrictive: it discourages the use of certain metric units and favors others. Although the SI has advantages over the older metric system, it also has significant disadvantages, and for this reason U.S. chemists have been very slow to adopt it. At this

"SI" is actually an abbreviation of the French title *Système Internationale.*

Figure **1.4** • A highway sign in Missouri. (Photograph by Beverly March.)

Figure **1.5**
Soft drinks in metric sizes. (Photograph by Beverly March.)

time, approximately 35 years after its introduction, not many U.S. chemists use the entire SI, although some of its preferred units are gaining ground.

In this book we shall use the metric system. Occasionally we shall mention the preferred SI unit.

Length

The key to the metric system (and the SI) is that there is one basic unit for each kind of measurement and that other units are related to the basic unit only by powers of ten. For example, let us look at measurements of length. In the English system we have the inch, the foot, the yard, and the mile (not to mention such older units as the league, furlong, ell, and rod). If you want to convert one to another, you must memorize or look up these conversion factors:

$$12 \text{ inches} = 1 \text{ foot}$$
$$3 \text{ feet} = 1 \text{ yard}$$
$$1760 \text{ yards} = 1 \text{ mile}$$

The official definition of the meter is the distance that light travels in $\frac{1}{299\,792\,458}$ of a second.

All this is unnecessary in the metric system (and the SI). In both systems the basic unit of length is the **meter (m)**. To convert to larger or smaller units we do not use arbitrary numbers like 12, 3, and 1760, but only 10, 100, 1/10, 1/100, or other powers of ten. This means that **to convert from one metric or SI unit to another, we only have to move the decimal point.** Furthermore, the other units are named by putting prefixes in front of "meter," and **these prefixes are the same throughout the metric system and the SI.**

The most important of these prefixes are shown in Table 1.2. If we put some of these prefixes in front of "meter," we have

$$1 \text{ kilometer (km)} = 1000 \text{ meters (m)}$$
$$1 \text{ centimeter (cm)} = 0.01 \text{ meter}$$
$$1 \text{ nanometer (nm)} = 10^{-9} \text{ meter}$$

Table 1.2 The Most Common Metric Prefixes

Prefix	Symbol	Value
giga	G	10^9 = 1 000 000 000 (one billion)
mega	M	10^6 = 1 000 000 (one million)
kilo[a]	k	10^3 = 1000 (one thousand)
deci	d	10^{-1} = 0.1 (one tenth)
centi	c	10^{-2} = 0.01 (one hundredth)
milli	m	10^{-3} = 0.001 (one thousandth)
micro	μ	10^{-6} = 0.000001 (one millionth)
nano	n	10^{-9} = 0.000000001 (1 billionth)

[a]The prefixes shown in color are those we encounter most frequently in this book.

Table 1.3 Some Conversion Factors Between the English and Metric Systems

Length	Mass	Volume
1 in. = 2.54 cm	1 oz (avdp) = 28.35 g	1 qt = 0.946 L
1 m = 39.37 in.	1 lb (avdp) = 453.6 g	1 gal = 3.785 L
1 mile = 1.609 km	1 kg = 2.205 lb	1 L = 33.81 fl oz
	1 g = 15.43 grains	1 fl oz = 29.57 mL
		1 L = 1.057 qt

For people who have grown up using English units, it is helpful to have some idea of the size of metric units. A meter is equal to 39.37 in., which makes it a little longer than a yard (36 in.). An inch is about 2.5 cm. Some conversion factors between the English and metric systems are shown in Table 1.3.

Now that we have discussed length, we shall look at the metric units for some other measurements important in chemistry.

Volume

Volume is space. The volume of a liquid, solid, or gas is the space occupied by that substance. The basic unit of volume in the metric system is the **liter** (**L**). This unit is a little larger than a quart (Table 1.3). The only other common metric unit is the milliliter (mL), which is of course equal to 10^{-3} L:

$$1000 \text{ mL} = 1 \text{ L}$$

One milliliter is exactly equal to one cubic centimeter (cc or cm^3):

$$1 \text{ mL} = 1 \text{ cc}$$

which means that there are 1000 cc in 1 L.

Volume measurements are very important in chemistry and in medicine, and there are many devices for this purpose (Fig. 1.6). Graduated cylinders are used for approximate measurements; burets, pipets, and volumetric flasks are used for more exact ones. In a hypodermic syringe, the means for injecting a drug (the needle) is combined with the means for measuring the volume injected (the barrel).

Mass

Mass is the quantity of matter in an object. The basic unit of mass in the metric system is the **gram** (g). As always in the metric system,

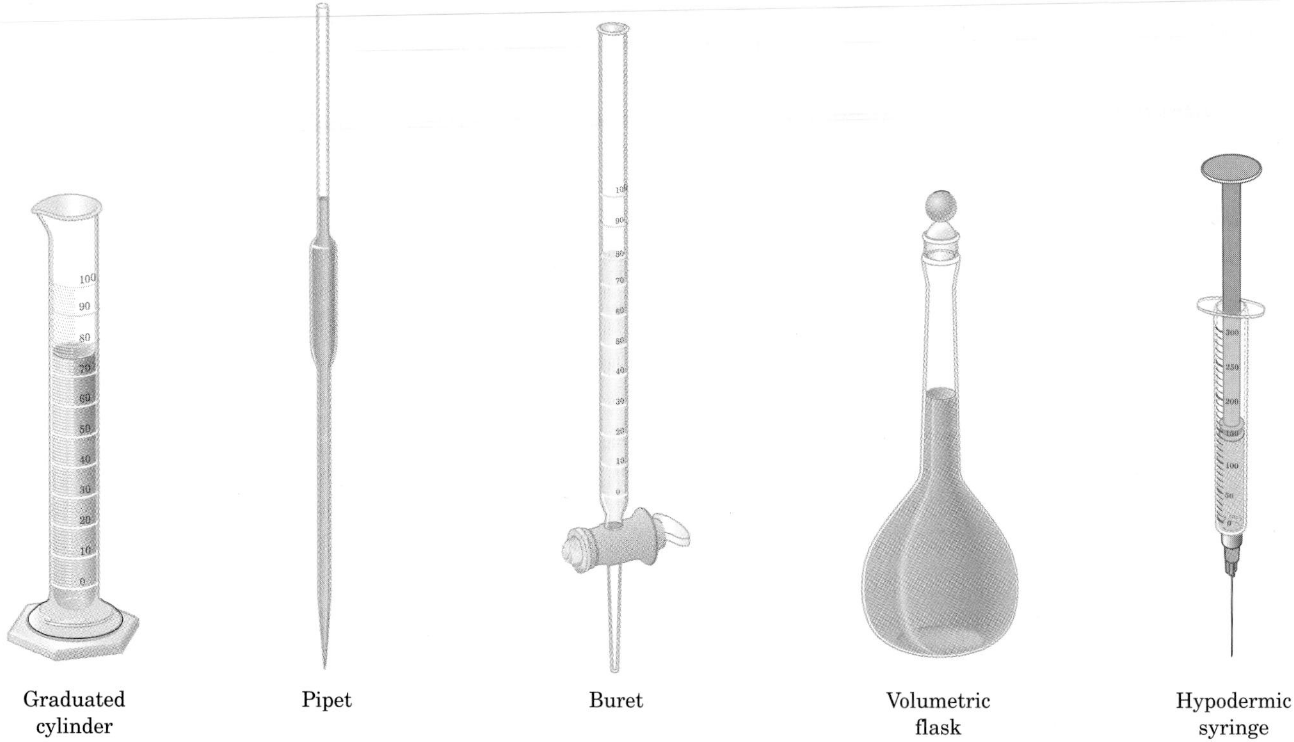

Figure **1.6** • Some instruments for measuring volume.

larger and smaller units are indicated by prefixes. The ones in common use are

The official kilogram is defined by the mass of a particular platinum-iridium bar kept at the International Bureau of Standards in Paris.

$$1 \text{ kilogram (kg)} = 1000 \text{ g}$$

$$1 \text{ milligram (mg)} = 0.001 \text{ g}$$

The gram is a small unit; there are 453.6 g in one pound (Table 1.3).

In chemistry we use balances to measure mass. As with devices for measuring volume, some balances are used for approximate measurements and others for more exact measurements. Some laboratory balances are shown in Figure 1.7. Most balances operate on the same very simple principle, shown in Figure 1.8. A beam is balanced on a pivot. The object whose mass is to be determined is placed on one side, and some known masses are placed on the other. When enough known masses have been placed to make the beam horizontal, the unknown mass equals the known mass.

There is a basic difference between mass and weight. Mass is independent of location. The mass of a stone, for example, is the same whether we measure it at sea level, on top of a mountain, or in the depths of a mine. In contrast, weight is not independent of location. **Weight is the force a mass experiences under the pull of gravity.** This was dramatically demonstrated when the astronauts walked on the surface of the moon. The moon, being a smaller body than the earth, exerts a weaker

Both mass and weight are independent of temperature.

Figure **1.7** • Two laboratory balances. (Courtesy of Brinkman Instruments Co.)

gravitational pull. Consequently, the astronauts, although they wore space suits and equipment that would be heavy on earth, felt lighter on the moon and could execute great leaps and bounces during their walks.

Although mass and weight are different concepts, they are related to each other by the force of gravity. We frequently use the words interchangeably because we weigh objects by comparing their masses to standard reference masses (weights) on a balance, and the gravitational pull is the same on the unknown object and on the standard masses (Fig. 1.8). Because the force of gravity is essentially constant, mass is always directly proportional to weight.

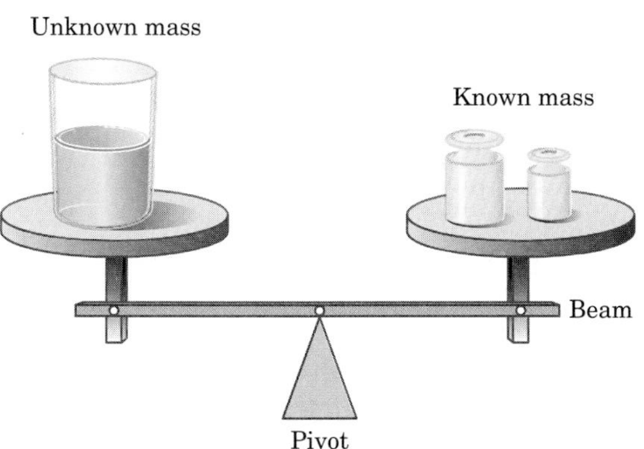

Figure **1.8**
Principle of the balance. When the beam is horizontal, the masses are equal.

Box 1B

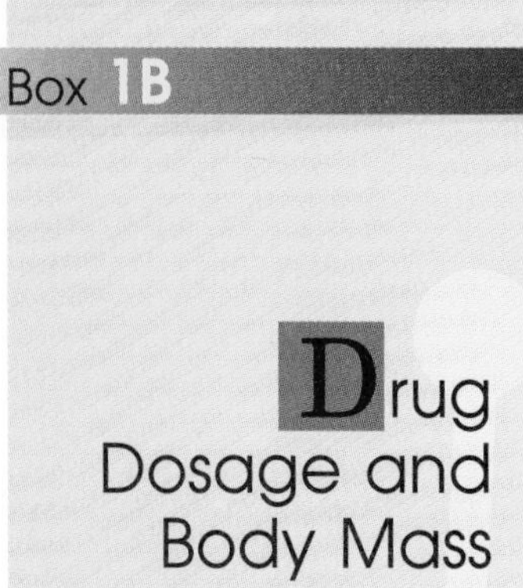

Drug Dosage and Body Mass

In many cases, drug dosages are prescribed on the basis of body mass. For example, the recommended dosage of a drug may be 3 mg of drug for each kilogram of body weight. In this case, a 50-kg (110-lb) woman would receive 150 mg and an 82-kg (180-lb) man would get 246 mg. This is especially important for children, since a dose suitable for an adult will generally be too much for a child, who has much less body mass. For this reason, manufacturers package and sell smaller doses of certain drugs, such as aspirin, for children.

Drug dosage may also vary with age. Occasionally, when an elderly patient has an impaired kidney or liver function, the clearance of a drug from the body is delayed, and the drug may stay in the body longer than is normal. This can cause dizziness, vertigo, and migraine-like headaches, resulting in falls and broken bones. Such delayed clearance must be monitored, and the drug dosage adjusted accordingly.

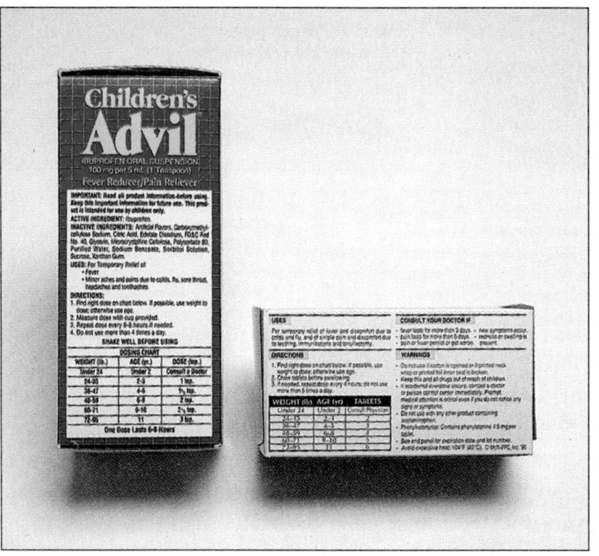

These packages of Advil™ and Tylenol™ have charts showing the proper dose for children of a given weight. (Photograph by Beverly March.)

Time

Time is the one quantity for which the units are the same in all systems: English, metric, and SI. The basic unit is the **second** (**s**), and

$$60 \text{ s} = 1 \text{ min}$$

$$60 \text{ min} = 1 \text{ h}$$

Temperature

Most people in the United States are familiar with the Fahrenheit scale of temperature. In the metric system the **centigrade,** or **Celsius,** scale is used. In this scale, the boiling point of water is set at 100°C and the freezing point at 0°C. We can convert from one scale to the other by using the formulas

$$°F = \tfrac{9}{5}°C + 32$$

$$°C = \tfrac{5}{9}(°F - 32)$$

The 32 in these equations is a defined number and is therefore treated as if it had an infinite number of zeros following the decimal point. (See p. A.4)

EXAMPLE

Normal body temperature is 98.6°F. Convert this to Celsius.

Answer

$°C = \tfrac{5}{9}(98.6 - 32) = \tfrac{5}{9}(66.6) = 37°C$

Problem 1.6

Convert: (a) 64.0°C to Fahrenheit (b) 47°F to Celsius.

The relationship between the Fahrenheit and Celsius scales is shown in Figure 1.9.

There is also another temperature scale, called the **Kelvin** (**K**) scale or the **absolute** scale. The *size* of the Kelvin degree is the same as that of the Celsius degree. The only difference is the zero point. The temperature −273°C is taken as the zero point on the Kelvin scale. This makes conversions between Kelvin and Celsius very easy. To go from Celsius to Kelvin, just *add* 273; to go from Kelvin to Celsius, *subtract* 273:

$$K = °C + 273$$

The relationship between the Kelvin and Celsius scales is also shown in Figure 1.9. Note that we don't use the degree symbol in the Kelvin scale: 100°C equals 373 K, not 373°K.

Why was −273°C chosen as the zero point on the Kelvin scale? The reason is that **−273°C, or 0 K, is the lowest possible temperature.** Because of this, 0 K is called **absolute zero.** Temperature depends on how fast molecules move. The more slowly they move, the colder it gets. At absolute zero molecules stop moving altogether, and it is not possible for them to move more slowly than that. Therefore, the temperature cannot get any lower. For some purposes it is convenient to have a scale that begins at the lowest possible temperature; the Kelvin scale fulfills this need.

Scientists in laboratories have been able to attain temperatures as low as 0.001 K, but a temperature of 0 K is impossible.

1.5 Unit Conversions. The Factor-Label Method

We frequently need to convert a measurement from one unit to another. The best and most foolproof way to do this is the **factor-label method.** In this method we follow the rule that *when multiplying numbers we also multiply units, and when dividing numbers we also divide units.*

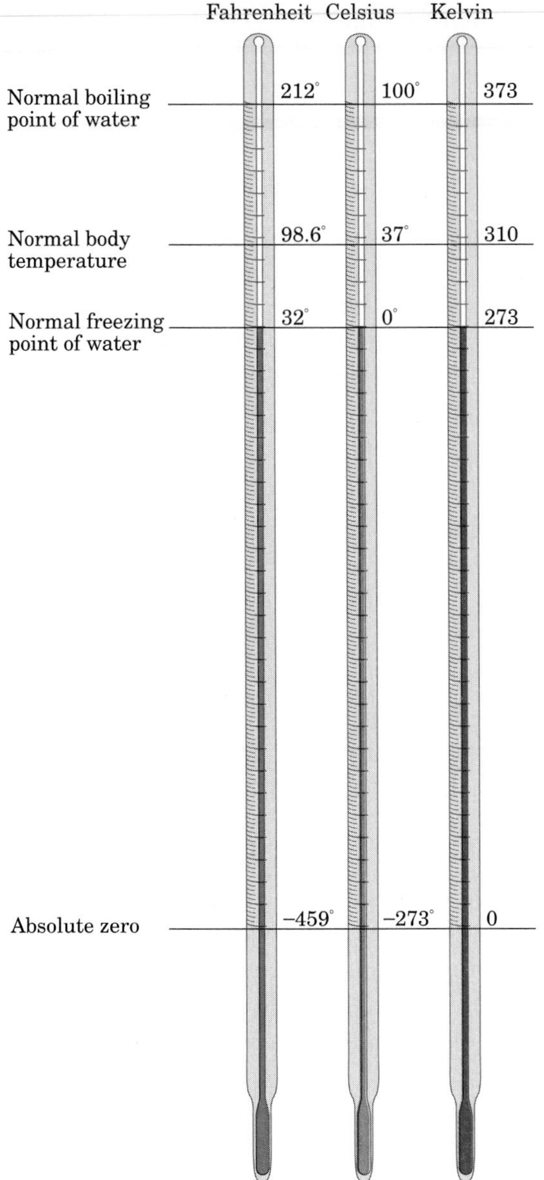

Figure **1.9** • Three temperature scales.

A conversion factor is a ratio of two different units.

For conversions between one unit and another, it is always possible to set up two fractions, called **conversion factors.** Suppose we wish to convert 381 g to pounds. We can see in Table 1.3 that there are 453.6 g in 1 lb. The conversion factors between grams and pounds therefore are

$$\frac{1 \text{ lb}}{453.6 \text{ g}} \quad \text{and} \quad \frac{453.6 \text{ g}}{1 \text{ lb}}$$

To convert 381 g to pounds, all we need do is multiply by the proper conversion factor. But which one? Let us try both and see what happens. First let us multiply by 1 lb/453.6 g:

$$381\ \cancel{g} \times \frac{1\ \text{lb}}{453.6\ \cancel{g}} = 0.840\ \text{lb}$$

Following the procedure of multiplying and dividing units when we multiply and divide numbers, we find that dividing grams by grams cancels out the grams and we are left with pounds, which is the answer we want. This is the correct method because it converts grams to pounds. Suppose we had done it the other way, multiplying by 453.6 g/1 lb:

$$381\ \text{g} \times \frac{453.6\ \text{g}}{1\ \text{lb}} = 173\ 000\ \frac{\text{g}^2}{\text{lb}}$$

When we multiply grams by grams, we get g^2 (grams squared). Dividing by pounds gives g^2/lb. Since this is not the unit we want, this must be the incorrect method.

The advantage of the factor-label method is that it lets us know when we have made a wrong calculation. *If the units of the answer are not the ones we are looking for, the calculation must be wrong.* Incidentally, this principle works not only in unit conversions but in all problems where we make calculations using measured numbers.

The factor-label method gives the correct mathematical solution for a problem. However, it is a mechanical technique and does not require you to think through the problem. Thus it may not provide a deeper understanding. For this reason and also to check your answer (because it is easy to make mistakes in arithmetic, for example by punching wrong numbers into a calculator), it is beneficial to ask yourself if the answer you have obtained is *reasonable.* For example, the question might ask the actual mass of a single oxygen atom. If your answer comes out 8.5×10^6 g, it is not reasonable. A single atom cannot weigh more than you do! In such a case, you have obviously made a mistake and should go back and take another look to see where you went wrong. Everybody makes mistakes at times, but if you check you can at least determine whether your answer is reasonable. If it is not, you will immediately know that there is a mistake and will be able to correct it.

You will get a deeper understanding of the problem because you will think through the relationship between the question and the answer. We will now give a few examples of unit conversions and then test the answers to see if they are reasonable. To save space, we will do this only in this chapter. But a similar approach should be used in problem solving in the later chapters as well. In conversion problems you will get two results. First, the process should tell you whether the number in the answer should be larger or smaller than the number in the question. Second, it should provide a number that has the same magnitude as the answer. For example, if the correct answer is 4.62 mg, an answer of 5 mg is reasonable, but 50 mg or 500 mg would not be reasonable, nor would 0.5 mg or 0.05 mg.

EXAMPLE

The distance between Rome and Milan (the two largest cities in Italy) is 358 miles. How many kilometers is this?

Answer • We want to convert miles to kilometers. Table 1.3 shows that 1 mile = 1.609 km. From this we get two conversion factors:

$$\frac{1 \text{ mile}}{1.609 \text{ km}} \quad \text{and} \quad \frac{1.609 \text{ km}}{1 \text{ mile}}$$

Which should we use? We use the one that gives the answer in kilometers:

$$358 \text{ miles} \times \text{conversion factor} = ? \text{ km}$$

This means that the miles must cancel, so the conversion factor 1.609 km/1 mile is appropriate.

$$358 \cancel{\text{miles}} \times \frac{1.609 \text{ km}}{1 \cancel{\text{mile}}} = 576 \text{ km}$$

Is it reasonable? We want to convert miles to kilometers. The conversion factor in Table 1.2 tells us that in a given distance the number of kilometers is larger than the number of miles. How much larger? The exact number is 1.609, which is approximately 1.5 times larger. Thus we expect that the answer in kilometers will be about 1.5 times greater than the number given in miles. The number given in miles is 358, which *for the purpose of getting an approximately reasonable answer* we can round off to, say, 400. Multiplying this by 1.5 gives an approximate answer of 600 kilometers. Since our actual answer, 576 km, was of the same order of magnitude as the estimated answer, we can say that it is reasonable. If the estimated answer had been 6 km, or 60 km, or 6000 km, we would suspect that we had made a mistake in calculating the actual answer.

Problem 1.7

How many kilograms are in 241 lb? Check your answer to see if it is reasonable.

EXAMPLE

The label on a container of olive oil says 1.844 gal. How many milliliters is this?

Answer • Table 1.3 shows no factor converting gallons to milliliters, but it does show that 1 gal = 3.785 L. Since we know that 1000 mL = 1 L, we can solve this problem by multiplying by two conversion factors, making certain that all units cancel except milliliters:

$$1.844\ \cancel{\text{gal}} \times \frac{3.785\ \cancel{\text{L}}}{1\ \cancel{\text{gal}}} \times \frac{1000\ \text{mL}}{1\ \cancel{\text{L}}} = 6980\ \text{mL}$$

Is it reasonable? The conversion factor in Table 1.3 tells us that there are more liters in a given volume than gallons. How much more? Approximately four times more. We also know that any volume in milliliters is one thousand times larger than the same volume in liters. Thus we expect that the volume expressed in milliliters will be 4 × 1000 or 4000 times more than in gallons. The estimated volume in milliliters will be approximately 1.8 × 4000 or 7000 mL. But we also expect that the actual answer should be somewhat less than the estimated figure because we overestimated the conversion factor (4 rather than 3.785). Thus the answer, 6980 mL, is quite reasonable.

Problem 1.8

Calculate the number of kilometers in 8.55 miles. Check your answer to see if it is reasonable.

EXAMPLE

The maximum speed limit on many roads in the United States is 65 mph. How many meters per second is this?

Answer • Here we have essentially a double conversion problem. We must convert miles to meters and hours to seconds. We use as many conversion factors as necessary, always making sure that we use them in such a way that the proper units cancel:

$$65\ \frac{\cancel{\text{miles}}}{\cancel{\text{h}}} \times \frac{1.609\ \cancel{\text{km}}}{1\ \cancel{\text{mile}}} \times \frac{1000\ \text{m}}{1\ \cancel{\text{km}}} \times \frac{1\ \cancel{\text{h}}}{60\ \cancel{\text{min}}} \times \frac{1\ \cancel{\text{min}}}{60\ \text{s}} = 29\ \frac{\text{m}}{\text{s}}$$

Is it reasonable? To estimate the 65 mph speed in meters per second, we must first establish the relationship between miles and meters. As in the first example, we know that there are more kilometers than miles in a given distance. How much more? Since there are approximately 1.5 km in a mile, there must be approximately 1500 times more meters. We also know that in 1 hour there are 60 × 60 = 3600 seconds. There are approximately half as many more meters in a mile than seconds in an hour: 1500/3600. Therefore, we estimate that the speed in meters per second will be about one half of that in miles per hour or 32 m/s. Once again, the actual answer, 29 m/s, is not far from the estimate of 32 m/s, so the answer is reasonable.

Estimating the answer is a good thing to do when working any mathematical problem, not just unit conversions.

As shown in these examples, when canceling units we do *not* cancel the numbers. The numbers are multiplied and divided in the ordinary way.

Problem **1.9**

Convert the speed of sound, 332 m/s, to miles per hour. Check your answer to see if it is reasonable.

1.6 The States of Matter

The most familiar gas is the air around us, which we depend on for life itself.

Matter can exist in three states: gas, liquid, and solid. *Gases* have no definite shape or volume. They expand to fill whatever container they are put into. On the other hand, they are highly compressible and can be forced into small containers. *Liquids* also have no definite shape. However, they do have a definite volume that remains the same when they are poured from one container to another. Liquids are only slightly compressible. *Solids* have definite shapes and definite volumes. They are essentially incompressible.

Whether a substance is a gas, a liquid, or a solid depends on its temperature. On a cold winter day a puddle of liquid water turns to *ice;* it becomes a solid. If we boil water, the liquid becomes a gas—we call it *steam.* Most substances can exist in the three states: they are gases at high temperature, liquefy at a lower temperature, and solidify when their temperature becomes low enough. Figure 1.10 shows a single substance in the three different states.

We do not regard a substance as having changed its identity when it is converted from one state to another. Water is still water whether in the form of ice, steam, or liquid water. We discuss the three states of matter, and the changes between one state and another, at greater length in Chapter 5.

(*a*)

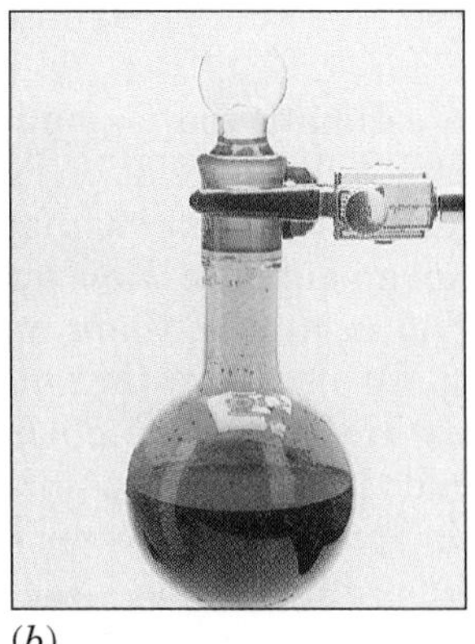
(*b*)

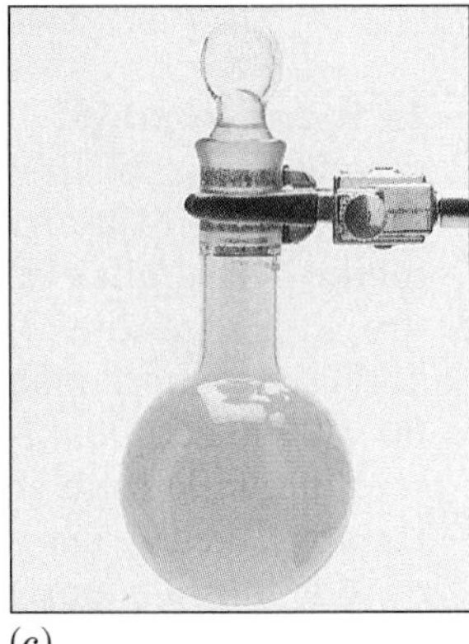
(*c*)

Figure **1.10** • The three states of matter for bromine: (a) bromine as a solid, (b) bromine as a liquid, and (c) bromine as a gas. (Photographs by C. D. Winters.)

1.7 Density and Specific Gravity

One of the many pollution problems that the world faces is the spillage of petroleum in the oceans from oil tankers or from offshore drilling. When oil spills into the ocean, it floats on top of the water. Why doesn't it sink? The reason is that water has a higher *density* than oil, and when two liquids are mixed (assuming that one does not dissolve in the other), the one of lower density floats on top (Fig. 1.11).

The **density** of any substance is defined as its **mass per unit volume.** Not only do all liquids have a density, but so do all solids and gases, too. Density is calculated by dividing the mass of a substance by its volume:

$$d = \frac{m}{V} \qquad d = \text{density}, m = \text{mass}, V = \text{volume}$$

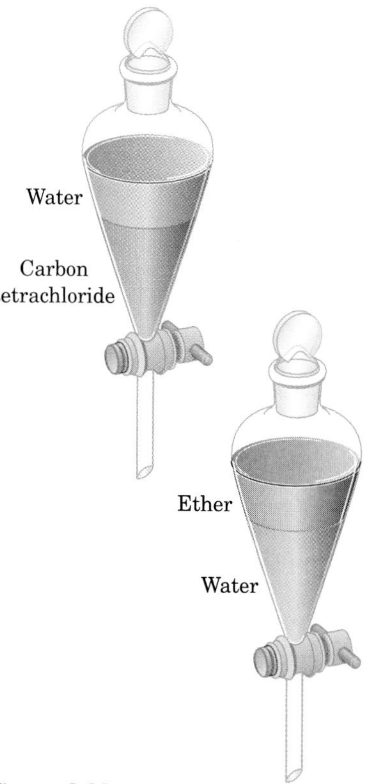

Figure **1.11**
Two separatory funnels containing water and another liquid. In each case the liquid with the lower density is on top.

EXAMPLE

If 73.2 mL of a liquid has a mass of 61.5 g, what is its density in g/mL?

Answer

$$d = \frac{m}{V} = \frac{61.5\text{ g}}{73.2\text{ mL}} = 0.840\,\frac{\text{g}}{\text{mL}}$$

EXAMPLE

The density of iron is 7.86 g/mL. What is the volume, in mL, of an irregularly shaped piece of iron that has a mass of 524 g?

Answer • Here we are given the mass. In this type of problem, it is useful to derive a conversion factor that we can obtain from the density. We know that the density is 7.86 g/mL. This means that 1 mL has a mass of 7.86 g. From this we can get two conversion factors:

$$\frac{1\text{ mL}}{7.86\text{ g}} \quad \text{and} \quad \frac{7.86\text{ g}}{1\text{ mL}}$$

As usual, we multiply the mass by whichever conversion factor results in the cancellation of all but the correct unit:

$$524\text{ g} \times \frac{1\text{ mL}}{7.86\text{ g}} = 66.7\text{ mL}$$

Is it reasonable? The density 7.86 g/mL shows us that the volume in mL of any piece of iron is always less than its mass in grams. How much less? Approximately eight times less. Thus we expect the volume to be approximately 500/8 = 63 mL. Since the actual answer is 66.7 mL, it is reasonable.

The spillage of more than 10 million gallons of petroleum in Prince William Sound, Alaska, in March 1989 caused a great deal of environmental damage. (Wide World Photos.)

Problem 1.10

The density of titanium is 4.54 g/mL. What is the mass, in g, of 17.3 mL of titanium? Check your answer to see if it is reasonable.

Problem 1.11

An unknown substance has a mass of 56.8 g and occupies a volume of 23.4 mL. What is its density, in g/mL? Check your answer to see if it is reasonable.

In old movies, villains frequently used a chloroform-wetted handkerchief to subdue their victims.

The density of any liquid or solid is a physical property that is constant, which means that it always has the same value at a given temperature. We use physical properties to help identify a substance. For example, the density of chloroform (a liquid formerly used as an anesthetic) is 1.483 g/mL at 20°C. If we want to find out if an unknown liquid is chloroform, one thing we might do is measure its density at 20°C. If the density is, say, 1.355 g/mL, we know the liquid isn't chloroform. If the density does come out to be 1.483 g/mL, we cannot be sure the liquid is chloroform, because other liquids might also have this density, but we can then measure other physical properties (the boiling point, for example). If all the physical properties we measure match those of chloroform, we can be reasonably sure the liquid *is* chloroform.

We have said that the density of a liquid or solid is a constant at a given temperature. Density does change when the temperature changes. Almost always, *density decreases with increasing temperature.* This is true because *mass* does not change when a substance is heated, but volume almost always increases because atoms and molecules tend to get farther apart as the temperature increases. Since $d = m/V$, if m stays the same and V gets larger, d must get smaller.

This is one of the many discoveries made by Benjamin Thompson (1753–1814), who was born in Massachusetts but later became Count Rumford in Munich.

The most common liquid, water, provides a partial exception to this rule. As we go from 4°C to 100°C, the density of water does decrease, but from 0°C to 4°C the density *increases.* That is, water has its maximum density at 4°C.

Since density is equal to mass divided by volume, it always has units, most commonly g/mL or g/cc (g/L for gases). **Specific gravity** is numerically the same as density, but it has no units (it is dimensionless). The reason is that specific gravity is defined as a comparison of the density of a substance with that of water, which is taken as a standard. For example, the density of copper at 20°C is 8.92 g/mL. The density of water at the same temperature is 1.00 g/mL. Therefore, copper is 8.92 times as dense as water, and its specific gravity at 20°C is 8.92. Because water is taken as the standard and because the density of water is 1.00 g/mL at 20°C, the specific gravity of any substance is always numerically equal to its density, provided that the density is measured in g/mL or g/cc.

EXAMPLE

The density of pure ethyl alcohol at 20°C is 0.789 g/mL. What is its specific gravity?

Answer

$$\text{Specific gravity} = \frac{0.789\ \cancel{\text{g/mL}}}{1.00\ \cancel{\text{g/mL}}} = 0.789$$

Problem **1.12**

The specific gravity of a urine sample is 1.016. What is its density, in g/mL?

1.8 Energy

Matter constantly undergoes changes. In order to understand these changes, we must know something about energy. **Energy** is defined as **the capacity to do work.** Energy can be described as being either kinetic energy or potential energy.

Kinetic energy is **the energy of motion.** Any object that is moving possesses kinetic energy. We can calculate how much energy by the formula $KE = \frac{1}{2}mv^2$, where m is the mass of the object and v is its velocity. This means that kinetic energy increases (1) when an object moves faster and (2) when it is a heavier object that is moving. When a truck and a bicycle are moving at the same velocity, the truck has more kinetic energy.

You impart kinetic energy to a baseball or a football when you throw it.

Potential energy is **stored energy.** The potential energy possessed by an object arises from its capacity to move or to cause motion. For example, a book balanced at the edge of a table contains potential energy—it is capable of doing work. If given a slight push, it will fall. The potential energy the book has while it is on the table is converted to the kinetic energy it has while it is falling. Another way in which potential energy is converted to kinetic energy is shown in Figure 1.12.

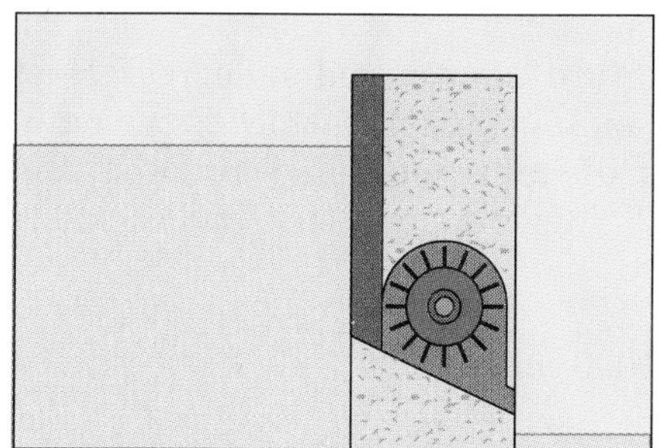

Potential energy ready to do work

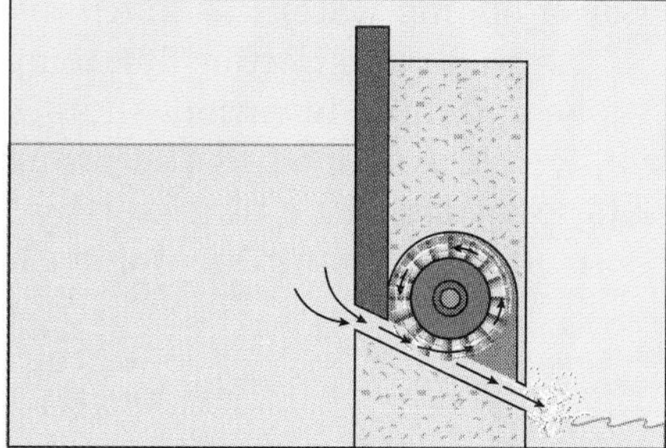

Kinetic energy doing work

Figure **1.12** • The water held back by the dam possesses potential energy, which is converted to kinetic energy when the water is released.

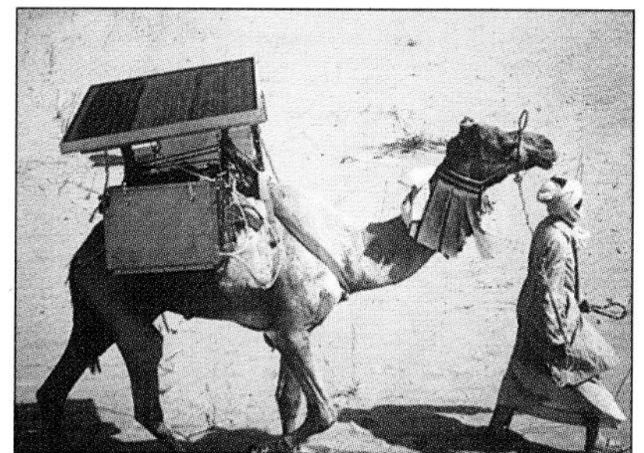

An example of energy conversion: Light energy from the sun is converted to electrical energy by solar cells. The electricity runs the refrigerator on the back of the camel, keeping the vaccines inside cool, so they can be delivered to remote locations. (Courtesy of Siemens Solar Industries, Camarillo, CA 93011.)

In chemistry the most important form of potential energy is **chemical energy.** This is the energy stored within chemical substances. The energy is given off when they react. For example, a piece of paper possesses chemical energy. When the paper is ignited, the chemical energy (potential) is turned into energy in the form of heat and light.

An important principle in nature is that things have a tendency to seek their lowest possible potential energy. We all know that water always flows downhill and not uphill.

Hydroelectric plants use the kinetic energy of falling water.

There are several forms of energy, of which the most important are mechanical energy (the kinetic energy possessed by moving objects), heat, light, electrical energy, chemical energy, and nuclear energy. The various forms of energy can be converted from one to another. In fact, we do this all the time. A power plant operates either on the chemical energy derived from burning fuel or on nuclear energy. This energy is converted to heat, which is converted to the electricity that is sent over transmission wires into houses and factories. Here we convert the electricity to light, heat (in an electrical heater, for example), or mechanical energy (in the motors of refrigerators, vacuum cleaners, and so on).

Although one form of energy can be converted to another, the *total amount* of energy in any system does not change. **Energy can neither be created nor destroyed.** This statement is called **the law of conservation of energy.***

1.9 Heat

One form of energy that is particularly important in chemistry is heat. This is the form of energy that most frequently accompanies chemical reactions. Heat is not the same as temperature. Heat is a form of energy, but temperature is not.

The difference between heat and temperature can be seen in the following example. If we have two beakers, one containing 100 mL of water and the other 1 L of water at the same temperature, the heat content of the water in the larger beaker is ten times that of the water in the smaller beaker, even though the temperature is the same in both. If you were to dip your hand accidentally into a liter of boiling water, you would be much more severely burned than if only one drop fell on your hand. Even though the water is at the same temperature in both cases, the liter of boiling water has much more heat.

As we saw in Section 1.4, temperature is measured in degrees. Heat can be measured in various units, the most common of which is the **calorie,** which is defined as **the amount of heat necessary to raise the temperature of 1 g of water by 1°C.** This is a small unit, and chemists more often use the **kilocalorie** (kcal):

$$1 \text{ kcal} = 1000 \text{ cal}$$

*This statement is not completely true. As discussed in Box 9F, it is possible to convert matter to energy, and vice versa. Therefore a more correct statement would be **matter-energy can neither be created nor destroyed.** However, the law of conservation of energy is valid for most purposes and is highly useful.

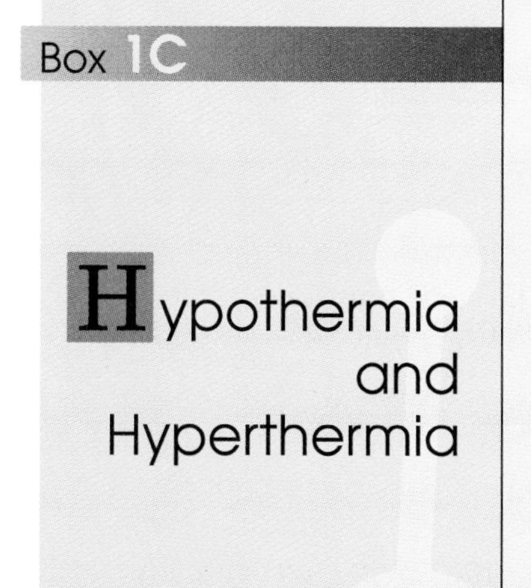

Box 1C

Hypothermia and Hyperthermia

The human body cannot tolerate temperatures that are too low. A person outside in very cold weather (say, −20°F, or −29°C) who is not protected by heavy clothing will eventually freeze to death because the body loses heat. Normal body temperature is 37°C, and when the outside temperature is lower than that, heat flows out of the body. When the air temperature is moderate (10 to 25°C), this poses no problem and is, in fact, necessary because the body produces more heat than it needs and must lose some. At extremely low temperatures, however, too much heat is lost and body temperature drops, a condition called **hypothermia.** A drop in body temperature of 1 or 2°C causes shivering, which is an attempt by the body to increase its temperature by the heat generated through muscular action. A greater drop than that results in unconsciousness and eventually death.

The opposite condition is **hyperthermia.** This can be caused either by high outside temperatures or by the body itself when an individual develops a high fever. A sustained body temperature as high as 41.7°C (107°F) is usually fatal.

Nutritionists use the word "Calorie" (with a capital C) to mean the same thing as "kilocalorie"; that is, 1 Cal = 1000 cal = 1 kcal. The calorie is not part of the SI. The official SI unit for heat is the **joule (J)**, which is about one fourth as big as the calorie:

Calories and joules are units not only of heat, but of other forms of energy as well.

$$1 \text{ cal} = 4.184 \text{ J}$$

Specific Heat

As we noted, it takes 1 cal to raise the temperature of 1 g of water by 1°C. **The amount of heat (in calories) necessary to raise the temperature of 1 g of any substance by 1°C** is called the **specific heat** of that substance. Each substance has its own specific heat, which is a physical property of that substance, like density or melting point. Specific heats for a few common substances are shown in Table 1.4. For ex-

Water has the highest specific heat of any common substance.

Table 1.4 Specific Heat Values for Some Common Substances

Substance	Specific Heat (cal/g · deg)	Substance	Specific Heat (cal/g · deg)
Water	1.00	Wood (typical)	0.42
Ice	0.48	Glass (typical)	0.12
Steam	0.48	Rock (typical)	0.20
Iron	0.11	Ethyl alcohol	0.59
Aluminum	0.22	Methyl alcohol	0.61
Copper	0.092	Ether	0.56
Lead	0.038	Carbon tetrachloride	0.21

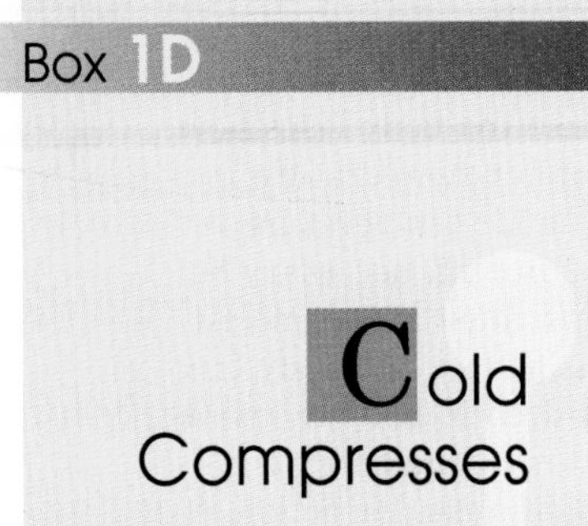

The high specific heat of water is useful in cold compresses and makes them last a long time. For example, consider two patients with cold compresses, one compress made by soaking a towel with water and the other made with ethyl alcohol, both at 0°C. Each gram of water in the water compress requires 25 cal to make the temperature of the compress rise to 25°C (after which it must be changed). Since the specific heat of ethyl alcohol is 0.59 cal/g · deg (Table 1.4), each gram of the alcohol requires only 15 cal to reach 25°C. If the two patients give off heat at the same rate, the alcohol compress is less effective because it will reach 25°C a good deal sooner than the water compress and will need to be changed sooner.

ample, the specific heat of copper is 0.092 cal/g · deg. Therefore, if we had 1 g of copper at 20°C, it would require only 0.092 cal to increase the temperature to 21°C. Note from Table 1.4 that ice and steam do not have the same specific heat as water.

It is easy to make calculations involving specific heats. The equation is

$$\text{Heat absorbed} = \text{specific heat} \times \text{mass} \times \text{change in temperature}$$

$$\text{Heat} = SH \times m \times (T_2 - T_1)$$

EXAMPLE

How many calories are required to heat 352 g of water from 23°C to 95°C?

Answer

$$\begin{aligned}\text{Heat} &= SH \times m \times (T_2 - T_1)\\ &= \frac{1.0\ \text{cal}}{\cancel{\text{g}} \cdot \cancel{\text{deg}}} \times 352\ \cancel{\text{g}} \times (95 - 23)\ \cancel{\text{deg}}\\ &= 2.5 \times 10^4\ \text{cal} = 25\ \text{kcal}\end{aligned}$$

Is it reasonable? Each gram of water requires 1 calorie to raise its temperature by one degree. We have approximately 350 g water. To raise its temperature by one degree would therefore require approximately 350 calories. But we are raising the temperature not by one degree but by approximately 70 degrees (from 23 to 95). Thus the total number of calories will be approximately 70 × 350 = 24,500 cal or 25 kcal. In this case the estimate comes out exactly the same as the calculated answer.

Problem **1.13**

How many calories are required to heat 731 g of water from 8 to 74°C? Check your answer to see if it is reasonable.

EXAMPLE

How many calories are required to heat 755 g of iron from 23 to 175°C?

Answer • According to Table 1.4, 0.11 cal/g · deg is the specific heat of iron.

$$\begin{aligned} \text{Heat} &= SH \times m \times (T_2 - T_1) \\ &= \frac{0.11 \text{ cal}}{\cancel{\text{g}} \cdot \cancel{\text{deg}}} \times 755 \cancel{\text{g}} \times (175 - 23) \cancel{\text{deg}} \\ &= 1.3 \times 10^4 \text{ cal} = 13 \text{ kcal} \end{aligned}$$

Is it reasonable? Once again, we may round off the numbers to get an approximate answer. The difference in temperature is about 150°C, and the specific heat of iron is about one tenth. Multiplying the two, we get 15. There are about 800 g of iron, so we should get about 15 × 800 = 12,000 cal or 12 kcal. The actual answer, 13 kcal, is reasonable.

Problem 1.14

How many calories are required to heat 26 g of aluminum from 20 to 115°C? Check your answer to see if it is reasonable.

EXAMPLE

If we add 450 cal of heat to 37 g of ethyl alcohol at 20°C, what is the final temperature?

Answer • The specific heat of ethyl alcohol is 0.59 cal/g · deg (Table 1.4).

$$\begin{aligned} \text{Heat} &= SH \times m \times (T_2 - T_1) \\ 450 \text{ cal} &= \frac{0.59 \text{ cal}}{\cancel{\text{g}} \cdot \text{deg}} \times 37 \cancel{\text{g}} \times (T_2 - T_1) \\ (T_2 - T_1) &= \frac{450 \cancel{\text{cal}}}{0.59 \cancel{\text{cal}}/\text{deg} \times 37} = 21 \text{ deg} \end{aligned}$$

Since the starting temperature is 20°C, the final temperature is 41°C.

Is it reasonable? The specific heat of ethyl alcohol is 0.59 cal/g · deg. This is close to 0.5, meaning that about half a calorie will raise the temperature of 1 g by 1 deg. 37 g of ethyl alcohol needs approximately 40 times as many calories for a 1-deg rise, and $40 \times \frac{1}{2} =$ 20 calories. But we are adding 450 calories, which is about 20 times as much. Thus, we expect the temperature to rise by about 20 deg, from 20 to 40 deg. The actual answer, 41 deg, is therefore quite reasonable.

Problem 1.15

It required 88.2 cal to heat 13.4 g of an unknown substance from 23 to 176°C. What is the specific heat of the unknown substance? Check your answer to see if it is reasonable.

SUMMARY

In order to understand modern medicine, it is necessary to know some chemistry. **Chemistry** is the science that deals with the structure of matter and the changes it can undergo. In **chemical changes** (**chemical reactions**) compounds or elements are used up and others are formed. In **physical changes** substances do not change their identity. There are three states of matter: **solid, liquid, and gas.**

The **scientific method** is a tool used in science and medicine. The heart of the scientific method is the testing of **hypotheses** and **theories.**

In chemistry we use the **metric system** for measurements. The main units are the meter for length, the liter for volume, and the gram for mass. Other units are indicated by prefixes that represent powers of ten. **Temperature** is measured in degrees Celsius or in kelvins. Conversions from one unit to another are best done by the **factor-label method,** in which units are multiplied and divided.

Density is mass per unit volume. **Specific gravity** is relative density and thus has no units. Density usually decreases with increasing temperature.

Kinetic energy is energy of motion; **potential energy** is stored energy. Energy can neither be created nor destroyed, but it can be converted from one form to another.

Heat is not the same as temperature. Heat is measured in calories. A calorie is the amount of heat necessary to raise the temperature of 1 g of water by 1°C. Every substance has a **specific heat,** which is a physical constant. The specific heat is the number of calories required to raise the temperature of 1 g of the substance by 1°C.

KEY TERMS

Chemical change (Sec. 1.1)
Chemical property (Sec. 1.1)
Chemical reaction (Sec. 1.1)
Chemistry (Sec. 1.1)
Coefficient (Sec. 1.3)
Conversion factor (Sec. 1.5)
Density (Sec. 1.7)
Energy (Sec. 1.8)
Exponential notation (Sec. 1.3)
Fact (Sec. 1.2)
Gas (Sec. 1.6)
Heat (Sec. 1.9)
Hyperthermia (Box 1C)
Hypothermia (Box 1C)
Hypothesis (Sec. 1.2)
Kinetic energy (Sec. 1.8)
Liquid (Sec. 1.6)
Mass (Sec. 1.4)
Matter (Sec. 1.1)
Metric system (Sec. 1.4)
Physical change (Sec. 1.1)
Physical property (Sec. 1.1)
Potential energy (Sec. 1.8)
Power (Sec. 1.3)
Scientific method (Sec. 1.2)
Scientific notation (Sec. 1.3)
SI (Sec. 1.4)
Solid (Sec. 1.6)
Specific gravity (Sec. 1.7)
Specific heat (Sec. 1.9)
Temperature (Secs. 1.4, 1.9)
Theory (Sec. 1.2)
Weight (Sec. 1.4)

PROBLEMS

Difficult problems are designated by an asterisk.

Chemistry and the Scientific Method

1.16 What subject does the science of chemistry deal with?
1.17 Why is the study of chemistry necessary for students planning a career in the health sciences?
1.18 Define (a) matter (b) chemistry
1.19 What is the difference between a hypothesis and a fact?
1.20 What does it take to get scientists to accept a theory? If they do accept it, does that make it correct?
1.21 Classify each as a chemical or physical change:
(a) burning gasoline
(b) making ice cubes
(c) boiling oil
(d) melting lead
(e) rusting iron
(f) making ammonia from nitrogen and hydrogen
(g) digesting food

Exponential Notation

1.22 Write in exponential notation:
(a) 37.5
(b) 62 900
(c) 0.091
(d) 0.00000028

1.23 Write out in full:
(a) 4.03×10^5 (c) 7.13×10^{-5}
(b) 3.2×10^3 (d) 5.55×10^{-10}

1.24 Multiply:
(a) $(4.73 \times 10^5)(1.37 \times 10^2)$
(b) $(2.7 \times 10^{-4})(5.9 \times 10^8)$
(c) $(6.49 \times 10^7)(7.22 \times 10^{-3})$
(d) $(3.4 \times 10^{-5})(8.2 \times 10^{-11})$

1.25 Divide:
(a) $\dfrac{6.02 \times 10^{23}}{3.01 \times 10^5}$ (d) $\dfrac{5.8 \times 10^{-6}}{6.6 \times 10^{-8}}$
(b) $\dfrac{3.14}{2.30 \times 10^{-5}}$ (e) $\dfrac{7.05 \times 10^{-3}}{4.51 \times 10^5}$
(c) $\dfrac{7.08 \times 10^{-8}}{300}$

1.26 Add:
(a) $(7.9 \times 10^4) + (5.2 \times 10^4)$
(b) $(8.73 \times 10^4) + (6.7 \times 10^3)$
(c) $(3.61 \times 10^{-4}) + (4.776 \times 10^{-3})$

1.27 Subtract:
(a) $(7.00 \times 10^5) - (4.20 \times 10^5)$
(b) $(5.82 \times 10^3) - (7.9 \times 10^2)$
(c) $(2.134 \times 10^{-7}) - (5.62 \times 10^{-8})$

***1.28** Solve:

$$\frac{(3.14 \times 10^3) \times (7.80 \times 10^5)}{(5.50 \times 10^2)}$$

***1.29** Solve:

$$\frac{(9.52 \times 10^4) \times (2.77 \times 10^{-5})}{(1.39 \times 10^7) \times (5.83 \times 10^{-3})}$$

Measurements

1.30 In the metric system, what is the basic unit of (a) mass (b) length (c) volume?

1.31 How many grams are there (a) in 1 kg (b) in 1 mg?

1.32 How many milliliters are there (a) in 1 L (b) in 1 cc?

1.33 For each of these, tell which answer is closest:
(a) a baseball bat has a length of 100 mm or 100 cm or 100 m
(b) a glass of milk holds 23 cc or 230 mL or 23 L
(c) a man weighs 75 mg or 75 g or 75 kg
(d) a tablespoon contains 15 mL or 150 mL or 1.5 L
(e) a paper clip weighs 50 mg or 50 g or 50 kg
(f) your hand has a width of 100 mm or 100 cm or 100 m
(g) an audiocassette weighs 40 mg or 40 g or 40 kg

1.34 What is the difference between mass and weight?

1.35 Convert to Celsius and to Kelvin:
(a) 320°F (c) 0°F
(b) 212°F (d) −250°F

1.36 Convert to Fahrenheit and to Kelvin:
(a) 25°C (c) 250°C
(b) 40°C (d) −273°C

Unit Conversions

1.37 Make the following conversions (conversion factors are given in Table 1.3):
(a) 42.6 kg to lb (f) 62 g to oz (avdp)
(b) 1.62 lb to g (g) 33.61 qt to L
(c) 34 in. to cm (h) 43.7 L to gal
(d) 37.2 km to miles (i) 1.1 miles to km
(e) 2.73 gal to L (j) 34.9 mL to fl oz

1.38 Make the following metric conversions:
(a) 96.4 mL to L (g) 0.044 L to mL
(b) 275 mm to cm (h) 711 g to kg
(c) 45.7 kg to g (i) 63.7 mL to cc
(d) 475 cm to m (j) 0.073 kg to mg
(e) 21.64 cc to mL (k) 83.4 m to mm
(f) 3.29 L to cc (l) 361 mg to g

1.39 What is your weight in kilograms? Your height in meters?

1.40 The speed limit in some European cities is 80 km/h. How many miles per hour is this?

1.41 The speed of light is 186 000 miles per second. What is it in cm/s?

Density and Specific Gravity

1.42 The density of a solid is 7.5 g/mL. What is its density in kg/L?

1.43 The density of manganese is 7.21 g/mL, that of calcium chloride is 2.15 g/mL, and that of sodium acetate is 1.528 g/mL. If you place these three solids in a liquid that has a density of 2.15 g/mL, which will sink to the bottom, which will stay on the top, and which will stay in the middle of the liquid?

1.44 The density of titanium is 4.54 g/mL. What is the volume, in mL, of 163 g of titanium?

1.45 A 335.0-cc sample of urine has a mass of 342.6 g. What is the density, in g/mL, to three decimal places?

1.46 The density of methanol at 20°C is 0.791 g/mL. What is the mass, in g, of a 280-mL sample?

1.47 The density of dichloromethane, a liquid insoluble in water, is 1.33 g/cc. If dichloromethane and water are placed in a separatory funnel, which will be the upper layer?

***1.48** A sample of 10.00 g of oxygen has a volume of 6702 mL. The same weight of carbon dioxide occupies 5058 mL. (a) What are the densities in g/L? (b) Carbon dioxide is used as a fire extinguisher to cut off the fire's supply of oxygen. Do the densities of these two gases explain the fire-extinguishing ability of carbon dioxide?

***1.49** Two 1-quart containers of butter are for sale. Butter A has a density of 1.2 g/mL and costs $5.00; butter B has a density of 0.96 g/mL and costs $4.00. Which is a better buy?

Energy

1.50 What is the difference between kinetic and potential energy?

1.51 How many calories are required to heat 39.0 g of water from 75 to 130°F?

1.52 How many calories are required to heat the following (specific heats are given in Table 1.4)?
(a) 52.7 g of aluminum from 100 to 285°C
(b) 93.6 g of methyl alcohol from −35 to 55°C
(c) 3.4 kg of lead from −33 to 730°C
(d) 71.4 g of ice from −77 to −5°C

*__1.53__ If 168 g of an unknown liquid requires 2750 cal of heat to raise its temperature from 26 to 74°C, what is the specific heat of the liquid?

Boxes

1.54 (Box 1A) Suppose that in examining the patient described in Box 1A the doctors had found normal lead levels in the blood. What would they conclude about their hypothesis of lead poisoning?

1.55 (Box 1A) How did the physicians mentioned in Box 1A test their hypothesis?

1.56 (Box 1B) If the recommended dose of a drug is 3.7 mg for each kilogram of body weight, how many mg should be prescribed for a 114-lb woman?

1.57 (Box 1B) If the recommended dose of a drug is 445 mg for a 180-lb man, what would be a suitable dose for a 135-lb man?

1.58 (Box 1B) The average lethal dose of heroin is 150 mg per kilogram of body weight. Estimate how many grams of heroin would be lethal for a 200-lb man.

1.59 (Box 1C) How does the body react to hypothermia?

1.60 (Box 1C) Low temperatures often cause people to shiver. What is the function of this involuntary body action?

1.61 (Box 1D) Which would make a more efficient cold compress, ethyl alcohol or methyl alcohol? (Refer to Table 1.4.)

Appendix: Significant Figures

1.62 How many significant figures are there in:
(a) 3216
(b) 0.007
(c) 0.0910
(d) 7.3004
(e) 16.104
(f) 0.0001
(g) 0.0000900

1.63 How many significant figures are there in:
(a) 5.71×10^{13}
(b) 4.4×10^{5}
(c) 3×10^{-6}
(d) 4.000×10^{-11}
(e) 5.5550×10^{-3}

1.64 Round off to two significant figures:
(a) 91.621
(b) 7.329
(c) 0.677
(d) 0.003249
(e) 5.88

1.65 Multiply these numbers, using the correct number of significant figures in your answer:
(a) 3630.15×6.8
(b) 512×0.0081
(c) $5.79 \times 1.85825 \times 1.4381$

1.66 Divide these numbers, using the correct number of significant figures in your answer:
(a) $\dfrac{3.185}{2.08}$
(b) $\dfrac{6.5}{3.0012}$
(c) $\dfrac{0.0035}{7.348}$

1.67 Add these groups of measured numbers:
(a)
37.4083
5.404
10916.3
3.94
0.0006

(b)
84
8.215
0.01
151.7

(c)
51.51
100.27
16.878
3.6817

1.68 A student had to solve the following equation, where all the quantities were measured numbers:

$$x = \frac{19.227 \times 4.300}{0.000450 \times 370.00}$$

Her calculator gave the answer $x = 496.5531532$. What answer should she report to her chemistry teacher?

Additional Problems

1.69 If 80 g of an unknown substance released 3000 cal of heat when it was cooled from 100 to 50°C, what was the specific heat of the substance?

1.70 The *meter* is a measure of length. Tell what each of the following units measures:
(a) cm^3
(b) mL
(c) kg
(d) cal
(e) g/cc
(f) joule
(g) °C
(h) cm/s

1.71 A brain weighing 1 pound occupied a volume of 620 mL. What is the specific gravity of the brain?

*__1.72__ If the density of air is 1.25×10^{-3} g/cc, what is the mass in kilograms of the air in a room that is 5.3 m long, 4.2 m wide, and 2.0 m high?

*__1.73__ Is it possible to convert 83 mL to grams?

*__1.74__ The kinetic energy possessed by an object with a mass of 1 gram moving with a velocity of 1 cm/s is called **1 erg.** What is the kinetic energy, in ergs, of an athlete with a mass of 127 lb running at a velocity of 14.7 miles/hr?

1.75 What will the final temperature be if, at room temperature (25°C), 1500 cal are added to a sample of 200.0 g each of: (a) water (b) ethyl alcohol (c) iron (d) lead?

1.76 Classify these as kinetic or potential energy:
(a) water held by a dam
(b) a speeding train
(c) a book on its edge before falling
(d) a falling book
(e) electric current in a light bulb

1.77 Make the following conversions (conversion factors are given in Table 1.3):
(a) 8854 inches to km
(b) 394 mL to quarts
(c) 2.66 kilograms to oz (avdp)
(d) 278 grains to pounds

1.78 When the astronauts walked on the moon they could make giant leaps in spite of their heavy gear.

(a) Why were their weights on the moon so low?
(b) Were their masses different on the moon than on the earth?

1.79 Which of the following is the largest mass and which the smallest?
(a) 41 g
(b) 3×10^3 mg
(c) 8.2×10^6 μg
(d) 4.1×10^{-8} kg

1.80 What is the difference between heat and temperature?

1.81 In Japan, high-speed bullet trains move with an average speed of 220 km/h. If Dallas and Los Angeles were connected by such a train, how long would it take to travel nonstop between these cities (1490 miles)?

*__1.82__ The specific heats of some elements at 25°C are: aluminum = 0.215 cal/g · deg; carbon (graphite) = 0.170 cal/g · deg; iron = 0.107 cal/g · deg; mercury = 0.0331 cal/g · deg. (a) Which element would require the smallest amount of heat to raise the temperature of 100 g of the element by 10°C? (b) If the same amount of heat needed to raise the temperature of 1 g of aluminum by 25°C were applied to 1 g of mercury, by how many degrees would its temperature be raised? (c) If a certain amount of heat is used to raise the temperature of 1.6 g of iron by 10°C, the temperature of 1 g of which element would also be raised by 10°C, using the same amount of heat?

1.83 The density of gasoline is 0.71 g/mL. What is this in lb/gal?

1.84 One quart of milk costs 80 cents and one liter costs 86 cents. Which is the better buy?

*__1.85__ (a) If 1.00 mL of butter, density 0.860 g/ml, is thoroughly mixed with 1.00 mL of sand, density 2.28 g/ml, what is the density of the mixture? (b) What would be the density of the mixture if 1.00 g of the same butter were mixed with 1.00 g of the same sand?

1.86 Which speed is the fastest?
(a) 70 miles/h
(b) 140 km/h
(c) 4.5 km/s
(d) 48 miles/min

Chapter **2**

Atoms

2.1 Introduction

What is matter made of? This question was discussed for thousands of years, long before human beings had any reasonable way of getting an answer. In ancient Greece two possible answers were given, each having its own group of supporters. One group, led by a scholar named Democritus (about 470–380 BC), believed that all matter was made of very small particles—much too small to see. Democritus called these particles *atoms*. Some of his followers developed the idea that there were different kinds of atoms, with different properties, and that the properties of the atoms caused ordinary matter to have the properties we all know.

They said that water is a liquid because its atoms are smooth and round and can easily glide over each other, whereas atoms of iron are rough and hard.

But these ideas were not accepted by all ancient thinkers. Another group, led by Zeno of Elea (born about 450 BC), did not believe in atoms at all. They insisted that matter was infinitely divisible; that is, if you took any object, such as a coin or a piece of wood, or a liquid such as water or olive oil, you could cut it or otherwise divide it into two parts, divide each of these into two more parts, and continue the process forever. According to Zeno and his followers, you would never reach a particle of matter that could no longer be divided.

Today we know that Democritus was right and Zeno was wrong, but there is a great difference in the way we look at this question. Today our ideas are based on *evidence*. Democritus had no evidence to prove that matter cannot be divided an infinite number of times, just as Zeno had no evidence for his claim that matter can be divided infinitely. Both claims were based not on evidence but on visionary beliefs: one in unity, the other in diversity.

In Section 2.3 we will discuss the evidence for the existence of atoms, but first we need to look at the diverse forms of matter.

2.2 Classifications of Matter

Matter can be divided into three classes: elements, compounds, and mixtures.

Elements

An **element** is **a pure substance that cannot be broken down into simpler substances by chemical reactions.** At this time, 112 elements are known. Of these, 88 occur in nature; the others have been made by chemists. The elements are the building blocks of all matter. A list of the known elements is given on the inside back cover of this book, along with their symbols. These symbols, consisting of one, two, or three letters, are the same in all countries. Most of the symbols correspond directly to the name in English (for example, C for carbon, H for hydrogen, Li for lithium), but a few come from the Latin or German names. Photographs of some elements are shown in Figure 2.1. Some of the more important symbols are listed in Table 2.1.

The elements technetium, promethium, astatine, francium, and all those in the periodic table higher than uranium do not occur naturally on earth, but have all been synthesized by chemists and physicists.

The symbol for tungsten, W, comes from the German *wolfram*. The symbol for iron, Fe, comes from the Latin *ferrum*.

Phosphorus (P)

Magnesium (Mg) (left), calcium (Ca) (right)

Clockwise from the top: lead (Pb), tin (Sn), carbon (C), silicon (Si)

Arsenic (As) (left), antimony (Sb) (right), bismuth (Bi) (top)

Nitrogen (N_2)

Zinc (Zn) (left), mercury (Hg) (right)

Figure **2.1** • Some elements. (Photographs by Charles D. Winters.)

Compounds

A **compound** is **a pure substance made up of two or more elements in a fixed proportion by weight.** For example, water is a compound made up of hydrogen and oxygen, and table salt is a compound made up of sodium and chlorine. There are millions of known compounds, many of which we will meet in this book.

Box 2A

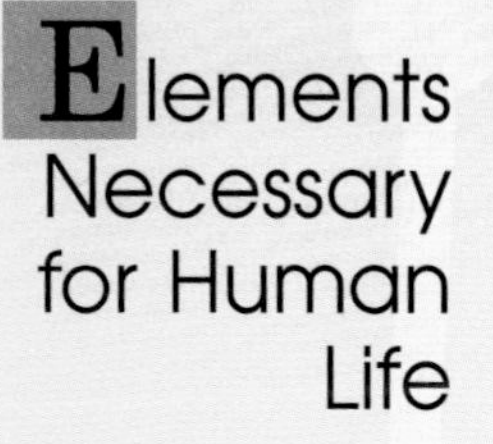

Elements Necessary for Human Life

Only a few elements are necessary for human life. The four most important of these—carbon, hydrogen, oxygen, and nitrogen—make up the subjects of organic chemistry and biochemistry (Chapters 10–26), so we need not consider them here. Seven other elements are also quite important, and besides these there are at least nine additional ones (trace elements) that our bodies use in very small quantities. The table shows these 20 elements and their functions in the human organism. Many of these are more fully discussed later in the book. For the average daily requirements, sources in foods, and symptoms of deficiencies, see Chapter 26.

Element	Function
The Big Four	
Carbon Hydrogen Oxygen Nitrogen	Discussed in Chapters 10–26
The Next Seven	
Calcium	Strengthens bones and teeth; aids in blood clotting
Phosphorus	Present in phosphates, which regulate neutrality of body fluids and are involved in energy transfer
Potassium	Helps regulate electrical balance of body fluids
Sulfur	An essential component of proteins; important in protein structure
Chlorine	Helps regulate electrical balance of body fluids
Sodium	Helps regulate electrical balance of body fluids
Magnesium	Helps nerve and muscle action; present in bones
The Trace Elements	
Chromium	Increases effectiveness of insulin
Cobalt	A part of vitamin B_{12}
Copper	Strengthens bones; assists enzyme action
Fluorine	Reduces incidence of dental cavities
Iodine	Necessary for thyroid function
Iron	An essential part of such proteins as hemoglobin, myoglobin, cytochromes, and FeS protein
Manganese	Present in bone-forming enzymes; aids fat and carbohydrate metabolism
Molybdenum	Necessary for activity of certain enzymes
Zinc	Necessary for normal growth

Mixtures

Elements and compounds are pure substances. The third category of matter consists of **mixtures** of pure substances. In our daily life most of the matter we meet (including our own bodies) consists of mixtures rather than pure substances. For example, blood, butter, gasoline, soap, the metal

Table 2.1 Names and Symbols of Some Important Elements

Symbol	Element	Symbol	Element
Al	Aluminum	Hg	Mercury
As	Arsenic	Mo	Molybdenum
B	Boron	Ne	Neon
Br	Bromine	N	Nitrogen
Ca	Calcium	O	Oxygen
C	Carbon	P	Phosphorus
Cl	Chlorine	Pt	Platinum
Cr	Chromium	Pu	Plutonium
Co	Cobalt	K	Potassium
Cu	Copper	Se	Selenium
F	Fluorine	Si	Silicon
Au	Gold	Ag	Silver
He	Helium	Na	Sodium
H	Hydrogen	Sr	Strontium
I	Iodine	S	Sulfur
Fe	Iron	Sn	Tin
Pb	Lead	U	Uranium
Mg	Magnesium	V	Vanadium
Mn	Manganese	Zn	Zinc

Students will do well to learn these symbols.

in a wedding ring, the air we breathe, and the earth we walk on are all mixtures of pure substances.

The difference between mixtures and compounds can be illustrated by a simple experiment. The elements zinc and sulfur are shown in Figure 2.2(a) and (b). If we mix them, Figure 2.2(c), we have a *mixture* of zinc and sulfur. We can put the mixture in a blender, so as to make it a very thorough mixture, with the particles evenly distributed, but it still consists of tiny pieces of sulfur and tiny pieces of zinc. On the other hand, if in this case we heat the mixture of zinc and sulfur, we get the *compound* zinc sulfide, shown in Figure 2.2(d). This compound is an entirely different substance, with properties that are completely different from those of either zinc or sulfur. An important difference between a compound and a mixture is that the proportions by weight of the elements in a compound are fixed (see Sec. 2.3), whereas, for example, in a mixture of zinc and sulfur, the two elements can be mixed in any proportion.

A mixture of iron and sulfur can be separated by a magnet, which attracts only the iron. A magnet applied to the compound iron sulfide will not separate the iron from the sulfur. (Photograph by Charles D. Winters.)

One of the most important tasks of chemists is the separation of the components of a mixture. Many methods are known, all of which rely on some difference in the properties of the components. For example, one such method is **distillation,** used to separate the components of a mixture of liquids. Here we take advantage of the difference in boiling points. The mixture is heated until the liquid with the lower boiling point evaporates. The vapor is then condensed to form a liquid. The liquid with the higher boiling point remains in the flask. A distillation apparatus is shown in Figure 2.3. You will probably encounter other methods of separation in the laboratory.

Note that elements present in a compound cannot be separated by such methods.

Figure **2.2**
(a) Zinc; (b) sulfur; (c) a mixture of zinc and sulfur. Both yellow sulfur and gray zinc particles are present; (d) the compound zinc sulfide. Neither gray nor yellow particles are present. The compound is white. (Photographs by Beverly March.)

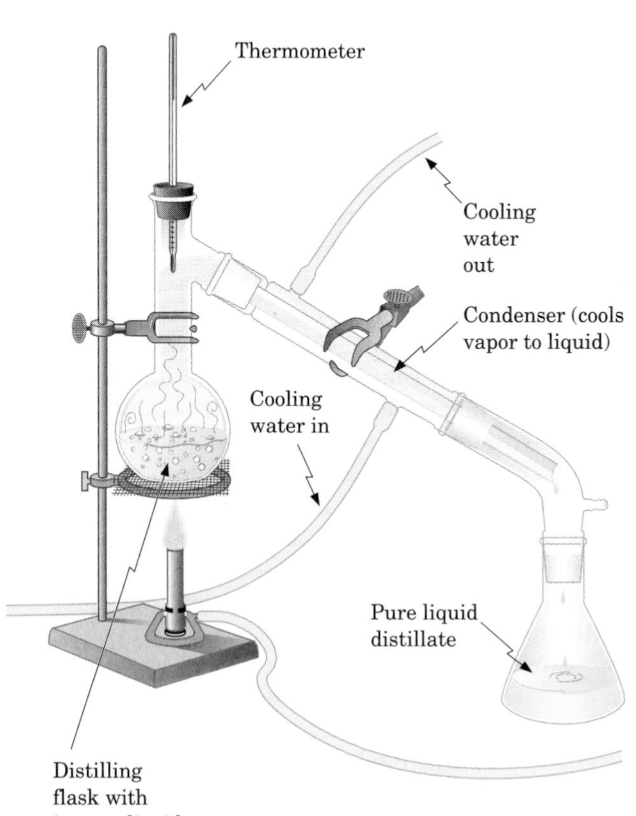

Figure **2.3**
A laboratory distillation apparatus. The liquid mixture in the flask on the left is heated until it boils. The component with the lower boiling point evaporates first, and the resulting vapor is condensed in the water-cooled condenser and collected in the flask on the right.

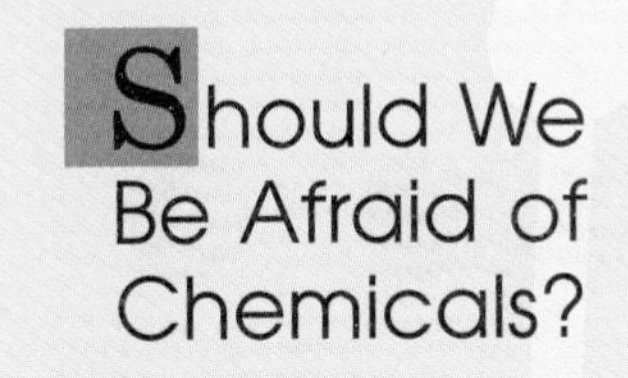

Box 2B

Should We Be Afraid of Chemicals?

Are chemicals dangerous? Many people, seeing something like "one tenth of one percent sodium sulfite," "carrageenan," or "carboxymethylcellulose" on a package of food will automatically react negatively, as if any chemical with a hard-to-pronounce name must be harmful. This reaction, while understandable in people who know little about chemistry, is quite mistaken. Actually, all the food we eat is made entirely of chemicals, and so are the trees and grass, the ground itself, all animals, and even our bodies. In fact all matter is chemical.

To answer our opening question, some chemicals are dangerous; others are not; still others are essential to life (oxygen is a chemical). In many cases, a small amount of a chemical is essential, while a larger amount of the same chemical may well be toxic.

The important thing is not to be frightened by words just because they sound unfamiliar. The names of chemicals are often long and, to the student, hard to pronounce. The reason is that millions of chemicals are known, and each must be given a different name, with the name in most cases carrying enough information for chemists to be able to tell the composition from the name. An example is the chemical 1,3,3-trimethyl-2-(9-hydroxy-3,7-dimethylnona-1,3,5,7-tetraenyl)cyclohexene. Far from being harmful, this chemical, also known as Vitamin A, is essential to human life (although excessive amounts can indeed be harmful; see Box 26B).

The intelligent approach to chemicals is to obtain an understanding of what the substances are and what they can do. In this book we will be doing just that.

2.3 Dalton's Atomic Theory

The atomic theory of Democritus (Sec. 2.1) was vague. Around 1805 an English chemist, John Dalton (1766–1844), put forth a scientific atomic theory. The big difference between Dalton's theory and that of Democritus was that Dalton based his on evidence. First let us state the theory and then see what kind of evidence supported it.

1. All matter is made up of very tiny indivisible particles, which Dalton called **atoms,** using the same name that Democritus had used.

An atom is the smallest particle of an element that shows the chemical behavior of that element.

2. All atoms of the same element are identical to each other.

3. Atoms of any one element are different from those of any other element.

4. Atoms combine to form molecules. A **molecule** is a tightly bound group of atoms that acts as a unit.

The Evidence for Dalton's Atomic Theory

A. The great French chemist Antoine Laurent Lavoisier (1743–1794) discovered the *law of conservation of mass.* This says that **matter can neither be created nor destroyed.** Lavoisier proved this by many experiments in which he showed that, no matter what chemical changes took place, the total mass of the matter at the end was exactly the same as at the beginning. Dalton's theory explained this fact very nicely. If all matter is made up of indestructible atoms, then any chemical change just changes the attachments between atoms but does not destroy the atoms themselves. For example, carbon monoxide reacts with lead oxide to give carbon dioxide and lead. This is how Dalton might have explained what happens:

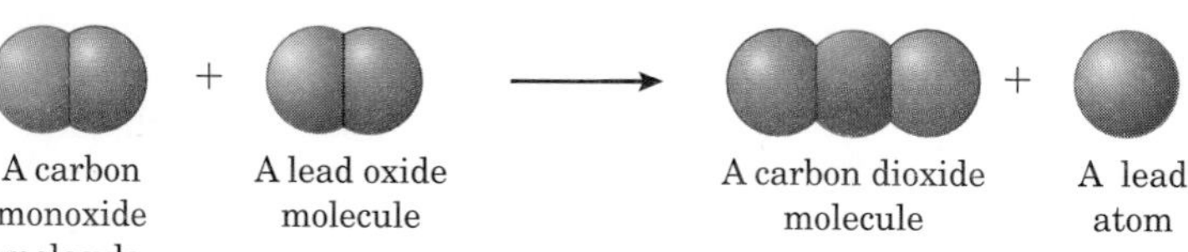

The real atoms, of course, do not have these colors.

In this figure, the carbon atom is shown as a red ball, the lead atom as a black ball, and the oxygen atoms as blue balls. All the original atoms are still there in the end; they have only changed partners. This explains why the total mass at the end is the same as at the beginning.

B. Before Dalton presented his theory, another French chemist, Joseph Proust (1754–1826), had demonstrated the *law of constant composition,* which states that **any compound is always made up of elements in the same proportion by mass.** For example, the compound water contains 89 percent oxygen and 11 percent hydrogen by mass. Proust said that all pure water contains these two elements in these exact proportions, whether the water comes from the Atlantic Ocean or the Seine River or is collected as rain, squeezed out of a watermelon, or distilled from urine. Water is never formed in any other proportions. This fact was also evidence for Dalton's theory. If a water molecule is made up of two atoms of hydrogen and one of oxygen, and if an oxygen atom has a mass of 16 times that of a hydrogen atom (as today we know it has), water must always contain 89 percent oxygen and 11 percent hydrogen and can never be found in any other proportions.

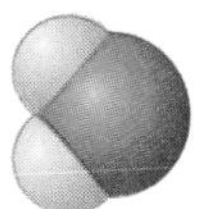

Monatomic, Diatomic, and Polyatomic Elements

Another form of oxygen is a gas called **ozone,** whose formula is O_3.

Some elements—for example, helium and neon—consist of single atoms not connected to each other. In contrast, oxygen, in its most common form, contains two atoms in each molecule, connected to each other by chemical bonds. We write this as O_2, the little number (subscript) telling us how many atoms are in the molecule. Other elements that occur as diatomic molecules are hydrogen (H_2), nitrogen (N_2), fluorine (F_2), chlorine (Cl_2), bromine (Br_2), and iodine (I_2). It is important to understand that under normal conditions the free atoms O, H, N, F, Cl, Br, and I do not exist. These elements occur only as diatomic molecules.

Some elements have even more atoms in each molecule. In one form of phosphorus, each molecule has four atoms (P_4); sulfur forms S_8 molecules. There are even elements whose molecules are much larger than this. For example, a molecule of carbon (a diamond) has millions of carbon atoms all bonded together in a gigantic cluster. Just the same, all these are elements because they all contain only one kind of atom.

2.4 Inside the Atom

Today we know that matter is more complex than Dalton believed. A wealth of evidence obtained over the last 100 years or so has convinced us that atoms are not indivisible, but are made up of even smaller particles. There are **three elementary particles that make up all atoms:** *protons, electrons,* and *neutrons.* The main things we know about these three particles are their charge and their mass (Table 2.2).

There are many other subatomic particles, but we do not deal with them in this book.

Protons and electrons carry electric charges. There are two kinds of charges, which we call *positive* and *negative.* The only way we can tell that they exist is that positive charges repel other positive charges, negative charges repel other negative charges, and positive charges attract negative charges. In order words, **like charges repel and unlike charges attract.**

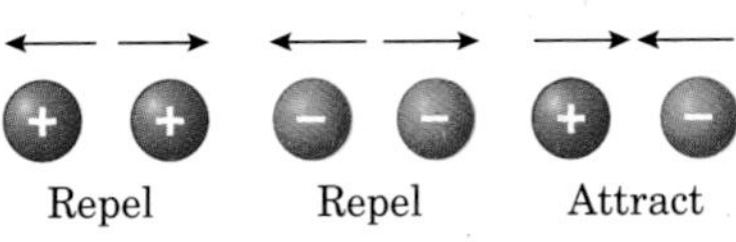

A **proton** has a positive charge. By convention we say that the magnitude of the charge is 1. Thus, the proton has a charge of +1 (two protons have a total charge of +2, and so forth). The mass of a proton is 1.673×10^{-24} g, but this number is so small that it is more convenient to use another unit, called the atomic mass unit (amu):

$$1 \text{ amu} = 1.67 \times 10^{-24} \text{ g}$$

The proton has a mass of approximately 1 amu.

To be more exact, the mass of the proton is 1.0073 amu.

The **electron** has a charge of −1, equal in magnitude to the charge on the proton but opposite in sign. However, the mass of the electron is much less, about 1/1835 that of the proton. It takes about 1835 electrons to equal the mass of one proton.

Neutrons have no charge at all. Therefore they do not attract or repel each other or any other particle. The mass of the neutron is slightly greater than that of the proton: 1.675×10^{-24} g. This mass is so close to that of the proton, however, that for most purposes we say that these two particles have the same mass: 1 amu.

The exact mass of the neutron is 1.0087 amu.

Table 2.2 Properties and Location of the Elementary Particles

	Mass (g)	Mass (amu)	Charge	Found in
Proton	1.673×10^{-24}	1	+1	Nucleus
Electron	9.110×10^{-28}	1/1835	−1	Outside of nucleus
Neutron	1.675×10^{-24}	1	0	Nucleus

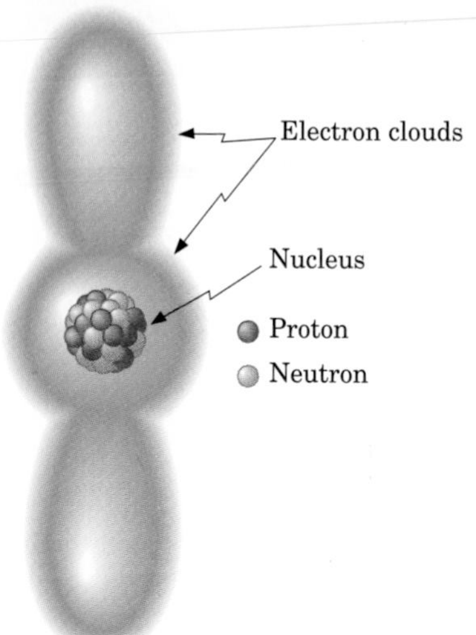

Figure **2.4**
A typical atom. Protons and neutrons make up the nucleus. The electrons are found as clouds outside the nucleus. This figure is not drawn to scale.

These are the three particles that make up atoms, but where are they found? Protons and neutrons are found in a tight cluster in the center of the atom (Fig. 2.4). This part of the atom is called the nucleus, and both protons and neutrons are called **nucleons.** Electrons are found as a diffuse cloud outside the nucleus. We will discuss the nucleus in greater detail in Chapter 9.

Mass Number

Each atom has a fixed number of protons, electrons, and neutrons. **The total number of protons and neutrons (nucleons) in any atom** is called the **mass number.** For example, an atom with 5 protons, 5 electrons, and 6 neutrons has a mass number of 11 (electrons are not counted in determining mass number).

EXAMPLE

What is the mass number of (a) an atom with 58 protons, 58 electrons, and 78 neutrons and (b) an atom with 17 protons, 17 electrons, and 20 neutrons?

Answer

(a) The mass number is $58 + 78 = 136$.
(b) The mass number is $17 + 20 = 37$.

In each case we add the number of protons and neutrons and ignore the electrons.

Problem **2.1**

What is the mass number of (a) an atom with 47 protons, 47 electrons, and 62 neutrons and (b) an atom with 91 protons, 91 electrons, and 139 neutrons?

Atomic Number

Atomic numbers for all the elements are given in the atomic weight table on the inside back cover.

The number of protons contained in any atom is very important. *It determines what element the atom is.* For example, any atom with 7 protons is a nitrogen atom; if an atom has 24 protons it is a chromium atom, and if there are 80 protons it is an atom of mercury. This is true no matter how many electrons or neutrons are present. The identity of an element is established only by the number of protons in the atom. We call this number the **atomic number** of the element. At the present time 112 elements have been discovered, and these elements have atomic numbers from 1 to 112. The smallest atomic number belongs to the element hydrogen, which has only 1 proton, and the largest (so far) to the element ununbium (Uub), with 112 protons.

EXAMPLE

Name the elements mentioned in the previous example.

Answer

(a) Since there are 58 protons, we see from the atomic weight table on the inside back cover that the atom is a cerium atom (Ce), atomic number 58.
(b) This atom has 17 protons, making it a chlorine (Cl) atom.

Problem **2.2**

Name the elements mentioned in Problem 2.1.

EXAMPLE

(a) What are the atomic numbers of the elements sodium and bromine? (b) How many protons does an atom of each have?

Answer

(a) Atomic numbers are most easily found in the list of elements on the inside back cover. A look at this table shows the atomic numbers of sodium (Na) and bromine (Br) to be 11 and 35, respectively.
(b) This means that an atom of Na has 11 protons, and an atom of Br has 35 protons.

Problem **2.3**

(a) What are the atomic numbers of zirconium (Zr) and mercury (Hg)?
(b) How many protons does an atom of each have?

Ions

Atoms are electrically neutral; that is, they have no charge. This means that the number of electrons in any atom must be equal to the number of protons. The mercury atom, for example, with 80 protons, also has 80 electrons. So why don't we say that the number of electrons establishes the identity of the atom? The reason is that atoms can lose or gain electrons but do not thereby change to other elements. If a mercury atom, with 80 protons and 80 electrons, loses two electrons, it becomes a mercury *ion,* with 80 protons and 78 electrons. We no longer call it an atom, but it is still mercury because it still has 80 protons. We call it an **ion,** which is defined as **a particle with unequal numbers of protons and electrons.** Atoms do not normally lose or gain protons or neutrons, only electrons. The mercury ion we just mentioned has 78 electrons and 80 protons. The total charge of the 78 electrons (−78) is canceled by the charge on 78 of the protons (+78), but there are no electrons present to cancel the other two positive charges. Therefore this particular mercury ion has a charge of +2, and we write it as Hg^{2+}.

A sulfide ion, S^{2-}, has 16 protons and 18 electrons.

If in an ion there are more protons than electrons, the ion has a positive charge. If there are more electrons than protons, the ion has a negative charge. We will discuss ions again in Section 3.2.

Isotopes

The fact that isotopes exist means that the second statement of Dalton's theory (Sec 2.3) is not correct.

Although we can say that a mercury atom or ion always has 80 protons and a neutral mercury atom has 80 electrons, we cannot say that a mercury atom (or ion) must have any particular number of neutrons. Some of the mercury atoms found in nature have 122 neutrons; the mass number of these atoms is 202, and they are designated as either ^{202}Hg or mercury-202. Other mercury atoms have only 120 neutrons (and therefore a mass number of 200). Still others have 118, 119, or 121 neutrons, and others as many as 124. **Atoms with the same number of protons but different numbers of neutrons** are called **isotopes.** All isotopes of mercury contain 80 protons (or they wouldn't be mercury). If they are neutral atoms, they all contain 80 electrons. Each isotope contains a different number of neutrons, however, and therefore has a different mass number.

EXAMPLE

How many neutrons are there in each of the following isotopes of oxygen: (a) oxygen-16 (b) oxygen-17 (c) oxygen-18?

Answer • Each oxygen atom has eight protons. The difference between the mass number and the number of protons gives the number of neutrons: (a) $16 - 8 = 8$; (b) $17 - 8 = 9$; (c) $18 - 8 = 10$.

Problem 2.4

How many neutrons has each of the following isotopes of selenium: (a) selenium-80 (b) selenium-82 (c) selenium-84?

The properties of isotopes of the same element are almost identical, and for most purposes we regard them as identical. (They differ in radioactivity properties, which we will discuss in Chapter 9.)

Atomic Weight

Strictly speaking, the atomic weight of an element is a ratio of the weighted average mass of the atom compared with the mass of the carbon isotope whose mass number is 12 and whose mass is taken to be 12.0000 amu. Thus the amu is defined as $\frac{1}{12}$ the mass of a ^{12}C atom.

Most elements are found on earth as mixtures of isotopes, in a more or less constant ratio. For example, any sample of the element chlorine contains 75.5 percent chlorine-35 (18 neutrons) and 24.5 percent chlorine-37 (20 neutrons). Silicon is composed of a fixed ratio of three isotopes, with 14, 15, and 16 neutrons. For some elements these ratios may vary slightly, but for most purposes the slight variations can be ignored. The **atomic weight** of an element is a **weighted average of the masses of the isotopes** (the mass of an isotope is approximately the same as its mass number). Thus the atomic weight of chlorine is 35.5 amu, which is a weighted average of the masses of the chlorine atoms, 75.5 percent of which have a mass number of 35 and 24.5 percent of which have a mass number of 37:

$$\left(\frac{75.5}{100} \times 35.0 \text{ amu}\right) + \left(\frac{24.5}{100} \times 37.0 \text{ amu}\right) = 35.5 \text{ amu}$$

This illustrates how to calculate a weighted average.

Some elements—for example, gold, fluorine, and aluminum—occur naturally as only one isotope. The atomic weights of these elements are of course close to whole numbers (gold, 196.97; fluorine, 18.998; aluminum, 26.98).

The atomic weight is a very important property of an element, and we shall make much use of it in this course. A table of atomic weights is found on the inside back cover of this book.

Some biochemists use **daltons** as a unit of atomic weight.

1 amu = 1 dalton

EXAMPLE

Magnesium has three stable isotopes, with abundances as follows: ^{24}Magnesium—78.6%; ^{25}magnesium—10.1%, and ^{26}magnesium—11.3%. Calculate the atomic weight of magnesium.

Answer • To calculate the weighted average of the masses of the isotopes, we must multiply each atomic mass by its abundance and add these together:

$$\left(\frac{78.6}{100} \times 24.0 \text{ amu}\right) + \left(\frac{10.1}{100} \times 25.0 \text{ amu}\right) + \left(\frac{11.3}{100} \times 26.0 \text{ amu}\right)$$

$$= 24.3 \text{ amu}$$

Problem **2.5**

The atomic weight of lithium is 6.941. Lithium has only two naturally occurring isotopes, lithium-6 and lithium-7. Estimate which one is in greater abundance.

2.5 The Periodic Table

In the 1860s a number of scientists, most notably the Russian Dmitri Mendeleev (1834–1907), produced periodic tables. In its modern form (see the inside front cover), the periodic table consists of all the elements arranged in order of atomic number. When this is done, we find that the elements fall into rows, in such a way that elements in the same vertical column (*group*) have similar properties. This is best understood by looking at specific examples in the table. The elements fluorine (atomic number 9), chlorine (17), bromine (35), iodine (53), and astatine (85) are all in the same column in the table. These elements are all called *halogens.* They are all colored substances, and the colors deepen as we go

Dmitri Mendeleev. (Courtesy of E. F. Smith Memorial Collection, Special Collection Department, Van Pelt Library, University of Pennsylvania.)

Note that hydrogen (H) appears in columns IA and VIIA but is neither a halogen nor an alkali metal.

Figure **2.5**
Three halogens: chlorine, a gas; bromine, a liquid; and iodine, a solid. (Photograph by Larry Cameron.)

"E" is a general symbol meaning element.

The elements of group 0 were not yet discovered when Mendeleev first made his table, so this column does not have a Roman numeral.

down the table (see Fig. 2.5). All form compounds with sodium that have the formula NaE (for example, NaCl, NaBr) and not NaE_2, Na_2E, Na_3E, or anything else. Elements not in this column do not share these properties.

At this point we must say a word about the numbering of the columns of the periodic table. Mendeleev gave them Roman numerals and added the letter A for some columns and B for others. These numbers have been used for many years. Recently, however, because U.S. and European chemists used the A and B in different ways, international and U.S. chemistry organizations have agreed to recommend a different set of numbers, 1 to 18, going from left to right. Thus, by Mendeleev's numbering, the halogens are in group VIIA, but in the new numbering they are in group 17. Although in this book we use the traditional numbering, both kinds are shown on the periodic table on the inside front cover.

Not only do the elements in any one column have similar properties, but the properties vary in some fairly regular way as we go up or down a column. As shown in Table 2.3, the melting and boiling points regularly increase as we go down a column. Another example is found in group IA. Once again, the elements here (called *alkali metals*) all share similar properties. All are metals with low melting points and are soft enough to be cut with a knife. All react with water (Fig. 2.6), and all form compounds with fluorine, whose formula is EF. Also, as we go down the column, the melting and boiling points of the alkali metals change in a reg-

Table **2.3** Melting and Boiling Points of Group VIIA Elements

Element	Melting Point (°C)	Boiling Point (°C)
Fluorine	−220	−188
Chlorine	−110	−35
Bromine	−7	59
Iodine	114	184
Astatine	302	337

Box 2C

Strontium-90

The similarity of properties shown by elements in the same column of the periodic table can have biological consequences. One important, and troubling, example comes about because the properties of the element strontium are fairly similar to those of calcium, which is just above it on the table. Calcium is an important element for humans, since our bones and teeth consist largely of calcium compounds. We need some of this mineral in our diet every day, and we get it mostly from milk, cheese, and other dairy products, where it is present in the form of calcium ions.

Unfortunately, one of the products released by nuclear explosions is an isotope of the element strontium, strontium-90. This isotope is radioactive, with a half-life of 28.1 years. (Half-life is discussed in Section 9.3.) Strontium-90, present in the fallout from above-ground nuclear test explosions, was carried all over the earth and slowly settled to the ground, where it was eaten by cows and other animals. By this means it got into milk and eventually into our bodies. If it were not similar to calcium, our bodies would eliminate all of it within a few days, but because it is similar to calcium, some of the strontium-90 ions were deposited in bones and teeth (especially of children), subjecting all of us to a small amount of radioactivity for long periods of time. Fortunately, above-ground nuclear testing was pretty much stopped some years ago by a treaty between the United States and the former Soviet Union, and although a few other countries still conduct occasional tests, there is reason to hope that such tests will be completely halted in the near future.

ular way, the violence of the reaction with water increases, and so does the softness of the metal. The elements in the last column (group 0) provide still another example. All are gases at room temperature and form either no compounds at all or very few.

The same type of regularity is found in all the columns of the periodic table. The table is so useful that it hangs in nearly all chemistry classrooms and in most chemical laboratories throughout the world. What makes it so useful is that it correlates a vast amount of data about the elements and their compounds and allows us to make many predictions about physical as well as chemical properties. For example, if you were told that the boiling point of germane (GeH_4) is −88°C and that of methane (CH_4) is −164°C, could you predict the boiling point of silane (SiH_4)? The position of silicon in the table, between germanium and carbon, might lead you to a prediction of about −120°C. The actual boiling point of silane is −112°C, not far from this prediction.

If we look at the periodic table on the inside front cover of this book, we see that columns IA and IIA extend above the rest, and so do columns IIIA to 0. The elements in these eight columns are called **representative elements.** The elements in the B columns and those in group VIII are called **transition elements.** Notice that elements 58 to 71 and 90 to 103 are not included in the main body of the table but are shown separately at the bottom. These elements (called **inner transition elements**) actually belong in the main body, between columns IIIB and IVB (the place is between La and Hf). As is customary, we put them outside the main body only in order to make a neater and more convenient pre-

Figure **2.6**
A piece of metallic sodium reacting with water. (Photograph by C. D. Winters.)

sentation. So, if you will, you may mentally take a pair of scissors, cut through the heavy line between columns IIIB and IVB, move them apart, and insert the inner transition elements. You will now have a table with 32 columns.

Metals, Nonmetals, and Metalloids

Another way to classify the elements is as metals or nonmetals. **Metals** are shiny substances whose atoms tend to give up electrons. They are usually good conductors of electricity and heat and are **malleable,** which means that they can be hammered into flat sheets, and **ductile,** which means that they can be drawn into thin wires. Most elements are metals, including all the transition and inner transition elements. In addition, the elements in columns IA and IIA are all metals, as are those in columns IIIA to VIA *below the heavy step-like line.*

Nonmetals do not conduct electricity, and their atoms do not give up electrons but rather tend to accept them. The nonmetals are the ones *above the heavy step-like line* in the periodic table. The elements right next to the heavy line share some of the properties of metals and nonmetals and are called **metalloids.** The following elements are classified as metalloids: B, Si, Ge, As, Sb, Te, Po, and At. In general, representative elements behave more like metals in going down a column of the periodic table, and in going from right to left.

The six elements in group 0 (called **noble gases**) are a special group, with properties different from those of the other nonmetals. They neither give nor take electrons. Apart from them, the nonmetals are H, C, N, O, F, P, S, Cl, Se, Br, and I. Although there are only 11 of these elements, they make up nearly all the compounds we shall study when we get to organic chemistry and biochemistry later in this book. Figure 2.7 shows the distribution of the elements in the periodic table.

Metals
Metalloids
Nonmetals
Noble gases

H																H	He
Li	Be											B	C	N	O	F	Ne
Na	Mg											Al	Si	P	S	Cl	Ar
K	Ca	Sc	Ti	V	Cr	Mn	Fe	Co	Ni	Cu	Zn	Ga	Ge	As	Se	Br	Kr
Rb	Sr	Y	Zr	Nb	Mo	Tc	Ru	Rh	Pd	Ag	Cd	In	Sn	Sb	Te	I	Xe
Cs	Ba	La	Hf	Ta	W	Re	Os	Ir	Pt	Au	Hg	Tl	Pb	Bi	Po	At	Rn
Fr	Ra	Ac	Rf	Db	Sg	Bh	Hs	Mt	Uun	Uuu	Uub						

Ce	Pr	Nd	Pm	Sm	Eu	Gd	Tb	Dy	Ho	Er	Tm	Yb	Lu
Th	Pa	U	Np	Pu	Am	Cm	Bk	Cf	Es	Fm	Md	No	Lr

Figure **2.7**
Classification of the elements.

Box 2D

The Use of Metals as Historical Landmarks

The malleability of metals played an important role in the development of human society. In the early days, tools were made from stone, which has no malleability. This period is called the Stone Age. Later it was discovered that copper found on the surface of the earth could be hammered into sheets, which made it suitable for vessels, utensils, and religious or artistic objects. This practice started about 9000 or 10 000 years ago, and this period is sometimes called the Copper Age. However, pure copper on the surface of the earth is scarce. Around 5000 BC human beings found a way to obtain copper from its ore by putting the green copper-containing stone, malachite, into a fire. Malachite yielded pure copper at the relatively low temperature of 200°C. Copper is a soft metal made of layers of copper crystals. It can easily be drawn into wires as the layers slip past each other. When hammered, the large crystals break into smaller ones with rough edges. The layers cannot slip, so hammered copper sheets were harder. As a result, the ancient profession of coppersmith was born, and beautiful plates, pots, and ornaments were produced.

Around 4000 BC, it was discovered that by mixing molten copper with tin a greater hardness could be achieved. Copper, even when hammered, does not take an edge, but a copper-tin mixture (an alloy called bronze) does take an edge, and knives and swords could be manufactured. Thus the Bronze Age was born somewhere in the Middle East and quickly spread to China and all over the world.

An even harder metal was soon to come. The first raw iron was found in meteorites. (The ancient Sumerian name of iron is "metal from heaven.") Later, around 2500 BC, it was smelted from its ores and the Iron Age began. Higher technology was needed because iron melts at a higher temperature (about 1500°C) than copper (about 1100°C). Thus, it took a longer time to perfect this process and to learn how to manufacture steel (about 90 to 95 percent iron and 5 to 10 percent carbon). Steel objects appeared first in India around 100 BC.

Modern historians look back at ancient cultures and use the discovery of a new metal as a landmark for that age.

King Solomon's copper mines, Timna, Israel. (Courtesy of Holy Views, Ltd., Jerusalem.)

2.6 The Electronic Structure of Atoms

Niels Bohr was awarded the Nobel prize for this work in 1922.

In discussing the electronic structure of atoms, we will begin with hydrogen because it is the simplest, with only one electron, but before we do that it is necessary to mention the work of Niels Bohr (1885–1962). In 1913 Bohr, a Danish physicist, found that the electron in a hydrogen atom could have one of a number of energies. An electron is always moving around the nucleus and so possesses energy. What Bohr proposed is that certain values were possible for this energy, but not others. This was a very surprising proposal. If you were told that you could drive your car at 23.4 mph or 28.9 mph or 34.2 mph, but never at any speed in between these values, you wouldn't believe it. Yet this is just what Bohr said about electrons in atoms. There is a lowest possible energy level; we call it the **ground state** (electrons in atoms cannot stop moving, so the ground-state energy is not zero). If an electron is to have more energy than in the ground state, however, there are only certain values allowed; values in between are not permitted. We can liken the situation to walking up a flight of stairs (Fig. 2.8). You can put your foot on any step, but you cannot stand anyplace between two steps. Bohr was unable to explain why these levels exist, but the accumulated evidence forced him to the conclusion that they do. We say that the energy of the electron is **quantized.**

Electrons in the lowest energy state of the hydrogen atom—the ground state—occupy a spherical cloud surrounding the nucleus (Fig. 2.9). Clouds of this type are called **orbitals.** They are designated by numbers and letters. This one is called the **1*s* orbital.** There are also orbitals of higher energy. Some of them (2*s* and 2*p*) are shown in Figure 2.9. There are three 2*p* orbitals (designated $2p_x$, $2p_y$, and $2p_z$), which have equal energy. These three orbitals look like dumbbells (but are still clouds) and are at right angles to each other.

There are also orbitals of still higher energy. After 2*p* come 3*s*, 3*p*, 4*s*, and others even higher. All *s* orbitals are spherical; all *p* orbitals look

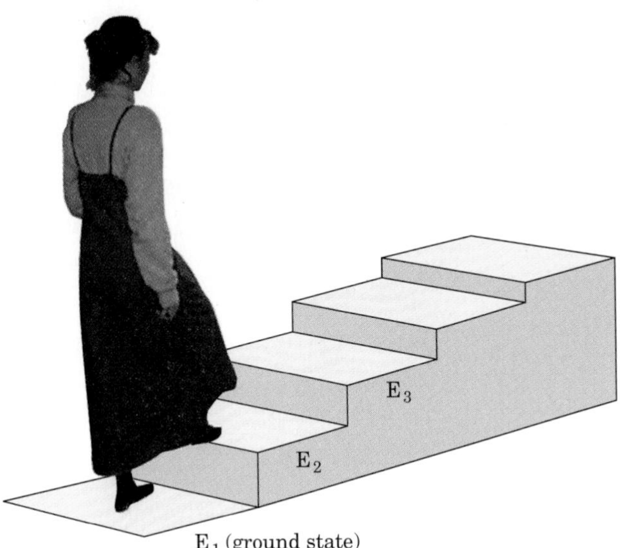

Figure **2.8**
The energy stairway. Unlike our stairways, the spaces between levels are not equal, but get smaller as they go up.

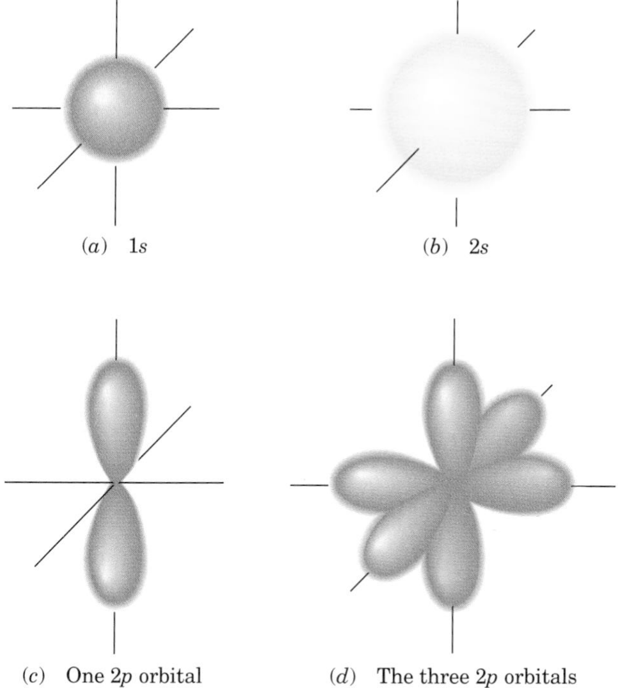

Figure **2.9** • The 1*s*, 2*s*, and 2*p* orbitals.

like dumbbells and come in sets of three. Besides *s* and *p* orbitals, there are two other kinds, called *d* and *f*. All *d* orbitals come in sets of five and *f* orbitals in sets of seven.

The *d* and *f* orbitals are less important to us, and we shall not discuss their shapes.

The number used in the designation of orbitals is called the **principal energy level.** The first principal energy level (the 1 level) contains only one orbital, the 1*s* orbital. The second principal energy level (the 2 level) contains four orbitals (one 2*s* and three 2*p*). The number of orbitals in each level increases as the principal energy level increases.

There are no 1*p*, 1*d*, or 1*f* orbitals.

All the orbitals in any principal energy level constitute a **shell.** Thus, the 1 shell contains only the 1*s* orbital, and the 2 shell contains one 2*s* and three 2*p* orbitals. A **subshell** consists of all the orbitals with the same number and letter. For example, there are two subshells in the 2 shell: One of them consists of the three 2*p* orbitals, and the other contains only the single 2*s* orbital.

A subshell can also be called a **sublevel.**

Note that the orbital designations show both the energy aspect and the spatial aspect of each orbital. The number tells the energy level and the size of the orbital; the letter tells its shape.

The Hydrogen Atom

We now have a good picture of the electronic structure of the hydrogen atom. In its lowest possible energy state (1*s*), the atom looks like the picture shown in Figure 2.9(a), with the nucleus at the center. This is the ground state, but it is not the only possible state. The electron can also exist in any one of the higher energy levels, called *excited states.* This is how it works: A hydrogen atom is in its ground state, looking like Figure 2.9(a). It then receives additional energy from some outside source (usually in the form of heat or light). This additional energy, when added to the energy it already had, equals the amount of energy the electron needs to reach some higher orbital—the $2p_x$, for example. The 1*s* orbital is now vacant, and the electron appears in the $2p_x$ orbital. The hydrogen atom now looks like Figure 2.9(c) and no longer like Figure 2.9(a). We can imagine a whole collection of hydrogen atoms, some having the electron in the 1*s* orbital, some in the $2p_x$, some in the 3*d*, and some in other orbitals. In any individual hydrogen atom, only one orbital is occupied at any time. At room temperature this orbital is overwhelmingly the 1*s*.

Rules for Other Atoms

Now that we have seen the electronic structure of hydrogen, what about other atoms? The situation is similar except that these atoms have more than one electron. The orbitals available to all atoms are the same as in hydrogen: 1*s*, 2*s*, 2*p*, and so on. In the ground state of each atom, only the orbitals of lowest energy are occupied; all higher ones are empty. The electrons fill in according to the following rules:

Electrons can spin either clockwise or counterclockwise.

1. Each orbital can hold two electrons, but no more. Furthermore, if there are two electrons in an orbital, they must have opposite *spin*. This is a statement of the **Pauli exclusion principle.** When there is only one electron in an orbital, we call the electron **unpaired** and say that the orbital is singly occupied.

2. When we have a set of orbitals of equal energy (such as the $2p_x$, $2p_y$, and $2p_z$ orbitals), each orbital becomes half filled before any is completely filled. This is called **Hund's rule.**

3. The orbitals can be likened to a set of boxes. The electrons in any atom go into whichever boxes have the lowest energies, until there are no more electrons left. If we want to know which boxes are occupied, we have to know the order of energies. This is shown in Figure 2.10. Note that the lowest *p* orbitals are 2*p*, the lowest *d* orbitals are 3*d*, and the lowest *f* orbitals are 4*f*. Within any shell the energy levels increase in the order *s*, *p*, *d*, *f*. That is, *s* is the lowest and *f* the highest.

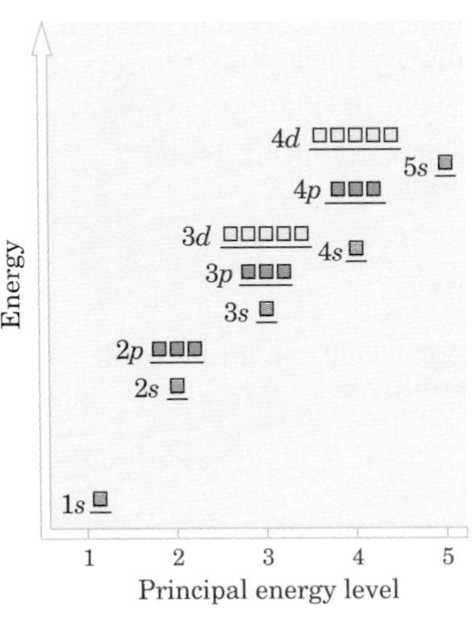

Figure **2.10**
Energy levels of atomic orbitals up to the 4*d* level.

The Electronic Buildup

Using the rules just given, we now determine the ground-state electronic configuration for the first 36 elements.

1. Hydrogen As we have already seen, the single hydrogen electron is in the 1*s* orbital in the ground state.

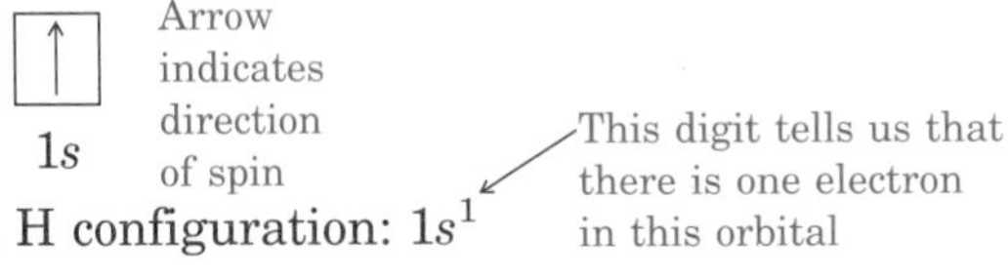

2. Helium There are two electrons in this atom. Since any orbital can hold two electrons, both are able to go into the lowest-energy orbital, the 1*s*.

↑↓

1*s*

He configuration: $1s^2$

The 2 tells us that there are two electrons in this orbital

All electrons at the 1 level (the first principal energy level) constitute the first *shell.* This shell can hold only two electrons, so helium *has a complete shell.*

3. Lithium The first two of lithium's three electrons fill the 1*s* orbital. The other must go into the orbital of next-lowest energy, the 2*s*.

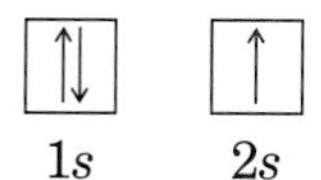

↑↓ ↑

1*s* 2*s*

Li configuration: $1s^22s^1$

In lithium the first shell (the 1 shell) is complete, and there is one electron in the 2 shell. The 1 shell is the *inner shell,* and the 2 shell is the incomplete *outer shell.*

4. Beryllium The first two electrons fill the 1*s* orbital (and the 1 shell). The remaining two fill the 2*s* orbital.

↑↓ ↑↓

1*s* 2*s*

Be configuration: $1s^22s^2$

5. Boron Now the 1*s* and 2*s* orbitals are filled and there is one electron left over, which must go into one of the three 2*p* orbitals. It doesn't matter which; we shall say $2p_x$.

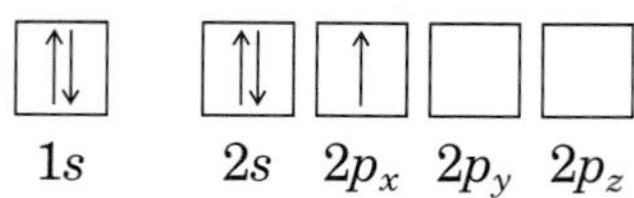

1*s* 2*s* $2p_x$ $2p_y$ $2p_z$

B configuration: $1s^22s^22p^1$

6. Carbon We now come to our first application of Hund's rule. The sixth electron of carbon obviously must go into one of the 2*p* orbitals, but which one? Hund's rule says that it cannot fill the $2p_x$ orbital while either of the other two 2*p* orbitals is vacant, and so it goes into the $2p_y$ or $2p_z$ (it doesn't matter which; we shall say $2p_y$). A carbon atom thus has *two* unpaired electrons.

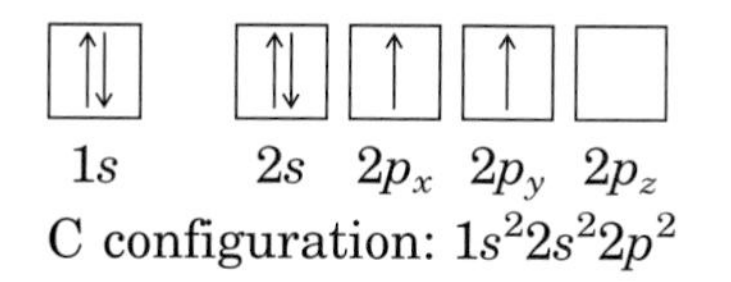

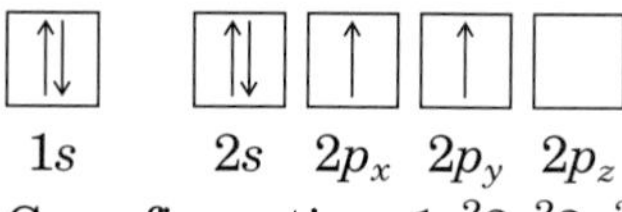

1*s* 2*s* $2p_x$ $2p_y$ $2p_z$

C configuration: $1s^22s^22p^2$

There is experimental proof for Hund's rule. Magnetic measurements show that a nitrogen atom has three unpaired electrons.

7. Nitrogen Hund's rule puts the seventh electron into the $2p_z$ orbital. A nitrogen atom has three unpaired electrons.

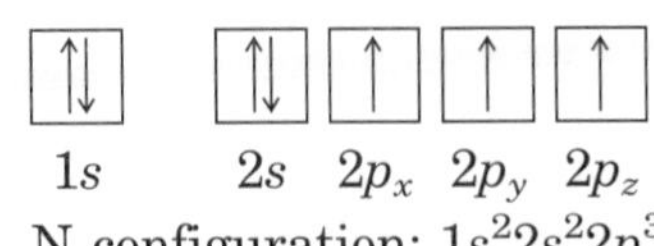

N configuration: $1s^2 2s^2 2p^3$

8–10. Oxygen, fluorine, neon Now that all three $2p$ orbitals are half filled, the next three elements fill these orbitals, one at a time.

8	O	$1s^2 2s^2 2p^4$
9	F	$1s^2 2s^2 2p^5$
10	Ne	$1s^2 2s^2 2p^6$

Oxygen has two unpaired electrons, and fluorine has one. The tenth electron of neon fills not only the $2p_z$ orbital but also the entire 2 shell. Neon is the second element with all its electrons in completely filled shells (helium was the first).

11. Sodium Since the 1 and 2 shells are now complete, the eleventh electron of sodium goes into the $3s$ orbital and begins the 3 shell. The 2 shell is now an inner shell, and the 3 shell becomes the outer shell.

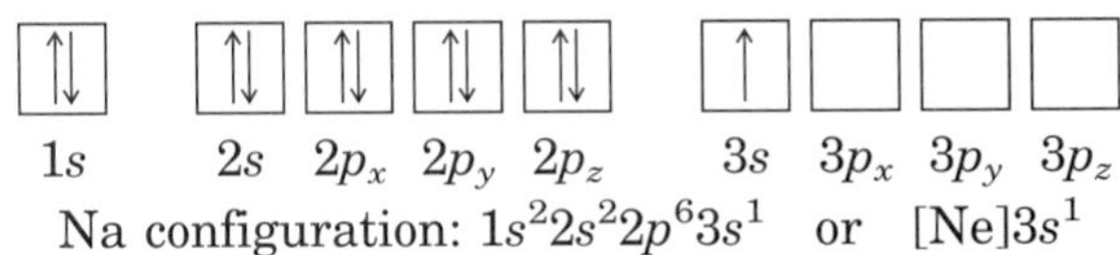

Na configuration: $1s^2 2s^2 2p^6 3s^1$ or $[Ne]3s^1$

Since the configuration of the two inner shells is the same as that of the neon atom, we use the abbreviation [Ne] to show the configuration of the two inner shells.

12–18. The next seven elements follow the same pattern as elements 4 through 10, but at the 3 rather than the 2 level.

12	Mg	$[Ne]3s^2$
13	Al	$[Ne]3s^2 3p^1$
14	Si	$[Ne]3s^2 3p^2$
15	P	$[Ne]3s^2 3p^3$
16	S	$[Ne]3s^2 3p^4$
17	Cl	$[Ne]3s^2 3p^5$
18	Ar	$[Ne]3s^2 3p^6$

Argon is the third element to have all its electrons in complete shells. This is true even though the $3d$ orbitals are vacant. We say this because the behavior of argon is very similar to that of helium and neon (none of the three form any compounds). This is what makes us regard argon as an element with no incomplete shells.

19–20. Potassium, calcium The next two electrons go into the $4s$ orbital.

19	K	$[Ar]4s^1$
20	Ca	$[Ar]4s^2$

21. Scandium If scandium followed the previous pattern, the twenty-first electron would go into the 4*p* orbital. It does not do so, however, because the 3*d* orbital has a lower energy (Fig. 2.10).

$$21 \quad \text{Sc} \quad [\text{Ar}]4s^2 3d^1$$

22–30. The next nine elements fill the five 3*d* orbitals; the last of these is zinc.

$$30 \quad \text{Zn} \quad [\text{Ar}]4s^2 3d^{10}$$

Zinc does not have a complete outer shell. The 3 shell *is* complete, but the 3 shell is an inner shell here. **The occupied shell with the highest number is always the outer shell.** In the case of zinc, the occupied shell with the highest number is the 4 shell, and that one is not complete since the 4*p* orbitals are vacant.

31–36. The next six elements fill the 4*p* orbitals.

$$36 \quad \text{Kr} \quad [\text{Ar}]4s^2 3d^{10} 4p^6$$

Krypton is the fourth element with a complete outer shell (even though the 4*d* and 4*f* orbitals are vacant).

EXAMPLE

Show how to find the electronic configuration of (a) iron (26) (b) rubidium (37).

Answer

(a) The previous element with a complete outer shell is argon (18), so we know that the first 18 electrons of iron will fill the 1*s*, 2*s*, 2*p*, 3*s*, and 3*p* orbitals, for a total of 18 electrons. Since iron has 26 electrons, there are 8 left over. Figure 2.10 shows that the subshell of next higher energy is the 4*s* followed, not by the 4*p*, but by the 3*d*. Therefore, 2 of the 8 electrons go into the 4*s*, and the remaining six go into the five 3*d* orbitals. Hund's rule tells us that one of these will get two electrons and the other four, one each.

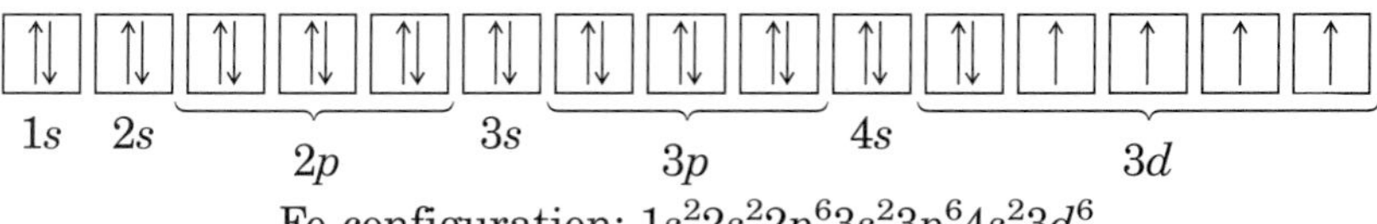

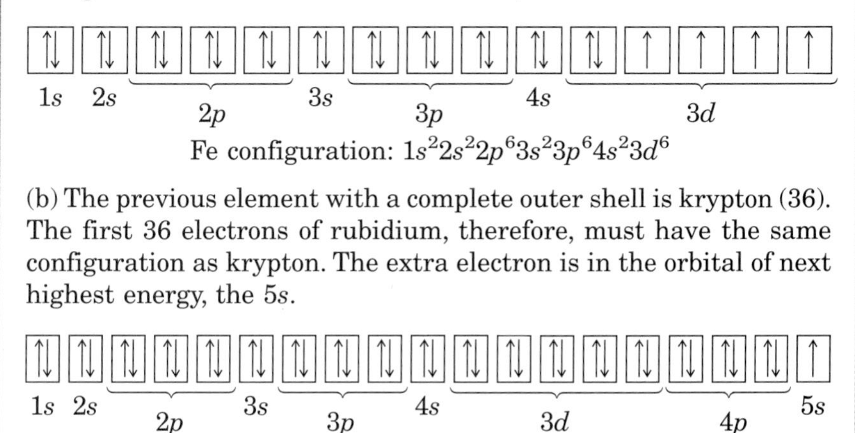

Fe configuration: $1s^2 2s^2 2p^6 3s^2 3p^6 4s^2 3d^6$

(b) The previous element with a complete outer shell is krypton (36). The first 36 electrons of rubidium, therefore, must have the same configuration as krypton. The extra electron is in the orbital of next highest energy, the 5*s*.

1*s* 2*s* 2*p* 3*s* 3*p* 4*s* 3*d* 4*p* 5*s*

Rb configuration: $1s^2 2s^2 2p^6 3s^2 3p^6 4s^2 3d^{10} 4p^6 5s^1$

Problem 2.6

Show how to find the electronic configuration of (a) vanadium (23) (b) arsenic (33).

We have now described the electronic configuration of the first 36 elements (they are summarized in Table 2.4). The remaining elements fill up the next orbitals (5*s*, 4*d*, and so on) in a similar manner, in the order shown in Figure 2.10.

Table 2.4 Electronic Configurations of the First 36 Elements

1	H	$1s^1$	
2	He	$1s^2$	
3	Li	$1s^22s^1$	
4	Be	$1s^22s^2$	
5	B	$1s^22s^22p^1$	
6	C	$1s^22s^22p^2$	
7	N	$1s^22s^22p^3$	
8	O	$1s^22s^22p^4$	
9	F	$1s^22s^22p^5$	
10	Ne	$1s^22s^22p^6$	
11	Na	$1s^22s^22p^63s^1$	or $[Ne]3s^1$
12	Mg	$1s^22s^22p^63s^2$	or $[Ne]3s^2$
13	Al	$1s^22s^22p^63s^23p^1$	or $[Ne]3s^23p^1$
14	Si	$1s^22s^22p^63s^23p^2$	or $[Ne]3s^23p^2$
15	P	$1s^22s^22p^63s^23p^3$	or $[Ne]3s^23p^3$
16	S	$1s^22s^22p^63s^23p^4$	or $[Ne]3s^23p^4$
17	Cl	$1s^22s^22p^63s^23p^5$	or $[Ne]3s^23p^5$
18	Ar	$1s^22s^22p^63s^23p^6$	
19	K	$1s^22s^22p^63s^23p^64s^1$	or $[Ar]4s^1$
20	Ca	$1s^22s^22p^63s^23p^64s^2$	or $[Ar]4s^2$
21	Sc	$1s^22s^22p^63s^23p^64s^23d^1$	or $[Ar]4s^23d^1$
22	Ti	$1s^22s^22p^63s^23p^64s^23d^2$	or $[Ar]4s^23d^2$
23	V	$1s^22s^22p^63s^23p^64s^23d^3$	or $[Ar]4s^23d^3$
24	Cr	$1s^22s^22p^63s^23p^64s^13d^5$[a]	or $[Ar]4s^13d^5$
25	Mn	$1s^22s^22p^63s^23p^64s^23d^5$	or $[Ar]4s^23d^5$
26	Fe	$1s^22s^22p^63s^23p^64s^23d^6$	or $[Ar]4s^23d^6$
27	Co	$1s^22s^22p^63s^23p^64s^23d^7$	or $[Ar]4s^23d^7$
28	Ni	$1s^22s^22p^63s^23p^64s^23d^8$	or $[Ar]4s^23d^8$
29	Cu	$1s^22s^22p^63s^23p^64s^13d^{10}$[a]	or $[Ar]4s^13d^{10}$
30	Zn	$1s^22s^22p^63s^23p^64s^23d^{10}$	or $[Ar]4s^23d^{10}$
31	Ga	$1s^22s^22p^63s^23p^64s^23d^{10}4p^1$	or $[Ar]4s^23d^{10}4p^1$
32	Ge	$1s^22s^22p^63s^23p^64s^23d^{10}4p^2$	or $[Ar]4s^23d^{10}4p^2$
33	As	$1s^22s^22p^63s^23p^64s^23d^{10}4p^3$	or $[Ar]4s^23d^{10}4p^3$
34	Se	$1s^22s^22p^63s^23p^64s^23d^{10}4p^4$	or $[Ar]4s^23d^{10}4p^4$
35	Br	$1s^22s^22p^63s^23p^64s^23d^{10}4p^5$	or $[Ar]4s^23d^{10}4p^5$
36	Kr	$1s^22s^22p^63s^23p^64s^23d^{10}4p^6$	

[a]The electronic configurations of chromium and copper are slightly irregular, because a filled (10 electrons) or half-filled (5 electrons) *d* subshell has a special stability.

2.7 Electronic Configuration and The Periodic Table

When Mendeleev published his first periodic table in 1869, he had no explanation for why it worked, and indeed nobody else had any good explanation either. It was not until after the discovery of electronic structure that chemists finally understood why the table works. The answer is very simple. *The periodic table works because elements in the same column have the same configuration of electrons in the outer shell.* For example, let us look at the elements in column IA. We already know the configuration for lithium, sodium, and potassium. We now add rubidium and cesium:

Li [He]$2s^1$
Na [Ne]$3s^1$
K [Ar]$4s^1$
Rb [Kr]$5s^1$
Cs [Xe]$6s^1$

All these atoms have complete inner shells and an outer shell consisting of one *s* electron. The properties of elements largely depend on the electronic configuration of their outer shell. This being so, it is not surprising that these elements, all of which have the same outer-shell configuration, should have such similar properties.

Because the highest-energy electrons of the elements in Groups IA and IIA are in *s* orbitals, these elements are often called ***s*-block elements.** Similarly, the ***p*-block elements** are those in columns IIIA to 0 (except hydrogen), the ***d*-block elements** are the transition elements (the eight B columns), and the ***f*-block elements** are the inner transition elements.

We could not expect *identical* properties because the number of protons in the nucleus is different and so is the number of inner shells. These numbers also have an effect on the properties and allow us to explain why properties change regularly as we move down a column of the periodic table. For example, all the elements in group IA are metals, which means they tend to give up electrons. In this case they give up only the single electron in the outer shell, but cesium gives it up most easily, rubidium next, and so on up to lithium. We explain this by saying that the 2*s* electron is closer to the positive nucleus in lithium than is the 3*s* electron in sodium, and so the 2*s* electron is more attracted to its nucleus. Furthermore, sodium has one more filled inner shell than lithium, and this filled inner shell (made up of negatively charged electrons) "shields" the 3*s* electron from the nucleus more effectively than the fewer inner electrons of lithium. Since atoms get bigger as we go down the table, these effects explain why loss of the outer electron becomes easier as we go down a column.

Ionization Energy

The energy necessary to remove an electron from an atom in the gaseous state is called the *ionization energy.* We can actually measure its numerical value for each atom. Table 2.5 shows the ionization energy for atoms of some of the representative elements. We can see from Table 2.5 that ionization energy decreases as we go down any column. We have al-

Table 2.5 Ionization Energy of Some Representative Elements

IA	IIA	IIIA	IVA	VA	VIA	VIIA	0
Li 124[a]	Be 215	B 191	C 260	N 335	O 314	F 402	Ne 497
Na 119	Mg 176	Al 138	Si 188	P 242	S 239	Cl 299	Ar 363
K 100	Ca 141	Ga 138	Ge 182	As 226	Se 225	Br 272	Kr 323

[a]All values are kilocalories/mole, a unit of energy.

ready seen why this is so: Electrons are easier to remove the farther they get from the nucleus. Table 2.5 also shows that, with a few exceptions, ionization energy increases regularly as we go from left to right across any row. This is consistent with the change in metallic character mentioned in Section 2.5. As we go from left to right in the table, elements become less metallic. Metals tend to give up electrons more readily than nonmetals, and the information in Table 2.5 largely bears this out. Figure 2.11, showing how ionization energy varies with increasing atomic number, shows it graphically.

In all cases these regularities exist because elements in the same column have the same outer-shell electronic configuration.

The regularity that we see in Table 2.5 is not unusual. We could construct similar tables for many other properties, and we would find similar regularities, going both down and across. Likewise, graphs of many other properties have shapes very much like that of Figure 2.11.

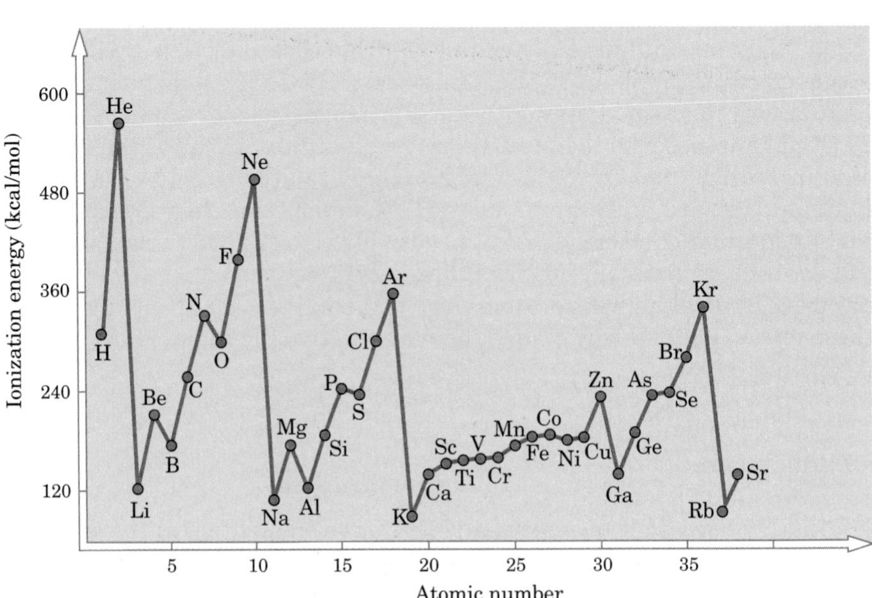

Figure **2.11** • A plot of ionization energy versus atomic number for the first 38 elements. Many other properties give similar plots.

Electronic configuration also explains the presence of the transition elements (p. 53). Let us look at the electronic structure of iron, a typical transition element:

Two incomplete shells

$$26 \quad Fe \quad [Ar]4s^2 3d^6$$

In this atom, the outer shell—the 4 shell—contains two electrons and is obviously incomplete because the $4p$ orbitals, which can hold six more electrons, are empty. The 3 shell, however, an inner shell, is also incomplete. There are six electrons in the five $3d$ orbitals, and these orbitals, of course, can hold a total of ten electrons. So iron has not only an incomplete outer shell but also an incomplete inner shell. We define a **transition element** as **an element whose atoms have one or more incomplete inner shells.** (As you might have guessed, inner transition elements have two incomplete inner shells.)

The presence of unfilled inner shells gives transition elements special properties. For example, most of them form more than one kind of ion.

2.8 How Small Are Atoms?

In the previous section we examined the electronic structure of atoms. It is useful to get some idea of how small atoms and their components really are.

The Mass of an Atom

A typical heavy atom (although not the heaviest) is lead-208, a lead atom with 82 protons, 82 electrons, and 126 neutrons. It has a mass of 3.5×10^{-22} g. You would need 1.3×10^{24} lead-208 atoms to make 1 lb of lead. This is a very large number. There are about five billion people on earth right now. If you divided 1 lb of these atoms among all the people on earth, each person would get about 2.6×10^{14} atoms. This is still a large number. If these atoms had a value of 1 cent per 1000 atoms, every person on earth could get more than two billion dollars by selling his or her share.

The Size of an Atom

An atom of lead has a diameter of about 3.5×10^{-8} cm. If you could line them up with the atoms just touching, it would take 73 million lead atoms to make a line 1 inch long. Despite their tiny size, we can actually see atoms, in certain cases, by the use of special microscopic techniques. An example is shown in Figure 2.12.

Figure **2.12**
Atoms of silicon on a specially reconstructed surface of a silicon crystal as seen by a scanning tunneling microscope. (Courtesy of Dr. Sang-Il Park, Applied Physics, Stanford University, Stanford, CA.)

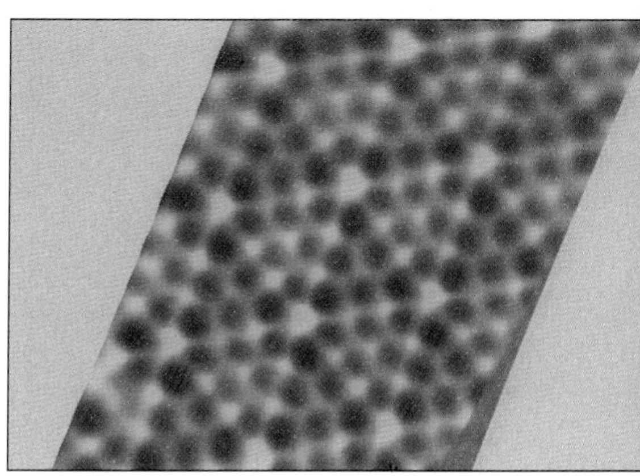

Box 2E

Abundance of Elements on Earth and in People

Table 2E shows the abundance of the elements present in the human body. Oxygen is the most abundant element by weight, followed by hydrogen, carbon, and nitrogen. If we go by number of atoms instead of by weight, hydrogen is even more abundant than oxygen.

Table 2E also shows the abundance of elements in the earth's crust. Although 88 elements are found in the earth's crust (we know very little about the interior of the earth because we have not been able to penetrate very far), they are not present in anything like equal amounts. The most abundant element is also oxygen, but otherwise the picture is greatly different from that of the human body. Silicon, aluminum, and iron, which are not major elements in the body (although iron is an important trace element, as mentioned in Box 2A), are the three next most abundant in the earth's crust, and only six other elements (Ca, Na, K, Mg, H, and Ti) make up even as much as one half of 1 percent by weight of the earth's crust.

Table 2E Relative Abundance of Elements in the Human Body and in the Earth's Crust, Including Atmosphere and Oceans

Element	Percent in Human Body: *By Number of Atoms*	Percent in Human Body: *By Weight*	Percent in Earth's Crust, by Weight
H	63.0	10.0	0.9
O	25.4	64.8	49.3
C	9.4	18.0	0.08
N	1.4	3.1	0.03
Ca	0.31	1.8	3.4
P	0.22	1.4	0.12
K	0.06	0.4	2.4
S	0.05	0.3	0.06
Cl	0.03	0.2	0.2
Na	0.03	0.1	2.7
Mg	0.01	0.04	1.9
Si	—	—	25.8
Al	—	—	7.6
Fe	—	—	4.7
Others	0.01	—	—

Only oxygen is a principal element in both the sand, gravel, and water, and in the water buffaloes. (Photograph by Beverly March.)

The Mass of an Electron and of a Nucleus

An electron has a mass of 9.1×10^{-28} g. The sun is very big; it weighs about 333 000 times as much as the earth. The mass of the sun is to the mass of a 3-lb cantaloupe as the mass of the cantaloupe is to the mass of an electron (Fig. 2.13). As shown in Table 2.2, the mass of an electron is very much smaller than that of a proton or neutron. Since protons and neutrons are in the nucleus and electrons are outside, this means that almost all the mass of an atom is in the nucleus. The total mass of the 82 electrons in a lead-208 atom is 7.5×10^{-26} g, and the mass of the whole atom is 3.5×10^{-22} g.

This means that 99.98 percent of the mass of this atom is in the nucleus. A similar percentage holds for all other atoms.

The Size and Density of the Nucleus

Although the mass of the nucleus is very great, the size of the nucleus is very small. The nucleus of a lead-208 atom has a diameter of about 1.6×10^{-12} cm. When we compare this with the diameter of the whole atom, which is 3.5×10^{-8} cm, we see that the nucleus occupies only a tiny fraction of the atom. If the nucleus of a lead-208 atom were the size

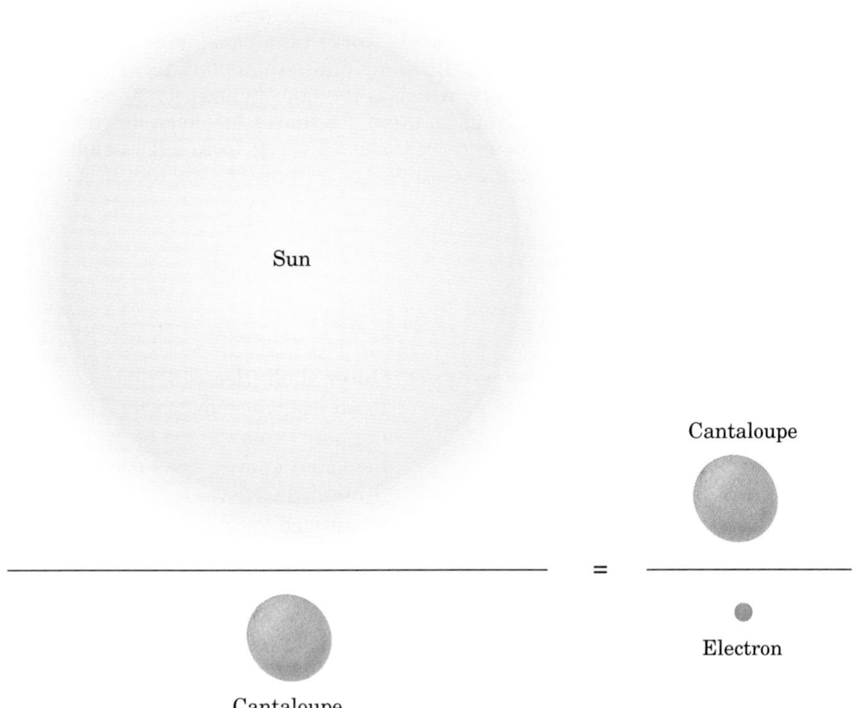

Figure **2.13** • The mass of the sun is to that of a cantaloupe as the mass of a cantaloupe is to that of an electron.

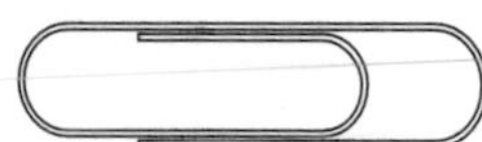

Figure **2.14**
If a paper clip this size were made entirely of atomic nuclei packed together, it would weigh about ten million tons.

of a baseball, then the whole atom would be much larger than a baseball stadium. In fact, it would be a sphere about one mile in diameter. With such a large mass in such a small space, it is obvious that the nucleus must be very dense. For example, the density of the lead-208 nucleus is 1.8×10^{14} g/cc. This is an extremely high density. Nothing in our daily life has a density anywhere near as high. If a paper clip had this density, it would weigh about ten million (10^7) tons (Fig. 2.14).

SUMMARY

Matter can be classified as **elements** (substances all of whose atoms are the same), **compounds** (substances made up of at least two elements), or **mixtures.** Dalton's atomic theory says that all atoms of a given element are identical. The theory is based on the law of conservation of mass (matter cannot be created or destroyed) and on the law of constant composition (any compound is always made up of elements in the same proportion by weight). An **atom** is the smallest unit of an element that shows the chemical behavior of the element. A **molecule** is a cluster of two or more atoms connected by chemical bonds.

Atoms consist of **protons** and **neutrons** inside the nucleus and **electrons** outside it. Electrons are very light and have a charge of −1. Protons and neutrons are much heavier; their masses are about equal. Protons have a charge of +1; neutrons have no charge. The **atomic number** of an element is the number of protons; this determines the identity of the element. The sum of the number of protons and neutrons is the **mass number.** Atoms have an equal number of electrons and protons; but **ions,** which are atoms that have lost or gained electrons, have an unequal number of electrons and protons and therefore are charged. **Isotopes** are atoms with the same number of protons but different numbers of neutrons. The **atomic weight** of an element is a weighted average of the masses of the isotopes as they occur in nature.

The **periodic table** arranges elements into columns of elements with similar properties, which gradually change as we move down a column. Metallic elements, which tend to lose electrons, are below the heavy line in the periodic table. Nonmetals are above the line.

Electrons in atoms can exist in a number of **energy levels.** The energy levels are clouds called **orbitals.** All *s* orbitals are spherical, and all *p* orbitals are shaped like dumbbells. Each orbital can hold a maximum of two electrons, of opposite spin. Electrons in the **ground state** (lowest energy state) of atoms fill up the available orbitals of lowest energy. Atoms are very tiny, with a very small mass, almost all of which is in the nucleus. The nucleus is extremely tiny, with a high density.

The periodic table works because elements in the same column have the same outer-shell configuration. **Ionization energy,** which is the energy necessary to remove an electron from an atom, decreases as we go down the periodic table and increases as we go from left to right. Transition elements have one or more incomplete inner shells.

KEY TERMS

Alkali metal (Sec. 2.5)
Atom (Sec. 2.3)
Atomic number (Sec. 2.4)
Atomic weight (Sec. 2.4)
Compound (Sec. 2.2)
Distillation (Sec. 2.2)
Electric charge (Sec. 2.4)
Electron (Sec. 2.4)
Element (Sec. 2.2)
Excited state (Sec. 2.6)
Ground state (Sec. 2.6)
Hund's rule (Sec. 2.6)
Inner transition element (Secs. 2.5, 2.7)
Ion (Sec. 2.4)
Ionization energy (Sec. 2.7)
Isotope (Sec. 2.4)
Mass number (Sec. 2.4)
Metal (Sec. 2.5)
Metalloid (Sec. 2.5)
Mixture (Sec. 2.2)
Molecule (Sec. 2.3)
Neutron (Sec. 2.4)
Noble gas (Sec. 2.5)
Nonmetal (Sec. 2.5)
Nucleon (Sec. 2.4)
Nucleus (Sec. 2.4)
Orbital (Sec. 2.6)
Outer shell (Sec. 2.6)
Pauli exclusion principle (Sec. 2.6)
Periodic table (Sec. 2.5)
Principal energy level (Sec. 2.6)
Proton (Sec. 2.4)
Quantized (Sec. 2.6)
Representative element (Sec. 2.5)
Shell (Sec. 2.6)
Subshell (Sec. 2.6)
Transition element (Secs. 2.5, 2.7)

PROBLEMS

2.7 Why are the theories of Democritus and Zeno considered nonscientific?

Classifications of Matter

2.8 Try to classify each of the following as an element, a compound, or a mixture:
(a) oxygen (g) diamond
(b) table salt (h) a pebble
(c) sea water (i) gasoline
(d) wine (j) milk
(e) air (k) carbon dioxide
(f) silver

2.9 Define (a) element (b) compound (c) mixture

2.10 Name these elements (try not to look at a table): (a) O (b) Pb (c) Ca (d) Na (e) C (f) S (g) Fe (h) H (i) K

2.11 Explain how a mixture of two liquids can be separated by distillation.

Dalton's Atomic Theory

2.12 How does Dalton's atomic theory explain (a) the law of conservation of mass (b) the law of constant composition?

2.13 When 2.16 g of mercury oxide is decomposed to yield 2.00 g of mercury and 0.16 g of oxygen, which law is supported by the experiment?

2.14 The compound carbon monoxide contains 42.9% carbon and 57.1% oxygen. The compound carbon dioxide contains 27.2% carbon and 72.7% oxygen. Does this disprove Proust's law of constant composition?

Inside the Atom

2.15 Where in the atom are these particles located:
(a) protons
(b) electrons
(c) neutrons

2.16 What is the mass number of an atom or ion with
(a) 24 protons, 24 electrons, 28 neutrons
(b) 9 protons, 9 electrons, 10 neutrons
(c) 34 protons, 36 electrons, 45 neutrons
(d) 83 protons, 83 electrons, 127 neutrons
(e) 88 protons, 86 electrons, 138 neutrons

2.17 What is the name of each element in Problem 2.16?

2.18 Given the mass number and the number of neutrons, what is the name of each of these elements?
(a) mass number 45; 24 neutrons
(b) mass number 48; 26 neutrons
(c) mass number 107; 60 neutrons
(d) mass number 246; 156 neutrons
(e) mass number 36; 18 neutrons

2.19 If each of the atoms in Problem 2.18 acquired two more neutrons, what would their names be?

2.20 How many neutrons are there in:
(a) a carbon atom of mass number 13
(b) a germanium atom of mass number 73
(c) an osmium atom of mass number 188
(d) an yttrium atom of mass number 89

2.21 How many protons and how many neutrons does each of these isotopes of radon contain: Rn-210, Rn-218, and Rn-222?

2.22 How many nucleons are there in:
(a) ^{22}Ne (d) tellurium-128
(b) ^{104}Pd (e) lithium-7
(c) ^{35}Cl (f) uranium-238

2.23 State the number of protons and neutrons in each isotope in Problem 2.22.

2.24 Define nucleon.

2.25 Tin-118 is one of the isotopes of tin. Name the other isotopes of tin that contain two, three, and six more neutrons than tin-118.

2.26 Define: (a) ion (b) isotope.

2.27 Write the symbol for each atom or ion, given the number of protons (p) and electrons (e) [example: 3p, 2e = Li^+]:
(a) 9p, 10e (e) 11p, 10e
(b) 19p, 18e (f) 16p, 18e
(c) 18p, 18e (g) 35p, 36e
(d) 1p, 0e (h) 25p, 23e

2.28 For each atom or ion, tell how many electrons, protons, and neutrons are present:
(a) Cl with a mass number of 37
(b) K^+ with a mass number of 39
(c) Pb^{2+} with a mass number of 207
(d) I^- with a mass number of 127
(e) Pd^{2+} with a mass number of 106
(f) W with a mass number of 186

2.29 If copper occurred in only two isotopic forms, ^{63}Cu and ^{65}Cu, and if the naturally occurring mixture contained 69.00% ^{63}Cu and 31.00% ^{65}Cu, what would be the atomic weight of copper?

2.30 If the three isotopes of hydrogen occurred in nature as follows: 98.00% ^1H, 1.800% ^2H, and 0.2000% ^3H, what would be the atomic weight of hydrogen?

*2.31 If there are only two naturally occurring isotopes of antimony, ^{121}Sb and ^{123}Sb, and the atomic weight of antimony is 121.75, what are the abundances of the two isotopes?

The Periodic Table

2.32 Which of the following elements would you expect to have fairly similar properties (look at the periodic table): As, I, Ne, F, Mg, K, Ca, Ba, Li, He, N, P.

2.33 Which are transition elements?
(a) Pd (d) Ce
(b) K (e) Br
(c) Co (f) Cr

2.34 Which element in each pair is more metallic?
(a) silicon or aluminum
(b) arsenic or phosphorus
(c) gallium or germanium
(d) gallium or aluminum

2.35 Classify as metal, nonmetal, or metalloid:
(a) argon (g) iodine
(b) boron (h) antimony
(c) lead (i) vanadium
(d) arsenic (j) sulfur
(e) potassium (k) nitrogen
(f) silicon

Electronic Structure of Atoms

2.36 What is the maximum number of electrons that can go into (a) an orbital (b) an outer shell (c) the 3 shell?
2.37 Write ground-state electronic configurations for (a) Li (b) Ne (c) Be (d) C (e) He (f) Mn (g) Cl (h) P
2.38 What is common and what is different in the electronic configuration of (a) Na and Cs (b) O and Te (c) C and Ge?
2.39 Positive ions are obtained from neutral atoms by removing the number of electrons indicated by the charge on the ion. Give the electronic configuration of (a) Mg and Mg^{2+} (b) Na and Na^+ (c) Al and Al^{3+}.
2.40 Negative ions are obtained from neutral atoms by adding the number of electrons indicated by the charge on the ion. Give the electronic configuration of (a) F and F^- (b) S and S^{2-} (c) P and P^{3-}.
2.41 The electronic configurations for the elements with atomic numbers higher than 36 follow the same rules as given in the text for the first 36. Write the ground-state electronic configuration for (a) Y (b) Ba (c) I
***2.42** The element radon (atomic number 86) has the electronic configuration $[Xe]6s^25d^{10}4f^{14}6p^6$. The element francium (atomic number 87) has the same configuration plus one more electron. Which orbital does this extra electron occupy?
2.43 What is the total number of electrons that can fit into the 6*s*, 6*p*, 6*d*, and 6*f* orbitals?

Electronic Configuration and the Periodic Table

2.44 Why does the periodic table work?
2.45 Why do the elements in column IA of the periodic table (the alkali metals) have similar but not identical properties?
2.46 List in order of increasing ionization energy: Al, Li, N, Ar, B, O, K, Cl, C.

Boxes

2.47 (Box 2A) Which four elements are the most important for human life?
2.48 (Box 2A) Why does the body need sulfur, calcium, and iron?
2.49 (Box 2B) Are all chemicals poisonous? Explain.
2.50 (Box 2C) Why is strontium-90 more dangerous to human beings than most other radioactive isotopes that are present in fallout?
2.51 (Box 2D) Bronze is an alloy of which two metallic elements?
2.52 (Box 2D) Historically the Copper Age preceded the Iron Age. Why was that so?
2.53 (Box 2E) Which are the two most abundant elements, by weight, in (a) the earth's crust (b) the human body?

Additional Problems

2.54 Give the designation of all subshells in:
(a) the 1 shell (c) the 3 shell
(b) the 2 shell (d) the 4 shell
2.55 Tell whether metals or nonmetals are more likely to have each of the following characteristics:
(a) conduct electricity and heat
(b) accept electrons
(c) be malleable
(d) be a gas at room temperature
(e) be a transition element
2.56 What is the outer-shell electronic configuration of the elements in (a) group IIIA (b) group VIIA (c) group VA?
2.57 How many protons, electrons, and neutrons are present in
(a) ^{35}Cl (d) $^{127}I^-$
(b) ^{98}Mo (e) ^{158}Gd
(c) $^{44}Ca^{2+}$ (f) $^{212}Bi^{3+}$
2.58 Define molecule.
2.59 How many neutrons are there in (a) polonium-210 (b) uranium-233 (c) radon-208 (d) californium-249?
2.60 What is the symbol for (try not to look at a table):
(a) phosphorus (f) silver
(b) potassium (g) calcium
(c) sodium (h) carbon
(d) nitrogen (i) tin
(e) bromine (j) zinc
2.61 This state of C

$$1s^22s^22p^13s^1$$

represents (a) a C^+ ion (b) a C^- ion (c) the ground state of a C atom (d) an excited state of a C atom.
***2.62** Write the electronic structure of an excited state of the Be atom.
***2.63** The natural abundance of titanium isotopes is as follows: titanium-46 = 7.95%, titanium-47 = 7.75%, titanium-48 = 73.45%, titanium-49 = 5.51%, and titanium-50 = 5.34%. Calculate the atomic weight of titanium.
2.64 Write the symbol for each atom or ion, given the number of protons (p) and electrons (e) [example: 3p, 2e = Li^+]:
(a) 10p, 10e (d) 56p, 54e
(b) 19p, 18e (e) 1p, 2e
(c) 35p, 36e (f) 21p, 18e
2.65 How many electrons are there in the outer shell of (a) Si (b) Br (c) P (d) K (e) He (f) Ca (g) Kr (h) Pb (i) Se (j) O?
***2.66** What percent of the mass of a hydrogen atom is in the nucleus?
***2.67** After the *s*, *p*, *d*, and *f* orbitals, the next higher series is the *g* series. (a) How many electrons does each *g* subshell hold? (b) What is the lowest principal energy level in which *g* orbitals can exist?

***2.68** The mass of a proton is 1.67×10^{-24} g. The mass of a grain of salt is 1.0×10^{-2} g. How many protons would it take to have the same mass as a grain of salt?

2.69 (a) What are the charges of an electron, a proton, and a neutron? (b) What are the masses (in amu) of an electron, a proton, and a neutron?

2.70 Define the excited state of an atom. Can an atom be in more than one unique excited state?

***2.71** Rubidium has two natural isotopes: rubidium-85 and rubidium-87. What is the natural abundance of each of these isotopes if the atomic weight of rubidium is 85.47?

Chapter 3

Chemical bonds

3.1 Introduction

In Chapter 2 we mentioned that molecules are tightly bound clusters of atoms. In this chapter we shall see what it is that holds the atoms together. Atoms are held together by powerful attractions called **chemical bonds.** There are two main types: *ionic bonds* and *covalent bonds.* In order to talk about ionic bonds, we must first discuss ions.

3.2 Ions

An ion is a particle with an unequal number of protons and electrons.

In Section 2.4 we defined *ions* and mentioned that there are two kinds.

Cations

The smallest cation is the hydrogen ion, H^+, which is a bare proton.

Many atoms have a tendency to lose one or more electrons from the electron cloud that surrounds the nucleus. An atom that loses one or more electrons becomes an ion. It now has a positive charge. **Positive ions** are called **cations.** Some examples are

$Li \longrightarrow e^- + Li^+$ (the lithium ion has three protons and two electrons; the charge is +1)

$Mg \longrightarrow 2e^- + Mg^{2+}$ (the magnesium ion has 12 protons and 10 electrons; the charge is +2)

$Al \longrightarrow 3e^- + Al^{3+}$ (the aluminum ion has 13 protons and 10 electrons; the charge is +3)

The naming of cations is usually very simple. Just give the name of the element followed by the word "ion." Thus the three ions mentioned here are the lithium ion, magnesium ion, and aluminum ion.

Anions

If an atom gains one or more electrons, it becomes an ion that contains more electrons than protons. It now has a negative charge because it has extra electrons. **Negative ions** are called **anions.** Some examples are

$Br + e^- \longrightarrow Br^-$ (the bromide ion has 35 protons and 36 electrons; the charge is −1)

$S + 2e^- \longrightarrow S^{2-}$ (the sulfide ion has 16 protons and 18 electrons; the charge is −2)

$N + 3e^- \longrightarrow N^{3-}$ (the nitride ion has seven protons and ten electrons; the charge is −3)

Anions formed from a single atom are named by replacing the ending in the name of the atom with **-ide.** For example, fluorine forms fluor**ide** ions.

Table 3.1 Some Common Cations and Anions[a]

Group	Ion Name	Ion Symbol
	Hydrogen	H^+
	Hydride	H^-
IA	Lithium	Li^+
	Sodium	Na^+
	Potassium	K^+
IIA	Magnesium	Mg^{2+}
	Calcium	Ca^{2+}
	Strontium	Sr^{2+}
	Barium	Ba^{2+}
IIIA	Aluminum	Al^{3+}
IVA	Lead	Pb^{2+}
VIA	Oxide	O^{2-}
	Sulfide	S^{2-}
VIIA	Fluoride	F^-
	Chloride	Cl^-
	Bromide	Br^-
	Iodide	I^-
Transition elements	Copper(I)	Cu^+
	Copper(II)	Cu^{2+}
	Iron(II)	Fe^{2+}
	Iron(III)	Fe^{3+}
	Mercury(II)	Hg^{2+}
	Silver(I)	Ag^+

[a]See also Table 3.3.

Table 3.1 lists the names of some of the more important ions that are formed from a single atom.

In Section 3.10 we will meet some polyatomic ions.

Atoms and Their Ions

As you might have guessed from our discussion of metals and nonmetals in Section 2.5, metals generally form cations, and nonmetals, if they form ions at all, form anions. Two important elements that do not form ions are carbon and boron. Also, nitrogen is reluctant to form N^{3-} ions, and those ions are not very stable.

It is important to understand that there is an enormous difference between the properties of atoms and the properties of their corresponding ions. For a common example, we can look at the elements sodium and chlorine. Sodium is a soft metal made of sodium atoms and reacts violently with water. Chlorine atoms are very unstable and even more reactive than sodium atoms. Both sodium and chlorine are poisonous. Yet common table salt, NaCl, is made up entirely of sodium *ions* and chloride *ions*. These two ions are quite stable and unreactive. Sodium ions do not react violently with water; in fact, they do not react with water at all. Neither do chloride ions. In all cases ions and atoms

The element chlorine exists not as free chlorine atoms but as chlorine molecules, Cl_2.

Box 3A Biologically Important Ions

The human body uses several ions to fulfill important functions. For example, electrical neutrality (Sec. 3.3) must be maintained both inside and outside body cells. Inside the cells the main positive ion used for this purpose is potassium (K^+) and the main negative ion is hydrogen phosphate (HPO_4^{2-}). In body fluids outside the cells, different ions are used to accomplish this task, chiefly sodium (Na^+) and chloride (Cl^-).

The calcium ion (Ca^{2+}) is a major component of bones and teeth. It also has other functions, including assistance in blood clotting, muscle action, and heartbeat control. About 90 percent of the calcium in the body is present in bones as calcium phosphates and carbonates. The iron(II) ion (Fe^{2+}) is a part of hemoglobin, which transports oxygen from the lungs to the cells (see Box 3C).

Another important ion is magnesium (Mg^{2+}), which is involved in the action of nerves and muscles and is necessary for the activity of certain enzymes. It is also present in bones and teeth, along with Ca^{2+}.

All of these ions, as well as certain others present in lesser amounts, must be obtained from the diet. Lack of some of them causes diseases; for example, a deficiency of Fe^{2+} results in anemia. However, too much of them can also be a problem. Too much table salt, NaCl, for example, can cause water retention (edema) and high blood pressure in some people. See also Section 25.8.

Biologically important ions are also considered in Section 26.6.

A **chemical species** is any type of particle: atom, ion, or molecule.

are completely different chemical species with completely different chemical and physical properties.

Sometimes people are not careful to distinguish between atoms and ions. One example is the drug many people call "lithium," which is used to treat manic-depressive symptoms (a form of mental illness). The element lithium, like sodium, is a soft metal that reacts with water. It would certainly be poisonous, probably fatal, if ingested. The drug is not lithium (Li) but lithium *ions* (Li^+), usually given in the form of lithium carbonate (Li_2CO_3) (Fig. 3.1). Another example comes from the fluoridation of water. Fluorine (F_2), an extremely reactive element and a deadly poison, is *not* used for this. Water is fluoridated with fluoride ions (F^-), which are unreactive and not poisonous in the low concentrations used.

Small amounts of tin(II) fluoride are added to water to prevent tooth decay.

Figure 3.1
Lithium (a) and lithium carbonate (b) are entirely different substances. (Photograph by Beverly March.)

The Octet Rule

In Table 3.1 we see that K^+ is listed, but not K^{2+} or K^{3+} or K^-. In fact, K^{2+}, K^{3+}, and K^- are extremely unstable and do not normally exist. Is there any way to predict which ions are stable? Yes, there is, although our predictions will not be perfect. We use the principle that **atoms and ions are most stable when they have a complete outer shell of electrons.** Since a complete outer shell of an atom or ion has eight electrons (Sec. 2.6) — except for the very first shell, which has only two — we call this the **octet rule.**

The outer shell of an atom is called the **valence shell.**

The octet rule readily explains why K^+ is stable but the other potassium ions are not. Potassium is in column IA of the periodic table, so a potassium atom has one electron in its outer or valence shell ($4s^1$). If it loses this electron

$$K \longrightarrow e^- + K^+$$

then there are no electrons left in the 4 shell, and the 3 shell becomes the outer shell in the K^+ ion. If the K^+ were to lose another electron (to form K^{2+}), the electron would have to come from the 3 shell, leaving only seven electrons in it. Because K^+ has a complete octet but K^{2+} does not, K^{2+} is much less stable than K^+ and is not found in normal matter. Once an atom or ion has formed a complete octet, it is usually difficult to get it to lose or gain any electrons.

That explains why K^{2+} and K^{3+} are not stable ions, but what about K^-? This ion does not exist because K^- has only two electrons in its outer shell, and this is a long way from an octet.

We can use the octet rule with a reasonable degree of success to predict the stable ions of the representative elements. All the elements in column IA, like potassium, have one electron in the valence shell, and so all form only E^+ ions. The elements in column IIA have two outer-shell electrons and lose these to form E^{2+} ions. For example,

As in Section 2.5, "E" is a general symbol for any element.

$$Ba \longrightarrow 2e^- + Ba^{2+}$$

The elements in column VIIA have seven electrons in the outer shell. They can achieve a complete octet by gaining an electron, but only one, so they form E^- ions. For example,

$$Cl + e^- \longrightarrow Cl^-$$

In column VIA each element has six electrons in the outer shell and so gains two electrons to form an E^{2-} ion. For example,

$$Se + 2e^- \longrightarrow Se^{2-}$$

One test of the octet rule is column 0. The elements in this column (the noble gases) already have a complete outer shell, so the octet rule predicts they will form no ions at all. This prediction is amply borne out: No stable ions of any group 0 elements have ever been found.

The beaker on the left contains Fe^{3+} ions; the one on the right contains Fe^{2+} ions. Note the color difference. (Photograph by Beverly March.)

The octet rule can help us to predict the charges of many ions. However, it is not perfect, for two reasons:

1. Concentrated charges are unstable. Boron has three valence-shell electrons. If it lost these three, it would, as B^{3+}, have a complete outer shell. It seems, however, that this is too concentrated a charge for such a tiny ion, and this ion is not found in normal matter. This also explains why the upper group IVA elements (C, Si) do not form any E^{4+} or E^{4-} ions and why an atom like chlorine, which forms Cl^- ions by gaining one electron, does not form Cl^{7+} ions by losing seven electrons.

2. The octet rule cannot be applied to transition elements because they are too far removed from the noble-gas structure. Table 3.1 shows that iron forms Fe^{2+} and Fe^{3+} ions. This behavior is typical of transition elements, which frequently form more than one type of ion.

In spite of these weaknesses, the octet rule is useful in those cases where it does apply.

3.3 Ionic Bonds

Since sodium chloride (NaCl), also known as table salt, is the most common ionic compound, let us look at its structure rather closely. It is made up of sodium ions, Na^+, and chloride ions, Cl^-. An **ionic bond** is **the attraction between positive and negative ions,** but this attraction is not on a one-to-one basis. In its solid (crystalline) form, sodium chloride consists of a three-dimensional array of Na^+ and Cl^- ions arranged as shown in Figure 3.2. Note that, for any particular Na^+ ion, there is no individual Cl^- ion that is closer than any other. Rather, each Na^+ has six Cl^- ions as "nearest neighbors," as shown in Figure 3.3(a). Since the six Cl^- ions are equidistant from the Na^+, each is attracted to the same degree. Likewise, each Cl^- ion is surrounded by six Na^+ ions, each an equal distance away, as shown in Figure 3.3(b). This situation is typical of ionic compounds in the solid state. There are no discrete molecules in these compounds. The entire crystal consists of ions arranged as shown in Figure 3.2.

Ionic compounds are compounds that contain ionic bonds.

There are many other ionic compounds, some of which are shown in Box 3B. All are solids at room temperature, and all consist of

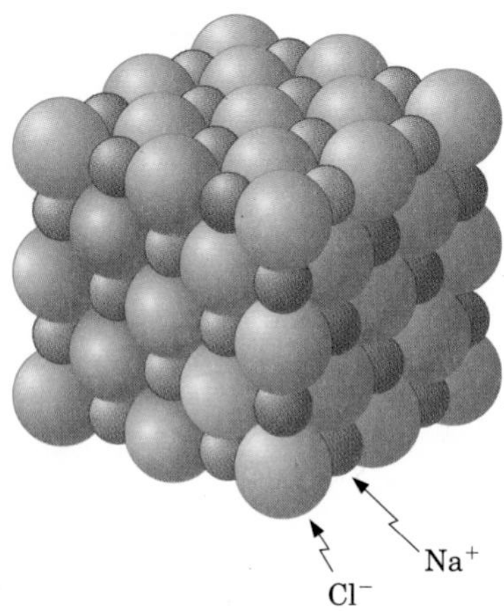

Figure **3.2**
The structure of the sodium chloride crystal.

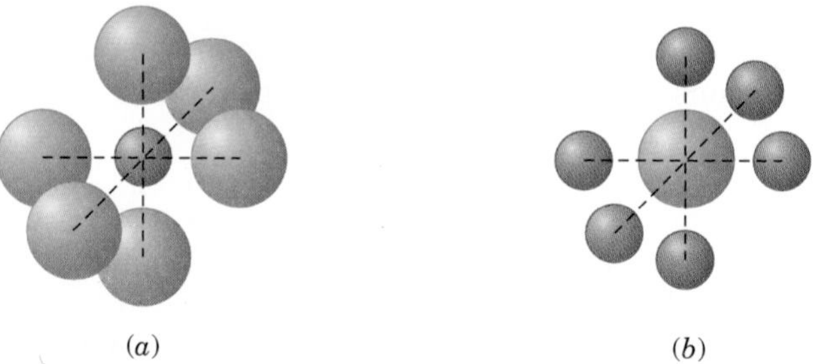

Figure **3.3** • (a) The environment around one Na^+ ion. (b) The environment around one Cl^- ion.

Box 3B

Ionic Compounds in Medicine

There are many ionic compounds that have medical uses. Some of them are shown here.

Formula	Name	Medical Use
$AgNO_3$	Silver nitrate	Astringent, styptic in veterinary medicine
$BaSO_4$	Barium sulfate	Radiopaque medium in x-ray work
$CaSO_4$	Calcium sulfate	Plaster casts
$FeSO_4$	Iron(II) sulfate	For iron deficiency
$KMnO_4$	Potassium permanganate	Anti-infective (external)
KNO_3	Potassium nitrate (saltpeter)	Diuretic
Li_2CO_3	Lithium carbonate	Antidepressive
$MgSO_4$	Magnesium sulfate (epsom salts)	Cathartic
$NaHCO_3$	Sodium bicarbonate (baking soda)	Antacid
NaI	Sodium iodide	Source of iodide for thyroid
NH_4Cl	Ammonium chloride	To acidify the digestive system
$(NH_4)_2CO_3$	Ammonium carbonate	Expectorant
SnF_2	Tin(II) fluoride	To strengthen teeth (external)
ZnO	Zinc oxide	Astringent (external)

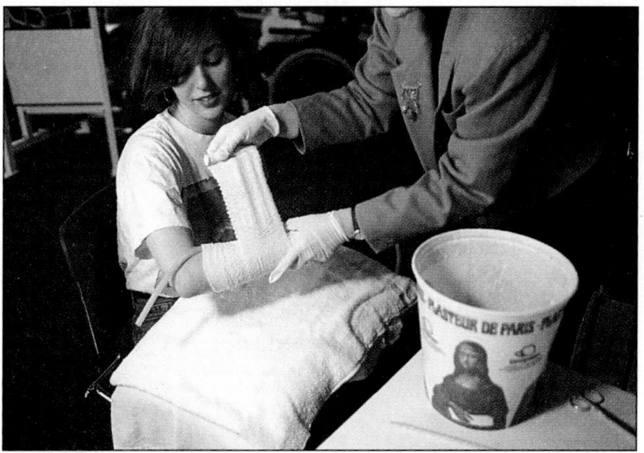

A plaster cast being applied to a patient's arm. (Photograph by C. D. Winters.)

ions arranged in some kind of regular three-dimensional array. The arrays are different in each case since they depend on the size of the ions and on the relative number of negative and positive ions, but in every case there is some regular array. No ionic compounds contain discrete molecules.

Ions are charged particles, but the matter we see all around us and deal with every day is electrically neutral (uncharged). Nature does not allow a large concentration of charge, either positive or negative, to build up in any one place. We can assume that all the substances we deal with are uncharged. If there are ions present in any sample of matter, the total number of positive charges must equal the total number of negative charges. Therefore, we cannot have a sample containing only Na^+ ions. Any sample that contains Na^+ ions must also contain negative ions, which may be Cl^-, Br^-, S^{2-}, or other anions, and the sum of the positive charges must equal the sum of the negative charges. This is called electrical neutrality.

Predicting Formulas of Ionic Compounds

The naming of ionic compounds is discussed in Section 3.11.

The principle of electrical neutrality allows us to predict the formulas of ionic compounds.

EXAMPLE

What is the formula for the compound formed from lithium ion and bromide ion?

Answer • The charge on Li^+ is +1; the charge on Br^- is −1. The formula for lithium bromide is LiBr (one to one).

Ionic charges are given in Tables 3.1 and 3.3.

EXAMPLE

What is the formula for the compound formed from barium ion and iodide ion?

Answer • The charge on Ba^{2+} is +2, and I^- has a charge of −1. Two I^- ions are required to neutralize one Ba^{2+} ion. Therefore, the formula for barium iodide is BaI_2.

EXAMPLE

What is the formula for the compound formed from aluminum ion and sulfide ion?

Answer • The charge on Al^{3+} is +3, and S^{2-} has a charge of −2. It takes three S^{2-} ions to neutralize the positive charge of two Al^{3+} ions, so the formula is Al_2S_3.

Problem 3.1

Write formulas for compounds formed from (a) potassium ion and chloride ion (b) calcium ion and fluoride ion (c) iron(III) ion and oxide ion.

Note that we can generally arrive at the formula for ionic compounds by "crossing" the charges; for example, for aluminum sulfide,

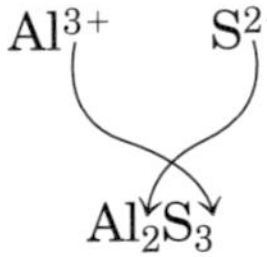

Al_2S_3, like all other ionic compounds, obeys the principle of electrical neutrality:

$$\begin{aligned} 2Al^{3+} &= +6 \\ 3S^{2-} &= -6 \\ \hline \text{Net charge} &= 0 \end{aligned}$$

Remember that the subscripts in the formulas for ionic compounds represent the *ratio* of the ions. Thus, a crystal of BaI_2 has twice as many I^- as Ba^{2+} ions. (This ratio is, of course, necessary to keep the substance neutral.) For ionic compounds, when both charges are 2, as in the compound formed from Pb^{2+} and O^{2-}, we must "reduce to lowest terms." That is, lead oxide is PbO, and not Pb_2O_2. The reason is that we are looking at ratios only, and the ratio of ions is 1:1. The same is true for the case where both charges are 3; thus aluminum nitride is AlN.

3.4 Covalent Bonds

The second type of chemical bond—the covalent bond—is very different from the ionic bond. Because of this the properties of compounds containing covalent bonds are very different from those of ionic compounds.

A covalent bond is one in which two atoms share a pair of electrons. Let us examine a typical case, that of the element fluorine. Each fluorine atom has seven outer-shell electrons and needs one more to complete the octet. Of course, the fluorine atom could gain an electron and thereby become an F^- ion, but it can do this only if there is a nearby atom (or molecule) from which to take the electron. It cannot get an electron from nowhere. If we have a sample that contains only fluorine atoms, then the only nearby atom would be another fluorine atom. Would a fluorine atom take an electron from another fluorine atom? A little thought shows that the answer is no. The first atom would then have a complete octet and be a stable F^- ion, but the second atom would have only six outer-shell electrons and be an extremely unstable F^+ ion. It turns out that fluorine atoms do not have to take electrons from each other. They can (and do) **share** electrons to form a covalent bond. When they do this, *both* atoms have a complete octet. We can write it this way (the dots stand for outer-shell electrons):

$$:\ddot{\underset{..}{F}}\cdot \;+\; \cdot\ddot{\underset{..}{F}}: \longrightarrow \;:\ddot{\underset{..}{F}}\cdot\cdot\ddot{\underset{..}{F}}:$$

F atom F atom F_2 molecule

It was Gilbert N. Lewis (1875–1946) who first proposed that two atoms can form a bond by sharing a pair of electrons.

G. N. Lewis. (AIP Niels Bohr Library; photograph by Francis Simon.)

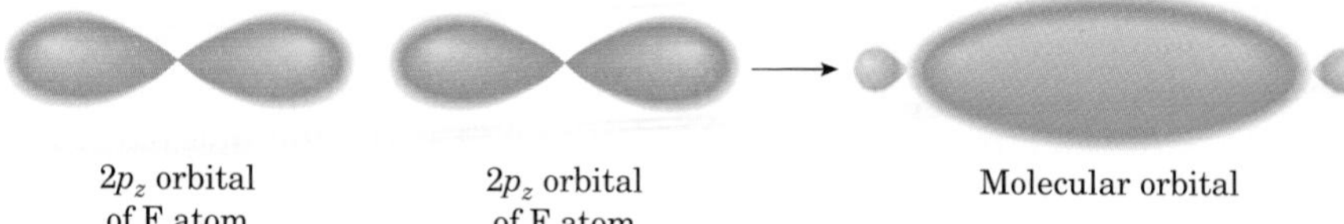

Figure **3.4** • When two F atoms form a molecule of F_2, their two $2p_z$ orbitals overlap to form a molecular orbital (called a sigma orbital). The other orbitals are not affected. (The 1*s* and 2*s* orbitals are not shown in this figure.)

As we can see, each fluorine atom still has possession of six of its original seven electrons. The other two (one contributed by each atom) are now *shared* between them. Each fluorine atom acts as if those two electrons are part of its outer shell, that is, as if each had a complete octet.

Another way to look at it is to remember that the electrons of fluorine (and all other) atoms are in orbitals, which are clouds of electrons (Sec. 2.6). A fluorine atom has the electron configuration $1s^22s^22p_x^{\;2}2p_y^{\;2}2p_z^{\;1}$, which means that its seventh outer-shell electron is in a 2*p* orbital, which is shaped like a dumbbell. When two fluorine atoms come together to form a covalent bond, these orbitals overlap to form a new cloud that surrounds both nuclei (Fig. 3.4). This new cloud is the covalent bond. We call it a **molecular orbital** because it binds two atoms together to form a molecule. The other orbitals (containing the 1*s*, 2*s*, $2p_x$, and $2p_y$ electrons) are not much affected by the bonding and remain virtually unchanged. Molecular orbitals are given Greek letters, and this type is called a **sigma (σ) orbital.** In Section 11.3 we will meet another type of molecular orbital, called pi (π).

Clouds in the sky can overlap, and so can electron clouds.

Any orbital, atomic or molecular, can hold a maximum of two electrons.

A fluorine atom can achieve an outer-shell octet by sharing *one* of its electrons. The same is true for the other elements in group VIIA. Because of this, each of these atoms forms only *one* covalent bond. The same principle holds for the other nonmetallic elements. **Each nonmetallic atom has a strong tendency to form a particular number of covalent bonds.** This number is equal to the number of electrons that must be shared to achieve a noble-gas electronic configuration. For example, carbon, which has four electrons in its outer shell, forms four covalent bonds; oxygen, with six electrons, forms two covalent bonds; and nitrogen, with five electrons, forms three covalent bonds.

Elements in groups VA, VIA, and VIIA (except O, N, and F) may form different numbers of bonds in different compounds. For example, phosphorus forms three bonds in some compounds (PCl_3), and five in others (PCl_5) (see pp. 86–87).

When an atom forms more than one bond, do all the bonds have to be to the same atom? The answer is no. The bonds can be formed in many different ways. As an example, let us look at carbon, which forms four bonds. Here are some examples of stable, well-known compounds that contain carbon:

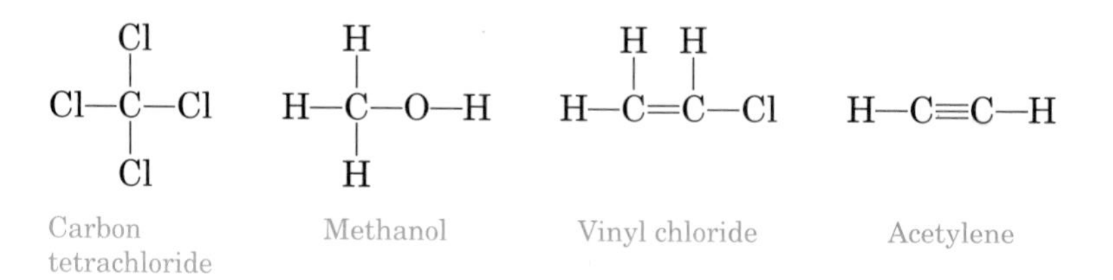

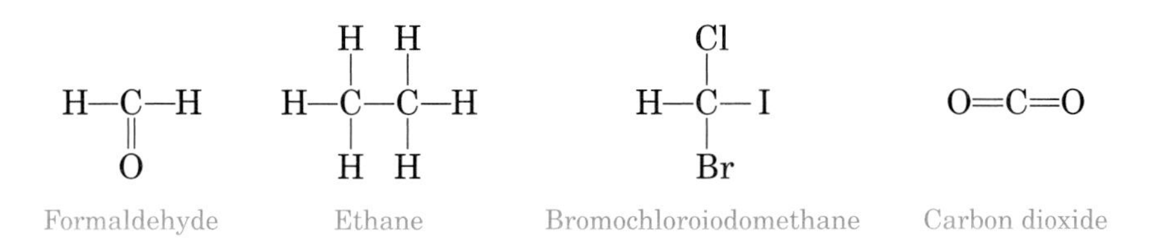

```
     Cl             H            H  H
     |              |            |  |
Cl—C—Cl        H—C—O—H      H—C=C—Cl      H—C≡C—H
     |              |
     Cl             H
```

Carbon tetrachloride | Methanol | Vinyl chloride | Acetylene

$$\underset{\text{Formaldehyde}}{H-\overset{\displaystyle O}{\underset{}{C}}-H} \qquad H-\underset{H}{\overset{H}{C}}-\underset{H}{\overset{H}{C}}-H \qquad H-\underset{Br}{\overset{Cl}{C}}-I \qquad O{=}C{=}O$$

Formaldehyde (H—C—H with C=O below) Ethane Bromochloroiodomethane Carbon dioxide

In these formulas, a covalent bond (a pair of electrons in a molecular orbital) is shown as a straight line connecting two atoms. We understand that, when we draw them this way, we are showing that the two atoms involved are sharing the electrons.

Double and Triple Bonds

An inspection of these formulas shows that a carbon atom can use its four bonds to connect with four atoms of the same element or with four different atoms, and can also form *double* and *triple* bonds. For example, in vinyl chloride the carbon on the left is bonded to two H atoms and, by a double bond, to another C atom. In the carbon-carbon double bond, *two* pairs of electrons are shared. The total number of bonds for each C atom must be four, but they can be made with any combination of single, double, or triple bonds adding up to four. In virtually all carbon compounds, every carbon atom has four bonds, and stable compounds containing carbon atoms with three or five bonds, or any other number, are practically unknown (though a few curiosities do exist).

The formulas given above are called structural formulas. A **structural formula** shows all the atoms in a molecule and all the bonds connecting them. A **molecular formula** gives less information, since it shows only the number of atoms and not the bonds. For example, the molecular formulas for methanol and vinyl chloride are CH_4O and C_2H_3Cl, respectively.

The structural formulas for these compounds are shown above.

Lewis Structures

We have now seen two kinds of formulas for molecules, structural and molecular. There is still another kind, one that gives even more information than a structural formula. This kind, called a **Lewis structure,** shows not only all atoms and covalent bonds but also all other outer-shell electrons, including those that are unshared.

Lewis structures can be drawn for atoms as well as molecules. They show all outer-shell electrons as dots. We have already shown the fluorine atom in this way on page 81. Some other examples are

$$Li\cdot \quad \cdot\dot{B}\cdot \quad \cdot\underset{\cdot}{\dot{C}}\cdot \quad \cdot\underset{\cdot}{\ddot{N}}\cdot \quad \cdot\underset{\cdot\cdot}{\ddot{O}}\cdot \quad :\underset{\cdot\cdot}{\ddot{Ne}}:$$

As an example of a Lewis structure of a molecule, consider ammonia, whose molecular formula is NH_3. Because we know that a hydrogen atom forms only one covalent bond, it is clearly not possible for a hydrogen atom *in this molecule* to be connected to another hydrogen atom.

Each hydrogen atom must be bonded to the nitrogen. Since nitrogen forms three bonds and since there are three hydrogen atoms, this works out all right.

$$\begin{array}{c}H\text{—}N\text{—}H\\ |\\ H\end{array}$$ Structural formula for ammonia

Unshared pairs are also called **nonbonding pairs** or **lone pairs.**

But a nitrogen atom has five outer-shell electrons (Sec. 2.6). It must use three of them to bond to the hydrogens (one each), and so two remain. We show the two unshared electrons as a pair of dots:

$$\begin{array}{c}H\text{—}\ddot{N}\text{—}H\\ |\\ H\end{array}$$ Lewis structure for ammonia

Similarly, the formula we gave for F_2 is also a Lewis structure:

$$:\ddot{\underset{..}{F}}\text{—}\ddot{\underset{..}{F}}:$$

Lewis structures are usually easy to draw if we follow a few steps:

Step 1. Find out how many atoms of each kind are in the molecule and which are connected to which. Sometimes we can figure out the connections (as we did for NH_3), but this is not always possible.

Step 2. Calculate the total number of outer-shell electrons brought into the molecule by all the atoms in it. For any given atom, this number is always the same—it is the number of outer-shell electrons in the free atom. For example, every time we write a Lewis structure for an oxygen-containing compound, we can be certain that each oxygen atom brings six outer-shell electrons.

Step 3. Insert these electrons in pairs (either bonds or unshared pairs) so that every atom has an octet if possible (two for hydrogen). Where necessary, use double or triple bonds.

The procedure is best illustrated by examples.

EXAMPLE

Draw the Lewis structure for hydrogen peroxide, H_2O_2.

O	$2 \times 6 =$	12
H	$2 \times 1 =$	2
Total		14

Answer • (Step 1) From the fact that a hydrogen atom forms only one covalent bond, we know the connections must be H—O—O—H. (Step 2) An oxygen atom has six outer-shell electrons, so the two O atoms bring in a total of 12 electrons. Each H atom brings in 1 electron, so the total is 14 outer-shell electrons, or 7 pairs. (Step 3) We can begin by putting in the three bonds we know are there:

$$H\text{—}O\text{—}O\text{—}H$$

This leaves four electron pairs to be accounted for. If we want to give each O atom an octet, we need to put the four pairs in as follows:

$$\text{H—}\ddot{\underset{..}{\text{O}}}\text{—}\ddot{\underset{..}{\text{O}}}\text{—H}$$ Lewis structure for hydrogen peroxide

Problem 3.2

Draw the Lewis structure for phosphorus tribromide, PBr_3.

EXAMPLE

Draw the Lewis structure for formaldehyde, H_2CO. The atoms are connected like this:

```
H  C  H

   O
```

Answer • (Step 1) Given. (Step 2) The C brings in four electrons, the O brings in six, and each H brings in one. The total is thus 12 electrons, or 6 pairs. (Step 3) As usual, we begin by putting in pairs to bond all the atoms:

C		4
O		6
H	2 × 1 =	2
Total		12

```
H—C—H
  |
  O
```

Three pairs remain. If we put one of them between C and O, making this a double bond, the carbon has a complete octet:

```
H—C—H
  ||
  O
```

Two pairs of electrons remain, and these go to fill the octet of the O:

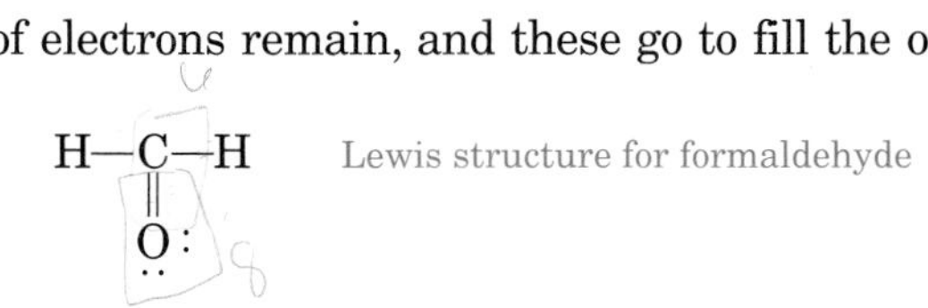

```
H—C—H
  ||
  O:
  ..
```

Lewis structure for formaldehyde

Problem 3.3

Draw the Lewis structure for acetic acid, $C_2H_4O_2$. The connections are

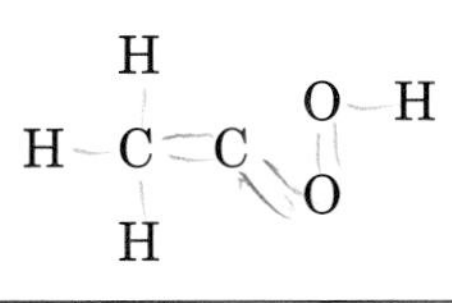

```
   H
        O  H
H  C  C
        O
   H
```

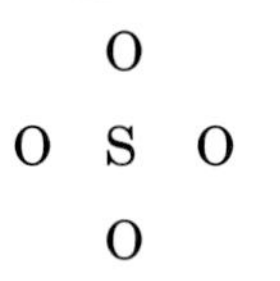

The -2 charge means that this ion has two more electrons than protons. It is clear that the atoms involved have not brought in any more electrons than protons, since they were neutral. The two extra electrons therefore came from some outside source, not from the atoms that went into it. The extra electrons are there, though, so we must count them.

EXAMPLE

Draw the Lewis structure for the sulfate ion, SO_4^{2-}. The connections are

```
    O
O   S   O
    O
```

Answer • This time we have an ion rather than a molecule, so our counting procedure has an extra step. A sulfur atom has 6 outer-shell electrons and each oxygen has 6, so the total brought in by these five atoms is 30 electrons. However, the SO_4^{2-} ion has a charge of -2, meaning that it must have two more electrons than this, or a total of 32 (16 pairs). Note that if the ion had a positive charge, we would subtract a number equal to the charge. (Step 3) As usual, we begin by putting in four electron pairs to bond all the atoms:

S		6
O	$4 \times 6 =$	24
		30
-2 charge		2
Total		32

```
    O
    |
O — S — O
    |
    O
```

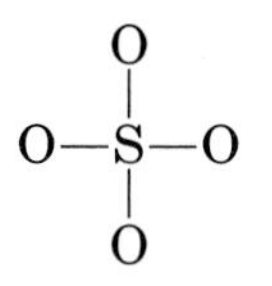

This leaves 12 pairs. If we put three pairs around each O, we give all atoms an octet and use up all the electrons:

When we write the Lewis structure of an ion, it is customary to use brackets and to show the total charge outside the brackets.

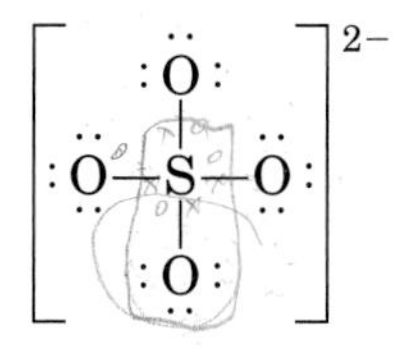

```
 ┌     ··      ┐ 2−
 |    :O:      |
 |  ··  |  ··  |
 | :O — S — O: |
 |  ··  |  ··  |
 |    :O:      |
 └     ··      ┘
```

Problem 3.4

Draw the Lewis structure for the sulfite ion, SO_3^{2-}. The connections are

```
O   S   O
    O
```

Exceptions to the Octet Rule

The octet rule is a useful guide to the drawing of Lewis structures because it is valid in the vast majority of cases. Nevertheless there are some exceptions. One such exception involves elements below the second row of the periodic table. The second-row atoms boron, carbon, nitrogen, oxygen, and fluorine can hold a maximum of 8 electrons in their outer shells, but atoms below this can hold more. For example, sulfur and phosphorus can hold 10 or 12, as shown by the Lewis structures for these compounds:

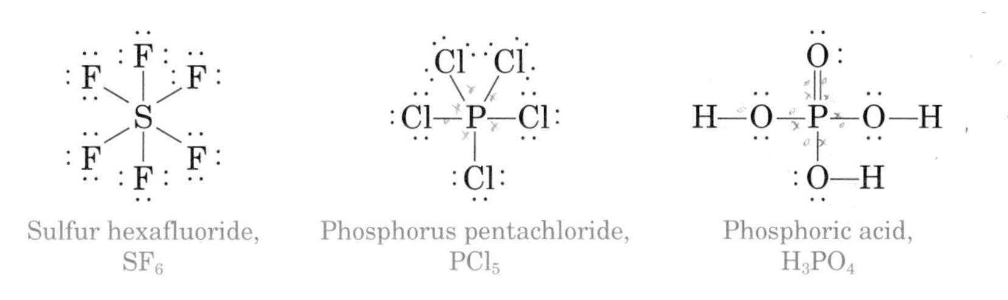

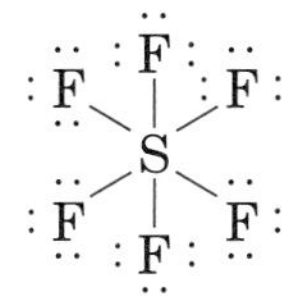

Sulfur hexafluoride, SF_6

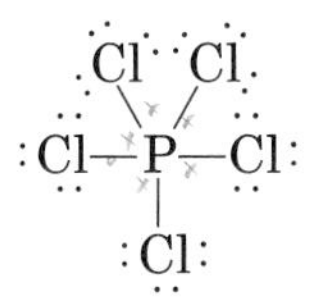

Phosphorus pentachloride, PCl_5

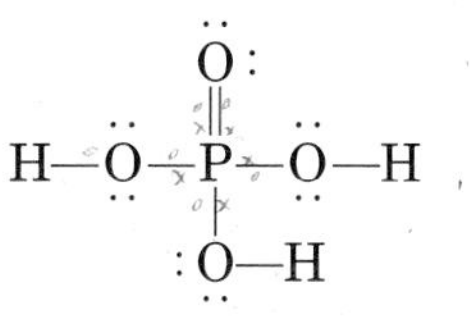

Phosphoric acid, H_3PO_4

In a different kind of exception, some perfectly stable compounds simply do not have enough electrons to give each atom an octet. One example is the simple molecule nitric oxide.

EXAMPLE

Draw the Lewis structure of nitric oxide, NO.

Answer • (Step 1) Since this molecule contains only two atoms, they must be connected to each other. (Step 2) The N brings in 5 electrons, the O brings in 6, for a total of 11. This is unusual, although it is not the only such case. In most molecules all electrons are paired, but the odd number here means one must be left over (any molecule with an unpaired electron is called a **free radical**; see Box 3D). It is therefore impossible to draw a Lewis structure in which both atoms have an octet. (Step 3) Putting in the electrons as best we can gives

$$:\ddot{N}=\ddot{O}\cdot$$

The oxygen atom has only seven electrons.

Problem 3.5

Draw the Lewis structure of nitrogen dioxide, NO_2. The connections are O N O.

An example of a molecule in which all electrons are paired, but which still has an atom with fewer than an octet is BF_3, discussed in Section 3.5. Unlike NO, most free radicals are so reactive that they cannot be isolated as pure compounds.

3.5 Coordinate Covalent Bonds

We have seen that a covalent bond is formed when a single electron from one atom and a single electron from another atom are both shared by the two atoms. There is another way a covalent bond can form, though it is less common. In this case one atom supplies two electrons and the other none. This type of bond is called a **coordinate covalent bond.** As an example, consider the compound ammonia, which has an unshared pair of electrons.

Note that atoms in coordinate covalent bonds do not form their normal number of bonds. The nitrogen atom in the adduct has four bonds rather than its usual three.

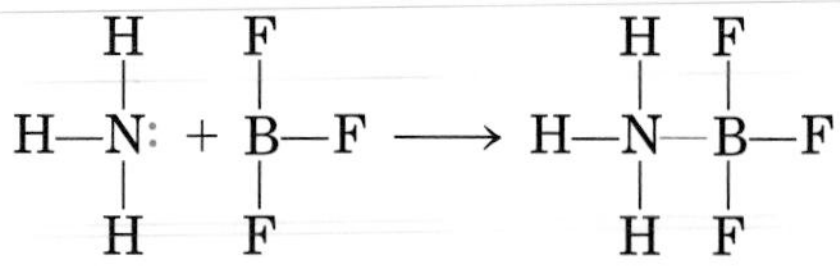

$$H_3N{:} + BF_3 \longrightarrow H_3N{-}BF_3$$

Ammonia Boron trifluoride Adduct

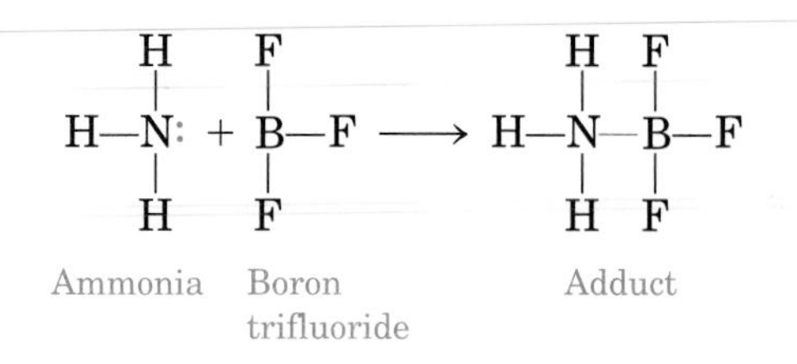

It can share this pair with a boron atom in the compound boron trifluoride to give a compound in which both the boron and the nitrogen have a complete octet. This type of bond can form only between two species that have the right characteristics: One of them must have an unshared pair of electrons, and the other must have space for a pair. In BF_3 the boron has only six electrons in its outer shell and thus lacks an octet. Therefore, it makes a fine partner for the ammonia because it needs two more to make eight.

Boron trifluoride is one of the compounds that violates the octet rule.

A coordinate covalent bond is just like any other covalent bond once it has been formed. The only difference is that both electrons are supplied by one atom rather than each atom supplying one electron.

Certain metal ions, most notably those of the transition metals, form large numbers of compounds that contain coordinate covalent bonds. Many of these ions can take two, four, or even six pairs of electrons. Such compounds are called **coordination compounds,** and thousands of them are known. Some examples follow (coordinate covalent bonds are shown in color).

The coordination number is the number of bonds connected to a central atom or ion.

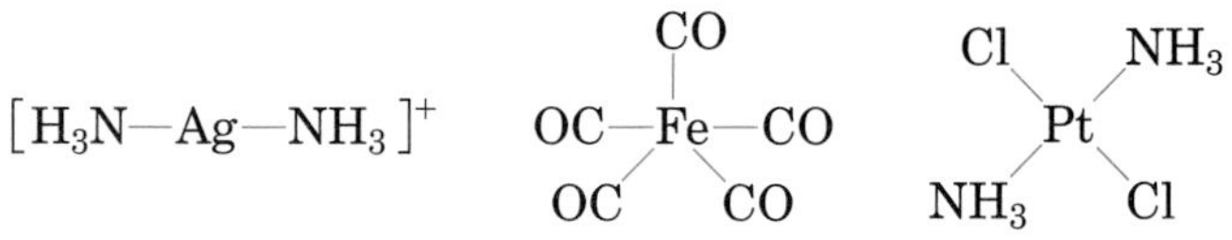

$$[H_3N{-}Ag{-}NH_3]^+ \qquad Fe(CO)_5 \qquad Pt(Cl)_2(NH_3)_2$$

3.6 The Shapes of Molecules

The atoms in a molecule are not arranged randomly. **Molecules have definite three-dimensional shapes.** These shapes are important because they help to determine the properties of the molecules. For molecules with only two atoms, only one shape is possible: The two atoms must be in a straight line. Thus diatomic molecules, such as F_2, HCl, and CO, can only be linear (two points determine a line). With three or more atoms, however, different possibilities arise. Three atoms could be *linear* (all three in a straight line) or *angular* (the three atoms forming an angle). An example of a linear three-atom molecule is carbon dioxide, CO_2; an angular one is water, where the H—O—H angle is about 105° (Fig. 3.5). These shapes can be determined in laboratories through the use of x-ray and electron diffraction.

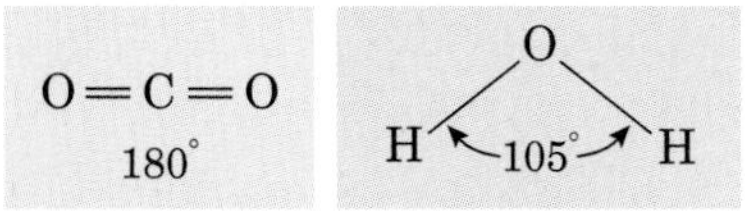

O=C=O 180° | H–O–H 105°

Carbon dioxide Water

Figure 3.5
Carbon dioxide is a linear molecule; water is an angular one.

Is there any way we can predict that CO_2 is linear and H_2O angular? The answer is yes. There is a simple method that allows us to predict accurately the shapes of almost all molecules. The method, called the *valence-shell electron pair repulsion theory* (VSEPR), is based on Lewis structures.

Box 3C

Hemoglobin

The function of the red cells of the blood is to carry oxygen, O_2, to the cells of the body. The actual carrier of the O_2 is a protein called **hemoglobin.** The O_2 attaches itself to the hemoglobin by a coordinate covalent bond. Hemoglobin is a very large molecule (molecular weight about 68 000) whose structure is discussed in Section 18.9, but of all the atoms in this molecule, the only ones that concern us here are four iron atoms. Each of these four atoms is imbedded in a portion of the hemoglobin called **heme,** which has the following structural formula:

H_3C $CH{=}CH_2$
H H
$CH_2{=}HC$ N CH_3
N—Fe—N
H_3C $CH_2CH_2CO_2H$
N
H H
H_3C $CH_2CH_2CO_2H$

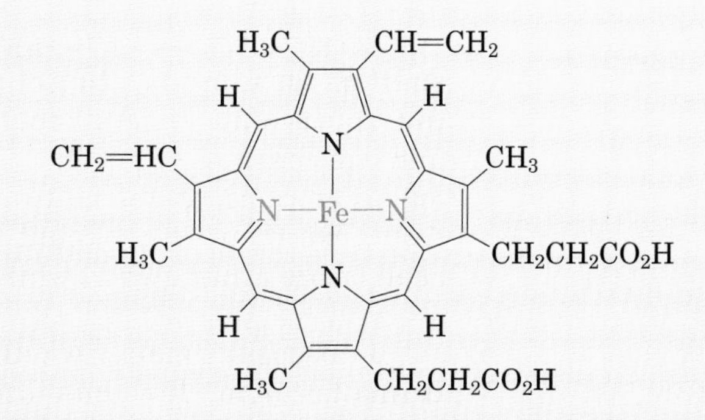

At this point we need not concern ourselves with the meaning of all these symbols. All that is relevant here is that the iron atom in the center is held in place by covalent bonds to two nitrogen atoms and by coordinate covalent bonds to two other nitrogen atoms (shown in color). Every hemoglobin molecule has four of these heme units, each containing one iron atom. When hemoglobin picks up oxygen in the lungs, each O_2 molecule bonds to one of the Fe atoms by a coordinate covalent bond, with both electrons supplied by the oxygen, because the Fe atom, though bonded to four N atoms, still has room for additional electrons.

$$:\ddot{O}{=}\ddot{O}{-}Fe$$

The hemoglobin now carries the O_2 to the cells, where it is released as needed.

The coordinate covalent bonding ability of the Fe in hemoglobin is not restricted to O_2. There are many other species that are also able to use an electron pair to form such a bond with the Fe in hemoglobin. Among these is the poison carbon monoxide, CO.

$$:O{\equiv}C{-}Fe$$

Carbon monoxide is poisonous because the bond it forms with the Fe in hemoglobin is stronger than the O_2—Fe bond. When a person breathes in CO the hemoglobin combines with this molecule rather than with O_2. The cells, deprived of O_2, can no longer function, and the person dies.

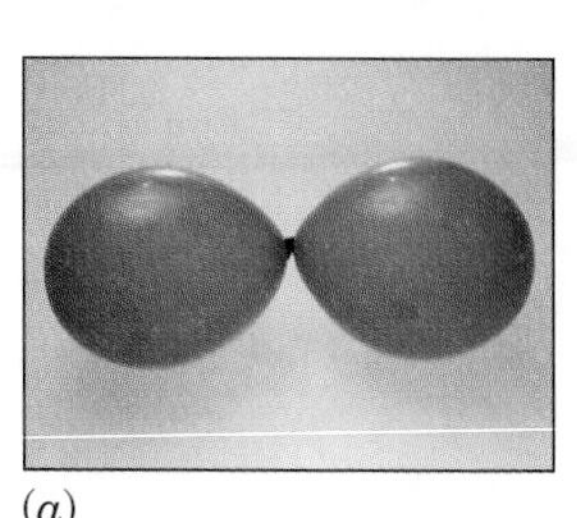
(*a*)

(*b*)

(*c*)

Figure **3.6**
When balloons are tied together, they naturally assume the proper shapes, because that is how they can keep farthest apart. (a) Two balloons are linear. (b) Three balloons are trigonal. (c) Four balloons are tetrahedral. (Photographs by Charles D. Winters.)

Very simply, **VSEPR** says that **the electron pairs in the outer shell of an atom try to get as far away from each other as possible.** This is quite understandable because they are all negatively charged, and like charges repel. What we do is draw the Lewis structure and look for an atom connected to two or more other atoms (this is the *central atom*). We then simply count the number of electron pairs in the outer shell of the central atom while keeping in mind these two rules:

1. It doesn't matter whether an electron pair is unshared or bonded to another atom. It counts just the same.

2. For the purposes of VSEPR, a double or triple bond counts as *one* pair because it occupies one region of space.

These are the simplest cases. Atoms with five or more pairs are not considered in this book.

The number of pairs tells us the shape. We will consider cases in which there are two, three, or four pairs. Each of these numbers corresponds to a different shape, and in each case this shape is adopted by the molecule because that is how the electrons can get farthest away from each other (Fig. 3.6). With *two pairs,* the shape is linear:

With *three pairs,* the shape is trigonal (planar triangular):

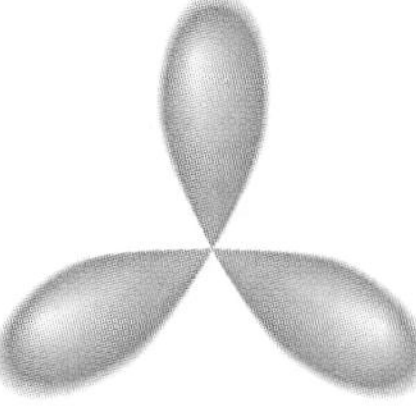

With *four pairs,* the shape is tetrahedral:

A regular tetrahedron is a geometric figure made up of four equilateral triangles. The central atom is *inside* the tetrahedron, and the four electron pairs point to the four corners.

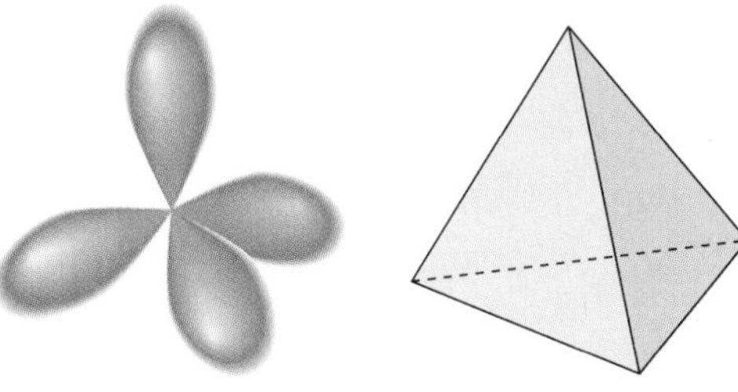

In compounds with more than one central atom, these shapes apply to each central atom. Let us look at some examples.

Two Pairs

There are not many examples in this category, because most molecules obey the octet rule, which means that they have four outer-shell pairs. However, remember that we count double or triple bonds as one pair, so that a molecule can obey the octet rule and still have two pairs for the purposes of VSEPR. Two important molecules in this category are carbon dioxide and acetylene, C_2H_2, whose Lewis structures are

:Ö=C=Ö: H—C≡C—H

In CO_2 the central atom, carbon, has two VSEPR pairs. In C_2H_2 there are two central atoms, each with two VSEPR pairs. The VSEPR theory predicts that both should be linear, and both are.

Three Pairs

As mentioned earlier, boron trifluoride, BF_3, does not obey the octet rule. The Lewis structure shows that there are only six electrons in the outer shell of the boron atom:

```
:F̤̈      F̤̈:
    \   /
     B      120°
     |
    :F̤̈:
```

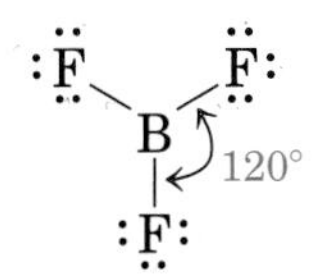

Since there are three electron pairs around the B atom, VSEPR says that this molecule should be triangular, with all four atoms in a plane and F—B—F angles of 120°. This is indeed the case. Another example is ethylene, C_2H_4, whose Lewis structure is

```
H  120°  H
 \      /
  C == C
 /      \
H        H
```

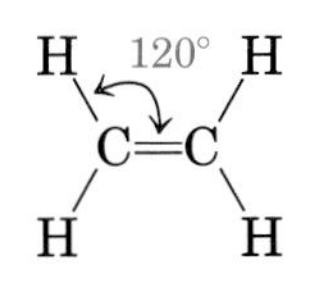

The octet rule is obeyed here, but we count the double bond as one, so each central atom has three pairs, and this compound too is planar, with 120° angles.

Four Pairs

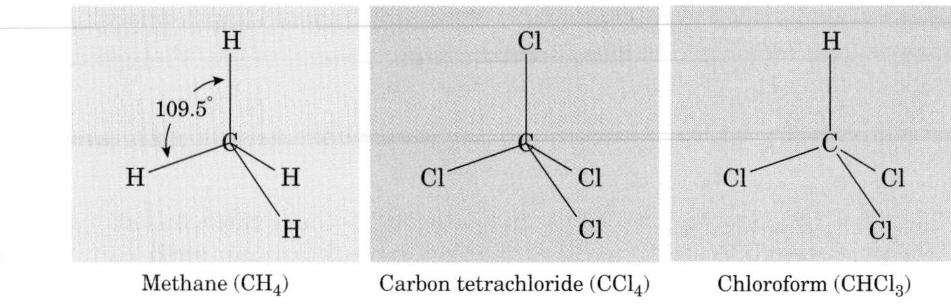

This is the most important case, since all molecules in which atoms have complete octets and no double or triple bonds are in this category. Some of these atoms have four bonds, some three, and some two. The most com-

Methane (CH_4) Carbon tetrachloride (CCl_4) Chloroform ($CHCl_3$)

Figure **3.7**
Some simple carbon compounds with tetrahedral shapes.

mon four-bonded atom is carbon, and VSEPR says that **all carbon atoms with four single bonds have tetrahedral shapes.** Some examples are CH_4, CCl_4, and $CHCl_3$ (Fig. 3.7). The angle between any two groups in this geometry is 109.5°.

Nitrogen, which forms three bonds, also has a complete octet (four pairs), and thus most nitrogen-containing compounds belong in this category. Ammonia is a typical example, as well as the simplest. VSEPR says that the four pairs point to the corners of a tetrahedron, even if one of these pairs is not connected to another atom:

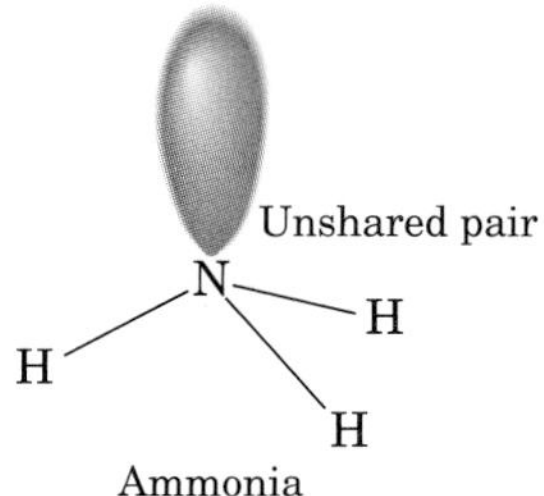

If we look only at the atoms and not at the unshared pair, we see that the ammonia molecule (as well as any other nitrogen-containing molecule with no double or triple bonds) has a *pyramidal* shape.

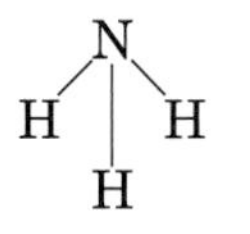

Finally, in this category, we may look at oxygen. In most of its compounds, oxygen has a complete octet (four pairs) but only two bonds. When these are single bonds, the four pairs again mean a tetrahedral structure. The simplest example is water:

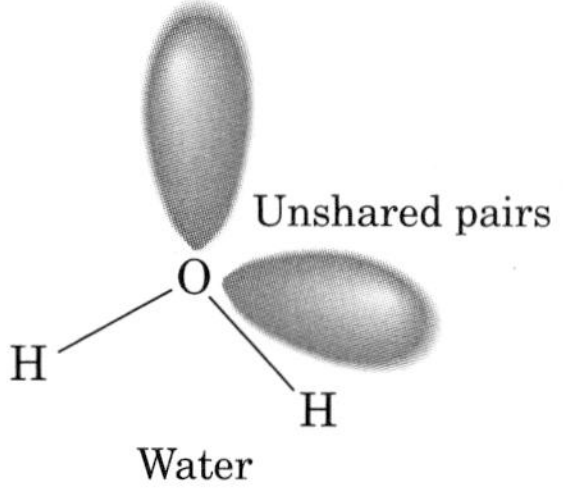

Again, if we look only at the three atoms (and ignore the unshared pairs), water is predicted by VSEPR to have an angle of 109.5° (the tetrahedral angle). The actual angle, 105°, is not very far off. Water is thus an angular molecule.

Though simple, VSEPR theory is very powerful. Without it we would have a hard time explaining what would otherwise be some very puzzling facts. Why should two such similar molecules as HOH and OCO have such different geometries, one angular and the other linear? The same question can be asked about NH_3 and BF_3. Why should one be pyramidal and the other planar triangular? VSEPR theory gives us simple and satisfying answers.

3.7 Electronegativity and Dipoles

In Section 3.4 we discussed the electronic configuration of F_2. We saw that the covalent bond is actually a cloud of electrons (containing two electrons), as shown in Figure 3.4. The electrons in this cloud are equally distributed around the two fluorine nuclei because there is no reason for them to prefer one side over the other. But that is not the case when a covalent bond connects two different atoms. For example, the Lewis structure for hydrogen fluoride, HF, is

$$\mathrm{H{-}\ddot{\underset{..}{F}}:}$$

In this case the two electrons in the bond do not remain equidistant from the two nuclei; they are closer to the F atom than to the H atom. That is, the electron cloud is distorted, as shown in Figure 3.8, because the F atom attracts the electron pair of the bond more than the H atom does.

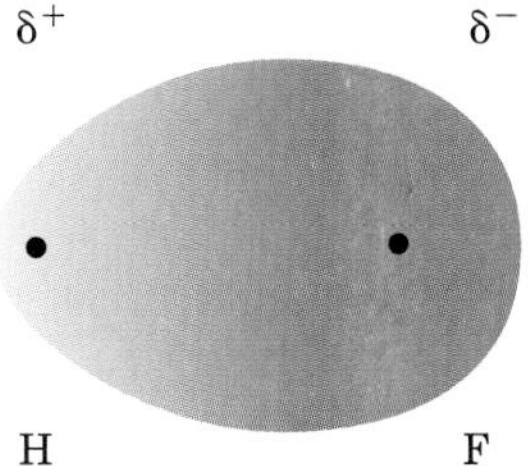

Figure **3.8**
The covalent bond in HF is distorted toward the fluorine atom.

The name given to this attraction is **electronegativity.** Fluorine has a higher electronegativity than hydrogen and so attracts the electrons of the bond more. Note that we are not talking about attracting anything outside the molecule. Electronegativity refers only to the attraction that an atom has for a pair of electrons *in a covalent bond to itself.*

We now know that fluorine has a higher electronegativity than hydrogen, but how much higher? Linus Pauling (1901–1994) set up a scale in which each element is assigned a number that measures its electronegativity. Part of this scale is shown in Table 3.2. Fluorine has the highest electronegativity, 4.0, and cesium the lowest, 0.7. With this scale we can tell in which direction the electrons of any covalent bond are shifted.

EXAMPLE

Toward which atom are the electrons shifted in (a) sulfur dichloride, SCl_2 (b) iodine bromide, IBr?

Answer • (a) The electronegativity of Cl is 3.0, and that of S is 2.5. The electrons are shifted toward the chlorine atoms. (b) The electrons are shifted toward the Br, but since the difference is small, the distortion is not very much.

Table 3.2 Electronegativities of Some Elements

H 2.1						
Li 1.0	Be 1.5	B 2.0	C 2.5	N 3.0	O 3.5	F 4.0
Na 0.9	Mg 1.2	Al 1.5	Si 1.8	P 2.1	S 2.5	Cl 3.0
Rb 0.8			Ge 1.8	As 2.0	Se 2.4	Br 2.8
Cs 0.7					Te 2.1	I 2.5

Problem **3.6**

Toward which atom are the electrons shifted in (a) oxygen difluoride, OF_2 (b) phosphorus tribromide, PBr_3?

Table 3.2 shows that electronegativity, like ionization energy (Sec. 2.7), changes with position in the periodic table. We can see two trends:

1. Electronegativity increases from left to right in the table (except for the noble gases).
2. Electronegativity decreases down any column.

Thus electronegativity correlates with metallic character: Nonmetals have high electronegativity; metals have low electronegativity.

Let us now return to HF. The electron cloud is distorted so that it covers the fluorine atom more than the hydrogen atom (Fig. 3.8). This means that there is a partial negative charge on the fluorine atom and a partial positive charge on the hydrogen, which we can show as

The Greek letter δ (delta) means "partial."

$$^{\delta+}\text{H—F}^{\delta-}$$

There are two things we should remember:

1. These are *partial* charges, that is, less than +1 or −1. If the electron pair were transferred *completely,* the charges would be +1 and −1. Since the cloud is not completely transferred, the charges are less than that. How much less varies from case to case and depends on the difference between the electronegativities of the two atoms (the greater the difference, the greater the partial charges).

2. Whatever the magnitude of $\delta+$, it must be the same as $\delta-$ because the sum of $\delta+$ and $\delta-$ must be zero (the total number of electrons in the whole molecule is equal to the total number of protons, and so the net charge must be zero).

What are the consequences of these partial charges? Because each end of the H—F molecule has a charge, the molecule behaves like a little bar magnet. If you put two bar magnets together, the north poles repel each other and attract the south poles (Fig. 3.9). The molecules behave exactly the same way. The H end of one repels the H end of the other and attracts the F end (Fig. 3.9).

Because the H—F molecule has two poles, like a magnet, we call it a **dipole** or a **polar molecule.** We have already seen how we can tell which is the negative and which the positive end. The *extent* of charge separation can be measured in laboratories by a special technique. For molecules with two atoms, the extent is related to the electronegativities. For example, we would expect HF to be more polar than HCl (and it is), whereas F_2 and H_2 should be completely nonpolar (and they are).

The electronegativity difference between H and F is 4.0 − 2.1 = 1.9; between H and Cl it is only 3.0 − 2.1 = 0.9. Thus, HF is more polar.

The polarity of a molecule depends not only on electronegativities but also on the three-dimensional geometry of a molecule. For example, let us consider H_2O, which has an angular geometry:

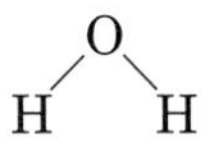

```
  O
 / \
H   H
```

The electronegativity of oxygen is 3.5 and that of hydrogen is 2.1. Thus the oxygen must have $\delta-$ and the hydrogen $\delta+$. Molecules behave like

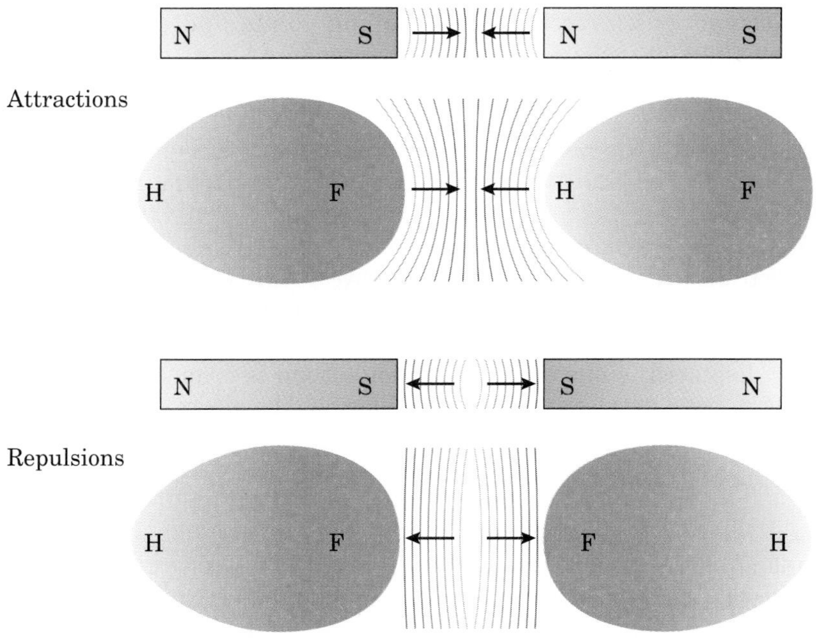

Attractions

N S → ← N S

H F → ← H F

Repulsions

N S ← → S N

H F ← → F H

Figure **3.9** • Dipoles, like magnets, attract and repel each other.

dipoles, however, not like tripoles. Each polar molecule has only one negative end and one positive end. We know where the negative end is: on the oxygen. But where is the positive end? The answer is, halfway between the hydrogens:

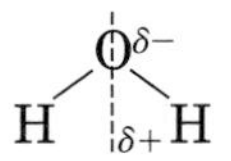

$$\overset{\delta-}{\text{O}}(\text{H})_2 \quad \delta+ \text{ between the H atoms}$$

Sometimes the geometry is such that the bond polarities completely cancel. For example, consider carbon dioxide, which, as we saw, is a linear molecule:

$$\overset{\delta-}{\text{O}}=\overset{\delta+}{\text{C}}=\overset{\delta-}{\text{O}}$$

Each C=O bond is polar. Since the electronegativities are 2.5 for carbon and 3.5 for oxygen, the electrons are shifted away from the central carbon and toward the two oxygens. Yet the entire molecule is nonpolar. Because the two polar bonds point in opposite directions (180° apart) and because the polarity of each bond is the same, the two polarities cancel.

Another such case is the tetrahedral molecule carbon tetrachloride, where four C—Cl bond polarities cancel:

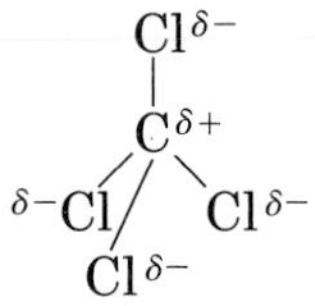

$$\text{C}^{\delta+}(\text{Cl}^{\delta-})_4$$

This compound is also nonpolar.

Whether a molecule is polar or not has a major effect on the behavior of compounds, especially their boiling points and solubilities. We shall see the consequences of this in Chapter 5 and in several subsequent chapters.

3.8 How to Predict the Kinds of Bonds that Form

In Section 3.7 we learned that, in a bond between two different atoms, the electrons are not usually halfway between the two but are shifted toward the atom of higher electronegativity. When the difference in electronegativities is great, we can say that the electron pair is (almost) completely displaced to one side, and an ionic bond is formed. For example, consider a bond between cesium and chlorine. The electronegativity of Cl is 3.0; that of Cs is 0.7.

$$\text{Cs}\cdot + \cdot\ddot{\underset{\cdot\cdot}{\text{Cl}}}\!: \longrightarrow \text{Cs}^+ \quad :\ddot{\underset{\cdot\cdot}{\text{Cl}}}\!:^-$$

This is a very large difference; it means that Cs has very little attraction for the electron pair and essentially allows the Cl to have it completely. The Cs, having contributed one electron to the bond, no longer

Box 3D

Nitric Oxide—Air Pollutant

Nitric oxide is a small, very reactive molecule whose Lewis structure is

$$:\ddot{N}{=}\ddot{O}\cdot$$

The high reactivity of the molecule is due to the unpaired electron (p. 87). The N—O bond is polar, but only to a small extent: The electronegativity difference is 0.5. The low polarity and the small size of the molecule allows NO to penetrate and exit from cells, by passing through nonpolar cell membranes (Box 24F).

The importance of NO was first realized when it was noticed that it is a by-product of fossil fuel burning that contributes to the formation of photochemical smog (as in warm-climate cities such as Los Angeles or Mexico City) and acid rain (Box 6A). Due to its reactivity, NO in the air rapidly reacts with O_2 forming NO_2, which is the major acidifying component of acid rain. When inhaled, NO passes from the lungs into the blood vessels and interacts with the iron of hemoglobin (Box 3C), decreasing its ability to carry oxygen.

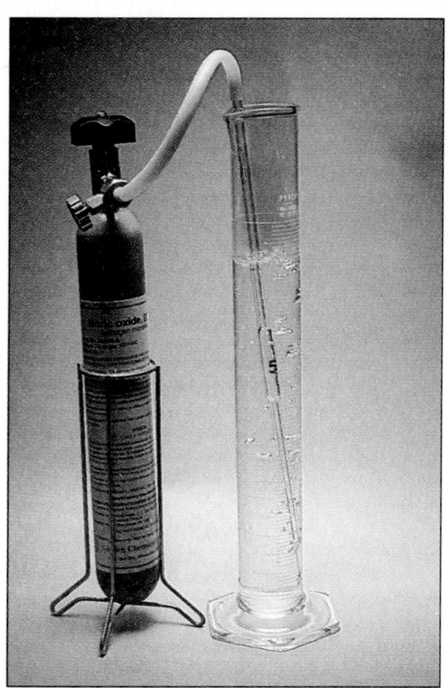

Colorless nitric oxide gas, coming from the tank, bubbles through the water. When it reaches the air it is oxidized to brown NO_2. (Photograph by Charles D. Winters.)

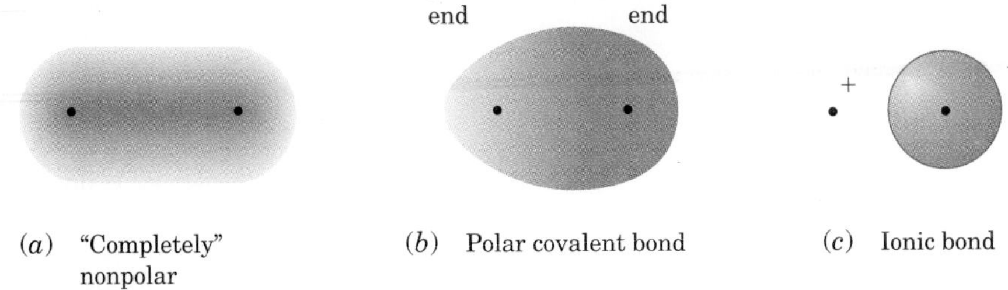

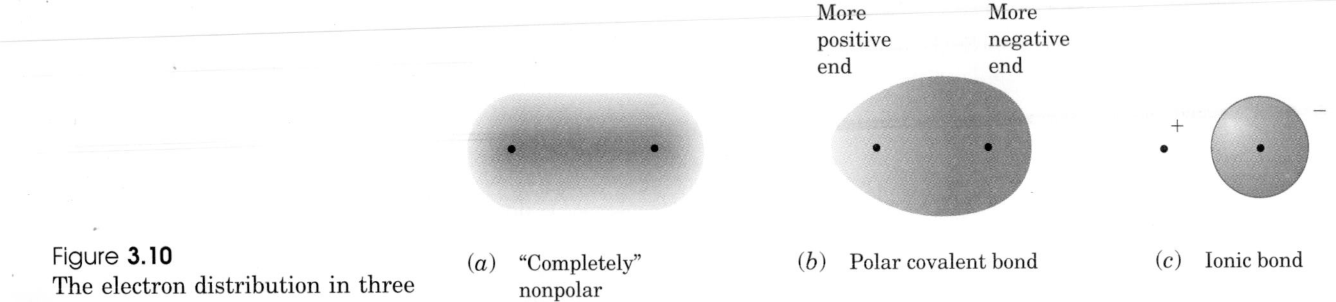

Figure **3.10**
The electron distribution in three types of bonds.

has it and has become a positive cesium ion. The chlorine atom has also contributed one electron but now has both, so that it has become a negative chloride ion. In this case we don't write $\delta+$ or $\delta-$, but + or −, because each ion has acquired a full charge. Cesium chloride, CsCl, is therefore an ionic compound. (We must remember that in ionic compounds the bonding is not one to one. Instead, in the solid state the ions are arranged in a lattice like the one shown in Figure 3.2.)

We can thus have basically three kinds of bonds between two atoms: a "completely" nonpolar covalent bond (when the electronegativities are identical or fairly similar), a polar covalent bond, or an ionic bond (when the electronegativities are very different) (see Fig. 3.10). We shall say that a difference of 1.8 or less gives a polar covalent bond, and a difference of 1.9 or more gives an ionic bond. Although this is a purely arbitrary rule, we shall not go very wrong in using it.

EXAMPLE

Classify as nonpolar, polar covalent, or ionic: (a) Cl—Cl (b) C—N (c) Rb—F.

Answer • To answer this question we look at the difference in electronegativities (Table 3.2). (a) This is a nonpolar bond because the electronegativity difference is zero. (b) The electronegativity difference is $3.0 - 2.5 = 0.5$, which is less than 1.9, so this is a polar covalent bond. (c) The electronegativity difference is $4.0 - 0.8 = 3.2$, which is greater than 1.8, so this is an ionic bond.

Problem **3.7**

Classify as nonpolar, polar covalent, or ionic: (a) N—N (b) S—Cl (c) Na—F.

3.9 What Bonds to What?

We have now learned to predict the polarity of bonds and the shape of molecules (Sec. 3.6). In this section, let us consider one more thing: What bonds to what? In Section 2.5 we saw that most elements are either metals or nonmetals (or somewhere between). With respect to bonding, the main tendencies of metals and nonmetals can be summarized:

Metals	Form **positive** ions	Usually reluctant to form covalent bonds
Nonmetals	Form **negative** ions	Form covalent bonds

We can then state the following:

1. Metals bond to nonmetals, usually by ionic bonds. If the difference in electronegativity is moderate, they use covalent bonds.
2. Metals do not form compounds with other metals.
3. Nonmetals bond to nonmetals, usually by covalent bonds.

These rules are broad and general and have many exceptions in specific cases. Still, they serve as a rough guide and are valid for most cases.

3.10 Polyatomic Ions

Many compounds have both ionic and covalent bonds. In such cases the compounds contain ions with more than one atom. These ions are called **polyatomic ions.** An example is the important compound sodium hydroxide, NaOH. This is an ionic compound containing the positive Na^+ ion and the negative OH^- ion. The structure is not very different from that of NaCl (Fig. 3.2). The difference is that Cl^- is a monatomic ion and OH^- is a polyatomic ion. The O and H are connected by a covalent bond, and the Lewis structure is

$$\left[:\ddot{\underset{..}{O}}—H\right]^-$$

Another notable polyatomic ion, one important in biochemistry, is the phosphate ion:

$$\left[\begin{array}{c} :\ddot{\underset{..}{O}}: \\ | \\ :\ddot{\underset{..}{O}}—P—\ddot{\underset{..}{O}}: \\ \| \\ \underset{..}{O}: \end{array}\right]^{3-}$$

A list of some important polyatomic ions is given in Table 3.3. As with OH^-, the atoms in each ion are connected by covalent bonds, and Lewis structures can be drawn for each (we did this for SO_4^{2-} in Sec. 3.4). Most of the important polyatomic ions are negatively charged, but there are some positively charged ones also, most notably the ammonium ion, NH_4^+:

$$\left[\begin{array}{c} H \\ | \\ H—N—H \\ | \\ H \end{array}\right]^+$$

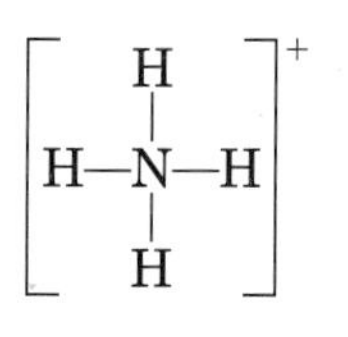

Table 3.3 Some Important Polyatomic Ions

Ion	Name
NH_4^+	Ammonium
OH^-	Hydroxide
NO_3^-	Nitrate
NO_2^-	Nitrite
CH_3COO^-	Acetate
CN^-	Cyanide
MnO_4^-	Permanganate
CO_3^{2-}	Carbonate
HCO_3^-	Bicarbonate
SO_3^{2-}	Sulfite
HSO_3^-	Bisulfite
SO_4^{2-}	Sulfate
PO_4^{3-}	Phosphate
HPO_4^{2-}	Hydrogen phosphate
$H_2PO_4^-$	Dihydrogen phosphate

3.11 Naming of Simple Inorganic Compounds

Inorganic compounds are those that do not contain the element carbon. The naming of organic compounds will be discussed beginning in Chapter 10.

There are a large number of inorganic compounds. It would be convenient to be able to name them just by looking at their formulas and, conversely, to be able to write their formulas by looking at their names. This is relatively easy to do for the simple ones, which include most of the inorganic compounds you are ever likely to come across.

Binary Ionic Compounds

A **binary** compound contains just two elements.

The rule for binary ionic compounds is that the positive ion (cation) is written first and then the negative ion (anion), in both the formula and the name. The name of the compound is simply the name of the positive ion followed by that of the negative ion.

EXAMPLE

Name these compounds: LiBr, Ag_2S, SrF_2.

Answer • Lithium bromide, silver sulfide, strontium fluoride.

Note that the *number* of ions of each type is *not* given in the name. SrF_2 is called strontium fluoride, and not strontium difluoride. Note also that negative ions of a single element usually end in **ide**; for example, sulfide, iodide, oxide.

Problem 3.8

Name these compounds: MgO, BaI_2, NaBr.

EXAMPLE

Give formulas for barium hydride and sodium sulfide.

Answer • BaH_2, Na_2S.

Problem 3.9

Give formulas for strontium sulfide and sodium fluoride.

It is mostly transition elements that form ions with more than one charge.

In Table 3.1 we showed that some metals can form positive ions with more than one charge (for example, Cu^+, Cu^{2+}). In such cases we use Roman numerals in the name to show the charge. We do this because we must have some way of distinguishing them.

EXAMPLE

Name these compounds: CuCl, $CuCl_2$.

Answer • Copper(I) chloride, copper(II) chloride.

An older way to do this was to use the suffix **ic** for the higher charge and **ous** for the lower one. By this method CuCl is cuprous chloride and $CuCl_2$ is cupric chloride.

The older method is still in widespread use. Note that the Latin forms *cupr* and *ferr* are used in the older method, but not in the newer one.

EXAMPLE

Name $FeBr_2$ and $FeBr_3$ by the two methods.

Answer • $FeBr_2$: ferrous bromide; iron(II) bromide. $FeBr_3$: ferric bromide; iron(III) bromide.

Problem 3.10

Name CuBr, $HgBr_2$, and Fe_2O_3 (see Table 3.1).

Ionic Compounds that Contain Polyatomic Ions

The rule for these cases is exactly the same (that is, the positive ion comes first), except that now one or both of the ions contain two or more elements.

EXAMPLE

Name these compounds: $NaNO_3$, $AlPO_4$, $(NH_4)_2SO_3$, NaH_2PO_4.

Answer • Sodium nitrate, aluminum phosphate, ammonium sulfite, sodium dihydrogen phosphate.

Problem 3.11

Name these compounds: K_2HPO_4, $Al_2(SO_4)_3$, $FeSO_3$.

Binary Molecular Compounds

Compounds that do not contain ions are called *molecular compounds*.

There are two rules we use for these compounds. (1) The less-electronegative element (see Table 3.2) comes first, in both the formula and the name; and (2) prefixes *di*, *tri*, and so on, are used to show the number of atoms of each element.

Prefixes up to 8 are:

1	mono	5	penta
2	di	6	hexa
3	tri	7	hepta
4	tetra	8	octa

EXAMPLE

Name these compounds: NO, SF_2, P_2O_3.

Answer • Nitrogen oxide (more commonly called nitric oxide), sulfur difluoride, diphosphorus trioxide.

Problem 3.12

Name these compounds: NO_2, PBr_3, SCl_2.

Note that the prefix *mono* is usually omitted from names, but there are exceptions; for example CO is carbon monoxide.

We will not deal with the names of molecular compounds that contain more than two elements in this chapter. Some of them will be given in the chapter on acids and bases (Chapter 8) and others in the organic and biochemistry chapters.

SUMMARY

Atoms form positive ions (**cations**) by losing electrons or negative ions (**anions**) by gaining electrons. An **ionic bond** is the attraction between positive and negative ions. Atoms usually lose or gain enough electrons to convert them to ions with complete octets. Atoms and ions have completely different properties. **Ionic compounds** are not composed of molecules. In the solid state they consist of regular arrays of ions.

Covalent bonds are formed when atoms share electrons. Atomic orbitals overlap to give **molecular orbitals.** Here too the atoms usually try to form octets and share enough electrons to do so, although there are some important exceptions. Each atom has a strong tendency to form a certain number of covalent bonds. Carbon always forms four bonds; nitrogen forms three; oxygen forms two. The four bonds of carbon can be made from single, double, or triple bonds. In **coordinate covalent bonds,** one atom supplies both electrons.

A **molecular formula** shows all the atoms in the molecule. A **structural formula** also shows all the bonds connecting them. A **Lewis structure** also shows all unshared outer-shell electrons.

Molecules have definite three-dimensional shapes, and these shapes can be determined by **VSEPR theory,** which is based on counting all electron pairs in the outer shell of each atom. In this theory, double and triple bonds are treated as single pairs. Atoms with two pairs are **linear,** and those with three pairs are **planar triangular.** With four pairs the shape is **tetrahedral.**

Electronegativity is a measure of how much an atom attracts the electrons in a covalent bond. When electronegativities of the two atoms in a bond differ, the atom with the higher electronegativity preferentially attracts the electron pair. This gives rise to **polar covalent bonds.** Molecules with polar covalent bonds may be **dipoles.** When a very large difference in electronegativities (1.9 or greater) exists, we say that the bond is ionic.

Simple ionic inorganic compounds are named by giving the name of the positive ion followed by the name of the negative ion. Binary molecular compounds are named by giving the less-electronegative element first, followed by the more-electronegative one. In this case, prefixes are used to show the number of atoms of each element.

KEY TERMS

Anion (Sec. 3.2)
Binary compound (Sec. 3.11)
Cation (Sec. 3.2)
Chemical bond (Sec. 3.1)
Chemical species (Sec. 3.2)
Coordinate covalent bond (Sec. 3.5)
Covalent bond (Sec. 3.4)
Dipole (Sec. 3.7)
Double bond (Sec. 3.4)
Electronegativity (Sec. 3.7)
Free radical (Sec. 3.4)
Inorganic compound (Sec. 3.11)
Ionic bond (Sec. 3.3)
Ionic compound (Sec. 3.3)
Lewis structure (Sec. 3.4)
Linear shape (Sec. 3.6)
Molecular compound (Sec. 3.11)
Molecular formula (Sec. 3.4)
Molecular orbital (Sec. 3.4)
Octet rule (Sec. 3.2)
Polar covalent bond (Sec. 3.8)
Polar molecule (Sec. 3.7)
Polyatomic ion (Sec. 3.10)
Pyramidal shape (Sec. 3.6)
Sigma (σ) orbital (Sec. 3.4)
Structural formula (Sec. 3.4)
Tetrahedral shape (Sec. 3.6)
Trigonal shape (Sec. 3.6)
Triple bond (Sec. 3.4)
Valence shell (Sec. 3.2)
VSEPR (Sec. 3.6)

PROBLEMS

Ions and Ionic Bonding

3.13 Calcium is a soft metal that reacts with water. When a physician prescribes calcium for osteoporosis, what species will the patient actually get?

3.14 How many electrons does each of these atoms have to gain or lose to reach a complete outer shell?
(a) Li (b) Cl (c) Ne (d) Al (e) Sr (f) S (g) Si (h) O (i) Kr (j) P

3.15 Write the formula for the most stable ion of (a) Mg (b) F (c) Al (d) S (e) K (f) Br.

3.16 Predict which of these ions are stable:
(a) I^- (b) Se^{2+} (c) Na^+ (d) S^{2-} (e) Li^{2+} (f) Ba^{3+} (g) Br^{2-} (h) C^{4-} (i) Ca^+ (j) Ar^+ (k) Na^- (l) Cs^+

3.17 Why is Li^- not a stable ion?

3.18 The ion O^{6+} has a complete outer shell. Why is this ion never found in normal matter?

3.19 Why are carbon and silicon reluctant to form ionic bonds?

3.20 Table 3.1 shows the following ions of Cu: Cu^+ and Cu^{2+}. Do these violate the octet rule? Explain.

3.21 Is there such a thing as a KF molecule? Explain.

3.22 Describe the structure of sodium chloride in the solid state.

3.23 Since sodium sulfate, Na_2SO_4, is an ionic compound, there are no discrete Na_2SO_4 molecules. What does the "2" in Na_2SO_4 mean?

3.24 Complete the chart by writing formulas:

	Br^-	ClO_4^-	O^{2-}	NO_3^-	SO_4^{2-}	PO_4^{3-}	OH^-
Li^+							
Ca^{2+}							
Co^{3+}							
K^+							
Cu^{2+}							

3.25 Write formulas for ionic compounds formed from these elements:
(a) sodium and bromine
(b) potassium and oxygen
(c) aluminum and chlorine
(d) barium and chlorine
(e) calcium and oxygen

3.26 The compound Na_2SO_4 is made up of Na^+ and SO_4^{2-} ions. Write the formulas for the ions in
(a) NaBr (b) $FeSO_3$ (c) $Mg_3(PO_4)_2$ (d) KH_2PO_4 (e) $NaHCO_3$ (f) $Ba(NO_3)_2$

3.27 What are the charges on the ions in each of the following? (a) CaS (b) MgF_2 (c) Cs_2O (d) $ScCl_3$ (e) Al_2S_3

Covalent Bonding

3.28 Explain how a covalent bond forms.

3.29 How many covalent bonds are normally formed by
(a) N (b) F (c) C (d) Br (e) O

3.30 What is (a) a double bond (b) a triple bond?

3.31 Draw the structural formula for
(a) CH_4 (b) C_2H_2 (c) C_2H_4 (d) BF_3 (e) CH_2O (f) C_3Cl_8

3.32 Some of the following formulas are incorrect (that is, they do not represent any real compound) because they contain atoms that do not have their normal number of covalent bonds. Which compounds are they, and which atoms have the incorrect number of bonds?

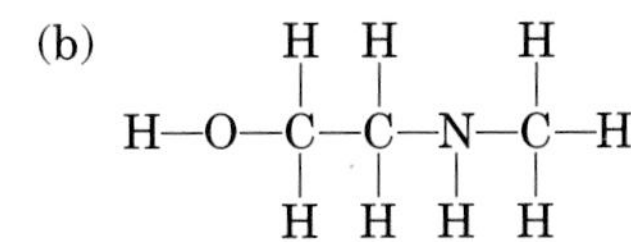

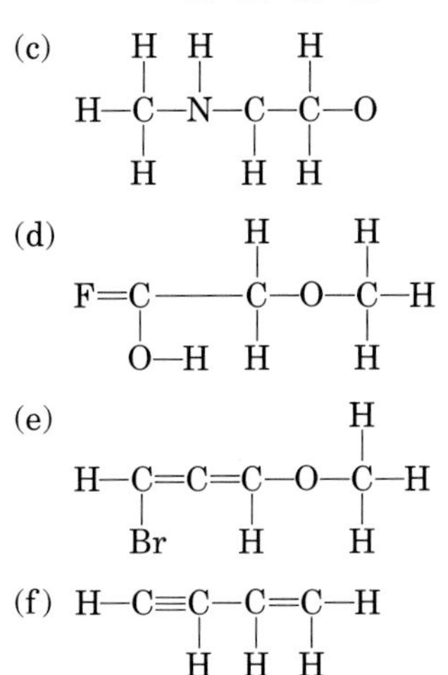

(a)
```
     H  H
     |  |
Cl—C=C—H
     |
     H
```

(b)
```
          H  H     H
          |  |     |
   H—O—C—C—N—C—H
          |  |  |  |
          H  H  H  H
```

(c)
```
      H  H     H
      |  |     |
   H—C—N—C—C—O
      |     |  |
      H     H  H
```

(d)
```
              H     H
              |     |
   F=C———C—O—C—H
      |       |     |
      O—H     H     H
```

(e)
```
                      H
                      |
   H—C=C=C—O—C—H
      |     |         |
      Br    H         H
```

(f)
```
   H—C≡C—C=C—H
         |  |  |
         H  H  H
```

3.33 Draw structural formulas for molecules in which a carbon atom is connected by a double bond to (a) another carbon atom (b) an oxygen atom (c) a nitrogen atom.

3.34 Why can't a carbon atom be connected to a chlorine atom by a double bond?

3.35 Explain why argon does not form (a) ionic bonds (b) covalent bonds.

Lewis Structures

3.36 What is the difference between a molecular formula, a structural formula, and a Lewis structure?

3.37 Draw the Lewis dot structure for (a) K (b) Se (c) N (d) I (e) Ar (f) Be (g) Cl.

3.38 Give the total number of outer-shell electrons in

(a) NH_3 (e) CCl_4
(b) C_3H_6 (f) NO_3^-
(c) $C_2H_4O_2$ (g) PO_4^{3-}
(d) C_2H_6O

3.39 Draw the Lewis structure for the following (in each case, there is only one way the atoms can be connected):

(a) Br_2 (d) CN^-
(b) H_2S (e) NH_4^+
(c) N_2H_2 (f) BF_4^-

3.40 Draw the Lewis structure for

(a) HCN H C N

(b) HNO_2 H O N O

(c) H_2SO_4

        O
H O S O H
        O

(d) $C_2H_3O_2^-$

  H
H C C O
  H O

(e) CO_3^{2-}

O C O
  O

3.41 What is the difference between a bromine atom, a bromine molecule, and a bromide ion? Write the formula for each.

***3.42** Which of these have an atom that does not obey the octet rule (not all of these are stable molecules)?

(a) BF_3 (e) CH_3
(b) CF_4 (f) N_2
(c) BeF_2 (g) NO
(d) C_2H_4

3.43 What is the maximum number of electrons that each of these atoms can hold in its outer shell when it is present in a molecular compound? (a) S (b) O (c) P (d) N (e) C

Coordinate Covalent Bonds

***3.44** Draw a Lewis structure for the cyanide ion, CN^-. Is a coordinate covalent bond present?

3.45 The reaction between ammonia, NH_3, and the hydrogen ion, H^+, produces the ammonium ion, NH_4^+. Show how this reaction involves the formation of a coordinate covalent bond.

The Shapes of Molecules

3.46 What is the shape of a molecule whose central atom contains (a) two (b) three (c) four electron pairs?

3.47 Predict the shape of

(a) CH_4 (g) C_2Cl_4
(b) PH_3 (h) NO_2^-
(c) BF_3 (i) H_3O^+
(d) C_2F_2 (j) NO_3^-
(e) SO_2 (k) SeO_2
(f) SO_3 (l) PCl_3

Electronegativity and Dipoles

3.48 Which atom are the electrons shifted toward in a bond between

(a) K and Cl (e) Na and O
(b) S and N (f) P and S
(c) C and O (g) H and O
(d) Cl and Br

3.49 Which of these bonds is the most polar? The least polar? (a) C—N (b) C—C (c) C—O

3.50 The H_2S molecule has an angular geometry similar to that of water. Is the H_2S molecule a dipole? Where are the partial charges located?

***3.51** Like ethylene, the chlorinated ethylenes are planar (flat). Which, if any, of these molecules is a dipole?

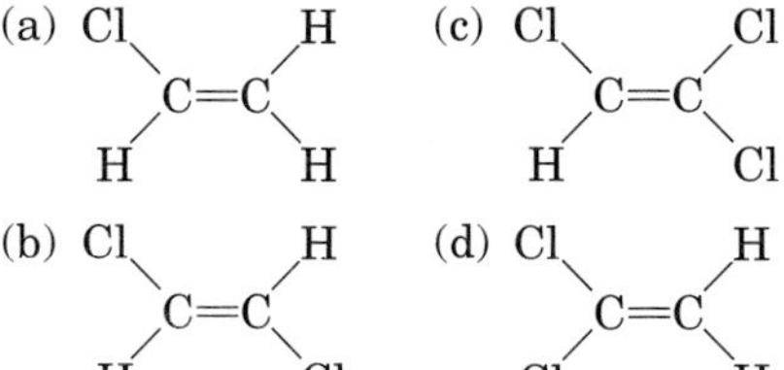

Kinds of Bonds

3.52 In each case, tell whether the bond is ionic, polar covalent, or nonpolar covalent:

(a) Br_2 (e) SiH_4
(b) BrCl (f) CO
(c) HCl (g) N_2
(d) SrF_2 (h) CsCl

3.53 Predict whether a bond will form and, if so, whether it will be ionic or covalent, between

(a) Cl and I (e) Na and Cr
(b) Li and K (f) Ar and S
(c) O and K (g) Na and Ca
(d) Al and N (h) Li and Ne

Polyatomic Ions

3.54 Draw Lewis structures for the following, each of which contains at least one ionic bond (show the ions as separate structures):

(a) Na_2SO_3 (d) NH_4OH
(b) KNO_3 (e) K_2HPO_4
(c) Cs_2CO_3

3.55 The perchlorate ion is ClO_4^-. (a) Draw its Lewis structure. (b) Write the molecular formula for calcium perchlorate.

Naming Inorganic Compounds

3.56 Write formulas for the following ionic compounds (see Tables 3.1 and 3.3):

(a) potassium bromide
(b) calcium oxide
(c) lead(II) hydroxide
(d) copper(II) phosphate
(e) lithium sulfate
(f) iron(III) sulfide
(g) ammonium bisulfite
(h) magnesium acetate
(i) strontium dihydrogen phosphate
(j) silver(I) carbonate
(k) strontium chloride
(l) barium permanganate
(m) mercury(II) oxide

3.57 Name these ionic compounds:

(a) NaF (g) $Sr_3(PO_4)_2$
(b) MgS (h) $Fe(OH)_2$
(c) Al_2O_3 (i) NaH_2PO_4
(d) $BaCl_2$ (j) $Pb(CH_3COO)_2$
(e) $Ca(HSO_3)_2$ (k) BaH_2
(f) KI (l) $(NH_4)_2HPO_4$

3.58 Write molecular formulas for these compounds:

(a) carbon monoxide
(b) bromine chloride
(c) phosphorus pentachloride
(d) silicon tetrabromide
(e) dinitrogen trioxide
(f) sulfur hexafluoride

3.59 Name these molecular compounds:

(a) H_2S (e) N_2O_3
(b) IF (f) SO_2
(c) IF_7 (g) SeO_3
(d) NO

Boxes

3.60 (Box 3A) What is the main use of Ca^{2+} in the body?

3.61 (Box 3A) What are the main ions inside the body cells?

3.62 (Box 3A) What are the main ions in the body fluids outside the cells?

3.63 (Box 3B) Why is sodium iodide often present in the table salt we buy in the supermarket?

3.64 (Box 3B) What is the medical use of barium sulfate?

3.65 (Box 3B) What is the medical use of $KMnO_4$?

3.66 (Box 3C) How is the central iron(II) ion bound to the nitrogens in heme?

3.67 (Box 3D) Why is NO considered to be a health hazard that pollutes the air?

Additional Problems

3.68 Write formulas for the compounds formed between

(a) potassium ion and sulfide ion
(b) aluminum ion and nitrate ion
(c) lead(II) ion and sulfite ion
(d) calcium ion and bicarbonate ion
(e) mercury(II) ion and dihydrogen phosphate ion
(f) barium ion and hydride ion

3.69 Show how BF_3 forms a coordinate covalent bond with phosphine, PH_3.

3.70 Name these ionic compounds:

(a) Na_2SO_3 (e) $(NH_4)_2CO_3$
(b) $KMnO_4$ (f) $LiNO_2$
(c) $Sr(CN)_2$ (g) $NaHCO_3$
(d) FeO

***3.71** What is the shape of the NH_4^+ ion?

***3.72** Predict which of these molecules will be polar:

(a) CBr_4 (d) CH_4
(b) $CHCl_3$ (e) CCl_2I_2
(c) CH_2Br_2 (f) CH_3Cl

3.73 Why is each of the following ions never found in normal matter?

(a) O^{3-} (b) C^{4-} (c) S^+

3.74 Which of these magnesium compounds are ionic compounds? (a) MgF_2 (b) Mg_2C (c) MgS

3.75 A covalent bond is directed in space from one particular atom to another. Can you say the same thing for an ionic bond? Explain.

***3.76** The two compounds $AlCl_3$ and NCl_3 would seem to be very similar in structure. Yet one has a flat planar geometry, while the other has a pyramidal geometry. Explain why these compounds have such different structures.

3.77 In analogy to Figure 3.4, show how the two 1*s* orbitals overlap to form a molecular orbital in the H_2 molecule.

3.78 Which type of element (metal or nonmetal) is more likely to form (a) anions (b) cations?

***3.79** The correct Lewis structure for ethylene is

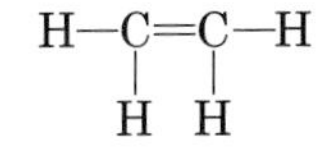

```
H—C=C—H
  |  |
  H  H
```

Two other structures that might be drawn are incorrect:

```
      ..          ..  ..
(a) H—C—C—H   (b) H—C—C—H
      |  |          |  |
      H  H          H  H
```

Explain why each of these is incorrect.

3.80 What type of bond is formed between BF_3 and NCl_3?

***3.81** In the ferricyanide ion $Fe(CN)_6^{3-}$ there are six coordinate covalent bonds. Draw the Lewis structure of the ferricyanide ion.

3.82 A book on pharmacology for nursing care contains the following sentence: "Suppression of aldosterone release [in human blood] further enhances secretion of sodium, while at the same time causing retention of potassium." Does this mean that sodium and potassium metals are present in human blood?

Chapter **4**

Chemical reactions

4.1 Introduction

In Chapter 1 we learned that chemistry is mainly concerned with two things: the structure of matter and the transformations of one form of matter to another. In Chapters 2 and 3 we discussed the first of these topics ("what is matter made of"), and now we are ready to turn our attention to the other. In a chemical change, also called a **chemical reaction,** one or more original substances (called the **starting materials** or **reactants**) are converted to one or more new substances (called the **products**). Chemical reactions are all around us. They fuel and keep alive the cells of living tissues; they occur when we light a match, cook dinner, start a car, listen to a portable radio, or watch television (Fig. 4.1). Most of the world's manufacturing processes involve chemical reactions (Fig. 4.2); they include petroleum refining, metal-smelting operations, and food processing, as well as the manufacture of drugs, plastics, artificial fibers, explosives, and many other materials.

(a)

Figure **4.1**
Examples of chemical reactions: (a) fireworks, (b) baking (photographs by Beverly March), (c) rocket blast-off (courtesy of NASA).

(b)

(c)

Figure **4.2** • A chemical factory that makes isopropyl alcohol, as well as other compounds. (Photograph by Beverly March.)

In this chapter we discuss three aspects of chemical reactions: (a) mass relationships, (b) types of reactions, and (c) heat gains and losses. In later chapters we will discuss many reactions in greater detail, especially those taking place in living organisms.

4.2 Formula Weight

We begin our study of mass relationships with formula weight. In Section 2.4 we learned that every atom has an atomic weight. The **formula weight** of any substance is simply **the sum of the atomic weights of all the atoms in the molecular formula.**

A table of atomic weights is given on the inside back cover. Atomic weights can also be found in the periodic table on the inside front cover.

From our previous discussions it is evident that molecular formulas mean different things for molecular compounds (those that contain only covalent bonds) and ionic compounds. Molecular compounds are made of molecules, and the molecular formula tells how many atoms of each element are present in that molecule. For example, a CCl_4 molecule contains one C and four Cl atoms. Since there are no molecules in an ionic compound, however, its "molecular" formula represents the ratio of ions. For example, Na_2SO_4 means that two Na^+ ions are present for every one SO_4^{2-} ion.

The term "formula weight" can be used for both ionic and molecular compounds and tells nothing about whether the compound is ionic or molecular. There is an older term, **molecular weight,** that is strictly correct only when used for molecular compounds. In this book we use it only for them. Strictly speaking, a more accurate term than "weight" would be "mass" (Sec. 1.4), but the expressions "molecular weight" and "formula weight" are more commonly used.

Many people use "molecular weight" for both molecular and ionic compounds.

In this book we take formula weights to one decimal place.

EXAMPLE

What is the formula weight of (a) CO_2 (b) $(NH_2)_2CO$ (urea)?

Answer

(a)

	AW (amu)	Total weight
C	$12.0 \times 1 =$	12.0
O	$16.0 \times 2 =$	32.0
	Formula weight of CO_2	44.0 amu

(b)

	AW (amu)	Total weight
N	$14.0 \times 2 =$	28.0
H	$1.0 \times 4 =$	4.0
C	$12.0 \times 1 =$	12.0
O	$16.0 \times 1 =$	16.0
	Formula weight of $(NH_2)_2CO$	60.0 amu

Note that in (b) we include all the atoms. The formula is $(NH_2)_2CO$. The $(NH_2)_2$ tells us that there are two nitrogen atoms and four hydrogen atoms.

Problem **4.1**

What is the formula weight of (a) aspirin, $C_9H_8O_4$ (b) barium phosphate, $Ba_3(PO_4)_2$?

4.3 The Mole

One-mole quantities of several compounds are shown in Figure 4.3.

Atoms and molecules are so tiny (Sec. 2.8) that chemists are seldom able to deal with them one at a time (very special and expensive apparatus is necessary for that). When a chemist weighs out even a very small quantity of a molecular compound, huge numbers of molecules (perhaps 10^{19}) are present. To overcome this problem, chemists long ago defined a unit called the mole. Quite simply, one **mole** of any substance is its **formula weight expressed in grams.** For instance, the formula weight of CO_2, as we learned in our last example, is 44.0 amu; therefore, 44.0 g of CO_2 is 1 mole of CO_2. Likewise, the formula weight of $(NH_2)_2CO$ is 60.0 amu, and so 60.0 g of $(NH_2)_2CO$ is 1 mole of this compound. For atoms, 1 mole is the atomic weight expressed in grams: 12 g of carbon is 1 mole of carbon; 32 g of sulfur is 1 mole of sulfur.

More than 70 years ago, scientists found out that 1 mole of any compound always contains 6.02×10^{23} molecules. In other words, 44.0 g of CO_2 always contains 6.02×10^{23} molecules of CO_2, and 60.0 g of $(NH_2)_2CO$ always contains 6.02×10^{23} molecules of $(NH_2)_2CO$. Thus we can say, as a secondary definition, that a **mole** is **6.02×10^{23} of anything:** A mole of hydrogen atoms is 6.02×10^{23} hydrogen atoms, a mole

Figure **4.3** • One-mole quantities of five compounds (clockwise from upper left): $CuSO_4 \cdot 5H_2O$ (249.68 g); $K_2Cr_2O_7$ (294.19 g); $(NH_2)_2CO$ (urea, 60.0 g); $KMnO_4$ (158.0 g); $C_2H_5NO_2$ (glycine, 75.1 g). (Photograph by Beverly March.)

of sugar molecules is 6.02×10^{23} sugar molecules, and a mole of apples is 6.02×10^{23} apples. As you can see, there is nothing mysterious about the mole. Just as we call 12 of anything a dozen, 20 a score, and 144 a gross, we call 6.02×10^{23} of anything a mole. The number 6.02×10^{23} is called **Avogadro's number,** after the Italian physicist Amedeo Avogadro (1776–1856).

One mole of pennies placed side by side would stretch for more than one million lightyears, a distance far outside our solar system and even our own galaxy. Six moles of freshman chemistry books would weigh as much as the earth.

Let us see why the fact that 1 mole always contains 6.02×10^{23} molecules is very useful information. An atomic weight table tells us that the mass of a silicon atom is 28 amu and the mass of a nitrogen atom is 14 amu; that is, a silicon atom has twice the mass of a nitrogen atom (to the nearest integer). Suppose we had ten of each. Ten Si atoms have a mass of 280 amu, and ten N atoms have a mass of 140 amu. The ratio is still 2:1. In fact, the mass ratio is 2:1 no matter how many atoms we have, as long as we have the same number of N and Si atoms. Now suppose we have a mole of each. Avogadro's number of N atoms have a mass of 14 g. Therefore, the same number of Si atoms *must* have a mass of 28 g.

The same argument can be made for any two atoms or molecules. In 1 mole each of any two substances, the mass ratios are the same as they are for the individual atoms or molecules. This fact allows chemists to measure out a given *number* of atoms or molecules by *weighing out* a sample. It is impossible for a chemist to weigh out one molecule or even 100 molecules, but it is very easy to weigh out one mole of molecules, or half a mole, or 2.7 moles, or any convenient fraction or multiple desired.

Even 100 molecules represent such a small mass that no balance is sensitive enough to detect it.

Let's look at some examples. In these examples we use the factor-label method discussed in Section 1.5.

EXAMPLE

We have 27.5 g of stannous fluoride, SnF_2. How many moles is this?

Note that such calculations can be performed for ionic compounds, such as SnF_2, as well as molecular compounds, such as CO_2 and urea.

Answer • The formula weight of $SnF_2 = 118.7 + 2(19.0) = 156.7$ amu. This means that each mole of SnF_2 has a mass of 156.7 g, allowing us to use the conversion factor 1 mole SnF_2 = 156.7 g SnF_2:

$$27.5 \text{ g } \cancel{SnF_2} \times \frac{1 \text{ mole } SnF_2}{156.7 \text{ g } \cancel{SnF_2}} = 0.175 \text{ mole } SnF_2$$

Problem 4.2

We have 756.3 g of $ZnCl_2$. How many moles is this?

EXAMPLE

We wish to weigh out 3.41 moles of ethyl alcohol, C_2H_5OH. How many grams is this?

Answer • The formula weight of C_2H_5OH is $2(12.0) + 6(1.0) + 16.0 = 46.0$ amu, and so the conversion factor is 1 mole C_2H_5OH = 46.0 g C_2H_5OH. In order for the proper units to cancel, we multiply as follows:

$$3.41 \cancel{\text{ moles } C_2H_5OH} \times \frac{46.0 \text{ g } C_2H_5OH}{1 \cancel{\text{ mole } C_2H_5OH}} = 157 \text{ g } C_2H_5OH$$

Problem 4.3

We wish to weigh out 2.84 moles of sodium sulfide, Na_2S. How many grams is this?

EXAMPLE

TNT (trinitrotoluene) is a powerful explosive.

How many moles of nitrogen atoms and oxygen atoms are there in 21.4 moles of TNT, $C_7H_5(NO_2)_3$?

Answer • The formula $C_7H_5(NO_2)_3$ tells us that each molecule of TNT contains three nitrogen atoms. Therefore each mole of TNT must contain 3 moles of nitrogen atoms. The number of moles of N atoms in 21.4 moles of TNT is therefore

$$21.4 \cancel{\text{ moles TNT}} \times \frac{3 \text{ moles N}}{1 \cancel{\text{ mole TNT}}} = 64.2 \text{ moles N}$$

Similarly, each mole of TNT contains 6 moles of oxygen atoms, so that the total number of moles of oxygen atoms is

$$21.4 \cancel{\text{ moles TNT}} \times \frac{6 \text{ moles O}}{1 \cancel{\text{ mole TNT}}} = 128.4 \text{ moles O}$$

Problem 4.4

How many moles of C atoms, H atoms, and O atoms are there in 2.5 moles of glucose, $C_6H_{12}O_6$?

EXAMPLE

How many moles of sodium ions, Na^+, are there in 5.63 g of Na_2SO_4?

Answer • First we find out how many moles of Na_2SO_4 we have. As in the SnF_2 example, we obtain the conversion factor by calculating the formula weight: 2(23.0) + 32.1 + 4(16.0) = 142.1 amu. Then

$$5.63\ \cancel{\text{g Na}_2\text{SO}_4} \times \frac{1\ \text{mole Na}_2\text{SO}_4}{142.1\ \cancel{\text{g Na}_2\text{SO}_4}} = 0.0396\ \text{mole Na}_2\text{SO}_4$$

The formula Na_2SO_4 tells us that there are two Na^+ ions in each mole of Na_2SO_4. The number of moles of Na^+ ions in 5.63 g of Na_2SO_4 is therefore

$$0.0396\ \cancel{\text{mole Na}_2\text{SO}_4} \times \frac{2\ \text{moles Na}^+}{1\ \cancel{\text{mole Na}_2\text{SO}_4}} = 0.0792\ \text{mole Na}^+$$

Problem 4.5

How many moles of copper(I) ions, Cu^+, are there in 0.062 g of copper(I) nitrate, $CuNO_3$?

EXAMPLE

A typical aspirin tablet, $C_9H_8O_4$, contains 0.36 g aspirin. How many aspirin molecules are present?

Answer • The formula weight of aspirin is 9(12.0) + 8(1.0) + 4(16.0) = 180.0 amu. First we find out how many moles of aspirin there are in 0.36 g:

$$0.36\ \cancel{\text{g aspirin}} \times \frac{1\ \text{mole aspirin}}{180.0\ \cancel{\text{g aspirin}}} = 0.0020\ \text{mole aspirin}$$

$$= 2.0 \times 10^{-3}\ \text{mole aspirin}$$

Each mole contains 6.02×10^{23} molecules, and so the number of molecules is

$$(2.0 \times 10^{-3}\ \cancel{\text{mole}})\left(6.02 \times 10^{23}\ \frac{\text{molecules}}{\cancel{\text{mole}}}\right) = 1.2 \times 10^{21}\ \text{molecules}$$

A single aspirin tablet contains 1.2×10^{21} molecules of $C_9H_8O_4$. Chemists seldom need to find out the number of molecules in anything. They work with grams and moles. If they ever do need this information, however, it is very easy to do, as the last example shows.

Problem **4.6**

How many molecules of water, H_2O, are there in a glass of water (235 g)?

4.4 Chemical Equations

A typical chemical reaction is the burning of propane (bottled gas). In this reaction, propane reacts with the oxygen in the air; both of these substances are converted to the products: carbon dioxide and water. We could write this

propane plus oxygen gives carbon dioxide and water

If we didn't know the molecular formulas, we would in fact be forced to write it this way. We do know the formulas, however, so a better way to write it is

The arrow indicates that the starting materials (on the left) are converted to the products (on the right).

$$\underset{\text{Propane}}{C_3H_8} + \underset{\text{Oxygen}}{O_2} \underset{\text{Yields}}{\longrightarrow} \underset{\text{Carbon dioxide}}{CO_2} + \underset{\text{Water}}{H_2O}$$

This kind of expression is called a **chemical equation.**

The equation we have written is incomplete, however. Even though it does tell us the formulas for all the starting materials and all the products (which every chemical equation must do), it does not give the amounts correctly. That is, it is qualitatively right but quantitatively wrong. *It is not balanced.* In chemical reactions atoms are never destroyed or created; they are only shifted from one substance to another. This means that all the atoms present at the start (on the left side of the arrow) must still be there at the end (on the right side), as seen in Figure 4.4. In the equation we have written, there are three carbon atoms on the left and only one on the right. This means the equation cannot be correct. In order to make it correct, we must balance it.

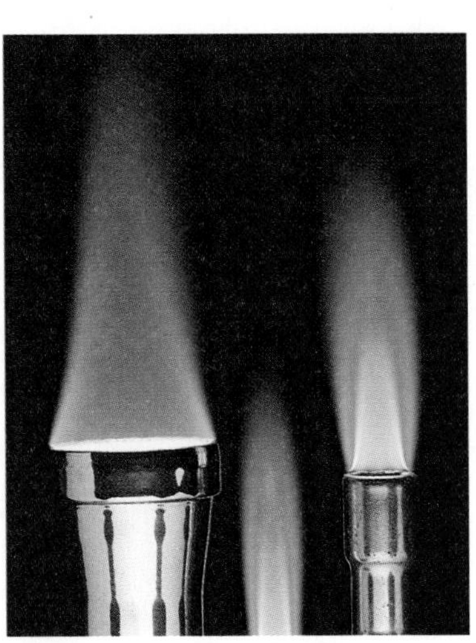

Propane burning in air. (Photograph by C. D. Winters.)

Balancing Equations

Most equations are easily balanced. It only takes a few minutes. Basically, all we do is place numbers in front of the formulas until the equation is balanced. These numbers are called **coefficients** and represent numbers of molecules. As an example, let us balance our propane equation:

$$C_3H_8 + O_2 \longrightarrow CO_2 + H_2O$$

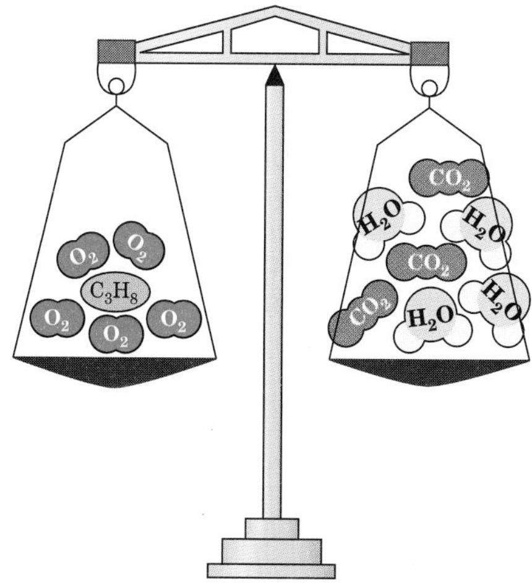

Figure **4.4**
In the combustion of propane, one C_3H_8 molecule and five O_2 molecules are converted to three CO_2 and four H_2O molecules. No atoms are created or destroyed. They merely change partners, and so the total mass is unchanged.

How do we choose our coefficients? Let us begin with carbon. There are three carbon atoms on the left and one on the right. If we put a 3 in front of the CO_2 (indicating that three CO_2 molecules are formed), there will be three carbons on each side and the carbons will be balanced:

$$C_3H_8 + O_2 \longrightarrow 3CO_2 + H_2O$$

Next we look at the hydrogens. There are eight on the left and two on the right. We therefore put a 4 in front of the H_2O to make a total of eight hydrogens on the right (4×2). We now have

$$C_3H_8 + O_2 \longrightarrow 3CO_2 + 4H_2O$$

The only thing still unbalanced is oxygen. There are two oxygen atoms on the left and ten on the right (make sure you can find ten). If we put the coefficient 5 in front of the oxygen on the left, we balance the oxygens and the whole equation:

$$C_3H_8 + 5O_2 \longrightarrow 3CO_2 + 4H_2O \qquad \text{Balanced equation}$$

At this point the equation ought to be balanced, but we should always check, just to make sure. **In a balanced equation there must be the same number of atoms of each element on both sides.** A check of our work shows three C, ten O, and eight H atoms on each side. The equation is indeed balanced.

Note that the numbers we put in to balance an equation are always coefficients. We are not permitted to change formulas. For example, when we started to balance $C_3H_8 + O_2 \rightarrow CO_2 + H_2O$, we saw that we needed three C atoms on the right. Why not change CO_2 to C_3O_2? We cannot because the product of the reaction is carbon dioxide and the formula for carbon dioxide is CO_2, not C_3O_2. In balancing equations we are permitted to say that three CO_2 molecules are formed (or whatever number is necessary for a balanced equation), but we are *not* permitted to write incorrect formulas. Here are two other examples.

EXAMPLE

Balance:

$$Ca(OH)_2 + HCl \longrightarrow CaCl_2 + H_2O$$

Answer • The calcium is already balanced: There is one on each side. There is one Cl on the left and two on the right. In order to balance them, we add the coefficient 2 to HCl. The equation now is

$$Ca(OH)_2 + 2HCl \longrightarrow CaCl_2 + H_2O$$

In the next step we note that there are four hydrogens on the left but only two on the right. Placing the coefficient 2 in front of H_2O remedies the situation:

$$Ca(OH)_2 + 2HCl \longrightarrow CaCl_2 + 2H_2O$$ Balanced equation

Note that the last step also balanced the oxygen atoms.

Problem **4.7**

Balance: $As_2O_5 + H_2O \rightarrow H_3AsO_4$

EXAMPLE

Balance:

$$C_3H_6 + O_2 \longrightarrow H_2O + CO_2$$

Propane, C_3H_8, and cyclopropane, C_3H_6, are different compounds.

Answer • This equation, for the combustion of cyclopropane, is very similar to the first one we did. As before, we put a 3 in front of the CO_2 and this time a 3 in front of the H_2O (since there are six hydrogens on the left):

$$C_3H_6 + O_2 \longrightarrow 3H_2O + 3CO_2$$

But now when we balance the oxygens, we find 2 on the left and 9 on the right. The only way to balance the equation is to put a $4\frac{1}{2}$ in front of the O_2:

$$C_3H_6 + 4\tfrac{1}{2}O_2 \longrightarrow 3H_2O + 3CO_2$$

Although there are times when chemists have good reasons to write equations with fractional coefficients, it cannot be completely correct because there is no such thing as half an oxygen molecule. We get around the difficulty by multiplying everything by 2:

$$2C_3H_6 + 9O_2 \longrightarrow 6H_2O + 6CO_2$$ Balanced equation

A check will show that the equation is balanced.

Problem **4.8**

Balance: $C_6H_{14} + O_2 \rightarrow H_2O + CO_2$

EXAMPLE

Balance:

$$Na_2SO_3 + H_3PO_4 \longrightarrow H_2SO_3 + Na_3PO_4$$

Answer • The key to balancing equations like this is to realize that units like SO_3 and PO_4 usually remain intact on both sides of the equation. We can begin by balancing the sodiums, which is done by putting a 3 in front of Na_2SO_3 and a 2 in front of Na_3PO_4, thus giving us six sodiums on each side. Our equation now looks like this:

$$3Na_2SO_3 + H_3PO_4 \longrightarrow H_2SO_3 + 2Na_3PO_4$$

There are three SO_3 units on the left and only one on the right, so we put a 3 in front of H_2SO_3 and now have

$$3Na_2SO_3 + H_3PO_4 \longrightarrow 3H_2SO_3 + 2Na_3PO_4$$

Since there are two PO_4 units on the right, but one on the left, we put a 2 in front of H_3PO_4:

$$3Na_2SO_3 + 2H_3PO_4 \longrightarrow 3H_2SO_3 + 2Na_3PO_4 \quad \text{Balanced equation}$$

In doing so, we not only balanced the PO_4 units, but also the hydrogens. Finally we check to make sure the equation is balanced.

Problem **4.9**

Balance: $K_2C_2O_4 + Ca_3(AsO_4)_2 \rightarrow K_3AsO_4 + CaC_2O_4$

One more thing. If

$$C_3H_8 + 5O_2 \longrightarrow 3CO_2 + 4H_2O$$

is a correctly balanced equation (and indeed it is), would it be correct if we doubled all the coefficients?

$$2C_3H_8 + 10O_2 \longrightarrow 6CO_2 + 8H_2O$$

The answer is that it is mathematically and scientifically correct, but chemists do not normally write equations with multiple coefficients. A correctly balanced equation is almost always written with the coefficients expressed "in lowest terms."

4.5 Mass Relationships in Chemical Reactions

"Stoichiometry" comes from the Greek *stoicheion,* element, and *metron,* measure.

As we saw in Section 4.4, a balanced chemical equation tells us not only which substances react and which are formed, but also in what proportions. This is very useful information when we want to carry out a chemical reaction. For example, we can easily calculate how much starting materials to weigh out if we want to produce a particular mass of a product. **The study of mass relationships in chemical reactions** is called **stoichiometry.**

Let us look once again at the balanced equation for the burning of propane:

$$C_3H_8 + 5O_2 \longrightarrow 3CO_2 + 4H_2O$$

In Section 4.4 we saw that the coefficients in an equation represent numbers of molecules. Since moles are proportional to molecules (Sec. 4.3), the coefficients in an equation also represent numbers of moles.

This equation not only tells us that propane and oxygen are converted to carbon dioxide and water; it also tells us that **1 mole** of propane combines with **5 moles** of oxygen to produce **3 moles** of carbon dioxide and **4 moles** of water; that is, we know the quantities involved. The same is true for any balanced equation.

This fact allows us to answer questions such as:

1. How many moles of any particular product are produced if we start with a given mass of any starting material?

2. How many grams of one starting material are necessary to react completely with a given number of grams of another starting material?

3. If we want to produce a certain number of grams (or moles) of a certain product, how many grams (or moles) of starting materials are needed?

4. If a certain amount of one product is produced, how much of another product is also produced?

It sounds as if we have four different types of problems here, and yet they are all done by the same very simple procedure, which involves three steps.

EXAMPLE

Consider the reaction

$$N_2 + 3H_2 \longrightarrow 2NH_3$$

Ammonia, NH_3, which is used to make fertilizer, is produced by this reaction (the Haber process).

How many grams of N_2 are necessary to produce 7.5 g of NH_3?

Answer

Step 1. The coefficients in an equation refer to moles, not grams. Therefore we must first find out how many moles of NH_3 there are in 7.5 g. As before, we multiply the 7.5 g by the proper conversion factor (1 mole NH_3 = 17.0 g NH_3). At this stage we don't do the actual multiplication, but just leave it set up:

$$7.5 \text{ g } \cancel{NH_3} \times \frac{1 \text{ mole } NH_3}{17.0 \text{ g } \cancel{NH_3}} = \text{number of moles of } NH_3$$

Step 2. Next we turn to the balanced equation, which tells us that *two* moles of NH_3 are produced from *one* mole of N_2. We multiply the number of moles of NH_3 by this fraction:

$$7.5 \text{ g } \cancel{NH_3} \times \frac{1 \text{ } \cancel{\text{mole } NH_3}}{17.0 \text{ g } \cancel{NH_3}} \times \frac{1 \text{ mole } N_2}{2 \text{ } \cancel{\text{moles } NH_3}} = \text{number of moles of } N_2$$

The fraction obtained from the balanced equation is shown in color.

This gives us the number of *moles* of N_2 that are necessary.

Step 3. We now find the number of grams of N_2, as usual, by multiplying by the proper conversion factor (28.0 g N_2 = 1 mole N_2):

$$7.5 \text{ g } \cancel{NH_3} \times \frac{1 \text{ } \cancel{\text{mole } NH_3}}{17.0 \text{ g } \cancel{NH_3}} \times \frac{1 \text{ } \cancel{\text{mole } N_2}}{2 \text{ } \cancel{\text{moles } NH_3}} \times \frac{28.0 \text{ g } N_2}{1 \text{ } \cancel{\text{mole } N_2}} = 6.2 \text{ g } N_2$$

At this point we do the arithmetic to get our answer, 6.2 g N_2.

In all such problems, we are given a mass (or number of moles) of one compound and asked to find the mass (or number of moles) of another compound. The two compounds can be on the left or the right, on the same side of the equation, or on opposite sides. All such problems can be done by the three simple steps we have just used:

1. If you are given a mass, convert to moles by multiplying by the proper conversion factor. If you are given the number of moles, just write it down.

2. Multiply by a fraction consisting of the coefficient of the compound whose mass is unknown divided by the coefficient of the compound whose mass is known. Since this information is obtained from the balanced equation, we cannot do such problems without a balanced equation.

Note that we look at the coefficients of the balanced equation only in step 2. These coefficients play no part in steps 1 and 3.

3. If you are asked for the mass of a compound, multiply by the appropriate conversion factor. If you are asked for moles, this step is unnecessary and the answer is already present at the end of step 2.

The simplicity of the method is shown in the following examples.

EXAMPLE

How many moles of Cl_2 are required to produce 11.6 moles of PCl_3 by the reaction

$$2P + 3Cl_2 \longrightarrow 2PCl_3$$

Answer

Step 1. Since we are given moles, we merely write the information down: 11.6 moles PCl_3.

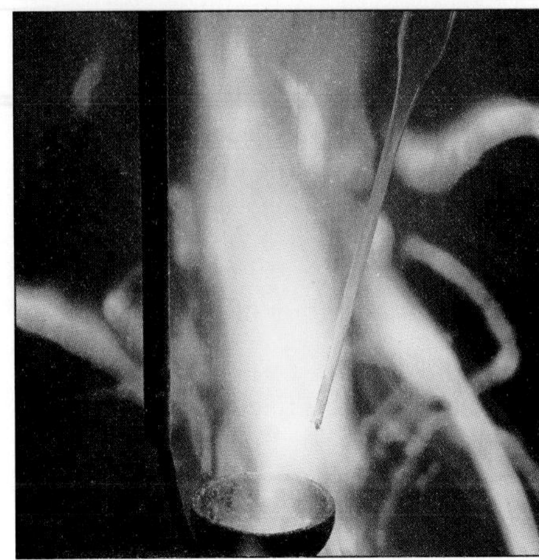

Phosphorus reacting with chlorine to give PCl_3 (Photograph by C. D. Winters.)

Step 2. Multiply by the fraction obtained from the equation:

$$11.6 \text{ moles PCl}_3 \times \frac{3 \text{ moles Cl}_2}{2 \text{ moles PCl}_3} = 17.4 \text{ moles Cl}_2$$

Step 3. Unnecessary, since we were asked for moles.

Problem **4.10**

In the reaction

$$CH_4 + 2O_2 \longrightarrow CO_2 + 2H_2O$$

how many moles of CH_4 are required to produce 16.6 moles of H_2O?

EXAMPLE

When urea, $(NH_2)_2CO$, is acted on by the enzyme urease in the presence of water, ammonia and carbon dioxide are produced:

$$(NH_2)_2CO + H_2O \longrightarrow 2NH_3 + CO_2$$

If excess water is present (more than necessary for the reaction), how many grams each of CO_2 and NH_3 are produced from 0.83 mole of urea?

Answer • For CO_2:

Step 1. Once again, we merely write down the number of moles: 0.83 mole urea.

Step 2. Multiply by the fraction obtained from the equation:

$$0.83 \text{ mole urea} \times \frac{1 \text{ mole CO}_2}{1 \text{ mole urea}}$$

Step 3. Multiply by the conversion factor ($44.0\text{ g } CO_2 = 1$ mole CO_2):

$$0.83 \text{ mole urea} \times \frac{1 \text{ mole } CO_2}{1 \text{ mole urea}} \times \frac{44.0 \text{ g } CO_2}{1 \text{ mole } CO_2} = 37 \text{ g } CO_2$$

For NH_3:
Step 1 is the same, but in step 2 the equation gives a different fraction:

$$0.83 \text{ mole urea} \times \frac{2 \text{ moles } NH_3}{1 \text{ mole urea}} \times \frac{17.0 \text{ g } NH_3}{1 \text{ mole } NH_3} = 28 \text{ g } NH_3$$

Problem **4.11**

In the reaction

$$C_2H_2 + 2Br_2 \longrightarrow C_2H_2Br_4$$

how many grams of $C_2H_2Br_4$ are produced from 7.24 moles of Br_2 (assuming excess C_2H_2)?

Excess means more than is necessary for complete reaction.

EXAMPLE

In the reaction

$$2CH_3OH + PCl_5 \longrightarrow 2CH_3Cl + POCl_3 + H_2O$$

how many grams of PCl_5 are necessary to react with 137 g of methanol, CH_3OH?

Answer • The formula weight of methanol is 32.0 amu and that of PCl_5 is 208.2 amu. Steps 1, 2, and 3 give us

$$137 \text{ g } CH_3OH \times \frac{1 \text{ mole } CH_3OH}{32.0 \text{ g } CH_3OH} \times \frac{1 \text{ mole } PCl_5}{2 \text{ moles } CH_3OH} \times \frac{208.2 \text{ g } PCl_5}{1 \text{ mole } PCl_5} = 446 \text{ g } PCl_5$$

Problem **4.12**

In the reaction

$$C + 2H_2 \longrightarrow CH_4$$

how many grams of H_2 are necessary to react with 51.4 g of C?

4.6 Percentage Yield

When carrying out chemical reactions in the laboratory, we often get less of a product than expected from the type of calculation we performed in Section 4.5. For example, if we start with 1 mole (32.0 g) of CH_3OH and add excess PCl_5 in the reaction

$$2CH_3OH + PCl_5 \longrightarrow 2CH_3Cl + POCl_3 + H_2O$$

Reactions that do not give the main product are called *side reactions*.

we should get 1 mole (50.5 g) of CH_3Cl. However, we often don't; we get less. Does this mean that the law of conservation of mass is being violated? No, it does not. We get less than 50.5 g of CH_3Cl either because some of the CH_3OH does not react or because some of it reacts in some other way, or perhaps because our laboratory technique is not perfect and we lose a little in transferring from one container to another. Whatever the reason, the mass of product we get in the laboratory is called the **actual yield.** The expected mass of the product—the amount calculated by the method of Section 4.5—is called the **theoretical yield.** The **percentage yield** is actual yield divided by theoretical yield multiplied by 100:

$$\% \text{ yield} = \frac{\text{actual yield}}{\text{theoretical yield}} \times 100$$

EXAMPLE

In an experiment using the reaction just described, the actual yield of CH_3Cl was 46.8 g (theoretical yield is 50.5 g). What is the percentage yield?

Answer

$$\frac{46.8 \cancel{g}}{50.5 \cancel{g}} \times 100 = 92.7\%$$

Most chemists consider this a pretty good percentage yield for this reaction.

Problem **4.13**

In one method for making aspirin, the theoretical yield is 153.7 g. If the actual yield obtained by a student was 124 g, what is the percentage yield?

An aqueous solution is a water solution.

4.7 Reactions Between Ions in Aqueous Solution

Many ionic compounds are soluble in water. As we saw in Section 3.3, ionic compounds always consist of both positive and negative ions. When they dissolve in water, positive and negative ions are separated from each other by the water molecules (Sec. 6.6). We can say that a **dissociation** has taken place; for example,

$$NaCl(s) \xrightarrow{H_2O} Na^+(aq) + Cl^-(aq)$$

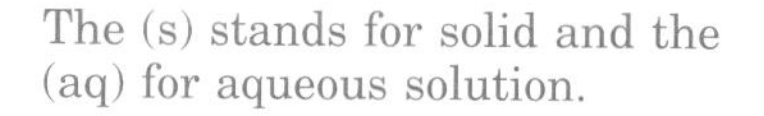

The (s) stands for solid and the (aq) for aqueous solution.

What happens when we mix two different ionic solutions? Does a reaction take place between the ions? The answer depends on what the ions are. If negative and positive ions can come together to form a compound, then a reaction takes place; otherwise it does not. For example, if we mix a solution of NaCl with one of silver nitrate, $AgNO_3$, four ions are present in the solution: Ag^+, Na^+, Cl^-, and NO_3^-. It so happens that two of these, Ag^+ and Cl^-, form a compound, AgCl (silver chloride), that is insoluble in water. A reaction therefore takes place,

$$Ag^+(aq) + Cl^-(aq) \longrightarrow AgCl(s)$$

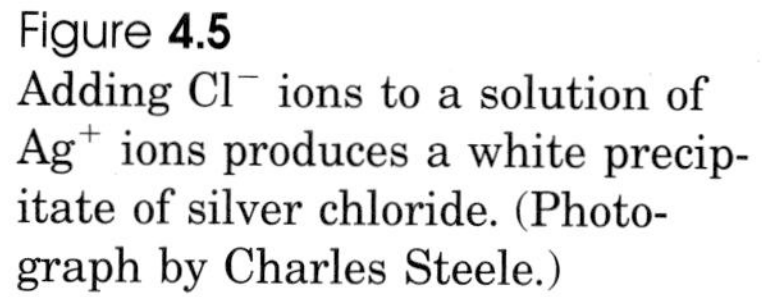

Figure **4.5**
Adding Cl^- ions to a solution of Ag^+ ions produces a white precipitate of silver chloride. (Photograph by Charles Steele.)

and a white precipitate forms that slowly sinks to the bottom of the container. What about the Na^+ and NO_3^- ions? They do nothing at all and merely remain dissolved in the water. Ions like these, which do not participate in a reaction, are called **spectator ions,** certainly an appropriate name.

In general, ions in solution react with each other only when one of these things can happen:

1. Two of them form a solid that is insoluble in water (AgCl is one example; Fig. 4.5).

2. Two of them form a gas. In this case the reaction takes place because the gas is not very soluble in water. An example (Fig. 4.6) is

$$\underset{\text{Bicarbonate ion}}{HCO_3^-(aq)} + H^+(aq) \longrightarrow CO_2(g) + H_2O(l)$$

The (g) stands for gas and the (l) for liquid.

(*a*) (*b*)

Figure **4.6** • (a) The left beaker contains a solution of HCO_3^- ions; the right beaker a solution of H^+ ions. (b) When the beakers are mixed the reaction between HCO_3^- ions and H^+ ions produces the gas CO_2, which can be seen as bubbles. (Photographs by Beverly March.)

(*a*) (*b*)

Figure **4.7** • (a) The beakers contain copper nitrate (blue) and potassium sulfate (colorless). (b) When they are mixed, no reaction takes place. The blue color gets lighter because the copper nitrate is less concentrated, but nothing else happens. (Photographs by Beverly March.)

3. An acid neutralizes a base. Acid-base reactions are so important that we devote a whole chapter to them (Chapter 8).

4. One of the ions can oxidize another. This type of reaction is also important enough to be discussed separately (Sec. 4.8).

In many cases no reaction takes place when we mix solutions of ions because none of these situations holds. For example, if we mix a solution of copper nitrate, $Cu(NO_3)_2$, and potassium sulfate, K_2SO_4, we merely have a mixture containing Cu^{2+}, K^+, NO_3^-, and SO_4^{2-} ions dissolved in water. None of these react with each other, and we will see nothing happening (Fig. 4.7).

The equations we write for ions in solution are called ionic equations. Like all other chemical equations, they must be balanced. We balance them the same way as any other equation, except that we must make sure that charges balance as well as atoms. For example, the precipitation of arsenic sulfide takes place according to this equation:

$$2As^{3+}(aq) + 3S^{2-}(aq) \longrightarrow As_2S_3(s)$$

Not only are there two arsenic and three sulfur atoms on each side, but the total charge on the left side is the same as the total charge on the right side: They are both zero. *Ionic equations show only the ions that react.* Spectator ions are not shown.

EXAMPLE

When a solution of barium nitrate, $Ba(NO_3)_2$, is added to a solution of potassium sulfate, K_2SO_4, a white precipitate of barium sulfate, $BaSO_4$, forms. Write the ionic equation for this reaction.

Answer • Because they are ionic compounds, both barium nitrate and potassium sulfate exist in water as the dissociated ions:

$$Ba^{2+}(aq) + 2NO_3^-(aq) + 2K^+(aq) + SO_4^{2-}(aq)$$

We are told that a precipitate of barium sulfate forms:

$$Ba^{2+}(aq) + 2NO_3^-(aq) + 2K^+(aq) + SO_4^{2-}(aq) \longrightarrow BaSO_4(s) + 2K^+(aq) + 2NO_3^-(aq)$$

Since K^+ and NO_3^- appear on both sides (nothing happens to them), we cancel them (they are spectator ions) and are left with

$$Ba^{2+}(aq) + SO_4^{2-}(aq) \longrightarrow BaSO_4(s) \quad \text{Ionic equation}$$

Problem **4.14**

When a solution of copper(II) chloride, $CuCl_2$, is added to a solution of potassium sulfide, K_2S, a black precipitate of copper(II) sulfide, CuS, forms. Write the ionic equation for the reaction.

Of the four ways for ions to react in water, one of the most common is the formation of an insoluble compound. We can predict when this will happen if we know the solubilities of the ionic compounds. There is no simple way to remember which ionic solids are soluble in water and which are not, but there are some useful generalizations.

The solubilities of compounds are found in handbooks.

1. All compounds containing Na^+, K^+, or NH_4^+ ions are soluble in water.

2. All nitrates (NO_3^-) and acetates (CH_3COO^-) are soluble in water.

3. Most chlorides (Cl^-) and sulfates (SO_4^{2-}) are soluble. Some important exceptions are silver chloride, AgCl, barium sulfate, $BaSO_4$, and lead sulfate, $PbSO_4$, which are insoluble.

4. Most carbonates (CO_3^{2-}), phosphates (PO_4^{3-}), sulfides (S^{2-}), and hydroxides (OH^-) are insoluble in water (Fig. 4.9). Important exceptions are those of Na^+, K^+, and NH_4^+, as well as barium hydroxide, $Ba(OH)_2$.

Some of these insoluble compounds are shown in Figure 4.8.

4.8 Oxidation–Reduction

One of the most important, as well as one of the most common, types of reaction is oxidation–reduction. We define **oxidation** as **the loss of electrons.** Reduction is defined as **the gain of electrons.** An oxidation–reduction reaction (often called a **redox reaction**) is the transfer of electrons from one species to another. An example is the oxidation of zinc by copper ions:

$$Zn(s) + Cu^{2+}(aq) \longrightarrow Cu(s) + Zn^{2+}(aq)$$

There must also be negative ions present, such as Cl^- or NO_3^-, for example, but we do not write them in the equation because they are spectator ions.

If you put a piece of zinc metal into a beaker containing copper ions in aqueous solution, you will see, after a short time, three things happening (Fig. 4.10):

1. Some of the zinc metal goes into solution.
2. Copper metal deposits on the zinc metal.
3. The blue color of Cu^{2+} ions gradually disappears.

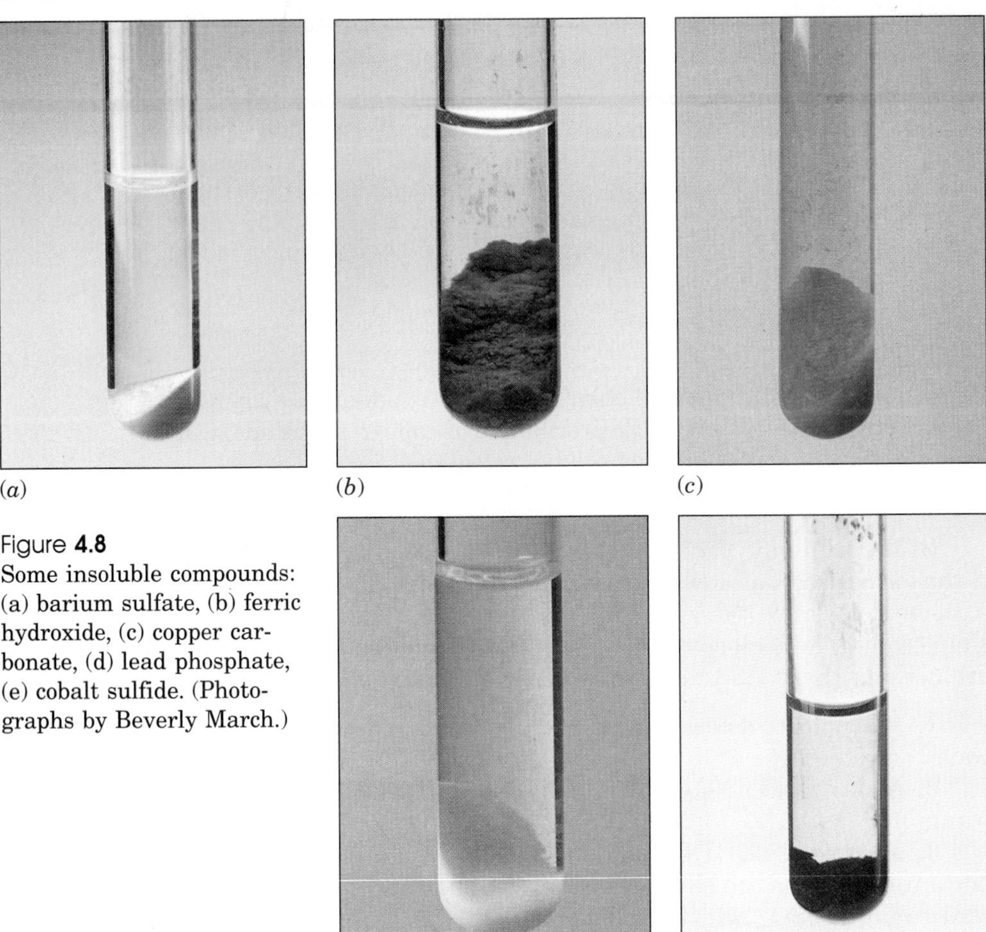

Figure **4.8**
Some insoluble compounds: (a) barium sulfate, (b) ferric hydroxide, (c) copper carbonate, (d) lead phosphate, (e) cobalt sulfide. (Photographs by Beverly March.)

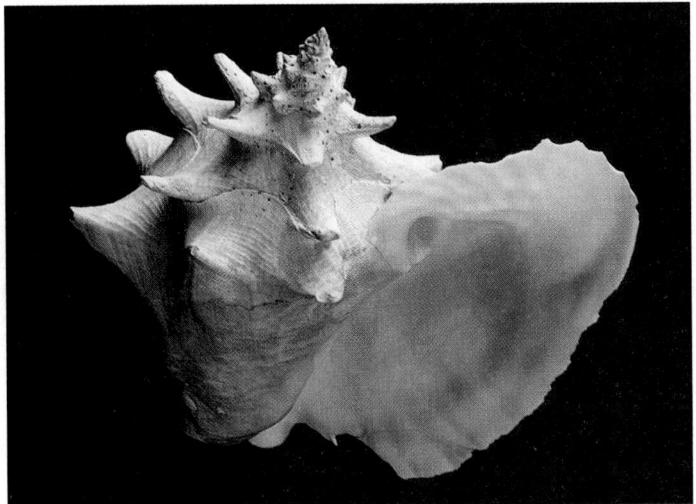

Figure **4.9** • A conch shell. Sea animals of the mollusc family often use insoluble $CaCO_3$ to construct their shells. (Photograph by Beverly March.)

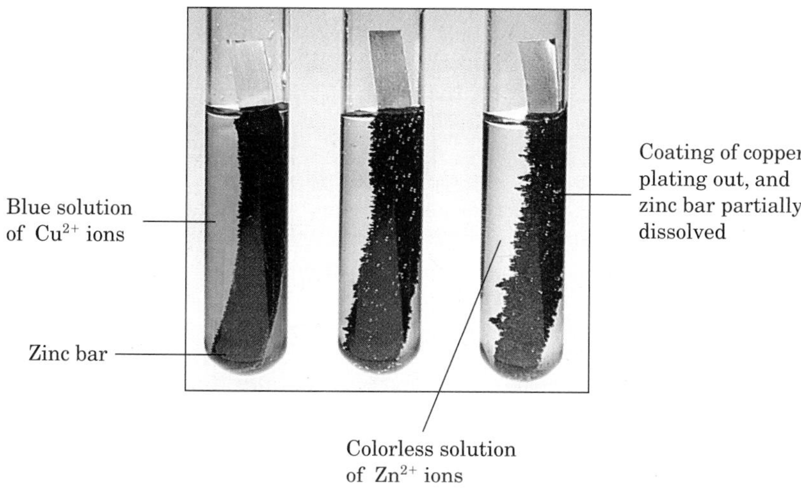

Figure **4.10**
When a piece of zinc is added to a solution containing Cu^{2+} ions, the zinc is oxidized by the Cu^{2+} ions. (Photograph by Charles D. Winters.)

The zinc metal is giving electrons to the copper ions. Since it is losing electrons, we say that the zinc is being oxidized:

$$Zn(s) \longrightarrow Zn^{2+}(aq) + 2e^- \quad \text{Oxidation of Zn}$$

At the same time, the Cu^{2+} ions are taking the electrons from the zinc. We say that they are being reduced:

$$Cu^{2+}(aq) + 2e^- \longrightarrow Cu(s) \quad \text{Reduction of } Cu^{2+}$$

It is evident that oxidation and reduction are not independent reactions. A species cannot take electrons from nowhere; a species cannot give electrons to nothing. **There can be no oxidation without an accompanying reduction, and vice versa.** In this reaction, the Cu^{2+} is oxidizing the Zn. Thus we call it an **oxidizing agent.** Similarly, the Zn is reducing the Cu^{2+}, and so we call it a **reducing agent.**

The oxidizing agent is the species that gets reduced. The reducing agent is the species that gets oxidized.

Although the definitions we have given for oxidation (loss of electrons) and reduction (gain of electrons) are easy to apply in many redox reactions, they are not so easy to apply in others. For example, another redox reaction is the combustion (burning) of methane, CH_4, in which the CH_4 is oxidized to CO_2 while O_2 is reduced to CO_2 and H_2O:

$$CH_4(g) + 2O_2(g) \longrightarrow CO_2(g) + 2H_2O(g)$$

It is not easy to see the electron loss and gain in cases like this, and so chemists have another definition of oxidation and reduction, one that is easier to apply in many cases, especially where organic compounds are involved:

Oxidation is the gain of oxygen or the loss of hydrogen.
Reduction is the loss of oxygen or the gain of hydrogen.

Box 4A

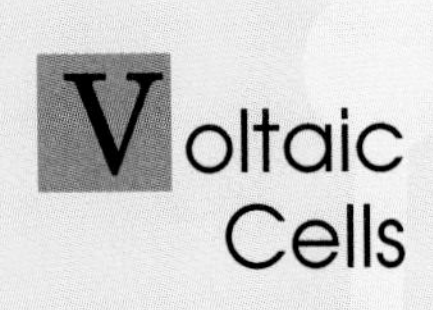

Voltaic Cells

In Figure 4.10 we see that when a piece of zinc metal is put in a solution containing Cu^{2+} ions, the zinc atoms give electrons to the Cu^{2+} ions. If we change the experiment by putting the zinc metal in one beaker and the Cu^{2+} ions in another and connect the two beakers by a piece of wire and a salt bridge (Fig. 4A), *the reaction still takes place.* That is, zinc atoms still give electrons to the Cu^{2+} ions, but now the electrons must flow through the wire to get from the Zn to the Cu^{2+}. **This flow of electrons is an electric current,** and the current keeps flowing until either the zinc or the Cu^{2+} is used up. This apparatus is therefore a device that generates an electric current by making use of a redox reaction; in other words, it is a **voltaic cell** or, as we commonly say, a **battery.**

It is obvious that this particular battery would not be useful for flashlights, portable radios, and so on. The batteries that we do use for such purposes use different redox reactions, in many cases ones that do not require aqueous solutions, but the principle is the same in every case: A reducing agent transmits electrons to an oxidizing agent in an apparatus that prevents it from doing so directly. The electrons must go through an outside circuit and in doing so produce the electric current of the battery. (Electricity is the flow of electrons.) To see why a salt bridge is necessary, we must look at the Cu^{2+} solution. Since we cannot have positive charges in any place without an equivalent number of negative charges, there must be negative ions in the beaker also, which may be sulfate, ${SO_4}^{2-}$, nitrate, ${NO_3}^-$, or some other ion. When electrons come over the wire, the Cu^{2+} is converted to Cu:

$$Cu^{2+}(aq) + 2e^- \longrightarrow Cu(s)$$

This diminishes the number of Cu^{2+} ions, but the number of negative ions is unaffected. The salt bridge is necessary to carry some of these negative ions to the other beaker, where they are needed to balance the Zn^{2+} ions being produced by the reaction

$$Zn(s) \longrightarrow 2e^- + Zn^{2+}(aq)$$

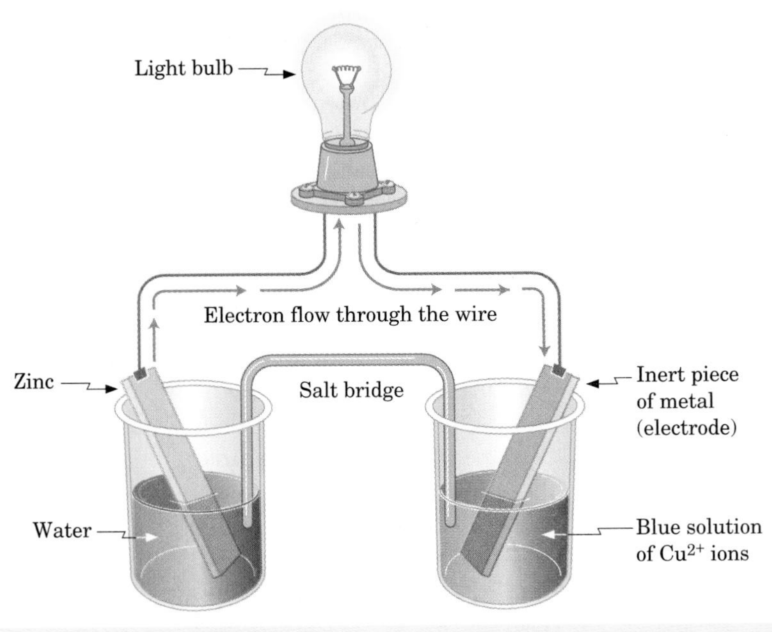

Figure **4.A** • A voltaic cell. The electron flow, over the wire from Zn to Cu^{2+}, is an electric current that causes the light bulb to glow.

In fact, this second definition is much older than the one involving electron transfer; it is the definition given by Lavoisier when he first discovered oxidation and reduction more than 200 years ago. Note that here, too, oxidation and reduction must occur together. Note also that we could not apply this definition to our zinc-copper example.

EXAMPLE

In each equation, identify the substance that is oxidized, the one that is reduced, and the oxidizing and reducing agents.

(a) $Al(s) + Fe^{3+}(aq) \longrightarrow Al^{3+}(aq) + Fe(s)$

(b) $CH_3OH(l) + O_2(g) \longrightarrow H{-}\underset{\underset{O}{\|}}{C}{-}OH(l) + H_2O(l)$

Methanol   Formic acid

Answer

(a) It is evident that aluminum has lost three electrons, since it went from an aluminum atom, which has no charge, to an aluminum ion, with a charge of +3. Since loss of electrons is oxidation, the Al has been oxidized. On the other hand, the iron, which went from Fe^{3+} to Fe must have gained three electrons, and has therefore been reduced. In every redox reaction, one substance is reduced and another oxidized. The substance that has been oxidized (Al in this case) must be the reducing agent, and Fe^{3+} is the oxidizing agent.

(b) It is not easy to look for electron loss or gain here, so we apply the second definitions. The methanol, which has one oxygen and four hydrogens, has been converted to a compound that has two oxygens and two hydrogens. It has therefore both gained oxygen and lost hydrogen, meeting the definition of a compound that has been oxidized. The O_2 molecule has been converted to H_2O, losing one O and gaining two hydrogens in the process. Again, the substance that has been oxidized (methanol) is the reducing agent, and O_2 is the oxidizing agent.

Problem **4.15**

In each equation, identify the substance that is oxidized, the one that is reduced, and the oxidizing and reducing agents:

(a) $Ni^{2+}(aq) + Cr(s) \longrightarrow Cr^{2+}(aq) + Ni(s)$

(b) $H_2CO(g) + H_2(g) \longrightarrow H_3COH(l)$

We have said that redox reactions are extremely common. Some important categories are the following:

1. Combustion All combustion (burning) reactions are redox reactions in which the compounds or mixtures that are burned are oxidized by the oxygen in the air. This includes the burning of gasoline, diesel oil,

Box 4B

Oxidizing Antiseptics

An **antiseptic** is a compound that kills bacteria. Antiseptics are used to treat wounds—not to heal them any faster but to prevent them from becoming infected by bacteria. Some antiseptics operate by oxidizing (hence destroying) compounds essential to the normal functioning of the bacteria. One example is iodine, I_2, which was used as a household antiseptic for minor cuts and bruises for many years, not in the pure form but as a dilute solution in alcohol, called a *tincture.* Pure I_2 is a steel-gray solid that gives a purple vapor when heated. The tincture is a brown liquid. Other examples of oxidizing antiseptics are dilute solutions of hydrogen peroxide, H_2O_2, and potassium permanganate, $KMnO_4$, both of which oxidize a large number of organic compounds. Sodium hypochlorite (a bleaching agent) was also once used as an antiseptic. It was one of the main ingredients of *Dakin's solution,* which was used to treat the wounded in World War I.

Oxidizing antiseptics are often regarded as too harsh, however. They not only kill bacteria but also harm skin and other normal tissues. For this reason, they have largely been replaced by phenolic antiseptics (Box 12D). *Disinfectants* are also used to kill bacteria, but on inanimate objects rather than on living tissues. Many disinfectants are oxidizing agents. Two important examples are chlorine, Cl_2, a pale green gas, and ozone, O_3, a colorless gas. Both of these gases are added in small quantities to municipal water supplies to kill any harmful bacteria that may be present. Both gases must be handled carefully because they are very poisonous.

These tablets, which produce chlorine when mixed to water, are added to a swimming pool filter. (© Yoav Levy/Phototake NYC.)

fuel oil, natural gas, coal, wood, and paper. All these materials contain carbon and most of them contain hydrogen also. If the combustion is complete, the carbon is oxidized to carbon dioxide and the hydrogen to water. In an incomplete combustion, these elements are oxidized to other compounds, many of which cause air pollution. It is unfortunate that much of the combustion that takes place in gasoline engines and in furnaces is incomplete and so contributes to air pollution.

An important product of incomplete combustion is carbon monoxide, CO, a significant air pollutant.

2. Respiration Humans and animals get their energy by respiration. As in combustion, the oxygen in the air we breathe oxidizes carbon-containing compounds in our cells to produce CO_2 and H_2O. In fact, this oxidation reaction is really the same as combustion, although it takes place more slowly and at a much lower temperature. We discuss these reactions more fully in Chapter 20. The important product of these reactions is not the CO_2 (which the body gets rid of anyhow) or the H_2O, but the energy.

3. Rusting We all know that when iron or steel objects are left out in the open air, they eventually rust (steel is mostly iron, but contains certain other elements also). In rusting (Fig. 4.11), iron is oxidized to a mixture of oxides. We can represent the main reaction by the equation

$$4Fe(s) + 3O_2(g) \longrightarrow 2Fe_2O_3(s)$$

Figure **4.11**
The rusting of iron and steel can be a serious problem in an industrial society. (Photograph by C. D. Winters.)

4. Bleaching Most bleaching is oxidation, and common bleaches are oxidizing agents. The colored compounds are usually organic compounds; oxidation converts them to colorless compounds.

5. Batteries A voltaic cell (Box 4A) is a device in which electricity is generated from a chemical reaction. Such cells are often called **batteries.** We are all familiar with them in our cars and in such portable devices as radios and flashlights (Fig. 4.12). In all cases, the reaction taking place in the battery is a redox reaction.

Figure **4.12**
(a) Dry cell batteries are used in a wide variety of applications. (Photograph by Charles D. Winters.) (b) A lead storage battery.

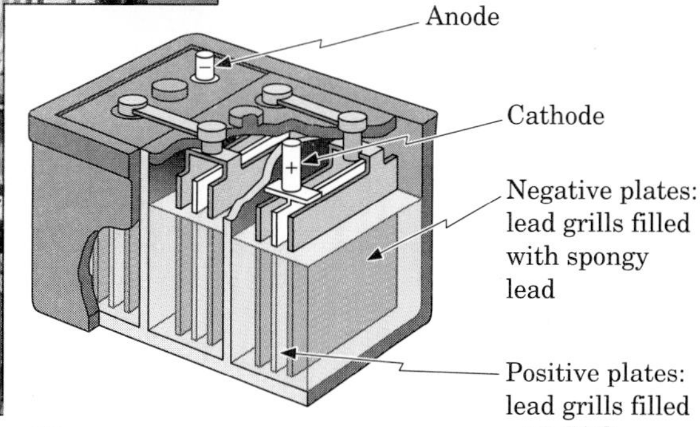

Anode
Cathode
Negative plates: lead grills filled with spongy lead
Positive plates: lead grills filled with PbO_2

(*a*) (*b*)

Box 4C

Photography

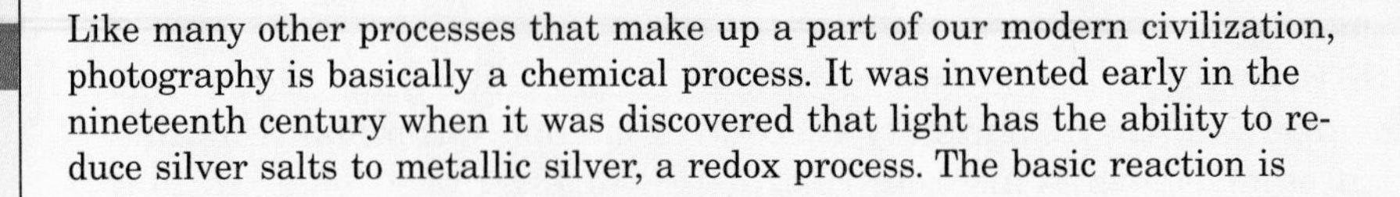

Like many other processes that make up a part of our modern civilization, photography is basically a chemical process. It was invented early in the nineteenth century when it was discovered that light has the ability to reduce silver salts to metallic silver, a redox process. The basic reaction is

$$2Ag^+ + 2Br^- + \text{light} \longrightarrow 2Ag + Br_2$$

In a photographic film tiny crystals of silver bromide, AgBr, are embedded in a gelatin emulsion. When light hits the emulsion, the silver ions are reduced to metallic silver, turning the film black in just those places where the light hits, but not where the light does not hit. The extent of reduction is proportional to the intensity and duration of the light exposure. The reduction is further aided by organic reducing agents in the developing process. The photographic image is provided by the dark silver metal atoms still embedded in the emulsion. In the last process, called fixation, the unreacted AgBr crystals are dissolved and washed away together with the gelatin emulsion. Both the development and the fixation must be done in a dark room. The negative, thus obtained, can be converted to the final photograph by the following process: The negative is used as a screen through which the light passes before it hits a piece of photographic paper in which AgBr crystals are embedded in a gelatin emulsion layer. The processing of the positive photo again involves development and fixation. What is obtained is a black-and-white photo. The description of color photography is more complicated, but it too involves chemical reactions.

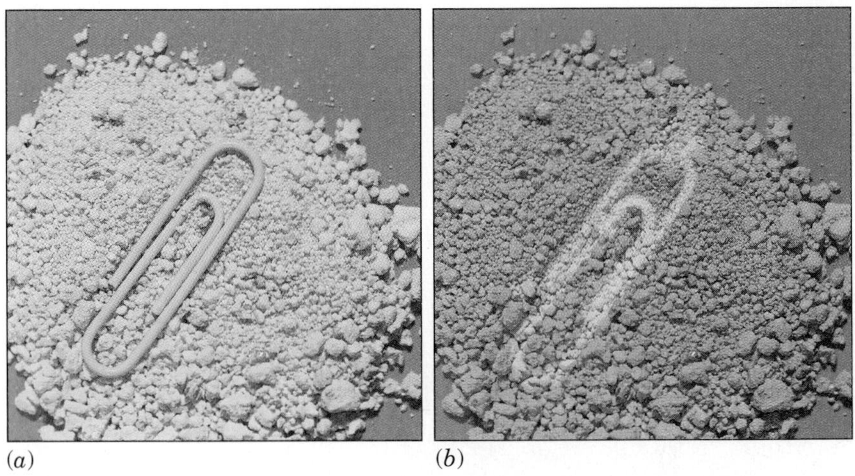

The principle on which photography is based. (a) Silver chloride crystals are exposed to light, which will darken them. (b) When the paper clip is removed, the non-darkened area stands out vividly. (© 1995 Richard Megna/Fundamental Photographs.)

4.9 Heat of Reaction

In almost all chemical reactions, not only are starting materials converted to products, but there is also a gain or loss of heat. For example, when 1 mole of pure carbon is oxidized by oxygen to give CO_2, 94 000 cal of heat are given off:

$$C(s) + O_2(g) \longrightarrow CO_2(g) + 94\,000 \text{ cal}$$

The heat given off increases the temperature of the reaction mixture.

The heat lost or gained in a reaction is called the **heat of reaction** (Fig. 4.13). A reaction that gives off heat is said to be **exothermic;** a reaction that requires heat is **endothermic** (Fig. 4.14). The *amount* of heat given off or taken in is proportional to the amount of material. For example, when two moles of pure carbon are oxidized, $2 \times 94\,000 = 188\,000$ cal of heat are given off.

An **exergonic** reaction is one that gives off energy in any form, not necessarily heat. An **endergonic** reaction is the opposite: It uses up energy.

The energy changes accompanying a chemical reaction are not limited to heat. In some reactions, as in voltaic cells (Box 4A), the energy

(*a*)

(*b*)

Figure **4.13** • A reaction that gives off a large quantity of heat. (a) The thermite reaction. After the reaction is started with a fuse of burning magnesium wire, iron(III) oxide reacts with aluminum powder. (b) The reaction generates so much heat that the product, iron, is produced in the molten state. For another example, see Figure 1.3. (Photographs by Charles D. Winters.)

Box 4D

The Caloric Value of Foods

All of us have seen the food package labels that say, for example, that one serving has 50 Calories (see Sec. 26.2). What does this mean? We have pointed out (p. 130) that combustion is essentially the same reaction as the metabolism of food. In both cases the starting materials are converted to CO_2 and H_2O, although metabolism takes place at 37°C and by a series of reactions, whereas combustion is a high-temperature process that takes place directly. Nevertheless, **the heat of reaction depends only on the starting compounds and the final products, not on the pathway.** (This is called **Hess's Law.**) This means that if a chemist takes a given weight of any food (such as sugar) and burns it in order to measure the amount of heat given off, then the chemist knows that the same amount of heat is given off in the body when that weight of that food is metabolized.

By this simple method, the energy content of many foods has been determined (Table 4D). Of course, we all know that it works two ways. If we don't take in enough calories in our food to replace the heat we lose or use for energy, our bodies begin to "burn" some of our normal tissues and we lose weight. On the other hand, if we eat food with more than enough calories to match our losses, our bodies turn that extra energy to fatty tissue and we gain weight. That is why many people carefully count the number of Calories they eat. Remember (Sec. 1.9) that a nutritionist's Calorie is equal to 1000 chemist's calories.

Table 4D Caloric Values of Some Foods

Food	Calories	Food	Calories
Milk, 8 oz	165	Apple, 5 oz	70
Swiss cheese, 1 oz	105	Banana, 5 oz	85
Egg, boiled, large	80	Grapefruit, $\frac{1}{2}$ medium	50
Hamburger patty, 3 oz	245	Orange juice, 8 oz	100
Beef, lean, 2 oz	115	Bread, rye, 1 slice	55
Lamb, shoulder, lean, 2 oz	125	Bread, white, 1 slice	60
		Fudge, 1 oz	115
Haddock, fried, 3 oz	135	Honey, 1 tablespoon	60
Shrimp, 3 oz	110	Cookie, 3-in.	110
Peanuts, roasted, 2 oz	210	Cornflakes, 1 oz	110
Carrot, raw, 8 oz	45	Noodles, 8 oz	200
Corn, 5-in. ear	65	Butter, 1 tablespoon	100
Potato, 5 oz	90	Margarine, 1 tablespoon	100
Tomato, 5 oz	30		

output is in the form of electricity. In other reactions, such as photosynthesis (the reaction whereby plants convert water and carbon dioxide to sugar and oxygen), the energy absorbed is in the form of light. In most cases, however, the energy change is in the form of heat. An example of an endothermic reaction is the decomposition of mercury(II) oxide:

$$2HgO(s) + 43\,400 \text{ cal} \longrightarrow 2Hg(l) + O_2(g)$$

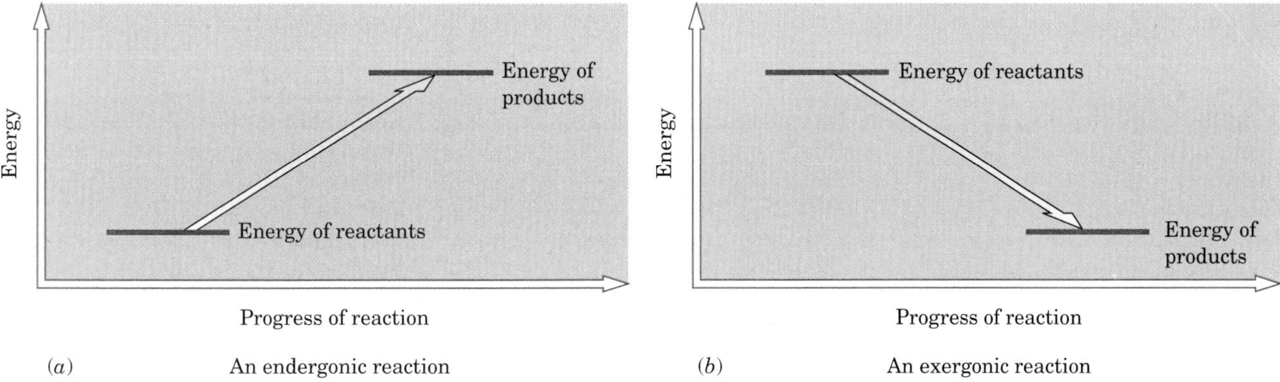

Figure **4.14**
In an endergonic reaction (a) the products have *more* energy than the starting compounds. Energy has been absorbed. In an exergonic reaction (b) the products have *less* energy than the starting compounds. Energy has been given off.

What this equation tell us is that if we want to break down 2 moles of mercury(II) oxide into the elements Hg and O_2, we must add 43 400 cal of energy.

Incidentally, the law of conservation of energy tells us that the reverse reaction, the oxidation of mercury, must give off exactly the same amount of heat:

$$2Hg(l) + O_2(g) \longrightarrow 2HgO(s) + 43\,400 \text{ cal}$$

Especially important are heats of reaction for combustion reactions. As we saw in the previous section, combustion reactions are the most important heat-producing reactions, since most of our energy is derived from them. All combustions are exothermic. The heat given off in a combustion reaction is called the **heat of combustion.** Accurate heat of combustion values are available for thousands of compounds.

SUMMARY

A **chemical equation** tells which starting materials are converted to which products. All equations must be balanced.

The **formula weight** of a compound is the sum of the atomic weights of all atoms in the molecular formula. A **mole** of a substance is defined as **Avogadro's number** (6.02×10^{23}) of particles. The mass of this number of molecules of any compound is numerically equal to the formula weight expressed in grams.

A **balanced equation** tells how many moles of each starting material are converted to how many moles of each product. By the use of balanced equations we can calculate the required number of grams (or moles) of any starting material or product if we know the number of grams (or moles) of any other starting material or product. **Percentage yield** equals actual yield divided by **theoretical yield** multiplied by 100.

When ions are mixed in water solution, they react with each other only if (a) a precipitate is formed, (b) a gas is formed, (c) an acid neutralizes a base, or (d) an oxidation–reduction takes place. Otherwise, no reaction takes place. Ions that do not react are called **spectator ions.** All compounds containing Na^+, K^+, NH_4^+, NO_3^-, or CH_3COO^- ions are soluble in water, as are most compounds containing Cl^- or SO_4^{2-} ions. Most compounds containing CO_3^{2-}, PO_4^{3-}, S^{2-}, or OH^- are insoluble. **Ionic equations** show only the ions that react. In an ionic equation the charges as well as the atoms must be balanced.

In an **oxidation,** electrons are lost; in a **reduction,** electrons are gained. These two processes must take place together. The joint process is often called a **redox** reaction. Oxidation can also be defined as the gain of oxygen or loss of hydrogen; reduction is the reverse. Important examples of oxidation are combustion, respiration, bleaching, and rusting. Electric batteries use redox reactions to generate electricity.

Almost all reactions are accompanied by a loss or gain of heat. This is called the **heat of reaction.** Reactions that give off heat are **exothermic;** those that absorb heat are **endothermic.** The heat of a combustion reaction is called the **heat of combustion.**

KEY TERMS

Antiseptic (Box 4B)
Avogadro's number (Sec. 4.3)
Chemical equation (Sec. 4.4)
Disinfectant (Box 4B)
Dissociation (Sec. 4.7)
Endergonic (Sec. 4.9)
Endothermic (Sec. 4.9)
Exergonic (Sec. 4.9)
Exothermic (Sec. 4.9)
Formula weight (Sec. 4.2)
Heat of combustion (Sec. 4.9)
Heat of reaction (Sec. 4.9)
Hess's law (Box 4D)
Ionic equation (Sec. 4.7)
Mole (Sec. 4.3)
Molecular weight (Sec. 4.2)
Oxidation (Sec. 4.8)
Oxidizing agent (Sec. 4.8)
Percentage yield (Sec. 4.6)
Products (Sec. 4.1)
Reactants (Sec. 4.1)
Redox (Sec. 4.8)
Reducing agent (Sec. 4.8)
Reduction (Sec. 4.8)
Spectator ion (Sec. 4.7)
Starting materials (Sec. 4.1)
Stoichiometry (Sec. 4.5)
Theoretical yield (Sec. 4.6)
Voltaic cell (Box 4A)
Yield (Sec. 4.6)

PROBLEMS

Formula Weight and Moles

4.16 Calculate the formula weight of (a) Cl_2 (b) Ar (c) P_4 (d) N_2 (e) He

4.17 Calculate the formula weight of
(a) KCl
(b) Na_3PO_4
(c) $Fe(OH)_2$
(d) sucrose, $C_{12}H_{22}O_{11}$
(e) $NaAl(SO_3)_2$
(f) glycine, $C_2H_7O_2N$
(g) $Al_2(SO_4)_3$
(h) $(NH_4)_2CO_3$
(i) DDT, $CCl_3CH(C_6H_4Cl)_2$

4.18 How many
(a) moles of CO_2 are there in 83.2 g of CO_2?
(b) moles of glycerol, $C_3H_8O_3$, are there in 428 g of glycerol?
(c) moles of NaH_2PO_4 are there in 14 g of NaH_2PO_4?
(d) moles of quinine, $C_{20}H_{24}N_2O_2$, are there in 51.6 g of quinine?

4.19 How many
(a) grams of nitrogen dioxide, NO_2, are there in 1.77 moles of NO_2?
(b) grams of 2-propanol, C_3H_8O (rubbing alcohol), are there in 0.84 mole of 2-propanol?
(c) grams of WCl_5 are there in 3.69 moles of WCl_5?
(d) grams of galactose, $C_6H_{12}O_6$, are there in 0.348 mole of galactose?
*(e) grams of vitamin C, $C_6H_8O_6$, are there in 4.9×10^{-2} mole of vitamin C?

4.20 How many
(a) moles of O atoms are there in 18.1 moles of formaldehyde, H_2CO?
(b) moles of Br atoms are there in 0.41 mole of bromoform, $CHBr_3$?
*(c) moles of O atoms are there in 3.5×10^3 moles of $Al_2(SO_4)_3$?
(d) moles of Hg atoms are there in 87 g of HgO?
*(e) moles of N atoms are there in 2.1×10^{-3} g of $K_3Fe(CN)_6$?

4.21 How many
(a) moles of Ag^+ ions are there in 3.71 moles of $AgNO_3$?
(b) moles of Na^+ ions are there in 1.44 g of Na_2CO_3?
*(c) moles of SO_4^{2-} ions are there in 84 g of $Al_2(SO_4)_3$?

4.22 A single atom of cerium weighs just about twice as much as a single atom of gallium. What is the weight ratio of 25 atoms of cerium to 25 atoms of gallium?

4.23 What is the mass in grams of (a) 100 (b) 3000 (c) 5.0×10^6 (d) 2.0×10^{24} molecules of formaldehyde, H_2CO?

4.24 What is the mass in grams of a single molecule of sucrose, $C_{12}H_{22}O_{11}$?

4.25 How many
(a) molecules of TNT, $C_7H_5N_3O_6$, are there in 2.9 moles of TNT?
(b) molecules are in one drop (0.0500 g) of water?

4.26 A typical cholesterol deposit in an artery might have a mass of 3.9 mg of cholesterol ($C_{27}H_{46}O$). How many molecules of cholesterol are in this mass?

4.27 Protein molecules are very large. The molecular weight of hemoglobin is about 68 000. What is the mass in grams of a single molecule of hemoglobin?

Balancing Equations

4.28 Balance:
(a) $H_2 + I_2 \longrightarrow HI$
(b) $Al + O_2 \longrightarrow Al_2O_3$
(c) $Na + Cl_2 \longrightarrow NaCl$
(d) $CaCO_3 \longrightarrow CO_2 + CaO$

4.29 Balance:
(a) $HI + NaOH \longrightarrow NaI + H_2O$
(b) $Ba(NO_3)_2 + H_2S \longrightarrow BaS + HNO_3$
(c) $CH_4 + O_2 \longrightarrow CO_2 + H_2O$
(d) $C_4H_{10} + O_2 \longrightarrow CO_2 + H_2O$
(e) $Fe + CO_2 \longrightarrow Fe_2O_3 + CO$
(f) $Al + HBr \longrightarrow AlBr_3 + H_2$
(g) $P_4 + O_2 \longrightarrow P_4O_{10}$

Mass Relationships in Chemical Reactions

4.30 Define:
(a) stoichiometry
(b) aqueous solution

4.31 For the reaction

$$2N_2(g) + 3O_2(g) \longrightarrow 2N_2O_3(g)$$

(a) how many moles of N_2 are required to react completely with 1 mole of O_2?
(b) how many moles of N_2O_3 are produced from the complete reaction of 1 mole of O_2?
(c) how many moles of O_2 are required to produce 8 moles of N_2O_3?

4.32 For the reaction

$$2Al(s) + 6HCl(g) \longrightarrow 2AlCl_3(s) + 3H_2(g)$$

how many moles of H_2 are produced by the complete reaction of 132 g of Al?

4.33 For the reaction

$$CH_4(g) + 3Br_2(l) \longrightarrow \underset{\text{Bromoform}}{CHBr_3(l)} + 3HBr(g)$$

how many grams of Br_2 are needed to produce 1.50 moles of bromoform?

4.34 Acetaldehyde, C_2H_4O, is produced commercially from acetylene by the reaction

$$C_2H_2(g) + H_2O(l) \longrightarrow C_2H_4O(g)$$

How many grams of C_2H_4O can be produced from 81.7 g of C_2H_2, assuming enough water is present?

4.35 Ethyl alcohol, C_2H_6O, can be added to gasoline to produce "gasohol," a fuel for automobile engines. The equation for the combustion of ethyl alcohol is

$$C_2H_6O(l) + 3O_2(g) \longrightarrow 2CO_2(g) + 3H_2O(g)$$

How many grams of O_2 are required for the combustion of 421 g of C_2H_6O?

4.36 In the process called photosynthesis, plants convert CO_2 and H_2O to glucose, $C_6H_{12}O_6$, by the reaction

$$6CO_2(g) + 6H_2O(l) \longrightarrow C_6H_{12}O_6(s) + 6O_2(g)$$

How many grams of CO_2 are required to produce 5.1 g of glucose?

4.37 Iron ore is converted to iron by a process that can be represented as

$$2Fe_2O_3(s) + 6C(s) + 3O_2(g) \longrightarrow 4Fe(s) + 6CO_2(g)$$

If the process is run until 3940 g of Fe is produced, how many grams of CO_2 will also be produced?

4.38 Given the reaction in Problem 4.37, how much C (in grams) is necessary to react completely with 0.58 g of Fe_2O_3?

4.39 Aspirin can be made in the laboratory by the reaction

$$\underset{\text{Salicylic acid}}{C_7H_6O_3(s)} + \underset{\text{Acetyl chloride}}{C_2H_3ClO(l)} \longrightarrow \underset{\text{Aspirin}}{C_9H_8O_4(s)} + HCl(g)$$

How many grams of aspirin are produced if 85.0 g of salicylic acid is treated with excess acetyl chloride?

4.40 The compound aniline reacts with chlorine as follows:

$$\underset{\text{Aniline}}{C_6H_7N(l)} + 3Cl_2(g) \longrightarrow \underset{\text{Trichloroaniline}}{C_6H_4Cl_3N(l)} + 3HCl(g)$$

How many grams of trichloroaniline are produced if 34.7 g of aniline is treated with excess Cl_2?

***4.41** If 29.7 g of N_2 is added to 3.31 g of H_2 in the reaction

$$N_2(g) + 3H_2(g) \longrightarrow 2NH_3(g)$$

(a) which reactant is completely used up?
(b) how many grams of the other reactant are left over?
(c) how many grams of NH_3 are formed if the reaction goes to completion?

Percentage Yield

4.42 A chemist calculated the theoretical yield in a particular reaction to be 60.4 g. The mass of the product actually obtained was 51.7 g. What is the percentage yield?

***4.43** Diethyl ether can be made by the reaction

$$2C_2H_5OH(l) \longrightarrow (C_2H_5)_2O(l) + H_2O(l)$$

If a chemist using 517 g of ethanol, C_2H_5OH, produced 391 g of ethyl ether, $(C_2H_5)_2O$, what is the percentage yield?

4.44 A chemist who ran this reaction

$$CH_3CH_2Cl(l) + KOH(aq) \longrightarrow CH_3CH_2OH(l) + KCl(aq)$$

by treating 47.2 g of CH_3CH_2Cl with excess KOH, obtained a yield of 28.7 g of CH_3CH_2OH. What was the percentage yield?

Reactions Between Ions in Aqueous Solution

4.45 In the equation

$$Na^+(aq) + CO_3^{2-}(aq) + Sr^{2+}(aq) + Cl^-(aq) \longrightarrow Na^+(aq) + SrCO_3(s) + Cl^-(aq)$$

(a) which are the spectator ions?
(b) Write the balanced ionic equation.

4.46 Balance:
(a) $Ag^+(aq) + Br^-(aq) \longrightarrow AgBr(s)$
(b) $Cd^{2+}(aq) + S^{2-}(aq) \longrightarrow CdS(s)$
(c) $Sc^{3+}(aq) + SO_4^{2-}(aq) \longrightarrow Sc_2(SO_4)_3(s)$
*(d) $Sn^{2+}(aq) + Fe^{2+}(aq) \longrightarrow Fe^{3+}(aq) + Sn(s)$
(e) $K(s) + H_2O(l) \longrightarrow K^+(aq) + OH^-(aq) + H_2(g)$

4.47 When a solution of ammonium chloride, NH_4Cl, is added to a solution of lead nitrate, $Pb(NO_3)_2$, a white precipitate, $PbCl_2$, is formed. Write a balanced ionic equation for this reaction. Both ammonium chloride and lead nitrate exist as dissociated ions in aqueous solution.

4.48 When a solution of iron(II) chloride, $FeCl_2$, is added to a solution of potassium hydroxide, KOH, iron(II) hydroxide, $Fe(OH)_2$, comes out of the solution as a pale green precipitate. Write a balanced ionic equation for this reaction. Both $FeCl_2$ and KOH exist as dissociated ions in aqueous solutions.

4.49 When a solution of hydrochloric acid, HCl, is added to a solution of sodium sulfite, Na_2SO_3, sulfur dioxide gas, SO_2, is released from the solution. Write an ionic equation for this reaction. An aqueous solution of HCl contains H^+ and Cl^- ions, and Na_2SO_3 exists as dissociated ions in aqueous solution.

4.50 When a solution of sodium hydroxide, NaOH, is added to a solution of ammonium carbonate, $(NH_4)_2CO_3$, ammonia gas, NH_3, is released when the solution is heated. Write an ionic equation for this reaction. Both NaOH and $(NH_4)_2CO_3$ exist as dissociated ions in aqueous solution.

4.51 Using the solubility generalizations given in Section 4.7, predict whether each of the following ionic compounds is soluble in water.

(a) KCl (d) Na_2SO_4
(b) NaOH (e) Na_2CO_3
(c) $BaSO_4$ (f) $Fe(OH)_2$

4.52 Using the solubility generalizations given in Section 4.7, predict whether each of the following ionic compounds is soluble in water.

(a) $CaCO_3$ (d) AgCl
(b) $Mg_3(PO_4)_2$ (e) $Fe(NO_3)_3$
(c) $(NH_4)_2CO_3$ (f) PbS

4.53 Predict whether or not a solid precipitate will come out of solution when aqueous solutions of the following compounds are mixed. If a precipitate will form, write its formula.

(a) $CaCl_2 + K_3PO_4$
(b) $KCl + Na_2SO_4$
(c) $(NH_4)_2CO_3 + Ba(NO_3)_2$
(d) $FeCl_2 + KOH$
(e) $Ba(NO_3)_2 + NaOH$
(f) $Na_2S + SbCl_3$
(g) $Pb(NO_3)_2 + K_2SO_4$

Oxidation–Reduction

4.54 Give two definitions of oxidation and two definitions of reduction.

4.55 Can a reduction take place without an oxidation? Explain.

4.56 In the reaction

$$Pb(s) + 2Ag^+(aq) \longrightarrow 2Ag(s) + Pb^{2+}(aq)$$

(a) which species gets oxidized and which gets reduced?
(b) which is the oxidizing agent and which the reducing agent?

4.57 In the reaction

$$C_7H_{12}(l) + 10O_2(g) \longrightarrow 7CO_2(g) + 6H_2O(g)$$

(a) which species gets oxidized and which reduced?
(b) which is the oxidizing agent and which the reducing agent?

Heat of Reaction

4.58 Define: (a) exothermic (b) endothermic.

4.59 Which of these reactions are endothermic and which are exothermic?

(a) $2NH_3(g) + 22\,000 \text{ cal} \longrightarrow N_2(g) + 3H_2(g)$
(b) $H_2(g) + F_2(g) \longrightarrow 2HF(g) + 124\,000 \text{ cal}$
(c) $C(s) + O_2(g) \longrightarrow CO_2(g) + 94\,000 \text{ cal}$
(d) $H_2(g) + CO_2(g) + 9800 \text{ cal} \longrightarrow H_2O(g) + CO(g)$
(e) $C_3H_8(g) + 5O_2(g) \longrightarrow 3CO_2(g) + 4H_2O(g) + 531\,000 \text{ cal}$

4.60 In the reaction

$$H_2(g) + CO_2(g) + 9800 \text{ cal} \longrightarrow H_2O(l) + CO(g)$$

9800 cal are used up. How much heat would be given off (or used up) if 2 moles of water were to react with 2 moles of carbon monoxide?

4.61 The equation for the combustion of acetone is

$$\underset{\text{Acetone}}{2C_3H_6O(l)} + 8O_2(g) \longrightarrow 6H_2O(g) + 6CO_2(g) + 853\,600 \text{ cal}$$

How much heat will be given off if 0.37 mole of acetone is completely burned?

***4.62** To convert one mole of iron(III) oxide to its elements requires 196 500 cal:

$$Fe_2O_3(s) + 196\,500 \text{ cal} \longrightarrow 2Fe(s) + 3O_2(g)$$

How many grams of iron can be produced if 156 000 calories of heat are absorbed by a large-enough sample of iron(III) oxide?

Boxes

4.63 (Box 4A) The equation

$$Fe(s) + Zn^{2+}(aq) \longrightarrow Zn(s) + Fe^{2+}(aq)$$

represents a redox reaction that can be used to make a voltaic cell. Draw a voltaic cell that uses this reaction, showing the direction of electron flow.

4.64 (Box 4B) Hydrogen peroxide is not only an antiseptic but also an oxidizing agent. A typical equation is

$$4H_2O_2(l) + CH_4(g) \longrightarrow CO_2(g) + 6H_2O(l)$$

In this reaction, what is oxidized and what is reduced? To what is it being reduced?

4.65 (Box 4C) Explain how an image is formed on a photographic film when light hits it in a camera.

4.66 (Box 4D) The caloric value of an apple is given in Table 4D. If this quantity of heat were added to 1.00 kg of water at 25°C, what would be the final temperature of the water? (Sec. 1.9.)

4.67 (Box 4D) How many ounces of milk are calorically equivalent to one slice of rye bread?

Additional Problems

4.68 Which has more mass, a mole of lead (Pb) or a mole of silver (Ag)?

4.69 In a certain reaction Cu^+ was converted to Cu^{2+}. What happened to the Cu^+ ion?

4.70 In the equation

$$Fe_2O_3(s) + 3CO(g) \longrightarrow 2Fe(s) + 3CO_2(g)$$

(a) how many moles of Fe_2O_3 are required to produce 38.4 moles of Fe?
(b) how many grams of CO are required to produce 38.4 moles of Fe?

4.71 Explain this statement: Our modern civilization depends on the heat obtained from chemical reactions.

4.72 Ammonium sulfide, $(NH_4)_2S$, and arsenic nitrate, $As(NO_3)_3$, are both ionic compounds and dissociate in water to give the corresponding ions. When aqueous solutions of these compounds are mixed, arsenic sulfide, As_2S_3, precipitates. Write the ionic equation and identify the spectator ions.

4.73 The active ingredient in one analgesic pill is 488 mg of aspirin, $C_9H_8O_4$. How many moles of aspirin does the pill contain? How many molecules?

***4.74** Chlorophyll, the compound responsible for the green color of leaves and grass, contains one atom of magnesium in each molecule. If the percentage by weight of magnesium in chlorophyll is 2.72%, what is the molecular weight of chlorophyll?

***4.75** The 1992 Clean Air Act mandates that an additive such as methyl *tert*-butyl ether, $CH_3—O—C(CH_3)_3$, be added to gasoline. (a) How many moles of CO_2 are produced in the complete combustion of 100 g of methyl *tert*-butyl ether? (b) To the extent that this ether is used in gasoline, it replaces, weight-for-weight, the alkanes normally present. A typical one is heptane, C_7H_{16}. How many moles of CO_2 are produced in the complete combustion of 100 g of heptane?

***4.76** If 7 g of N_2 is added to 11 g of H_2 in the Haber reaction,

$$N_2(g) + 3H_2(g) \longrightarrow 2NH_3(g)$$

which reactant is in excess?

***4.77** In the reaction

$$3Pb(NO_3)_2 + 2AlCl_3 \longrightarrow 3PbCl_2 + 2Al(NO_3)_3$$

8.00 grams of lead nitrate interact with 2.67 grams of aluminum chloride. The reaction yields 5.55 g of lead chloride. (a) Which reagent is in excess? (b) What is the percentage yield?

4.78 Define (a) exergonic (b) endergonic.

***4.79** A test tube contains some calcium oxide, CaO. The total weight of the test tube and the CaO is 10.860 g. The CaO reacts with water vapor in the air to give calcium hydroxide, by this equation:

$$CaO(s) + H_2O(g) \longrightarrow Ca(OH)_2(s)$$

After completely reacting with the water vapor, the test tube and its contents weigh 11.149 g. What is the weight of the empty test tube?

4.80 Define spectator ion.

***4.81** Assume that the average red blood cell has a mass of 2×10^{-8} g and that 20 percent of its content is hemoglobin (a protein whose molecular weight is 68 000). How many molecules of hemoglobin are there in one red blood cell?

4.82 (a) Balance

$$C_4H_8 + O_2 \longrightarrow CO_2 + H_2O$$

(b) In this reaction, what is oxidized and what is reduced?
(c) What is the oxidizing agent and what is the reducing agent?

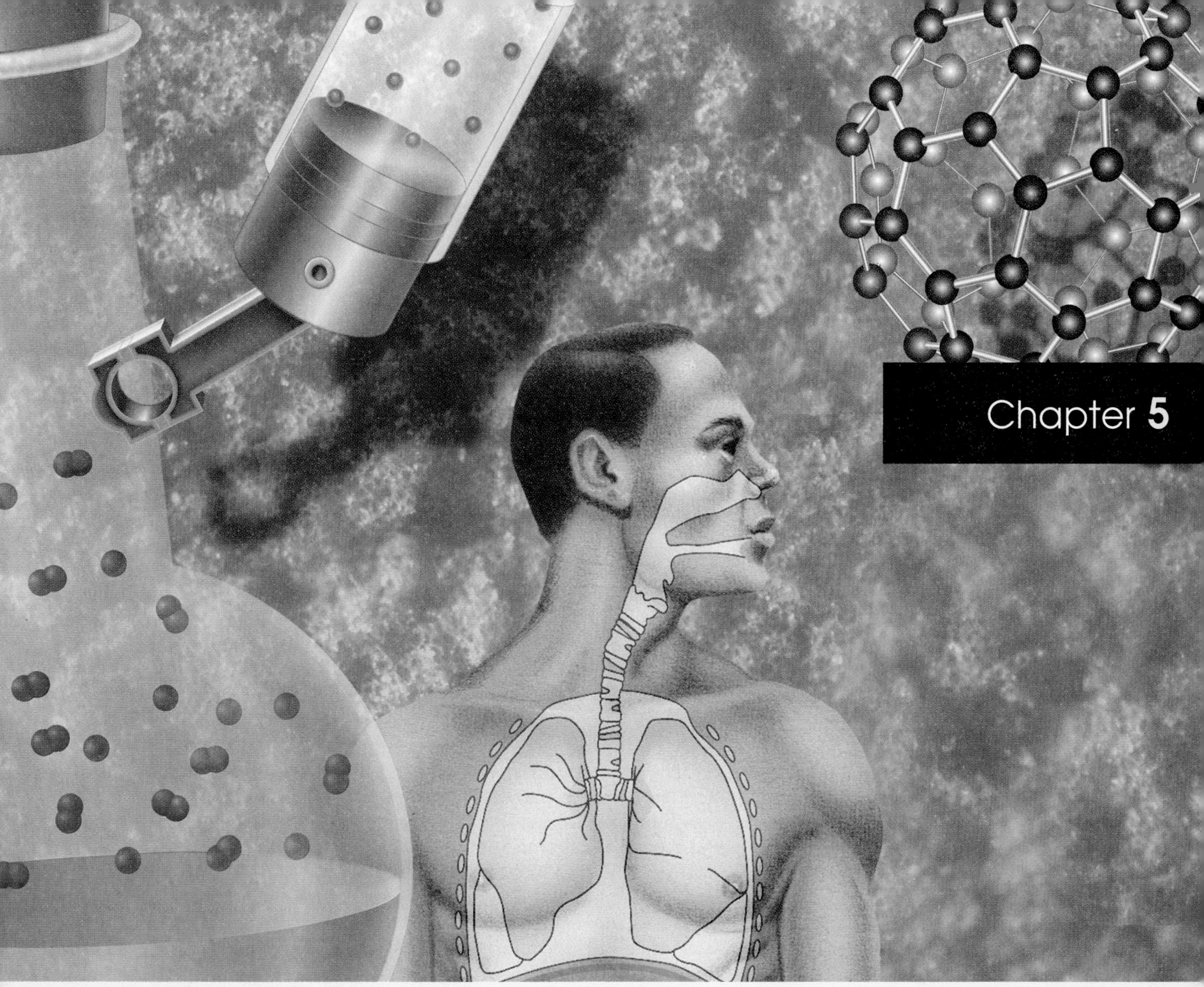

Chapter 5

Gases, liquids, and solids

5.1 The Organization of Matter

There are various forces that hold matter together and cause it to take different forms. In the atomic nucleus, very strong forces keep the protons and neutrons together (Chapter 9). In the atom itself, there are attractions between the positive nucleus and the negative electrons. Within molecules, atoms are attracted to each other by covalent bonds, the arrangement of which causes the molecules to assume a particular shape. Within ionic crystals, three-dimensional shapes are brought about by electrostatic attractions between ions.

In addition to all these forces, there are also attractive forces *between* molecules. These forces are weaker than any already mentioned, but they nevertheless help to determine whether a particular compound or element is a solid, a liquid, or a gas at room temperature. Because these are **forces between one molecule and another** (rather than within a molecule), they are called **intermolecular forces.**

All these attractive forces provide the potential energy that holds matter together and keeps it from falling apart, by counteracting another form of energy, kinetic energy, which tends to disorganize matter. In the absence of attractive forces, the kinetic energy that particles possess keeps them constantly moving, mostly in a random, disorganized way. This kinetic energy increases with increasing temperature. Therefore, **the higher the temperature, the greater the tendency of particles to fly randomly in every direction.**

The physical state of matter thus depends on a balance between the kinetic energy of particles and the attractive forces between them. At high temperatures, molecules possess a large amount of kinetic energy and move so fast that intermolecular forces are too weak to hold them together. This is the **gaseous state.** At lower temperatures, the molecules slow down. When the temperature is sufficiently reduced, the gas condenses to form the **liquid state.** The molecules are still moving past each other, but much more slowly. The attractive forces now become important. When the temperature goes even lower, the molecules no longer have enough velocity to move past each other. This is the **solid state** (Fig. 5.1). In this state each molecule has a certain number of nearest neighbors, and these neighbors do not change.

The attractive forces between molecules in the solid state are the same as in the gaseous and liquid states. The difference is that in the gaseous (and to a lesser degree in the liquid) state the kinetic energy of the molecules is great enough to overcome the attractive forces.

Most substances can exist in any of the three states. Typically a solid, when heated to a sufficiently high temperature, melts and becomes a liquid. This temperature, called the **melting point,** is different for each solid. Further heating causes the temperature to rise to the point at which the liquid boils and becomes a gas. This is the boiling point (see also p. 163). However, not all substances can exist in all three states. For example, wood and paper cannot be melted. On heating, they decompose or burn (depending on whether or not air is present) but they do not melt. Another example is sugar, which does not melt when heated, but forms a dark substance called caramel.

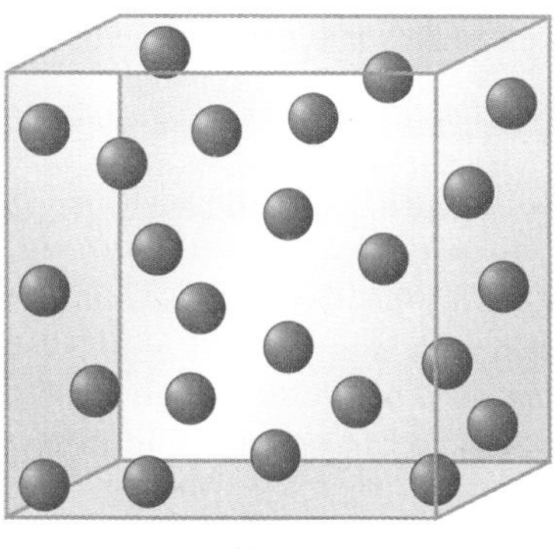

Gas

Molecules far apart and disordered
Negligible interactions between molecules

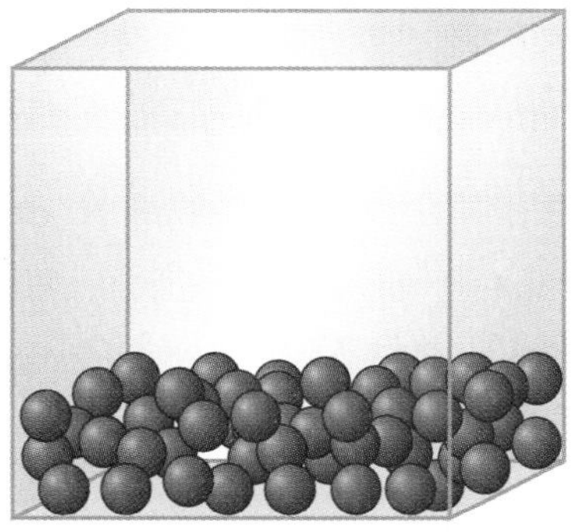

Liquid

Intermediate situation

Solid

Molecules close together and ordered
Strong interactions between molecules

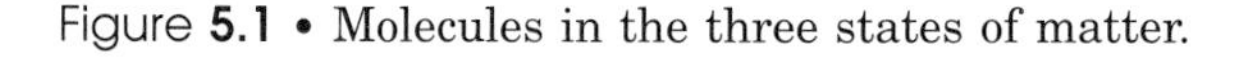

Figure **5.1** • Molecules in the three states of matter.

5.2 Gases

Of the three states of matter, the gaseous state is the one most studied and best understood. The most important characteristic of a gas is that it completely fills whatever container it is in. In doing so, the gas exerts a pressure on the walls of the container. This behavior, and all other properties of gases, can be explained by the **kinetic molecular theory,** which assumes that

The kinetic molecular theory explains the behavior of ideal gases.

1. Gases consist of molecules that are constantly moving through space in straight lines, randomly, and with varying speeds.

2. The average kinetic energy of the gas molecules is proportional to the absolute (Kelvin) temperature. This means that as the temperature increases the gas molecules move faster.

3. The gas molecules collide with each other, as billiard balls do, bouncing off each other and changing direction. Each time they collide they also may exchange kinetic energies (one moves faster, the other slower than before), but the total kinetic energy is conserved (assuming that they do not react with each other).

4. The molecules of a gas are very tiny compared to the distances between the molecules, so tiny that the kinetic molecular theory assumes that the molecules have no volume at all. The **volume of a gas** is thus the **volume of the container** it is in.

5. The molecules do not stick together after collision because there are no attractions between them.

6. The molecules collide with the walls of the container. The collisions with the walls constitute the pressure of the gas. The greater the number of collisions per unit time, the greater the pressure.

These six assumptions of the kinetic molecular theory obviously give an idealized picture of what is going on. In real gases there are inter-

Box 5A

Entropy, a Measure of Disorder

Section 5.2 states that the molecules of a gas move randomly and that the higher the temperature, the faster they move. In general, random motion means disorder. When the temperature is lowered, the molecules slow down and more and more order is apparent. When all the molecules of a system become motionless and they line up perfectly, we achieve the greatest possible order. (Of course, at this point the substance is a solid—it can no longer be a gas, or even a liquid.) There is a measure of such order, called **entropy.** When the order is perfect the entropy of the system is zero. If the system has the slightest impurity, the perfect order is broken and the entropy increases. When molecules rotate or move from one place to another, the disorder increases and so does the entropy. Thus when a crystal melts, there is an increase in entropy; when a liquid vaporizes, the disorder, and hence the entropy, increases. When we combine two pure substances and a mixture is formed, disorder increases and so does the entropy. When the temperature increases, the entropy of any substance or mixture of substances increases—not just for gases but also for liquids and solids—because an increase in temperature means increased molecular motions. Conversely, when a system is cooled, its entropy decreases.

We learned about the absolute, or Kelvin, temperature scale in Section 1.4. Absolute zero or −273°C, is the lowest possible temperature. At this temperature, all molecular motions cease, and perfect order reigns. No such perfect order can be reached in reality, although scientists have been able to produce temperatures much less than one degree above absolute zero. Of course, any substance at a temperature this low must have extremely little entropy.

High entropy

Low entropy

The concept of entropy does not apply only to molecules. The general tendency in nature is towards disorder. Keeping the entropy low requires energy, as demonstrated in these photos. (Photographs by Charles D. Winters.)

molecular attractions, and the molecules really do occupy some volume. Because of this, a gas described by these six assumptions is called an **ideal gas.** There cannot actually be an ideal gas; all gases are real. However, under normal environmental conditions, the behavior of real gases does approximate that of an ideal gas, and we can safely use these concepts.

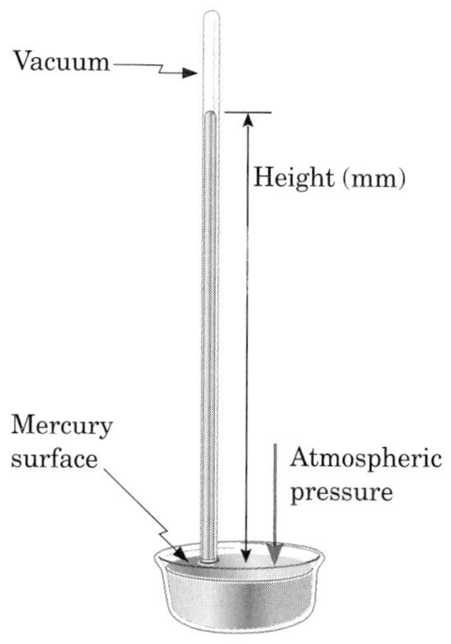

Figure **5.2**
The mercury barometer. A long glass tube completely filled with mercury is inverted into a pool of mercury in a dish. There is no air at the top of the mercury column in the tube because there is no way air could get in. Thus there is no gas pressure exerted on the mercury column. However, on the open dish of mercury the whole atmosphere exerts its pressure. The difference in the heights of the two mercury levels is the atmospheric pressure.

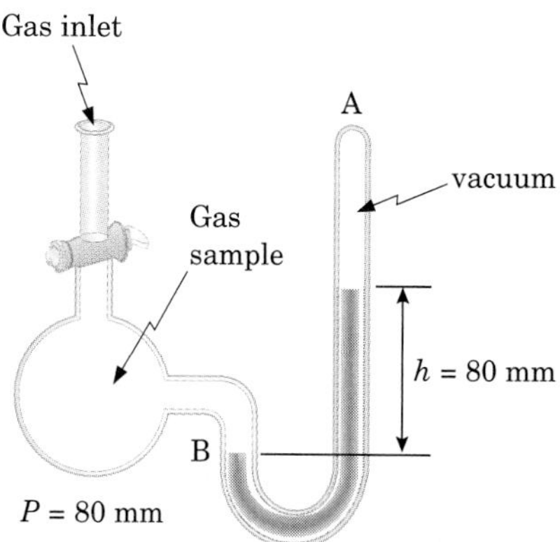

Figure **5.3** • A manometer.

5.3 Pressure

On the earth we live under a blanket of air that presses down on us, on the ground, and on everything around us. As we all know from weather reports, the pressure of the atmosphere is not constant, but varies from day to day. A device called a **barometer** is used to measure atmospheric pressure. In this instrument, as shown in Figure 5.2, there is a vacuum inside the tube, and the atmosphere presses down on the liquid in the dish. Since the vacuum has zero pressure, there is no pressure on the mercury inside the tube. On the other hand, there is a pressure (the pressure of the air) on the mercury in the dish. The atmospheric pressure is measured as the difference in height between the two liquid levels. This difference in height is used directly as a unit of pressure. **Pressure is measured in millimeters of mercury** (mm Hg). (The weather bureau usually gives it in inches of mercury.) In honor of Torricelli, we define another unit, the **torr,** as being equal to 1 mm Hg. At sea level the average pressure of the atmosphere is 760 mm Hg, and we use this number to define still another unit of pressure, the **atmosphere** (atm). There are several other units used for pressure. The official SI unit is the pascal, defined as 1 atm = 101 325 pascals. In this book we use only mm Hg and atm.

The barometer was invented in 1643 by Evangelista Torricelli (1608–1647).

1 atm = 760 torr = 760 mm Hg

A barometer is fine for measuring the pressure of the atmosphere, but to measure the pressure of a gas in a container, we use a simpler instrument called a **manometer.** One form of manometer consists of a U-shaped tube containing mercury (Fig. 5.3) in which one arm (A) has been evacuated and sealed. This arm has zero pressure. The other arm (B) is connected to the container in which the gas sample is enclosed. The pressure of the gas depresses the level of the mercury in the B arm of the U tube, and **the difference between the two mercury levels gives the pressure directly in mm Hg.** If more gas is added to the bulb in Figure 5.3, the mercury level in B will be pushed down and that in A will rise, because the pressure in the bulb has now increased.

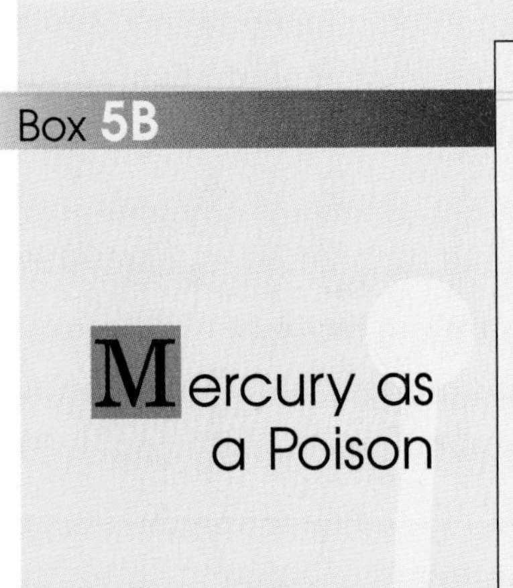

Box 5B

Mercury as a Poison

Mercury is used in most manometers and barometers because it is a liquid over a wide temperature range and so can be used for pressure measurements between −30 and +300°C. However, mercury is poisonous. Mercury spills should be collected and stored under water. Frequent and prolonged contact with mercury should be avoided (use gloves). Prolonged inhalation of mercury vapor can cause teeth to fall out; later, nervous disorders develop.

Industrial mercury poisoning was prevalent among people who made felt hats, beginning in the middle of the seventeenth century. Felt hats are made of the fine hair of rabbits and hares. The workers treated the hairs with mercury salts, which twisted them and made them limp, aiding the felting process. This contaminated the air with mercury vapors, which the hat makers breathed in. As a consequence they developed twitches, a symptom of nervous disorders. The term "mad hatter" (as in *Alice in Wonderland*) originates from this industrial disease.

5.4 Gas Laws

These gas laws hold not only for pure gases but also for any mixture of gases.

In observing the behavior of gases, a number of relationships have been established:

1. Boyle's law At constant temperature, the volume of any sample of gas decreases as its pressure increases, and vice versa. This means that for a fixed mass of gas at a constant temperature, the volume is inversely proportional to the pressure. This is what happens in the cylinder of a car engine; as the pistons move up, the volume of gas decreases and the pressure increases (Fig. 5.4).

Although Jacques Charles (1746–1823) had earlier formulated a crude version of his law, it was first stated mathematically by Joseph Gay-Lussac (1778–1850) in 1802.

2. Charles's law At constant pressure, the volume of any sample of gas increases as its temperature increases, and vice versa. Putting it more precisely, the volume of a fixed mass of gas at a constant pressure is directly proportional to the absolute temperature. This is the basis of the hot-air balloon (Fig. 5.5). The air in the balloon expands as it is heated

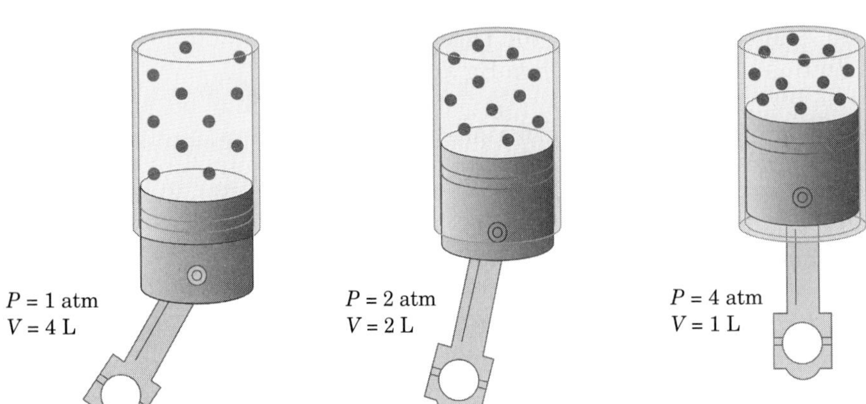

Figure **5.4**
Boyle's law illustrated in an automobile cylinder. When the piston moves up, the volume occupied by the gas (a mixture of air and gasoline vapor) decreases. Since the temperature is constant, the pressure must increase.

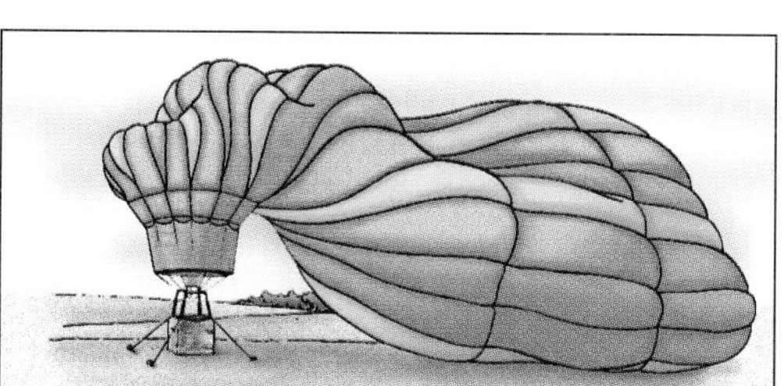

Figure **5.5**
Charles's law illustrated in a hot-air balloon. Because the balloon can stretch, the pressure inside it is constant. When the air in the balloon is heated, its volume increases, expanding the balloon. Since the volume of the air has increased but not the mass, its density has decreased. The air inside the balloon being less dense than the air outside provides the lift, and the balloon can ascend.

from below; this makes the air inside the balloon less dense than the surrounding air, providing the lift (Charles was one of the first balloonists).

3. Gay-Lussac's law At constant volume, as the temperature increases, the pressure of any sample of gas increases, and vice versa. For a fixed mass of gas at constant volume, the pressure is directly proportional to the absolute temperature. This is what happens inside an autoclave. Autoclaves are used to sterilize hospital and laboratory equipment (Fig. 5.6). Steam generated at 1 atm pressure has a temperature of 100°C. When this steam is further heated in an enclosed vessel (autoclave) at constant volume, the increased temperature of the steam increases its pressure. A valve controls the maximum pressure for which the autoclave is designed; if the pressure exceeds this, the valve opens and the steam is released. At maximum pressure the temperature may be as high as 120 to 150°C. At these high temperatures all microorganisms in the equipment are destroyed.

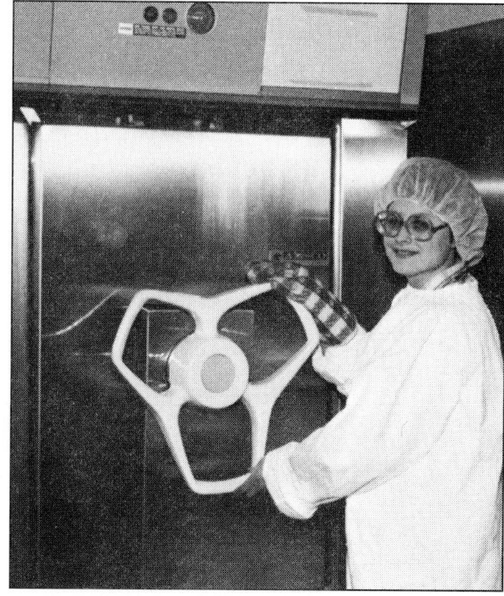

Figure **5.6**
An autoclave used to sterilize hospital equipment.

Mathematical expressions of these three laws are shown in Table 5.1. The three gas laws can be combined and expressed by a mathematical equation called the **combined gas law:**

$$\frac{P_1V_1}{T_1} = \frac{P_2V_2}{T_2}$$

Table **5.1** Mathematical Expressions of the Three Gas Laws for a Fixed Mass of Gas

Name	Expression	Constant
Boyle's law	$P_1V_1 = P_2V_2$	T
Charles's law	$\frac{V_1}{T_1} = \frac{V_2}{T_2}$	P
Gay-Lussac's law	$\frac{P_1}{T_1} = \frac{P_2}{T_2}$	V

Box 5C

Blood Pressure Measurement

Blood pressure is the result of the blood's pushing against the walls of the blood vessels. When the heart ventricles contract, pushing blood out into the arteries, the blood pressure is high **(systolic blood pressure)**; when the ventricles relax, the blood pressure is low **(diastolic pressure).**

Blood pressure varies with age. In young adults the normal range is 100 to 120 mm Hg systolic and 60 to 80 mm Hg diastolic. In older people the corresponding normal pressures are 115 to 135 and 75 to 85. Blood pressure is usually expressed as a fraction showing systolic over diastolic pressure, for instance, 120/80. Normal blood pressure is lower for women than for men. Blood pressures higher than normal may lead to heart attack, stroke, or kidney failure. Early diagnosis is important because blood pressure can be reduced by proper diet (low in sodium ions) and drugs. The measurement for blood pressure is routinely performed by nurses and physicians.

The *sphygmomanometer,* or aneroid blood pressure measurement device, consists of a bulb, a cuff, a manometer, and a stethoscope. The cuff is wrapped around a bare arm and the bulb repeatedly squeezed. This increases the amount of air in the cuff and thus the pressure. This increased pressure is transmitted to the cuff, which transmits it to the upper arm. The pressure applied can be read on the manometer. At the beginning, the cuff is inflated by repeated squeezing of the bulb to a pressure of 200 mm Hg. Then a release valve is opened slowly and the pressure is allowed to drop 2 or 3 mm Hg per second. While the pressure is dropping, a stethoscope is placed over the brachial artery above the elbow, and one listens for a clear tapping sound. At the start no sound is heard because, in most cases, 200 mm Hg applied pressure is greater than the blood pressure and so no pulsating blood flows under the stethoscope. When the first faint tapping sound is heard, the pressure on the manometer is read. This is the systolic pressure. The applied pressure just matches the blood pressure when the ventricle contracts, allowing pulsating blood to flow into the lower arm. The clear tapping sound becomes louder as the applied cuff pressure drops and eventually fades away. When the last tapping sound is heard, the

In this equation, P_1, V_1, and T_1 are the pressure, volume, and **absolute (Kelvin) temperature** of any sample of gas. When any change is made to the sample, P_2, V_2, and T_2 are the new pressure, volume, and temperature. The following examples illustrate calculations using the combined gas law equation.

EXAMPLE

A gas occupies 3.00 L at 1.50 atm pressure. What is its volume at 10.00 atm at the same temperature?

In this type of problem, we first identify the quantities that remain constant, and then solve the equation for the property (P, V, or T) we are interested in and substitute the numbers.

Answer • First we identify the known quantities:

$P_1 = 1.50$ atm; $V_1 = 3.00$ L; $P_2 = 10.00$ atm; $T_1 = T_2$

$$\frac{P_1V_1}{T_1} = \frac{P_2V_2}{T_2}$$

manometer is read again. This is the diastolic pressure. The applied cuff pressure now matches the blood pressure when the ventricle is relaxed, allowing continuous blood flow into the lower arm. Repeated blood pressure measurements can be taken, but the cuff should be deflated completely for two or three minutes before attempting the next reading.

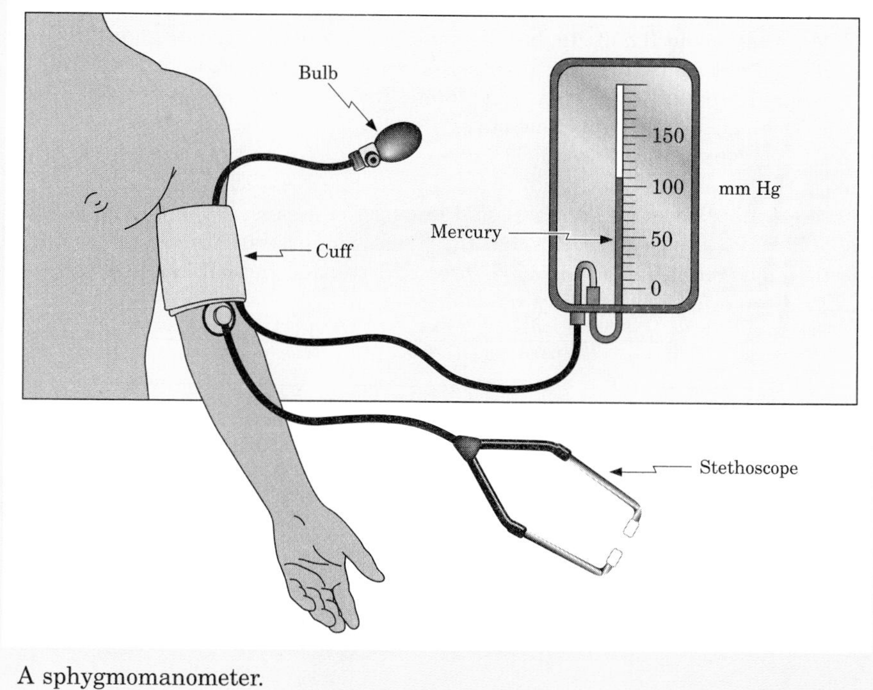

A sphygmomanometer.

Since we want the new volume, we solve our equation for V_2. Because T_1 and T_2 are the same in this example and consequently cancel each other, we do not even have to know what the temperature is. Thus,

$$V_2 = \frac{P_1V_1\cancel{T_2}}{\cancel{T_1}P_2} = \frac{(1.50\ \cancel{\text{atm}})(3.00\ \text{L})}{10.00\ \cancel{\text{atm}}} = 0.45\ \text{L}$$

The new volume is 0.45 L, or 450 mL.

Problem **5.1**

A gas occupies 3.8 L at 0.7 atm pressure. If we expand the volume at constant temperature to 6.5 L, what is the new pressure?

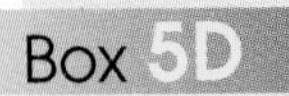

Box 5D

Breathing and Boyle's Law

Robert Boyle (1627–1691) discovered the pressure-volume relationship in 1662, but humans and animals were using the process described by the law long before then, in order to breathe. We breathe about 12 times a minute, each time taking in and exhaling about 500 mL of air. When we breathe, we lower the diaphragm or raise the rib cage (Fig. 5D). Either of these motions increases the volume of the chest cavity. In accord with Boyle's law, the pressure is therefore decreased and becomes lower than the outside pressure. Air thus flows from the higher-pressure area into the lungs. The difference is only about 3 mm Hg, but this is enough to make the air go in. In breathing out we reverse the process: We raise the diaphragm or lower the rib cage. The resulting decrease in volume increases the pressure inside the chest cavity, causing air to flow out.

In certain diseases the chest becomes paralyzed, and the patient cannot move the diaphragm or the rib cage. In such cases artificial respirators are used, pushing down on the chest and forcing the air inside the patient to go out the only way possible, through the lungs. The pressure in the respirator is then lowered below atmospheric pressure, causing the patient to breathe in, and the cycle is constantly repeated.

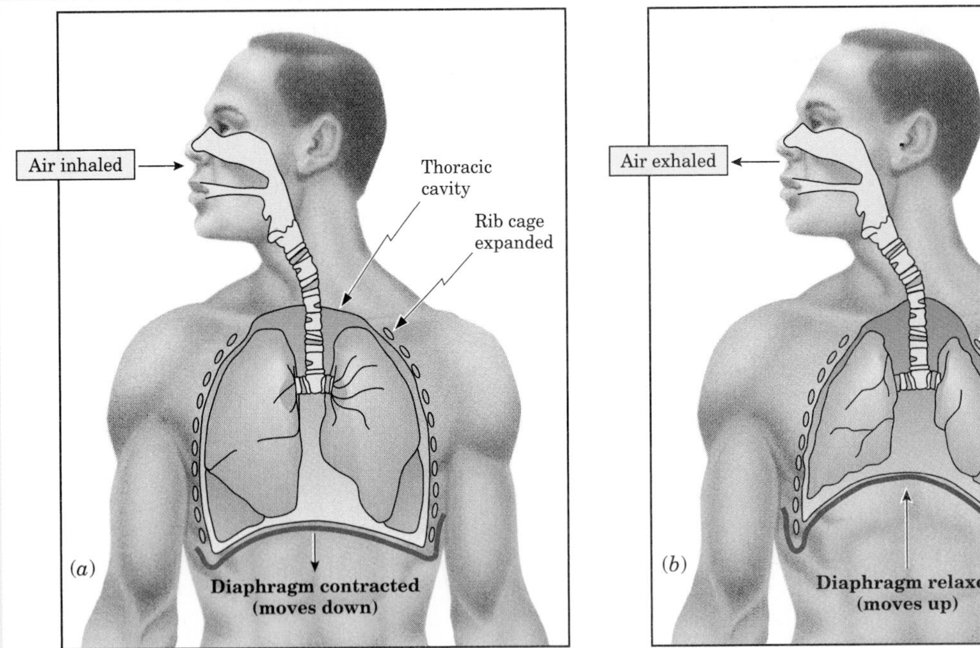

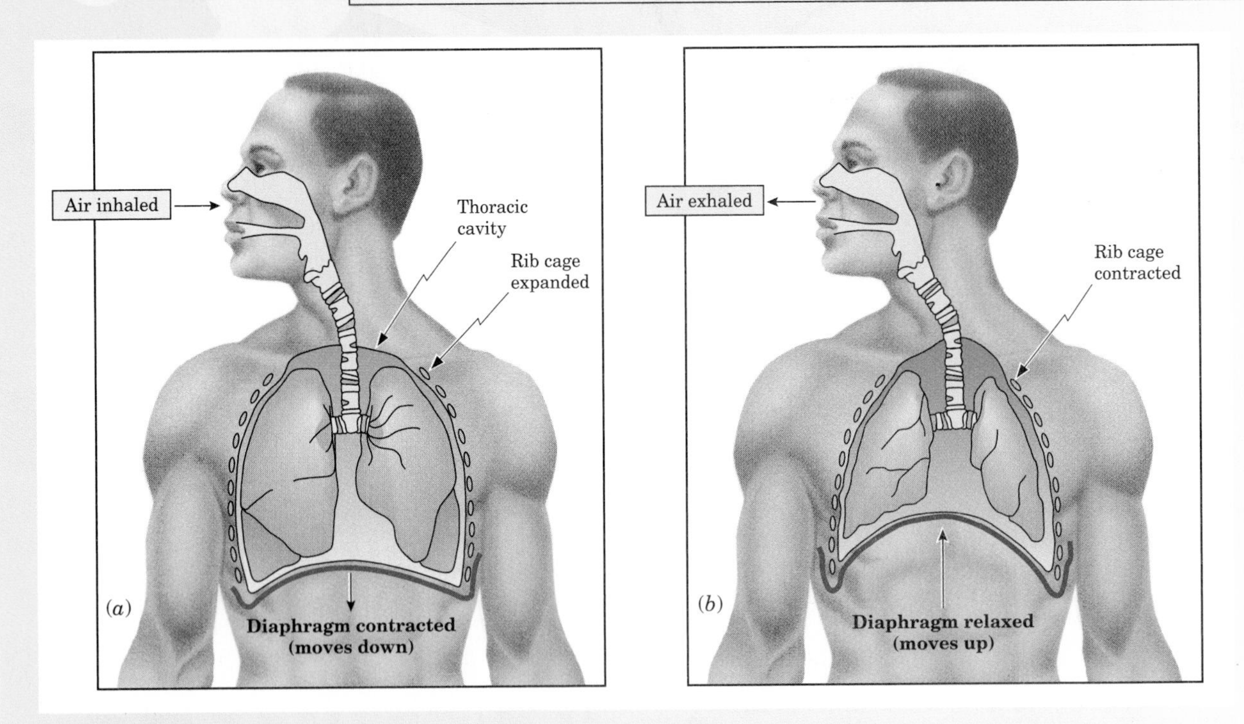

Figure **5D** • Schematic drawings of the chest cavity. In (a) the lungs fill with air. In (b) air empties from the lungs.

EXAMPLE

In an autoclave, steam at 100°C is generated at 1.00 atm. After the autoclave is closed, the steam is heated at constant volume until the pressure gauge indicates 1.13 atm. What is the new temperature in the autoclave?

Answer • Since all temperatures in gas law calculations are Kelvin, we must first convert the Celsius temperature to Kelvin:

$$100°C = 100 + 273 = 373\ K$$

We then identify the known quantities:

$$P_1 = 1.00\ atm; \quad T_1 = 373\ K; \quad P_2 = 1.13\ atm; \quad V_1 = V_2$$

$$\frac{P_1V_1}{T_1} = \frac{P_2V_2}{T_2}$$

Since we want a new temperature, we solve this equation for T_2:

$$T_2 = \frac{P_2\cancel{V_2}T_1}{P_1\cancel{V_1}}$$

$$T_2 = \frac{(1.13\ \cancel{atm})(373\ K)}{(1.00\ \cancel{atm})} = 421\ K$$

The new temperature is 421 K, or 421 − 273 = 148°C. In this example the two volumes cancel out.

Problem 5.2

A constant volume of oxygen gas is heated from 120°C to 212°C. The final pressure is 20.3 atm. What was the initial pressure?

EXAMPLE

A gas is under 1.0 atm pressure in a flexible container with a volume of 0.50 L at a temperature of 393 K. When the gas is heated to 500 K and the volume expanded to 3.0 L, what will the new pressure be?

Answer • The known quantities are

$$P_1 = 1.0\ atm; \quad V_1 = 0.50\ L; \quad T_1 = 393\ K; \quad V_2 = 3.0\ L; \quad T_2 = 500\ K$$

$$\frac{P_1V_1}{T_1} = \frac{P_2V_2}{T_2}$$

Solving for P_2, we get

$$P_2 = \frac{P_1V_1T_2}{T_1V_2} = \frac{(1.0\ atm)(0.50\ \cancel{L})(500\ \cancel{K})}{(3.0\ \cancel{L})(393\ \cancel{K})} = 0.21\ atm$$

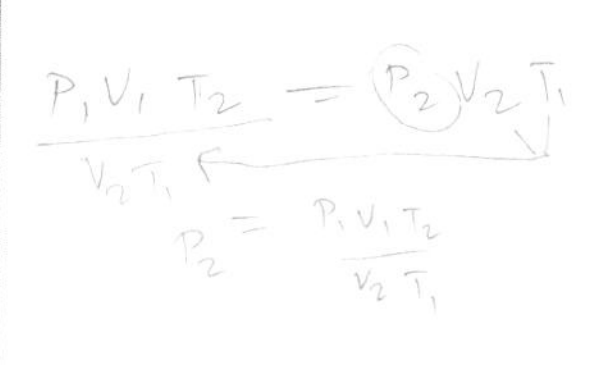

The new pressure will be 0.21 atm, or

$$0.21\ \cancel{atm} \times \frac{760\ mm\ Hg}{1\ \cancel{atm}} = 161\ mm\ Hg$$

Problem 5.3

If 3.0 L of a gas has an initial pressure of 2.8 atm at 280 K, what will be its final pressure if it is compressed to 0.466 L and cooled to 200 K?

Box 5E

Hyperbaric Medicine

Ordinary air contains 20.9 percent oxygen. Under certain conditions, the cells of tissues are starved for oxygen (hypoxia), and there is a need for quick oxygen delivery. Increasing the percentage of oxygen in the air supplied to the patient is one way to remedy the situation, but sometimes even breathing pure (100%) oxygen may not be enough. For example, in carbon monoxide poisoning, the hemoglobin in the blood, which usually carries most of the O_2 from the lungs to the tissues, binds CO and cannot take up any O_2 in the lungs. Without any help, the tissues would soon be starved for oxygen and the patient would die. If oxygen is administered, especially under high pressure (2 or 3 atm), oxygen dissolves in the plasma to such a degree that the tissues receive enough to recover. Many large hospitals now have special *hyperbaric units* where pure oxygen under pressure is given to patients. The entire unit is under high pressure, and the patient and the accompanying medical staff must undergo gradual compression and, at the end of the treatment, decompression.

Besides treating carbon monoxide poisoning, hyperbaric units are also used along with other treatment. For example, in radiation necrosis, when tissues are damaged by x-rays during cancer treatment or accidents, hyperbaric medicine is beneficial in 80 to 100 percent of the cases. Chronic refractory osteomyelitis, a bone infection that does not heal, also responds well to hyperbaric medicine. Other conditions for which hyperbaric medicine is used are gas gangrene, smoke inhalation, cyanide poisoning, skin grafts, and thermal burns.

However, breathing pure oxygen for prolonged periods is toxic. For example, if O_2 is administered at 2 atm for more than six hours, the lung tissues begin to be destroyed, and the central nervous system can also be damaged. Therefore, recommended exposures of O_2 are two hours at 2 atm and 90 minutes at 3 atm.

The benefits of hyperbaric medicine must be carefully weighed against the contraindications. Reports have been published showing that this treatment can cause nuclear cataract formation, thus necessitating postrecovery eye surgery.

5.5 Avogadro's Law and the Ideal Gas Law

The combined gas law is valid only for a given constant sample of gas. That is, it does not work if any more gas is added or any taken away. However, it is often necessary to work with changing amounts of gas. Fortunately, there is an easy way to do this. The relationship between the amount of gas present and its volume is described by **Avogadro's law,** which states that **equal volumes of gases at the same temperature and pressure contain equal numbers of molecules.** For example, Figure 5.7 shows two tanks of equal volume, one containing hydrogen and the other carbon dioxide. Avogadro's law tells us that, if the temperature and pressure are the same, the two tanks contain the same number of molecules, regardless of what the gases are. The actual temperature and pressure don't matter. Avogadro's law is valid no matter what the gases are, as long as they don't become liquids. However, it is convenient to select one temperature and one pressure as standard, and chemists have chosen 1 atm to be the standard pressure and 0°C (273 K) to be the standard temperature. These conditions are called **STP (standard temperature and pressure).**

Avogadro's law holds only for gases, not liquids or solids.

STP = 1 atm and 273 K

Just as at any other temperature and pressure, all gases at STP contain the same number of molecules in any given volume. But how many is that? In Chapter 4 we saw that the *mole* (which contains 6.02×10^{23} molecules) is a convenient quantity of matter. What volume of gas at STP contains 1 mole of molecules? This quantity is very easy to measure, and it was done long ago. It turns out to be 22.4 L:

1 mole of any gas at STP occupies 22.4 L.

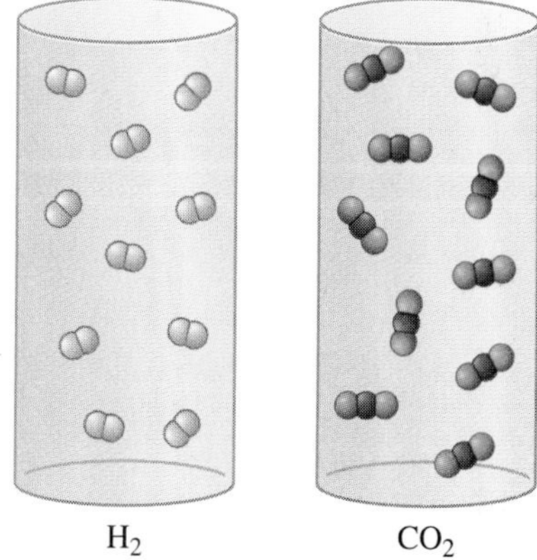

Figure **5.7**
Two tanks of gas of equal volume at the same temperature and pressure contain the same number of molecules.

Avogadro's law allows us to write a gas law that is valid not only for any pressure, volume, and temperature but also for any quantity of gas. This law, called the **ideal gas law,** is

$$PV = nRT$$

where P = pressure of the gas in atm
V = volume of the gas in L
n = number of moles of the gas
T = temperature of the gas in K
R = a constant for all gases, for all conditions of n, P, V, and T.

We can find out what the constant is by using the fact that 1 mole at STP occupies 22.4 L:

$$PV = nRT$$

$$R = \frac{PV}{nT} = \frac{(1\ \text{atm})(22.4\ \text{L})}{(1\ \text{mole})(273\ \text{K})} = 0.0821\ \frac{\text{L} \cdot \text{atm}}{\text{mole} \cdot \text{K}}$$

Thus R, called the **universal gas constant,** is equal to 0.0821 L · atm/mole · K.

The **ideal gas law, $PV = nRT$, is a universal law.** It holds for all gases, including all mixtures of gases, at any temperature, pressure, volume, or amount, as long as they are ideal gases. As we mentioned before, real gases, while not exactly ideal, in most cases behave as ideal gases, and so we can use this equation with little trouble. Using $PV = nRT$, we can calculate any one quantity—P, V, T, or n—if we know the other three.

Real gases behave most like ideal gases at low pressures (1 atm or less) and high temperatures (300 K or higher).

Furthermore, since the number of moles of any substance is its mass in grams divided by its molecular weight (Sec. 4.3),

MW stands for molecular weight.

$$n = \frac{m}{MW}$$

we can replace n in the ideal gas equation to give

$$PV = \frac{mRT}{MW}$$

This equation allows us to calculate the molecular weight of a gas if we know the other quantities. Some examples of the use of the ideal gas law follow.

EXAMPLE

One mole of CH_4 gas occupies 20.0 L at 1.00 atm pressure. What is the temperature of the gas in K?

Answer • Solve the ideal gas law for T:

$$T = \frac{PV}{nR}$$

$$T = \frac{(1.00\ \cancel{\text{atm}})(20.0\ \cancel{\text{L}})}{(1\ \cancel{\text{mole}})(0.0821\ \cancel{\text{L}} \cdot \cancel{\text{atm}}/\cancel{\text{mole}} \cdot \text{K})} = 244\ \text{K}$$

Note that the answer would be the same if the gas were CO_2, N_2, NH_3, or *any other gas.*

Problem 5.4

If 2.00 moles of NO occupy 10.0 L at 295 K, what is the pressure in atm?

EXAMPLE

If there is 5.0 g of CO_2 gas in a 10-L vessel at 350 K, what is the pressure?

Answer • Here we are given a mass, and so we use $PV = mRT/MW$. The molecular weight of CO_2 = 44:

$$P = \frac{mRT}{V(MW)} = \frac{(5.0\ \cancel{\text{g}})(0.0821\ \cancel{\text{L}} \cdot \text{atm}/\cancel{\text{mole}} \cdot \cancel{\text{K}})(350\ \cancel{\text{K}})}{(10\ \cancel{\text{L}})(44\ \cancel{\text{g}}/\cancel{\text{mole}})}$$

$$= 0.33\ \text{atm}$$

Problem 5.5

A certain quantity of neon gas is under 1.05 atm pressure at 303 K in a 10.0-L vessel. How many moles of neon are present?

EXAMPLE

If 1.0 L of an unknown gas has 1.15 atm pressure at 40°C and the mass of the gas is 3.3 g, what is the molecular weight?

Answer • Convert 40°C to absolute temperature: 273 + 40 = 313 K.

$$MW = \frac{mRT}{PV} = \frac{(3.3\text{ g})(0.0821\ \not{L}\cdot\not{\text{atm}}/\text{mole}\cdot\not{K})(313\ \not{K})}{(1.15\ \not{\text{atm}})(1.0\ \not{L})}$$

$$= 74\ \frac{\text{g gas}}{\text{mole gas}}$$

Therefore, the molecular weight is 74 amu.

Problem 5.6

An unknown gas weighs 5.0 g and occupies 1.50 L at 1.00 atm and 300 K. What is its molecular weight?

5.6 Dalton's Law and Graham's Law

In a mixture of gases each molecule acts independently of all the others, assuming that the gases do not react with each other. Unlike liquids, where molecules are close together and strongly influence each other, the molecules in a gas are so far apart that they have little or no effect on each other. This is why the ideal gas law works for mixtures as well as for pure gases. In particular, the collisions with the walls of the container made by each molecule contribute independently to the total pressure. **Dalton's law of partial pressures** states that the total pressure of a mixture of gases is the sum of the individual partial pressures:

$$P_T = P_1 + P_2 + P_3 + \cdots$$

The **partial pressure of a single gas** in a mixture is defined as **the pressure the gas would exert if it were alone in the container.** The equation $PV = nRT$ holds separately for each gas in the mixture, as well as for the total mixture.

Box 5F

Breathing and Dalton's Law

A gas naturally flows from an area of higher pressure to one of lower pressure. This fact is responsible for respiration, though we must look not at the total pressure of air but only at the partial pressure of O_2 and CO_2. Our cells use O_2 to oxidize carbon compounds to CO_2 and thus obtain energy. Respiration is the process by which the blood carries O_2 from the lungs to the tissues and collects the CO_2 manufactured by the cells and brings it to the lungs, where it is breathed out.

The air we breathe contains about 21 percent O_2; the partial pressure of O_2 is about 159 mm Hg. The partial pressure of O_2 in the alveoli (tiny air sacs) in the lungs is about 100 mm Hg. Because the partial pressure is higher in the inhaled air, O_2 flows to the alveoli. At this point, venous blood (blood from the veins) flows past the alveoli. The partial pressure of O_2 in the venous blood is only 40 mm Hg, and so O_2 flows from the alveoli to the venous blood. This new supply of O_2 raises the partial pressure to about 100 mm Hg, and the blood that now has had its O_2 pressure increased becomes arterial blood and flows to the body tissues. There it finds that the partial pressure of O_2 is about 30 mm Hg or less (because the cells of the body tissues have used up most of their oxygen to provide energy). Oxygen now flows from the arterial blood to the body tissues, giving them a fresh supply. This naturally lowers the partial pressure of O_2 in the blood (to about 40 mm Hg), so that the blood now becomes venous blood again and returns to the lungs for a fresh supply of O_2. At each step of the cycle, O_2 flows from a region of higher partial pressure to one of lower partial pressure.

Meanwhile, the CO_2 goes the opposite way, for the same reason. The partial pressure of CO_2 in oxidized body tissue is about 60 mm Hg. Thus, CO_2 flows from the tissues to the arterial blood, which of course is changed by the process to venous blood and now has a CO_2 pressure of 46 mm Hg. When the venous blood reaches the lungs, it finds that the CO_2 pressure in the alveoli is 40 mm Hg, and so it flows to the alveoli. Since the partial pressure of CO_2 in air is only 0.3 mm Hg, CO_2 flows from the alveoli and is exhaled.

EXAMPLE

To a tank already containing N_2 at 2.0 atm and O_2 at 1.0 atm, we add an unknown quantity of CO_2 until the total pressure is 4.6 atm. What is the partial pressure of the CO_2?

Answer • Dalton's law tells us that the addition of CO_2 does not affect the N_2 and O_2 that are already there. Thus the partial pressures of N_2 and O_2 remain at 2.0 atm and 1.0 atm, respectively, even after the CO_2 is added. The sum of the partial pressures of N_2 and O_2 is

$$2.0 \text{ atm} + 1.0 \text{ atm} = 3.0 \text{ atm}$$

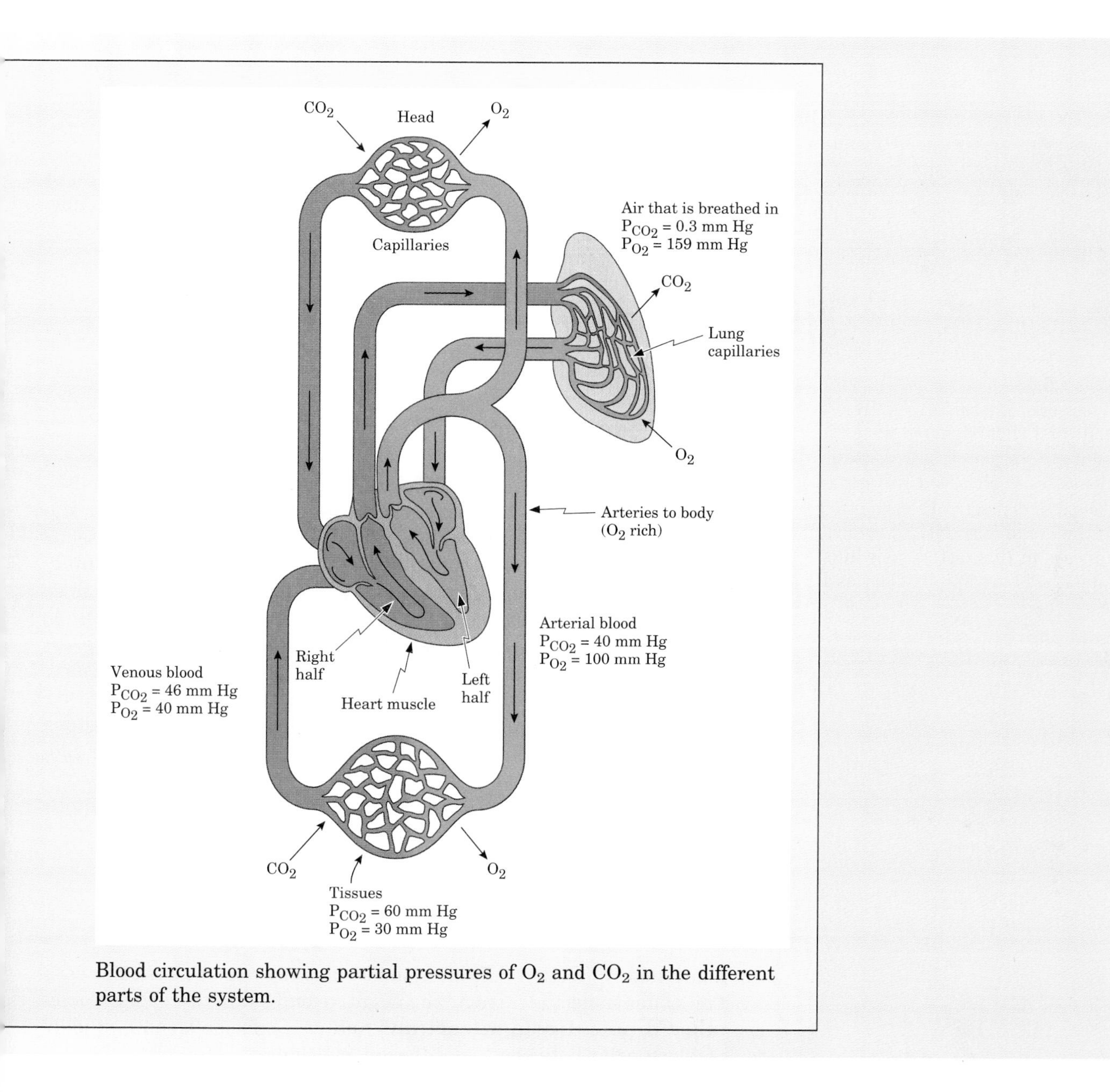

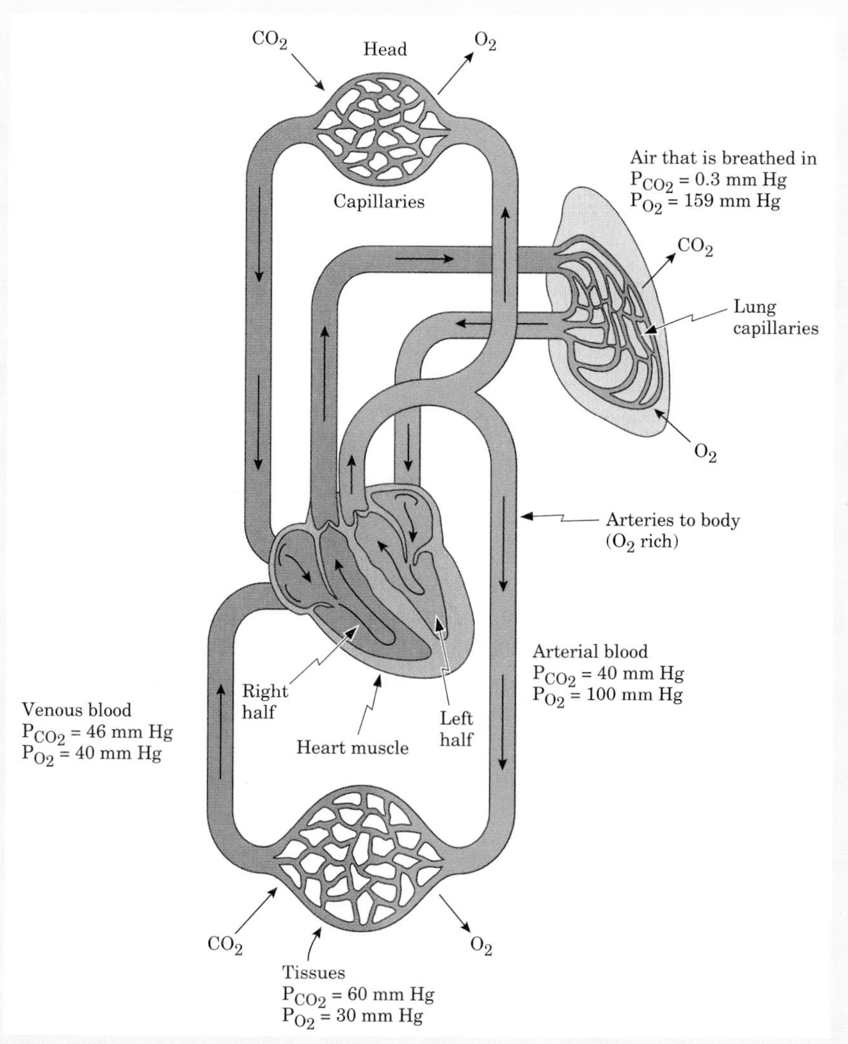

Blood circulation showing partial pressures of O_2 and CO_2 in the different parts of the system.

Since the rest of the total pressure of 4.6 atm must be caused by the CO_2, we can determine it by subtracting:

4.6 atm	− 3.0 atm	= 1.6 atm
Total pressure	Pressure of N_2 and O_2	Pressure of CO_2

Problem 5.7

A vessel under 2.015 atm pressure contains nitrogen and water vapor. The partial pressure of the nitrogen is 1.908 atm. What is the partial pressure of the water vapor?

Figure **5.8**
When open containers of NH_3 (left) and HCl (right) in water solution are placed side by side, molecules of NH_3 and HCl gases escape and react with each other in the air to form NH_4Cl, which appears in this picture as white fumes. (Photograph by Charles D. Winters.)

When a bottle of ammonia, NH_3, is opened in one corner of a room, within minutes its odor spreads to other parts of the room (Fig. 5.8). The process whereby two gases mix is called **diffusion.** In this case, the ammonia molecules diffuse into the air. When a bottle of perfume is opened, its fragrance also fills the room, but this takes longer than with ammonia. Thomas Graham (1805–1869) observed that molecules with small masses diffuse faster than heavy molecules and formulated what is now known as **Graham's law:**

$$\frac{(\text{Diffusion rate of A})^2}{(\text{Diffusion rate of B})^2} = \frac{MW \text{ of B}}{MW \text{ of A}}$$

5.7 Intermolecular Forces

If the temperature of a gas is sufficiently decreased, the gas condenses and becomes a liquid. If the temperature is lowered still more, the liquid freezes and becomes a solid. The great difference between the gaseous state and condensed states (liquid and solid) is the effect of **intermolecular forces.** In gases the individual molecules are far apart, so much so that we can ignore intermolecular attraction and treat most gases as ideal. This is valid when the temperature is high (room temperature and above) and the pressure low (1 atm or less), so that the molecules are far apart. When the temperature decreases and/or the pressure increases, distances between molecules become less, and intermolecular forces can no longer be ignored. As a matter of fact, it is these forces that cause condensation and solidification.

Therefore, before discussing the properties and structures of liquids and solids, we must look at the nature of intermolecular forces.

Dipole-Dipole Attractions

As mentioned in Section 3.7, many molecules are dipoles, and the positive end of one molecule attracts the negative end of another. These forces are called **dipole-dipole attractions** and can exist between two identical molecules or between two different molecules:

$$^{\delta+}(\text{H—Cl})^{\delta-}\ \ ^{\delta+}(\text{H—Cl})^{\delta-}$$

$$^{\delta+}(\text{H—Br})^{\delta-}\ \ ^{\delta+}(\text{H—Cl})^{\delta-}$$

Hydrogen Bonds

An especially strong attraction exists between molecules of water. The polarity of the O—H bond shifts the electrons toward the oxygen within the molecule so much that the hydrogen acquires a partial positive charge. Such a hydrogen is very strongly attracted toward the electron cloud of an oxygen in a neighboring water molecule. This attraction, indicated by the dotted line in Figure 5.9, is called a **hydrogen bond.**

Hydrogen bonds are not restricted to water. They form whenever two molecules, the same or different, come together, provided that

1. One molecule has a hydrogen atom attached by a covalent bond to an atom of oxygen, nitrogen, or fluorine.

2. The other molecule has an oxygen, nitrogen, or fluorine atom.

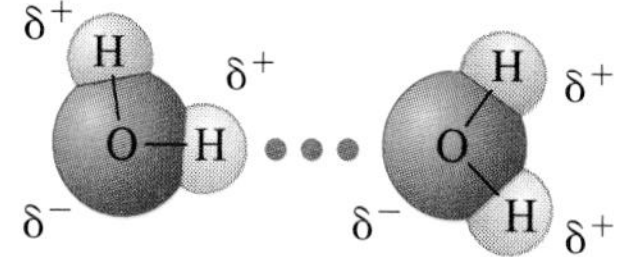

Figure **5.9**
Two water molecules joined by a hydrogen bond.

Therefore, we expect hydrogen bonds to form here:

```
H
 \
  O···H—F
 /
H
```

but not here:

```
H                H
 \               |
  O          H—C—H
 /               |
H                H
```

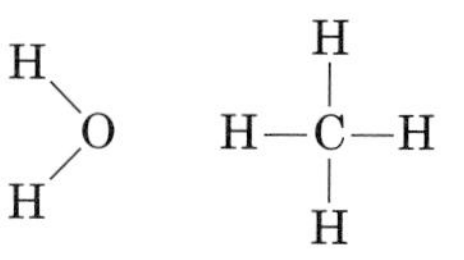

Although hydrogen bonds are among the strongest of the intermolecular forces, it must be kept in mind that they are still much weaker than the covalent bonds within the molecules. Covalent bonds are about ten times stronger than hydrogen bonds (Table 5.2). As we shall see in later chapters, hydrogen bonds play an important role in biological molecules.

Table **5.2** Approximate Energy of Atomic and Molecular Interactions

Interaction	Energy (kcal/mole)
Covalent bond	50–100
Hydrogen bond	5–10
Dipole-dipole attraction	0.1–1
London dispersion force	0.001–0.2

EXAMPLE

Can a hydrogen bond form (a) between two molecules of methanol, CH_3OH? (b) between two molecules of formaldehyde, H—C(=O)—H? (c) between one molecule of methanol and one of formaldehyde?

Answer

(a) Yes. One molecule has a hydrogen atom attached to an oxygen atom (actually, both have, but only one is necessary), and the other molecule has an oxygen atom:

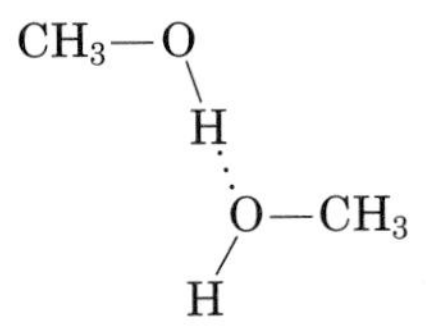

```
CH3—O
     \
      H
       :
        O—CH3
       /
      H
```

(b) No. Neither molecule has a hydrogen attached to an oxygen, nitrogen, or fluorine.

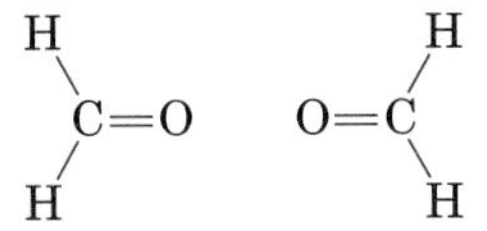

```
H              H
 \            /
  C=O    O=C
 /            \
H              H
```

(c) Yes. One molecule (methanol) has a hydrogen atom attached to an oxygen atom; the other (formaldehyde) has an oxygen atom.

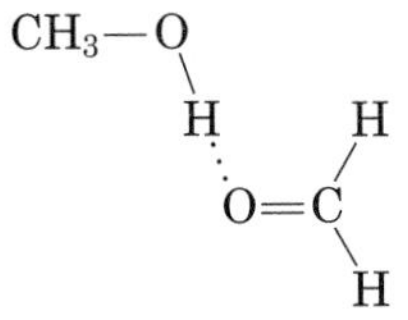

```
CH3—O
     \
      H     H
       :   /
        O=C
           \
            H
```

Problem 5.8

(a) Will a molecule of water form a hydrogen bond with a molecule of methanol? (b) Will two molecules of methane (CH_4) form hydrogen bonds with each other?

London Dispersion Forces

London forces were first explained by Fritz London (1900–1954) in 1930.

Not all gases contain polar molecules. Yet even nonpolar gases, such as helium and hydrogen, liquefy if the temperature is lowered sufficiently. Are there intermolecular forces even between molecules that are not dipoles? The answer is yes, though such forces, called **London dispersion forces,** are *very* weak (Table 5.2). Even though London forces are very weak, in large molecules they contribute significantly to the attractive forces between the molecules, because they act over large surface areas.

Even London forces have their origin in electrostatic interaction. We can explain the attraction between nonpolar atoms by assuming that, when the two atoms get close to each other, the motion of the electron clouds becomes synchronized. For a very small fraction of a second, the negative charge is located on one side of each spherical cloud. If the motion of the electrons in two helium clouds becomes synchronized during that tiny fraction of a second, the charges will be at the same relative position, as shown in Figure 5.10. Thus, the electrons of atom B attract the nucleus of atom A during this short interval. A fraction of a second later, when the electrons have moved to the 3 o'clock position, the electrons of A interact with the nucleus of B. Thus there are very short-lived and constantly changing interactions between the two atoms—the London dispersion forces.

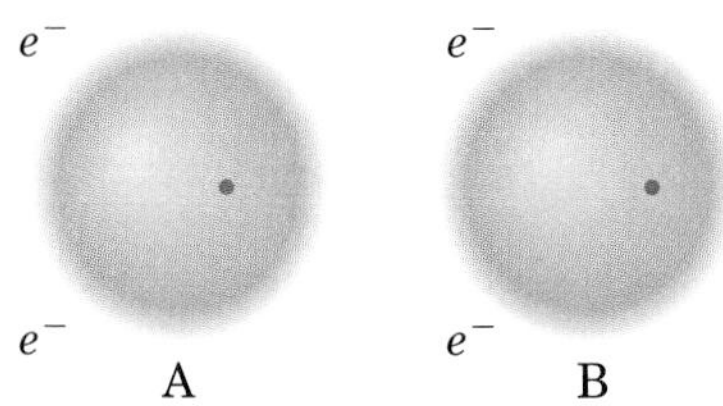

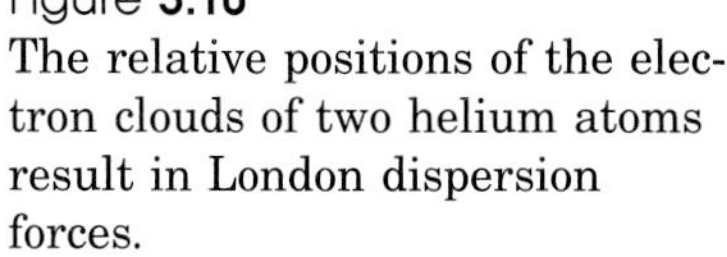

Figure **5.10**
The relative positions of the electron clouds of two helium atoms result in London dispersion forces.

5.8 Liquids

We have seen that the behavior of gases under most circumstances can be described by the ideal gas law. As pressure increases, however, the molecules of a gas are squeezed into a smaller space. The distance the average molecule travels between collisions becomes smaller. Over short distances intermolecular forces become significantly stronger, and so molecules have a greater tendency to stick together.

If the contact between molecules increases to the extent that almost all the molecules touch (or almost touch) each other, the gas becomes a liquid. In contrast to gases, liquids do not fill all the available space. They have a definite volume, irrespective of the container. Because gases have a lot of empty space between molecules, it is easy to compress them into a smaller volume. In liquids there is very little empty space; consequently, liquids are hard to compress. A great increase in pressure is needed to cause a very small decrease in the volume of a liquid. Liquids are, for all practical purposes, incompressible. The density of liquids is much greater than that of gases because the same mass occupies a much smaller volume in liquids than in gases.

The position of the molecules in the liquid phase is random, and there is some irregular empty space into which molecules can slide. The molecules thus constantly change their position with respect to neighboring molecules. This property causes the *fluidity* of liquids; that is, they do not have a constant shape but only a constant volume.

5.9 Evaporation and Condensation. Boiling Point

An important property of liquids is that they evaporate. A few hours after a heavy rain, most of the puddles have dried up. The water has evaporated, gone into the air. The same thing occurs if we leave a glass of water out in the open. Why does this happen?

There is a distribution of velocities among molecules. In our water example, some of the molecules have high energy and are moving rapidly. Others have low energy and are moving slowly. Whether fast or slow, a molecule in the center or at the bottom of the liquid cannot go very far before it hits another molecule and has its speed and direction changed by the collision. A molecule at the surface, however, is in a different sit-

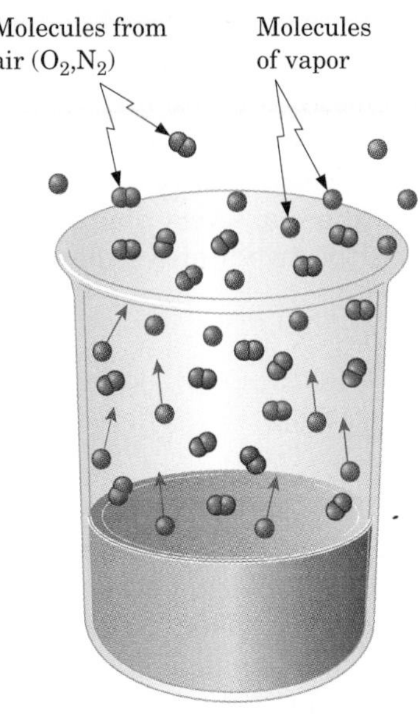

Figure **5.11**
Some molecules at the surface of a liquid are moving fast enough to escape into the gas phase.

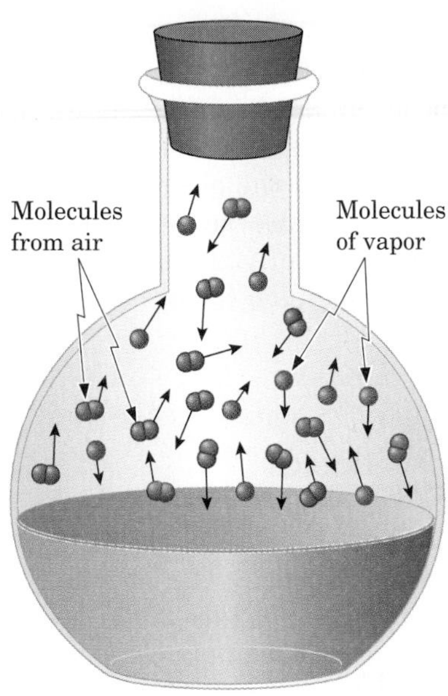

Figure **5.12**
In a closed container, molecules of liquid escape into the vapor phase, and vapor molecules are recaptured by the liquid.

uation (Fig. 5.11). If it is moving slowly, it cannot escape from the liquid because of the attractions of neighboring molecules. If, however, it is moving rapidly (a high-energy molecule) and is moving upward, it can escape from the liquid and enter the gaseous space above it.

If the container is open, this process continues until eventually all the molecules have left the liquid and it has dried up. However, if the liquid is in a closed container, as in Figure 5.12, this does not happen because the water vapor molecules in a closed container cannot diffuse away (as they do if the container is open). They remain in the air space above the liquid, where they behave, of course, like any other gas molecules, meaning that they move rapidly in straight lines until they collide with something. Some of these water vapor molecules move *downward* and thus can be recaptured by the liquid.

At equilibrium the rate of vaporization is equal to the rate of liquefaction.

At this point we have an **equilibrium.** The amount of water in the vapor no longer increases or decreases, as long as the temperature does not change. (We will meet other types of equilibria in Chapters 6 and 7.) After equilibrium is reached, the space above the liquid shown in Figure 5.12 contains air and water vapor, and it is possible to measure the partial pressure of the water vapor. We call this the **vapor pressure of the liquid.** Note that we measure the partial pressure *of a gas* but call it the vapor pressure *of the liquid.*

The vapor pressure of a liquid is a physical property of the liquid at a given temperature. A liquid has the same vapor pressure no matter what gas is above it (air or any other gas) and no matter what the pressure of that gas is. In fact, the vapor pressure is the same even if a vacuum exists in the space above the gas. Vapor pressure is also independent of the volume of the liquid.

The molecules that evaporate are the more energetic molecules. When they go into the gas phase, the molecules left behind are the less energetic ones. Therefore, the temperature of the liquid drops as a result of evaporation. This is what causes the cooling effect of perspiration. The liquid that comes out of the pores evaporates, so the skin temperature drops. The same cooling effect is felt when you come out of a swimming pool and the layer of water starts to evaporate from your skin.

The vapor pressure of a liquid depends on the temperature. Figure 5.13 shows that **vapor pressure increases as temperature increases.** Eventually the liquid boils. The **boiling point** of a liquid is the **temperature at which its vapor pressure is equal to the pressure of the atmosphere in contact with the surface of the liquid.** The boiling point when the atmospheric pressure is 1 atm is called the **normal boiling point.** Thus, 100°C is the normal boiling point of water because that is the temperature at which water boils at 1 atm pressure.

Boiling is a special form of evaporation during which bubbles form inside the liquid so that the surface of the liquid is greatly increased.

The use of a pressure cooker is an example of boiling at higher temperatures. In this type of pot, food is cooked at, say, 2 atm, where the boiling point of water is 121°C. Since the food has been raised to such a high temperature, it cooks much faster than it would in an open pot, in which boiling water cannot get hotter than 100°C. Conversely, at low pressures water boils at lower temperatures. For example, at the top of a mountain, the boiling point might be 95°C.

Different liquids have different normal boiling points. There are two factors that cause liquids to have high or low boiling points:

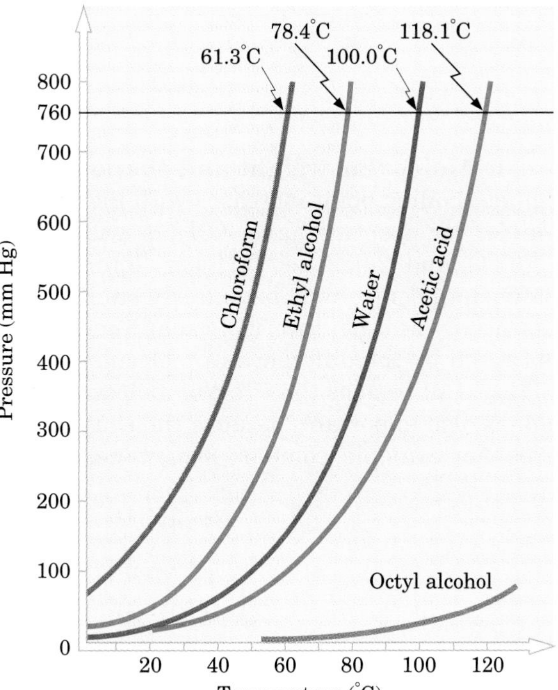

Figure **5.13**
The change in vapor pressure with temperature for five liquids.

Box 5G

The Densities of Ice and Water

The hydrogen-bonded superstructure of ice (Fig. 5G.1) has empty spaces in the middle. Therefore the molecules in ice are not so closely packed as those in liquid water. Because of this, ice has a lower density than liquid water. In the latter, more of the hydrogen bonds are broken, and the hexagonal superstructure is collapsed into a more densely packed organization (Fig. 5G.2). This is why ice floats on top of water instead of sinking to the bottom. Such behavior is highly unusual. Most solids are denser than the corresponding liquids. The lower density of ice keeps fish and microorganisms alive in many rivers and lakes that would freeze solid each winter if the ice sank to the bottom. The presence of ice on top insulates the remaining water and keeps it from freezing.

The fact that ice has a lower density than water means it has a greater molar volume (that is, any particular mass of ice takes up more space than the same mass of water). This explains the damage done to biological tissues by freezing. When parts of the body (usually fingers, toes, noses, and ears) are subjected to extreme cold, they develop a condition called **frostbite.** The water in the cells freezes despite the blood's attempt to keep the temperature at 37°C. The frozen water expands and in doing so ruptures the cell walls, causing damage. In some cases, frostbitten fingers or toes must be amputated.

Cold weather can also damage plants in a similar way. Many plants are killed when the air temperature drops below the freezing point of water for several hours. Trees can survive cold winters because they have a low water content inside their trunks and branches.

Many times, slow freezing is more damaging to plant and animal tissues than quick freezing. In slow freezing only a few crystals are formed, and these can grow to large sizes, rupturing the cells. In quick freezing, such as can be achieved in liquid nitrogen (at a temperature of −196°C), many tiny crystals are formed. Since these do not grow much, tissue damage may be much less.

1. Molecular weight and shape Where intermolecular forces are similar, boiling point increases with increasing molecular weight. For example, consider methane, CH_4, a gas whose normal boiling point is −161°C, and *n*-hexane, C_6H_{14}, a liquid with a normal boiling point of 69°C. Both compounds have fairly similar London dispersion forces and essentially no other intermolecular forces. The difference in boiling points is caused by the difference in molecular weights (16 for CH_4 and 86 for C_6H_{14}). But besides the effect of molecular weight, the shape of the molecule is also important, because the shape helps to determine the surface area. For example, compare *n*-pentane, a chain-like molecule

```
     H   H   H   H   H
     |   |   |   |   |
 H — C — C — C — C — C — H
     |   |   |   |   |
     H   H   H   H   H
```

(text continues on page 166)

Box 5G (continued)

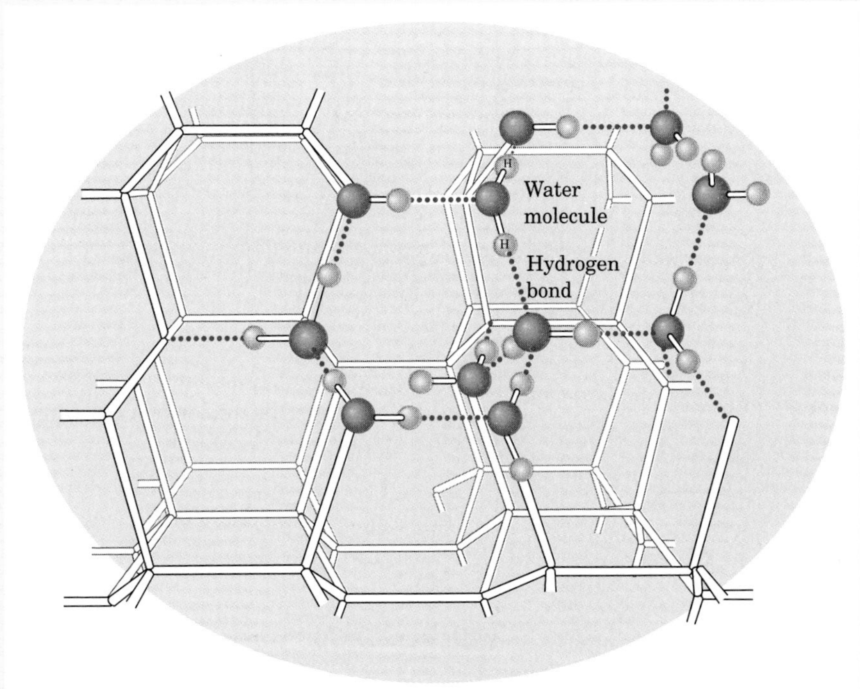

Figure **5G.1** • The structure of ice.

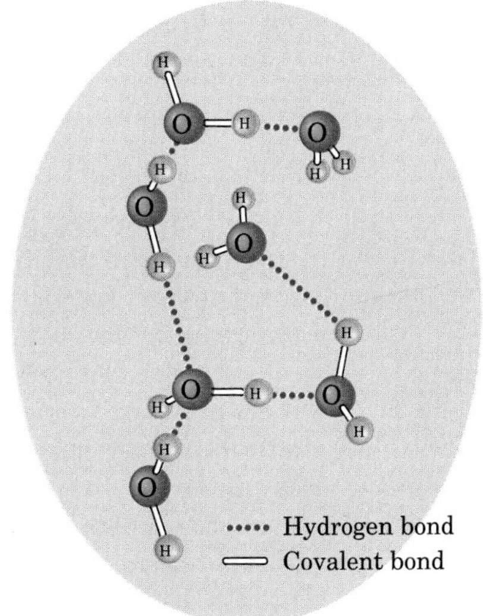

Figure **5G.2** • The structure of liquid water.

(*a*) (*b*)

Figure **5.14** • Molecular models of (a) *n*-pentane and (b) dimethylpropane. (Photograph by Beverly March.)

Although all of the carbon bonds in dimethylpropane point to the corners of a tetrahedron, the 12 hydrogens on the outside of the molecule can be thought of as being on the surface of a sphere (Fig. 5.14).

with dimethylpropane, a spherically shaped molecule (Fig. 5.14).

```
             H
             |
           H—C—H
        H    |    H
        |    |    |
      H—C————C————C—H
        |    |    |
        H    |    H
           H—C—H
             |
             H
```

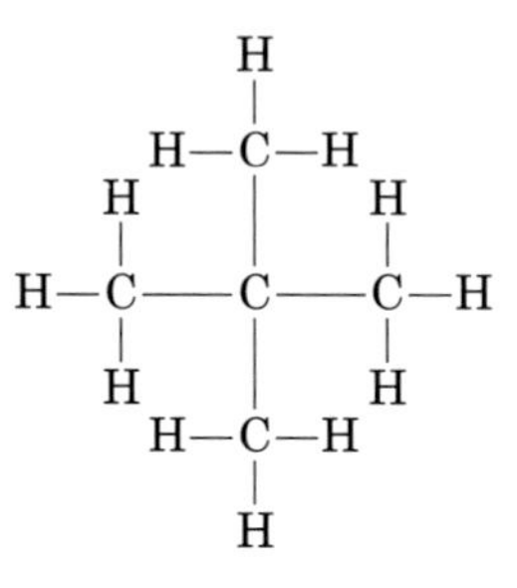

These compounds have the same molecular weight (they are both C_5H_{12}), but the boiling point of *n*-pentane is 36.2°C, significantly higher than that of dimethylpropane, 9.5°C. In the chain-like molecule, the London dispersion forces act over a larger molecular surface than in the spherically shaped molecule of the same molecular weight. Thus, the *n*-pentane molecules need a higher temperature than the dimethylpropane molecules in order to move them from the liquid to the gas phase.

2. Intermolecular forces Water (MW = 18) and methane (MW = 16) have about the same molecular weight. The normal boiling point of water is 100°C and that of methane is −161°C. The difference in boiling points is due to the fact that the CH_4 molecules must overcome only London dispersion forces in order to go into the vapor phase (low boiling point), whereas water molecules, being hydrogen-bonded to each other, need more kinetic energy (and a higher boiling temperature) to escape into the vapor phase.

Ethyl chloride spray is used to alleviate pain from minor muscular injuries and bruises. This compound evaporates very quickly so that the skin is cooled rapidly. The net result is a freezing effect. This minimizes swelling from bruises and reduces the pain suffered in the injury by deadening nerves. Athletes routinely use ethyl chloride spray.

5.10 Solids

When liquids are cooled and/or compressed, the intermolecular forces stop the random motion of the molecules, and a solid is formed. In the solid state, not only has each molecule or ion stopped moving through space, but the entire three-dimensional array of molecules or ions is usually orderly. A regular three-dimensional array of atoms, ions, or molecules is called a **crystal lattice.** The formation of solid from liquid is called **crystal formation** or **crystallization.**

Even in the solid state, molecules and ions do not stop moving completely. They vibrate around fixed points.

All crystals have a regular shape that in many cases is obvious to the eye (Fig. 5.15). This regular shape often reflects the arrangement of the atoms within the crystal. For example, table salt is made of small crystalline cubes, and the Na^+ and Cl^- ions in the crystal are also arranged in a cubic system. Metals also consist of particles in a regular crystal lattice (generally not cubic), but here the particles are atoms rather than ions. Because molecules or ions in a solid are usually closer together, solids almost always have a higher density than the corresponding liquids.

As can be seen in Figure 5.15, crystals have characteristic shapes and symmetries. We are familiar with the cubic nature of table salt and the hexagonal ice crystals in snow flakes. It is less well known, however, that **the same compound can have more than one type of solid state.** The best known example of this is the element carbon. Carbon has a crystalline form, diamond, that occurs when solidification takes place under very high pressure (thousands of atmospheres). Another crystalline form of carbon is the graphite in a pencil. The carbon atoms are packed differently in high-density, hard diamonds and in low-density, soft graphite (Fig. 5.16[a]).

When two or more forms of a solid are possible, one is more stable than the others.

Recently, a third form of carbon has been discovered in which each molecule contains 60 carbon atoms arranged in a structure having 12 pentagons and 20 hexagons as faces, resembling a soccer ball (Fig. 5.16[c]). Because the famous architect Buckminster Fuller (1895–1983) invented domes of a similar structure (he called them *geodesic domes*), the C-60 substance was named buckminsterfullerene, or bucky balls for short. This discovery has generated a whole new area of carbon chemistry. Similar cage-like carbon structures, containing 72, 80, and even larger numbers of carbon have been synthesized.

Richard E. Smalley (1943–   ), Robert F. Curl, Jr. (1933–   ), and Harold Kroto (1939–   ) were awarded the 1996 Nobel Prize in Chemistry for the discovery of these compounds.

(*a*) Garnet (*b*) Sulfur (*c*) Quartz (*d*) Pyrite

Figure **5.15** • Some crystals. (Photographs by Beverly March.)

Soot is still a fourth form of solid carbon. This substance solidifies directly out of carbon vapor and is an *amorphous solid;* that is, its atoms have no set pattern and are arranged randomly (Fig. 5.16[d]). Another example of an amorphous solid is glass. In essence, glass is a frozen liquid.

As we have now seen, some crystalline solids consist of orderly arrays of ions (Fig. 3.2) and others of molecules (molecular solids). Ions are held in the crystal lattice by ionic bonds (Sec. 3.3). Molecules are held only by intermolecular forces, which are much weaker than ionic bonds. Therefore, molecular solids generally have much lower melting points than ionic solids.

Besides these, there are other types of solids. Some are extremely large molecules; each molecule may have 10^{22} or 10^{23} atoms, all connected by covalent bonds. In such a case, the entire crystal is one big molecule. We call such molecules **network solids** or **network crystals.** A good example is diamond (Fig. 5.16[b]). When you hold a diamond in your hand, you are holding a single gigantic molecule. Like ionic crystals, network crystals have very high melting points—if they can be liquefied at all. In many cases they cannot be. We can be fairly confident, then, that any solid with a relatively low melting point (less than 300°C) can only be a molecular solid and not an ionic crystal or a network crystal. The types of solid are summarized in Table 5.3.

There are other very large molecules, called polymers and containing about 10^4 to 10^7 atoms, that do not form network crystals (Sec. 11.7).

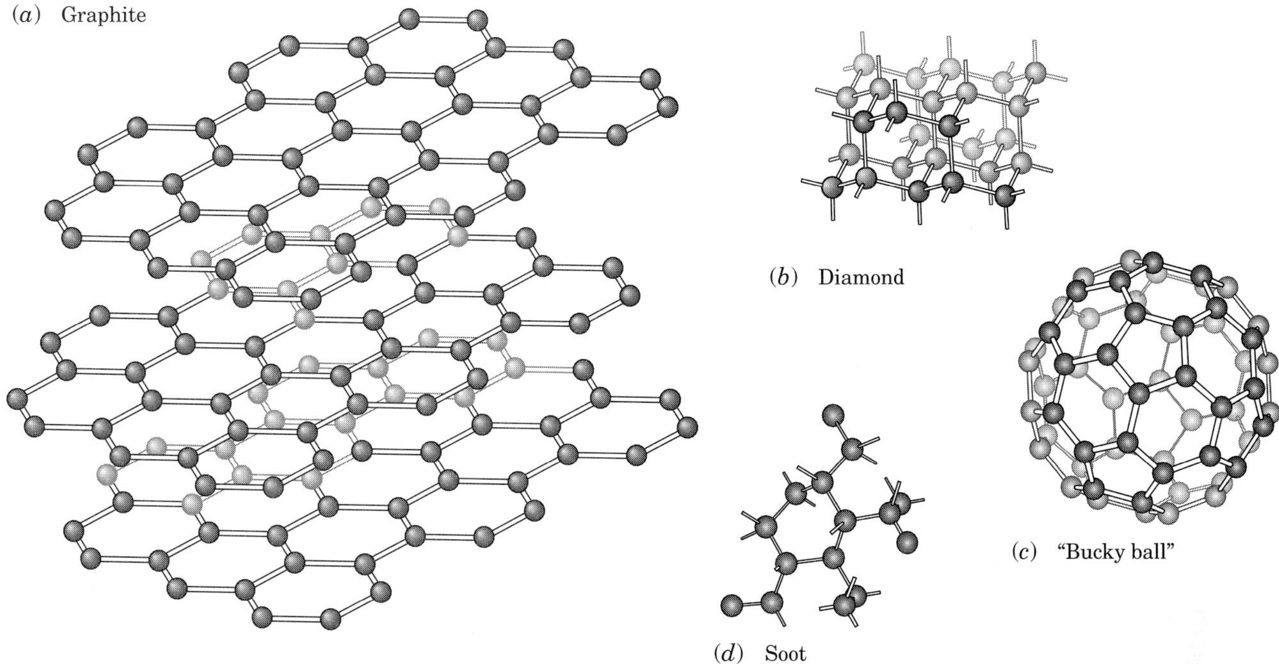

Figure **5.16** • Solid carbon structures: (a) graphite, (b) diamond, (c) buckminsterfullerene, (d) soot.

Table 5.3 Types of Solids

Type	Made up of	Characteristics	Example
Ionic	ions in a crystal lattice	high-melting	NaCl, K_2SO_4
Molecular	molecules in a crystal lattice	low-melting	sugar, ice
Metallic	metal atoms in a crystal lattice	shiny, low-to-high melting; soft-to-hard	sodium, iron, gold
Polymeric	giant molecules; can be crystalline, semi-crystalline, or amorphous	low-melting or cannot be melted; soft-to-hard	rubber, plastics, proteins
Network	a very large number of atoms connected by covalent bonds	very hard; very high melting points or cannot be melted	diamond, quartz
Amorphous	randomly arranged atoms or molecules	mostly soft, can be made to flow, but no melting points	soot, tar, glass

The element phosphorus is one of those that exist in more than one solid state. Like the forms of carbon, white and red phosphorus have different structures. (Photograph by Charles D. Winters.)

5.11 Freezing and Boiling. Phase Changes

The **melting point** is the temperature at which a solid changes to a liquid. The temperature at which the reverse process occurs is the **freezing point.** For any substance, the melting point is the same as the freezing point.

Imagine the following experiment: We heat a piece of ice that is initially at, let's say, −20°C. At first we don't see any change in the physical state of the ice. The temperature is increasing, but the appearance of the ice does not change. At 0°C, the ice melts and liquid water appears. While we continue the heating, more and more ice melts, but *the temperature stays constant* at 0°C until all the ice disappears. After all the ice has changed to liquid, the temperature of the water again begins to increase as more heat is added. These changes in the state of matter are called **phase changes.** A **phase** is any part of a system that looks uniform (homogeneous) throughout. Ice (solid) is one phase, and water (liquid) is another.

The criterion of uniformity is the way it appears to our eyes and not as it is on the molecular level.

The diagram describing this experiment (Fig. 5.17) is called a heating curve. It shows that from −20 to 0°C added heat raises the temperature of the ice. After the ice reaches 0°C, **additional heat does not increase the temperature.** Instead, it melts the ice by breaking some of the hydrogen bonds that provide the rigid hexagonal structure (Fig. 5G.1) of solid ice. The heat necessary to melt 1 g of any solid is called its **heat of fusion.** For ice, this is a rather large value: 80 cal/g. During melting, the two phases (solid and liquid) coexist side by side. Only after the ice has completely melted does the temperature of the water rise, assuming that we continue to add heat. For every gram of water, it takes 100 cal to raise the temperature from 0 to 100°C, the normal boiling point. (Contrast this with the 80 cal that only melted the gram of ice but did not raise the temperature.)

This was discussed in Section 1.9.

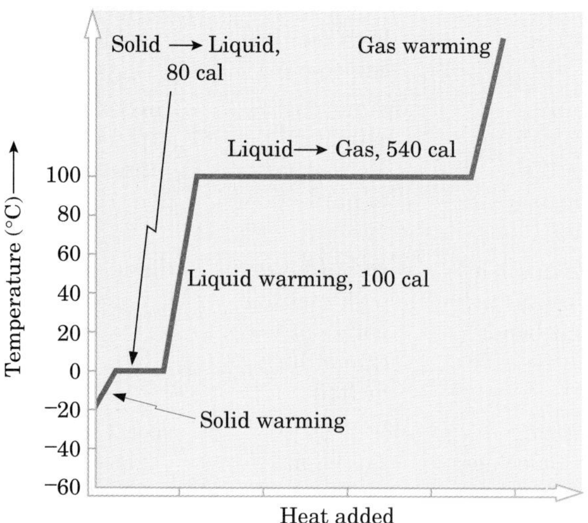

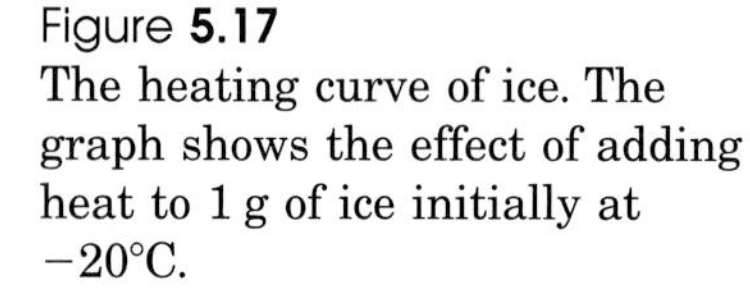

Figure **5.17**
The heating curve of ice. The graph shows the effect of adding heat to 1 g of ice initially at −20°C.

The important aspect of this phase transition is that it is reversible. If we start with liquid water, let's say at room temperature, and cool it by immersing the container in a Dry Ice bath, the same process is observed, but in reverse. The temperature drops until it reaches 0°C, and then the ice starts to crystallize. During the crystallization (phase transition) the temperature stays constant, but heat is given off. The amount of heat given off is exactly the same as the amount of heat absorbed during melting.

Similar considerations apply to the remaining part of the heating curve in Figure 5.17. When the temperature of the liquid reaches the boiling point, another phase change takes place: The liquid is vaporized. While vaporization is taking place, the temperature stays constant at the boiling point. The heat supplied at the boiling point does not increase the temperature; it changes the liquid to a vapor. The amount of heat necessary to vaporize 1 g of any liquid is called its **heat of vaporization.** For water this value is 540 cal/g (as with the heat of fusion, this is a very high value). As long as any liquid water remains, the temperature does not go above 100°C (at 1 atm pressure).

Transition from the solid state directly into the vapor state without going through the liquid state is called **sublimation** (Fig. 2.6[c]). Solids usually sublime only at reduced pressures (less than 1 atm). At high altitudes, where the atmospheric pressure is low, snow sublimes. Solid CO_2 (Dry Ice) sublimes at −78.5°C under 1 atm pressure. At 1 atm pressure, CO_2 can exist only as a solid or as a gas, never as a liquid.

EXAMPLE

To melt 1.0 g of ice requires 80 cal. How many calories are required to melt 1.0 mole of ice?

Answer • Since the molecular weight of ice (H_2O) is 18.0 g/mole, 1.0 mole of ice has a mass of 18.0 g. Using the factor-label method:

$$\frac{80\text{ cal}}{\cancel{\text{g ice}}} \times 18.0\ \cancel{\text{g ice}} = 1.4 \times 10^3 \text{ cal or } 1.4 \text{ kcal}$$

Problem 5.9

What mass of water at 100°C can be vaporized by the addition of 45.0 kcal of heat?

EXAMPLE

What will the final temperature be if we add 1000 cal to 10.0 g of ice at 0°C?

Answer • The first thing the added heat will do is melt the ice. This will use up 10.0 g × 80 cal/g = 800 cal. The remaining 200 cal will be used to heat the liquid water. The number of calories required for this is calculated by the equation on page 32:

$$\begin{aligned} \text{Heat} &= SH \times m \times (T_2 - T_1) \\ 200 \text{ cal} &= 1 \text{ cal/g} \cdot \text{degree} \times 10.0 \text{ g} \times (T_2 - T_1) \\ (T_2 - T_1) &= 200 \cancel{\text{cal}}/(1 \cancel{\text{cal}}/\cancel{\text{g}} \cdot \text{deg})(10.0 \cancel{\text{g}}) \\ &= 20 \text{ deg} \end{aligned}$$

Thus the temperature of the liquid water will have risen by 20 deg from 0°C, and it will now be 20°C.

Problem **5.10**

The specific heat of iron is 0.11 cal/g · deg. The heat of fusion of iron is 63.7 cal/g. Iron melts at 1530°C. How much heat must be added to 1.0 g of iron at 25°C to completely melt it?

SUMMARY

Matter can exist in three different states: gaseous, liquid, and solid. Attractive forces between molecules tend to hold matter together, whereas kinetic energy tends to disorganize matter. At high temperatures (high kinetic energy), matter is in the gaseous state. Molecules in the gaseous state move randomly. Gases fill all the available space of their container. Gas molecules, in their random motion, collide with the walls of the container and thereby exert **pressure.** The **kinetic molecular theory** explains the behavior of gases. The pressure of the atmosphere is measured with a **barometer.** The units of pressure are 1 mm Hg = 1 torr and 760 mm Hg = 1 atm.

Boyle's, Charles's, and **Gay-Lussac's laws** describe the behavior of gases under different conditions. These laws are combined and expressed as the **combined gas law:**

$$\frac{P_1V_1}{T_1} = \frac{P_2V_2}{T_2}$$

Avogadro's law states that, under constant temperature and pressure, equal volumes of different gases contain the same number of molecules. The **ideal gas law,** $PV = nRT$, incorporates Avogadro's law into the combined gas law. Different gases in a mixture exert their own pressures independently. The total pressure is the sum of the partial pressures of the gases (**Dalton's law**).

Intermolecular forces are responsible for the condensation of gases into the liquid and solid states. The intermolecular forces discussed in the chapter are, in increasing order of magnitude, **London dispersion forces, dipole-dipole interactions,** and **hydrogen bonds.**

Liquids have **vapor pressures.** The most energetic molecules of liquids escape from the surface into the vapor phase and exert their own pressure. The vapor pressure of a liquid increases with increasing temperature. When the temperature of a liquid is raised so that the vapor pressure becomes equal to the atmospheric pressure, the liquid boils. The temperature at which this occurs is the **boiling point.** The boiling point of a liquid is determined by its molecular weight and shape and by intermolecular forces.

Solids crystallize in well-formed geometrical shapes that often reflect the patterns in which the atoms are arranged within the crystals. Amorphous solids have in essence a frozen liquid structure. When a solid is heated, its temperature is raised until the melting point is reached; at the melting point there is a **phase change:** The solid becomes a liquid. After all the solid has melted, further heating raises the temperature of the liquid. At the boiling point there is a second phase change: The liquid becomes a gas; further heating can only increase the temperature of the gas. The heat necessary to convert 1 g of any liquid to vapor is the **heat of vaporization** of the liquid. The heat necessary to melt 1 g of any solid to liquid is its **heat of fusion.**

KEY TERMS

Amorphous solid (Sec. 5.10)
Avogadro's law (Sec. 5.5)
Barometer (Sec. 5.3)
Boiling point (Sec. 5.9)
Boyle's law (Sec. 5.4)
Charles's law (Sec. 5.4)
Combined gas law (Sec. 5.4)
Crystal lattice (Sec. 5.10)
Dalton's law (Sec. 5.6)
Diffusion (Sec. 5.6)
Dipole-dipole attraction (Sec. 5.7)
Freezing point (Sec. 5.11)
Gay-Lussac's law (Sec. 5.4)
Graham's law (Sec. 5.6)
Heat of fusion (Sec. 5.11)
Heat of vaporization (Sec. 5.11)
Hydrogen bond (Sec. 5.7)
Ideal gas (Sec. 5.2)
Ideal gas law (Sec. 5.5)
Intermolecular forces (Secs. 5.1, 5.7)
Kinetic molecular theory (Sec. 5.2)
London dispersion forces (Sec. 5.7)
Manometer (Sec. 5.3)
Melting point (Sec. 5.11)
Network solid (Sec. 5.10)
Normal boiling point (Sec. 5.9)
Partial pressure (Sec. 5.6)
Phase (Sec. 5.11)
Pressure (Sec. 5.3)
STP (Sec. 5.5)
Sublimation (Sec. 5.11)
Vapor pressure (Sec. 5.9)
Volume (Sec. 5.2)

PROBLEMS

5.11 What forces hold matter together?

Gases and Pressure

5.12 In terms of the kinetic molecular theory, what causes (a) the pressure (b) the temperature of gases?

5.13 A sphygmomanometer (Box 5C, page 148) reads a pressure of 120 mm Hg. Draw a diagram indicating the level of Hg in the evacuated reference arm and in the arm connected to the inflating cuff.

5.14 Express in atm:
(a) 7835 mm Hg
(b) 25 mm Hg
(c) 22.7 in. Hg
(d) standard pressure

***5.15** Under certain weather conditions (just before rain), the air becomes less dense. How does this affect the barometric pressure reading?

Gas Laws

5.16 If a gas undergoes a change in which both the pressure and the Kelvin temperature are doubled, does the volume of the gas increase, decrease, or stay the same? Explain.

5.17 At constant temperature a sample of xenon gas at 2.0 atm is expanded from 1.0 L to 4.0 L. What is the new pressure?

5.18 A sample of 23.0 L of NH_3 gas at 10.0°C is heated at constant pressure until it fills a volume of 50.0 L. What is the new temperature in °C?

5.19 If a sample of 4.17 L of C_2H_6 gas at 725°C is cooled to 175°C at constant pressure, what is the new volume?

5.20 A sample of SO_2 gas has a volume of 5.2 L. It is heated at constant pressure from 30 to 90°C. What is the new volume?

5.21 A sample of B_2H_6 gas in a 35-mL container is at a pressure of 450 mm Hg and a temperature of 625°C. If the gas is allowed to cool at constant volume until the pressure is 375 mm Hg, what is the new temperature in °C?

***5.22** A certain quantity of helium gas is at a temperature of 27°C and a pressure of 1.00 atm. What will the new temperature be if its volume is doubled at the same time as its pressure is dropped to one-half of its original value?

5.23 A 480-mL balloon is filled with helium at 730 mm Hg barometric pressure. The balloon is released and climbs to an altitude where the barometric pressure is 395 mm Hg. What will the volume of the balloon be if, during the ascent, the temperature drops from 20 to 3°C?

5.24 An ideal gas occupies 56.44 L at 2.00 atm and 310.0 K. If the gas is compressed to 23.52 L and the temperature lowered to 281.3 K, what is the new pressure?

5.25 A sample of 30.0-mL Kr gas is at 756 mm Hg at 25.0°C. What is the new volume if the pressure is decreased to 325 mm Hg and the temperature to −12.5°C?

5.26 A sample of helium gas has a volume of 180.0 L at 4.40 atm and −138°C. What is its volume at STP?

5.27 A 26.4-mL sample of C_2H_4 gas has a pressure of 2.50 atm at 2.5°C. If the volume is increased to 36.2 mL and the temperature raised to 10°C, what is the new pressure?

5.28 The inside pressure of an automobile tire is 1.80 atm at a temperature of 20°C. What will the pressure in the tire be if, after 10 miles of driving, the temperature of the tire increases to 47°C?

5.29 In an autoclave, a constant amount of steam is generated at a constant volume. Under 1.00 atm pressure the steam temperature is 100°C. What pressure setting should be used to obtain a 165°C steam temperature for the sterilization of surgical instruments?

Avogadro's Law and the Ideal Gas Law

5.30 A sample of a gas at 77°C and 1.33 atm occupies a volume of 50.3 L.
(a) How many moles of the gas are present?

(b) Does your answer depend on knowing what gas it is?

5.31 What volume is occupied by 2.5 g O_2 gas at 25°C and at 600 mm Hg pressure?

5.32 What volume is occupied by 5.8 g of propane gas (C_3H_8) at 23°C and 1.15 atm pressure?

5.33 If 5.50 moles of BrCl gas occupies 20.2 L at 200.0°C, what is the pressure in mm Hg?

5.34 At a mountain top where the pressure is 650 mm Hg, 0.500 mole of F_2 gas occupies 8.51 L. What is its temperature in °C?

5.35 What volume, in milliliters, does 0.275 g of UF_6 gas occupy at 425°C and 365 torr?

5.36 A hyperbaric chamber has a volume of 200 L. (a) How many moles of oxygen are needed to fill the chamber at room temperature (300 K) to 3.00 atm pressure? (b) How many grams of oxygen?

5.37 What is the volume of 32 g of argon gas at 0°C and a pressure of 1.5 atm?

***5.38** At STP, 1 mole of every ideal gas occupies 22.4 L. What are the densities, in g/L, of SO_2, CH_4, and F_2 at STP?

5.39 A sample of cyanogen gas weighing 6.802 g occupies 3.74 L at 25.0°C and 650.0 mm Hg. Calculate the molecular weight of cyanogen.

5.40 How many molecules of CO are there in 100 L of CO at STP?

5.41 The density of C_2H_2 gas in a 4-L container at 0°C and 2 atm pressure is 0.02 g/mL. What would be the density of the gas under identical temperature and pressure if the container is partitioned into two 2-L compartments?

Dalton's and Graham's Laws

5.42 Three gases in open vessels are all at one end of a room. One is a perfume with the molecular formula $C_{10}H_{18}O_2$, the second is NH_3, and the third is HCl. If you sit in the corner farthest from the samples, which odor reaches you first? Which last?

5.43 The pressure of CH_4 over water registered 342 mm Hg. If the water had a partial pressure of 18.2 mm Hg, what was the partial pressure of CH_4?

***5.44** The rate of diffusion of an unknown gas is three times lower than that of helium. If helium has a molecular weight of 4.0 and a diffusion rate of 60 cm/min, what is the molecular weight of the unknown gas?

Intermolecular Forces

5.45 Under which condition does water vapor behave most ideally:
(a) 0.5 atm, 400 K (c) 0.01 atm, 500 K
(b) 4 atm, 500 K

5.46 Which molecule has the strongest intermolecular interaction: (a) CO (b) HF (c) Cl_2?

5.47 Explain how helium can become a liquid at a sufficiently low temperature even though the atoms of helium are nonpolar.

5.48 What kind of intermolecular interactions are there in (a) NH_3 (b) Cl_2 (c) HBr

Evaporation, Condensation, and Boiling Point

5.49 Can dipole-dipole interactions ever be weaker than London dispersion forces? Explain.

5.50 Which compound has a higher boiling point: butane, C_4H_{10} or hexane, C_6H_{14}?

5.51 Two compounds, *n*-octane and tetramethylbutane, have the same molecular weight, and the London forces acting between the molecules are the same. Yet *n*-octane has a boiling point of 126°C, while tetramethylbutane boils at 107°C. Explain.

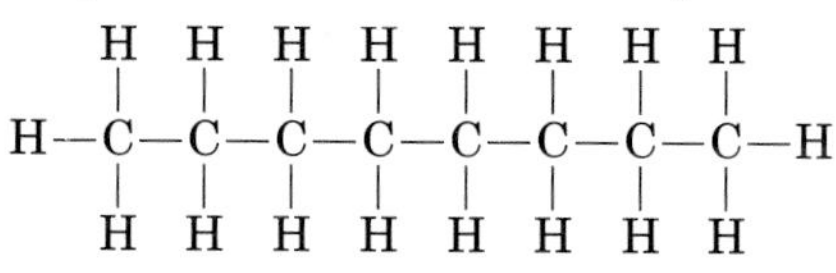

Octane

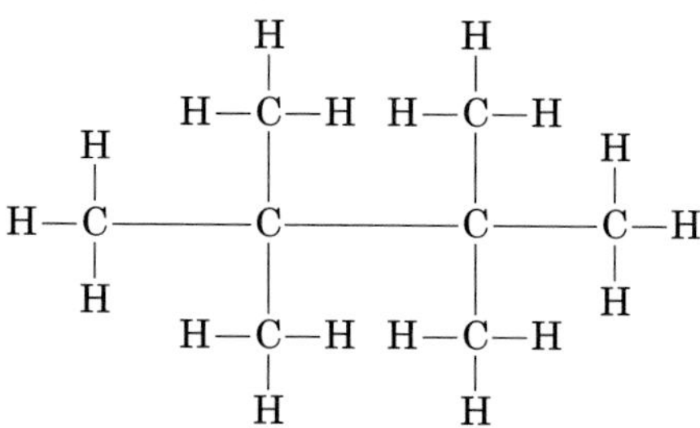

Tetramethylbutane

5.52 Using Figure 5.13 estimate the vapor pressure of ethyl alcohol: (a) at 30°C (b) at 40°C (c) at 60°C.

5.53 Explain, in terms of the behavior of the molecules, why a puddle of water dries up after it stops raining.

5.54 The normal boiling point of a substance depends on both the mass of the molecule and the intermolecular interactions. Considering these, rank in order of increasing boiling point and explain your answer: (a) HCl, HBr, HI (b) O_2, HCl, H_2O_2

***5.55** Kinetic energy increases with increasing temperature, whereas potential energy stays constant. On the basis of this simple principle, explain how a gas condenses to a liquid when the temperature is decreased.

Solids and Phase Changes

5.56 Define: (a) crystal (b) amorphous solid.

5.57 Considering their structures, explain why diamond is harder than buckminsterfullerene.

5.58 When iodine vapor hits a cold surface, iodine crystals are formed. Name the phase change that is the reverse of this condensation.

5.59 Draw the heating curve for acetic acid if its melting point is 17°C and its normal boiling point 118°C.

5.60 Define sublimation.

Boxes

5.61 (Box 5A) Which has lower entropy, a gas at 100°C or at 200°C? Explain.

5.62 (Box 5A) When a liquid vaporizes, does the entropy increase or decrease? Explain.

5.63 (Box 5B) Why should you collect the mercury from a broken thermometer and store it under water?

5.64 (Box 5C) In a sphygmomanometer one listens to the first tapping sound as the constrictive pressure of the arm cuff is slowly released. What is the significance of this tapping sound?

5.65 (Box 5D) What happens when we lower the diaphragm in our chest cavity?

5.66 (Box 5E) In carbon monoxide poisoning, the hemoglobin is incapable of transporting oxygen to the tissues. How does the oxygen get delivered to the cells when the patient is put into a hyperbaric chamber?

5.67 (Box 5F) In circulating blood, where is the pressure of CO_2 the highest? The lowest?

5.68 (Box 5G) Why is the damage by severe frost bite irreversible?

5.69 (Box 5G) If you fill a bottle with water, cap it, and freeze to −10°C, the bottle will crack. Explain.

Additional Problems

5.70 Differentiate between evaporation and sublimation.

5.71 An ideal gas occupies 387 mL at 275 mm Hg and 75°C. If the pressure is changed to 1.36 atm and the temperature increased to 105°C, what is the new volume?

5.72 Distinguish among a molecular crystal, a network crystal, and an ionic crystal.

5.73 Which compound has greater intermolecular interactions, CO or CO_2? Assume that O═C═O is linear.

5.74 On the basis of what you have learned of intermolecular forces, predict which liquid has the highest boiling point:
(a) pentane, C_5H_{12}
(b) chloroform, $HCCl_3$
(c) water, H_2O

***5.75** Your automobile tire contains 10 L of N_2 under 35 in. Hg pressure. The car manufacturer recommends a tire pressure of 60 in. Hg. How many moles of N_2 do you have to pump into your tire to raise the pressure as recommended? Assume a constant temperature of 27°C.

***5.76** Explain why gases are transparent.

***5.77** The density of a gas is 0.00300 g/cc at 100°C and 1.00 atm. What is the molecular weight of the gas?

***5.78** Because natural gas (methane) has no odor, gas companies add a warning gas so that leaks can be detected. Two candidates for this are NH_3 and CH_3SH, because of their strong odors. Which could be used to warn the household faster of an impending disaster? How much faster?

5.79 The normal boiling point of hexane is 69°C, and that of pentane is 36°C. Predict which of these has a higher vapor pressure at 20°C.

5.80 Which gas molecules diffuse faster, HCl or C_3H_8?

***5.81** What is the density of CH_4 at 1 atm and at 35°C?

5.82 If 60.0 g of NH_3 occupies 35.1 L under 77.2 in. Hg pressure, what is the temperature of the gas, in °C?

5.83 Given that 2.0 L of diborane gas weighs 2.48 g at STP, calculate its molecular weight.

5.84 Why does the temperature of a liquid drop as a result of evaporation?

Chapter 6

Solutions and colloids

Making a homogeneous solution. A green solid, nickel nitrate, is stirred into water, where it dissolves with stirring to form a homogeneous solution. (Photograph by Charles D. Winters.)

6.1 Introduction

In Chapter 5, we discussed pure substances—systems having only one component. Such systems are the easiest to study, so it was convenient to begin with them. In our daily lives, however, we more frequently encounter systems having more than one component. Such systems—for example, air, smoke, seawater, milk, blood, rocks—are **mixtures.**

If a mixture is uniform throughout, we call it a **homogeneous mixture.** In a homogeneous mixture the molecules (or ions) are thoroughly mixed. A homogeneous mixture is called a **solution.**

In contrast, in most rocks we can see distinct regions separated from each other by well-defined boundaries. Such rocks are **heterogeneous mixtures.** For another example, a mixture of sand and sugar is heterogeneous. Even though sand and sugar are both present, we can easily distinguish between the two; the mixing is not at the molecular level (see also Fig. 2.2[c]).

Mixtures are classified on the basis of how they look to the unaided eye. There are systems, however, that fall between homogeneous and heterogeneous mixtures. Cigarette smoke, muddy water, and blood plasma may look homogeneous, but they do not have the transparency of air or seawater. These mixtures are classified as **colloidal dispersions.** We will deal with such systems in Section 6.7.

Although mixtures can contain many components, we shall restrict our discussion mainly to two-component systems, with the understanding that everything we say can be extended to multicomponent systems.

6.2 Types of Solutions

When we think of a solution, we normally think of a liquid. Liquid solutions are the most common kind, but there are also solutions that are gases or solids. In fact, all mixtures of gases are solutions. Because gas molecules are far apart from each other and there is much empty space between them, two or more gases can mix with each other in any proportions. Because the mixing is at the molecular level, a true solution always forms. That is, there are no heterogeneous mixtures of gases.

Many alloys are solid solutions; one example is stainless steel, which is mostly iron but also contains carbon, chromium, and other elements.

With solids, we are at the other extreme. Anytime we mix solids we get a heterogeneous mixture. Since even microscopic pieces of solid still contain many billions of particles (molecules, ions, or atoms), there is no way to obtain mixing at the molecular level. Homogeneous mixtures of solids (that is, true solutions) do exist, but we make them by first melting the solids, then mixing the molten components and allowing the mixture to solidify.

We normally do not use the terms "solute" and "solvent" in talking about solutions of gases in gases or solids in solids.

The five most common types of solutions are shown in Table 6.1. Examples of other types (gas in solid, liquid in gas, and so on) are also known but are much less important. In this chapter, we deal almost entirely with liquid solutions (the first three types listed in Table 6.1).

When a solution consists of a solid or a gas dissolved in a liquid, the liquid is called the **solvent** and the solid or gas the **solute.** A solvent

Table 6.1 Types of Solutions

Solute		Solvent	Appearance of Solution	Example
Gas	in	Liquid	Liquid	Carbonated water
Liquid	in	Liquid	Liquid	Wine
Solid	in	Liquid	Liquid	Salt water (saline solution)
Gas	in	Gas	Gas	Air
Solid	in	Solid	Solid	14-carat gold

Beer is a solution in which a liquid (alcohol), a solid (malt), and a gas (CO_2) are dissolved in the solvent water. (Photograph by Beverly March.)

may have several solutes dissolved in it, even of different types. A common example is beer, in which a liquid (alcohol), a gas (carbon dioxide), and a solid (malt) are dissolved in the solvent, water.

When a liquid is dissolved in another liquid, there is a question about which is the solvent and which the solute. In most cases the one present in the greater amount is called the solvent, but there is no rigid rule about this.

6.3 Characteristics of Solutions

The following are some properties of true solutions:

1. The distribution of particles in a solution is uniform. Every part of the solution has exactly the same composition and properties as every other part. That, in fact, is the definition of homogeneous. This means that we cannot usually tell a solution from a pure solvent simply by looking at it. A glass of pure water looks the same as a glass of water containing dissolved salt or sugar. In some cases we can tell by looking, for example, if the solution is colored and we know that the solvent is colorless.

2. The components of a solution do not separate on standing. A solution of vinegar (acetic acid in water), for example, can remain in the closet for many years without separating out.

3. A solution cannot be separated into its components by filtration. Both the solvent and the solute pass through the filter paper.

4. For any given solute and solvent, it is possible to make solutions of many different compositions. For example, we can easily make a solution of 1 g of glucose in 100 g of water, or 2 g, or 6, or 8.7, or any amount at all, up to the solubility limit (Sec. 6.4).

5. Solutions are almost always transparent. They may be colorless or colored, but you can usually see through them.

We use the word "clear" to mean transparent. A solution of copper sulfate in water is blue, but clear.

6. Solutions can be separated into pure components, not by filtration but by such other methods as distillation (Sec. 2.2) and chromatography, which you may learn about in the laboratory portion of this course. The separation of a solution into its components is a physical, not a chemical, change.

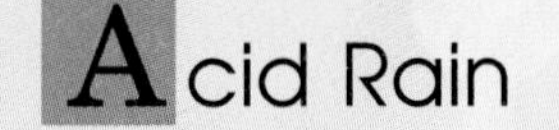

Box 6A

Acid Rain

The water vapor evaporated by the sun from oceans, lakes, and rivers condenses and forms clouds of water vapor that eventually fall as rain. The raindrops contain small amounts of CO_2, O_2, and N_2. Table 6A shows that, of these gases, CO_2 is the most soluble in water. When CO_2 dissolves in water, it forms carbonic acid, H_2CO_3.

Table 6A Solubility of Some Gases in Water

Gas	Solubility (g per kg of H_2O at 20°C and 1 atm)
O_2	0.0434
N_2	0.0190
CO_2	1.688
H_2S	3.846
SO_2	112.80
NO_2	0.0617

The acidity caused by the CO_2 is not harmful. However, a serious acid rain problem is caused by contaminants that result from industrial pollution. Burning coal or oil that contains sulfur generates sulfur dioxide, which has a high solubility in water. Sulfur dioxide in the air is oxidized to sulfur trioxide. The combination of water and sulfur dioxide results in sulfurous acid, and sulfur trioxide and water give sulfuric acid. Smelting industries produce other soluble gases as well. The result is that in many parts of the world, especially downwind from heavily industrial areas, acid rain pours down on forests and lakes. This damages vegetation and kills fish. Such is the situation in the northeastern United States, where in the Adirondacks and New England areas, as well as in eastern Canada, acid rain has been observed with increasing frequency.

Trees killed by acid rain in eastern Europe. (Simon Fraser/SPL/Photo Researchers.)

6.4 Solubility

Suppose we wish to make a solution of table salt (NaCl) in water. We take some water, add a few grams of salt, and stir. At first we see the particles of salt suspended in the water, but soon all the salt dissolves. Now let us add more salt to the same solution and continue to stir. Again the salt all dissolves. Can we repeat this process indefinitely? The answer is no; there is a limit. The solubility of table salt at 25°C is 36.2 g per 100 g of water. If we add more salt than that, the excess solid does not dissolve but remains suspended as long as we keep stirring, and it sinks to the bottom after we stop.

The **solubility** of a solid in a liquid is **the maximum amount of that solid that dissolves in a given amount of liquid at a given temperature.** Solubility is a *physical constant,* like melting point or boiling point. Each solid has a different solubility in every liquid. Some solids have a very low solubility in a particular solvent. We often call these solids insoluble. Others have a much higher solubility. We call these soluble, but even for soluble solids, there is always a solubility limit. The same is true for gases dissolved in liquids.

Tables of solubility values are found in handbooks.

For solutions of liquids in liquids, however, the situation may be different. Some liquids are essentially insoluble in other liquids (gasoline in water), and others are soluble to a limit. For example, 100 g of water dissolves about 4 g of ethyl ether (another liquid). If we add more ether than that, we see two layers (Fig. 6.1). Some liquids, however, are *completely* soluble in other liquids, no matter how much is present. The most important example is ethyl alcohol and water, which form a solution no matter what quantities of each are mixed. We say that the water and ethyl alcohol are *miscible* in all proportions.

We use the word "miscible" to refer to a liquid dissolving in a liquid.

When a solvent contains all the solute it can hold at a given temperature, we call the solution **saturated.** Any solution containing a lesser amount of solute is **unsaturated.** It may seem surprising, but there are also solutions in which the solvent holds *more* solute than it can normally hold at a given temperature! Such solutions are called **supersaturated** and are fairly common (p. 182).

If, to a saturated solution at a constant temperature, an additional amount of solute is added, it does not appear that any of the additional solid will dissolve, since the solution already holds all the solute that it can. But there is actually an equilibrium here, similar to the one discussed in Section 5.9. Some particles of the additional solute will dissolve, while an equal quantity of dissolved solute will come out of solution. So, while the concentration of dissolved solute does not change, the solute particles themselves are constantly going in and out of solution.

Whether a particular solute dissolves in a particular solvent depends on several factors.

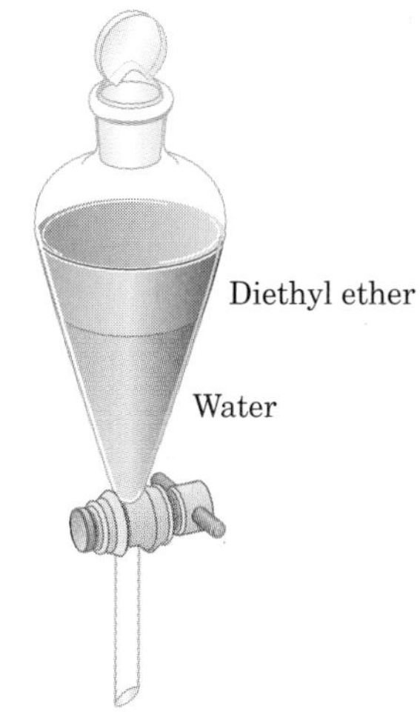

Figure **6.1**
Diethyl ether and water form two layers. A separatory funnel permits the bottom layer to be drawn off.

Nature of the Solute and Solvent

Here the rule is **"like dissolves like."** The more similar two compounds are, the more likely it is that one is soluble in the other. This is not an absolute rule, but it does apply in most cases.

When we say "like," we mostly mean alike in *polarity.* That is, polar compounds dissolve in polar compounds and nonpolar compounds dissolve in nonpolar compounds. For example, the liquids benzene and carbon tetrachloride are nonpolar compounds. They dissolve in each other, and other nonpolar materials, such as gasoline and camphor, dissolve in them. On the other hand, ionic compounds (such as NaCl) and polar compounds (such as table sugar) are insoluble in all of these solvents.

Polar compounds dissolve in polar compounds because the positive end of one molecule attracts the negative end of the other.

The most important polar solvent is water, and we have already seen that most ionic compounds are soluble in water, as are small covalent compounds that can form hydrogen bonds with water. It is worth noting that even polar molecules are usually insoluble in water if they cannot either react with the water or form hydrogen bonds with water molecules. Water as a solvent is discussed in Section 6.6.

Temperature

For most solids and liquids that dissolve in liquids, the rule is **solubility increases with increasing temperature.** Sometimes the increase is great. For example, the solubility of glycine, a building block of proteins, is 52.8 g in 100 g of water at 80°C but only 33.2 g at 30°C.

If we prepare a saturated solution of glycine in 100 g of water at 80°C, it will hold 52.8 g of glycine. If we now allow it to cool to 30°C, where the solubility is 33.2 g, we might expect the excess glycine, 19.6 g, to come out of solution and form crystals. It often does, but on many occasions it does not. Even though the solution contains more glycine than the water can normally hold at 30°C, the excess may well stay in solution. This is an example of the supersaturated solution referred to earlier. The excess glycine stays in solution because the molecules need a surface on which to begin crystallizing, and no such surface is present.

The difference between saturated, unsaturated, and supersaturated solutions is illustrated in Figure 6.2.

Supersaturated solutions are not indefinitely stable, however. If we shake or stir the solution, we may find that all of the excess solid precipitates out at once. Another way to crystallize the excess solute is to add a grain of glycine crystal. This is called **seeding.** The seed crystal provides the surface onto which the solute molecules can converge.

For gases, **solubility in liquids almost always decreases with increasing temperature.** This explains why a bottle of beer or a carbonated soft drink foams up when opened at room temperature. When the beverage is cold, the CO_2 it contains is more soluble. At room temperature it is less soluble, and more of it comes out of solution. This additional gaseous CO_2 increases the pressure. When the pressure is released, the gas pushes out some of the liquid.

The effect of temperature on the solubility of gases in water can have important consequences for fish. Oxygen is only slightly soluble in water, but fish need that oxygen to live. When the temperature of the water is increased, perhaps by the output from a power plant, the solubility of the oxygen decreases and may become so low that fish die. This is called *thermal pollution.*

Box 6B

The Bends

Deep-sea divers encounter high pressures. In order for them to breathe properly under such conditions, oxygen must be supplied under pressure. At one time this was achieved with compressed air. As pressure increases, the solubility of the gases in the blood increases. This is especially true for nitrogen, which constitutes almost 80 percent of the air.

When a diver comes up and the pressure on the body decreases, the solubility of nitrogen in the blood decreases as well. As a consequence, the previously dissolved nitrogen in the blood and in the tissues starts to form small bubbles, especially in the veins. The formation of gas bubbles, called **the bends,** can prevent blood circulation. If allowed to develop uncontrolled, the resulting pulmonary embolism can be fatal.

If the ascent is gradual, the dissolved gases are removed by regular exhalation and diffusion through the skin. Divers use decompression chambers, where the high pressure is gradually reduced to normal pressure.

Because of the problem caused by nitrogen, divers often use a helium-oxygen mixture instead of air. The solubility of helium in blood is less affected by pressure than is that of nitrogen. The sudden decompression and ensuing bends are important not only in deep-sea diving but also in high-altitude flight, especially orbital flight.

Pressure

Pressure has little effect on the solubility of liquids or solids. For gases, however, **the higher the pressure, the greater the solubility of a gas in a liquid.** This is the basis of the hyperbaric medicine discussed in Box 5E. When the pressure is increased, more oxygen dissolves in the blood plasma and reaches the tissues at higher-than-normal pressures (2 to 3 atm).

Henry's law states that the solubility of a gas in a liquid is directly proportional to the pressure.

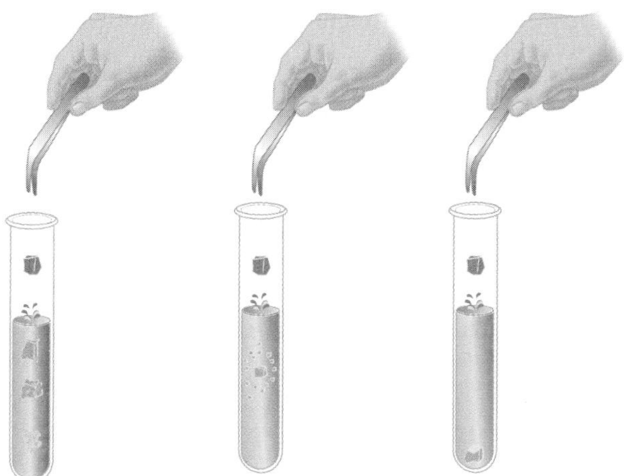

Figure **6.2**
The difference in behavior of unsaturated, saturated, and supersaturated solutions when a crystal of the solute is added.

6.5 Concentration Units

The amount of a solute dissolved in a given quantity of solvent is called the **concentration.** This can be expressed in a number of ways. Some concentration units are better suited than others for some purposes. Sometimes qualitative terms are good enough. We say a solution is **dilute** or **concentrated.** This tells us little about the concentration, but we know that a concentrated solution contains more solute than a dilute solution.

For most purposes, we need quantitative concentrations. For example, a nurse must know how much glucose to give to a patient. Many methods of expressing concentration exist, but in this chapter we deal only with the three most important: percent concentration, molarity, and parts per million (ppm).

Percent Concentration

Percent concentration (% w/v) is the number of grams of solute in 100 mL of solution.

There are three different ways of representing percent concentration. The most common is *weight* of solute per *volume* of solution. If 10 g of sugar is dissolved in enough water so that the total volume is 100 mL, the concentration is 10 percent w/v. We need to know the total volume of the solution, not the volume of the solvent. For this purpose we use **volumetric flasks,** which come in different sizes. Volumetric flasks of different sizes are shown in Figures 6.3 and 6.4.

EXAMPLE

How would we make 500 mL of a 1.5 percent w/v solution of KCl in water?

Answer • We find out how many grams of KCl there are in this solution by taking the percentage:

$$\frac{1.5 \text{ g KCl}}{100 \cancel{\text{mL}}} \times 500 \cancel{\text{mL}} = 7.5 \text{ g KCl}$$

We then put 7.5 g of KCl into a beaker, add some water, and stir until the KCl is all dissolved. We then transfer the solution to a 500-mL volumetric flask and fill with water to the mark. We don't know exactly how much water we have added, but we do know that the total volume is 500 mL.

Problem **6.1**

How would you make 250 mL of a 4.4 percent w/v KBr solution in water? Assume that a 250-mL volumetric flask is available.

EXAMPLE

If 3.7 g of sodium nitrate is dissolved in enough water to make 250.0 mL of solution, what is the percentage of $NaNO_3$ w/v?

Answer • We divide the weight of the solute by the volume of solution and multiply by 100:

$$\frac{3.7 \text{ g}}{250.0 \text{ mL}} \times 100 = 1.5\% \text{ w/v}$$

The percentage is 1.5% w/v.

Problem 6.2

If 7.7 g of lithium iodide is dissolved in enough water to make 400.0 mL of solution, what is the percentage of LiI w/v?

A second type of percentage unit is *weight* of solute per *weight* of solution (w/w). Calculations are essentially the same as with w/v except that the *weight* of solution is used instead of the volume. A volumetric flask is not used for these solutions. (Why not?)

Finally, there is *volume* of solute per *volume* of solution (v/v). This unit is used only for solutions of liquids in liquids, most notably alcoholic beverages. For example, 40 percent v/v ethyl alcohol in water means that 40 mL of ethyl alcohol has been added to enough water to make 100 mL of solution.*

A 40% v/v solution of ethyl alcohol in water is 80 proof. Proof in the United States is twice the percent concentration (v/v) of ethyl alcohol in water.

Molarity

For many purposes it is easiest to express concentration by the weight or volume percentage methods just discussed, but when we want to focus on the *number of molecules* present we need another way. For example, a 5 percent solution of sugar in water does not contain the same concentration of solute molecules as a 5 percent solution of alcohol in water. That is why chemists often use molarity. The **molarity** of a solution is defined as the **number of moles of solute dissolved in 1 L of solution.** Thus, in the same volume of solution, a 0.2-molar solution of sugar in water contains the same number of molecules of solute as a 0.2-molar solution of alcohol. In fact, this holds true for equal volumes of any solutions, as long as the molarities are the same. The units of molarity are moles per liter, and the symbol is M.

$$\text{Molarity } (M) = \frac{\text{moles } (n)}{\text{liters } (V)}$$

*A 40 percent v/v ethyl alcohol solution contains not 60 mL of water for every 40 mL of alcohol, but somewhat less than that. *Volumes of liquids are not always additive.* When one liquid is dissolved in another, the new volume may be more or less than the sum of the volumes of the individual liquids because the molecules move either closer to each other (because of greater attraction) or farther apart.

We can prepare a given volume of a given molarity in essentially the same way we prepare a solution of given w/v concentration, except that we use moles instead of grams in our calculations. We can always find out how many moles of solute there are in any solution of known molarity by the formula

$$\text{Molarity} \times \text{volume in liters} = \text{number of moles}$$

$$M \times V = \text{moles}$$

The solution is then prepared as shown in Figure 6.3.

EXAMPLE

How do we prepare 2.0 L of a 0.15 M solution of NaOH?

Answer • First we find out how many moles of NaOH there will be in this solution:

$$M \times V = \text{moles}$$

$$\frac{0.15 \text{ mole NaOH}}{1 \not{L}} \times 2.0 \not{L} = 0.30 \text{ mole NaOH}$$

We need 0.30 mole of NaOH to make up this solution. How many grams is this? To convert moles to grams, we multiply by the appropriate conversion factor (Sec. 4.3). The formula weight of NaOH = 40.0, so

$$0.30 \cancel{\text{mole}} \text{ NaOH} \times \frac{40.0 \text{ g NaOH}}{1 \cancel{\text{mole}} \text{ NaOH}} = 12 \text{ g NaOH}$$

We therefore weigh out 12 g of NaOH, put it into a beaker, add some water, stir until the solid has dissolved, and then transfer the solution to a 2-L volumetric flask and fill with water to the 2-L mark.

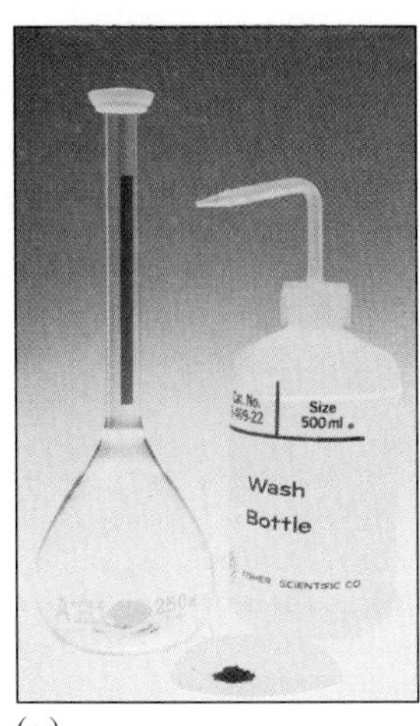

(*a*)

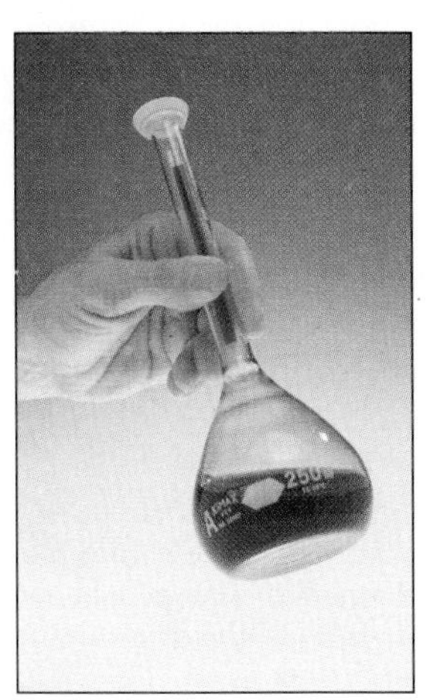

(*b*)

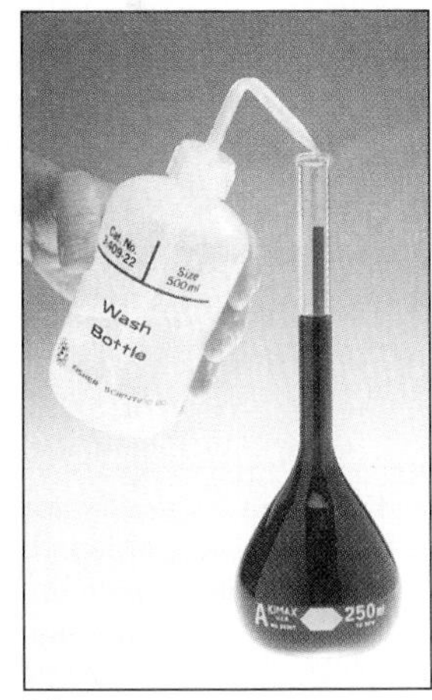

(*c*)

Figure **6.3**
(a) A 0.0100 M solution of $KMnO_4$ is made by adding enough water to 0.395 g of $KMnO_4$ to make 0.250 L of solution. (b) To ensure the correct solution volume, the $KMnO_4$ is placed in a volumetric flask and dissolved in a small amount of water. (c) After dissolving is complete, sufficient water is added to fill the flask to the mark. The flask contains 0.250 L of solution. (Photographs by Charles D. Winters.)

Problem 6.3

How would you prepare 2.0 L of a 1.06 *M* aqueous solution of KCl?

EXAMPLE

What is the molarity of a solution made by dissolving 75 g of glucose, $C_6H_{12}O_6$, in enough water to have 550 mL of solution?

Answer • We are looking for molarity, that is, moles per liter. First we find the number of moles of glucose, then the number of liters of solution, and then divide. To find moles of glucose, we multiply grams of glucose by the conversion factor 1 mole glucose/180.0 g glucose:

$$75\ \cancel{\text{g glucose}} \times \frac{1\ \text{mole glucose}}{180.0\ \cancel{\text{g glucose}}} = 0.42\ \text{mole glucose}$$

To find liters of solution, we divide mL by 1000:

$$550\ \cancel{\text{mL}} \times \frac{1\ \text{L}}{1000\ \cancel{\text{mL}}} = 0.55\ \text{L}$$

To find moles of glucose per liter of solution, we divide:

$$\frac{0.42\ \text{mole glucose}}{0.55\ \text{L solution}} = 0.76\ M$$

Problem 6.4

If 2.3 g of KI is dissolved in enough solvent to make 250 mL of solution, what is the molarity of the solution?

EXAMPLE

The concentration of sodium chloride in blood serum is approximately 0.14 *M*. What volume of blood serum contains 2.0 g of NaCl?

Blood serum is the liquid part of the blood that remains after the cellular particulates and the fibrinogen have been removed.

Answer • Since we know the concentration in *moles* per liter, we must first find out how many moles there are in 2.0 g of NaCl (formula weight = 58.5) by multiplying by the proper conversion factor:

$$2.0\ \cancel{\text{g NaCl}} \times \frac{1\ \text{mole NaCl}}{58.5\ \cancel{\text{g NaCl}}} = 0.034\ \text{mole NaCl}$$

We next find liters:

$$0.034\ \cancel{\text{mole NaCl}} \times \frac{1\ \text{L}}{0.14\ \cancel{\text{mole NaCl}}} = 0.24\ \text{L} = 240\ \text{mL}$$

Problem 6.5

If 0.300 M glucose ($C_6H_{12}O_6$) solution is available for intravenous infusion, how many milliliters are needed to deliver 10.0 g of glucose?

EXAMPLE

How many grams of HCl are there in 225 mL of concentrated HCl, whose concentration is 6.00 M?

Answer • First we find out how many moles of HCl there are:

$$M \times V = \text{moles}$$

$$\frac{6.00 \text{ moles}}{1 \cancel{\text{L}}} \times 0.225 \cancel{\text{L}} = 1.35 \text{ moles}$$

We get grams by multiplying by the conversion factor 1 mole = 36.5 g:

$$1.35 \cancel{\text{moles HCl}} \times \frac{36.5 \text{ g HCl}}{1 \cancel{\text{mole HCl}}} = 49.3 \text{ g HCl}$$

Problem 6.6

How many grams of H_2SO_4 are there in 50.0 mL of a 3.0 M H_2SO_4 solution?

Dilution

We frequently prepare solutions by diluting more concentrated solutions rather than by weighing out pure solute (Fig. 6.4). Since we are adding only solvent, the number of moles of solute remains unchanged. Before we dilute, the equation that applies is

$$M_1V_1 = \text{moles}$$

After we dilute, the volume and molarity have both changed and we have

$$M_2V_2 = \text{moles}$$

but the number of moles of solute is the same before and after, so we can say that

This is a handy equation that we can use for dilution problems.

$$M_1V_1 = M_2V_2$$

(*a*) (*b*) (*c*)

Figure **6.4** • Making a solution by dilution. (a) A 100-mL volumetric flask is filled to the mark with 0.100 *M* $K_2Cr_2O_7$. (b) This is transferred to a 1.00-L volumetric flask. (c) The 1.00-L volumetric flask is filled to the mark with distilled water. The concentration of the now diluted $K_2Cr_2O_7$ is 0.0100 *M*. (Photographs by Charles D. Winters.)

EXAMPLE

How do we prepare 200 mL of a 3.5 *M* solution of acetic acid if we have a bottle of concentrated acetic acid (6.0 *M*)?

Answer • We use the equation

$$M_1V_1 = M_2V_2$$

$$\frac{6.0 \text{ moles}}{1 \text{ L}} \times V_1 = \frac{3.5 \text{ moles}}{1 \text{ L}} \times 0.200 \text{ L}$$

$$V_1 = \frac{(3.5 \cancel{\text{moles/L}})(0.200 \text{ L})}{6.0 \cancel{\text{moles/L}}} = 0.12 \text{ L}$$

We put 0.12 L, or 120 mL, of the concentrated acetic acid in a 200-mL volumetric flask, add some water and mix, and then fill to the mark with water.

Problem **6.7**

How would you prepare 200.0 mL of 0.200 *M* HCl from a 10.0 *M* stock HCl solution?

A similar equation,

$$\%_1V_1 = \%_2V_2$$

can be used for dilution problems involving percent concentrations.

EXAMPLE

How do we prepare 500 mL of a 0.5 percent w/v solution of NaOH if we have a stock solution of 50 percent w/v NaOH on hand?

Answer

$$(50\%)V_1 = (0.5\%)(500 \text{ mL})$$

$$V_1 = \frac{(0.5\%)(500 \text{ mL})}{50\%} = 5 \text{ mL}$$

We add 5 mL of the concentrated solution to a 500-mL volumetric flask, add some water and mix, and then fill to the mark with water.

Problem **6.8**

A concentrated solution of 15 percent w/v KOH solution is available. How would you prepare 20.0 mL of a 0.10 percent w/v KOH solution?

Parts per Million

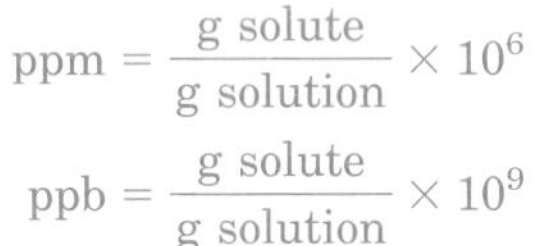

$$\text{ppm} = \frac{\text{g solute}}{\text{g solution}} \times 10^6$$

$$\text{ppb} = \frac{\text{g solute}}{\text{g solution}} \times 10^9$$

Sometimes we need to deal with very dilute solutions, for example, 0.0001 percent. In such cases it is more convenient to use **parts per million (ppm).** For example, if drinking water is polluted with lead ions to the extent of 1 ppm, it means that there is 1 mg of lead ions in 1 kg (1 L) of the water. Some solutions are so dilute that we use **parts per billion (ppb).** Modern methods of analysis allow us to detect concentrations this small. Some substances are harmful even at concentrations measured in ppb. One such substance is dioxin, an impurity in the 2,4,5-T herbicide sprayed by the United States as a defoliant in Vietnam.

6.6 Water as a Solvent

Water has many unusual properties. Among these are very low vapor pressure, high heat of vaporization, and high boiling point. All of these are caused by the hydrogen-bonded structure of water. In Box 5G we noted that ice has a lower density than liquid water, a property that causes it to float on top of water.

Water covers about 75 percent of the earth's surface in the form of oceans, ice caps, glaciers, lakes, and rivers. Water vapor is always present in the atmosphere. Life evolved in water, and without it life as we know it could not exist. The human body is about 60 percent water. This water is located both inside the cells of the body (intracellular) and outside (extracellular). Most of the important chemical reactions in living tissue occur in aqueous solution; water serves as a solvent to transport reactants and products from one place in the body to another. Water is also itself a reactant or product in many biochemical reactions.

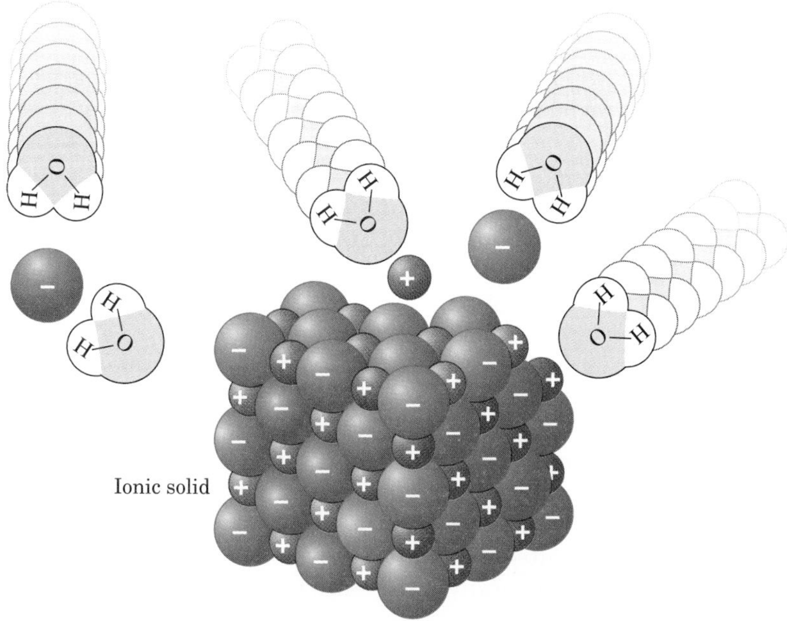

Figure **6.5** • Water molecules remove anions and cations from the surface of an ionic solid.

The properties that make water such a good solvent are its polarity and its hydrogen-bonding capacity. Water dissolves many compounds because it **solvates** ions and molecules. Solutions in which water is the solvent are called **aqueous solutions.**

How Water Dissolves Ionic Compounds

We learned in Section 3.3 that ionic compounds in the solid state are composed of a regular array of ions. The crystal is held together by ionic bonds, which are attractions between positive and negative ions. Water, of course, is a polar molecule. When a solid ionic compound is added to water, the ions at the surface of the crystal become surrounded by water molecules. The negative ions (anions) attract the positive ends of water molecules, and the positive ions (cations) attract the negative ends of water molecules (Fig. 6.5). Each ion attracts two or three water molecules, and the combined force of attraction is usually enough for the ion to be completely dislodged from its position in the crystal. Once the ion is removed from the crystal, it is *completely* surrounded by water molecules (Fig. 6.6). We call these ions **hydrated.** A more general term is **solvated.** The **solvation layer**—that is, the surrounding shell of solvent molecules—acts as a cushion. It prevents a solvated anion from colliding directly with a solvated cation and thus keeps the solvated ions in solution.

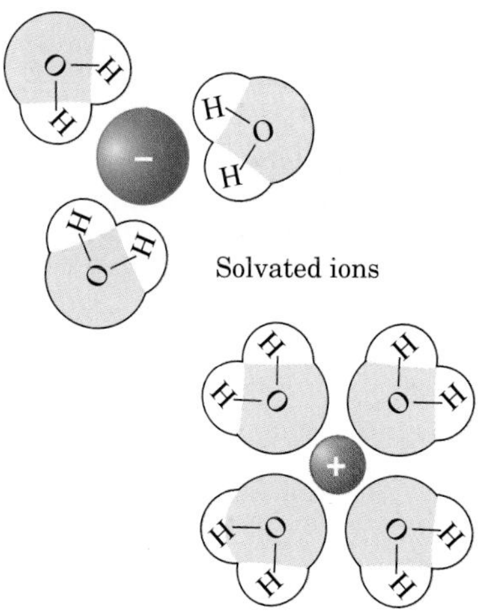

Figure **6.6**
Anions and cations solvated by water.

Not all ionic solids are soluble in water. Some—for example, AgCl and $Ca_3(PO_4)_2$—are insoluble in water. However, even "insoluble" solids have at least a tiny solubility. Some rules for predicting solubilities were given in Section 4.7.

Solid Hydrates

The attraction between ions and water molecules is so strong in some cases that water molecules are an integral part of the crystal structure of many solids. Water molecules in a crystal are called **water of hydration.** The substances that contain water in their crystals are themselves called **hydrates.** For example, gypsum and plaster of Paris are hydrates of calcium sulfate: Gypsum is $CaSO_4 \cdot 2H_2O$, calcium sulfate dihydrate, and plaster of Paris is $(CaSO_4)_2 \cdot H_2O$, calcium sulfate monohydrate.

The dot in the formula indicates that the H_2O is present in the crystal, but it is not covalently bonded to the ions of Ca^{2+} or SO_4^{2-}.

Some of these crystals hold their water tenaciously. To remove it, the crystals must be heated for some time at a high temperature (Fig. 6.7). The crystal without its water is called **anhydrous.** In many cases the anhydrous crystals are so strongly attracted to water that they take it from the water vapor in the air. That is, some anhydrous crystals become hydrated on standing in the open air. Crystals that do this are called **hygroscopic.**

If we want hygroscopic compounds to remain anhydrous, they must be placed in sealed containers that contain no water vapor.

Hydrated crystals often look different from the anhydrous forms. For example, copper sulfate pentahydrate, $CuSO_4 \cdot 5H_2O$, is blue, but the anhydrous form is white (see Fig. 6.7).

The difference between hydrated and anhydrous crystals can sometimes have an effect in the body. For example, the compound sodium urate in the anhydrous form exists as spherical crystals, but in the monohydrate the crystals are needle-shaped (Fig. 6.8). The deposition of sodium urate monohydrate in the joints (mostly in the big toe) causes gout.

Figure **6.7**
When blue hydrated copper sulfate is strongly heated in a crucible, it changes to the white anhydrous salt. (Photograph by Charles D. Winters.)

Electrolytes

Ions in water migrate from one place to another, maintaining their charge. Because of this, solutions of ions conduct electricity. They can do this because the ions in the solution migrate independently; as shown in Figure 6.9, the cations migrate to a negative electrode, called the **cathode,** and the anions migrate to a positive electrode, the **anode.** In doing so, they carry the electric current and complete the circuit initiated by the battery, and the electric bulb lights up.

Substances that conduct an electric current when dissolved in water or when in the molten state are called **electrolytes.** Sodium chloride is an electrolyte. Hydrated Na^+ ions carry positive charge and hydrated Cl^- ions carry negative charge. Substances that do not conduct electricity are called **nonelectrolytes.** Distilled water is a nonelectrolyte. The light bulb shown in Figure 6.9 does not light up if only distilled water is placed in the container. However, when tap water is placed in the beaker, the bulb lights dimly. Tap water contains enough ions to carry electricity, but the concentration of these ions is so low that only a small amount of electricity is conducted.

This experiment shows that electric conductance depends on the concentration of ions. **The higher the ion concentration, the greater the electric conductance.** However, there are differences in electrolytes. If we take a 0.1 *M* aqueous solution of NaCl and compare it with

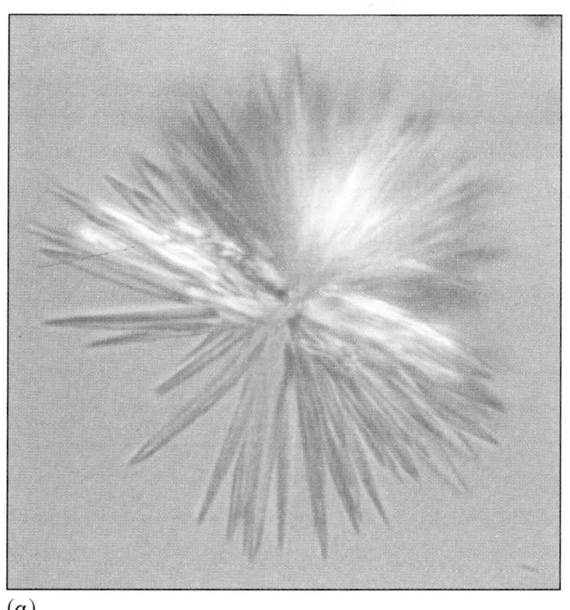

(*a*)

(*b*)

Figure **6.8** • (a) The needle-shaped sodium urate monohydrate crystals that cause gout and (b) the pain of gout as depicted by a cartoonist. (a) Courtesy of P. A. Dieppe, P. A. Bacon, A. N. Bamji, and I. Watt and the Gower Medical Publishing Co., Ltd., London, (b) courtesy of National Library of Medicine.

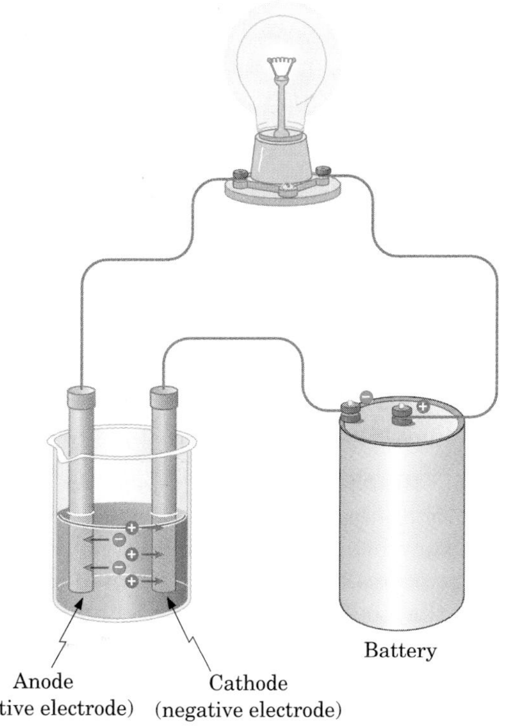

Figure **6.9** • Conductance by an electrolyte.

Box 6C

Hydrates and Air Pollution: The Decay of Buildings and Monuments

Many buildings and monuments in urban areas throughout the world are decaying, ruined by air pollution. The main culprit in this process is gypsum. The most common stones used for buildings and stone monuments are limestone and marble, both of which are largely calcium carbonate. In the absence of polluted air these can last for thousands of years, and many statues and buildings from ancient times (Babylonian, Egyptian, Greek, and others) have survived until recently with little change, and indeed still do, in rural areas. But in urban areas the air is polluted with SO_2 and SO_3, which mostly come from the combustion of coal and petroleum products containing small amounts of sulfur compounds as impurities (Box 6A). They react with the calcium carbonate at the surface of the stones to form calcium sulfate. When calcium sulfate interacts with water in the rain it forms the dihydrate gypsum. The problem is that gypsum has a larger volume than the original marble or limestone, causing the surface of the stone to expand, resulting in flaking. This is how statues such as those in the Parthenon (in Athens, Greece) become noseless and later faceless.

Effects of air pollution on statues. The photo at the left was taken at the Lincoln Cathedral in England in 1910; the one on the right in 1984. (Dean and Chapter of Lincoln.)

a 0.1 *M* HF solution, we find that the NaCl solution lights a bulb brightly, but the HF solution lights it only dimly. We might have expected the two solutions to behave similarly. They have the same concentration, 0.1 *M*, and each "molecule" provides two ions, a cation and an anion (Na^+ and Cl^-; H^+ and F^-). The reason they behave differently is that, whereas Na^+Cl^- dissociates completely to two ions (each hydrated and each moving independently), in the case of HF only a few molecules are dissociated into ions. Most of the HF molecules do not dissociate, and undisso-

Box 6D

Hypoglycemia and Potassium Balance

In the body, potassium ions are mainly concentrated inside the cells, where the concentration is about 0.158 *M*. The K^+ concentration in the blood and in the interstitial fluids is much lower (about 0.004 *M*). Among the functions of K^+ in the heart cells is maintenance of the proper contractions, and the K^+ ions in the extracellular fluids help to control nerve transmission to the muscles.

In a condition called **hypoglycemia,** the body output of insulin is elevated and blood sugar is depleted. This may suddenly shift the already small amount of K^+ from the extracellular media into the cells. The general result is lack of nerve impulses going to the muscles and to the extremities of the body. Muscular weakness and numbness of fingers and toes are symptoms of low K^+ content. Since the heartbeat is also influenced, later symptoms may include tachycardia (fast heartbeat) and, still later, weak pulse and falling blood pressure. Potassium may be obtained from food (bananas, orange and pineapple juice, veal, chicken, and pork) or, if necessary, intravenously as KCl solution. The latter is used to prevent severe lack of K^+ from causing cardiac arrest.

Sweating causes loss of K^+ ions. In hot weather, when athletes perspire heavily, they lose a large amount of K^+. That is why football and baseball players, long-distance runners, and other athletes often suffer serious muscle cramps in such weather.

ciated molecules do not conduct electricity. Compounds that dissociate completely are called **strong electrolytes,** and those that dissociate into ions only partially are called **weak electrolytes.**

Electrolytes are important components of the body because they help to maintain the acid-base balance and the water balance. The most important cations in body tissues are Na^+, K^+, Ca^{2+}, and Mg^{2+}. The most abundant anions in the body are bicarbonate, HCO_3^-; chloride, Cl^-; monohydrogen phosphate, HPO_4^{2-}; and dihydrogen phosphate, $H_2PO_4^-$.

How Water Dissolves Covalent Compounds

Water is a good solvent not only for ionic compounds; it also dissolves many covalent compounds. In a few cases, the covalent compounds dissolve because they *react* with the water. An important example is the covalent compound HCl. HCl is a gas (with a penetrating, choking odor) that attacks the eyes. When dissolved in water, the HCl molecules react with water to give ions:

$$HCl(g) + H_2O(l) \longrightarrow Cl^-(aq) + \underbrace{H_3O^+(aq)}_{\text{Hydronium ion}}$$

In aqueous solution, H^+ does not exist; it combines with water and forms hydronium ion, H_3O^+.

Another example is the gas sulfur trioxide, which reacts as follows:

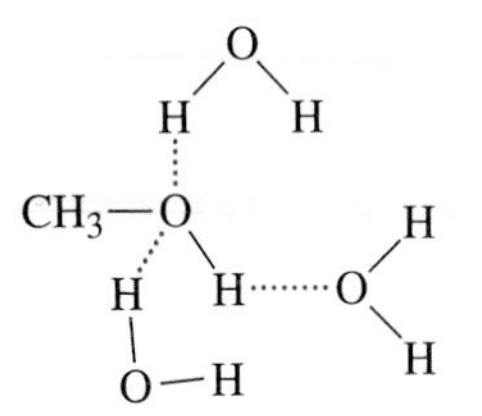

$$SO_3(g) + 2H_2O(l) \longrightarrow HSO_4^-(aq) + H_3O^+(aq)$$

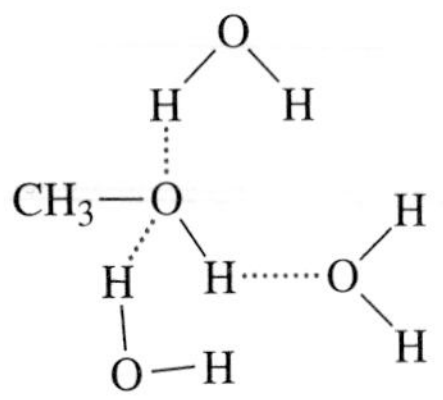

Figure **6.10**
Solvation of a covalent compound by water. The dotted lines represent hydrogen bonds.

Because HCl and SO_3 are completely converted to ions in dilute aqueous solution, these solutions are ionic solutions and behave just as other electrolytes do (they conduct a current) even though HCl and SO_3 are themselves covalent compounds, unlike salts such as NaCl.

However, most covalent compounds that dissolve in water do not react with water. They dissolve because the water molecules surround the entire covalent molecule and solvate it. For example, when methanol, CH_3OH, is dissolved in water, the methanol molecules are solvated by the water molecules (Fig. 6.10).

There is a simple way to predict which covalent compounds dissolve in water and which do not. **Covalent compounds dissolve in water if they can form hydrogen bonds with water, provided that the molecules are fairly small.** You may recall from Section 5.7 that hydrogen bonding is possible between two molecules if one of them contains an O, N, or F atom and the other contains an O—H, N—H, or F—H bond. Every water molecule contains an O atom and O—H bonds. Therefore, **water can form hydrogen bonds to any molecule that contains an O, N, or F atom or an O—H, N—H, or F—H bond.** If these molecules are small enough, they will be soluble in water. How small? In general, they should have no more than about three carbon atoms for each O, N, or F.

For example, acetic acid, CH_3COOH, is soluble in water, but benzoic acid, C_6H_5COOH, is not. Similarly, ethyl alcohol, C_2H_5OH, is soluble, but dipropyl ether, $C_6H_{14}O$, is not. Table sugar, $C_{12}H_{22}O_{11}$, is very soluble in water. Though it contains a large number of carbon atoms, it has so many oxygen atoms that it forms many hydrogen bonds with water molecules, and thus this molecule is very well solvated. **Covalent molecules that do not contain O, N, or F atoms are almost always insoluble in water,** except in the rare cases where they react with water. For example, methyl alcohol, CH_3OH, is infinitely soluble in water, but methyl chloride, CH_3Cl, is not.

Water in the Body

Water is important in the body not only because it dissolves ionic substances as well as some covalent compounds, but also because it hydrates all polar molecules in the body. It thus serves as a vehicle to carry most of the organic compounds, nutrients, and fuels the body uses, as well as waste material. Blood and urine are but two examples of aqueous body fluids.

Beyond that, the hydration by water of macromolecules such as proteins, nucleic acids, and polysaccharides allows the proper motions within these molecules.

6.7 Colloids

$1\ nm = 1 \times 10^{-9}\ m$

Up to now, we have been discussing solutions. In a true solution the maximum diameter of the solute particles is about 1 nm. If the diameter of

Table 6.2 Types of Colloidal Systems

Type	Example
Gas in gas	None
Gas in liquid	Whipped cream
Gas in solid	Marshmallows
Liquid in gas	Clouds, fog
Liquid in liquid	Milk, mayonnaise
Liquid in solid	Cheese, butter
Solid in gas	Smoke
Solid in liquid	Jelly, plasma extender
Solid in solid	Dried paint

the solute particles is larger than this, we no longer have a true solution—we have a colloid. In a **colloid** (or **colloidal dispersion**), the diameter of the solute particles ranges from about 1 to 1000 nm. Particles of this size usually have a very large surface area, and that accounts for the two basic characteristics of colloidal systems:

1. They scatter light and therefore appear turbid, cloudy, or milky.

2. Although colloidal particles are large, they form stable dispersions; that is, they do not form separate phases that settle out.

Like true solutions, colloids can exist in a variety of phases (Table 6.2).

All colloids exhibit the following characteristic effect. When we shine light through a colloidal system and look at the system from a 90° angle, we see the pathway of the light without seeing the individual colloidal particles (Fig. 6.11). This is called the **Tyndall effect** and is due to light scattering. Smoke, starch solutions, and fog, to name a few examples, all exhibit the Tyndall effect. We are all familiar with the sunbeams that can be seen when sunlight passes through dusty air. This, too, is an example of the Tyndall effect. Ordinary solutions do not show this effect.

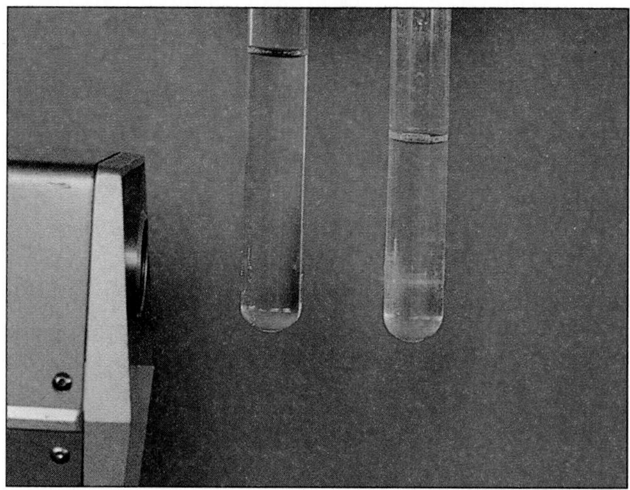

Figure **6.11**
The Tyndall effect. The test tube on the left contains a true solution; the one on the right, a colloidal solution of starch. A laser beam, generated by the apparatus at the far left, passes through both tubes. In the colloidal solution the beam leaves a trail of scattered light. In the true solution it is invisible. (Photograph by Beverly March.)

Ordinary household dust particles, magnified 2200 times. (Courtesy of David Scharf.)

Colloidal systems are stable. Mayonnaise stays emulsified and does not separate into oil and water. When the size of colloidal particles is larger than about 1000 nm, the system is unstable and separates into phases. Such systems are called **suspensions.** For example, if we take a lump of soil and disperse it in water, we get a muddy suspension. The soil particles are anywhere from 10^3 to 10^6 nm in diameter. The muddy mixture scatters light and therefore appears turbid, but it is not a stable system. If left alone, the soil particles soon settle to the bottom, with clear water above the sediment. Therefore, soil in water is a suspension, not a colloidal system.

The properties of true solutions, suspensions, and colloids are summarized in Table 6.3.

What makes a colloidal dispersion stable? To answer this question, we must first realize that the colloidal particles are in constant motion. Just look at the dust particles dancing in a ray of sunlight that enters your room. Actually you are not seeing the dust particles themselves; they are too small to see. What you are seeing are flashes of scattered light. The motion of the dust particles dispersed in air is a random, chaotic motion. It is the motion of any colloidal particle suspended in a solvent. This is called **Brownian motion** (Fig. 6.12).

Brownian motion is named for Robert Brown (1773–1858), the Scottish botanist who first observed it in 1827.

The cause of Brownian motion is the buffeting and collisions the colloidal particles suffer from the solvent molecules (in the case of the dust particles, the solvent is air). This constant Brownian motion creates favorable conditions for collisions between colloidal particles. When such large particles collide, they stick together, combine to give larger particles, and finally settle out of the solution.

So why do colloidal particles remain in solution? For two reasons:

1. Most colloidal particles carry a large *solvation layer.* If the solvent is water, as in the case of protein molecules in the blood, the colloidal particles are surrounded by a large number of water molecules, which move together with the colloidal particles and cushion them. When two colloidal particles collide as a result of Brownian motion, they do not actually touch each other; only the solvent layers collide. This way, the particles do not stick together and thus stay in solution.

2. The large surface area of colloidal particles acquires *charges* from the solution. All colloids in a particular solution acquire the same kind of charge, for example, negative charge. This leaves a net positive charge

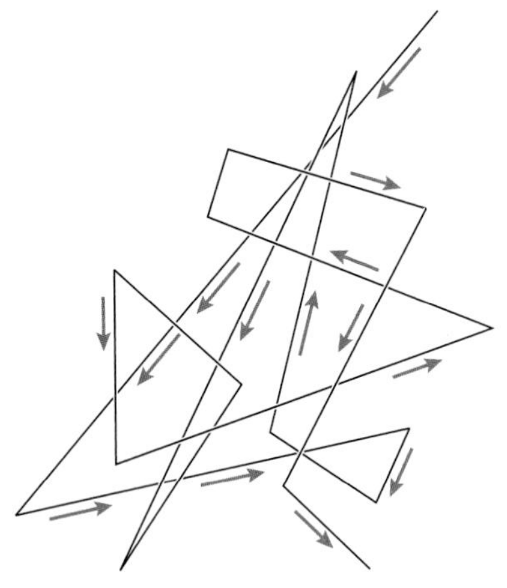

Figure **6.12**
Brownian motion.

Table 6.3 Properties of Three Types of Mixtures

Property	**Solutions**	**Colloids**	**Suspensions**
Particle size (nm)	0.1–1.0	1–1000	>1000
Filterable with ordinary paper	No	No	Yes
Homogeneous	Yes	Borderline	No
Settles on standing	No	No	Yes
Behavior to light	Transparent	Tyndall effect	Translucent or opaque

Box 6E

Emulsions and Emulsifying Agents

Oil and water do not mix. Even when we stir them vigorously and the oil droplets are dispersed in the water, the two phases separate as soon as we stop stirring. However, there are a number of stable colloidal systems made of oil and water; these are **emulsions.** For example, the oil droplets in milk are dispersed in aqueous solution. This is possible because in milk there is a protective colloid, the milk protein called casein. The casein molecules surround the oil droplets, and since they are polar and carry a charge, they protect and stabilize the oil droplets. Casein is thus an **emulsifying agent.** A similar emulsifying agent is egg yolk, which is the ingredient in mayonnaise that coats the oil droplets and prevents them from separating.

in the dispersion medium. When a charged colloidal particle encounters another colloidal particle, the two repel each other because of their like charge.

Thus, the combined effect of solvation layer and surface charge keeps colloidal particles in a stable dispersion. By using these effects, chemists can either increase or decrease the stability of a colloidal system. If we want to get rid of a colloidal dispersion, we can remove either the solvation layer or the surface charge, or both. For example, proteins in the blood form a colloidal dispersion. If we want to isolate a protein from blood, we may want to precipitate it. We can accomplish this in two ways: either remove the hydration layer or remove the surface charges. If we add solvents such as ethanol or acetone, which have great affinity for water, the water will be removed from the solvation layer of the protein, and then the unprotected protein molecules will stick together when they collide and form a sediment. Similarly, by adding electrolytes such as NaCl to the solution, we can remove the charges from the surface of the proteins (by a mechanism that is too complicated to discuss here). Without their protective charges, two protein molecules no longer repel each other. Instead, when they collide they stick together and precipitate from the solution.

6.8 Colligative Properties. Freezing-Point Depression

Certain properties of solutions depend only on the *number* of solute particles dissolved in a given amount of solvent and not on the *nature* of these particles. Such properties are called **colligative properties.** For example, one such property is **freezing-point depression.** One mole (6.02×10^{23} particles) of *any* solute dissolved in 1000 g of water lowers the freezing point of the water by 1.86°C. The *nature* of the solute does not matter, only the number of particles.

Note that in preparing a solution for this purpose we do not use molarity. That is, we do not need to measure the total volume of the solution.

This principle is used in a number of practical ways. In the winter we use salts (sodium chloride and calcium chloride) to melt snow and ice on our streets. The salts dissolve in the melting snow and ice, and this lowers the freezing point of the water. Another application is the use of antifreeze in automobile radiators. Since water expands on freezing (Box 5G), the water in a car's cooling system can crack the engine block of a parked car when the outside temperature falls below 0°C. The addition of antifreeze prevents this because it makes the water freeze at a much lower temperature. The most common antifreeze is ethylene glycol, $C_2H_6O_2$.

EXAMPLE

If we add 275 g of ethylene glycol per 1000 g of water in a car radiator, what will the freezing point of the solution be?

Answer • Since $C_2H_6O_2$ has a molecular weight of 62.0, 275 g contains

$$275\ \cancel{g}\ C_2H_6O_2 \times \frac{1\ \text{mole}\ C_2H_6O_2}{62.0\ \cancel{g}\ C_2H_6O_2} = 4.44\ \text{moles}\ C_2H_6O_2$$

Each mole lowers the freezing point by 1.86°C, so the freezing point is lowered by

$$\frac{1.86°\text{C}}{1\ \cancel{\text{mole}}} \times 4.44\ \cancel{\text{moles}} = 8.26°\text{C}$$

The freezing point of the water will be lowered from 0°C to −8.26°C, and the radiator will not crack if the outside temperature remains above −8.26°C (17.1°F).

Problem **6.9**

If we add 215 g of CH_3OH to 1000 g of water, what will be the freezing point of the solution?

If a solute is ionic, then each mole of solute dissociates (breaks up) to more than 1 mole of particles. For example, if we dissolve 1 mole (58.5 g) of NaCl in 1000 g of water, the solution contains 2 moles of solute particles: 1 mole each of Na^+ and Cl^-. The freezing point of water will be lowered by *twice* 1.86°C, or 3.72°C.

EXAMPLE

What will be the freezing point of the solution if we dissolve 1 mole of K_2SO_4 in 1000 g of water?

Answer • One mole of K_2SO_4 dissociates to produce 3 moles of ions: 2 moles of K^+ and 1 mole of SO_4^{2-}. The freezing point will be lowered by 3 × 1.86°C = 5.58°C, and the solution will freeze at −5.58°C.

Problem **6.10**

What will be the freezing point of the solution if 2.5 moles of $Al(NO_3)_3$ is added to 1000 g of water?

Besides freezing-point depression, there are several other colligative properties, including vapor-pressure lowering, boiling-point elevation, and osmotic pressure. We discuss only the last of these because biologically it is the most important.

6.9 Osmotic Pressure

In order to understand osmotic pressure, let us consider the beaker shown in Figure 6.13(b). In this beaker two compartments are separated by an osmotic semipermeable membrane. A **semipermeable membrane** is a thin slice of some material, such as cellophane, that contains very tiny holes (far too small for us to see) that are big enough to let small solvent molecules pass through, but not big enough to let large solute molecules pass. In the right-hand compartment of the beaker is pure solvent—in this example, water; in the left-hand compartment is a solution of sugar in water.

An osmotic membrane is a very selective semipermeable membrane that allows only solvent molecules, and nothing else, to pass through.

The sugar molecules are too big to go through the membrane and thus are forced to remain in the left-hand compartment. The water molecules, however, easily go back and forth; as far as they are concerned, there is no membrane. Therefore, water molecules go from the right to the left side because the concentration of water in the right-hand compartment is higher (100%) than in the left-hand compartment. Molecules

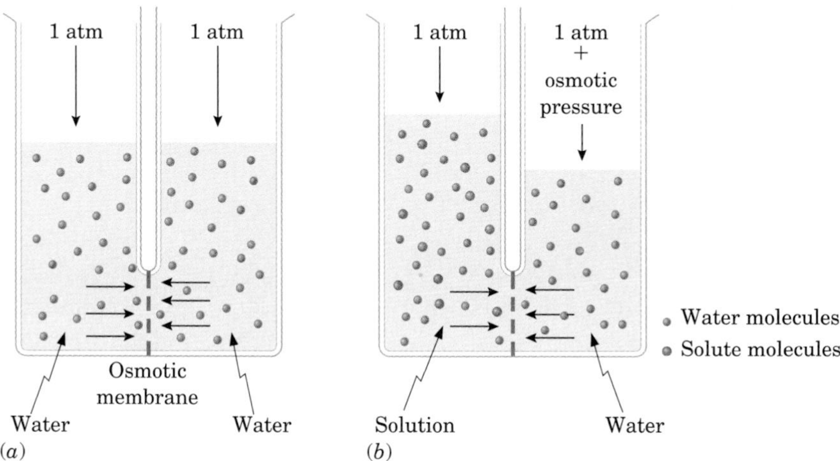

Figure **6.13** • Osmotic pressure. (a) Two compartments separated by an osmotic semipermeable membrane both contain only solvent molecules that can pass through the membrane. (b) The compartment on the right contains only solvent, the one on the left both solute and solvent. Solute molecules cannot pass through the membrane. The solvent molecules move to the left compartment in an effort to dilute the solution, raising the liquid level on that side.

will always diffuse from an area of higher concentration to one that is lower, and that is what the water molecules do. If there were no membrane separating the two solutions, the sugar molecules would also follow this rule, moving from the higher concentration area (in this case the left side) to the lower one. The membrane, of course, prevents the movement of the sugar molecules, but does not hinder the water molecules from moving into the left compartment. **This passage of solvent molecules from the dilute to the concentrated side of a semipermeable membrane** is called **osmosis.** When a significant number of water molecules have moved from the right to the left side, the liquid level on the right side goes down and that on the left side goes up, as shown in Figure 6.13(b).

The process cannot continue indefinitely because gravity prevents the difference in levels from becoming too great. Eventually the process stops with the levels unequal. The levels can be made equal again if we apply an external pressure to the higher side. **The amount of external pressure necessary to equalize the levels** is called the **osmotic pressure.**

Some ions are small but still do not go through the membrane because they are solvated by a shell of water molecules (see Fig. 6.6).

Although this discussion assumes that one compartment contained pure solvent, the same principle applies if both compartments contain solutions, as long as the concentrations are different. The solution of higher concentration always has a higher osmotic pressure than the one of lower concentration, which means that *the flow of solvent is always from the more dilute to the more concentrated side.* Of course, it is the number of particles that matters, so we must remember that, in ionic solutions, each mole of solvent gives rise to more than 1 mole of particles. For convenience in calculation, we define a new term, **osmolarity (osmol),** as **molarity multiplied by the number of particles produced by each molecule of solute.**

EXAMPLE

A 0.89 percent w/v NaCl solution is referred to as a *physiological saline solution* because it has the same concentration of salts as normal human blood, though blood contains several salts and saline solution has only NaCl. What is the osmolarity of this solution?

Answer • 0.89% w/v NaCl = 8.9 g NaCl in 1 L of solution

$$\frac{8.9\ \text{g NaCl}}{1\ \text{L solution}} \times \frac{1\ \text{mole NaCl}}{58.5\ \text{g NaCl}} = \frac{0.15\ \text{mole NaCl}}{\text{L solution}} = 0.15\ M$$

Osmolarity = 0.15 × 2 = 0.30 osmol

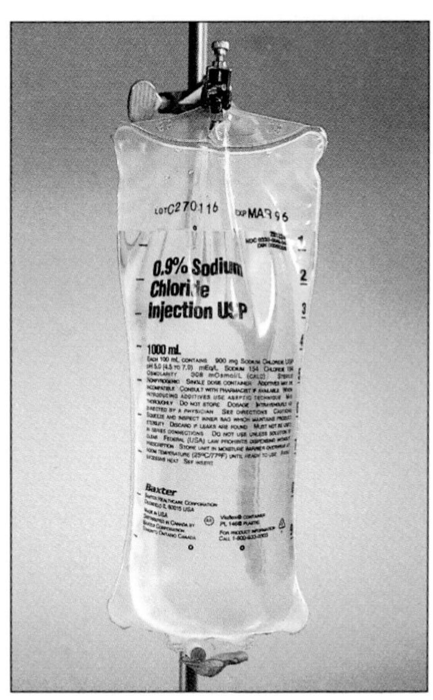

An isotonic saline solution. (Photograph by Charles D. Winters.)

Problem **6.11**

What is the osmolarity of a 3.3 percent w/v Na_3PO_4 solution?

Osmotic pressure is a colligative property. The osmotic pressure generated by a solution across the semipermeable membrane—the difference between the heights of the two columns in Figure 6.13(b)—depends on

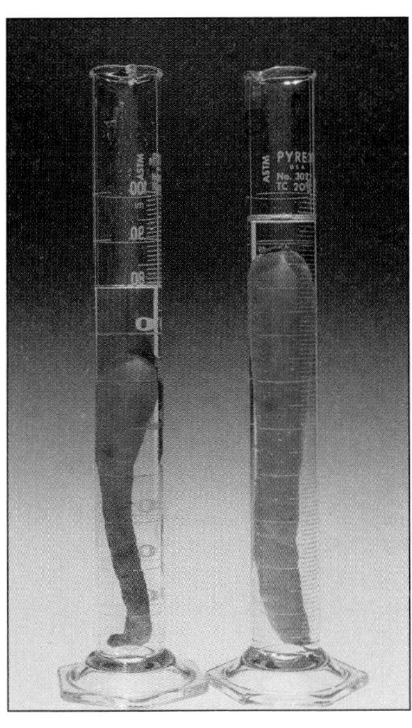

The effect of osmosis on a carrot. The carrot on the left is in a solution that contains a high concentration of NaCl. Water flowing out of the carrot has caused it to shrink. The carrot on the right, in pure water, is slightly swollen. (Photograph by Charles D. Winters.)

the osmolarity of the solution. If the osmolarity increases by a factor of 2, the osmotic pressure will also increase by a factor of 2.

Osmotic pressure is very important in biological organisms because cell membranes are semipermeable membranes. Therefore, biological fluids must have the proper osmolarity. For example, the red blood cells in the body are suspended in a medium called plasma that must have the same osmolarity as the red blood cells. **Two solutions with the same osmolarity** are called **isotonic,** and plasma is said to be isotonic with the red blood cells. This means that no osmotic pressure is generated across the cell membrane.

What would happen if we suspended red blood cells in distilled water instead of in plasma? Inside the red blood cells the osmolarity is about the same as in a physiological saline solution, 0.30 osmol. Distilled water has zero osmolarity. As a consequence, water flows into the red blood cells. The volume of the cells increases, and the cells swell, as shown in Figure 6.14(b). The membrane cannot resist the osmotic pressure, and the red blood cells eventually burst, spilling their contents into the water. We call this process **hemolysis.**

Solutions with an osmolarity (and hence osmotic pressure) lower than that of suspended cells are called **hypotonic solutions.** Obviously, it is very important that in intravenous feeding and blood transfusion we always use *isotonic* solutions and never *hypotonic* solutions. The latter would simply kill the red blood cells by hemolysis.

Iso-osmotic, hypo-osmotic, and hyper-osmotic are recent synonyms for isotonic, hypotonic, and hypertonic, respectively.

Equally important, we should not use *hypertonic* solutions either. A **hypertonic solution** has a greater osmolarity (and greater osmotic pressure) than the reference cells. If red blood cells are placed in a hypertonic solution (for example, 0.5 osmol glucose solution), water flows from the cells into the glucose solution through the semipermeable cell membrane. This process, called **crenation,** shrivels the cells, as shown in Figure 6.14(c).

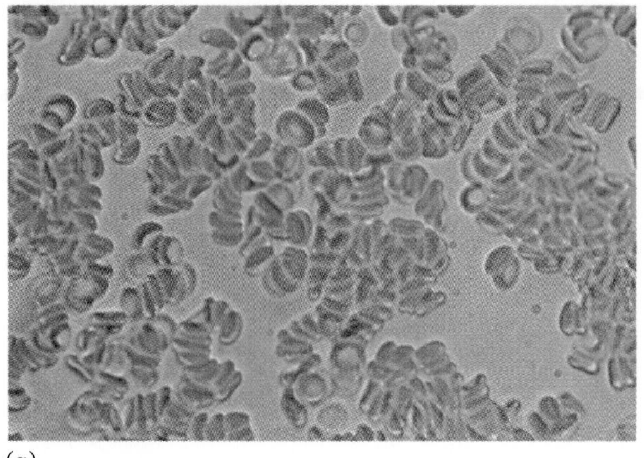

(*a*)

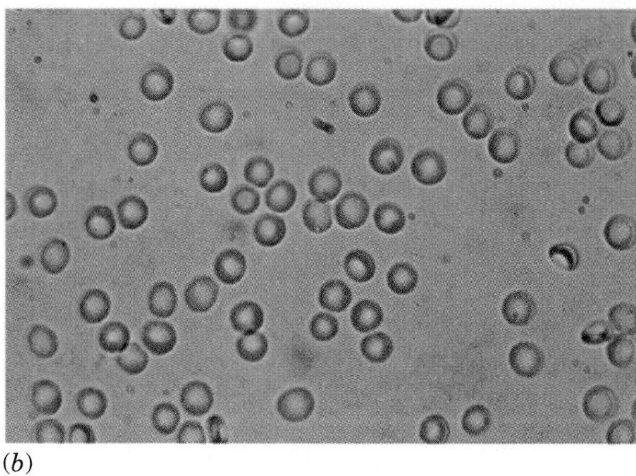

(*b*)

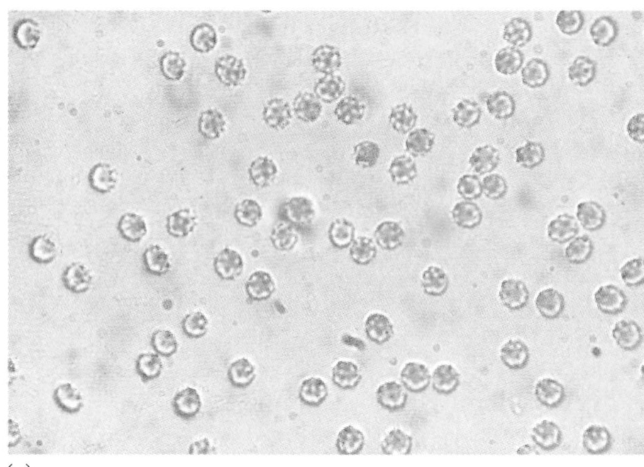

(*c*)

Figure **6.14** • Red blood cells: (a) in an isotonic solution, (b) in a hypotonic solution, (c) in a hypertonic solution. (Peter Arnold, Inc.,/© Philip A. Harrington.)

A 5.5 percent glucose solution is also isotonic and is used in intravenous feeding.

As already mentioned, 0.89 percent NaCl (physiological saline) is isotonic with red blood cells and is used in intravenous injections.

EXAMPLE

Is a 0.50% solution of KCl (a) hypertonic (b) hypotonic or (c) isotonic, compared to red blood cells?

Answer • 0.50% KCl contains 5.0 g KCl in 1.0 L of solution.

$$\frac{5.0\text{ g }\cancel{\text{KCl}}}{1.0\text{ L soln}} \times \frac{1.0\text{ mole KCl}}{74.6\text{ g }\cancel{\text{KCl}}} = \frac{0.067\text{ mole KCl}}{\text{L}} = 0.067\ M$$

The osmolarity is 0.067 × 2 = 0.13 osmol; this is smaller than the osmolarity of the red blood cells which is 0.30 osmol. Therefore, the KCl solution is hypotonic.

Problem **6.12**

Which solution is isotonic compared to red blood cells: (a) 0.1 (b) 1.0 (c) 0.2 M Na_2SO_4?

6.10 Dialysis

In osmosis there is an osmotic semipermeable membrane that allows only solvent and not solute molecules to pass. If, however, the openings in the membrane are somewhat larger, then *small* solute molecules can also get through, but large solute molecules, such as macromolecular and colloidal particles, cannot. This process is called **dialysis.** For example, ribonucleic acid is an important biological molecule that we will study in Chapter 23. When biochemists prepare ribonucleic acid solutions, they must remove small molecules, such as NaCl and ethyl alcohol, from the solution in order to get a pure nucleic acid preparation. This is done by placing the nucleic acid solution in a dialysis bag (cellophane) of sufficient pore size to allow all the small molecules to diffuse and retain only the large nucleic acid molecules. If the dialysis bag is suspended in flowing distilled water, all the NaCl and small molecules leave the bag and, after a certain amount of time, it contains only the pure nucleic acids dissolved in water.

This is also the way our kidneys work. The millions of nephrons, or kidney cells, have very large surface areas in which the capillaries of the blood vessels are in contact with the nephrons. The kidneys serve as a gigantic filtering machine. The waste products of the blood dialyse out through semipermeable membranes in the glomeruli and enter collecting tubes that carry the urine to the ureter. Large protein molecules and cells are retained in the blood.

The glomeruli of the kidney are fine capillary blood vessels in which the body's waste products are removed from the blood.

Box 6F

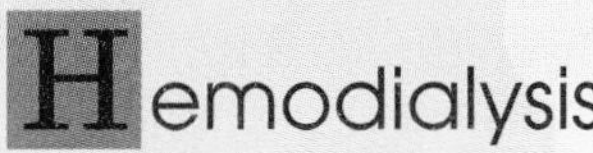

The kidneys' main function is to remove toxic waste products from the blood. When the kidneys are not functioning properly, these waste products may threaten life. Hemodialysis is an artificial process that performs the same filtration function (see Figure 25.5).

The patient's blood is circulated through a long tube of cellophane membrane suspended in an isotonic solution and then returned to the patient's vein. The cellophane membrane retains all the large molecules (for example, proteins) but allows the small ones, including the toxic wastes, to pass through. Thus, the wastes are removed from the blood by dialysis.

If the cellophane tube were suspended in distilled water, other small molecules and ions, such as Na^+, Cl^-, and glucose, would also be removed from the blood. The isotonic solution used in hemodialysis consists of 0.6 percent NaCl, 0.04 percent KCl, 0.2 percent $NaHCO_3$, and 0.72 percent glucose (all w/v). In this manner, no glucose or Na^+ is lost from the blood.

A patient is usually on an artificial kidney machine for 4 to 7 hours. During this time the isotonic bath is changed every 2 hours. Kidney machines allow people with kidney failure to lead a fairly normal life, though they must take these hemodialysis treatments regularly.

SUMMARY

Systems containing more than one component are **mixtures. Homogeneous mixtures** are uniform throughout. **Heterogeneous mixtures** exhibit well-defined boundaries between phases. When a solution consists of a solid or gas dissolved in a liquid, the liquid is the **solvent** and the solid or gas is the **solute.** The **solubility** of a substance is the maximum amount of substance that dissolves in a given amount of solvent at a given temperature.

"Like dissolves like" means that polar molecules are soluble in polar solvents and nonpolar molecules in nonpolar solvents. The solubility of solids and liquids in liquids usually increases with temperature; the solubility of gases usually decreases with an increase in temperature. Percent concentration is given in either weight per unit volume (**w/v**) or weight per unit weight (**w/w**). **Molarity** is the number of moles of solute in 1 L of solution. Water is the most important solvent because it dissolves polar compounds and ions through hydrogen bonding and dipole interactions. Hydrated ions are surrounded by water molecules (hydration layer) that move together with the ion and cushion it from collisions with other ions. Aqueous solutions of ions and molten salts are **electrolytes** and conduct electricity.

Colloids exhibit a scattering of light called the **Tyndall effect.** Colloids are stable mixtures in spite of the large size of the colloidal particles (1 to 1000 nm). The stability is due to the solvation layer that cushions the colloids from direct collisions and to an electric charge on the surface of colloidal particles.

Colligative properties are properties that depend only on the number of solute particles present. **Osmotic pressure** is one of the most important colligative properties. Osmotic pressure operates across an osmotic semipermeable membrane that allows only solvent molecules to pass but screens out all other molecules. In osmotic pressure calculations, concentration is measured in **osmolarity,** which is the molarity of the solution multiplied by the number of particles produced by dissociation of the solute. Red blood cells in **hypotonic solution** swell and burst; this is called **hemolysis.** Red blood cells in **hypertonic solution** shrink; this is called **crenation.** Some semipermeable membranes allow small solute molecules to pass along with solvent molecules. In **dialysis,** such membranes are used to separate large molecules from smaller ones.

KEY TERMS

Acid rain (Box 6A)
Anhydrous (Sec. 6.6)
Anode (Sec. 6.6)
Aqueous solution (Sec. 6.6)
Brownian motion (Sec. 6.7)
Cathode (Sec. 6.6)
Colligative property (Sec. 6.8)
Colloid (Sec. 6.7)
Concentrated (Sec. 6.5)
Crenation (Sec. 6.9)
Dialysis (Sec. 6.10)
Dilute (Sec. 6.5)
Electrolyte (Sec. 6.6)
Emulsion (Box 6E)
Freezing-point depression (Sec. 6.8)
Hemodialysis (Box 6F)
Hemolysis (Sec. 6.9)
Henry's law (Sec. 6.4)
Heterogeneous mixture (Sec. 6.1)
Homogeneous mixture (Sec. 6.1)
Hydrate (Sec. 6.6)
Hygroscopic (Sec. 6.6)
Hypertonic (Sec. 6.9)
Hypoglycemia (Box 6D)
Hypotonic (Sec. 6.9)
Isotonic (Sec. 6.9)
Miscible (Sec. 6.4)
Molarity (Sec. 6.5)
Osmolarity (Sec. 6.9)
Osmosis (Sec. 6.9)
Osmotic membrane (Sec. 6.9)
Osmotic pressure (Sec. 6.9)
Parts per million (Sec. 6.5)
Percent concentration (Sec. 6.5)
Saturated solution (Sec. 6.4)
Seeding (Sec. 6.4)
Semipermeable membrane (Sec. 6.9)
Solubility (Sec. 6.4)
Solute (Sec. 6.2)
Solution (Sec. 6.1)
Solvated ions (Sec. 6.6)
Solvent (Sec. 6.2)
Supersaturated solution (Sec. 6.4)
Suspension (Sec. 6.7)
Tyndall effect (Sec. 6.7)
Unsaturated solution (Sec. 6.4)
Water of hydration (Sec. 6.6)

PROBLEMS

Solutions and Their Characteristics

6.13 What is the difference between a homogeneous and a heterogeneous mixture?

6.14 A solution is made by dissolving glucose in water. Which is the solvent and which the solute?

6.15 In each of the following tell whether the solutes and solvents are gases, liquids, or solids:
(a) bronze (see Box 2D) (c) car exhaust
(b) cup of coffee (d) champagne

6.16 Give a familiar example of solutions of each of these types:
(a) liquid in liquid (c) gas in liquid
(b) solid in liquid (d) gas in gas

6.17 Are mixtures of gases true solutions or heterogeneous mixtures? Explain.

Solubility

6.18 If two liquids, such as benzene and toluene, are completely soluble in each other, can they form a supersaturated solution? Explain.

6.19 The solubility of a compound is 2.5 g in 100 mL of aqueous solution at 25°C. If you put 1.12 g of the compound in a 50-mL volumetric flask at 25°C and add sufficient water to fill it to the 50-mL mark, what kind of solution do you get, saturated or unsaturated?

6.20 To a separatory funnel with two layers, the nonpolar diethyl ether and the polar water, is added a small amount of solid. After shaking the separatory funnel, in which layer will you find each of the following solids? (a) NaCl (b) camphor (c) KOH

6.21 On the basis of polarity and of hydrogen-bonding capacity, list the following in order of increasing solubility in water:
(a) $CH_3—O—CH_3$
(b) $CH_3—CH_2—CH_2—CH_3$
(c) $(CH_3)_2S{=}O$
(d) $HO—CH_2—CH_2—OH$

6.22 Suppose you are trying to dissolve 2.0 g of a certain solid drug in water. You add it to 100 mL of water and part, but not all, of it dissolves. Give two procedures you might try to dissolve the entire 2.0 g.

6.23 How can you convert a saturated solution at room temperature to a supersaturated one?

6.24 The solubility of aspartic acid in water is 0.500 g in 100 mL at 25°C. If you dissolve 0.251 g aspartic acid in 50.0 mL water at 50°C and let the solution cool to 25°C without stirring, shaking, or otherwise disturbing the solution, would you get a saturated, unsaturated, or supersaturated solution?

6.25 Near power plants, warm water is discharged into a river. Sometimes, dead fish are observed in the area. Why do fish die from the warm water?

6.26 If a bottle of beer, after being opened, is allowed to stand for several hours, it becomes "flat" (it loses CO_2). Explain.

6.27 Would you expect the solubility of ammonia gas in water at 2 atm pressure to be: (a) greater (b) the same or (c) smaller than at 0.5 atm pressure?

Concentration Units

6.28 Describe exactly how you would make
(a) 250.0 g of a 34.6 percent w/w solution of KCl in water
(b) 156.00 g of a 4.00 percent w/w solution of NaOH in water
(c) 300.00 g of a 0.32 percent w/w solution of formic acid (HCOOH) in water

6.29 Describe exactly how you would make
(a) 280 mL of a 27 percent v/v solution of ethyl alcohol, C_2H_5OH, in water
(b) 435 mL of a 1.8 percent v/v solution of ethyl acetate, $C_4H_8O_2$, in water
(c) 1.65 L of an 8.00 percent v/v solution of benzene C_6H_6, in chloroform, $HCCl_3$

6.30 Describe exactly how you would make
(a) 250 mL of a 3.6 percent w/v solution of NaCl in water
(b) 625 mL of a 4.9 percent w/v solution of glycine, $C_2H_5NO_2$, in water
(c) 43.5 mL of a 13.7 percent w/v solution of Na_2SO_4 in water
(d) 518 mL of a 2.1 percent w/v solution of acetone, C_3H_6O, in water

6.31 Calculate the w/v percentage of each of these solutes:
(a) 15.3 g NaCl in 825 mL of solution
(b) 72 g sucrose in 1.39 L of solution
(c) 2.35 g $Ba(NO_3)_2$ in 415 mL of solution

6.32 Describe how you would prepare 250 mL of 0.10 *M* NaOH from solid NaOH and water.

6.33 Assuming that the appropriate volumetric flasks are available, describe exactly how you would make
(a) 175 mL of a 1.14 *M* solution of NH_4Br in water
(b) 1.35 L of a 0.825 *M* solution of NaI in water
(c) 330 mL of a 0.16 *M* solution of ethanol, C_2H_5OH, in water

6.34 What is the molarity of each of these solutions?
(a) one made by dissolving 47 g of KCl in enough water to give 375 mL of solution
(b) one made by dissolving 82.6 g of sucrose, $C_{12}H_{22}O_{11}$, in enough water to give 725 mL of solution
(c) one made by dissolving 9.3 g of $(NH_4)_2SO_4$ in enough water to give 2.35 L of solution

6.35 A teardrop with a volume of 0.5 mL contains 5.0 mg NaCl. What is the molarity of the NaCl in the teardrop?

6.36 What is the molarity of a 2.0 percent w/v HCl solution?

6.37 If 79.49 g of H_2SO_4 is dissolved in enough solvent to make 2.00 L of solution, what is the molarity?

6.38 If 3.18 g of $BaCl_2$ is dissolved in enough solvent to make 500.0 mL of solution, what is the molarity?

6.39 In the reaction

$$CO_2(g) + H_2O(l) \longrightarrow H_2CO_3(aq)$$

1.25 g of CO_2 was added to enough water to make 50.0 mL of solution. What is the molarity of the H_2CO_3 formed? Assume the reaction goes to completion.

6.40 If 0.456 g of NaF is dissolved in enough water to make 100 mL of solution, what is the molarity of the solution?

***6.41** A student has a bottle labeled 0.750% albumin solution. The bottle contained exactly 5.00 mL. How much water must the student add to make the concentration of albumin 0.125%?

6.42 How many grams of solute are present in each of the following aqueous solutions?
(a) 575 mL of a 2.00 *M* solution of HNO_3
(b) 1.65 L of a 0.286 *M* solution of alanine, $C_3H_7NO_2$
(c) 320 mL of a 0.0081 *M* solution of $CaSO_4$

6.43 A student has a stock solution of 30.0 percent w/v H_2O_2. Describe how she should prepare 250 mL of a 0.25 percent H_2O_2 (hydrogen peroxide) solution.

6.44 To make 5.0 L of a fruit punch that will contain 10% v/v ethyl alcohol how much 95 percent ethyl alcohol v/v must be mixed with how much fruit juice?

6.45 Describe exactly how you would prepare each of the following aqueous solutions, in each case by diluting a more concentrated solution:
(a) 550 mL of a 0.90 *M* solution of HNO_3, starting with 4.0 *M* HNO_3
(b) 1.75 L of a 2.5 *M* solution of KOH, starting with 12.0 *M* KOH
(c) 385 mL of a 0.0095 *M* solution of glycine, $C_2H_5NO_2$, starting with 2.0 *M* glycine

6.46 Analysis of a batch of concentrated HCl has shown it to contain 0.50 mg Fe in 200 g of acid. How much is this in ppm?

***6.47** Dioxin is considered to be poisonous in concentrations above 2 ppb. If a lake containing 1×10^7 L has been contaminated by 0.1 g of dioxin, did the concentration reach a dangerous level?

Water as a Solvent

6.48 How is an anion such as Cl^- solvated by water?

6.49 Define:
(a) hygroscopic
(b) water of hydration
(c) electrolyte
(d) anhydrous

6.50 Considering polarities, electronegativities, and similar concepts learned in Chapter 3, classify the following as strong electrolyte, weak electrolyte, or nonelectrolyte:
(a) KCl
(b) C_2H_5OH (ethanol)
(c) NaOH
(d) HF
(e) $C_6H_{12}O_6$ (glucose)

6.51 Which would produce the brightest light in the conductance apparatus in Figure 6.9?
(a) 0.1 *M* KCl
(b) 0.1 *M* $(NH_4)_3PO_4$
(c) 0.5 *M* sucrose

6.52 Ethanol, C_2H_6O, is very soluble in water. Describe how water dissolves ethanol.

6.53 Predict which of these covalent compounds is soluble in water:
(a) C_2H_6
(b) CH_3OH
(c) HF
(d) NH_3
(e) CCl_4

Colloids

6.54 If the dust in the air of a room shows the Tyndall effect, what is the average size of the dust particles?

6.55 On the basis of Tables 6.1 and 6.2, classify the following systems as homogeneous, heterogeneous, or colloidal mixtures:
(a) physiological saline solution
(b) orange juice
(c) a cloud
(d) wet sand
(e) suds
(f) milk

6.56 Table 6.2 shows no examples of a gas-in-gas colloidal system. Considering the definition of a colloid, explain why this is so.

6.57 What makes the water molecules in the solvation layer of a colloid different from the water molecules in the bulk?

6.58 Define Brownian motion. On the basis of Brownian motion, explain why many colloidal systems become unstable upon heating.

Colligative Properties

6.59 Calculate the freezing points of solutions made by dissolving 1 mole of each of the following ionic solutes in 1000 g of H_2O:
(a) NaCl
(b) $MgCl_2$
(c) $(NH_4)_2CO_3$
(d) $Al(HCO_3)_3$

6.60 If we add 175 g of ethylene glycol ($C_2H_6O_2$) per 1000 g of water to a car radiator, what will be the freezing point of the solution?

6.61 Methanol is used as an antifreeze. How many grams of methanol, CH_3OH, do you need per 1000 g of water for an aqueous solution to stay liquid at −20°C?

6.62 In winter, after a snow storm, salt, NaCl, is spread to melt the ice on roads. How many grams of salt per 1000 g of ice is needed to make it liquid at −5.0°C?

6.63 How many moles of glycerol, $C_3H_8O_3$, which does not dissociate in aqueous solution, must you add to 1000 g of water to make a solution that freezes at −10°C?

Osmosis

6.64 What is the difference between an osmotic semipermeable membrane and a semipermeable membrane?

6.65 In each case, tell which side (if either) rises and why. The solvent is water.

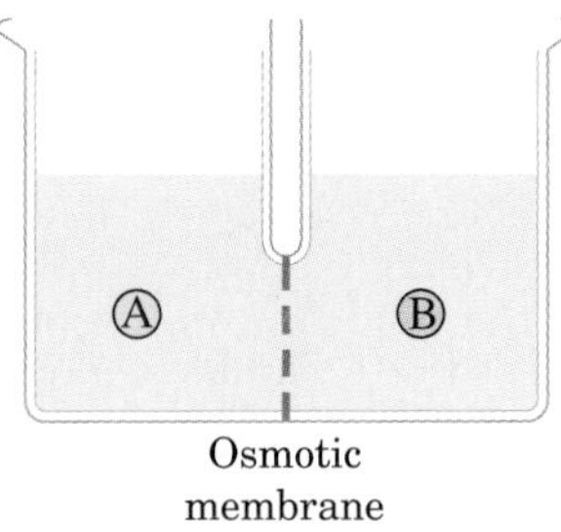

	Ⓐ	Ⓑ
(*a*)	1% glucose	5% glucose
(*b*)	1 *M* glucose	5 *M* glucose
(*c*)	1 *M* NaCl	1 *M* glucose
(*d*)	1 *M* NaCl	1 *M* K_2SO_4
(*e*)	3% NaCl	3% NaI
(*f*)	1 *M* NaBr	1 *M* NaCl

6.66 An osmotic semipermeable membrane that allows only water to pass separates two compartments, A and B. Compartment A contains 0.9 percent NaCl and B contains 3 percent glycerol, $C_3H_8O_3$. (a) In which compartment will the level of solution rise? (b) Which compartment has the higher osmotic pressure (if either)?

6.67 Calculate the osmolarity of
(a) 3.9 M Na_2CO_3 (c) 4.2 M LiBr
(b) 0.62 M $Al(NO_3)_3$ (d) 0.009 M K_3PO_4

6.68 Two compartments are separated by a semipermeable osmotic membrane through which only water molecules can pass. Compartment A contains a 0.3 M KCl solution; compartment B contains a 0.2 M Na_3PO_4 solution. Predict from which compartment the water will flow to the other compartment.

***6.69** A 0.9 percent NaCl solution is isotonic with blood plasma. Which solution would crenate red blood cells?
(a) 0.3% NaCl
(b) 0.9 M glucose
(c) 0.9% glucose (MW = 180)

Boxes

6.70 (Box 6A) In 1992 a volcano in the Philippines emitted large quantities of CO_2 and SO_2 gases as high as 2 to 3 miles into the atmosphere. Would you expect an increase in acid rain locally or globally?

6.71 (Box 6A) Oxides of nitrogen (NO, NO_2, N_2O_3) are also responsible for acid rain. Which acids do you suppose can be formed from these oxides?

6.72 (Box 6A) What makes normal rain water slightly acidic?

***6.73** (Box 6A) In a carbonated drink of 1 L volume, 3.456 g CO_2 was dissolved under pressure. When you remove the cap, CO_2 gas escapes. How many grams of CO_2 were lost when you opened the bottle at 20°C?

6.74 (Box 6B) Why do deep-sea divers use a helium-oxygen mixture in the tank instead of air?

6.75 (Box 6C) What is the chemical formula for the main component of limestone and marble?

***6.76** (Box 6C) Write balanced equations (two steps) for the conversion of marble to gypsum dihydrate.

6.77 (Box 6D) Why do long-distance runners suffer muscle cramps in hot and humid weather?

6.78 (Box 6D) Explain how hypoglycemia causes a lack of nerve impulses to the muscles.

6.79 (Box 6E) Which is the emulsifying agent in mayonnaise: oil, vinegar, or egg yolk?

***6.80** (Box 6F) The artificial kidney machine uses a solution containing 0.6 percent NaCl, 0.04 percent KCl, 0.2 percent $NaHCO_3$, and 0.72 percent glucose (all w/v). Show that this is an isotonic solution.

Additional Problems

6.81 If 10.0 g of ethyl alcohol and 30.0 g of water are mixed and the solution has a final volume of 38.0 mL, what is the concentration of ethyl alcohol (a) in % w/v and (b) in % w/w?

6.82 When a cucumber is put into a saline solution to pickle it, the cucumber shrinks; when a prune is put into the same solution, the prune will swell. Explain what happens in each case.

6.83 A solution of As_2O_3 has a molarity of 2×10^{-5} M. What is this concentration in ppm? (Assume the density of the solution is 1.00 g/mL.)

6.84 What effect does increasing the temperature have on the solubility of most substances?

***6.85** Both methanol, CH_3OH, and ethylene glycol, $C_2H_6O_2$, are used as antifreeze. Which is more efficient; that is, which produces a lower freezing point if equal weights of each are added to the same weight of water?

6.86 You know that a 0.89% saline (NaCl) solution is isotonic with blood. In a real-life emergency you run out of physiological saline solution and have only KCl as a salt, and distilled water. Would it be all right to make a 0.89% aqueous KCl solution and use it for intravenous infusion?

***6.87** Carbon dioxide and sulfur dioxide are soluble in water because they react with the water. Write possible equations for these reactions.

6.88 A reagent label shows that it contains 0.05 ppm lead as a contaminant. How many micrograms of lead are in 5.0 g of the reagent?

6.89 You have a 20 percent w/v stock solution of NaOH. Describe how you would prepare 250 mL of a 0.50 M NaOH solution.

6.90 Which will have greater osmotic pressure: (a) a 0.9 percent w/v NaCl solution or (b) a 25 percent w/v solution of a nondissociating dextran with a molecular weight of 15 000?

6.91 Government regulations permit a 6 ppb concentration of a certain pollutant. How many grams of pollutant are allowed in one ton (1016 kg) of water?

***6.92** The average osmolarity of sea water is 1.18. How much pure water would have to be added to 1.0 mL of sea water for it to achieve the osmolarity of blood (0.30 osmol)?

6.93 How can we make a solution of a solid in a solid?

Chapter 7

Reaction rates and equilibrium

7.1 Introduction

Some chemical reactions take place rapidly; others are very slow. For example, hydrogen gas and oxygen gas react with each other to form water:

$$2H_2(g) + O_2(g) \longrightarrow 2H_2O(l)$$

This reaction is extremely slow. A tank containing a mixture of H_2 and O_2 shows no measurable change even after many years.

In the course of several years, a few molecules of H_2 and O_2 will react, but not enough for us to detect.

In contrast to such a slow reaction, consider what happens when you take one or two aspirin tablets for a slight headache. Very often, the pain disappears in half an hour or so. This shows that the aspirin must have reacted with compounds in the body within that time.

Many known reactions are much faster still. For example, if we add a solution of silver ions to a solution of chloride ions (Sec. 4.7)

$$Ag^+(aq) + Cl^-(aq) \longrightarrow AgCl(s)$$

a precipitate of silver chloride forms almost instantaneously. This reaction is essentially complete in considerably less than one second.

The study of reaction rates is called **chemical kinetics.** The **rate** of a reaction is **the change in concentration of a reactant (or product) per unit time.** For example, consider the reaction

$$CH_3{-}Cl + I^- \xrightarrow[\text{acetone}]{} CH_3{-}I + Cl^-$$

This is a net ionic reaction. Spectator ions are not shown.

carried out in the solvent acetone. To determine the reaction rate, we can measure the concentration of iodomethane, CH_3I, in the acetone at periodic time intervals, say every ten minutes. The rate of the reaction is the increase in the concentration of iodomethane divided by the time interval. For example, the concentration might increase from 0 to 0.12 mole/L over a period of 30 minutes. If so, the rate of the reaction during that time is

The rate could also be determined by following the decrease in concentration of CH_3Cl, if that is more convenient.

$$\frac{0.12 \text{ mole/L} - 0.0 \text{ mole/L}}{30 \text{ min}} = \frac{0.12 \text{ mole/L}}{30 \text{ min}} = 0.004 \frac{\text{mole/L}}{\text{min}}$$

This unit is read "0.004 mole per liter per minute." During each minute of the reaction, an average of 0.004 mole of chloromethane, CH_3Cl, has been converted to iodomethane, for each liter of solution. Every reaction has its own rate, which must be measured in the laboratory.

Chemical reactions are very important to us, both the ones that we carry out in the laboratory and the ones that take place inside our bodies. The rates of these reactions are also important. A reaction that goes more slowly than we need may be useless, whereas a reaction that goes too fast may be dangerous. We would like to have some idea as to what causes the enormous variety in reaction rates. In the next three sections we examine this question.

7.2 Molecular Collisions

In order for two molecules or ions to react with each other, they must first collide. As we saw in Chapter 5, molecules in gases and liquids are constantly in motion and frequently collide with each other. If we want a reaction to take place between two compounds, A and B, we generally mix them together, either by dissolving them in the same solvent or, if they are gases, by simply allowing them to mix. In either case, we can be sure that the constant motion of all the molecules will cause frequent collisions between molecules of A and B. In fact, we can even calculate how many such collisions will take place within a given period of time.

When we make such calculations, we generally get a surprising result. There are so many collisions between A and B molecules that most reactions should be over in considerably less than one second. Since the actual reactions are generally much slower, we must conclude that **most collisions do not result in a reaction.** In most cases, when a molecule of A collides with a molecule of B, they simply bounce apart without reacting. **A collision that results in a reaction** is called an **effective collision.**

Why are some collisions effective and others not? There are two main reasons.

1. In most cases, for a reaction to take place between A and B, one or more covalent bonds must be broken in A or B or both, and energy is required for this to happen. This energy comes from the collision between A and B. If the energy of the collision is large enough, the bonds can break and the reaction will take place. If the collision energy is too low, the molecules will bounce apart without reacting. **The minimum energy necessary for the reaction to happen** is called the **activation energy.**

The energy of any collision depends on the relative speed of the colliding objects and on the angle of approach. Much greater damage is done in a head-on collision of two cars both going 40 miles per hour than in a collision in which a car going 20 miles per hour sideswipes one going 10 miles per hour. It is the same with molecules. There is a much greater collision energy when two fast-moving molecules collide head-on, for example, than when two slow-moving molecules collide at an angle (see Fig. 7.1).

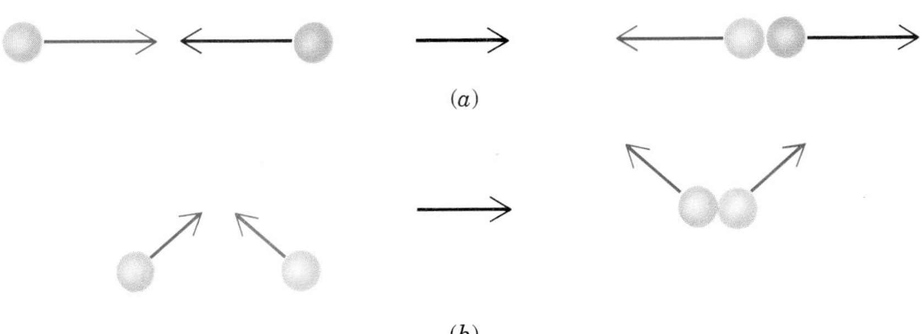

Figure **7.1**
The energy of molecular collisions varies. Two fast-moving molecules colliding head-on (a) have a higher collision energy than two slower-moving molecules colliding at an angle (b).

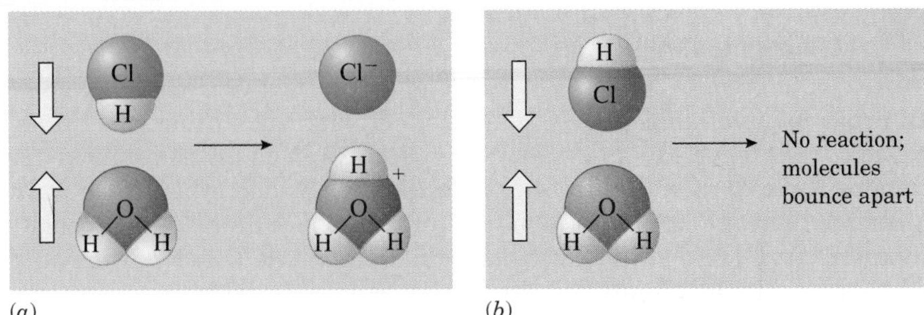

Figure **7.2** • (a) The two molecules are properly oriented for a reaction to take place. (b) The orientation is such that a reaction cannot take place, because the Cl, not the H, collides with the O. The colored arrows show the path of the molecules.

2. Even if two molecules collide with an energy greater than the energy of activation, a reaction still may not take place if the molecules are not oriented properly when they collide. Consider, for example, the reaction between H_2O and HCl:

$$H_2O + HCl \longrightarrow H_3O^+ + Cl^-$$

For this reaction to take place, the molecules must collide in such a way that the H of the HCl hits the O of the water, as shown in Figure 7.2(a). A collision in which the Cl of the HCl hits the O, as shown in Figure 7.2(b), cannot lead to reaction even if there is sufficient energy.

Coming back to the example at the beginning of this chapter, we can now see why the reaction between H_2 and O_2 is so slow. The O_2 and H_2 molecules are constantly colliding, but the percentage of effective collisions is extremely tiny at room temperature.

7.3 Activation Energy and Energy Diagrams

Molecules tend to go spontaneously from higher to lower energy states.

Figure 7.3 shows a typical energy diagram for a reaction. The products have a lower energy than the reactants; we might therefore expect the reaction to take place rapidly. As the curve shows, however, the reactants cannot be converted to products without the necessary activation energy. The activation energy is like a hill. If we are in a mountainous region, we may find that the only way to go from one point to another is to climb over a hill. It is the same in a chemical reaction. Even though the products may have a lower energy than the reactants (H_2O has a lower energy than $H_2 + O_2$), they cannot be formed unless the reactants "go over the hill," that is, gain energy of activation.

Let us look into this more closely. In a typical reaction, bonds are broken *and* new bonds are formed. For example, when H_2 reacts with N_2

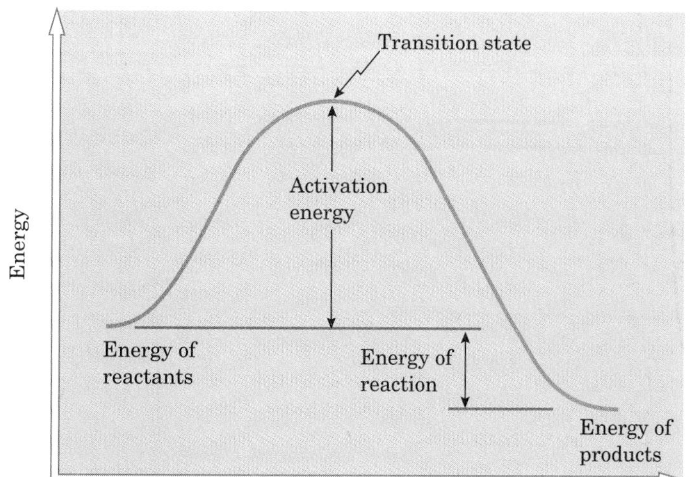

Figure **7.3**
Energy diagram for a typical reaction.

to give NH_3, six covalent bonds must break and six new covalent bonds must be formed (Fig. 7.4). Breaking a bond requires energy, but forming a bond releases energy. In a "downhill" reaction, of the type shown in Figure 7.3, the amount of energy released in making the new bonds is greater than that required to break the original bonds. That is, the reaction is exothermic. Yet it may well have a substantial activation energy, or energy barrier, because, in most cases, at least one bond must break *before* any new bonds can form. This means energy must be put in before we get any back. This is like the situation you might find yourself in if somebody offered to let you buy into a business from which, for an investment of \$100 000, you could get an income of \$40 000 per year, beginning in one year. In the long run you would do very well, but first you need to put up the \$100 000 right now (the activation energy).

"Downhill" reactions are exothermic. "Uphill" reactions are endothermic.

Every reaction has a different energy diagram. In some cases the energy of the products is higher than that of the reactants ("uphill" reactions, Fig. 7.5), but in almost all cases there is an energy hill—the activation energy. The activation energy is inversely related to the rate of the reaction. **The lower the activation energy, the faster the reaction; the higher the activation energy, the slower the reaction.**

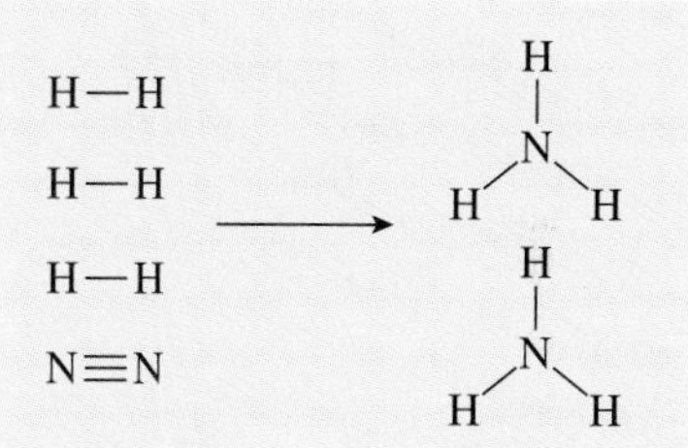

Figure **7.4**
In the reaction $3H_2 + N_2 \rightarrow 2NH_3$, six covalent bonds (three single and one triple) break, and six new bonds form.

The top of the energy hill is called the **transition state.** When the reacting molecules reach this point, one or more original bonds are partially broken and one or more new bonds may be in the process of formation. A typical situation is shown in Figure 7.6.

Let us look once more at the reaction between H_2 and O_2. The activation energy is so high that at room temperature no measurable reaction occurs even after many years. But if we light a match, the situation suddenly changes. The heat from the match causes the reaction not only to take place, but to take place so rapidly that the whole mixture explodes as soon as the match is lit. What has happened?

The heat of the match has supplied activation energy to some of the molecules in its vicinity, causing them to react. (It has done this by mak-

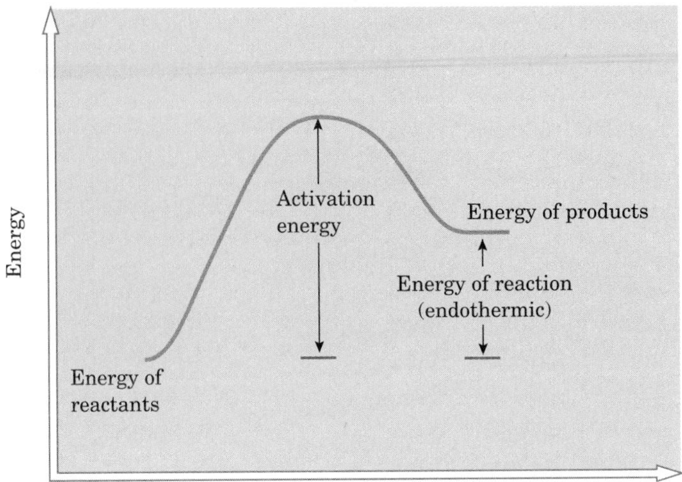

Figure 7.5
Energy diagram for an endothermic reaction, in which the energy of the products is greater than that of the reactants.

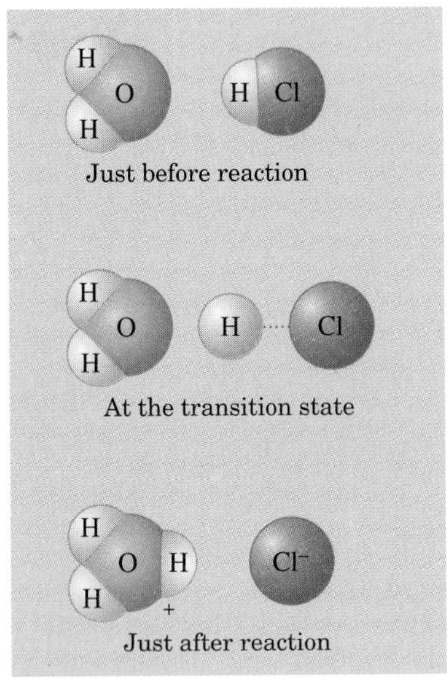

Figure 7.6
The position of two typical reacting molecules, before, at, and after the transition state.

ing the molecules move faster—see the next section.) Once they react, they give off additional energy (the heat of the reaction, which is 137 000 cal for each mole of O_2 used up). This additional energy becomes the activation energy for the neighboring molecules, which in turn give off still more. The result is that the small amount of energy from the lighted match is enough to start a process that becomes self-sustaining so rapidly that the mixture explodes as soon as the match is lit.

This is a dramatic example of the importance of activation energy. In most cases explosions do not take place, but activation energy is required for the large majority of reactions.

7.4 Factors Affecting Rates of Reaction

In the previous section we saw that reactions occur as a result of collisions between fast-moving molecules possessing a certain minimum energy (the activation energy). In this section we examine some of the factors that affect activation energies and reaction rates.

Nature of the Reactants

As we have already pointed out, every reaction has its own rate and its own activation energy. In general, reactions that take place between ions in aqueous solution (Sec. 4.7) are extremely rapid, almost instantaneous. Activation energies for these reactions are very low because usually no covalent bonds need to be broken. As we might expect, reactions between covalent molecules, whether in aqueous solution or not, are much slower. Many of these require anywhere from 15 minutes to 24 hours or more for most of the starting compounds to be converted to the products. There are, of course, reactions that take a good deal longer, but such reactions are seldom useful.

Box 7A

Why High Fever Is Dangerous

Box 1C points out that a sustained body temperature of 41.7°C (107°F) is invariably fatal. We can now see why a high fever is dangerous. Normal body temperature is 37°C (98.6°F), and all the many reactions in the body—including respiration, digestion, and the synthesis of various compounds—take place at that temperature. If an increase of 10°C causes the rates of most reactions approximately to double, then an increase of even 1°C makes them go significantly faster.

Fever is a protective mechanism, and a small increase in temperature allows the body to kill germs faster by mobilizing the immune defense mechanism. But this increase must be small: An increase of 1°C brings the temperature to 100.4°F; an increase of 3°C brings it to 104°F. A temperature higher than 104°F increases reaction rates to the danger point.

One can easily detect the increase in reaction rates when a patient has a high fever. The pulse rate increases and breathing is faster in an effort to supply increased amounts of oxygen for the fast reactions. A marathon runner may become overheated on a hot and humid day. After a time perspiration can no longer cool his body effectively, and he may suffer hyperthermia or heat stroke, which, if not treated properly, can cause brain damage.

Concentration

Consider a reaction $A + B \rightarrow C + D$. In most cases, a reaction rate increases when the concentration of either or both reactants (A or B) is increased. For many reactions, though by no means all, there is a direct relationship between concentration and rate; that is, when the concentration of a reactant is doubled, the rate also doubles. All of this is easily understandable on the basis of the collision theory. If we double the concentration of A, there are twice as many molecules of A in the same volume, so the molecules of B in that volume now collide with twice as many A molecules per second as before. Since the reaction rate depends on the number of collisions per second, the rate is doubled (Fig. 7.7).

For reactions in the gas phase, an increase in pressure usually increases the rate.

In the case where one of the reactants is a solid, the rate is affected by the surface area of the solid, so a substance in powder form reacts faster than the same substance in the form of large chunks.

Temperature

In virtually all cases, reaction rates increase with increasing temperature. An approximate rule for many reactions is that every time the temperature goes up by 10°C, the rate of reaction doubles. This rule is far from exact, but it is not far from the truth in many cases. As you can see, this is quite a large effect and says, for example, that if we run a reaction at 90°C instead of at room temperature (20°C), the reaction will go about 128 times faster. Putting it another way, if it takes 20 hours to convert 100 g of A to product C at 20°C, then it would take only about 10 minutes at 90°C. Temperature, therefore, is a powerful tool that lets us increase the rates of reactions that are inconveniently slow. It also lets us decrease the rates of reactions that are inconveniently fast. In some cases we choose to run reactions at low temperatures because explosions would result or the reactions would otherwise be out of control at room temperature.

There are seven 10° increments between 20° and 90°, and $2^7 = 128$.

Figure **7.7**
When heated in air, steel wool glows but does not burn rapidly, because the O_2 concentration of the air is only about 20%. When the glowing steel wool is put into 100% O_2 (in the photo) it burns vigorously. (Photograph by Leon Lewandowski.)

What causes reaction rates to increase with increasing temperature? Once again, we turn to collision theory. Here temperature has two effects:

1. In Section 5.2 we learned that temperature is related to the average kinetic energy of molecules. When the temperature increases, molecules move faster, which means they collide more frequently. More frequent collisions mean higher reaction rates. However, this factor is much less important than the second factor.

2. Recall from Section 7.2 that a reaction between two molecules takes place only if there is an *effective collision*—a collision with an energy greater than the activation energy. However, when the temperature

Box 7B

The Effects of Lowering Body Temperature

As with a significant increase in body temperature, a substantial decrease below 37°C is also harmful because reaction rates are lower than they should be. However, it is sometimes possible to take advantage of this effect. In some heart operations, it is necessary to stop the flow of oxygen to the brain for a considerable time. At 37°C, the brain cannot survive without oxygen for more than about five minutes without permanent damage. When the patient's body temperature is deliberately lowered to about 28 to 30°C (82.4 to 86°F), however, the oxygen flow can be stopped for a considerable time without causing damage because reaction rates are lowered. At 25.6°C (78°F) the body's oxygen consumption is reduced by 50 percent.

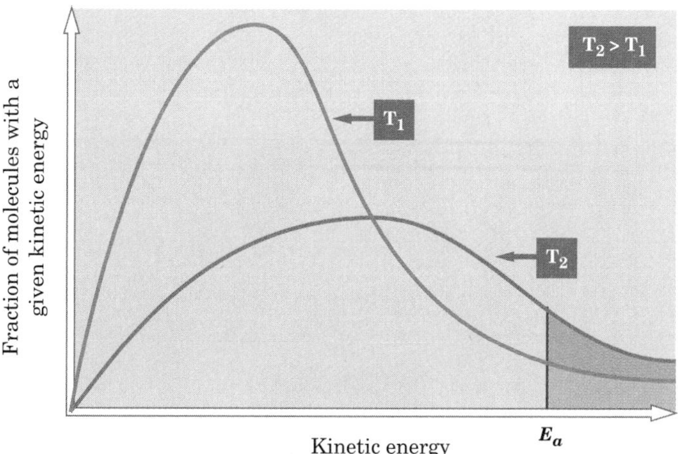

Figure **7.8**
Distribution of kinetic energies (molecular velocities) at two temperatures. The vertical line at the right indicates the energy (velocity) necessary to pass through the activation energy barrier. The shaded areas represent the fraction of molecules that have kinetic energies (velocities) greater than the activation energy.

increases, not only is the average speed of the molecules greater, but there is also a different *distribution* of speeds. The number of very fast molecules increases much more than the average speed (Fig. 7.8). This means that the number of effective collisions rises even more than the total number of collisions. Not only are there more collisions, but the percentage of collisions that have an energy greater than the activation energy also rises. It is this factor that is mainly responsible for the sharp increase in reaction rates with increasing temperature.

In this dish Cl^- ions act as a catalyst in the decomposition of NH_4NO_3. (Photograph by Charles D. Winters.)

Presence of a Catalyst

Any substance that increases the rate of a reaction without itself being used up is called a **catalyst.** Many catalysts are known, some that increase the rate of only one reaction, others that can affect several reactions. Although we have seen that reactions can be speeded up by increasing the temperature, in some cases they are still too slow even at the highest temperatures we can conveniently reach. In other cases it is not feasible to increase the temperature, for example, because other, unwanted, reactions would also be speeded up. In such cases a catalyst, if we can find the right one for a given reaction, can prove very valuable. Many important industrial processes rely on catalysts (see Box 7E), and virtually all reactions that take place inside living organisms are catalyzed by enzymes (Chapter 19).

Catalysts work by allowing the reaction to take a different pathway, one with a lower activation energy (Fig. 7.9). Without the catalyst, the reactants would have to get over the energy hill. The catalyst provides a lower hill. As we have seen, a lower activation energy means a higher reaction rate. Each catalyst has its own way of providing an alternate pathway. Many catalysts do it by providing a surface on which the reactants can meet. Thus, the reaction between ethylene gas, C_2H_4, and hydrogen gas, H_2, to give ethane gas, C_2H_6,

$$C_2H_4(g) + H_2(g) \xrightarrow[\text{Pt}]{} C_2H_6(g)$$

We often write the catalyst over or under the arrow.

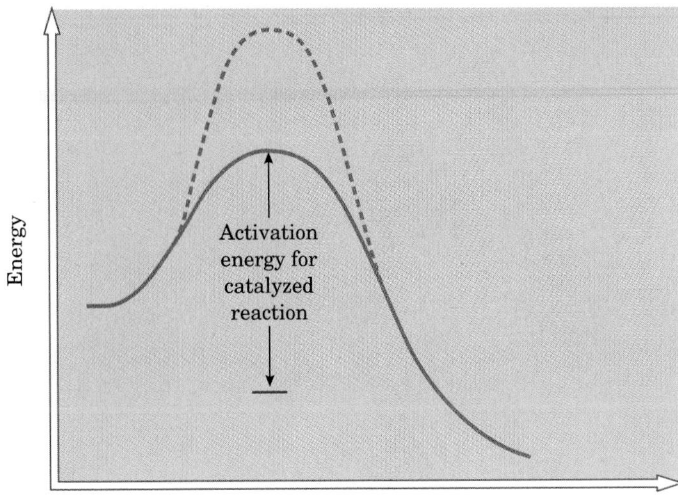

Figure **7.9**
Energy diagram for a catalyzed reaction. The dashed line shows the energy curve for the uncatalyzed process. The catalyst provides an alternate pathway whose activation energy is lower.

Catalysts can be homogeneous or heterogeneous. A **homogeneous catalyst** is in the same phase as the reactants, for example, enzymes in body tissues. A **heterogeneous catalyst** is in a separate phase from the reactants, for example, the solid platinum in the reaction between the gases C_2H_4 and H_2.

goes so slowly without a catalyst that it is not practical even if we increase the temperature to any reasonable level. But if the mixture of gases is shaken with finely divided solid platinum, the reaction takes place at a convenient rate. The C_2H_4 molecules and the H_2 molecules meet each other on the surface of the platinum where the proper bonds can be broken and the reaction can proceed.

7.5 Reversible Reactions and Equilibrium

Many reactions are irreversible. When a piece of paper is completely burned, the products are CO_2 and H_2O. Anybody who took pure CO_2 and H_2O and tried to make them react to give paper and oxygen would not succeed. A tree, of course, turns CO_2 and H_2O into wood and oxygen, and we, in sophisticated factories, make paper from the wood, but this is not the same as taking CO_2, H_2O, and energy and directly combining them in a single process to get paper and oxygen. We can certainly consider the burning of paper an irreversible reaction.

Other reactions, however, are reversible. A **reversible reaction** is **one that can be made to go in either direction.** For example, if carbon monoxide is mixed with water in the gas phase at a high temperature, hydrogen and carbon dioxide are produced:

$$CO(g) + H_2O(g) \longrightarrow CO_2(g) + H_2(g)$$

But, if we desire, we can also make the reaction take place the other way: We can mix hydrogen and carbon dioxide to get carbon monoxide and water vapor:

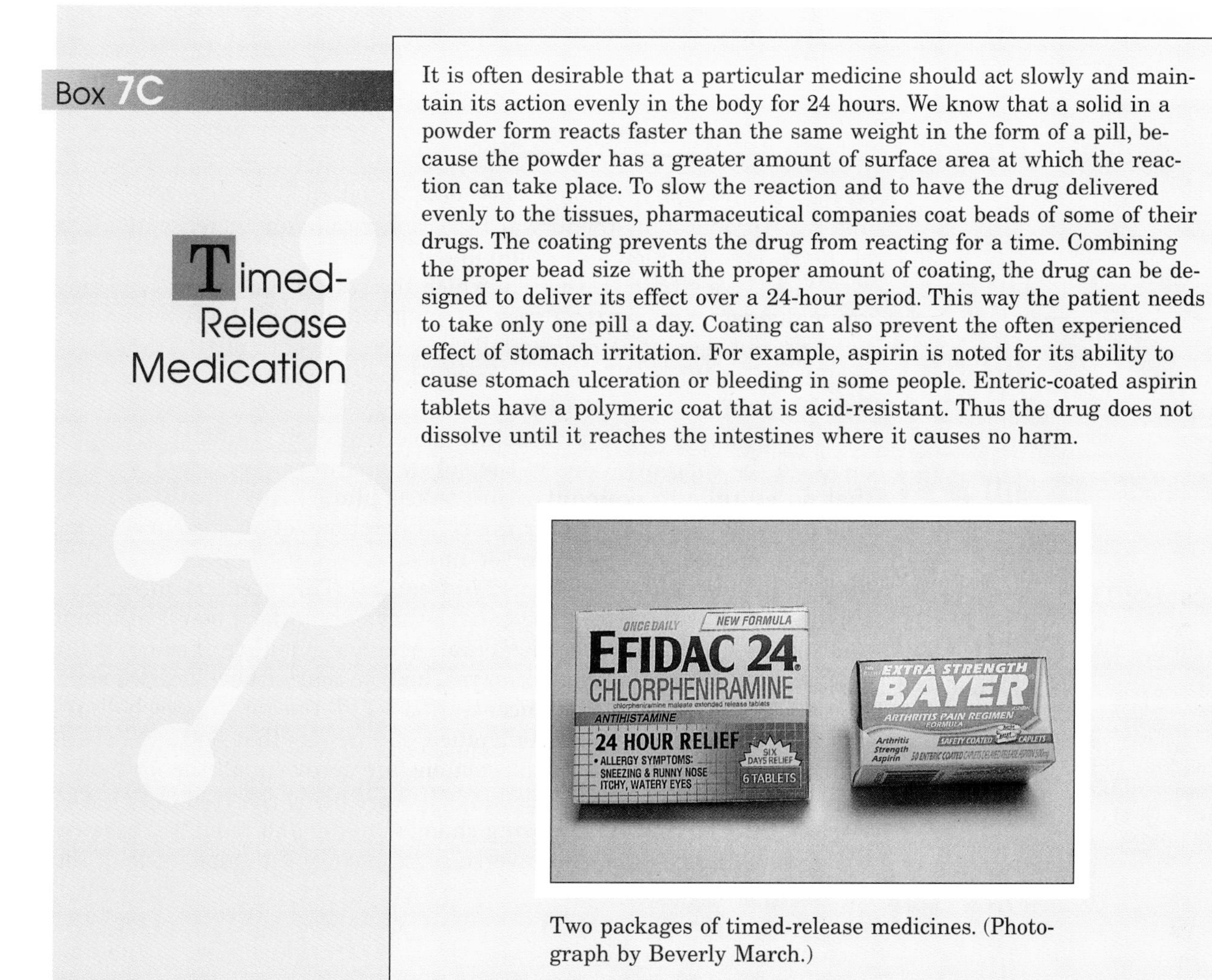

$$CO_2(g) + H_2(g) \longrightarrow CO(g) + H_2O(g)$$

Box 7C

Timed-Release Medication

It is often desirable that a particular medicine should act slowly and maintain its action evenly in the body for 24 hours. We know that a solid in a powder form reacts faster than the same weight in the form of a pill, because the powder has a greater amount of surface area at which the reaction can take place. To slow the reaction and to have the drug delivered evenly to the tissues, pharmaceutical companies coat beads of some of their drugs. The coating prevents the drug from reacting for a time. Combining the proper bead size with the proper amount of coating, the drug can be designed to deliver its effect over a 24-hour period. This way the patient needs to take only one pill a day. Coating can also prevent the often experienced effect of stomach irritation. For example, aspirin is noted for its ability to cause stomach ulceration or bleeding in some people. Enteric-coated aspirin tablets have a polymeric coat that is acid-resistant. Thus the drug does not dissolve until it reaches the intestines where it causes no harm.

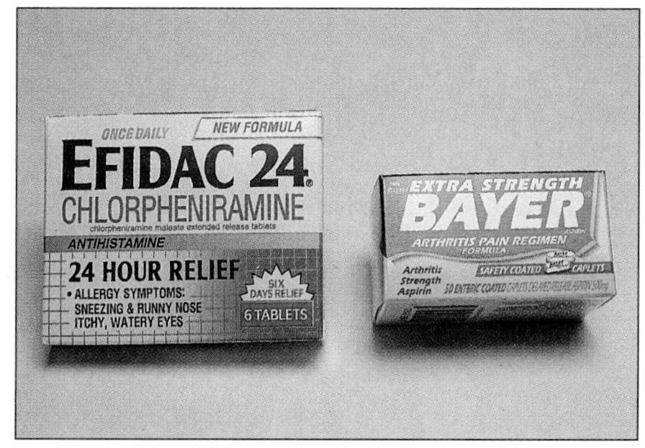

Two packages of timed-release medicines. (Photograph by Beverly March.)

This reaction, like many others, is reversible. Let us see what happens when we run a reversible reaction. We will add some carbon monoxide to water vapor in the gas phase. The two compounds begin to react at a certain rate (the *forward* reaction):

$$CO(g) + H_2O(g) \longrightarrow CO_2(g) + H_2(g)$$

As the reaction proceeds, the concentrations of CO and H_2O gradually decrease, because they are being used up. This means that the *rate* of the reaction also gradually decreases because the rate of a reaction depends on the concentration of the reactants (Sec. 7.4).

The units of concentration in the gas phase are the same as in solution: moles per liter.

But what is happening in the other direction? Before we added the carbon monoxide, there was no carbon dioxide or hydrogen present. As soon as the forward reaction begins, however, it produces small amounts of these substances, and we now have some CO_2 and H_2. These two compounds will now, of course, begin reacting with each other (the *reverse* reaction):

$$CO_2(g) + H_2(g) \longrightarrow CO(g) + H_2O(g)$$

At first the reverse reaction is very slow, but as the concentrations of H_2 and CO_2 (produced by the forward reaction) gradually increase, the rate of the reverse reaction also gradually increases.

We have a situation, then, in which the rate of the forward reaction gradually decreases as time goes on while the rate of the reverse reaction (which began at zero) gradually increases. Eventually the two rates become equal. When this point is reached, we call the process a **dynamic equilibrium** or just an **equilibrium.**

What happens in the container once we reach equilibrium? If we measure the concentrations of the substances in the container, we find that **no change in concentration takes place after equilibrium is reached** (Fig. 7.10). Whatever the concentrations of all the substances, they will remain the same forever unless something is done to disturb the equilibrium (this is discussed in Sec. 7.8). This does not mean that all the concentrations must be equal—they can, in fact, be all different and usually are—but it does mean that, whatever they are, they no longer change once equilibrium has been reached, no matter how long we wait.

Given the fact that the concentrations of all the substances, both reactants and products, no longer change, can we say that nothing is happening? No, we know that both reactions are going on; all the molecules are constantly reacting—the CO and H_2O are being changed to CO_2 and H_2, and the CO_2 and H_2 are being changed to CO and H_2O. But because the rates of the forward and reverse reactions are the same, none of the concentrations can change.

Another way to look at it is that the concentration of carbon monoxide (and of the other three compounds) does not change because it is being used up as fast as it is being formed.

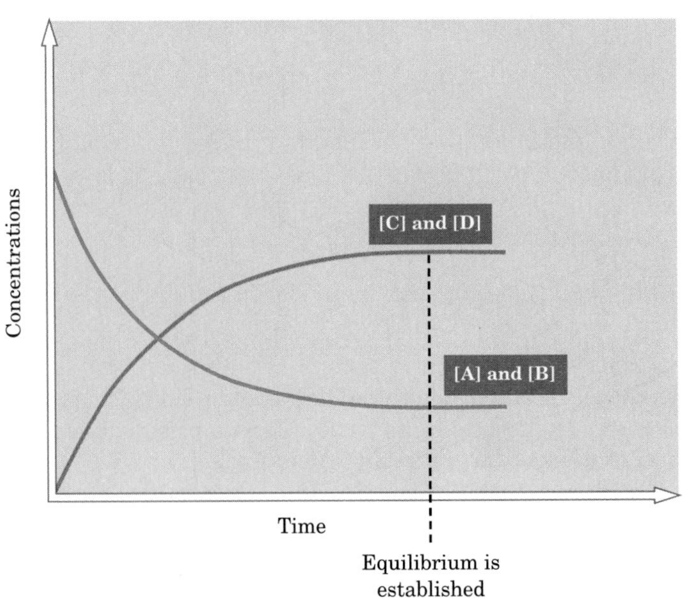

Figure **7.10**
Changes in the concentrations of A, B, C, and D in the $A + B \rightleftharpoons C + D$ system, as equilibrium is approached beginning with A and B only. This equilibrium lies far to the right.

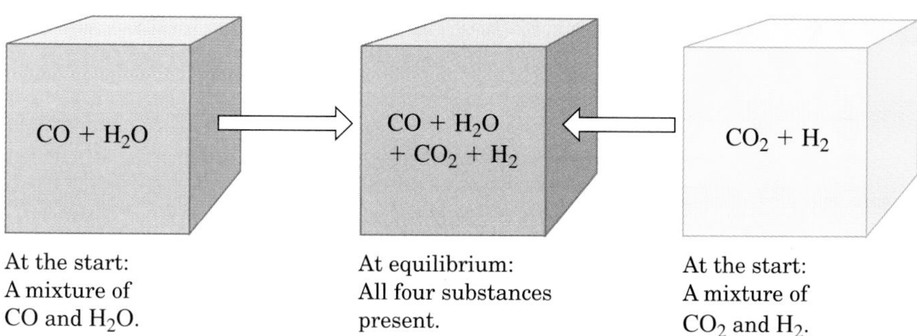

Figure **7.11**
An equilibrium can be approached from either direction.

In the example just discussed, we approached equilibrium by adding carbon monoxide to water vapor. We could also have done it by adding carbon dioxide to hydrogen. In either case, we eventually get an equilibrium mixture containing the same four compounds (Fig. 7.11).

It is not necessary to begin with equal amounts. We could, for example, take 10 moles of carbon monoxide and 0.2 mole of water vapor. We would still arrive at an equilibrium mixture of all four compounds.

Reversible chemical reactions are not the only places in which we find dynamic equilibria. They are very common in the body and elsewhere in science.

We have already met two examples of dynamic equilibrium in this book: the equilibrium between a liquid and its vapor (Sec. 5.9), and the equilibrium involving solute particles in saturated solutions (Sec. 6.4).

7.6 Equilibrium Constants

Chemical equilibria can be treated by a simple mathematical expression. First let us write

$$aA + bB + \cdots \rightleftharpoons cC + dD + \cdots$$

as the general equation for all reversible reactions. In this equation, the capital letters stand for substances—H_2O and CO_2, for instance—and the small letters are the coefficients of the balanced equation. The double arrow $\rightleftharpoons$ is used to show that the reaction is reversible. The dots mean that more than two substances can be present on either side.

Once equilibrium is reached, the following equation is valid:

$$K = \frac{[C]^c[D]^d \cdots}{[A]^a[B]^b \cdots} \quad \text{The equilibrium expression}$$

This expression was first derived in 1863 by Cato Guldberg (1836–1902) and Peter Waage (1833–1900), two Norwegian chemists.

The brackets [ ] are a shorthand way of saying "concentration." Let us examine this equation. The K is a constant called the **equilibrium constant.** What this expression tells us is that, when we multiply the concentrations of the substances on the right-hand side of the chemical equation and divide this product by the concentrations of the substances on the left-hand side (after raising each number to the appropriate power), *we get a number that does not change:* the equilibrium constant.

It is understood that the concentration of the species within the brackets is always expressed in moles per liter.

Let us look at examples of how to set up equilibrium expressions.

The concentration terms in brackets in the expression for K are for species in solution, either liquid or gaseous.

EXAMPLE

Write the equilibrium expression for the reaction

$$CO(g) + H_2O(g) \rightleftharpoons CO_2(g) + H_2(g)$$

Answer

$$K = \frac{[CO_2][H_2]}{[CO][H_2O]}$$

This expression tells us that, at equilibrium, the concentration of carbon dioxide multiplied by the concentration of hydrogen and divided by the concentrations of water and carbon monoxide is a constant (K) and will not change. Mathematically, it would be just as correct to write the left-hand compounds on top, but the universal custom is to write them as shown here, with the products on top and the reactants on the bottom.

EXAMPLE

Write the equilibrium expression for

$$PCl_5 \rightleftharpoons Cl_2 + PCl_3$$

Answer

$$K = \frac{[Cl_2][PCl_3]}{[PCl_5]}$$

Problem 7.1

Write the equilibrium expression for the reaction

$$SO_2 + H_2O \rightleftharpoons H_2SO_3$$

EXAMPLE

Write the equilibrium expression for

$$O_2(g) + 4ClO_2(g) \rightleftharpoons 2Cl_2O_5(g)$$

Answer

$$K = \frac{[Cl_2O_5]^2}{[O_2][ClO_2]^4}$$

In this case the chemical equation has coefficients other than unity, so the equilibrium expression contains powers.

Problem 7.2

Write the equilibrium expression for the reaction

$$2NH_3(g) \rightleftharpoons N_2(g) + 3H_2(g)$$

Now let us see how K is calculated.

EXAMPLE

Some H_2 is added to I_2 at 427°C and the reaction

$$I_2(g) + H_2(g) \rightleftharpoons 2HI(g)$$

is allowed to come to equilibrium. When this point is reached, the concentrations are found to be $[I_2] = 0.42$ mole/L, $[H_2] = 0.025$ mole/L, and $[HI] = 0.76$ mole/L. Calculate K at 427°C.

Answer • The equilibrium expression is

$$K = \frac{[HI]^2}{[I_2][H_2]}$$

Substituting the concentrations, we get

$$K = \frac{(0.76)^2}{(0.42)(0.025)} = 55$$

Problem 7.3

What is the equilibrium constant for the reaction

$$PCl_3 + Cl_2 \rightleftharpoons PCl_5$$

if the equilibrium concentrations are $[PCl_3] = 1.66\ M$, $[Cl_2] = 1.66\ M$, and $[PCl_5] = 1.66\ M$?

The last example shows us that the reaction between I_2 and H_2 to give HI has an equilibrium constant of 55. What does this mean? At constant temperature, equilibrium constants remain the same no matter what concentrations we have. That is, at 427°C, if we begin by adding, say, 5 moles of H_2 to 5 moles of I_2, the two reactions take place and equilibrium is reached. At that point, the concentration of HI, squared, divided by the concentrations of I_2 and H_2 equals 55. If we begin instead, at 427°C, with 7 moles of H_2 and 2 of I_2, or 0.4 mole of H_2 and 17 moles of I_2, or *any* number of moles of HI, once equilibrium is reached, the value of $[HI]^2/[I_2][H_2]$ is again 55. It makes no difference what concentrations of the three substances we begin with. At 427°C, as long as all three substances are present and equilibrium has been reached, the concentrations of the three substances so adjust themselves that the value of the equilibrium expression is 55.

The equilibrium constant is different for every reaction. Some reactions have a large *K*, others a small *K*. A reaction with a very large *K* proceeds almost to completion (to the right). For example, *K* for

This symbol $\rightleftharpoons$ means that the equilibrium lies far to the right.

$$HCl(aq) + H_2O(l) \rightleftharpoons H_3O^+(aq) + Cl^-(aq)$$

is about 10 000 000, or 10^7, at 25°C, meaning that at equilibrium $[H_3O^+]$ and $[Cl^-]$ must be very large and $[HCl]$ and $[H_2O]$ very small, in order that $[H_3O^+][Cl^-]/[HCl][H_2O]$ be equal to 10^7. This means that if we add HCl to H_2O we can be certain that, when equilibrium has been reached, an essentially complete reaction has taken place.

On the other hand, a reaction with a very small *K*, say 10^{-7}, hardly goes at all. Equilibrium effects are most obvious in reactions with *K* values between 10^3 and 10^{-3}. In such cases, the reaction goes part way, and significant concentrations of all substances are present at equilibrium. An example is the reaction between carbon monoxide and water, discussed in Section 7.5, for which *K* is equal to 10 at 600°C.

The equilibrium constant for a given reaction remains the same no matter what happens to the concentrations, but this is not true for changes in temperature. The value of *K* does change when the temperature is changed.

7.7 Equilibrium and Reaction Rates

As pointed out in the previous section, the equilibrium expression is valid only after equilibrium has been reached. Before that, there is no equilibrium, and the equilibrium expression is not valid. But how long does it take for a reaction to reach equilibrium? For this there are no easy answers. Some reactions, if the reactants are well mixed, reach equilibrium in less than one second; others will still not get there even after millions of years.

There is no relationship between the rate of a reaction (how long it takes to reach equilibrium) and the value of *K*. It is possible to have a large *K* and a low rate, as in the reaction between H_2 and O_2 to give H_2O, which does not reach equilibrium for millions of years (Sec. 7.1), or a small *K* and a high rate, as well as reactions in which both the rate and *K* are large or both are small.

7.8 Le Chatelier's Principle

When a reaction has reached equilibrium, the forward and reverse reactions are taking place at the same rate, and the concentrations of all the components do not change as long as we don't do anything to the system. But what happens if we do? In 1888 Henri Le Chatelier (1850–1936) put forth the statement known as **Le Chatelier's principle: If an external stress is applied to a system in equilibrium, the system reacts in such a way as to partially relieve the stress.** Let us look at four types of stress that can be put on chemical equilibria.

Addition of Reaction Components

Suppose that this reaction has reached equilibrium:

$$\underset{\text{Acetic acid}}{CH_3COOH} + \underset{\text{Ethyl alcohol}}{C_2H_5OH} \overset{HCl}{\rightleftharpoons} \underset{\text{Ethyl acetate}}{CH_3COOC_2H_5} + H_2O$$

The HCl is a catalyst, but this does not affect the equilibrium.

This means that the reaction flask contains all these substances and that the concentrations no longer change. We now disturb the system by adding some acetic acid from outside. The result is that the concentration of acetic acid suddenly increases, which increases the rate of the forward reaction. As a consequence, the concentrations of products (ethyl acetate and water) begin to increase. At the same time the concentrations of reactants are decreasing. Now, an increase in the concentrations of products causes the rate of the reverse reaction to increase, but the rate of the forward reaction is decreasing, so eventually the two rates will be equal again and a new equilibrium is established.

When that happens, the concentrations once again remain constant, but they are not the same as they were before. The concentrations of ethyl acetate and water are higher now, and the concentration of ethyl alcohol is lower. The concentration of acetic acid is higher *because we added some,* but it is less than it was immediately after we made the addition.

The equilibrium constant K of course remains the same.

This always happens when we add more of any component to a system in equilibrium. The addition constitutes a stress; the system relieves the stress by increasing the concentrations of the components on the other side. We say that the equilibrium shifts in the opposite direction:

$$\text{Adding } CH_3COOH \curvearrowright \; CH_3COOH + C_2H_5OH \overset{HCl}{\rightleftharpoons} CH_3COOC_2H_5 + H_2O \quad (\longrightarrow)$$

The addition of acetic acid causes the reaction to move toward the right: More $CH_3COOC_2H_5$ and H_2O are formed and some of the CH_3COOH and C_2H_5OH are used up. The same thing happens if we add C_2H_5OH.

On the other hand, if we add H_2O or $CH_3COOC_2H_5$, the reaction shifts to the left:

$$CH_3COOH + C_2H_5OH \overset{HCl}{\rightleftharpoons} CH_3COOC_2H_5 + H_2O \; \curvearrowleft \text{Adding } CH_3COOC_2H_5 \quad (\longleftarrow)$$

We can summarize by saying that **the addition of any component causes the equilibrium to shift to the opposite side.**

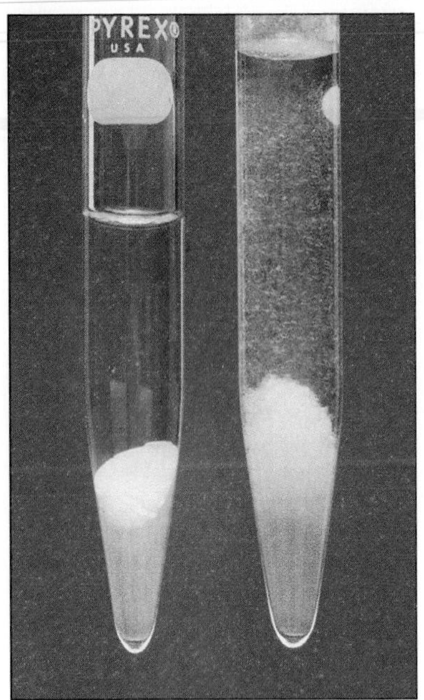

The tube at the left contains a saturated solution of silver acetate (Ag^+ ions and CH_3COO^- ions) in equilibrium with solid silver acetate. When silver ions, in the form of silver nitrate solution, are added the equilibrium

$$Ag^+(aq) + CH_3COO^-(aq) \rightleftharpoons CH_3COOAg(s)$$

shifts to the right, producing more solid silver acetate, as can be seen in the tube on the right. (Photograph by Charles D. Winters.)

EXAMPLE

When dinitrogen tetroxide, N_2O_4, a colorless gas, is enclosed in a vessel a color will appear indicating the formation of brown nitrogen dioxide, NO_2 (see Fig. 7.12). The intensity of the brown color indicates the amount of nitrogen dioxide formed. The equilibrium reaction is

$$N_2O_4(g) \rightleftharpoons 2NO_2(g)$$

When more N_2O_4 is added to the equilibrium mixture, the brownish color becomes darker. Explain what happened.

Answer • The darker color indicates that more nitrogen dioxide was formed. This happened because the addition of the reactant shifted the equilibrium to the right, forming more product, NO_2.

Problem 7.4

What happens to the equilibrium reaction

$$2NOBr(g) \rightleftharpoons 2NO(g) + Br_2(g)$$

when Br_2 gas is added to the equilibrium mixture?

Removal of a Reaction Component

It is not always as easy to remove a component from a reaction mixture as it is to add one, but there are often ways to do it. The removal of a component, or even a decrease in its concentration, lowers the corresponding reaction rate and changes the position of the equilibrium. In this case, the reaction is shifted toward the side from which the reactant was removed.

In the case of the acetic acid–ethyl alcohol equilibrium, ethyl acetate has the lowest boiling point of the four components and can be removed by distillation. If this is done, the equilibrium shifts to that side:

$$CH_3COOH + C_2H_5OH \overset{HCl}{\rightleftharpoons} CH_3COOC_2H_5 + H_2O \quad \text{Removing } CH_3COOC_2H_5$$

The concentrations of CH_3COOH and C_2H_5OH decrease and that of H_2O increases. The effect of removing a component is thus the opposite of adding one: **The removal of a component causes the equilibrium to shift to the side the component was removed from.**

No matter what happens to the individual concentrations, **the value of the equilibrium constant remains unchanged.**

EXAMPLE

When acid rain containing sulfuric acid attacks a marble statue made of calcium carbonate, the following equilibrium reaction can be written

$$CaCO_3(s) + H_2SO_4(aq) \rightleftharpoons CaSO_4(s) + CO_2(g) + H_2O(l)$$

How does the fact that CO_2 is a gas influence the equilibrium?

Answer • The gaseous CO_2 diffuses away from the reaction site. In essence this product is removed from the equilibrium mixture. The equilibrium shifts to the right.

Problem **7.5**

In a power plant using coal, SO_2 is generated from pyrite, FeS_2, small amounts of which are present in the coal, according to the reaction:

$$4FeS_2(s) + 11O_2(g) \rightleftharpoons 2Fe_2O_3(s) + 8SO_2(g)$$

SO_2 is much more soluble in water than O_2 (Table 6A) and reacts with water forming H_2SO_3. What is the effect on the equilibrium if the above reaction is performed in a moist atmosphere?

Change in Temperature

The effect of a change in temperature on a reaction that has reached equilibrium depends on whether the reaction is exothermic (gives off heat) or endothermic (requires heat). Let us look at an exothermic reaction:

$$2H_2(g) + O_2(g) \rightleftharpoons 2H_2O(l) + 137\,000 \text{ cal}$$

Box 7D

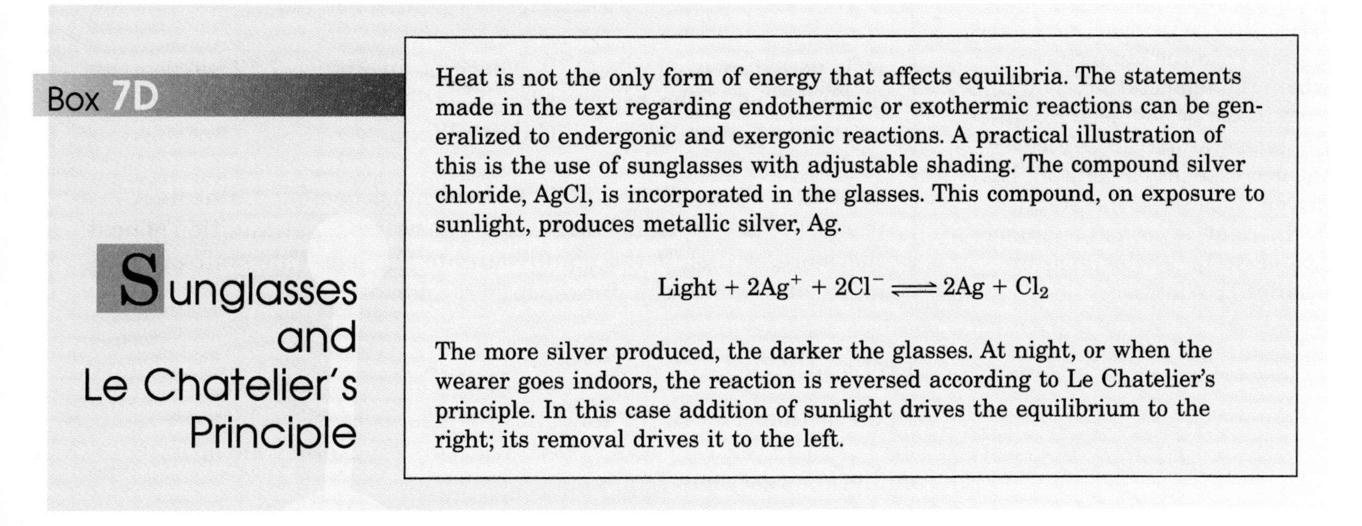

Sunglasses and Le Chatelier's Principle

Heat is not the only form of energy that affects equilibria. The statements made in the text regarding endothermic or exothermic reactions can be generalized to endergonic and exergonic reactions. A practical illustration of this is the use of sunglasses with adjustable shading. The compound silver chloride, AgCl, is incorporated in the glasses. This compound, on exposure to sunlight, produces metallic silver, Ag.

$$\text{Light} + 2Ag^+ + 2Cl^- \rightleftharpoons 2Ag + Cl_2$$

The more silver produced, the darker the glasses. At night, or when the wearer goes indoors, the reaction is reversed according to Le Chatelier's principle. In this case addition of sunlight drives the equilibrium to the right; its removal drives it to the left.

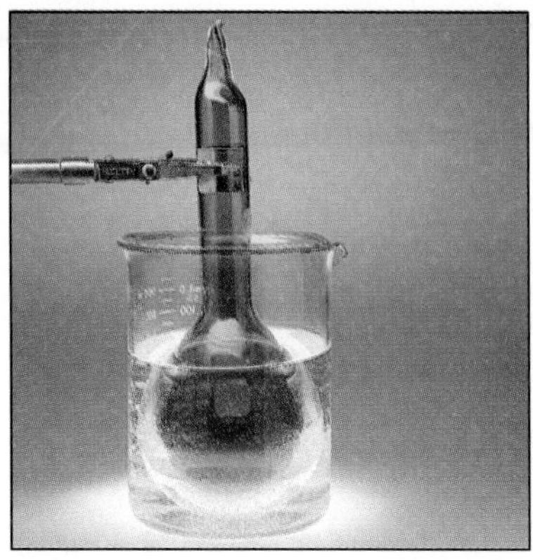

Figure **7.12**
Effect of temperature on the equilibrium:

$$2NO_2(g) \rightleftharpoons N_2O_4(g) + 13\ 700\ cal$$

NO_2 is a brown gas; N_2O_4 is colorless. The concentration of NO_2 is greater in the top picture, where the temperature is 50°C, than in the bottom picture, where the temperature is 0°C, because the decrease in temperature has driven this exothermic reaction to the right, favoring the production of N_2O_4. (Photographs by Charles D. Winters.)

This equation tells us that 2 moles of H_2 react with 1 mole of O_2 to give 2 moles of H_2O and heat. If we look upon the heat as a product of this reaction, then we can use the same type of reasoning as we did before. An increase in temperature means that we are adding heat. Since heat is a product, its addition pushes the equilibrium to the opposite side. We can therefore say that, if the reaction is at equilibrium and we increase the temperature, the reaction goes to the left—the concentrations of H_2 and O_2 increase and that of H_2O decreases. This is true of all exothermic reactions.

An **increase** in temperature drives an **exothermic** reaction toward the **reactants** (to the left).

A **decrease** in temperature drives an **exothermic** reaction toward the **products** (to the right).

For an endothermic reaction, of course, the opposite is true:

An **increase** in temperature drives an **endothermic** reaction toward the **products.**

A **decrease** in temperature drives an **endothermic** reaction toward the **reactants.**

Remember that *a change in temperature changes not only the position of equilibrium but the value of* K *as well.*

EXAMPLE

The conversion of nitrogen dioxide to dinitrogen tetroxide is an exothermic reaction:

$$2NO_2(g) \rightleftharpoons N_2O_4(g) + 13\ 700\ cal$$

NO_2 is a brown gas; N_2O_4 is colorless. In Figure 7.12 we see that at 50°C (right) the brown color is darker; hence, the NO_2 concentration is higher than at 0°C (left). Why?

Answer • To go from 0 to 50°C, heat must be added. But heat is a product of this equilibrium reaction, as written. The addition of heat, therefore, will shift the equilibrium to the left. This shift produces more NO_2, hence the darker brown color.

Problem **7.6**

In a reaction $A \rightleftharpoons B$, the amount of B has increased when the temperature of the reaction was raised. Was this an exothermic or an endothermic reaction? Explain.

Box 7E

The Haber Process

Human beings and animals need proteins and other nitrogen-containing compounds in order to live. Ultimately, the nitrogen in these compounds comes from the plants that we eat. There is plenty of N_2 in the atmosphere, but the only way that nature converts this N_2 to compounds usable by biological organisms is by certain bacteria that have the ability to "fix" atmospheric N_2, that is, convert it to ammonia, NH_3. Most of these bacteria live in the roots of certain plants such as clover, alfalfa, peas, and beans. However, the amount of N_2 fixed by such bacteria each year is far less than the amount necessary to feed all the humans and animals in the world.

The world today can support its population only by artificial fixing, primarily by the **Haber process,** which converts N_2 to NH_3 by the reaction

$$N_2(g) + 3H_2(g) \rightleftharpoons 2NH_3(g) + 22\,000 \text{ cal}$$

As indicated above, this is an exothermic equilibrium process. Early workers on the problem of fixing nitrogen were troubled by a conflict between equilibrium and rate. Since this is an exothermic reaction, an increase in temperature drives the equilibrium to the left, so the best results (largest possible yield of NH_3) should be obtained at low temperatures. But at low temperatures, the *rate* is too low to produce any meaningful amounts of NH_3. In 1908 Fritz Haber (1868–1934) solved this problem by discovering a catalyst that permits the reaction to take place at a convenient rate at 500°C.

The NH_3 produced by the Haber process is converted to fertilizers, which are used all over the world. Without these fertilizers, food production would diminish so much that widespread starvation would result.

An industrial plant for the manufacture of ammonia. (M. W. Kellogg Company.)

Addition of a Catalyst

As we saw in Section 7.4, a catalyst increases the rate of a reaction without itself being changed. For a reversible reaction, catalysts always increase the rates of both the forward and reverse reactions to the same extent. Therefore, **the addition of a catalyst has no effect on the position of equilibrium.** However, adding a catalyst to a system not yet at equilibrium causes it to reach equilibrium faster than it would without the catalyst.

SUMMARY

Some reactions are fast; others are slow. The **rate of a reaction** is the change in concentration of a reactant or a product per unit time. Reactions take place when molecules or ions collide. The rate of a reaction increases with an increasing number of **effective collisions,** that is, ones that lead to a reaction. The energy necessary for a reaction to begin is the **activation energy.** Effective collisions are those that have (a) more than the activation energy and (b) the proper orientation in space.

The lower the activation energy, the faster the reaction. An energy diagram shows the progress of a reaction. The position at the top of the curve is the **transition state.**

Reaction rates generally increase with increasing concentration and temperature and also depend on the nature of the reactants. The rate of some reactions can be increased by the addition of a **catalyst,** which lowers the activation energy.

Many reactions are reversible and eventually reach equilibrium—some slowly, some quickly. At **equilibrium,** the forward and reverse reactions take place at equal rates, and concentrations do not change. Every equilibrium has an **equilibrium expression** and an **equilibrium constant,** K, which does not change when concentrations change but does change when the temperature changes. The equilibrium constant is large for some reactions and small for others. There is no relationship between K and the rate of a reaction.

Le Chatelier's principle tells us what happens when we put stress on a system in equilibrium. Addition of a component causes the equilibrium to shift to the opposite side. Removal of a component causes the equilibrium to shift to the side the component is taken from. Increasing the temperature drives an exothermic equilibrium to the side of the reactants. Addition of a catalyst has no effect on the position of equilibrium.

KEY TERMS

Activation energy (Sec. 7.2)
Catalyst (Sec. 7.4)
Effective collision (Sec. 7.2)
Equilibrium (Sec. 7.5)
Equilibrium constant (Sec. 7.6)
Equilibrium expression (Sec. 7.6)
Haber process (Box 7E)
Heterogeneous catalyst (Sec. 7.4)
Homogeneous catalyst (Sec. 7.4)
Kinetics (Sec. 7.1)
Le Chatelier's principle (Sec. 7.8)
Rate of a reaction (Sec. 7.1)
Reversible reaction (Sec. 7.5)
Transition state (Sec. 7.3)

PROBLEMS

Reaction Rates

7.7 Define reaction rate. What might be a typical unit for reaction rate?

7.8 In the reaction

$$CH_3Br + I^- \longrightarrow CH_3I + Br^-$$

we start the reaction with an *initial* I^- concentration of 0.450 M. After ten minutes the I^- concentration is 0.268 M. What is the rate of reaction?

7.9 If the concentration of a product increases from 0.320 mole/L to 0.750 mole/L over a period of 1 hour and 10 minutes, what is the average rate per minute?

Molecular Collisions and Activation Energy

7.10 Give two reasons why most collisions between N_2 and O_2 molecules do not result in a reaction to give NO.

7.11 In the reaction

$$H_2(g) + Br_2(g) \longrightarrow 2HBr(g)$$

how many covalent bonds are broken and how many new ones are formed?

7.12 Is the breaking of chemical bonds an exothermic or an endothermic process, or is no energy released or absorbed? Explain.

7.13 Why are reactions between ions in aqueous solution generally much faster than reactions between covalent molecules?

7.14 A certain reaction is exothermic by 9 kcal/mole and has an activation energy of 14 kcal/mole. Draw an energy curve for this reaction, and label the transition state.

Factors Affecting Rates of Reaction

7.15 A quart of milk quickly spoils if left at room temperature but keeps for several days in a refrigerator. Explain.

7.16 Explain why the rates of most reactions increase when the concentration of reactants is increased.

7.17 If a certain reaction takes 16 hours to go to completion at 10°C, at about what temperature should we run it if we want it to go to completion in 1 hour?

***7.18** In most cases when we run a reaction by mixing a fixed quantity of A with a fixed quantity of B, the rate of the reaction begins at a maximum and then decreases as time goes by. Explain.

7.19 If you were running a reaction and wished it to go faster, what three things might you try to accomplish this?

7.20 What factors determine whether a reaction run at a given temperature will be fast or slow?

7.21 Explain how a catalyst increases the rate of a reaction.

7.22 If you add a piece of marble ($CaCO_3$) to a 6 *M* HCl solution at room temperature you will see some bubbles form around the marble as H_2 gas slowly rises. If you crush another piece of marble and add it to the same solution at the same temperature, you will see vigorous gas formation, so much so that the solution appears to be boiling. Explain.

Reversible Reactions and Equilibrium

7.23 Burning a piece of paper is an irreversible reaction. Can you give some other examples of irreversible reactions?

7.24 If the reaction $PCl_3 + Cl_2 \rightleftharpoons PCl_5$ is at equilibrium, are the concentrations of PCl_3, Cl_2, and PCl_5 necessarily equal? Is the concentration of PCl_3 necessarily equal to that of Cl_2? Explain.

Equilibrium Constants

7.25 Write equilibrium expressions for
(a) $COCl_2 \rightleftharpoons CO + Cl_2$
(b) $2H_2O + 2SO_2 \rightleftharpoons 2H_2S + 3O_2$
(c) $2C_2H_6 + 7O_2 \rightleftharpoons 4CO_2 + 6H_2O$

7.26 Write the chemical equations corresponding to the following equilibrium expressions:

(a) $K = \dfrac{[H_2CO_3]}{[CO_2][H_2O]}$

(b) $K = \dfrac{[P_4][O_2]^5}{[P_4O_{10}]}$

(c) $K = \dfrac{[F_2]^3[PH_3]}{[HF]^3[PF_3]}$

(d) $K = \dfrac{[Ca]^2[O_2]}{[CaO]^2}$

7.27 When a mixture containing H_2, CO_2, H_2O, and CO at 987°C reached equilibrium, the concentrations were

$[H_2] = 3.37\ M \quad [CO_2] = 0.133\ M$

$[H_2O] = 0.72\ M \quad [CO] = 0.933\ M$

What is the equilibrium constant at 987°C for the reaction

$$H_2(g) + CO_2(g) \rightleftharpoons H_2O(g) + CO(g)$$

7.28 When a mixture containing N_2O_4 and NO_2 at 25°C reached equilibrium, the concentrations were

$[N_2O_4] = 0.877\ M \quad [NO_2] = 0.0714\ M$

What is the equilibrium constant at 25°C for the reaction

$$2NO_2(g) \rightleftharpoons N_2O_4(g)$$

7.29 When a mixture containing NOCl, NO, and Cl_2 at 25°C reached equilibrium, the concentrations were

$[NO] = 1.4\ M \quad [NOCl] = 2.6\ M$

$[Cl_2] = 0.34\ M$

What is the equilibrium constant at 25°C for the reaction

$$2NOCl(g) \rightleftharpoons 2NO(g) + Cl_2(g)$$

7.30 The reaction

$$NH_3(aq) + H_2O(l) \rightleftharpoons NH_2^-(aq) + H_3O^+(aq)$$

is very fast but has an equilibrium constant of 10^{-34}. Would you be wise to try to run it?

7.31 Equilibrium constants for several reactions are
(a) 4.5×10^8 (d) 3×10^{-7}
(b) 32 (e) 0.0032
(c) 4.5
Which of them favor the formation of products and which favor the formation of reactants?

7.32 A particular reaction has an equilibrium constant of 1.13 under one set of conditions and 1.72 under a different set of conditions. Which conditions would be more advantageous in an industrial process to obtain the maximum amount of products?

Equilibrium and Reaction Rates

7.33 (a) If a reaction is very exothermic—that is, the products have a much lower energy than the reactants—can we be reasonably certain that it will take place rapidly? (b) If a reaction is very endothermic—that is, the products have a much higher energy than the reactants—can we be reasonably certain that it will take place extremely slowly or not at all? (Think carefully about this one.)

Le Chatelier's Principle

7.34 The reaction

$$H_2(g) + I_2(g) \rightleftharpoons 2HI(g)$$

is exothermic. If the reaction is at equilibrium, tell whether the equilibrium will be shifted to the right or the left if we
(a) remove some HI
(b) add some H_2
(c) remove some I_2
(d) increase the temperature
(e) add a catalyst

7.35 The reaction

$$3O_2(g) \rightleftharpoons 2O_3(g)$$

is endothermic. If the reaction is at equilibrium, tell whether the equilibrium will be shifted to the right or the left if we
(a) remove some O_3
(b) remove some O_2
(c) add some O_3
(d) decrease the temperature
(e) add a catalyst

7.36 The reaction

$$C(s) + H_2O(g) \rightleftharpoons CO(g) + H_2(g)$$

is exothermic. If the reaction is at equilibrium, tell whether the equilibrium will be shifted to the right or to the left if we
(a) add some H_2
(b) remove some CO
(c) remove some H_2O
(d) decrease the temperature
(e) add a catalyst

7.37 The reaction

$$4NO_2(g) + 6H_2O(g) \rightleftharpoons 7O_2(g) + 4NH_3(g)$$

is an endothermic process. What happens to the position of equilibrium if we raise the reaction temperature from 350 to 420°C?

7.38 Is there any change in conditions that changes the equilibrium constant K of a given reaction?

***7.39** The equilibrium constant at 1127°C for the endothermic reaction

$$2H_2S(g) \rightleftharpoons 2H_2(g) + S_2(g)$$

is 571. If the mixture is at equilibrium, what happens to K if we
(a) add some H_2S
(b) add some H_2
(c) lower the temperature to 1000°C?

Boxes

7.40 (Box 7A) In a bacterial infection body temperature may rise to 101°F. Does this body defense kill the bacteria directly by heat or by another mechanism? If so, which mechanism?

7.41 (Boxes 7A and 7B) Why is a high fever dangerous? Why is a low body temperature dangerous?

7.42 (Box 7B) Why do surgeons sometimes lower body temperatures during heart operations?

7.43 (Box 7C) How do time-release medicines work?

7.44 (Box 7D) What reaction takes place when sunlight hits the compound silver chloride?

7.45 (Box 7E) If the equilibrium for the Haber process is unfavorable at high temperatures, why do the factories nevertheless use high temperatures?

Additional Problems

7.46 In the reaction

$$H_2(g) + Cl_2(g) \longrightarrow 2HCl(g)$$

a 10°C increase in temperature doubles the rate of reaction. If the rate of reaction at 15°C is 2.8 moles of HCl per liter per second, what is the rate at −5°C and at 45°C?

***7.47** Draw an energy diagram for an exothermic reaction that yields 75 kcal/mole. The energy of activation is 30 kcal/mole.

7.48 Write equilibrium expressions for these ionic equilibria:
(a) $2OH^-(aq) + H_2PO_4^-(aq) \rightleftharpoons PO_4^{3-}(aq) + 2H_2O(l)$
(b) $H_2O(l) + HNO_3(aq) \rightleftharpoons H_3O^+(aq) + NO_3^-(aq)$
(c) $2H_2O(l) \rightleftharpoons OH^-(aq) + H_3O^+(aq)$

7.49 A mixture of hydrogen and oxygen gases can be kept for many years without reacting, but if we light a match the two react so rapidly that an explosion results. Explain how the lighted match causes such a large change in the reaction rate.

7.50 The equilibrium constant for

$$COCl_2(g) \rightleftharpoons CO(g) + Cl_2(g)$$

is 25. A measurement made on the equilibrium mixture found that the concentrations of CO and Cl_2 were each 0.80 M. What is the concentration of $COCl_2$ at equilibrium?

7.51 In the reaction

$$N_2O_4(g) \longrightarrow 2NO_2(g)$$

the concentration of N_2O_4 was measured at the end of the times shown:

Time(s)	$[N_2O_4]$
0	0.200
10	0.180
20	0.160
30	0.140

What is the rate of reaction?

***7.52** How could you increase the rate of a gaseous reaction without adding more reactants or a catalyst and without changing the temperature?

***7.53** The equilibrium constant for the reaction α-D-glucose $\rightleftharpoons$ β-D-glucose is 1.5 at 30°C. This reaction takes place only in aqueous solution. If you begin with a fresh solution of 1 *M* α-D-glucose in water, what will be its concentration when equilibrium is reached?

***7.54** In an endothermic reaction, A → B, the activation energy was 100 kcal/mole. Would the activation energy of the reverse reaction, B → A, also be 100 kcal/mole, or would it be more or less? Explain with the aid of a diagram.

7.55 The reaction

$$C_2H_4(g) + Cl_2(g) \rightleftharpoons C_2H_4Cl_2(g)$$

has an equilibrium constant of 10. What is the equilibrium concentration of $C_2H_4Cl_2$ if the equilibrium concentration of C_2H_4 is 0.2 *M* and that of Cl_2 is 0.4 *M*?

7.56 For the reaction

$$H_2(g) + Br_2(g) \rightleftharpoons 2HBr(g)$$

the equilibrium concentrations are $[H_2] = 0.37$ *M*, $[Br_2] = 0.70$ *M*, $[HBr] = 3.6$ *M*. What is the equilibrium constant?

***7.57** Assume that there are two different reactions taking place at the same temperature. In A, two different spherical molecules collide to yield a product. In B, the shape of the colliding molecules is rod-like. Each reaction has the same number of collisions per second and the same activation energy. Which reaction goes faster?

***7.58** Is it possible for an endothermic reaction to have zero activation energy?

***7.59** In the reaction $2HI(g) \rightarrow H_2(g) + I_2(g)$ the rate of appearance of I_2 is measured. The following concentrations were found at the end of the times shown:

Time (min)	$[I_2]$
0	0
10	0.30
20	0.60
30	0.90

What is the rate of reaction?

Chapter 8

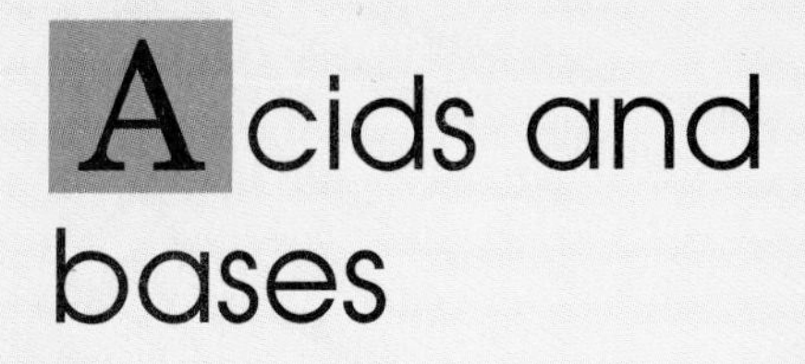

Acids and bases

8.1 Introduction

The word "alkali" is an older term for "base"; it means the same thing.

We frequently encounter acids and bases in our daily lives. Oranges, lemons, and vinegar are examples of acidic foods. Sulfuric acid is in our automobile batteries. As for bases, we take antacid tablets for heartburn and use household ammonia as a cleaning agent. What do these substances have in common, and why are acids and bases usually discussed together?

In 1884 a young Swedish chemist named Svante Arrhenius (1859–1927) proposed an answer to the first of these questions by giving what was then a new definition of acids and bases. In modern terms, his definitions are

An **acid** is a **substance that produces H_3O^+ ions in aqueous solution.**

A **base** is a **substance that produces OH^- ions in aqueous solution.**

This definition of acid is a slight modification of what Arrhenius himself said, which was that an acid produces hydrogen ions, H^+. Today we know that H^+ ions cannot exist in water. A H^+ ion is a bare proton, and a charge of +1 is too concentrated for such a tiny particle (Sec. 3.2). Because of this, any H^+ ion in water immediately combines with an H_2O molecule to give a **hydronium ion, H_3O^+:**

Although all chemists know that acidic aqueous solutions do not contain H^+ ions, they frequently use the terms "H^+" and "proton" when they really mean "H_3O^+." The three terms are generally interchangeable.

$$H^+(aq) + H_2O(l) \longrightarrow H_3O^+(aq)$$

Apart from this modification, the Arrhenius definitions of acid and base are still valid and useful today, after more than 100 years, as long as we are talking about aqueous solutions. When we dissolve an acid in water, it reacts with the water to produce H_3O^+. For example, hydrogen chloride, HCl, in its pure state is a poisonous, choking gas. When dissolved in water, it reacts as follows:

$$\text{H—}\ddot{\underset{..}{\text{Cl}}}\text{:} + \text{H—}\ddot{\underset{..}{\text{O}}}\text{—H} \longrightarrow \left[\begin{array}{c}\text{H—}\ddot{\text{O}}\text{—H} \\ | \\ \text{H}\end{array}\right]^+ + \text{:}\ddot{\underset{..}{\text{Cl}}}\text{:}^-$$

The bottle on the shelf labeled "HCl" is actually not HCl but a solution of H_3O^+ and Cl^- ions in water.

With bases the situation is slightly different. Many bases are solid metallic hydroxides, such as KOH, NaOH, $Mg(OH)_2$, and $Ca(OH)_2$. These compounds are ionic even in the solid state. When dissolved in water, the ions merely separate, and each ion is solvated by water molecules (Sec. 6.6). For example,

$$NaOH(s) \xrightarrow{H_2O} Na^+(aq) + OH^-(aq)$$

Box 8A

Some Important Acids and Bases

Strong acids. Sulfuric acid, H_2SO_4, is used in a great many industrial processes. More tons of sulfuric acid are manufactured in the United States than any other chemical, organic or inorganic. It is a strong acid and a powerful oxidizing agent, as well as a dehydrating agent. The dissolving of H_2SO_4 in water gives off a lot of heat. The acid should always be added to water, and *not* water to it, because of the danger of splashing.

Hydrochloric acid, HCl, is an important acid in chemistry laboratories. Pure HCl is a gas, and the HCl in laboratories is an aqueous solution. HCl is the acid in the gastric fluid in your stomach, where it is secreted at a strength of about 5 percent.

Nitric acid, HNO_3, is a strong oxidizing agent. A drop of this acid causes the skin to turn yellow. This is due to a reaction with skin protein. A yellow color with nitric acid has long been a test for proteins.

Weak acids. Acetic acid, CH_3COOH, is present in vinegar (about 5 percent). Vinegar is made by allowing wine to oxidize in air, whereupon the ethyl alcohol in the wine is oxidized to acetic acid (Sec. 12.3). Pure acetic acid is called *glacial* because of its melting point of 17°C, which means that it freezes on a moderately cold day.

Boric acid, H_3BO_3, is a solid. Solutions of boric acid in water were once used as antiseptics, especially for eyes. Boric acid is toxic when swallowed.

Phosphoric acid, H_3PO_4, is one of the strongest of the weak acids. The ions produced from it—$H_2PO_4^-$, HPO_4^{2-}, and PO_4^{3-}—are important in biochemistry (see also Sec. 14.12). Phosphates are used in detergents.

Strong bases. Sodium hydroxide, NaOH, also called lye, is the most important of the strong bases. It is a solid whose aqueous solutions are used in many industrial processes, including the manufacture of glass and soap.

Potassium hydroxide, KOH, also a solid, is used for many of the same purposes as NaOH.

Weak bases. Ammonia, NH_3, the most important weak base, is a gas with many industrial uses. One of its chief uses is for fertilizers. A 5 percent solution is sold in supermarkets as a cleaning agent, and weaker solutions are used as "spirits of ammonia" to revive people who have fainted.

Magnesium hydroxide, $Mg(OH)_2$, is a solid that is insoluble in water. A suspension of about 8 percent $Mg(OH)_2$ in water is called milk of magnesia and is used as a laxative.

Copper (*left*) and zinc (*center*) react with concentrated nitric acid. (From Metcalfe, Williams, and Castka, *Modern Chemistry*, Holt, Rinehart, and Winston, 1986.)

The boiling point of ammonia is −33°C.

Other bases, though, are not hydroxides. These bases produce OH^- ions in water by reacting with the water. The most important example of this kind of base is ammonia, NH_3. Like HCl, ammonia is a poisonous, choking gas. When dissolved in water, it reacts as follows:

We show the equation with equilibrium arrows because it is reversible.

$$NH_3(aq) + H_2O(l) \rightleftharpoons NH_4^+(aq) + OH^-(aq)$$

Ammonia produces OH^- ions by taking H^+ from water molecules and leaving OH^- behind. This equilibrium lies well to the left. In a 1 *M* solution of NH_3 in water, only about 4 molecules of NH_3 out of every 1000 have reacted to form NH_4^+. Nevertheless, some OH^- ions are produced, so NH_3 is therefore a base.

Another reason not to use these labels is that NH_4OH cannot be isolated away from water as a pure substance.

Bottles of NH_3 in water are sometimes labeled "ammonium hydroxide" or "NH_4OH" but this gives a false impression of what is really in the bottle. Most of the NH_3 molecules have not reacted with the water, so the bottle contains mostly NH_3 and H_2O and only a little NH_4^+ and OH^-.

8.2 Acid and Base Strength

All acids are not equally strong. According to the Arrhenius picture of acids, a **strong acid** is one that **reacts almost completely with water to form H_3O^+ ions.** A **weak acid** reacts much less completely. There are six common acids that react completely, or almost completely, when dissolved in water. They are shown in Table 8.1. Like HCl, the other five acids listed exist in water almost entirely as H_3O^+ and the corresponding anion. All other acids are weaker and produce a much smaller concentration of H_3O^+ ions. For example, acetic acid in water exists primarily as acetic acid molecules. The equilibrium

$$\underset{\text{Acetic acid}}{CH_3COOH(aq)} + H_2O(l) \rightleftharpoons \underset{\text{Acetate ion}}{CH_3COO^-(aq)} + H_3O^+(aq)$$

lies well to the left, and only a few acetic acid molecules are converted to acetate ions. Many acids are even weaker than acetic acid.

Table 8.1 The Six Strong Acids and Four Strong Bases

Acids		Bases	
Formula	***Name***	***Formula***	***Name***
$HClO_4$	Perchloric acid	NaOH	Sodium hydroxide
HI	Hydroiodic acid	KOH	Potassium hydroxide
HBr	Hydrobromic acid	LiOH	Lithium hydroxide
H_2SO_4	Sulfuric acid	$Ba(OH)_2$	Barium hydroxide
HNO_3	Nitric acid		
HCl	Hydrochloric acid		

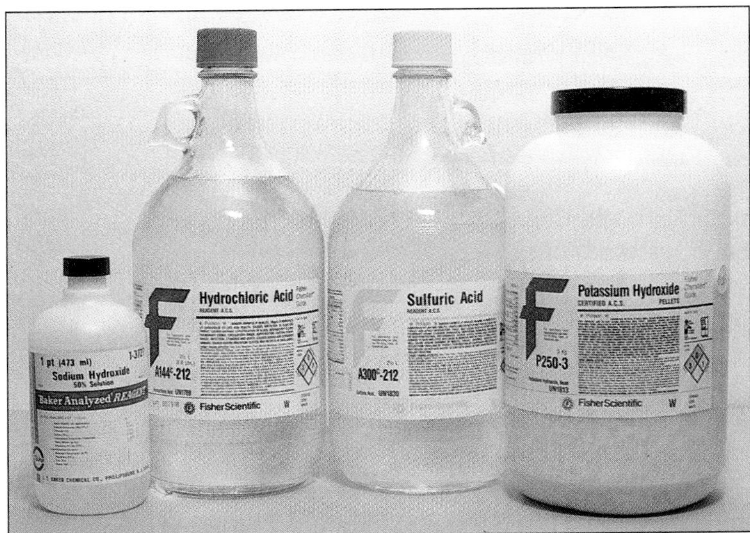

Some strong acids and strong bases. (Photograph by Beverly March.)

A similar situation exists with bases. There are only four common strong bases (Table 8.1), all of the metal hydroxide type. As we saw in Section 8.1, ammonia is a weak base because the equilibrium

These bases are strong because they are soluble in water and dissociate completely to OH^-.

$$NH_3(aq) + H_2O(l) \rightleftharpoons NH_4^+(aq) + OH^-(aq)$$

lies to the left.

It is important to understand that the *strength* of an acid or base is not related to its *concentration*. HCl is a strong acid, whether it is concentrated or dilute, because in its reaction with water the equilibrium lies well to the right:

$$HCl(aq) + H_2O(l) \rightleftharpoons H_3O^+(aq) + Cl^-(aq)$$

Acetic acid is a weak acid, whether it is concentrated or dilute, because in its reaction with water the equilibrium lies well to the left.

In Section 6.6 we saw that electrolytes (substances that produce ions in solution) can be strong or weak. The strong acids and bases in Table 8.1 are strong electrolytes. Almost all other acids and bases are weak electrolytes.

8.3 Brønsted-Lowry Acids and Bases

The Arrhenius definitions of acid and base are very useful in aqueous solutions. But what if water is not involved? In 1923 new definitions of acids and bases were proposed simultaneously by Brønsted and Lowry. These definitions are more general and do not require that the solvent be water. They are:

Johannes Brønsted (1879–1947) was a Danish chemist, and Thomas Lowry (1874–1936) was an English chemist.

An **acid** is **a proton donor.**
A **base** is **a proton acceptor.**
An **acid-base reaction** is **the transfer of a proton.**

As noted in Section 8.1, this is precisely how acids react with water. An acid gives a proton, H^+, to water:

$$\underset{\text{Acid}}{HCl} + \underset{\text{Base}}{H_2O} \rightleftharpoons \underset{\text{Acid}}{H_3O^+} + \underset{\text{Base}}{Cl^-}$$

We have, in this equation, two new ideas:

1. The water, because it accepts the proton, is a base by the Brønsted-Lowry definition.
2. The hydronium ion is an acid since it can now, at least in theory, give a proton back to the Cl^-. Similarly, the Cl^- is a base because it can take the proton back.

To repeat, an acid, in the Brønsted-Lowry picture, is a proton donor. When it gives up its proton, it always becomes a base. This base is called the **conjugate base** of the acid. Some examples are given in Table 8.2.

Similarly, a base is a proton acceptor. When it accepts a proton, it always becomes an acid called the **conjugate acid** of that base. The examples in Table 8.2 illustrate this also. Every acid and the species that remains when it loses its proton are called a **conjugate acid-base pair.**

Table 8.2 Some Acids and Their Conjugate Bases, in Decreasing Order of Acid Strength

	Acid		Conjugate Base		
Strong Acids	HI	Hydroiodic acid	I^-	Iodide ion	**Weak Bases**
	H_2SO_4	Sulfuric acid	HSO_4^-	Hydrogen sulfate ion	
	HCl	Hydrochloric acid	Cl^-	Chloride ion	
	HNO_3	Nitric acid	NO_3^-	Nitrate ion	
	H_3O^+	Hydronium ion	H_2O	Water	
	HSO_4^-	Hydrogen sulfate ion	SO_4^{2-}	Sulfate ion	
	H_3PO_4	Phosphoric acid	$H_2PO_4^-$	Dihydrogen phosphate ion	
	CH_3COOH	Acetic acid	CH_3COO^-	Acetate ion	
	H_2CO_3	Carbonic acid	HCO_3^-	Bicarbonate ion	
	H_2S	Hydrogen sulfide	HS^-	Hydrogen sulfide ion	
	$H_2PO_4^-$	Dihydrogen phosphate ion	HPO_4^{2-}	Hydrogen phosphate ion	
	NH_4^+	Ammonium ion	NH_3	Ammonia	
	C_6H_5OH	Phenol	$C_6H_5O^-$	Phenylate ion	
Weak Acids	HCO_3^-	Bicarbonate ion	CO_3^{2-}	Carbonate ion	**Strong Bases**
	HPO_4^{2-}	Hydrogen phosphate ion	PO_4^{3-}	Phosphate ion	
	H_2O	Water	OH^-	Hydroxide ion	
	C_2H_5OH	Ethyl alcohol	$C_2H_5O^-$	Ethoxide ion	
	NH_3	Ammonia	NH_2^-	Amide ion	

Thus HCl and Cl^- are a conjugate acid-base pair, as are H_3O^+ and H_2O, and CH_3COOH and CH_3COO^-.

Note the following points from the examples in Table 8.2.

1. An acid can be positively or negatively charged, or it can be neutral. A base can be negatively charged, or it can be neutral.

2. Some acids can give up two or three protons (other acids, not shown in Table 8.2, can give up even more). Such acids are called **diprotic** or **triprotic.** For example, H_2SO_4 can give up two protons to become SO_4^{2-} and H_3PO_4 can give up three to become PO_4^{3-}. In the Brønsted-Lowry picture, however, each acid is considered to give up only one proton **(monoprotic).** Thus H_2SO_4 is treated as if it gives up a single proton to become HSO_4^-, and HSO_4^- is regarded as a separate acid in its own right (note that HSO_4^- is listed separately in Table 8.2).

3. Some substances appear on *both* sides of Table 8.2. For example, HCO_3^-, the bicarbonate ion, can give up a proton to become CO_3^{2-} (thus it is an acid) or take a proton to become H_2CO_3 (thus it is a base). A substance that can act as either an acid or a base is called **amphoteric.** The most important amphoteric substance in Table 8.2 is of course water, which takes a proton to become H_3O^+ or loses one to become OH^-.

4. A substance cannot be a Brønsted-Lowry acid unless it contains a hydrogen atom, but not all hydrogen atoms can in fact be given up. For example, acetic acid, CH_3COOH, has four hydrogens but is monoprotic. It gives up only one of them. Similarly, phenol, C_6H_5OH, gives up only one of its six hydrogens.

An original box of Arm & Hammer baking soda (sodium bicarbonate). Sodium bicarbonate is composed of Na^+ and HCO_3^-, the amphoteric bicarbonate ion. (Arm & Hammer.)

We have seen that HCl reacts with water as follows:

$$\underset{\text{Acid}}{HCl} + \underset{\text{Base}}{H_2O} \rightleftharpoons \underset{\text{Acid}}{H_3O^+} + \underset{\text{Base}}{Cl^-}$$

The equilibrium is far to the right. Out of every 1000 HCl molecules dissolved in water, more than 990 are converted to Cl^- ions. However, H_3O^+ is an acid and Cl^- a base. Why doesn't H_3O^+ give a proton to Cl^- to produce HCl and H_2O? What we are really asking here is, why does the equilibrium lie far to the right in this case?

We can answer this question by using the concept of acid and base strength. In the Brønsted-Lowry picture, **acid strength** is **the willingness to give up the proton** and **base strength** is **the willingness to accept the proton.** Because HCl is more willing to give up its proton, it is a stronger acid than H_3O^+. Looking at it the other way, H_2O is a stronger base than Cl^- and is therefore more willing to accept the proton. **In the battle for the proton, the stronger base always wins.** Since H_2O is a stronger base than Cl^-, it gets the proton.

It should be obvious that the stronger an acid is, the weaker must be its conjugate base, since the more willing an acid is to donate the proton, the less willing is the conjugate base to take it back.

The acids in Table 8.2 are listed in order of *decreasing* strength (strongest at the top, weakest at the bottom). It follows that the conjugate bases are listed in order of *increasing* strength (strongest at the bottom, weakest at the top). We can therefore use this table (or any other table that lists acids with the strongest at the top) to predict whether or not any acid will donate its proton to any base. The rule is that **any acid in Table 8.2 donates its proton to the conjugate base of any acid *below* it but does not donate its proton to the conjugate base of any acid above it.**

These reactions are all equilibria, and what we are really doing is predicting in which direction the equilibria lie.

EXAMPLE

Which of these reactions proceed from left to right?

(a) $H_3O^+(aq) + CH_3COO^-(aq) \rightleftharpoons H_2O(l) + CH_3COOH(aq)$
(b) $H_3O^+(aq) + I^-(aq) \rightleftharpoons HI(aq) + H_2O(l)$
(c) $H_2CO_3(aq) + OH^-(aq) \rightleftharpoons HCO_3^-(aq) + H_2O(l)$
(d) $HPO_4^{2-}(aq) + NH_3(aq) \rightleftharpoons PO_4^{3-}(aq) + NH_4^+(aq)$

Answer • All we need to do is find the acids and bases in Table 8.2. Reactions (a) and (c) proceed because H_3O^+ is above CH_3COOH and H_2CO_3 is above H_2O. Reactions (b) and (d) do not proceed because H_3O^+ is below HI and HPO_4^{2-} is below NH_4^+.

Problem **8.1**

Which of these reactions proceed from left to right?

(a) $CH_3COO^-(aq) + NH_4^+(aq) \rightleftharpoons CH_3COOH(aq) + NH_3(aq)$
(b) $H_3PO_4(aq) + HS^-(aq) \rightleftharpoons H_2PO_4^-(aq) + H_2S(aq)$

8.4 Acid Dissociation Constants

In Section 8.2 we learned that acids vary in the extent to which they produce H_3O^+ when added to water. The six strong acids (see Table 8.1) are completely converted to H_3O^+ and the corresponding anions, whereas weak acids undergo this reaction to a lesser extent.

Since these reactions are all equilibria, we can use equilibrium constants to tell us quantitatively just how strong any weak acid actually is. The reaction that takes place when a weak acid is added to water is

HA is the general formula for an acid.

$$HA(aq) + H_2O(l) \rightleftharpoons H_3O^+(aq) + A^-(aq)$$

The equilibrium expression (Sec. 7.6) is

$$K = \frac{[H_3O^+][A^-]}{[HA][H_2O]} \tag{8.1}$$

Notice that this expression contains the concentration of water. Since water is the solvent, its concentration changes very little when we add HA, so we can treat $[H_2O]$ as constant and rearrange Equation 8.1 to bring the constant terms together on the left side:

$$K[H_2O] = \frac{[H_3O^+][A^-]}{[HA]} \tag{8.2}$$

Box 8B

Acid and Alkali Burns of the Cornea

In the laboratory every student must wear safety glasses to prevent the accidental splashing of chemicals into the eyes. The outermost part of the eye is the cornea. This transparent tissue is very sensitive to chemical burns, whether caused by acids or by bases. Acids and bases in small amounts and in low concentrations that would not seriously damage other tissues, such as skin, may cause severe corneal burns. If not treated promptly, these can lead to permanent loss of vision.

Acids are less dangerous than bases.

For acid in the eyes, first aid should be provided by immediately washing the eyes with a steady stream of cold water. The fast removal of the acid is of utmost importance. Time should not be wasted looking for some mild alkaline solution to neutralize the acid because the removal of the acid is more helpful than the neutralization. Even if it is sulfuric acid in the eyes, washing with copious quantities of water is *the* remedy. The heat generated by the interaction of sulfuric acid with water (lots of water) will be removed by the washing. The extent of the burn can be assessed after all the acid has been washed out.

Not so with alkali (base) burns. Bases are dangerous chemicals, not only in the laboratory and in industry but also in the home. Concentrated ammonia or sodium hydroxide of the "liquid plumber" type can cause severe damage when splattered in the eyes. Again, immediate washing is of utmost importance, but alkali burns can later develop ulcerations in which the healing wound deposits scar tissue that is not transparent, so that vision is impaired. Fortunately, the cornea is a tissue with no blood vessels and therefore can be transplanted without immunological rejection problems. Today, corneal grafts are common.

Cleaning products like these contain bases. Care must be taken to make sure that they are not splashed into anyone's eyes. (Photograph by Charles D. Winters.)

Table 8.3 Some K_a and pK_a Values[a] for Weak Acids, in Decreasing Order of Strength

Name	Formula	K_a	pK_a
Phosphoric acid	H_3PO_4	7.5×10^{-3}	2.1
Hydrofluoric acid	HF	3.5×10^{-4}	3.45
Formic acid	HCOOH	1.8×10^{-4}	3.75
Lactic acid	$CH_3CH(OH)COOH$	1.4×10^{-4}	3.85
Acetic acid	CH_3COOH	1.8×10^{-5}	4.75
Carbonic acid	H_2CO_3	4.3×10^{-7}	6.37
Dihydrogen phosphate ion	$H_2PO_4^-$	6.2×10^{-8}	7.21
Boric acid	H_3BO_3	7.3×10^{-10}	9.14
Ammonium ion	NH_4^+	5.6×10^{-10}	9.25
Phenol	C_6H_5OH	1.3×10^{-10}	9.9
Bicarbonate ion	HCO_3^-	5.6×10^{-11}	10.25
Hydrogen phosphate ion	HPO_4^{2-}	2.2×10^{-13}	12.66

[a]The pK_a is explained in Section 8.8.

When we multiply one constant, K, by another, $[H_2O]$, we get a third constant, which we call K_a:

$$K_a = K[H_2O] = \frac{[H_3O^+][A^-]}{[HA]} \tag{8.3}$$

K_a is also called the **acidity constant** and the **ionization constant.**

The K_a is called the **acid dissociation constant,** and every weak acid has one. Table 8.3 lists some weak acids and their dissociation constants. The larger the K_a, the stronger the acid. We can see this by looking at the equation

$$HA(aq) + H_2O(l) \rightleftharpoons H_3O^+(aq) + A^-(aq)$$

A stronger acid produces more H_3O^+ ions (and A^- ions also) than a weaker acid. This increases the numerator of Equation 8.3 so that K_a is larger for the stronger acid.

The importance of K_a is that it immediately tells us how strong an acid is. For example, Table 8.3 shows us that although acetic acid, formic acid, and phenol are all weak acids, their strengths as acids are not the same. The table gives K_a for acetic acid as 1.8×10^{-5}. Thus formic acid, with a K_a of 1.8×10^{-4}, is stronger than acetic acid while phenol, whose K_a is 1.3×10^{-10}, is much weaker than acetic acid.

8.5 Some Properties of Acids and Bases

Chemists stopped tasting chemicals when some of them got sick or died from doing this.

Chemists of today do not taste the substances they work with, but 150 and 200 years ago it was routine to do so. That is how we know that acids taste sour and bases taste bitter. The sour taste of lemons, vinegar, and many other foods is due to the acids they contain.

Both strong acids and strong bases damage living tissue. Externally, they cause chemical burns. Internally, they are even more harmful and

All these fruits and fruit drinks contain organic acids. (Photograph by C. D. Winters.)

can cause severe damage to the digestive tract. One way they cause internal damage is by breaking down the organism's protein molecules.

The most important property of acids and bases is that they react with each other in a process called **neutralization.** This is an appropriate name because, when a strong corrosive acid such as hydrochloric acid reacts with a strong corrosive base such as sodium hydroxide, the product (a solution of ordinary table salt in water) has neither acidic nor basic properties. We call such a solution **neutral.** Neutralization reactions are further discussed in the next section.

In the laboratory, you must be careful not to get any strong acids or bases on your skin or, even worse, in your eyes, where they can cause blindness.

8.6 Reactions of Acids

Acids have several characteristic reactions. In all of them, they react as proton donors, giving their protons to a base. In this section we look at three types.

Reactions with Metals, Metal Hydroxides, and Oxides

Acids react with certain metals (called *active metals*) to produce hydrogen (H_2) and a salt. An example is (Fig. 8.1):

$$\underset{\text{Active metal}}{Mg(s)} + \underset{\text{Acid}}{2HCl(aq)} \longrightarrow \underset{\text{A salt}}{MgCl_2(aq)} + \underset{\text{Hydrogen}}{H_2(g)}$$

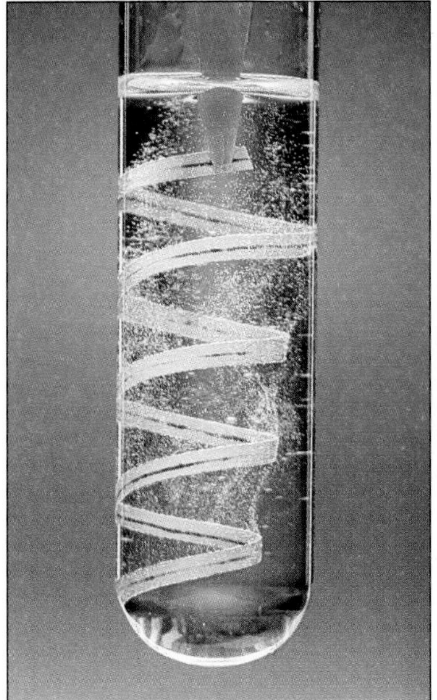
Figure **8.1**
A ribbon of magnesium metal reacts with aqueous HCl to give H_2 gas and aqueous $MgCl_2$. (Photograph by Charles D. Winters.)

Acids react with metal hydroxides to give water and a salt:

$$\underset{\text{Acid}}{HNO_3} + \underset{\text{Base}}{KOH} \longrightarrow H_2O + \underset{\text{A salt}}{KNO_3}$$

A **salt** is **an ionic compound.** We may consider all salts to be composed of a negative ion from an acid and a positive ion from a base. For example, potassium nitrate, KNO_3, is made of nitrate ions, NO_3^-, from nitric acid, HNO_3, and potassium ions, K^+, from potassium hydroxide, KOH. The *actual* equation for the reaction between HNO_3 and KOH is therefore (Sec. 4.7)

$$H_3O^+ + NO_3^- + K^+ + OH^- \rightleftharpoons H_2O + H_2O + NO_3^- + K^+$$

This can be simplified by omitting the spectator ions (K^+ and NO_3^-):

$$H_3O^+ + OH^- \rightleftharpoons H_2O + H_2O$$

This is the actual ionic equation when *any* strong acid reacts in water with *any* strong base. Note that even in this view an acid and a base are converted to a salt and water, and the salt ends up dissolved in the water; an example is KNO_3 in the case mentioned above.

Strong acids also react with metal oxides; for example,

$$2H_3O^+(aq) + CaO(s) \longrightarrow 3H_2O(l) + Ca^{2+}(aq)$$

Reactions with Carbonates and Bicarbonates

When a strong acid is added to a carbonate or bicarbonate, such as sodium carbonate or potassium bicarbonate, bubbles of gas are rapidly given off. The gas, carbon dioxide, is formed in this manner:

$$2H_3O^+(aq) + CO_3^{2-}(aq) \rightleftharpoons H_2CO_3(aq) + 2H_2O(l)$$

$$H_2CO_3(aq) \rightleftharpoons CO_2(g) + H_2O(l)$$

or

$$H_3O^+(aq) + HCO_3^-(aq) \rightleftharpoons H_2CO_3(aq) + H_2O(l)$$

$$H_2CO_3(aq) \rightleftharpoons CO_2(g) + H_2O(l)$$

In either case the direct product of the proton transfer is carbonic acid, H_2CO_3, but this compound is very unstable and breaks down to CO_2 and H_2O. As shown in Tables 8.2 and 8.3, carbonic acid is a weak acid. By the rule given in Section 8.3, not only the six strong acids but any acid stronger than H_2CO_3 reacts with carbonates or bicarbonates to give CO_2 gas.

Reactions with Ammonia and Amines

Any acid stronger than NH_4^+ (see Table 8.2) is strong enough to react with NH_3 in this way:

$$HCl + NH_3 \rightleftharpoons NH_4^+ + Cl^-$$

In Section 15.2 we will meet a family of compounds called amines, which are similar to ammonia except that one or more of the three hydrogen

Box 8C

Drugstore Antacids

Stomach fluid is normally quite acidic because of its HCl content, but sometimes we get "heartburn" that may be caused by excess stomach acidity. Then many of us take an antacid, which, as the name implies, is a substance that neutralizes acids—in other words, a base. The word "antacid" is a medical term, not one used by chemists. It is, however, found on the labels of many medications available in drugstores and supermarkets. Almost all of them use weak bases (hydroxides and/or carbonates) to decrease the acidity of the stomach. The active ingredients of some of these antacids are as follows (some brands are sold in more than one formulation, but we list only one in each case):

Alka-Seltzer: sodium bicarbonate, $NaHCO_3$, citric acid, and aspirin

Brioschi: sodium bicarbonate

Chooz: calcium carbonate, $CaCO_3$

Di-gel: $CaCO_3$, magnesium hydroxide, $Mg(OH)_2$, and simethicone (an anti-flatulent)

Gaviscon: aluminum hydroxide, $Al(OH)_3$, and magnesium carbonate, $MgCO_3$

Gelusil: $Al(OH)_3$, $Mg(OH)_2$, and simethicone

Maalox: $Mg(OH)_2$, $Al(OH)_3$, and simethicone

Mylanta: $Al(OH)_3$, $Mg(OH)_2$, and simethicone

Phillips' Milk of Magnesia: $Mg(OH)_2$

Riopan: magnesium aluminum hydrate (magaldrate)

Rolaids: $CaCO_3$, $Mg(OH)_2$

Tempo: $CaCO_3$, $Al(OH)_3$, $Mg(OH)_2$, and simethicone

Tums: $CaCO_3$

Also in the supermarkets and drugstores are other drugs, labeled "acid reducers," like the antacids available without prescription, that lower stomach acidity without being bases themselves. Among these are Zantac, Tagamet, Pepcid, and Axid. Instead of neutralizing acidity, these compounds restrict the secretion of acid in the stomach. In larger doses (sold only with a prescription) some of these drugs are used in the treatment of stomach ulcers.

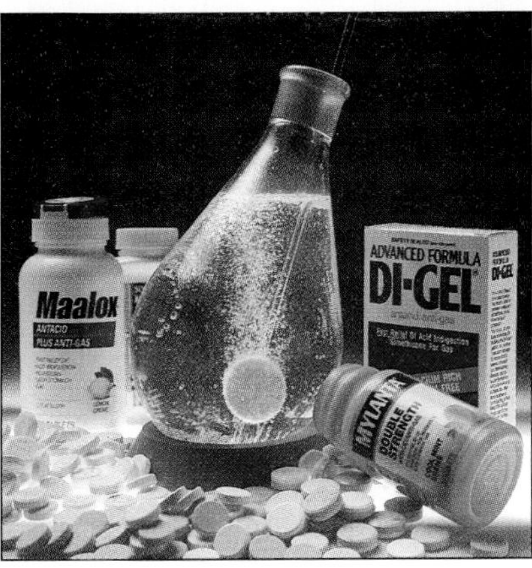

Commercial remedies for excess stomach acid. (Photograph by Charles D. Winters.)

Box 8D

Rising Cakes and Fire Extinguishers

When an acid is added to the bicarbonate ion, CO_2 gas is given off:

$$HCO_3^-(aq) + H_3O^+(aq) \rightleftharpoons H_2O(l) + H_2CO_3(aq) \rightleftharpoons CO_2(g) + H_2O(l)$$

The acid protonates HCO_3^-, and the resulting H_2CO_3, being unstable, breaks down to CO_2 and water. This reaction is used in a number of ways, of which we discuss two.

Cake rising. Baking powder consists of sodium bicarbonate, $NaHCO_3$, and a solid acidic compound. When the powder is dry, little or no reaction takes place, but in the presence of water the acid and bicarbonate ion react to produce CO_2 gas, which is what causes a cake to rise. Potassium acid tartrate, $KC_4H_5O_6$, called cream of tartar, was the earliest acidic compound used for this purpose and is still one of the best, but because of its high cost it has been largely replaced in recent years by monocalcium phosphate or similar compounds. Because of its use in baking powders, $NaHCO_3$ is often called baking soda. Baking powder is seldom used for bread. Bread rising is accomplished by yeast, which produces CO_2 by fermenting the sugars present in the dough.

Fire extinguishers. There are several types of fire extinguishers, but one type uses the reaction between $NaHCO_3$ and an acid to produce CO_2. The pressure of the CO_2 forces water out of the fire extinguisher. In another type of extinguisher CO_2 itself, in the form of a foam, is used to extinguish the flames.

This extinguisher used CO_2 to put out the fire. (Photograph by C. D. Winters.)

atoms are replaced by carbon. A typical example is methylamine, $CH_3—NH_2$. The base strength of most amines is similar to that of NH_3, which means that amines react with acids in the same way:

$$HCl + CH_3—NH_2 \rightleftharpoons CH_3—NH_3^+ + Cl^-$$

Neither NH_3 nor amines survive as such in the presence of an acid. They are converted to the corresponding positive ions (conjugate acids). This is very important in the chemistry of the body, as we shall see in later chapters.

8.7 Self-Ionization of Water

We have seen that an acid produces H_3O^+ ions in water and a base produces OH^- ions. Suppose we have absolutely pure water, with no added acid or base. Surprisingly enough, even pure water contains a very small number of H_3O^+ and OH^- ions. They come from the reaction

$$\underset{\text{Acid}}{H_2O} + \underset{\text{Base}}{H_2O} \rightleftharpoons \underset{\text{Acid}}{H_3O^+} + \underset{\text{Base}}{OH^-}$$

This reaction is called the **self-ionization of water.**

in which one water molecule acts as an acid and gives a proton to a second water molecule, which is therefore acting as a base.

What is the extent of this reaction; how far does it go? The answer is, very little. As shown by the arrows, it is an equilibrium lying far to the left. We shall soon see exactly how far, but first let us write the equilibrium expression:

$$K = \frac{[H_3O^+][OH^-]}{[H_2O]^2}$$

This expression contains the concentration of water, but since only a very few water molecules react, the concentration of water remains essentially constant. Since this is so, it is convenient to change the expression to

$$K[H_2O]^2 = [H_3O^+][OH^-]$$

As we did with the dissociation constant for acids (Sec. 8.4), we can replace $K[H_2O]^2$ by a new constant, K_w. Our equilibrium expression is now

$$K_w = [H_3O^+][OH^-]$$

The K_w is called the **ion product of water.** At room temperature, K_w has the value of 1.0×10^{-14}. Since in pure water H_3O^+ and OH^- are formed by the same process (the self-ionization of water), their concentrations must be equal. This means that, in pure water,

$$[H_3O^+] = 1 \times 10^{-7} \text{ moles/L}$$
$$[OH^-] = 1 \times 10^{-7} \text{ moles/L}$$

Thus, in pure water, each liter of H_2O contains only 1×10^{-7} moles of H_3O^+ and 1×10^{-7} mole of OH^-. These are very small concentrations, not enough to make pure water a conductor of electricity. Water is not an electrolyte.

The equation $K_w = [H_3O^+][OH^-]$ is important because it applies not only to pure water; **it is valid in any water solution.** It says that the product of $[H_3O^+]$ and $[OH^-]$ in any aqueous solution is equal to 1×10^{-14}. If, for example, we add 0.01 mole of HCl to 1 L of pure water, it reacts essentially completely to give H_3O^+ ions. The concentration of H_3O^+ will be 0.01 M, or 1×10^{-2} M. This means that the $[OH^-]$ *must be* 1×10^{-12} M because the product *must be* 1×10^{-14}.

EXAMPLE

The $[OH^-]$ of an aqueous solution is 1×10^{-14} M. What is the $[H_3O^+]$?

Answer • We substitute into the equation:

$$K_w = [H_3O^+][OH^-] = 1 \times 10^{-14}$$
$$[H_3O^+](10^{-4}) = 1 \times 10^{-14}$$
$$[H_3O^+] = \frac{1 \times 10^{-14}}{1 \times 10^{-4}} = 1 \times 10^{-10}\ M$$

Problem **8.2**

The $[OH^-]$ of an aqueous solution is 1×10^{-12} M. What is the $[H_3O^+]$?

Aqueous solutions can have a very high $[H_3O^+]$, but the $[OH^-]$ must then be very low, and vice versa. Any solution with a $[H_3O^+]$ higher than 1×10^{-7} M is **acidic.** In such solutions, of course, $[OH^-]$ must be lower than 1×10^{-7} M. The higher the $[H_3O^+]$, the more acidic the solution is. Similarly, any solution with an $[OH^-]$ higher than 1×10^{-7} M is **basic.**

Pure water, in which $[H_3O^+]$ and $[OH^-]$ are equal (they are both 1×10^{-7} M) is **neutral,** that is, neither acidic nor basic.

Not only pure water, but *any* aqueous solution in which $[H_3O^+] = [OH^-] = 1 \times 10^{-7}$ M is a neutral solution.

8.8 pH

Because the $[H_3O^+]$ of any aqueous solution is such an important characteristic, chemists, workers in the health sciences, and other scientists must frequently talk about it. When they do so, it is inconvenient to keep saying, for example, "6×10^{-11} *M*, 2×10^{-2} *M*," or other numbers of this form. Fortunately, a shorter way exists. Chemists have defined a number, called the pH, to express these concentrations in a more convenient way. The mathematical definition of **pH** is

$$\mathbf{pH = -log[H_3O^+]}$$

When the $[H_3O^+]$ is of the form 1×10^{-A} *M*, where the A stands for any number, we don't have to worry about the mathematical definition—the pH is simply A. Instead of saying that a $[H_3O^+]$ is 1×10^{-4} *M*, we simply say that the pH is 4. As you can see, all we do is take the exponent of the $[H_3O^+]$ and drop the minus sign.

EXAMPLE

The $[H_3O^+]$ of a certain liquid soap is 1×10^{-9} *M*. What is the pH?

Answer • The pH is 9.

Problem **8.3**

The $[H_3O^+]$ of an acidic solution is 1×10^{-3} *M*. What is the pH?

EXAMPLE

The pH of black coffee is 5. What is the $[H_3O^+]$?

Answer • The $[H_3O^+]$ is 1×10^{-5} *M*.

Problem **8.4**

The pH of tomato juice is 4. What is the $[H_3O^+]$?

EXAMPLE

The $[OH^-]$ in a strongly alkaline solution is 1×10^{-1} *M*. What is the pH?

Answer • Here we are given $[OH^-]$, so we must first calculate $[H_3O^+]$ before we can find pH:

$$[H_3O^+][OH^-] = K_w$$

$$[H_3O^+](10^{-1}) = 10^{-14}$$

$$[H_3O^+] = \frac{10^{-14}}{10^{-1}} = 10^{-13}$$

Since $[H_3O^+] = 10^{-13}$, the pH is 13.

Problem 8.5

The $[OH^-]$ of an acidic solution is 1×10^{-11}. What is the pH?

The convenience of this system is readily apparent. It is much easier to say "the pH is 12" than to say "the $[H_3O^+] = 1 \times 10^{-12}$ *M*," and yet both mean the same thing.

Suppose that the $[H_3O^+]$ is 7×10^{-4}. What do we do then? That's when we need to use logarithms. Now that inexpensive scientific calculators are readily available, this is no problem at all. We simply take the logarithm of 7×10^{-4} with the calculator and get the answer -3.15. Just drop the minus sign (because $pH = -\log[H_3O^+]$) and you have the pH: 3.15.

EXAMPLE

What is the pH of a solution whose $[H_3O^+]$ is 3×10^{-2} *M*?

Answer • Enter 3×10^{-2} into the calculator, and press log. The answer is -1.52. Since $pH = -\log[H_3O^+]$, we drop the minus sign, and the pH is 1.52.

Problem 8.6

What is the pH of a solution whose $[H_3O^+]$ is 6×10^{-9} *M*?

EXAMPLE

A solution has a pH of 11.70. What is the $[H_3O^+]$?

Answer • Since this is the reverse of the method just described, we first convert 11.70 to -11.70 and then find the antilogarithm on the calculator ("inverse log" or 10^x on some calculators). When we do this we get $[H_3O^+] = 2.0 \times 10^{-12}$ *M*.

Problem 8.7

The pH of blood is 7.4. What is the $[H_3O^+]$?

As already indicated, pH is universally used to express the acidity of aqueous solutions (you probably had heard of pH even before you took this course). **The lower the pH, the more acidic the solution; the higher the pH, the more basic the solution.** Pure water, with a $[H_3O^+]$ of 1×10^{-7} *M*, has a pH of 7 and is neutral.

Although the pH of pure water is 7, tap water usually has a lower pH, generally about 6. The reason is that tap water normally contains dissolved CO_2 (from the atmosphere), which reacts with water to give the dissolved acid H_2CO_3.

Acidic solutions have a pH below 7.
Basic solutions have a pH above 7.

All fluids in the human body are aqueous; that is, the only solvent is water. Consequently, all body fluids have a pH value. Some of them have a narrow pH range, others a wide range. The pH of blood must be between 7.35 and 7.45 (slightly basic). If it goes outside these limits, illness and even death will result (see Box 8E). In contrast, the pH of urine can vary from 5.5 to 7.5. Table 8.4 gives the pH values of some body fluids, as well as of some other common materials (see Fig. 8.2).

One thing we must remember when we see pH values is that a difference of one pH unit means a multiplication by 10 of the $[H_3O^+]$. For example, a pH of 3 does not sound very different from a pH of 4, but the first means a $[H_3O^+]$ of 10^{-3} *M* and the second means a $[H_3O^+]$ of 10^{-4} *M*. The $[H_3O^+]$ of the pH 3 solution is ten times the $[H_3O^+]$ of the pH 4 solution.

There are two ways to measure the pH of an aqueous solution. In one method pH paper is used (Fig. 8.3). When a drop of solution is placed on this paper, the paper turns a certain color. The pH is determined by comparing the color of the paper with the colors on a chart supplied with the paper.

pH paper is made by soaking plain paper with a mixture of indicators. An **indicator** is **a substance that changes color at a certain pH.** An example is the compound methyl orange. When a drop of methyl orange solution is added to any aqueous solution with a pH of 3.2 or lower it turns red (and turns the whole solution red). Added to an aqueous solution with a pH of 4.4 or higher, it turns yellow (and turns the whole solution yellow). These are the particular limits and colors for this indicator. Other indicators have other limits and colors. Phenolphthalein, for example, is colorless below pH 8.2 but pink above pH 10.0. The colors and pH values of these and some other indicators are shown in Figure 8.4.

Table **8.4** pH Values for Common Materials

Material	pH
Battery acid	0.5
Gastric juice	0.9 – 1.8
Soft drinks	2.0 – 4.0
Lemon juice	2.2 – 2.4
Vinegar	2.4 – 3.4
Tomatoes	4.0 – 4.4
Black coffee	5.0 – 5.1
Urine	5.5 – 7.5
Rain (unpolluted)	6.2
Milk	6.3 – 6.6
Saliva	6.5 – 7.5
Pure water	7.0
Blood	7.35 – 7.45
Bile	7.6 – 8.6
Pancreatic fluid	7.8 – 8.0
Seawater	8 – 9
Soap	8 – 10
Milk of magnesia	10.5
Household ammonia	11.7
Lye (1 *M* NaOH)	14

Litmus paper turns red in acidic solution and blue in basic solution. This indicator is often used as an approximate guide to acidity or basicity.

Figure **8.2**
The indicators show that vinegar has a pH of about 3, a carbonated beverage has a pH of about 4 to 5, and a household cleaner is strongly basic. (Photograph by Charles D. Winters.)

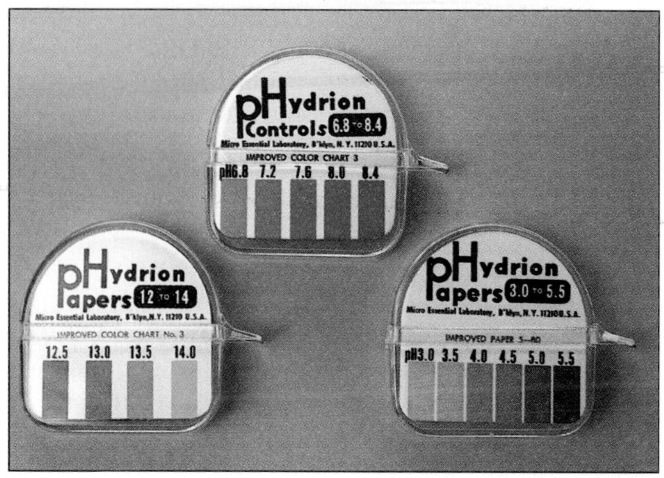

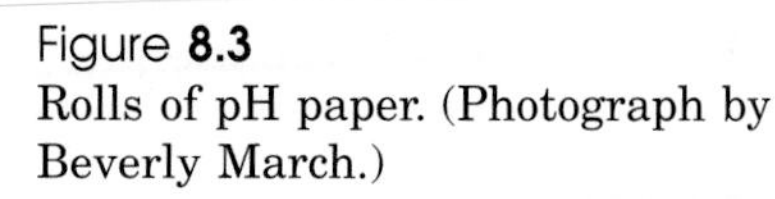

Figure **8.3**
Rolls of pH paper. (Photograph by Beverly March.)

The other method for determining pH is more accurate. In this method a pH meter is used (Fig. 8.5). The electrode of the meter is dipped into the unknown solution and the pH, to the nearest tenth of a pH unit, is read on a dial.

In Section 8.4 we saw that K_a values for weak acids are given in exponential notation. Because of this, pK_a values are often used instead. The pK_a is defined as $-\log K_a$ and is calculated from K_a values in exactly the same way pH is calculated from $[H_3O^+]$. Table 8.3 lists pK_a as well as K_a values for some weak acids.

The higher the pK_a, the weaker the acid.

EXAMPLE

The K_a for benzoic acid is 6.5×10^{-5}. What is the pK_a of this acid?

Answer • This is done exactly the same way as the calculation of pH from $[H_3O^+]$. The number 6.5×10^{-5} is entered into a scientific calculator, and the log key is pressed. The answer is -4.19. We drop the minus sign, and the pK_a is 4.19.

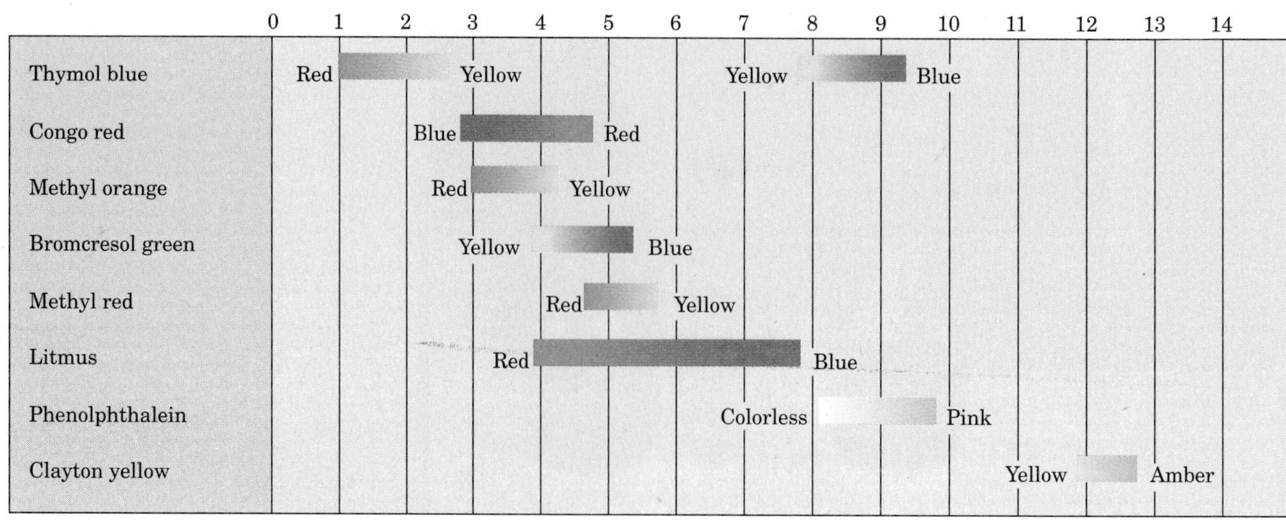

Figure **8.4** • Some acid-base indicators. Note that some have two color changes.

Problem 8.8

The K_a for hydrocyanic acid, HCN, is 4.9×10^{-10}. What is the pK_a?

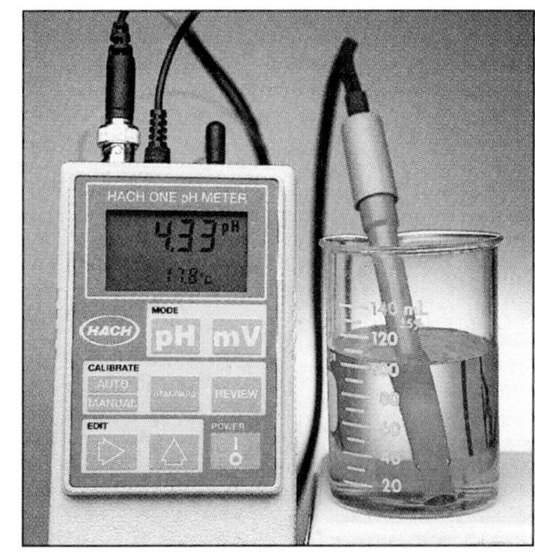

Figure **8.5**
A pH meter. The blue solution is 0.1 *M* $CuSO_4$. This is the salt of a strong acid and a weak base, so the pH is 4.3 (Sec. 8.9). (Photograph by Charles D. Winters.)

8.9 The pH of Aqueous Salt Solutions

When an acid is added to water, the solution becomes acidic (pH below 7); when a base is added to water, the solution becomes basic (pH above 7). Suppose we add a salt to water? What happens to the pH then? At first thought, you might suppose that all salts are neutral and that addition of any salt to pure water leaves the pH at 7. This is in fact true for some salts. They are neutral and do not change the pH of pure water. Many salts are acidic or basic, however, and cause the pH of pure water to change. In order to see why, let us look at sodium acetate, a typical basic salt.

Like almost all salts, sodium acetate is 100 percent ionized in water, so that when we add CH_3COONa to water we get a solution containing Na^+ and CH_3COO^- ions. The Na^+ ions do not react with the water, but the CH_3COO^- ions do:

$$CH_3COO^-(aq) + H_2O(l) \rightleftharpoons CH_3COOH(aq) + OH^-(aq)$$

Because this is a reaction with water, it is sometimes called **hydrolysis.**

This is an equilibrium reaction lying well to the left. In a 0.1 *M* solution of sodium acetate, only about one out of every 13 500 acetate ions is converted to acetic acid and OH^- ions at equilibrium. Nevertheless, OH^- ions *are* being produced, enough to cause the pH to rise to about 8.9. Because this is higher than 7, the solution is basic.

We can now understand why sodium acetate is a basic salt. Some of the acetate ions react with water to produce OH^- ions, but none of the Na^+ ions react with water at all. Because OH^- ions are being produced, the solution is basic. It is true that acetic acid is also being produced, but the pH of an aqueous solution depends only on the concentration of H_3O^+ (and of OH^-) and *not* on the concentration of any other acid or base.

How can we predict whether any particular salt will produce an acidic, basic, or neutral solution when added to water? This is easily done if we can remember the six strong acids and four strong bases listed in Table 8.1. Acetate ions, CH_3COO^-, react with water because the corresponding acid, acetic acid, is a weak acid. Sodium ions do not react with water because the corresponding base, NaOH, is a strong base.

The four strong bases are NaOH, KOH, LiOH, and $Ba(OH)_2$. All other bases are considered weak. The six strong acids are HI, HBr, HCl, HNO_3, H_2SO_4, and $HClO_4$.

All salts can be thought of as being combinations of an acid and a base. We can predict whether a salt will be acidic, neutral, or basic in water solution when we know something about the strength of the acid and base. We have four cases:

1. *A salt of a strong base and a weak acid.* As we see from the sodium acetate example, such salts are **basic salts** and raise the pH of pure water.

Box 8E

Acidosis and Alkalosis

The pH of blood is normally between 7.35 and 7.45. If the pH goes lower than this, the condition is called **acidosis.** A blood pH higher than 7.45 is called **alkalosis.** Both of these are abnormal conditions. Acidosis leads to depression of the acute nervous system. Mild acidosis can result in fainting; a more severe case can cause coma. Alkalosis leads to overstimulation of the nervous system, muscle cramps, and convulsions. If the acidosis or alkalosis persists for a sufficient period of time, or if the pH gets too far away from 7.35 to 7.45, the patient dies.

Acidosis has several causes. One type, called *respiratory acidosis,* is caused by difficulty in breathing **(hypoventilation).** An obstruction in the windpipe or diseases such as pneumonia, emphysema, asthma, or congestive heart failure diminish the amount of CO_2 that leaves the body through the lungs. (You can even produce mild acidosis by holding your breath.) The pH of the blood is decreased because the CO_2, unable to escape fast enough, remains in the blood, where it decreases the $[HCO_3^-]/[H_2CO_3]$ ratio.

Respiratory alkalosis is the reverse. It arises from rapid or heavy breathing, called **hyperventilation.** This can come about from fever, infection, the action of certain drugs, or even hysteria. Here the excessive loss of CO_2 raises the $[HCO_3^-]/[H_2CO_3]$ ratio and the pH.

Acidosis caused by other factors is called *metabolic acidosis.* Two causes of this condition are starvation or fasting and heavy exercise. When the body doesn't get enough food, it burns its own fat, and the products of this reaction are acidic compounds that get into the blood. This sometimes happens to people on fad diets. Heavy exercise causes the muscles to produce excessive amounts of lactic acid—this is what makes you feel tired and sore. Metabolic acidosis is also caused by a number of metabolic irregularities. For example, the disease diabetes mellitus produces acidic compounds called ketone bodies (Sec. 21.7).

Metabolic alkalosis can also result from various metabolic irregularities and in addition can be caused by excessive vomiting. The contents of the stomach are strongly acidic (pH $\approx$ 1), and loss of substantial amounts of this acidic material raises the pH of the blood.

2. *A salt of a strong acid and a weak base.* We have the opposite behavior here. An example is ammonium chloride, NH_4Cl. The Cl^- ion does not react with H_2O, but the NH_4^+ ion does:

$$NH_4^+(aq) + H_2O(l) \rightleftharpoons H_3O^+(aq) + NH_3(aq)$$

The newly formed H_3O^+ lowers the pH, as it does for all salts in this category. Such salts are **acidic** (See Fig. 8.5).

3. *A salt of a strong acid and a strong base.* In this case, neither ion reacts with water, and the solution remains neutral. These are **neutral salts.** Examples are NaCl, KNO_3, and Na_2SO_4.

4. *A salt of a weak acid and a weak base.* Both ions react with water. Sometimes the effects cancel each other and sometimes not. We will not try to make predictions in such cases.

8.10 Buffers

In Section 8.8 we saw that the body must keep the pH of blood between 7.35 and 7.45 (Box 8E). Yet we frequently eat acidic foods such as oranges, lemons, sauerkraut, and tomatoes, and this eventually adds considerable quantities of H_3O^+ to the blood. How does the body manage to keep the pH of blood so constant? It does it with buffers. A **buffer solution** is **a solution whose pH does not change very much when H_3O^+ or OH^- ions are added to it.**

How do we make a buffer solution, and how does it work? All we need to do to make a buffer solution is to add to water approximately equal quantities of a weak acid and its conjugate base. For example, if we dissolve 1 mole of acetic acid (a weak acid) and 1 mole of its conjugate base (in the form of sodium acetate) in 1 L of water, we have a very good buffer solution.

How does it work? It works because we have substantial concentrations of *both* CH_3COOH and CH_3COO^-. If H_3O^+ is added, the CH_3COO^- ions remove it by reacting with it. If OH^- is added, the CH_3COOH molecules perform the same function

$$H_3O^+(aq) + CH_3COO^-(aq) \rightleftharpoons CH_3COOH(aq) + H_2O(l)$$
$$OH^-(aq) + CH_3COOH(aq) \rightleftharpoons CH_3COO^-(aq) + H_2O(l)$$

(Note from Table 8.2 that both equilibria lie well to the *right.*) The important thing here is that when the CH_3COO^- ions remove H_3O^+, they are converted to CH_3COOH molecules. Since a substantial amount of CH_3COOH is already present, all that happens is that the $[CH_3COO^-]$ goes down a bit and the $[CH_3COOH]$ goes up a bit.

We saw in Section 8.4 that the following equilibrium expression holds for acetic acid (a similar one holds for any weak acid):

$$K_a = \frac{[H_3O^+][CH_3COO^-]}{[CH_3COOH]}$$

If we rearrange this, we get

$$[H_3O^+] = K_a \frac{[CH_3COOH]}{[CH_3COO^-]}$$

This equation tells us that the $[H_3O^+]$ is equal to the constant K_a multiplied by the ratio of $[CH_3COOH]$ to $[CH_3COO^-]$. We've just seen, however, that H_3O^+ ions added to the buffer solution make only a slight change in this ratio: the $[CH_3COOH]$ goes up a bit and the $[CH_3COO^-]$ down a bit. Therefore the $[H_3O^+]$, and the pH, are only slightly changed.

The same thing happens if we add OH^- to the buffer, but in the opposite direction. The OH^- ions are removed by the CH_3COOH, which in the process is converted to CH_3COO^-. Thus the $[CH_3COO^-]$ goes up a bit and the $[CH_3COOH]$ down a bit, so that the $[H_3O^+]$, and the pH, are only slightly changed.

In essence, a buffer solution works because it is able to neutralize, and hence remove from the solution, most of any OH^- or H_3O^+ ions that are added. This effect can be quite powerful. For example, pure water has a pH of 7. If we add 0.1 mole of HCl to 1 L of pure water, with no buffer, the pH drops to 1. If we add 0.1 mole of NaOH to 1 L of pure water, the pH rises to 13. A buffer solution containing 1 *M* $H_2PO_4^-$ ion and 1 *M* HPO_4^{2-} ion (its conjugate base) has a pH of 7.21. If we add 0.1 mole of HCl to 1 L of this solution, the pH drops only to 7.12. If we add 0.1 mole of NaOH, the pH rises only to 7.30. In summary,

Pure H_2O + HCl	pH 7 $\longrightarrow$ 1
Buffer + HCl	pH 7.21 $\longrightarrow$ 7.12
Pure H_2O + NaOH	pH 7 $\longrightarrow$ 13
Buffer + NaOH	pH 7.21 $\longrightarrow$ 7.30

It is obvious from this example that a buffer really does the job: It holds the pH relatively constant (Fig. 8.6).

Every buffer solution has two characteristics, its pH and its capacity.

Buffer pH

In the example just given, the pH of the $H_2PO_4^-/HPO_4^{2-}$ buffer solution is 7.21 as long as the concentrations of $H_2PO_4^-$ and HPO_4^{2-} are equal. If you look at Table 8.3, you see that 7.21 is the pK_a of the $H_2PO_4^-$ ion. This is not a coincidence. If we make a buffer solution by mixing equimo-

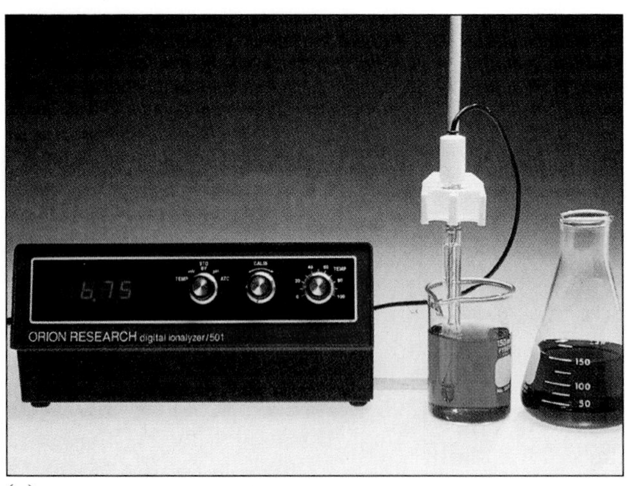

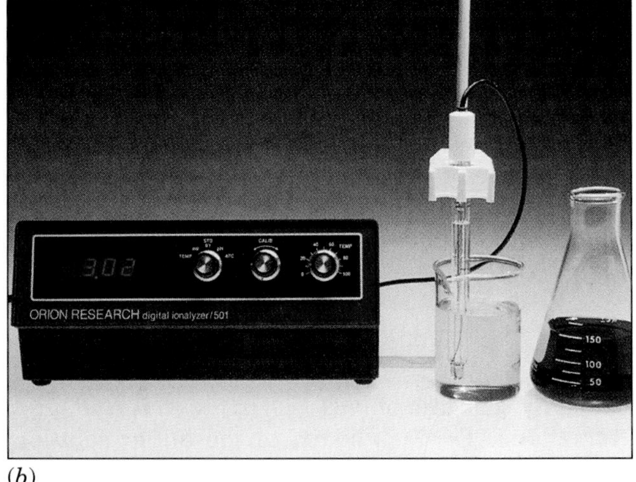

(*a*) (*b*)

Figure **8.6** • Buffer solutions. (a) The pH meter shows the pH of water that contains a trace of acid (and bromcresol green indicator, see Fig. 8.4). The solution at the right is a buffer that has the same pH as human blood, 7.4, and contains the same indicator. (b) When 5 mL of 0.10 *M* HCl is added to each solution, the pH of the water drops several units, while the pH of the buffer stays constant. (Photographs by Charles D. Winters.)

lar concentrations of any weak acid and its conjugate base, the pH of the solution is always equal to the pK_a of the acid.

This allows us to prepare buffer solutions to maintain almost any pH. For example, if we want to maintain a pH of 9.14, we could make a buffer solution from boric acid and its conjugate base (see Table 8.3).

EXAMPLE

What is the pH of buffer systems containing equimolar quantities of (a) H_3PO_4 and $H_2PO_4^-$ (b) H_2CO_3 and HCO_3^-?

Answer • The pH is equal to the pK_a of the acid, which we find in Table 8.3: (a) pH = 2.1 (b) pH = 6.37.

Problem 8.9

What is the pH of buffer systems containing equimolar quantities of (a) H_3BO_3 and $H_2BO_3^-$ (b) HCO_3^- and CO_3^{2-}?

Suppose the concentrations of the weak acid and its conjugate base are not equal. In that case, the pH of the buffer will be a bit higher or lower (but not by very much) as long as significant amounts of both acid and conjugate base are present (Box 8F). The most efficient buffers do contain equimolar concentrations, and buffer efficiency decreases as the ratio becomes more unequal. If the acid-to-conjugate-base ratio is more than about 15 to 1 or less than about 1 to 15, the solution is essentially no longer a buffer.

Buffer Capacity

We could make a buffer solution by dissolving 1 mole each of CH_3COONa and CH_3COOH in 1 L of H_2O, or we could use only 0.1 mole of each. Both solutions have the same pH: 4.75. However, the former has a **buffer capacity** ten times that of the latter. If we add 0.2 mole of HCl to the former solution, it performs the way we expect—the pH drops only a little (to 4.57). If we add 0.2 mole of HCl to the latter solution, however, the pH drops all the way to 1.0 because the buffer has been **swamped out.** The amount of H_3O^+ added has exceeded the buffer capacity. The first 0.1 mole of HCl completely neutralizes essentially all the CH_3COO^- present. After that, the solution contains only CH_3COOH and is no longer a buffer, so that the second 0.1 mole of HCl is able to lower the pH to 1.0.

Blood Buffers

To hold the pH of the blood at close to 7.4, the body uses three buffer systems: carbonate, phosphate, and proteins (proteins are discussed in

Box 8F

The Henderson-Hasselbalch Equation

Whenever we work with buffers, a useful equation is

$$pH = pK_a + \log \frac{[A^-]}{[HA]}$$

where [HA] is the concentration of the acid and $[A^-]$ that of its conjugate base. This equation, called the **Henderson-Hasselbalch equation,** is easily derived from Equation 8.3, but uses pK_a instead of K_a. Suppose that we wish to make a buffer solution whose pH does not correspond exactly to the pK_a of any acid we may have on hand. We can make up such a solution by choosing an acid with a pK_a that is not too far from the desired pH and adjusting the concentrations of the weak acid (HA) and its conjugate base (A^-) to obtain the desired pH. The Henderson-Hasselbalch equation allows us to calculate the proper ratio.

EXAMPLE

How can we make a buffer solution with a pH of 7.99, using an acid-base pair from Table 8.3?

Answer • The closest pK_a value in Table 8.3 is 7.21, the value for $H_2PO_4^-$. We therefore use $H_2PO_4^-/HPO_4^{2-}$ as our buffer and adjust the $[HPO_4^{2-}]/[H_2PO_4^-]$ ratio to get a pH of 7.99:

$$pH = pK_a + \log \frac{[HPO_4^{2-}]}{[H_2PO_4^-]}$$

$$7.99 = 7.21 + \log \frac{[HPO_4^{2-}]}{[H_2PO_4^-]}$$

$$0.78 = \log \frac{[HPO_4^{2-}]}{[H_2PO_4^-]}$$

On a calculator we find that the antilogarithm of 0.78 is 6. Taking the antilogarithm of both sides, we have

$$6 = \frac{[HPO_4^{2-}]}{[H_2PO_4^-]}$$

Thus, in order to get a buffer solution of pH 7.99, we must use a molar concentration of HPO_4^{2-} that is six times that of $H_2PO_4^-$.

Chapter 18). The most important of these is the carbonate system. The acid is carbonic acid, H_2CO_3; the base is the bicarbonate ion, HCO_3^-. The pK_a of H_2CO_3 is 6.37 (see Table 8.3). Since the pH of an equimolar mixture of acid and conjugate base is equal to its pK_a, a buffer made of equal concentrations of H_2CO_3 and HCO_3^- has a pH of 6.37.

Blood, however, has a pH of 7.4. The carbonate buffer can maintain this pH only if $[H_2CO_3]$ and $[HCO_3^-]$ are not equal. In fact, the necessary $[HCO_3^-]/[H_2CO_3]$ ratio is about 10 to 1. The normal concentrations in blood are about 0.025 *M* HCO_3^- and 0.0025 *M* H_2CO_3. The buffer works, as all buffers do, because any added H_3O^+ is neutralized by the HCO_3^- and any added OH^- is neutralized by the H_2CO_3.

Problem 8.10

How can we make a buffer solution of pH 4.15, using an acid-base pair from Table 8.3?

We can also use the Henderson-Hasselbalch equation to calculate the pH if we know the $[A^-]/[HA]$ ratio.

EXAMPLE

What is the pH of an $H_2PO_4^-/HPO_4^{2-}$ buffer system in which $[HPO_4^{2-}]$ is twice $[H_2PO_4^-]$?

Answer

$$pH = pK_a + \log \frac{[HPO_4^{2-}]}{[H_2PO_4^-]}$$

$$pH = 7.21 + \log \frac{2}{1}$$

The log of 2 (from a calculator) is 0.30. Thus

$$pH = 7.21 + 0.30 = 7.51$$

Problem 8.11

What is the pH of a buffer solution in which $[CH_3COO^-] = 0.20\ M$ and $[CH_3COOH] = 0.60\ M$?

The fact that the $[HCO_3^-]/[H_2CO_3]$ ratio is 10 to 1 means that this is a better buffer for acids, which lower the ratio and thus improve buffer efficiency, than for bases, which raise the ratio. This is in harmony with the actual functioning of the body, since under normal conditions larger amounts of acidic than basic substances enter the blood. The 10-to-1 ratio is easily maintained under normal conditions because the body can very quickly increase or decrease the amount of CO_2 entering the blood.

The second buffering system of the blood is made up of hydrogen phosphate ion, HPO_4^{2-}, and dihydrogen phosphate ion, $H_2PO_4^-$. In this case, a 1.6-to-1 ratio of $[HPO_4^{2-}]$ to $[H_2PO_4^-]$ is necessary to maintain a pH of 7.4. This is well within the limits of good buffering action.

8.11 Titration, Equivalents, and Normality

In laboratories, the question frequently arises as to how much acid or base is present in a given amount of solution. To measure this, we use a method called titration. If a solution is acidic, **titration** consists of **adding base to it until all the acid is neutralized.** To do this, we need two things: (1) a means of measuring how much base is added and (2) a means of telling just when the acid is completely neutralized. Titration is shown in Figure 8.7. The **buret** is the means of measuring the amount of base. An **indicator** in the solution tells us when the acid is exactly neutralized (we could also use a pH meter) because at that point there is a sudden large change in pH. This is called the **end point.**

It is important to understand that *titration is not a method of determining the acidity (or basicity) of a solution.* If we want to do that, we measure the pH, which is the only measurement of solution acidity or basicity. Titration is a method for determining the total acid or base concentration of a solution, which is not the same as the acidity. For example, a 0.1 *M* solution of HCl in water has a pH of 1, but a 0.1 *M* solution of acetic acid has a pH of 2.9. The solutions have very different acidities, but the total concentration of acid is the same and each of these solutions neutralizes the same amount of NaOH solution.

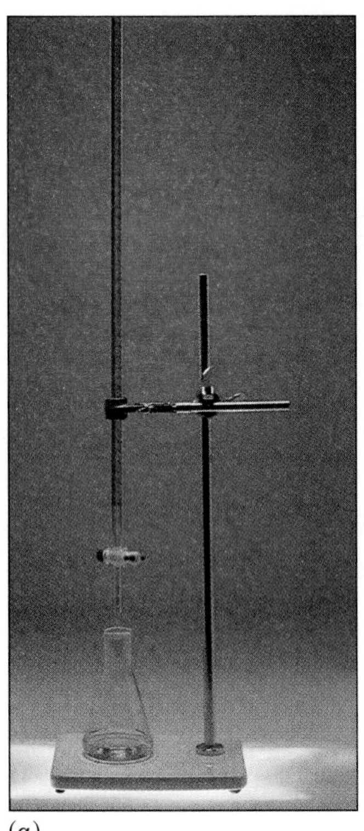
(*a*)

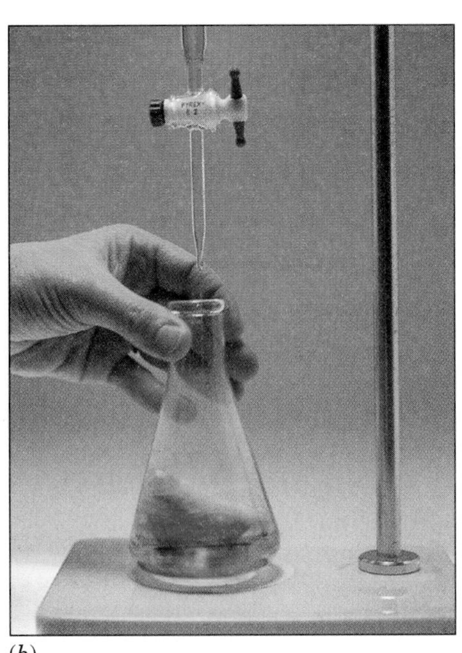
(*b*)

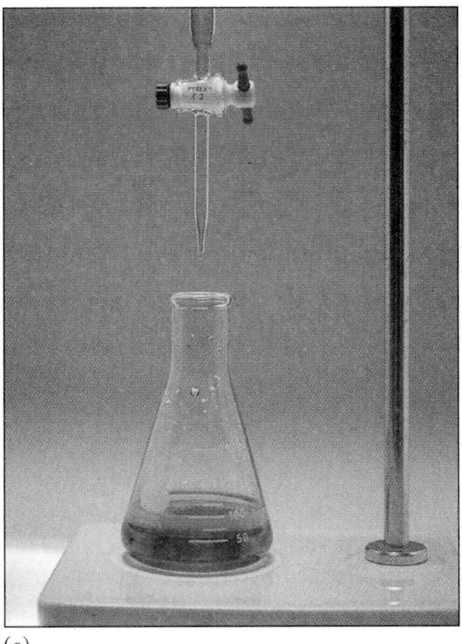
(*c*)

Figure **8.7** • Titration. An acid of unknown concentration is in the Erlenmeyer flask. When base from the buret is added, the acid is neutralized. The end point is reached when the indicator color change becomes permanent. (Photographs by Charles D. Winters.)

When we have reached the end point in a titration, we can easily read, directly on the buret, exactly how many milliliters of NaOH solution were required to neutralize the acid. How do we then calculate the amount of acid? Before we can answer that, we must first define equivalents and normality.

We have seen that some acids are diprotic (H_2SO_4, H_2CO_3) or triprotic (H_3PO_4). Similarly, there are some bases that produce two or three OH^- ions in solution—for example, $Ba(OH)_2$, $Mg(OH)_2$, and $Al(OH)_3$. For acids and bases, we define the **equivalent weight (EW)** as **the formula weight (FW) divided by the number of H^+ or OH^- ions produced.**

$Al(OH)_3$ yields 3 OH^- ions only under optimal reaction conditions.

EXAMPLE

What is the equivalent weight of these acids and bases: (a) HNO_3, FW = 63 (b) NaOH, FW = 40 (c) H_2SO_4, FW = 98 (d) $Mg(OH)_2$, FW = 58?

Answer • In each case we divide the formula weight by the number of H^+ or OH^- ions produced by the acid or base:

(a) HNO_3	EW = 63/1 = 63
(b) NaOH	EW = 40/1 = 40
(c) H_2SO_4	EW = 98/2 = 49
(d) $Mg(OH)_2$	EW = 58/2 = 29

Problem 8.12

What is the equivalent weight of these acids and bases: (a) CH_3COOH, FW = 60.0 (b) $Ba(OH)_2$, FW = 171.3 (c) H_3PO_4, FW = 98.0?

We define an **equivalent (eq)** of any substance as its **equivalent weight expressed in grams.** Thus 1 eq of HNO_3 is 63 g, and 49 g is 1 eq of H_2SO_4. This definition is similar to that of the mole, which is the *formula* weight expressed in grams, and we deal with equivalents in a manner similar to the way we deal with moles. That is, if 49 g is 1 eq of H_2SO_4, then 4.9 g = 0.1 eq, 98 g = 2 eq, and so forth.

Because we often deal with very small quantities, we use **milliequivalents:** 1 meq = 1/1000 eq.

EXAMPLE

How many equivalents are there in (a) 18 g of H_2SO_4 (b) 51 g of $Mg(OH)_2$?

Answer • We multiply the number of grams by the appropriate conversion factor:

$$\text{(a)}\quad 18\ \text{g}\ \cancel{H_2SO_4} \times \frac{1\ \text{eq}\ H_2SO_4}{49\ \text{g}\ \cancel{H_2SO_4}} = 0.37\ \text{eq}\ H_2SO_4$$

$$\text{(b)}\quad 51\ \text{g}\ \cancel{Mg(OH)_2} \times \frac{1\ \text{eq}\ Mg(OH)_2}{29\ \text{g}\ \cancel{Mg(OH)_2}} = 1.8\ \text{eq}\ Mg(OH)_2$$

Problem **8.13**

How many equivalents are there in (a) 63.5 g of H_2SO_3 (b) 3.68 g of $Ba(OH)_2$?

The normalities discussed in this section are limited to acid-base reactions. Other systems of normality, not used in this book, apply to other kinds of reactions.

Why have chemists gone through the trouble of defining equivalent weight and equivalent? The answer to this is present in the word used: "equivalent." In an acid-base reaction, **one equivalent of any acid exactly neutralizes one equivalent of any base.** Since most acid-base reactions occur in aqueous solution, we can go one step further and define a new concentration unit, **normality (*N*),** as **equivalents of solute per liter of solution.**

EXAMPLE

What is the normality of a solution made by dissolving 8.5 g of H_2SO_4 in enough water to make 500 mL of solution?

Answer • First we need the number of equivalents of H_2SO_4 per liter of solution, and we know that the equivalent weight of H_2SO_4 is 49. Thus,

$$8.5\ \cancel{g} \times \frac{1\ \text{eq}}{49\ \cancel{g}} = 0.17\ \text{eq}\ H_2SO_4$$

To find normality we divide equivalents of H_2SO_4 by liters of solution:

$$\frac{0.17\ \text{eq}}{0.500\ \text{L}} = 0.34\ \frac{\text{eq}}{\text{L}} = 0.34\ N$$

Problem **8.14**

What is the normality of a solution made by dissolving 3.4 g of $Mg(OH)_2$ in enough water to make 1250 mL of solution?

Concentrations of acids and bases in many laboratories are often expressed in normalities.

There is a simple relationship between the normality and the molarity (Sec. 6.5) of an acidic or basic solution. Normality is simply the molarity multiplied by the number of H^+ or OH^- ions produced by one molecule of acid or base. For example, in HCl only one H^+ ion is produced. This means that, for all HCl solutions, the normality is equal to the molarity. Sulfuric acid produces two H^+ ions, so for H_2SO_4 solutions the normality is always twice the molarity; thus a 6 *M* solution of H_2SO_4 is 12 *N*. *Normality is always equal to or greater than molarity.*

Now that we have defined normality, we can come back to titration and see how an unknown acid concentration is calculated. We know that, at the end point, the number of equivalents of acid equals the number of equivalents of base:

$$\mathrm{Eq}_{\mathrm{acid}} = \mathrm{Eq}_{\mathrm{base}}$$

Since normality multiplied by volume gives the number of equivalents, the following equation is valid:

$$V_{\mathrm{acid}}N_{\mathrm{acid}} = V_{\mathrm{base}}N_{\mathrm{base}}$$

If we are titrating an acid with a base, we begin by putting the acid in an Erlenmeyer flask. In doing so, we must carefully measure the volume of acid added (because this is V_{acid}). At the end point, we read the buret to find the volume of base added (V_{base}). The normality of the base (N_{base}) must be known before we do the titration. It is then simple to calculate the unknown (N_{acid}).

A pipet is often used for this.

EXAMPLE

If 50.0 mL of an acid solution of unknown concentration is titrated with 0.32 *N* base and it takes 24.6 mL of base to reach the end point, what is the normality of the acid?

Answer

$$V_{\mathrm{acid}}N_{\mathrm{acid}} = V_{\mathrm{base}}N_{\mathrm{base}}$$

$$(50.0\ \mathrm{mL})(N_{\mathrm{acid}}) = (24.6\ \mathrm{mL})(0.32\ N)$$

$$N_{\mathrm{acid}} = \frac{(24.6\ \cancel{\mathrm{mL}})(0.32\ N)}{50.0\ \cancel{\mathrm{mL}}} = 0.16\ N$$

The concentration of the acid is 0.16 *N*.

Problem 8.15

A sample of 50.00 mL of an acid solution of unknown concentration was titrated with 0.152 *N* base, and it took 27.5 mL of base to reach the end point. What was the normality of the acid?

Note that this answer is correct no matter which acid is in the flask and no matter which base is in the buret. The advantage of using normalities is that the simple equation $V_{\mathrm{acid}}N_{\mathrm{acid}} = V_{\mathrm{base}}N_{\mathrm{base}}$ works for any acid and any base, no matter how many H^+ or OH^- ions they produce.

The same procedure and calculations can be used to determine an unknown base concentration. In this case, a known amount of the base is put in the flask, and the buret is filled with an acid solution of known concentration.

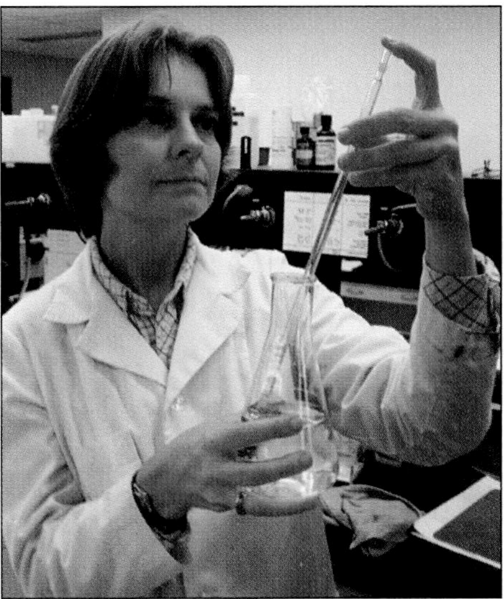

Laboratory worker measuring a volume with a pipet. (FourBy-Five.)

SUMMARY

By the **Arrhenius** definitions, **acids** are substances that produce H_3O^+ ions in water and **bases** are substances that produce OH^- ions in water. A **strong acid or base** produces high concentrations of these ions. The **Brønsted-Lowry** definitions expand this concept beyond water: An acid is a proton donor, and a base is a proton acceptor. Every acid has a **conjugate base.** An **amphoteric** substance, such as water, can act as either an acid or a base. Table 8.2 shows which Brønsted-Lowry acid-base reactions will proceed; the favored direction is always toward the weaker acid and base.

The strengths of weak acids are expressed by **K_a values;** the higher the K_a, the stronger the acid. Strong acids and bases are corrosive. Acids neutralize metals, metal hydroxides, and oxides to give **salts,** which are made up of positive and negative ions. Acids also react with carbonates, bicarbonates, ammonia, and amines.

In pure water, a small percentage of molecules undergo the reaction $2H_2O \rightleftharpoons H_3O^+ + OH^-$, enough to produce a concentration of 10^{-7} M each of H_3O^+ and OH^-. The **ion product,** $K_a = [H_3O^+][OH^-]$, is equal to 10^{-14} in any aqueous solution. Hydronium ion concentrations are generally expressed in **pH units,** with $pH = -\log[H_3O^+]$. Solutions with pH less than 7 are **acidic;** those with pH higher than 7 are **basic. Neutral** solutions have a pH of 7. The pH is measured with indicators or with a pH meter.

Many salts hydrolyze in water. Salts of strong acids and weak bases are acidic; salts of weak acids and strong bases are basic; salts of strong acids and strong bases are neutral.

A **buffer solution** does not change its pH very much when a strong acid or base is added. Buffers are made up of approximately equal concentrations of a weak acid and its conjugate base. Every buffer solution has a pH and a capacity. The most important buffers for blood are carbonate and phosphate.

The concentration of aqueous solutions of acids and bases can be measured by **titration,** in which a base of known concentration is added to an acid of unknown concentration (or vice versa) until an end point is reached, at which point the solution is completely neutralized. By the use of **normalities,** the unknown concentration can be calculated.

KEY TERMS

Acid (Sec. 8.1)
Acid dissociation constant (Sec. 8.4)
Acidic salt (Sec. 8.9)
Acidic solution (Sec. 8.8)
Acidity constant (Sec. 8.4)
Acidosis (Box 8E)
Alkali (Sec. 8.1)
Alkalosis (Box 8E)
Amphoteric (Sec. 8.3)
Base (Sec. 8.1)
Basic salt (Sec. 8.9)
Basic solution (Sec. 8.8)
Brønsted-Lowry acid and base (Sec. 8.3)
Buffer (Sec. 8.10)
Buffer capacity (Sec. 8.10)
Conjugate acid and base (Sec. 8.3)
Diprotic acid (Sec. 8.3)
End point (Sec. 8.11)
Equivalent (Sec. 8.11)
Equivalent weight (Sec. 8.11)
Henderson-Hasselbalch equation (Box 8F)
Hyperventilation (Box 8E)
Hypoventilation (Box 8E)
Indicator (Sec. 8.8)
Ion product (Sec. 8.7)
Ionization constant (Sec. 8.4)
K_a (Sec. 8.4)
K_w (Sec. 8.7)
Monoprotic acid (Sec. 8.3)
Neutral salt (Sec. 8.9)
Neutral solution (Secs. 8.5, 8.8)
Neutralization (Sec. 8.5)
Normality (Sec. 8.11)
pH (Sec. 8.8)
Salt (Sec. 8.6)
Self-ionization of water (Sec. 8.7)
Strong acid or base (Sec. 8.2)
Titration (Sec. 8.11)
Triprotic acid (Sec. 8.3)
Weak acid or base (Sec. 8.2)

PROBLEMS

Arrhenius Acids and Bases

8.16 Write equations for the reactions that take place when each of the following acids is added to water:
(a) HF
(b) HBr
(c) H_2SO_3
(d) H_2SO_4
(e) HCO_3^-

Which of these reactions proceed nearly to completion?

8.17 Write equations for the reactions that take place when the bases LiOH and $(CH_3)_2NH$ are each added to water.

Brønsted-Lowry Theory

8.18 Write formulas for the conjugate bases of these acids:
(a) H_2SO_4 (f) NH_3
(b) H_3BO_3 (g) HPO_4^{2-}
(c) HI (h) $H_2PO_4^-$
(d) H_2O (i) H_2S
(e) H_3O^+ (j) HCO_3^-

8.19 Write formulas for the conjugate acids of these bases:
(a) OH^- (f) HCO_3^-
(b) NH_2^- (g) H_2O
(c) NH_3 (h) HPO_4^{2-}
(d) $C_6H_5O^-$ (i) $CH_3{-}NH_2$
(e) CO_3^{2-} (j) PO_4^{3-}

8.20 Tell whether each of these acids is monoprotic, diprotic, or triprotic:
(a) H_2SO_4 (d) CH_3COOH
(b) HNO_3 (e) H_2CO_3
(c) H_3PO_4

8.21 Is there a contradiction between the Brønsted-Lowry and Arrhenius definitions of acid strength? Explain.

8.22 Which of these reactions proceed from left to right?
(a) $H_3PO_4 + OH^- \rightleftharpoons H_2PO_4^- + H_2O$
(b) $H_2O + Cl^- \rightleftharpoons HCl + OH^-$
(c) $HCO_3^- + OH^- \rightleftharpoons H_2O + CO_3^{2-}$
(d) $C_6H_5OH + C_2H_5O^- \rightleftharpoons C_6H_5O^- + C_2H_5OH$
(e) $HCO_3^- + H_2O \rightleftharpoons CO_3^{2-} + H_3O^+$
(f) $CH_3COOH + H_2PO_4^- \rightleftharpoons CH_3COO^- + H_3PO_4$

Acid Dissociation Constants

8.23 Looking at the values in Table 8.3, indicate which is the stronger acid in each of the following pairs:
(a) $H_2PO_4^-$ and HCOOH
(b) HF and H_3BO_3
(c) H_2CO_3 and CH_3COOH

Acid and Base Properties

8.24 What is the most important chemical property of acids and bases?

8.25 Write equations for the reaction of HCl with
(a) Na_2CO_3 (d) NH_3
(b) Mg (e) $CH_3{-}NH_2$
(c) NaOH (f) $NaHCO_3$

Self-Ionization of Water

8.26 Calculate the $[OH^-]$ of each of the following aqueous solutions, given that $[H_3O^+] =$
(a) $10^{-11}\,M$ (c) $10^{-7}\,M$
(b) $10^{-4}\,M$ (d) $10^{1}\,M$

8.27 Determine the $[H_3O^+]$ of each of the following aqueous solutions, given that $[OH^-] =$
(a) $10^{-10}\,M$ (c) $10^{-7}\,M$
(b) $10^{-3}\,M$ (d) $10^{0}\,M$

pH and pK_a

8.28 What is the pH of a solution having a $[H_3O^+]$ of
(a) $10^{-8}\,M$ (d) $10^{0}\,M$
(b) $10^{-10}\,M$ (e) $10^{-7}\,M$
(c) $10^{-2}\,M$
Which of these solutions are acidic, which basic, and which neutral?

8.29 What is the pH of a solution having a $[OH^-]$ of
(a) $10^{-5}\,M$ (d) $10^{-7}\,M$
(b) $10^{-9}\,M$ (e) $10^{-2}\,M$
(c) $10^{-13}\,M$
Which of these solutions are acidic, which basic, and which neutral?

8.30 What is the pH of a solution having a $[H_3O^+]$ of
(a) 3×10^{-9} (c) 8×10^{-12}
(b) 6×10^{-2} (d) 5×10^{-7}
Which of these solutions are acidic, which basic, and which neutral?

8.31 What is the $[H_3O^+]$ of an aqueous solution having a pH of
(a) 12 (e) 8.40
(b) 7 (f) 12.10
(c) 2 (g) 3.52
(d) 13
Which of these solutions are acidic, which basic, and which neutral?

8.32 What is the $[OH^-]$ of an aqueous solution having a pH of
(a) 12 (d) 14
(b) 1 (e) 7
(c) 4
Which of these solutions are acidic, which basic, and which neutral?

***8.33** Three acids have the following pK_a values: butyric acid, 4.82; barbituric acid, 5.00; lactic acid, 3.85. (a) What are the K_a values? (b) Which is the strongest of the three and which the weakest?

The pH of Aqueous Salt Solutions

8.34 Give three examples of acidic salts and three of basic salts.

8.35 If we dissolve some solid sodium carbonate, Na_2CO_3, in pure water, the resulting solution is basic, that is, the pH rises above 7. Explain how this happens.

8.36 Tell whether each of these salts is acidic, basic, or neutral:
(a) Na_3PO_4 (e) NH_4Br
(b) $Al_2(SO_4)_3$ (f) Li_2CO_3
(c) MgI_2 (g) $BaCl_2$
(d) K_2SO_4 (h) $Fe(ClO_4)_3$

Buffers

***8.37** What is the connection between buffer action and Le Chatelier's principle?

8.38 Write equations to show what happens when, to a buffer solution containing equimolar amounts of CH_3COOH and CH_3COO^-, we add (a) H_3O^+ (b) OH^-.

8.39 Write equations to show what happens when, to a buffer solution containing equimolar amounts of HPO_4^{2-} and $H_2PO_4^-$, we add (a) H_3O^+ (b) OH^-.

8.40 What is the pH of a buffer solution made by dissolving 0.1 mole of formic acid, HCOOH, and 0.1 mole of sodium formate, HCOONa, in 1 L of water?

8.41 What is meant by buffer capacity?

8.42 The pH of a solution made by dissolving 1 mole of propionic acid and 1 mole of sodium propionate in 1 L of water is 4.85. (a) What would the pH be if we used 0.1 mole of each (in 1 L of water) instead of 1 mole? (b) With respect to buffer action, how would the two solutions differ?

8.43 What are the two main buffers in blood? Describe how each works.

Titration, Equivalents, and Normality

8.44 What is the purpose of titration?

8.45 What is the equivalent weight of
(a) HBr
(b) H_2CO_3
(c) LiOH
(d) $Ca(OH)_2$
(e) $Al(OH)_3$
(f) NH_3
(g) H_3AsO_4

8.46 How many equivalents are there in
(a) 72.5 g of KOH
(b) 37 g of oxalic acid $H_2C_2O_4$ (a diprotic acid)
(c) 124 g of H_3PO_4
(d) 2.45 g of $Al(OH)_3$

8.47 What is the normality of a solution made by dissolving 12.7 g of HCl in enough water to make 1 L of solution?

8.48 What is the normality of a solution made by dissolving 3.4 g of $Ba(OH)_2$ in enough water to make 450 mL of solution?

8.49 What is the normality of a solution made by dissolving 73.2 g of H_3PO_4 in enough water to make 400 mL of solution?

8.50 Describe exactly how you would prepare each of the following solutions (in each case assume you have the solid bases):
(a) 400.0 mL of a 0.75 *N* solution of NaOH
(b) 1.0 L of a 0.071 *N* solution of $Ba(OH)_2$

8.51 What is the normality of
(a) 3.0 *M* HNO_3
(b) 4.5 *M* H_2SO_4
(c) 0.34 *M* KOH
(d) 2.7 *M* $Ca(OH)_2$

8.52 If 25.0 mL of an aqueous solution of H_2SO_4 requires 19.7 mL of 0.72 *M* NaOH to reach the end point, what is the normality of the H_2SO_4 solution?

8.53 A sample of 27.0 mL of a 0.310 *N* solution of NaOH is titrated with 0.740 *N* H_2SO_4 solution. How many mL of the H_2SO_4 solution are required to reach the end point?

8.54 A 0.300 *M* solution of H_2SO_4 was used to titrate 10.00 mL of an unknown base; 15.00 mL of acid was required to neutralize the basic solution. What was the normality of the base?

8.55 A solution of an unknown base was titrated with 0.150 *N* HCl, and 22.0 mL of acid was needed to reach the end point of the titration. How many equivalents of the unknown base were in the solution?

8.56 What is the molarity of
(a) 1.5 *N* HNO_3
(b) 32 *N* H_2SO_4
(c) 0.035 *N* LiOH
(d) 2.8 *N* NH_3
(e) 0.0618 *N* $Ca(OH)_2$

***8.57** The usual concentration of HCO_3^- ions in blood plasma is approximately 24 meq/L. How would you make up 1 L of a solution containing this concentration of HCO_3^- ions?

Boxes

8.58 (Box 8A) Which weak base is used as a laxative?

8.59 (Box 8B) What is the most important *immediate* first aid in any chemical burn of the eyes?

8.60 (Box 8B) With respect to corneal burns, which are more dangerous, strong acids or strong bases?

8.61 (Box 8C) Which of the antacids listed in Box 8C contain no hydroxides?

8.62 (Box 8D) What compound in baking powder produces the sponginess of a cake?

8.63 (Box 8D) What compounds generate the CO_2 in fire extinguishers?

8.64 (Box 8E) What causes (a) hypoventilation (b) respiratory alkalosis (c) metabolic alkalosis?

8.65 (Box 8F) Calculate the pH of an aqueous solution containing
(a) 0.8 *M* lactic acid and 0.4 *M* lactate ion
(b) 0.3 *M* NH_3 and 1.50 *M* NH_4^+
(c) 0.42 *M* HF and 1.38 *M* F^-

***8.66** (Box 8F) You wish to make a buffer whose pH is 8.21. You have available a stock solution of 1.00 L of 0.100 *M* NaH_2PO_4, as well as some solid Na_2HPO_4. How many grams of the solid Na_2HPO_4 must be added to the stock solution to accomplish this task? (Assume the volume remains at 1.00 L.)

8.67 (Box 8F) Describe how you would make a buffer solution of pH 4.55, using CH_3COOH/CH_3COO^- as the buffer system.

***8.68** (Box 8F) You want to make a CH_3COOH/CH_3COO^- buffer solution with a pH of 5.60. The acetic acid concentration is to be 0.10 *M*. What should the acetate ion concentration be?

Additional Problems

***8.69** Assume that you have a dilute solution of HCl (0.1 *M*) and a concentrated solution of acetic acid (5 *M*). Which solution is more acidic? Explain.

8.70 If the $[OH^-]$ of a solution is 1×10^{-14}, (a) what is the pH of the solution? (b) What is the $[H_3O^+]$?

8.71 (a) What are the Arrhenius definitions of acid and base? (b) What are the Brønsted-Lowry definitions?

8.72 What is the $[H_3O^+]$ of a solution that has a pH of 1.1?

8.73 What is the normality of a solution made by dissolving 0.583 g of the diprotic acid oxalic acid, $H_2C_2O_4$, in enough water to make 1.75 L of solution?

8.74 The pK_a value of barbituric acid is 5.00. If the H_3O^+ and the barbiturate ion concentrations are each 0.00300 *M*, what is the concentration of the undissociated barbituric acid?

8.75 If pure water self-ionizes to give H_3O^+ and OH^- ions, why doesn't pure water conduct an electric current?

***8.76** A scale of K_b values for bases could be set up in a manner similar to that for the K_a scale for acids. However, this is generally considered unnecessary. Can you explain why?

8.77 Do a 1 *M* CH_3COOH solution and a 1 *M* HF solution have the same pH? Explain.

8.78 What is the pH of a solution that has a $[H_3O^+]$ of 6×10^{-12}?

***8.79** 4.00 grams of an unknown acid were dissolved in 1.00 L of solution. This was titrated with 0.600 *N* NaOH, and 38.7 mL of the NaOH solution were needed to neutralize the acid. What is the equivalent weight of the acid?

8.80 Write equations to show what happens when, to a buffer solution containing equal amounts of HF and F^-, we add (a) H_3O^+ (b) OH^-.

***8.81** If a solution of NH_3 in water (aqueous ammonia) is added to a solution of sodium ethoxide (containing ethoxide ions $C_2H_5O^-$), will an acid-base reaction take place? If so, write an ionic equation for the reaction.

8.82 If we add 0.1 mole of NH_3 to 0.5 mole of HCl dissolved in enough water to make 1 L of solution, what happens to the NH_3? Will any of it remain? Explain.

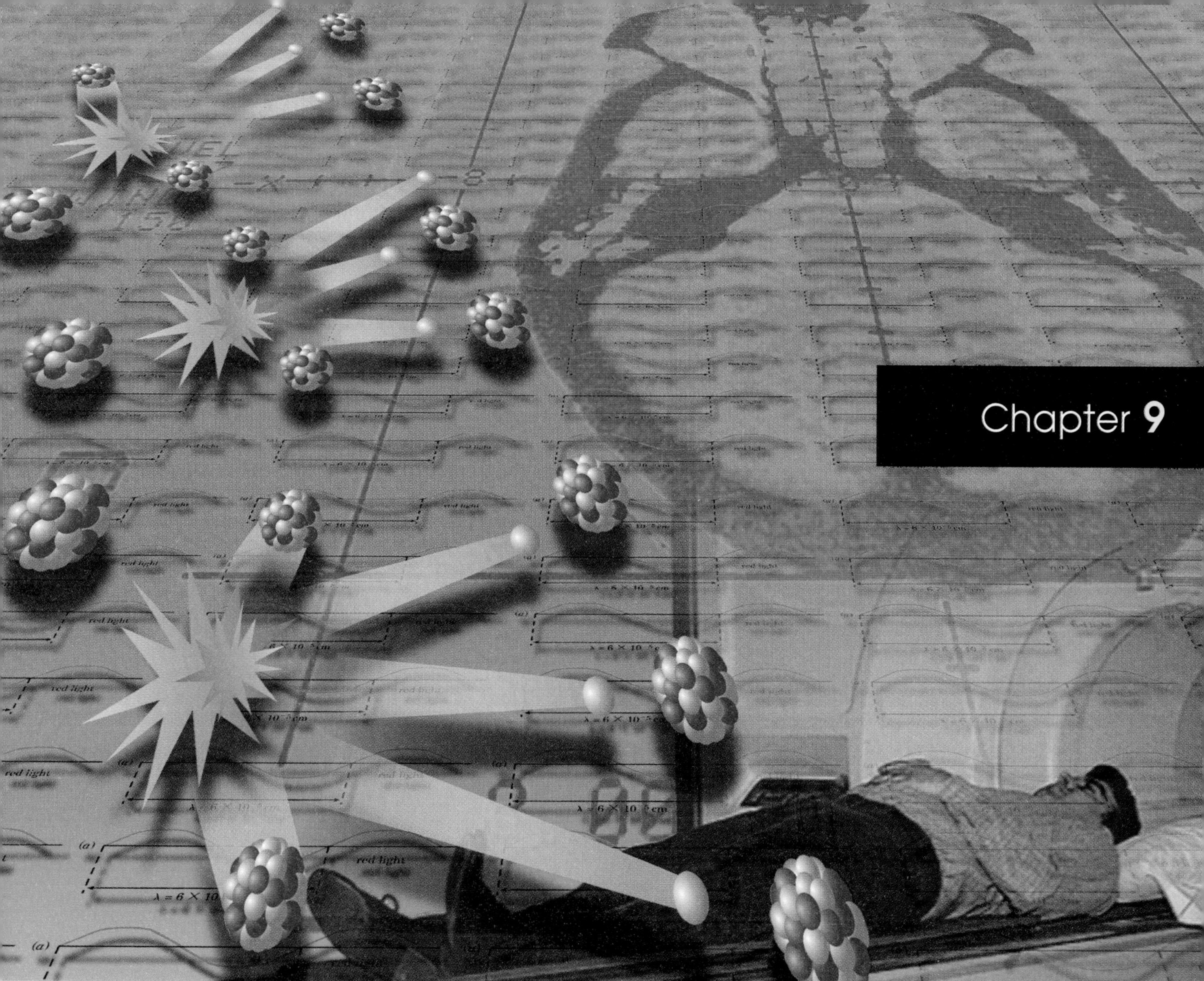

Chapter 9

Nuclear chemistry

9.1 Introduction

Every so often, a scientist makes the kind of discovery that changes the future of the world in some significant way. In 1896 a French scientist, Henri Becquerel (1852–1908), made one of the most important of these discoveries. Completely by accident he found that a certain rock was giving off mysterious radiation. The discovery was totally unexpected; nobody had ever had the faintest idea that anything of the kind was happening, and yet that rock, and others like it, had been giving off radiation since long before there were humans on the earth. It was simply that, before 1896, nobody knew it.

Becquerel's rock contained uranium. The radiation given off by the uranium caused photographic plates to "fog" even though they were wrapped in black paper. This meant that the radiation coming from the rock must be more energetic than visible light because it could pass through the paper.

Becquerel and the Curies were awarded the Nobel Prize in 1903 for this discovery. Rutherford won the Nobel Prize in 1908.

Shortly after Becquerel's discovery, other scientists, most notably Pierre (1859–1906) and Marie (1867–1934) Curie and Ernest Rutherford (1871–1937), began to study this radiation, which Marie Curie called *radioactivity.* They soon learned that there were three types, which they named alpha, beta, and gamma, after the first three letters of the Greek alphabet.

Alpha particles contain two protons and two neutrons and are thus the same as helium nuclei. They have an atomic number of 2.

Beta particles are electrons. Beta "radiation" is just a stream of electrons.

Gamma rays are not particles; they are a high-energy form of electromagnetic radiation.

Gamma rays are only one form of **electromagnetic radiation.** There are many others, including visible light. All consist of waves (Fig. 9.1). The only difference between one form of electromagnetic radiation and another is the **wavelength** (λ), which is the distance from one wave crest to the next. All electromagnetic radiation travels at the same speed in vacuum (the speed of light, which is 3.0×10^{10} cm/s). Because of this, the different forms of electromagnetic radiation have different frequencies. The **frequency** (ν) of a radiation is the number of crests that pass a given point in one second. **The longer the wavelength, the lower the frequency.** Mathematically they are related by the equation

$$\lambda = \frac{c}{\nu}$$

where c is the speed of light. There is also a direct relationship between the frequency of electromagnetic radiation and its energy: **The higher**

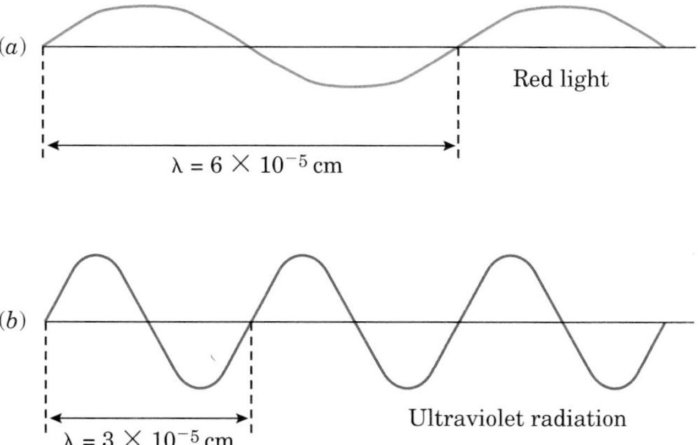

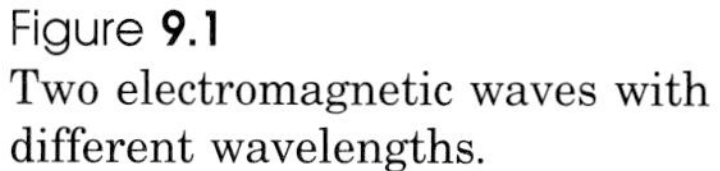

Figure **9.1**
Two electromagnetic waves with different wavelengths.

the frequency, the higher the energy. Electromagnetic radiation comes in packets; the smallest units are called **photons.**

Gamma rays are electromagnetic radiation of very high frequency. We cannot see them because our eyes are not sensitive to waves of this frequency, but they can be detected by instruments (Sec. 9.4).

The only radiation we know of with even higher frequency (and energy) than gamma rays are cosmic rays.

Another kind of radiation, called **x-rays,** has frequencies (and energies) higher than those of visible light but less than those of gamma rays. Figure 9.2 shows the position of all these waves on the **electromagnetic spectrum.**

Materials that emit radiation (alpha, beta, or gamma) are called **radioactive. Radioactivity comes from the atomic nucleus** and not from the electron cloud. Table 9.1 summarizes the properties of the particles and rays that come out of radioactive nuclei, along with the properties of some other particles and rays. Note that x-rays are not considered to be a form of radioactivity. They do not come out of the nucleus but are generated in other ways.

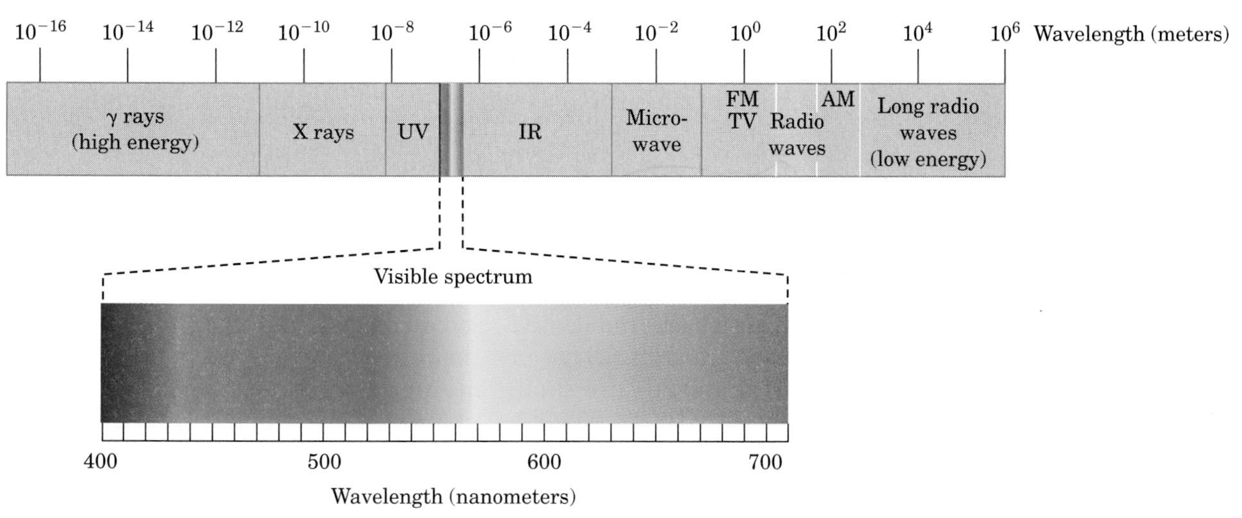

Figure **9.2** • The electromagnetic spectrum.

Table 9.1 Particles and Rays Frequently Encountered in Radiation

Particle or Ray	Common Name of Radiation	Symbol	Charge	Atomic Mass Units	Penetrating Power[a]	Energy Range[b]
Proton	Proton beam	${}^{1}_{1}H$	+1	1	1–3 cm	60 MeV
Electron	Beta particle	${}^{0}_{-1}e$ or β	−1	0.00055 $\left(\frac{1}{1835}\right)$	0–4 mm	1–3 MeV
Neutron	Neutron beam	${}^{1}_{0}n$	0	1	—	—
Positron	—	${}^{0}_{+1}e$ or β^+	+1	0.00055	—	—
Helium nucleus	Alpha particle	${}^{4}_{2}He$ or α	+2	4	0.02–0.04 mm	3–9 MeV
Energetic radiation	Gamma ray	γ	0	0	1–20 cm	0.01–10 MeV
	X-ray		0	0	0.01–1 cm	0.1–10 keV

[a] Distance at which half the radiation has been stopped in water.
[b] $1\ \text{MeV} = 1.602 \times 10^{-13}\ \text{J} = 3.829 \times 10^{-14}$ cal.

Characteristics of Radioactivity

We have said that humans cannot see gamma rays. This is also true for the other kinds of radioactivity. We cannot see alpha or beta particles either. Not only that, but we also cannot hear them, smell them, or feel them. They are undetectable by our senses, which is why they were undiscovered for most of human history. Radioactivity can be detected only by instruments. We will discuss this subject in Section 9.4.

9.2 What Happens When a Nucleus Emits Radioactivity? Natural Transmutation

As mentioned in Section 2.4, different nuclei are made up of different numbers of protons and neutrons. It is customary to indicate these numbers with subscripts and superscripts placed to the left of the atomic symbol. The number of protons is shown as a subscript and the mass number (total number of protons and neutrons) as a superscript. For example, ${}^{1}_{1}H$, ${}^{2}_{1}H$, and ${}^{3}_{1}H$ are the symbols for the three isotopes of hydrogen. We call them hydrogen-1, hydrogen-2, and hydrogen-3, respectively.

Hydrogen-2 is also called deuterium, and hydrogen-3 is called tritium.

Radioactive and Stable Nuclei

A radioactive isotope is often called a **radioisotope.**

Not all nuclei are radioactive. Some isotopes (Sec. 2.4) are radioactive and others are stable. There are more than 300 naturally occurring isotopes. Of these, 264 are stable, meaning that these nuclei never give off any radioactivity. As far as we can tell, they will last forever. The remainder are radioactive—they do give off radioactivity. Furthermore, more than 1000 artificial isotopes have been made in laboratories.

All artificial isotopes are radioactive.

Isotopes in which the number of protons and neutrons are balanced seem to be stable. In the lighter elements, this occurs when the numbers of protons and neutrons are approximately equal. For example, ${}^{12}_{6}C$ is a stable nucleus (six protons and six neutrons) and so are ${}^{16}_{8}O$ (eight protons and eight neutrons) and ${}^{20}_{10}Ne$ (ten protons and ten neutrons). Among the heavier elements, stability requires more neutrons than protons. Lead-206, one of the most stable isotopes of lead, contains 82 protons and 124 neutrons.

The role of neutrons seems to be to provide binding energy to overcome the repulsion between protons.

Beta Emission

If there is a serious imbalance in the proton-to-neutron ratio, a nucleus will emit an alpha or beta particle and thereby make the ratio more favorable and the nucleus more stable. For example, if a nucleus has an excess number of neutrons, it can stabilize itself by converting a neutron to a proton and an electron:

$$ {}^{1}_{0}n \longrightarrow {}^{1}_{1}H + {}^{0}_{-1}e $$

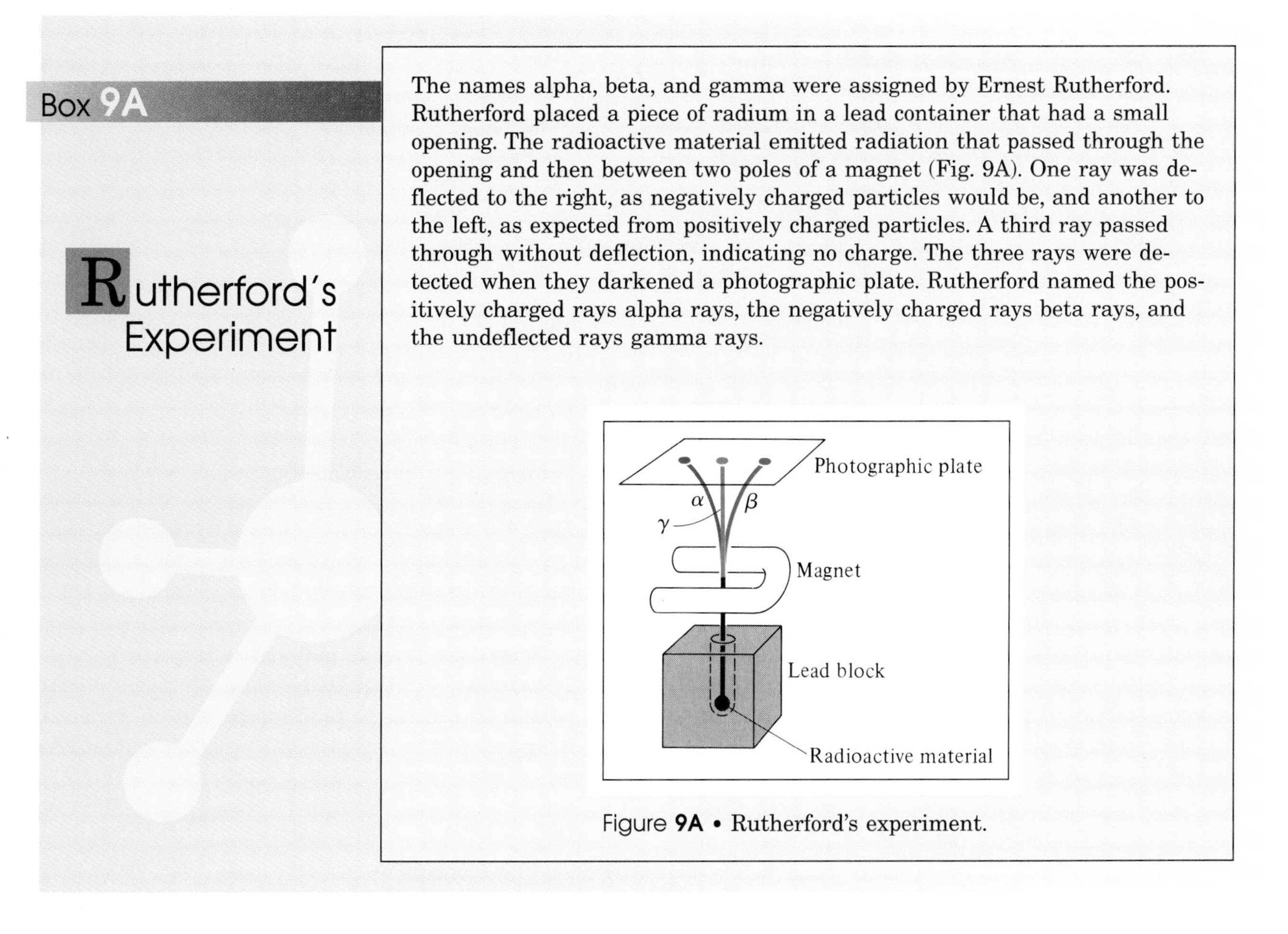

Box 9A

Rutherford's Experiment

The names alpha, beta, and gamma were assigned by Ernest Rutherford. Rutherford placed a piece of radium in a lead container that had a small opening. The radioactive material emitted radiation that passed through the opening and then between two poles of a magnet (Fig. 9A). One ray was deflected to the right, as negatively charged particles would be, and another to the left, as expected from positively charged particles. A third ray passed through without deflection, indicating no charge. The three rays were detected when they darkened a photographic plate. Rutherford named the positively charged rays alpha rays, the negatively charged rays beta rays, and the undeflected rays gamma rays.

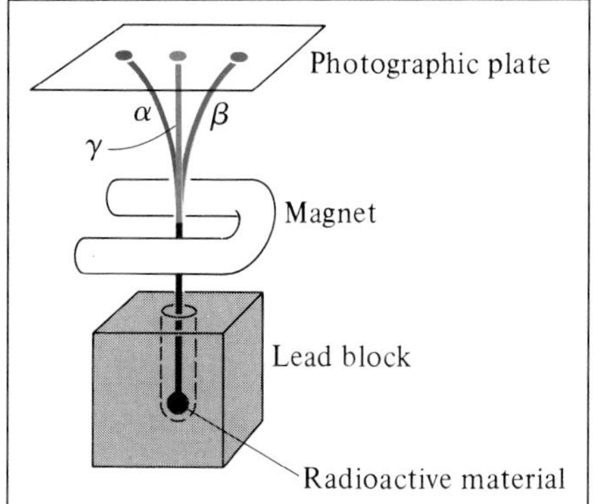

Figure **9A** • Rutherford's experiment.

The electrons of the beta emission come from the nucleus and not from the surrounding electron cloud.

This process is called **beta emission** because the electrons (beta particles) that are produced are emitted from the nucleus. For example, phosphorus-32 is a beta emitter:

$$^{32}_{15}\text{P} \longrightarrow {}^{32}_{16}\text{S} + {}^{0}_{-1}\text{e}$$

When the unstable phosphorus-32 is converted to sulfur-32, stability is achieved (16 protons and 16 neutrons).

Every time a nucleus emits a beta particle, it is transformed into another nucleus with the same mass number but an atomic number one unit greater.

In balancing nuclear equations, **the sum of the masses (superscripts) must be equal on both sides of the equation. The same is true for the atomic numbers (subscripts).** To find the proper symbol for a nucleus corresponding to a certain atomic number, use the periodic table on the inside cover of this book.

Note that for balancing purposes the electron is given an atomic number of −1.

EXAMPLE

What nucleus is $^{14}_{6}\text{C}$, a beta emitter, converted to?

$$^{14}_{6}\text{C} \longrightarrow ? + {}^{0}_{-1}\text{e}$$

Answer • The $^{14}_{6}\text{C}$ nucleus has six protons and eight neutrons. One of its neutrons is converted to a proton and an electron. The electron is emitted (this is the beta particle), but the proton remains. The nucleus now has seven protons (therefore it must be nitrogen) and seven neutrons. The equation is

$$^{14}_{6}\text{C} \longrightarrow {}^{14}_{7}\text{N} + {}^{0}_{-1}\text{e}$$

Note that this equation is balanced: the sum of the masses is 14 on both sides; and the sum of the atomic numbers is 6 on both sides.

Problem **9.1**

Iodine-139 is a beta emitter. What is the reaction product?

Alpha Emission

The changing of one element into another is called **transmutation.** It happens naturally every time an element gives off beta particles. What about alpha emission? Transmutation takes place here also. For heavy elements, the loss of alpha particles is an especially important stabilization process. For example,

$$^{238}_{92}\text{U} \longrightarrow {}^{234}_{90}\text{Th} + {}^{4}_{2}\text{He}$$

$$^{210}_{84}\text{Po} \longrightarrow {}^{206}_{82}\text{Pb} + {}^{4}_{2}\text{He} + \gamma$$

The symbol γ, the Greek letter gamma, represents a gamma ray.

In the second example, stable lead-206 is obtained after alpha emission. Note that the polonium-210 emits both alpha particles and gamma rays.

Both alpha and beta emission can be either "pure" or mixed with gamma rays.

Can you deduce, from the examples given, the general rule for alpha emission? **The new nucleus always has a mass number four units lower and an atomic number two units lower than the original.** Like beta emission, alpha emission always means that natural transmutation has taken place.

EXAMPLE

Polonium-218 is an α emitter. Write an equation for this α decay and identify the product.

Answer • The atomic number of polonium is 84. The equation is

$$^{218}_{84}\text{Po} \longrightarrow {}^{4}_{2}\text{He} + ?$$

Subtract 4 from the mass number 218 to get the new mass number, 214. Subtract 2 from atomic number 84 to get the new atomic number, 82. In the periodic table we find that the element with an atomic number of 82 is lead, Pb. Therefore the product is $^{214}_{82}\text{Pb}$.

Problem **9.2**

Thorium-223 is an α emitter. Identify the isotope formed.

Positron Emission

A positron is a particle that has the same mass as an electron, but a charge of +1, rather than −1. Its symbol is β^+ or $_{+1}^{0}\text{e}$. Positron emission is much less frequent than beta emission, but some nuclei do emit positrons. When it happens the nucleus is transformed into another nucleus with the same mass number, but an atomic number one unit less. An example is

$$^{11}_{6}\text{C} \longrightarrow {}_{+1}^{0}\text{e} + {}^{11}_{5}\text{B}$$

EXAMPLE

Nitrogen-13 is a positron emitter. Write an equation for this nuclear reaction and identify the product.

Answer • The mass number of the new isotope will still be 13, since a positron has no appreciable mass. To get the atomic number we subtract 1 from 7, the atomic number of nitrogen. The new atomic number is therefore 6. The element with an atomic number of 6 is carbon. Therefore, the equation is

$$^{13}_{7}\text{N} \longrightarrow {}_{+1}^{0}\text{e} + {}^{13}_{6}\text{C}$$

Problem **9.3**

Nitrogen-15 is found to be the product in a certain experiment. There was evidence that this was the result of positron emission. What was the original positron emitter?

Gamma Emission

Besides alpha, beta, and positron emitters, one occasionally encounters pure gamma emitters:

$$^{11}_{5}\text{B}^* \longrightarrow \, ^{11}_{5}\text{B} + \gamma$$

In this equation, the $^{11}_{5}\text{B}^*$ symbolizes the boron nucleus in a high-energy (excited) state. In this case, there is no transmutation. The boron is still boron, but it is in a lower-energy form (more stable) after the emission of excess energy in the form of gamma rays.

9.3 Half-Life

When a nucleus gives off radiation, it is said to decay.

Suppose we have 40 g of a radioactive isotope, say $^{90}_{38}\text{Sr}$. It contains many atoms, of course (in this case, about 2.7×10^{23}). These atoms are unstable and decay. In this case they give off beta particles and thereby become $^{90}_{39}\text{Y}$ atoms. But do they all decay at once? No, they do not. They decay one at a time, at a fixed rate. For strontium-90, the decay rate is such that one half of our sample (about 1.35×10^{23} atoms) will have decayed by the end of 28.1 years. **The time it takes for one half of any sample of radioactive material to decay** is called the **half-life** ($t_{\frac{1}{2}}$).

It does not matter how big or small a sample is. For example, in the case of our 40 g of strontium-90, 20 g will be left at the end of 28.1 years (the rest will have been converted to yttrium-90). It will then take another 28.1 years for half of the remainder to decay, so that at the end of that time we will have 10 g. If we wait still a third period of 28.1 years, then 5 g will be left. If we had begun with 100 g, then 50 g would be left after the first 28.1-year period.

Figure 9.3 shows the radioactive decay curve of $^{131}_{53}\text{I}$. Inspection of this graph shows that, at the end of eight days, half of the original $^{131}_{53}\text{I}$ has disappeared. Thus, the half-life of iodine-131 is eight days. It would take a total of 16 days, or two half-lives, for three fourths of the original iodine-131 to decay.

EXAMPLE

If 10 mg of fresh $^{131}_{53}\text{I}$ is administered to a patient, how much is left in the body after 32 days?

Answer • We know that 32 days corresponds to four half-lives because $t_{\frac{1}{2}}$ for this isotope is eight days. After one half-life, 5 mg is left; after two half-lives, 2.5 mg; after three half-lives, 1.25 mg; and after four half-lives, 0.63 mg.

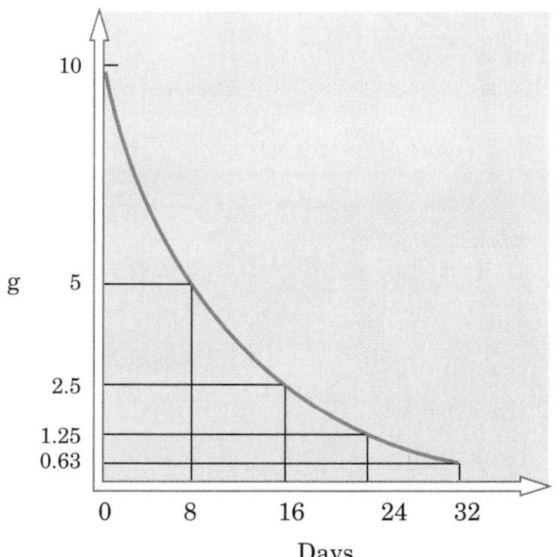

Figure **9.3**
The decay curve of iodine-131.

Problem **9.4**

Barium-122 has a half-life of 2 minutes. You just obtained a fresh sample weighing 10 g. It takes 10 minutes to set up the experiment in which barium-122 will be used. How many grams of barium-122 will be left when you begin the experiment?

It must be noted that, in theory, it would take an infinite amount of time for all of a radioactive sample to disappear. However, most of the radioactivity disappears after five half-lives; by that time, only 3 percent of the original is left. After ten half-lives less than 0.1 percent of the activity remains.

It is important to note that the half-life of an isotope is independent of temperature and pressure, and indeed of all other conditions, and is a property of the particular isotope only. It does not depend on what kind of other atoms surround the particular nucleus, that is, what kind of molecule the nucleus is part of. **We do not know any way to speed up radioactive decay or slow it down.**

Some half-lives are given in Table 9.2. Even this brief sampling indicates that there are tremendous differences in half-lives. Some isotopes, such as technetium-99m, decay and disappear in a day; others, such as uranium-238, will be radioactive for billions of years. Very short-lived isotopes, especially the artificial heavy elements (Sec. 9.7) with atomic numbers greater than 100, have half-lives of the order of seconds. The usefulness, or the inherent danger, in some of these radioactive isotopes is related to their half-lives.

The "m" in technetium-99m means *metastable.*

In assessing long-range health effects of atomic-bomb damage or of nuclear power plant accidents, such as what happened at Three Mile Island, Pennsylvania, in 1979 or Chernobyl (in the former Soviet Union) in 1986 (see Box 9G), we can see that radioactive isotopes with long half-lives, such as $^{85}_{36}Kr$ ($t_{\frac{1}{2}}$ = 10 years) or $^{60}_{27}Co$ ($t_{\frac{1}{2}}$ = 5.2 years), are more important than the short-lived ones. On the other hand, when a radioactive isotope is used in medical diagnosis or therapy, short-lived isotopes

Table 9.2 Half-lives of Some Radioactive Nuclei

Name	Symbol	Half-Life	Radiation
Hydrogen-3 (tritium)	$^{3}_{1}H$	12.26 years	Beta
Carbon-14	$^{14}_{6}C$	5730 years	Beta
Phosphorus-28	$^{28}_{15}P$	0.28 seconds	Positrons
Phosphorus-32	$^{32}_{15}P$	14.3 days	Beta
Potassium-40	$^{40}_{19}K$	1.28×10^9 years	Beta + gamma
Scandium-42	$^{42}_{21}Sc$	0.68 seconds	Positrons
Cobalt-60	$^{60}_{27}Co$	5.2 years	Gamma
Strontium-90	$^{90}_{38}Sr$	28.1 years	Beta
Technetium-99m	$^{99m}_{43}Tc$	6.0 hours	Gamma
Indium-116	$^{116}_{49}In$	14 seconds	Beta
Iodine-131	$^{131}_{53}I$	8 days	Beta + gamma
Mercury-197	$^{197}_{80}Hg$	65 hours	Gamma
Polonium-210	$^{210}_{84}Po$	138 days	Alpha
Radon-205	$^{205}_{86}Rn$	1.8 minutes	Alpha
Radon-222	$^{222}_{86}Rn$	3.8 days	Alpha
Uranium-238	$^{238}_{92}U$	4×10^9 years	Alpha

are more useful because they disappear faster from the body. Among the isotopes listed in Table 9.2, those with half-lives of only hours or days, such as $^{99m}_{43}Tc$, $^{32}_{15}P$, $^{131}_{53}I$, and $^{197}_{80}Hg$, find much use in medicine.

9.4 Characteristics of Radiation

As already noted, radioactivity is not detectable by our senses. We cannot see it, hear it, feel it, or smell it. Then how do we know it is there? Alpha, beta, gamma, and x-rays, as well as proton beams, all have a prop-

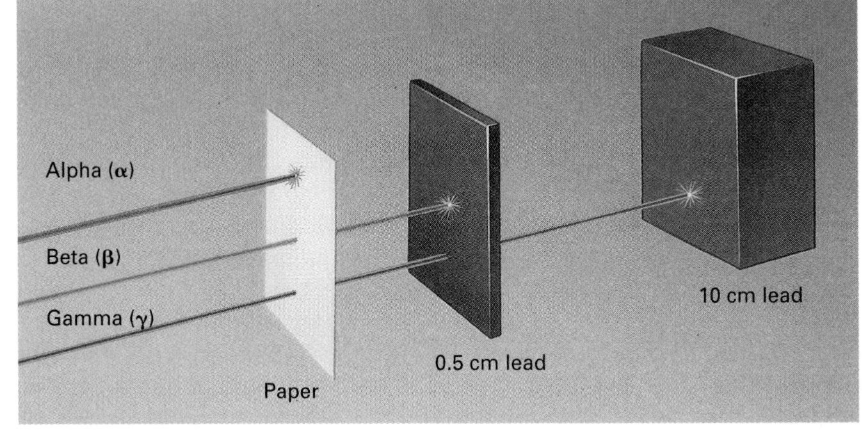

Alpha particles penetrate the least. They are stopped by a piece of paper. Beta particles easily penetrate paper, but cannot go through a 0.5 cm sheet of lead. Gamma rays have the greatest penetrating power. It takes 10 cm of lead to stop them.

erty we can use to detect them: When they interact with matter, they usually knock electrons out of the electron cloud surrounding the atomic nucleus and thereby create positively charged ions from neutral atoms. Because of this, we call all of them **ionizing radiation.**

Radiation is characterized by two physical measurements: (a) its **intensity,** which is the number of particles or photons emerging per unit time, and (b) the **energy** of each particle or photon.

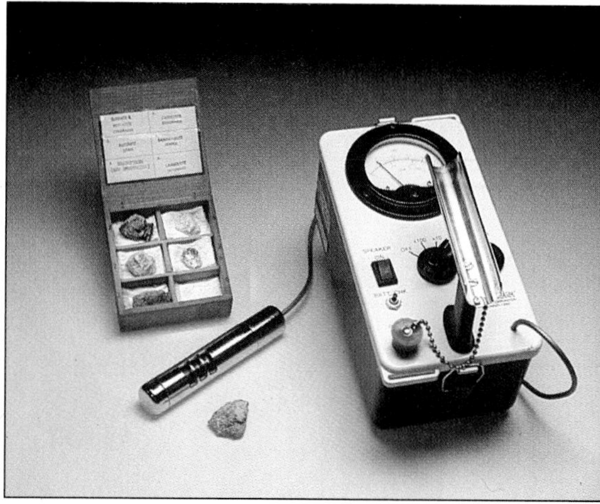

A Geiger-Müller counter. (Photograph by Charles D. Winters.)

Intensity

To measure intensity we take advantage of the ionizing property of radiation. Devices such as the **Geiger-Müller counter** (Fig. 9.4) or the **proportional counter** contain a gas such as helium or argon. When a radioactive nucleus emits beta particles (electrons), these particles ionize the gas, and the instrument registers this by indicating that an electric current is passing between two electrodes. In this way, the instrument counts particle after particle.

Other measuring devices, such as **scintillation counters,** have a material called a phosphor that emits a unit of light when an alpha or beta particle or gamma ray strikes it. Again the particles are counted one by one. The quantitative measure of radioactivity can be reported in counts/minute or counts/second.

A television picture tube works on a similar principle.

One easy way to protect against ionizing radiation is to wear lead aprons, covering sensitive organs. This is done routinely when diagnostic x-rays are taken. Another way to lessen the damage from ionizing radiation is to move further away from the source. The intensity of any radiation decreases with the square of the distance. Thus,

$$\frac{I_1}{I_2} = \frac{d_2^{\,2}}{d_1^{\,2}}$$

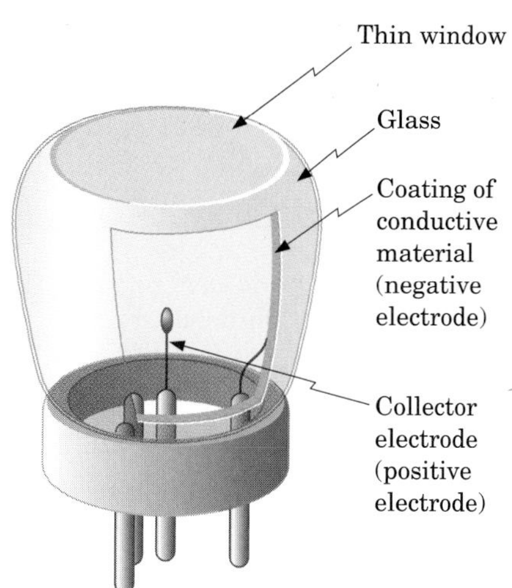

Figure **9.4**
The Geiger-Müller counter. The counter is made of a glass cylinder. The inner surface is coated with a material that conducts electricity. In the center is the collector electrode, made of tungsten wire. This is positively charged to about 1200 volts. The space between is filled with helium or argon gas at a low pressure. A thin Mylar plate (window) at the end keeps the gas in and at the same time allows the ionizing radiation to pass through. When the radiation enters the tube, it converts an argon or helium atom to a positive ion, which goes to the negative electrode and is counted.

Box 9B

Radioactive Dating

Carbon-14, with a half-life of 5730 years, can be used to date archeological objects as old as 60 000 years. The dating is based on the principle that an organism, plant or animal, as long as it lives, maintains a constant carbon-12/carbon-14 ratio. When the organism dies, the carbon-12 level remains constant but the carbon-14 decays through the reaction

$$^{14}_{6}C \longrightarrow ^{14}_{7}N + ^{0}_{-1}e$$

Thus, the changing carbon-12/carbon-14 ratio can be used to determine the date of an artifact.

For example, in charcoal made from a tree that has recently died, the carbon-14 gives a radioactive count of 13.70 disintegrations per minute per gram of carbon. In a piece of charcoal found in a cave in France near some ancient cave paintings, the carbon-14 count was 1.71 disintegrations/min for each gram of carbon. From this information the cave paintings can be dated. After one half-life, the number of disintegrations per minute per gram is 6.85; after two half-lives it is 3.42, and after three half-lives it is 1.71. Therefore, three half-lives have passed since the paintings were done. Since carbon-14 has a half-life of 5730 years, the paintings are about $3 \times 5730 = 17\ 190$ years old.

The famous Shroud of Turin, a piece of linen cloth with the image of a man's head on it, was believed by many to be the original cloth that was wrapped around Jesus Christ. However, radioactive dating showed, with 95 percent certainty, that the plants from which the linen was obtained

A prehistoric cave painting (Hall of Bulls) from Lascaux, Dordogne, France.

were alive sometime between AD 1260 and 1380, proving that it could not have been the shroud of Christ. It was not necessary to destroy the shroud in order to make the tests. Scientists used just a few square centimeters of cloth from its edge.

Rock samples can be dated on the basis of their lead-206 and uranium-238 content. The assumption is that all of the lead-206 has come from the decay of uranium-238,* which has a half-life of 4.5 billion years. The oldest rock found on the earth is a granite outcrop in Greenland, dated to be 3.7×10^9 years old.

On the basis of dating meteorites, the estimated age of the solar system is 4.6×10^9 years.

*Uranium-238 does not decay *directly* to lead-206. It first goes to thorium-234, which then also decays to radium-230, and so on. Lead-206 is the final result of a series of such transmutations.

(*a*)

(*b*)

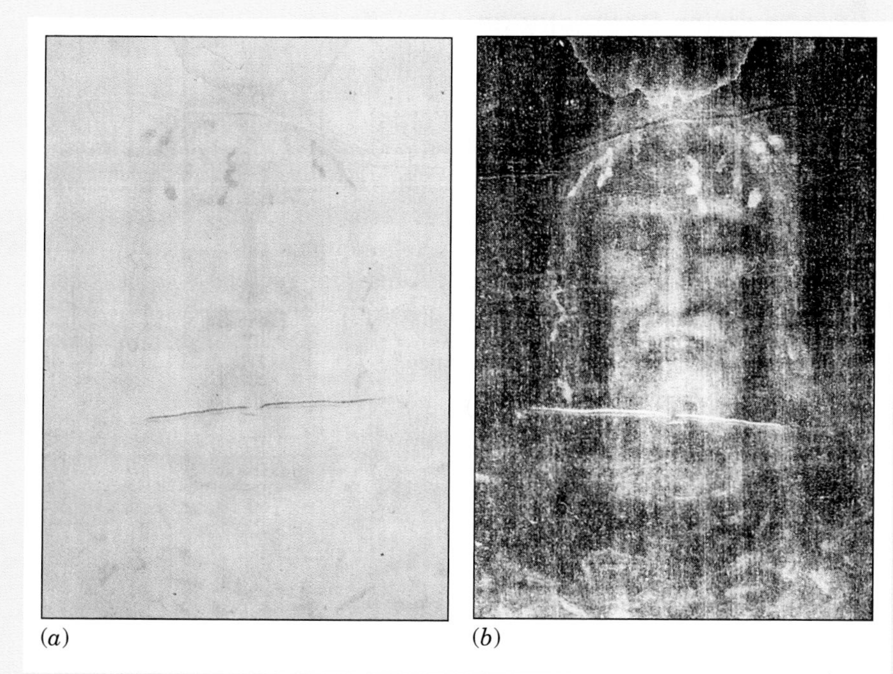

The Shroud of Turin: (a) as it appears to the eye, and (b) photographic negative, which brings out the image more clearly. (Santa Visalli/The IMAGE BANK.)

mCi and other units of radiation are defined in the next section.

EXAMPLE

If the intensity of radiation is 28 mCi at a distance of 1.0 m what is the intensity at a distance of 2.0 m?

Answer • From the preceding equation we have

$$\frac{28 \text{ mCi}}{I_2} = \frac{2.0^2}{1.0^2}$$

$$I_2 = \frac{28 \text{ mCi}}{4.0} = 7.0 \text{ mCi}$$

Problem **9.5**

If the intensity of radiation 1.0 cm from a source is 300 mCi, what is the intensity at 3.0 m?

9.5 Radiation Dosimetry and the Effects of Radiation on Human Health

A common unit of radiation activity or intensity is the **curie** (Ci), named in honor of Marie Curie, whose lifelong work with radioactive materials greatly helped our understanding of nuclear phenomena. One curie is defined as 3.7×10^{10} disintegrations per second (or counts per second). This is radiation of very high intensity, the amount a person would get from 1 g of pure $^{286}_{88}Ra$. This activity is too high for regular medical use, and the most common units used in the health sciences are small fractions of this:

1 millicurie (mCi) = 3.7×10^7 counts/s

1 microcurie (μCi) = 3.7×10^4 counts/s

EXAMPLE

A radioactive isotope with an activity of 100 mCi per vial is sold to a hospital. The vial contains 10 mL of liquid. The instruction is to administer 2.5 mCi intravenously. How much of the liquid should one use?

Answer • The activity of a sample is directly proportional to the amount present, so

$$2.5 \cancel{\text{mCi}} \times \frac{10 \text{ mL}}{100 \cancel{\text{mCi}}} = 0.25 \text{ mL}$$

Box 9C

The Indoor Radon Problem

Most of our exposure to ionizing radiation comes from natural sources (Table 9.4). Radon gas is the main cause of this exposure. Radon has more than 20 isotopes, all radioactive, but the most important is radon-222, an alpha particle emitter. Radon-222 is a natural decay product of uranium-238, which is widely distributed in the earth's crust. Radon poses a particular health hazard among radioactive elements because it is a gas at normal temperatures and pressures. This means that radon enters our lungs with the air we breathe and gets trapped in the mucous lining of the lung. Radon-222 has a half-life of 3.8 days. It decays naturally and produces, among other isotopes, two harmful alpha emitters: polonium-218 and polonium-214. Radon-222 and its decay products can cause lung cancer over a long exposure.

The Environmental Protection Agency sets a standard of 4 pCi/L (one picocurie, pCi, is 10^{-12} curie) as a safe exposure level. A survey of single-family homes in the United States showed that 7 percent exceeded this level. Most radon seeps into dwellings through cracks in cement foundations and around pipes. It accumulates mostly in basements. The remedy is to ventilate basements as well as houses enough to reduce the radiation levels. In a notorious case, in Grand Junction, Colorado, houses were built from bricks made from uranium tailings. Obviously the radiation levels in these buildings were unacceptably high. They could not be controlled and the buildings had to be destroyed. In our modern radiation-conscious age more and more home buyers request a certification of radon levels before buying a house.

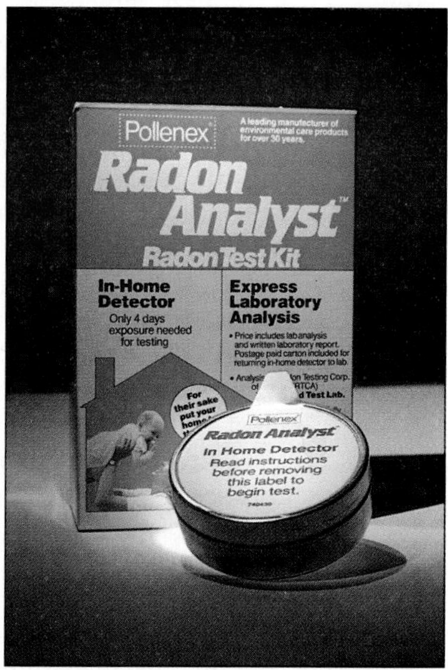

Testing devices are available to determine whether radon is building up in a home. (Photograph by Charles D. Winters.)

Problem 9.6

A radioactive isotope in a 9.0-mL vial has an activity of 300 mCi. A patient is required to take 50 mCi intravenously. How much liquid should be used for the injection?

Another unit of radiation activity is the **becquerel** (Bq). One becquerel is one disintegration per second.

In studying the effect of radiation on the body, neither the energy of the radiation (in kcal/mole) nor its activity (in Ci) alone or in combination is of particular importance. The critical question is what kind of effects such radiation produces in the body. Three different units are used to describe the **effects of radiation:** roentgens, rads, and rems.

Wilhelm Roentgen (1845–1923) discovered x-rays in 1895.

Roentgens. Roentgens measure the ability of the radiation to affect matter, and in particular human beings and animals. One **roentgen** is the amount of radiation that produces ions having 2.58×10^{-4} coulombs per kilogram (a coulomb is a unit of electrical charge). The number of roentgens is the **effective dose delivered** by the radiation and is therefore a measure of exposure to a particular form of radiation.

Rads. However, exposure, or delivered energy, does not take into account the effect on tissue. Radiation damages body tissue by causing ionization. In order for ionization to occur, the delivered energy must be absorbed by the tissue. The measure of the amount of **radiation absorbed** is the **rad,** which stands for **r**adiation **a**bsorbed **d**ose. The SI unit is the **gray** (Gy). One Gy equals 100 rad.

The relationship between delivered dose in roentgens and absorbed dose in rads is that exposure to 1 roentgen yields 0.97 rad of absorbed radiation in water, 0.96 rad in muscle, and 0.93 rad in bone. This relationship holds for photons with high energies. For lower-energy photons, such as "soft" x-rays, each roentgen yields 3 rads of absorbed dose in bone. This is the principle of diagnostic x-rays, wherein soft tissue lets the radiation through to strike a photographic plate but the bones absorb the radiation and cast a shadow on the plate.

Rems. The most important biological measure of radiation is the **rem,** which stands for **r**oentgen **e**quivalent **m**an. One rem is a measure of the effect of the radiation when 1 roentgen is absorbed by a person. Other units are the **millirem** (mrem) (1 mrem $= 1 \times 10^{-3}$ rem) and the **sievert** (Sv) (1 Sv = 100 rem). The sievert is the SI unit. The need for such units arises because 1 rad of absorbed energy from different sources can cause different damage to the tissues. One rad from alpha rays or neutrons causes ten times more damage than 1 rad from x-rays or gamma rays.

The various radiation units are summarized in Table 9.3.

Although alpha particles cause more damage than x-rays or gamma rays, they have a very low penetrating power and cannot pass through the skin (see Table 9.1). Consequently, they are not harmful to humans or animals as long as they do not get into the body some other way. If they do get in, however, they are quite harmful. They can get inside, for example, if a person swallows or inhales a small particle of a substance that emits alpha particles.

Table 9.3 Radiation Dosimetry

Measurement	International Unit	Conventional Unit
Activity	1 becquerel (Bq) = 1 disintegration/s	1 curie (Ci) = 3.7×10^{10} disintegrations/s
Production of ions	1 roentgen (R) = 2.58×10^{-4} coulomb/kg	1 roentgen (R) = 2.58×10^{-4} coulomb/kg
Absorbed dose	1 gray (Gy) = 1 joule/kg	1 rad = 0.01 Gy
Absorbed dose by humans	1 sievert (Sv) = 1 joule/kg	1 rem = 0.01 Sv

Beta particles are less damaging to tissue than alpha particles but penetrate farther and so are generally more harmful. But gamma rays, which can completely penetrate the skin, are by far the most dangerous and harmful form of radiation.

Therefore, for comparison purposes and for determining exposure from all kinds of sources, the **equivalent dose** is an important measure. If an organ receives radiation from different sources, the total effect can be summed up in rems (or mrem or Sv). For example, 10 mrem of alpha particles and 15 mrem of gamma radiation give a total of 25 mrem absorbed equivalent dose. Table 9.4 shows the amount of radiation exposure an average person obtains yearly from both natural and artificial sources. People who work in nuclear medicine are of course exposed to greater amounts. In order to assure that exposures do not get too high,

Table 9.4 Average Exposure to Radiation from Common Sources[a]

	Dose (mrem/year)[b]
Naturally Occurring Radiation	
Cosmic rays	27
Terrestrial radiation (rocks, buildings)	28
Inside human body (K-40 and Ra-226 in the bones)	39
Radon in the air	200
Total	294
Man-made (Artificial) Radiation	
Medical x-rays[c]	39
Nuclear medicine	14
Consumer products	10
Nuclear power plants	0.5
All others	1.5
Total	65.0
Grand total	359[d]

[a]Source: *National Council on Radiation Protection and Measurements* NCRP Report No. 93 (1987).
[b]100 mrem = 1 mSv.
[c]Individual medical procedures may expose certain parts of the body to much higher doses. For instance, one chest x-ray gives 27 mrem, and a diagnostic GI series gives 1970 mrem.
[d]The federal safety standard for allowable occupational exposure is about 5000 mrem/year. However, it has been suggested that this be lowered to 4000 mrem/yr or even lower, to reduce the risk of cancer stemming from low levels of radiation.

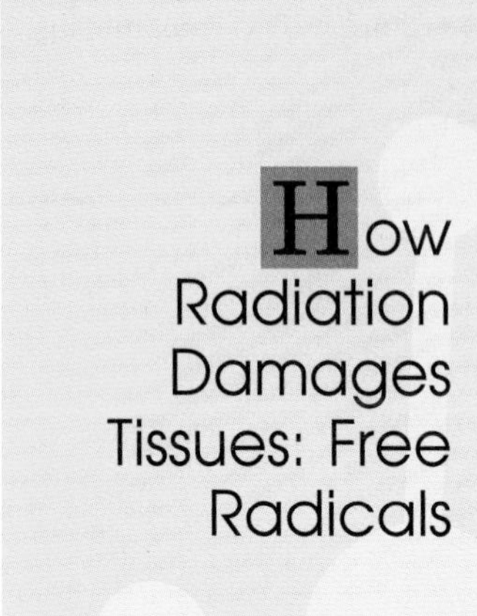

Box 9D How Radiation Damages Tissues: Free Radicals

As mentioned in the text, high-energy radiation damages tissue by causing ionization. This means that the radiation knocks electrons out of molecules that make up the tissue (generally one electron per molecule), thereby forming unstable ions. For example, interaction of high-energy radiation with water forms an unstable cation

$$\text{Energy} + \text{H—O—H} \longrightarrow \text{H—O—H}^+ + e^-$$

The + charge on the H_2O^+ means that one of the electrons that is normally present in the water molecule, either in a covalent bond or an unshared pair, is missing in this cation; it has been knocked out. Once formed, the cation is unstable and decomposes:

$$\text{H—O—H}^+ \longrightarrow \text{H}^+ + \cdot\text{O—H}$$

The O—H has an unpaired electron (indicated by the dot), which makes it extremely reactive. As we saw in Section 3.5, compounds that have unpaired electrons are called free radicals. Because they are so reactive, they rapidly interact with other molecules, causing chemical reactions that damage the tissues. These reactions are all the more serious if they occur inside the cell nucleus and damage genetic material. They affect rapidly dividing cells more than they do stationary cells. Thus the damage is greater to embryonic cells, cells of the bone marrow and intestines, and in the lymph.

they wear radiation badges (Fig. 9.5). A single whole-body irradiation of 25 rem is noticeable in the blood count, and 100 rem causes the typical symptoms of radiation sickness. A dose of 400 rem causes death within one month in 50 percent of exposed persons, and 600 rem is almost invariably lethal within a short time. It should be noted that up to 50 000 rem is needed to kill bacteria and up to 10^6 rem to inactivate viruses.

Symptoms of radiation sickness are nausea, vomiting, a decrease in the white blood cell count, and loss of hair.

Most of us, fortunately, never get a single dose of more than a few rem and so never suffer from any form of radiation sickness. This does not mean, however, that small doses are totally harmless. The harm comes in two ways:

1. Small doses of radioactivity over a period of years can cause cancer, especially blood cancers such as leukemia. Nobody knows how many cancers have been caused this way because the doses are so small and continue for so many years that they cannot generally be measured. Also, because there seem to be so many other causes of cancer, it is difficult or impossible to decide if any particular case is caused by radiation.

2. If any form of radiation strikes an egg or sperm cell, it can cause a change in the genes (Box 23E). Such changes are called **mutations.** If this particular egg or sperm cell mates, grows, and becomes a new individual, that individual may have mutated characteristics, which are usually harmful and frequently lethal.

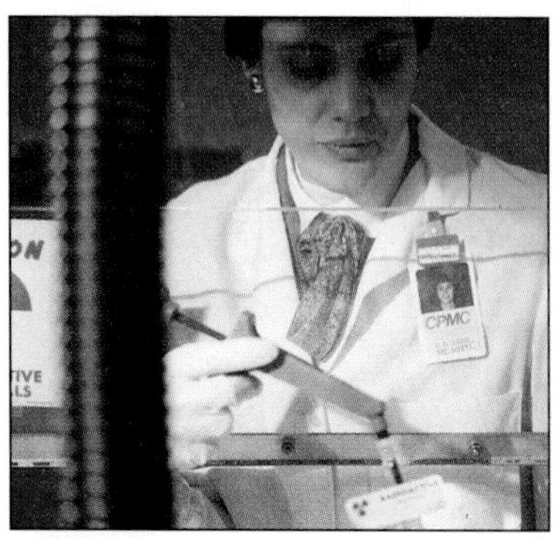

Figure **9.5**
Worker with a radiation badge. (Phototake/© Yoav Levy.)

Since radiation carries so much potential for harm, it would be nice if we could totally escape it. But can we? Table 9.4 shows that this is impossible. Naturally occurring radiation, which we call **background radiation,** is present everywhere on earth. As Table 9.4 shows, this background radiation is much more than the average radiation level from artificial sources (which are mostly diagnostic x-rays). Therefore, if we stopped all forms of artificial radiation, including medical uses, we would still be exposed to the background radiation.

9.6 Medical Uses of Radioactive Materials

Radioactive isotopes have two main uses in medicine: diagnosis and therapy (Table 9.5).

Diagnosis

This is the more widely used aspect of nuclear medicine.

Table 9.5 Some Medically Useful Radioactive Isotopes

	Isotope	Use
$^{3}_{1}H$	Tritium (hydrogen-3)	Measure water content of body
$^{11}_{6}C$	Carbon-11	Brain scan with positron emission transverse tomography (PET) to trace glucose pathway
$^{14}_{6}C$	Carbon-14	Radioimmunoassay
$^{32}_{15}P$	Phosphorus-32	Detection of eye tumors
$^{51}_{24}Cr$	Chromium-51	Albumism diagnosis, size and shape of spleen, gastrointestinal disorders
$^{59}_{26}Fe$	Iron-59	Bone marrow function, diagnosis of anemias
$^{60}_{27}Co$	Cobalt-60	Treatment of cancer
$^{67}_{31}Ga$	Gallium-67	Whole-body scan for tumors
$^{75}_{34}Se$	Selenium-75	Pancreas scan
$^{81m}_{36}Kr$	Krypton-81m	Lung ventilation scan
$^{85}_{38}Sr$	Strontium-85	Bone scan for bone diseases, including cancer
$^{99m}_{43}Tc$	Technetium-99m	Brain, liver, kidney, bone marrow scans; diagnosis of damaged heart muscles
$^{131}_{53}I$	Iodine-131	Diagnosis of thyroid malfunction; treatment of hyperthyroidism and thyroid cancer
$^{197}_{80}Hg$	Mercury-197	Kidney scan

Chemically and metabolically, a radioactive isotope in the body behaves exactly the same way as the nonradioactive isotopes of the same element. In the simplest form of diagnosis, a radioactive isotope is injected intravenously and a technician uses the various types of detectors to monitor how the radiation is distributed.

The use of iodine-131 to diagnose malfunctioning thyroid glands is a good example. The thyroid glands in the neck produce a hormone, thyroxine, that controls the overall rate of metabolism (use of food) in the body. Thyroxine contains iodine atoms. When radioactive iodine-131 is administered into the blood stream, the thyroid glands take it up and incorporate it into the thyroxine (see Box 12I). A normally functioning thyroid absorbs about 12 percent of the administered iodine within a few hours. An overactive thyroid (hyperthyroidism) absorbs and localizes iodine-131 in the glands faster, and an underactive thyroid (hypothyroidism) does this much more slowly than normal. Thus, by counting the radioactive particles emitted from the neck, one can determine the rate of uptake of iodine-131 into the glands and diagnose hyper- or hypothyroidism.

Most organ scans are similarly based on the preferential uptake of some radioactive isotopes by a particular organ (Fig. 9.6).

Another important use of radioactive isotopes is to learn what happens to ingested material. The foods and drugs swallowed or otherwise taken in by the body are transformed, decomposed, and excreted. To understand the pharmacology of a drug, it is important to know how these processes occur. For example, a certain drug may be effective in treating

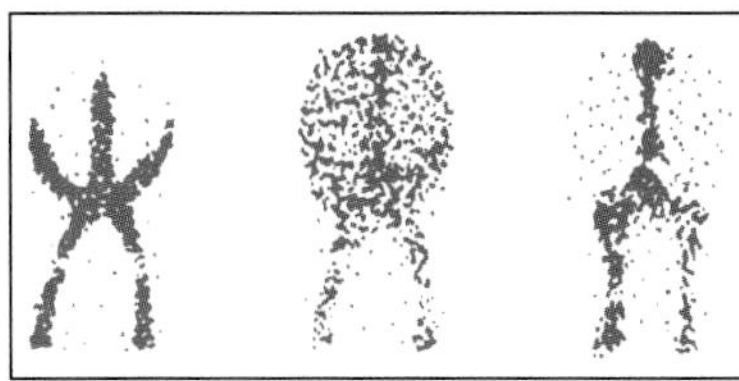

Normal

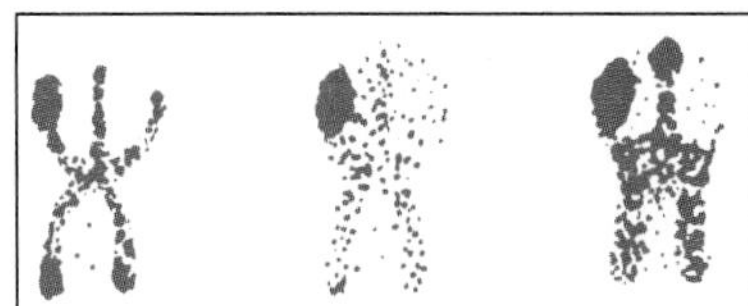

Meningioma (brain tumor)

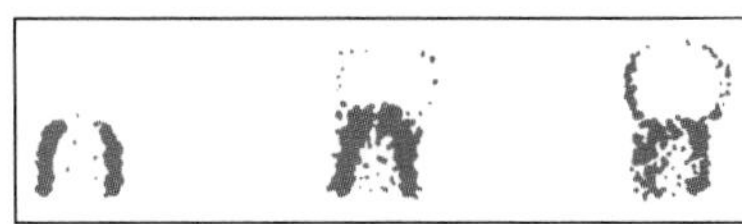

"Brain death"

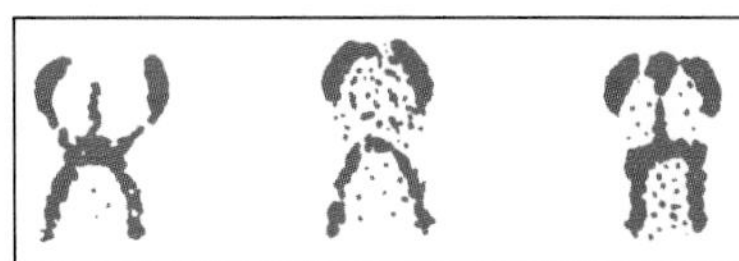

Scalp tumor

Figure **9.6**
A comparison of dynamic scan patterns for normal and pathological brains. The studies were performed by injecting technetium-99m into blood vessels. (From *CRC Handbook in Clinical Laboratory Science,* Vol. 1, Nuclear Medicine. CRC Press, Inc.)

certain bacterial infections. Before this drug can be used in a clinical trial, the manufacturer must prove that the drug is not harmful to humans. In a typical case this is done first in animal studies. The drug in question is synthesized, and some radioactive isotope, such as hydrogen-3, carbon-14, or phosphorus-32, is incorporated into its structure. The drug is administered to the animals, and after a certain period the animals are sacrificed. The fate of the drug is then followed by isolating from the body any radioactive compounds formed.

This is called *tagging* a drug.

A typical pharmacological experiment was the study of the effects of tetracycline. This powerful antibiotic tends to accumulate in bones and is not given to pregnant women because it is transferred to the bones of the fetus. A particular tetracycline was tagged with radioactive hydrogen-3, and its uptake in rat bones was monitored in the presence and absence of a sulfa drug. With the aid of a scintillation counter, the activity of the maternal and fetal bones was determined, and it was found that the sulfa drug helped to minimize this undesirable side effect of tetracycline, that is, its accumulation in the fetal bones.

The metabolic fate of essential chemicals in the body can also be followed with radioactive tracers. A number of normal and pathological body functions have been elucidated by the use of radioactive isotopes.

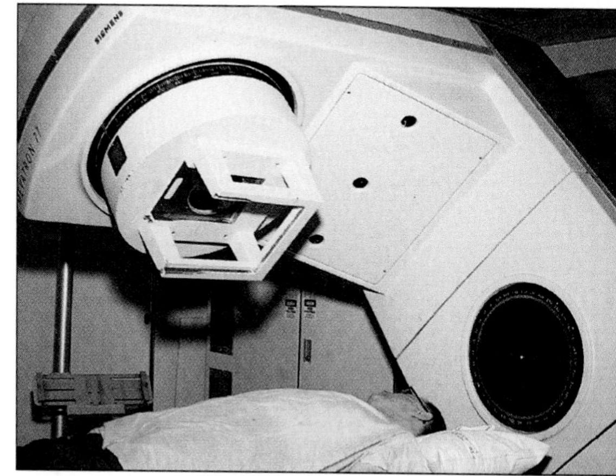

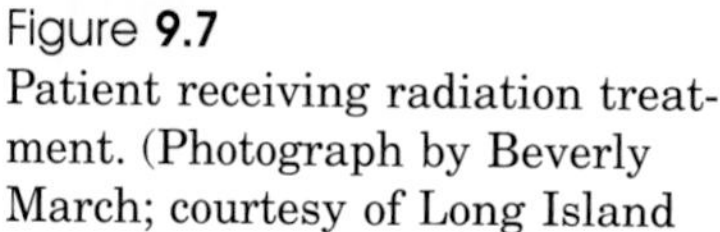

Figure **9.7**
Patient receiving radiation treatment. (Photograph by Beverly March; courtesy of Long Island Jewish Hospital.)

Therapy

The main use of radioactive isotopes in therapy is the selective destruction of pathological cells and tissues. Remember that radiation, whether from gamma rays or other sources, is detrimental to cells. Ionizing radiation damages cells, especially those that divide rapidly. The damage may be great enough to destroy the cells or to sufficiently alter the genes so that multiplication of the cells slows down.

Cancerous cells are the main target for such ionizing radiation. Radiation is usually used when the cancer is not well localized and therefore surgical removal is not the complete answer, and also when the cancerous cells are in a metastatic state. It is also used for preventive purposes when one wants to eliminate any possible remaining cancerous cells after surgery has been performed. The idea, of course, is to kill cancerous but not normal cells. Therefore, radiation such as x-rays or gamma rays from cobalt-60 is highly *collimated,* that is, aimed at a small part of the body where the cancerous cells are suspected. Besides the gamma rays from cobalt-60, other ionizing radiation is also used to treat inoperable tumors. Proton beams from cyclotrons have been used to treat ocular melanoma and tumors of the skull base and spine.

A metastatic state exists when the cancerous cells have broken off from their primary site(s) and are moving to other parts of the body.

Despite this pinpointing, the radiation kills normal cells along with the cancerous cells. Because the radiation is most effective against rapidly dividing cancer cells, rather than normal cells, and because the radiation is aimed at a specific location, the damage to healthy tissues is minimized. Figure 9.7 shows a patient receiving radiation treatment.

Because of this, radiation sickness is a side effect of radiation therapy.

Another way to localize radiation damage in therapy is to use specific radioactive isotopes. In the case of thyroid cancer, large doses of iodine-131 are given, which are taken up by the glands. The isotope,

Box 9E

CAT, mri, and PET Scans

One particular x-ray diagnostic technique, **computer assisted tomography (CAT scan)** has seen a phenomenal growth in the more than 20 years since its invention. Allan M. Cormack of Tufts University and Godfrey N. Hounsfield, an English scientist, shared the 1979 Nobel prize in medicine for this invention.

CAT scans can reveal small differences in the density of tissues because different tissues slow down x-rays to different extents. A modern CAT scan needs one rotation of the x-ray source around the body. It has a large number of detectors to measure simultaneously the x-rays passing through the body. A computer then clearly reconstructs the internal structure of the body.

CAT scans are used mainly to detect brain tumors because no other technique is so accurate in locating tissues within the skull. Not only are internal structures displayed (Fig. 9E.1), but one can even distinguish among the various fluid-filled cavities in the brain and between gray and white matter. CAT scans are also extensively used to detect the damage caused by strokes. Their use is not restricted to the brain—any part of the body can be examined with this tool.

Another diagnostic tool, **magnetic resonance imaging (mri),** is especially good for the soft tissues that contain many hydrogen atoms, mostly in

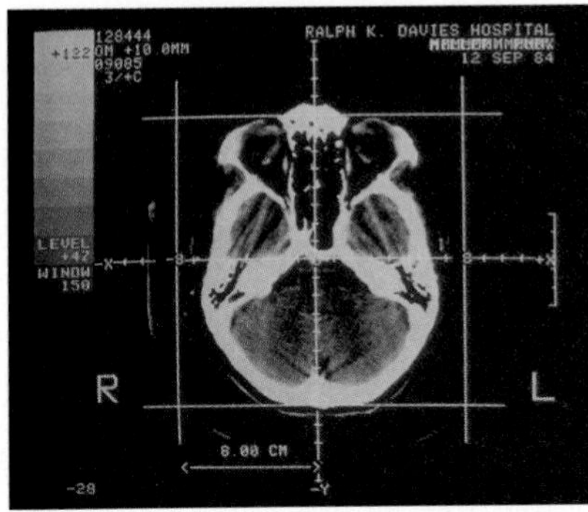

Figure **9E.1** • A CAT scan of a human brain. (The IMAGE BANK/© Bill Varie.)

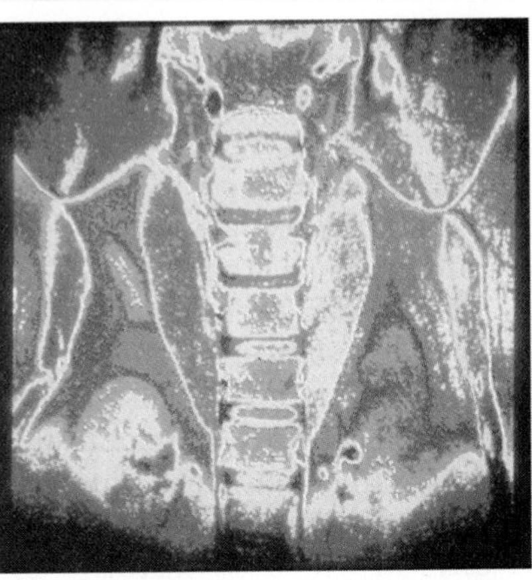

Figure **9E.2** • An mri scan. (Phototake/© Debbie Breen.)

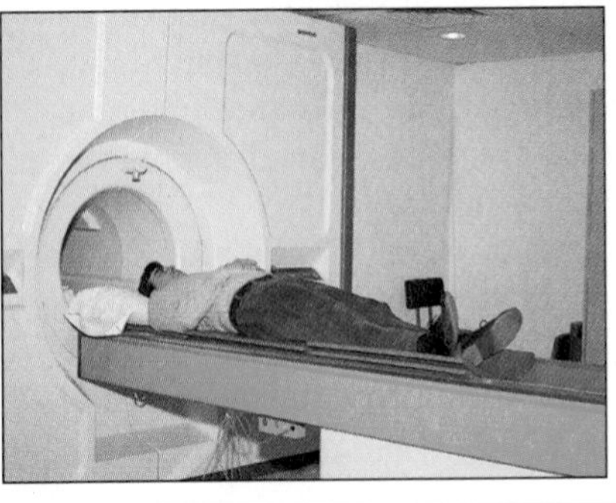

Figure **9E.3** • A medical mri imager. (Photograph by Beverly March; courtesy of Long Island Jewish Hospital.)

the form of water or fat. It probes the nuclear spins of atoms, mainly protons, which at a certain frequency invert their spins when subjected to radio-frequency waves. An advantage of mri is that bones contain few hydrogens, so that they are almost transparent in this imaging. Thus the soft tissues with their differing water contents stand out in the image (Fig. 9E.2). Furthermore, mri has the advantage that it does not use ionizing radiation (Fig. 9E.3), as CAT (and PET) scans do.

The third important type of imaging is called **positron emission tomography (PET scans).** It is based on the property that certain isotopes (such as carbon-11) have of emitting positrons (Sec. 9.2). Positrons have very short lives because when they collide with an electron, they annihilate each other, resulting in gamma rays. Because electrons are present in every atom, there are always lots of them around, so positrons generated in the body cannot live long. For example, if one is fed glucose labeled with carbon-11, the glucose gets into the blood and from there into the brain. Detectors for the gamma rays can pick up the signals that come from the areas where the glucose is used up. In this way one can see which areas of the brain are involved when we process, for example, visual information (Fig. 9E.4). PET scans can diagnose early stages of epilepsy and other diseases that involve abnormal glucose metabolism, such as schizophrenia.

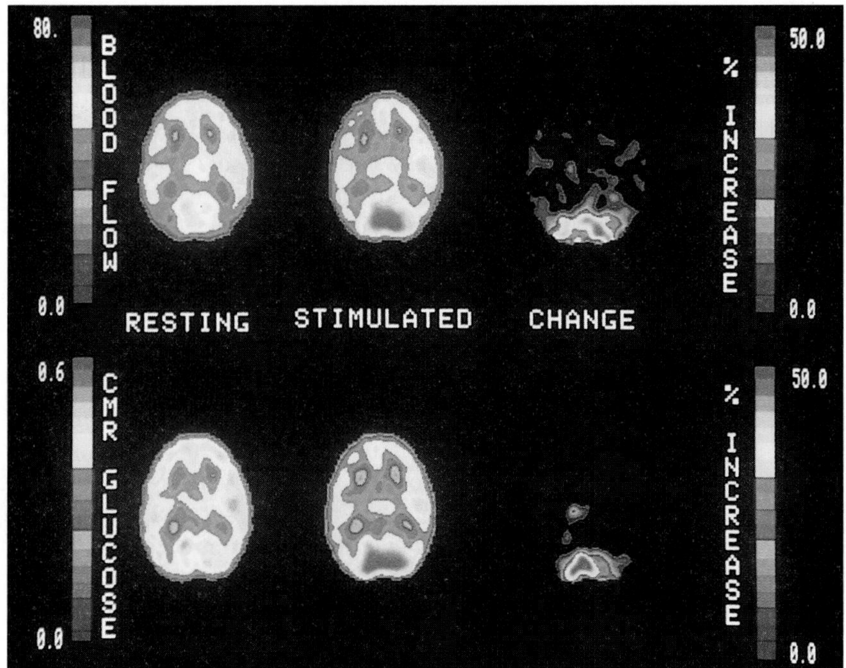

Figure **9E.4** • Positron emission tomography brain scans showing that glucose can cross the blood-brain barrier. The second scan in each case shows the result of visual stimulation. Both the local cerebral blood flow (48 percent increase) and the glucose metabolism (56 percent increase) go up during visual stimulation. (Courtesy of Dr. Peter T. Fox, Washington University School of Medicine.)

which has high radioactivity, kills all the cells of the gland, the cancerous as well as the healthy ones, but does not appreciably damage other organs. In a similar manner, localization can be achieved by implanting a tiny grain of yttrium-90 in the pituitary gland. This beta emitter destroys the gland but nothing much beyond that. The hope is that this treatment will slow the growth of tumors elsewhere in the body because the pituitary gland stimulates cell reproduction, and without the growth hormone of the pituitary, the cell proliferation of tumors will cease.

Still another way to localize radiation damage is boron neutron capture therapy (BNCT). This treatment is based on the fact that boron absorbs neutrons selectively. A boron-containing compound, such as boronophenylalanine, is administered to the patient. This drug concentrates in the tumor. When a high-energy neutron beam is applied, it is selectively absorbed by the boron and destroys the tumor tissues at the point where the boron is concentrated. Inoperable melanomas have been treated in this manner. (See also the use of boron control rods in Sec. 9.8).

9.7 Nuclear Fusion. Artificial Transmutation

It is estimated that 98 percent of all matter in the universe is made up of hydrogen and helium. This is explained by the "big bang" theory of the formation of the universe. This theory postulates that our universe started with an explosion (big bang) in which matter was formed out of energy, and, at the beginning, only the lightest element, hydrogen, was in existence. Later, as the universe expanded, stars were born when the hydrogen clouds collapsed under gravitational forces. In the cores of these stars hydrogen nuclei fused together and formed helium.

The transformation of hydrogen nuclei into a helium nucleus liberates a very large amount of energy in the form of photons (Box 9F), largely by this equation:

$$^{2}_{1}\text{H} + ^{3}_{1}\text{H} \longrightarrow ^{4}_{2}\text{He} + ^{1}_{0}\text{n} + \text{energy}$$

The reactions occurring in the sun are essentially the same as those in the hydrogen bomb.

This process, called **fusion,** is how the sun makes its energy. If we can ever achieve a controlled version of this fusion reaction, we shall be able to solve our energy problems. However, this may not occur for a long time.

All the **transuranium elements** (elements with atomic numbers greater than 92) from 93 to 112 are artificial and have been prepared by a process in which heavy nuclei are bombarded with light ones. Many, as their names indicate, were first prepared at the Lawrence Laboratory of the University of California, Berkeley, by Glenn Seaborg (1912–) (Nobel laureate in chemistry, 1951) and his colleagues:

$$^{244}_{96}\text{Cm} + ^{4}_{2}\text{He} \longrightarrow ^{245}_{97}\text{Bk} + ^{1}_{1}\text{H} + 2\,^{1}_{0}\text{n} \quad \text{Berkelium}$$

$$^{238}_{92}\text{U} + ^{12}_{6}\text{C} \longrightarrow ^{246}_{98}\text{Cf} + 4\,^{1}_{0}\text{n} \quad \text{Californium}$$

$$^{252}_{98}\text{Cf} + ^{10}_{5}\text{B} \longrightarrow ^{257}_{103}\text{Lr} + 5\,^{1}_{0}\text{n} \quad \text{Lawrencium}$$

$c = 3.0 \times 10^8$ m/s

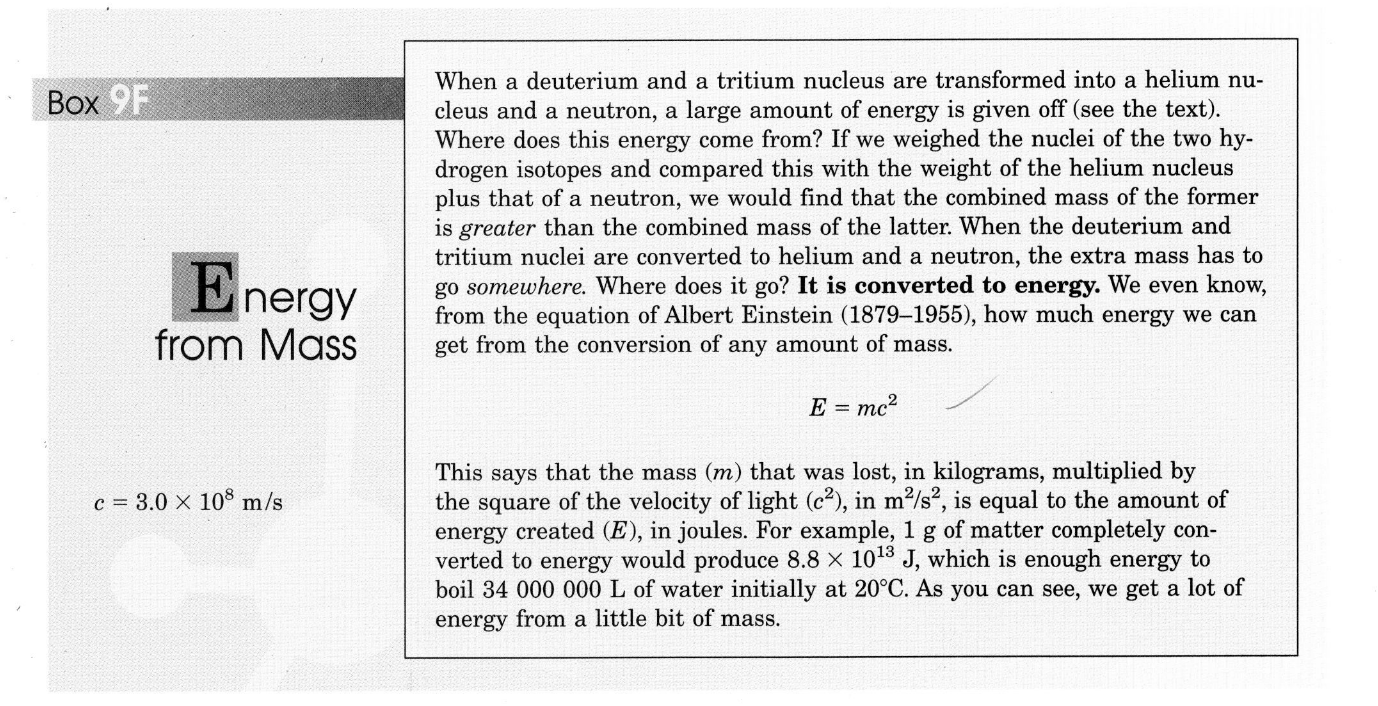

When a deuterium and a tritium nucleus are transformed into a helium nucleus and a neutron, a large amount of energy is given off (see the text). Where does this energy come from? If we weighed the nuclei of the two hydrogen isotopes and compared this with the weight of the helium nucleus plus that of a neutron, we would find that the combined mass of the former is *greater* than the combined mass of the latter. When the deuterium and tritium nuclei are converted to helium and a neutron, the extra mass has to go *somewhere.* Where does it go? **It is converted to energy.** We even know, from the equation of Albert Einstein (1879–1955), how much energy we can get from the conversion of any amount of mass.

$$E = mc^2$$

This says that the mass (m) that was lost, in kilograms, multiplied by the square of the velocity of light (c^2), in m^2/s^2, is equal to the amount of energy created (E), in joules. For example, 1 g of matter completely converted to energy would produce 8.8×10^{13} J, which is enough energy to boil 34 000 000 L of water initially at 20°C. As you can see, we get a lot of energy from a little bit of mass.

These transuranium elements are unstable, and most have very short half-lives. For example, that of $^{257}_{103}Lr$ is 0.65 second.

9.8 Nuclear Fission. Atomic Energy

In the 1930s Enrico Fermi (1901–1954) and his colleagues in Rome, and Otto Hahn (1879–1968), Lisa Meitner (1878–1968), and Fritz Strassman (1902–) in Germany, tried to produce new transuranium elements by bombarding uranium-235 with neutrons. To their surprise, Hahn and his co-workers found that, rather than fusion, they obtained **nuclear fission** (fragmentation of large nuclei into smaller pieces):

$$^{235}_{92}U + ^{1}_{0}n \longrightarrow ^{139}_{56}Ba + ^{94}_{36}Kr + 3\,^{1}_{0}n + \gamma$$

In this reaction the uranium-235 nucleus is broken into smaller particles, but the most important product is energy, which is produced because the products have less mass than the starting materials (Box 9F). This form of energy is called **atomic energy** and has been used for both war (in the atomic bomb) and peace.

In the fission of uranium-235, each fission produces three neutrons, which in turn can generate more fission by colliding with other uranium-235 nuclei. If even one of these neutrons produces new fission, the process

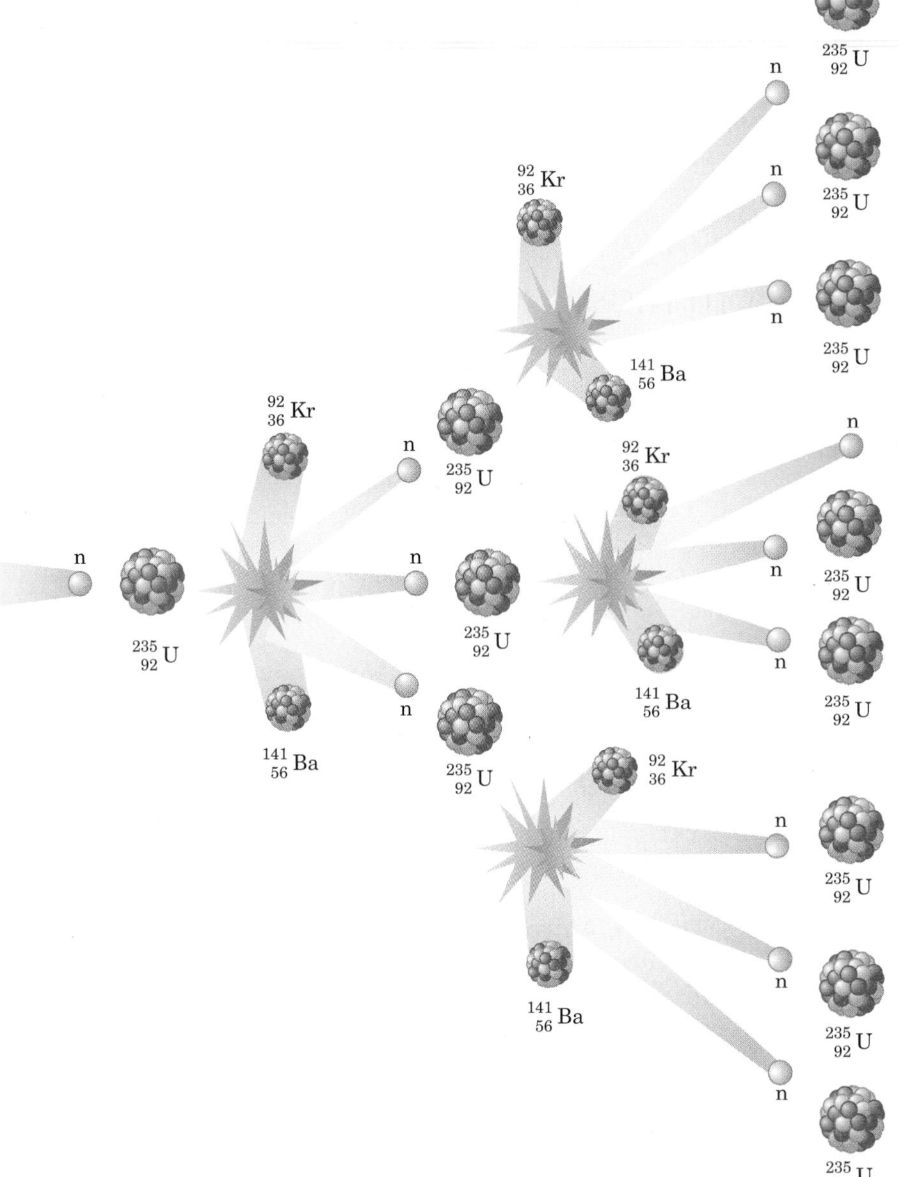

Figure **9.8** • A chain reaction is begun when a neutron collides with a nucleus of $^{235}_{92}U$.

Figure **9.9**
Haddam Neck nuclear reactor (NUS). (Courtesy of Atomic Industrial Forum.)

becomes a self-propagating **chain reaction** (Fig. 9.8) that continues at a constant rate. If all three neutrons are allowed to produce new fission, the rate of the reaction increases constantly and eventually culminates in a nuclear explosion. The rate of reaction can be controlled by inserting boron control rods to absorb neutrons.

In nuclear plants, the energy produced by fission is sent to heat exchangers and used to generate steam, which drives a turbine to produce electricity (Fig. 9.9). Today about 10 percent of the electrical energy in the United States is supplied by nuclear plants. The opposition to nu-

Figure **9.10**
The nuclear power plant at Chernobyl after the accident. (Novosti/Gamma Liaison)

clear plants is based on safety considerations and on the unsolved problems of waste disposal. Nuclear plants in general have had good safety records, but accidents such as those at Chernobyl (Box 9G; Fig. 9.10) and Three Mile Island have caused concern. The location of reactors in highly populated areas is definitely objectionable.

Waste disposal is a long-range problem. The fission products of nuclear reactors are highly radioactive themselves, with long half-lives. Thus, storing them underwater or burying them underground (Fig. 9.11) just postpones the disposal for generations to come. The fear is that corrosion and leakage from such storage tanks may contaminate water supplies. Remember that, up to now, no one has ever discovered a way to speed up or slow down radioactive decay.

It is true that most other ways of generating large amounts of electrical power also have their environmental problems. For example, burning coal or oil contributes to the accumulation of CO_2 in the atmosphere (Box 10F) and to acid rain (see Box 6A).

Figure **9.11**
Storing nuclear wastes in a storage room carved out of an underground salt bed. (Courtesy U.S. Department of Energy.)

Box 9G

Radioactive Fallout from Nuclear Accidents

On April 26, 1986, an accident occurred at the nuclear reactor in the town of Chernobyl in the former Soviet Union (Fig. 9.10). It was a clear reminder of the dangers involved in this industry and of the far-reaching contamination that such accidents create. In Sweden, more than 500 miles away from the accident, the radioactive cloud increased the background radiation from 4 to 15 times normal. The radioactive cloud reached England, about 1300 miles away, a week later. As a consequence the natural background radiation was increased by 15 percent.

Radioactivity from iodine-131 was 400 Bq/L in milk and 200 Bq/kg in leafy vegetables. Even some 4000 miles away in Spokane, Washington, iodine-131 activity of 242 Bq/L was found in rainwater, and smaller activities—1.03 Bq/L of ruthenium-103 and 0.66 Bq/L of cesium-137—were also recorded. These levels are not harmful.

Closer to the source of the nuclear accident, however, in neighboring Poland, potassium iodide pills were given to children. This was done to prevent radioactive iodine-131 (which might come from contaminated food) from concentrating in their thyroid glands, which could lead to cancer. Soon after the accident about 115 000 people were evacuated from the immediate area and nearby fallout zones. But even that was not sufficient, and more than three years later many additional villages in the fallout path were evacuated.

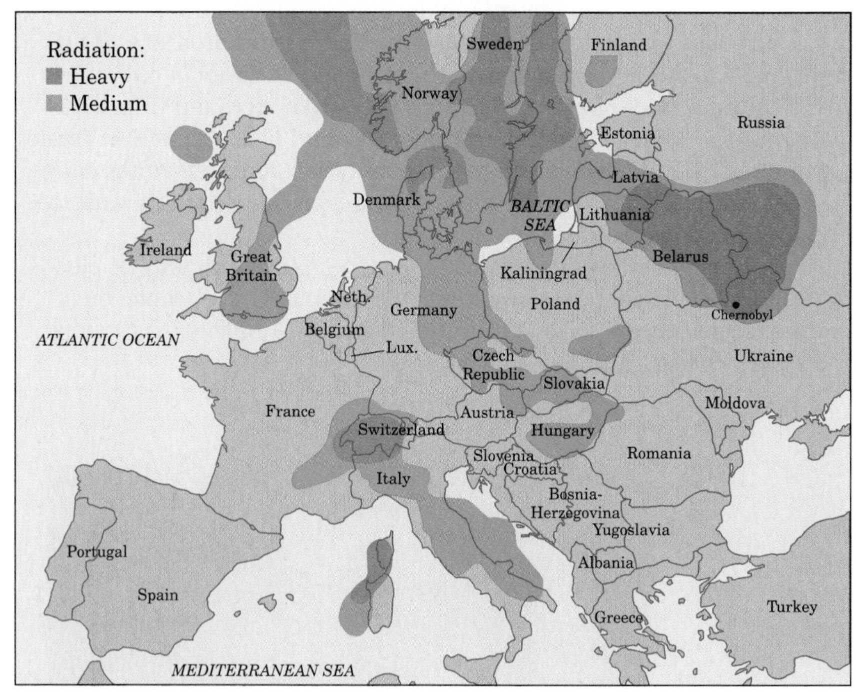

Map showing areas most greatly affected by the Chernobyl accident.

SUMMARY

Certain isotopes are radioactive; others are stable. The three major types of radioactivity are **alpha particles** (helium nuclei), **beta particles** (electrons), and **gamma rays** (high-energy radiation). **Radioactivity** comes from the nucleus, not from the electron cloud. When a nucleus emits an alpha or beta particle, it is changed to the nucleus of another element. When an alpha particle is emitted, the new element has an atomic number two units lower and a mass number four units lower. When a beta particle is emitted, the new element has the same mass number and an atomic number one unit higher. **Positrons** (positive electrons) are emitted by fewer isotopes, but in this case the new element has an atomic number one unit lower.

Each radioactive isotope decays at a fixed rate described by its **half-life,** which is the time required for half the sample to decay. Radiation is detected and counted by devices such as Geiger-Müller counters. The main unit of intensity of radiation is the **curie,** which is 3.7×10^{10} disintegrations per second. Radiation can be harmful in several ways. For medical purposes and to measure potential radiation damage, the absorbed dose, measured in **rads,** is an important factor. Different particles damage body tissues differently. The **rem** is a measure of the damage. Radioactive nuclei are used in medicine for both diagnosis and therapy.

Helium is synthesized in the interiors of stars by **fusion** of hydrogen nuclei. The energy released in this process is the energy of our sun. Artificial elements are created by bombarding atomic nuclei with high-energy nuclei. With certain nuclei, such bombardment may cause the nucleus to split (**fission**). Nuclear fission releases large amounts of energy, which can be either controlled (nuclear reactors) or uncontrolled (nuclear weapons).

KEY TERMS

Alpha emission (Sec. 9.2)
Alpha particle (Sec. 9.1)
Atomic energy (Sec. 9.8)
Background radiation (Sec. 9.5)
Becquerel (Sec. 9.5)
Beta emission (Sec. 9.2)
Beta particle (Sec. 9.1)
CAT scan (Box 9E)
Chain reaction (Sec. 9.8)
Curie (Sec. 9.5)
Electromagnetic radiation (Sec. 9.1)
Frequency (Sec. 9.1)
Gamma emission (Sec. 9.2)
Gamma rays (Sec. 9.1)
Gray (Sec. 9.5)
Half-life (Sec. 9.3)
Ionizing radiation (Sec. 9.4)
Magnetic resonance imaging (mri) (Box 9E)
Nuclear fission (Sec. 9.8)
Nuclear fusion (Sec. 9.7)
Photon (Sec. 9.1)
Positron (Sec. 9.2)
Positron emission tomography (PET) scans (Box 9E)
Rad (Sec. 9.5)
Radioactive dating (Box 9B)
Radioactive isotope (Sec. 9.2)
Radioactivity (Sec. 9.1)
Radioisotope (Sec. 9.2)
Rem (Sec. 9.5)
Sievert (Sec. 9.5)
Transmutation (Sec. 9.2)
Transuranium element (Sec. 9.7)
Wavelength (Sec. 9.1)
X-ray (Sec. 9.1)

PROBLEMS

Radioactivity and Electromagnetic Radiation

9.7 What is the smallest unit of electromagnetic radiation?

9.8 How did Becquerel discover radioactivity?

***9.9** Microwaves are a form of electromagnetic radiation that is used for the rapid heating of foods. What is the frequency of a microwave with a wavelength of 5.8 cm?

***9.10** In each case given the frequency, give the wavelength in cm or nm (10^{-9} m) and tell what kind of radiation it is:
(a) 7.5×10^{14}/s (c) 1.1×10^{15}/s
(b) 1.0×10^{10}/s (d) 1.5×10^{18}/s

9.11 Why are x-rays not a form of radioactivity?

9.12 Which has the longest wavelength: (a) infrared (b) ultraviolet (c) x-rays? Which has the highest energy?

Radioactive and Stable Isotopes

9.13 Write the symbol for a nucleus with
(a) 9 protons and 10 neutrons
(b) 15 protons and 17 neutrons
(c) 37 protons and 50 neutrons

9.14 In each pair, tell which isotope is more stable:
(a) ${}^{1}_{1}H$ and ${}^{3}_{1}H$ (c) ${}^{40}_{20}Ca$ and ${}^{43}_{20}Ca$
(b) ${}^{18}_{8}O$ and ${}^{16}_{8}O$ (d) ${}^{26}_{14}Si$ and ${}^{28}_{14}Si$

9.15 How many artificial stable (nonradioactive) nuclei are known?

Natural Transmutation and Nuclear Reactions

9.16 Samarium-151 is a beta emitter. What is the transmutation product of the nuclear reaction?

9.17 The following nuclei turn into new nuclei by emitting beta particles. In each case write the formula for the new nucleus:

(a) $^{159}_{63}Eu$ (c) $^{242}_{95}Am$

(b) $^{141}_{56}Ba$

***9.18** Chromium-51 is used in diagnosing pathology of the spleen. The nucleus of this isotope captures an electron according to the equation

$$^{51}_{24}Cr + {}^{0}_{-1}e \longrightarrow ?$$

What is the transmutation product?

9.19 The following nuclei turn into new nuclei by emitting alpha particles. In each case write the formula for the new nucleus:

(a) $^{210}_{83}Bi$ (c) $^{174}_{72}Hf$

(b) $^{238}_{94}Pu$

9.20 When curium-244 is bombarded by alpha particles, the emission products are a proton, two neutrons, and one other product. What is the other product?

9.21 Phosphorus-29 is a positron emitter. Write the balanced nuclear equation.

9.22 For each of the following, write a balanced nuclear equation and identify the radiation emitted:
(a) beryllium-10 changes to boron-10
(b) europium-151* changes to europium-151
(c) thallium-195 changes to mercury-195
(d) plutonium-239 changes to uranium-235

9.23 In the radioactive decay of uranium-238, the first two steps are (1) alpha emission and (2) beta emission. Write the balanced nuclear equations for these two steps.

9.24 What kind of emission does not result in transmutation?

9.25 Complete the following nuclear reactions:

(a) $^{16}_{8}O + {}^{16}_{8}O \longrightarrow ? + {}^{4}_{2}He$

(b) $^{235}_{92}U + {}^{1}_{0}n \longrightarrow {}^{90}_{38}Sr + ? + 3\ {}^{1}_{0}n$

(c) $^{13}_{6}C + {}^{4}_{2}He \longrightarrow {}^{16}_{8}O + ?$

(d) $^{210}_{83}Bi \longrightarrow {}^{0}_{-1}e + ?$

(e) $^{12}_{6}C + {}^{1}_{1}H \longrightarrow ? + \gamma$

Half-Life

9.26 What is a decay curve?

9.27 The half-life of uranium-232 is 70 years. How long will it take 5.0 g of this element to be reduced to 1.25 g?

***9.28** A rock containing 1 mg plutonium-239/kg rock is found in a glacier. The half-life of plutonium-239 is 25 000 years. If this rock was deposited 100 000 years ago during an Ice Age, how much plutonium-239, per kg of rock, was in the rock at that time?

9.29 The element radium is extremely radioactive. If you converted a piece of radium metal to radium chloride (the weight of radium remaining the same), would it become less radioactive?

9.30 In what ways can we increase the rate of radioactive decay? Decrease it?

9.31 50.0 mg of potassium-45, a beta emitter, was isolated in pure form. After one hour only 3.1 mg of the radioactive material was left. What is the half-life of potassium-45?

***9.32** A patient receives 200 mCi of iodine-131, which has a half-life of eight days. (a) If 12 percent of this is taken up by the thyroid gland after two hours, what will be the activity of the thyroid after two hours, in mCi and in counts per minute? (b) After 24 days how much activity will remain in the thyroid gland?

Characteristics of Radiation

9.33 Why is it necessary to use instruments to detect radioactivity?

9.34 What is the difference between a Geiger-Müller counter and a scintillation counter? Do they measure the same properties?

9.35 What do Geiger-Müller counters measure: (a) intensity or (b) energy of radiation?

***9.36** It is known that radioactivity is being emitted with an intensity of 175 mC at a distance of 1.0 m from the source. How far, in meters, from the source should you stand if you wish to be subjected to no more than 0.20 mC?

Radiation Dosimetry

9.37 State clearly the properties that are measured in
(a) rads
(b) rems
(c) roentgens
(d) curies
(e) grays
(f) becquerels
(g) sieverts

9.38 A radioactive isotope with an activity of 80.0 mCi per vial is delivered to a hospital. The vial contains 7.0 cc of liquid. The instruction is to administer 7.2 mCi intravenously. How many cubic centimeters of liquid should be used for one injection?

***9.39** Why is it that when a hand is exposed to alpha rays the damage to the person is not serious, but if the alpha emitter gets into the lung as an aerosol, the damage to the person's health is very serious?

9.40 What is the source of the largest amount of radiation our bodies are exposed to? Can we control this exposure?

9.41 Assuming the same amount of effective radiation, in rads, from three sources, which would be the most damaging to the tissues: alpha particles, beta particles, or gamma rays?

Medical Uses of Radioactive Materials

9.42 Which radioactive isotopes are used for
(a) cancer treatment
(b) pancreas scans
(c) bone scans
(d) brain scans
(e) kidney scans

9.43 In 1986 the nuclear reactor in Chernobyl had an accident and spewed radioactive nuclei that was carried by the winds for hundreds of miles. Today one finds that among the child survivors the most common damage is thyroid cancer. What radioactive nucleus do you expect to be responsible for these cancers?

9.44 Radioactive isotopes can be used for cancer therapy, especially if they are going to be concentrated in the diseased organ. Name two radioactive isotopes that are used in cancer therapy because of this property.

9.45 Cobalt-60, with a half-life of 5.26 years, is used in cancer therapy. The energy of the radiation from cobalt-62 (half-life, 14 minutes) is even higher. Why isn't this isotope also used for cancer therapy?

9.46 Match the radioactive isotope with its proper use:

______	(a) $^{14}_{6}C$	1. Cancer therapy
______	(b) $^{60}_{27}Co$	2. Brain scan
______	(c) $^{99m}_{43}Tc$	3. Thyroid scan
______	(d) $^{131}_{53}I$	4. Artifact dating

Fusion and Fission

9.47 What is the most abundant nucleus in the universe?

9.48 What is the product of the fusion of hydrogen nuclei?

9.49 Assuming that one proton and two neutrons will be produced in an alpha bombardement fusion reaction, what target nucleus would you use to obtain berkelium-249?

9.50 Element 109 was first prepared in 1982. A single atom of this element, with a mass number of 266 ($^{266}_{109}Une$), was made by bombarding a bismuth-209 nucleus with an iron-58 nucleus. What other products, if any, must have been formed besides $^{266}_{109}Une$?

9.51 What nuclear particles cause chain reactions? How do we control chain reactions?

9.52 Why are some people opposed to nuclear reactors?

Boxes

9.53 (Box 9A) Explain why, in Rutherford's experiment, the three rays were deflected in different directions.

9.54 (Box 9B) In a recent archeological dig in the Amazon region of Brazil, charcoal paintings were found in a cave. The carbon-14 content of the charcoal was one fourth of what is found in charcoal prepared from this year's tree harvest. How long ago was the cave settled?

***9.55** (Box 9B) Carbon-14 dating of the Shroud of Turin indicated that the plant from which the shroud was made was alive around AD 1350. How many half-lives does this correspond to?

9.56 (Box 9B) The half-life of carbon-14 is 5730 years. The wrapping of an Egyptian mummy gave off 7.5 counts per minute per gram of carbon. A piece of linen purchased today would give an activity of 15 counts per minute per gram of carbon. How old is the mummy?

9.57 (Box 9C) How does radon-222 produce polonium-218?

9.58 (Box 9D) Explain how high energy damages tissues.

9.59 (Box 9E) What are the advantages of an mri scan over a CAT scan?

9.60 (Box 9E) Can one use CAT scans to detect cerebral hemorrhages (bleeding in the brain)? Explain.

9.61 (Box 9E) Why do positrons have a very short lifetime?

9.62 (Box 9F) Explain how the fusion of hydrogen nuclei to produce a helium nucleus gives off energy.

9.63 (Box 9G) In the nuclear accident, one of the radioactive nuclei that people are concerned about is iodine-131. This is so because iodine is easily vaporized, can be carried by the winds, and can cause radioactive fallout hundreds, even thousands, of miles away. Why is iodine-131 especially harmful?

Additional Problems

9.64 Which of these isotopes of nitrogen is the most stable: nitrogen-12, nitrogen-14, or nitrogen-17?

9.65 Neon-19 and sodium-20 are positron emitters. What are the products in each case?

9.66 The half-life of nitrogen-16 is 7 seconds. How long does it take for 100 mg of nitrogen-16 to be reduced to 6.25 mg?

9.67 Do curie and becquerel measure the same or different properties of radiation?

***9.68** Selenium-75 has a half-life of 120.4 days. This means that it would take 602 days (five half-lives) to diminish to 3 percent of the original quantity. Yet this isotope is used for pancreatic scans without any fear that the radioactivity will cause undue harm to the patient. Suggest a possible explanation.

9.69 Use Table 9.4 to determine the percentage of annual radiation we receive from
(a) naturally occurring sources
(b) diagnostic medical sources
(c) nuclear power plants

9.70 A beta emitter sends an electron out of the nucleus, yet there are no electrons in the nucleus. Explain where the electron comes from.

9.71 Which radiation will cause more ionization, x-rays or radar?

9.72 You have an old wristwatch that still has radium paint on its dial. Measurement of the radioactivity of the watch shows a beta-ray count of 0.50 counts/s. 1.0 microcurie radiation of this sort produces 1000 mrem/year. How much radiation, in mrem, do you expect from the wristwatch if you wear it for a year?

9.73 Californium-244 was first produced in 1950 by bombarding curium-242 with alpha particles. What were the other reaction products?

9.74 On rare occasions a nucleus, instead of emitting a beta particle, captures one. Berkelium-246 is such a nucleus. What is the product of this nuclear transmutation?

9.75 Where does the energy of the atomic bomb come from?

***9.76** What is the ground state of a nucleus?

9.77 Explain why (a) it is impossible to have a completely pure sample of any radioactive isotope, (b) beta emission of a radioactive isotope creates a new isotope with an atomic number one unit higher than that of the radioactive isotope.

9.78 Yttrium-90, which emits beta particles, is used in radiotherapy. What is the decay product of yttrium-90?

***9.79** The half-lives of some oxygen isotopes are: oxygen-14 = 71 s; oxygen-15 = 124 s; oxygen-19 = 29 s; and oxygen-20 = 14 s. Oxygen-16 is the stable nonradioactive isotope. Do the half-lives indicate anything about the stability of the other oxygen isotopes?

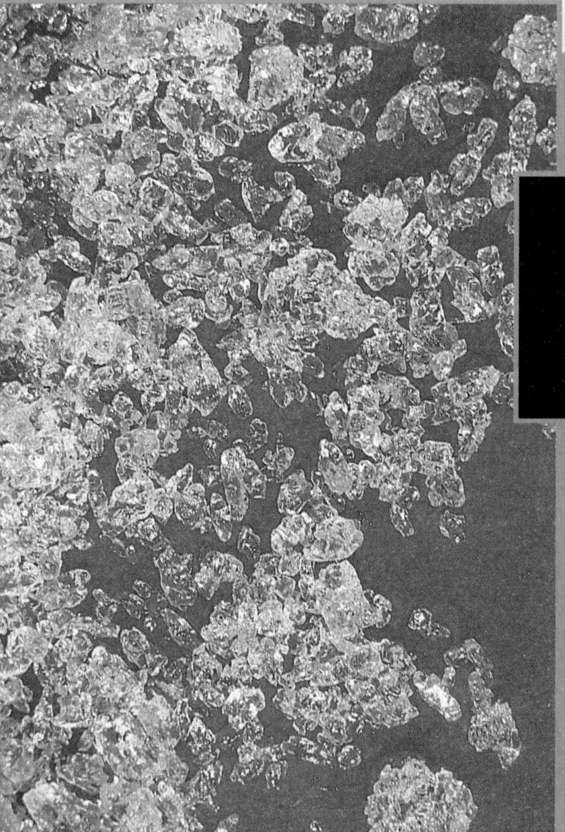

Photograph by Charles D. Winters.

Part **two**

Organic chemistry

Carl Djerassi

Carl Djerassi, Professor of Chemistry at Stanford University, is a man of many achievements, in science and in letters. Born in Vienna in 1923, he was forced to leave Europe in 1939 in the face of the Nazi occupation. Although he had not completed high school, he was allowed to enroll at Newark Junior College. Following this he went to Tarkio College and then to Kenyon College, where he graduated summa cum laude *in 1942.*

Still short of his nineteenth birthday, he joined the CIBA Corporation, where he and Charles Huttrer synthesized pyribenzamine, one of the first two antihistamines—drugs that act to control allergies. He left CIBA for the University of Wisconsin, where he received his Ph.D. in 1945, at the age of 22. For the next seven years he worked first at CIBA (until 1949) and then at Syntex, S. A., a fledgling pharmaceutical corporation then located in Mexico City. Here he was part of the research team that produced the first synthesis of cortisone, an important hormone, and later synthesized norethindrone, the first oral contraceptive and one that is still widely used all over the world.

In 1952 he moved to Wayne University (now Wayne State) in Detroit, as Associate Professor, and in 1959 accepted a Professorship at Stanford University, where he has remained ever since. Both at Wayne and at Stanford he maintained an association with Syntex and served as president of Syntex Research from 1968 to 1972. In 1968 he helped to found the Zoecon Corporation, of which he also became president and which pioneered in developing hormonal methods of insect control that do not rely on conventional insecticides.

Dr. Djerassi has always been involved in steroid chemistry, but his early interest in synthesis turned to the equally important area of the structural determination of naturally occurring steroids, of which there are many thousands. When he began his career, determining the structure of a single naturally occurring steroid was a difficult process that would often take years. As an illustration of the difficulty, it may be mentioned that two German chemists, Heinrich Wieland and Adolph Windaus, were awarded Nobel Prizes in 1927 and 1928, partly for determining the structures of a few steroids, and these structures were later shown (by Wieland) to be wrong!

Beginning approximately in the 1950s instrumental methods of structure determination slowly began to replace the laborious older chemical methods, and Dr. Djerassi was one of the pioneers in developing and using techniques such as optical rotatory dispersion, circular dichroism, and mass spectrometry for steroids and other organic compounds. Owing to the work of Dr. Djerassi and other pioneers, most unknown structures now can be determined in a few days or less.

Dr. Djerassi has had a prolific career, publishing more than 1000 research papers and seven books dealing with the topics already mentioned as well as with alkaloids, antibiotics, lipids, and terpenoids, and with chemical applications of computer artificial intelligence. For this work he has received many awards, including the National Medal of Science (1973), the National Medal of Technology (1991), the first Wolf Prize in Chemistry (1978), the Perkin Medal of the Society for Chemical Industry (1975), The American Chemical Society's Award in Pure Chemistry (1958), the Roussel Prize (1988), the Priestley medal, the highest award given by the American Chemical Society (1992), and many others. In 1978 he was inducted into the National Inventor's Hall of Fame. He has also received a dozen honorary degrees and is a member of the U.S. National Academy of Sciences as well as the Academies of Science of several other countries.

In recent years Dr. Djerassi, without abandoning his interest in chemistry, has turned to literature. He has published a collection of short stories, three novels (Cantor's Dilemma, The Bourbaki Gambit, *and* Marx, deceased)*, a collection of poems, and two books of autobiography, one entitled* Steroids Made it Possible *and the other, intended for a more general audience, called* The Pill, Pygmy Chimps, and Degas's Horse. *Under the auspices of the Djerassi Foundation, he has established an artist's colony near Woodside, California, that provides residences and studio space for visual, literary, choreographic, and musical artists.*

Photograph by Beverly March.

Why did you choose a career in organic chemistry?

I really didn't choose it; it sort of happened to me. I was the child of two physicians and it was always assumed that I would go to medical school and become a practicing physician. When I arrived in this country at the age of 16 I had taken no chemistry and had not even graduated from high school. Fortunately, I got straight into a Junior College in New Jersey and began a pre-med curriculum that included first-year chemistry. I had a first-rate teacher there, named Nathan Washton, who got me interested in chemistry. After one semester at Tarkio College in Missouri (the alma mater of Wallace Carothers, the inventor of nylon), I went to Kenyon College, a small (at that time) college in Ohio, that had only two faculty members in chemistry. Chemistry classes were very small, but I got a first-class education, and that is where I decided to become a chemist, rather than go to medical school.

Could you tell us about your work in developing antihistamines?

When I graduated from Kenyon I needed to earn some money, so I got a job as a junior chemist at CIBA Pharmaceutical Corporation, in New Jersey. Just at that time the company became interested in antihistamines, and I was one of only two chemists working on this project. So despite my youth, I was involved in the synthesis of one of the first two antihistamines produced in this country, pyribenzamine. This compound, which was synthesized during the first year after I graduated from college, turned out to make a significant contribution in the treatment of allergies.

What got you interested in the chemistry of steroids?

While working at CIBA, I began taking classes at NYU and Brooklyn Polytechnic Institute at night, so as to get an advanced degree, but commuting from New Jersey after a day of work was murderous, and, after a year of night school, I decided I would go to graduate school full time. CIBA was involved in steroid projects, and even though I was working on antihistamines, I started reading books on steroids, especially Louis Fieser's *Natural Products Related to Phenanthrene,* a superb book. This turned me on to steroid chemistry, so when I went full time to the University of Wisconsin, I was prepared to work in this field. Fortunately, Wisconsin had two young Assistant Professors in this area, and I did my Ph.D. with one of them, A. L. Wilds, on the conversion of androgens to estrogens, which at that time was a tough problem.

How did you get the idea to try to synthesize an oral contraceptive?

We did not set out with that objective. We set out with the objective of developing an orally active progestational hormone; in other words, a compound that would mimic the biological properties of progesterone. At that time progesterone was clinically used for menstrual disorders and infertility, but there were ideas about using it as a contraceptive, because it is progesterone that naturally stops further ovulation after an ovum is fertilized. However, progesterone itself is not active by mouth, and daily injections would be needed. By this time I was Associate Director of Chemical Research at Syntex Corporation, located in Mexico City. By combining ideas discovered by previous investigators, we set out to synthesize a steroid that would not only be active by mouth, but would also have enhanced progestational activity. This compound was 19-nor-17α-ethynyltestosterone (norethindrone), whose synthesis we completed on 15 October 1951. It was first tested for menstrual disorders and fertility problems and then as an oral contraceptive. Forty-five years after its synthesis, it is still the active ingredient of about a third of all the oral contraceptives used throughout the world.

How do you feel about the social impact of your contraceptive work?

On the whole, it was enormously positive, and if I could do it over again, there is no question that I would proceed to do so. By now about 13–14 million women in the United States and 50–60 million in the world use the pill, making it the most widely used method of reversible birth control. Population growth is probably the biggest problem facing us in the world, assuming that the possibility of nuclear warfare is now greatly diminished, and the widespread use of oral contraceptives helps in controlling that. I think the development of these contraceptives is one of the most important contributions that chemistry has made to society. New drugs cure diseases of individuals during their lifetimes, but the use of contraceptives has implications for generations, because if you do not control the production of offspring, you do not control future generations. Furthermore, these compounds have had an enormous impact on women, empowering them to be in control of their own fertility.

Do you think there will be further research in oral contraception, for example, the development of a pill for men?

I think there will be very little fundamental new research in that area over the next couple of decades, and a pill for men is completely out of the question for the next 15 to 20 years. The reason is that the pharmaceutical industry has turned its back on this field for a number of very complex reasons, not only in the United States, but in other industrialized countries as well. The only really new approach in chemical birth control in the last 25 years is the French "morning-after pill," RU486—at present used almost exclusively as an early abortifacient, although it can also be employed as a method of birth control.

Tell us about your work on insect control.

Conceptually there was a relationship to our work on oral contraceptives, because in a way you could say that steroid oral contraceptives were true biorational methods of human birth control, since progesterone—our conceptional lead compound—is really nature's contraceptive. That was a model on which insect control could be based. At this time (the late 1960s) I was in charge of research at Syntex in addition to being Professor at Stanford University. Governments and the public realized that conventional methods of insect control—largely spraying with chlorinated hydrocarbons such as DDT—were damaging the environment. DDT and similar compounds were now being banned, and a new approach was needed. We formed a new company called Zoecon. Two insect hormones—the moulting hormone ecdysone, which is a complicated steroid, and the juvenile hormone, based on a much simpler sesquiterpene skeleton—had been discovered in the 1960s, and we decided to focus on the latter. Insects pass through a juvenile stage controlled by the juvenile hormone, whose production is later shut off by another hormone so that the insect can then mature. Our biorational approach was to synthesize an artificial juvenile hormone which would continue to be applied to immature insects, so that the insect would never reach the stage at which it could reproduce. This turned out to be a new biorational approach to controlling mosquitoes, fleas, cockroaches, and other insects that do their damage as adults, and it was approved by the Environmental Protection Agency for public use.

You have been a leader in both the synthetic and structure determination areas in organic chemistry. Which of these activities gave you more satisfaction?

In my early days—the 1950s and 1960s—undoubtedly structure determination was more satisfying, especially because I was involved in the development of new physical methods for structure determination. However, these methods have now made structural determination much more routine and automatic. For example, mass spectrometry enables us to identify compounds on a micro scale, which was unheard of before.

When you embarked on a research project, what proportion of your driving force was provided by (a) pure intellectual curiosity, (b) consideration of potential benefits to society, and (c) material rewards?

In my own work, material rewards never played any initial role, because they always came much later. To a large extent it was initially pure intellectual curiosity, but very soon I started to look for the potential benefits to society. My interest has always been in the biological areas, and I was never interested in war-related research. I never had the slightest question about the societal appropriateness of the work I was doing.

What is your opinion about the interplay between academia and industry in developing new inventions, including chemical inventions?

I am a great believer in this, and have always been connected with a University and with chemical companies. This made me a much better professor, because I became aware of the many steps needed to take a laboratory discovery up to practical realization. Conversely, I was a much better industrial research director because of what I learned in the university. For example, should scientists be concerned with societal implications of their work? Obviously yes, but it is not easy for a scientist who stays only in the ivory tower. You ought to have a responsibility for taking your work a step further. There should, however, be guidelines to prevent actual or potential conflicts of interest.

Interview conducted by Dr. Jerry March.

Chapter **10**

Organic chemistry. Alkanes

10.1 Introduction

Organic chemistry is **the study of compounds that contain carbon.** What is there about the element carbon that justifies our making it the basis of a whole branch of chemistry, whereas compounds that may contain any of the other elements, but not carbon, are all classified as inorganic compounds? One answer is that organic and inorganic compounds, in general, differ in many properties, some of which are shown in Table 10.1. Most of these differences stem from the fact that all the chemical bonds in most organic compounds are covalent, while many inorganic compounds have ionic bonding. Figure 10.1 shows some organic and inorganic compounds.

Organic and inorganic compounds differ in properties because they differ in structure, not because they obey different natural laws. There is only one set of natural laws for all compounds.

It is important to note that there are many exceptions to the statements in Table 10.1—some organic compounds behave like inorganic compounds, and vice versa—but still, the generalizations in Table 10.1 are true for the vast majority of compounds in both groups.

But how did the name "organic" come about? In the early days of chemistry, scientists thought that there were two classes of compounds. One class was produced by living organisms—these compounds they called "organic." The other class, found in minerals or rocks, they called "inorganic." They also believed that chemists working in laboratories could not synthesize any organic compound starting only from inorganic compounds. They thought that a "vital force," possessed only by living organisms, was necessary to produce organic compounds. This kind of theory is very easy to disprove if indeed it is wrong. All it takes is one experiment in which a chemist makes an organic compound starting only from inorganic ones. Such an experiment was carried out by Friedrich Wöhler (1800–1882) in 1828. He heated an aqueous solution of two inorganic compounds, ammonium chloride and silver cyanate, and to his surprise obtained urea,

$$NH_4Cl + AgNCO \xrightarrow[\text{heat}]{} \underset{\text{Urea}}{NH_2CONH_2} + AgCl$$

In this experiment, Wöhler was trying to prepare ammonium cyanate, then an unknown compound. Shortly afterward he did prepare ammonium cyanate, which yielded urea when heated:

$$NH_4NCO \xrightarrow[\text{heat}]{} NH_2CONH_2$$

Table 10.1 A Comparison of Properties of Organic and Inorganic Compounds

Organic Compounds	Inorganic Compounds
Bonding is almost entirely covalent.	Many compounds have ionic bonds.
Compounds may be gases, liquids, or solids with low melting points (less than 360°C).	Mostly high-melting solids (above 350°C).
Mostly insoluble in water.	Many soluble in water.
Mostly soluble in organic solvents such as gasoline, benzene, carbon tetrachloride.	Almost entirely insoluble in organic solvents.
Solutions in water or any other solvent do not conduct electricity.	Water solutions conduct electricity.
Almost all burn.	Very few burn.
Reactions are usually slow.	Reactions are often very fast.

(a)

(b)

Figure **10.1** • (a) Some organic compounds. Solids: on the left, *meta*-nitroaniline; on the right, aspartic acid. Liquids: cyclohexane and methanol. (b) Some inorganic compounds. Clockwise from the top: copper(II) nitrate, ammonium sulfate, potassium dichromate, and iron(III) sulfate. (Photographs by Beverly March.)

which definitely fit the old definition of "organic" because it had been isolated from human urine.

This single experiment of Wöhler's was enough to disprove the "doctrine of vital force," even though it took several years and a number of additional experiments for the entire chemical world to accept the fact. This meant that the words "organic" and "inorganic" no longer had real definitions, since, for example, urea could be obtained from both sources. A few years later Friedrich Kekulé (1829–1896) assigned the modern definition—**organic compounds are those containing carbon**—and this has been accepted ever since. The chemistry of living organisms is no longer called organic chemistry; today we call that branch of science **biochemistry.**

Kekulé put forth the structural theory of organic chemistry (Sec. 10.3) and also solved the difficult question of the structure of benzene (Sec. 11.9).

10.2 Sources of Organic Compounds

You might suppose that, with 111 elements other than carbon to choose from, there would be many more inorganic compounds known than organic, but in fact the opposite is true. There are more than eight million known organic compounds, but only about 200 000 to 300 000 known inorganic compounds. Furthermore, more than 100 000 new organic compounds are becoming known every year. This is another reason a whole branch of chemistry is devoted to carbon compounds.

What is there about carbon that permits the formation of so many compounds? The answer is

1. Carbon atoms form stable bonds with other carbon atoms, so that both long and short chains of carbon atoms, and even whole networks, are possible.

2. Carbon atoms also form stable bonds with certain other atoms, including hydrogen, oxygen, nitrogen, the halogens, and sulfur.

3. A carbon atom forms four bonds, and these can be made up in many ways, so that many combinations and arrangements of atoms are possible.

We have used the expression "known compound." What does this mean? A compound becomes "known" when a chemist obtains that compound, determines its formula, and measures some of its properties. He or she then publishes the results of the work in a chemical journal, describing the properties and how the compound was obtained. The compound is then "known." In many cases, the chemist then uses the compound to make some other compound, or it decomposes or is discarded. In any event, it no longer exists. Still, it remains "known" because any chemist can read that journal and obtain the compound once more by following the original procedure.

It is likely that only a few hundred thousand of the millions of known compounds actually exist at any one time.

There are two principal ways of obtaining known as well as new organic compounds: isolation from nature and synthesis.

Isolation from Nature

Living organisms are chemical factories. Every plant and animal makes hundreds of organic compounds; even microorganisms such as bacteria synthesize organic compounds. This process is called **biosynthesis.** One way to get organic compounds is by extraction from a biological source. In this book we will meet many compounds that have been obtained in this way. Some important examples are ethyl alcohol, acetic acid, cholesterol, cane sugar (sucrose), nicotine, quinine, and ATP (adenosine triphosphate). Many thousands of organic compounds have been and still are being extracted from nature. The process is a long way from being finished. The number of animal and especially plant species is so large that in many cases nobody has even tried to extract the compounds they contain. Besides this, nature also supplies us with two other important sources of organic compounds: petroleum (see Box 10D) and coal.

Coal and petroleum were formed from the remains of organisms that lived millions of years ago.

Synthesis

Ever since Wöhler synthesized urea, organic chemists have been constantly making the same compounds that nature also makes. In recent years the methods for doing this have been so greatly improved that there are few natural organic compounds, no matter how complicated, that cannot be synthesized by chemists (Fig. 10.2). It must be emphasized that *the compounds made by chemists are identical to those in nature.* The ethyl alcohol made by a chemist is exactly the same as the ethyl alcohol we get by distilling wine. The molecules are the same and so are all the properties; thus there is no way that anyone can tell whether a given sample of ethyl alcohol was made by a chemist or obtained from nature. Therefore, there is no advantage in paying more money for, say, vitamin C obtained from a natural source than for synthetic vitamin C, since they are exactly the same thing.

Figure **10.2**
A modern laboratory for synthetic organic chemistry. (Visuals Unlimited.)

Organic chemists have not rested with duplicating nature's compounds, however. They also synthesize compounds not found in nature. In fact, the majority of the more than 8 000 000 known organic compounds are purely synthetic and do not exist in living organisms. Many of them are used only for research, but a large number find more practical use. A few examples are acetylene, carbon tetrachloride, diethyl ether, DDT, and TNT. A very important area in which synthetic organic compounds find much use is drugs. Many modern drugs—for example, aspirin, Valium, Vasotec, Prozac, Zantac, Cardizem, Inderal, Lasix, sulfadiazine, and Enovid—are synthetic organic compounds not found in nature.

10.3 Structures of Organic Compounds

The molecular formula for ethyl alcohol is C_2H_6O. As we learned in Section 3.4, a *molecular formula* shows which atoms, and how many of each, are present in a molecule. A *structural formula* shows more than this: It shows not only all the atoms present in the molecule but also all the bonds that connect the atoms to each other. The structural formula for ethyl alcohol is

$$\begin{array}{ccccccccc} & & H & & H & & & & \\ & & | & & | & & & & \\ H & - & C & - & C & - & O & - & H \\ & & | & & | & & & & \\ & & H & & H & & & & \end{array}$$

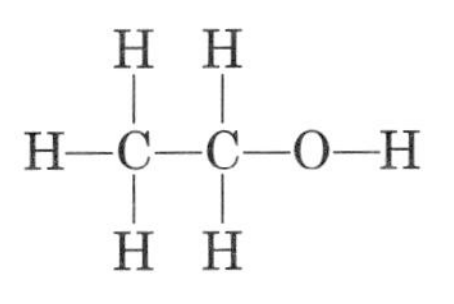

Remember that a straight line indicates a pair of electrons in a covalent bond.

Box 10A

Combinatorial Synthesis

Since the beginning of synthetic organic chemistry, about 200 years ago, its goal has been to produce highly pure organic compounds with high percentage yields (Sec. 4.6). More often than not, a particular use for a new compound was discovered accidentally. For example, this was the case with aspirin (Box 14D) and the barbiturates (Box 15E). Later, when the mode of action of certain drugs was discovered, it became possible to design molecules that would have a reasonable probability of being more effective or with fewer side effects than some existing drug. Even in such a "rational" approach (as compared to a "shotgun" approach in which compounds were synthesized at random), the aim was still to obtain a single pure compound, no matter how many steps were involved in the synthesis or how difficult the steps. In the last few years this strategy has changed, mostly in pharmaceutical research.

Combinatorial synthesis is a process in which the chemist tries to produce as many compounds from a few building blocks as is possible. For example, if 20 compounds (building blocks) are allowed to combine at random, and a method can be found to restrict the combination in such a way that only molecules containing three of these building blocks are produced, there would be obtained a mixture containing 8000 different compounds (20^3), each with three building blocks. Of course, in such a mixture the concentration of any one compound is extremely small. But perhaps in this mixture, there are two or three (or even just one) that might have the potential to be a useful drug. So how does the chemist find out which ones they are? The task is to select from this synthetic soup those molecules that promise to be biologically active. This is accomplished by specific binding to certain receptors (proteins found in the body). The few active ones will bind; the many others will not. In essence, it is cheaper and faster to produce a large "library" of combinatorial synthetic compounds and from such an assembly to "fish out" and isolate biologically active compounds, than to pursue the old way of randomly synthesizing natural products and then screening them, one at a time.

However, ethyl alcohol is not the only compound whose molecular formula is C_2H_6O. An entirely different compound, dimethyl ether, is also C_2H_6O. The structural formula for this compound is

```
    H     H
    |     |
H — C — O — C — H
    |     |
    H     H
```

This kind of isomerism is called **structural isomerism** or **constitutional isomerism.** We will meet other kinds in Sections 10.9 and 16.2.

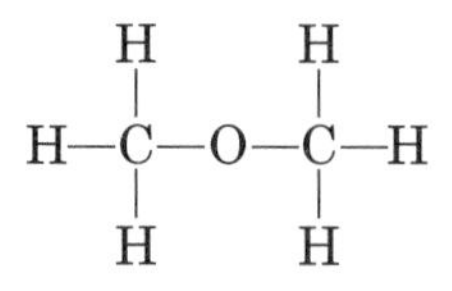

Compounds that have the same molecular formula but different structural formulas are called **isomers.** Ethyl alcohol and dimethyl ether, both being C_2H_6O, are therefore isomers. Isomers are not at all unusual; in fact, they are very common. There are three isomers of C_5H_{12}, five of C_4H_8, and in many cases dozens and even hundreds of different isomers for a given molecular formula. The various isomers sharing a given molecular formula all have different properties because **the**

Table 10.2 Some Properties of Ethyl Alcohol and Dimethyl Ether

Property	Ethyl Alcohol (CH_3-CH_2-O-H)	Dimethyl Ether (CH_3-O-CH_3)
Physical state at room temperature	Liquid	Gas
Boiling point (°C)	78	−23
Melting point (°C)	−117	−138
Reacts with sodium	Yes	No
Poisonous (in moderate amounts)	No	Yes
Anesthetic (in small amounts)	No	Yes

properties of a compound depend on its structural formula and not on its molecular formula. That is why in organic chemistry we almost always use structural rather than molecular formulas. For example, the isomers we just mentioned, dimethyl ether and ethyl alcohol, have entirely different properties (Table 10.2). If we just said "C_2H_6O" no one would know which compound we meant.

Although organic chemists use structural formulas all the time, there are times when even these are not enough to let us understand the behavior of organic compounds. We need pictures, or better yet, three-dimensional models, that show the shapes of organic molecules. Such models for ethyl alcohol and dimethyl ether are shown in Figure 10.3. Models of this type are very useful in the study of organic chemistry and biochemistry. Of course, the proper angles (approximately) must be used in these models. In Section 3.6 we saw that carbon points its four bonds to the corners of a tetrahedron and that the bond angles are about 110°.

Model sets are often sold in college book stores, but you can make your own by using marshmallows or gumdrops for atoms and toothpicks or matchsticks for bonds.

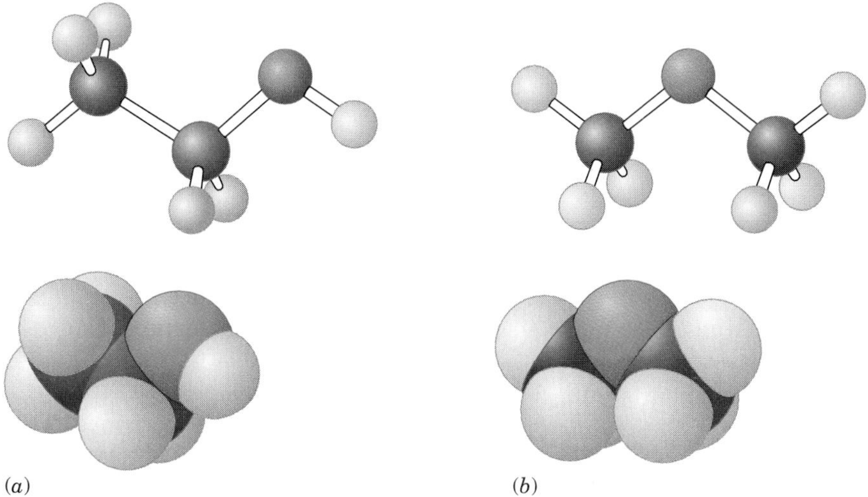

Figure **10.3** • Models of (a) ethyl alcohol and (b) dimethyl ether.

Nearly all of the millions of known organic compounds are made up of only a few elements. More than 95 percent of them contain no elements other than C, H, O, N, S, and the four halogens. These elements form the following numbers of bonds (see Sec. 3.4):

This statement does not hold for atoms with charges. For example, as we already saw, the nitrogen in the ammonium ion, NH_4^+, forms four bonds.

H	1	C	4	N	3	O	2	F	1
						S	2	Cl	1
								Br	1
								I	1

These numbers are very important, so you should memorize them as soon as possible. One reason that so many organic compounds can be made from so few elements is that stable compounds can be formed from almost any combination of these elements, as long as each atom has the correct number of bonds. This means that if we can draw the structural formula of a molecule, it is usually possible to go to the laboratory and make it (chances are it has been made already).

Remember (Sec. 3.4) that carbon and nitrogen can form double and triple bonds as well as single bonds and that oxygen can form both double and single bonds (Table 10.3). All these possibilities give rise to a vast number of possible organic compounds, for example,

Table 10.3 Possible Double and Triple Bonds

>C=C<	—C≡C—
>C=N—	—C≡N
>C=O	
>C=S	Less important than the others
—N=N—	

```
     H          H       H                            
     |           \     /                             
   H—C—H          C=C          H—C—H        H—C≡N    
     |           /     \         ‖                   
     H          H       H        O                   
  Methane        Ethene      Formaldehyde  Hydrogen cyanide

               H             H  H              H
               |             |  |              |
  O=C=O      H—C—C≡C—H     H—C——N—H         H—C—O—H
               |             |                 |
               H             H                 H
Carbon dioxide  Propyne     Methylamine       Methanol
```

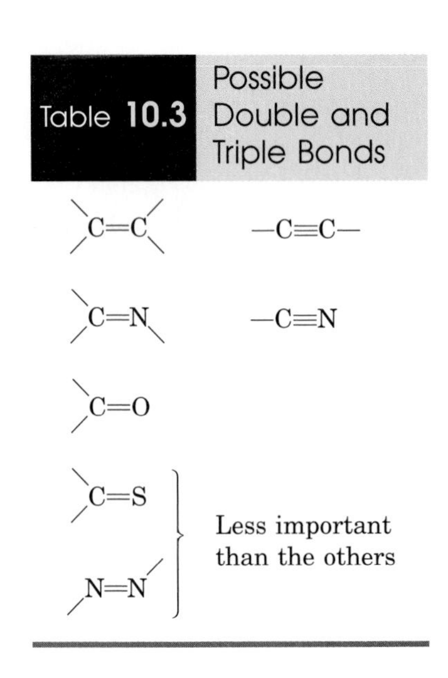

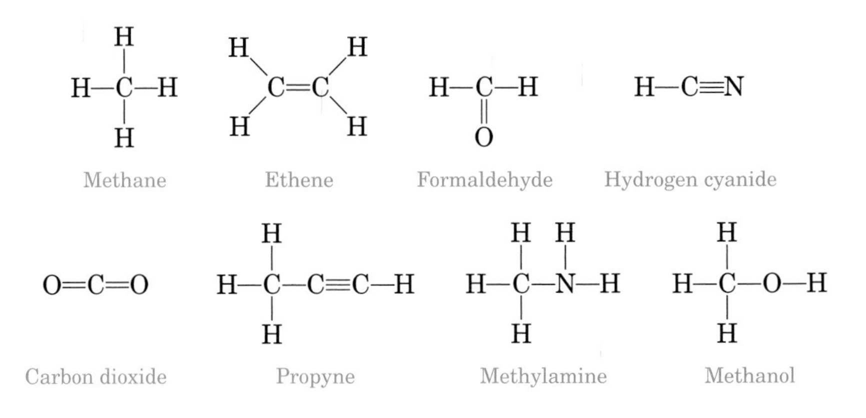

10.4 Hybrid Orbitals

The concept of Lewis structures and the VSEPR model (Sec. 3.6) give us an insight into chemical bonding and allow us to predict the shapes of molecules. But these models are insufficient to predict certain important characteristics of organic molecules. For example, the Lewis model cannot explain why a carbon-carbon double bond is more reactive than a carbon-carbon single bond. Neither is it clear why a carbon atom forms four equal bonds with four hydrogens in methane (CH_4) when its valence electrons are in different atomic orbitals ($2s^2\ 2p^2$).

In order to explain these and other properties, theoretical chemists came up with the idea of **hybrid orbitals.** Let's take the case of methane, which is made of four C—H bonds, all of which have the same properties. The Lewis structure is simple:

```
   H
   ..
H : C : H
   ..
   H
```

The four covalent bonds are formed by overlap between the atomic orbitals of the carbon atom and the atomic orbitals (1*s*) of four hydrogen atoms. But the valence electrons in a carbon atom do not occupy four orbitals. Two electrons are in a 2*s* orbital and the other two are in 2*p* orbitals that are 90° to each other (see Fig. 2.9).

We can schematically show it like this:

↑↓	↑	↑	
2s	$2p_x$	$2p_y$	$2p_z$

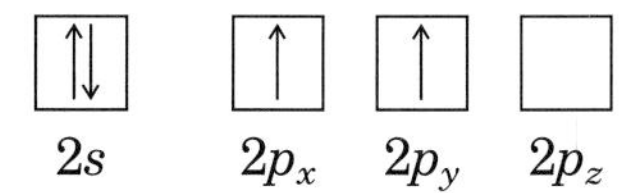

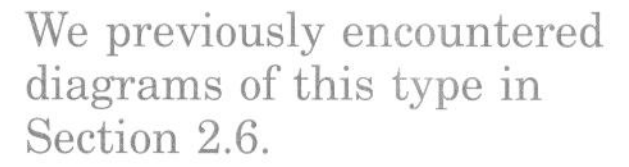

We previously encountered diagrams of this type in Section 2.6.

From this picture we should expect carbon to form only two bonds. The two bonds should form by overlap of the two 2*p* electrons with orbitals from outside atoms. The 2*s* orbital, being filled, should not form any bonds at all. But we know that a carbon atom forms four bonds, and that the formula of methane is CH_4.

No stable compound with the formula CH_2 has ever been found.

What happens is that the carbon atom promotes one of its electrons from the 2*s* orbital into the empty $2p_z$ orbital:

↑	↑	↑	↑
2s	$2p_x$	$2p_y$	$2p_z$

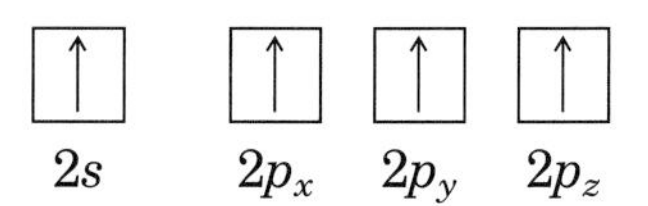

This promotion requires the expense of a certain amount of energy, but the ability to form four rather than two bonds returns the energy investment manyfold.

We can now explain why all four bonds in methane are equal and also why the bond angles are 109.5°. We say that the valence orbitals of a carbon atom are neither *s* nor *p* but a *hybrid* or *mixture* of orbitals. Since three *p* orbitals are mixed with one *s* orbital, we call the hybrid orbitals $\boldsymbol{sp^3}$, meaning that each of them has one-fourth *s* character and three-fourths *p* character. An sp^3 orbital is neither spherical nor dumbbell-shaped but has a mixture of the two shapes (Fig. 10.4).

Figure **10.4**
The four sp^3 orbitals.

Each of the four hybrid sp^3 orbitals contains one electron, so that four equivalent C—H bonds are formed simply by the overlap between the four carbon sp^3 orbitals and four hydrogen *s* orbitals. This overlap allows the sharing of two electrons between a carbon and a hydrogen atom. The bond formed between them is a σ bond.

The four equivalent sp^3 orbitals get as far away from each other as they can. As the VSEPR theory showed us, this means that the bond angles will be 109.5°.

The sp^3 orbitals of a carbon atom can similarly overlap with other orbitals such as the *p* orbital of F or Cl or indeed any *s* or *p* orbital of any atom, or even the sp^3 orbital of another carbon atom.

EXAMPLE

The compound chloroethane can be represented as follows:

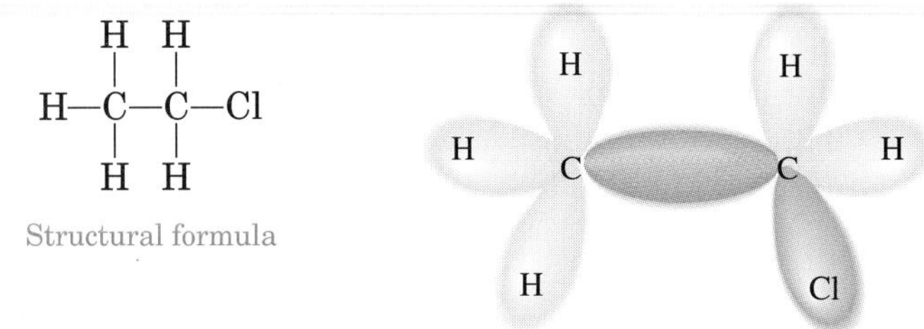

Each of the molecular orbitals shown in color is made up by the overlap of an sp^3 orbital from a carbon atom with an s, p, or sp^3 orbital of another atom.

In fact we can assume that in any case where a carbon atom bonds with four other atoms, it uses sp^3 hybrid orbitals.

Problem 10.1

Describe the orbitals that make up all the bonds in (a) CH_3F (b) $CH_3—CH_2—F$.

10.5 Hydrocarbons

We begin our study of organic compounds with the **hydrocarbons,** which are **compounds that contain only carbon and hydrogen.** Even with only two elements, there are many thousands of known hydrocarbons and a much larger number of possible ones that have not yet been synthesized.

There are several families of hydrocarbons:

Alkanes have only single bonds.
Alkenes have C=C double bonds.
Alkynes have C≡C triple bonds.
Aromatic hydrocarbons have benzene rings (Sec. 11.9).

We discuss alkanes in this chapter and reserve the other types for Chapter 11.

10.6 Alkanes

In Section 11.1 we will see why alkanes are called saturated hydrocarbons.

Alkanes are **hydrocarbons with only single bonds.** They are also called **saturated hydrocarbons.** The simplest one, *methane,* has only one carbon atom:

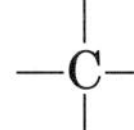

The clouds of Jupiter contain methane. (NASA.)

Since carbon forms four bonds and hydrogen forms one, methane must have this formula:

$$\begin{array}{c} \text{H} \\ | \\ \text{H—C—H} \\ | \\ \text{H} \end{array} \qquad \text{CH}_4$$

Methane

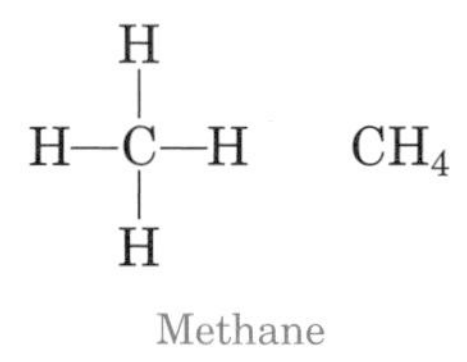

In a similar way, we can construct the next two alkanes, ethane and propane:

$$\begin{array}{c} \text{H}\ \ \text{H} \\ |\quad | \\ \text{H—C—C—H} \\ |\quad | \\ \text{H}\ \ \text{H} \end{array} \qquad \text{C}_2\text{H}_6 \qquad\qquad \begin{array}{c} \text{H}\ \ \text{H}\ \ \text{H} \\ |\quad |\quad | \\ \text{H—C—C—C—H} \\ |\quad |\quad | \\ \text{H}\ \ \text{H}\ \ \text{H} \end{array} \qquad \text{C}_3\text{H}_8$$

Ethane   Propane

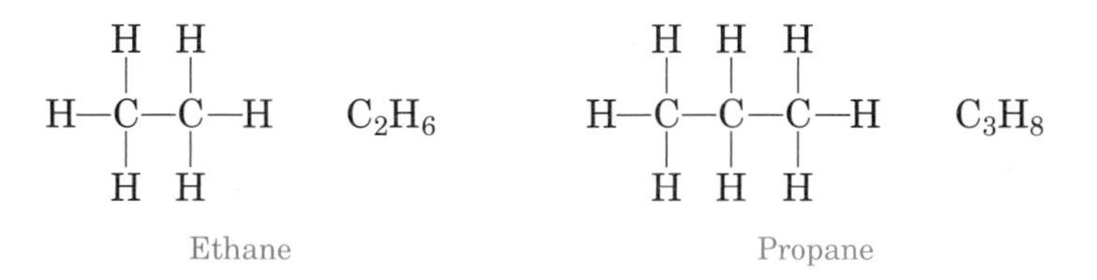

All three of these compounds are gases. Methane, also called *marsh gas,* is the main constituent of *natural gas,* which is used as a fuel in many homes and factories. Propane is also used as a fuel.

Propane is the fuel sold as "bottled gas."

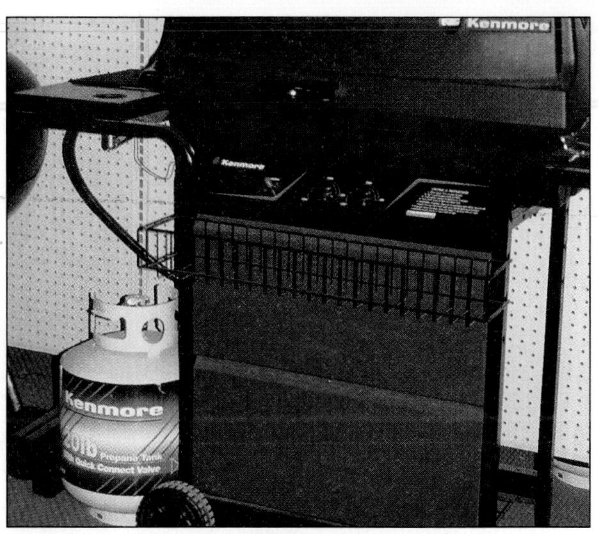

An outdoor barbecue grill is fueled by propane. (Photograph by Beverly March.)

When written out as just shown, these formulas occupy a lot of space. Therefore, chemists generally write them in a shorthand way:

Methane	CH_4		
Ethane	$CH_3—CH_3$	or	CH_3CH_3
Propane	$CH_3—CH_2—CH_3$	or	$CH_3CH_2CH_3$

Occasionally the hydrogen atoms are written before the carbons, $H_3C—CH_3$, to emphasize the C—C bond.

These shorthand formulas are called **condensed structural formulas.** Though they do not explicitly show all the bonds (or even any of the bonds), we must not forget that the bonds are there. Of course, as we learned in Section 10.3, none of these formulas shows the real shape of the molecules. For that we need models.

Figure 10.5 shows models for the first three alkanes, of two different types. As we could predict from the fact that carbon atoms use sp^3 hybrid orbitals to form their bonds (see Fig. 10.4), the carbon atoms in all alkanes point their four bonds to the corners of a tetrahedron, so that all angles are approximately 110°. This means that when we write, for example,

$$CH_3—CH_2—CH_3$$

for the structure of propane (as chemists often do), we must remember that, in the real molecule, the three carbons are not really in a straight line, as the formula suggests. They are actually connected by an angle of about 110°:

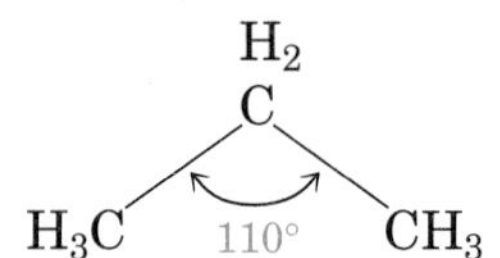

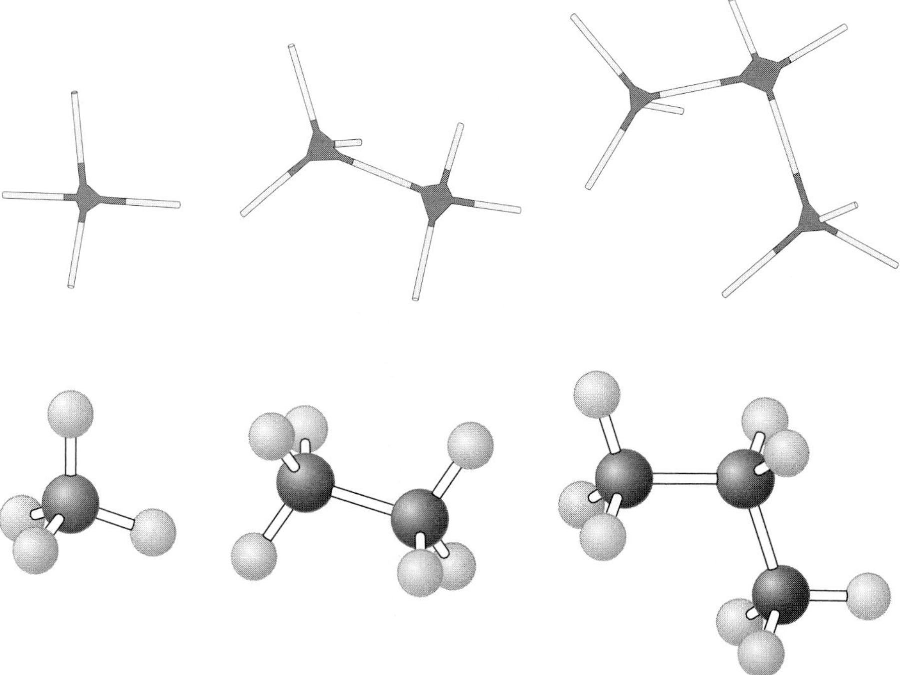

Figure **10.5** • Models of methane, ethane, and propane.

We write the condensed formula in a straight line because it is convenient to do so, in spite of the fact that it is geometrically inaccurate.

As you might suppose, it is possible to make carbon chains containing many more than three carbon atoms. In fact, alkanes of this type can contain any number of carbon atoms. Table 10.4 lists the name and condensed formula for all those up to ten carbons. It is important that you

In Section 11.7 we shall see that such chains can be many thousands of carbon atoms long and still be perfectly stable.

Table 10.4 Molecular and Condensed Structural Formulas, and Boiling and Melting Points, for the First Ten Normal Alkanes

Molecular Formula[a]	Condensed Structural Formula	Name	Boiling Point (°C)	Melting Point (°C)
CH_4	CH_4	Methane	−164	−182
C_2H_6	CH_3CH_3	Ethane	−89	−183
C_3H_8	$CH_3CH_2CH_3$	Propane	−42	−190
C_4H_{10}	$CH_3CH_2CH_2CH_3$	Butane	0	−138
C_5H_{12}	$CH_3CH_2CH_2CH_2CH_3$	Pentane	36	−130
C_6H_{14}	$CH_3CH_2CH_2CH_2CH_2CH_3$	Hexane	69	−95
C_7H_{16}	$CH_3CH_2CH_2CH_2CH_2CH_2CH_3$	Heptane	98	−91
C_8H_{18}	$CH_3CH_2CH_2CH_2CH_2CH_2CH_2CH_3$	Octane	126	−57
C_9H_{20}	$CH_3CH_2CH_2CH_2CH_2CH_2CH_2CH_2CH_3$	Nonane	151	−51
$C_{10}H_{22}$	$CH_3CH_2CH_2CH_2CH_2CH_2CH_2CH_2CH_2CH_3$	Decane	174	−30

[a]The general formula for all alkanes, normal or branched, is C_nH_{2n+2}, where n = the number of carbons.

memorize these names because they are the basis for naming most of the other organic compounds. Note that the names of all alkanes end in "-ane."

In all the alkanes in Table 10.4, the carbons form a continuous chain; in other words, there is no **branching.** Such alkanes are called *normal* alkanes, abbreviated *n*-alkanes. Branched alkanes are also possible. For example, consider this compound:

$$\begin{array}{c} CH_3{-}\underset{\displaystyle CH_3}{\underset{|}{C}}H{-}CH_3 \end{array} \quad \text{Isobutane}$$

There are four carbons and ten hydrogens in this molecule, so the molecular formula is C_4H_{10}. This is the same as that of butane (Table 10.4), which means, of course, that butane and this new compound are isomers. Because of this, we call it *isobutane.* As in other cases of isomerism, butane and isobutane have different structures and different properties.

Notice that isobutane has a carbon atom (the one in the middle) that is connected to *three* other carbon atoms, but butane has no such carbon. We define four types of carbon atoms:

Why can't there be a quintary carbon?

> A **quaternary carbon** is attached to four other carbons.
> A **tertiary carbon** is attached to three other carbons.
> A **secondary carbon** is attached to two other carbons.
> A **primary carbon** is attached to one other carbon or to no other carbons.

Note that normal alkanes have only primary or secondary carbons and branched alkanes have at least one tertiary or quaternary carbon.

We have seen that there are two alkanes containing four carbons: butane and isobutane. What about five-carbon alkanes? Of course, there is one normal compound, pentane (there can never be more than one normal alkane of any particular chain length), but there are *two* branched compounds:

$$CH_3{-}\underset{\displaystyle CH_3}{\underset{|}{C}}H{-}CH_2{-}CH_3 \quad \text{Tertiary carbon} \qquad CH_3{-}\overset{\displaystyle CH_3}{\overset{|}{\underset{\displaystyle CH_3}{\underset{|}{C}}}}{-}CH_3 \quad \text{Quaternary carbon}$$

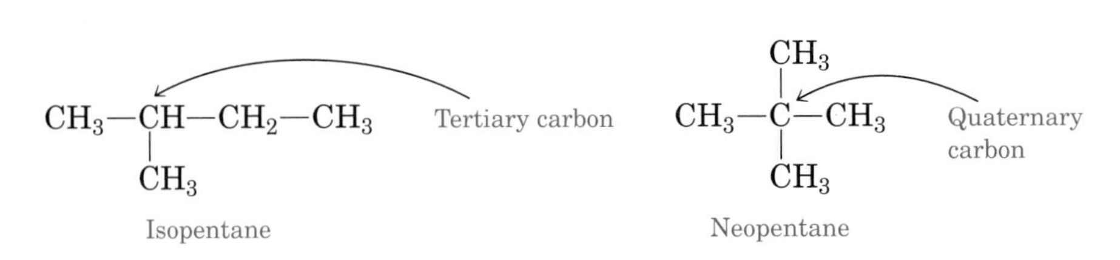

Isopentane Neopentane

If we go to six carbons, we find that, besides *normal*-hexane, there are four branched alkanes:

$$CH_3{-}\underset{\displaystyle CH_3}{\underset{|}{C}}H{-}CH_2{-}CH_2{-}CH_3 \qquad CH_3{-}CH_2{-}\underset{\displaystyle CH_3}{\underset{|}{C}}H{-}CH_2{-}CH_3$$

A B

$$\begin{array}{rcl} & CH_3 & \\ & | & \\ CH_3-CH_2- & C & -CH_3 \\ & | & \\ & CH_3 & \end{array}$$

C

$$\begin{array}{lcccl} CH_3- & CH & - & CH & -CH_3 \\ & | & & | & \\ & CH_3 & & CH_3 & \end{array}$$

D

for a total of five isomers (all having the molecular formula C_6H_{14}). If we were to continue, we would find the number of isomers growing steadily. There are 75 alkanes with 10 carbons (all $C_{10}H_{22}$), 366 319 with 20 carbons (all $C_{20}H_{42}$), and more than four billion with 30 carbons (all $C_{30}H_{62}$). Only a few of this vast number of alkanes are known compounds, but almost all of them could be made fairly easily, if we really wanted to. Why haven't we done it? Simply because there are more important things to do. We don't generally synthesize new compounds without a good reason.

There are some important points to notice about the formulas we just gave. Take a look at the formula marked **A.** First we note that **A** is not the same as **B.** Both have a continuous chain of five carbons (the **main chain**) and a $-CH_3$ attached to the main chain (this $-CH_3$ constitutes a **side chain**), but in **A** the side chain is attached to the *second* carbon of the main chain, while in **B** it is attached to the *third* carbon of the main chain. This means that **A** and **B** are different compounds (isomers). Another way we can see the difference is to look at the tertiary carbons (both **A** and **B** have one tertiary carbon). In **A** the tertiary carbon is connected to *two* $-CH_3$ groups and one 3-carbon chain. In **B** the tertiary carbon is connected to *one* $-CH_3$ and two 2-carbon chains.

Now let us look once more at compound **A.** We wrote it like this:

$$\begin{array}{lcl} CH_3- & CH & -CH_2-CH_2-CH_3 \\ & | & \\ & CH_3 & \end{array}$$

A

that is, a five-carbon main chain with an extra $-CH_3$ on the second carbon of the chain, pointing down. But we could have shown it pointing up, or on the fourth carbon (up or down), or we could even have twisted the chain:

$$\begin{array}{lcl} & CH_3 & \\ & | & \\ CH_3- & CH & -CH_2-CH_2-CH_3 \end{array}$$

A

$$\begin{array}{rcl} & CH_3 & \\ & | & \\ CH_3-CH_2-CH_2- & CH & -CH_3 \end{array}$$

A

$$\begin{array}{lclc} CH_3- & CH & - & CH_2 \\ & | & & | \\ & CH_3 & & CH_2-CH_3 \end{array}$$

A

$$\begin{array}{rcl} CH_3-CH_2-CH_2- & CH & -CH_3 \\ & | & \\ & CH_3 & \end{array}$$

A

$$\begin{array}{lclc} CH_3 & & CH_3 & \\ | & & | & \\ CH_2 & -CH_2- & CH & -CH_3 \end{array}$$

A

$$\begin{array}{rl} & CH_3 \\ & | \\ & CH-CH_3 \\ & | \\ & CH_2 \\ & | \\ CH_3- & CH_2 \end{array}$$

A

All these are equally correct ways of writing the *same compound* (2-methylpentane) because they all show the *same connections.* In all these formulas there is one tertiary carbon connected to two $—CH_3$ groups and one 3-carbon chain. Looking at it another way, in all of them there is a main chain of five carbons and a $—CH_3$ side chain connected to the second carbon. Of course, the real molecule does not look like any of these. The model does get much closer, though; at least the angles are correct (Fig. 10.6).

EXAMPLE

Which of these represent the same compound?

(a)
$$\begin{array}{ccccccccc} & & CH_3 & & & & & & \\ & & | & & & & & & \\ CH_3 & - & CH & - & CH & - & CH_2 & - & CH_3 \\ & & & & | & & & & \\ & & & & CH_3 & & & & \end{array}$$

(b)
$$\begin{array}{ccccccccc} & & CH_3 & & & & & & \\ & & | & & & & & & \\ CH_3 & - & CH & - & CH_2 & - & CH & - & CH_3 \\ & & & & & & | & & \\ & & & & & & CH_3 & & \end{array}$$

(c)
$$\begin{array}{ccccccc} & & & & CH_3 & & \\ & & & & | & & \\ & & & & CH & - & CH_3 \\ & & & & | & & \\ CH_3 & - & CH_2 & - & CH & - & CH_3 \end{array}$$

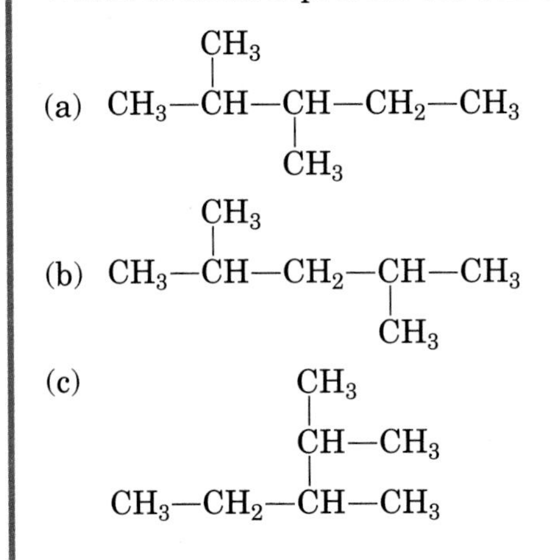

Answer • Pay no attention to the positions of the atoms on the paper. The important thing is the connections. In all three structures the longest chain has five carbons, and there are two $—CH_3$ side chains. Begin by identifying the chain of five carbons in all three formulas. It will then become clear that in (a) and (c) the two $—CH_3$ groups are on the second and third carbon, whereas in (b) the two

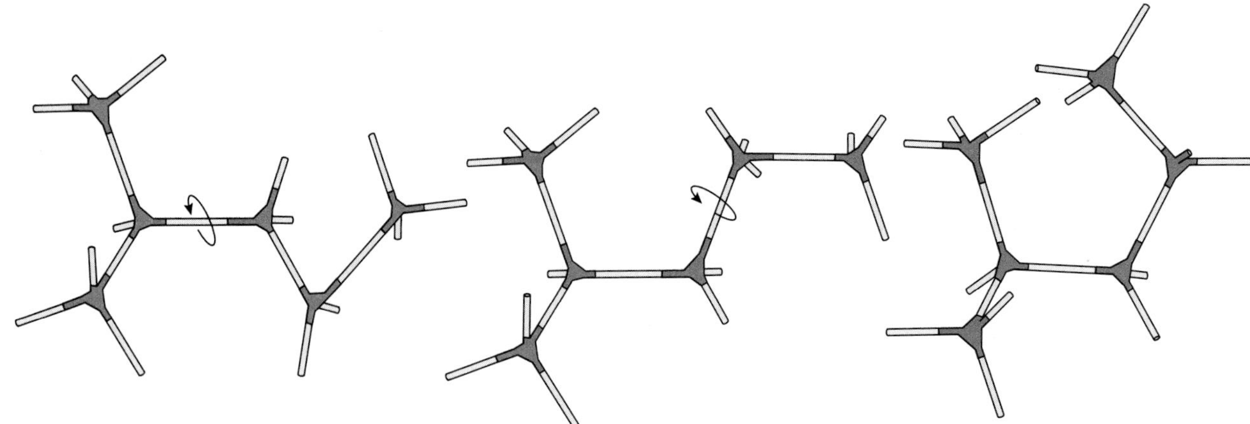

Figure **10.6** • Some of the shapes of the 2-methylpentane molecule (**A** in the text). The curved arrows show the rotations that change one shape to another.

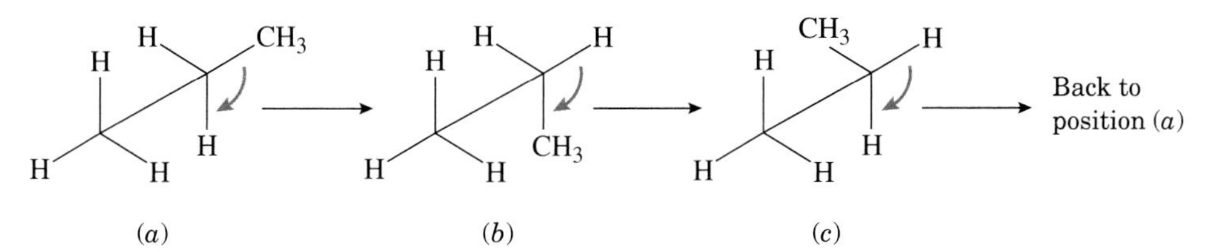

(a) (b) (c) Back to position (a)

Figure **10.7** • Free rotation in the propane molecule. Three conformations are shown.

groups are on the second and fourth carbons. Thus, (a) and (c) are different ways of writing the same compound, but (b) is a different compound. Note that if we start counting from the other end, we could also say that in (a) and (c) the two —CH_3 groups are on the third and fourth carbons, but (a) and (c) are still the same compound.

There are also other ways to tell the difference; for example, (a) and (c) have two CH carbons directly connected to each other, but in (b) the two CH carbons have a CH_2 between them, proving that (b) must be a different compound. (This fact, *by itself,* does not prove that (a) is the same as (c), but all the facts together prove this statement.)

Problem **10.2**

Which of these represent the same compound?

```
(a)        CH3
           |
           CH—CH3
           |
    CH3—CH
           |
           CH2—CH3
(b) CH3—CH—CH—CH2—CH3
           |    |
           CH3 CH3
(c)        CH3
           |
    CH3—C—CH2—CH2—CH3
           |
           CH3
```

As Figure 10.6 shows, the atoms in **A** are free to rotate around the bonds, so that the molecule may assume any of a number of shapes (these shapes are called **conformations**), though the 110° angles are present in all of them. There is *free rotation* about the C—C bonds of all the alkanes, as shown in Figure 10.7.

10.7 IUPAC Nomenclature of Alkanes and Alkyl Halides

How do we name all the compounds we just met in Section 10.6? The International Union of Pure and Applied Chemistry (IUPAC) is an organization that, among other things, issues rules for naming chemical compounds, so that all the chemists of the world can use the same names. Most people do use the IUPAC names for most compounds, but unfortunately some compounds are still called by their "common"

The IUPAC system for naming organic compounds was made official in 1892.

names, which are not part of the IUPAC system. Since it is important for students to know the names that people actually use, we will be using some of these common names in this book. A major advantage of the IUPAC names is that they are *systematic*. This means that a person can derive the name by looking at the structural formula, and vice versa. It is not necessary to memorize a different name for every structure. Common names, which are often shorter, do not have this advantage.

An important part of the IUPAC naming system is the idea of the **group.** If we take a hydrogen away from methane, we are left with a CH_3—.

$$\begin{array}{c}\text{H}\\|\\\text{H—C—H}\\|\\\text{H}\end{array}\xrightarrow{-\text{H}}\begin{array}{c}\text{H}\\|\\\text{H—C—}\\|\\\text{H}\end{array}$$

Methyl group

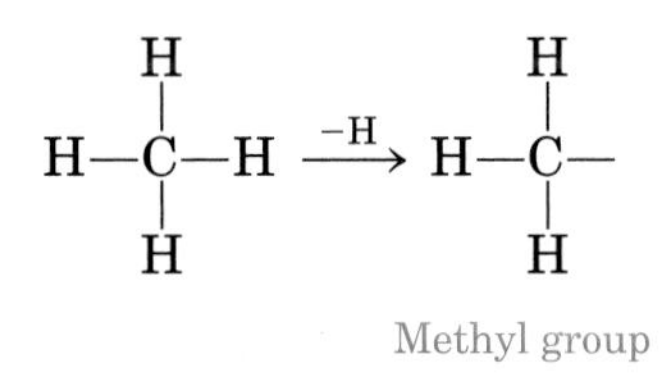

This is called the **methyl group.** Note that we changed the ending of methane from "-ane" to "-yl." In a similar way we can have the **ethyl group,** the **propyl group,** and so on:

$CH_3—CH_2—$ Ethyl group

$CH_3—CH_2—CH_2—$ Propyl group

The propyl group is also called *n*-propyl.

All of these are called **alkyl groups** because they are derived from **alk**anes. Alkyl groups are not independent compounds. They do not exist by themselves. They are *parts* of molecules, not whole molecules themselves. We are now ready to give the IUPAC rules for naming alkanes and alkyl halides.

1. Find the longest continuous chain of carbon atoms. Assign a parent name from Table 10.4 based on this number.

As an example, let us try this on our compound **A:**

$$\begin{array}{l}\text{CH}_3\text{—CH—CH}_2\text{—CH}_2\text{—CH}_3\\\qquad\quad|\\\qquad\ \ \text{CH}_3\end{array}$$

The longest continuous chain has *five* carbons, so the parent name (from Table 10.4) is *pentane.*

2. Find whatever groups are *not* part of the longest continuous chain. Name these as prefixes.

If we look at compound **A,**

$$\begin{array}{l}CH_3{-}CH{-}CH_2{-}CH_2{-}CH_3\\ \qquad\;\; |\\ \qquad\; CH_3\end{array}$$

we see that the only group that is not part of the chain is CH_3—, the methyl group. We therefore put *methyl* in front of pentane to get methylpentane.

We are almost at the end of the naming process for this compound, but not quite. If we apply the above two rules to compound **B** on page 322, we also get methylpentane (try it). Since we can't have the same name for two different compounds, we need another rule:

> **3.** Assign numbers to groups by counting from one end of the chain. A chain has two ends, and the end we start from is the one that gives the lowest possible numbers to the groups.

$$\begin{array}{l}\overset{1}{CH_3}{-}\overset{2}{CH}{-}\overset{3}{CH_2}{-}\overset{4}{CH_2}{-}\overset{5}{CH_3}\\ \qquad\;\; |\\ \qquad\; CH_3\end{array}\quad \text{2-Methylpentane}$$

A

If we count from left to right, we arrive at the number 2 for the position of the methyl group. Our final name for compound **A** is 2-methylpentane. If we had begun at the other end we would have said 4-methylpentane, but this is not a correct name because 2 is lower than 4. It doesn't matter whether we begin at the left or the right. *All that matters is that we come out with the lowest numbers.* If we use these rules to name the other formulas on page 323 (try it), we find that the name 2-methylpentane applies to all of them, which is another indication that they are all different ways of drawing the same compound. As for compound **B,**

$$\begin{array}{l}\overset{1}{CH_3}{-}\overset{2}{CH_2}{-}\overset{3}{CH}{-}\overset{4}{CH_2}{-}\overset{5}{CH_3}\\ \qquad\qquad\quad |\\ \qquad\qquad\; CH_3\end{array}\quad \text{3-Methylpentane}$$

B

this is 3-methylpentane no matter which end we start from.

Problem **10.3**

Name the following compounds by the IUPAC system:

(a) $CH_3{-}CH_2{-}CH_2{-}CH_3$

(b) $\begin{array}{l}CH_3{-}CH_2{-}CH{-}CH_3\\ \qquad\qquad\quad |\\ \qquad\qquad\; CH_3\end{array}$

(c) $CH_3{-}CH_2{-}CH_2{-}CH_2{-}\underset{\substack{|\\ CH_2\\ |\\ CH_3}}{CH}{-}CH_2{-}CH_3$

Problem 10.4

Draw structural formulas for (a) 2-methylbutane (b) 3-methylhexane.

Let us now look at some other cases.

4. If there are two or more identical groups, we use these prefixes:

Number of Identical Groups	Prefix	Number of Identical Groups	Prefix
2	di	5	penta-
3	tri-	6	hexa-
4	tetra-		

and so on. For example, let's name compounds **C** and **D:**

$$CH_3{-}CH_2{-}\overset{\substack{CH_3\\ |}}{\underset{\substack{|\\ CH_3}}{C}}{-}CH_3 \qquad CH_3{-}\underset{\substack{|\\ CH_3}}{CH}{-}\underset{\substack{|\\ CH_3}}{CH}{-}CH_3$$

C D

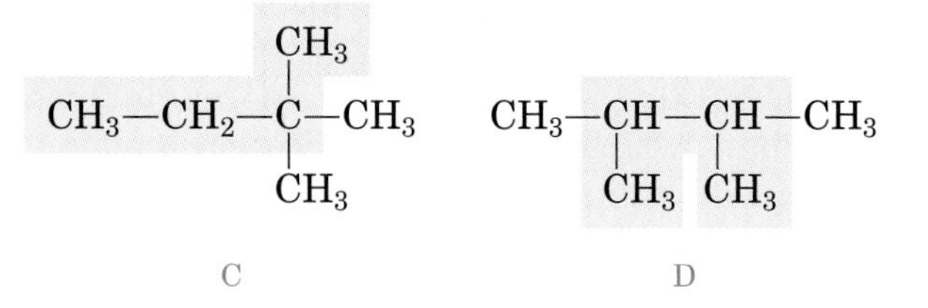

The longest continuous chain in each has four carbons, so the parent name in both cases is butane. In both cases there are two methyl groups, so both compounds are dimethylbutane. As required by rule 3, we use numbers to tell them apart:

Why isn't compound **C** called 3,3-dimethylbutane?

Compound **C** is 2,2-dimethylbutane.

Compound **D** is 2,3-dimethylbutane.

The comma is used to separate position numbers; the hyphen connects the number to the name.

Note that we must use a separate number for every prefix, even if two of the same groups are on the same carbon. Compound **C** is 2,2-dimethylbutane, and it would be incorrect to call it 2-dimethylbutane. It is also incorrect to say 2,2-methylbutane. If there are two identical groups we must say "di-." Here are examples showing the use of "tri-" and "tetra-":

```
    CH3  CH3
     |    |
CH3—CH—CH—CH—CH3
              |
              CH2
              |
              CH3
```

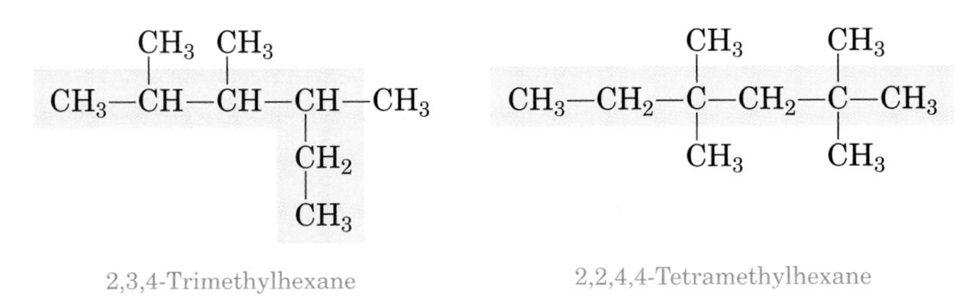

2,3,4-Trimethylhexane

```
              CH3          CH3
               |            |
CH3—CH2—C—CH2—C—CH3
               |            |
              CH3          CH3
```

2,2,4,4-Tetramethylhexane

Problem 10.5

Name:

```
(a) CH3—CH—CH—CH2—CH3
         |    |
        CH3  CH3

                     CH3
                      |
(b) CH3—CH—CH2—C—CH3
         |            |
        CH3          CH3
```

Problem 10.6

Draw structural formulas for (a) 2,2-dimethylpropane (b) 3,3,4-triethylheptane.

5. If there are two or more different groups, we put all of them into the prefix in alphabetical order.

```
CH3—CH2—CH2—CH—CH—CH—CH3
             |    |    |
            CH3  CH2  CH3
                  |
                 CH3
```

The longest continuous chain in this molecule has seven carbons, so the parent name is heptane. When we look for the groups, we find two methyl groups and one ethyl group. Since ethyl comes before methyl alphabetically, the prefixes will be ethyl- followed by dimethyl-. Numbering to give the lowest numbers, we come out with

3-ethyl-2,4-dimethylheptane

Note that we ignore "di-" in determining alphabetical order. Ethyl- comes before *di*methyl-.

Note that we must put the methyl groups together. It is wrong to say 2-methyl-3-ethyl-4-methylheptane.

Problem 10.7

Name:

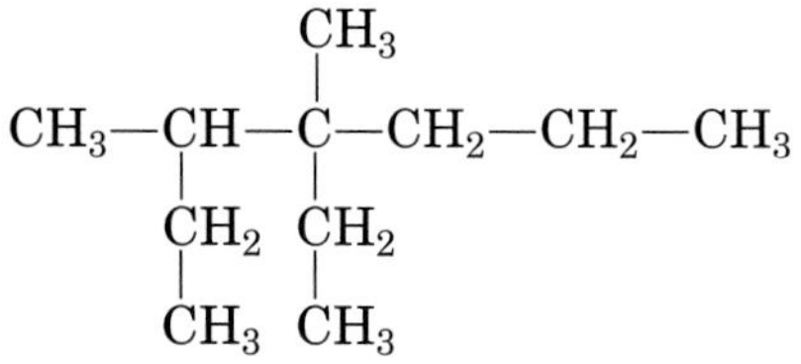

$$\begin{array}{ccccccccccc} & & & & \mathrm{CH_3} & & & & & & \\ & & & & | & & & & & & \\ \mathrm{CH_3} & - & \mathrm{CH} & - & \mathrm{C} & - & \mathrm{CH_2} & - & \mathrm{CH_2} & - & \mathrm{CH_3} \\ & & | & & | & & & & & & \\ & & \mathrm{CH_2} & & \mathrm{CH_2} & & & & & & \\ & & | & & | & & & & & & \\ & & \mathrm{CH_3} & & \mathrm{CH_3} & & & & & & \end{array}$$

Problem 10.8

Draw the structural formula for 4-ethyl-3,4-dimethylheptane.

In rule 6 we extend this system to **alkyl halides,** which are compounds containing carbon and halogens and only single bonds (most also contain hydrogen):

> **6.** The four halogens are named by prefixes:
>
> | | | | |
> |---|---|---|---|
> | **F** | fluoro- | **Br** | bromo- |
> | **Cl** | chloro- | **I** | iodo- |

All other rules are the same. For example,

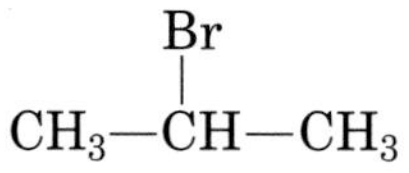

$$\begin{array}{ccccc} & & \mathrm{Br} & & \\ & & | & & \\ \mathrm{CH_3} & - & \mathrm{CH} & - & \mathrm{CH_3} \end{array}$$

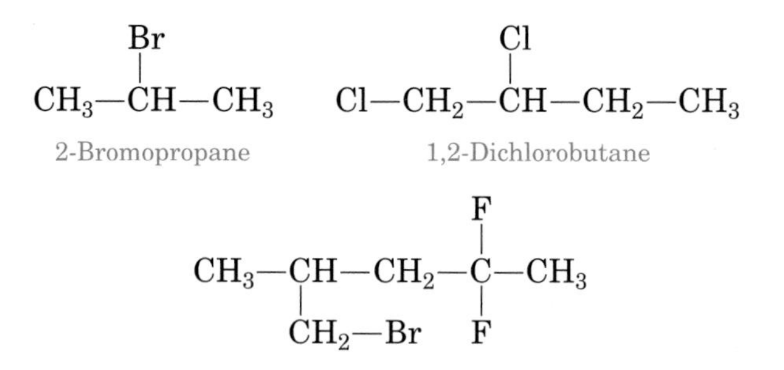

2-Bromopropane

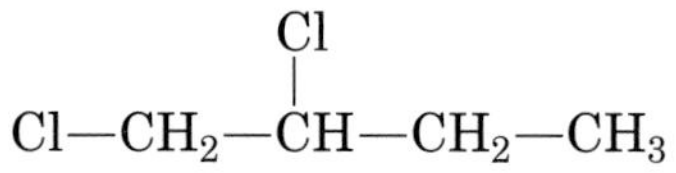

$$\begin{array}{ccccccccc} & & & & \mathrm{Cl} & & & & \\ & & & & | & & & & \\ \mathrm{Cl} & - & \mathrm{CH_2} & - & \mathrm{CH} & - & \mathrm{CH_2} & - & \mathrm{CH_3} \end{array}$$

1,2-Dichlorobutane

$$\begin{array}{ccccccccc} & & & & & & \mathrm{F} & & \\ & & & & & & | & & \\ \mathrm{CH_3} & - & \mathrm{CH} & - & \mathrm{CH_2} & - & \mathrm{C} & - & \mathrm{CH_3} \\ & & | & & & & | & & \\ & & \mathrm{CH_2}-\mathrm{Br} & & & & \mathrm{F} & & \end{array}$$

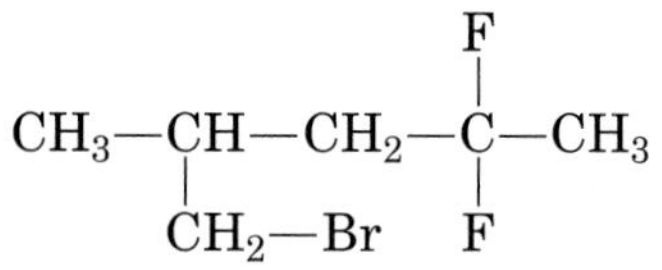

1-Bromo-4,4-difluoro-2-methylpentane

The prefixes bromo-, fluoro-, and methyl- are listed in alphabetical order. Let us look at the last example. There are two ways of taking the longest continuous chain, but we choose the one shown in color

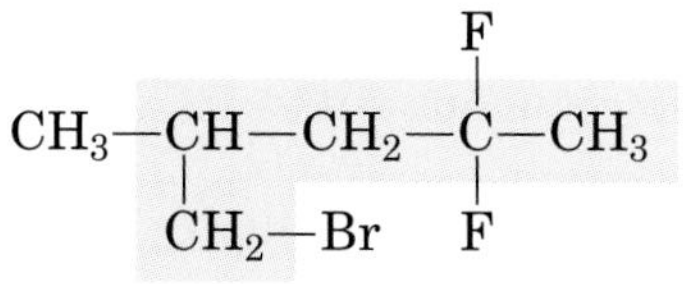

$$\begin{array}{ccccccccc} & & & & & & \mathrm{F} & & \\ & & & & & & | & & \\ \mathrm{CH_3} & - & \mathrm{CH} & - & \mathrm{CH_2} & - & \mathrm{C} & - & \mathrm{CH_3} \\ & & | & & & & | & & \\ & & \mathrm{CH_2}-\mathrm{Br} & & & & \mathrm{F} & & \end{array}$$

because that way the bromine is a substituent on the parent chain instead of on a side chain.

Problem **10.9**

Name:

```
                Cl   Cl
                |    |
(a) CH3—CH2—CH—CH—CH3
```

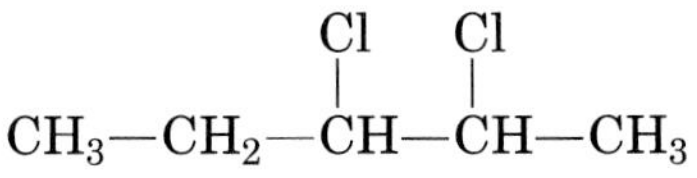

```
                    Br  I
                    |   |
(b) CH3—CH2—CH2—C—C—CH3
                    |   |
                    I   Cl
```

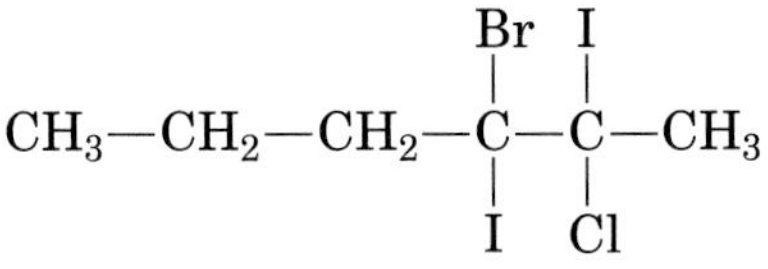

On page 326 we met a series of alkyl groups, none of which are branched. In some compounds, however, we need to use branched alkyl groups as well. The most important of these are shown in Table 10.5. Isopropyl is a **secondary group** because **the bond by which it is connected to the parent is on a secondary carbon.** The unbranched alkyl groups we have already met are all primary groups. Except for methyl and ethyl, the unbranched ones are often called *normal* alkyl groups. Here is an example of the use of a branched group in the naming of an alkane:

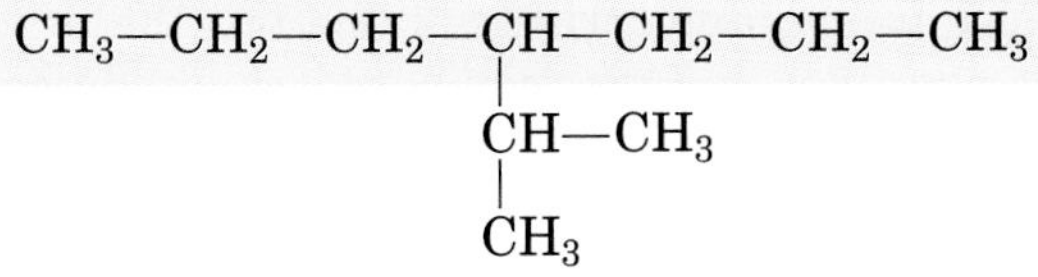

```
CH3—CH2—CH2—CH—CH2—CH2—CH3
              |
              CH—CH3
              |
              CH3
```

4-Isopropylheptane

Table 10.5 The Most Important Branched Alkyl Groups

Name	Structure	Type of Group
Isopropyl	CH_3 | CH_3-CH-	Secondary
Isobutyl	CH_3 | $CH_3-CH-CH_2-$	Primary
sec-Butyl	CH_3 | CH_3-CH_2-CH-	Secondary
tert-Butyl	CH_3 | CH_3-C- | CH_3	Tertiary
Neopentyl	CH_3 | CH_3-C-CH_2- | CH_3	Primary

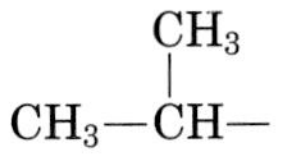

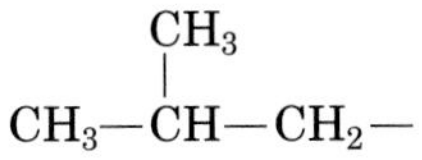

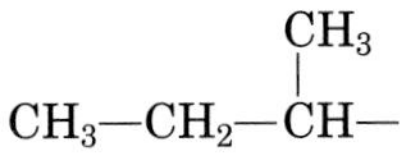

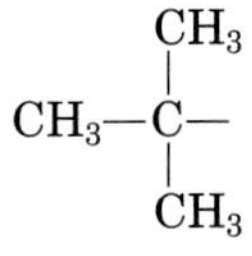

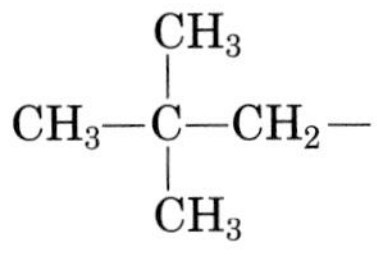

Branched groups are also often used in the naming of alkyl halides:

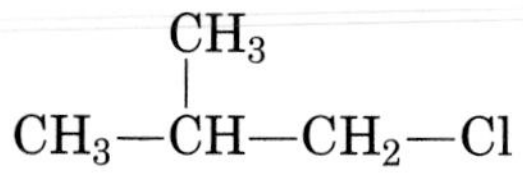

$$\begin{array}{c} CH_3 \\ | \\ CH_3{-}CH{-}CH_2{-}Cl \end{array}$$

IUPAC name: 1-chloro-2-methylpropane
Common name: isobutyl chloride

Problem 10.10

Give the IUPAC and common name for

(a) 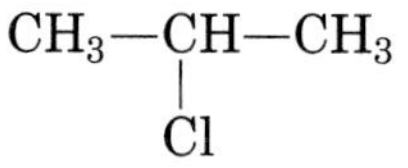 $\begin{array}{c} CH_3{-}CH{-}CH_3 \\ | \\ Cl \end{array}$ (b) 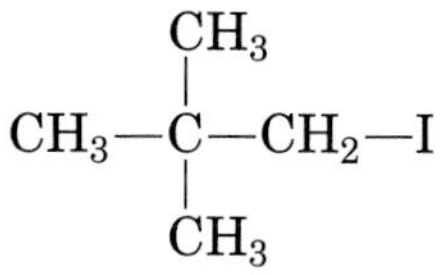 $\begin{array}{c} CH_3 \\ | \\ CH_3{-}C{-}CH_2{-}I \\ | \\ CH_3 \end{array}$

10.8 Cycloalkanes

Carbon atoms can be connected to each other in rings as well as in chains, giving rise to a series of hydrocarbons called **cycloalkanes.** The rings may be of any size from three carbons on up, but the most common and most important are five- and six-membered rings. Some examples are shown in Table 10.6.

For cyclic compounds it is customary to use the very condensed formulas shown on the right in Table 10.6. These formulas show only the

Table 10.6 Three Common Cycloalkanes

Name	Structural Formula	Very Condensed Structural Formula	Boiling Point (°C)	Melting Point (°C)
Cyclopropane	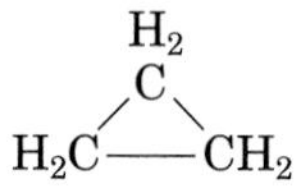 H_2C–CH_2–CH_2 ring (three CH_2 groups)	 △	−33	−128
Cyclopentane	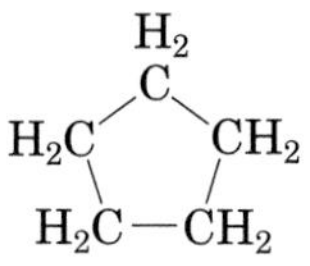 ring of five CH_2 groups	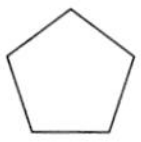 ⬠	49	−94
Cyclohexane	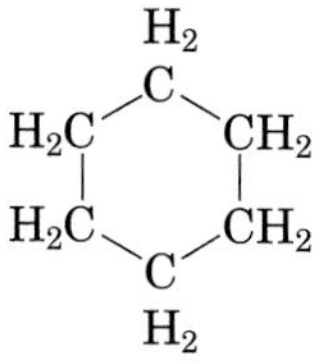 ring of six CH_2 groups	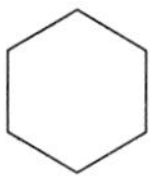 ⬡	81	7

C—C bonds. Although the carbon and hydrogen atoms are omitted from the formulas, we must not forget that they are there.

Cycloalkanes are named by the same IUPAC rules we discussed in Section 10.7, except that we need a new rule for numbering because unlike a chain, a ring has *no* ends. The rule is

> **7.** We may begin at any position and may proceed either clockwise or counterclockwise, but we must end up with the lowest possible numbers.

EXAMPLE

Name these compounds:

(a) $-CH_3$ (b) H_3C CH_3 (c) Br Br Br Br

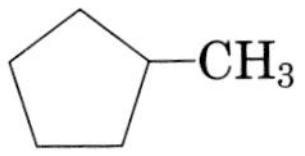

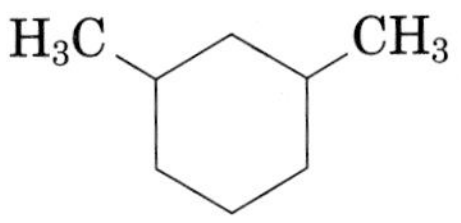

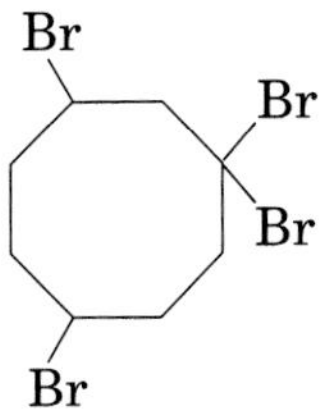

(d) CH_3 Br

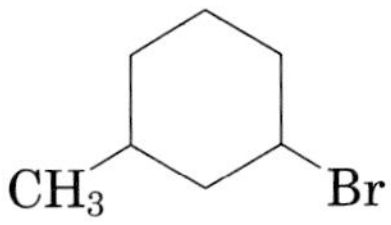

Answer

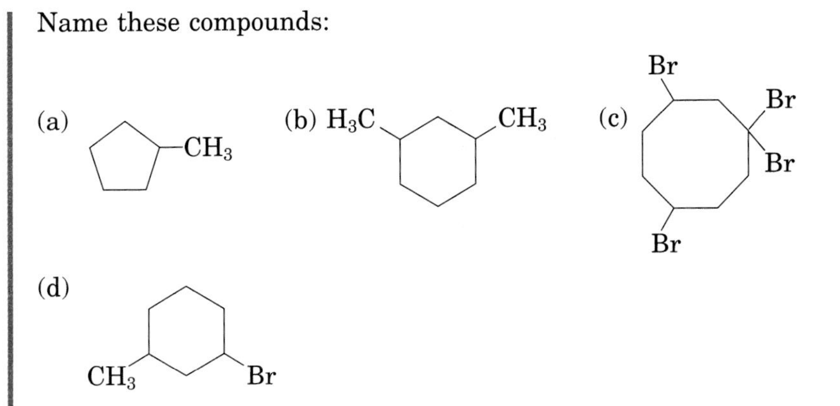

(a) This compound is methylcyclopentane. No number is needed.

(b) To get the lowest numbers, we can begin at either methyl group and count in the direction of the other. Either way, we get 1,3-dimethylcyclohexane:

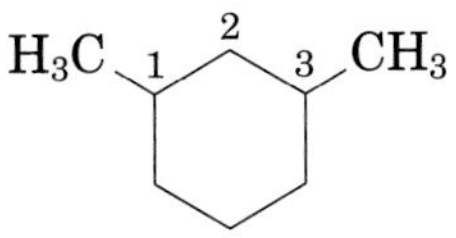

H_3C 1 2 3 CH_3

(c) We get the lowest numbers only by starting at the position with two bromines. The correct name is 1,1,3,6-tetrabromocyclooctane. If we went clockwise instead of counterclockwise, we would get 1,1,4,7. This would be wrong because it is a higher set of numbers. The lowest is always correct.

(d) When we have two or more different groups on a ring, the lower number goes to the group that is cited first. Since bromo comes before methyl alphabetically, it is cited first and gets the lower number, so this is 1-bromo-3-methylcyclohexane. Note that it makes no difference whether the groups are written at the top, bottom, left, or right side of the ring.

Problem **10.11**

Name:

(a) 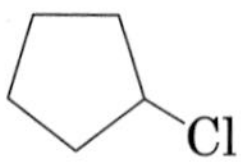 Cl

(b) 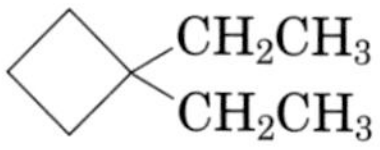 CH_2CH_3 CH_2CH_3

(c) 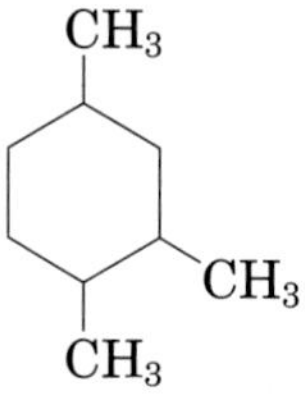 CH_3 CH_3 CH_3

(d) 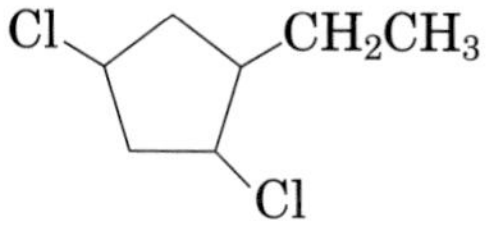 Cl CH_2CH_3 Cl

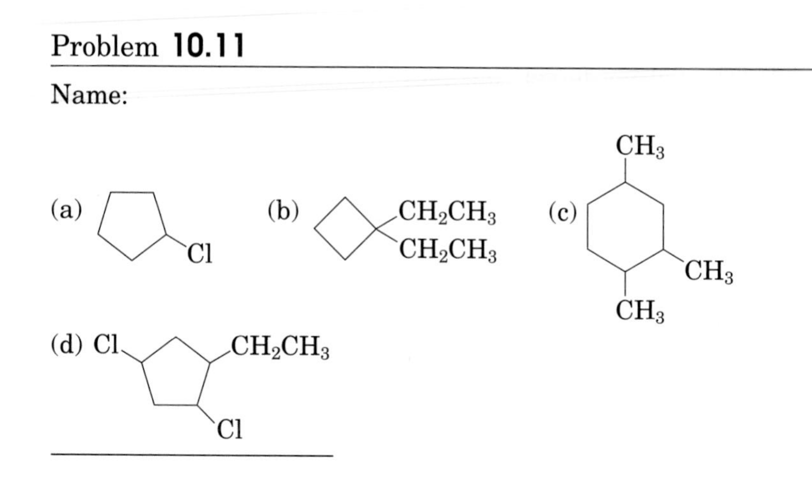

10.9 Stereoisomerism in Cyclic Compounds

There is a major difference between cycloalkane structures and the structures of the branched and unbranched alkanes we considered before (we may call those *acyclic* alkanes). In the acyclic compounds there is free rotation about all the C—C bonds. This free rotation cannot exist in the cyclic compounds because the ring prevents it. Because of this we have a new kind of isomerism here, which can be illustrated by the models shown in Figure 10.8. These two compounds not only have the same molecular formula, C_7H_{14} (and are therefore isomers), but also the same structural formula:

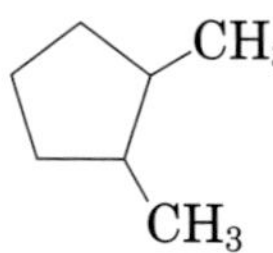

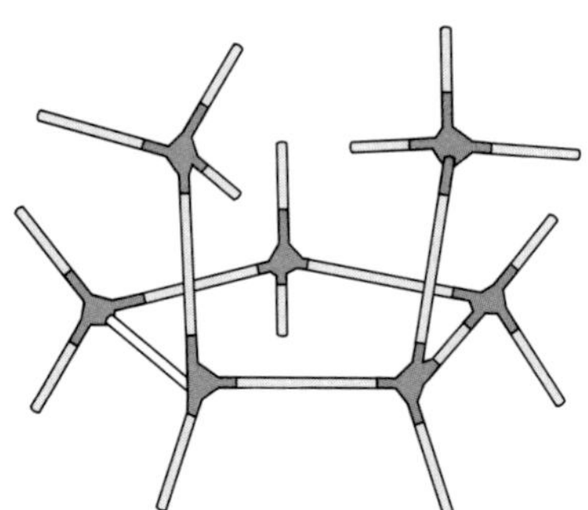

cis-1,2-Dimethylcyclopentane

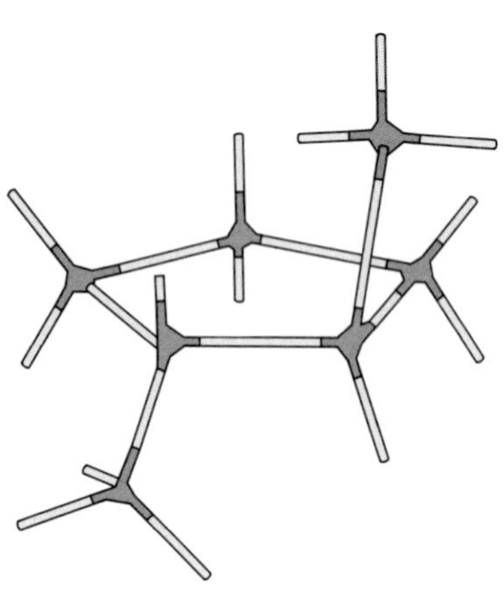

trans-1,2-Dimethylcyclopentane

Figure **10.8**
The two stereoisomers of 1,2-dimethylcyclopentane.

By our naming system, both are called 1,2-dimethylcyclopentane. Yet they are different compounds, with different structures and different properties. How can we tell that two molecules have different structures? There is a very simple test. Mentally try to superimpose the two structures—slide one on top of the other and see if it fits. If it does, the molecules are identical; if not, they are different. When we try this with the models in Figure 10.8, we find that they *don't* fit. One molecule has both methyl groups on the same side of the ring, and the other has one methyl above and the other below. If we could twist the second molecule so as to put both methyl groups on the same side, we might be able to superimpose it on the first molecule, but the lack of rotation about the C—C bonds prevents us from doing this without breaking the ring.

Compounds that have the same structural formulas but different three-dimensional shapes (cannot be superimposed) are called **stereoisomers.** Since they are different compounds we cannot give them exactly the same name. Both compounds in Figure 10.8 are called 1,2-dimethylcyclopentane, but we use the prefixes *cis* and *trans* to

Box 10B

The Actual Shape of the Cyclohexane Ring

In the text, rings are drawn as if they are flat (planar). Most rings, however, are not planar. Since six-membered rings are the most important, let us look at them. The cyclohexane ring has this shape:

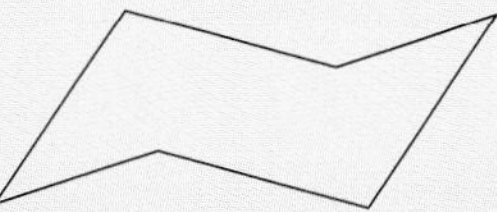

We call it the **chair** conformation. (Can you see a chair in it?) The ring adopts this shape because doing so allows all the carbon angles to be about 110° (the normal carbon angle). If this molecule were a planar hexagon, the angles would have to be 120°.

Let us look at the 12 hydrogens in cyclohexane:

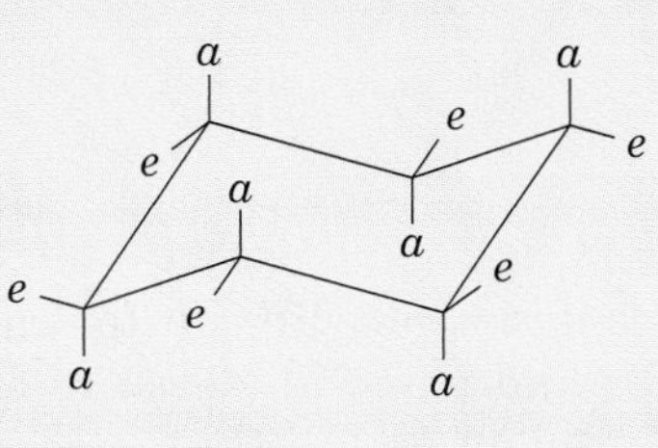

The lines show the C—H bonds.

Six of the hydrogens (marked *a*) are pointing up or down (three up, three down), and the other six (marked *e*) are pointing approximately in the "plane" of the ring. The six up-and-down hydrogens are called **axial** hydrogens; the other six are **equatorial.** You might conclude that there should be two stereoisomers of, say, bromocyclohexane: one with the bromine axial and the other with it equatorial. There is only one, however, because the two molecules rapidly interconvert by going through an intermediate stage called the **boat** conformation:

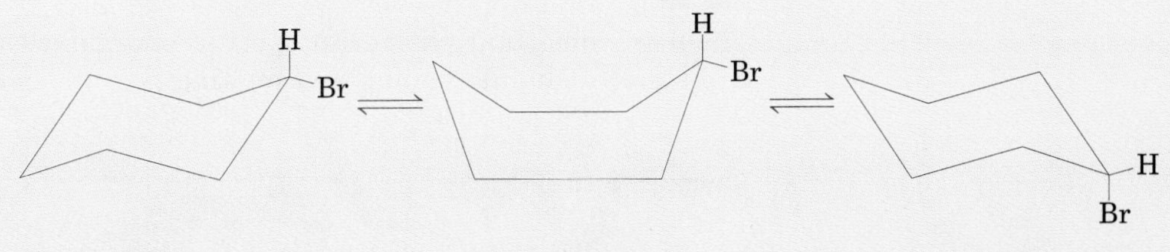

Equatorial bromine (more stable) | Boat conformation | Axial bromine

It is true that the two molecules have different structures (nonsuperimposable), but there is no stereoisomerism here because we cannot separate the two forms. When we try to make only one of them, we get an equilibrium mixture of both.

In general, however, molecules with groups other than hydrogen in equatorial positions are more stable than those with groups in axial positions.

tell them apart. **Cis** means **the two groups are on the same side; trans** means **they are on opposite sides:**

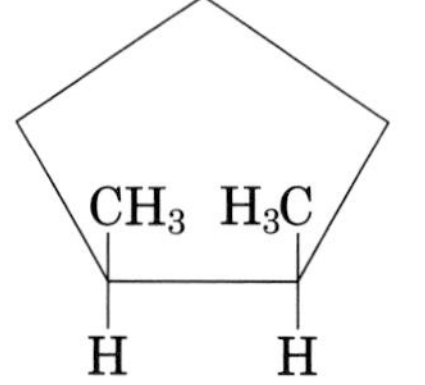

cis-1,2-Dimethylcyclopentane

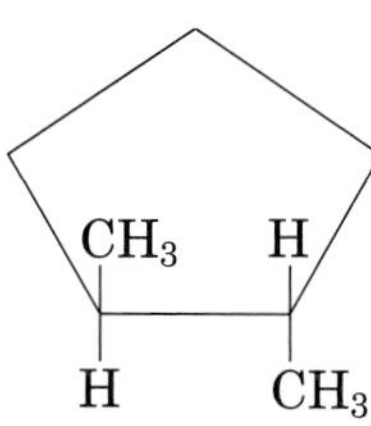

trans-1,2-Dimethylcyclopentane

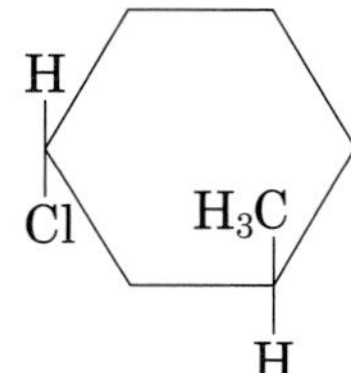

trans-1-Chloro-3-methylcyclohexane

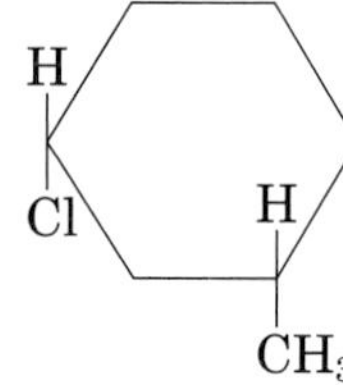

cis-1-Chloro-3-methylcyclohexane

Note that it is not difficult to draw these isomers.

Cis and trans isomers are found in rings of all sizes, from three-membered on up. All that is necessary is that at least two groups, the same or different, be substituted for H atoms on different carbons of a cycloalkane ring. In general, the properties of cis and trans isomers, both physical and chemical, are different. They may be similar, but they are not identical.

Cis-trans isomers are sometimes called **geometrical isomers.**

10.10 Physical Properties

The most important physical properties of organic compounds are physical state (including boiling and melting points), solubility, and density.

Physical State

The properties of cycloalkanes are very similar to those of the acyclic alkanes.

In general, both the melting and the boiling points of hydrocarbons increase with increasing molecular weight (see Tables 10.4 and 10.6). The low-molecular-weight alkanes, including cycloalkanes, up to about 5 carbons, are gases at room temperature. From about 5 to 17 carbons, they are liquids. Larger alkanes are white, waxy solids. Solid paraffin is a mixture of high-molecular-weight alkanes.

In any state, alkanes are colorless, odorless, and tasteless.

Solubility

Even though carbon has a slightly higher electronegativity than hydrogen (Sec. 3.7), alkanes are nonpolar compounds. Therefore *they are not*

Because they are so unreactive, alkanes in general are not particularly poisonous. Gasoline, which is a mixture of alkanes, should not be swallowed because it generally contains additives, some of which can be quite harmful. Liquid alkanes can cause damage if they get into the lungs, though by a physical rather than a chemical process. They dissolve the lipid molecules in the cell membranes and cause pneumonia-like symptoms. For this reason, anyone who swallows liquid alkanes should *not* be induced to vomit them up because some might be forced into the lungs in that manner.

Liquid alkanes can also harm the skin by a similar physical process. Human skin is kept moist by natural body oils. Liquid alkanes dissolve these oils and cause the skin to dry out. However, mixtures of high-molecular-weight liquid alkanes, sold in drugstores under the name *mineral oil,* soften and moisten skin. *Petroleum jelly* ("Vaseline" is one brand) is a mixture of solid and liquid hydrocarbons also used to protect skin. Because alkanes are not soluble in water, a coating of petroleum jelly protects skin from too much contact with water. For example, it can protect babies from diaper rash caused by the skin's contact with urine.

soluble in water, which dissolves only ions and polar compounds. Alkanes are soluble in each other, however—an example of "like dissolves like"—as well as in other nonpolar organic compounds, such as benzene, diethyl ether, and carbon tetrachloride.

Density

Each liquid alkane has a slightly different density, but all are less dense than water. Since none of them are soluble in water, this means that alkanes always float on water.

10.11 Chemical Properties

Combustion

The most important chemical property of alkanes is that they burn. Much of our current civilization depends on this property, called *combustion.* Petroleum (Box 10D) is a mixture of organic compounds consisting mostly of alkanes. From petroleum we get gasoline, diesel fuel, and fuel oil, which we use to power our cars, trucks, airplanes, and ships and to heat our houses. Natural gas, another important fuel, is chiefly methane, with a small amount of other low-molecular-weight alkanes. Other sources of energy, such as coal, hydroelectric, and nuclear power, provide us with less than half our energy needs.

Box 10D

Petroleum

Petroleum, a thick black liquid found underground, is a mixture of hundreds, probably thousands, of different alkanes, ranging from methane up to alkanes with 40 or more carbons. It also contains varying amounts of sulfur-, oxygen-, and nitrogen-containing compounds. Because it is such a complex mixture, petroleum must be refined before use. The most important refinery process is distillation, in which the crude petroleum is separated into fractions, each with a different boiling range (hence a different average molecular weight). The major fractions, each a mixture of many compounds, are

Gasoline (about C_4 to C_{12}, boiling range 20 to 200°C)
Kerosene (about C_{10} to C_{14}, boiling range 200 to 275°C)
Fuel oil and diesel fuel (about C_{14} to C_{18}, boiling range 275 to 350°C)
Lubricating oil (about C_{16} to C_{20}, boiling higher than 350°C)
Residue, used for greases and asphalt (more than C_{20})

The most important fraction is gasoline, and it would be convenient if a very large proportion of the alkanes in petroleum had molecular weights that put them in this fraction. However, nature has not made petroleum like that; more than half of it consists of C_{13} and higher alkanes.

Another disadvantage of petroleum is that a high percentage of the molecules are normal (unbranched) alkanes. This is a disadvantage because unbranched alkanes give rise to much more "knocking" when used in automobile engines. Knocking takes place when the gasoline-air mixture in the cylinder explodes before the spark plug ignites it. The tendency for a fuel to cause knocking can be measured; the *octane number* is used for this. The higher the octane number of a gasoline, the less knocking it will cause. Branched alkanes have higher octane numbers than unbranched alkanes. Another way to increase octane number is to add the compound lead tetraethyl, but this practice has been almost entirely phased out in the United States because, when burned, the lead fouls up the catalytic converters designed to reduce air pollution. (Also, lead itself is a poisonous pollutant.)

Though most petroleum is eventually used for fuel, about 2 percent is used to synthesize organic compounds, as mentioned in Section 10.2. This sounds like a small percentage, but so much petroleum is refined that even 2 percent is a very large volume indeed. More than 50 percent of all industrial synthetic organic compounds are made from this source. These chemicals are eventually turned into dyes, explosives, drugs, plastics, artificial fibers, detergents, insecticides, and other materials.

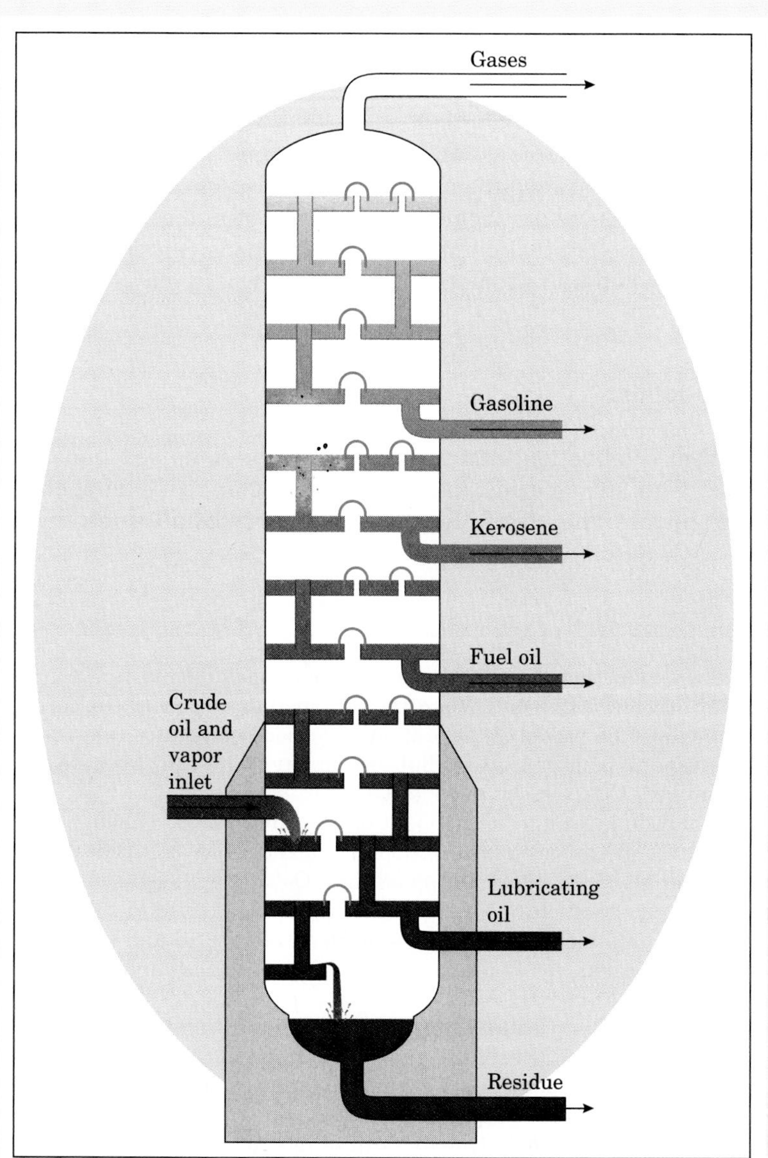

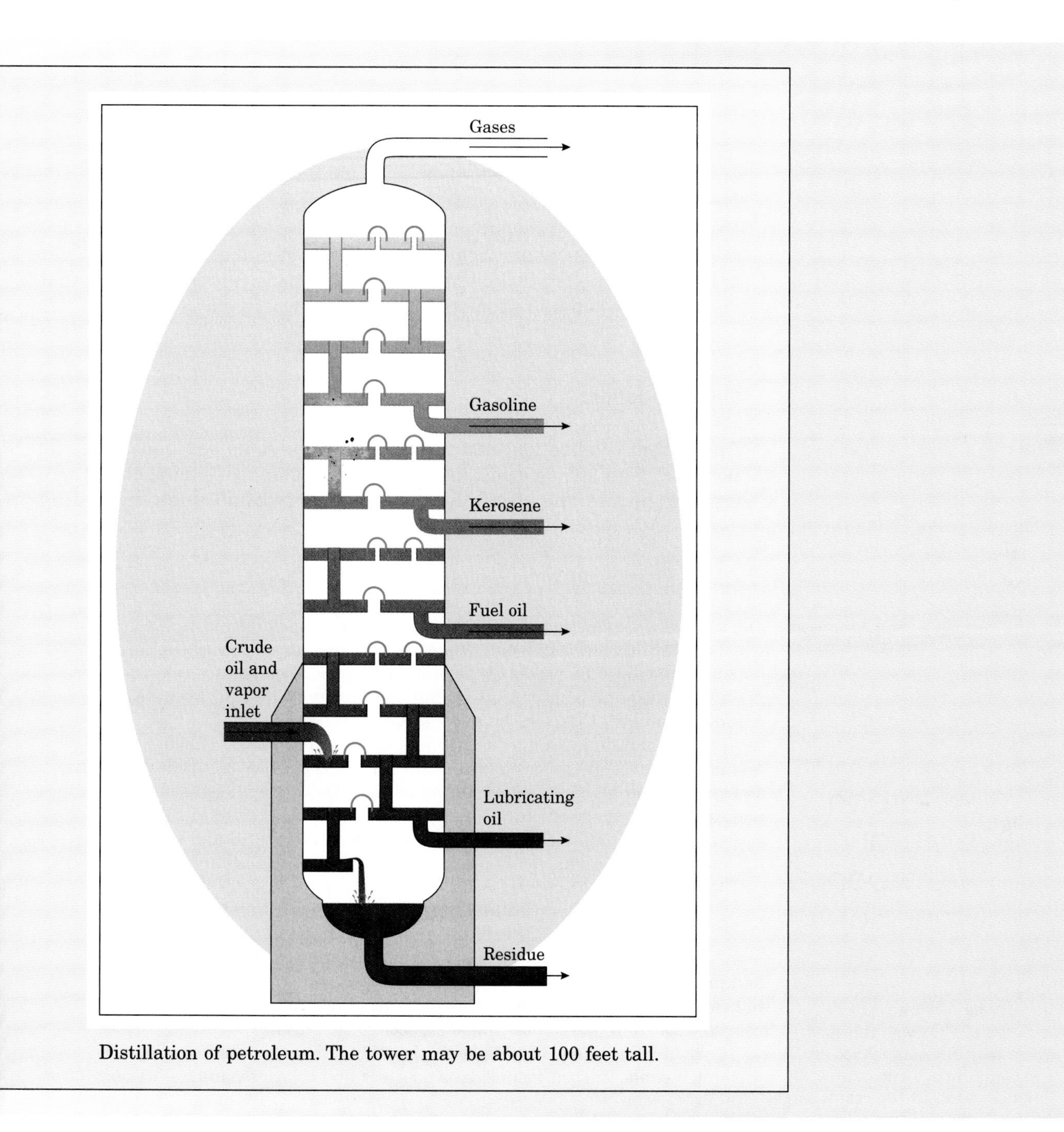

Distillation of petroleum. The tower may be about 100 feet tall.

The complete combustion of any alkane gives only two products: carbon dioxide and water. For example, this is the equation for the combustion of propane (bottled gas):

Combustion reactions are always exothermic.

$$C_3H_8 + 5O_2 \longrightarrow 3CO_2 + 4H_2O + \text{heat}$$

However, complete combustion takes place only under ideal conditions. Most combustion of gasoline and other fuels is not complete, and a variety of other products are formed, including soot and carbon monoxide. It is because of the production of the very poisonous carbon monoxide that one must not run an automobile engine in a closed garage.

Other Reactions

Apart from combustion, alkanes undergo almost no reactions. They do not react with strong acids, bases, or oxidizing or reducing agents. To an organic chemist, **the chief chemical property of alkanes (including**

Box 10E

Carbon Dioxide and Carbon Monoxide

In August 1986 a huge volume of CO_2 was released from beneath Lake Nyos (in Cameroon, Africa), a lake situated in the crater of a volcano, killing more than 1700 local inhabitants. In less dramatic fashion, seeping CO_2 gas from an underground volcano near Mammoth Lake in California destroyed hundreds of acres of woodland in 1996.

The complete combustion of fuels produces carbon dioxide. This occurs when enough oxygen is present. When the combustion is incomplete, carbon monoxide may also be produced. Oxidation of carbon compounds takes place not only in furnaces and in our cars but also in our bodies. In this case the product is entirely CO_2, which we exhale.

Carbon dioxide is a minor component of our atmosphere, although it is produced in large quantities by the action of volcanos (Fig. 10E.1) as well as by the burning of fossil fuels. However, the CO_2 in the air is used in photosynthesis of carbohydrates by plants. It is also converted to solid carbonates, first by dissolving in water to form carbonic acid and later by interacting with cations, for example, with Ca^{2+} to form insoluble calcium carbonate (Fig. 10E.2).

Carbon monoxide is produced in incomplete combustion. For example, 4 to 7 percent of automobile exhaust gas and of chimney gases is CO. It is highly poisonous because it strongly combines with hemoglobin (Box 3C) and prevents the transport of oxygen to the tissues. Carbon monoxide binds to hemoglobin 200 times more strongly than oxygen does and is therefore very difficult to remove once it is absorbed in the blood.

Carbon monoxide poisoning is particularly lethal because it has no warning signals. Carbon monoxide is odorless and colorless, and its action is rapid.

Carbon monoxide itself is also produced in the body by the breakdown of heme, but this amount of CO blocks only about 1 percent of the hemoglobin activity and produces no symptoms. When 10 to 30 percent of the hemoglobin molecules are combined with carbon monoxide, headache and dizziness develop, combined with dimmed vision and nausea. A greater amount of blocked hemoglobin results in unconsciousness, depressed heart action, coma, and death. Treatment involves administration of a mixture of oxygen and carbon dioxide (see also Box 5E) and blood transfusion if necessary.

cycloalkanes) is that they are chemically inert. There are a few exceptions (for example, alkanes react with Cl_2 at high temperatures or under the influence of ultraviolet light), but these are unimportant for our purposes. Actually, the inertness of alkanes is quite useful, as we shall see in the next section.

10.12 Functional Groups

Let us look at the structural formula of the compound 1-butanol:

$$CH_3—CH_2—CH_2—CH_2—OH$$

The only difference between this molecule and the alkane butane is the —OH group (called a **hydroxy group**). The molecule may be thought of as consisting of two parts: an alkane-like portion and the —OH group. We use the term **functional group** to refer to an atom or group of atoms that substitutes for a hydrogen atom. In 1-butanol the —OH is

Figure **10E.1** • Erupting volcanoes, like Mount St. Helens shown here, produce CO_2 among other gases. (David Weintraub/Photo Researchers, Inc.)

Figure **10E.2** • These stalactites, in Mammoth Cave, Kentucky, are made of $CaCO_3$ that has precipitated out of water solution. (Photograph by Beverly March.)

a functional group. Functional groups can consist of any combination of N, O, S, H, halogen, and even C atoms, but some of them are more important than others. The most important functional groups are shown in Table 10.7.

Just as there are many alkanes, so are there a large number of compounds containing any given functional group, because in general each functional group can replace hydrogen at any position of any alkane. For example, see how many compounds can be made from alkanes of up to four carbons if we substitute —OH for —H (note that we do this *mentally*):

Since alkanes are inert, we cannot make —OH compounds in this way in practice.

CH_4 CH_3OH

Methane

Table 10.7 The Most Important Functional Groups

Group	Type of Compound	Typical Example	
$>C{=}C<$	Alkene	$CH_2{=}CH_2$	Ethylene
$-C{\equiv}C-$	Alkyne	$HC{\equiv}CH$	Acetylene
R—F (or Cl, Br, I)	Alkyl halide	$CH_3{-}Cl$	Chloromethane
R—OH	Alcohol	$CH_3{-}OH$	Methanol
R—O—R′	Ether	$CH_3CH_2{-}O{-}CH_2CH_3$	Diethyl ether
R—SH	Thiol (mercaptan)	$CH_3CH_2{-}SH$	Ethanethiol
$R{-}NH_2$	Amines	$CH_3{-}NH_2$	Methylamine
R_2NH	Amines	$CH_3{-}NH{-}CH_3$	Dimethylamine
R_3N	Amines	$CH_3{-}N(CH_3){-}CH_3$	Trimethylamine
$R{-}C({=}O){-}H$	Aldehyde	$CH_3{-}C({=}O){-}H$	Acetaldehyde
$R{-}C({=}O){-}R'$	Ketone	$CH_3{-}C({=}O){-}CH_3$	Acetone
$R{-}C({=}O){-}OH$	Carboxylic acid	$H{-}C({=}O){-}OH$	Formic acid
$R{-}C({=}O){-}OR'$	Carboxylic ester	$CH_3{-}C({=}O){-}O{-}CH_2CH_3$	Ethyl acetate
$R{-}C({=}O){-}NH_2$	Amides	$CH_3{-}C({=}O){-}NH_2$	Acetamide
$R{-}C({=}O){-}NHR'$	Amides	$CH_3{-}C({=}O){-}NH{-}CH_3$	*N*-Methylacetamide
$R{-}C({=}O){-}NR'R''$	Amides	$CH_3{-}C({=}O){-}N(CH_3){-}CH_3$	*N*, *N*-Dimethylacetamide

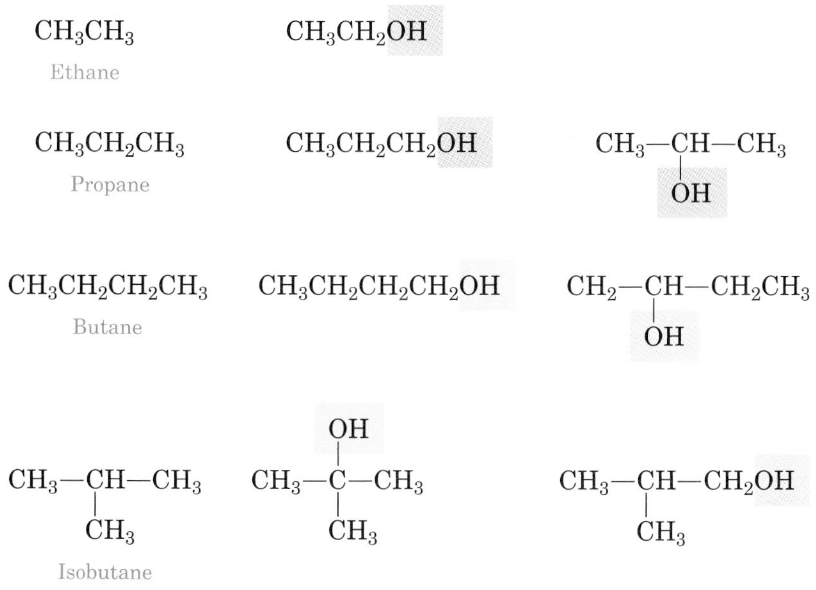

$$CH_3CH_3 \qquad CH_3CH_2OH$$

Ethane

$$CH_3CH_2CH_3 \qquad CH_3CH_2CH_2OH \qquad CH_3{-}\underset{\displaystyle OH}{\underset{|}{CH}}{-}CH_3$$

Propane

$$CH_3CH_2CH_2CH_3 \qquad CH_3CH_2CH_2CH_2OH \qquad CH_2{-}\underset{\displaystyle OH}{\underset{|}{CH}}{-}CH_2CH_3$$

Butane

$$CH_3{-}\underset{\displaystyle CH_3}{\underset{|}{CH}}{-}CH_3 \qquad CH_3{-}\overset{\displaystyle OH}{\overset{|}{\underset{\displaystyle CH_3}{\underset{|}{C}}}}{-}CH_3 \qquad CH_3{-}\underset{\displaystyle CH_3}{\underset{|}{CH}}{-}CH_2OH$$

Isobutane

The five alkanes give rise to eight hydroxy-containing compounds. All of these compounds are called alcohols, an **alcohol being a carbon compound containing an —OH group.** The general formula for an alcohol is

$$R{-}OH$$

where the **R** is a symbol that stands for **any alkyl group.** The alcohols shown above contain eight of these R groups, and we have already learned their names: methyl, ethyl, *n*-propyl, isopropyl, and so forth. These are all the possible noncyclic alkyl groups with up to four carbons, but there are many others with five or more carbons. Note that R groups may be branched or unbranched.

The family of alcohols, symbolized R—OH, have many properties in common. We can see why by looking at their structures. All of them have an alkyl group connected to an —OH, for example:

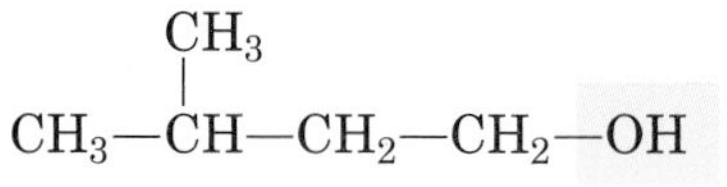

$$CH_3{-}\overset{\displaystyle CH_3}{\overset{|}{CH}}{-}CH_2{-}CH_2{-}OH$$

We saw in the previous section that alkanes are inert. *Therefore, so are the alkane-like portions of other molecules (alkyl groups).* Any compound that reacts with an alcohol reacts not with the alkyl part but only with the —OH. Since the —OH group is the same in all alcohols, it follows that all alcohols have many properties in common. This does not mean that the properties of alcohols are *identical.* Each alcohol has its own set of specific properties, and anyone who works with any particular alcohol must be aware of the properties of that alcohol. For example, methanol, CH_3OH, is poisonous, but ethanol, CH_3CH_2OH, is present in wine, beer, and whisky. Still, the properties of an entire family are similar enough to make it convenient to study organic compounds by family. We will do just that in the next five chapters.

Methanol is also called methyl alcohol, and ethanol is also called ethyl alcohol.

Box 10F

The Greenhouse Effect

The world gets most of its energy by burning fossil fuels—not only petroleum but also coal and natural gas. If the combustion is complete, all the carbon in these materials is converted to carbon dioxide. The carbon in the fossil fuels has been stored in the earth for hundreds of millions of years. In the last century or two, large amounts of it have been converted to carbon dioxide in this way. Not all the carbon dioxide remains in the atmosphere (some of it dissolves in the oceans and lakes, and some is converted to rocks in the form of calcium and magnesium carbonates), but measurements show that the CO_2 content of the atmosphere is slowly increasing every year and has been doing so for decades.

This increase in the CO_2 content of the atmosphere poses a potential problem for the following reason. Carbon dioxide has the property of allowing visible light to pass through it but absorbing infrared light (see Sec. 9.1). In order for the earth to maintain a more or less constant average temperature, it must give off about as much energy as it receives. The energy it receives, from the sun, is mostly in the form of visible light. Since the CO_2 in the atmosphere allows this to pass through, the energy goes right through to the earth's surface. But the energy reemitted by the earth is mostly in the form of infrared light, not visible light, so it is absorbed by the atmospheric CO_2. Once a CO_2 molecule absorbs this energy, it does not keep it, but reemits it in all directions, sending some of it back down to the surface of the earth. The net consequence is that the atmospheric CO_2 does nothing to stop the sun's energy from reaching the surface of the earth, but it does stop some energy from going back into space. This is called the **greenhouse effect** (Fig. 10F).

We might therefore expect a gradual increase in the average temperature of the earth's surface, as the amount of CO_2 in the atmosphere increases every year. It would not take a very large increase in the average temperature to cause substantial effects on the earth's climate. A rise of about 4°C would probably cause enough Antarctic ice to melt to flood most of the world's coastal cities. But has the average temperature really been rising? Measurements show that earth's average temperature did rise, by a total of about 0.6°C, from 1880 to 1940, but then fell, by about 0.3°C, from 1940 to about 1975, even though the concentration of atmospheric CO_2 was still increasing all that time. Since about 1975 the average temperature has been slowly rising again. Consequently, we cannot be certain at this time that the greenhouse effect will really have major effects on the climate. There are several other factors involved, and much research must still be carried out.

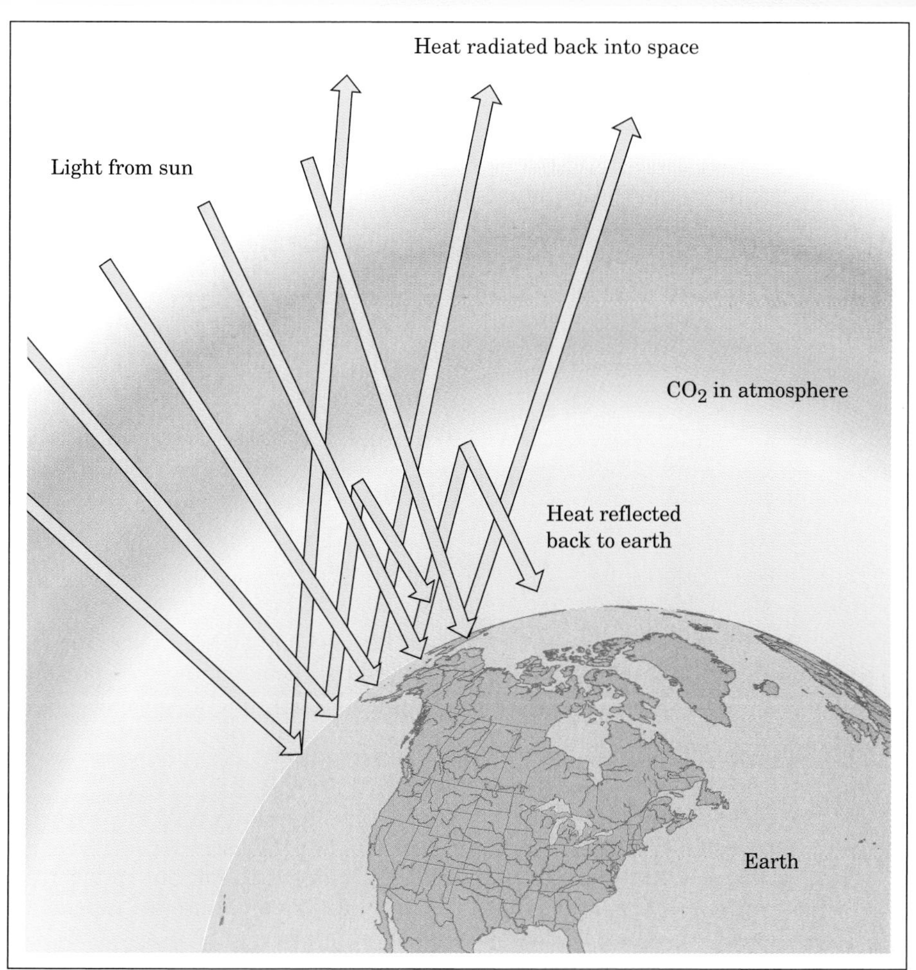

Figure **10F** • The greenhouse effect.

Box 10G

Taxol, a New Anticancer Drug

Many naturally-occurring organic compounds have a multiplicity of functional groups. An example is paclitaxel (trade name **Taxol**), which can be extracted from the bark of the Pacific Yew tree found in the old-growth forests of the Pacific Northwest. As can be seen in the formula, this one compound belongs to six of the families in Table 10.7: alkene, alcohol, ether, ketone, carboxylic ester, and amide.

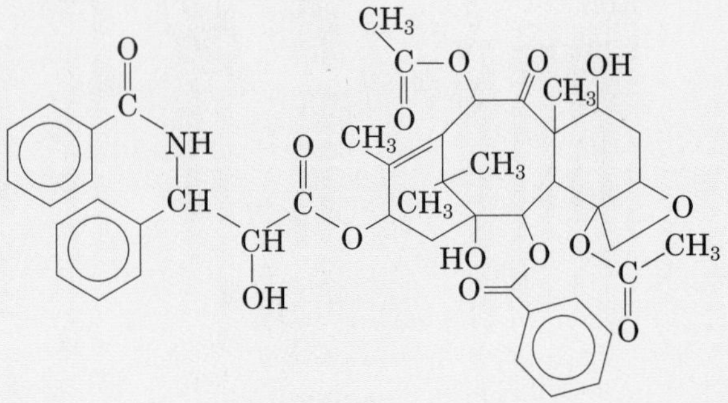

Taxol

The reason Taxol is important is that it shows promise in the treatment of ovarian and breast cancer even in cases where chemotherapy fails. Taxol acts by preventing the division of cancer cells. It does this by blocking the disassembly of microtubules within the cells. Before dividing, the cell must disassemble these units, and Taxol prevents this.

Originally, the entire supply of the drug came from Pacific Yew bark, but to get 10 grams of the drug it was necessary to cut down 25 trees. Obviously, it would be better to obtain it some other way. What about synthesis? Despite its complicated structure, early in 1994 chemists working in laboratories did succeed in synthesizing Taxol, but their methods are too expensive to supply the amount needed.

Fortunately, a new way of obtaining Taxol has been found. The needles of a related plant, *Taxus baccata,* contains a compound closely related to taxol, which can be chemically converted to Taxol rather easily. Because only the needles are used, and these are renewed each year, it is not necessary to cut the trees down.

SUMMARY

Organic chemistry is the study of carbon compounds. The properties of organic compounds are generally very different from those of inorganic compounds. Many organic compounds are obtained from natural sources, but many others, which do not exist in nature, are synthesized in laboratories. Organic chemists use structural formulas and models to show three-dimensional shapes of molecules. The number of bonds formed by C, N, O, H, and the halogens are four, three, two, one, and one, respectively.

In order to explain the differing reactivities of double and single covalent bonds, chemists developed the concept of **hybrid orbitals.** A carbon atom that bonds to four other atoms uses orbitals that are neither *s* nor *p* but a mixture of these called sp^3 hybrid orbitals. An $\boldsymbol{sp^3}$ **orbital** is neither spherical nor dumbbell-shaped but a mixture of the two. In forming a single covalent bond, an sp^3 orbital of the carbon atom overlaps with an *s* or *p* orbital of any atom or with an sp^3 orbital of another carbon atom.

Hydrocarbons contain only carbon and hydrogen. **Alkanes,** which are hydrocarbons containing only single bonds, are named by the **IUPAC system,** in which (1) the longest continuous chain is the parent, (2) other groups are named as prefixes, and (3) chains are numbered to produce the lowest numbers for the groups. There is free rotation about the C—C bonds of alkanes. **Alkyl halides** are named the same way as alkanes. **Cycloalkanes** have the carbon atoms connected in a ring. Cyclic compounds with two substituents on different carbons can exist as **cis and trans isomers. Stereoisomers** are compounds with the same structural formula but different three-dimensional shapes.

The melting and boiling points of alkanes increase with increasing molecular weight. Alkanes are insoluble in water and soluble in each other. They are less dense than water. Though alkanes burn, they are otherwise chemically inert and undergo few reactions.

Most organic molecules that are not alkanes contain **functional groups.** These groups, of which about a dozen are very important, are used to divide organic compounds into families. The letter **R** stands for any alkyl group.

KEY TERMS

Alkane (Sec. 10.6)
Alkyl group (Sec. 10.7)
Alkyl halide (Sec. 10.7)
Axial hydrogen (Box 10B)
Boat conformation (Box 10B)
Branched alkane (Sec. 10.6)
Chair conformation (Box 10B)
Cis isomer (Sec. 10.9)
Combinatorial synthesis (Box 10A)
Condensed structural formula (Sec. 10.6)
Conformation (Sec. 10.6)
Constitutional isomerism (Sec. 10.3)
Cycloalkane (Sec. 10.8)
Equatorial hydrogen (Box 10B)
Functional group (Sec. 10.12)
Geometrical isomers (Sec. 10.9)
Greenhouse effect (Box 10F)
Group (Sec. 10.7)
Hybrid orbital (Sec. 10.4)
Hydrocarbon (Sec. 10.5)
Isomers (Sec. 10.3)
IUPAC system (Sec. 10.7)
Main chain (Sec. 10.6)
Normal alkane (Sec. 10.6)
Octane number (Box 10D)
Organic chemistry (Sec. 10.1)
Parent name (Sec. 10.7)
Primary carbon (Sec. 10.6)
Quaternary carbon (Sec. 10.6)
Saturated hydrocarbon (Sec. 10.6)
Secondary carbon (Sec. 10.6)
Side chain (Sec. 10.6)
sp^3 orbital (Sec. 10.4)
Stereoisomerism (Sec. 10.9)
Structural isomerism (Sec. 10.3)
Tertiary carbon (Sec. 10.6)
Trans isomer (Sec. 10.9)

PROBLEMS

Organic and Inorganic Compounds

10.12 Which are organic compounds?
(a) CH_4 (d) C_2Cl_6
(b) NaCl (e) $KMnO_4$
(c) $C_6H_{12}O_6$ (f) C_7H_{10}

10.13 Do liquid organic compounds conduct electricity?

10.14 In each case pick out the compound with the higher melting point:
(a) CH_3Cl and KCl
(b) NaOH and CH_3CH_2OH
(c) $CH_3CH_2CH_3$ and $CH_3(CH_2)_5CH_3$

10.15 An unknown solid has a melting point of 91°C, is insoluble in water, and burns. Is it likely to be an inorganic or an organic compound?

Sources of Organic Compounds

10.16 (a) What is meant by the expression "known compound"? (b) What are the two ways in which compounds become known?

10.17 Is there any difference between vanillin made synthetically and vanillin extracted from vanilla beans, assuming that both are chemically pure?

10.18 What important experiment was carried out by Wöhler in 1828?

Isomers

10.19 For each of the following pairs, tell whether the structures shown are (1) different formulas for the same compound, (2) isomers, or (3) different compounds that are not isomers:

```
(a) CH3—CH—CH2—CH—CH3   and
        |         |
        CH3       CH3

                               CH3
                               |
                               CH—CH2
                               |   |
                               CH3 CH—CH3
                                   |
                                   CH3

(b) CH2—CH2     and     CH3—CH—CH3
    |    |                  |
    CH2—CH2                 CH3
```

(c)
```
CH3—CH—CH3
    |
    CH2  CH2—CH3      and
    |    |
    O————CH2
```
```
CH3—CH2—CH2—O—CH2—CH—CH3
                  |
                  CH3
```

(d)
```
             CH3
             |
CH3—CH2—CH—CH2—CH3     and
```
```
           CH3
           |
CH3—CH—CH—CH3
    |
    CH3
```

(e)
```
       CH3
       |
CH2=C—CH=CH2     and
```
```
          CH3
          |
HC≡C—CH—CH3
```

(f)
```
         NH2                      CH3
        /                        /
  H2C—CH          and      H2C—CH
  /     \                  /     \
H2C      CH2             H2C      CH2
  \     /                  \     /
   CH2                       NH
```

(g)
```
                      CH3
                      |
CH3—CH2—CH2—NH     and
```
```
CH3—CH—CH2—NH2
    |
    CH3
```

(h)
```
  H2C—CH2
  /     \          and     CH3—C—CH2—CH3
H2C      CH2                   ||
  \     /                      O
     O
```

(i)
```
  H2C—CH2
  /     \          and
H2C      CH2
  \     /
     O
```
```
CH3—CH2—O—CH2—CH3
```

10.20 Draw structural formulas for
(a) two different compounds of molecular formula $C_2H_4Cl_2$
(b) three different compounds of molecular formula C_5H_{12}

10.21 Draw structural formulas for all compounds whose molecular formula is C_3H_8O.

Bonding in Organic Compounds

10.22 List the principal elements that make up organic compounds and the number of bonds they form.

10.23 Complete these structural formulas by supplying the missing hydrogens:

(a)
```
C—C=C—C—C
      |
      C
```

(b)
```
                 C
                 |
N—C—C—O—C—C—Br
```

(c)
```
Cl—C=C—C≡C—N—C
            |
            C
```

10.24 Some of the following formulas are incorrect (that is, they do not represent any real compound) because they have atoms with an incorrect number of bonds. Which are they, and which atoms have an incorrect number of bonds?

(a)
```
   H  H
   |  |
H—C—N—H
   |  |
   H  H
```

(b)
```
   H  H
   |  |
H—C=C—H
   |  |
   H  Cl
```

(c)
```
       H  H
       |  |
H—N—C—C—O—H
   |  |  |
   H  H  H
```

(d)
```
   H
   |
H—C—O
   |
   H
```

(e)
```
   H     H
   |     |
H—C≡C—C—H
         |
         H
```

(f)
```
          H
          |
H—C=C—C=N—Br
      |  |
      Br H
```

(g)
```
H—C=C=C=N—O—H
   |
   H
```

(h)
```
      H     H
      |     |
F—C—C—N—C=O
   || |
   O  H
```

Hybrid Orbitals

10.25 What is the shape of (a) an *s* orbital (b) a *p* orbital (c) an sp^3 orbital?

10.26 The promotion of an electron from a 2*s* to a $2p_z$ orbital
(a) requires energy input
(b) goes without energy gain or loss
(c) produces energy

10.27 Show the overlap of orbitals in (a) CH_2Cl_2 (b) CH_3CH_2Br.

Structure of Alkanes

10.28 The condensed formula for n-butane is $CH_3CH_2CH_2CH_3$. Explain why this formula does not show the geometry of the real molecule.

10.29 Draw condensed or very condensed structural formulas for

(a)
$$\begin{array}{ccccccc} & H & & H & & H & H \\ & | & & | & & | & | \\ H- & C & - & C & - & C- & C-H \\ & | & & | & & | & | \\ & H & & H-C-H & & H & H \\ & & & | & & & \\ & & & H & & & \end{array}$$

(b)
$$\begin{array}{ccccc} & & & H & \\ & & & | & \\ & H & & H-C-H & & H \\ & | & & | & & | \\ H- & C & - & C & - & C-H \\ & | & & | & & | \\ & H & & H-C-H & & H \\ & & & | & & \\ & & & H & & \end{array}$$

(c)
$$\begin{array}{ccccccc} & & & H & & H & \\ & H & H & \diagdown & C & \diagup & H \\ & | & \diagdown & \diagup & & \diagdown & \diagup \\ H- & C & - & C & & & C-H \\ & | & & | & & & | \\ & H & & H-C & - & C-H & \\ & & & | & & | & \\ & & & H & & H & \end{array}$$

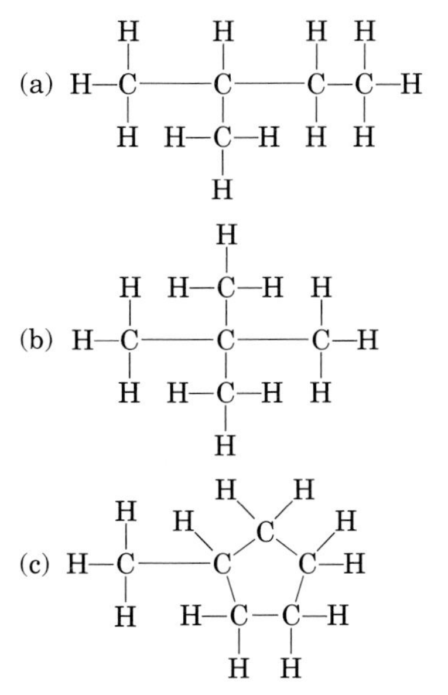

10.30 Pick out all the tertiary carbons and all the quaternary carbons:

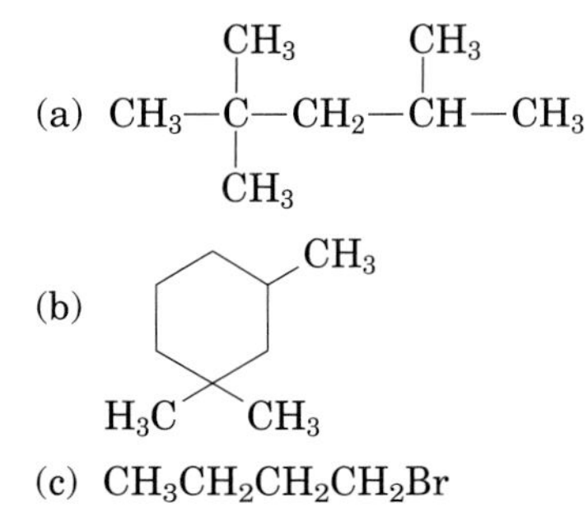

(a)
$$\begin{array}{ccccc} & CH_3 & & CH_3 & \\ & | & & | & \\ CH_3- & C & -CH_2- & CH & -CH_3 \\ & | & & & \\ & CH_3 & & & \end{array}$$

(b) cyclohexane ring bearing CH_3 on one carbon and H_3C, CH_3 on another carbon

(c) $CH_3CH_2CH_2CH_2Br$

Nomenclature of Alkanes and Alkyl Halides

10.31 (a) Draw structural formulas for all compounds whose molecular formula is C_6H_{14}. (b) Give the IUPAC name of each.

10.32 Draw and name all compounds whose molecular formula is (a) C_4H_9Br (b) $C_5H_{11}F$.

10.33 Name each of the following by the IUPAC system:

(a) $CH_3CH_2CH_2CH_2CH_2CH_2CH_3$

(b)
$$\begin{array}{cc} CH_3-CH_2-CH_2- & CH-CH_3 \\ & | \\ & CH_3 \end{array}$$

(c)
$$\begin{array}{ccccc} CH_3-CH_2- & CH & -CH_2- & CH & -CH_3 \\ & | & & | & \\ & CH_3 & & CH_3 & \end{array}$$

(d)
$$\begin{array}{ccccc} & & & CH_3 & \\ & & & | & \\ CH_3- & CH & -CH_2-CH_2- & CH & -CH_3 \\ & | & & & \\ & CH_2 & & & \\ & | & & & \\ & CH_3 & & & \end{array}$$

(e)
$$\begin{array}{cccc} & CH_3 & CH_3 & \\ & | & | & \\ CH_3- & CH & -C & -CH_3 \\ & & | & \\ & & CH_2 & \\ & & | & \\ & & CH_3 & \end{array}$$

(f)
$$\begin{array}{ccccc} & CH_3 & & & \\ & | & & & \\ & CH_2 & & & \\ & | & & & \\ CH_3- & CH & -CH_2- & CH & -CH_3 \\ & & & | & \\ & & & CH_2 & \\ & & & | & \\ & & & CH_3 & \end{array}$$

(g)
$$\begin{array}{ccccccc} & & & CH_3 & & & \\ & & & | & & & \\ CH_3 & & & CH_2 & & & \\ | & & & | & & & \\ CH_2- & CH & - & C & -CH_2- & CH & -CH_2-CH_3 \\ & | & & | & & | & \\ & CH_3 & & CH_2 & & CH_2 & \\ & & & | & & | & \\ & & & CH_2 & & CH-CH_3 & \\ & & & | & & | & \\ & & & CH_3 & & CH_3 & \end{array}$$

10.34 Draw the structural formula for
(a) 3-methylheptane
(b) dimethylpropane
(c) 2,3,4-trimethylheptane
(d) octachloropropane
(e) 5-isopropyldecane
(f) 2-ethyl-1,1,2-trimethylcyclobutane

10.35 Name each of the following halides by the IUPAC system:

(a) $CH_3-CH_2-CH_2-Br$

(b)
$$\begin{array}{ccccc} CH_3- & CH & -CH_2- & CH & -CH_3 \\ & | & & | & \\ & F & & CH_2F & \end{array}$$

(c)
$$\begin{array}{cc} Br-CH_2-CH_2- & CH-CH_2-Br \\ & | \\ & Br \end{array}$$

(d)
$$\begin{array}{cccccc} CH_3- & CH & - CH & - CH & - CH & -CH_3 \\ & | & | & | & | & \\ & Cl & CH_3 & I & Cl & \end{array}$$

(e)
$$\begin{array}{cccc} & & I & \\ & & | & \\ CH_3- & CH & -C & -CH_3 \\ & | & | & \\ & I & CH_2I & \end{array}$$

(f)
$$\begin{array}{ccccc} CH_3- & CH & -CH_2- & CH & -Cl \\ & | & & | & \\ & Cl & & Cl & \end{array}$$

10.36 Name these groups:

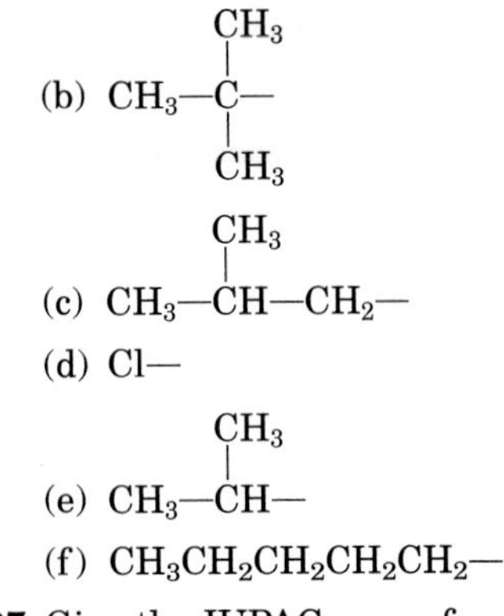

(a) CH_3CH_2-

(b) 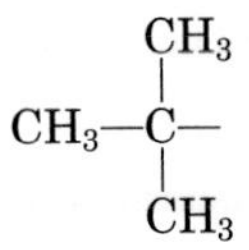 CH_3—C(CH_3)(CH_3)—

(c) CH_3—CH(CH_3)—CH_2—

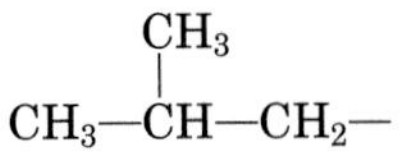

(d) Cl—

(e) 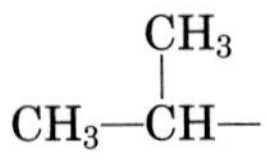 CH_3—CH(CH_3)—

(f) $CH_3CH_2CH_2CH_2CH_2$—

10.37 Give the IUPAC name for
(a) isobutane
(b) isopropyl chloride
(c) neopentyl iodide
(d) *tert*-butyl chloride
(e) *sec*-butyl bromide

Cycloalkanes

10.38 Name each of the following by the IUPAC system (ignore stereoisomerism):

(a)

(b) 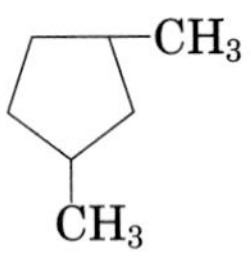 CH_3 CH_3

(c) 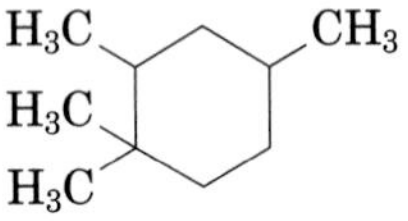 H_3C H_3C H_3C CH_3

(d) 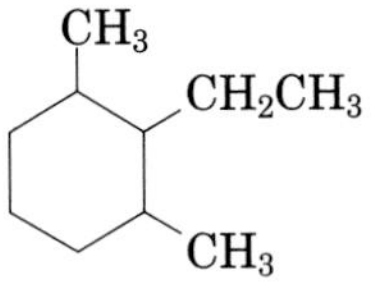 CH_3 CH_2CH_3 CH_3

(e) 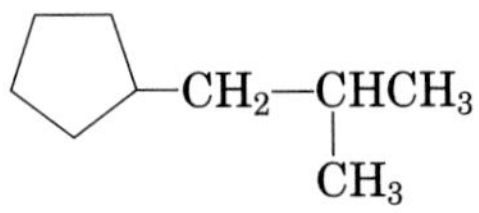 —CH_2—CH(CH_3)CH_3

10.39 Name the following cyclic halogen-containing compounds by the IUPAC system (ignore stereoisomerism):

(a) 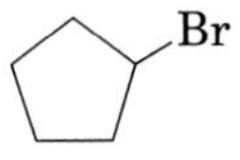 Br

(b) 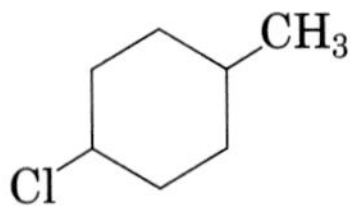 CH_3 Cl

(c) 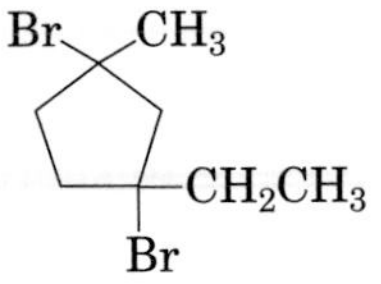 Br CH_3 CH_2CH_3 Br

(d) 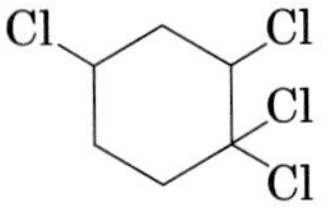 Cl Cl Cl Cl

(e) 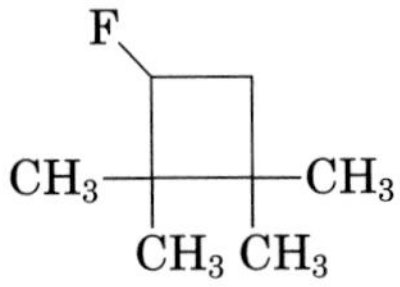 F CH_3 CH_3 CH_3 CH_3

10.40 There are six *cycloalkanes* with the molecular formula C_5H_{10}. Draw all of them and give the IUPAC names.

10.41 Draw the condensed or very condensed structural formulas for

(a)

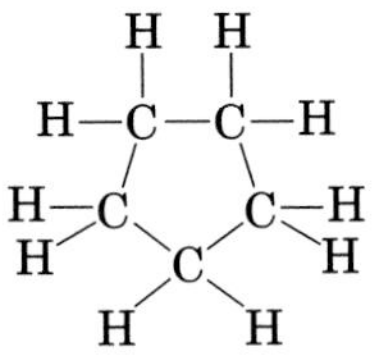

(b) 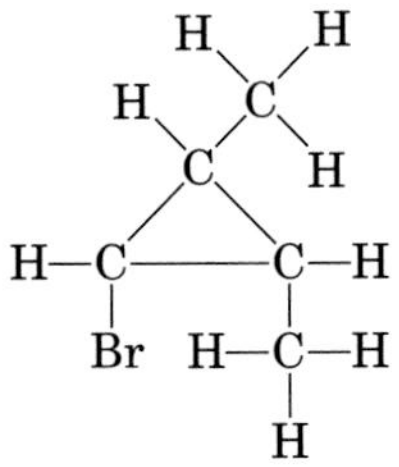 Br

(c)

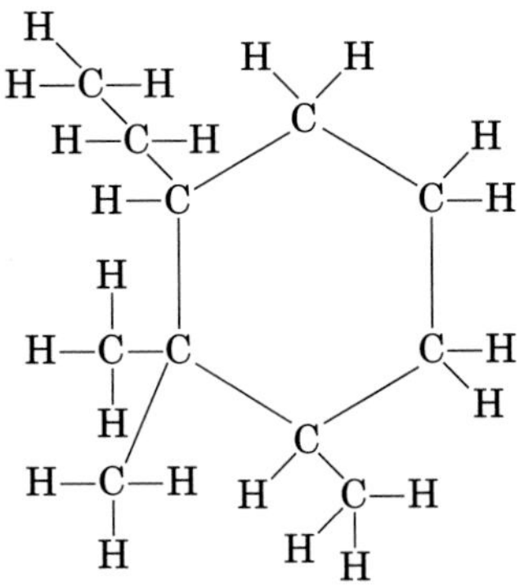

10.42 Give IUPAC names:

(a) 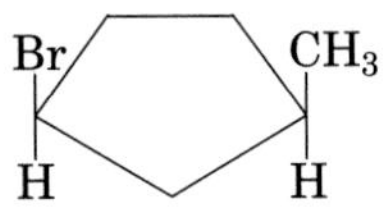 Br H CH_3 H

(b) 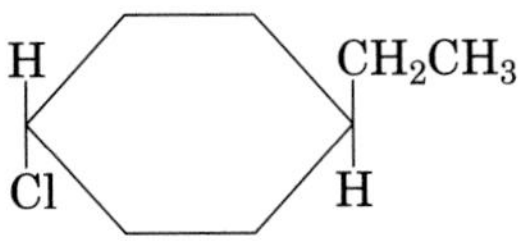 H Cl CH_2CH_3 H

(c) CH_3 CH_3 H H

10.43 Draw structural formulas for
(a) *trans*-1,3-dimethylcyclobutane
(b) *cis*-1,4-dichlorocyclohexane
(c) *trans*-1,3-diethyl-4,4-dimethylcyclopentane

Properties of Alkanes

10.44 The compound *n*-decane, $C_{10}H_{22}$, is an unbranched alkane. Predict the following:
(a) Does it dissolve in water?
(b) Does it dissolve in *n*-hexane?
(c) Will it burn when ignited?
(d) Is it a liquid, solid, or gas at room temperature?
(e) Is it more or less dense than water?

10.45 What are the chief chemical properties of the alkanes and cycloalkanes?

10.46 Balance the complete combustion reaction of
(a) cyclopropane
(b) 2,2-dimethylpentane
(c) *n*-octane

Functional Groups

10.47 State which family each belongs to:

(a) $CH_3-CH_2-CH(OH)-CH_3$

(b) $CH_3-CH(CH_3)-C(=O)-OH$

(c) cyclopentyl$-C(=O)-CH_3$

(d) $CH_3-CH(CH_3)-CH(CH_3)-CH_3$

(e) $CH_3-N(CH_3)-CH_2-CH_3$

(f) $CH_3-C(CH_3)_2-Cl$

(g) $CH_3-C{\equiv}C-CH_2-CH_3$

10.48 List all the functional groups in Table 10.7 that contain oxygen atoms.

10.49 Draw the structural formula for one member of each of the following classes (other than the one shown in Table 10.7):

(a) alkene	(f) ether
(b) carboxylic acid	(g) thiol
(c) amine	(h) aldehyde
(d) amide	(i) carboxylic ester
(e) alcohol	(j) ketone

10.50 Each of the following compounds has two or three functional groups. Put each into as many families as is appropriate:

(a) $CH_3-O-CH_2CH_2-NH-CH_3$

(b) $CH_2{=}CH-CH_2-Cl$

(c) $CH_2(OH)-CH(OH)-CH_2(OH)$

(d) $CH_3-CH(F)-CH_2CH_2-OH$

(e) $CH_3-C(=O)-CH_2-C(=O)-O-CH_3$

(f) cyclohexyl$-C(=O)-CH_2-C(=O)-H$

10.51 Tell which functional groups are present in

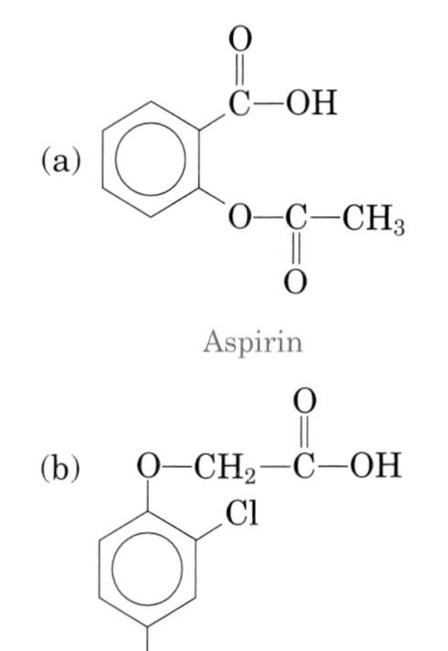

(a) benzene ring bearing $-C(=O)-OH$ and, on the adjacent carbon, $-O-C(=O)-CH_3$

Aspirin

(b) benzene ring bearing $-O-CH_2-C(=O)-OH$, with Cl on the adjacent (2-) carbon and Cl at the 4-position

2,4-D (a herbicide)

(c) $HO-C(=O)-CH_2-CH_2-CH(NH_2)-COOH$

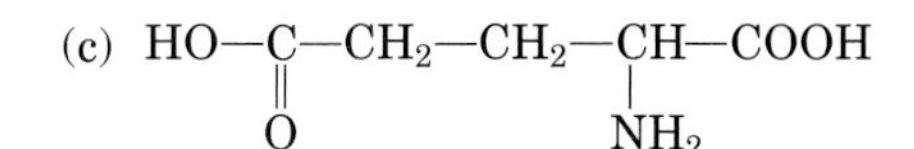

Glutamic acid

(d) benzene ring bearing NH_2 and, para to it, $-C(=O)-OH$

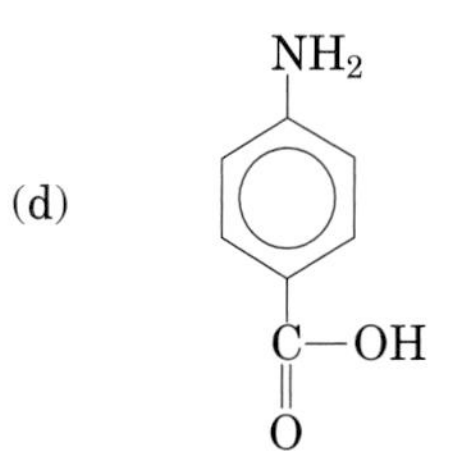

para-Aminobenzoic acid

Boxes

10.52 (Box 10A) What are the advantages of combinatorial synthesis, compared to the standard methods?

10.53 (Box 10B) Explain why it is not possible to isolate separate isomers of axial and equatorial chlorocyclohexane.

10.54 (Box 10C) The solubility of body oils in alkanes decreases with increasing molecular weight of the alkanes. On the basis of this principle, explain how petroleum jelly protects the skin but gasoline causes chapping.

10.55 (Box 10D) Why is it necessary to distill petroleum?

10.56 (Box 10E) Carbon dioxide is a product of the burning of fossil fuels, as well as of the breathing of humans and animals. Explain why carbon dioxide is still a minor component of the atmosphere.

10.57 (Box 10F) Certain gases, among them CO_2, generate the greenhouse effect. What property of these gases promotes an increase in the temperature of the atmosphere?

10.58 (Box 10G) How many carboxylic ester groups are present in the Taxol molecule?

***10.59** (Box 10G) How many OH groups in Taxol are located at primary, secondary, and tertiary carbons?

Additional Problems

10.60 Classify all the carbons as primary, secondary, tertiary, or quaternary:

(a) $CH_3—CH_2—CH(CH_3)—CH_3$

(b) $CH_3—C(CH_3)_2—CH_2Cl$

(c) Cyclopentane ring bearing Br on one carbon, and CH_3 and Br on the adjacent carbon

(d) $CH_3—CH_2—CH_3$

***10.61** Write structural formulas for two different amines, each of which has two carbons.

10.62 Draw structural formulas and give IUPAC names for all compounds whose molecular formula is C_7H_{16}.

10.63 Describe the orbitals that overlap, making up all bonds in the following compounds:

(a) $F—C(H)(H)—C(F)(F)—F$ (H and H on the first carbon; F and F on the second)

(b) $H—C(Cl)(H)—C(H)(H)—C(H)(Cl)—H$ (Cl at top carbon; H at bottom)

10.64 Define:

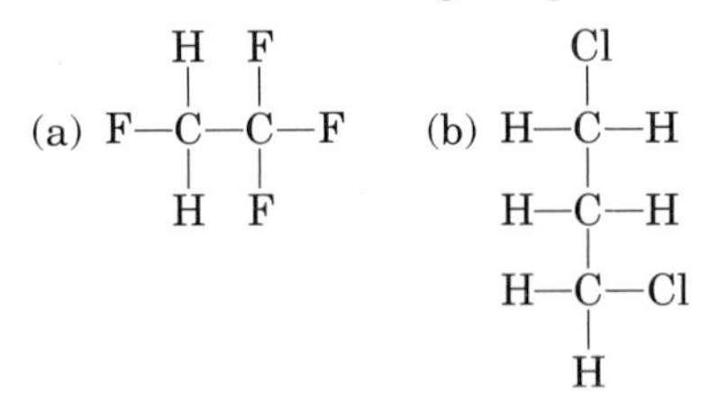

(a) organic chemistry
(b) hydrocarbon
(c) alkane
(d) functional group

10.65 The following names are all incorrect. State what is wrong with each (do not try to give correct names)

(a) 4-methylpentane
(b) 1-methylbutane
(c) 2-dimethylbutane
(d) 2,3-methylhexane
(e) 2-methylcyclohexane
(f) 2-ethyl-4,4-dimethylpentane
(g) 2-ethylpentane
(h) cycloethane

10.66 Write structural formulas for

(a) An alcohol that has five carbons
(b) A carboxylic acid that has three carbons
(c) A ketone that has three carbons
(d) An ether that has four carbons
*(e) A cyclic alcohol that has four carbons

10.67 Suppose you have a sample of 2-methylheptane and a sample of 2-methyloctane. Could you tell the difference by looking at them? What color would they be? How could you tell which is which?

***10.68** As can be seen in Table 10.4, an increase in the carbon chain by one carbon increases both the melting point and the boiling point. This increase is greater going from CH_4 to C_2H_6 and from C_2H_6 to C_3H_8 than it is in going from C_8H_{18} to C_9H_{20} or from C_9H_{20} to $C_{10}H_{22}$. What do you think is the reason for this?

10.69 Draw the structural formula for

(a) 2,3-dichlorohexane
(b) 1,2,4-trichlorocyclohexane
(c) 1,1-dimethylcyclobutane
(d) 2-*sec*-butyl-1-isopropyl-3-neopentylcycloheptane

***10.70** Draw structural formulas and give IUPAC names for all the dibromopropanes ($C_3H_6Br_2$).

10.71 Identify the functional groups:

(a) $CH_2OH—C(=O)—CH_3$

(b) $HO—C(=O)—C(=O)—O—CH_3$

(c) $H—C(=O)—C(H)(OH)—CH_3$

Chapter 11

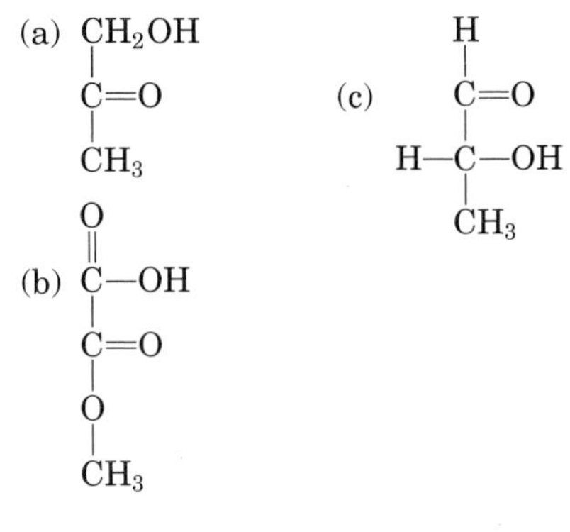

Alkenes, alkynes, and aromatic compounds

11.1 Introduction

In Chapter 10 we discussed the alkanes, including the cycloalkanes. In this chapter we deal with the other families of hydrocarbons—alkenes, alkynes, and aromatic hydrocarbons.

Alkenes are also called **olefins,** an older name that is still widely used.

Alkenes contain **carbon-carbon double bonds; alkynes** contain **carbon-carbon triple bonds.** Because compounds with double and triple bonds can add more hydrogen (see Sec. 11.6), alkenes and alkynes are called **unsaturated hydrocarbons.** Alkanes, which cannot add any more hydrogen, are called saturated hydrocarbons.

11.2 Nomenclature of Alkenes

Alkenes are named by the same IUPAC rules we met in Section 10.7, but with several additional rules to handle the double bonds.

1. The suffix "-ene" is used when double bonds are present.

Because an alkene must have a C=C bond, the simplest alkene has two carbon atoms:

$$H_2C{=}CH_2$$

Ethene or ethylene

In this molecule, the longest chain has two carbon atoms. If this were an alkane, it would be ethane. Because it has a double bond, the suffix "-ane" is changed to "-ene" and we call it **ethene.** This compound is more frequently called by its common name, ethylene. Similarly, $CH_3{-}CH{=}CH_2$ is called propene or propylene.

$CH_3{-}CH{=}CH_2$ is the condensed formula for

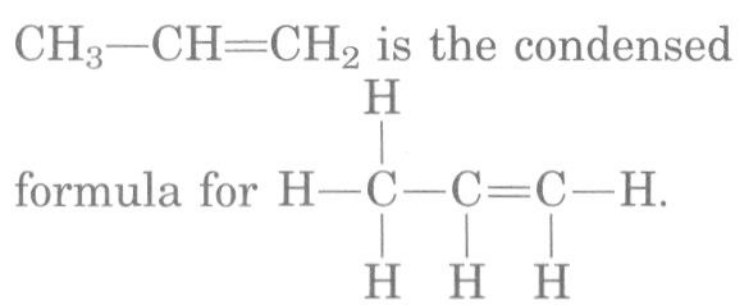

When the chain has four or more carbons, there is structural isomerism because the double bond can be in more than one position. Consider these compounds:

$$CH_2{=}CH{-}CH_2{-}CH_3 \qquad CH_3{-}CH{=}CH{-}CH_3$$

The rules we know so far say that both these compounds are butene, but they are obviously different compounds, with different structural formulas (though they are isomers because both are C_4H_8). We need a new rule.

2. When a chain has four or more carbons, **we number the position of the double bond, using only the lower of the possible numbers.**

Box 11A

Terpenes

Terpenes are a family of compounds found widely distributed throughout the plant world. Many are alkenes, but some contain other functional groups as well. Many are cyclic or bicyclic (two rings). Hundreds of different ones are known. Some have important biological uses; others we use for various commercial purposes. Among the latter are compounds called *essential oils,* which are what give plants their pleasant odors. For thousands of years, essential oils have been extracted from flowers and other parts of plants and used in perfumes and cosmetics. All terpenes have two things in common: They are made by plants (some are made by animals as well) and they follow the **isoprene rule.** Isoprene is a common name for 2-methyl-1,3-butadiene:

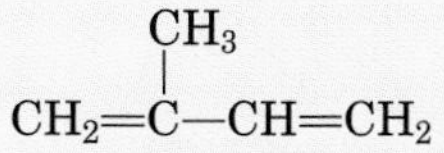

$$CH_2{=}C(CH_3){-}CH{=}CH_2$$

Isoprene

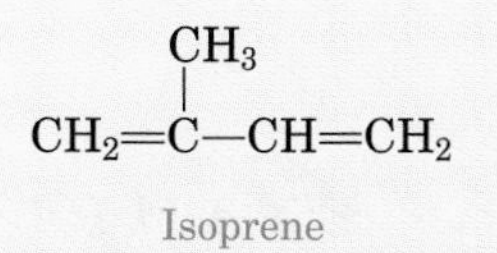

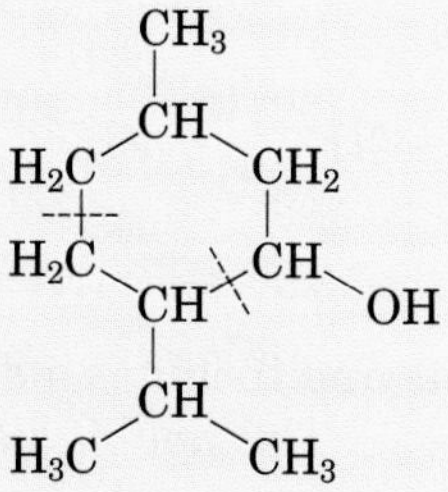

Isoprene itself is not a terpene, but the isoprene rule says that all terpenes are made up of isoprene units—that is, the carbon atoms of any terpene molecule can be divided into pieces, each having the C—C(—C)—C—C skeleton.

Because of this, the number of carbon atoms in a terpene is always a multiple of five. Terpenes can have 10 carbon atoms, or 15 (sesquiterpenes), or 20, or 30, or 40. They are found in nature in a bewildering array of structures, of which we can look at only a few.

We will see how the isoprene rule works by using dashed lines to divide the molecules into isoprene units. One of the simplest terpenes is menthol. Menthol is one of the chief constituents of mint, especially Japanese peppermint oil.

Menthol

Two larger terpenes are

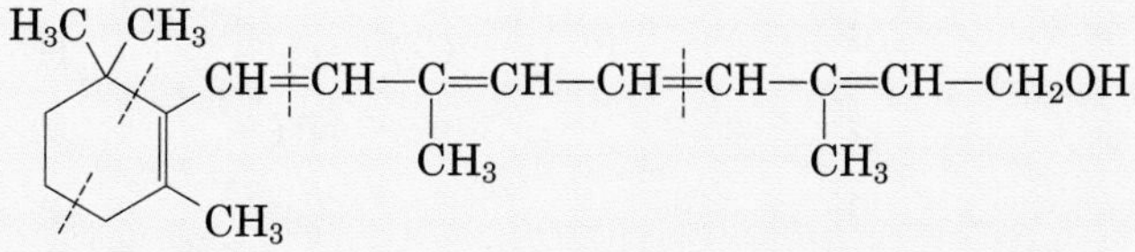

$$\text{(ring)}{-}CH{=}CH{-}C(CH_3){=}CH{-}CH{=}CH{-}C(CH_3){=}CH{-}CH_2OH$$

Vitamin A

$$\text{(ring)}{-}CH{=}CH{-}C(CH_3){=}CHCH{=}CHC(CH_3){=}CHCH{=}CHCH{=}C(CH_3)CH{=}CHCH{=}C(CH_3)CH{=}CH{-}\text{(ring)}$$

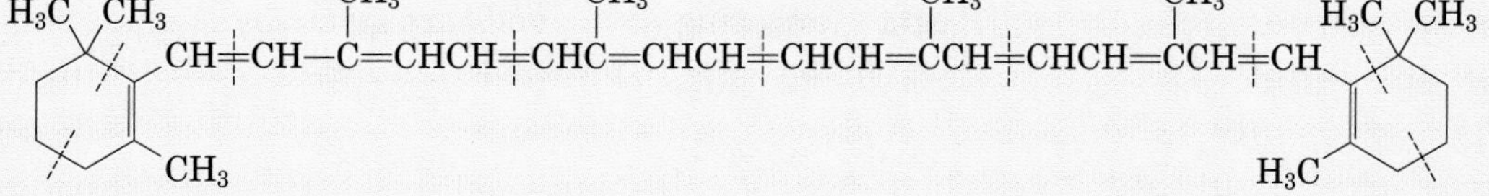

β-Carotene

β-Carotene, a tetraterpene, has 40 carbons (and is therefore made up of eight isoprene units) and 11 double bonds. It is the compound that gives the orange color to carrots. As you can see, *β*-carotene consists of two vitamin A units joined together. Because the body can break the carotene down, carrots are a good source of vitamin A.

Thus these two compounds are

$$\overset{1}{C}H_2=\overset{2}{C}H-\overset{3}{C}H_2-\overset{4}{C}H_3 \qquad \overset{1}{C}H_3-\overset{2}{C}H=\overset{3}{C}H-\overset{4}{C}H_3$$

1-Butene — 2-Butene

In 1-butene the double bond is between carbons 1 and 2, but in the name we use only the 1. It is understood that the double bond goes from the digit shown to the next higher digit.

As in all the cases of isomerism we have seen so far, 1-butene and 2-butene have different physical and chemical properties.

3. Branches and halogen substituents are named as in the rules for alkanes.

EXAMPLE

Name this compound:

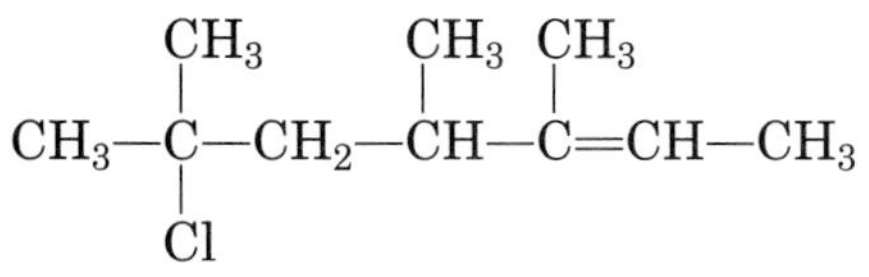

$$CH_3-\underset{\underset{Cl}{|}}{\overset{\overset{CH_3}{|}}{C}}-CH_2-\overset{\overset{CH_3}{|}}{C}H-\overset{\overset{CH_3}{|}}{C}=CH-CH_3$$

Answer • The longest chain has seven carbons and includes a double bond. Therefore the parent name is heptene. The double bond goes from the second to the third carbon, so the parent is 2-heptene. (If we counted from the other end, it would be 5-heptene, but 2 is lower than 5.) When we add the other groups, we get

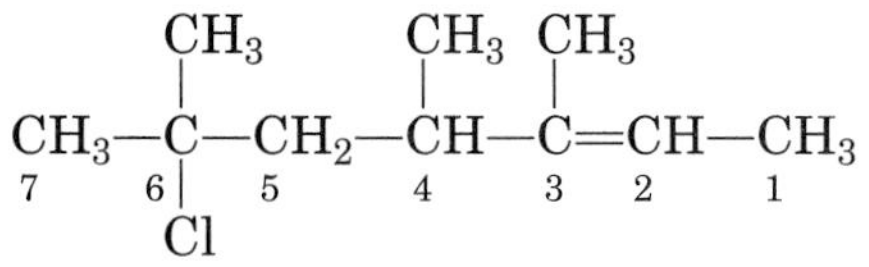

$$\underset{7}{CH_3}-\underset{\underset{Cl}{6|}}{\overset{\overset{CH_3}{|}}{C}}-\underset{5}{CH_2}-\underset{4}{\overset{\overset{CH_3}{|}}{C}H}-\underset{3}{\overset{\overset{CH_3}{|}}{C}}=\underset{2}{CH}-\underset{1}{CH_3}$$

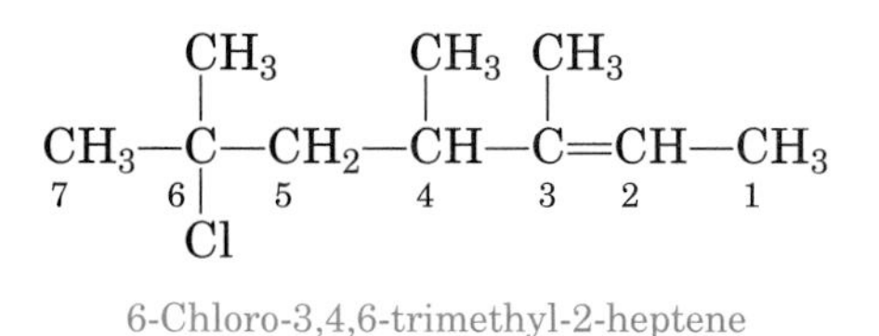

6-Chloro-3,4,6-trimethyl-2-heptene

As this example shows, the double bond takes precedence over alkyl and halo groups in determining which end to start numbering from. We begin numbering at the end that gives the double bond the lowest number, no matter where the alkyl or halo groups are.

Problem **11.1**

Name:

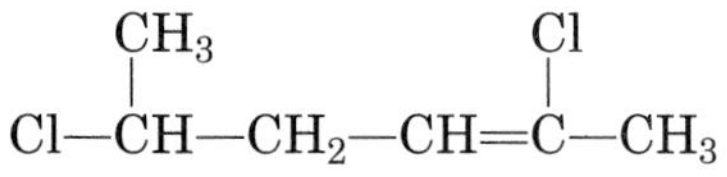

$$Cl-\overset{\overset{CH_3}{|}}{C}H-CH_2-CH=\overset{\overset{Cl}{|}}{C}-CH_3$$

4. The "longest chain" must include the double bond.

$$\overset{}{CH_3}-CH_2-\overset{2}{C}(=\overset{1}{C}H_2)-\overset{3}{C}H_2-\overset{4}{C}H_3$$

2-Ethyl-1-butene

This compound contains a five-carbon chain, but that chain does not contain the double bond. The longest chain with the double bond contains four carbons, so the parent is *but* and not *pent.*

Problem 11.2

Name:

$$CH_3-CH_2-CH_2-C(-CH_2-CH_3)=CH-CH_3$$

5. Cyclic compounds containing double bonds are called cycloalkenes. When there is only one double bond in a ring, it always goes from the 1 to the 2 position, so it is not necessary to show the number in the name.

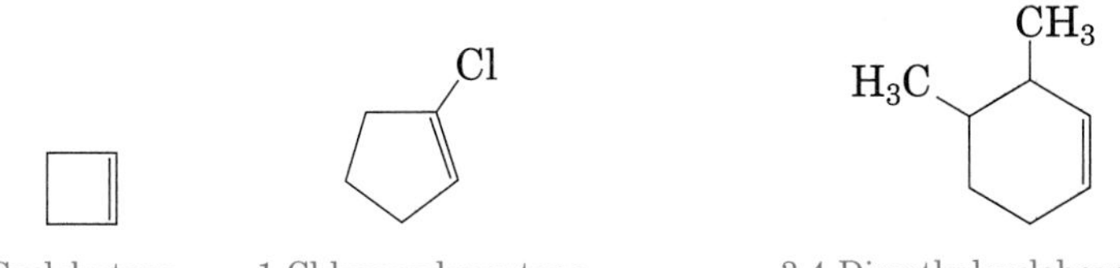

Cyclobutene   1-Chlorocyclopentene (*not* 1-chloro-1-cyclopentene)   3,4-Dimethylcyclohexene

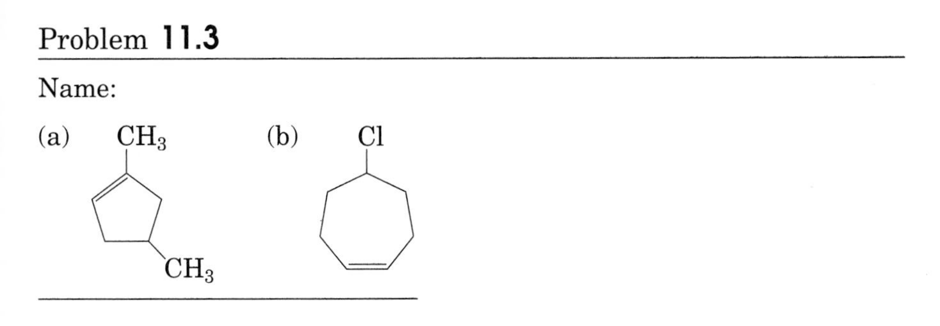

Problem 11.3

Name:

(a) CH_3 CH_3

(b) Cl

6. Compounds with two or more double bonds are given the suffix "-diene," "-triene," and so on.

$$CH_2{=}CH{-}CH{=}CH_2$$

1,3-Butadiene

$$CH_2{=}C(CH_3){-}CH{=}CH{-}CH_2{-}CH{=}CH{-}CH_3$$

2-Methyl-1,3,6-octatriene

Box 11B

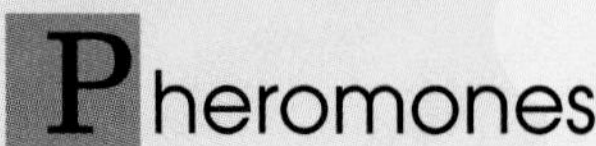

Pheromones

Insects emit compounds called **pheromones** to transmit messages to each other. These compounds are remarkably species-specific; that is, each insect is sensitive only to the pheromones of its own species. One of the main uses for these compounds is as sex attractants. Normally, the female emits the pheromone and the male is attracted by its scent. Only a very small quantity is necessary, in some cases as little as 10^{-12} g.

Pheromones are often very simple compounds, though usually with long chains. Two examples are 9-tricosene, the sex attractant of the common housefly, and bombykol, secreted by the female silkworm moth:

$$\begin{matrix} CH_3(CH_2)_7 & & (CH_2)_{12}CH_3 \\ & C{=}C & \\ H & & H \end{matrix}$$

cis-9-Tricosene

$$\begin{matrix} & & H & & (CH_2)_9OH \\ CH_3(CH_2)_2 & & & C{=}C & \\ & C{=}C & & & H \\ H & & H & & \end{matrix}$$

Bombykol

cis double bond   trans double bond

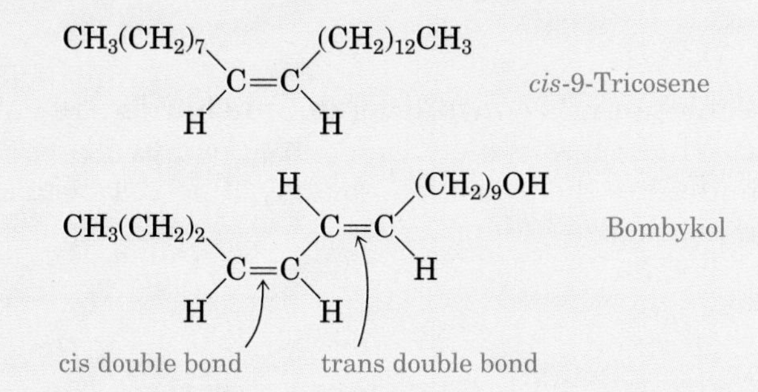

The 9-tricosene secreted by the female housefly is the cis isomer. The trans isomer does not attract male houseflies and is not secreted by the female. Similarly, the bombykol of the silkworm moth is the trans-cis compound, and none of the three other stereoisomers work. (This is a general rule in nature: In most cases where more than one stereoisomer is possible, nature uses only one.)

The discovery of pheromones has given us a possible way to control harmful insects without polluting the environment (as happens with chlorinated insecticides; see Box 12H). Because pheromones have simple structures, they are easily synthesized in laboratories. A small quantity of a synthetic pheromone can be used to lure male insects into a trap, where they can be killed. Without males, the females cannot propagate. Some of this has already been done—for example, with Japanese beetles—and it is likely that its use will increase.

When it was discovered that insects use pheromones, the question was raised as to whether there are human pheromones also, but no such compound has as yet been positively identified.

The group $H_2C{=}CH{-}$ is called the **vinyl group.** Thus the IUPAC name of $H_2C{=}CH{-}Cl$ is chloroethene, but it is also frequently called vinyl chloride.

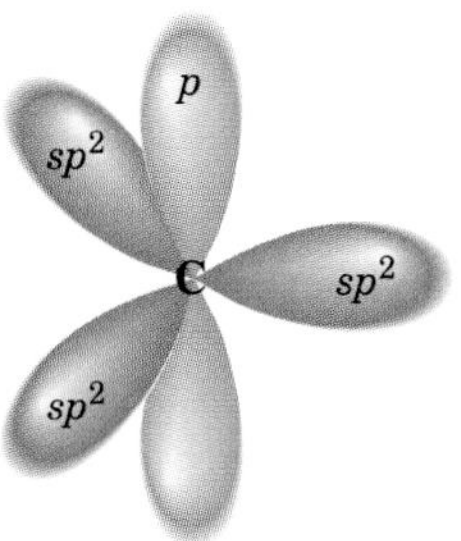

Figure 11.1
The three sp^2 orbitals and the unchanged p orbital.

11.3 sp^2 Hybridization and π Orbitals

In Section 10.4 we learned that a carbon atom uses sp^3 orbitals when it is connected to four other atoms. The carbon atom of a double bond, however, is connected to only three other atoms, not four. It hybridizes its orbitals in a different way. The hybridization takes place only among three orbitals; the fourth orbital remains unchanged.

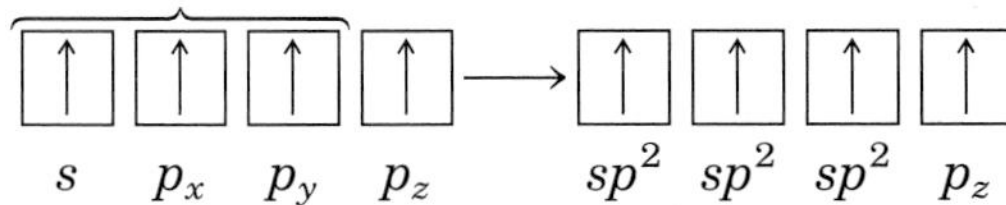

Since the mixing or hybridization of three orbitals involves one s and two p orbitals, we call the hybrid orbitals **sp^2 orbitals.** Each of them has one-third s character and two-thirds p character.

Three sp^2 orbitals are formed.

The three sp^2 orbitals all lie in one plane, and the angle between them is 120°. The fourth orbital, the unchanged p orbital, forms a 90° angle with the sp^2 orbitals (Fig. 11.1).

When two carbon atoms, each having sp^2 orbitals, approach each other, they form a sigma bond by sharing the two electrons from the overlapping sp^2 orbitals (Fig. 11.2).

The two p orbitals also form a bond. This bond is not the result of direct overlap, as in the sigma bond, but of the overlap of two parallel p orbitals (sideways overlap). An orbital formed by the overlap of two parallel p orbitals is called a π (Greek letter pi) **orbital,** and the bond is called a **π bond.** A carbon-carbon double bond is therefore made of one σ bond and one π bond. Since the π bond is weaker than the σ bond, it is more reactive and easier to break.

There is less overlap here, so the π bond is weaker than the σ bond.

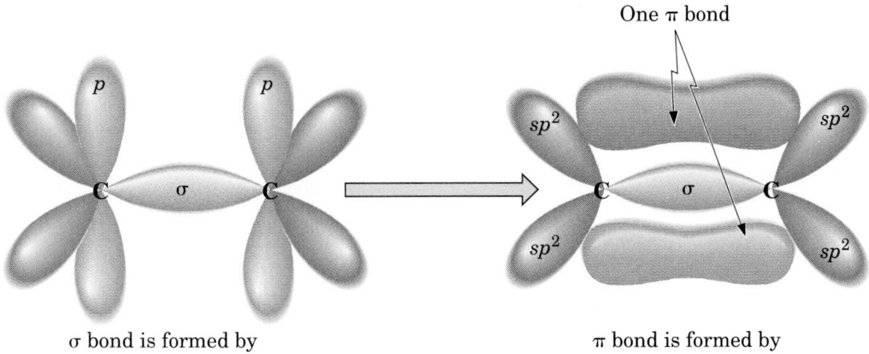

Figure 11.2
Two sp^2 orbitals overlap to form a σ bond and two p orbitals overlap sideways to form a π bond.

Pi bonds can exist only in double or triple (Sec. 11.8) bonds. The parallel alignment and the sideways overlap of two *p* orbitals in the absence of a σ bond is simply not strong enough to form a stable single bond.

Pi bonds and the double bonds that contain them are not restricted to carbon-carbon double bonds. Carbon-oxygen and carbon-nitrogen double bonds are also made of a σ bond and a π bond.

11.4 Geometry and Stereoisomerism of Alkenes

The fact that a double bond is made of one σ bond and one π bond has an important consequence for the structure of alkenes. The π bond can only exist if the two *p* orbitals are parallel to each other. If one of the *p* orbitals were to rotate, so that it was no longer parallel to the other (Fig. 11.3), the bond would be broken. This does not normally happen, because the molecule is more stable with two bonds (a σ and a π) rather than with just a σ bond.

The bond can be broken by the addition of heat or light.

The fact that the *p* orbitals remain parallel (to produce the π bond by sideways parallel overlap) also affects the sp^2 orbitals. We saw in Section 11.3 that the sp^2 orbitals are all perpendicular to the *p* orbitals. Therefore, a molecule like ethylene is planar, with angles of 120° (the angles between sp^2 orbitals):

$$\begin{array}{c} H \quad\quad H \\ C{=}C \quad 120^\circ \\ H \quad\quad H \end{array}$$

Other alkenes also have this geometry. In each case, six atoms—the two C=C atoms and the four atoms directly attached to them—are all in one plane. Furthermore, **there is no rotation about the double bond.** This is in sharp contrast to the case of single bonds, where free rotation takes place (Sec. 10.6, Fig. 10.7).

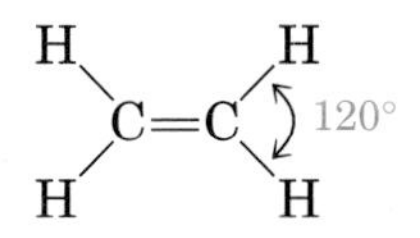

C=C double bonds are shorter than C—C single bonds.

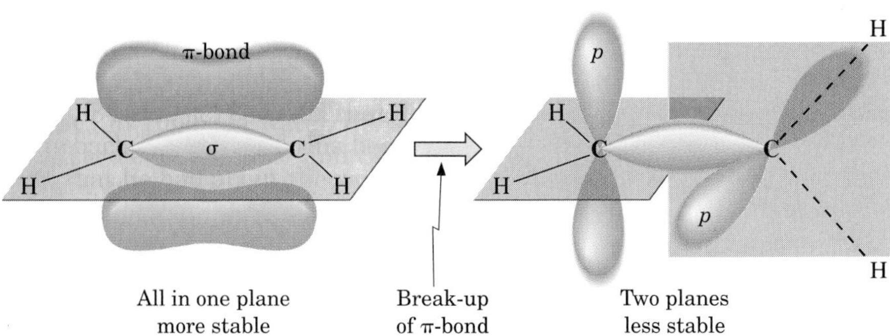

Figure **11.3** • The *p* orbitals of a C=C double bond remain parallel because if one rotated, it would break the π bond.

Because there is no rotation about the double bond, we can have the same type of stereoisomerism we saw for cycloalkanes in Section 10.9. There are two different compounds called 2-butene. This is easily seen if we draw the formulas this way, remembering that the molecules are planar, with 120° angles:

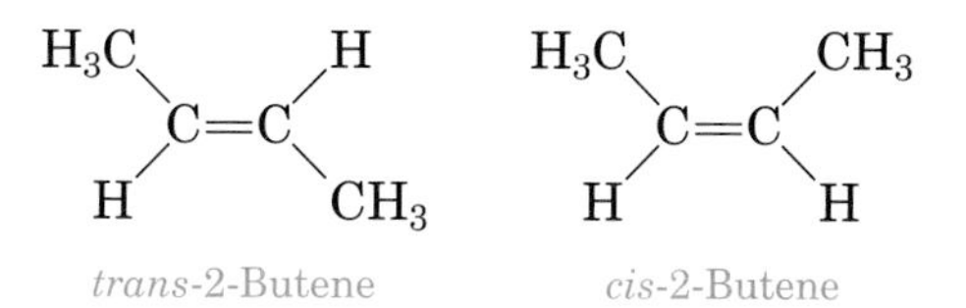

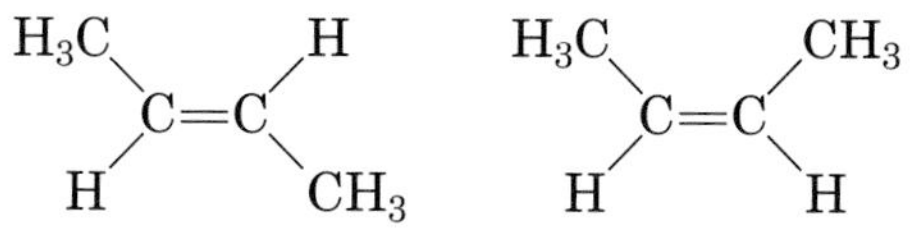

trans-2-Butene   *cis*-2-Butene

As in the case of cycloalkanes, we use the prefix *cis* for the isomer in which the two similar or identical groups are on the same side and *trans* for the one in which they are on opposite sides. As usual, the boiling points, freezing points, reactivities, and all other physical and chemical properties of the two isomers are different. Because their structures are similar, most properties are usually similar (see Table 11.1) *but not identical.* The fact that there is no free rotation about the C=C bond prevents them from interconverting. A bottle containing, say, the trans isomer will continue to hold only the trans isomer no matter how long it remains on the shelf. (When the compounds are heated to very high temperatures or treated with certain catalysts, rotation can take place, and then cis-trans conversion does occur.)

Not all double-bond compounds show cis-trans stereoisomerism. The test for its presence is very simple. **In order for there to be cis-trans stereoisomers, both double-bonded carbons must be connected to two different groups.** If *either* carbon is connected to two identical groups, there are no stereoisomers. It is easy to see why:

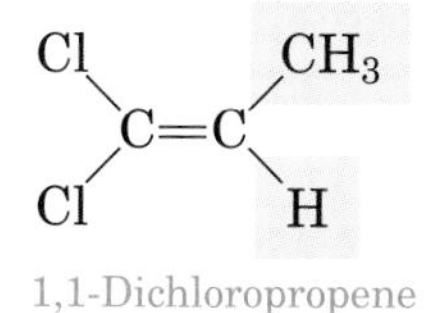

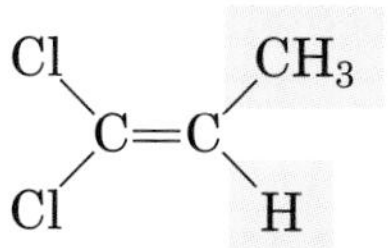

1,1-Dichloropropene

In this case, one of the two C=C carbons is connected to two chlorines. Thus, if we interchange the groups on either carbon, we get the same compound and not a different one. On the other hand, when both carbons of the double bond are attached to two different groups, we do have cis-trans isomers:

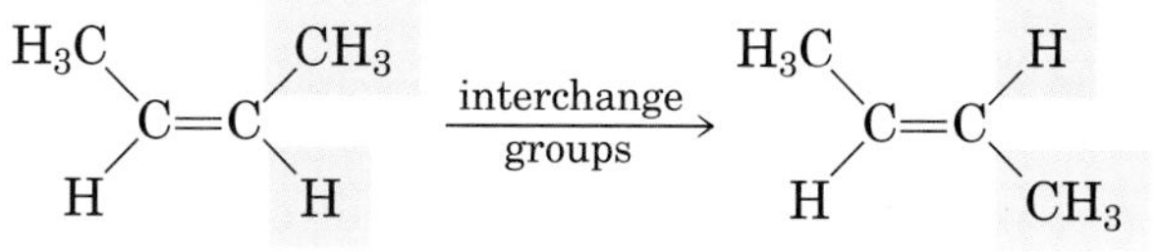

interchange groups →

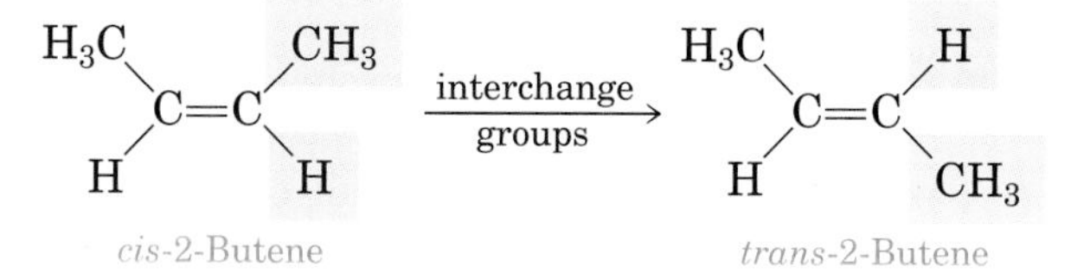

cis-2-Butene   *trans*-2-Butene

EXAMPLE

In each case decide whether cis-trans isomers exist: (a) 1-butene (b) 2-methyl-2-butene (c) 1-chloro-2-methyl-2-butene

Answer

(a) $(H)_2C{=}CH{-}CH_2{-}CH_3$ 1-Butene

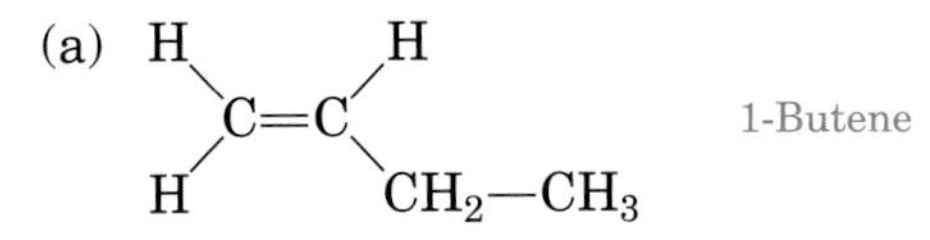

Because one double-bonded carbon is connected to two hydrogens, there are no cis-trans isomers.

(b) $(H_3C)_2C{=}C(CH_3)H$ 2-Methyl-2-butene

Here one double-bonded carbon is connected to two methyl groups, so again there are no cis-trans isomers.

(c) $(H_3C)(Cl{-}CH_2)C{=}C(CH_3)(H)$ *cis* ; $(H_3C)(Cl{-}CH_2)C{=}C(H)(CH_3)$ *trans* 1-Chloro-2-methyl-2-butene

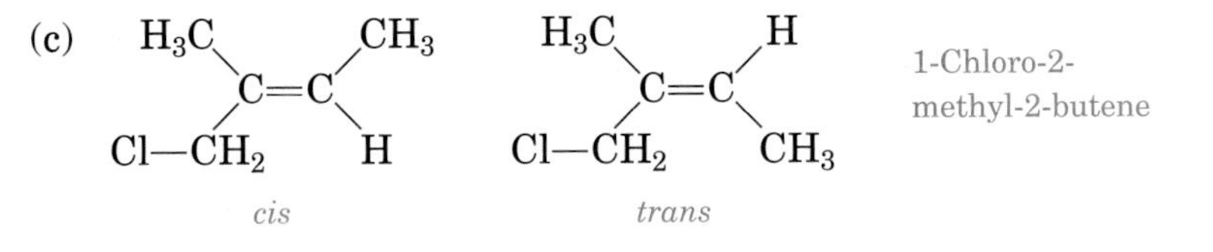

This case is very similar to (b); but the presence of the Cl makes the two groups on the left-hand carbon different. Since the carbon on the right is also connected to two different groups, cis-trans isomers do exist.

Problem 11.4

In each case, decide whether cis-trans isomers exist: (a) 2-pentene (b) 1-pentene (c) 3-ethyl-2-pentene.

11.5 Physical Properties of Alkenes

The physical properties of alkenes are very similar to those of the corresponding alkanes (Sec. 10.10). Up to about 5 carbons, alkenes are gases at room temperature. From about 5 to 17 carbons they are liquids, and then solids above that. Table 11.1 shows boiling and melting points for some typical compounds. Note that virtually all alkenes are insoluble in water and have densities much lower than that of water (Table 11.1).

Following the "like dissolves like" principle, alkenes are generally soluble in alkanes and in each other.

11.6 Chemical Properties of Alkenes: Addition Reactions

Unlike alkanes, which are inert to almost all chemical reagents (Sec. 10.11), alkenes undergo many reactions. Since the only difference between an alkane and an alkene is the double bond, we are not surprised

Table 11.1 Physical Properties of Some Alkenes

Name	Structural Formula	BP(°C)	MP(°C)	Density (g/mL)
Ethene (ethylene)	$CH_2{=}CH_2$	−104	−169	—
Propene	$CH_2{=}CHCH_3$	−47	−185	—
1-Butene	$CH_2{=}CHCH_2CH_3$	−6	−185	0.595
cis-2-Butene	$(H_3C)(H)C{=}C(CH_3)(H)$	4	−139	0.621
trans-2-Butene	$(H)(H_3C)C{=}C(CH_3)(H)$	1	−105	0.604
Methylpropene (isobutylene)	$CH_2{=}C(CH_3)_2$	−7	−140	0.594
1-Pentene	$CH_2{=}CHCH_2CH_2CH_3$	30	−138	0.641
1-Hexene	$CH_2{=}CHCH_2CH_2CH_2CH_3$	63	−140	0.673
Cyclohexene	(cyclohexene ring)	83	−104	0.810

to learn that most of the reactions of alkenes take place at the double bond. These reactions follow the pattern

$$-\overset{|}{\underset{|}{C}}{=}\overset{|}{\underset{|}{C}}- + AB \longrightarrow -\overset{A}{\underset{|}{\overset{|}{C}}}-\overset{B}{\underset{|}{\overset{|}{C}}}-$$

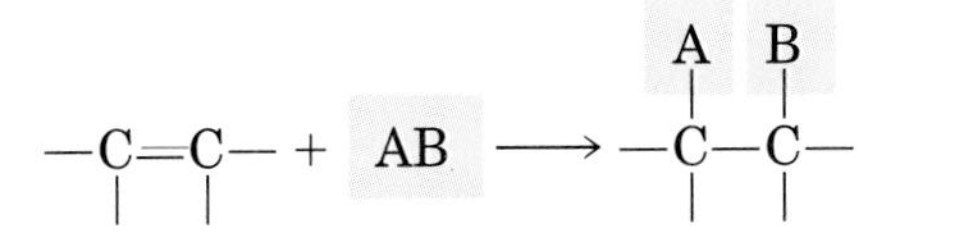

AB is a general symbol to show something adding to a double bond.

They are called **addition reactions** because something is added to the double bond. One of the two bonds of the double bond (the one shown in color) is broken, and two new bonds are formed. The original double bond is converted to a single bond. We will discuss only a few of the many known reactions of this type.

Addition of H_2

General reaction

$$-\underset{|}{C}{=}\underset{|}{C}- + H_2 \xrightarrow{\text{catalyst}} -\overset{H}{\underset{|}{\overset{|}{C}}}-\overset{H}{\underset{|}{\overset{|}{C}}}-$$

Specific example

$$\underset{\text{Propene}}{CH_3CH{=}CH_2} + H_2 \xrightarrow{Pt} \underset{\text{Propane}}{CH_3CH_2CH_3}$$

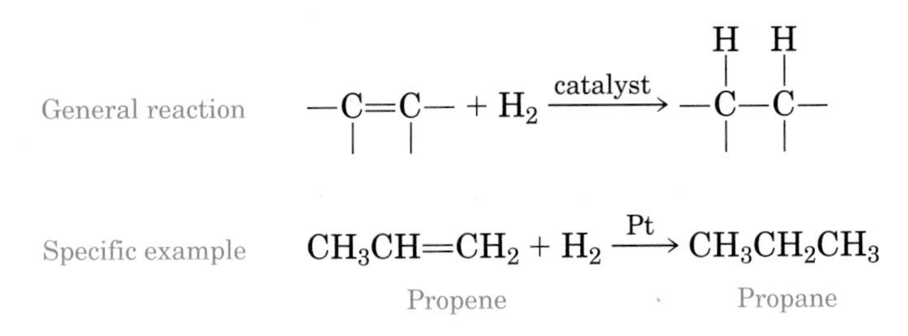

Ethene is converted to ethane, cyclohexene to cyclohexane, and so forth.

Any alkene can be converted to the corresponding alkane by reaction with hydrogen gas and a catalyst, which may be finely divided platinum,

Box 11C

Cis-Trans Isomerism in Vision

The retina, the light-detector layer in the back of our eyes, contains colored compounds called *visual pigments.* They are insoluble in water and can be extracted from the retina with aqueous detergents. In the dark these pigments are reddish (their name, rhodopsin, comes from a Greek word meaning rose-colored), but the color fades upon exposure to light.

The rhodopsin molecules contain a protein called opsin plus a derivative of vitamin A (see Box 11A) called 11-*cis*-retinal:

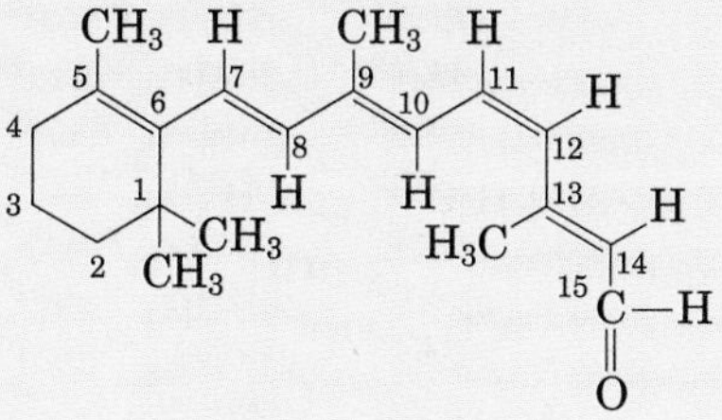

11-*cis*-Retinal

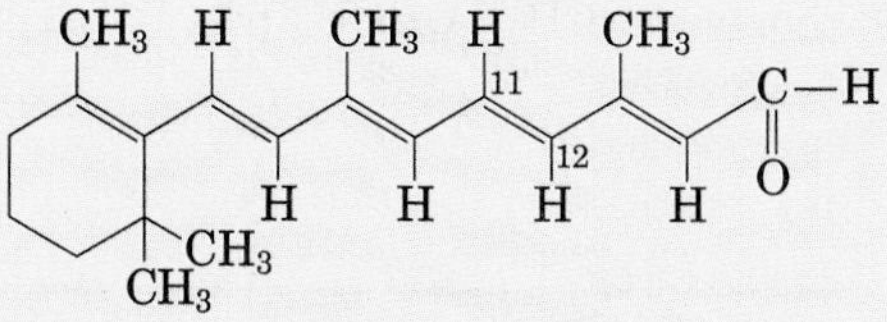

all-trans-Retinal

In the dark, 11-*cis*-retinal is stable and fits nicely into the folds of the surrounding opsin. When light hits the rhodopsin, the 11-*cis*-retinal becomes *all-trans*-retinal and no longer fits into the cavity of opsin. The opsin and the *all-trans*-retinal separate. The change in rhodopsin conformation is eventually transmitted to the nerve cells and then the brain. The stereoisomerism of retinal is thus an important part of the vision process.

nickel, rhodium, or some other transition metal or transition metal compound. The alkene and the catalyst are shaken in a bottle containing hydrogen gas at a pressure slightly above 1 atm. In this quick and convenient reaction, called **catalytic hydrogenation,** the unsaturated hydrocarbon (alkene) is *saturated* with hydrogen to become a saturated hydrocarbon (alkane).

In Section 17.3 we shall see how catalytic hydrogenation is used to solidify liquid oils.

Problem 11.5

Write the structural formula of the product:

$$CH_3CH_2CH_2CH{=}CH_2 + H_2 \xrightarrow{Pt}$$

Note that only one of the five double bonds is affected in this transformation, but when this one changes from cis to trans, the shape of the entire molecule is changed (Fig. 11C). An enzyme later catalyzes the change of *all-trans*-retinal back to 11-*cis*-retinal so that it can once again bind the opsin and wait for the next exposure to light.

(*a*)

(*b*)

Figure **11C** • Molecular models of (a) 11-*cis*-retinal and (b) *all-trans*-retinal show the different shapes of the two isomers. (Photographs by Beverly March.)

The retinas of vertebrates have two kinds of cells that contain rhodopsin. These cells are distinguished by their shapes: rods and cones. The cones, which function in bright light and are used in color vision, are concentrated in the central portion of the retina, called the macula, and are responsible for the greatest visual acuity. The remaining area of the retina consists mostly of rods, which are used for peripheral and night vision. 11-*cis*-Retinal is present both in cones and rods. However, the opsin is somewhat different in the two kinds of cells, and the cones have three different opsins, one kind each for perception of blue, green, and red colors.

Addition of Br_2 and Cl_2

General reaction

$$-\overset{|}{\underset{|}{C}}=\overset{|}{\underset{|}{C}}- + X_2 \longrightarrow -\overset{X}{\overset{|}{\underset{|}{C}}}-\overset{X}{\overset{|}{\underset{|}{C}}}-$$

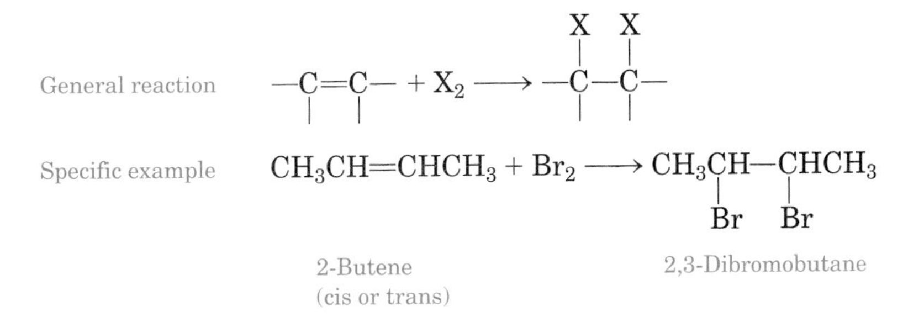

The letter X is frequently used to stand for a halogen atom.

Specific example

$$CH_3CH{=}CHCH_3 + Br_2 \longrightarrow CH_3\underset{\underset{Br}{|}}{C}H-\underset{\underset{Br}{|}}{C}HCH_3$$

2-Butene (cis or trans) — 2,3-Dibromobutane

Bromine adds to almost all alkenes to give a compound that has two bromines on adjacent carbons. Chlorine reacts in the same way. The reac-

Box 11D

Ethylene

Because of their reactivity, simple alkenes serve as starting materials for many products of the chemical industry. Drugs, paints, and plastics (Sec. 11.7) are the major products. Ethylene, the most important starting material, is obtained from petroleum. But ethylene is also a natural product, being produced in many fruits and vegetables upon ripening. It acts as a hormone, enhancing the softening process. Commercial marketers of these products take advantage of this property. The application of ethylene gas in warehouses produces a controlled ripening of fruits and vegetables that were picked green. In animal cells, ethylene acts as a messenger, in a manner similar to that of nitric oxide (Box 24F).

Green tomatoes. On their way to market these tomatoes may be ripened by exposure to ethylene gas. (Jan Halaska/Photo Researchers.)

tion is very similar to the one with H_2, and again the double bond of an alkene is converted to a single bond. The reaction is not useful for F_2 or I_2.

Problem 11.6

Write the structural formula of the product:

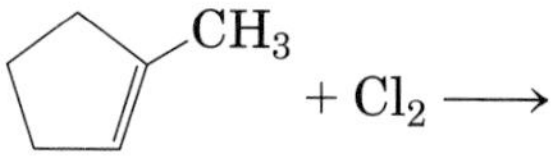

$$CH_3 \quad + Cl_2 \longrightarrow$$

Because almost all alkenes add bromine, this reaction is often used as a test for the presence of unsaturation (a double or triple bond). Organic chemists have at their disposal many such simple laboratory tests. What we want in a laboratory test is something that will give a quick answer (less than five minutes) and is simple to run. This test for unsaturation is an excellent example. All we do is add a drop of bromine (usually dissolved in carbon tetrachloride) to our unknown liquid and we get our answer at once. If the deep red color of Br_2 disappears, a double or triple bond is present (the color disappears because the Br_2 is reacting with the double or triple bond) (Fig. 11.4). Chlorine gives the same reaction but is not used as a test for unsaturation because it is difficult to see the pale green chlorine disappearing.

Addition of HX

General reaction

$$-\overset{|}{\underset{|}{C}}=\overset{|}{\underset{|}{C}}- + HX \longrightarrow -\overset{H}{\overset{|}{\underset{|}{C}}}-\overset{X}{\overset{|}{\underset{|}{C}}}-$$

Specific example

Cyclohexene + HBr ⟶ Bromocyclohexane (H and Br on adjacent carbons)

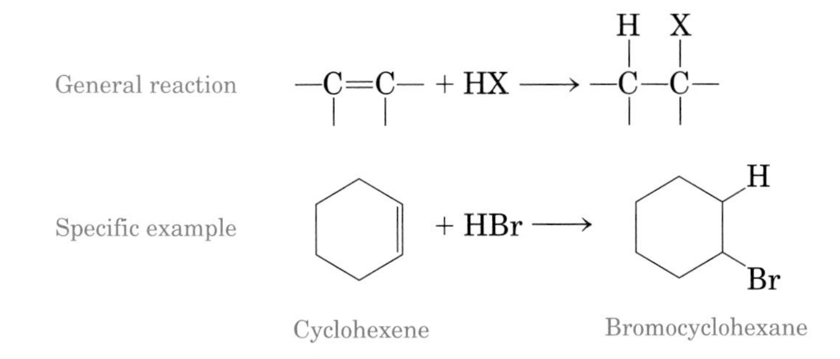

Any of the four hydrogen halides—HF, HCl, HBr, and HI—add to the double bond of an alkene to give the corresponding alkyl halide. The reaction is very similar to H_2 and X_2 addition. Once again, a double bond is converted to a single bond. However, we do encounter one new factor in this reaction. In H_2 and X_2 addition, the *same* group adds to each carbon atom (H and H or X and X). Here a *different* group is adding to each carbon (H and X). This means that, for most alkenes, two different products are possible. For example, when HCl adds to propene we can get either 1-chloropropane or 2-chloropropane:

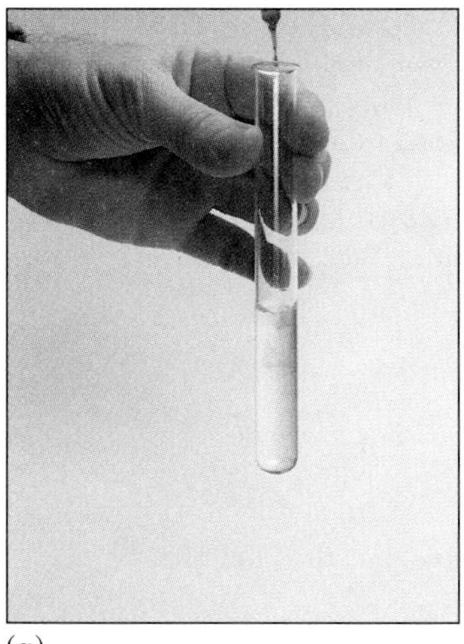
(*a*)

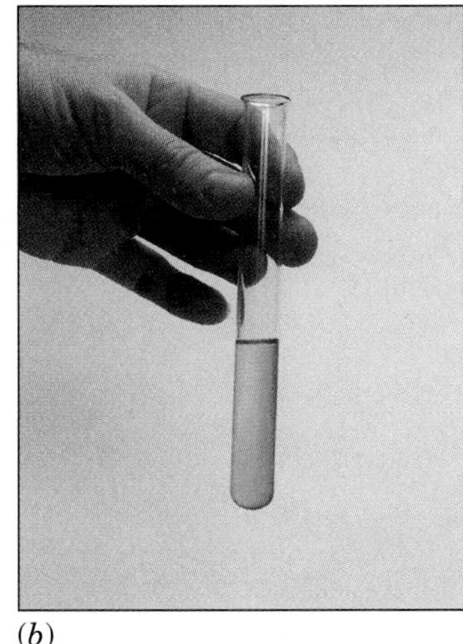
(*b*)

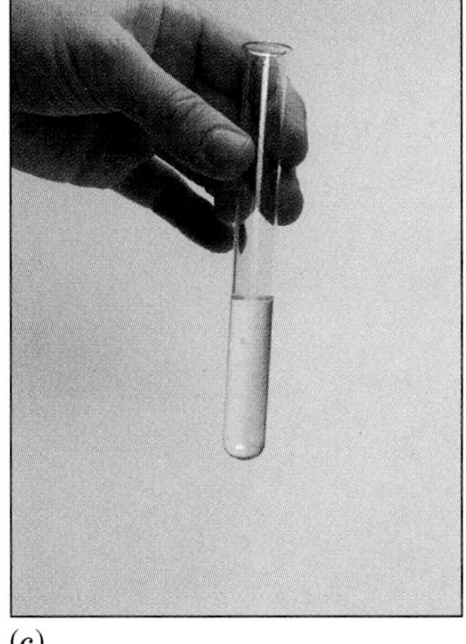
(*c*)

Figure **11.4** • (a) A drop of Br_2 dissolved in CCl_4 is added to an unknown liquid. If the color remains after stirring (b), it indicates the absence of unsaturation. If the color disappears (c), the unknown is unsaturated. (Photographs by Beverly March.)

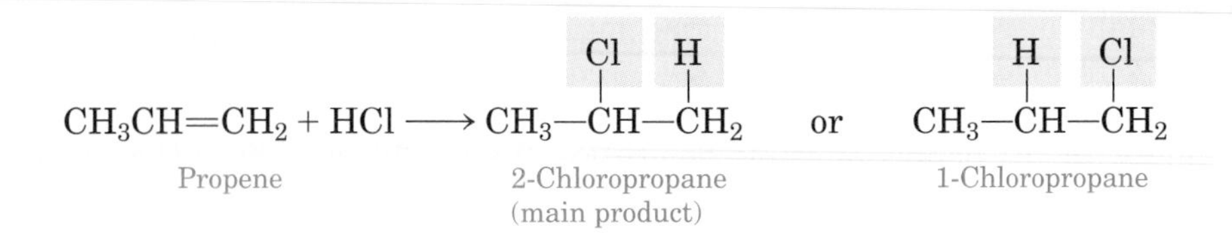

$$\underset{\text{Propene}}{CH_3CH{=}CH_2} + HCl \longrightarrow \underset{\substack{\text{2-Chloropropane}\\ \text{(main product)}}}{CH_3{-}\overset{\overset{\large Cl}{|}}{C}H{-}\overset{\overset{\large H}{|}}{C}H_2} \quad \text{or} \quad \underset{\text{1-Chloropropane}}{CH_3{-}\overset{\overset{\large H}{|}}{C}H{-}\overset{\overset{\large Cl}{|}}{C}H_2}$$

Markovnikov's rule is frequently stated as "the rich get richer."

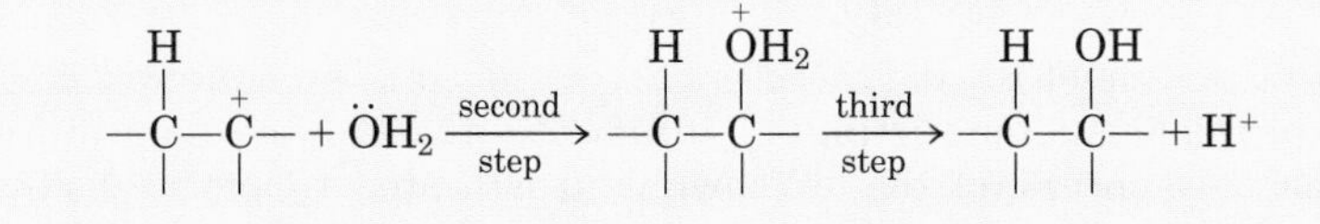

More than 100 years ago, the Russian chemist Vladimir Markovnikov (1838–1904), after studying many such cases, formulated a rule that predicts which product will be exclusively or predominantly formed. **Markovnikov's rule** states that, **in the addition of HX, the hydrogen goes to the carbon that already has more hydrogens.** The X

Box 11E

The Mechanism of HX Addition

Organic chemists spend a great deal of time investigating the mechanisms of reactions. A **mechanism** describes exactly how a reaction takes place: which bonds break, in what order, and so forth. Additions of HBr, HCl, HF, HI, and H_2O to double bonds all have essentially the same mechanism. The mechanism begins when an H^+ ion approaches the double bond and forms a bond by using one of the electron pairs of the double bond:

$$-\underset{|}{C}{=}\underset{|}{C}- + H^+ \longrightarrow \underset{\text{Carbocation}}{-\underset{|}{\overset{\overset{H}{|}}{C}}-\underset{|}{\overset{+}{C}}-} \qquad \text{First step of the mechanism}$$

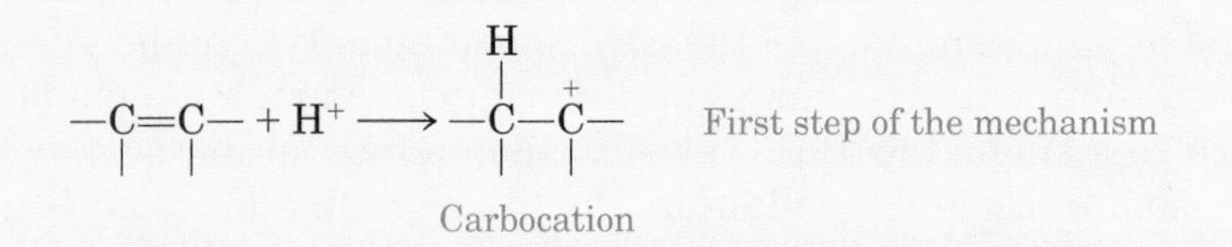

It is only natural that a double bond, which has a high electron density (because there are four electrons), should be attractive to a positive ion like H^+. One of the ways we know that it is H^+ attacking (and not, say, HBr) is that H_2O alone does not add to an alkene. An acid catalyst is necessary to supply H^+ ions.

Let us now look at the ion formed from the alkene and the H^+. This ion has a positive charge on a carbon atom. Such ions are called **carbocations.** A carbocation is extremely reactive. It cannot last for more than a tiny fraction of a second, and so of course it is impossible to isolate it in a stable salt, as we can isolate the Na^+ ion as NaCl, for example. As soon as it is formed, the carbocation reacts with any species in its vicinity that can supply a pair of electrons. In the case of HX this is a halide ion, so the second step of the mechanism is

$$-\underset{|}{\overset{\overset{H}{|}}{C}}-\underset{|}{\overset{+}{C}}- + Br^- \longrightarrow -\underset{|}{\overset{\overset{H}{|}}{C}}-\underset{|}{\overset{\overset{Br}{|}}{C}}-$$

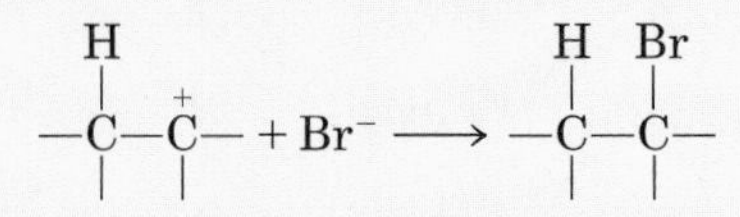

In the case of water, there is no negative ion available, but the O of H_2O has two unshared electron pairs, one of which is used:

$$-\underset{|}{\overset{\overset{H}{|}}{C}}-\underset{|}{\overset{+}{C}}- + \ddot{O}H_2 \xrightarrow[\text{step}]{\text{second}} -\underset{|}{\overset{\overset{H}{|}}{C}}-\underset{|}{\overset{\overset{\overset{+}{O}H_2}{|}}{C}}- \xrightarrow[\text{step}]{\text{third}} -\underset{|}{\overset{\overset{H}{|}}{C}}-\underset{|}{\overset{\overset{OH}{|}}{C}}- + H^+$$

goes to the carbon with fewer hydrogens. The only hydrogens we count are those *directly* attached to the carbons of the double bond. Applying this rule to propene, we find

$$\overset{3}{CH_3}—\overset{2}{CH}=\overset{1}{CH_2}$$

Two hydrogens (C-1); One hydrogen (C-2)

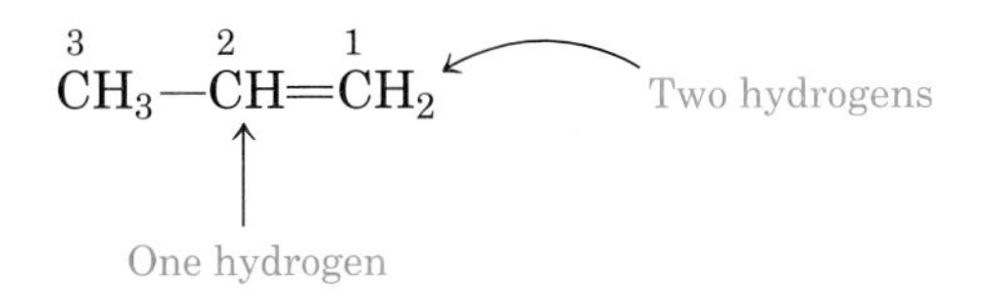

Therefore, the hydrogen of HX goes to the end carbon, and the main product is 2-chloropropane. The other compound, 1-chloropropane, is formed only in small quantities or not at all.

In this case a third step follows: loss of H^+ to give the alcohol. Note that the catalyst, H^+, that started the reaction is recovered unchanged.

A knowledge of mechanisms helps us to understand the facts about a reaction: why some go but others don't, and why those that go proceed the way they do. For example, understanding the mechanism of HX addition allows us to understand why Markovnikov's rule works (which is something Markovnikov himself didn't understand because the mechanism was unknown in his time). If we add HX to propene, the first step is attack by H^+. This could lead to two different carbocations:

$$\underset{\text{Propene}}{\overset{3}{C}H_3\overset{2}{C}H=\overset{1}{C}H_2} + H^+ \xrightarrow{\text{attack at C-1}} CH_3—\overset{+}{C}H—CH_3 \quad \text{Secondary carbocation}$$

$$\underset{\text{Propene}}{\overset{3}{C}H_3\overset{2}{C}H=\overset{1}{C}H_2} + H^+ \xrightarrow{\text{attack at C-2}} CH_3—CH_2—\overset{+}{C}H_2 \quad \text{Primary carbocation}$$

We know from much other evidence that secondary carbocations (where the positive charge is on a secondary carbon) are more stable than primary carbocations (tertiary carbocations are even more stable). Therefore, the H^+ is much more likely to attach to C-1 rather than C-2 because that gives a secondary rather than a primary carbocation. Once the secondary carbocation is formed, it must give the secondary halide or alcohol:

$$CH_3—\overset{+}{C}H—CH_3 + X^- \longrightarrow CH_3—\underset{}{\overset{X}{\overset{|}{C}H}}—CH_3 \quad \text{Second step of the mechanism}$$

A carbocation is an example of an **intermediate,** which is a species that is formed in a reaction but then is used up so quickly that it cannot be isolated.

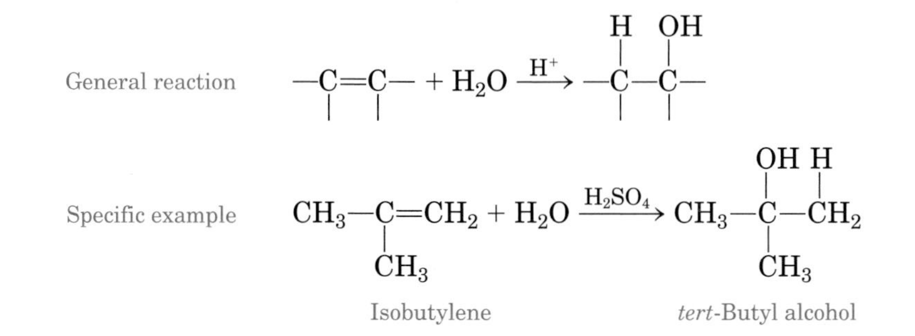

Problem 11.7

Write the structural formula of the principal product:

$$CH_3CH_2CH_2-\underset{\displaystyle CH_3}{\underset{|}{C}}=CH_2 + HBr \longrightarrow$$

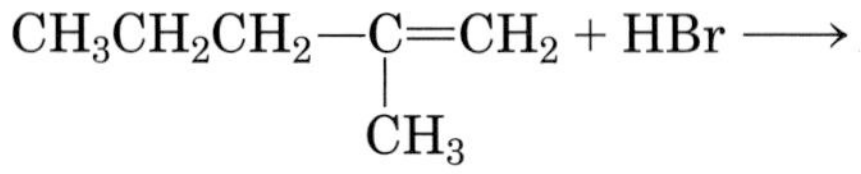

Addition of H_2O

General reaction

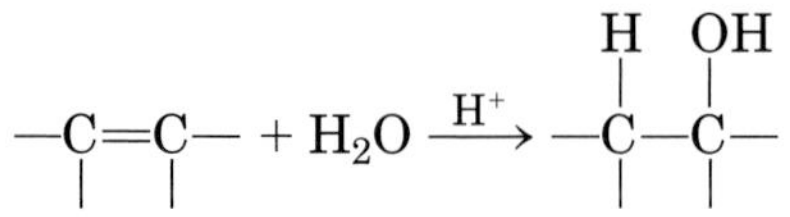

$$-\underset{|}{C}=\underset{|}{C}- + H_2O \xrightarrow{H^+} -\underset{|}{\overset{\overset{\displaystyle H}{|}}{C}}-\underset{|}{\overset{\overset{\displaystyle OH}{|}}{C}}-$$

Specific example

$$CH_3-\underset{\displaystyle CH_3}{\underset{|}{C}}=CH_2 + H_2O \xrightarrow{H_2SO_4} CH_3-\underset{\displaystyle CH_3}{\underset{|}{\overset{\overset{\displaystyle OH}{|}}{C}}}-\overset{\overset{\displaystyle H}{|}}{CH_2}$$

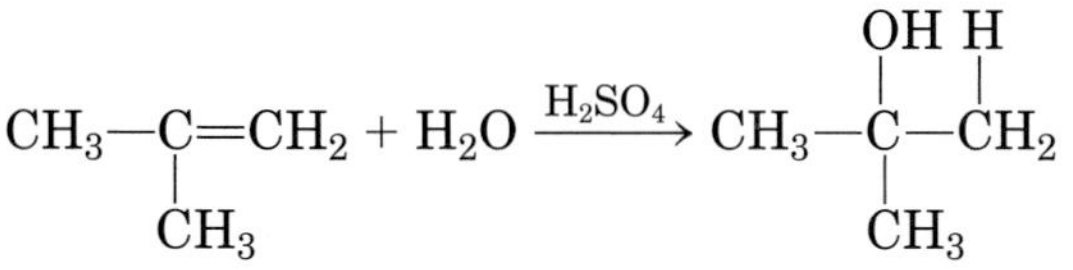

Isobutylene — *tert*-Butyl alcohol

In the absence of a catalyst, water does not react with alkenes. But if an acid catalyst—usually sulfuric acid—is added, water does add to C=C double bonds to give alcohols. The reaction is called **hydration.** As shown in the isobutylene reaction, Markovnikov's rule is followed: The H of the H_2O goes to the carbon that already has more hydrogens.

Problem 11.8

Write the structural formula of the principal product:

$$CH_3-CH=CH_2 + H_2O \xrightarrow{H_2SO_4}$$

Most industrial ethyl alcohol, CH_3CH_2OH, is made by the hydration of ethylene, $CH_2=CH_2$.

The conversion of alkenes to alcohols is important not only in the organic chemist's laboratory and industrially but also in the body, where the hydration of double bonds is catalyzed by enzymes rather than by acids. An important example is the hydration of fumaric acid to give malic acid, catalyzed by the enzyme fumarase:

This is one of the steps in the metabolism of carbohydrates, fats, and proteins (see Sec. 20.4).

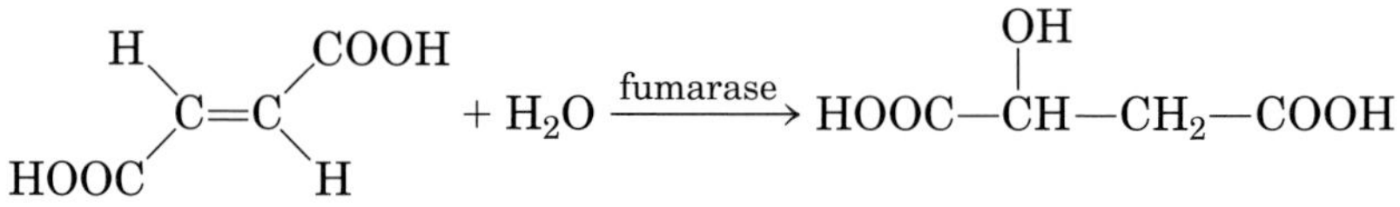

$$\begin{matrix} H & & COOH \\ & C=C & \\ HOOC & & H \end{matrix} + H_2O \xrightarrow{\text{fumarase}} HOOC-\overset{\overset{\displaystyle OH}{|}}{CH}-CH_2-COOH$$

Fumaric acid — Malic acid

Besides the addition reactions just discussed, two other chemical properties of double-bond compounds are worthy of mention. First, like alkanes and all other hydrocarbons, alkenes burn; the products of complete combustion are CO_2 and H_2O:

$$CH_2=CH_2 + 3O_2 \longrightarrow 2CO_2 + 2H_2O$$

Second, double-bond compounds can be polymerized to give long chains. This topic is treated in the following section.

11.7 Addition Polymers

Possibly the most important reaction of alkenes and other compounds that contain C=C bonds is that, when treated with appropriate catalysts, they add to each other to give long chains. For example, the simplest alkene, ethylene, polymerizes as follows:

$$5\,\mathrm{H_2C{=}CH_2} \xrightarrow{\text{catalyst}} \mathrm{{-}CH_2{-}CH_2{-}CH_2{-}CH_2{-}CH_2{-}CH_2{-}CH_2{-}CH_2{-}CH_2{-}CH_2{-}}$$

Polyethylene

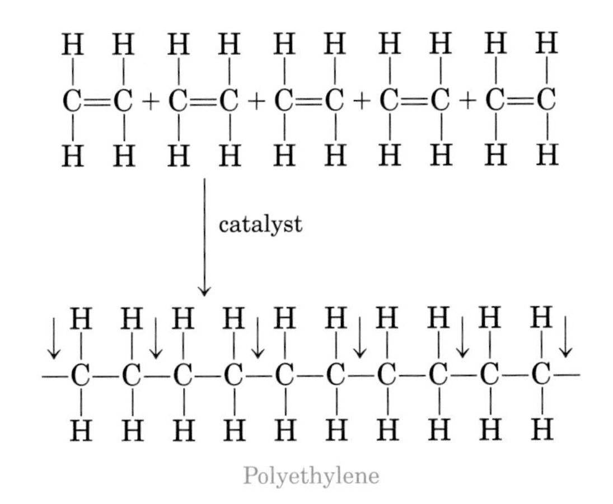

Arrows point to newly formed bonds.

Under the influence of the catalyst, each double bond becomes a single bond, and the extra electrons form new single bonds that tie the molecules together into long chains. **A long chain made up of repeating units** is called a **polymer,** and the process of making it is **polymerization.** Later in the book we will meet many biological polymers. The compounds made by polymerizing alkenes do not exist in nature; they are synthetic polymers. The simple compound that is polymerized (in the above case, ethylene) is called a **monomer.**

"Poly," from the Greek *polys,* means many.

There are two main types of synthetic polymers. The ones we are discussing in this section are called **addition polymers** because they are made by adding double-bonded molecules to each other. (The other kind, **condensation polymers,** are discussed in Boxes 14E and 15D.) The polymer made by polymerizing ethylene is called polyethylene. The molecules of this polymer are essentially unbranched alkanes thousands of units long. When another group is present in the monomer, the polymer will have this group on every second carbon atom. For example, polymerization of vinyl chloride gives the polymer polyvinyl chloride (PVC):

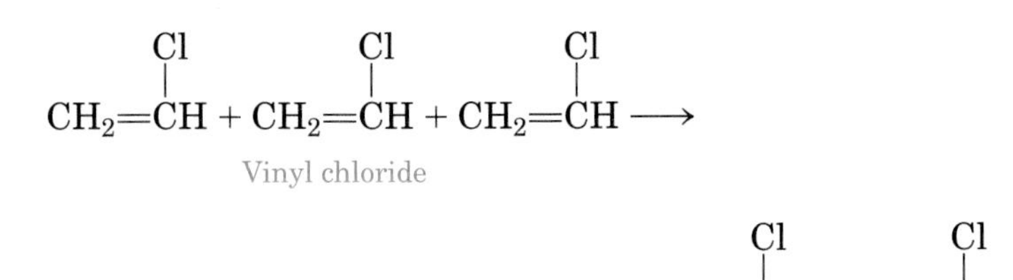

$$\mathrm{CH_2{=}CHCl + CH_2{=}CHCl + CH_2{=}CHCl} \longrightarrow$$

Vinyl chloride

$$\mathrm{{-}CH_2{-}CH(Cl){-}CH_2{-}CH(Cl){-}CH_2{-}CH(Cl){-}}$$

Polyvinyl chloride

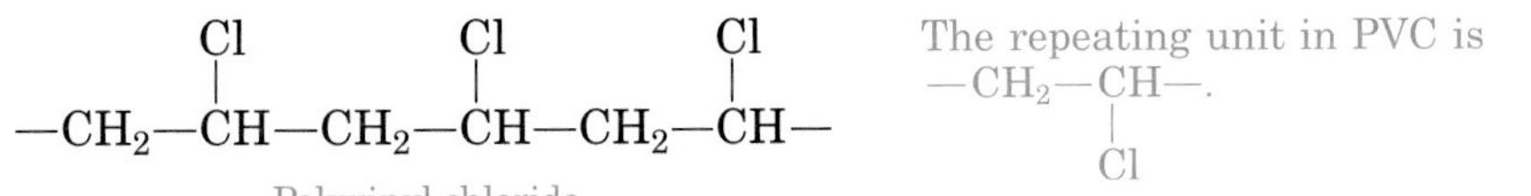

The repeating unit in PVC is $\mathrm{{-}CH_2{-}CH(Cl){-}}$.

Box 11F

Rubber

An important example of a naturally occurring addition polymer is rubber. Natural rubber is obtained upon coagulation of the sap (called *latex*) of the *Hevea* tree. Rubber differs from the other polymers discussed in this section in that the monomer has *two* double bonds, so the polymer still contains double bonds:

$$CH_2{=}\underset{}{\overset{CH_3}{\overset{|}{C}}}{-}CH{=}CH_2 + CH_2{=}\overset{CH_3}{\overset{|}{C}}{-}CH{=}CH_2 + CH_2{=}\overset{CH_3}{\overset{|}{C}}{-}CH{=}CH_2$$

$$\downarrow$$

$$\downarrow{-}CH_2{-}\overset{CH_3}{\overset{|}{C}}{\overset{\downarrow}{=}}CH{-}CH_2\overset{\downarrow}{-}CH_2{-}\overset{CH_3}{\overset{|}{C}}{\overset{\downarrow}{=}}CH{-}CH_2\overset{\downarrow}{-}CH_2{-}\overset{CH_3}{\overset{|}{C}}{\overset{\downarrow}{=}}CH{-}CH_2{-}\downarrow$$

Arrows point to newly formed bonds

The monomer is 2-methyl-1,3-butadiene, which has the common name **isoprene.** All the double bonds in natural rubber have the cis configuration, so the chemical name is *all-cis*-polyisoprene. The most important property of rubber is its elasticity; it can be extended a considerable amount, and then, when released, it snaps back to its original dimensions. However, raw rubber is too soft and tacky to be of much use, and it must be hardened by heating it with sulfur, a process called **vulcanization.** This process creates sulfur *cross links,* which hold the polymer chains together:

Rubber. The sap that comes from this tree is a natural polymer of isoprene.

Two to three percent sulfur gives rubber that is used for such things as shoe soles, erasers (which is where rubber got its name; it was used to rub out pencil marks), and rubber bands, while more sulfur (5 to 8 percent) gives rubber that can be used for tires.

Although natural rubber is a highly useful material, it is nevertheless not the best material for all applications where an elastic substance is needed. Chemists have been able to polymerize dienes other than isoprene to obtain **synthetic rubbers,** some of which are copolymers. Worldwide, nearly two thirds of all rubber used today is synthetic. The largest use of rubber is in tires, and most tires contain both natural rubber (which withstands the heat better) and one or more kinds of synthetic rubber.

Problem **11.9**

Draw the structural formula of polyacrylonitrile, made by polymerizing cyanoethene, $CH_2{=}CH{-}CN$.

Some of the more important commercial addition polymers are shown in Table 11.2. Saran is an example of a **copolymer,** which is a polymer made up of two or more different monomers.

As you can tell by looking at some of the familiar names in Table 11.2, synthetic polymers are very important in modern life, and we would find it hard to get along without them. Among the synthetic polymers are plastics. Some of these are rigid and are used for such things as combs, ballpoint pens, toys, and bottles for soft drinks and other liquids. Other plastics are produced as thin sheets and are used to wrap foods

Table 11.2 Some Important Addition Polymers

Monomer		**Polymer**	
Name	***Structure***	***Structure***	***Name***
Ethylene	$CH_2{=}CH_2$	$-CH_2-CH_2-CH_2-CH_2-$	Polyethylene
Propylene	$CH_2{=}CH(CH_3)$	$-CH_2-CH(CH_3)-CH_2-CH(CH_3)-$	Polypropylene
Styrene	$CH_2{=}CH(C_6H_5)$	$-CH_2-CH(C_6H_5)-CH_2-CH(C_6H_5)-$	Polystyrene
Vinyl chloride (chloroethene)	$CH_2{=}CH(Cl)$	$-CH_2-CH(Cl)-CH_2-CH(Cl)-$	Polyvinyl chloride (PVC)
Acrylonitrile (cyanoethene)	$CH_2{=}CH(CN)$	$-CH_2-CH(CN)-CH_2-CH(CN)-$	Polyacrylonitrile
Methyl methacrylate	$CH_2{=}C(CH_3)(COOCH_3)$	$-CH_2-C(CH_3)(COOCH_3)-CH_2-C(CH_3)(COOCH_3)-$	Acrylic plastics
Tetrafluoroethene	$CF_2{=}CF_2$	$-CF_2-CF_2-CF_2-CF_2-$	Teflon
Vinylidine chloride + Vinyl chloride	$CH_2{=}C(Cl)(Cl)$ + $CH_2{=}CH(Cl)$	$-CH_2-CCl_2-CH_2-CHCl-$	Saran (copolymer)

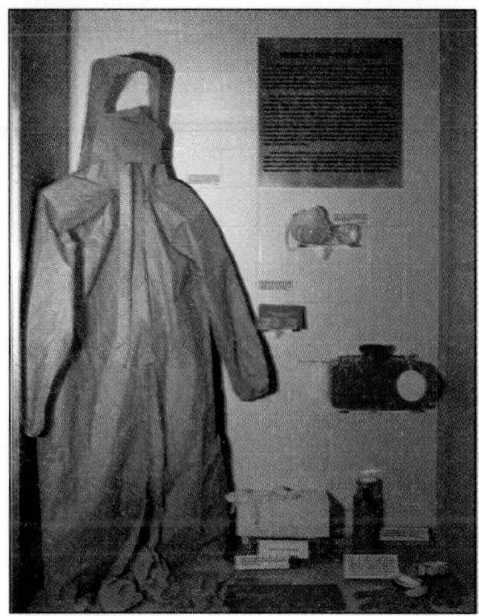

Figure **11.5**
A plastic garment and other materials used as protection when treating people with AIDS. (Courtesy of the National Museum of Health and Medicine, Armed Forces Institute of Pathology, Washington, D.C.)

and other merchandise. Others are used as artificial fibers (nylon and Dacron, Boxes 14E and 15D), synthetic rubbers and leathers, adhesives, and coatings (for example, the glazed surface of refrigerators), as well as other uses. Figure 11.5 shows a plastic garment used by surgeons as protection when treating patients with AIDS.

In polymers the chemical structure is less important in defining physical properties than are other attributes such as molecular weight or the orientation of the molecules. The polymer polyethylene supplies an example. Low-molecular-weight polyethylene (MW 500 to 600) has the consistency of grease and is used for lubrication. The ordinary polyethylene bag (low-density polyethylene) used in packaging owes its flexibility to short branches along the chain (the branches prevent the molecules from getting very close to each other, so that crystals cannot be formed), while high-density linear polyethylene (without branches) is more rigid and is used in manufacturing pipes, bottles, toys, and a host of houseware items. Both have molecular weights up to a few hundred thousand. Polyethylene that is solidified (crystallized) while being extended exhibits even more interesting properties. Here the molecular

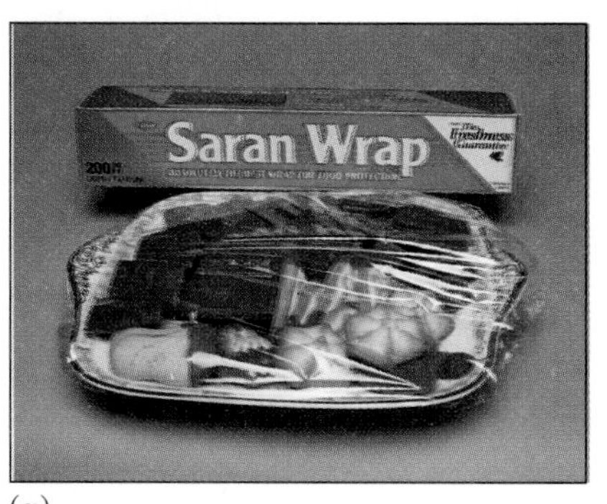

(*a*)

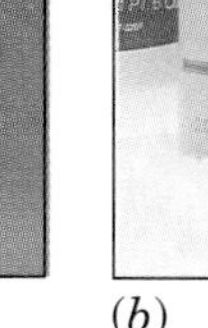

(*b*)

(*c*)

(*d*)

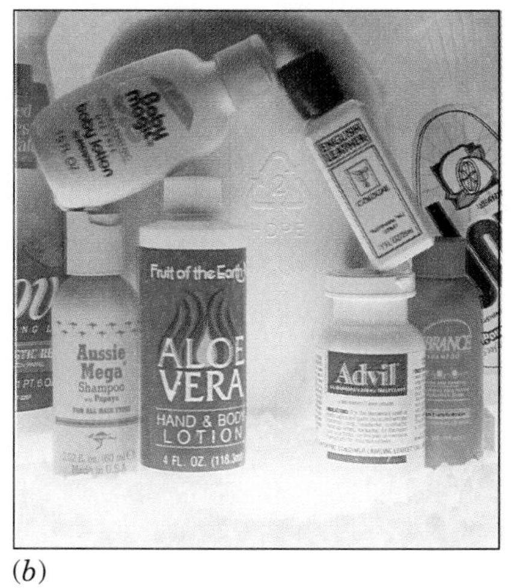

Some articles made from synthetic addition polymers.
(a) Saran wrap, a copolymer of vinyl chloride and vinylidene chloride. (b) Plastic containers for various supermarket products, mostly made of polyethylene and polypropylene. (c) Teflon-coated kitchenware. (d) Articles made from polystyrene. (Photographs (a) & (c) by Beverly March. Photographs (b) & (d) by C. D. Winters.)

weights are up to millions, and the molecules are stretched out, unfolded, and oriented in one direction. Such polyethylene films and fibers (manufactured under the name of Spectra) are so strong that they are used in manufacturing bullet-proof vests, riot helmets, and cut-resistant gloves for surgeons to wear under surgical latex gloves.

The importance of polymers can be judged by the fact that they make up more than half the output of our chemical factories.

11.8 Alkynes

Alkynes are named by the rules used for alkenes except that the suffix "-yne" is used for the triple bond:

$$HC{\equiv}CH \qquad CH_3{-}C{\equiv}CH \qquad CH_3{-}\overset{\displaystyle CH_3}{\overset{|}{C}H}{-}C{\equiv}C{-}CH_3$$

Ethyne (acetylene) — Propyne — 4-Methyl-2-pentyne

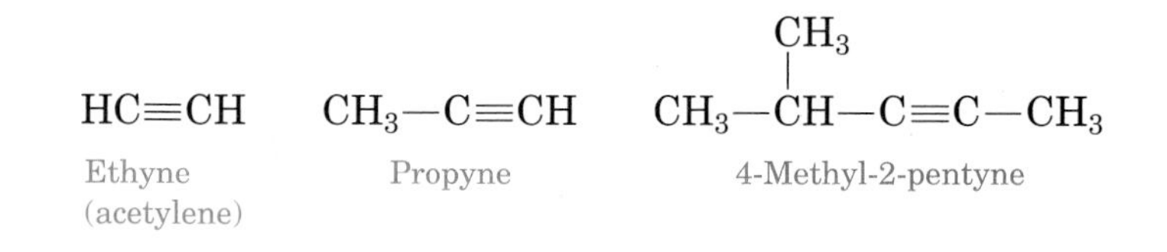

Ethyne, the simplest and by far the most important alkyne, is much more often called by its common name, acetylene.

The high temperature of an acetylene flame makes it useful for torches that can cut and weld steel.

We can see the geometry of triple bonds by examining the hybridization. A triple-bond carbon is attached to only two other atoms, so it hybridizes as follows:

$$\underbrace{\boxed{\uparrow}\,\boxed{\uparrow}}\,\boxed{\uparrow}\,\boxed{\uparrow} \longrightarrow \boxed{\uparrow}\,\boxed{\uparrow}\,\boxed{\uparrow}\,\boxed{\uparrow}$$

$s \quad p_x \quad p_y \quad p_z \qquad sp \quad sp \quad p_y \quad p_z$

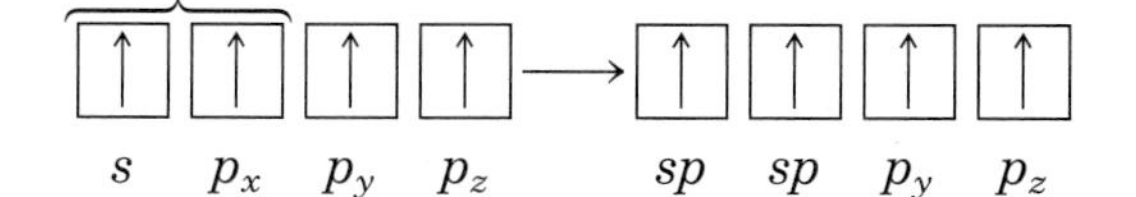

We now have two hybrid orbitals, each called *sp*. The ***sp* orbitals** have 50 percent *s* and 50 percent *p* character, and the angle between the two is 180°. Both of the remaining *p* orbitals are perpendicular to the *sp* orbitals and to each other (Fig. 11.6). Thus two carbon atoms that have *sp*

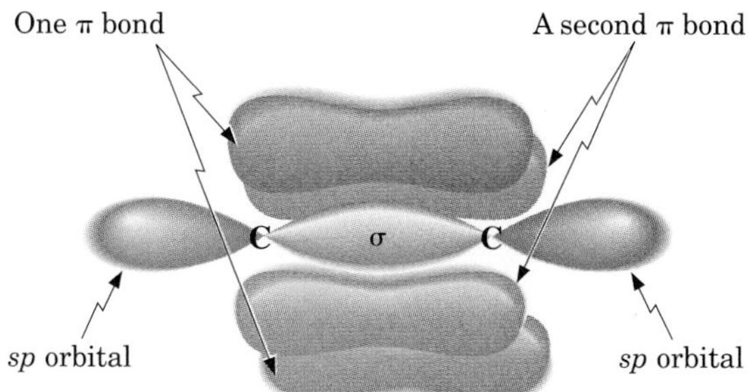

Figure **11.6** • A triple bond is made of one σ and two π bonds.

hybrid orbitals form a triple bond: one σ bond formed by the direct overlap of *sp* hybrid orbitals, and two π bonds formed by the parallel sideways overlap of the p_y and p_z orbitals of both carbons (Fig. 11.6). The remaining *sp* orbitals (one from each carbon) form σ bonds with the *s* orbitals of hydrogen atoms. This causes acetylene to be a linear molecule—all four atoms are in a straight line:

$$H{-}C{\equiv}C{-}H$$

The same geometry is found for all other alkynes: The two carbon atoms and the two atoms connected to them are all in a straight line.

Alkynes also undergo addition reactions similar to those undergone by alkenes.

11.9 Aromatic Hydrocarbons

The physical properties of aromatic hydrocarbons—boiling points, solubilities, densities, etc.—are similar to those of the aliphatic hydrocarbons.

All the hydrocarbons discussed so far—alkanes, alkenes, and alkynes (including the cyclic ones)—are called **aliphatic hydrocarbons.** More than 150 years ago, it became apparent to the early organic chemists that there was another kind of compound, one whose chemical properties are quite different from those of the aliphatic compounds. Because some of them have pleasant odors, they were named **aromatic compounds.** Today we know that most aromatic compounds do not have pleasant odors (in fact, some are downright unpleasant).

Benzene, the simplest and most important aromatic hydrocarbon, was discovered by Michael Faraday (1791–1867) in 1825.

If aromatic compounds were unsaturated, they would add Br_2 the same way alkenes and alkynes do.

The most common aromatic hydrocarbon is benzene. Between 1860 and 1880, benzene and the other aromatic compounds known at that time presented a major challenge to the theories of organic structure. The problem was that benzene has a molecular formula of C_6H_6, and a compound with so few hydrogens for its six carbons (compare hexane, C_6H_{14}) should be quite unsaturated, with several double and/or triple bonds. But benzene does not *behave* like an unsaturated compound. It does not give a positive Br_2 test; in fact, it does not react with Br_2 at all except in the presence of a catalyst (Sec. 11.11). When benzene does undergo reactions, they are *substitution* reactions, not *addition* reactions:

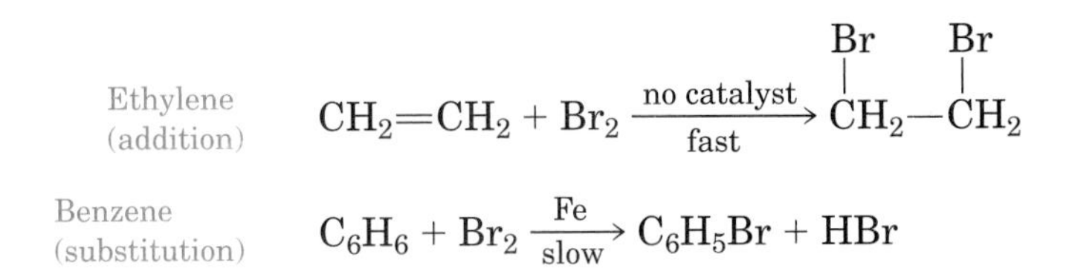

Benzene is an important industrial compound, but it must be handled carefully because it is toxic. Not only is it poisonous if ingested in the liquid form, but the vapor form is also toxic and can be absorbed either by breathing or through the skin. Long-term inhalation can cause liver damage and cancer.

Ethylene (addition)
$$CH_2{=}CH_2 + Br_2 \xrightarrow[\text{fast}]{\text{no catalyst}} \underset{\displaystyle Br}{\overset{}{C}}H_2{-}\underset{\displaystyle Br}{C}H_2$$

(Product: $BrCH_2{-}CH_2Br$, with Br on each carbon.)

Benzene (substitution)
$$C_6H_6 + Br_2 \xrightarrow[\text{slow}]{\text{Fe}} C_6H_5Br + HBr$$

This is a major difference in chemical properties and shows that aromatic compounds, despite their low hydrogen content, cannot be regarded as unsaturated molecules. The full story of how the dilemma was finally resolved is too long to tell here. We merely say that the man who did the most to resolve it was Kekulé, who proposed that the six carbons of benzene are connected in a ring. Today we know that benzene consists of a ring of six carbon atoms, each connected to a hydrogen atom (structure **A**):

```
          H
          |
   H      C      H
    \   /   \   /
      C       C
      |       |
      C       C
    /   \   /   \
   H      C      H
          |
          H
```

A

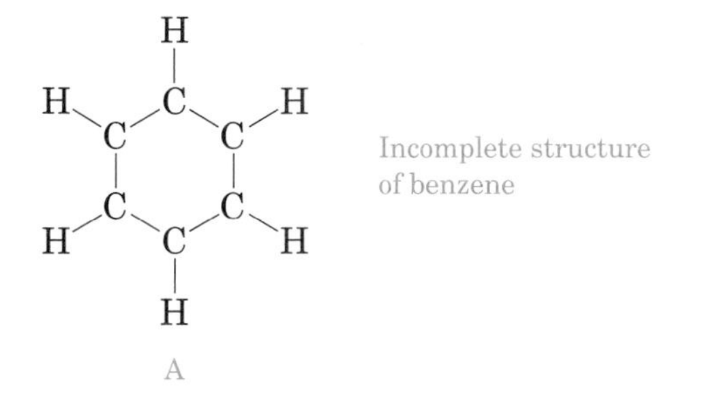

Incomplete structure of benzene

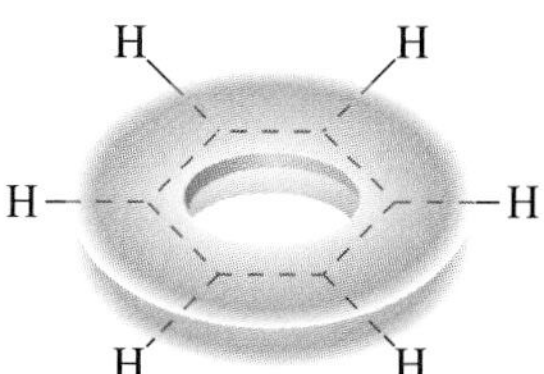

Figure **11.7**
Benzene. The two donut-shaped clouds hold a total of six electrons.

This gives each carbon three bonds. Carbon, of course, forms four bonds, and so each carbon has one remaining electron. These six electrons are located in a cloud (made up of three molecular orbitals) that looks like two donuts (Fig. 11.7). All twelve atoms—six carbons and six hydrogens—are in the same plane (benzene is a *flat* molecule); one donut lies just above the plane, the other just below it.

These three orbitals are also called π orbitals, because they result from the sideways overlap of p orbitals.

The picture of benzene shown as structure **A** does not violate the rule that carbon forms four bonds. Each carbon is connected to three other atoms by ordinary covalent bonds. The fourth bond can be thought

Friedrich August Kekulé. (Courtesy of E. F. Smith Memorial Collection, Center for History of Chemistry, University of Pennsylvania.)

of as being shared by all six carbons to create the donut-shaped clouds (the **aromatic sextet, or ring,** of electrons). It is true that Figure 11.7 is not a Lewis structure (Sec. 3.4), but benzene and other aromatic compounds simply cannot be represented by a single Lewis structure.

However, there is a way to represent benzene by Lewis structures. In this method we must draw not one, but two, Lewis structures:

The double-headed arrow is used to indicate resonance. It has no other use.

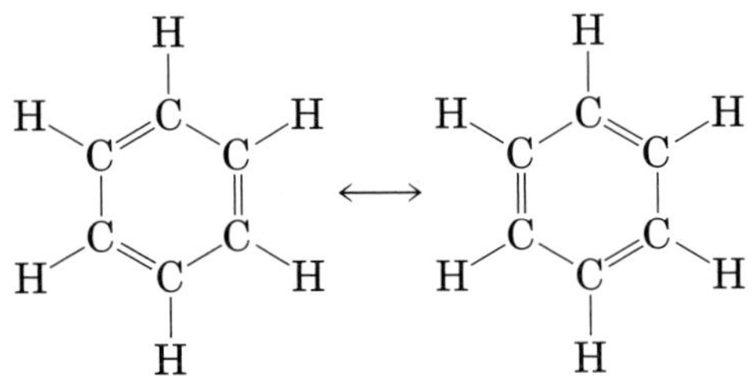

B ⟷ C

We start with the three ordinary covalent bonds for each carbon (structure **A**) and put in the extra six electrons as three double bonds. Because there are two ways to do this, we get two Lewis structures, **B** and **C.** Both are valid Lewis structures. However, neither of these structures is the correct formula for benzene. (As we have already said, benzene cannot be represented by a single Lewis structure.) We know that neither can be the correct structure because if either one were, then benzene would have three double bonds and would undergo addition reactions. **When we write these structures, we understand that the true benzene is neither of them but has a structure halfway between them.** In other words, *real* benzene is an *average* of **B** and **C.**

Chemists often use fictitious resonance structures because they obey the rules for Lewis structures, and chemists are comfortable with them.

The use of fictitious structures to represent a real structure that is the average of them is called **resonance.** It must always be understood that structures like **B** and **C** are fictitious—they don't exist. The real molecule is an average, or **hybrid,** of the two fictitious ones.

Wherever we find resonance, we find stability. The real structure is always more stable than the fictitious structures. (If any fictitious structure were more stable, it would exist.) The benzene ring is greatly stabilized by resonance. That is why benzene does not undergo addition reactions. If, say, Br_2 added to benzene, it would have to pull two electrons out of the aromatic sextet, resulting in a molecule without an aromatic sextet.

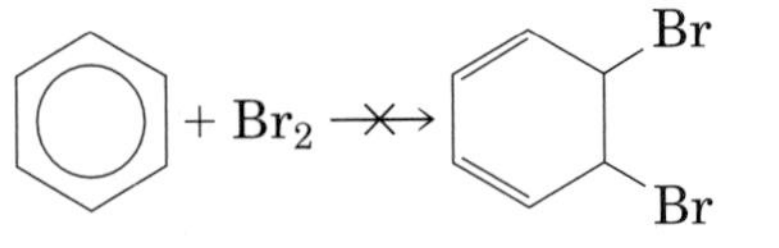

+ Br_2 ⟶ (does not occur) D

A reaction that doesn't happen

This does not happen because molecules with aromatic sextets are more stable than similar molecules (such as **D**) without aromatic sextets.

Both the resonance picture and the aromatic sextet picture shown in Figure 11.7 predict that all six C—C bonds in benzene should be the same length. Bond distances can be measured by x-ray crystallography, and such measurements show that the six bonds are indeed all the same length.

We can now define an **aromatic compound** as **one that contains an aromatic loop (sextet) of electrons,** just as benzene itself does. Because of the aromatic sextet, which is donut-shaped, we use formulas like **E,** with a circle, to represent benzene and most other aromatic compounds. However, many chemists use the Kekulé type formulas **F** or **G:**

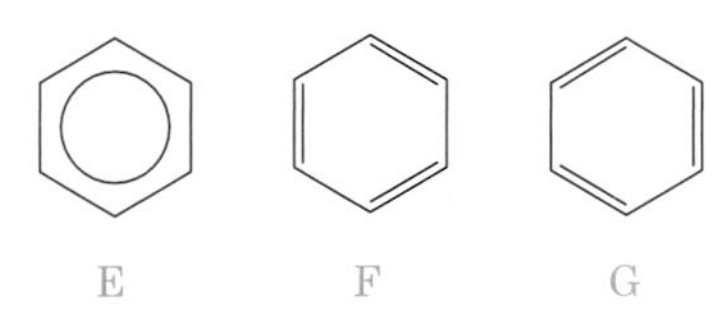

E F G

When you see formulas like these, don't forget that each carbon is connected to a hydrogen. We don't usually show the six hydrogens, but they are there just the same.

This is strictly a matter of preference. Whichever you choose, you should realize that they all mean the same thing. The real structure of benzene is that shown in Figure 11.7.

11.10 Nomenclature of Benzene Derivatives

A benzene ring may have one or more of its six hydrogens replaced by other groups. They are named as follows.

One Group

Some monosubstituted benzenes are named by giving the name of the group followed by the parent "benzene":

Cl NO_2 CH_2CH_3 Br

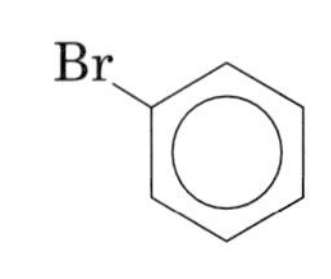

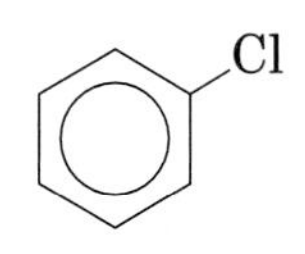

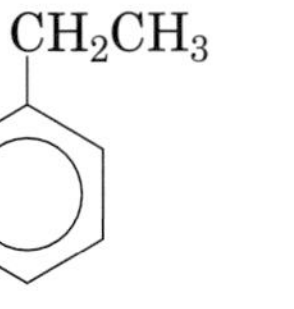

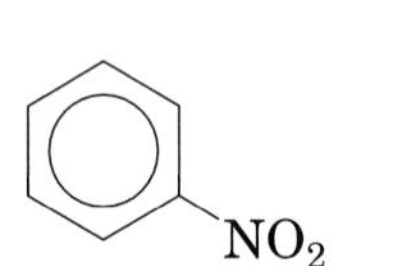

Chlorobenzene Nitrobenzene Ethylbenzene Bromobenzene

Others, however, are always named by common names:

CH_3 HO NH_2 C—OH ‖ O

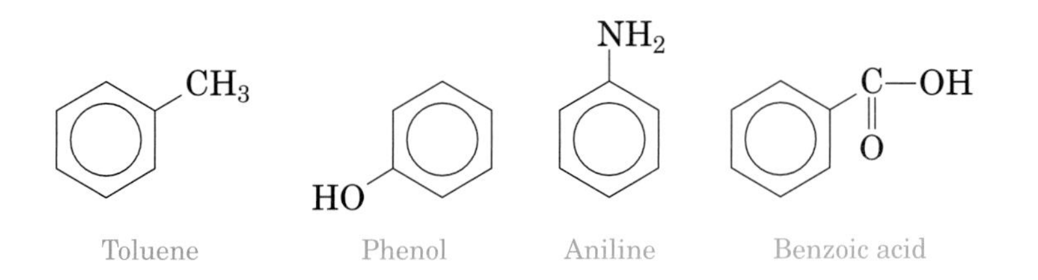

Toluene Phenol Aniline Benzoic acid

These four compounds are so important that their names should be memorized.

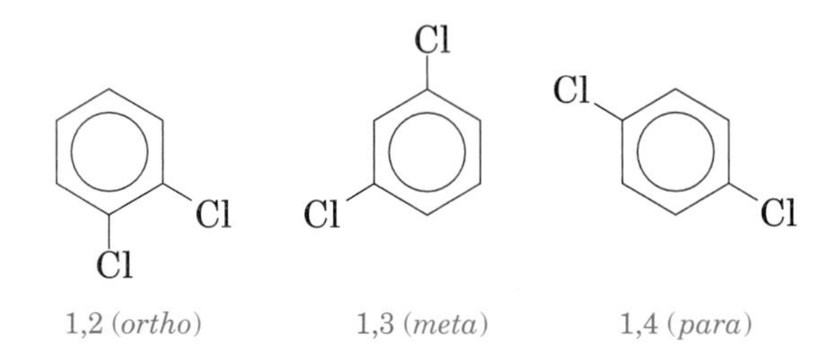

Note that all six benzene hydrogens are equivalent so it makes no difference whether we write the group up, down, right, or left.

Two Groups

When there are two groups on a benzene ring (disubstituted benzenes), three isomers are possible, whether the groups are the same or different:

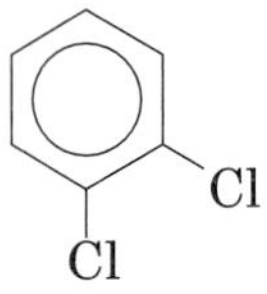

1,2 (*ortho*) 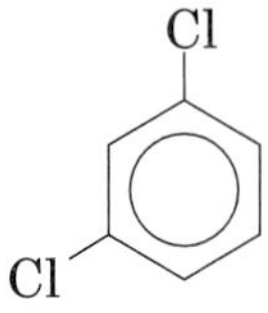 1,3 (*meta*) 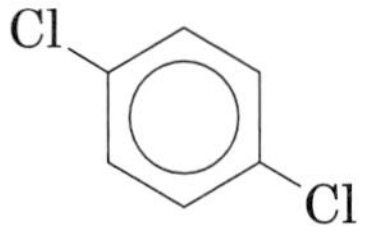 1,4 (*para*)

The letters *o*, *m*, and *p*, are often used as abbreviations for *ortho, meta,* and *para.*

All these are dichlorobenzene, and, as usual, we distinguish the *positions* of the groups by prefixes. There are two ways to do this. We can use numbers, following the rule for cyclohexanes (rule 7 in Sec. 10.8). However, a much more common way is to use prefixes: *ortho* for 1,2; *meta* for 1,3; and *para* for 1,4.

When one of the two groups produces a compound that has a common name, we use that name as the parent:

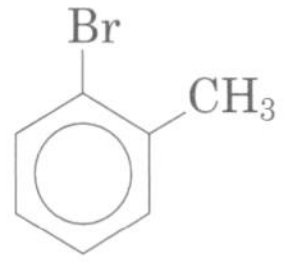

o-Bromotoluene 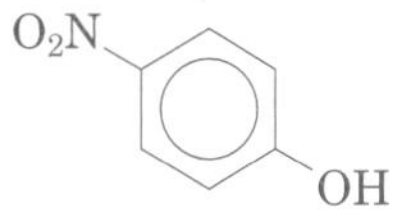 *p*-Nitrophenol

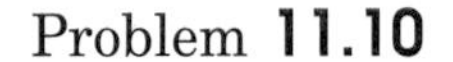

Problem 11.10

Name:

(a) 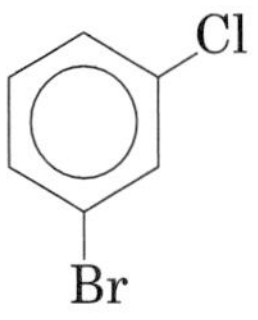 (b) 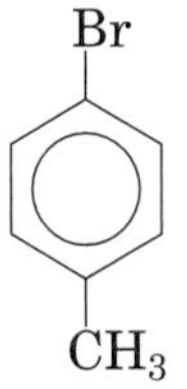 (c)

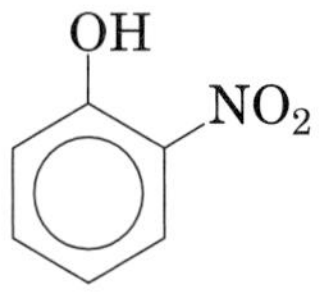

Three or More Groups

The rules do not permit the use of *ortho, meta,* and *para* when more than two groups are present. For these cases we must use numbers, following rule 7 in Section 10.8:

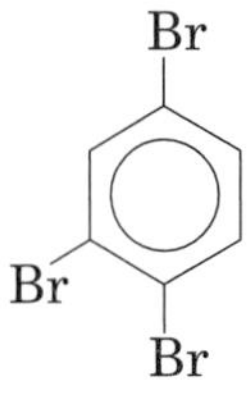

1,2,4-Tribromobenzene 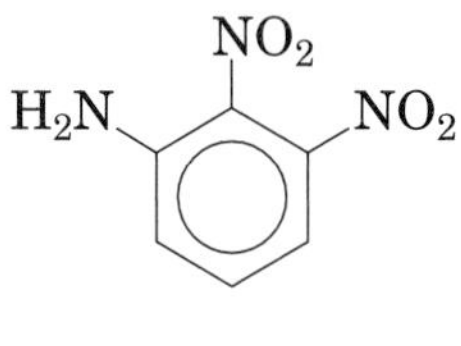 2,3-Dinitroaniline 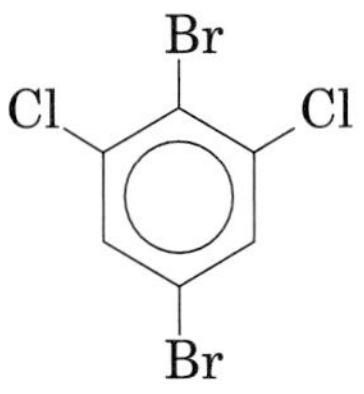 1,4-Dibromo-2,6-dichlorobenzene

Note in the example on the right that one of the bromos is given the number 1, because it comes first in alphabetical order [see Example (d) on p. 333].

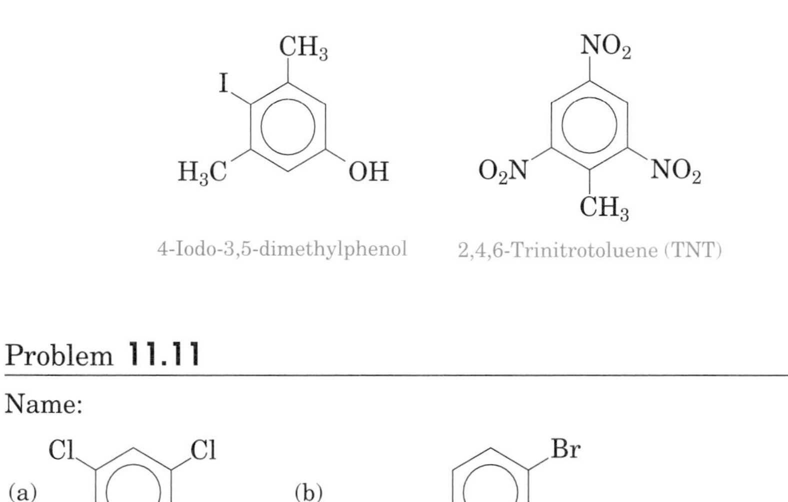

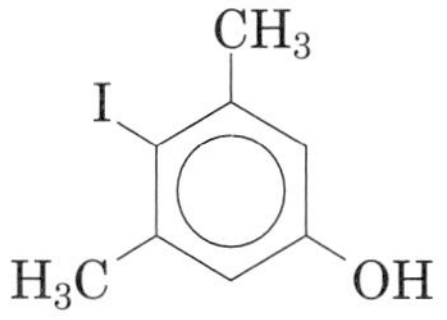

4-Iodo-3,5-dimethylphenol

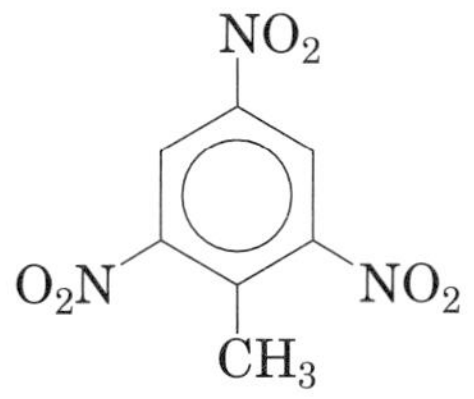

2,4,6-Trinitrotoluene (TNT)

Problem **11.11**

Name:

(a) (b)

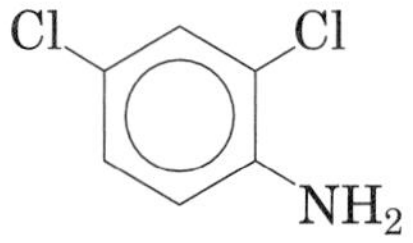

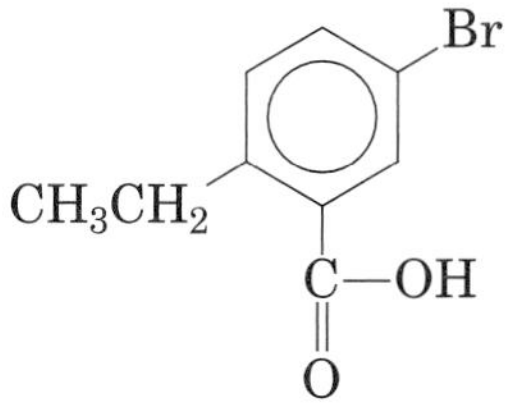

The Phenyl Group

In many cases it is necessary to name the benzene ring as a group. Although you might suppose that this is called the benzyl group, the name used is not that, but **phenyl.** For example:

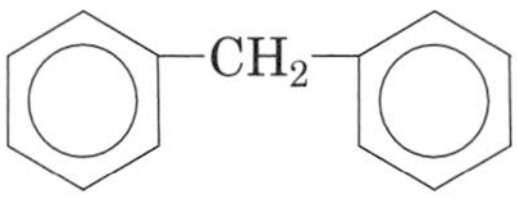

Diphenylmethane

As you can see, compounds like this would be difficult to name using the benzene ring as a parent. Remember that the phenyl group is $C_6H_5—$; that is, one of the six benzene hydrogens has been removed to make it a group (just as the methyl group, $CH_3—$, has one less hydrogen than methane).

Another example of the use of the phenyl group in naming compounds is

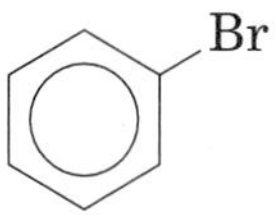

Phenyl bromide
(bromobenzene)

11.11 Reactions of Aromatic Compounds

We have stressed that aromatic compounds undergo substitution reactions rather than additions.

Most of the known reactions of aromatic compounds are **aromatic substitutions.** We will discuss three of them.

Nitration

$$C_6H_6 + HNO_3 \xrightarrow{H_2SO_4} C_6H_5\text{—}NO_2 + H_2O$$

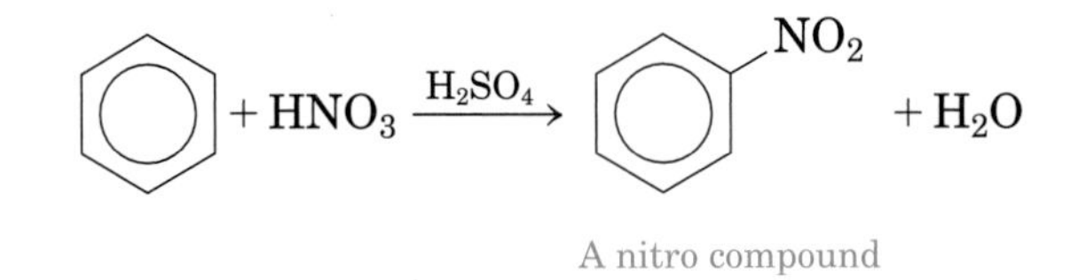

A nitro compound

When an aromatic compound is heated with a mixture of concentrated nitric and sulfuric acids, one of the hydrogens of the ring is replaced by a nitro group, $—NO_2$. Because this is such a convenient reaction, aromatic nitro compounds are very common.

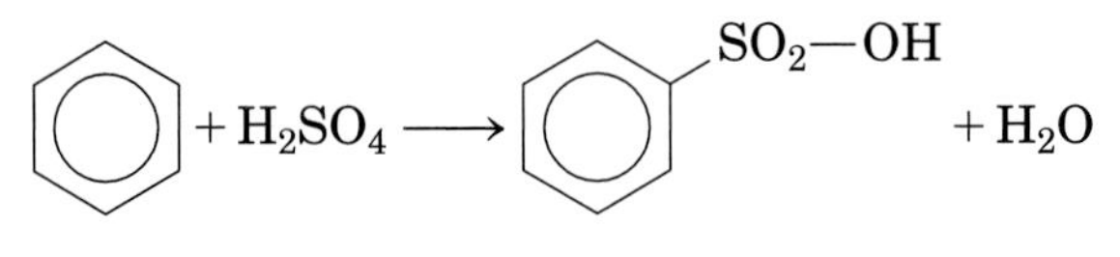

Sulfonation

$$C_6H_6 + H_2SO_4 \longrightarrow C_6H_5\text{—}SO_2\text{—}OH + H_2O$$

A sulfonic acid

Heating an aromatic compound with concentrated sulfuric acid alone gives a sulfonic acid. The one shown here is called benzenesulfonic acid. Sulfonic acids are strong acids, about as strong as sulfuric acid.

Halogenation

$$C_6H_6 + Br_2 \xrightarrow{Fe} C_6H_5\text{—}Br + HBr$$

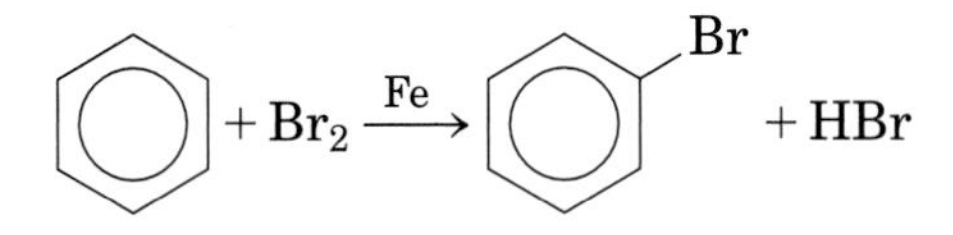

Aromatic rings react with bromine, in the presence of an iron catalyst, to give brominated aromatic compounds. The bromine replaces a hydrogen of the ring. Chlorine behaves in the same way and produces a chlorinated aromatic compound.

11.12 Fused Aromatic Rings

All the aromatic compounds we have mentioned up to now possess a single intact benzene ring. We have devoted so much space to these compounds because they are so common, both in laboratories and in our bodies. There are also compounds with two or more benzene rings fused together. The three simplest of these are

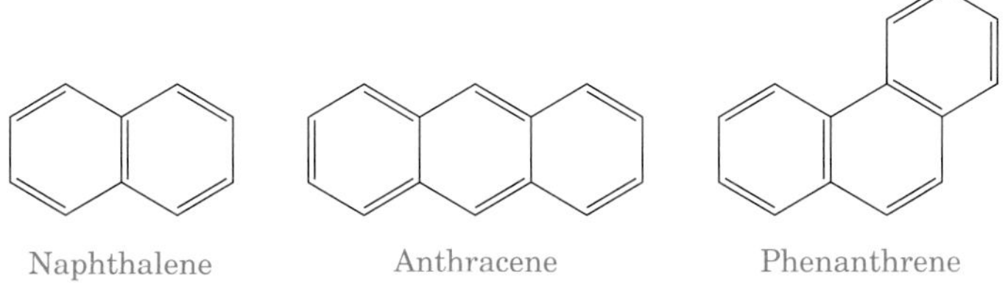

Naphthalene Anthracene Phenanthrene

Note that naphthalene does not have two *complete* benzene rings (nor do anthracene or phenanthrene have three complete benzene rings). Two complete benzene rings would require 12 carbon atoms, but naphthalene has only 10. Because it does not have two complete aromatic sextets, we do not draw circles inside the rings; instead, we represent these compounds by single resonance forms (other resonance forms can also be drawn).

Unlike benzene, which is a liquid, these compounds are all solids at room temperature. Naphthalene is used to make moth balls.

Fused aromatic compounds undergo aromatic substitutions, just as single-ring benzenes do.

11.13 Heterocyclic Compounds

A **heterocyclic compound** is **one that contains a ring in which at least one atom is not carbon.** Heterocyclic compounds are very important in biochemistry, and we will meet many of them in later chapters. Some heterocyclic compounds are aromatic; others are not. Two examples of nonaromatic heterocyclic compounds are

H_2C, CH_2, H_2C, CH_2, H_2C, O or (ring with O) Tetrahydropyran

H_2C, CH_2, H_2C, CH_2, H_2C, N–H or (ring with N–H) Piperidine

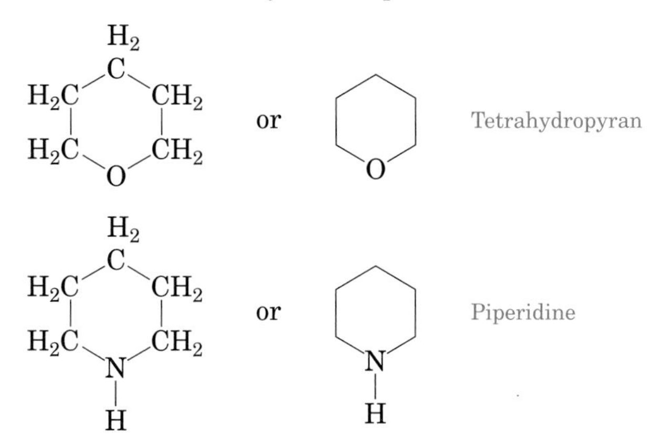

These compounds are nonaromatic because they do not have an aromatic sextet.

Two important aromatic heterocyclic compounds are

HC, CH, HC, CH, H–C, N or (ring with N) Pyridine

HC, N, HC, CH, H–C, N or (ring with N, N) Pyrimidine

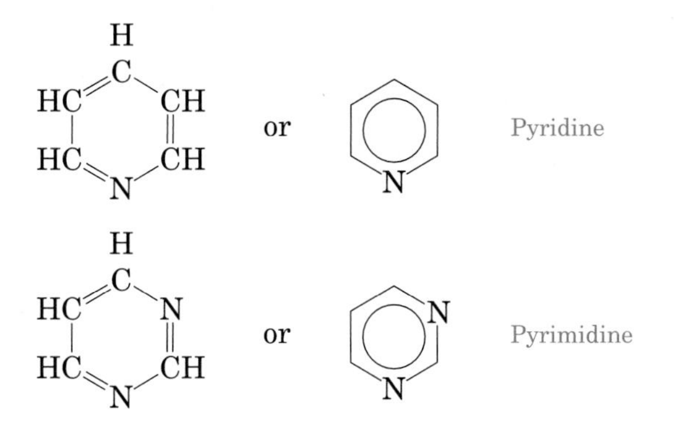

In pyrimidine, two nitrogens (in the 1 and 3 positions) replace CH groups of benzene. As we shall see in Section 23.2, pyrimidine rings are important constituents of nucleic acids.

Note that pyridine has the same formula as benzene except that one CH has been replaced by an N. There is no hydrogen on the N because nitrogen forms three bonds and not four.

Box 11G

Carcinogenic Fused Aromatics and Smoking

Some years ago it was discovered that certain organic compounds, when rubbed on the skin of rats, caused skin cancers. **A compound that causes cancer** is called a **carcinogen.** Although many compounds have now been found to be carcinogenic, the first to be discovered were a group of fused aromatic hydrocarbons, all of which have at least four rings and at least one angular junction (that is, structures resembling phenanthrene rather than anthracene). Some of these are

Dibenz[*a*,*h*]anthracene   Benzo[*a*]pyrene   Benzo[*e*]pyrene

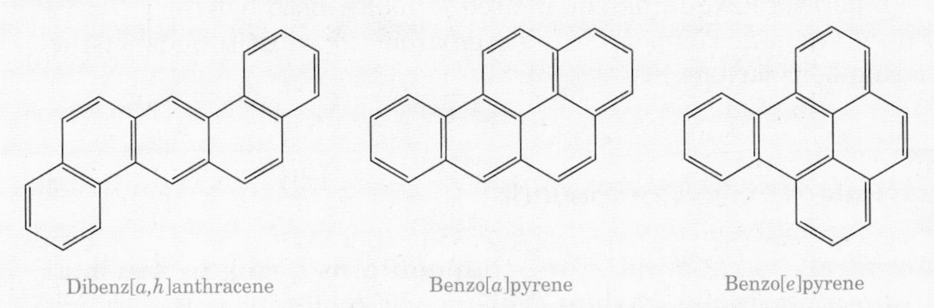

These carcinogens are produced by automobile exhausts and are found in cigarette smoke. If you smoke, you greatly increase your chances of getting cancer. A sizable proportion of cancers, especially lung cancers, but other cancers also, are caused by cigarette smoking.

About 200 years ago a London surgeon, Percivall Pott, found that chimney sweeps (boys employed to clean chimneys) were especially prone to cancer of the scrotum and other parts of the body. Today it is known that these cancers were caused by fused aromatic hydrocarbons present in the chimney soot.

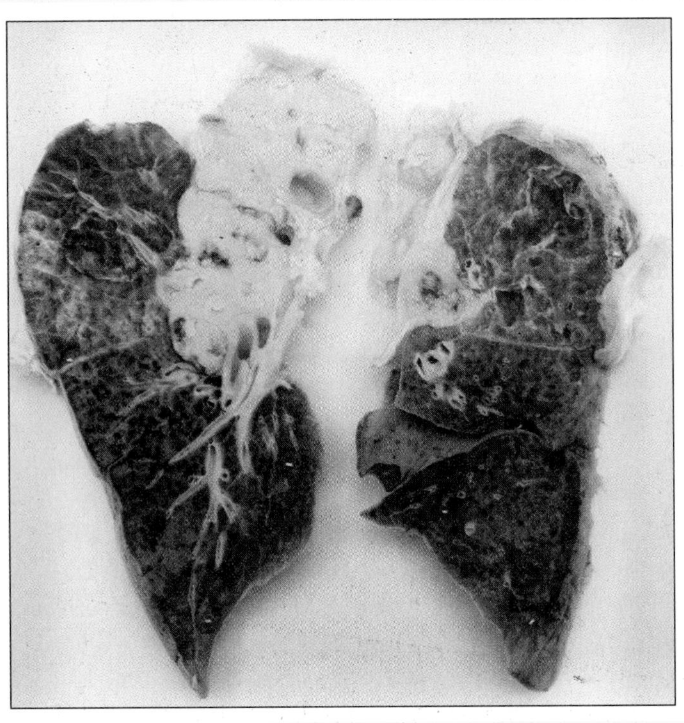

A smoker's lung showing carcinoma. (Courtesy of Anatomical Collections, National Museum of Health and Medicine, Armed Forces Institute of Pathology, Washington, D.C.)

SUMMARY

Alkenes are hydrocarbons containing double bonds. They are named the same way alkanes are except that the suffix "-ene" is used and the double bond is given the lower of the two carbon numbers.

Carbons forming double bonds use sp^2 orbitals. One of the two bonds in a double bond is σ, the other, π. All this causes alkenes to be planar molecules. There is no rotation about a double bond, so that **cis and trans isomers** occur in suitable cases.

The chief chemical property of alkenes is that they undergo **addition reactions,** for example with H_2 (in the presence of a catalyst), Br_2, Cl_2, HBr, HCl, and H_2O. Unsymmetrical reagents like HBr add according to **Markovnikov's rule:** The hydrogen goes to the carbon that already has more hydrogens. Compounds containing double bonds can be **polymerized** to give **addition polymers,** which are used for plastics, artificial fibers, synthetic rubbers, and coatings. The **polymers** consist of long chains of repeating units.

Alkynes are hydrocarbons containing triple bonds, which consist of one σ and two π bonds. They are linear molecules. The most important is acetylene (ethyne).

All the preceding hydrocarbons are **aliphatic. Aromatic hydrocarbons** and their derivatives have an aromatic ring of electrons like the one in benzene. The benzene ring is especially stable and is found in many naturally occurring compounds. Aromatic compounds undergo **substitution** (rather than addition) reactions, including nitration, sulfonation, and halogenation. The prefixes *ortho, meta,* and *para* are usually used to name disubstituted aromatic compounds.

Many aromatic compounds (such as naphthalene or anthracene) have two or more **fused rings.** Others are **heterocyclic**—they have one or more atoms in the ring that are not carbon.

Summary of Reactions

1. Addition of H_2 to alkenes (Sec. 11.6):

$$-\overset{|}{C}=\overset{|}{C}- + H_2 \xrightarrow{\text{catalyst}} -\underset{H}{\overset{|}{\underset{|}{C}}}-\underset{H}{\overset{|}{\underset{|}{C}}}-$$

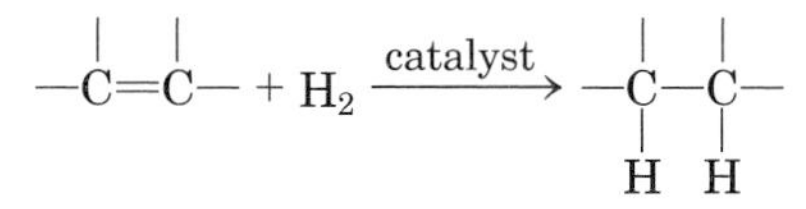

2. Addition of Br_2 or Cl_2 to alkenes (Sec. 11.6):

$$-\overset{|}{C}=\overset{|}{C}- + X_2 \longrightarrow -\underset{X}{\overset{|}{\underset{|}{C}}}-\underset{X}{\overset{|}{\underset{|}{C}}}- \qquad X = \text{Br or Cl}$$

3. Addition of HX to alkenes (Sec. 11.6):

$$-\overset{|}{C}=\overset{|}{C}- + HX \longrightarrow -\underset{H}{\overset{|}{\underset{|}{C}}}-\underset{X}{\overset{|}{\underset{|}{C}}}- \qquad X = \text{F, Cl, Br, or I}$$

Reactions 1 through 3 can also be done with alkynes.

4. Addition of H_2O to alkenes (Sec. 11.6):

$$-\overset{|}{C}=\overset{|}{C}- + H_2O \xrightarrow{H^+} -\underset{H}{\overset{|}{\underset{|}{C}}}-\underset{OH}{\overset{|}{\underset{|}{C}}}-$$

5. Nitration of aromatic rings (Sec. 11.11):

$$C_6H_6 + HNO_3 \xrightarrow{H_2SO_4} C_6H_5NO_2$$

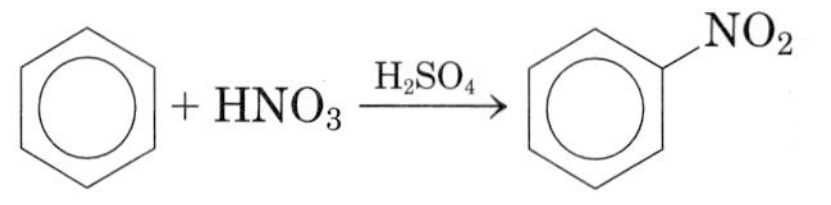

6. Sulfonation of aromatic rings (Sec. 11.11):

$$C_6H_6 + H_2SO_4 \longrightarrow C_6H_5SO_2OH$$

7. Halogenation of aromatic rings (Sec. 11.11):

$$C_6H_6 + X_2 \xrightarrow{Fe} C_6H_5X \qquad X = \text{Br or Cl}$$

KEY TERMS

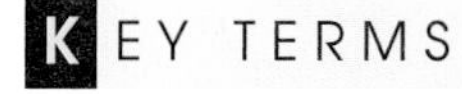

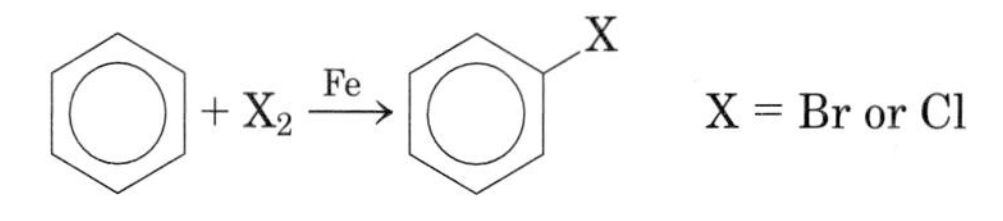

Addition polymer (Sec. 11.7)
Addition reaction (Sec. 11.6)
Aliphatic hydrocarbon (Sec. 11.9)
Alkene (Sec. 11.1)
Alkyne (Sec. 11.8)
Aromatic compound (Sec. 11.9)
Aromatic sextet (Sec. 11.9)
Aromatic substitution (Sec. 11.11)
Carbocation (Box 11E)
Carcinogen (Box 11G)
Catalytic hydrogenation (Sec. 11.6)
Copolymer (Sec. 11.7)
Fused aromatic ring (Sec. 11.12)
Heterocyclic compound (Sec. 11.13)
Hybrid (Sec. 11.9)
Hydration (Sec. 11.6)
Intermediate (Box 11E)
Isoprene rule (Box 11A)
Markovnikov's rule (Sec. 11.6)
Mechanism (Box 11E)
Monomer (Sec. 11.7)
Olefin (Sec. 11.1)
Pheromone (Box 11B)
Pi bond (Sec. 11.3)
Pi orbital (Sec. 11.3)
Polymer (Sec. 11.7)
Polymerization (Sec. 11.7)
Resonance (Sec. 11.9)
sp orbital (Sec. 11.8)
sp^2 orbital (Sec. 11.3)
Unsaturated hydrocarbon (Sec. 11.1)

PROBLEMS

11.12 Pick out the unsaturated compounds and classify each as an alkene or alkyne:

(a) $CH_3—CH═CH—CH_3$

(b) $CH_3—CH—CH_3$ with CH_3 on the middle C

(c) $CH_3—C≡CH$

(d)

(e) $CH_3—CH—Br$ with CH_3 on the CH

(f) CH_2CH_3

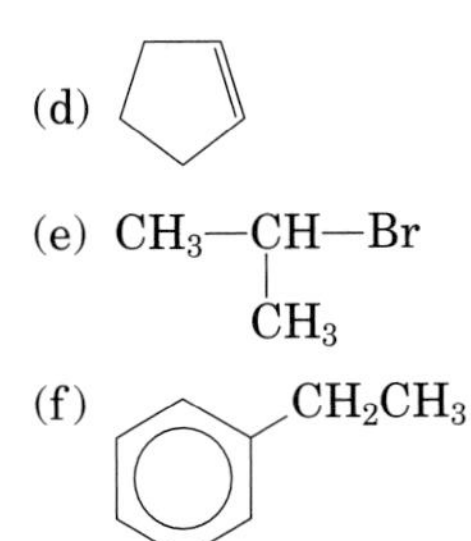

Nomenclature of Alkenes

11.13 Draw the structural formula for

(a) ethylene
(b) 2-pentene
(c) 2,4-dimethyl-1-pentene
(d) 1,2-dimethylcyclohexene
(e) 2,4-dibromo-6-methyl-3-octene
(f) 1-vinylcyclohexene
(g) 2,5-heptadiene
(h) 1,4-cyclohexadiene

11.14 Give the IUPAC name (ignore cis-trans isomerism):

(a) $CH_3—CH═CH_2$

(b) $CH_3—C═C—CH_3$ with Br on each double-bonded C

(c) $Cl—C—CH_2—CH═CH_2$ with Cl above and below the first C

(d) $CH_3CH_2—C—CH_3$ with $═CH_2$ on the C

(e) $CH_2═C—CH_2—CH—CH_3$ with $CH_2—CH_3$ below the second C and CH_3 above the CH

(f) H_3C

(g) Cl

11.15 Give the IUPAC name:

(a) $CH_2═CH—CH═CH_2$

(b) $CH_3CH_2—CH═CH—CH_2—CH═CH_2$

(c) $CH_3—CH═C═C—CH_2CH_3$ with Cl on the fourth C

(d) CH_3

11.16 The compound 1,1-dichloropropene is shown on page 361. (a) Why isn't this compound called 3,3-dichloropropene? (b) Is there a compound with this name? If so, draw its structural formula. (c) Is there a compound 2,2-dichloropropene?

sp^2 Hybridization

11.17 The angle between sp^2 orbitals is 120°. What is the angle between an sp^2 orbital and the remaining p orbital?

11.18 If a carbon atom uses sp^2 orbitals to form bonds, how many sp^2 orbitals does it use?

Stereoisomerism of Alkenes

11.19 In each case tell whether cis-trans isomers exist. If they do, draw the two and label them cis and trans:

(a) 2-butene
(b) 2-methylpropene
(c) 1,1-dichloro-1-hexene
(d) 3-hexene
(e) 3-methyl-3-hexene
(f) 3-ethyl-3-hexene
(g) 1-chloro-3-ethyl-3-hexene

11.20 Name the following cis or trans isomers:

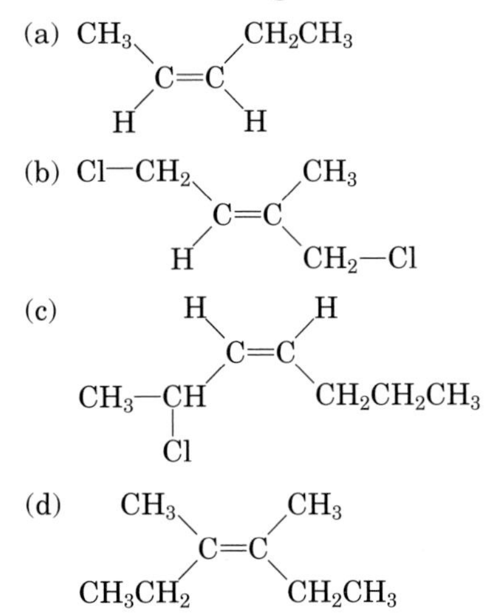

(a) C═C with CH_3 (upper left), CH_2CH_3 (upper right), H (lower left), H (lower right)

(b) C═C with $Cl—CH_2$ (upper left), CH_3 (upper right), H (lower left), $CH_2—Cl$ (lower right)

(c) C═C with H (upper left), H (upper right), $CH_3—CH$ bearing Cl (lower left), $CH_2CH_2CH_3$ (lower right)

(d) C═C with CH_3 (upper left), CH_3 (upper right), CH_3CH_2 (lower left), CH_2CH_3 (lower right)

Addition Reactions

11.21 Define addition reaction. Write the equation for any addition reaction to propene.

11.22 Write structural formulas for the principal reaction products when ethylene, $CH_2{=}CH_2$, reacts with

(a) Cl_2 (d) $H_2 + Pt$

(b) HBr (e) Br_2

(c) $H_2O + H_2SO_4$

11.23 Write structural formulas for the principal reaction products.

(a) $CH_3{-}CH{=}CH{-}CH_3 + H_2 \xrightarrow{Pt}$

(b) $CH_3{-}\underset{\underset{CH_3}{|}}{C}{=}CH{-}CH_3 + HBr \longrightarrow$

(c) CH_3 $+ Cl_2 \longrightarrow$

(d) $CH_2{=}CH{-}\underset{\underset{CH_3}{|}}{CH}{-}CH_3 + H_2O \xrightarrow{H_2SO_4}$

(e) $CH_3{-}\underset{\underset{CH_3}{|}}{C}{=}CH{-}CH_2CH_3 + HCl \longrightarrow$

(f) CH_3 $+ HBr \longrightarrow$

11.24 Show how to make the following compounds starting from propene, $CH_3{-}CH{=}CH_2$ (write equations):

(a) $CH_3CH_2CH_3$ (c) $CH_3\underset{\underset{Br}{|}}{CH}\underset{\underset{Br}{|}}{CH_2}$

(b) $CH_3\underset{\underset{Br}{|}}{CH}CH_3$ (d) $CH_3\underset{\underset{OH}{|}}{CH}CH_3$

11.25 Write the structural formula for the principal reaction product, if a reaction takes place. Where there is no reaction, say so.

(a) $CH_3{-}CH{=}CH{-}CH_3 + Cl_2 \longrightarrow$

(b) $CH_3{-}CH{=}CH{-}CH_3 + I_2 \longrightarrow$

(c) $CH_3{-}\underset{\underset{CH_3}{|}}{C}{=}CH_2 + H_2 \xrightarrow{\text{no catalyst}}$

(d) CH_3 $+ HI \longrightarrow$

11.26 Write structural formulas for all possible addition products when HBr is added to each of the following. In each case predict which of the possible products is actually formed, in accord with Markovnikov's rule:

(a) $CH_3{-}CH{=}CH_2$

(b) $CH_3{-}\underset{\underset{CH_3}{|}}{C}{=}CH{-}CH_3$

(c) $CH_3CH_2{-}\underset{\underset{CH_3}{|}}{C}{=}CH_2$

(d) $CH_3CH_2{-}CH{=}CH{-}CH_2CH_3$

(e) CH_3

11.27 What reagents and/or catalysts are necessary to make each of the following conversions in the laboratory?

(a) $CH_3{-}CH{=}CH{-}CH_3 \longrightarrow CH_3{-}CH_2{-}\underset{\underset{Br}{|}}{CH}{-}CH_3$

(b) $CH_3{-}\overset{\overset{CH_3}{|}}{C}{=}CH_2 \longrightarrow CH_3{-}\overset{\overset{CH_3}{|}}{\underset{\underset{OH}{|}}{C}}{-}CH_3$

(c) $\longrightarrow$ I

(d) $CH_2{=}CH{-}CH{=}CH_2 \longrightarrow CH_3CH_2CH_2CH_3$

(e) $CH_3{-}\underset{\underset{CH_3}{|}}{C}{=}CH_2 \longrightarrow CH_3{-}\overset{\overset{Cl}{|}}{\underset{\underset{CH_3}{|}}{C}}{-}\overset{\overset{Cl}{|}}{CH_2}$

11.28 Could you make 1-chloropropane from propene? Explain.

11.29 Assume that you are given a liquid that could be either cyclohexane or cyclohexene. What simple test could you perform to tell which it is? Describe exactly what you would do and what you would see in each case.

Addition Polymers

11.30 Draw structural formulas for the polymers made from
(a) $CH_2{=}CH{-}F$
(b) $Cl{-}CH{=}CH{-}Cl$

***11.31** Which monomers would you use to make the following polymers?

(a) $-\overset{\overset{Cl}{|}}{CH}{-}\overset{\overset{CH_3}{|}}{CH}{-}\overset{\overset{Cl}{|}}{CH}{-}\overset{\overset{CH_3}{|}}{CH}{-}\overset{\overset{Cl}{|}}{CH}{-}\overset{\overset{CH_3}{|}}{CH}{-}$

(b) $-CH_2{-}\overset{\overset{COOCH_3}{|}}{CH}{-}CH_2{-}\overset{\overset{COOCH_3}{|}}{CH}{-}CH_2{-}\overset{\overset{COOCH_3}{|}}{CH}{-}$

Alkynes

11.32 Name the three orbitals that make up a $C{\equiv}C$ triple bond.

11.33 Give the IUPAC name:

(a) $CH_3—CH_2—C{\equiv}C—CH_3$

(b) $CH_3—\underset{}{\overset{Cl}{|}}{CH}—C{\equiv}C—Cl$

(c) $Br—CH_2—C{\equiv}C—\overset{CH_3}{\overset{|}{C}}H—CH_3$

11.34 Draw structural formulas for (a) propyne (b) 1,4-dibromo-2-butyne (c) 3-bromo-1-iodo-4,4-dimethyl-1-pentyne.

11.35 Why is cis-trans isomerism not possible for alkynes?

Aromatic Compounds

11.36 Draw the two resonance forms of benzene. Is the real benzene represented by either of them? Explain.

11.37 Define aromatic compound.

11.38 The compound *para*-dichlorobenzene,

Cl

Cl

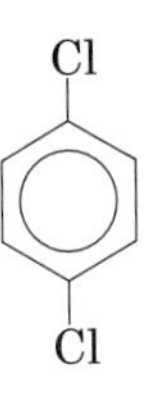

has a rigid geometry that allows no free rotation. Yet there are no cis-trans isomers for this structure (or any other benzene ring). Explain why.

11.39 Draw structural formulas for
(a) *ortho*-dinitrobenzene
(b) *para*-ethyltoluene
(c) *meta*-chloroaniline
(d) phenylcyclohexane
(e) *meta*-diphenylbenzene
(f) *ortho*-isopropylbenzoic acid

11.40 What kind of damage is caused when benzene is ingested in the liquid or vapor form?

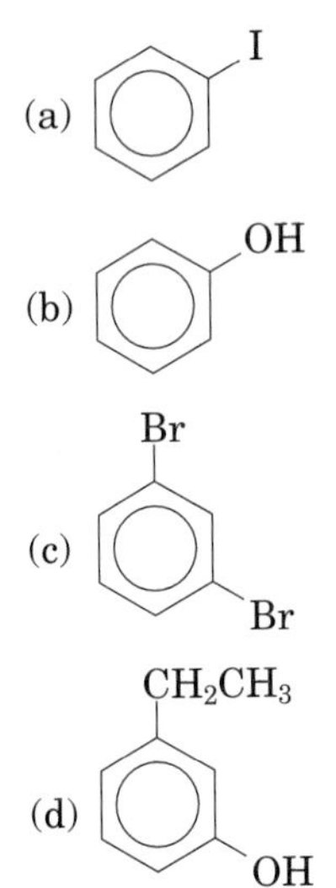

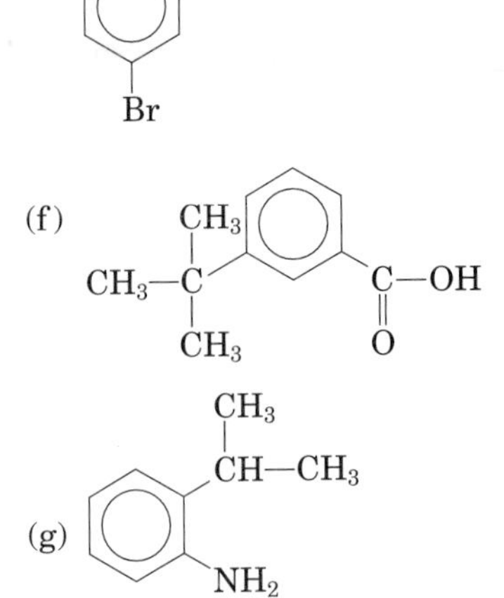

11.41 Name:

(a) I

(b) OH

(c) Br, Br

(d) CH_2CH_3, OH

(e) NO_2, Br

(f) $CH_3—\overset{CH_3}{\underset{CH_3}{C}}$, $\underset{O}{\overset{}{C}}—OH$

(g) $\overset{CH_3}{CH}—CH_3$, NH_2

11.42 Name:

(a) Cl, Cl, Cl

(b) CH_3, Cl, Cl, Cl

(c) I, I, I, I, OH

(d) CH_3, CH_3, $\underset{O}{C}—OH$

(e) $CH{=}CH_2$, $\underset{O}{C}—OH$, Cl

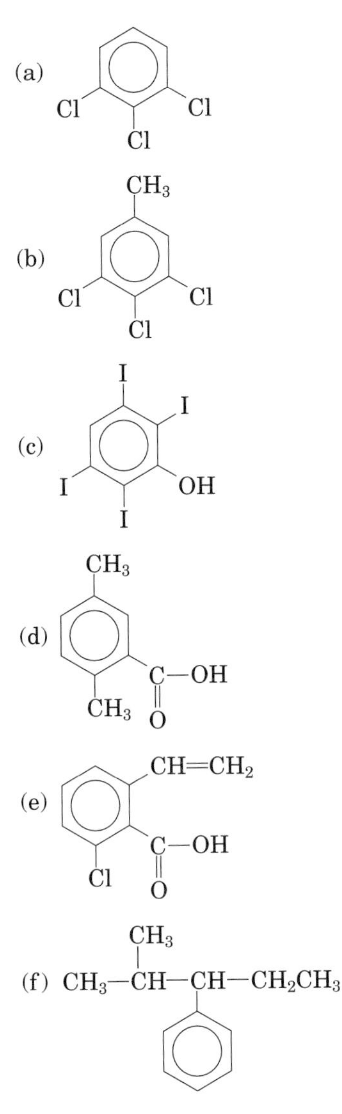

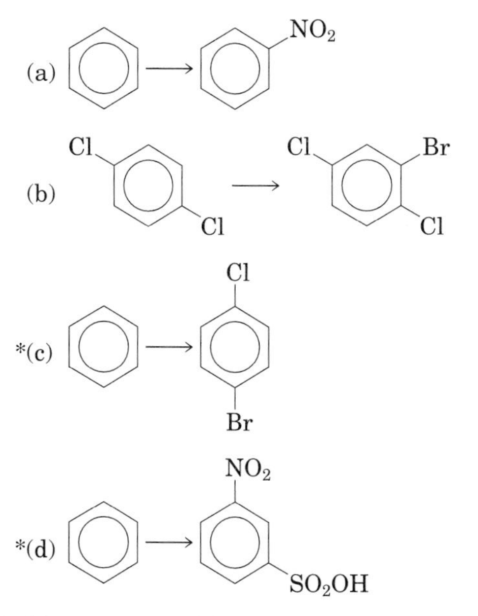

(f) $CH_3—\overset{CH_3}{CH}—CH—CH_2CH_3$

(g) $C_6H_5CH_2CH_2C_6H_5$

11.43 Why do aromatic compounds undergo substitution rather than addition reactions?

11.44 What reagents and/or catalysts are necessary to make each of the following conversions in the laboratory?

(a) benzene ⟶ nitrobenzene (NO_2)

(b) 1,4-dichlorobenzene (Cl, Cl) ⟶ 2-bromo-1,4-dichlorobenzene (Cl, Br, Cl)

*(c) benzene ⟶ 1-bromo-4-chlorobenzene (Cl, Br)

*(d) benzene ⟶ 3-nitrobenzenesulfonic acid (NO_2, SO_2OH)

11.45 Write structural formulas for the principal reaction products:

(a) benzene $+ HNO_3 \xrightarrow{H_2SO_4}$

(b) p-xylene (CH_3, CH_3) $+ Br_2 \xrightarrow{Fe}$

(c) 1,4-dibromobenzene (Br, Br) $+ H_2SO_4 \longrightarrow$

11.46 Three possible products can be formed in the chlorination of bromobenzene. Draw the three structures.

11.47 Draw structural formulas for the principal reaction products, if a reaction takes place. Where there is no reaction, say so.

(a) benzene $+ Br_2 \xrightarrow{\text{no catalyst}}$

(b) p-xylene (CH_3, CH_3) $+ Cl_2 \xrightarrow{Fe}$

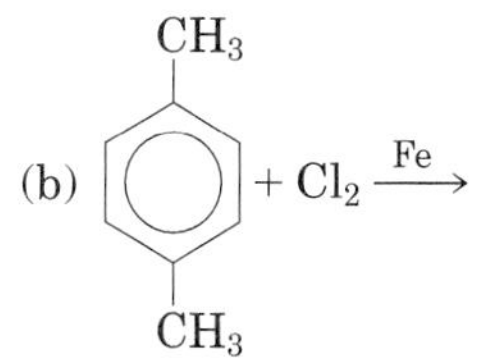

Fused and Heterocyclic Compounds

11.48 Define: (a) heterocyclic compound (b) heterocyclic aromatic compound.

11.49 The structure we have given for naphthalene, [naphthalene structure], is only one of three resonance forms. Draw the other two (note that each carbon atom must have one double bond in each form).

11.50 Aromatic substitution can also be done on naphthalene. When naphthalene is sulfonated with concentrated H_2SO_4, two (and only two) different naphthalenesulfonic acids are produced. Draw the two structures.

11.51 There are three different methylpyridines. Draw the three structural formulas.

Boxes

11.52 (Box 11A) Phytol is a terpene that is part of the chlorophyll molecule. Show where the isoprene units are in this molecule:

$$CH_3-\underset{}{\overset{CH_3}{\overset{|}{C}H}}-CH_2CH_2CH_2-\overset{CH_3}{\overset{|}{C}H}-CH_2$$
$$HO-CH_2-CH{=}\underset{CH_3}{\underset{|}{C}}-CH_2-CH_2$$

Phytol

11.53 (Box 11A) Limonene is the chief terpene in orange peel. Show where the isoprene units are in this molecule:

Ring: $CH(CH_3)$, H_2C, CH, H_2C, CH, CH (with $CH{=}CH$ double bond); substituent on CH: $C(=CH_2)CH_3$ (H_3C, CH_2)

Limonene

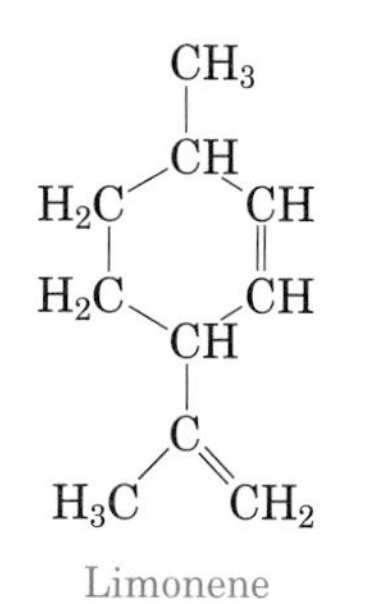

11.54 (Box 11A) How many carbon atoms are there in the β-carotene molecule? In vitamin A?

11.55 (Box 11B) Assume that 1×10^{-12} g of *cis*-9-tricosene is the amount secreted by a single housefly. How many molecules is this? (The molecular formula is $C_{23}H_{46}$.)

11.56 (Box 11C) What different functions are performed by the rods and cones in the eye?

11.57 (Box 11C) (a) In which isomer is the end-to-end distance longer, *all-trans-* or 11-*cis*-retinal? (b) Which does not fit into the opsin cavity?

11.58 (Box 11D) What are the uses of ethylene in industry?

11.59 (Box 11E) Show the mechanism of the reaction between H_2O and isobutene, $CH_3-\underset{\displaystyle CH_3}{\underset{|}{C}}=CH_2$, catalyzed by H_2SO_4.

***11.60** (Box 11E) Polymerization reactions can occur through a mechanism involving carbocation intermediates, for example:

$$CH_2=CH_2 + H^+ \longrightarrow CH_3-CH_2^+$$

$$CH_3-CH_2^+ + CH_2=CH_2 \longrightarrow CH_3CH_2CH_2-CH_2^+$$

Complete the mechanism for the preparation of polyisobutylene:

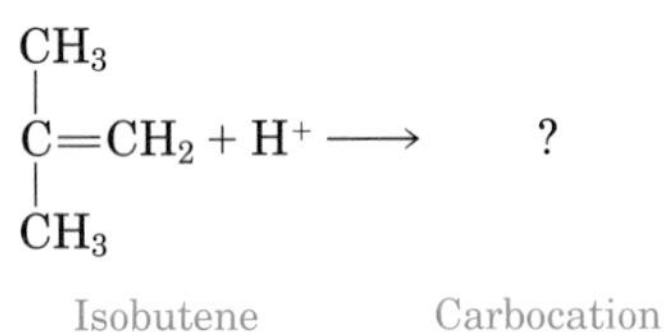

$$\underset{\text{Isobutene}}{\overset{\displaystyle CH_3}{\underset{\displaystyle CH_3}{\overset{|}{\underset{|}{C}}}}=CH_2} + H^+ \longrightarrow \underset{\text{Carbocation}}{?}$$

$$\underset{\text{Carbocation}}{?} + \overset{\displaystyle CH_3}{\underset{\displaystyle CH_3}{\overset{|}{\underset{|}{C}}}}=CH_2 \longrightarrow \underset{\text{Dimer}}{?}$$

***11.61** (Box 11F) *Neoprene* is an important synthetic rubber that is made by polymerizing the monomer 2-chloro-1,3-butadiene (called chloroprene). Draw the structural formula of neoprene.

11.62 (Box 11F) Why is it necessary to vulcanize rubber?

11.63 (Box 11G) What compounds caused the scrotum cancers found in London chimney sweeps 200 years ago?

11.64 (Box 11G) What is a carcinogen? What kind of carcinogens are found in cigarette smoke?

Additional Problems

11.65 Do aromatic rings have double bonds? Are they unsaturated? Explain.

11.66 Write structural formulas for all the different mononitro products that are possible in the nitration of
(a) *meta*-bromotoluene
(b) naphthalene

11.67 There are two different bromonaphthalenes. Draw the two structural formulas.

11.68 The following names are all incorrect. State what is wrong with each (do not try to give correct names):
(a) *cis*-2-methyl-2-pentene
(b) 2-bromo-*meta*-dichlorobenzene
(c) 2,2-dimethyl-1-pentene
(d) 3-pentene
(e) 1-bromo-3-butene
(f) *para*-bromomethylbenzene
(g) 2-bromopropyne
(h) 3-bromo-2,3-pentadiene
(i) 1-bromobutene

11.69 Draw the structures of and name all the isomers of dibromopropene (there are seven, including cis and trans isomers).

***11.70** Cyclohexyne, , is not a stable compound (it is unknown), but cyclohexene, , is a perfectly stable, known compound. Suggest a reason for the instability of cyclohexyne.

11.71 Show how tetrafluoroethene, $F_2C=CF_2$, polymerizes to give the polymer Teflon.

11.72 What is the difference between a saturated and an unsaturated compound?

11.73 Write the structural formula and name for the simplest (a) alkene (b) alkyne (c) aromatic hydrocarbon

11.74 (a) Why is it incorrect to call this compound 1,2- or 2,3-dimethylcyclohexene?

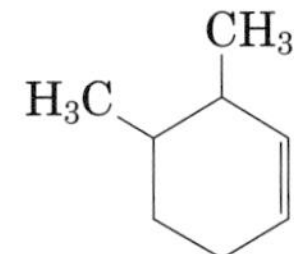

(b) What is the correct numbering?

11.75 Write a balanced equation for the combustion of toluene, C_7H_8.

***11.76** For each carbon atom in this compound,

$$\overset{1}{C}H_3-\overset{2}{C}H=\overset{3}{C}H-\overset{4}{C}\equiv\overset{5}{C}H,$$

tell whether the hybridization is *sp*, sp^2, or sp^3.

11.77 Because alkynes have triple bonds, they can add one mole of reagent or two. Draw structural formulas for the principal products after one mole of Br_2 reacts with 2-butyne, and after two.

$$CH_3-C\equiv C-CH_3 + Br_2 \longrightarrow$$

11.78 Explain why a π bond is weaker than a σ bond.

Chapter **12**

Alcohols, phenols, ethers, and halides

Table 12.1 Types of Compounds Derived from H_2O, H_2S, and HX

Formula	Name
H—O—H	Water
R—O—H	Alcohols
R—O—R′	Ethers[a]
H—S—H	Hydrogen sulfide
R—S—H	Thiols (mercaptans)
R—S—R′	Thioethers[a]
H—X	Hydrogen halide
R—X	Alkyl halides

[a]The R′ can be the same as R or different.

12.1 Introduction

In Chapters 10 and 11 we studied the hydrocarbons: alkanes, alkenes, alkynes, and aromatic compounds. Section 10.12 introduced the concept of **functional group.** A particular functional group, such as the hydroxyl group, —OH, can replace a hydrogen at almost any position of any hydrocarbon of any of the classes we have studied.

In this and the next three chapters, we will study the most important types of organic compounds—those listed in Table 10.7. All these functional groups contain either O, N, S, or halogen atoms. In this chapter we study those compounds that are simple derivatives of H_2O, H_2S, and the hydrogen halides, HX (Table 12.1).

12.2 Nomenclature of Alcohols

As already noted, **alcohols** are **compounds containing —OH groups connected to an alkyl carbon.** If the —OH is connected directly to an aromatic ring, the compound is not an alcohol but a **phenol** (see Sec. 12.5).

Alcohols are classified as **primary, secondary,** or **tertiary** depending on whether the —OH is connected to a primary, a secondary, or a tertiary carbon atom (Table 12.2). Recall that a primary carbon is connected to one (or no) other carbon, a secondary to two other carbons, and a tertiary to three other carbons.

Table 12.2 Classification, Common Names, and IUPAC Names for Some Alcohols

Structural Formula	Classification	Common Name	IUPAC Name
CH_3—OH	Primary	Methyl alcohol	Methanol
CH_3CH_2—OH	Primary	Ethyl alcohol	Ethanol
$CH_3CH_2CH_2$—OH	Primary	*n*-Propyl alcohol	1-Propanol
CH_3—CH(OH)—CH_3	Secondary	Isopropyl alcohol	2-Propanol
CH_3CH_2—CH(OH)—CH_3	Secondary	*sec*-Butyl alcohol	2-Butanol
CH_3—C(CH_3)(CH_3)—OH	Tertiary	*tert*-Butyl alcohol	2-Methyl-2-propanol

Problem 12.1

Classify these alcohols as primary, secondary, or tertiary:

(a) 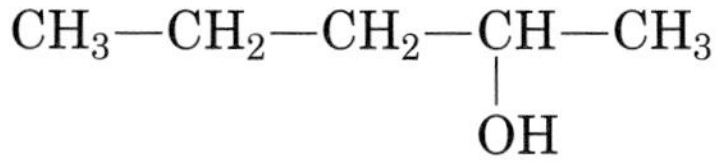 $CH_3—CH_2—CH_2—\underset{\displaystyle OH}{\underset{|}{C}H}—CH_3$

(b) 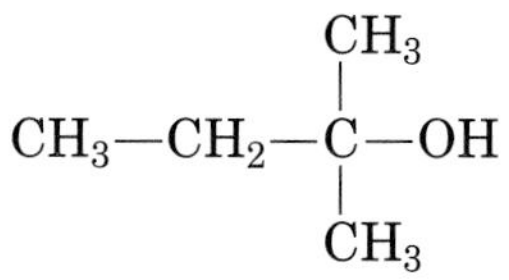 $CH_3—CH_2—\overset{\displaystyle CH_3}{\overset{|}{\underset{\displaystyle CH_3}{\underset{|}{C}}}}—OH$

(c) $CH_3—CH_2—CH_2—CH_2—OH$

There are two ways to name alcohols. Simple alcohols are often named by giving the name of the alkyl group (Sec. 10.7) followed by the word "alcohol"; Table 12.2 gives examples. The other way to name alcohols is by the IUPAC rules we use for alkanes, alkyl halides (Sec. 10.7), and alkenes (Sec. 11.2), except that we need two additional rules:

1. The "longest" chain must include the carbon bearing the —OH group.

2. The suffix "-ol" is used to represent the —OH group, and the lowest possible number is assigned to the carbon bearing this group.

When we assign the lowest number, the —OH takes precedence over all alkyl and halo groups.

Table 12.2 shows how the IUPAC system applies to some simple alcohols. Note that the prefixes iso, normal, tertiary, and so on are not used in the IUPAC names.

For many complicated molecules there are no simple names; IUPAC names must be used.

EXAMPLE

Give the IUPAC name for

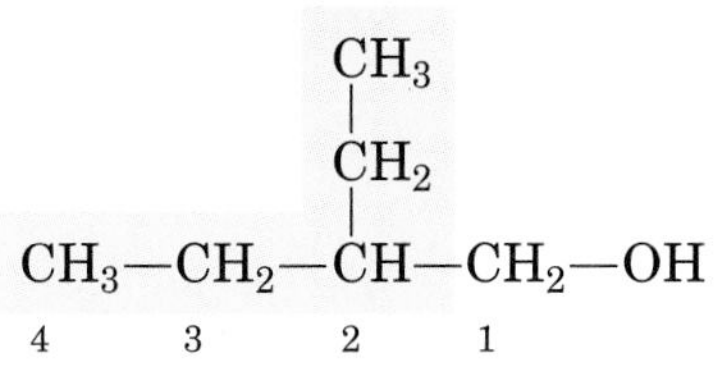

$$\underset{4}{CH_3}—\underset{3}{CH_2}—\underset{2}{\overset{\displaystyle CH_3}{\overset{|}{\overset{\displaystyle CH_2}{\overset{|}{C}}}}H}—\underset{1}{CH_2}—OH$$

2-Ethyl-1-butanol

Answer • This molecule has a five-carbon chain (shown in color), but this chain does not contain the carbon bearing the —OH group. The longest chain that contains this carbon has four carbons, and so the parent is butane. Because there is an —OH group, the suffix is "-ol." Thus the name is 2-ethyl-1-butanol, with the OH-bearing carbon given the lowest possible number.

EXAMPLE

Give the IUPAC name for

$$\overset{1}{CH_3}-\overset{2}{CH_2}-\underset{\large OH}{\overset{3}{CH}}-\overset{4}{CH_2}-\underset{\large CH_3}{\overset{\large Br}{\overset{5|}{C}}}-\overset{6}{CH_3}$$

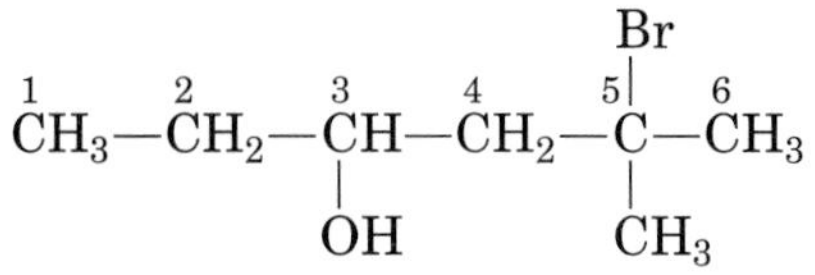

5-Bromo-5-methyl-3-hexanol

Answer • The longest chain has six carbons, and it contains the —OH group; thus the parent is hexanol. The chain is numbered as shown, because the —OH carbon becomes 3. If the chain were numbered the other way, this would be 4, and the rule is that the —OH group must be given the lowest number, even though this means that the methyl and the bromo groups must be numbered 5 and 5, respectively.

EXAMPLE

Give the IUPAC name for

(cyclohexane ring, positions numbered 1–6; HO on C-1, CH_3 on C-3)

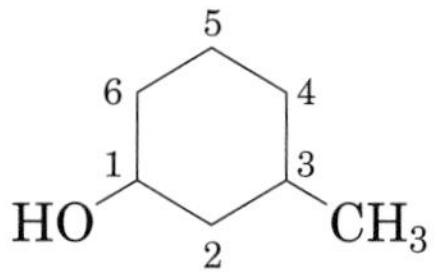

3-Methylcyclohexanol

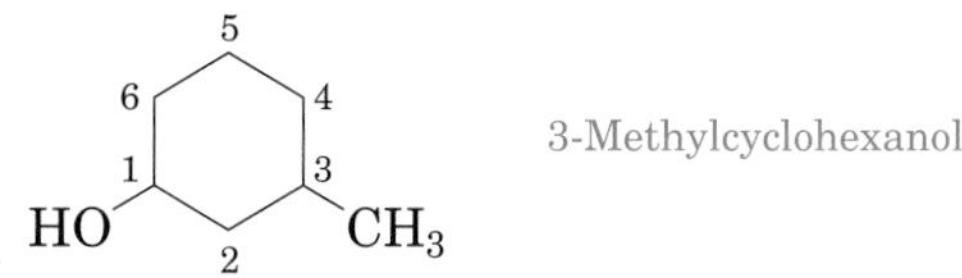

Because the —OH is always at the 1 position in a ring, we do not show the 1 in the name.

Answer • Since the —OH group is given preference, the carbon to which it is attached is C-1, meaning that the $-CH_3$ group is on C-3 (since 3 is lower than 5, which we would get by counting the other way).

Problem **12.2**

Give the IUPAC name for

(a) 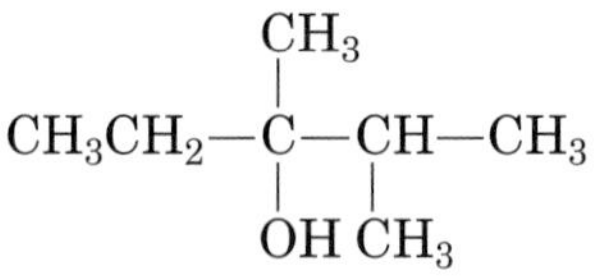 $CH_3CH_2-\underset{\large OH}{\overset{\large CH_3}{C}}-\underset{\large CH_3}{CH}-CH_3$

(b) 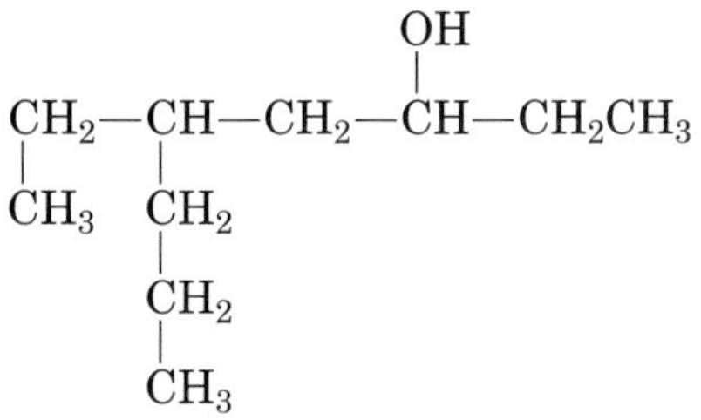 $\underset{\large CH_3}{CH_2}-\underset{\large CH_2CH_2CH_3}{CH}-CH_2-\overset{\large OH}{CH}-CH_2CH_3$

Compounds with two —OH groups on the same carbon are usually not stable.

Many compounds contain more than one —OH group in a molecule (though not generally two on the same carbon atom). Molecules of this type are called diols, triols, and so forth. Two of the most important are (common names) ethylene glycol and glycerol (also called glycerin):

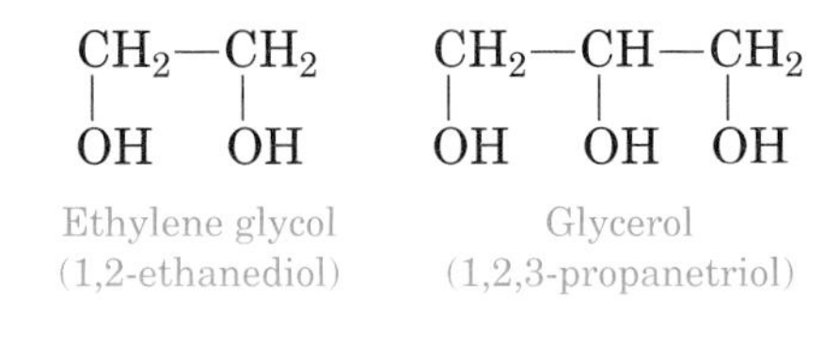

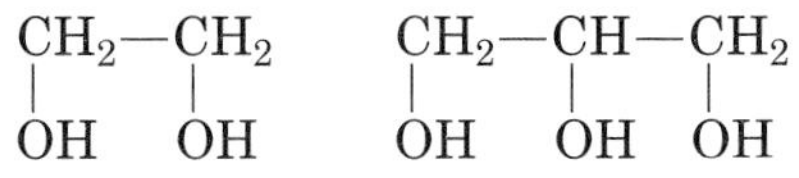

$$\underset{\text{OH}}{\underset{|}{\text{CH}_2}}-\underset{\text{OH}}{\underset{|}{\text{CH}_2}} \qquad \underset{\text{OH}}{\underset{|}{\text{CH}_2}}-\underset{\text{OH}}{\underset{|}{\text{CH}}}-\underset{\text{OH}}{\underset{|}{\text{CH}_2}}$$

Ethylene glycol (1,2-ethanediol) — Glycerol (1,2,3-propanetriol)

12.3 Chemical Properties of Alcohols

Metallic hydroxides, such as NaOH, $Mg(OH)_2$, and $Al(OH)_3$, are bases, some strong and some weak, because they produce OH^- ions in water. Alcohols do not behave this way. The —OH in an alcohol is connected to the carbon by a covalent bond and not by an ionic bond (as in the metal hydroxides). Therefore, alcohols have no basic properties. Nor are they acidic, despite the hydrogen connected to the oxygen (they are weaker acids than water). When an alcohol is dissolved in pure water, the pH remains 7.

This was mentioned in Section 8.1.

Alcohols are neutral compounds.

We give two reactions of alcohols in this section (we will see some others in Chapters 13 to 15).

Dehydration to Give an Alkene

General reaction

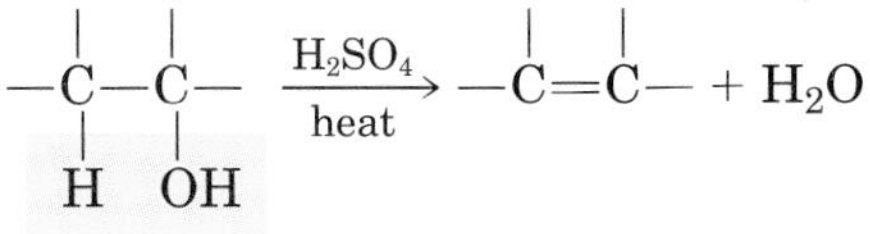

$$-\underset{\text{H}}{\underset{|}{\overset{|}{\text{C}}}}-\underset{\text{OH}}{\underset{|}{\overset{|}{\text{C}}}}- \xrightarrow[\text{heat}]{H_2SO_4} -\overset{|}{\text{C}}=\overset{|}{\text{C}}- + H_2O$$

Specific example

$$\text{CH}_3-\underset{\text{OH}}{\underset{|}{\text{CH}}}-\text{CH}_3 \xrightarrow[\text{heat}]{H_2SO_4} \text{CH}_3-\text{CH}=\text{CH}_2 + H_2O$$

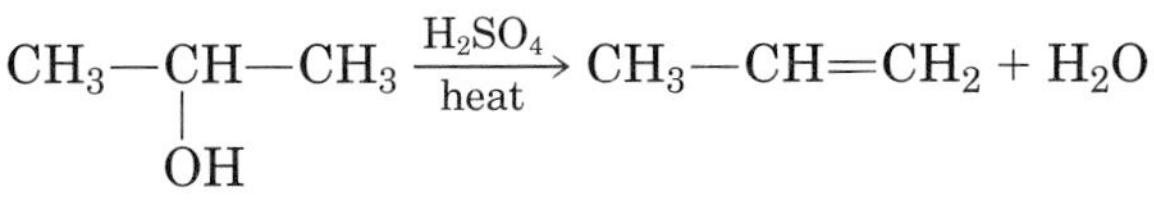

2-Propanol — Propene

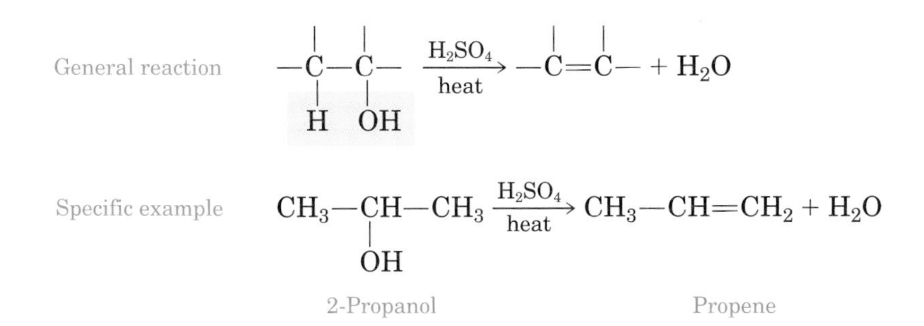

When heated with sulfuric or phosphoric acid as catalyst, an alcohol loses an OH group and an H atom on the adjacent carbon to give an alkene. This reaction can be performed only on alcohols that have at least one hydrogen on the adjacent carbon. (Methanol, CH_3OH, for example, cannot undergo this reaction.) Note that this is the reverse of the addition of water to a double bond (Sec. 11.6). That is, we can heat 2-propanol with sulfuric acid and get propene. If desired, we can then take the propene and convert it back to 2-propanol, as shown in Section 11.6. In this case the catalyst is the same, and we can shift the equilibrium either way by using Le Chatelier's principle (Sec. 7.8).

When two groups are removed from adjacent carbons to give a double bond, the reaction is called **elimination.**

Many reactions go both forward and backward, but in most cases different reagents are required for each direction.

Problem **12.3**

Draw the structural formula of the product:

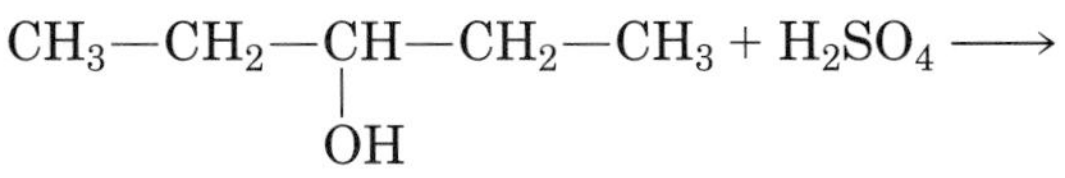

$$\text{CH}_3-\text{CH}_2-\underset{\text{OH}}{\underset{|}{\text{CH}}}-\text{CH}_2-\text{CH}_3 + H_2SO_4 \longrightarrow$$

Note that in Problem 12.3 the catalyst is not written over the arrow. In organic chemistry we often write catalysts as if they were reagents, and vice versa.

With some alcohols, dehydration can give only one product. In the preceding example, the only possible dehydration product from 2-propanol is propene. But with other alcohols, two products are possible. An example is dehydration of 2-butanol:

This carbon has 3 hydrogens (→ CH_3); This carbon has 2 hydrogens (→ CH_2)

$$\underset{\text{2-Butanol}}{CH_3-\underset{\displaystyle OH}{\underset{|}{CH}}-CH_2-CH_3} \xrightarrow[\text{heat}]{H_2SO_4} \underset{\text{2-Butene (major product)}}{CH_3-CH{=}CH-CH_3}$$

$$\xrightarrow[\text{heat}]{H_2SO_4} \underset{\text{1-Butene}}{CH_2{=}CH-CH_2-CH_3}$$

In such cases it is found that **the major product (or even the only product) is the one in which the H is lost from the carbon atom that had fewer hydrogens.** This is known as **Zaitsev's rule** (often given the German spelling Saytzeff). In the case of 2-butanol this means that the major product is 2-butene and not 1-butene, since 2-butene is formed by loss of a hydrogen from a CH_2 group, whereas to form 1-butene requires that a hydrogen be lost from a CH_3 group.

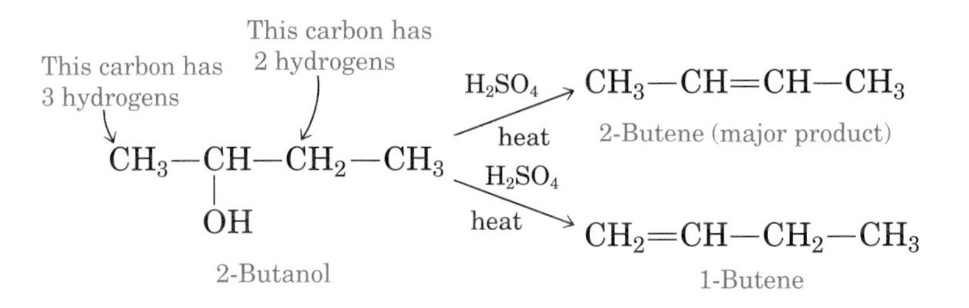

Zaitsev's rule is similar to Markovnikov's rule (Sec. 11.6) but applies to dehydration rather than addition.

Problem 12.4

Draw the structural formula of the principal product

$$CH_3-\underset{\displaystyle OH}{\underset{|}{CH}}-CH_2-CH_2-CH_3 \xrightarrow[\text{heat}]{H_2SO_4}$$

In Section 11.6 we saw that the hydration reaction is also carried out in the body, which uses enzymes as catalysts rather than sulfuric or phosphoric acid. The same is true for the dehydration reaction; for example, the dehydration of citric acid,

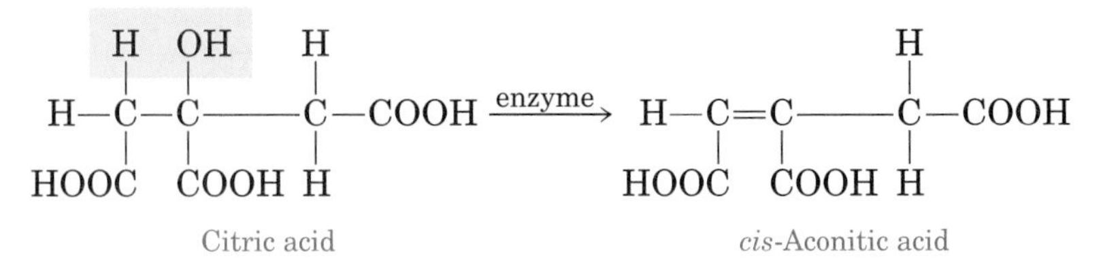

$$\underset{\text{Citric acid}}{\begin{array}{ccccc} & H & OH & & H \\ & | & | & & | \\ H- & C- & C & \text{——} & C-COOH \\ & | & | & & | \\ & HOOC & COOH & & H \end{array}} \xrightarrow{\text{enzyme}} \underset{\textit{cis}\text{-Aconitic acid}}{\begin{array}{ccccc} & & & & H \\ & & & & | \\ H- & C{=} & C & \text{——} & C-COOH \\ & | & | & & | \\ & HOOC & COOH & & H \end{array}}$$

is part of the citric acid cycle (Sec. 20.4). Note that the dehydration of citric acid involves exactly the same changes in bonding as the dehydration of 2-propanol (that is why we classify them as the same type of reaction). In both cases an H and an OH are removed from adjacent carbons, leaving a C=C double bond. The rest of the bonds in both

There are only a limited number of reactions possible for any class of compounds, and it is not surprising that the same reactions are found in the body and in the laboratory.

molecules do not change. It is important that you be able to recognize that the same thing is happening in both cases, despite the considerable differences in the structures of citric acid and 2-propanol. In fact, almost every molecule with OH and H on adjacent carbon atoms is capable of giving this reaction under the proper conditions.

Zaitsev's rule does not apply to reactions catalyzed by enzymes. When an alcohol is dehydrated with sulfuric acid, Zaitsev's rule operates because the more stable product forms, but enzyme-catalyzed reactions do not happen this way. The action of enzymes is highly specific (Chapter 19), and each enzyme produces the product it is designed to produce, whether it is the more stable or the less stable product.

Oxidation

Recall (Sec. 4.8) that one definition of oxidation is the gain of oxygen or loss of hydrogen. The three classes of alcohols behave differently toward oxidizing agents.

Primary Alcohols

General reaction

$$\underset{\text{Primary alcohol}}{R{-}CH_2{-}OH} \xrightarrow{[O]} \underset{\text{Aldehyde}}{R{-}\overset{\overset{\displaystyle O}{\|}}{C}{-}H} \xrightarrow{[O]} \underset{\text{Carboxylic acid}}{R{-}\overset{\overset{\displaystyle O}{\|}}{C}{-}OH}$$

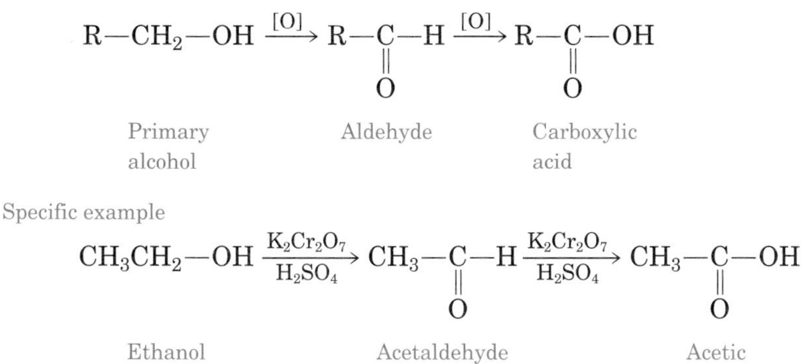

The [O] represents an unspecified oxidizing agent. Recall that R stands for any alkyl group.

Specific example

$$\underset{\text{Ethanol}}{CH_3CH_2{-}OH} \xrightarrow[H_2SO_4]{K_2Cr_2O_7} \underset{\text{Acetaldehyde}}{CH_3{-}\overset{\overset{\displaystyle O}{\|}}{C}{-}H} \xrightarrow[H_2SO_4]{K_2Cr_2O_7} \underset{\text{Acetic acid}}{CH_3{-}\overset{\overset{\displaystyle O}{\|}}{C}{-}OH}$$

Primary alcohols are oxidized by many oxidizing agents. The immediate product is an aldehyde, which results from the removal of two hydrogens:

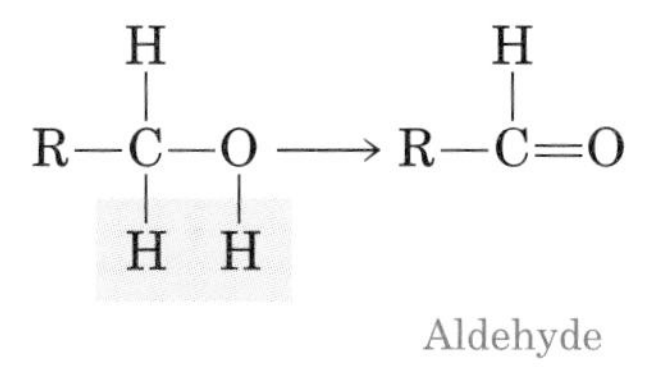

$$R{-}\underset{H}{\overset{H}{\underset{|}{\overset{|}{C}}}}{-}\underset{H}{\underset{|}{O}} \longrightarrow \underset{\text{Aldehyde}}{R{-}\overset{H}{\overset{|}{C}}{=}O}$$

During the oxidation water is produced (not shown in the equation). The two hydrogens lost by the alcohol combine with an oxygen from the oxidizing agent.

An important biological example of this reaction is oxidation of ethanol to acetic acid, which is catalyzed by enzymes. This is what happens when wine is oxidized to vinegar. The oxidizing agent in this case is the O_2 in the air.

Aldehydes are themselves readily oxidized by the same oxidizing agents (Sec. 13.5) to give carboxylic acids. Therefore, the usual product of oxidation of a primary alcohol is the corresponding carboxylic acid. It is often possible to stop the reaction at the aldehyde stage by distilling the mixture so that the aldehyde (which usually has a lower boiling point than either the alcohol or the carboxylic acid) is removed before it can be oxidized further. A common oxidizing agent in this reaction is a mixture of potassium dichromate, $K_2Cr_2O_7$, and sulfuric acid, but potassium permanganate, $KMnO_4$, and bromine are also commonly used.

Problem **12.5**

Draw the structural formula of the first and second products:

$$CH_3-\overset{\displaystyle CH_3}{\overset{|}{C}H}-CH_2-OH \xrightarrow[H_2SO_4]{K_2Cr_2O_7}$$

Secondary Alcohols

General reaction

$$\underset{\text{Secondary alcohol}}{R-\underset{\displaystyle OH}{\underset{|}{C}H}-R'} \xrightarrow{[O]} \underset{\text{Ketone}}{R-\underset{\displaystyle O}{\underset{\|}{C}}-R'}$$

Specific example

$$\underset{\text{2-Propanol}}{CH_3-\underset{\displaystyle OH}{\underset{|}{C}H}-CH_3} \xrightarrow[H_2SO_4]{K_2Cr_2O_7} \underset{\text{Acetone}}{CH_3-\underset{\displaystyle O}{\underset{\|}{C}}-CH_3}$$

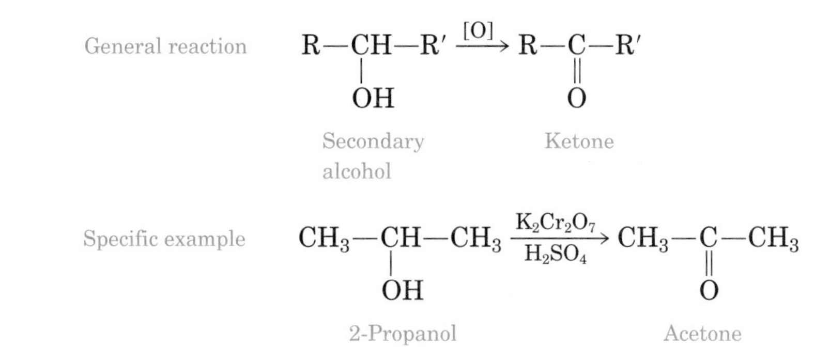

Secondary alcohols are oxidized to ketones in exactly the same way primary alcohols are oxidized to aldehydes, and by the same oxidizing agents. The difference is that, unlike aldehydes, ketones are resistant to further oxidation, so this reaction is an excellent method for the preparation of ketones.

Problem **12.6**

Draw the structural formula of the product:

$$C_6H_5-CH_2-\underset{\displaystyle OH}{\underset{|}{\overset{\displaystyle H}{\overset{|}{C}}}}-CH_3 \xrightarrow[H_2SO_4]{K_2Cr_2O_7}$$

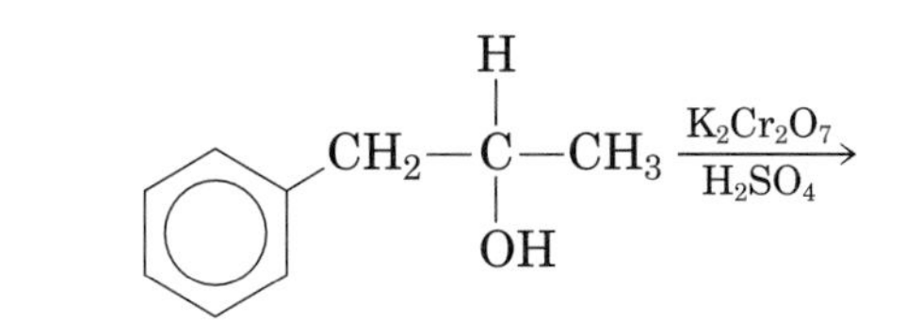

Tertiary Alcohols

$$R-\underset{\displaystyle R''}{\underset{|}{\overset{\displaystyle R'}{\overset{|}{C}}}}-OH \xrightarrow{[O]} \text{no reaction}$$

Tertiary alcohols, not having a hydrogen on the —OH carbon, do not react with oxidizing agents.

Note that some of the equations we have written in this section are not balanced; some are not even complete. It is difficult to balance these oxidation reactions, though it could be done if necessary. Most of the time, organic chemists do not find it necessary. Their focus is on the starting compound, reagent, and product and not on the balanced equation. This is the custom among organic chemists, and we shall follow that custom. This does not mean that the reactions are not balanced by nature. All reactions that actually take place follow the conservation of mass law and are balanced.

Organic reactions rarely produce only one product. Most often, side reactions yield a variety of products in small quantities. In contrast, most biochemical reactions are very specific and yield only one product.

12.4 Some Important Alcohols

Ethanol

When people say "alcohol" and mean a specific compound, ethanol, CH_3CH_2OH, is almost always the one that is meant. The production of ethanol by fermentation is as old as civilization. For the entire span of recorded history, human beings have been making ethanol-containing beverages by this process. Ethanol is sometimes called *grain alcohol* because large quantities are made by fermentation of such grains as corn, wheat, and rye, but fruits (especially grapes), vegetables (e.g., potatoes), and molasses (from sugar cane) are also used.

Fermentation of molasses. This mixture will be distilled to give rum. (FourByFive.)

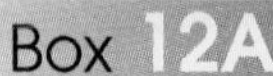

Box 12A

Alcoholic Beverages

Despite the great differences in alcoholic beverages, virtually all of them consist of at least 99 percent water and ethanol. They differ from one another in two ways: the relative percentages of water and ethanol, and the composition of the other 1 percent. Beer contains 3 to 4 percent ethanol, and table wine contains 10 to 13 percent. These beverages are made by direct fermentation. The yeast cells that catalyze the fermentation cannot live when the ethanol concentration gets higher than about 13 percent. The only way to get a higher ethanol concentration is to distill the wine. This process was not discovered until about AD 1200. Thus in all human history up to that time, people had nothing stronger to drink than table wine.

After 1200, distillation became a common process, and today we have spirits and fortified wines. Spirits are the direct products of distillation, often aged in bottles or casks, or given various other treatments. The ethanol content of spirits varies from 35 to 50 percent. Fortified wines (such as sherry and port) are produced in the normal way, but extra ethanol is added to bring the concentration to about 20 percent. In the United States, the ethanol concentration of beverages is expressed as **proof,** which is twice the percent ethanol (v/v). Thus 86-proof whiskey contains 43 percent ethanol.

The difference between one bottle of spirits and another of the same proof is in the composition of the last 1 percent. The compounds that make up this final 1 percent, called **congeners,** may be higher alcohols (such as 1-butanol), aldehydes, ketones, carboxylic esters, or other compounds. They get into the spirits either by being carried over in the original distillation or by absorption during the aging process (for example, from wooden casks). Alcohol itself is tasteless; the congeners provide the taste. Vodka is pure ethanol and water and has no congeners. Gin is made by adding juniper berries to vodka and has only a small amount of congeners. Rye, scotch,

In fermentation, sugars (represented here as $C_6H_{12}O_6$) are converted to ethanol and carbon dioxide by enzymes present in yeast cells:

$$C_6H_{12}O_6 \xrightarrow{\text{zymase}} 2CH_3CH_2OH + 2CO_2$$

The sugar content of grains and vegetables is very low, but they do contain starch, which is hydrolyzed to sugars during the fermentation process.

Sugars and starches are discussed in Chapter 16.

Of all alcohols, ethanol is the least toxic (except for glycerol), and it is the only alcohol present in alcoholic beverages (Box 12A), except perhaps for tiny amounts of others.

By law, the ethanol used in all alcoholic beverages and drugs must be made by fermentation, but industrial ethanol is mostly made by the hydration of ethylene (Sec. 11.6),

$$CH_2{=}CH_2 + H_2O \xrightarrow{H_2SO_4} CH_3CH_2{-}OH$$

because this process is cheaper. The ethanol made by the two processes is of course identical.

rum, bourbon, and brandy have greater amounts of congeners. Some alcoholic beverages also contain carbon dioxide (beer, champagne), which causes the alcohol to get into the blood faster, increasing the speed of intoxication.

The United States government puts a high tax on beverage ethanol. The tax on a 1-L bottle of 86-proof spirits is about $3.00. Laboratories and hospitals are permitted to use ethanol without paying the tax, but the government rigidly controls such usage with frequent inspections. The tax can also be avoided by using denatured alcohol (see text), but this can be done only when the denaturants are not harmful in the intended use.

Table wine contains about 10–13% ethanol. (Photograph by Charles D. Winters.)

Apart from its beverage use, ethanol is used as a solvent for medicines, perfumes, and varnishes, as a body rub (isopropyl alcohol is also used for this purpose—see below), and, in a 70 percent solution, as an antiseptic (Box 18H) used to prepare patients for surgery and to clean surgical instruments.

Solutions in which ethanol is a solvent are called **tinctures.**

Pure ethanol is called **absolute alcohol.** It is often found in laboratories in this form and in the form of a solution containing 5 percent water (95% alcohol). **Denatured alcohol** is ethanol that has been made unfit to drink by the addition of small quantities of a poison such as methanol (see Box 12B) or benzene. Denatured alcohol is used in some chemical laboratories and factories (where the addition of the poison does not affect its use) because no tax has to be paid on it.

Methanol

Sometimes called *wood alcohol* because at one time it was made from wood, methanol, CH_3OH, is now made by subjecting hydrogen and carbon monoxide to high temperatures and pressures in the presence of a

Box 12B

Alcohols as Drugs and Poisons

In the United States and many other countries, alcoholism is a major social problem. Pregnant women in particular must be careful about their alcohol consumption. Ethanol can cause brain damage in a developing fetus, and women who consume excessive amounts of alcohol during pregnancy endanger the well being of their offspring. Alcohol is a psychologically addictive drug, and many alcoholics remain intoxicated for weeks at a time. Alcohol causes deterioration of the liver (cirrhosis of the liver, caused almost entirely by alcohol intake, is a major cause of death) and loss of memory. People (not necessarily alcoholics) who drive while intoxicated are responsible for an estimated 40 percent of traffic deaths in the United States. Ethanol is not a stimulant, but a depressant. The individual may feel stimulated, but sensory perception is impaired and the reflexes are slowed.

The first step in normal ethanol metabolism is oxidation to acetaldehyde,

$$\underset{\text{Ethanol}}{CH_3CH_2OH} \xrightarrow{[O]} \underset{\text{Acetaldehyde}}{CH_3-\overset{\displaystyle O}{\overset{\|}{C}}-H} \xrightarrow{[O]} \text{other products}$$

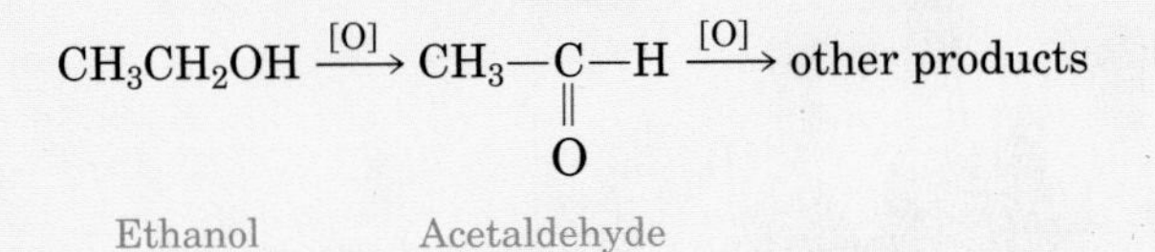

which is then oxidized further. One treatment for alcohol addiction involves use of the drug disulfuram (trade name Antabuse),

$$(CH_3CH_2)_2N-\overset{\displaystyle S}{\overset{\|}{C}}-S-S-\overset{\displaystyle S}{\overset{\|}{C}}-N(CH_2CH_3)_2$$

Disulfuram

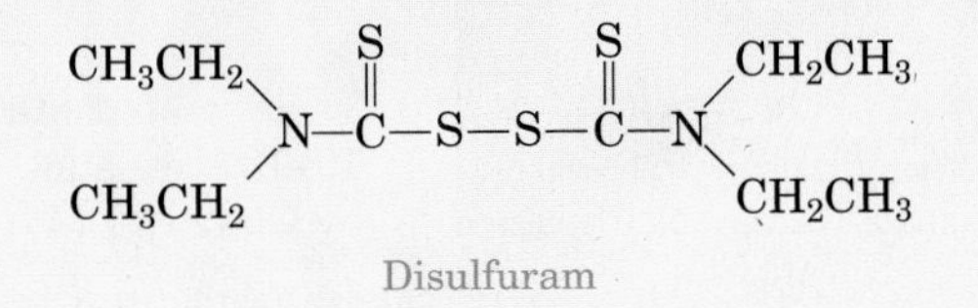

which interferes with the second step of the metabolism process. The ethanol is oxidized to acetaldehyde, but the acetaldehyde, which can no longer be oxidized, builds up in the blood, causing nausea, sweating, and vomiting. Knowing that this will happen if he or she drinks after taking disulfuram often keeps the person from drinking, though this is obviously not a permanent cure for alcoholism in all cases.

Methanol is highly toxic. Ingestion of 15 mL causes blindness and 30 mL, death. Methanol in the body is oxidized to formaldehyde and formic acid. We are not certain whether formic acid or formaldehyde poisons the cells of the retina and causes blindness, but we do know that, as with methanol, the ingestion of formic acid also causes acidosis in the blood, which can lead to death.

catalyst. Methanol is very toxic (Box 12B). It is used as a solvent for paints, shellacs, and varnishes as well as to make formaldehyde:

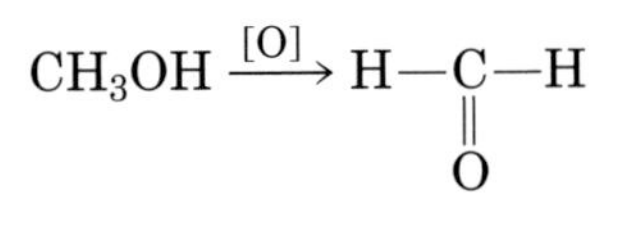

$$CH_3OH \xrightarrow{[O]} \underset{\text{Formaldehyde}}{H-\overset{\displaystyle O}{\overset{\|}{C}}-H}$$

Isopropyl Alcohol (2-Propanol)

Often called *rubbing alcohol,* isopropyl alcohol, $CH_3—\underset{}{\overset{OH}{\overset{|}{C}}}H—CH_3$, is useful for this purpose because it cools the skin by evaporation. It thus helps to lower fever. It is also an astringent—it hardens the skin and decreases the size of the pores, limiting secretions. It is used as a solvent for cosmetics, perfumes, and skin creams.

Isopropyl alcohol has a large heat of vaporization (though not as large as that of water), so when it evaporates it removes a lot of heat (see Sec. 5.11).

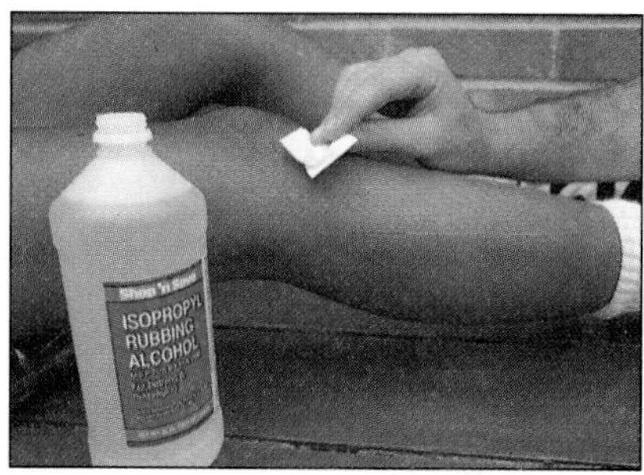

Isopropyl alcohol is also used to disinfect cuts and scrapes. (Photograph by Charles D. Winters.)

Menthol, Ethylene Glycol, and Glycerol

Menthol is a terpene alcohol (Box 11A):

[Structure: cyclohexane ring bearing CH_3 at the top, OH on an adjacent lower-right carbon, and $CH(CH_3)_2$ at the bottom]

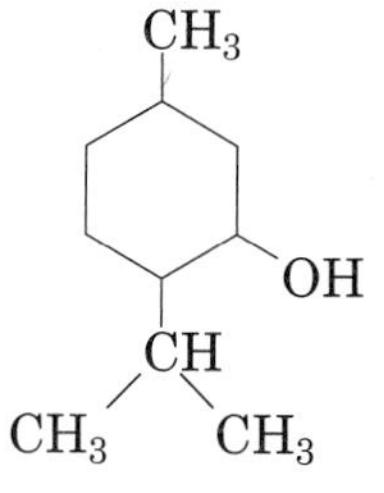

It has a pleasant, minty odor and is used in shaving creams, cough drops ("mentholated"), cigarettes, and toothpastes.

Menthol in throat sprays and lozenges soothes the respiratory tract.

Ethylene glycol, $HO—CH_2CH_2—OH$, is a viscous (thick), toxic liquid. Its most important use is as antifreeze for automobile radiators, where it lowers the freezing point of the water. Although any solute

See Section 6.8 for a discussion of freezing-point lowering.

Antifreeze being poured into an automobile radiator. (Photograph by Beverly March.)

Box 12C

Nitroglycerine, an Explosive and a Drug

In 1847 an Italian chemist named Ascanio Sobrero (1812–1888) discovered that glycerol reacts with a mixture of nitric and sulfuric acids to give a liquid called nitroglycerine (or glyceryl trinitrate):

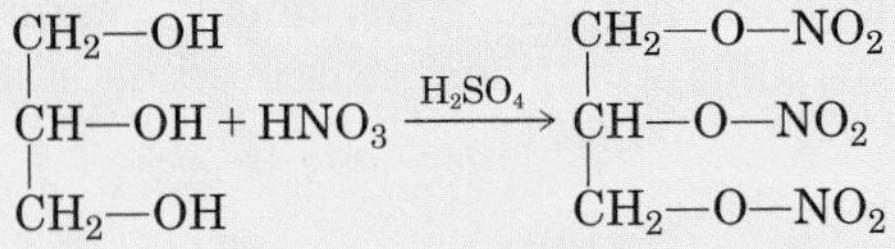

$$\begin{array}{l} CH_2\text{—}OH \\ | \\ CH\text{—}OH \\ | \\ CH_2\text{—}OH \end{array} + HNO_3 \xrightarrow{H_2SO_4} \begin{array}{l} CH_2\text{—}O\text{—}NO_2 \\ | \\ CH\text{—}O\text{—}NO_2 \\ | \\ CH_2\text{—}O\text{—}NO_2 \end{array}$$

Glycerol — Nitroglycerine

When Sobrero heated a small quantity of this liquid, it exploded. At that time there were few chemical explosives known, so nitroglycerine factories sprang up in many countries to make a compound that has many uses both for war and for peace (in digging canals and tunnels, blasting hills for road building, and mining, for example).

A problem soon arose because nitroglycerine is so explosive and difficult to handle that these factories would occasionally blow up. This problem was essentially solved by a Swedish chemist, Alfred Nobel (1833–1896), whose brother had been killed when Nobel's nitroglycerine factory exploded in 1864. In 1866 Nobel found that a clay-like substance called diatomaceous earth would absorb liquid nitroglycerine so that it would not explode without a fuse. He called the new material **dynamite,** and this is still one of our most important explosives.

Surprising as it may seem, nitroglycerine is also used in medicine, to treat angina pectoris. This condition, the symptoms of which are sharp chest pains, is caused by a reduced flow of blood to the heart. Nitroglycerine, available either in liquid form (diluted with alcohol to render it nonexplo-

will lower the freezing point, ethylene glycol is preferred because it is relatively cheap, infinitely soluble in water, noncorrosive to metal, and has a higher boiling point than water (thus it won't evaporate in summer).

Ethylene glycol is colorless; the color of most antifreezes comes from additives.

Glycerol (also called glycerin),

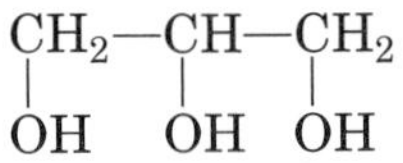

$$\begin{array}{ccc} CH_2\text{—} & CH\text{—} & CH_2 \\ | & | & | \\ OH & OH & OH \end{array}$$

is an even more viscous liquid than ethylene glycol. It is sweet, nontoxic, and infinitely soluble in water. These properties make it a good sweetening agent, solvent for medicines, and lubricant for suppositories and in chemical laboratories (for putting glass tubes through rubber stoppers). It is also used as a moistening agent (humectant) in cosmetics, skin creams, lotions, tobacco, and food. Another important use of glycerol is in the manufacture of nitroglycerine (Box 12C).

Glycerol is relatively inexpensive because it is a by-product in the production of soap from fats (Sec. 17.3).

sive), or as tablets, or in paste form, relaxes the smooth muscles of the blood vessels and dilates the arteries, allowing more blood to reach the heart and reducing or eliminating the chest pains.

Another compound used for the same purpose is isoamyl nitrite (also called amyl nitrite):

$$CH_3-\underset{}{\overset{\displaystyle CH_3}{\overset{|}{C}}}H-CH_2-CH_2-O-NO$$

However, this compound has an unpleasant odor.

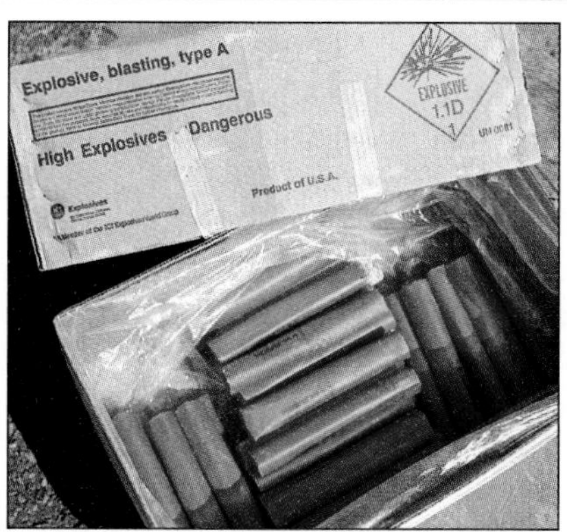

Nitroglycerine is used in the manufacture of dynamite. (Photograph by Charles D. Winters.)

12.5 Phenols

As we have said, phenols are compounds containing an —OH connected directly to an aromatic ring. The simplest phenol is phenol itself. Other phenols are generally named as derivatives of phenol (Sec. 11.10), though some have common names. As is evident, the suffix "-ol" is used for phenols as well as for alcohols (with a few exceptions, such as hydroquinone):

The same name, *phenol,* is used for a single compound and for a whole class of compounds. There are, unfortunately, several examples of this practice in the nomenclature of organic compounds.

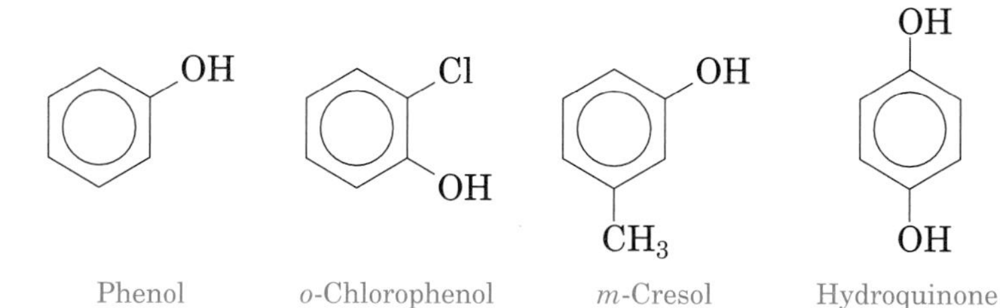

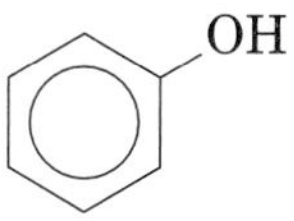

Phenol 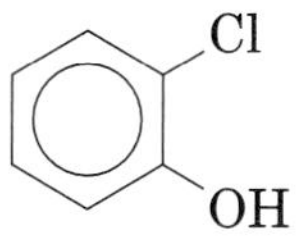 | *o*-Chlorophenol 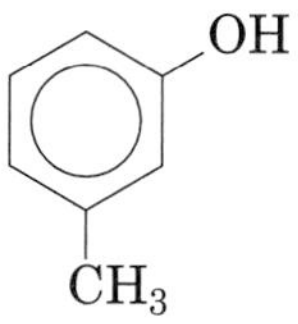 | *m*-Cresol 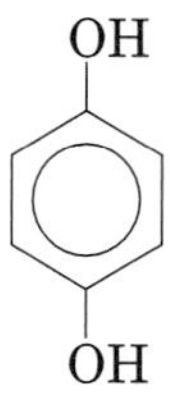 | Hydroquinone

Some phenols are important natural compounds. Examples are vanillin, which gives flavor to vanilla; tetrahydrourushiol, a principal irritant of poison ivy; and thymol, which is used in medicine to kill fungi and hookworms:

Vanillin contains three functional groups: It is not only a phenol but also an ether and an aldehyde. Its IUPAC name is 4-hydroxy-3-methoxybenzaldehyde.

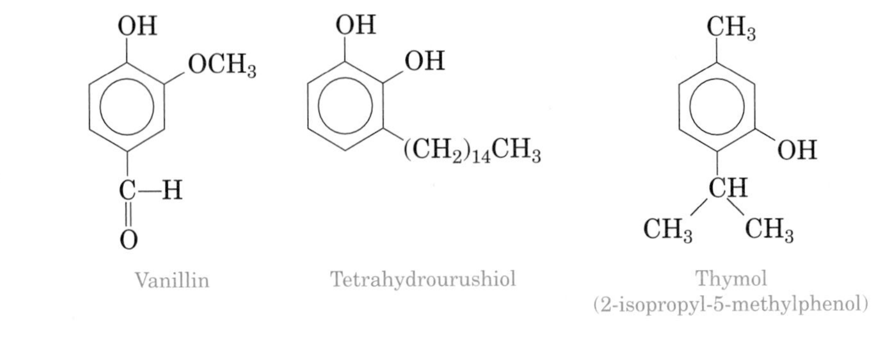

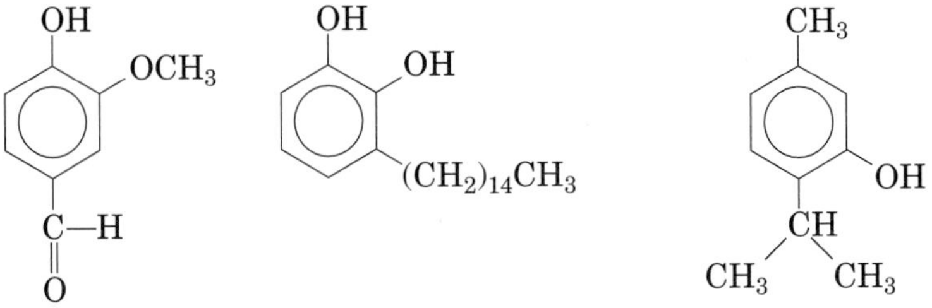

OH, OCH_3, C—H, O

Vanillin

OH, OH, $(CH_2)_{14}CH_3$

Tetrahydrourushiol

CH_3, OH, CH, CH_3, CH_3

Thymol
(2-isopropyl-5-methylphenol)

Phenol itself is a solid that is fairly soluble in water. An aqueous solution of phenol is called *carbolic acid.* This solution was one of the earliest antiseptics because it is toxic to bacteria (Box 12D). Today it is no longer used directly on patients because it burns the skin, but it is still used to clean surgical and medical instruments.

Because phenols are acidic, they dissolve in aqueous NaOH solution. This enables us to distinguish them from other, nonacidic, organic compounds.

The most important chemical property of phenols is that, unlike alcohols, they are acidic. As shown in Table 8.3 (p. 246), phenols have K_a values around 10^{-10} (pK_a = about 10), which makes them weak acids, considerably stronger than water but much weaker than carboxylic acids (pK_a about 5).

Another chemical property of phenols is that many of them are easily oxidized. This is not important in chemical synthesis, because the oxidation products are usually mixtures of products that are difficult to handle. But the property does make certain phenols useful as **antioxidants,** compounds added to foods or other products that must be protected from oxidation. When an oxidizing agent (most often the O_2 in air) approaches, the antioxidant is oxidized first, thus protecting the food or other product. The phenols whose trade names are BHT (butylated hydroxytoluene) and BHA (butylated hydroxyanisole) are added to many products, including rubber, fats and oils, food wrappings, and lubricating oils.

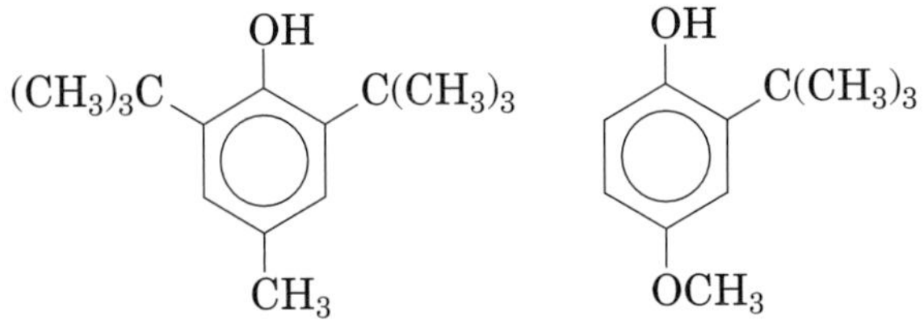

OH, $(CH_3)_3C$, $C(CH_3)_3$, CH_3

BHT

OH, $C(CH_3)_3$, OCH_3

BHA

12.6 Ethers

Ethers are compounds whose formula is R—O—R′, where R and R′ can be the same or different. Ethers are generally named by giving the names of the groups and adding the word “ether”:

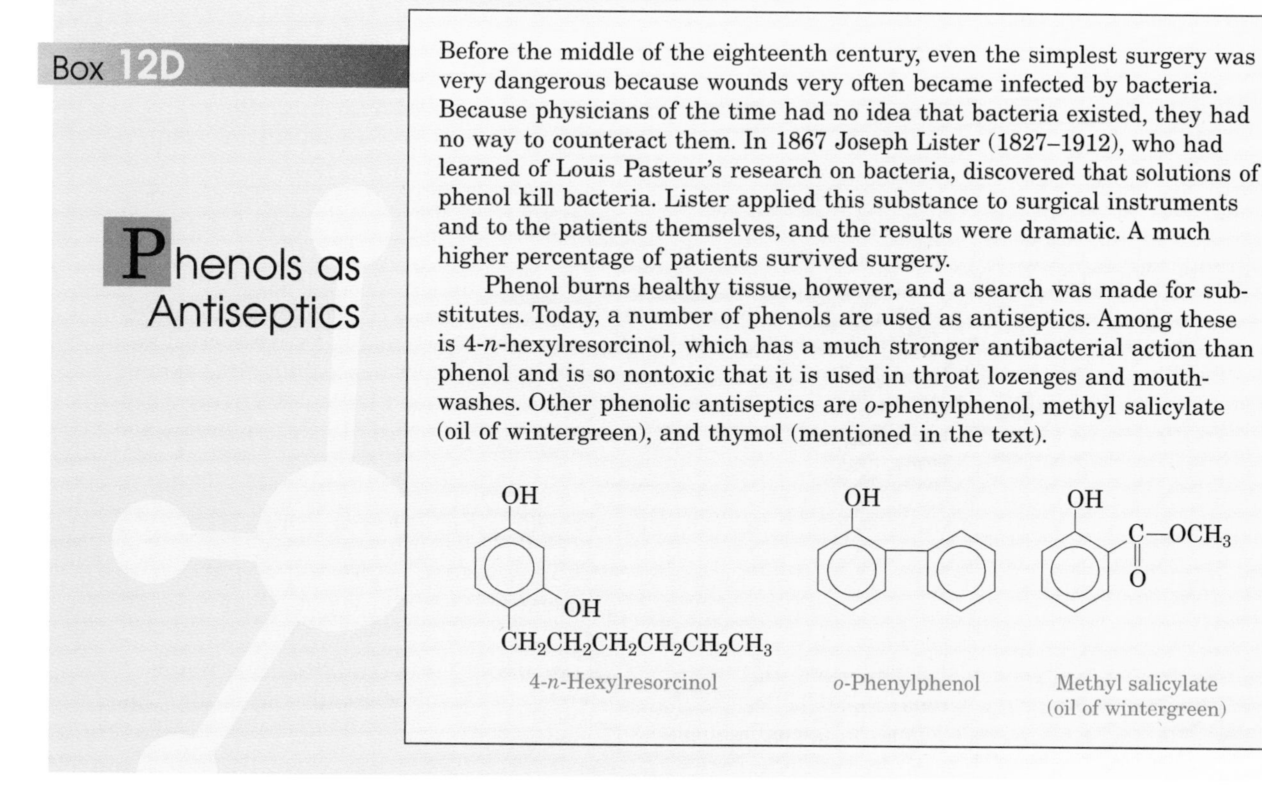

Box 12D

Phenols as Antiseptics

Before the middle of the eighteenth century, even the simplest surgery was very dangerous because wounds very often became infected by bacteria. Because physicians of the time had no idea that bacteria existed, they had no way to counteract them. In 1867 Joseph Lister (1827–1912), who had learned of Louis Pasteur's research on bacteria, discovered that solutions of phenol kill bacteria. Lister applied this substance to surgical instruments and to the patients themselves, and the results were dramatic. A much higher percentage of patients survived surgery.

Phenol burns healthy tissue, however, and a search was made for substitutes. Today, a number of phenols are used as antiseptics. Among these is 4-*n*-hexylresorcinol, which has a much stronger antibacterial action than phenol and is so nontoxic that it is used in throat lozenges and mouthwashes. Other phenolic antiseptics are *o*-phenylphenol, methyl salicylate (oil of wintergreen), and thymol (mentioned in the text).

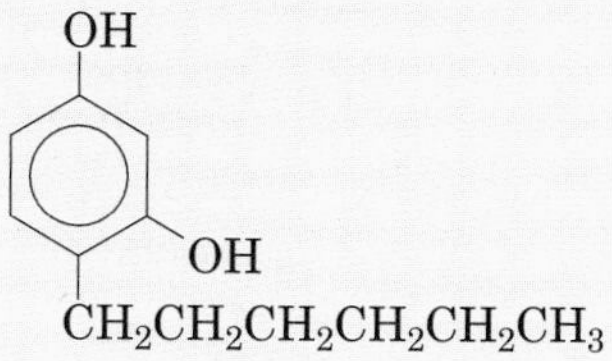

OH, OH, $CH_2CH_2CH_2CH_2CH_2CH_3$

4-*n*-Hexylresorcinol

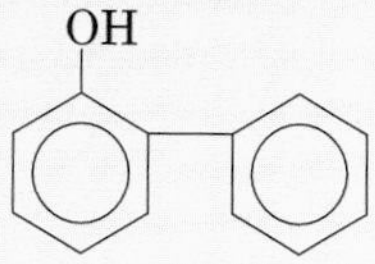

OH

o-Phenylphenol

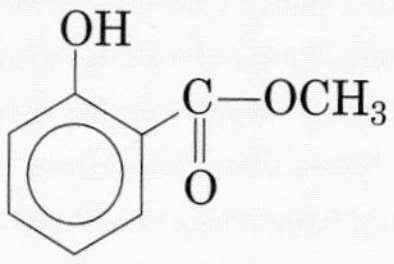

OH, $C(=O)—OCH_3$

Methyl salicylate (oil of wintergreen)

$CH_3—O—CH_3$ Dimethyl ether

$CH_3—O—CH_2CH_3$ Ethyl methyl ether

$CH_3CH_2—O—CH_2CH_3$ Diethyl ether

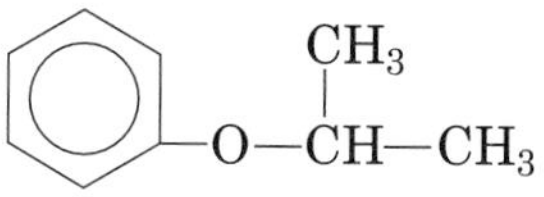

$C_6H_5—O—CH(CH_3)—CH_3$ Isopropyl phenyl ether

If the groups are the same, we use the prefix di-, as in diethyl ether, but many people omit the di- and simply say "ethyl ether."

Problem 12.7

Name:

(a) 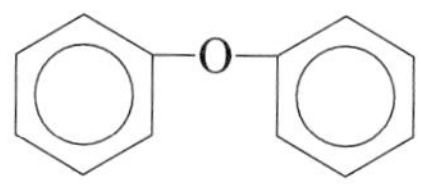 $C_6H_5—O—C_6H_5$ (b) $CH_3CH(CH_3)—O—CH_2CH_3$

Certain heterocyclic ethers (Sec. 11.13) are important in the laboratory, among them

Box 12E

Anesthetics

Before the mid-nineteenth century, not only did many patients die from surgery because of infection (Box 12D), but there were no anesthetics, so surgery was extremely painful; many patients died from shock. Physicians did use alcohol and such drugs as opium, but they were usually reluctant to operate except in extreme circumstances because the risks to the patient were so great. These problems were resolved by the discovery of antiseptics (Box 12D) and anesthetics. Diethyl ether was first used as a general anesthetic in 1842.

Diethyl ether has many advantages as an anesthetic. It is safe to the patient and easy to administer. However, it also has disadvantages: It is relatively slow acting, irritates the respiratory tract, and has a nauseating aftereffect. It is also highly flammable, and its vapor mixed with air is explosive. Because of these disadvantages, a number of other anesthetics are now used. Some of these are also ethers, the most important being enflurane and isoflurane.

$$F_2CH{-}O{-}CF_2{-}\underset{\large Cl}{\underset{|}{CH}}{-}F \qquad F_3C{-}O{-}\underset{\large Cl}{\underset{|}{CH}}{-}CF_3$$

Enflurane (Ethrane) — Isoflurane

Other anesthetics are not ethers at all. One of the most important of today's anesthetics is halothane, a simple halogen derivative of ethane. This compound is nonflammable and nonexplosive.

$$H{-}\overset{\large Br}{\overset{|}{\underset{|}{\underset{\large Cl}{C}}}}{-}\overset{\large F}{\overset{|}{\underset{|}{\underset{\large F}{C}}}}{-}F$$

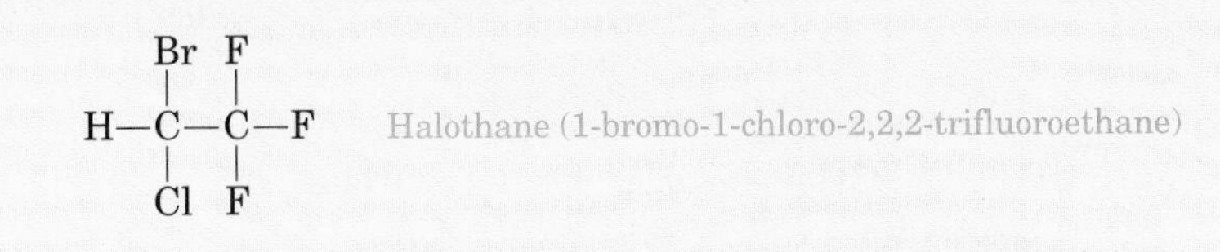

Halothane (1-bromo-1-chloro-2,2,2-trifluoroethane)

Cyclic ethers don't have "ether" in their names.

Anesthesiologists administer gas mixtures to patients undergoing surgery from a gas-mixing manifold like this one. (Photograph by Charles D. Winters.)

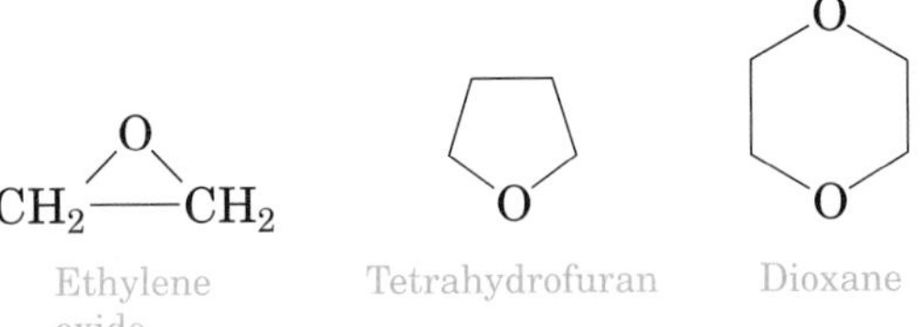

By far the most important ether is diethyl ether, which is often simply called "ether." This was formerly the most important general anesthetic, although it has now been largely replaced by other compounds (Box 12E). Diethyl ether is still an important laboratory solvent, as are tetrahydrofuran and dioxane. Ethylene oxide, a gas, is used in the manufacture of plastics and other compounds.

The chief chemical property of ethers is that, like the alkanes, they are virtually inert and do not react with most reagents. It is this property that makes some of them such valuable solvents. There are, how-

It does not cause discomfort and is considered safe, though it does depress respiratory and cardiovascular action. Another important anesthetic is the simple inorganic compound nitrous oxide, N_2O (also called laughing gas), which is used by dentists because its effects wear off quickly.

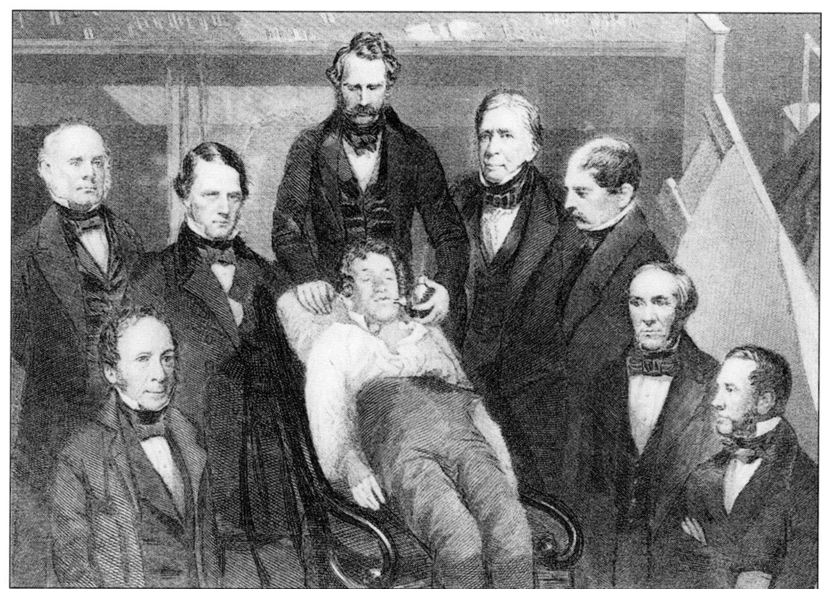

Early demonstration of the use of ether as an anesthetic. (Courtesy of the National Library of Medicine.)

ever, two chemical properties of ethers that laboratory workers must be aware of:

1. Like many organic compounds, ethers are flammable. Diethyl ether, the most common, is especially dangerous to work with because of its low boiling point (35°C), only a few degrees above room temperature. This means that the air in a room where ether is being used contains substantial amounts of ether vapor, which can quickly travel to a flame elsewhere in the room. It is therefore important that no source of flame (including pilot lights) or electric spark be present in any room where diethyl ether is being used.

2. On standing for long periods of time, diethyl ether (and some other ethers) is oxidized by air to give high-boiling explosive compounds called peroxides. To prevent this, diethyl ether must be stored in the dark in sealed containers away from any source of heat.

Box 12F

Deodorants

Body odor is caused by the action of certain skin bacteria on natural secretions that are part of perspiration. The secretions themselves have no odor, but the bacteria react with compounds in the secretion to produce compounds with unpleasant odors. Throughout most of history people dealt with this by using perfumes and/or soap and water, or else many simply chose to smell bad. Today in the United States and other developed countries the use of personal deodorants is widespread. These deodorants operate by killing the bacteria that react with the secretions. The most common antibacterial agent in deodorants today is called *triclosan* (chemical name 2,4,4′-trichloro-2′-hydroxydiphenyl ether).

Triclosan   Hexachlorophene

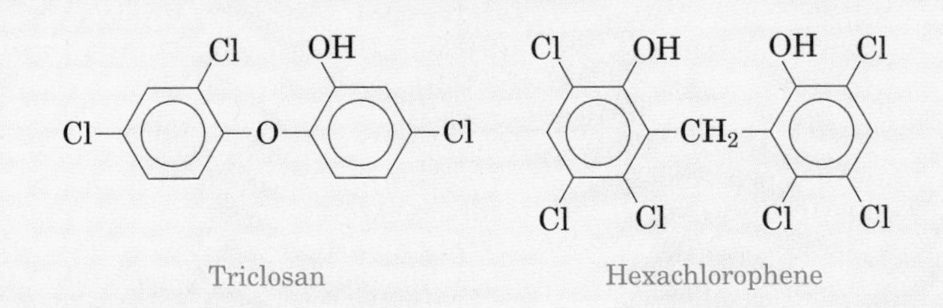

This compound is both a phenol and an ether, as well as a halide. A typical solid deodorant has less than 1 percent triclosan, the rest being inactive ingredients (that is, they don't kill bacteria), mostly ethanol (or 1,2-propanediol) and water. A fragrance is also usually added. At one time most deodorants contained the compound hexachlorophene (this was also used in soap and even in baby lotion), but this compound proved too toxic and is no longer used in any cosmetic products.

MTBE is also used medicinally, to dissolve gallstones.

Since 1992 the Clean Air Act has mandated the use of "oxygenated" gasoline in metropolitan areas of the United States during the winter months because of carbon monoxide pollution. By law, the oxygen content of this new gasoline must be at least 2.7% by weight. Since gasoline ordinarily contains mostly hydrocarbons, with only very small amounts of oxygen-containing compounds (Box 10D), gasoline companies achieve this goal by adding certain alcohols or ethers, mostly either methyl *tert*-butyl ether (MTBE) or ethanol, to the gasoline.

Oxygen-containing compounds are routinely added to gasoline in certain parts of the country in the winter months. (Photograph by Beverly March.)

$$CH_3—O—C(CH_3)_2—CH_3 \qquad CH_3CH_2—OH$$

Methyl *tert*-butyl ether   Ethanol

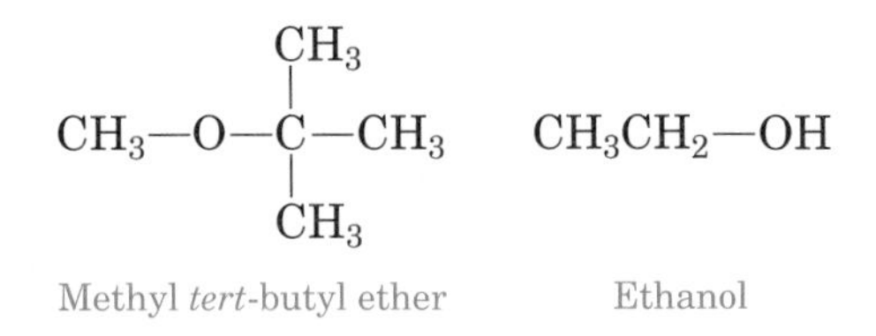

In certain parts of the United States and other countries, even more ethanol is added (10–20%), to give a product called *gasohol*.

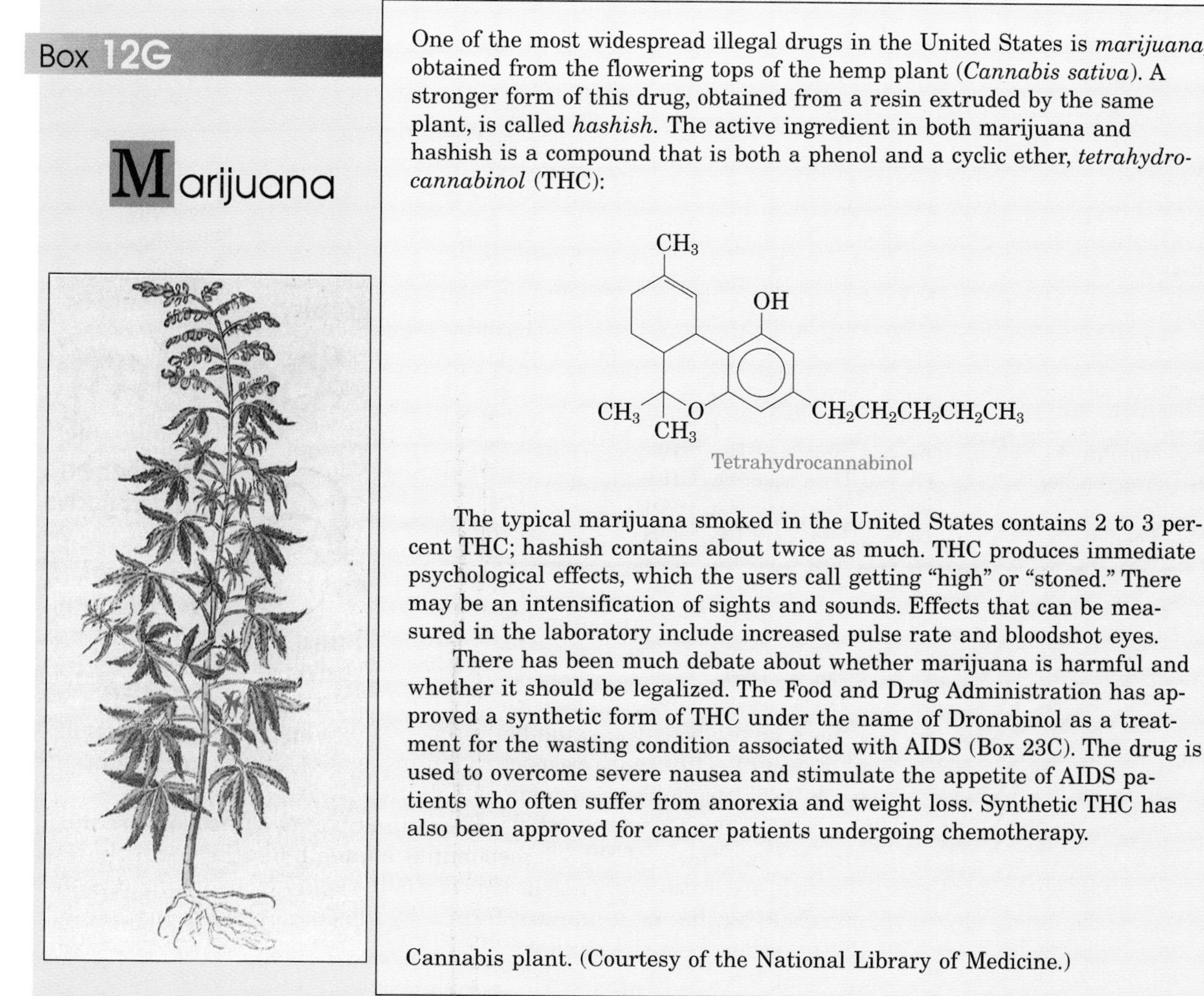

Box 12G

Marijuana

One of the most widespread illegal drugs in the United States is *marijuana,* obtained from the flowering tops of the hemp plant (*Cannabis sativa*). A stronger form of this drug, obtained from a resin extruded by the same plant, is called *hashish.* The active ingredient in both marijuana and hashish is a compound that is both a phenol and a cyclic ether, *tetrahydrocannabinol* (THC):

Tetrahydrocannabinol

The typical marijuana smoked in the United States contains 2 to 3 percent THC; hashish contains about twice as much. THC produces immediate psychological effects, which the users call getting "high" or "stoned." There may be an intensification of sights and sounds. Effects that can be measured in the laboratory include increased pulse rate and bloodshot eyes.

There has been much debate about whether marijuana is harmful and whether it should be legalized. The Food and Drug Administration has approved a synthetic form of THC under the name of Dronabinol as a treatment for the wasting condition associated with AIDS (Box 23C). The drug is used to overcome severe nausea and stimulate the appetite of AIDS patients who often suffer from anorexia and weight loss. Synthetic THC has also been approved for cancer patients undergoing chemotherapy.

Cannabis plant. (Courtesy of the National Library of Medicine.)

12.7 Physical Properties of Alcohols, Phenols, and Ethers

All the hydrocarbons considered in Chapters 10 and 11 have similar physical properties (this also includes the halides). Basically, the properties we are most interested in are boiling points (and melting points), which increase fairly regularly with increasing molecular weight, and solubilities: They are all insoluble in water but generally are soluble in each other.

When we come to alcohols, phenols, and ethers, we find a big difference in physical properties, as shown in Table 12.3.

Table 12.3 Molecular Weights, Boiling Points, and Solubilities of Some Simple Alkanes, Alcohols, and Ethers

Name	Structural Formula	MW	BP (°C)	Solubility in Water
Ethane	$CH_3—CH_3$	30	−88	Insoluble
Methanol	$CH_3—OH$	32	65	Soluble
Propane	$CH_3—CH_2—CH_3$	44	−42	Insoluble
Dimethyl ether	$CH_3—O—CH_3$	46	−23	Soluble
Ethanol	$CH_3—CH_2—OH$	46	78	Soluble
Butane	$CH_3—CH_2—CH_2—CH_3$	58	0	Insoluble
Ethyl methyl ether	$CH_3—CH_2—O—CH_3$	60	11	Soluble
1-Propanol	$CH_3—CH_2—CH_2—OH$	60	97	Soluble
Ethylene glycol	$HO—CH_2—CH_2—OH$	62	198	Soluble
Pentane	$CH_3—CH_2—CH_2—CH_2—CH_3$	72	36	Insoluble
Diethyl ether	$CH_3—CH_2—O—CH_2—CH_3$	74	35	Slightly soluble
Methyl propyl ether	$CH_3—CH_2—CH_2—O—CH_3$	74	39	Slightly soluble
1-Butanol	$CH_3—CH_2—CH_2—CH_2—OH$	74	117	Slightly soluble
1,3-Propanediol	$HO—CH_2—CH_2—CH_2—OH$	76	214	Soluble

Boiling Point

Ethers have about the same boiling points as alkanes of similar molecular weight. Thus, with respect to boiling points, ethers are essentially no different from hydrocarbons or halides.

This is not true for alcohols. The boiling points of alcohols are much higher than those of the corresponding alkane and ether, and those of the diols are much higher still. This is easily explained by hydrogen bonding. Each alcohol molecule can form a hydrogen bond with another one, but an ether molecule cannot:

Review Section 6.6 for a discussion of which molecules can form hydrogen bonds.

$$\underset{\text{Alcohol}}{R—\underset{\diagdown H}{O}}\cdots\underset{\text{Alcohol}}{H—\underset{\diagdown R}{O}} \qquad \underset{\text{Ether}}{R—\underset{\diagdown R'}{O}} \quad \underset{\text{Ether}}{R—\underset{\diagdown R'}{O}}$$

(no hydrogen bond possible)

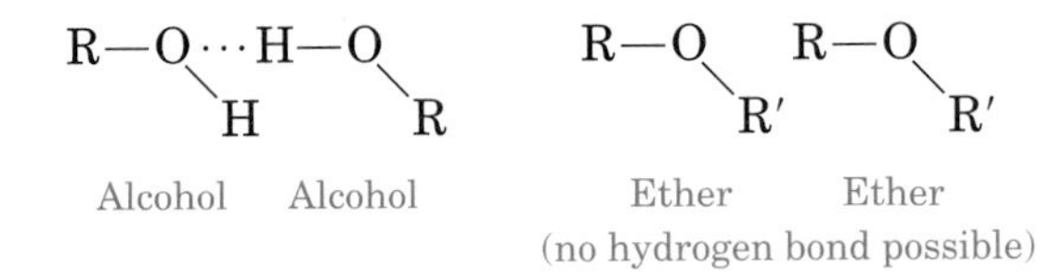

Alcohols have higher boiling points because hydrogen bonds must be broken during boiling, and this requires energy, available only at higher temperatures. Diols, with two —OH groups, have hydrogen bonds on two sides, so their boiling points are higher still. Hydrogen bonding also accounts for the viscosity of such liquids as ethylene glycol and glycerol.

Hydrogen bonding accounts for ethylene glycol's viscosity and its solubility in water. (Photograph by Charles D. Winters.)

Solubility in Water

A look at Table 12.3 shows that both alcohols *and* ethers are soluble in water, up to about three or four carbons. This behavior is, of course, completely different from that of the hydrocarbons and halides. Again, the

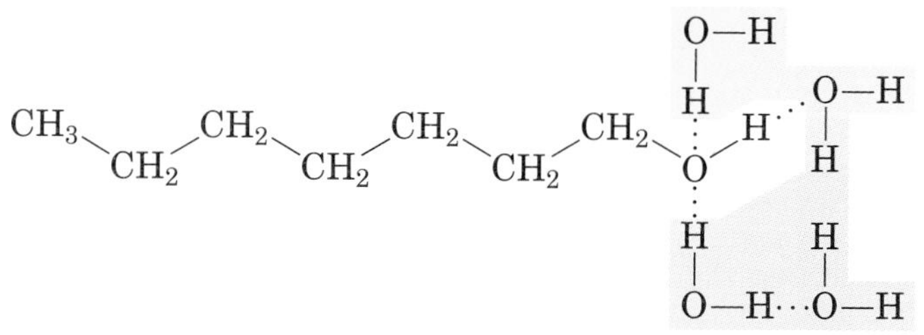

$CH_3CH_2CH_2CH_2CH_2CH_2CH_2$—O—H ··· (O—H, H—O—H, H—O—H ··· O—H)

Figure **12.1**
Water molecules surround the —OH end of 1-heptanol but do not come near the alkyl end. Because the alkyl end is unsolvated, 1-heptanol is insoluble in water.

reason is hydrogen bonding. Both alcohols and ethers have an oxygen atom, so both can form hydrogen bonds with water and so can dissolve.

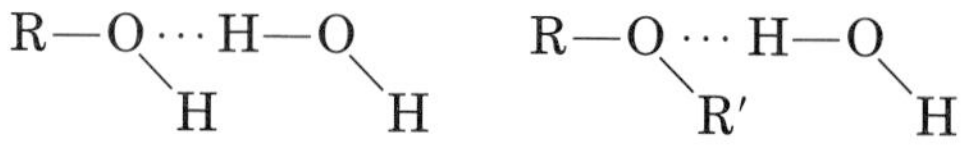

R—O(—H) ··· H—O—H   R—O(—R′) ··· H—O—H

Alcohol   Water   Ether   Water

Ethers can form hydrogen bonds with water but not with other ether molecules, because water has the group —OH but an ether molecule does not.

Alcohols and ethers of higher molecular weight, however, do not dissolve in water because the water molecules cannot solvate the large R groups. For example, consider 1-heptanol. This molecule consists of an alkyl chain of seven carbons and an —OH group. As shown in Figure 12.1, the —OH forms hydrogen bonds with and is surrounded by water molecules, but the alkyl portion of the molecule has *no* attraction for water molecules. Because this part of the molecule cannot be surrounded by water molecules, 1-heptanol is insoluble in water.

We call this part of the molecule *hydrophobic,* meaning water-hating.

As you might expect, the physical properties of phenols are much like those of alcohols of similar molecular weight.

12.8 Thiols, Thioethers, and Disulfides

Thiols and thioethers are sulfur analogs of the corresponding alcohols, phenols, and ethers (Table 12.1),

Thiols are also called mercaptans, and thioethers are also called sulfides.

R—SH   R—S—R′

Thiols   Thioethers

but are much less important chemically. Three properties of thiols are worth mentioning here.

1. Thiols have foul odors. You generally don't expect organic compounds to have pleasant odors (although some do), but thiols smell so bad that chemical companies put the word "stench" on the labels. The liquid squirted by skunks is a mixture of thiols and closely related compounds.
2. Thiols are weak acids, somewhat weaker than phenols.
3. Thiols are easily oxidizable to disulfides:

Gas companies take advantage of the odor of thiols. Natural gas (methane) has no odor, so the companies add a tiny amount of methanethiol, CH_3SH, so that gas leaks can be detected before a spark or a match sets off an explosion.

$$2RSH \xrightarrow{[O]} R—S—S—R$$

This reaction can be accomplished by many oxidizing agents. Note that the disulfides have an S—S bond. Disulfides are, in turn, easily reducible to thiols by several different reducing agents:

The [H] represents an unspecified reducing agent.

$$R—S—S—R \xrightarrow{[H]} 2R—SH$$

The easy conversion between thiols and disulfides is very important in protein chemistry, as we shall see in Section 18.9.

Thioethers are not very important, and we shall not discuss them further.

12.9 Alkyl and Aryl Halides

An aryl halide is an aromatic halide.

We have already discussed the naming of simple alkyl halides (Sec. 10.7) and aryl halides (Sec. 11.10). These compounds are not important in living tissues, but they are important in chemical laboratories and in industry because they undergo many reactions and so can be converted to other compounds. We will be chiefly interested in looking at a few of the simpler ones.

Chloromethane, CH_3Cl, also called methyl chloride, is a gas at room temperature but can be liquefied under pressure. The pressurized liquid is used as a local anesthetic, whose effect is derived from its rapid evaporation. Ethyl chloride (chloroethane), CH_3CH_2Cl, is used for the same purpose (Box 5H).

Dichloromethane, CH_2Cl_2, also called methylene chloride, is a common laboratory solvent.

Trichloromethane, $CHCl_3$, more commonly known as chloroform, was one of the earliest anesthetics but is no longer used for this purpose because it is too toxic. It is still used (carefully) as a laboratory solvent.

Tetrachloromethane, CCl_4, better known as carbon tetrachloride, was once the most important dry-cleaning solvent. Because of its toxicity, it is no longer used for that purpose. Today most dry cleaning is done with perchloroethylene (tetrachloroethene), $Cl_2C{=}CCl_2$, which is as good a solvent but has a much lower toxicity.

Triiodomethane, CHI_3, also called iodoform, is a yellow solid. In the past it was used in hospitals as a general disinfectant, so much so that its characteristic odor has become known as "hospital odor."

Dichlorodifluoromethane, CCl_2F_2, known by the trade name Freon-12, is the most important member of a group of compounds called chlorofluorocarbons (CFCs), compounds that contain carbon, chlorine, and fluorine. Freon-12, a nontoxic, noncorrosive, nonreactive gas, has been used for many years as the refrigerant in virtually all refrigerators and air conditioners. Other major uses of Freon-12 and other CFCs have been as the propellant gas in aerosol sprays and in the manufacture of plastic foams used as insulators in refrigerators and other appliances. However, it is now known that Freon-12 and the other CFCs are a threat to the ozone layer in the earth's upper atmosphere, which shields us from the harmful rays of the sun (Fig. 12.2). The use of these compounds in aerosol sprays was discontinued several years ago, and use of them for other purposes, especially as refrigerants, is now also being discontinued, by international agreement. The compound that is largely replacing Freon-12 as a refrigerant is 1,1,1,2-tetrafluoroethane, F_3CCH_2F, which is considered to be much less of a threat to the ozone layer.

The compound *para*-dichlorobenzene,

$$\mathrm{Cl}-\bigcirc-\mathrm{Cl}$$

sold under the trade name Para, is a moth repellent.

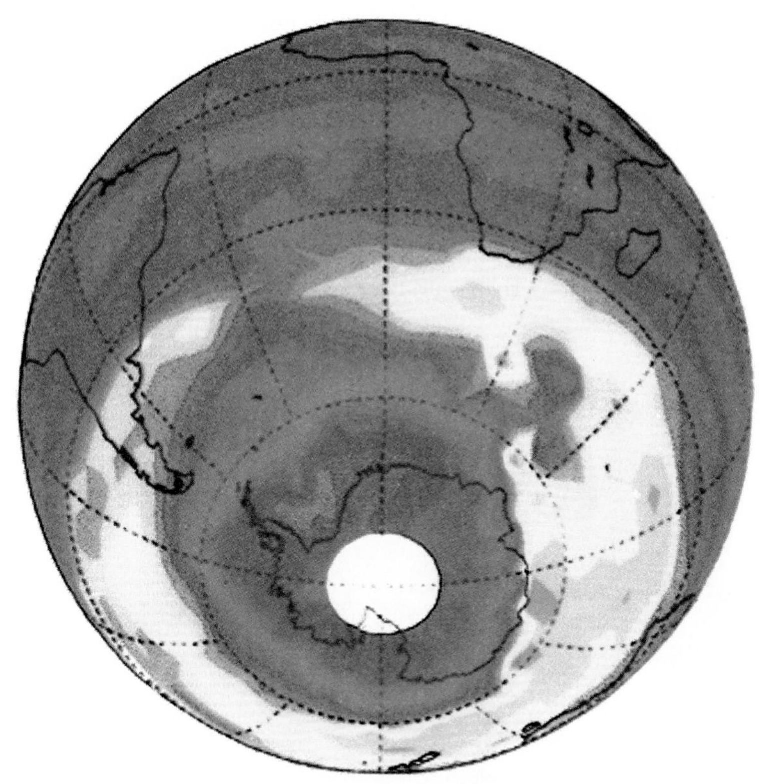

Figure **12.2** • A chart showing the density of the earth's ozone layer in September 1992. As the scheme at the bottom shows, a blue color indicates a low density of ozone. The area around the south pole is entirely blue, showing very low ozone concentration. (The colorless circle directly at the pole indicates an area where no measurements were made.) (Courtesy of NASA.)

Box 12H

Chlorinated Hydrocarbon Insecticides

Certain chlorinated hydrocarbons are widely used and effective insecticides. Two of the most important ones are DDT and chlordane:

1,1,1-Trichloro-2,2-bis(*p*-chlorophenyl)ethane (DDT)

Chlordane

In many places chlorinated hydrocarbons have been so effective in killing crop-eating insects that crop yields have increased enormously. Chlorinated hydrocarbons have also wiped out, in large areas of the world, a number of insects that carry such diseases as malaria and typhus.

However, chlorinated hydrocarbons are a two-edged sword. Despite their enormous benefits, they also have enormous disadvantages. They are toxic not just to insects but to all animal life, including human beings. They remain in the environment (soil and water) for years because they are not *biodegradable* (that is, there are no natural processes by which they decompose). Because of this, it is estimated that the tissues of the average person contain about 5 to 10 ppm of DDT. There is DDT in mother's milk. As a result of all these problems, some governments have banned some or all chlorinated hydrocarbons, and their worldwide use is declining.

An airplane spraying insecticide on a field of crops. (Courtesy of U.S. Department of Agriculture.)

Box 12I

Iodine and Goiter

In the disease called goiter, the thyroid gland in the neck becomes enlarged. The relationship between this disease and iodine intake from food was established 130 years ago. Centuries before, goiter was treated by feeding patients such sources of iodine as burnt sea sponges. The iodine interacts with a protein called thyroglobulin, and the aromatic rings of the protein become iodinated. Two iodinated molecules interact, forming a thyroxine unit that is bound to a protein. The aromatic unit is then released to become the powerful thyroid hormone thyroxine.

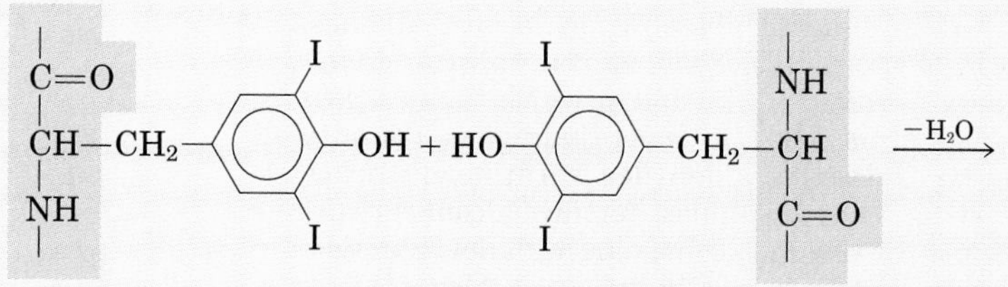

$-H_2O$

Protein part

Protein part

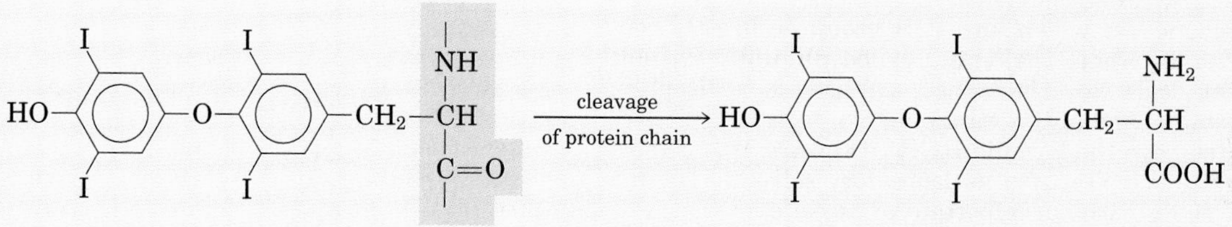

cleavage of protein chain

Thyroxine

Young mammals require thyroid hormones for normal growth and development. Lack of thyroid hormones during fetal development results in mental retardation. Low thyroid hormone levels in adults results in hypothyroidism, commonly called goiter, the symptoms of which are lethargy, obesity, and dry skin.

The iodine we get from our food and water is in the form of iodide ion, and the amount of it depends on the iodide content of the soil. Low-iodine soils the world over are responsible for what are called "goiter belts," large areas of a country where the average diet is very low in iodine. In the United States, approximately half of the table salt used is enriched with NaI (100 micrograms of iodide per gram of table salt). It is this iodized salt that has substantially reduced the incidence of goiter in this country, although it has not eliminated it completely.

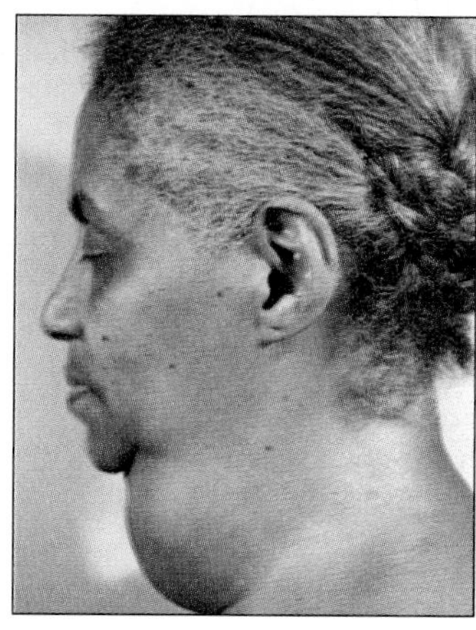

A patient suffering from goiter. (Phototake.)

SUMMARY

Alcohols have the formula **ROH,** where R is an alkyl group. Alcohols are classified as **primary, secondary,** or **tertiary** depending on whether the —OH is connected to a primary, secondary, or tertiary carbon. They can be named either by the IUPAC system or by common names. Alcohols are neither acidic nor basic. When treated with sulfuric acid they undergo dehydration to give alkenes. Where two products are possible, **Zaitsev's rule** is followed: the major product is the one in which the H is lost from the carbon atom that had fewer hydrogens. Primary alcohols can be oxidized to aldehydes or carboxylic acids, and secondary alcohols to ketones. Tertiary alcohols resist oxidation. Some important alcohols are ethanol (the alcohol of alcoholic beverages), methanol (wood alcohol), isopropyl alcohol (rubbing alcohol), menthol, ethylene glycol (antifreeze), and glycerol.

Compounds in which an —OH group is directly connected to an aromatic ring are called **phenols.** Phenols are weak acids. Several of them are used as antiseptics, disinfectants, and antioxidants.

Ethers have the formula **ROR′.** They are rather inert chemically and are used as laboratory solvents. Diethyl ether was formerly an important anesthetic.

Alcohols and phenols have high boiling points as a result of hydrogen bonding. Alcohols, phenols, and ethers are soluble in water if the molecules are small, also because of hydrogen bonding.

Thiols, RSH, have bad odors, are weak acids, and are easily oxidized to **disulfides, RSSR.** Some important alkyl halides are CH_3Cl, CH_2Cl_2, $CHCl_3$, and CCl_4, used mostly as solvents.

Summary of Reactions

1. Dehydration of alcohols to give alkenes (Sec. 12.3):

$$-\underset{\mid}{\overset{\mid}{\underset{H}{C}}}-\underset{\mid}{\overset{\mid}{\underset{OH}{C}}}- \xrightarrow[\text{heat}]{H_2SO_4} -\overset{\mid}{C}=\overset{\mid}{C}- + H_2O$$

2. Oxidation of alcohols (Sec. 12.3):

Primary

$$RCH_2OH \xrightarrow{[O]} R-\underset{\underset{O}{\|}}{C}-H \xrightarrow{[O]} R-\underset{\underset{O}{\|}}{C}-OH$$

Secondary

$$R-\underset{\underset{OH}{\mid}}{CH}-R' \xrightarrow{[O]} R-\underset{\underset{O}{\|}}{C}-R'$$

Tertiary

$$R-\overset{\overset{R'}{\mid}}{\underset{\underset{R''}{\mid}}{C}}-OH \xrightarrow{[O]} \text{no reaction}$$

3. Oxidation of thiols (Sec. 12.8):

$$2RSH \xrightarrow{[O]} R-S-S-R$$

4. Reduction of disulfides (Sec. 12.8):

$$R-S-S-R \xrightarrow{[H]} 2RSH$$

KEY TERMS

Alcohol (Sec. 12.2)
Antioxidant (Sec. 12.5)
Antiseptic (Sec. 12.5)
CFCs (Sec. 12.9)
Congener (Box 12A)
Dehydration (Sec. 12.3)
Denatured alcohol (Sec. 12.4)
Disinfectant (Sec. 12.5)
Disulfide (Sec. 12.8)
Elimination (Sec. 12.3)
Ether (Sec. 12.6)
Gasohol (Sec. 12.6)
Mercaptan (Sec. 12.8)
Phenol (Sec. 12.5)
Primary alcohol (Sec. 12.2)
Proof (Box 12A)
Secondary alcohol (Sec. 12.2)
Tertiary alcohol (Sec. 12.2)
Thioether (Sec. 12.8)
Thiol (Sec. 12.8)
Zaitsev's rule (Sec. 12.3)

PROBLEMS

Nomenclature of Alcohols

12.8 What is the difference between an alcohol and a phenol?

12.9 Give the IUPAC name and, where possible, the common name for these alcohols:

(a) $CH_3CH_2CH_2—OH$

(b) 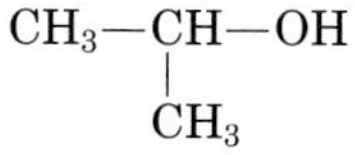
$$\begin{array}{l} CH_3—CH—OH \\ \quad\quad\;\; | \\ \quad\quad CH_3 \end{array}$$

(c) 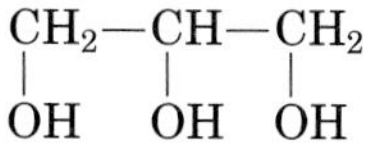
$$\begin{array}{ccccc} CH_2 & — & CH & — & CH_2 \\ | & & | & & | \\ OH & & OH & & OH \end{array}$$

(d) 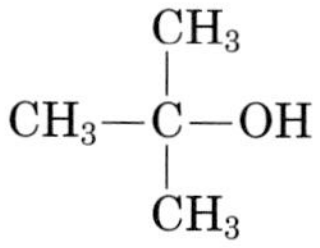
$$\begin{array}{ccc} & & CH_3 \\ & & | \\ CH_3 & — & C—OH \\ & & | \\ & & CH_3 \end{array}$$

(e) 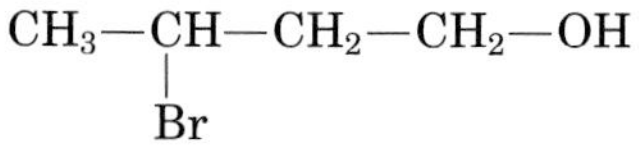
$$\begin{array}{l} CH_3—CH—CH_2—CH_2—OH \\ \quad\quad\;\; | \\ \quad\quad Br \end{array}$$

(f) 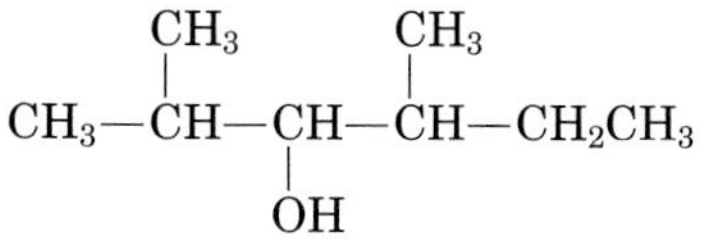
$$\begin{array}{ccccccccc} & & CH_3 & & & & CH_3 & & \\ & & | & & & & | & & \\ CH_3 & — & CH & — & CH & — & CH & — & CH_2CH_3 \\ & & & & | & & & & \\ & & & & OH & & & & \end{array}$$

(g) 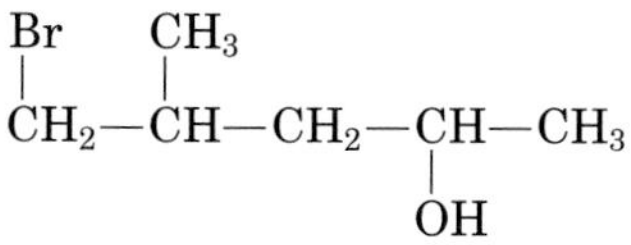
$$\begin{array}{ccccccccc} Br & & CH_3 & & & & & & \\ | & & | & & & & & & \\ CH_2 & — & CH & — & CH_2 & — & CH & — & CH_3 \\ & & & & & & | & & \\ & & & & & & OH & & \end{array}$$

(h) 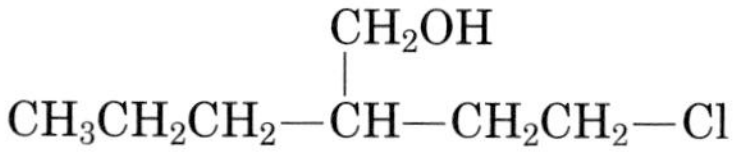
$$\begin{array}{ccccc} & & CH_2OH & & \\ & & | & & \\ CH_3CH_2CH_2 & — & CH & — & CH_2CH_2—Cl \end{array}$$

(i) 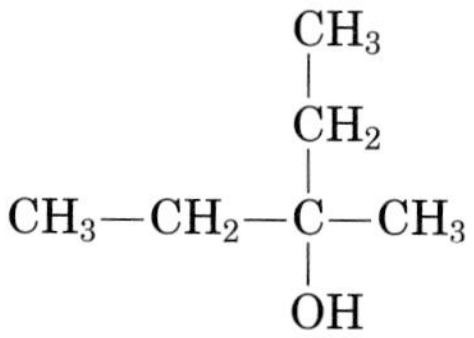
$$\begin{array}{ccccccc} & & & & CH_3 & & \\ & & & & | & & \\ & & & & CH_2 & & \\ & & & & | & & \\ CH_3 & — & CH_2 & — & C & — & CH_3 \\ & & & & | & & \\ & & & & OH & & \end{array}$$

(j) 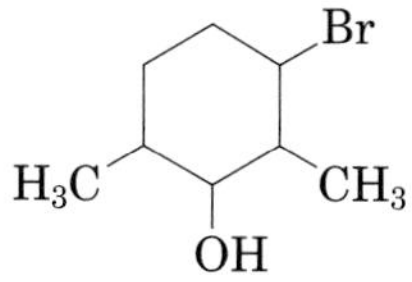 [cyclohexane ring bearing Br, CH₃, OH and H₃C substituents]
Br, H_3C, CH_3, OH

(k) 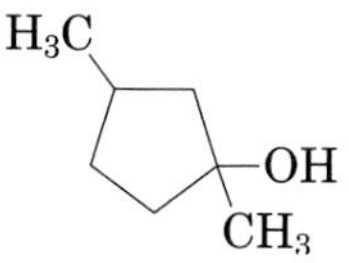 [cyclopentane ring bearing H_3C, and OH and CH_3 on the same carbon]
H_3C, OH, CH_3

12.10 Tell whether each alcohol in Problem 12.9 is primary, secondary, or tertiary.

12.11 What is the difference between a primary, a secondary, and a tertiary alcohol?

12.12 Draw the structural formula for (a) any primary alcohol (b) any secondary alcohol (c) any tertiary alcohol.

12.13 Draw the structural formula for
(a) ethanol
(b) *sec*-butyl alcohol
(c) cyclopentanol
(d) *para*-chlorobenzyl alcohol
(e) ethylene glycol
(f) wood alcohol

12.14 Draw the structural formula for
(a) 3-methyl-1-pentanol
(b) 2-methyl-3-hexanol
(c) 3-methyl-2-hexanol
(d) 2-chloro-3-heptanol
(e) 3,3-dibromocyclobutanol
(f) 1,2,3-pentanetriol

Chemical Properties of Alcohols

12.15 Show the structural formula of the principal organic product when ethanol is treated with
(a) H_2SO_4
(b) excess $K_2Cr_2O_7$ and H_2SO_4

12.16 Write the structural formula for the principal organic product (if no reaction, say so):

(a) $CH_3OH + H_2SO_4 \longrightarrow$

(b) 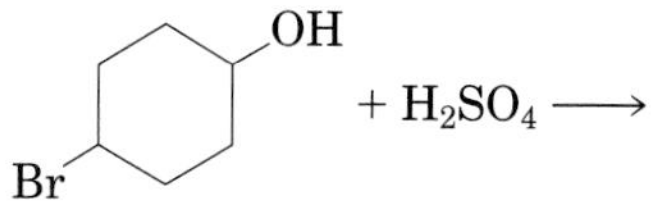 [cyclohexane ring with OH and Br substituents] $+ H_2SO_4 \longrightarrow$

(c) 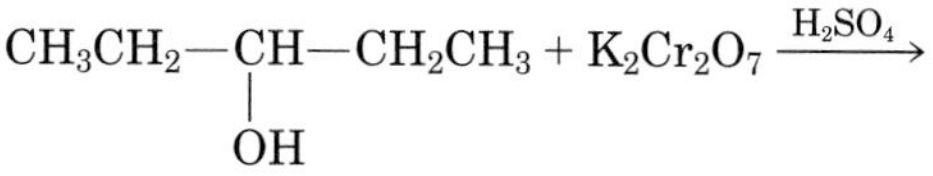
$$\begin{array}{l} CH_3CH_2—CH—CH_2CH_3 + K_2Cr_2O_7 \xrightarrow{H_2SO_4} \\ \quad\quad\quad\quad\;\; | \\ \quad\quad\quad\quad OH \end{array}$$

(d) $CH_3CH_2CH_2CH_2—OH + \overset{\text{excess}}{K_2Cr_2O_7} \xrightarrow{H_2SO_4}$

(e) 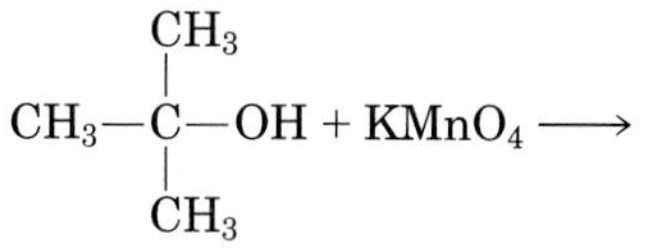
$$\begin{array}{ccl} & & CH_3 \\ & & | \\ CH_3 & — & C—OH + KMnO_4 \longrightarrow \\ & & | \\ & & CH_3 \end{array}$$

(f) 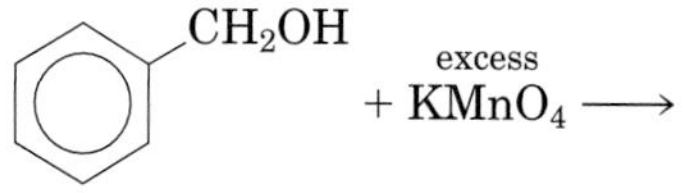 [benzene ring with CH_2OH substituent] $+ \overset{\text{excess}}{KMnO_4} \longrightarrow$

12.17 Write structural formulas for the two possible products when each of the following is heated with H_2SO_4. In each case predict which of the two products is actually formed, in accordance with Zaitsev's rule:

(a) 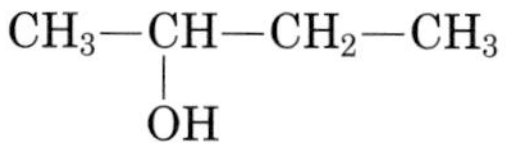
$$\begin{array}{l} CH_3—CH—CH_2—CH_3 \\ \quad\quad\;\; | \\ \quad\quad OH \end{array}$$

(b) 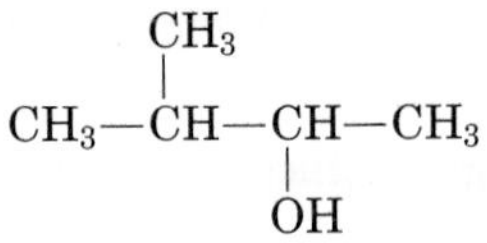 $CH_3-CH(CH_3)-CH(OH)-CH_3$

(c) 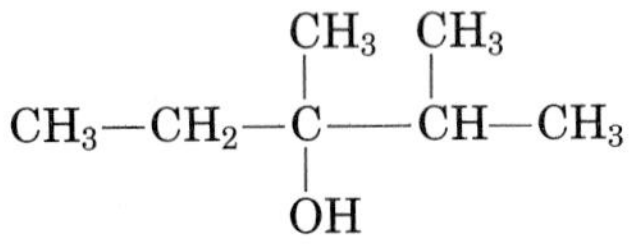 $CH_3-CH_2-C(CH_3)(OH)-CH(CH_3)-CH_3$

(d) 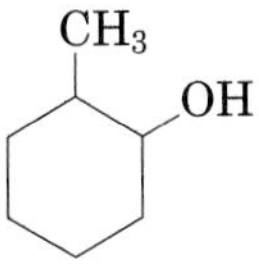 2-methylcyclohexanol structure: cyclohexane ring with CH_3 and OH on adjacent carbons

Some Important Alcohols

12.18 Define (a) denatured alcohol (b) absolute alcohol (c) tincture.

12.19 Which alcohol is
(a) used as an antifreeze
(b) found in fats
(c) used for hospital rubdowns
(d) oxidized to formaldehyde
(e) the alcohol of alcoholic beverages
(f) a cause of blindness and death when ingested in moderate quantities
(g) used to make nitroglycerine

12.20 Name these phenols:

(a) 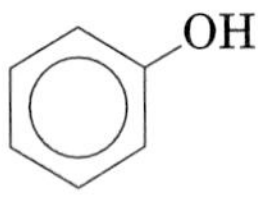 benzene ring–OH

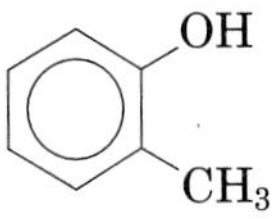

(b) 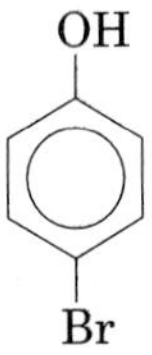 benzene ring with OH and Br para

(c) benzene ring with OH and CH_3 ortho

(d) 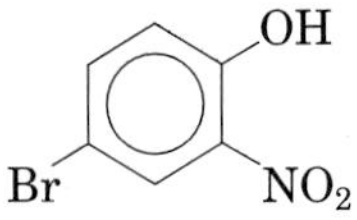 benzene ring with OH, NO_2 and Br

12.21 Complete these reactions:

(a) 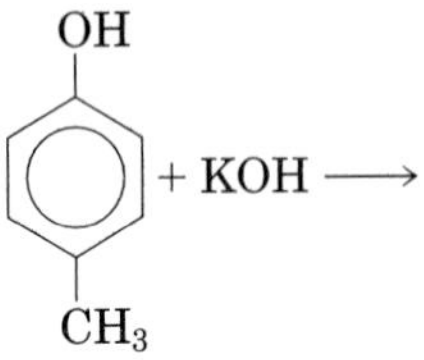 p-$CH_3C_6H_4OH$ + KOH $\longrightarrow$

(b) 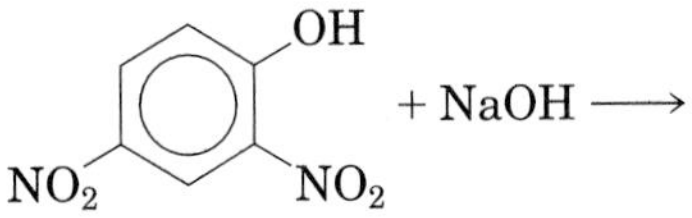 benzene ring with OH and two NO_2 groups + NaOH $\longrightarrow$

12.22 Which phenol mentioned in this chapter is
(a) a former antiseptic that is now considered too strong for use on patients
(b) a flavoring agent
(c) used commercially as an antioxidant

Ethers

12.23 Draw the structural formulas for
(a) dimethyl ether
(b) ethyl propyl ether
(c) diisopropyl ether
(d) phenyl propyl ether
(e) ethyl vinyl ether

12.24 Name these ethers:

(a) $CH_3-O-CH_2CH_3$

(b) $CH_3CH_2-O-CH_2CH_3$

(c) 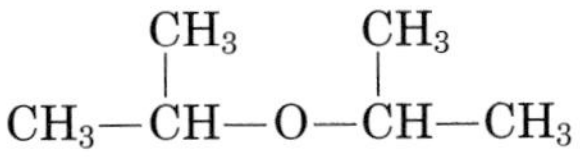 $CH_3-CH(CH_3)-O-CH(CH_3)-CH_3$

(d) 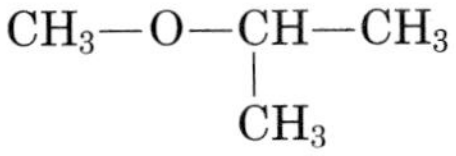 $CH_3-O-CH(CH_3)-CH_3$

(e) 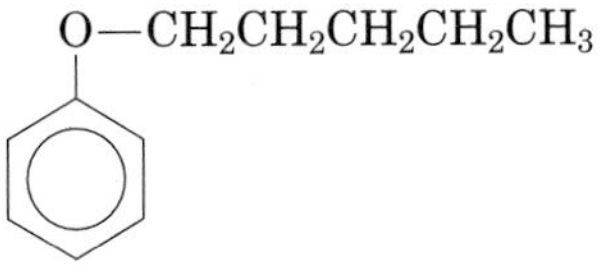 benzene ring–$O-CH_2CH_2CH_2CH_2CH_3$

(f) 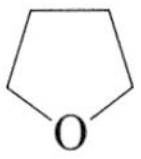 five-membered ring containing O

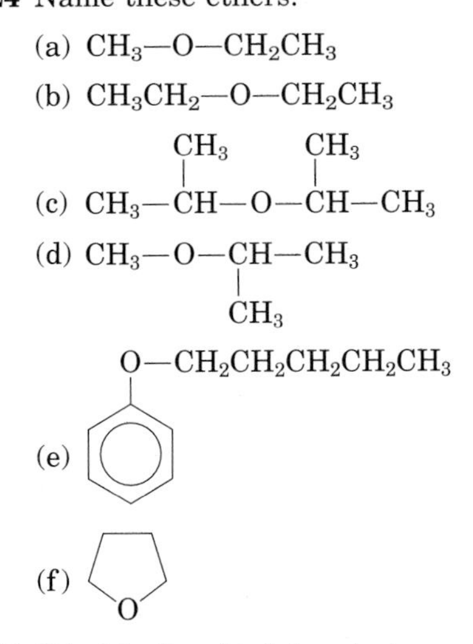

12.25 What is the chief chemical property of ethers?

12.26 Classify each of the following as (1) primary alcohol, (2) secondary alcohol, (3) tertiary alcohol, (4) phenol, (5) ether, or (6) none of these:

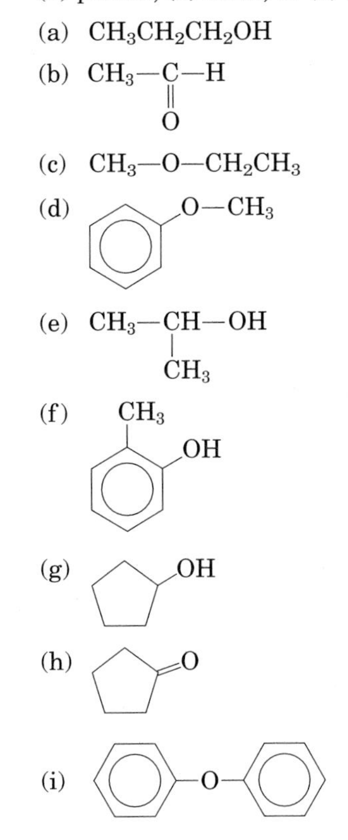

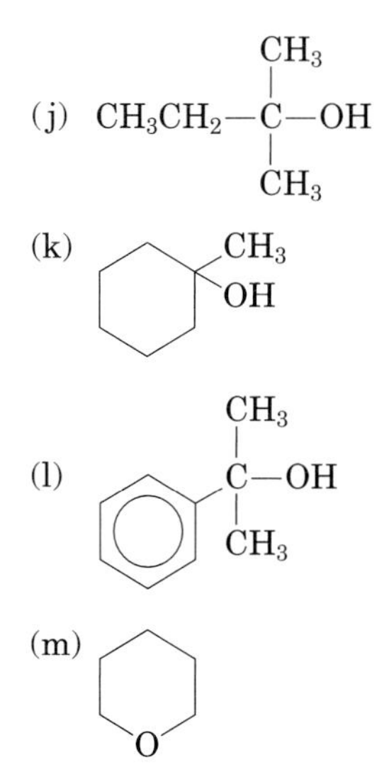

(a) $CH_3CH_2CH_2OH$

(b) 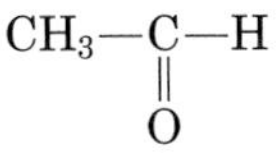 $CH_3-C(=O)-H$

(c) $CH_3-O-CH_2CH_3$

(d) 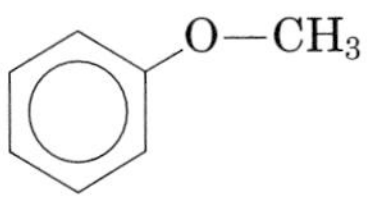 benzene ring–$O-CH_3$

(e) 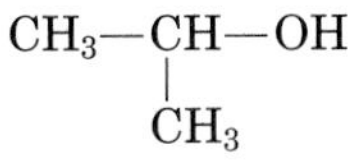 $CH_3-CH(CH_3)-OH$

(f) 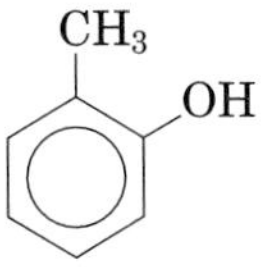 benzene ring with CH_3 and OH ortho

(g) 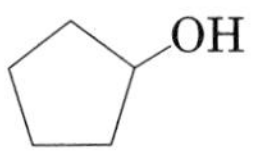 cyclopentane ring–OH

(h) 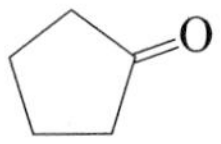 cyclopentane ring=O

(i) 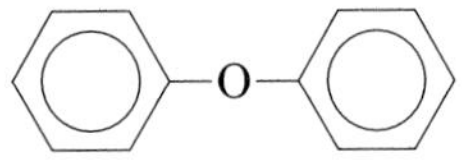 benzene ring–O–benzene ring

(j) $CH_3CH_2-C(CH_3)_2-OH$

(k) 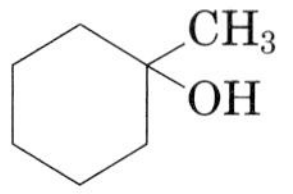 CH_3 OH

(l) 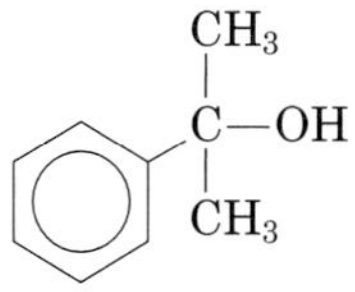 $C_6H_5-C(CH_3)_2-OH$

(m) 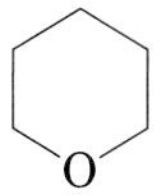 O

12.27 Each of these compounds belongs to two or more of the first five classes mentioned in Problem 12.26. Tell which classes:

(a) $CH_2(OH)-CH(OH)-CH_2(OH)$

(b) $CH_3-O-CH_2CH_2CH_2-OH$

(c) 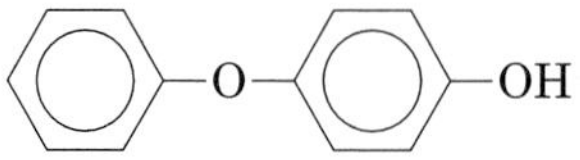 $C_6H_5-O-C_6H_4-OH$

(d) 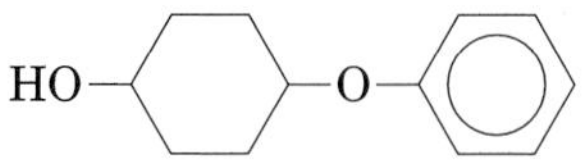 HO–(cyclohexane ring)–O–C_6H_5

(e) 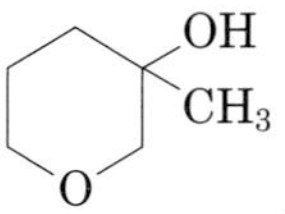 ring with O, bearing OH and CH_3

(f) 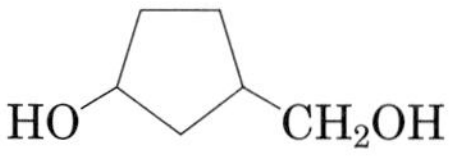 HO–(cyclopentane ring)–CH_2OH

(g) 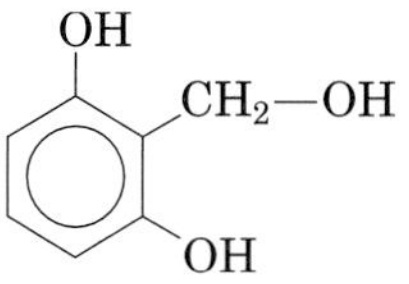 benzene ring with OH, CH_2-OH, OH

(h) 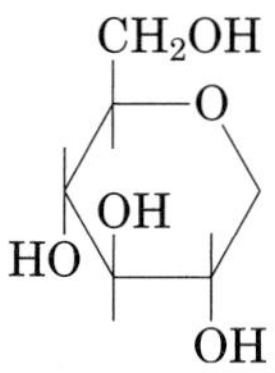 ring with O, bearing CH_2OH, OH, HO, OH

12.28 What is the chief chemical property of phenols?

12.29 Draw the structural formulas for all seven possible compounds whose molecular formula is $C_4H_{10}O$. Some are alcohols and some are ethers.

Physical Properties of Alcohols, Phenols, and Ethers

12.30 Which of these molecules can form a hydrogen bond with another molecule of itself?
(a) CH_3CH_2OH
(b) CH_3CH_2Cl
(c) $CH_3CH_2-O-CH_3$
(d) 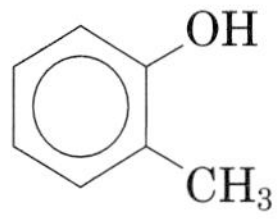 benzene ring with OH and CH_3
(e) 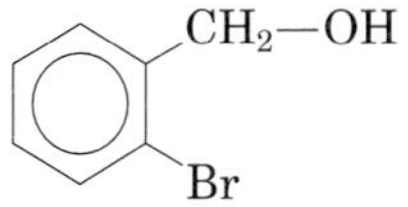 benzene ring with CH_2-OH and Br
(f) $CH_3CH_2-O-CH_2CH_2-F$

12.31 Predict which compound in each pair has the higher boiling point:
(a) CH_3OH and CH_3CH_2OH
(b) $CH_3CH_2CH_2OH$ and $CH_3CH_2CH_2CH_3$
(c) $CH_3CH_2CH_2OH$ and $CH_3CH_2-O-CH_3$

12.32 Predict whether each of the following is soluble in water:
(a) CH_3OH
(b) $CH_3CH_2CH_2CH_2CH_2CH_2OH$
(c) $CH_3OCH_2CH_3$
(d) CH_3Cl
(e) $CH_3CH_2CH_2CH_2-O-CH_2CH_3$
(f) 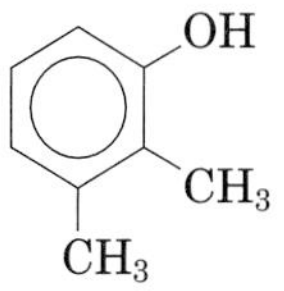 benzene ring with OH, CH_3, CH_3

12.33 1-Pentanol is only slightly soluble in water. Predict the solubility in water of 1,5-pentanediol and 1,3,6-hexanetriol. Explain.

12.34 Explain why the boiling point of ethylene glycol (198°C) is so much higher than that of 1-propanol (97°C) even though their molecular weights are about the same.

Thiols and Disulfides

12.35 Draw the structural formula for the principal organic product:
(a) $2CH_3CH_2-SH \xrightarrow{[O]}$
(b) $CH_3-S-S-CH_3 \xrightarrow{[H]}$

12.36 Write an equation for
(a) the oxidation of ethanethiol, CH_3CH_2SH
(b) the reduction of the disulfide $CH_3CH_2CH_2-S-S-CH_2CH_2CH_3$

12.37 The amino acid cysteine,

$$HS-CH_2-CH(NH_2)-COOH$$

is oxidized in the body to a disulfide. Draw its structure.

Halides

12.38 Which halogenated compound
(a) was formerly an anesthetic but is no longer used because it is too toxic
(b) was formerly a major dry-cleaning agent but is no longer used because it is too toxic
(c) is used as a local anesthetic
(d) is the cause of "hospital odor"
(e) has been an important refrigerant

12.39 Which halogenated compound (a) is used as a dry-cleaning agent today (b) is likely to become an important refrigerant?

12.40 Why is Freon-12 no longer being used as a refrigerant?

Boxes

12.41 (Box 12A) What is the proof in each of the following:
(a) a table wine that contains 12 percent alcohol
(b) a cordial that contains 19 percent alcohol
(c) a rum that contains 48 percent alcohol

12.42 (Box 12A) Why does the average person get drunk faster from champagne than from table wine even if both contain the same percentage of alcohol?

12.43 (Box 12A) How do we get an 86-proof whiskey from a fermentation broth that contains only 13 percent alcohol?

12.44 (Box 12B) How does disulfuram remedy alcohol addiction?

12.45 (Box 12B) Why must pregnant women be particularly careful about alcohol consumption?

12.46 (Box 12B) Distillation of wood yields wood alcohol (methanol). Why should this product never be used for drinking?

12.47 (Box 12C) Do nitroglycerine and isoamyl nitrite have the same functional group? If not, how do they differ?

12.48 (Box 12D) What phenolic compound is present in mouthwashes and throat lozenges?

***12.49** (Box 12E) Name the compound enflurane according to the method given in the text for naming ethers.

12.50 (Box 12E) What are the advantages and disadvantages of diethyl ether as an anesthetic?

12.51 (Box 12F) What causes body odor?

12.52 (Box 12F) Why has triclosan replaced hexachlorophene in deodorants?

12.53 (Box 12G) Find the ether linkage and the phenol group in tetrahydrocannabinol.

12.54 (Box 12G) Why is hashish a stronger drug than marijuana, even though both have the same active ingredient?

12.55 (Box 12H) What is meant by the term "biodegradable"?

12.56 (Box 12I) In the absence of iodine in the diet, goiter develops. Explain why goiter is a regional disease.

Additional Problems

12.57 Does Zaitsev's rule apply to the enzymatic dehydration of citric acid shown in the text? If not, why not?

12.58 List in order of increasing solubility in water:
(a) $CH_3CH_2CH_2OH$
(b) $CH_3CH_2CH_2SH$
(c) $CH_3CH_2—O—CH_2CH_3$

12.59 Draw the structural formulas for
(a) tetrahydrofuran
*(b) *para*-cresol
(c) ethylene oxide

12.60 Explain why methanethiol, CH_3SH, has a lower boiling point (6°C) than methanol, CH_3OH (65°C), even though it has a higher molecular weight.

12.61 Explain why the boiling point of 1-butanol (117°C) is so much higher than that of methyl propyl ether (39°C), even though they have the same molecular weight (they are isomers).

***12.62** 2-Propanol is commonly used as a rubbing alcohol, to cool the skin. 2-Hexanol, also a liquid, is not suitable for this purpose. Why not?

***12.63** The pK_a values of some phenols are: phenol = 9.89; *o*-chlorophenol = 8.49; *o*-methylphenol = 10.20. What is the effect of Cl and CH_3 substitution on the benzene ring on the acidity of the OH group?

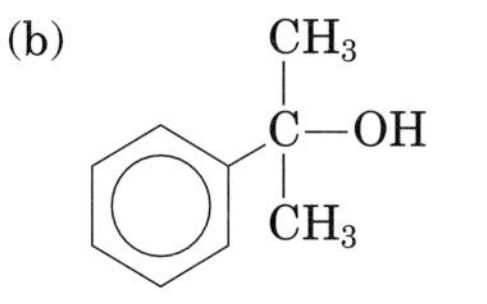

12.64 Draw the structural formulas for the principal organic products when each of these compounds is treated with $K_2Cr_2O_7$ and H_2SO_4 (if no reaction, say so):
(a) $CH_3CH_2CH_2—OH$
(b) $C_6H_5—C(CH_3)_2—OH$ (benzene ring bonded to C bearing two CH_3 groups and OH)
(c) $CH_3CH_2—O—CH_3$
(d) $CH_3CH_2—CH(OH)—CH_2CH_3$

12.65 Write a balanced equation for the reaction in which glucose, $C_6H_{12}O_6$, is fermented to beverage alcohol.

12.66 What is the difference in structure between thymol and menthol? Is either one acidic? Explain.

12.67 Draw the structural formulas for the principal organic products (if no reaction, say so):
(a) $CH_3CH_2—SH + NaOH \longrightarrow$
(b) cyclopentyl—OH $\xrightarrow{H_2SO_4}$
(c) $CH_3CH_2—O—CH_2CH_3 + NaOH \longrightarrow$

12.68 Explain why glycerol is much thicker (more viscous) than ethylene glycol, which in turn is much thicker than ethanol.

12.69 (a) Draw the structural formula for the two possible products when 1-butanol is oxidized. (b) Which of the two would you expect to form in greater amount? (c) How could you get the other one, if that was what you wanted?

12.70 Of the chemical families studied in this chapter, which are acidic and which are neutral?

12.71 Decide which family of organic compounds is referred to in each case:

(a) weak acids; very bad odors; contain sulfur
(b) neutral compounds; contain oxygen; chemically almost inert
(c) weak acids; many have been used as antiseptics
(d) neutral compounds; oxidizable, low-molecular-weight compounds soluble in water

Chapter 13

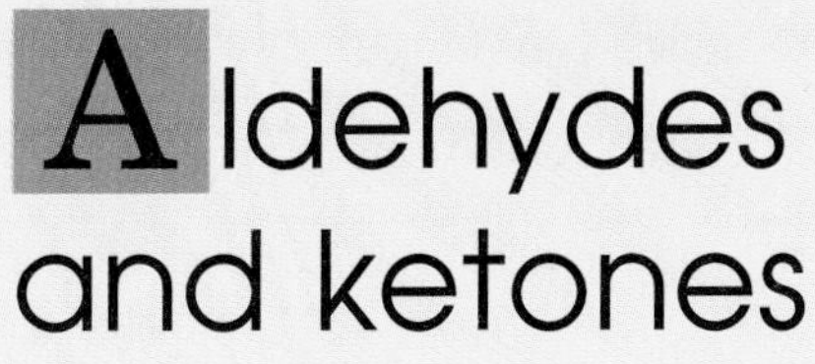

Aldehydes and ketones

13.1 Introduction

One of the most important functional groups in organic chemistry is the **carbonyl group:**

$$\underset{\underset{\textstyle O}{\|}}{-C-}$$

The carbonyl group

Fats, carbohydrates, proteins, nucleic acids, urea, and many other biological compounds contain the carbonyl group.

This group is present in aldehydes and ketones, which we study in this chapter, as well as in other compounds, which will be discussed in Chapters 14 and 15. The carbonyl group is also very important in biochemistry.

The carbon of the carbonyl group must be connected to two other atoms or groups. When it is connected to a hydrogen and an alkyl group or aromatic ring (or to two hydrogens) the compound is called an **aldehyde;** when it is connected to two alkyl groups or aromatic rings (or one of each) the compound is a **ketone:**

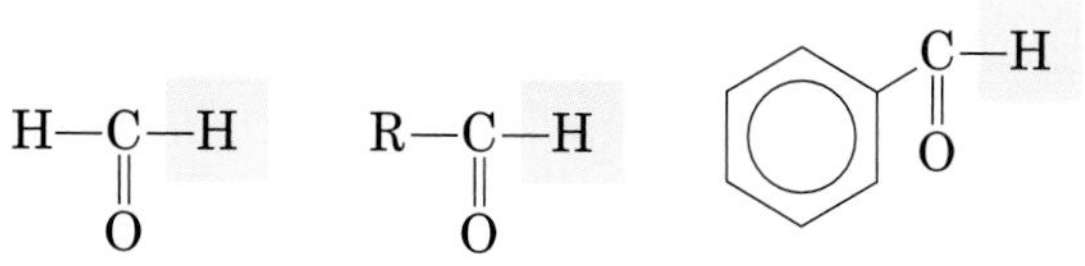

H—C(=O)—H R—C(=O)—H C_6H_5—C(=O)—H

Aldehydes

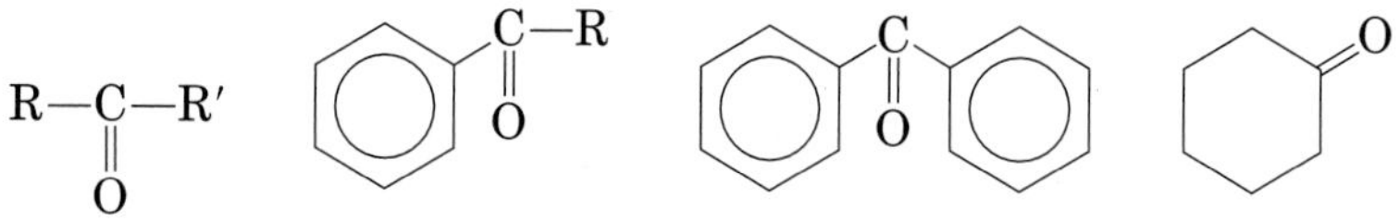

R—C(=O)—R′ C_6H_5—C(=O)—R C_6H_5—C(=O)—C_6H_5 (cyclohexanone)

Ketones

Do not confuse RCHO with alcohols, which are ROH.

Because they always contain at least one H connected to the C=O, aldehydes are often written RCHO to save space. Similarly, ketones may be written as RCOR′.

13.2 Nomenclature

Both aldehydes and ketones can be named by the IUPAC rules. The system is the same as we have seen before, except that the suffix "-al" is used for aldehydes and "-one" for ketones.

EXAMPLE

Name the following compounds.

(a) 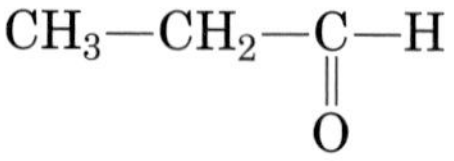 $CH_3—CH_2—\underset{\underset{\textstyle O}{\|}}{C}—H$ Propanal

The longest chain has three carbons; and the suffix for aldehydes is "-al." Therefore the name is propanal. Note that no number is needed because the aldehyde group is always at the end of the chain.

$$\text{(b)}\quad \overset{1}{CH_3}-\overset{2}{CH_2}-\underset{\underset{\displaystyle O}{\|}}{\overset{3}{C}}-\overset{4}{CH_2}-\overset{5}{CH_2}-\overset{6}{CH_3} \qquad \text{3-Hexanone}$$

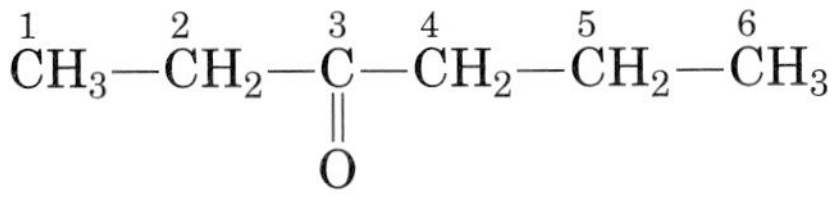

There are six carbons in the chain; and the suffix is "-one." The compound is therefore 3-hexanone. In this case we need a number to specify the position of the ketone group, remembering, of course, that we start numbering at the end that gives the lowest possible number to the functional group.

$$\text{(c)}\quad Cl-\overset{5}{CH_2}-\overset{\overset{\displaystyle OH}{|}}{\overset{4}{C}H}-\overset{3}{CH_2}-\overset{2}{CH_2}-\underset{\underset{\displaystyle O}{\|}}{\overset{1}{C}}-H \qquad \text{5-Chloro-4-hydroxypentanal}$$

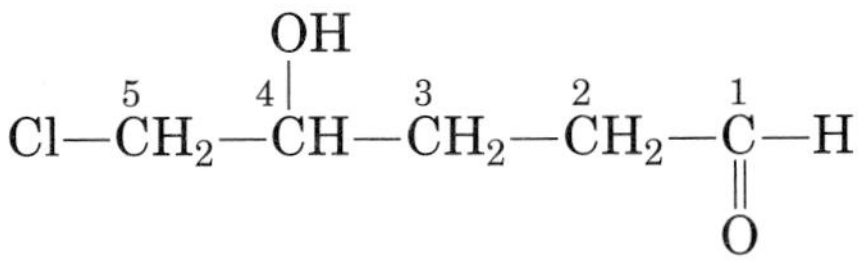

$$\text{(d)}\quad \overset{5}{CH_2}=\overset{4}{CH}-\overset{3}{CH_2}-\underset{\underset{\displaystyle O}{\|}}{\overset{2}{C}}-\overset{1}{CH_2}-Cl \qquad \text{1-Chloro-4-penten-2-one}$$

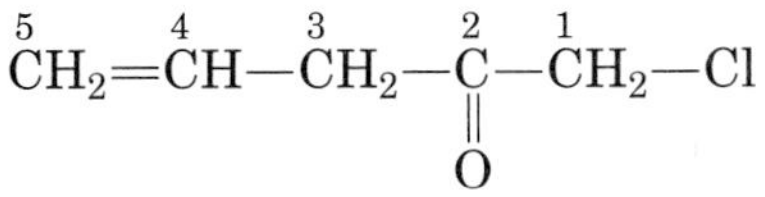

In this case both "-en-" and "-one" are placed in the suffix, and both are given numbers. The 2 goes just before the "-one," because if it were put anywhere else it would be confusing.

Problem **13.1**

Name these compounds by the IUPAC system:

$$\text{(a)}\quad CH_3CH_2-\underset{\underset{\displaystyle Cl}{|}}{CH}-\underset{\underset{\displaystyle O}{\|}}{C}-H \qquad\qquad \text{(b)}\quad CH_3-\underset{\underset{\displaystyle O}{\|}}{C}-CH_2CH_2-Cl$$

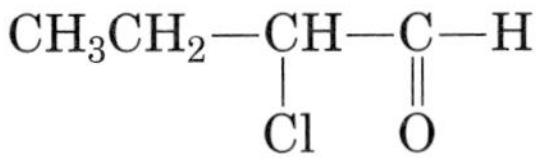

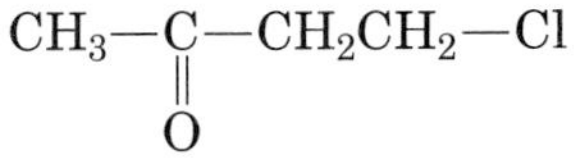

As these examples show, the aldehyde or ketone carbon takes precedence over any previously considered group in numbering the chain. That is, we number the chain so as to give the lowest possible number to the carbon of the aldehyde or ketone group, even if such groups as C=C or —OH must get higher numbers. See Table 13.1 for the names of some other aldehydes and ketones.

This means that aldehyde and ketone groups take precedence over OH, halogen, and alkyl groups, as well as double and triple bonds.

Simple aldehydes often have common names, some of which are shown in Table 13.1. These names, which end in "aldehyde," are derived from the names of the corresponding carboxylic acids, which we will consider in Section 14.2. The compounds HCHO and CH_3CHO are almost always called formaldehyde and acetaldehyde, respectively; the names methanal and ethanal are almost never used, perhaps because of possible confusion with methanol and ethanol.

Ketones also have common names, which simply consist of the word "ketone" preceded by the two group names (Table 13.1), though the simplest ketone, CH_3COCH_3, is almost always called *acetone.* There is a special system for aryl ketones. Thus, the compound that we might call methyl phenyl ketone is more often called acetophenone (Table 13.1). Cyclic ketones are very simply named (see cyclohexanone in Table 13.1).

Table 13.1 Names and Boiling Points of Some Aldehydes and Ketones

Formula	IUPAC Name	Common Name	Boiling Point (°C)
$H-C(=O)-H$	Methanal	Formaldehyde	−21
$CH_3-C(=O)-H$	Ethanal	Acetaldehyde	21
$CH_3CH_2-C(=O)-H$	Propanal	Propionaldehyde	49
$CH_3CH_2CH_2-C(=O)-H$	Butanal	Butyraldehyde	76
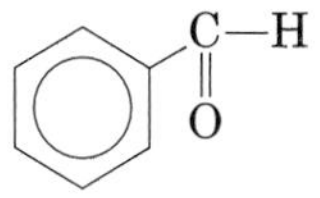 $C_6H_5-C(=O)-H$	Benzaldehyde	Benzaldehyde	178
$CH_3-C(=O)-CH_3$	Propanone	Acetone	56
$CH_3CH_2-C(=O)-CH_3$	Butanone	Ethyl methyl ketone	80
$CH_3CH_2-C(=O)-CH_2CH_3$	3-Pentanone	Diethyl ketone	102
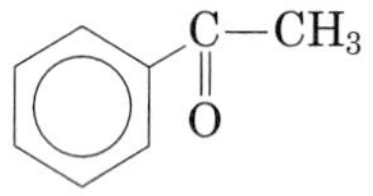 $C_6H_5-C(=O)-CH_3$	1-Phenylethanone	Acetophenone	203
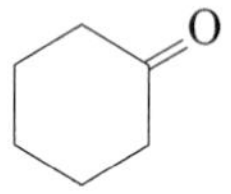 cyclohexane ring with =O	Cyclohexanone	Cyclohexanone	156

13.3 Physical Properties

Remembering what we learned in Section 12.7 about the boiling points and solubilities of alcohols and ethers, what would we expect for aldehydes and ketones? From their structures we can see these things:

1. Two molecules of an aldehyde or ketone cannot form hydrogen bonds with each other because neither molecule contains an O—H, N—H, or F—H bond (Sec. 6.6):

$$R-\underset{\underset{O}{\|}}{C}-H \qquad O=\underset{\underset{H}{|}}{C}-R$$

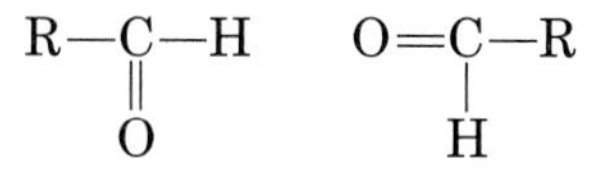

No hydrogen bond possible

We therefore expect their boiling points to be lower than those of alcohols, the molecules of which can form hydrogen bonds with each other. Table 13.2 shows that both propanal and acetone have much lower boiling points than 1-propanol, which has about the same molecular weight.

2. Aldehydes and ketones are polar molecules because the C=O bond is a polar bond:

$$\overset{\delta+}{C}=O_{\delta-}$$

The electrons in the C=O bond are closer to the more electronegative oxygen than to the carbon. This causes aldehyde and ketone molecules to behave as dipoles that attract each other. These attractions are not as strong as hydrogen bonds, but they do cause aldehydes and ketones to boil at higher temperatures than the nonpolar alkanes or the less polar ethers (Table 13.2).

Electronegativity and dipoles are discussed in Section 3.7.

3. Because aldehydes and ketones contain oxygen atoms, they do form hydrogen bonds with water:

$$R{-}C({=}O{\cdots}H{-}O{-}H){-}H \qquad R{-}C({=}O{\cdots}H{-}O{-}H){-}R'$$

Aldehyde Water Ketone Water

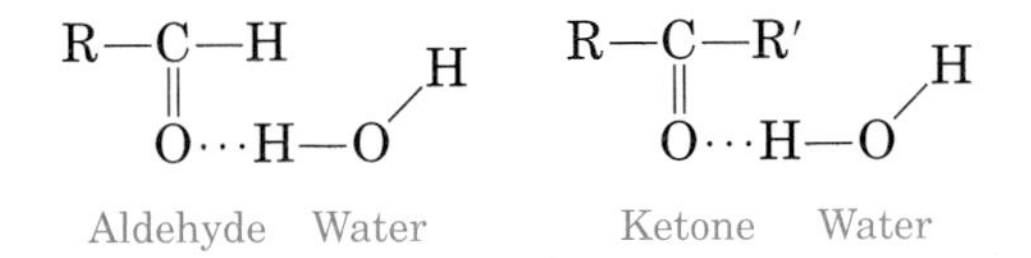

Therefore, their solubility in water is about the same as that of alcohols and ethers: The small molecules are soluble (Table 13.2), but the larger ones are not, for the reasons discussed in Section 12.7 for alcohols and ethers (see Fig. 12.1).

Most aldehydes and ketones have strong odors. The odors of ketones are generally pleasant, and many of them are used in perfumes and as flavoring agents (see Box 13D). The odors of aldehydes vary. You may be familiar with the smell of formaldehyde. If so, you know that it is not pleasant. However, many higher aldehydes are also used in perfumes.

Formaldehyde is a gas at room temperature, but we often encounter it as an aqueous solution.

Table 13.2 Molecular Weights, Boiling Points, and Solubilities of Some Compounds

Name	Formula	MW	BP (°C)	Solubility in Water
Butane	$CH_3{-}CH_2{-}CH_2{-}CH_3$	58	0	Insoluble
Ethyl methyl ether	$CH_3{-}CH_2{-}O{-}CH_3$	60	11	Soluble
Propanal	$CH_3{-}CH_2{-}C({=}O){-}H$	58	49	Soluble
Acetone	$CH_3{-}C({=}O){-}CH_3$	58	56	Soluble
1-Propanol	$CH_3{-}CH_2{-}CH_2{-}OH$	60	97	Soluble

13.4 Preparation of Aldehydes and Ketones by Oxidation

Aldehydes and ketones are most often made by the oxidation of primary and secondary alcohols, respectively:

$$\underset{\text{Primary alcohol}}{RCH_2OH} \xrightarrow{[O]} \underset{\text{Aldehyde}}{R-\overset{\displaystyle O}{\overset{\|}{C}}-H}$$

$$\underset{\text{Secondary alcohol}}{R-\underset{\displaystyle OH}{\underset{|}{C}H}-R'} \xrightarrow{[O]} \underset{\text{Ketone}}{R-\overset{\displaystyle O}{\overset{\|}{C}}-R'}$$

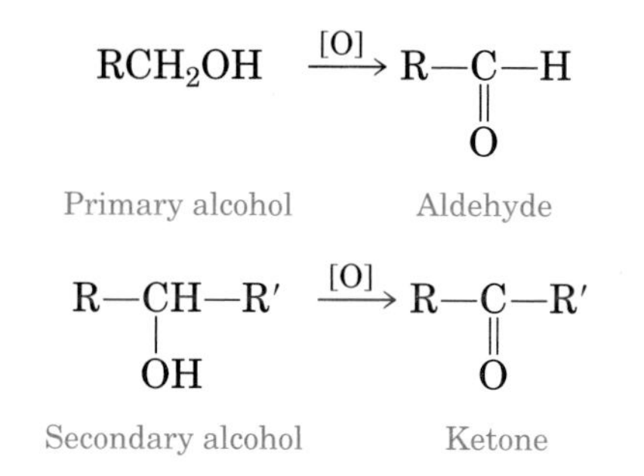

We discussed these reactions in Section 12.3 and pointed out that it is not always easy to stop the oxidation of a primary alcohol at the aldehyde stage because aldehydes are easily oxidizable to carboxylic acids. We will discuss this reaction in the next section.

13.5 Oxidation of Aldehydes

We have already mentioned that aldehydes are easily oxidized to the corresponding carboxylic acids.

$$R-\overset{\displaystyle O}{\overset{\|}{C}}-H \xrightarrow{[O]} R-\overset{\displaystyle O}{\overset{\|}{C}}-OH$$

Many oxidizing agents will do this, even the oxygen in the air. Unless a bottle containing an aldehyde is kept tightly stoppered, the air will oxidize the aldehyde, and the bottle will soon contain the carboxylic acid as well as the aldehyde.

Ketones, on the other hand, resist oxidation:

$$R-\overset{\displaystyle O}{\overset{\|}{C}}-R' \xrightarrow[\text{mild conditions}]{[O]} \text{no reaction}$$

When ketones are oxidized, C—C bonds are broken, leading to complex mixtures.

They can be oxidized by heating with very strong oxidizing agents, but no useful products are obtained.

Problem 13.2

Draw the structural formula for each product (if there is no reaction, say so):

(a) $\mathrm{CH_3{-}\underset{\substack{\| \\ O}}{C}{-}CH_2CH_2CH_3 \xrightarrow[\text{mild conditions}]{[O]}}$

(b) $\mathrm{CH_3{-}\underset{\substack{\| \\ O}}{C}{-}CH_2CH_2CH_3 \xrightarrow[\text{mild conditions}]{[O]}}$

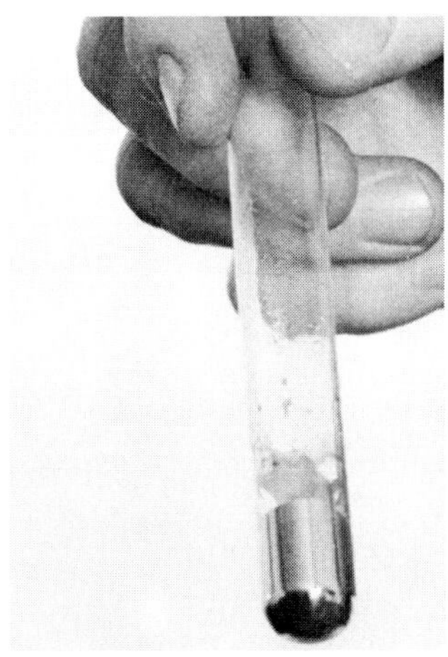

Figure **13.1**
A positive Tollens' test (silver mirror). (Photograph by Beverly March.)

The fact that aldehydes are easy to oxidize and ketones are not allows us to use simple tests to tell these compounds apart in the laboratory. Suppose we have a compound that we know is either an aldehyde or a ketone. All we have to do is treat it with a mild oxidizing agent. If it can be oxidized it is an aldehyde; otherwise it is a ketone. There are two tests commonly used for this: Tollens' and Benedict's tests.

Tollens' Test

Tollens' reagent contains silver nitrate and ammonia in water. When these reagents are mixed, the silver ion combines with NH_3 to form the ion $Ag(NH_3)_2^+$. When this solution is added to an aldehyde, the complexed silver ion is reduced to silver metal. If the test tube is clean, the silver metal deposits on the inside of the test tube, forming a mirror (Fig. 13.1). The equation is

The $Ag(NH_3)_2^+$ ion is a coordination compound (Sec. 3.5), also called a complex ion.

$$\mathrm{\underset{\text{Aldehyde}}{R{-}\underset{\substack{\| \\ O}}{C}{-}H} + 2[Ag(NH_3)_2]^+ + 3OH^- \longrightarrow \underset{\text{Carboxylic acid (as the ion)}}{R{-}\underset{\substack{\| \\ O}}{C}{-}O^-} + \underset{\text{Mirror}}{2Ag(s)} + 4NH_3 + 2H_2O}$$

Ketones are not oxidized under these conditions, so that no silver mirror forms.

Benedict's Test

This is similar to Tollens' test in that a metal ion is reduced, and we know that a positive test has taken place when we see the product come out of solution. But here the oxidizing agent is copper(II) ion, which is reduced to copper(I) oxide by an aldehyde:

$$\mathrm{\underset{\text{Aldehyde}}{R{-}\underset{\substack{\| \\ O}}{C}{-}H} + 2Cu^{2+} + 5OH^- \longrightarrow \underset{\text{Carboxylic acid (as the ion)}}{R{-}\underset{\substack{\| \\ O}}{C}{-}O^-} + \underset{\text{Orange-to-red precipitate}}{Cu_2O} + 3H_2O}$$

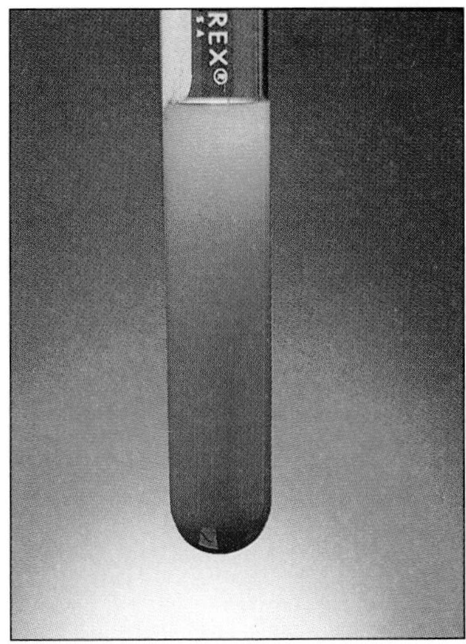

Figure **13.2**
The orange-to-red precipitate is a positive Benedict's test. (Photograph by Beverly March.)

The formation of an orange-to-red precipitate of copper(I) oxide and the disappearance of the blue color of Cu^{2+} are the positive test (Fig. 13.2). Benedict's solution is made by dissolving copper sulfate, sodium citrate, and sodium carbonate in water.

There is one type of ketone that does give positive tests with Tollens' and Benedict's reagents. This type of ketone has a hydroxy group on the carbon next to the C=O group:

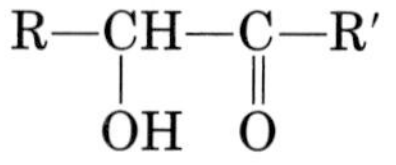

These compounds are called α-hydroxy ketones.

General formula

$$\mathrm{R{-}\underset{\displaystyle OH}{\underset{|}{CH}}{-}\underset{\displaystyle O}{\underset{\|}{C}}{-}R'}$$

Specific example

$$\mathrm{CH_3{-}\underset{\displaystyle OH}{\underset{|}{CH}}{-}\underset{\displaystyle O}{\underset{\|}{C}}{-}C_6H_5}$$

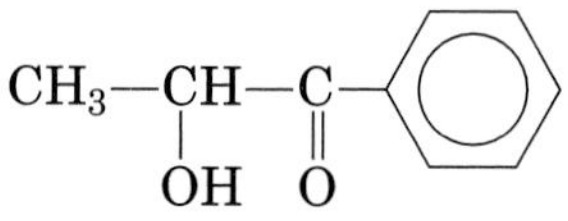

It must therefore be borne in mind that an unknown compound that gives a positive Tollens' or Benedict's test might not be an aldehyde, but an α-hydroxy ketone instead. These tests are especially important in carbohydrate chemistry (Sec. 16.8).

13.6 Other Chemical Properties of Aldehydes and Ketones

In Section 11.6 we saw that compounds containing C=C bonds undergo addition reactions. The same is true for C=O bonds. Unlike the C=C bond, the C=O bond is polar (Sec. 13.3):

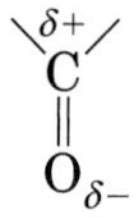

$$\mathrm{\diagdown\overset{\delta+}{C}\diagup \atop \overset{\|}{O_{\delta-}}}$$

so that negatively charged species are attracted to the C and positively charged species to the O. There are many known reactions of this type, but we look at only two (see also Box 13A).

Addition of H_2. Reduction to Alcohols

General reactions:

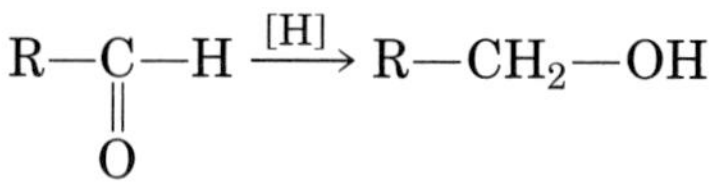

$$\underset{\text{Aldehyde}}{\mathrm{R{-}\underset{\displaystyle O}{\underset{\|}{C}}{-}H}} \xrightarrow{[H]} \underset{\text{Primary alcohol}}{\mathrm{R{-}CH_2{-}OH}}$$

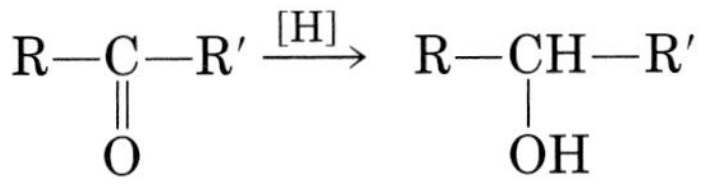

$$R-\underset{\underset{O}{\|}}{C}-R' \xrightarrow{[H]} R-\underset{\underset{OH}{|}}{CH}-R'$$

Ketone — Secondary alcohol

Specific examples:

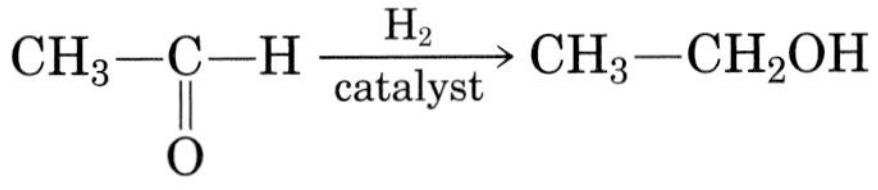

$$CH_3-\underset{\underset{O}{\|}}{C}-H \xrightarrow[\text{catalyst}]{H_2} CH_3-CH_2OH$$

Acetaldehyde — Ethanol

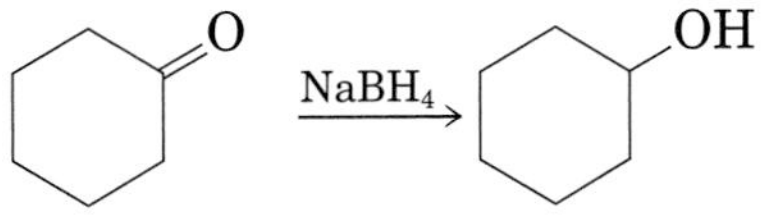

Cyclohexanone $\xrightarrow{NaBH_4}$ Cyclohexanol

In Section 11.6 we saw that C=C can be reduced to CH—CH. The C=O bonds of aldehydes and ketones undergo the same reaction, though the reducing agents are not always the same. Hydrogen gas and a catalyst reduce $>C=O$ to $>\underset{\underset{H}{|}}{C}-OH$ and are frequently used for this purpose, but the reaction is slower than the one with C=C bonds. Thus, if there is a C=C bond in the same molecule, it is reduced first. More often the reducing agent used for this purpose is sodium borohydride, $NaBH_4$, or lithium aluminum hydride, $LiAlH_4$. Neither of these reagents usually affects C=C double bonds, so a C=O double bond can be reduced while leaving a C=C double bond intact.

No matter what the reducing agent, *aldehydes are always reduced to primary alcohols and ketones to secondary alcohols.* (Can you see why?) These reactions are the opposite of the oxidation reactions discussed in Section 13.4.

Problem **13.3**

Draw the structural formula for each product:

(a) $C_6H_5-\underset{\underset{O}{\|}}{C}-H \xrightarrow{[H]}$

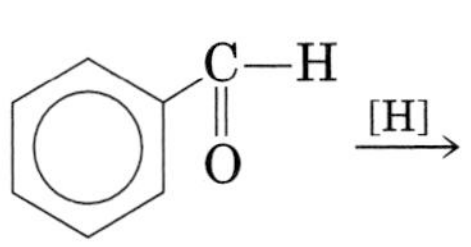

(b) $CH_3CH_2-\underset{\underset{O}{\|}}{C}-CH_2CH_3 \xrightarrow{[H]}$

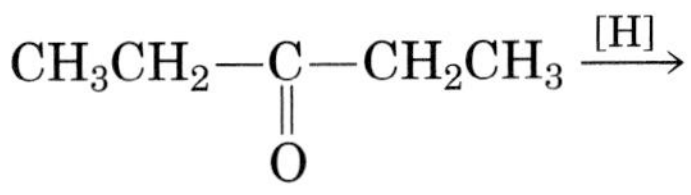

Reduction of aldehydes and ketones also takes place in biological organisms, in which case organic reducing agents are used. One of the most important is the reduced form of **nicotinamide adenine dinucleotide, NADH** (Sec. 20.3).

The body doesn't have H_2, $NaBH_4$, or $LiAlH_4$ available; it uses the reducing agent NADH instead.

Addition of ROH. Formation of Acetals

General reaction:

$$\underset{\text{Aldehyde}}{R{-}\overset{\displaystyle O}{\overset{\|}{C}}{-}H} + \underset{\text{Alcohol}}{R'OH} \rightleftharpoons \underset{\substack{\text{Hemiacetal}\\\text{(not isolated)}}}{R{-}\underset{OH}{\overset{OR'}{C}}{-}H} \xrightarrow[R'OH]{H^+} \underset{\text{Acetal}}{R{-}\underset{OR'}{\overset{OR'}{C}}{-}H} + H_2O$$

Specific example:

$$\underset{\text{Acetaldehyde}}{CH_3{-}\overset{\displaystyle O}{\overset{\|}{C}}{-}H} + \underset{\text{Ethanol}}{CH_3CH_2OH} \rightleftharpoons \underset{\text{(not isolated)}}{CH_3{-}\underset{OH}{\overset{OCH_2CH_3}{C}}{-}H} \xrightarrow[CH_3CH_2OH]{H^+} CH_3{-}\underset{OCH_2CH_3}{\overset{OCH_2CH_3}{C}}{-}H + H_2O$$

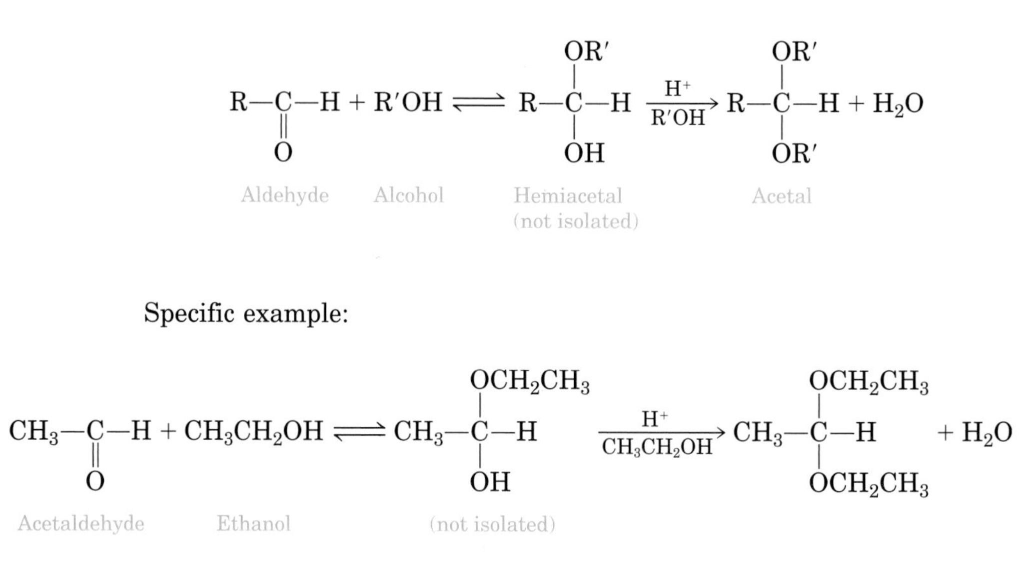

When an alcohol (R′OH) is added to an aldehyde, the positive part of the alcohol (H^+) adds to the O of the carbonyl group, and the negative part of the alcohol ($R'O^-$) adds to the C:

$$R{-}\overset{\delta+}{C}(=\overset{\delta-}{O}){-}H \quad \underset{\delta-}{R'O}{-}\underset{\delta+}{H} \rightleftharpoons \underset{\substack{\text{Hemiacetal}\\\text{(not stable)}}}{R{-}\underset{OH}{\overset{OR'}{C}}{-}H} \qquad \text{An addition reaction}$$

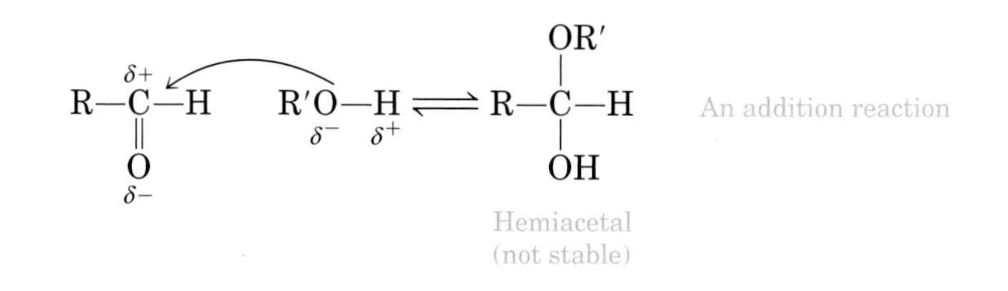

The compound produced in this way is both an alcohol and an ether. **A compound with an —OH and an —OR on the same carbon** is called a **hemiacetal.** Simple hemiacetals are not stable and cannot be isolated. However, if a small amount of a strong acid (such as HCl) is present in the mixture, the hemiacetal reacts with a second molecule of alcohol (the —OR′ replaces the —OH) to produce a compound that has two ether groups on the same carbon:

$$R'{-}OH + \underset{\text{Hemiacetal}}{R{-}\underset{OH}{\overset{OR'}{C}}{-}H} \overset{H^+}{\rightleftharpoons} \underset{\text{Acetal}}{R{-}\underset{OR'}{\overset{OR'}{C}}{-}H} + H_2O \qquad \text{A substitution reaction}$$

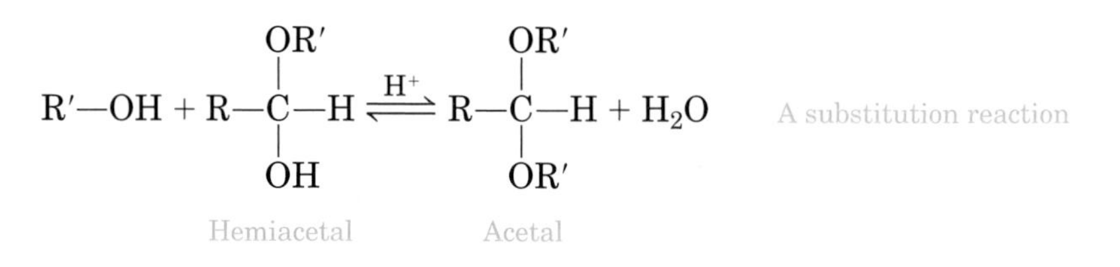

A compound containing two —OR groups on the same carbon is called an **acetal.**

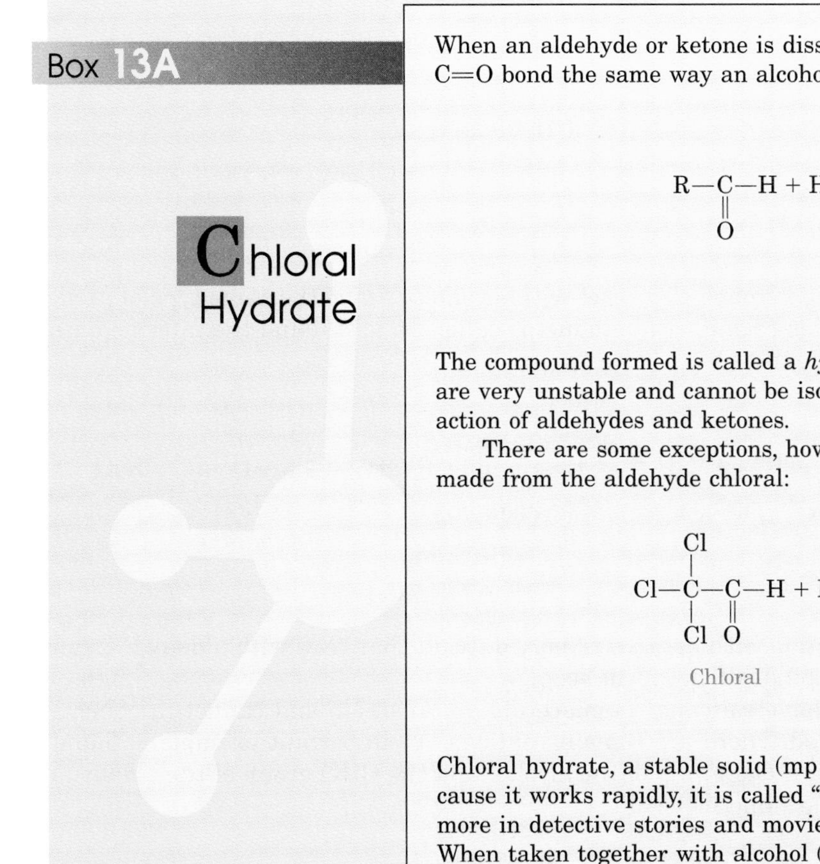

Although in simple cases we cannot isolate hemiacetals, simple acetals are quite stable and are easily prepared by treating an aldehyde

Box 13A

Chloral Hydrate

When an aldehyde or ketone is dissolved in water, the water can add to the C=O bond the same way an alcohol would:

$$R-\underset{\underset{O}{\parallel}}{C}-H + HOH \rightleftharpoons R-\underset{\underset{OH}{|}}{\overset{\overset{OH}{|}}{C}}-H$$

Hydrate (unstable)

The compound formed is called a *hydrate.* Like simple hemiacetals, hydrates are very unstable and cannot be isolated. Therefore, this is not a general reaction of aldehydes and ketones.

There are some exceptions, however. One of these is chloral hydrate, made from the aldehyde chloral:

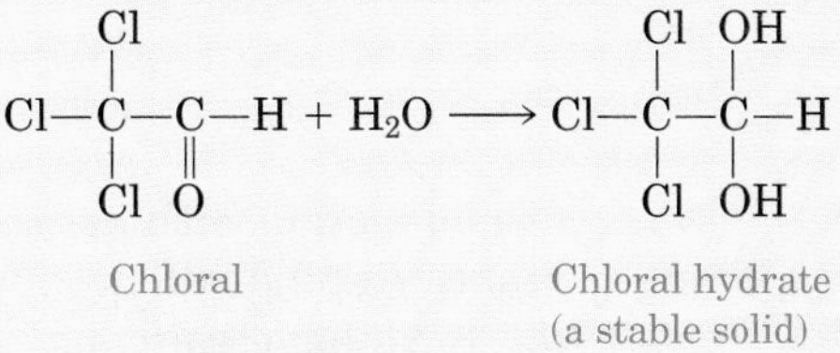

$$Cl-\underset{\underset{Cl}{|}}{\overset{\overset{Cl}{|}}{C}}-\underset{\underset{O}{\parallel}}{C}-H + H_2O \longrightarrow Cl-\underset{\underset{Cl}{|}}{\overset{\overset{Cl}{|}}{C}}-\underset{\underset{OH}{|}}{\overset{\overset{OH}{|}}{C}}-H$$

Chloral — Chloral hydrate (a stable solid)

Chloral hydrate, a stable solid (mp 57°C), is used as a sleeping tablet. Because it works rapidly, it is called "knockout drops." Another name, used more in detective stories and movies than in real life, is "Mickey Finn." When taken together with alcohol (for example, if added to a drink), chloral hydrate is dangerous and can easily be fatal.

with an alcohol in the presence of HCl. This reaction is an equilibrium, and acetals can be cleaved to produce the alcohol and aldehyde by treating with aqueous HCl:

$$R-\underset{\underset{OR'}{|}}{\overset{\overset{OR'}{|}}{C}}-H + H_2O \overset{H^+}{\rightleftharpoons} R-\underset{\underset{O}{\parallel}}{C}-H + 2R'-OH$$

A reaction with water is called **hydrolysis.**

As with any other equilibrium, we can make this one go in the direction we want by using Le Chatelier's principle. If we want to hydrolyze an acetal, there should be a large amount of water present; this drives the equilibrium to the side of the aldehyde and alcohol. If we want to form an acetal, there should be very little water present (the best would be none at all). Acetals are generally made by adding the pure alcohol to the pure aldehyde and passing dry HCl gas through the mixture. The HCl is a catalyst for both the forward and the reverse reaction.

Bases do not catalyze either reaction. Acetals are quite stable in the presence of a base.

Problem 13.4

Draw the structural formula for the hemiacetal intermediate and for the acetal product:

$$C_6H_5\text{—}CH_2CH_2\text{—}\underset{\underset{O}{\|}}{C}\text{—}H + CH_3OH \xrightarrow{H^+}$$

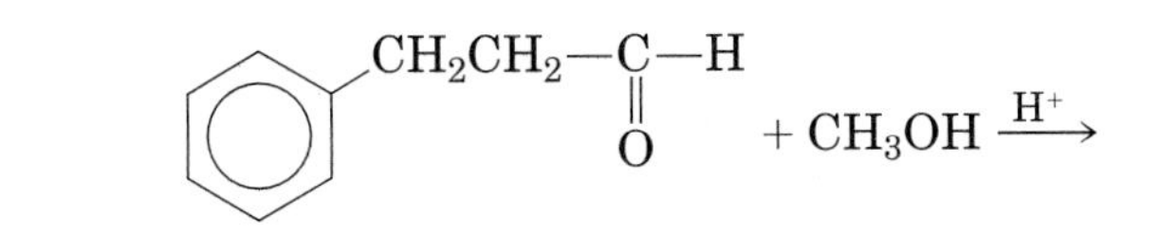

When the reaction takes place with ketones, the acetal is sometimes called a **ketal,** and the intermediate a **hemiketal,** though these names are not sanctioned by IUPAC.

Some ketones also react with alcohols to give the same type of product. In this case too the product is called an acetal and the intermediate a hemiacetal:

$$\underset{\text{Ketone}}{R\text{—}\underset{\underset{O}{\|}}{C}\text{—}R''} + R'OH \rightleftharpoons \underset{\substack{\text{Hemiacetal}\\\text{(unstable)}}}{R\text{—}\overset{\overset{OR'}{|}}{\underset{\underset{OH}{|}}{C}}\text{—}R''} + R'OH \rightleftharpoons \underset{\text{Acetal}}{R\text{—}\overset{\overset{OR'}{|}}{\underset{\underset{OR'}{|}}{C}}\text{—}R''} + H_2O$$

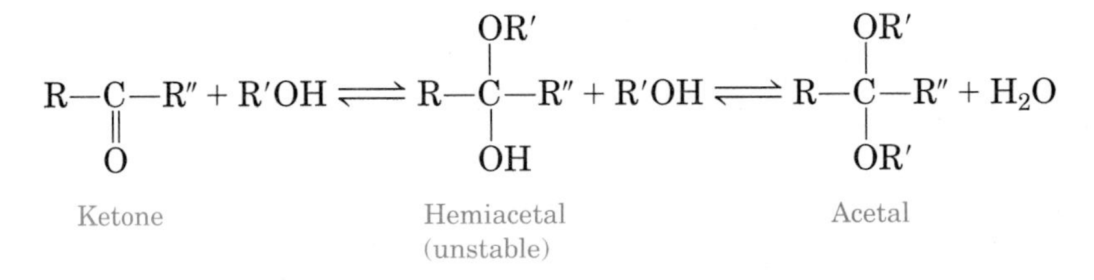

The reaction with ketones is more difficult than that with aldehydes, and many ketones do not form acetals.

We have said that hemiacetals are unstable and cannot in general be isolated. There are some exceptions. An important example is found when the —OH group of the alcohol and the CHO group of the aldehyde are in the same molecule:

This is an addition reaction.

$$\underset{\underset{OH}{|}}{\overset{5}{C}H_2}\text{—}\overset{4}{C}H_2\text{—}\overset{3}{C}H_2\text{—}\overset{2}{C}H_2\text{—}\overset{1}{\underset{\underset{O}{\|}}{C}}\text{—}H \text{ or } \begin{array}{c} \text{(open-chain form drawn as ring: } {}^5H_2C\text{—}O\text{—}H \cdots O{=}{}^1C\text{—}H,\ {}^4H_2C,\ {}^3C H_2,\ {}^2CH_2) \end{array} \longrightarrow \underset{\text{Cyclic hemiacetal}}{\begin{array}{c} H_2C\text{—}O\text{—}C(OH)H \\ | \qquad\qquad | \\ H_2C\text{—}CH_2\text{—}CH_2 \end{array}}$$

Cyclic hemiacetal

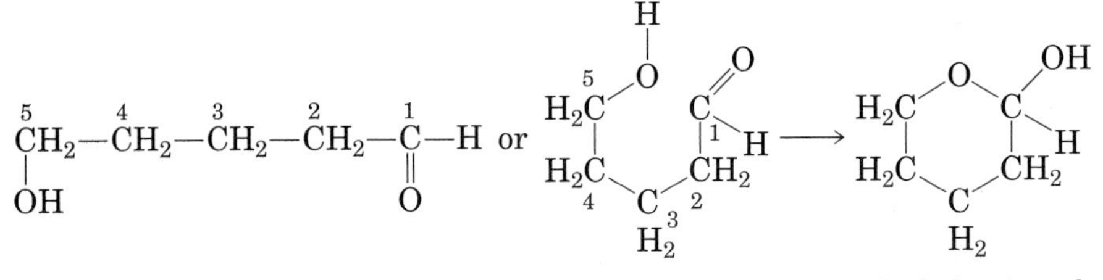

In this case, the H of the —OH group adds to the C=O *in the same molecule* (you can see that it is not very far away), and a six-membered ring is produced. The most stable rings contain five or six atoms, and we are not surprised to learn that aldehydes that contain an —OH group on the fourth or fifth carbon of the chain spontaneously produce cyclic hemiacetals. These cyclic hemiacetals react with alcohols in the presence of H^+ to produce cyclic acetals:

This is a substitution reaction.

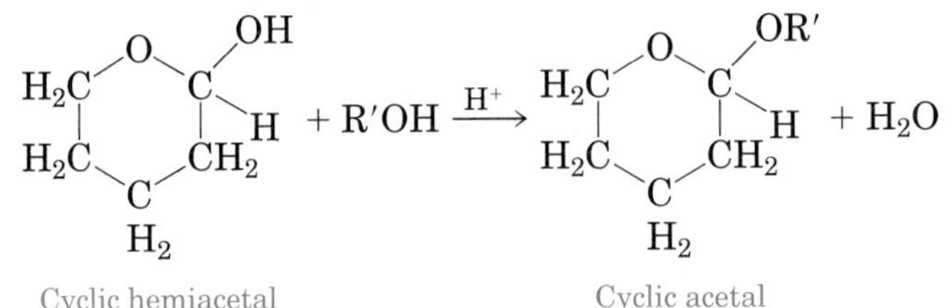

$$\underset{\text{Cyclic hemiacetal}}{\begin{array}{c} H_2C\text{—}O\text{—}C(OH)H \\ | \qquad\qquad | \\ H_2C\text{—}\underset{H_2}{C}\text{—}CH_2 \end{array}} + R'OH \xrightarrow{H^+} \underset{\text{Cyclic acetal}}{\begin{array}{c} H_2C\text{—}O\text{—}C(OR')H \\ | \qquad\qquad | \\ H_2C\text{—}\underset{H_2}{C}\text{—}CH_2 \end{array}} + H_2O$$

Cyclic hemiacetal　　Cyclic acetal

We shall see examples of cyclic hemiacetals and acetals when we study carbohydrates in Chapter 16. Because these groups are so important in carbohydrate chemistry, it is important that you be able to recognize them.

EXAMPLE

In the following structures, identify the acetals and hemiacetals and tell whether they are formed from aldehydes or ketones:

(a) $CH_3CH_2—O—CH_2CH_2—OH$

(b) $CH_3—CH(OCH_2CH_3)—OH$

(c) $CH_3CH_2—O—CH_2CH_2—O—CH_3$

(d) $CH_3CH_2—C(OCH_3)(OCH_2CH_3)—CH_3$

(e) five-membered ring: H_2C—C(H_2)—O—CH(—OH)—H_2C—H_2C

(f) five-membered ring: H_2C—C(H_2)—O—CH_2—CH(—O—CH_3)—H_2C; written $CH_3—O—CH—CH_2$

(g) six-membered ring: H_2C—C(H_2)—O—CH(—OCH_3)—C(H_2)—H_2C—H_2C

(h) six-membered ring: H_2C—C(H_2)—O—C(OH)(CH_3)—C(H_2)—H_2C—H_2C

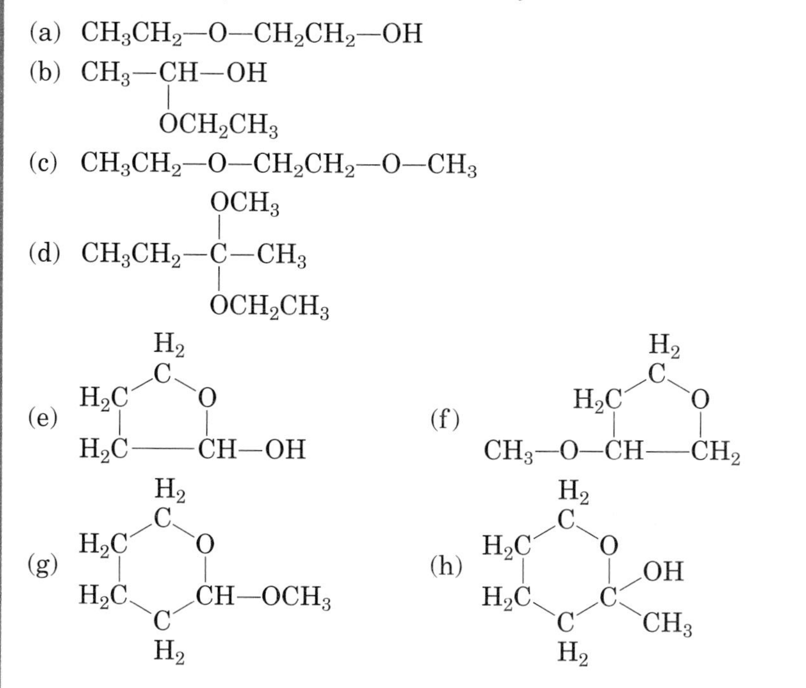

Answer • Both acetals and hemiacetals have two oxygen atoms connected to the same carbon. In compounds (a), (c), and (f), there is no carbon connected to two oxygen atoms. Compounds (c) and (f) are ethers; compound (a) is an ether and an alcohol. None of them is an acetal or hemiacetal. Compounds (b), (e), and (h) are hemiacetals. Each of them has one carbon atom connected to an —OH group and an —OR group (shown below in color). Compounds (b) and (e) are derived from aldehydes (there is an H on the colored carbon); compound (h) from a ketone (there is no H on the colored carbon).

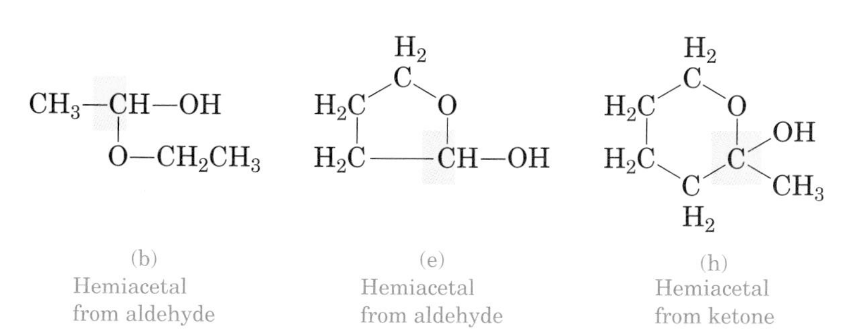

$CH_3—CH(O—CH_2CH_3)—OH$

(b) Hemiacetal from aldehyde

Five-membered ring: H_2C—C(H_2)—O—CH(—OH)—H_2C—H_2C

(e) Hemiacetal from aldehyde

Six-membered ring: H_2C—C(H_2)—O—C(OH)(CH_3)—C(H_2)—H_2C—H_2C

(h) Hemiacetal from ketone

Compounds (d) and (g) have two —OR groups on the same carbon, so they are acetals. Because compound (g) also has an H on this carbon, it is derived from an aldehyde. Compound (d), which has no H on this carbon, is derived from a ketone.

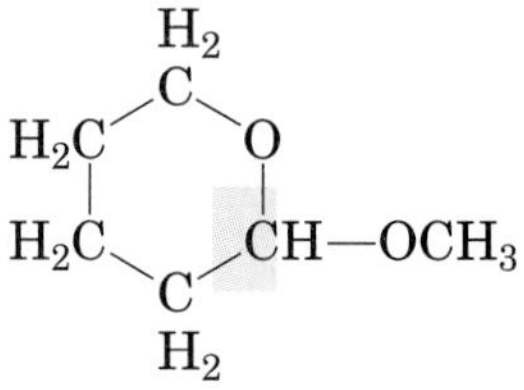

(g) Acetal from aldehyde: six-membered ring H_2C–CH_2–CH_2–CH_2–O–CH(OCH_3)

$$CH_3CH_2-\overset{OCH_3}{\underset{OCH_2CH_3}{C}}-CH_3$$

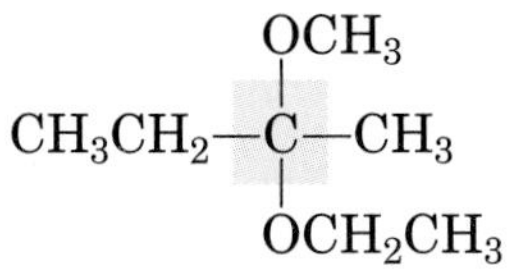

(d) Acetal from ketone

Problem **13.5**

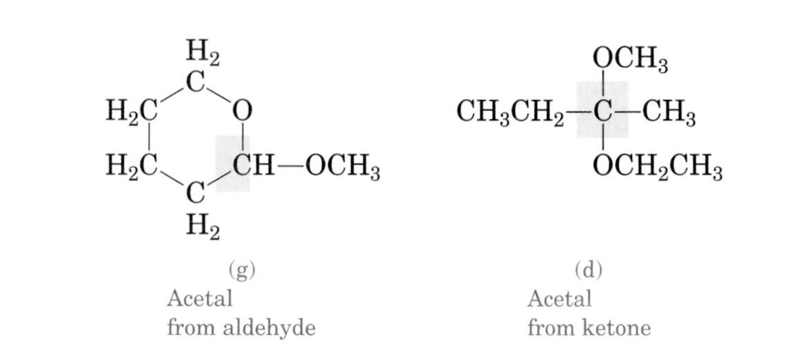

In the following structures, identify the acetals and hemiacetals and tell whether they are formed from aldehydes or ketones:

(a) 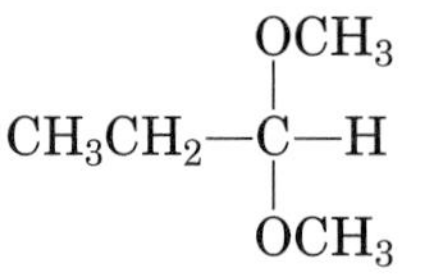 $CH_3CH_2-\overset{OCH_3}{\underset{OCH_3}{C}}-H$

(b) $CH_3O-\overset{OH}{CH}-CH_2CH_3$

(c) cyclobutane ring (CH_2, CH_2, CH_2, C) with C bearing OH and OCH_2CH_3

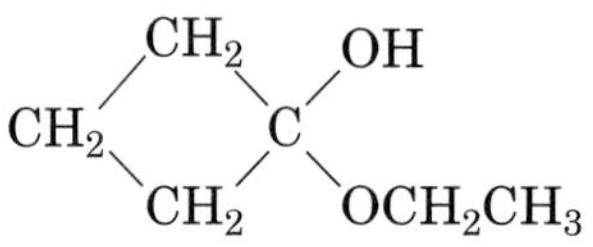

(d) 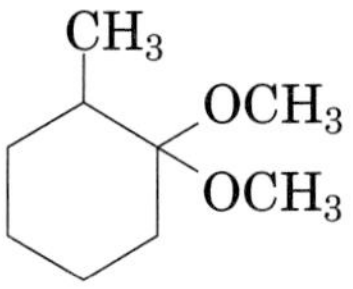 cyclohexane ring with CH_3 on one carbon and OCH_3, OCH_3 on the adjacent carbon

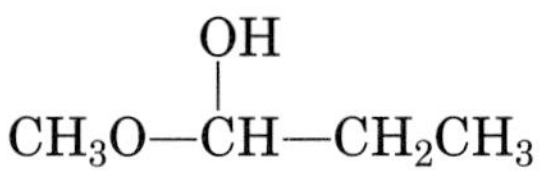

EXAMPLE

Draw the structures of the aldehyde (or ketone) and alcohol(s) produced when compounds b, e, d, g, and h in the previous example are hydrolyzed.

Answer • Pick out the carbon connected to the two oxygen atoms (the one shown in color). This becomes the aldehyde or ketone carbon, the one connected to the =O:

(b) 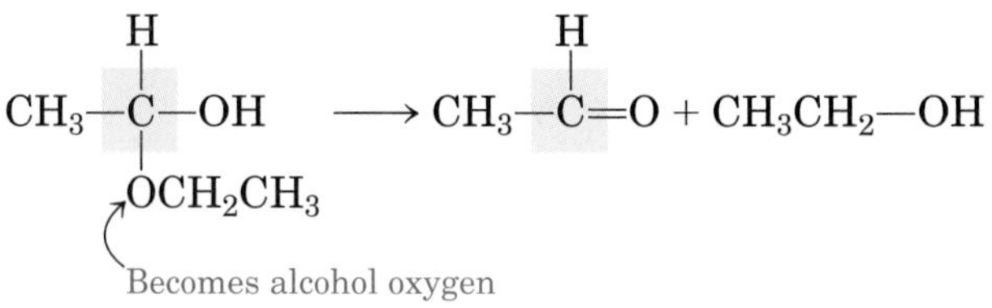 $CH_3-\overset{H}{\underset{OCH_2CH_3}{C}}-OH \longrightarrow CH_3-\overset{H}{C}{=}O + CH_3CH_2-OH$

Becomes alcohol oxygen

Becomes alcohol oxygen

(e) 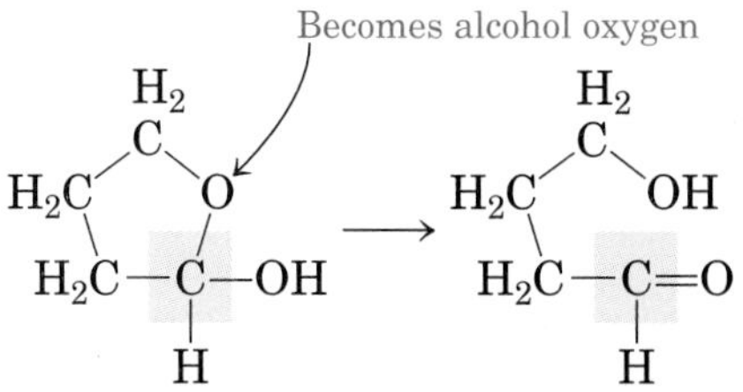 five-membered ring H_2C–CH_2–O–C(H)(OH)–CH_2 $\longrightarrow$ H_2C–CH_2–OH ... $H_2C-\underset{H}{C}{=}O$ (open chain: $HO-CH_2-CH_2-CH_2-CHO$)

(d) $CH_3CH_2-\underset{\substack{|\\ OCH_2CH_3}}{\overset{\substack{OCH_3\\ |}}{C}}-CH_3 \longrightarrow$ (Becomes alcohol oxygen; Becomes alcohol oxygen)

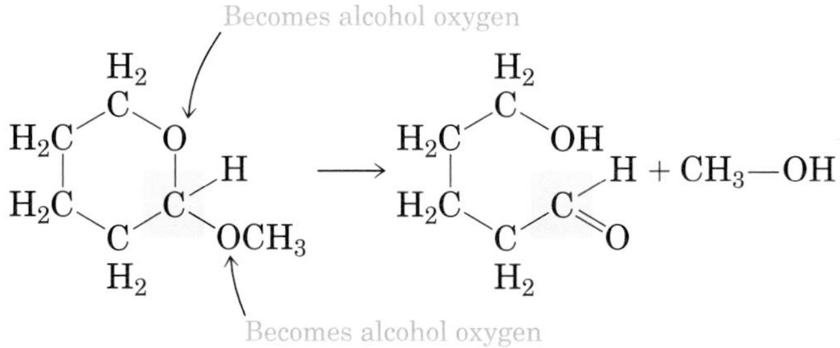

$CH_3CH_2-\underset{\substack{\|\\ O}}{C}-CH_3 + CH_3-OH + CH_3CH_2-OH$

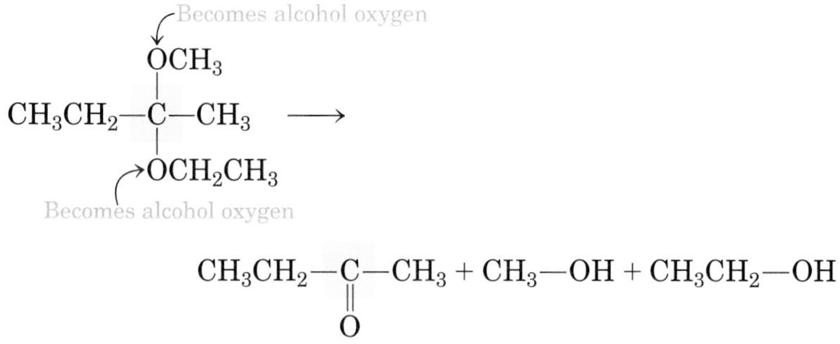

(g) Cyclic acetal: ring H_2C–$C(H_2)$–O–C(H)(OCH₃)–$C(H_2)$–H_2C $\longrightarrow$ $H_2C(CH_2OH)$…H_2C–$C(H_2)$–$C(=O)H$ + CH_3-OH (ring O: Becomes alcohol oxygen; OCH_3 oxygen: Becomes alcohol oxygen)

(h) Cyclic hemiketal: ring H_2C–$C(H_2)$–O–C(CH₃)(OH)–$C(H_2)$–H_2C $\longrightarrow$ $HO-CH_2-CH_2-CH_2-CH_2-C(=O)-CH_3$ (ring O: Becomes alcohol oxygen)

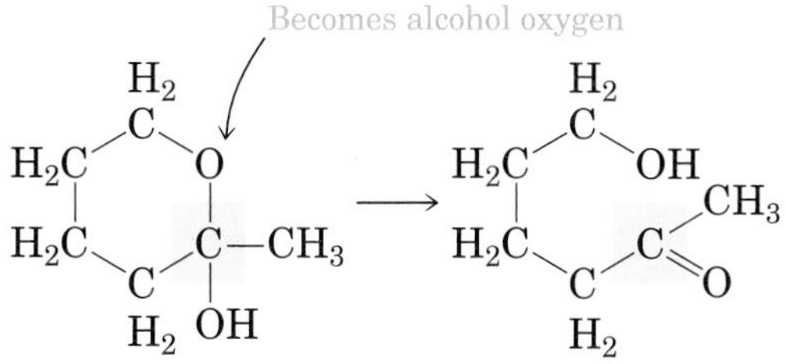

Problem 13.6

Draw the structure of the aldehyde or ketone and alcohol produced upon breakdown of the four compounds in Problem 13.5.

Keto–Enol Tautomerism

One other property of many aldehydes and ketones is worth noting. If the aldehyde or ketone possesses at least one hydrogen on the atom adjacent to the C=O group (called an α-hydrogen), this hydrogen can migrate to the oxygen. An example is

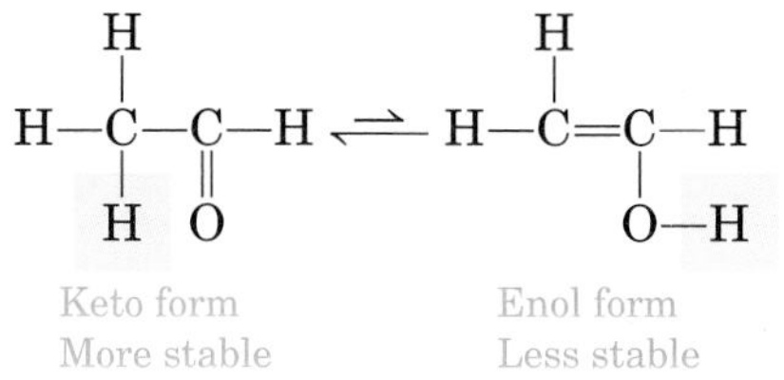

$$H-\underset{\substack{|\\ H}}{\overset{\substack{H\\ |}}{C}}-\underset{\substack{\|\\ O}}{C}-H \rightleftharpoons H-\overset{\substack{H\\ |}}{C}=\underset{\substack{|\\ O-H}}{C}-H$$

Keto form More stable — Enol form Less stable

This is an equilibrium, and in any such case both forms are present. The phenomenon is called **tautomerism.** For most simple aldehydes

Simple enols (compounds with a double bond and an OH group on the same carbon) are usually not stable enough to be isolated, because they tautomerize to the keto form.

and ketones the **keto form** is much more stable than the **enol form,** and much less than 1 percent of the molecules are in the enol form at any time. But there are some cases in which the enol is more stable than the keto form, one such being the diketone 2,4-pentandione:

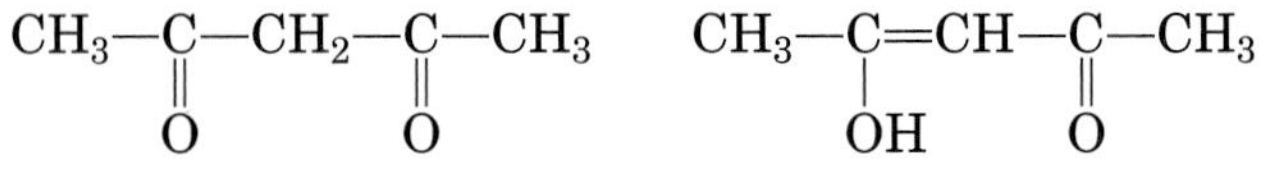

$$\underset{\text{Keto form, Less stable}}{CH_3-\underset{\underset{O}{\|}}{C}-CH_2-\underset{\underset{O}{\|}}{C}-CH_3} \qquad \underset{\text{Enol form, More stable}}{CH_3-\underset{\underset{OH}{|}}{C}=CH-\underset{\underset{O}{\|}}{C}-CH_3}$$

Keto form
Less stable

Enol form
More stable

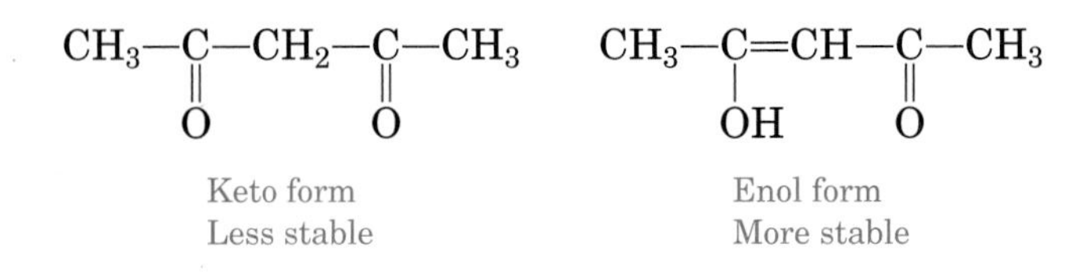

Another compound in which the enol form is more stable is vitamin C (Sec. 26.5).

Before leaving this section, we should mention that **aldehydes and ketones are neither basic nor acidic;** they are neutral compounds like alcohols and ethers.

Box 13B

Formaldehyde in Plastics and Plywood

In 1909 Leo Baekeland (1863–1944) announced the invention of the first completely synthetic plastic, made by combining formaldehyde with phenol under heat and pressure. These two compounds react to form a giant molecule (polymer) with a three-dimensional structure.

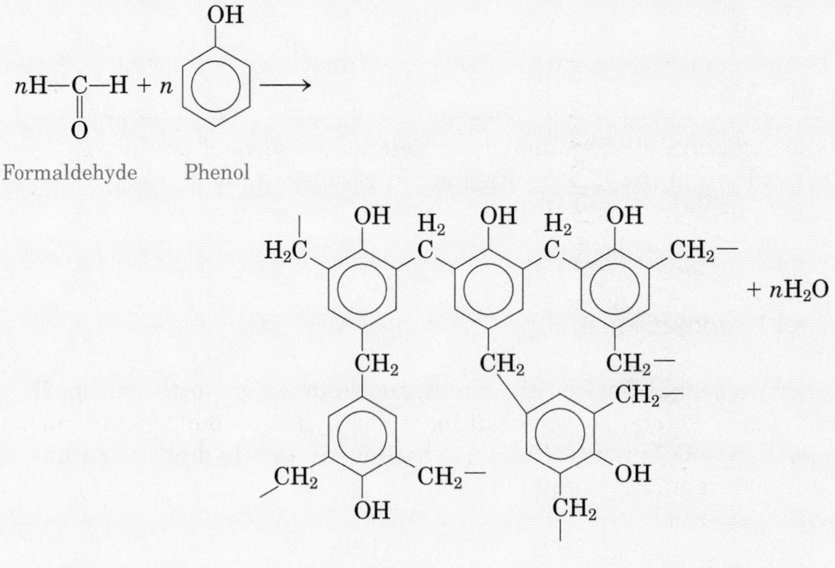

$$n\,H-\underset{\underset{O}{\|}}{C}-H + n\,C_6H_5OH \longrightarrow$$

Formaldehyde Phenol

[Phenol-formaldehyde polymer structure: phenol rings (OH groups) linked by CH_2 bridges] $+ \; nH_2O$

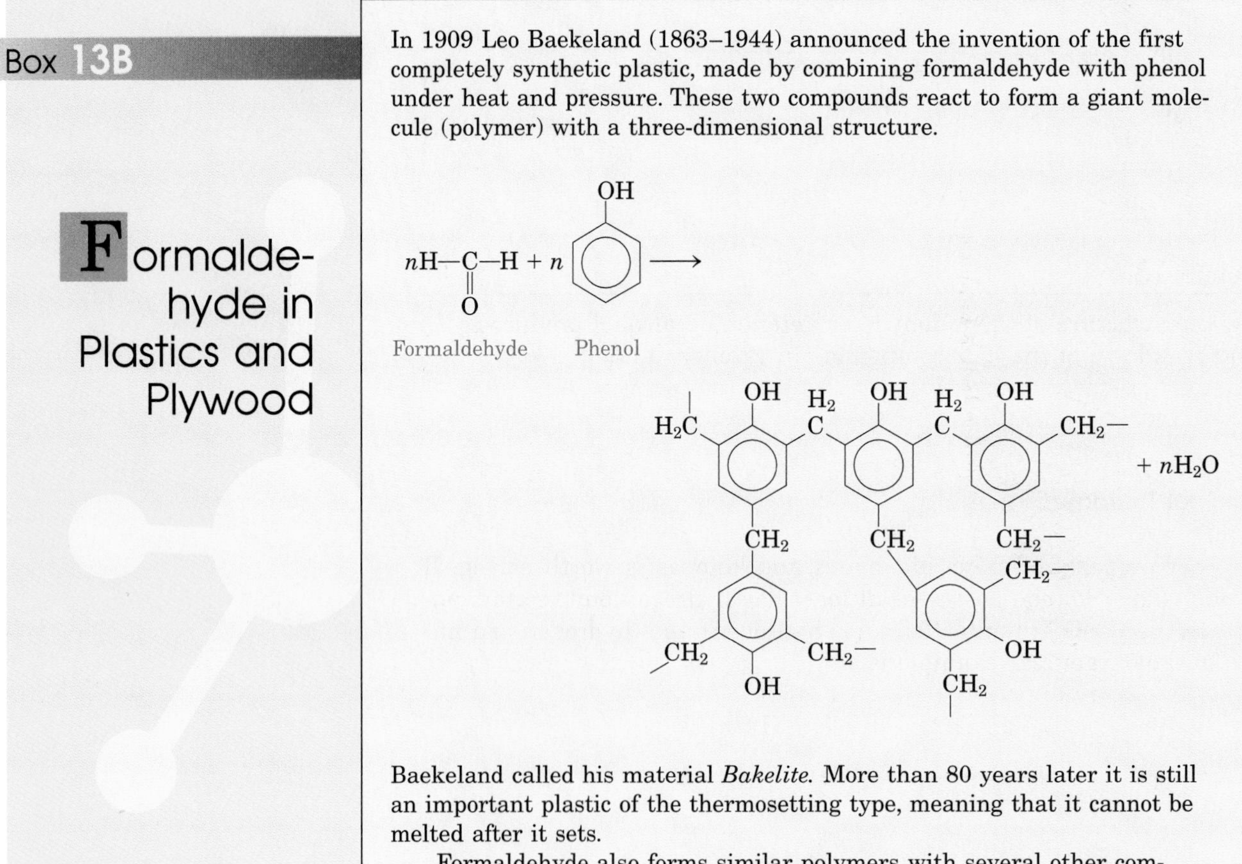

Baekeland called his material *Bakelite.* More than 80 years later it is still an important plastic of the thermosetting type, meaning that it cannot be melted after it sets.

Formaldehyde also forms similar polymers with several other compounds, the most important being urea and melamine. Phenol-formalde-

13.7 Some Important Aldehydes and Ketones

Formaldehyde, H—CHO

Although formaldehyde is a gas at room temperature (bp −21°C), it is soluble in water and so is normally found in laboratories as a 37 percent aqueous solution called *formalin*. This solution kills bacteria and is used to sterilize surgical instruments.

You may be familiar with formalin as a preservative for biological specimens. It does this by denaturing proteins (Sec. 18.10), hardening them. It is also used to embalm cadavers. Its use as a preservative and as an embalming liquid has declined, however, because it has been determined that formaldehyde is carcinogenic.

Industrially, formaldehyde is the most important aldehyde. Its principal use is in the production of polymers (Box 13B).

hyde, urea-formaldehyde, and melamine-formaldehyde polymers are used not only as plastics but even more importantly as adhesives and coatings. Plywood consists of thin sheets of wood glued together by one of these polymers. They are also combined with wood chips or sawdust to make fiberboard. Besides Bakelite, the trade names Formica and Melmac are used for some of these materials. Some of these polymers are produced as porous foams and are used for building insulation. In recent years many of these products have been removed from the market because they contain tiny amounts of free formaldehyde, which can get into the air and possibly cause cancer.

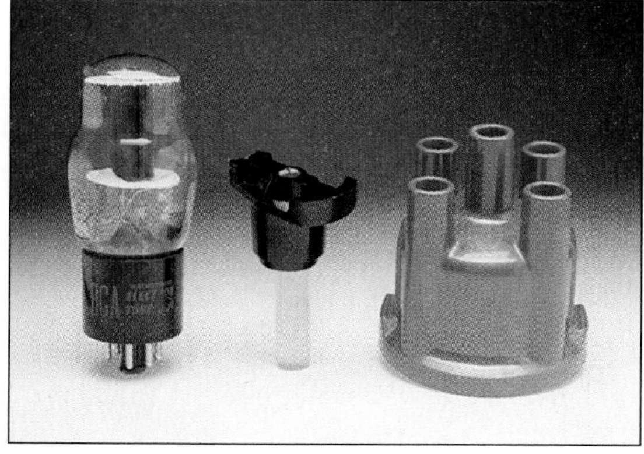

Some products that contain bakelite. (Photograph by Charles D. Winters.)

Acetaldehyde, CH_3—CHO

Because its boiling point, 21°C, is so close to room temperature, acetaldehyde is not easy to handle, either as a liquid or as a gas. For this reason, it is often converted to a cyclic trimer (a molecule made of three units) called *paraldehyde:*

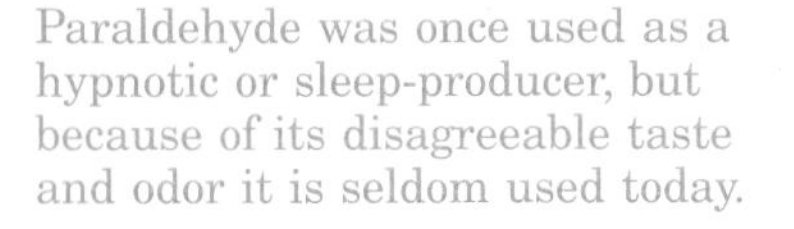

Paraldehyde was once used as a hypnotic or sleep-producer, but because of its disagreeable taste and odor it is seldom used today.

$$3CH_3\text{—}CHO \longrightarrow$$

Paraldehyde

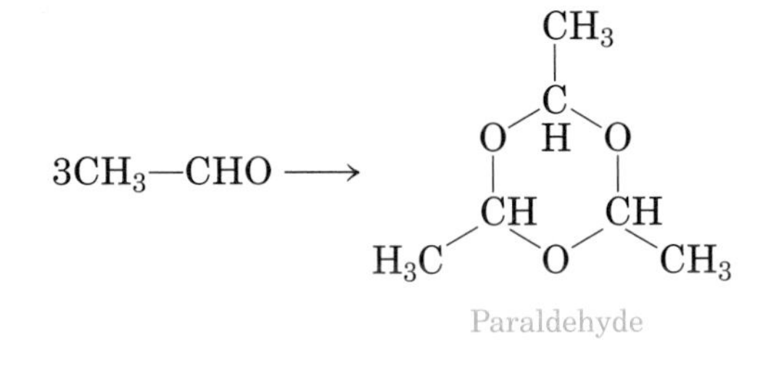

Acetone, CH_3—CO—CH_3

A liquid with a moderately low boiling point (56°C), acetone is one of the few organic compounds that is infinitely soluble in water and also dissolves a great many organic compounds. For this reason it is one of the most important industrial solvents. It is used in paints, varnishes, resins, coatings, and nail polish, among other things.

Acetone is produced in the body as a product of one pathway of lipid metabolism (Sec. 21.7). Normally it does not accumulate because the small amount produced is oxidized to CO_2 and H_2O. However, in diabetic patients a larger quantity of acetone is produced than the body can oxidize completely, so much of it is excreted in the urine. A positive test for acetone in the urine is one way in which diabetes is diagnosed. In severe cases of diabetes acetone is exhaled, and its sweet odor can be detected in the patient's breath.

α-Chloroacetophenone, C_6H_5—CO—CH_2—Cl

A number of chlorine- and bromine-containing organic compounds are **lachrymators,** compounds that cause the eyes to tear. α-Chloroacetophenone has long been used as a tear gas, by police and other government officials to control unruly mobs. More recently, this compound has been used as the active ingredient in Mace, a handheld aerosol tear gas projector, used by individuals to protect themselves from an attacker. Mace has a range of about 15 feet.

Many aldehydes and ketones are found in nature. Some of these are shown in Box 13D.

Box 13C

Toxicity and LD_{50}

Different compounds have different effects on the human body. Whether taken as food or medication or taken in as environmental pollutants, some compounds can exert a toxic effect. But it is the dosage that counts. For example, table salt, NaCl, is a common (and essential) ingredient of our diets. Taken in large enough quantities, however, it can cause death.

A crude measure of the relative toxicity of a compound is its LD_{50} value (LD stands for lethal dose). This is the single dosage, which, when administered to a group of experimental rats, kills 50% of them within 14 days. LD_{50} values are normally given as mg of substance per kg body weight. For sodium chloride this number is 3000. (This means, for example, that if a group of rats whose average weight is 0.2 kg is fed a dose of 600 mg of NaCl, half of them would die within 14 days.) For arsenic, a favorite poison in many novels, and sometimes in real life, the LD_{50} is 48 mg/kg.

Acetaldehyde, which is a common intermediate in alcohol production by fermentation (Path ⑩ in Fig. 21.4) and a key component in alcoholism (Box 12B), is somewhere between NaCl and arsenic in lethality. Its LD_{50} value is 1900 mg/kg. Nicotine (Sec. 15.9) is 50 times more toxic than arsenic, with an LD_{50} of 1 mg/kg, and the botulinum toxin of many food poisonings (Box 24C) is 10,000 times more toxic than nicotine, having an LD_{50} of 1×10^{-5} mg/kg.

As we said previously, the LD_{50} is only a crude indicator of toxicity. For one thing, the lethal dose determined in rats may be completely different in other species, including humans (and for obvious reasons, the LD_{50} test cannot be carried out on humans). Varying modes of administration, by oral intake, inhalation, or topical application to the skin, yield different numbers. Most importantly, however, *death is not the only type of toxicity that must be avoided.* U.S. government regulations set the acetaldehyde exposure limit by inhalation to 200 ppm over an 8-hr exposure period. The regulatory limit for formaldehyde exposure is even lower: 1.5 ppm. Setting such limits is often controversial. The case of formaldehyde is instructive. In experimental rats it was found that 14.3 ppm of inhaled formaldehyde caused nasal cancers. There is no evidence that formaldehyde has caused cancer in humans. Nevertheless, in all industrial countries there are limits set to formaldehyde exposure, though these limits vary from country to country.

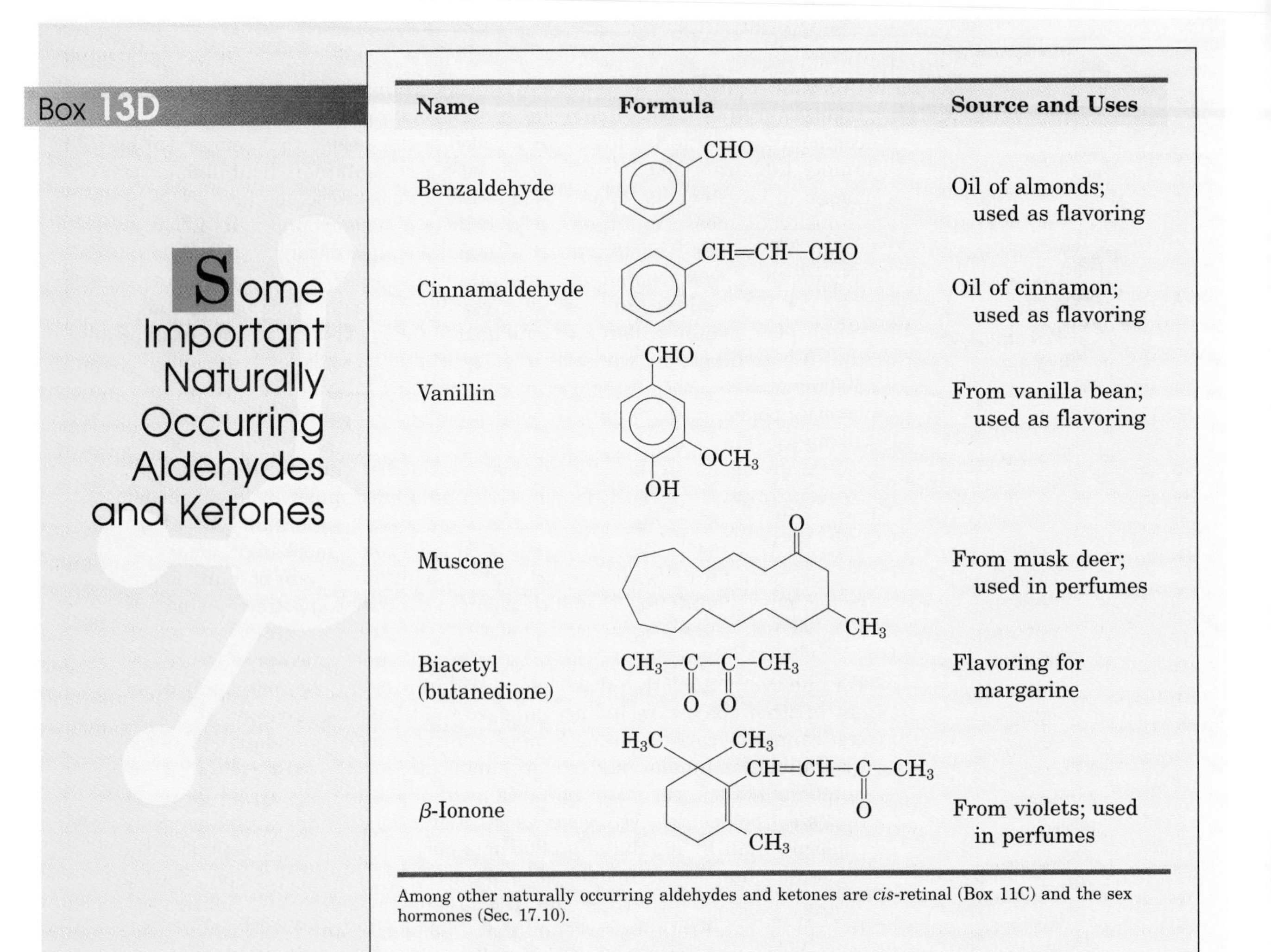

Box 13D

Some Important Naturally Occurring Aldehydes and Ketones

Name	Formula	Source and Uses
Benzaldehyde	C_6H_5–CHO	Oil of almonds; used as flavoring
Cinnamaldehyde	C_6H_5–CH=CH–CHO	Oil of cinnamon; used as flavoring
Vanillin	4-hydroxy-3-methoxybenzaldehyde (ring with CHO, OCH_3, OH)	From vanilla bean; used as flavoring
Muscone	3-methylcyclopentadecanone (large ring with C=O and CH_3)	From musk deer; used in perfumes
Biacetyl (butanedione)	CH_3–C(=O)–C(=O)–CH_3	Flavoring for margarine
β-Ionone	cyclohexene ring with H_3C, CH_3, CH_3 and CH=CH–C(=O)–CH_3	From violets; used in perfumes

Among other naturally occurring aldehydes and ketones are *cis*-retinal (Box 11C) and the sex hormones (Sec. 17.10).

Cinnamon, almonds, and vanilla beans all contain aldehydes. (Photograph by Beverly March.)

SUMMARY

Aldehydes and ketones contain the **carbonyl group,** $-\overset{|}{C}=O$. **Aldehydes** are **RCHO: ketones** are **RCOR′.** Both types of compounds can be named by IUPAC rules or by common names. Aldehydes and ketones are polar and hence have higher boiling points than hydrocarbons, but their molecules cannot form hydrogen bonds with each other, so their boiling points are lower than those of the alcohols. Their solubility in water is about the same as that of alcohols.

Aldehydes are easily oxidized to carboxylic acids; ketones resist oxidation. Several tests for aldehydes depend on this difference. Aldehydes can be reduced to primary alcohols, ketones to secondary alcohols. Aldehydes react with alcohols, in the presence of an acid catalyst, to produce **acetals. Hemiacetals** are intermediates. Ketones undergo this reaction less readily. These reactions are reversible, and acetals can be easily hydrolyzed to the aldehyde or ketone and the alcohol.

Aldehydes and ketones that possess an α-hydrogen exist in equilibrium with small amounts of their enol forms, a phenomenon called **tautomerism.** Important aldehydes are formaldehyde and acetaldehyde. The most important ketone is acetone.

Summary of Reactions

1. Oxidation (Sec. 13.5):

$$\underset{\text{Aldehyde}}{R-\underset{\underset{O}{\|}}{C}-H} \xrightarrow{[O]} \underset{\text{Carboxylic acid}}{R-\underset{\underset{O}{\|}}{C}-OH}$$

$$\underset{\text{Ketone}}{R-\underset{\underset{O}{\|}}{C}-R'} \xrightarrow[\text{Mild conditions}]{[O]} \text{no reaction}$$

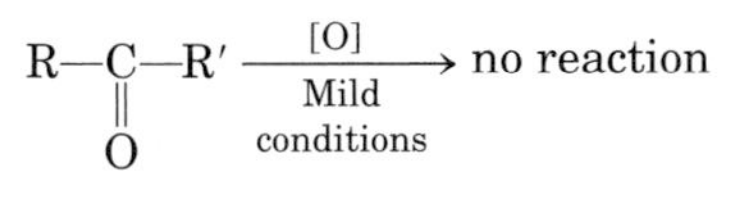

2. Reduction (Sec. 13.6):

$$\underset{\text{Aldehyde}}{R-\underset{\underset{O}{\|}}{C}-H} \xrightarrow{[H]} \underset{\text{Primary alcohol}}{R-CH_2-OH}$$

$$\underset{\text{Ketone}}{R-\underset{\underset{O}{\|}}{C}-R'} \xrightarrow{[H]} \underset{\text{Secondary alcohol}}{R-\overset{\overset{H}{|}}{\underset{\underset{OH}{|}}{C}}-R'}$$

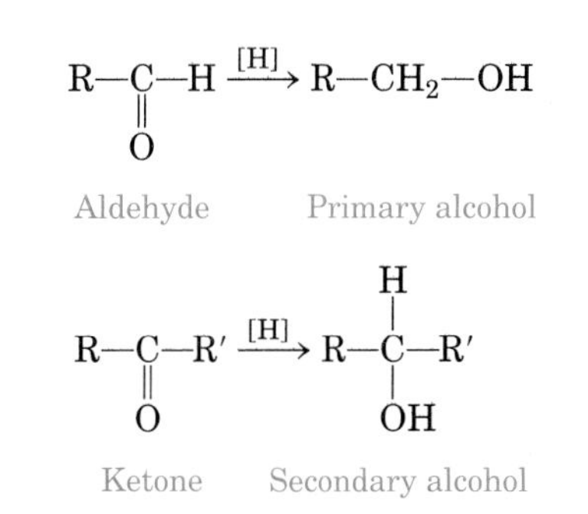

3. Formation of acetals (Sec. 13.6):

$$\underset{\text{Aldehyde}}{R-\underset{\underset{O}{\|}}{C}-H} + \underset{\text{Alcohol}}{R'OH} \longrightarrow \underset{\text{Hemiacetal}}{R-\overset{\overset{OR'}{|}}{\underset{\underset{OH}{|}}{C}}-H} \xrightarrow[R'OH]{H^+} \underset{\text{Acetal}}{R-\overset{\overset{OR'}{|}}{\underset{\underset{OR'}{|}}{C}}-H}$$

$$\underset{\text{Ketone}}{R-\underset{\underset{O}{\|}}{C}-R''} + R'OH \longrightarrow R-\overset{\overset{OR'}{|}}{\underset{\underset{OH}{|}}{C}}-R'' \xrightarrow[R'OH]{H^+} R-\overset{\overset{OR'}{|}}{\underset{\underset{OR'}{|}}{C}}-R''$$

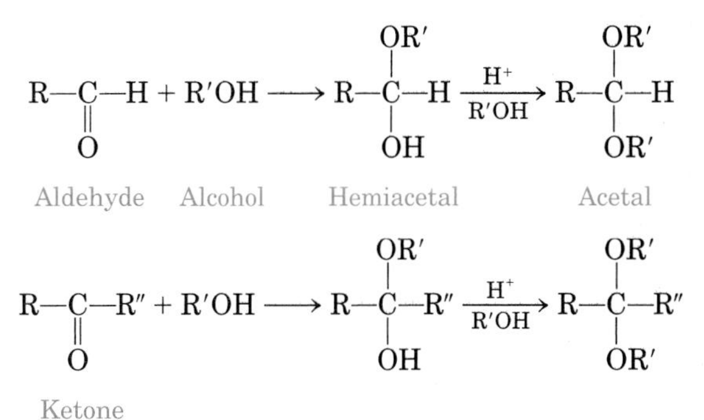

4. Hydrolysis of acetals (Sec. 13.6):

$$\underset{\text{Acetal}}{R-\overset{\overset{OR'}{|}}{\underset{\underset{OR'}{|}}{C}}-H} + H_2O \xrightarrow{H^+} \underset{\text{Aldehyde}}{R-\underset{\underset{O}{\|}}{C}-H} + \underset{\text{Alcohol}}{2R'OH}$$

KEY TERMS

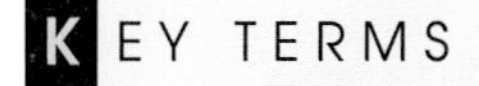

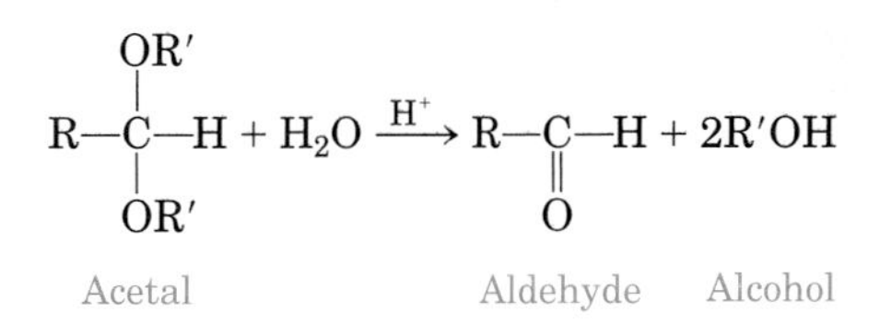

Acetal (Sec. 13.6)
Aldehyde (Sec. 13.1)
Benedict's test (Sec. 13.5)
Carbonyl group (Sec. 13.1)
Enol form (Sec. 13.6)
Hemiacetal (Sec. 13.6)
Hydrolysis (Sec. 13.6)
Keto form (Sec. 13.6)
Ketone (Sec. 13.1)
Lachrymator (Sec. 13.7)
Tautomerism (Sec. 13.6)
Tollens' test (Sec. 13.5)

PROBLEMS

Aldehydes and Ketones

13.7 Which of these compounds have a carbonyl group?

(a) $CH_3-CH(OH)-CH_3$
(b) $CH_3CH_2-C(=O)-H$
(c) $C_6H_5-C(=O)-NH_2$
(d) O
(e) tetrahydrofuran ring with OH on the carbon adjacent to O
(f) $CH_3CH_2-C(=O)-OH$

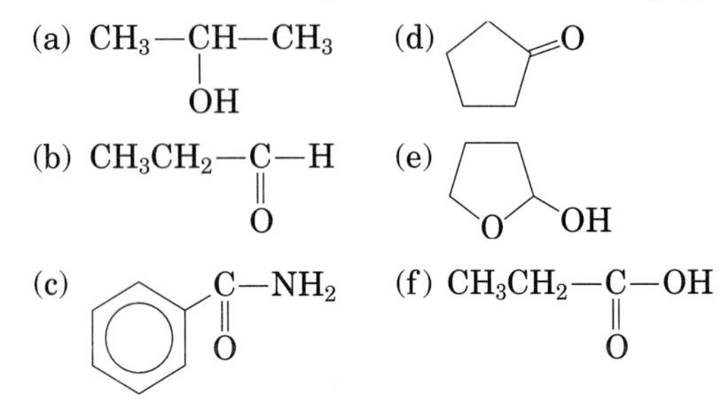

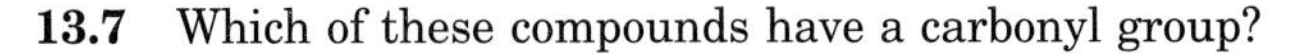

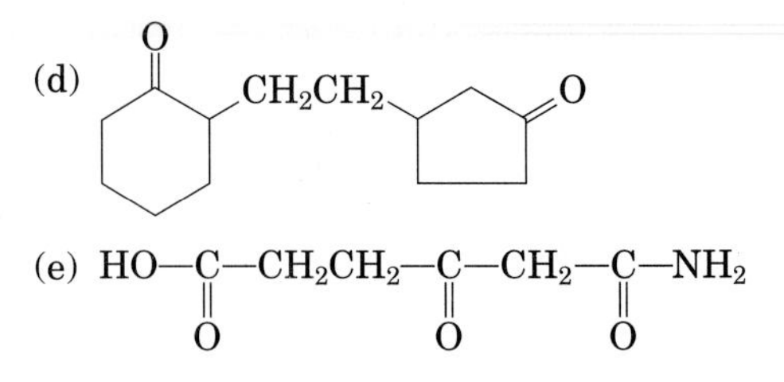

13.8 What is the difference in structure between an aldehyde and a ketone?

13.9 Which compounds are aldehydes, ketones, both, or neither?

(a) $CH_3CH_2CH_2-C(=O)-CH_2CH_3$
(b) $C_6H_5-C(=O)-CH_3$
(c) $CH_3-C(=O)-O-CH_2CH_3$
(d) benzene ring bearing $-C(=O)-H$ and, on the adjacent carbon, $-C(=O)-CH_2CH_3$
(e) $H-C(=O)-CH_2CH_2-C(=O)-O-C_6H_5$

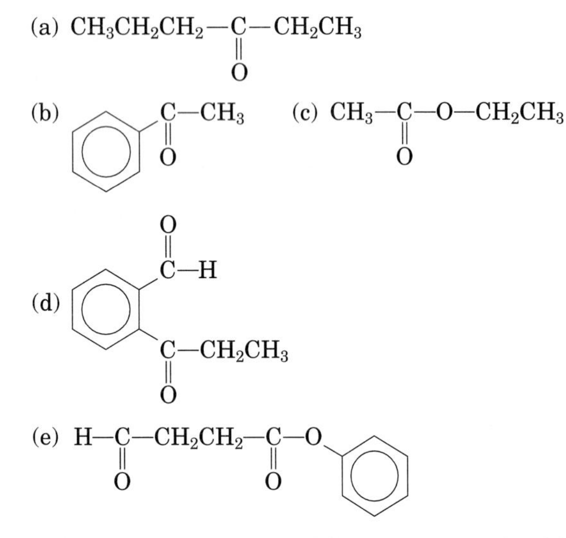

13.10 Indicate the aldehyde and keto groups in the following compounds:

(a) $H-C(=O)-CH_2CH_2-C(=O)-C(=O)-CH_3$
(b) cyclohexane ring bearing $-C(=O)-H$, with C=O on the adjacent ring carbon
(c) benzene ring bearing OH, with OCH_3 on the adjacent carbon, and $-C(=O)-H$ para to OH

(d) O CH_2CH_2 O
(e) $HO-C(=O)-CH_2CH_2-C(=O)-CH_2-C(=O)-NH_2$

Nomenclature of Aldehydes and Ketones

13.11 Name these compounds by the IUPAC system:

(a) $CH_3CH_2-CH(Cl)-CH_2CH_2-C(=O)-H$
(b) $CH_3CH_2-C(=O)-CH(OH)-CH_2CH_3$
(c) CH_3 O CH_3
(d) $CH_2=CH-CH_2-C(=O)-H$

13.12 Give common names:

(a) C_6H_5-CHO
(b) $CH_3-C(=O)-CH_3$
(c) $CH_3CH_2CH_2-C(=O)-CH_3$
(d) HCHO
(e) CH_3-CHO
(f) $C_6H_5-C(=O)-CH_3$

13.13 Draw the structural formulas for the following aldehydes:
(a) formaldehyde
(b) propanal
(c) 3,4-dimethylheptanal
(d) butyraldehyde

13.14 Draw structural formulas for the following ketones:
(a) isobutyl ethyl ketone
(b) 3-chlorocyclohexanone
(c) acetophenone
(d) hept-6-en-3-one
(e) diisopropyl ketone

Physical Properties of Aldehydes and Ketones

13.15 In each pair, pick the compound you would expect to have the higher boiling point:

(a) (1) CH_3CHO vs. (2) CH_3CH_2OH

(b) (1) $CH_3{-}C({=}O){-}CH_3$ vs.

(2) $CH_3CH_2{-}C({=}O){-}CH_3$

(c) (1) $CH_3CH_2CH_2CHO$ vs. (2) $CH_3CH_2CH_2CH_3$

(d) (1) $CH_3CH_2{-}C({=}O){-}CH_3$ vs.

(2) $CH_3CH_2{-}CH(OH){-}CH_3$

(e) (1) $HO{-}CH_2CH_2CH_2{-}OH$ vs.

(2) $H{-}C({=}O){-}CH_2{-}C({=}O){-}H$

13.16 Acetone, CH_3COCH_3, is completely soluble in water, but 4-heptanone,

$$CH_3CH_2CH_2COCH_2CH_2CH_3$$

is almost completely insoluble in water. Explain.

13.17 Why does acetone have a higher boiling point (56°C) than ethyl methyl ether (11°C), though their molecular weights are almost the same?

13.18 Why does acetone have a lower boiling point (56°C) than 2-propanol (82°C), though their molecular weights are almost the same?

13.19 Show how acetaldehyde and acetone can form hydrogen bonds with water molecules.

Oxidation of Aldehydes

13.20 Draw the structural formulas for the principal organic products. If no reaction, say so:

(a) $CH_3CH_2{-}C({=}O){-}H$ + Tollens' solution $\longrightarrow$

(b) $CH_3{-}C({=}O){-}CH_3$ + Tollens' solution $\longrightarrow$

(c) 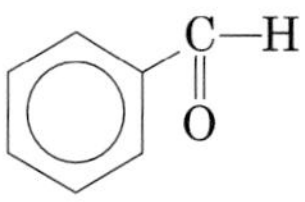 $C_6H_5{-}C({=}O){-}H$ (benzaldehyde) + Benedict's solution $\longrightarrow$

(d) 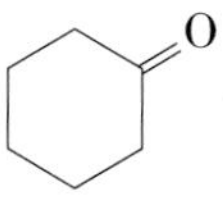 cyclohexanone (ring with =O) + Benedict's solution $\longrightarrow$

13.21 If you were given a compound that could be either $CH_3CH_2CH_2CH_2CHO$ or

$$CH_3CH_2{-}C({=}O){-}CH_2CH_3$$

what simple laboratory tests could you use to tell which it is?

13.22 Which kind of ketone gives a positive Benedict's test?

13.23 Show by means of an equation how Benedict's solution reacts with an aldehyde.

13.24 Which of the following compounds will give positive tests with Tollens' or Benedict's reagents?

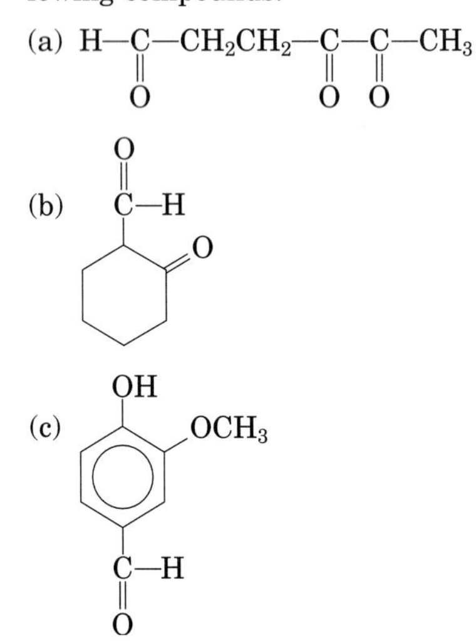

(a) $CH_3CH_2{-}C({=}O){-}CH_2CH_3$

(b) 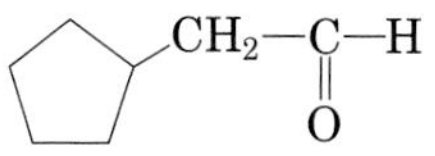 cyclopentyl${-}CH_2{-}C({=}O){-}H$

(c) $CH_3CH_2{-}CH(OH){-}C({=}O){-}CH_3$

(d) $CH_3{-}CH(OH){-}CH_2{-}C({=}O){-}CH_3$

(e) 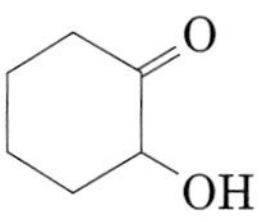 cyclohexane ring with =O and adjacent OH (2-hydroxycyclohexanone)

Reduction of Aldehydes and Ketones

13.25 Write the structural formulas for the principal organic products. If no reaction, say so:

(a) $CH_3{-}C({=}O){-}CH_3 + H_2 \xrightarrow{\text{catalyst}}$

(b) $CH_3CH_2{-}C({=}O){-}H + NaBH_4 \longrightarrow$

(c) 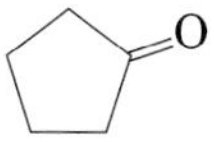 cyclopentanone (ring with =O) $+ H_2 \xrightarrow{\text{catalyst}}$

(d) 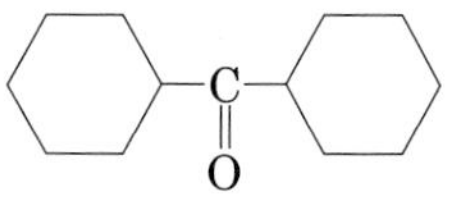 cyclohexyl${-}C({=}O){-}$cyclohexyl $+ LiAlH_4 \longrightarrow$

(e) $CH_3CH_2CH_2{-}C({=}O){-}H + NaBH_4 \longrightarrow$

13.26 Explain why reduction of an aldehyde always gives a primary alcohol and reduction of a ketone always gives a secondary alcohol.

13.27 Draw the structure of the aldehyde or ketone that can be reduced to produce each of these alcohols. If none exists, say so.

(a) $CH_3{-}CH(OH){-}CH_3$

(b) 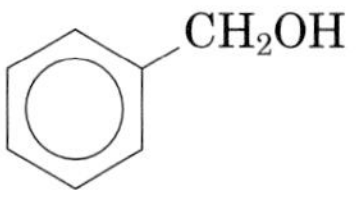 $C_6H_5{-}CH_2OH$ (benzene ring with CH_2OH)

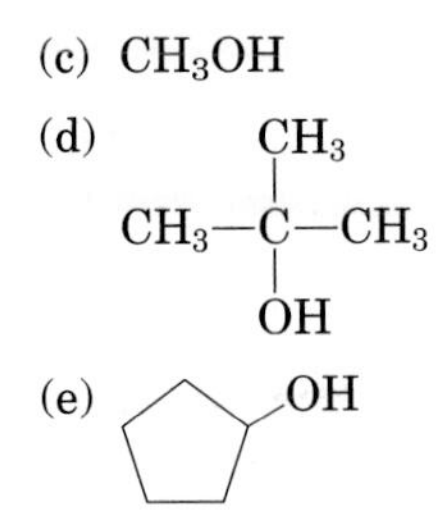

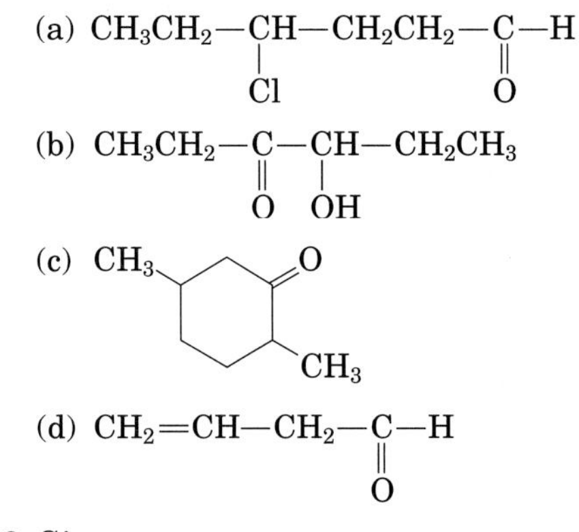

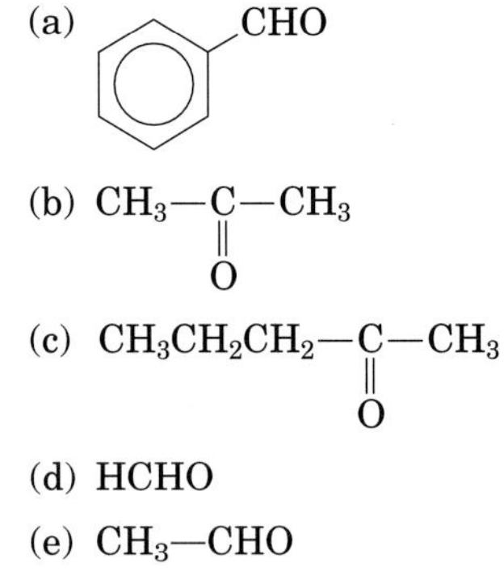

(c) CH_3OH

(d) $CH_3-C(CH_3)(OH)-CH_3$

(e) cyclopentane with OH

Other Chemical Properties of Aldehydes and Ketones

13.28 Which is the positive end of the C=O group and which is the negative end?

13.29 Aldehydes undergo addition reactions with HCN. Show the product when HCN adds to acetaldehyde. (The positive part of the HCN is H^+; the negative part is CN^-.)

Acetals and Hemiacetals

13.30 Tell which are acetals, hemiacetals, or neither:

(a) benzene ring–$CH(OCH_2CH_3)-OCH_2CH_3$

(b) $CH_3CH_2-CH(OH)-OCH_3$

(c) $CH_3O-CH_2CH_2-OCH_3$

(d) five-membered ring: H_2C–O–$CH-CH_3$, H_2C–O

(e) tetrahydrofuran ring (O) with OH on the carbon adjacent to O

(f) tetrahydrofuran ring (O) with OH on the carbon not adjacent to O

13.31 Tell which are acetals, hemiacetals, or neither:

(a) $CH_3-C(OCH_3)(OH)-CH_2CH_2CH_3$

(b) benzene ring–$C(OCH_3)(OCH_3)-CH_2CH_3$

(c) benzene ring–$C(OCH_3)(OCH_2CH_3)-CH_3$

(d) benzene ring–$C(CH_3)(OCH_2CH_3)-CH_2-OCH_3$

(e) cyclopentane with OCH_3 and OCH_3 on the same carbon

(f) cyclohexane with OCH_2CH_3 and OH on the same carbon

(g) five-membered ring lactone (O, C=O) with OCH_3

13.32 Tell which are acetals, hemiacetals, or neither.

(a) five-membered ring: H_2C–$C(H_2)$–O–$CH-OH$–H_2C

(b) six-membered ring: CH_2OH–C(H)–O–$CH-OH$, $HO-HC$, $HC(HO)-CH(OH)$

(c) benzene ring with OCH_3 and OH

(d) $CH_3CH_2-O-CH_2-OH$

13.33 Write the structural formulas for the principal organic products. If no reaction, say so:

(a) $CH_3-\underset{\parallel}{\overset{}{C}}(=O)-H + CH_3CH_2CH_2OH \xrightarrow{H^+}$

(b) cyclopentanone $+ CH_3OH \xrightarrow{H^+}$

(c) benzene ring–$C(=O)-H$ $+ CH_3OH \xrightarrow{H^+}$

*(d) tetrahydrofuran ring (O) with OCH_3 on the carbon adjacent to O $+ H_2O \xrightarrow{H^+}$

13.34 Draw the structural formulas for the principal organic products. If no reaction, say so:

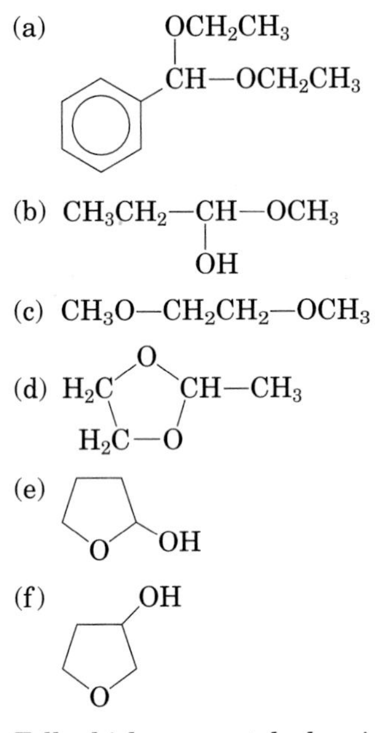

(a) benzene ring–$C(OCH_3)(OCH_3)-CH_3$ $+ H_2O \xrightarrow{H^+}$

(b) 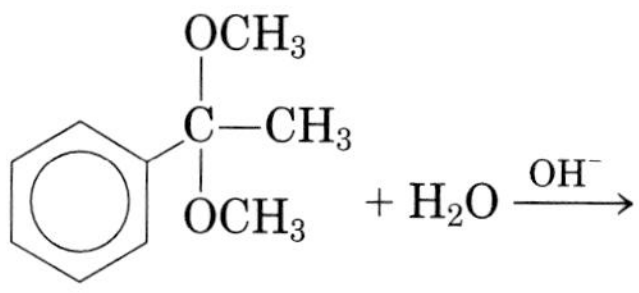 C_6H_5–$C(OCH_3)_2$–CH_3 $+ H_2O \xrightarrow{OH^-}$

13.35 Show the hemiacetal and acetal that would form in each case:

(a) $CH_3CH_2-\underset{\underset{O}{\|}}{C}-H + CH_3CH_2OH \xrightarrow{H^+}$

(b) 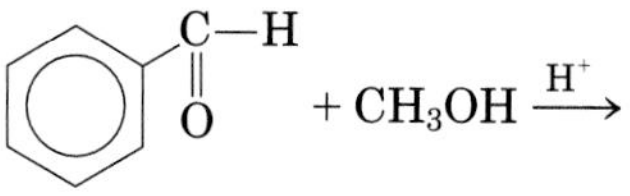 $C_6H_5-\underset{\underset{O}{\|}}{C}-H + CH_3OH \xrightarrow{H^+}$

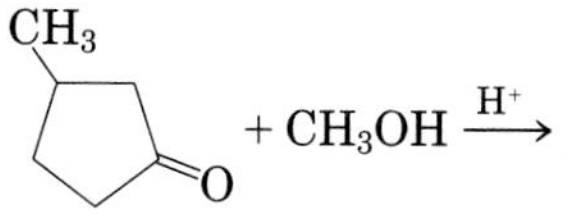

(c) 3-methylcyclopentanone $+ CH_3OH \xrightarrow{H^+}$

(d) (hemiacetal only)

$\underset{OH}{CH_2}CH_2CH_2CH_2-\underset{\underset{O}{\|}}{C}-H \xrightarrow{H^+}$

13.36 Draw the structures of the aldehyde or ketone and alcohol produced when these acetals or hemiacetals are hydrolyzed:

(a) $CH_3CH_2-\overset{OCH_2CH_3}{\overset{|}{CH}}-OCH_2CH_3$

(b) $CH_3CH_2-\overset{OCH_3}{\overset{|}{\underset{OH}{\underset{|}{C}}}}-CH_3$

(c) 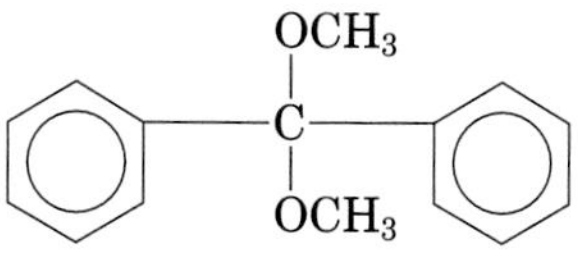 $C_6H_5-\overset{OCH_3}{\overset{|}{\underset{OCH_3}{\underset{|}{C}}}}-C_6H_5$

(d) 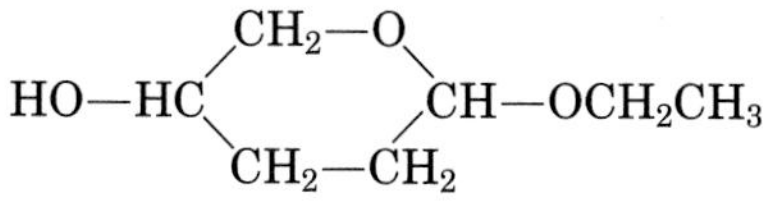 HO—HC(—CH_2—O—)(—CH_2—CH_2—)CH—OCH_2CH_3 (ring)

(e) 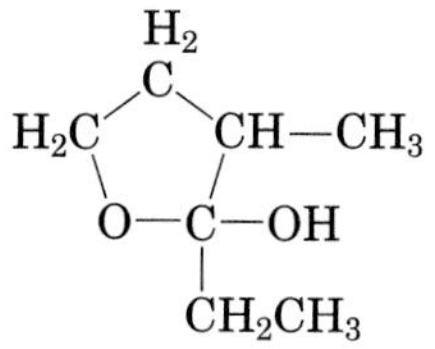 H_2C—$C(H_2)$—CH—CH_3; H_2C—O—C(—OH)(—CH_2CH_3) (five-membered ring)

(f) 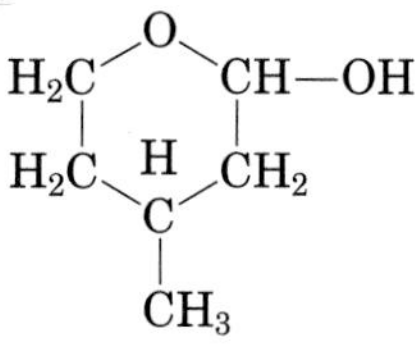 H_2C—O—CH—OH; H_2C—C(H)(—CH_3)—CH_2 (six-membered ring)

Tautomerism

13.37 Which of these compounds undergo tautomerism?

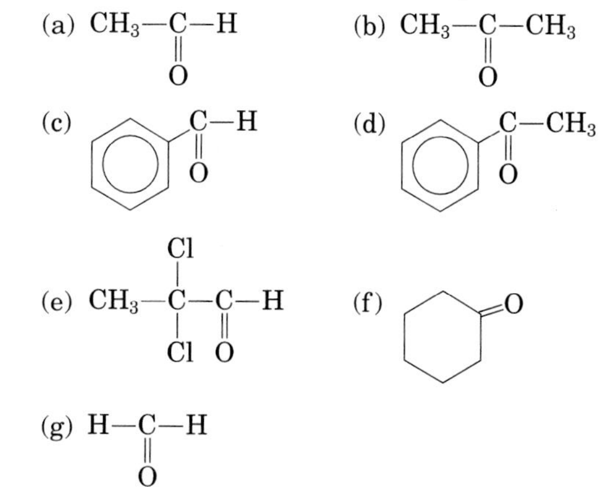

(a) $CH_3-\underset{\underset{O}{\|}}{C}-H$ (b) $CH_3-\underset{\underset{O}{\|}}{C}-CH_3$

(c) 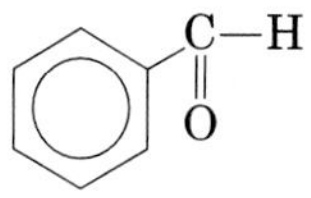 $C_6H_5-\underset{\underset{O}{\|}}{C}-H$ (d) 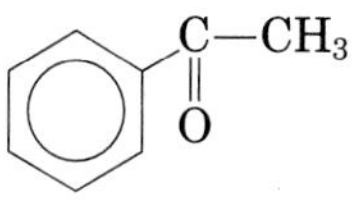 $C_6H_5-\underset{\underset{O}{\|}}{C}-CH_3$

(e) $CH_3-\overset{Cl}{\overset{|}{\underset{Cl}{\underset{|}{C}}}}-\underset{\underset{O}{\|}}{C}-H$ (f) 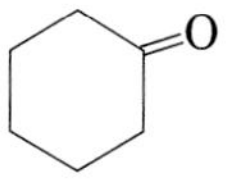 cyclohexanone

(g) $H-\underset{\underset{O}{\|}}{C}-H$

13.38 Draw the enol forms of these aldehydes and ketones:

(a) CH_3CH_2CHO

(b) $CH_3CH_2-\underset{\underset{O}{\|}}{C}-CH_3$

(c) 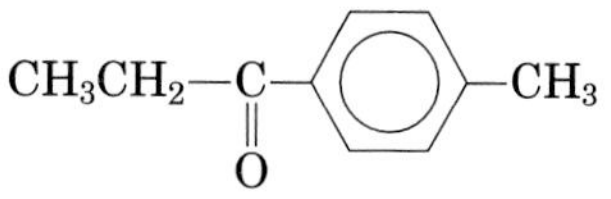 $CH_3CH_2-\underset{\underset{O}{\|}}{C}-C_6H_4-CH_3$

(d) 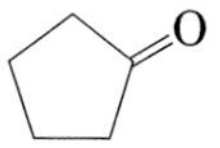 cyclopentanone

13.39 Draw the structural formula for the keto form of each of these enols:

(a) $CH_2{=}CH-OH$

(b) $CH_3-CH{=}\underset{\underset{OH}{|}}{C}-CH_3$

(c) $HO-CH{=}CH-CH{=}CH-OH$

(d) 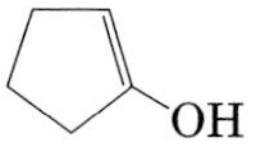 cyclopentenol (OH on the double-bonded carbon)

Some Important Aldehydes and Ketones

13.40 What is the most important industrial use for formaldehyde?

***13.41** Show how three molecules of acetaldehyde can add to each other to give paraldehyde (draw the bonds and show how they change in the course of the reaction).

13.42 Why is the sweet odor of acetone on the breath of a patient a bad clinical symptom?

13.43 What is the active ingredient in Mace?

13.44 How should you handle a compound known to be a lachrymator?

Boxes

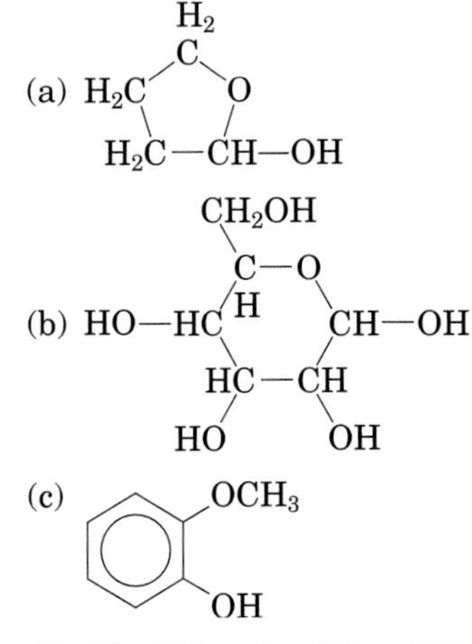

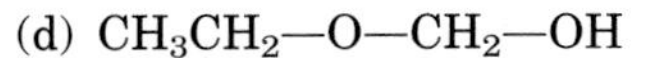

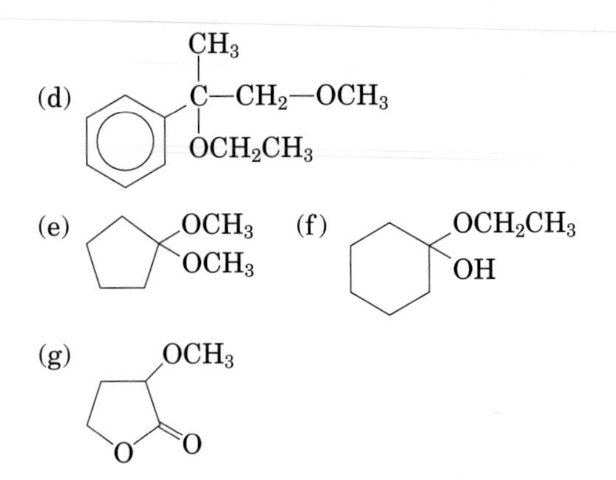

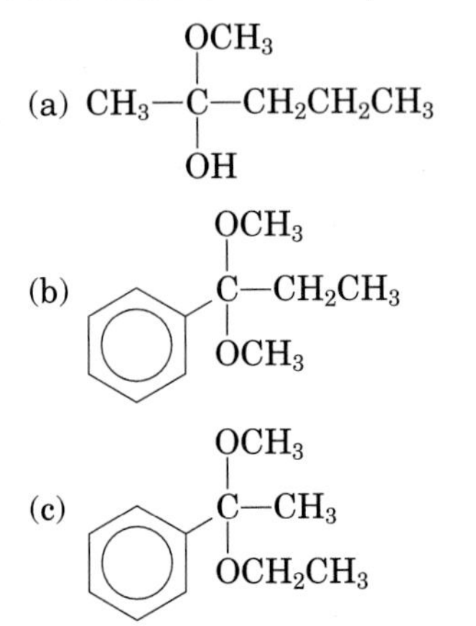

13.45 (Box 13A) Chloral is one of the few aldehydes that form stable hemiacetals. Draw the structure of the hemiacetal that forms when chloral reacts with ethanol:

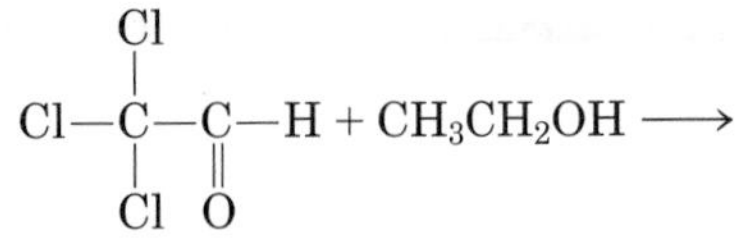

$$Cl_3C-\overset{\displaystyle O}{\overset{\|}{C}}-H + CH_3CH_2OH \longrightarrow$$

13.46 (Box 13A) What is the medical use of chloral hydrate?

13.47 (Box 13B) What is a thermosetting resin?

13.48 (Box 13B) Could a thermosetting resin be made by treating HCHO with *m*-cresol instead of phenol?

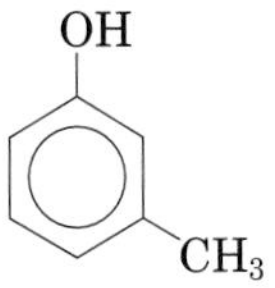

OH, CH_3 (benzene ring)

m-Cresol

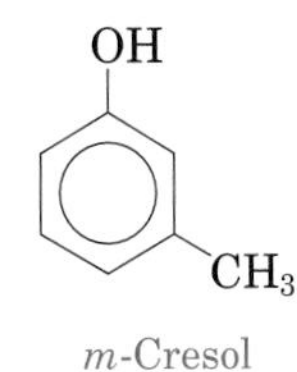

13.49 (Box 13C) The LD_{50} for arsenic is 48 mg/kg. What dose, per rat, would kill, within 14 days, half of a group of rats whose average weight is 0.35 kg?

13.50 (Box 13C) Why is the LD_{50} not a completely clear indication of the toxicity of a substance?

13.51 (Box 13D) What naturally occurring ketones are used in the manufacture of perfumes?

13.52 (Box 13D) Classify β-ionone into as many chemical families as is appropriate.

Additional Problems

13.53 Why can't two molecules of acetone form a hydrogen bond with each other?

13.54 Name these compounds by IUPAC or common names:

(a) 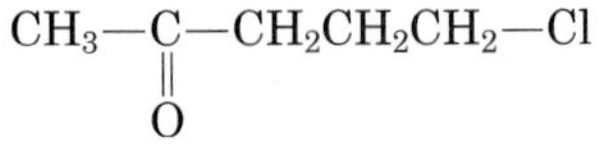 $CH_3-\overset{\displaystyle O}{\overset{\|}{C}}-CH_2CH_2CH_2-Cl$

(b) $C_6H_5-CH_2CH_2-\overset{\displaystyle O}{\overset{\|}{C}}-H$

(c) $CH_3-CH{=}CH-\overset{\displaystyle O}{\overset{\|}{C}}-CH_2CH_3$

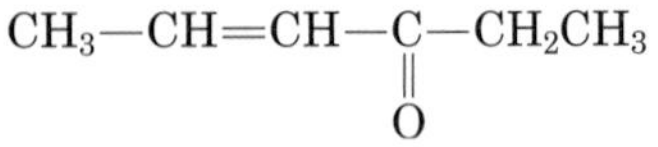

(d) 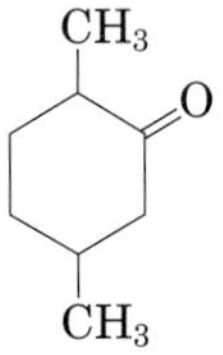 cyclohexanone ring with CH_3 at C-2 and CH_3 at C-4 (O on ring carbon)

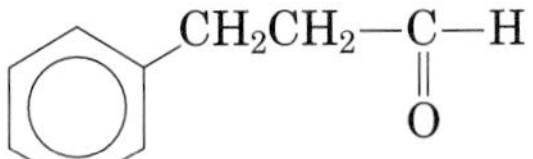

(e) 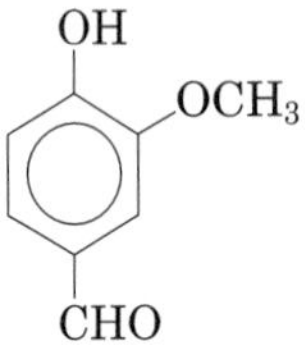 benzene ring with OH, OCH_3, and CHO

***13.55** A student took a bottle of butanal from the shelf and discovered that the contents turned litmus red (that is, it was acidic). Yet aldehydes are neutral compounds. Could you explain to the student what caused the acidic behavior?

***13.56** Acetone is an important laboratory solvent. For similar compounds, solvation ability increases as molecular weight decreases. Can you give a reason why formaldehyde and acetaldehyde are not important laboratory solvents?

13.57 What is formalin and what is it used for?

***13.58** Can you give a reason why 2,4-pentandione (p. 440) is more stable in the enol than in the keto form?

13.59 Propanal, CH_3CH_2CHO, and 1-propanol, $CH_3CH_2CH_2OH$, have about the same molecular weight (Table 13.2), yet their boiling points differ by almost 40°C. Explain this fact.

13.60 What kind of odors do aldehydes and ketones have?

***13.61** Write the stuctural formula for

(a) any cyclic hemiacetal that has a six-membered ring

(b) any cyclic acetal that has a five-membered ring

(c) any noncyclic hemiacetal formed from a ketone.

***13.62** Which, if any, of the three compounds in Problem 13.61 can be isolated?

***13.63** Write the name and structure of the compound that is formed by the reduction of dihydroxyacetone, $HOCH_2-\overset{\displaystyle O}{\overset{\|}{C}}-CH_2OH$.

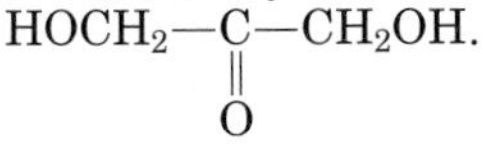

***13.64** Predict the product of this reaction:

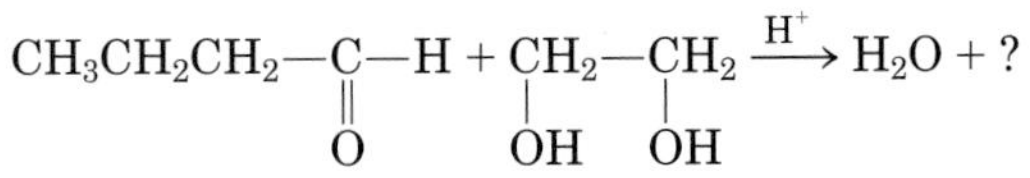

$$CH_3CH_2CH_2-\overset{\displaystyle O}{\overset{\|}{C}}-H + \underset{\displaystyle OH}{CH_2}-\underset{\displaystyle OH}{CH_2} \xrightarrow{H^+} H_2O + ?$$

What kind of compound is this?

13.65 Draw the structural formulas for the principal organic products. If no reaction, say so.

(a) 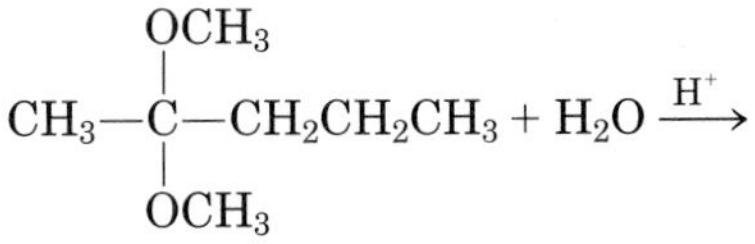 $CH_3-\overset{\displaystyle OCH_3}{\underset{\displaystyle OCH_3}{C}}-CH_2CH_2CH_3 + H_2O \xrightarrow{H^+}$

(b) cyclohexanone + Tollens' solution $\longrightarrow$

(c) $CH_3-\overset{\displaystyle O}{\overset{\|}{C}}-H + CH_3CH_2-OH \xrightarrow{H^+}$

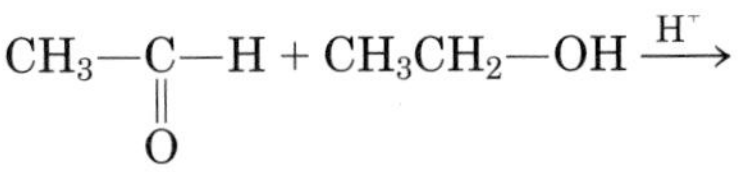

*(d) One mole

$$CH_2{=}CH-CH_2CH_2-\overset{\displaystyle O}{\overset{\|}{C}}-CH_3$$

$$+ \text{ one mole } H_2 \xrightarrow{\text{catalyst}}$$

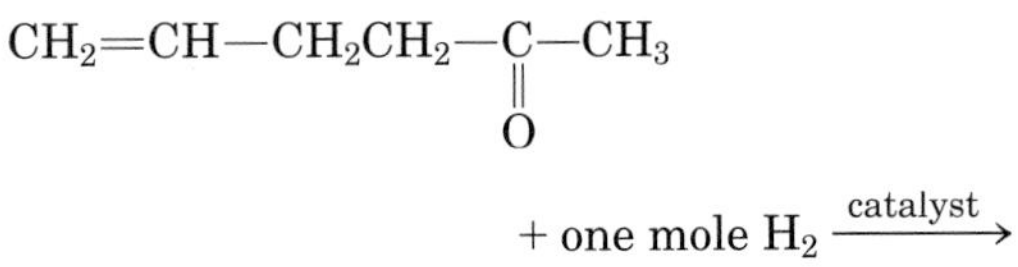

13.66 Explain why the compound prop-1-en-2-ol has never been isolated.

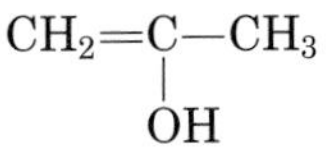

$$CH_2{=}\underset{\displaystyle OH}{C}-CH_3$$

***13.67** Write the structures of cyclic hemiacetals formed by ring closure of (a) 6-hydroxyhexanal (b) 4-hydroxy-5-iodoheptanal (c) 4,5-dihydroxypentanal (two different ones in this case).

Chapter 14

Carboxylic acids and esters

14.1 Introduction

Anytime we eat foods with a sour or tart taste, it is likely that the taste is due to one or more carboxylic acids. Vinegar contains acetic acid, lemons contain citric acid, and the tart taste of apples is caused by malic acid. Carboxylic acids are very important in biochemistry and in organic chemistry, as we shall see in later chapters. The free acids themselves are present in biological systems less often than their salts and derivatives. **Derivatives of carboxylic acids** include carboxylic esters, studied in this chapter, and amides, discussed in Chapter 15.

Acyl chlorides and anhydrides, discussed in Section 14.10, are also acid derivatives.

Like aldehydes and ketones, carboxylic acids and their derivatives all contain the carbonyl group, C=O, but in carboxylic acids, the carbonyl carbon is always connected to a hydroxyl group, so all carboxylic acids have the formula

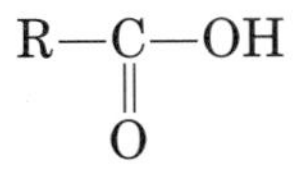

$$\mathrm{R{-}\underset{\underset{\displaystyle O}{\|}}{C}{-}OH}$$

An abbreviated form is either —COOH or —CO_2H, but we must not forget that the carbonyl group is there.

The $\mathbf{-\underset{\underset{\displaystyle O}{\|}}{C}-OH}$ group is called the **carboxyl group** (because it is a combination of **carb**onyl and hydr**oxyl**). As we shall see, the properties of carboxylic acids are mostly quite different from those of aldehydes and alcohols. Just as R groups, derived from alkanes, RH, are called alkyl groups, groups derived from carboxylic acids, $\mathrm{R{-}\underset{\underset{\displaystyle O}{\|}}{C}{-}}$, are called **acyl groups.**

14.2 Nomenclature of Carboxylic Acids

Carboxylic acids are easily named by the IUPAC system. The suffix is "-oic acid"; otherwise the rules are the same as before. Because it is always at the end of the chain, the —COOH carbon is always number 1, although this number is understood and is not stated as part of the name.

In numbering chains, the —COOH group takes precedence over all groups we have met up to now.

EXAMPLE

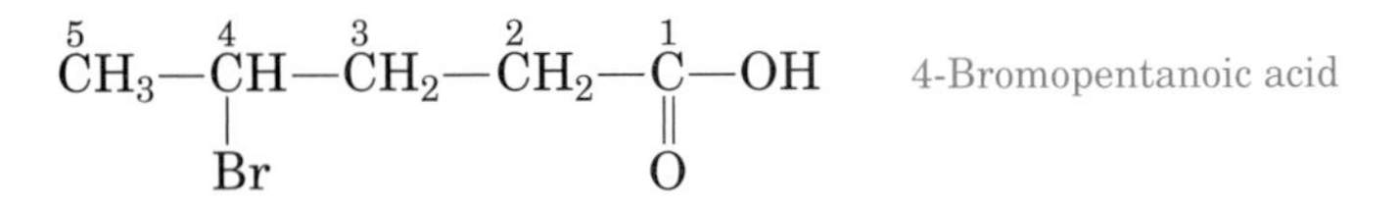

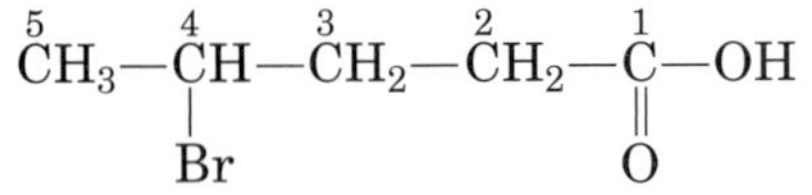

$$\overset{5}{C}H_3{-}\underset{\underset{\displaystyle Br}{|}}{\overset{4}{C}H}{-}\overset{3}{C}H_2{-}\overset{2}{C}H_2{-}\underset{\underset{\displaystyle O}{\|}}{\overset{1}{C}}{-}OH \qquad \text{4-Bromopentanoic acid}$$

The longest chain has five carbons, so the parent is pentanoic acid. The —COOH carbon is number 1, so the bromo is in the 4 position.

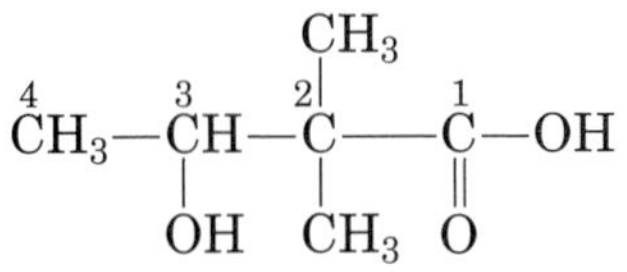

$$\overset{4}{C}H_3{-}\underset{\underset{\displaystyle OH}{|}}{\overset{3}{C}H}{-}\underset{\underset{\displaystyle CH_3}{|}}{\overset{\overset{\displaystyle CH_3}{|\,2}}{C}}{-}\underset{\underset{\displaystyle O}{\|}}{\overset{1}{C}}{-}OH \qquad \text{3-Hydroxy-2,2-dimethylbutanoic acid}$$

Here the parent is *buta-* because the longest chain has four carbons.

Problem 14.1

Give the IUPAC name for

(a) $CH_3(CH_2)_8COOH$

(b) $CH_3—\underset{\displaystyle OH}{\underset{|}{C}H}—CH_2—\underset{\displaystyle Br}{\underset{|}{C}H}—COOH$

Although the IUPAC system is, as usual, easy to apply, it is generally used only for relatively complicated carboxylic acids. Those with simple structures almost always have common names, and these are almost always used in preference to the IUPAC names. Table 14.1 shows the most important of these, but there are hundreds of other carboxylic acids with common names.

Table 14.1 Names and Physical Properties of Some Carboxylic Acids[a]

Formula	Common Name	BP(°C)	MP(°C)	Solubility in Water (g/100 g of H_2O)
H—COOH	Formic acid	101	8	Infinite
CH_3—COOH	Acetic acid	118	17	Infinite
CH_3—CH_2—COOH	Propionic acid	141	−21	Infinite
CH_3—$(CH_2)_2$—COOH	Butyric acid	162	−8	Infinite
CH_3—$(CH_2)_3$—COOH	Valeric acid	184	−35	5
CH_3—$(CH_2)_4$—COOH	Caproic acid	205	−2	1
CH_3—$(CH_2)_6$—COOH	Caprylic acid	240	16	Insoluble
CH_3—$(CH_2)_8$—COOH	Capric acid	268	31	Insoluble
CH_3—$(CH_2)_{10}$—COOH	Lauric acid	—	44	Insoluble
CH_3—$(CH_2)_{12}$—COOH	Myristic acid	—	54	Insoluble
CH_3—$(CH_2)_{14}$—COOH	Palmitic acid	—	63	Insoluble
CH_3—$(CH_2)_{16}$—COOH	Stearic acid	—	70	Insoluble
HOOC—COOH	Oxalic acid	—	190	9
HOOC—CH_2—COOH	Malonic acid	—	136	73
HOOC—$(CH_2)_2$—COOH	Succinic acid	—	190	8
HOOC—$(CH_2)_3$—COOH	Glutaric acid	—	99	65
HOOC—$(CH_2)_4$—COOH	Adipic acid	—	154	2
HOOC(H)C=C(H)COOH (cis: HOOC and COOH on same side)	Maleic acid	—	130	4
HOOC(H)C=C(H)COOH (trans: HOOC and COOH on opposite sides)	Fumaric acid	—	286	55
C_6H_5—COOH (benzene ring with COOH)	Benzoic acid	—	122	Insoluble
benzene ring with OH and COOH on adjacent carbons	Salicylic acid	—	159	Insoluble

[a]See also Table 17.1.

Formic acid is the compound chiefly responsible for the pain and blistering caused by the sting of red ants and bees.

The Formica used in kitchen counter tops is made from a formaldehyde resin (Box 13B). Formaldehyde got its name from formic acid, its oxidation product.

The unbranched acids having between 3 and 20 carbons are known as **fatty acids.** We will study them in Chapter 17.

Why do common names exist? The reason is that these acids were discovered more than 100 years ago, before the IUPAC system existed and, in most cases, before their structures were known. Since these compounds needed names, they were generally named according to their sources, using Latin word roots, as was customary. For example, formic acid was first isolated from red ants, and the Latin word for ant is *formica*. Similarly, acetic acid comes from vinegar (Latin, *acetum*) and butyric acid from rancid butter (Latin, *butyrum*). Three acids shown in Table 14.1 (caproic, caprylic, capric) have similar names. All of these were first isolated from goat fat (the Latin word for goat is *caper*).

Note that in Table 14.1 we have given common names for unbranched carboxylic acids up to 18 carbons long (stearic acid). These long-chain acids are important because esters of them constitute the fats, as well as some other lipids.

As Table 14.1 shows, **dicarboxylic acids** also have common names. These acids, from oxalic to adipic, are also important in biochemistry and in organic chemistry.

The names of all the acid derivatives, as well as those of the aldehydes, are based on the names of the acids. Therefore, if you know the name of any carboxylic acid, you can name its derivatives as well.

14.3 Physical Properties of Carboxylic Acids

When compared with other compounds of about the same molecular weights, carboxylic acids have the highest boiling points of any compounds we have met so far. The simplest one, formic acid, HCOOH (MW = 46), has a boiling point of 101°C, considerably higher than that of ethanol (78°C), which has the same molecular weight.

As with alcohols (Sec. 12.7), the high boiling points of carboxylic acids are caused by hydrogen bonding, but the hydrogen bonding is stronger here because each molecule forms two hydrogen bonds with another molecule:

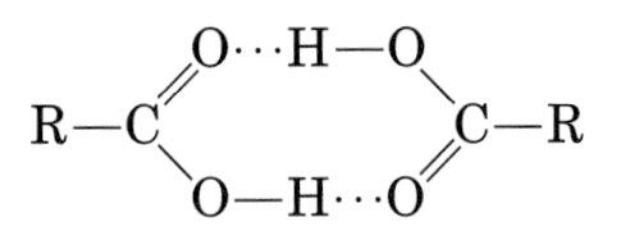

```
       O···H—O
      //       \
R—C              C—R
      \        //
       O—H···O
```

Carboxylic acid dimer

A **dimer** is a species made up of two identical units.

From the discussion in Section 12.7 we would expect the smaller carboxylic acids to be quite soluble in water (why?), and Table 14.1 shows that this is so. The first four are infinitely soluble in water, but then the solubility drops rapidly and becomes insignificant after caprylic acid (C_8).

The higher fatty acids might also have foul odors, but their volatility is very low. You cannot smell a compound unless molecules of it reach your nose.

One other physical property of carboxylic acids must be mentioned. The liquid fatty acids from propionic (C_3) to capric (C_{10}) have extremely foul odors, about as bad as (though different from) those of the thiols (Sec. 12.8). Butyric acid is found in stale perspiration and is an important component of "locker room odor." Valeric acid smells even worse, and goats, which excrete the C_6, C_8, and C_{10} acids, are not famous for their pleasant odors.

14.4 Some Important Carboxylic Acids

Acetic Acid, CH_3COOH

This acid is discussed in Box 8A.

Oxalic Acid, HOOCCOOH

The simplest dicarboxylic acid, oxalic acid is present as the monopotassium salt, $HOOCCOO^- K^+$, in certain leafy vegetables, including spinach and rhubarb.

Hydroxy Acids

Many hydroxy acids are present in various foods. Among the most important are lactic acid, present in sour milk, yogurt, sauerkraut, and pickles, and citric acid, found in citrus fruits:

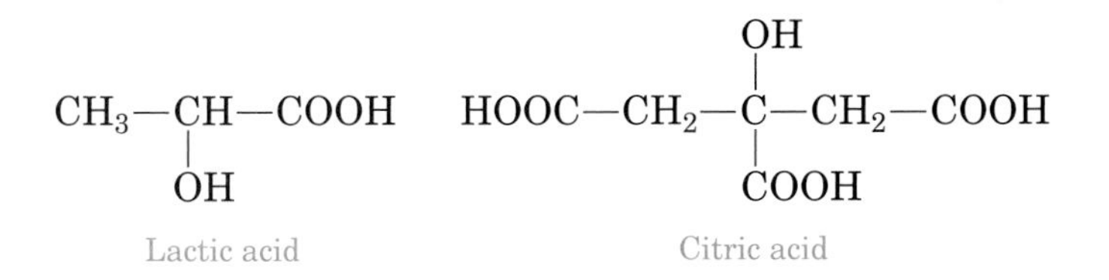

$$\underset{\text{Lactic acid}}{CH_3{-}\underset{\underset{\displaystyle OH}{|}}{C}H{-}COOH} \qquad \underset{\text{Citric acid}}{HOOC{-}CH_2{-}\overset{\overset{\displaystyle OH}{|}}{\underset{\underset{\displaystyle COOH}{|}}{C}}{-}CH_2{-}COOH}$$

Both of these acids are important in body processes. Lactic acid, in its salt form, is produced in the muscles in the breakdown of glucose (Box 21A). Citric acid, also in its salt form, is part of the Krebs cycle (Sec. 20.4). Both of these acids are normally present in blood. The intestines of human infants contain high concentrations of lactic acid due to the fermentation of mother's milk by the bacteria *Lactobacillus lipidus.* The lactic acid helps to prevent infections by suppressing the growth of harmful bacteria.

Spinach and rhubarb contain a salt of oxalic acid. (Photograph by Charles D. Winters.)

Trichloroacetic Acid, CCl_3COOH

A 50 percent aqueous solution of trichloroacetic acid is used by dentists to cauterize gums. This strong acid denatures the bleeding of diseased tissue and allows the growth of healthy gum. Trichloroacetic acid solution is also used to cauterize canker sores.

Box 14A

Retinoic Acid

Retinoic acid is the carboxylic acid corresponding to retinal (Box 11C). The all-trans form is a natural product that helps the development of limbs in embryos. It is a vertebrate morphogen, a substance that helps the development of differentiated tissues.

all-trans-Retinoic acid is also sold, under the name tretinoin (brand name, Retin-A), as a cream to be applied externally. The only use for this drug that is currently approved by the Food and Drug Administration is as a treatment for severe cystic acne, for which it generally works well. However, the impression has gotten around that it can remove wrinkles and other signs of aging, and some people have used it for this purpose, with varying results. For pregnant women there is the possibility of harm to the fetus if the drug is overused.

An isomer of *all-trans*-retinoic acid, 13-*cis*-retinoic acid, was also found beneficial in treatment of severe cystic acne. However, when taken orally by pregnant women, it induces birth defects in the form of cranial deformities.

It is interesting to compare three related compounds: The alcohol form, retinol or vitamin A, is essential for the normal development of epithelial tissues (Box 11A, Sec. 26.5); the aldehyde form, retinal, is essential to vision (Box 11C); the carboxylic acid form is a natural morphogen and a drug used to treat severe acne.

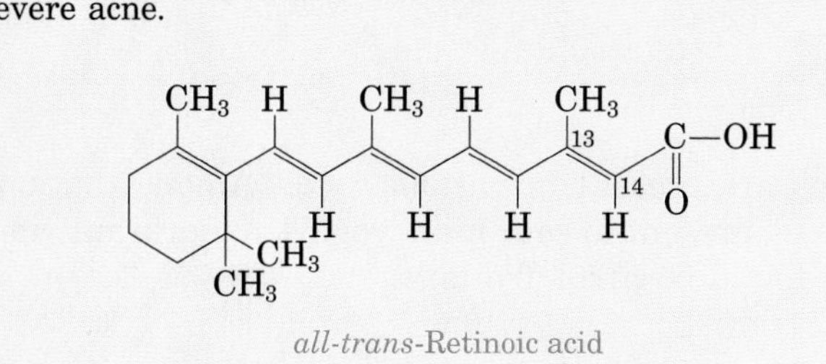

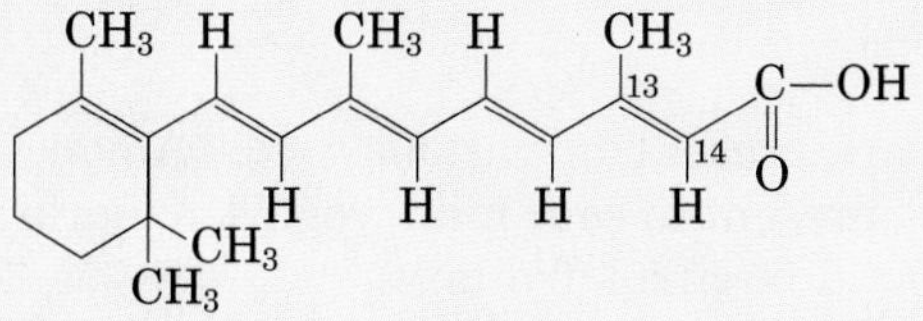

CH_3 H CH_3 H CH_3 (13) C—OH (14) O; H H H H; CH_3 CH_3

all-trans-Retinoic acid

14.5 Preparation of Carboxylic Acids

There are several ways to prepare carboxylic acids in the laboratory. We have already discussed the oxidation of primary alcohols and aldehydes (Secs. 12.3 and 13.5):

$$R{-}CH_2{-}OH \xrightarrow{[O]} R{-}\underset{\underset{O}{\|}}{C}{-}OH$$

$$R{-}\underset{\underset{O}{\|}}{C}{-}H \xrightarrow{[O]} R{-}\underset{\underset{O}{\|}}{C}{-}OH$$

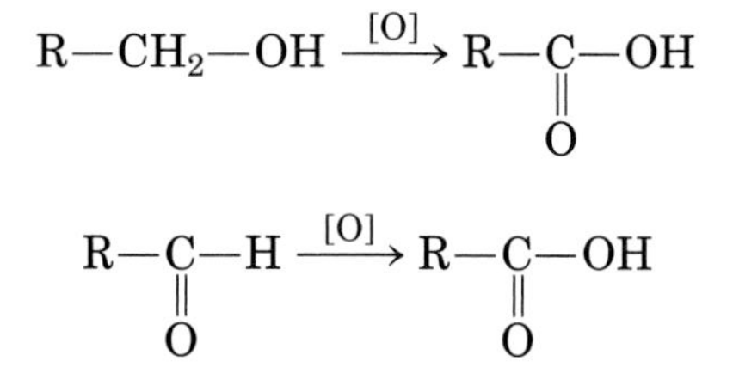

Other important methods for carboxylic acid preparation are the hydrolysis of carboxylic esters (Sec. 14.11) and of amides (Sec. 15.8).

The oxidation of a primary alcohol to a carboxylic acid is one of the oldest chemical reactions known, since even prehistoric humans found that their wine often turned to vinegar. The ethanol in the wine is oxidized to acetic acid by the oxygen in the air.

14.6 The Acidity of Carboxylic Acids

By far the most important property of carboxylic acids is the one implied by their name: They are acidic. Apart from the sulfonic acids (Sec. 11.11), carboxylic acids are the strongest organic acids. Still, compared with the strong inorganic acids (Table 8.1, p. 240), almost all carboxylic acids are weak. In Section 8.2 we saw that a strong acid reacts almost completely with water to form H_3O^+. The extent of this reaction for a typical unsubstituted carboxylic acid is only about 1 percent or less. That is, the equilibrium

$$\underset{\text{about 99\%}}{R-\underset{\underset{O}{\|}}{C}-OH} + H_2O \rightleftharpoons \underset{\text{about 1\%}}{R-\underset{\underset{O}{\|}}{C}-O^-} + H_3O^+$$

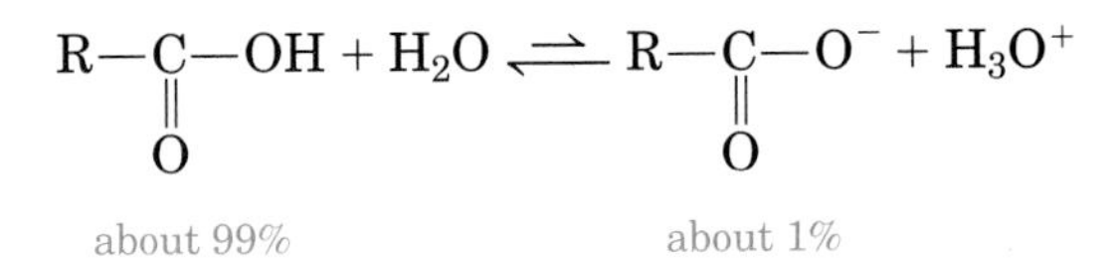

lies well to the left.

Although carboxylic acids are weak, they are strong enough to react with many bases stronger than water. Among these are hydroxide ion, OH^-, carbonate ion, CO_3^{2-}, bicarbonate ion, HCO_3^-, and ammonia (Sec. 8.6):

$$R-\underset{\underset{O}{\|}}{C}-OH + OH^- \longrightarrow R-\underset{\underset{O}{\|}}{C}-O^- + H_2O$$

$$R-\underset{\underset{O}{\|}}{C}-OH + CO_3^{2-} \longrightarrow R-\underset{\underset{O}{\|}}{C}-O^- + HCO_3^-$$

$$R-\underset{\underset{O}{\|}}{C}-OH + HCO_3^- \longrightarrow R-\underset{\underset{O}{\|}}{C}-O^- + H_2CO_3 \longrightarrow [H_2O + CO_2]$$

$$R-\underset{\underset{O}{\|}}{C}-OH + NH_3 \longrightarrow R-\underset{\underset{O}{\|}}{C}-O^- + NH_4^+$$

In all these reactions, the conversion of the carboxylic acid to its conjugate base (called a *carboxylate ion*) is essentially complete.

Problem **14.2**

Complete the following reaction:

$$CH_3CH_2CH_2-\underset{\underset{O}{\|}}{C}-OH + OH^- \longrightarrow$$

At this point it is useful to compare the acidity of carboxylic acids with that of other oxygen-containing organic compounds. In Section 8.4 we learned that the acidity of an acid can be expressed by its equilibrium

Table 14.2 The K_a and pK_a Values for Some Organic Compounds

Compound	Formula	K_a	pK_a
Formic acid	HCOOH	1.8×10^{-4}	3.74
Acetic acid	CH_3COOH	1.8×10^{-5}	4.74
Benzoic acid	C_6H_5—COOH (benzene ring–COOH)	6.5×10^{-5}	4.19
Phenol	C_6H_5—OH (benzene ring–OH)	1.3×10^{-10}	9.89
Ethanethiol	CH_3CH_2—SH	2.5×10^{-11}	10.6
Ethanol	CH_3CH_2—OH	Neutral	
Acetaldehyde	CH_3—C(=O)—H	Neutral	

As discussed in Section 8.8, p$K_a = -\log K_a$.

constant, K_a. Table 14.2 shows K_a and pK_a values for some organic compounds. As can be seen, carboxylic acids are much stronger acids than phenols, which in turn are somewhat stronger than thiols. Alcohols and aldehydes, on the other hand, are so weak (they are weaker acids than water) that we generally consider them to be neutral compounds, with no acidic properties at all.

14.7 Properties of Carboxylic Acid Salts

When a carboxylic acid reacts with a base, the acid is converted to the carboxylate ion, $RCOO^-$. Negative ions, of course, do not exist by themselves—there must also be positive ions present. As we learned in Chapter 8, the combination of a negative and a positive ion is called a salt, and the reaction of a carboxylic acid with a base always produces a salt. These salts can be isolated as pure compounds by evaporating the water from the solution. Salts of carboxylic acids are ionic compounds, for example,

$$R\text{—}\underset{\substack{\|\\ O}}{C}\text{—}O^-\ Na^+$$

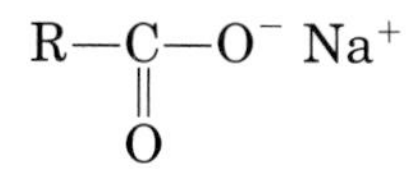

Note that the carboxylic acids themselves are often liquids, and many are insoluble in water.

and we are not surprised to find that they are always solids and are usually soluble in water (Sec. 6.6).

To name a carboxylate salt, always begin by naming the corresponding acid. Then change "-ic acid" to "-ate" and place the name of the positive ion before this word.

EXAMPLE

Name the salts shown.

(a) 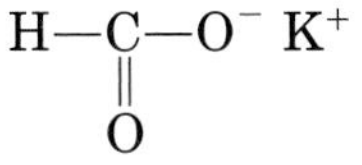 $H—\overset{\displaystyle O}{\overset{\|}{C}}—O^- K^+$ Potassium formate

The corresponding acid, HCOOH, is form*ic acid* (Table 14.1), and we change this to form*ate*. The positive ion is potassium, so the name is potassium formate.

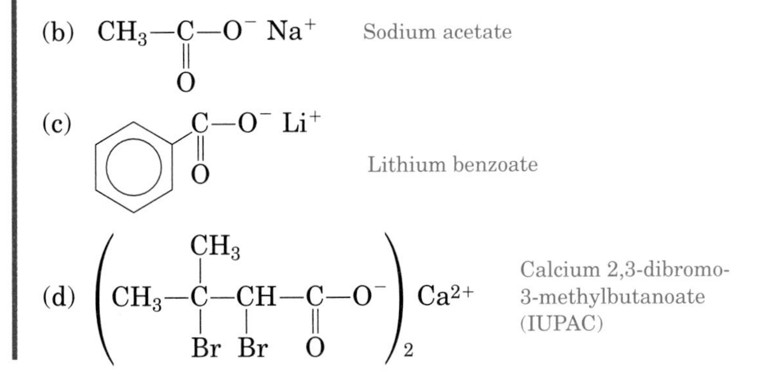

(b) 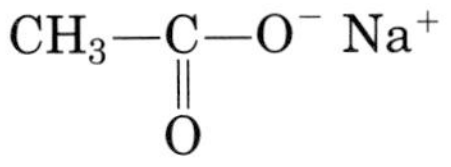 $CH_3—\overset{\displaystyle O}{\overset{\|}{C}}—O^- Na^+$ Sodium acetate

(c) 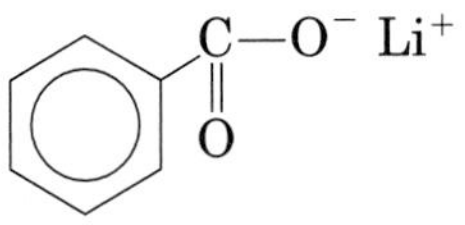 $C_6H_5—\overset{\displaystyle O}{\overset{\|}{C}}—O^- Li^+$ Lithium benzoate

(d) 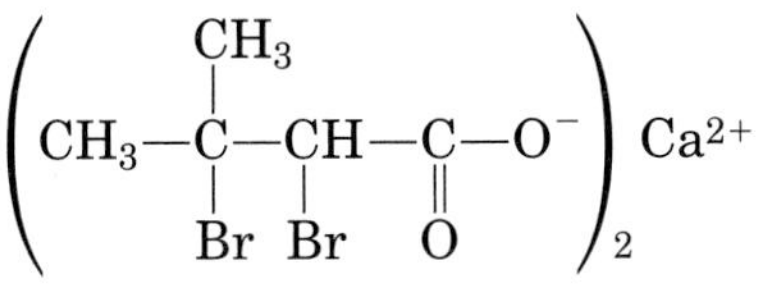 $\left(CH_3—C(CH_3)(Br)—CH(Br)—\overset{\displaystyle O}{\overset{\|}{C}}—O^-\right)_2 Ca^{2+}$ Calcium 2,3-dibromo-3-methylbutanoate (IUPAC)

The process is the same whether a common name or an IUPAC name is used for the acid.

Problem 14.3

Name these salts:

(a) 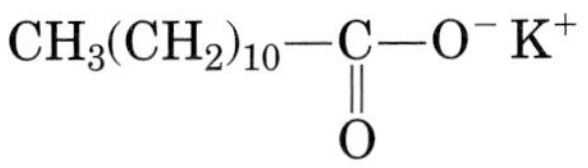 $CH_3(CH_2)_{10}—\overset{\displaystyle O}{\overset{\|}{C}}—O^- K^+$

(b) 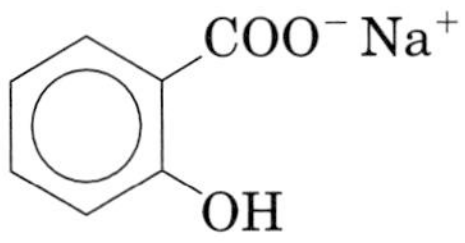 $C_6H_4(COO^- Na^+)(OH)$

Box 14B

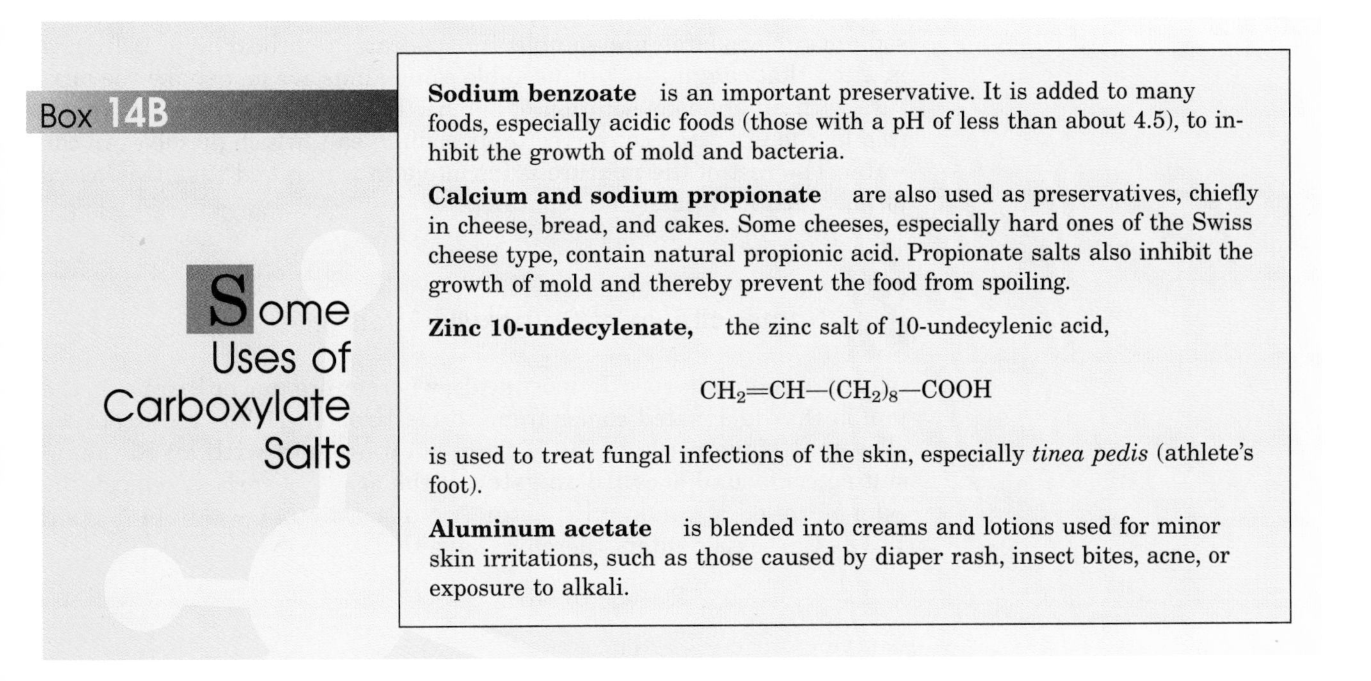

Some Uses of Carboxylate Salts

Sodium benzoate is an important preservative. It is added to many foods, especially acidic foods (those with a pH of less than about 4.5), to inhibit the growth of mold and bacteria.

Calcium and sodium propionate are also used as preservatives, chiefly in cheese, bread, and cakes. Some cheeses, especially hard ones of the Swiss cheese type, contain natural propionic acid. Propionate salts also inhibit the growth of mold and thereby prevent the food from spoiling.

Zinc 10-undecylenate, the zinc salt of 10-undecylenic acid,

$$CH_2{=}CH—(CH_2)_8—COOH$$

is used to treat fungal infections of the skin, especially *tinea pedis* (athlete's foot).

Aluminum acetate is blended into creams and lotions used for minor skin irritations, such as those caused by diaper rash, insect bites, acne, or exposure to alkali.

Because carboxylic acids are acidic, their structure in water depends on the pH, unlike the case for neutral compounds such as alcohols. When an alcohol, ROH, is dissolved in water, its structure remains ROH no matter what the pH. This is not the case for carboxylic acids. When a carboxylic acid is dissolved in water, the pH is always below 7. If we add enough base to the solution, (1) it will neutralize the acid (convert it to its salt), and (2) it will raise the pH. These two things always go together, and we can make the following useful statement:

> A carboxylic acid can exist in water only at a low pH; a carboxylate ion (salt) can exist in water only at a moderate or high pH.

The latter part of this statement is true because the carboxylate ion is always converted to the acid when the pH is lowered:

$$\underset{\text{Low pH}}{R-\underset{\underset{O}{\|}}{C}-OH} \underset{H_3O^+}{\overset{OH^-}{\rightleftharpoons}} \underset{\text{High pH}}{R-\underset{\underset{O}{\|}}{C}-O^-}$$

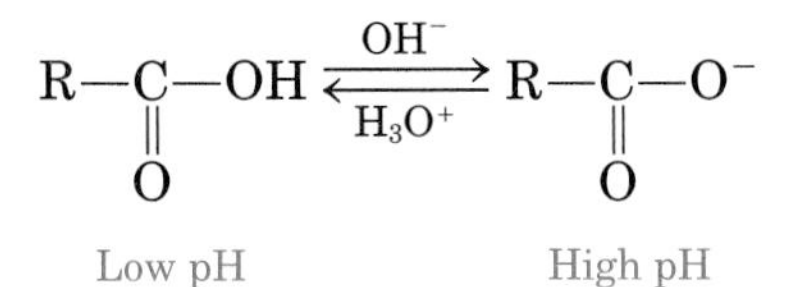

These interconversions are so easy and common that organic chemists consider carboxylic acids and their salts to be essentially interchangeable.

All this allows us to go back and forth easily. If we have a carboxylic acid, all we have to do is add NaOH and we get the salt. If we have the salt, we just add HCl and get the acid.

Most carboxylic acids are almost completely converted to the ionic form at a pH of 7. For example, acetic acid exists as more than 99% acetate ion at a pH of 7.

The acidity of carboxylic acids often allows us to separate them from other organic compounds. As we have seen, most organic compounds, including most carboxylic acids, are insoluble in water, but almost all sodium carboxylates are soluble. To separate a carboxylic acid from a mixture that contains other insoluble compounds, we just shake the mixture with an aqueous solution of NaOH. The carboxylic acid is neutralized by the OH^- and converted to its sodium salt, which dissolves in the water. The rest of the mixture is insoluble in water, and thus two layers form, which may easily be separated.

14.8 Nomenclature of Carboxylic Esters

Acids are proton donors. In most acids, whether organic or inorganic, the proton that is donated comes from an O—H bond (HCl is an exception, of course). **If we replace the H from such an acid with an R,** the resulting compound is called an **ester.** If the acid is a carboxylic acid, the ester is called a **carboxylic ester** (we discuss other esters in Section 14.12). Carboxylic esters therefore have the structure

$$R-\underset{\underset{O}{\|}}{C}-O-R' \quad \text{(abbreviated RCOOR}'\text{)}$$

As usual, we use R′ to show that the two R groups can be the same or different.

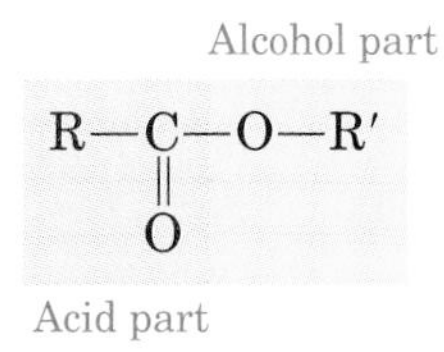

It is useful to think of a carboxylic ester as being derived from a carboxylic acid (the RCO part) and an alcohol (the OR′ part):

Alcohol part

$$\mathrm{R{-}\overset{\displaystyle O}{\overset{\|}{C}}{-}O{-}R'}$$

Acid part

We will see in Section 14.10 that carboxylic esters can in fact be made from a carboxylic acid and an alcohol. In naming a carboxylic ester, it is not wise to start by trying to name the ester as a whole. Esters are named as derivatives of the corresponding carboxylic acids, and the best way to begin is by naming the acid, using either the IUPAC name or the common name. After that, the procedure is the same as for naming salts. The "-ic acid" is changed to "-ate," and this word is the second word in the ester name. The first word is the name of the R′ group.

When a molecule contains both a carboxyl and a hydroxyl group, a cyclic ester, called a **lactone,** can be formed:

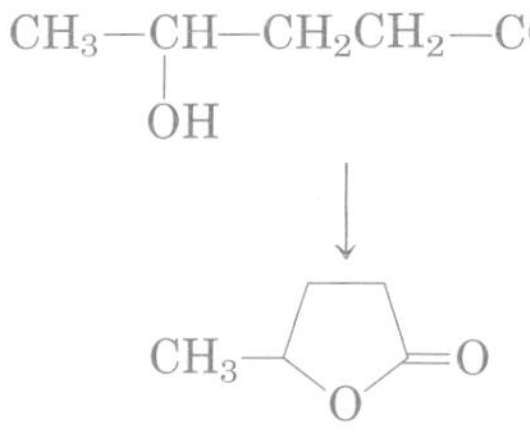

$$\mathrm{CH_3{-}\underset{\displaystyle OH}{\underset{|}{CH}}{-}CH_2CH_2{-}COOH}$$

$$\downarrow$$

CH$_3$– (five-membered ring with ring O) =O

EXAMPLE

Name the esters shown.

(a) $\mathrm{CH_3{-}\underset{\displaystyle O}{\underset{\|}{C}}{-}O{-}CH_2CH_3}$ Ethyl acetate

The corresponding acid, CH_3COOH, is acet*ic acid.* We change this to acet*ate.* The R group is ethyl, so the ester is called ethyl acetate. (The IUPAC name is ethyl ethanoate.)

(b) $\mathrm{CH_3CH_2CH_2{-}\underset{\displaystyle O}{\underset{\|}{C}}{-}O{-}CH_2CH_2CH_2CH_3}$ *n*-Butyl butyrate

The acid is butyric acid, and the alcohol is *n*-butyl alcohol. Thus the ester is *n*-butyl butyrate.

(c) (benzene ring with OH) $\mathrm{{-}\underset{\displaystyle O}{\underset{\|}{C}}{-}O{-}\overset{\displaystyle CH_3}{\overset{|}{\underset{\displaystyle CH_3}{\underset{|}{C}}}}{-}CH_3}$ *tert*-Butyl salicylate

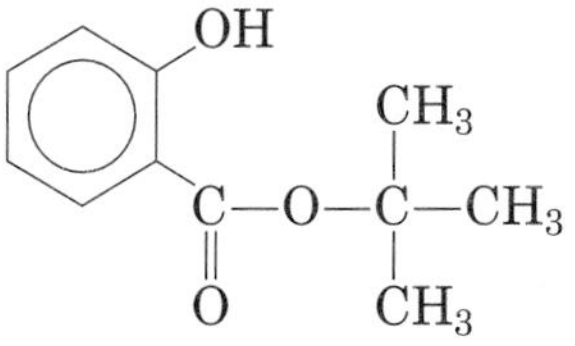

(d) $\mathrm{CH_3{-}\underset{\displaystyle OH}{\underset{|}{CH}}{-}\underset{\displaystyle O}{\underset{\|}{C}}{-}O}$–(benzene ring) Phenyl lactate

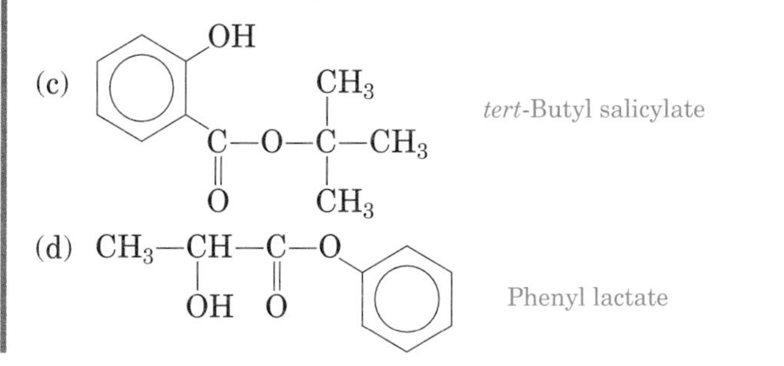

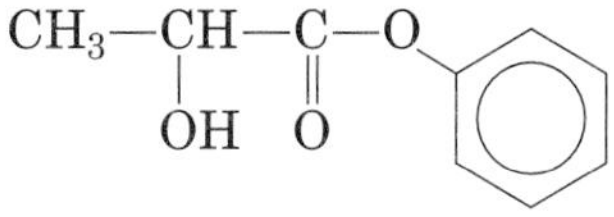

The last example shows that the R′ of a carboxylic ester can be an aromatic ring. In such a case the "alcohol" portion of the ester is a phenol. This type of ester is called a *phenolic ester.*

Problem 14.4

Name these compounds:

(a) $CH_3CH_2-\overset{\overset{O}{\|}}{C}-O-CH_2CH_3$ (b) $C_6H_5-\overset{\overset{O}{\|}}{C}-O-\underset{\underset{CH_3}{|}}{\overset{\overset{CH_3}{|}}{C}}-CH_3$

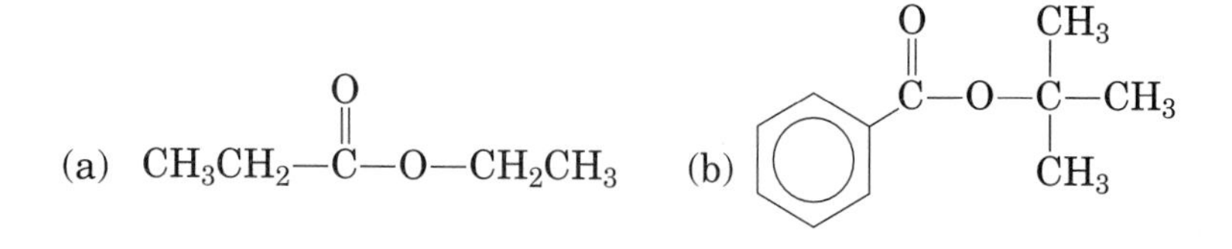

14.9 Physical Properties of Carboxylic Esters

The boiling points of carboxylic esters are not only lower than those of the corresponding carboxylic acids, but even lower than those of the alcohols.

Because ester molecules cannot form hydrogen bonds with each other, esters have lower boiling points than compounds of comparable molecular weight that do form hydrogen bonds (Table 14.3):

$$R-C(=O)-O-R' \qquad R'-O-C(=O)-R$$

Carboxylic ester — Carboxylic ester — No hydrogen bond possible

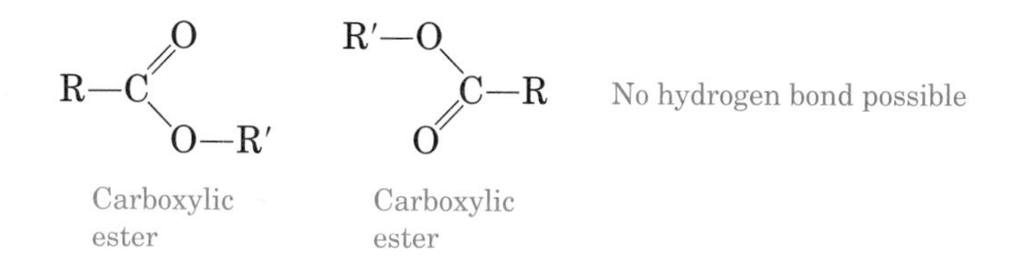

Since carboxylic esters can form hydrogen bonds with water:

$$R-C(-O-R')=O\cdots H-O-H$$

Carboxylic ester — Water

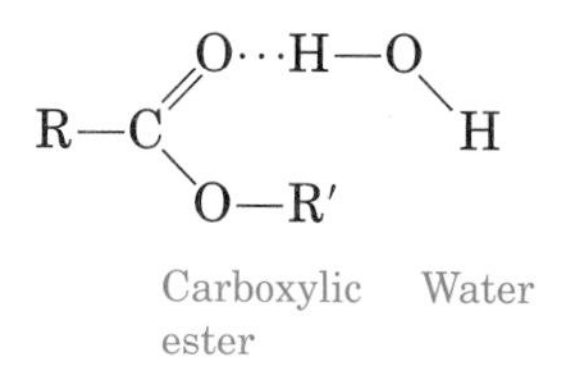

The low-molecular-weight esters, such as methyl acetate, are completely soluble in water. Solubility decreases rapidly, however, as molecular weight (and hence size of the nonpolar portion of the molecule) increases, and larger esters are insoluble in water.

we are not surprised to find that their solubility in water is about the same as that of carboxylic acids of the same molecular weight.

In Section 14.3 we learned that carboxylic acids have very disagreeable odors. In sharp contrast, the odors of the simple carboxylic esters

Table 14.3 Boiling Points of Some Esters Compared with Compounds of Similar Molecular Weight

Compound		MW	BP(°C)
Butane	$CH_3CH_2CH_2CH_3$	58	0
1-Propanol	$CH_3CH_2CH_2-OH$	60	97
Acetic acid	CH_3COOH	60	118
Methyl formate	$HCOOCH_3$	60	32
Hexane	$CH_3(CH_2)_4CH_3$	86	69
1-Pentanol	$CH_3(CH_2)_4-OH$	88	137
Butyric acid	$CH_3(CH_2)_2-COOH$	88	165
Ethyl acetate	$CH_3COOCH_2CH_3$	88	77

Box 14C

Carboxylic Esters as Flavoring Agents

The sweet and pleasant odors and tastes of many foods are due to complex mixtures of organic compounds, of which carboxylic esters generally are the most prevalent. Manufacturers of such products as soft drinks, ice cream, and maraschino cherries would like these products to taste and smell as much like the natural flavors as possible. In some cases they use extracts from the natural foods, but these are often too expensive, and artificial flavoring agents are very common. However, to reproduce synthetically the exact mixture found in a pineapple, for example, would be even more costly than using natural pineapple extract, and manufacturers have found that this is usually not necessary. Often the addition of only one or a few compounds, usually esters, makes the ice cream or soft drink taste natural.

The following are some carboxylic esters used as flavoring agents.

Structure	Name	Flavor
$HCOOCH_2CH_3$	Ethyl formate	Rum
$CH_3COOCH_2CH_2CH(CH_3)_2$	Isopentyl acetate	Banana
$HCOO(CH_2)_7CH_3$	Octyl formate	Orange
$CH_3(CH_2)_4COOCH_2CH{=}CH_2$	Allyl hexanoate	Pineapple
$CH_3(CH_2)_3COOCH_2CH_3$	Ethyl valerate	Apple
C_6H_5—CH=CH—$COOCH_3$ (benzene ring)	Methyl cinnamate	Strawberry
benzene ring with NH_2 and $COOCH_3$ (ortho)	Methyl anthranilate	Grape
benzene ring with OH and $COOCH_3$ (ortho)	Methyl salicylate	Wintergreen

are generally regarded as very pleasant. As a class, the esters have the most agreeable odors of all organic compounds. Many of the odors of fruits and flowers result from mixtures of carboxylic esters, and many esters (either naturally derived or synthetic) are used in perfumes and food flavorings (Box 14C).

Some products that contain ethyl acetate, a common carboxylic ester. (Photograph by Charles D. Winters.)

14.10 Preparation of Carboxylic Esters

We have learned that a carboxylic ester can be considered as a combination of a carboxylic acid and an alcohol (or a phenol). In fact, carboxylic esters can be synthesized by heating a mixture of a carboxylic acid and an alcohol, provided that an acid catalyst (usually H_2SO_4) is present.

General reaction:

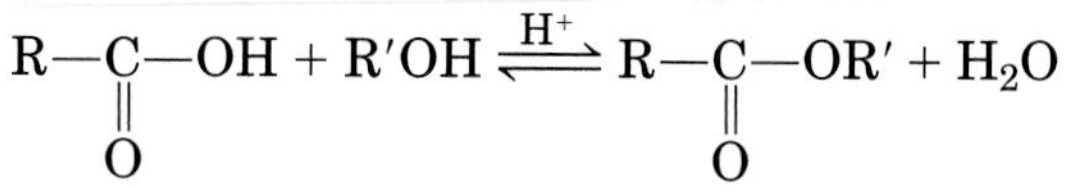

$$R{-}\underset{\underset{O}{\|}}{C}{-}OH + R'OH \underset{}{\overset{H^+}{\rightleftharpoons}} R{-}\underset{\underset{O}{\|}}{C}{-}OR' + H_2O$$

Specific example:

$$CH_3{-}\underset{\underset{O}{\|}}{C}{-}OH + CH_3CH_2OH \overset{H^+}{\rightleftharpoons} CH_3{-}\underset{\underset{O}{\|}}{C}{-}OCH_2CH_3 + H_2O$$

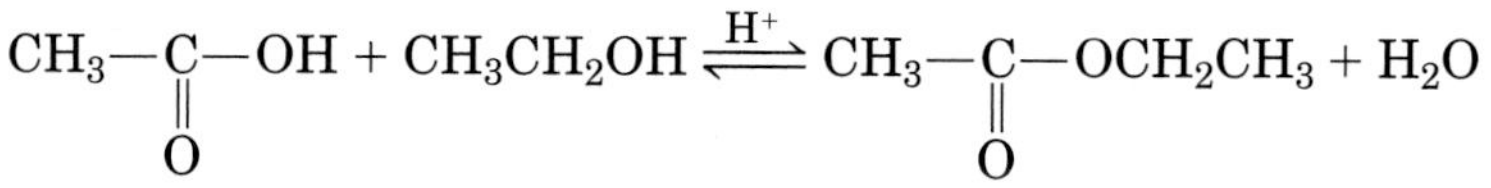

Acetic acid — Ethanol — Ethyl acetate

In this reaction, called **esterification,** a molecule of water comes out, and the carboxylic acid and alcohol combine to form the ester:

$$R{-}\underset{\underset{O}{\|}}{\overset{\overset{O-H}{|}}{C}} \quad \overset{\overset{H}{|}}{O}{-}R' \rightleftharpoons R{-}\underset{\underset{O}{\|}}{C}{-}O{-}R' + H{-}O{-}H$$

It is known that in most cases, the O that comes out comes from the acid and not from the alcohol (note that the same ester would be formed even if it happened the other way).

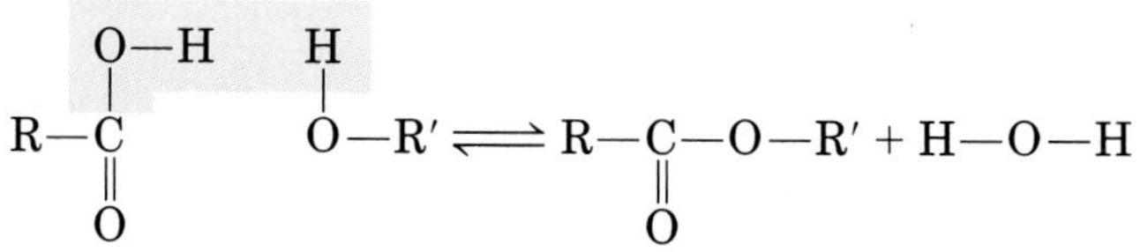

EXAMPLE

Draw the structural formula of the carboxylic ester obtained from each combination:

(a) $CH_3COOH + CH_3OH$

(b) $CH_3CH_2COOH +$ C_6H_5OH (phenol)

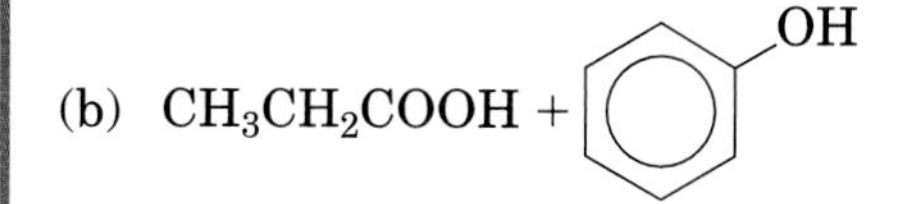

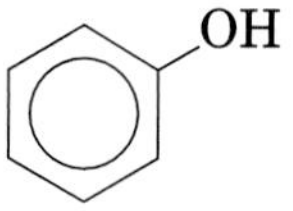

(c) $HOOC{-}CH_2CH_2{-}COOH + CH_3CH_2OH$

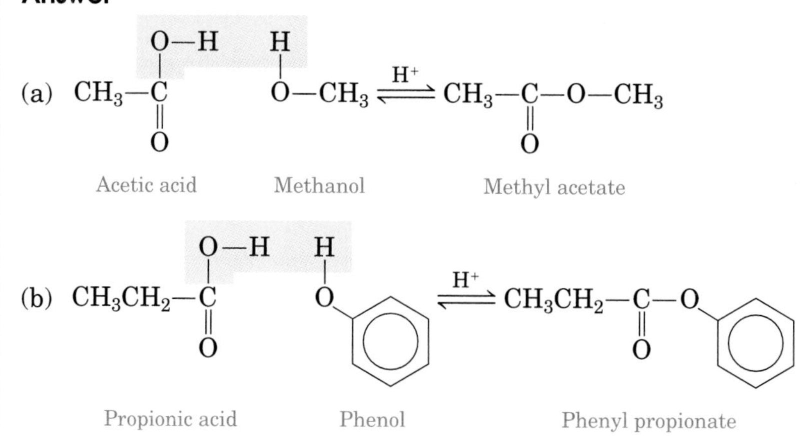

Answer

(a) 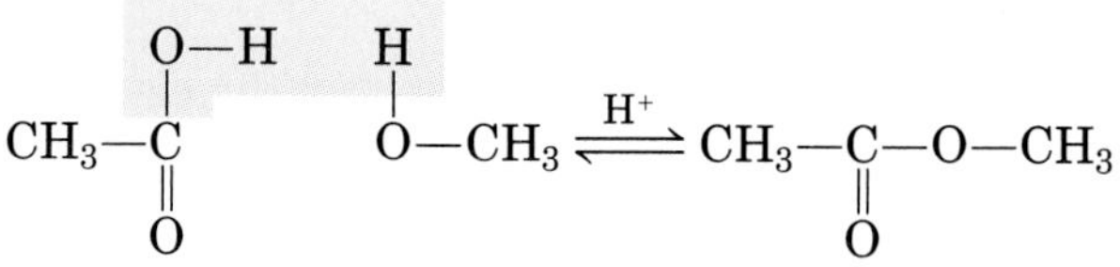 $$CH_3{-}\underset{\underset{O}{\|}}{\overset{\overset{O-H}{|}}{C}} \quad \overset{\overset{H}{|}}{O}{-}CH_3 \overset{H^+}{\rightleftharpoons} CH_3{-}\underset{\underset{O}{\|}}{C}{-}O{-}CH_3$$

Acetic acid — Methanol — Methyl acetate

(b) 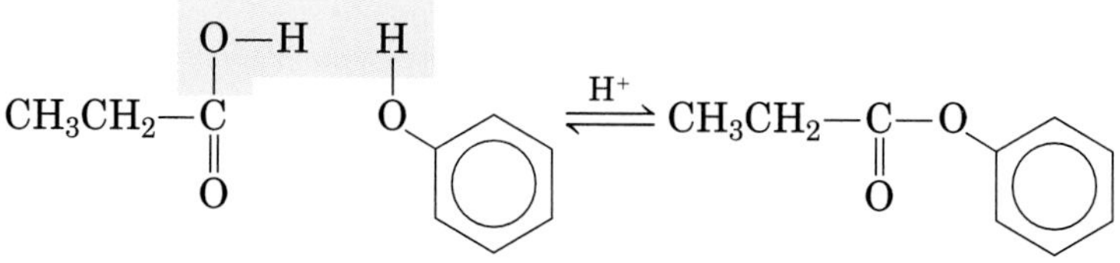 $$CH_3CH_2{-}\underset{\underset{O}{\|}}{\overset{\overset{O-H}{|}}{C}} \quad \overset{\overset{H}{|}}{O}{-}C_6H_5 \overset{H^+}{\rightleftharpoons} CH_3CH_2{-}\underset{\underset{O}{\|}}{C}{-}O{-}C_6H_5$$

Propionic acid — Phenol — Phenyl propionate

(c) Monoester:

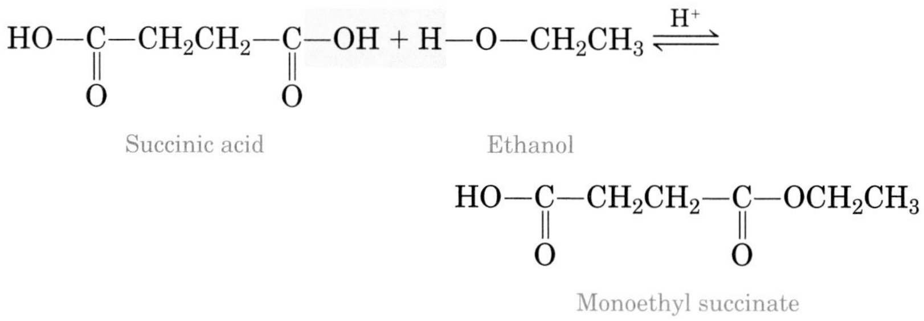

$$HO-\underset{\underset{O}{\|}}{C}-CH_2CH_2-\underset{\underset{O}{\|}}{C}-OH + H-O-CH_2CH_3 \overset{H^+}{\rightleftharpoons}$$

Succinic acid　　Ethanol

$$HO-\underset{\underset{O}{\|}}{C}-CH_2CH_2-\underset{\underset{O}{\|}}{C}-OCH_2CH_3$$

Monoethyl succinate

(d) Diester:

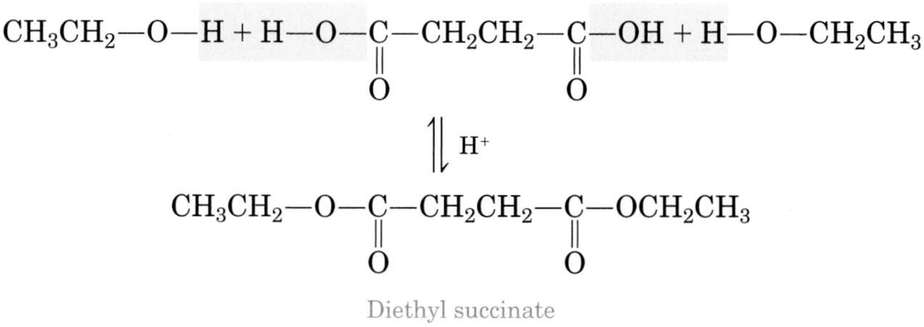

$$CH_3CH_2-O-H + H-O-\underset{\underset{O}{\|}}{C}-CH_2CH_2-\underset{\underset{O}{\|}}{C}-OH + H-O-CH_2CH_3$$

$$\Big\downarrow\!\!\uparrow\ H^+$$

$$CH_3CH_2-O-\underset{\underset{O}{\|}}{C}-CH_2CH_2-\underset{\underset{O}{\|}}{C}-OCH_2CH_3$$

Diethyl succinate

Problem 14.5

Show the structural formulas of the carboxylic esters obtained from a combination of these carboxylic acids and alcohols:

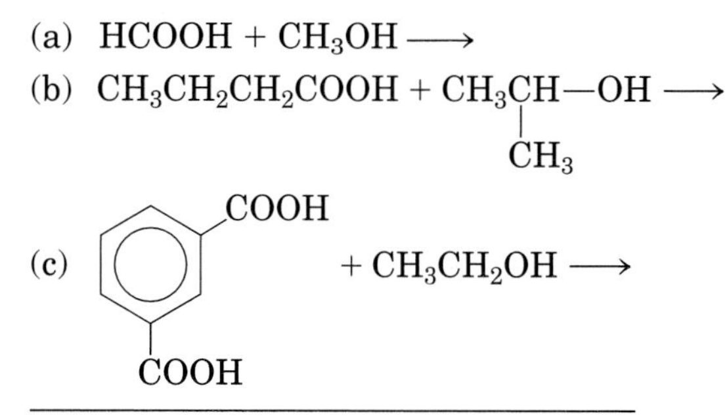

(a) $HCOOH + CH_3OH \longrightarrow$

(b) $CH_3CH_2CH_2COOH + CH_3\underset{\underset{CH_3}{|}}{C}H-OH \longrightarrow$

(c) 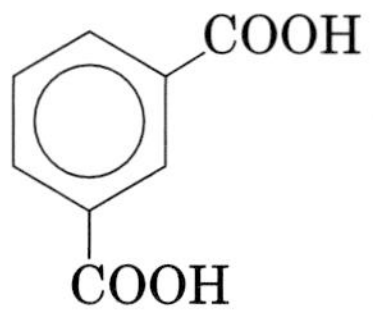 (benzene ring with COOH groups at positions 1 and 3) $+ CH_3CH_2OH \longrightarrow$

COOH
COOH

EXAMPLE

Identify the acid and alcohol (or phenol) portions of these carboxylic esters:

(a) $CH_3CH_2-\underset{\underset{O}{\|}}{C}-O-CH_2-\underset{\underset{CH_3}{|}}{C}H-CH_3$

(b) 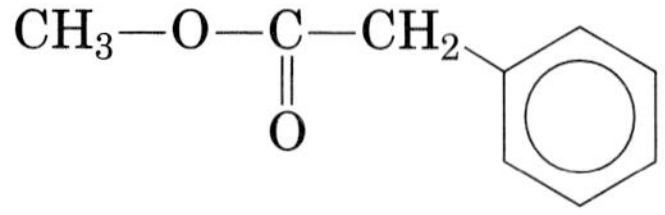 $CH_3-O-\underset{\underset{O}{\|}}{C}-CH_2-C_6H_5$

(c) Oil of wintergreen (benzene ring bearing OH and $C(=O)-O-CH_3$)

(d) Aspirin (benzene ring bearing $O-C(=O)-CH_3$ and $C(=O)-OH$)

Answer • The alcohol or phenol portions are shown in color:

(a) $CH_3CH_2-C(=O)-\boxed{O-CH_2-CH(CH_3)-CH_3}$

(b) $\boxed{CH_3-O}-C(=O)-CH_2-C_6H_5$

(c) $C_6H_4(OH)-C(=O)\boxed{-O-CH_3}$

(d) $\boxed{C_6H_4(COOH)-O}-C(=O)-CH_3$ (phenol portion: ring, O, and C(=O)—OH)

Problem **14.6**

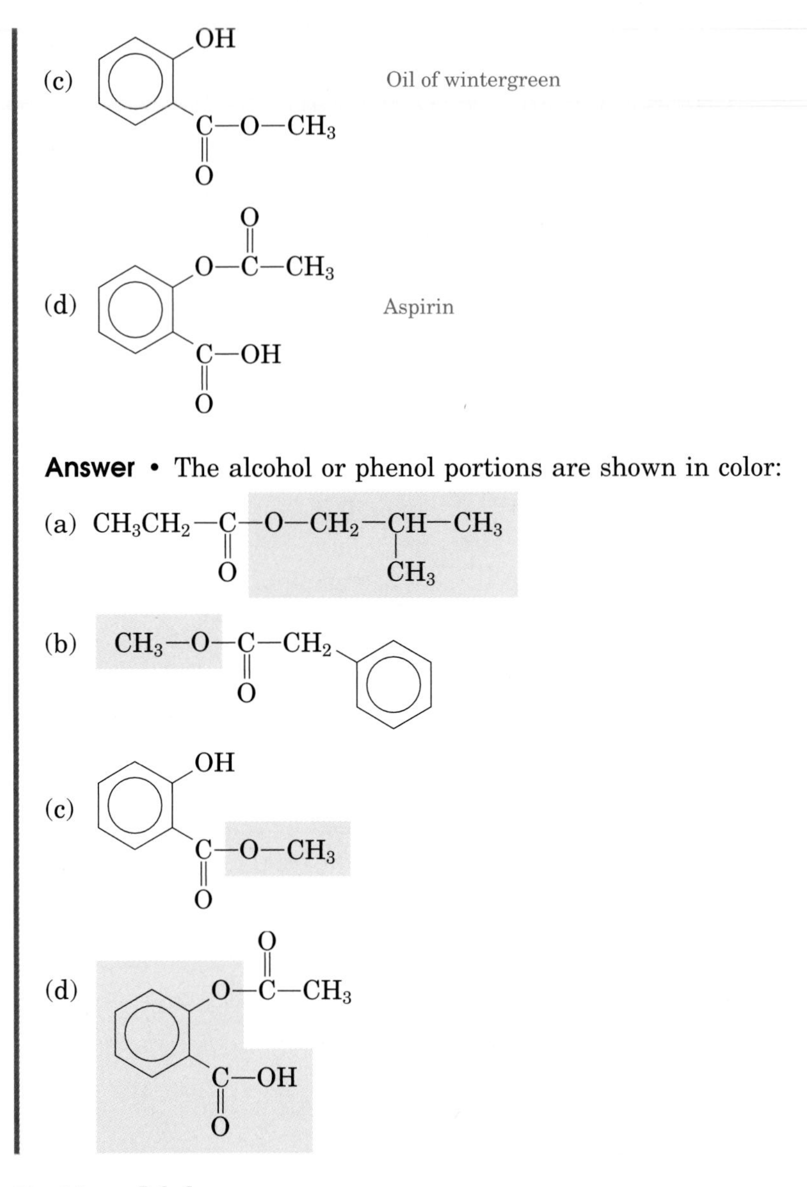

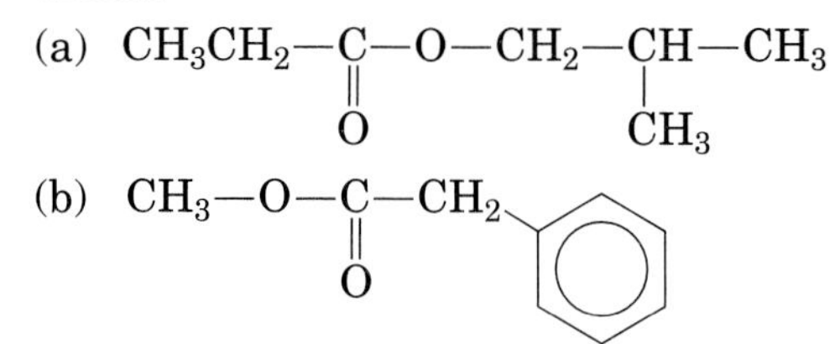

Identify the acid and the alcohol or phenol portions of these carboxylic esters:

(a) $CH_3CH_2CH_2-C(=O)-O-C_6H_5$ (b) $C_5H_9-O-C(=O)-CH_2-CH(CH_3)-CH_3$ (cyclopentyl ester)

Although a great many carboxylic esters have been synthesized simply by heating a carboxylic acid with an alcohol in the presence of a strong acid catalyst, the reaction does have its problems: It is an equilibrium.

Box 14D

Aspirin

For centuries, many people knew that they could relieve pain by chewing willow bark (Hippocrates recommended it around 400 BC). In 1860 an organic chemist extracted the compound salicylic acid from willow bark and soon discovered that it was this compound that was responsible for pain relief. Physicians immediately began to prescribe salicylic acid because it worked well, but it did have serious drawbacks: It had a bad (sour) taste and irritated the mouth and stomach lining.

Salicylic acid is both a carboxylic acid and a phenol. Because it is a phenol, it forms an ester if treated with acetic anhydride:

$$C_6H_4(OH)(COOH) + CH_3{-}\overset{\overset{\displaystyle O}{\|}}{C}{-}O{-}\overset{\overset{\displaystyle O}{\|}}{C}{-}CH_3 \longrightarrow C_6H_4(O{-}\overset{\overset{\displaystyle O}{\|}}{C}{-}CH_3)(COOH) + CH_3COOH$$

Salicylic acid — Acetic anhydride — Aspirin (acetylsalicylic acid)

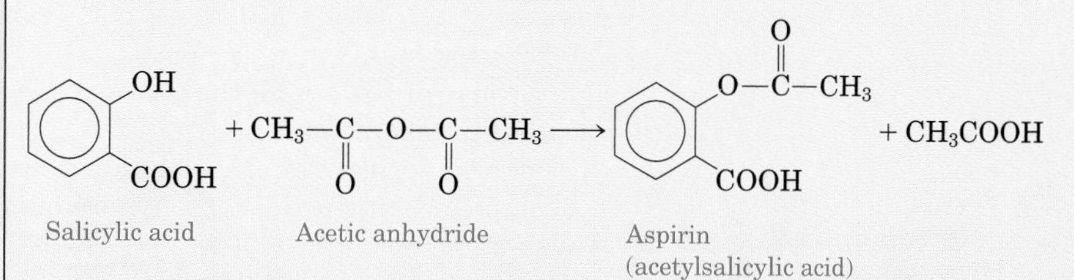

In 1893 a chemist at the Bayer Company in Germany did just that and converted salicylic acid to an ester that Bayer named "aspirin" (the chemical name is acetylsalicylic acid).

Aspirin turned out to be as good a pain-reliever as salicylic acid, or even better, and it did not have the same drawbacks. Its use grew very quickly, and it soon became the most widely used pain-reliever in the world, in fact, the most widely used drug of any kind.

Aspirin also reduces fever and the swelling caused by injuries and rheumatism. It is the most important drug in the treatment of arthritis, although it only relieves symptoms and does not cure the disease. It is one of the safest known drugs, and side effects are few.

However, as with most drugs, aspirin does have some side effects. It can prevent synthesis of stomach mucus, leading to ulceration by stomach acid of the unprotected stomach walls. It also prevents the aggregation of platelets whose function is to form a gel-like plug when a blood vessel is cut (see Box 25B). This side effect is now being turned to beneficial preventative use. Hundreds of thousands of patients at risk, and even many people not at risk, take small daily doses of aspirin to reduce or prevent the occurrence of strokes and heart attacks.

The drug is sold not only under the name aspirin, but is also the pain-relieving ingredient in such products as Anacin and Bufferin. Since all aspirin-containing products reduce pain equally well, the wise consumer buys the cheapest. See Box 17J for a discussion of how aspirin acts as an anti-inflammatory agent. For other nonprescription pain-relievers, see Box 15C.

Willow trees are a source of salicylic acid. (Photograph by Charles D. Winters.)

To make it go to completion, the equilibrium must be shifted to the right. In most cases this is done by boiling the mixture. When the mixture of four components is boiled, the one with the lowest boiling point evaporates first and can be removed by distillation. If this is either the carboxylic ester or the water, the equilibrium shifts to the right (Le Chatelier's principle), and the reaction goes to completion. Since carboxylic esters often have lower boiling points than either acids or alcohols, this is usually a feasible procedure. However, if the alcohol or acid has the lowest boiling point, boiling will not produce the ester, and some other way must be found.

Both acyl chlorides and anhydrides can be made from carboxylic acids.

Because of the equilibrium problem, carboxylic esters are more often prepared in laboratories not from the carboxylic acid but from two other types of acid derivatives: acyl chlorides and anhydrides. An **acyl chloride** is a compound in which a carbonyl group is connected to a chlorine:

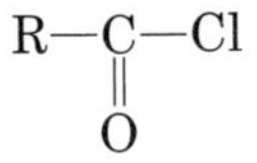

$$R{-}\underset{\underset{O}{\|}}{C}{-}Cl \qquad \text{An acyl chloride}$$

An **anhydride** can be thought of as a molecule formed when two carboxylic acids combine and lose water:

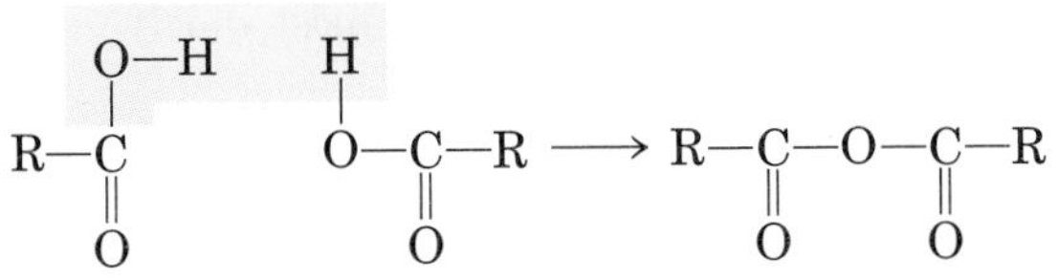

$$R{-}\underset{\underset{O}{\|}}{\overset{\overset{O-H}{|}}{C}} \quad \overset{\overset{H}{|}}{O}{-}\underset{\underset{O}{\|}}{C}{-}R \longrightarrow R{-}\underset{\underset{O}{\|}}{C}{-}O{-}\underset{\underset{O}{\|}}{C}{-}R$$

An anhydride

Both acyl chlorides and anhydrides react easily and quickly with alcohols or phenols to give carboxylic esters. These reactions are not reversible.

Since these reactions are not reversible, yields are higher than with carboxylic acids.

General reactions:

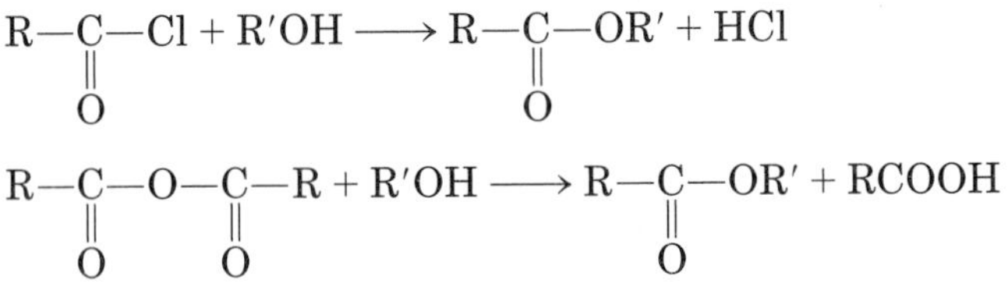

$$R{-}\underset{\underset{O}{\|}}{C}{-}Cl + R'OH \longrightarrow R{-}\underset{\underset{O}{\|}}{C}{-}OR' + HCl$$

$$R{-}\underset{\underset{O}{\|}}{C}{-}O{-}\underset{\underset{O}{\|}}{C}{-}R + R'OH \longrightarrow R{-}\underset{\underset{O}{\|}}{C}{-}OR' + RCOOH$$

Specific examples:

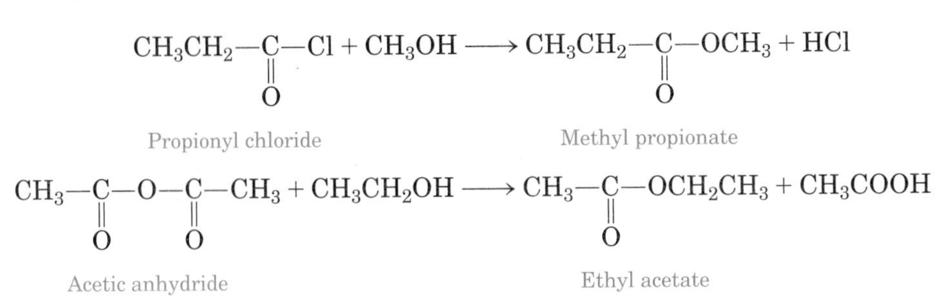

$$CH_3CH_2{-}\underset{\underset{O}{\|}}{C}{-}Cl + CH_3OH \longrightarrow CH_3CH_2{-}\underset{\underset{O}{\|}}{C}{-}OCH_3 + HCl$$

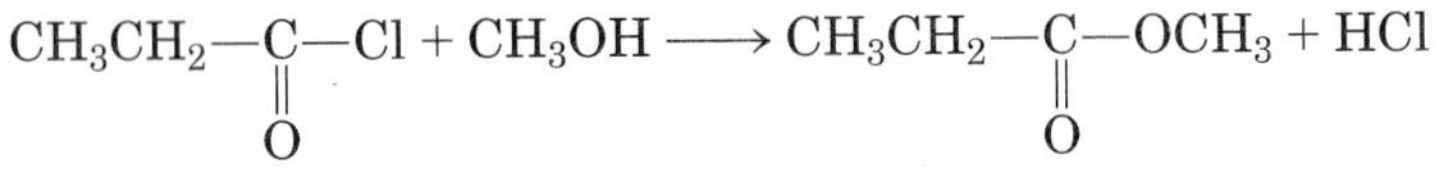

Propionyl chloride — Methyl propionate

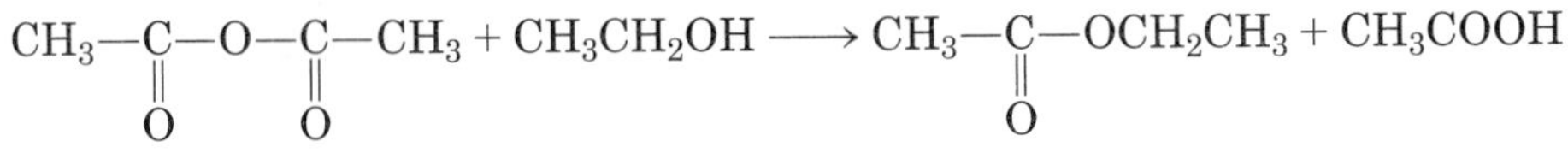

$$CH_3{-}\underset{\underset{O}{\|}}{C}{-}O{-}\underset{\underset{O}{\|}}{C}{-}CH_3 + CH_3CH_2OH \longrightarrow CH_3{-}\underset{\underset{O}{\|}}{C}{-}OCH_2CH_3 + CH_3COOH$$

Acetic anhydride — Ethyl acetate

Box 14E

Polyesters

There are two major types of polymers. In Section 11.7 we discussed one of these: addition polymers. The other type, **condensation polymers,** are constructed from molecules that have two (or more) functional groups. For example, we saw in the text that an alcohol can be connected to a carboxylic acid to form an ester. The ester group "ties" the two molecules together (this is why it is often called an ester *linkage*). If instead of using an ordinary carboxylic acid and alcohol we run the same reaction with a dicarboxylic acid and a diol, the molecules link at both ends, and a long chain (a polymer) is formed.

In the most important polyester, the two monomers are terephthalic acid and ethylene glycol:

$$—CH_2—OH + HO—\underset{\underset{O}{\|}}{C}—C_6H_4—\underset{\underset{O}{\|}}{C}—OH + HO—CH_2CH_2—OH + HO—\underset{\underset{O}{\|}}{C}—C_6H_4—\underset{\underset{O}{\|}}{C}—OH + HO—CH_2CH_2OH$$

Terephthalic acid   Ethylene glycol

$$\downarrow -H_2O$$

$$—CH_2—O—\underset{\underset{O}{\|}}{C}—C_6H_4—\underset{\underset{O}{\|}}{C}—O—CH_2CH_2—O—\underset{\underset{O}{\|}}{C}—C_6H_4—\underset{\underset{O}{\|}}{C}—O—CH_2CH_2—O—$$

Polyester (Dacron)

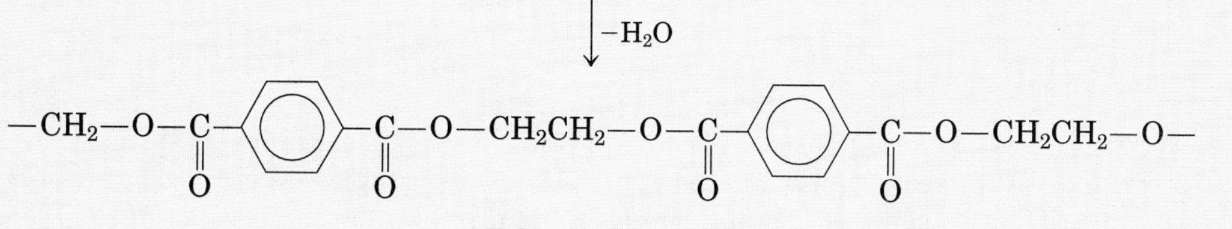

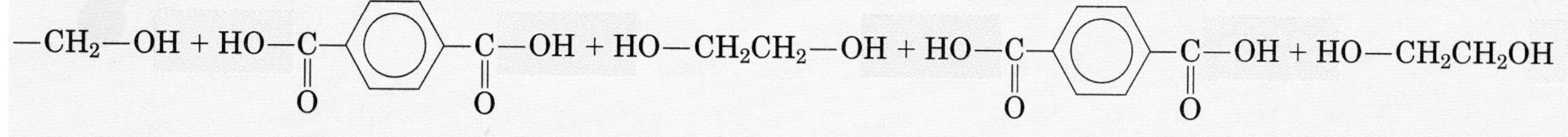

As you can see from the equation, a molecule of water is removed at each place a linkage is formed, and a chain in which all the links are ester groups has been formed. This particular polyester is called Dacron. It is widely used in clothing of all kinds (often just called polyester), as well as in tire cords, carpets, and many other products. The same polyester can also be produced in thin sheets, in which case it is called Mylar. Mylar is used to make recording and computer tape, among other things.

Dacron is also used to replace human arteries. In coronary bypass surgery, physicians prefer to use a section of the saphenous vein from the patient's leg to replace the diseased coronary artery. However, in nearly one third of patients this is unavailable because of previous surgery or previous disease (for example, phlebitis). In these cases synthetic arterial replacement is called for. Dacron and also Teflon (Table 11.2) have been used to make synthetic arteries. Dacron is also used to patch defective hearts.

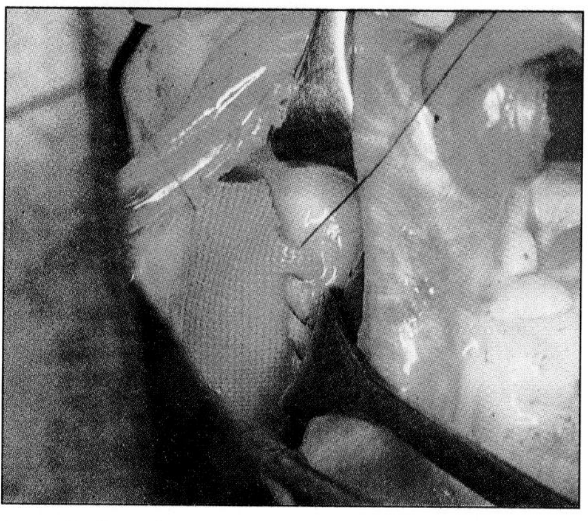

A Dacron patch is used to close an atrial septal defect in a heart patient. (Courtesy of Drs. James L. Monro and Gerald Shore and the Wolfe Medical Publications, London, England.)

Box 14F

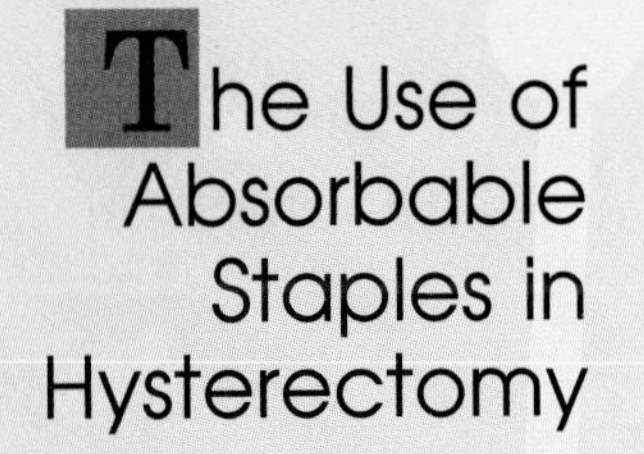

The Use of Absorbable Staples in Hysterectomy

Another type of polyester (see Box 14E), a copolymer of lactic acid and glycolic acid, provides an improved method for surgical closures of the vagina. Staples are manufactured from this polyester, which goes under the trade name Lactomer. Both lactic acid and glycolic acid have an OH group, so they form a copolymer as follows:

$$\underset{\text{Lactic acid}}{HO-\underset{\displaystyle CH_3}{\underset{|}{C}H}-COOH} + \underset{\text{Glycolic acid}}{HO-CH_2-COOH} \longrightarrow$$

$$HO-\underset{\displaystyle CH_3}{\underset{|}{C}H}-\underset{\displaystyle O}{\underset{\|}{C}}-O\left[CH_2-\underset{\displaystyle O}{\underset{\|}{C}}-O-\underset{\displaystyle CH_3}{\underset{|}{C}H}-\underset{\displaystyle O}{\underset{\|}{C}}-O\right]_n CH_2-\underset{\displaystyle O}{\underset{\|}{C}}-OH + nH_2O$$

The polyester starts to hydrolyze in the body after six to eight weeks, forming a mixture of the two acids, and the staples dissolve. Both acids are normal compounds in the body, so the staples do not produce a foreign material that would have to be removed by the phagocytes in the blood. The staples retain 50 percent of their initial strength for at least two months, which is more than adequate to heal the wounds in the vagina or uterine ligaments. They are also used in operations on the bladder and on the intestines.

The advantage of absorbable staples over sutures (absorbable or otherwise) is threefold: Stapling is a fast process and thus reduces the operating time and the ensuing trauma. It also greatly reduces the blood flow and consequent loss of blood during surgery. Finally, in suturing, each time the needle hits the tissue it produces a trauma. In stapling, the multiple trauma is reduced to one single shock, and healing is more rapid.

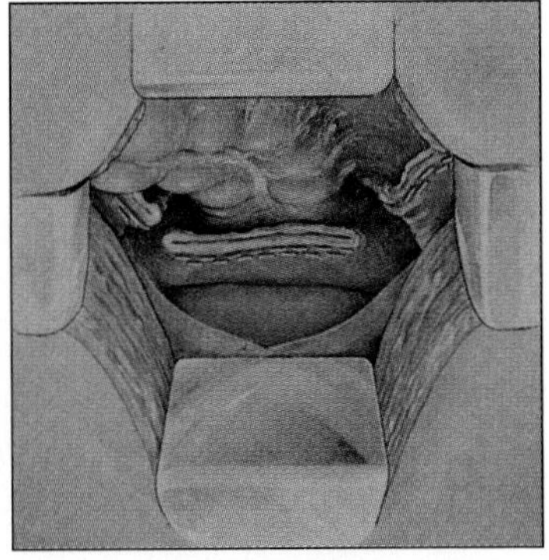

Staples used to close the vaginal vault in resection of the uterus. (Courtesy of United States Surgical Corporation.)

Problem **14.7**

Draw the structural formulas for the products:

(a) $CH_3{-}\underset{\underset{O}{\|}}{C}{-}Cl + CH_3CH_2OH \longrightarrow$

(b) 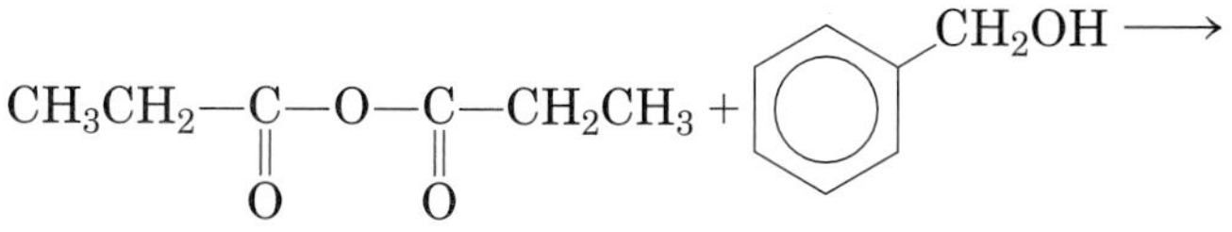 $CH_3CH_2{-}\underset{\underset{O}{\|}}{C}{-}O{-}\underset{\underset{O}{\|}}{C}{-}CH_2CH_3 + C_6H_5{-}CH_2OH \longrightarrow$

14.11 Hydrolysis of Carboxylic Esters

Carboxylic esters undergo a number of reactions, but we consider only one of them in this chapter, the one that is the most important, both in the laboratory and in the body. This reaction, called **hydrolysis,** is the reverse of the esterification reaction discussed in Section 14.10. In hydrolysis, a carboxylic ester reacts with water to produce the carboxylic acid and the alcohol or phenol from which the ester was originally formed.

Carboxylic esters are neutral compounds because the —COOH proton has been replaced by an R group.

To hydrolyze a carboxylic ester we usually need an acid or a base.

Acid-Catalyzed Hydrolysis

Strong acids are frequently used as catalysts to hydrolyze esters.

General reaction:

$$R{-}\underset{\underset{O}{\|}}{C}{-}OR' + H_2O \xrightleftharpoons{H^+} R{-}\underset{\underset{O}{\|}}{C}{-}OH + R'OH$$

Specific example:

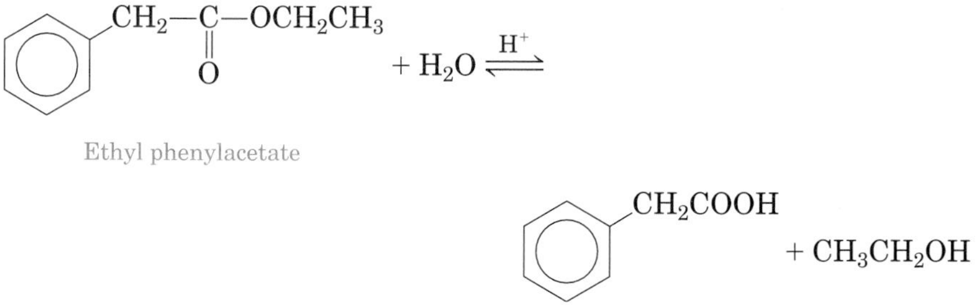

$$C_6H_5{-}CH_2{-}\underset{\underset{O}{\|}}{C}{-}OCH_2CH_3 + H_2O \xrightleftharpoons{H^+} C_6H_5{-}CH_2COOH + CH_3CH_2OH$$

Ethyl phenylacetate

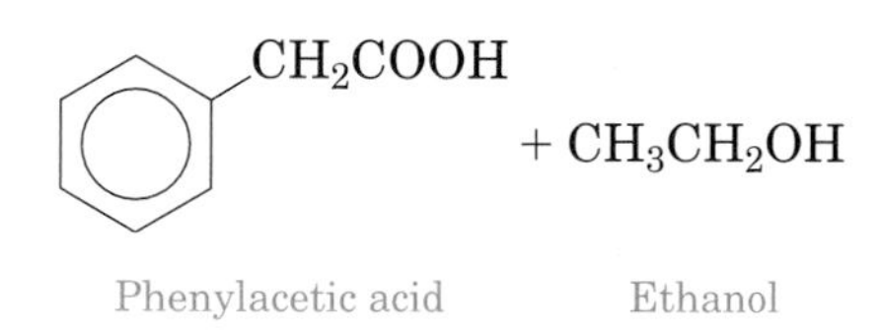

Phenylacetic acid

Ethanol

This, of course, is precisely the reverse of the formation of esters, discussed in Section 14.10, which means that it is also an equilibrium reaction (the same equilibrium, approached from the other direction). This time we desire to shift the equilibrium the other way, and that isn't always easy. One way is to use a large excess of water.

Problem 14.8

Draw the structural formulas for the products:

$$CH_3-\underset{\underset{O}{\parallel}}{C}-OCH_2CH_3 + H_2O \overset{H^+}{\rightleftharpoons}$$

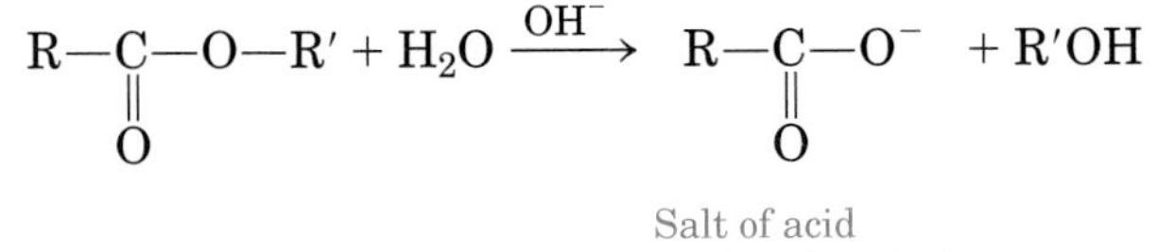

Hydrolysis with a Base

Esters can also be hydrolyzed by treatment with water and a base (typically, NaOH):

General reaction:

$$R-\underset{\underset{O}{\parallel}}{C}-O-R' + H_2O \xrightarrow{OH^-} R-\underset{\underset{O}{\parallel}}{C}-O^- + R'OH$$

Salt of acid (carboxylate ion)

Specific example:

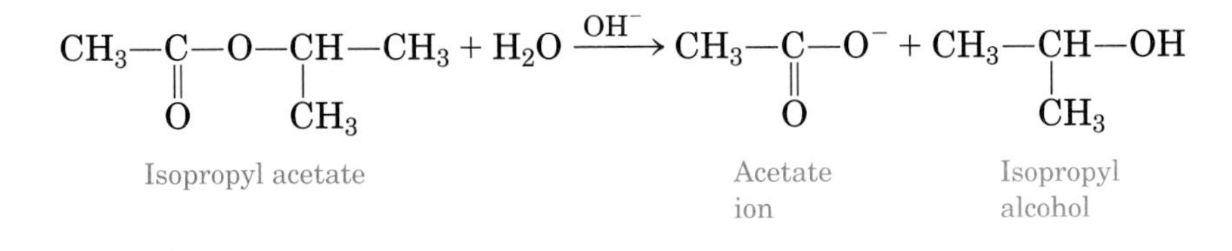

$$CH_3-\underset{\underset{O}{\parallel}}{C}-O-\underset{\underset{CH_3}{|}}{CH}-CH_3 + H_2O \xrightarrow{OH^-} CH_3-\underset{\underset{O}{\parallel}}{C}-O^- + CH_3-\underset{\underset{CH_3}{|}}{CH}-OH$$

Isopropyl acetate — Acetate ion — Isopropyl alcohol

Problem 14.9

Draw the structural formulas for the products:

$$C_6H_5-\underset{\underset{O}{\parallel}}{C}-O-CH_3 + H_2O \xrightarrow{OH^-}$$

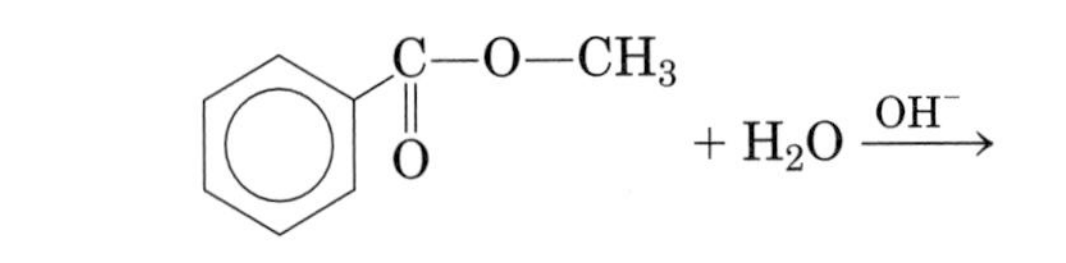

Since carboxylic acids cannot exist in basic solutions, it is only natural that in the presence of OH^- the product should be the carboxylate ion instead of the free acid.

Salts of carboxylic acids do not react with alcohols.

As with acid catalysis, the alcohol R′OH is a product. In this case, however, the other product is not the carboxylic acid itself, but its salt, the carboxylate ion. However, this certainly does not present any problem, since, once the hydrolysis is over, the carboxylate ion can easily be converted to the free acid (if that is what is wanted) simply by adding HCl (Sec. 14.7).

Because we get the salt rather than the free acid, basic hydrolysis of carboxylic esters is *not* reversible. Consequently, we needn't worry about shifting any equilibrium. Therefore, chemists usually prefer the basic to the acid-catalyzed method.

The **basic hydrolysis of carboxylic esters** has a special name, **saponification.** This word is derived from the soapmaking process, and, as we shall see in Section 17.3, soap is indeed made by saponification of fats, which are a type of carboxylic ester.

Enzyme-Catalyzed Hydrolysis

Biological organisms contain many carboxylic esters, and the body is constantly forming and hydrolyzing them. When the body hydrolyzes carboxylic esters, it uses neither acids nor bases, but enzyme catalysts, which are much more efficient. We will discuss the nature and action of enzymes in Chapter 19.

For example, the enzyme **lipase** hydrolyzes fats, which are esters of fatty acids and glycerol.

14.12 Esters and Anhydrides of Phosphoric Acid

We have defined an ester as the compound resulting when the H of an —OH acid is replaced by an R group. Most inorganic acids are —OH acids and can be converted to esters. The most important of these in biochemistry are the esters of phosphoric acid.

Phosphoric acid, H_3PO_4, is a triprotic acid and gives up its protons one at a time to sufficiently strong bases:

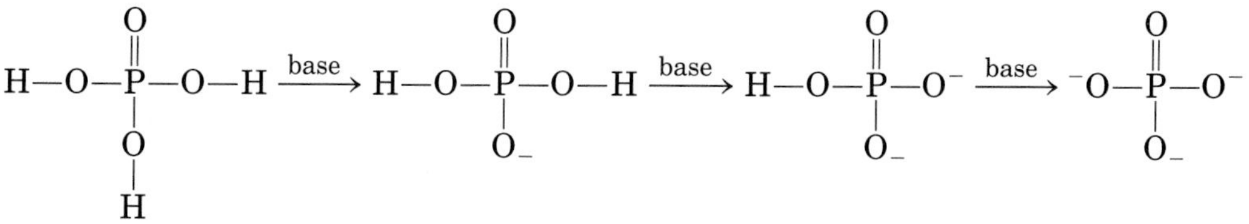

$$\underset{\text{Phosphoric acid}}{H{-}O{-}\overset{\overset{\displaystyle O}{\|}}{\underset{\underset{\displaystyle H}{|}}{\underset{\displaystyle O}{\underset{|}{P}}}}{-}O{-}H} \xrightarrow{\text{base}} \underset{\text{Dihydrogen phosphate ion}}{H{-}O{-}\overset{\overset{\displaystyle O}{\|}}{\underset{\displaystyle O^-}{\underset{|}{P}}}{-}O{-}H} \xrightarrow{\text{base}} \underset{\text{Hydrogen phosphate ion}}{H{-}O{-}\overset{\overset{\displaystyle O}{\|}}{\underset{\displaystyle O^-}{\underset{|}{P}}}{-}O^-} \xrightarrow{\text{base}} \underset{\text{Phosphate ion}}{{}^-O{-}\overset{\overset{\displaystyle O}{\|}}{\underset{\displaystyle O^-}{\underset{|}{P}}}{-}O^-}$$

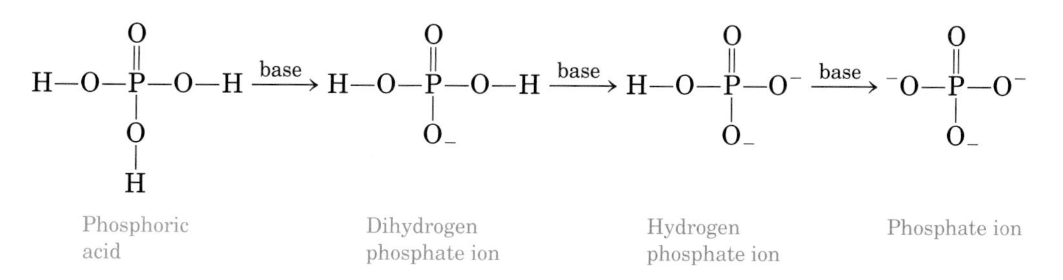

Because H_3PO_4 has three —OH groups, it can form mono-, di-, and triesters:

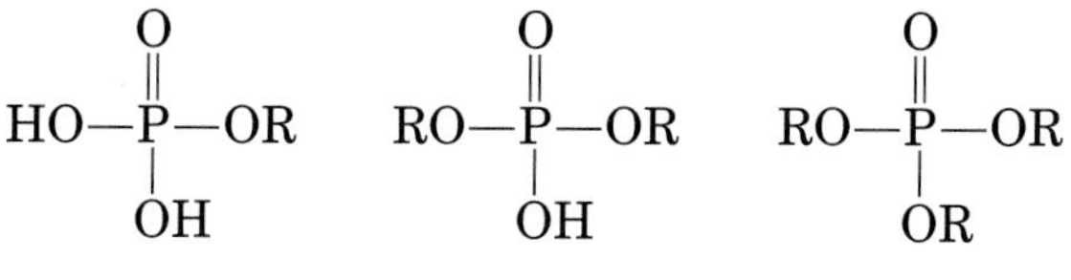

$$\underset{\text{Monoester}}{HO{-}\overset{\overset{\displaystyle O}{\|}}{\underset{\displaystyle OH}{\underset{|}{P}}}{-}OR} \qquad \underset{\text{Diester}}{RO{-}\overset{\overset{\displaystyle O}{\|}}{\underset{\displaystyle OH}{\underset{|}{P}}}{-}OR} \qquad \underset{\text{Triester}}{RO{-}\overset{\overset{\displaystyle O}{\|}}{\underset{\displaystyle OR}{\underset{|}{P}}}{-}OR}$$

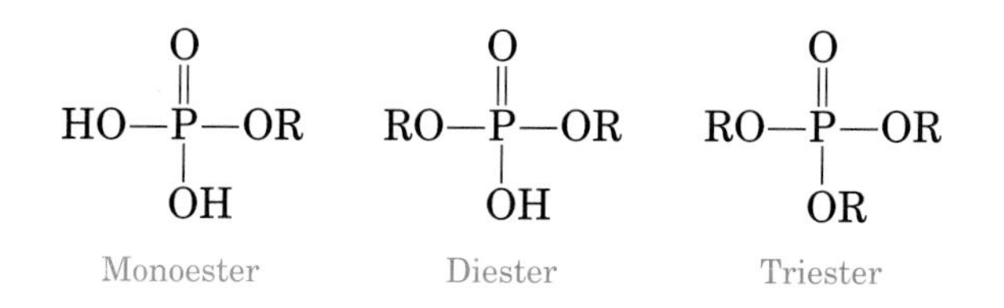

Specific example:

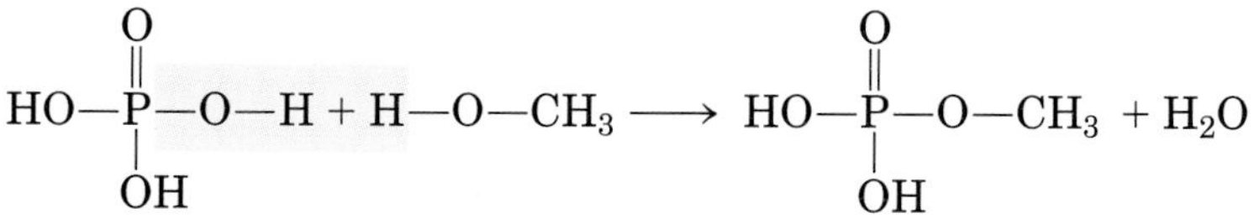

$$HO{-}\overset{\overset{\displaystyle O}{\|}}{\underset{\displaystyle OH}{\underset{|}{P}}}{-}O{-}H + H{-}O{-}CH_3 \longrightarrow \underset{\text{Monomethyl phosphate}}{HO{-}\overset{\overset{\displaystyle O}{\|}}{\underset{\displaystyle OH}{\underset{|}{P}}}{-}O{-}CH_3} + H_2O$$

Most formations of phosphate esters are catalyzed by enzymes.

Note that the mono- and diesters still have —OH groups. Therefore they are still acidic and can give their protons to a base.

Esters of phosphoric acid, especially the mono- and diesters, are important biological compounds. We shall see many examples of them in later chapters, including phospholipids (Sec. 17.4), glucose 6-phosphate (Sec. 21.2), and nucleic acids (Chap. 23). A typical example is glycerol 1-phosphate:

At body pH values, mono- and diphosphate esters usually exist in their ionic forms, as shown for glycerol 1-phosphate.

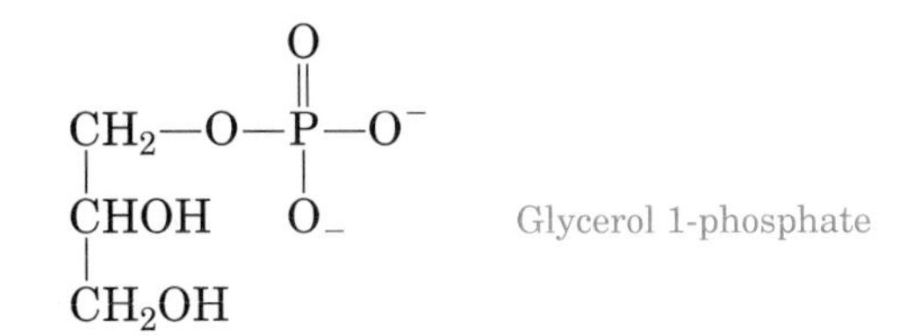

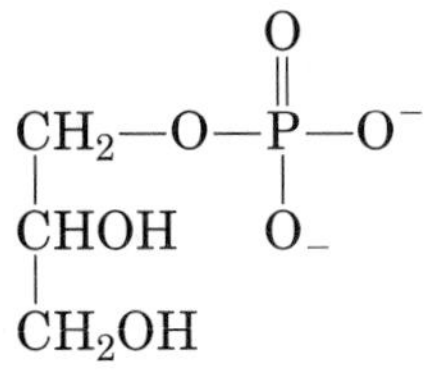

$$\begin{array}{l} CH_2{-}O{-}\overset{\displaystyle O}{\overset{\|}{P}}{-}O^- \\ \;|\qquad\quad\;\; | \\ CHOH \quad\; O^- \\ \;| \\ CH_2OH \end{array}$$

Glycerol 1-phosphate

Like carboxylic acids, phosphoric acid can also be converted to an anhydride, called pyrophosphoric acid, but unlike the anhydrides formed from carboxylic acids, pyrophosphoric acid still has —OH groups and so can be converted to another anhydride, triphosphoric acid:

At physiological pH values, pyrophosphoric acid exists as the pyrophosphate ion and triphosphoric acid exists as the triphosphate ion.

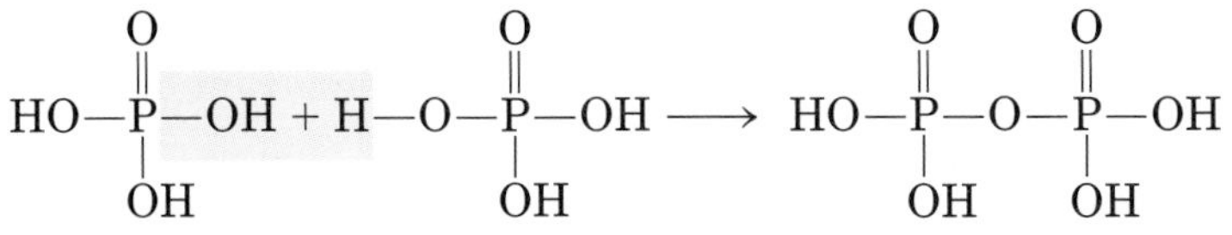

$$HO{-}\underset{OH}{\overset{O}{\overset{\|}{\underset{|}{P}}}}{-}OH + H{-}O{-}\underset{OH}{\overset{O}{\overset{\|}{\underset{|}{P}}}}{-}OH \longrightarrow HO{-}\underset{OH}{\overset{O}{\overset{\|}{\underset{|}{P}}}}{-}O{-}\underset{OH}{\overset{O}{\overset{\|}{\underset{|}{P}}}}{-}OH$$

Pyrophosphoric acid, $H_4P_2O_7$

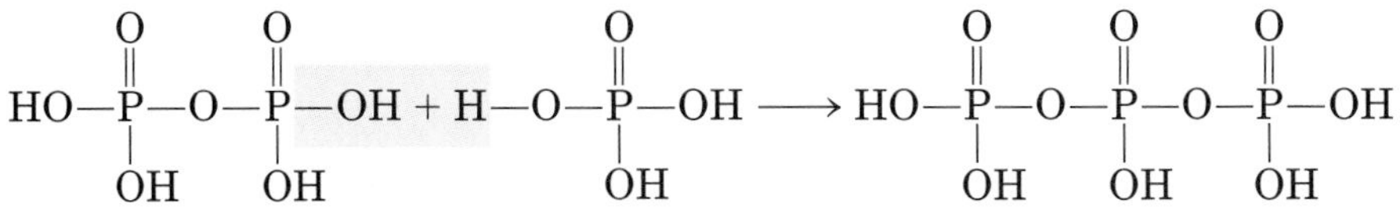

$$HO{-}\underset{OH}{\overset{O}{\overset{\|}{\underset{|}{P}}}}{-}O{-}\underset{OH}{\overset{O}{\overset{\|}{\underset{|}{P}}}}{-}OH + H{-}O{-}\underset{OH}{\overset{O}{\overset{\|}{\underset{|}{P}}}}{-}OH \longrightarrow HO{-}\underset{OH}{\overset{O}{\overset{\|}{\underset{|}{P}}}}{-}O{-}\underset{OH}{\overset{O}{\overset{\|}{\underset{|}{P}}}}{-}O{-}\underset{OH}{\overset{O}{\overset{\|}{\underset{|}{P}}}}{-}OH$$

Triphosphoric acid, $H_5P_3O_{10}$

These two anhydrides of phosphoric acid are extremely important in biochemistry. Or rather, it is their esters that are important. Both pyrophosphoric acid and triphosphoric acid still have —OH groups (which is why they are acids as well as anhydrides) and so form salts and esters. In the body these are generally the monoesters, which we show in the salt form because that is how they exist at body pH values:

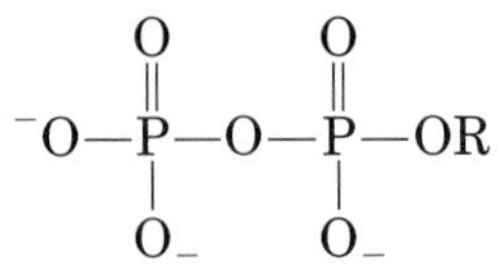

$$^-O{-}\underset{O^-}{\overset{O}{\overset{\|}{\underset{|}{P}}}}{-}O{-}\underset{O^-}{\overset{O}{\overset{\|}{\underset{|}{P}}}}{-}OR \qquad\qquad ^-O{-}\underset{O^-}{\overset{O}{\overset{\|}{\underset{|}{P}}}}{-}O{-}\underset{O^-}{\overset{O}{\overset{\|}{\underset{|}{P}}}}{-}O{-}\underset{O^-}{\overset{O}{\overset{\|}{\underset{|}{P}}}}{-}OR$$

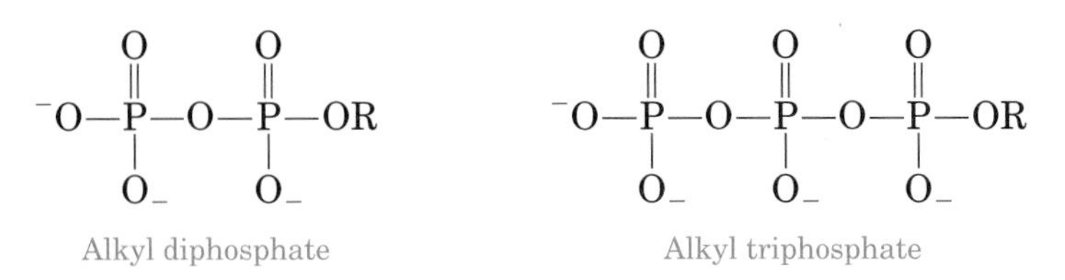

Alkyl diphosphate 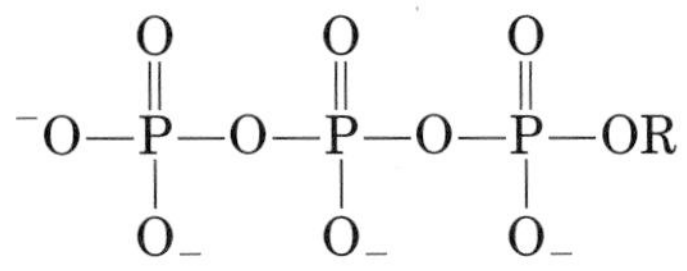 — Alkyl triphosphate

The most important diphosphate and triphosphate in the body are called ADP and ATP, respectively. The R group is rather complicated, and we need not look at it now. The complete structures of ADP and ATP are given in Section 20.3.

Anhydrides of phosphoric acid are important in the body because the anhydride bonds are used to transfer energy. The P—O—P linkage is a

high-energy linkage, which is just another way of saying that a lot of energy is given off when it is hydrolyzed. The body stores energy in part by forming P—O—P bonds. When it needs to use the energy, it hydrolyzes the linkage, so that the energy is given off, for example:

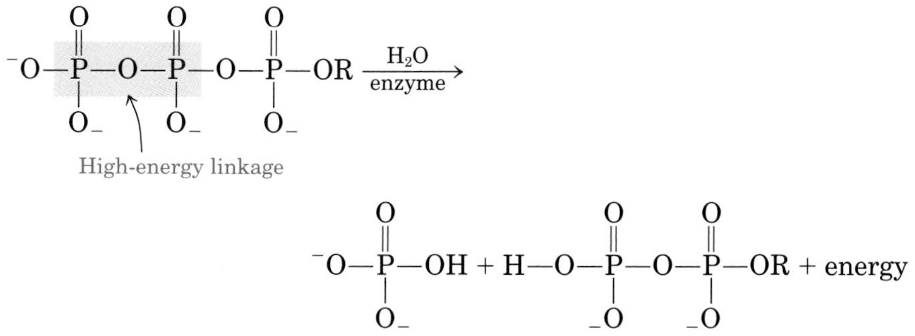

These phosphoric anhydrides do not react with water in the absence of a catalyst. In the body specific enzymes catalyze the reaction.

$$^{-}O-\overset{\overset{O}{\|}}{\underset{\underset{O^{-}}{|}}{P}}-O-\overset{\overset{O}{\|}}{\underset{\underset{O^{-}}{|}}{P}}-O-\overset{\overset{O}{\|}}{\underset{\underset{O^{-}}{|}}{P}}-OR \xrightarrow[\text{enzyme}]{H_2O}$$

High-energy linkage

$$^{-}O-\overset{\overset{O}{\|}}{\underset{\underset{O^{-}}{|}}{P}}-OH + H-O-\overset{\overset{O}{\|}}{\underset{\underset{^{-}O}{|}}{P}}-O-\overset{\overset{O}{\|}}{\underset{\underset{^{-}O}{|}}{P}}-OR + \text{energy}$$

Like the anhydrides of carboxylic acids (Sec. 14.10), anhydrides of phosphoric acid react with alcohols to give phosphate esters, for example:

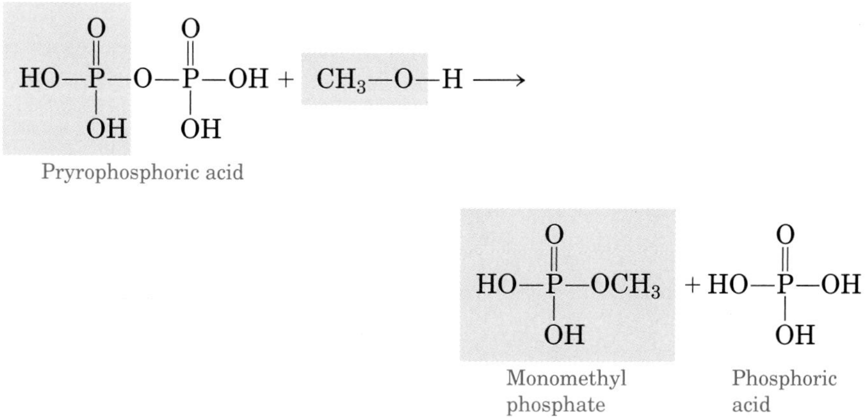

$$HO-\overset{\overset{O}{\|}}{\underset{\underset{OH}{|}}{P}}-O-\overset{\overset{O}{\|}}{\underset{\underset{OH}{|}}{P}}-OH + CH_3-O-H \longrightarrow$$

Pryrophosphoric acid

$$HO-\overset{\overset{O}{\|}}{\underset{\underset{OH}{|}}{P}}-OCH_3 + HO-\overset{\overset{O}{\|}}{\underset{\underset{OH}{|}}{P}}-OH$$

Monomethyl phosphate — Phosphoric acid

This is an important reaction in biochemistry. The body typically makes phosphate esters by this process. Because a phosphate group (shown in color) is transferred from the anhydride to the alcohol, this process is called **phosphorylation.** We shall see specific examples of phosphorylation in Chapters 20, 21, and 22.

Other Inorganic Esters

Although this section has discussed only esters of phosphoric acid, all other —OH acids also form esters (the general term is **inorganic ester**). In Box 12C we met nitroglycerine, an ester of nitric acid, and isoamyl nitrite, an ester of nitrous acid. Esters of sulfuric acid are also important in biochemistry, especially in acidic polysaccharides (see Sec. 16.12). Many detergents are esters of sulfuric acid (see Box 17C). Esters made from thiols, RSH, rather than alcohols, ROH, are important in biochemistry (see Secs. 20.3 and 22.3).

SUMMARY

Compounds that contain the **carboxyl group, —COOH,** are called **carboxylic acids.** Carboxylic acids can be named by the IUPAC system but are more often given common names. Because they form **dimers** by hydrogen bonding, they have high boiling points. They also have unpleasant odors. Their solubilities in water are about the same as those of other oxygen-containing organic compounds.

The most important chemical property of carboxylic acids is their acidity. They form salts when treated with strong bases like OH^- and even with weaker bases such as HCO_3^- and NH_3. These salts, called **carboxylates,** are soluble in water even when the acids themselves are not. A carboxylic acid can exist in water only at a low pH; a carboxylate ion can exist in water only at a moderate or high pH. Thus a carboxylic acid can always be converted to its salt by adding NaOH, and a carboxylate ion can always be converted to the acid by adding HCl.

An **ester** is made by replacing the H of an —OH acid with an R group. **Carboxylic esters** derived from carboxylic acids therefore have the formula **RCOOR′.** They are named by changing the "-ic acid" suffix to "-ate" and placing this new word after the name of the R′ group. **Carboxylate salts** are named the same way, except that the name of the metal ion becomes the first word. Because carboxylic ester molecules cannot form hydrogen bonds with each other, their boiling points are low, but their solubilities are similar to those of the carboxylic acids. They generally have very pleasant odors.

Carboxylic esters, RCOOR′, are prepared by the reaction between an alcohol (R′OH) and either an acyl halide (RCOCl), an anhydride (R—CO—O—CO—R), or a carboxylic acid (RCOOH). In the last case, a strong acid catalyst (such as H_2SO_4) is needed, and an equilibrium mixture of reactants and products is formed, which must be driven to completion. The most important reaction of carboxylic esters is their **hydrolysis** (cleavage by water), which may be carried out with water and either acids or bases. For synthetic purposes, the basic reaction is often preferred because it is not reversible. Basic ester hydrolysis is called **saponification.** In biological systems, ester hydrolysis is catalyzed by enzymes.

Esters and anhydrides of phosphoric acid, H_3PO_4, are very important in biochemistry. Phosphoric acid forms three series of esters: $(HO)_2PO(OR)$, $(HO)PO(OR)_2$, and $PO(OR)_3$. The P—O—P bond is a **high-energy bond,** and conversions between **alkyl diphosphates** (monoesters of pyrophosphoric acid, $H_4P_2O_7$) and **alkyl triphosphates** (monoesters of triphosphoric acid, $H_5P_3O_{10}$) are the major way the body transfers energy from one molecule to another. **Phosphorylation,** the transfer of a phosphate group from the —OH of one molecule to the —OH of another, is an important biological process.

Summary of Reactions

1. Reaction of carboxylic acids with bases (Sec. 14.6):

$$R-\underset{\displaystyle \overset{\|}{O}}{C}-OH + OH^- \longrightarrow R-\underset{\displaystyle \overset{\|}{O}}{C}-O^- + H_2O$$

$$R-\underset{\displaystyle \overset{\|}{O}}{C}-OH + NH_3 \longrightarrow R-\underset{\displaystyle \overset{\|}{O}}{C}-O^- + NH_4^+$$

2. Neutralization of carboxylate salts (Sec. 14.7):

$$R-\underset{\displaystyle \overset{\|}{O}}{C}-O^- + HCl \longrightarrow R-\underset{\displaystyle \overset{\|}{O}}{C}-OH + Cl^-$$

3. Preparation of carboxylic esters (Sec. 14.10):

a. From carboxylic acids and alcohols:

$$R-\underset{\displaystyle \overset{\|}{O}}{C}-OH + R'OH \underset{}{\overset{H^+}{\rightleftharpoons}} R-\underset{\displaystyle \overset{\|}{O}}{C}-OR' + H_2O$$

b. From acyl chlorides and alcohols:

$$R-\underset{\displaystyle \overset{\|}{O}}{C}-Cl + R'OH \longrightarrow R-\underset{\displaystyle \overset{\|}{O}}{C}-OR' + HCl$$

c. From anhydrides and alcohols:

$$R-\underset{\displaystyle \overset{\|}{O}}{C}-O-\underset{\displaystyle \overset{\|}{O}}{C}-R + R'OH \longrightarrow R-\underset{\displaystyle \overset{\|}{O}}{C}-OR' + R-\underset{\displaystyle \overset{\|}{O}}{C}-OH$$

4. Hydrolysis of carboxylic esters (Sec. 14.11):

a. Acid-catalyzed:

$$R-\underset{\displaystyle \overset{\|}{O}}{C}-OR' + H_2O \overset{H^+}{\rightleftharpoons} R-\underset{\displaystyle \overset{\|}{O}}{C}-OH + R'OH$$

b. With a base (saponification):

$$R-\underset{\displaystyle \overset{\|}{O}}{C}-OR' + H_2O \xrightarrow{OH^-} R-\underset{\displaystyle \overset{\|}{O}}{C}-O^- + R'OH$$

KEY TERMS

Acyl chloride (Sec. 14.10)
Acyl group (Sec. 14.1)
Anhydride (Sec. 14.10)
Carboxyl group (Sec. 14.1)
Carboxylate ion (Sec. 14.6)
Carboxylic acid (Sec. 14.1)
Carboxylic acid salt (Sec. 14.7)
Carboxylic ester (Sec. 14.8)
Condensation polymer (Box 14E)
Dimer (Sec. 14.3)
Ester (Sec. 14.8)
Ester hydrolysis (Sec. 14.11)
Esterification (Sec. 14.10)
Fatty acid (Sec. 14.2)
High-energy linkage (Sec. 14.12)
Hydrolysis (Sec. 14.11)
Inorganic ester (Sec. 14.12)
Lactone (Sec. 14.8)
Phenolic ester (Sec. 14.8)
Phosphate ester (Sec. 14.12)
Phosphoric anhydride (Sec. 14.12)
Phosphorylation (Sec. 14.12)
Polyester (Boxes 14E, 14F)
Saponification (Sec. 14.11)

PROBLEMS

Nomenclature of Carboxylic Acids

14.10 Give IUPAC names for

(a) $CH_3CH_2CH_2—COOH$

(b) $CH_3—CH(Cl)—CH_2—COOH$

(c) 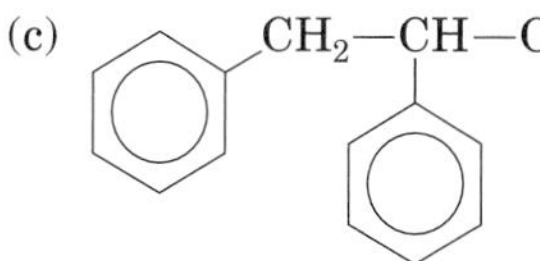
$CH_2—CH—COOH$

(d) $Cl—CH(Cl)—CH(Br)—CH(CH_3)—CH_2—COOH$

14.11 Draw structural formulas for
(a) 2-hydroxypentanoic acid
(b) 3-bromo-2-methylbutanoic acid
(c) 5-cyclohexylhexanoic acid
*(d) 3-methyl-3-heptenoic acid

14.12 Give common names for

(a) $H—COOH$

(b) $CH_3—(CH_2)_4—COOH$

(c) 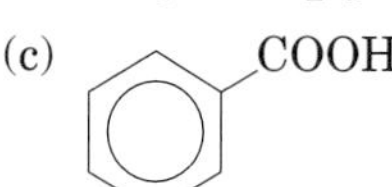
COOH

(d) $HOOC—CH_2CH_2—COOH$

(e) 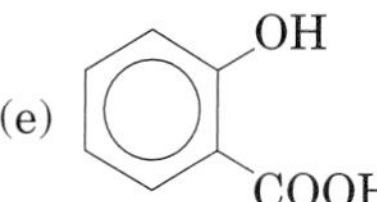
OH
COOH

(f) $CH_3—(CH_2)_{16}—COOH$

14.13 Draw the structural formulas for
(a) acetic acid
(b) propionic acid
(c) oxalic acid
(d) malonic acid
(e) salicylic acid
(f) caproic acid

14.14 Give the IUPAC names (common names given) for

(a) caproic acid, $CH_3(CH_2)_4COOH$

(b) capric acid, $CH_3(CH_2)_8COOH$

(c) pivalic acid, $CH_3—C(CH_3)_2—COOH$

14.15 Why was formic acid given that name?

14.16 (a) Define fatty acid.
(b) Why are the fatty acids given that name?

Physical Properties of Carboxylic Acids

14.17 Acetic acid has a boiling point of 118°C, considerably higher than that of 1-propanol (97°C), though they have the same molecular weight. Explain why.

14.18 Draw the structure of the dimer formed when two molecules of acetic acid form hydrogen bonds with each other.

***14.19** Malonic acid forms an internal hydrogen bond (the H of one —COOH group forms a hydrogen bond with an O of the other —COOH group). Draw a structural formula that shows this.

14.20 Rank the following compounds in decreasing order of solubility in water:
(a) $CH_3CH_2CH_2CH_2CH_2CH_2COOH$
(b) $CH_3CH_2CH_2CH_2CH_3$
(c) CH_3COOH

14.21 Of the organic compounds studied so far, which have particularly bad odors? Which have particularly pleasant odors?

14.22 Caproic acid has a solubility in water of 1 g per 100 g of H_2O (Table 14.1). Which part of the molecule contributes to the solubility in water, and which part prevents greater solubility?

Some important Carboxylic Acids

14.23 Draw structural formulas for (a) lactic acid (b) citric acid.

14.24 Lactic acid is a monocarboxylic acid. What would you call citric acid?

14.25 What is a medical use for Cl_3CCOOH?

14.26 Which carboxylic acid is present (as its salt) in spinach?

Preparation of Carboxylic Acids

***14.27** Write an equation to show how wine turns to vinegar if allowed to stand in the open air.

Acidity of Carboxylic Acids; Carboxylate Salts

14.28 Name these salts:

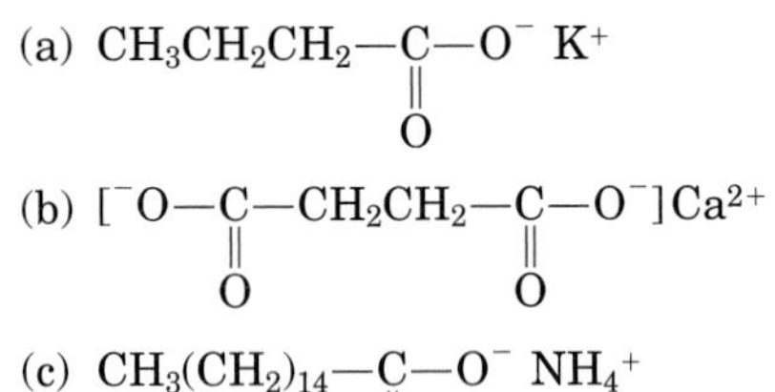

(a) $CH_3CH_2CH_2-\overset{\overset{\displaystyle O}{\|}}{C}-O^-\ K^+$ (C=O)

(b) $[^-O-C(=O)-CH_2CH_2-C(=O)-O^-]Ca^{2+}$

(c) $CH_3(CH_2)_{14}-C(=O)-O^-\ NH_4^+$

(d) $[CH_3(CH_2)_8-C(=O)-O^-]_2\,Mg^{2+}$

(e) $[^-O-C(=O)-C_6H_4-C(=O)-O^-]2Na^+$

(see Box 14E)

14.29 Draw structural formulas for
(a) potassium formate
(b) ammonium valerate
(c) monosodium oxalate
(d) iron(III) benzoate
(e) magnesium caproate
(f) dilithium maleate

14.30 Draw the structural formulas of malonic acid (a) at a very low pH, (b) at a medium pH, and (c) at a very high pH.

14.31 Complete these equations:

(a) $CH_3COOH + OH^- \longrightarrow$

(b) $HCOOH + NH_3 \longrightarrow$

(c) $HOOC-CH_2-COOH + \text{excess } OH^- \longrightarrow$

(d) $CH_3-CH(OH)-COOH + HCO_3^- \longrightarrow$

(e) $CH_3COO^- + HCl \longrightarrow$

(f) $C_6H_5-COO^- + H_2SO_4 \longrightarrow$

14.32 A small quantity of sodium acetate, $CH_3COO^-\ Na^+$, is added to an aqueous solution whose pH is 2. What happens to the salt? Draw its formula at this pH.

14.33 (a) Butyric acid is dissolved in water. Draw its structural formula. (b) Enough NaOH is added to raise the pH to 12. What happens to the butyric acid? Draw the formula now. (c) Enough HCl is added to lower the pH to 2. Draw the formula now.

14.34 You are given a mixture of caprylic acid and 1-heptanol, neither of which is soluble in water. Based on the information in this chapter, give a simple method for separating them. Describe in detail exactly what you would do.

Structure and Nomenclature of Carboxylic Esters

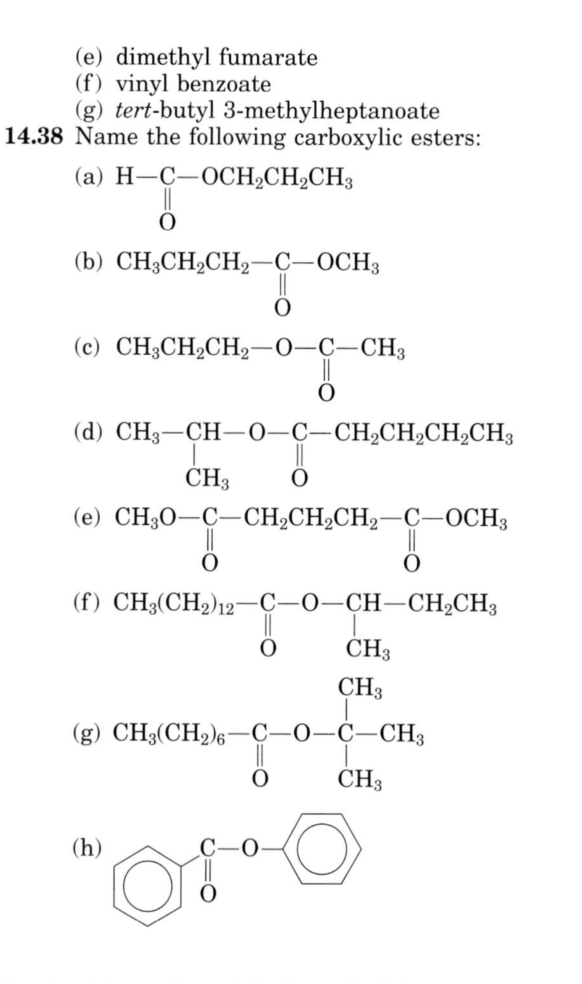

14.35 Define ester.

14.36 Tell whether each compound is a carboxylic ester and, if so, identify the alcohol (or phenol) and acid portions:

(a) $CH_3CH_2-C(=O)-OCH_3$

(b) $CH_3-CH(OCH_2CH_3)-OCH_2CH_3$

(c) cyclopentyl$-O-C(=O)-H$

(d) cyclohexyl$-CH_2-C(=O)-CH_2CH_3$

(e) phenyl$-O-C(=O)-CH_2CH_3$

(f) $HO-C(=O)-C_6H_4-C(=O)-O-$cyclohexyl

(g) $CH_3-C(=O)-CH_2-C(=O)-CH_3$

(h) $CH_3O-C(=O)-CH_2-C(=O)-OCH_3$

(i) five-membered ring lactone: ring O, C=O, with H_3C substituent on the carbon adjacent to the carbonyl

14.37 Draw structural formulas for
(a) ethyl formate
(b) methyl valerate
(c) isopropyl butyrate
(d) phenyl stearate

(e) dimethyl fumarate
(f) vinyl benzoate
(g) *tert*-butyl 3-methylheptanoate

14.38 Name the following carboxylic esters:

(a) $H-\underset{\underset{\displaystyle O}{\|}}{C}-OCH_2CH_2CH_3$

(b) $CH_3CH_2CH_2-\underset{\underset{\displaystyle O}{\|}}{C}-OCH_3$

(c) $CH_3CH_2CH_2-O-\underset{\underset{\displaystyle O}{\|}}{C}-CH_3$

(d) $CH_3-\underset{\underset{\displaystyle CH_3}{|}}{CH}-O-\underset{\underset{\displaystyle O}{\|}}{C}-CH_2CH_2CH_2CH_3$

(e) $CH_3O-\underset{\underset{\displaystyle O}{\|}}{C}-CH_2CH_2CH_2-\underset{\underset{\displaystyle O}{\|}}{C}-OCH_3$

(f) $CH_3(CH_2)_{12}-\underset{\underset{\displaystyle O}{\|}}{C}-O-\underset{\underset{\displaystyle CH_3}{|}}{CH}-CH_2CH_3$

(g) $CH_3(CH_2)_6-\underset{\underset{\displaystyle O}{\|}}{C}-O-\overset{\overset{\displaystyle CH_3}{|}}{\underset{\underset{\displaystyle CH_3}{|}}{C}}-CH_3$

(h) $C_6H_5-\underset{\underset{\displaystyle O}{\|}}{C}-O-C_6H_5$

Physical Properties of Carboxylic Esters

14.39 Arrange in order of increasing boiling point:

(a) CH_3CH_2COOH

(b) $CH_3CH_2CH_2CH_2CH_3$

(c) $CH_3CH_2COO^- \ Na^+$

(d) $CH_3CH_2COOCH_3$

(e) $CH_3CH_2CH_2CH_2OH$

14.40 Would you predict $CH_3CH_2CH_2COOH$ to be more soluble in water than $CH_3CH_2COOCH_3$, to be less soluble, or to have about the same solubility? Explain.

Preparation of Carboxylic Esters; Acyl Halides; Anhydrides

14.41 Show the structural formulas of the carboxylic esters obtained from a combination of

(a) $CH_3COOH + CH_3OH$

(b) $HCOOH + CH_3-\overset{\overset{\displaystyle CH_3}{|}}{CH}-OH$

(c) $CH_3CH_2OH + CH_3-\underset{\underset{\displaystyle CH_3}{|}}{CH}-COOH$

(d) $CH_3-\underset{\underset{\displaystyle OH}{|}}{CH}-COOH + C_6H_5-CH_2OH$

(e) $HOOC-(CH_2)_4-COOH$ + excess CH_3OH

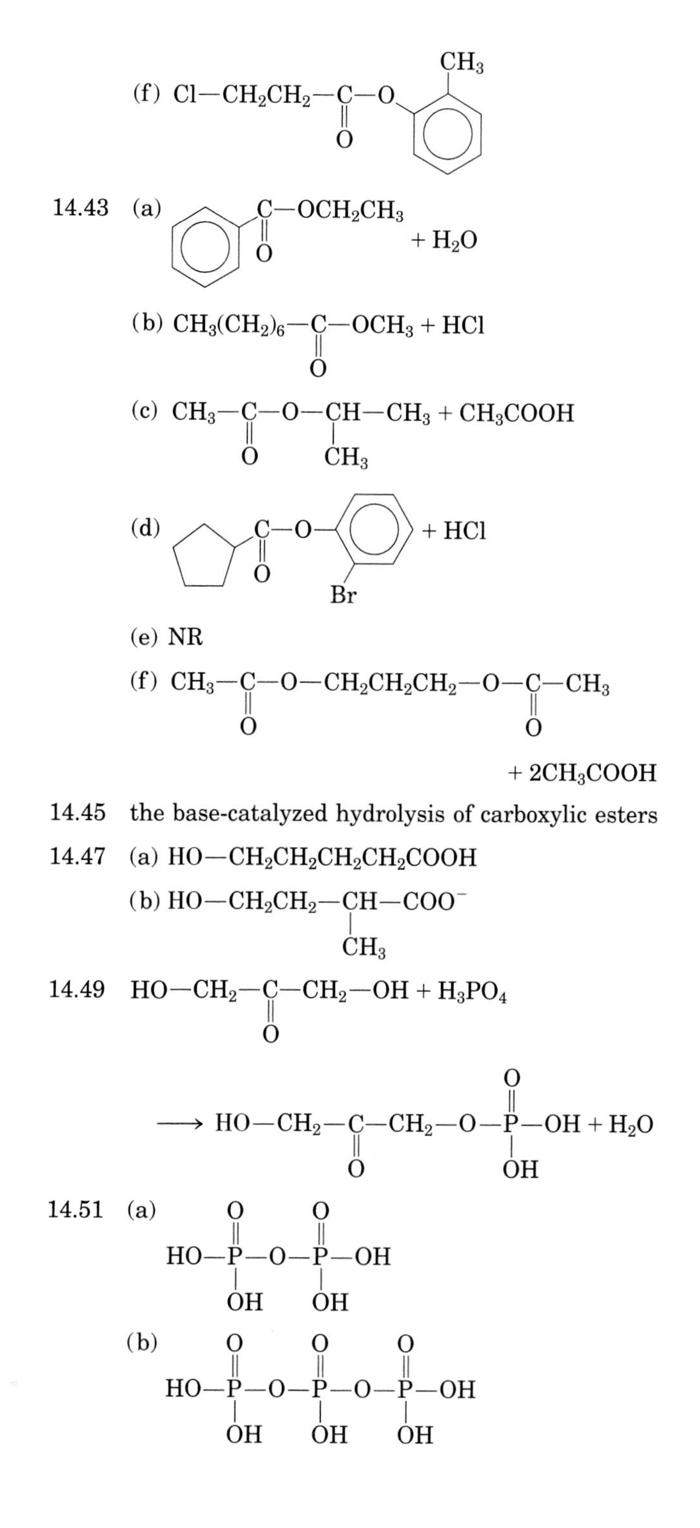

(f) $Cl-CH_2CH_2-COOH +$ (benzene ring bearing CH_3 and OH on adjacent carbons)

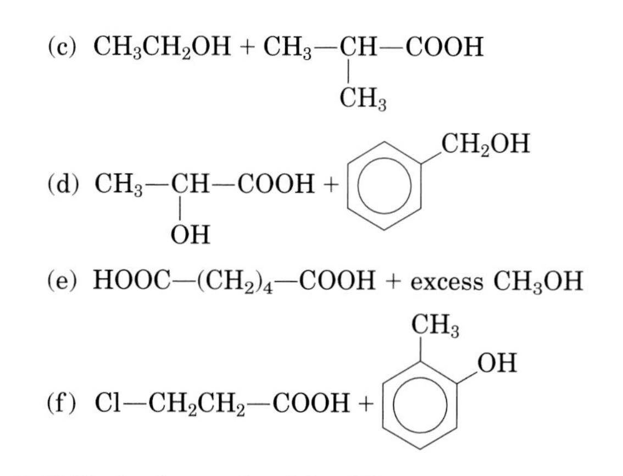

14.42 Tell whether each of the following compounds is a carboxylic acid, a carboxylic ester, an acyl chloride, an anhydride, or none of these:

(a) CH_3COOH

(b) $CH_3-\underset{\underset{\displaystyle O}{\|}}{C}-CH_2-Cl$

(c) $CH_3CH_2-\underset{\underset{\displaystyle O}{\|}}{C}-Cl$

(d) $C_6H_5-O-\underset{\underset{\displaystyle O}{\|}}{C}-CH_3$

(e) $CH_3-\underset{\underset{\displaystyle O}{\|}}{C}-O-\underset{\underset{\displaystyle O}{\|}}{C}-CH_3$

(f) $CH_3-\underset{\underset{\displaystyle OCH_3}{|}}{CH}-CH_2-OCH_3$

(g) $CH_3-\underset{\underset{\displaystyle Cl}{|}}{CH}-COOH$

(h) (benzene ring fused to a five-membered ring: $C(=O)-O-C(=O)$)

(i) (six-membered ring: H_2C, CH_2 (H_2C), CH_2, O, $C(=O)$, H_2C)

(j) (benzene ring bearing OH and, on the adjacent carbon, $\underset{\underset{\displaystyle O}{\|}}{C}-O-CH_2CH_3$)

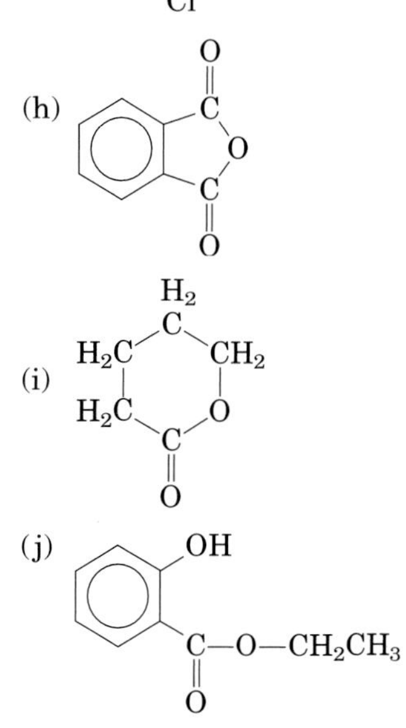

(k) $HOOC—CH_2CH_2—COOCH_3$

14.43 Complete the following equations:

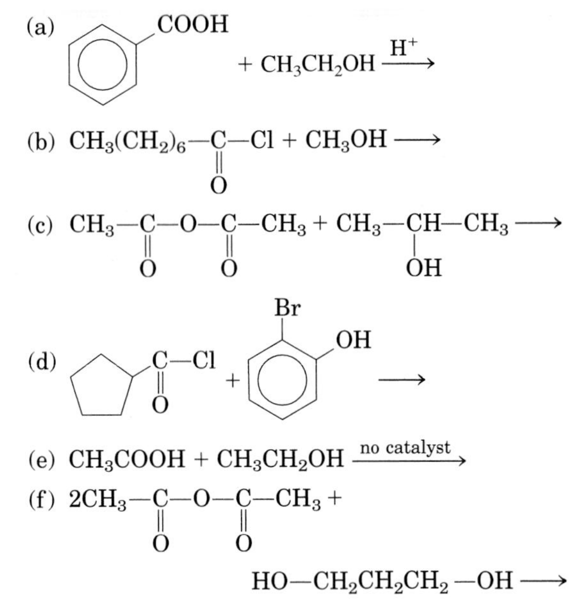

(a) $C_6H_5—COOH + CH_3CH_2OH \xrightarrow{H^+}$

(b) $CH_3(CH_2)_6—\underset{\underset{O}{\|}}{C}—Cl + CH_3OH \longrightarrow$

(c) $CH_3—\underset{\underset{O}{\|}}{C}—O—\underset{\underset{O}{\|}}{C}—CH_3 + CH_3—\underset{\underset{OH}{|}}{CH}—CH_3 \longrightarrow$

(d) $C_5H_9—\underset{\underset{O}{\|}}{C}—Cl + $ (2-bromophenol, $Br—C_6H_4—OH$) $\longrightarrow$

(e) $CH_3COOH + CH_3CH_2OH \xrightarrow{\text{no catalyst}}$

(f) $2CH_3—\underset{\underset{O}{\|}}{C}—O—\underset{\underset{O}{\|}}{C}—CH_3 +$

$HO—CH_2CH_2CH_2—OH \longrightarrow$

Hydrolysis of Carboxylic Esters

14.44 Write equations for the hydrolysis of ethyl butyrate, $CH_3CH_2CH_2COOCH_2CH_3$, under (a) acidic conditions (b) basic conditions.

14.45 Define saponification.

14.46 Draw the structural formula for each product:

(a) $CH_3—\underset{\underset{O}{\|}}{C}—OCH_3 + H_2O \xrightarrow{OH^-}$

(b) $CH_3—\underset{\underset{O}{\|}}{C}—OCH_3 + H_2O \xrightarrow{H^+}$

(c) $C_6H_5—O—\underset{\underset{O}{\|}}{C}—CH_2CH_3 + H_2O \xrightarrow{OH^-}$

(d) $C_6H_5—\underset{\underset{O}{\|}}{C}—O—CH_2CH_3 + H_2O \xrightarrow{H^+}$

*(e) $CH_3—O—\underset{\underset{O}{\|}}{C}—CH_2—\underset{\underset{O}{\|}}{C}—OCH_3 + H_2O \xrightarrow{OH^-}$

*(f) $CH_3CH_2—\underset{\underset{O}{\|}}{C}—O—CH_2CH_2—O—\underset{\underset{O}{\|}}{C}—CH_2CH_3$

$+ H_2O \xrightarrow{OH^-}$

(g) $CH_3(CH_2)_8—\underset{\underset{O}{\|}}{C}—O—CH_2—\underset{\underset{CH_3}{|}}{CH}—CH_3$

$+ H_2O \xrightarrow{OH^-}$

***14.47** A lactone is an internal ester. Show the product when the following lactones are hydrolyzed:

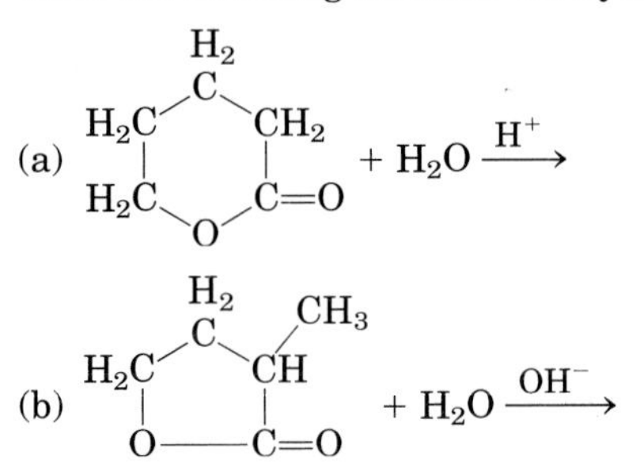

(a) six-membered ring: $H_2C—CH_2—CH_2—CH_2—C(=O)—O$ (ring) $+ H_2O \xrightarrow{H^+}$

(b) five-membered ring: $H_2C—CH_2—CH(CH_3)—C(=O)—O$ (ring) $+ H_2O \xrightarrow{OH^-}$

Esters and Anhydrides of Phosphoric Acid

14.48 (a) Draw the structural formulas for the mono-, di-, and triethyl esters of phosphoric acid.
(b) How many protons can each of these esters give up?

***14.49** Dihydroxyacetone, $\underset{\underset{OH}{|}}{CH_2}—\underset{\underset{O}{\|}}{C}—\underset{\underset{OH}{|}}{CH_2}$, and phosphoric acid form an important monoester called dihydroxyacetone phosphate. Write the equation for the formation of dihydroxyacetone phosphate.

14.50 Show how triphosphoric acid is formed from three molecules of phosphoric acid. How many H_2O molecules are split out?

14.51 Draw the structural formulas of (a) pyrophosphoric acid (b) triphosphoric acid.

14.52 Write equations showing the hydrolysis of (a) methyl triphosphate to methyl diphosphate (b) methyl diphosphate to methyl monophosphate.

14.53 Draw the formulas for the acid form and the salt form of the monoethyl ester of triphosphoric acid.

14.54 Write an equation to show how pyrophosphoric acid transfers a phosphate group to ethanol.

Boxes

14.55 (Box 14B) Draw the structural formula for calcium propionate.

14.56 (Box 14C) Show how to prepare the carboxylic esters that give orange flavor and apple flavor.

14.57 (Box 14D) Why is aspirin, a synthetic compound, a more satisfactory pain-reliever than salicylic acid, a natural product?

14.58 (Box 14D) Why do people who do not have headaches and pains regularly take small doses of aspirin?

14.59 (Box 14E) Draw the structure of a polyester obtained from the condensation of fumaric acid and 1,6-hexanediol:

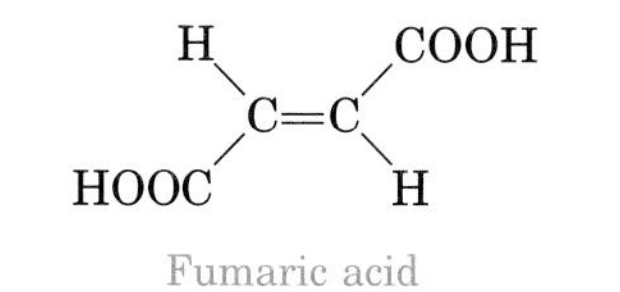

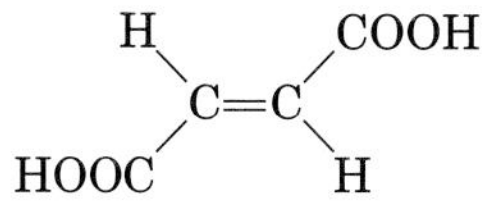

$$\begin{array}{c} H \quad\quad COOH \\ \diagdown \quad \diagup \\ C{=}C \\ \diagup \quad \diagdown \\ HOOC \quad\quad H \end{array}$$

Fumaric acid

$$HO{-}CH_2CH_2CH_2CH_2CH_2CH_2{-}OH$$

1,6-Hexanediol

14.60 (Box 14E) How does a condensation polymer differ from an addition polymer?

14.61 (Box 14F) (a) Could you form an internal ester (a lactone) from lactic acid or glycolic acid? (b) What would be their structures? (c) If such compounds were formed, how could they be separated from the polymers?

14.62 (Box 14F) Why do Lactomer staples disappear from the body after three to four months?

Additional Problems

14.63 Write the full equations for preparing the following esters from alcohols and anhydrides or acyl chlorides:

(a) 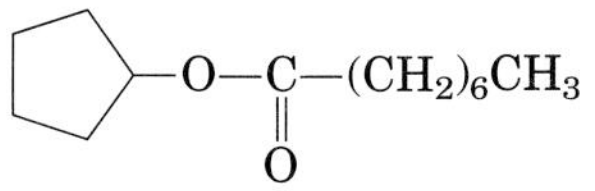 cyclopentyl$-O-\underset{\underset{O}{\|}}{C}-(CH_2)_6CH_3$

(b) 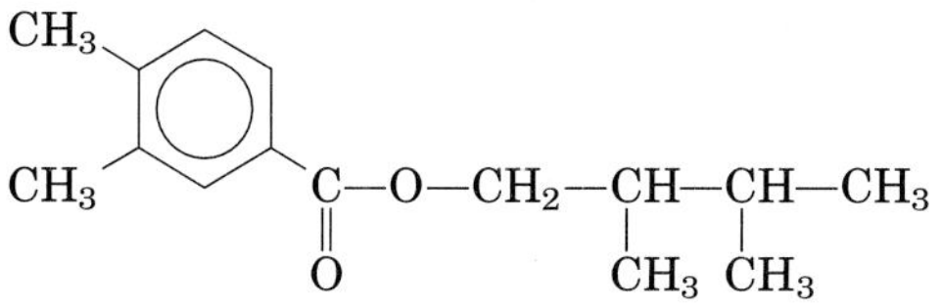 3,4-dimethylphenyl (ring bearing two CH_3 groups) $-\underset{\underset{O}{\|}}{C}-O-CH_2-\underset{\underset{CH_3}{|}}{CH}-\underset{\underset{CH_3}{|}}{CH}-CH_3$

14.64 You are given a solid that could be either a carboxylic acid or a carboxylic ester. Describe a simple test (less than 5 minutes) for telling which of these classes it belongs to.

***14.65** Draw the structural formula for dimethyl sulfate.

14.66 When synthesizing a carboxylic ester from an alcohol, why is it usually better to use the anhydride or the acyl halide rather than the corresponding carboxylic acid?

14.67 Of the organic compounds we have studied so far, which classes are acidic? Rank them in order of increasing acidity.

14.68 What is the most important property of carboxylic acids?

14.69 Which of these types of compounds will produce CO_2 bubbles when added to a solution of sodium bicarbonate in water? (a) a carboxylic acid (b) a carboxylate salt (c) a carboxylic ester.

14.70 Draw the structural formula for each of the following derivatives of acetic acid: (a) the ethyl ester (b) the anhydride (c) the acyl chloride.

***14.71** Write the structural formula for any compound that is both an ester and an anhydride of phosphoric acid.

14.72 If someone wanted to hydrolyze the ester methyl propionate, $CH_3CH_2\underset{\underset{O}{\|}}{C}-O-CH_3$, would it be better to do it under acidic or basic conditions?

***14.73** Would you expect a molecule of an anhydride (Sec. 14.10) to react with two molecules of an alcohol to produce two carboxylic ester molecules?

14.74 Is $CH_3(CH_2)_{10}COOH$ more or less soluble in water than its Li^+ salt? Explain.

Chapter **15**

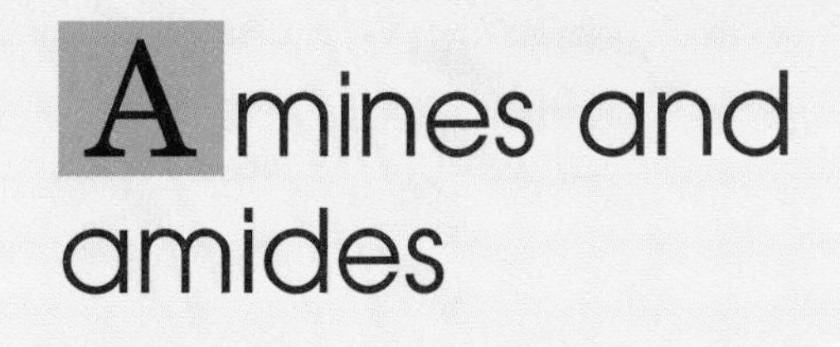

Amines and amides

15.1 Introduction

In this chapter we deal with organic compounds containing nitrogen. The two most important classes of nitrogen-containing organic compounds are amines and amides.

15.2 Nomenclature of Amines

Amines are **derivatives of ammonia.** Ammonia has three hydrogens, and we classify amines according to the number of hydrogens replaced by alkyl or aryl groups:

The R groups in secondary and tertiary amines may be the same or different.

NH_3	Ammonia
$R{-}NH_2$	**Primary amines** have one hydrogen replaced by R (*two* Hs remain)
$R{-}NH{-}R'$	**Secondary amines** have two hydrogens replaced by R (*one* H remains)
$R{-}\underset{\displaystyle R''}{\underset{\vert}{N}}{-}R'$	**Tertiary amines** have all three hydrogens replaced by R (*no* H remains)

Note that the use of primary, secondary, and tertiary for amines is not quite the same as for alcohols. For alcohols we look at how many R groups are on a *carbon;* for amines we look at how many are on the *nitrogen.* Thus, for example, *tert*-butylamine is a *primary* amine because there is only one R group on the nitrogen, even though the R group is a tertiary alkyl group:

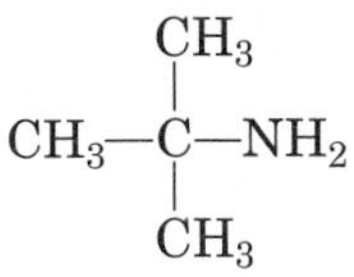

$$CH_3{-}\overset{\displaystyle CH_3}{\overset{\vert}{\underset{\displaystyle CH_3}{\underset{\vert}{C}}}}{-}NH_2$$

tert-Butylamine (a primary amine)

Problem 15.1

Classify each as a primary, secondary, or tertiary amine:

(a) $CH_3CH_2{-}NH{-}CH_3$

(b) $CH_3CH_2{-}NH_2$

(c) $C_6H_5{-}\overset{\displaystyle CH_3}{\overset{\vert}{N}}{-}CH_3$

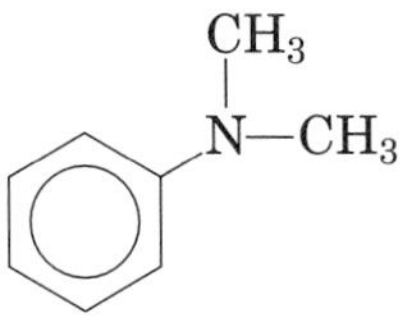

Simple primary amines are named by adding the suffix "-amine" to the name of the group:

CH_3NH_2 Methylamine

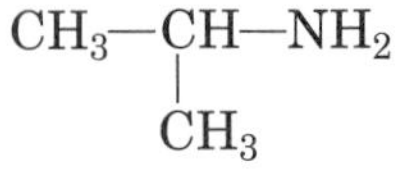

$CH_3{-}\underset{\displaystyle CH_3}{\underset{\vert}{CH}}{-}NH_2$ Isopropylamine

Recall (Sec. 11.10) that the simplest aromatic amine is aniline:

$C_6H_5-NH_2$ Aniline

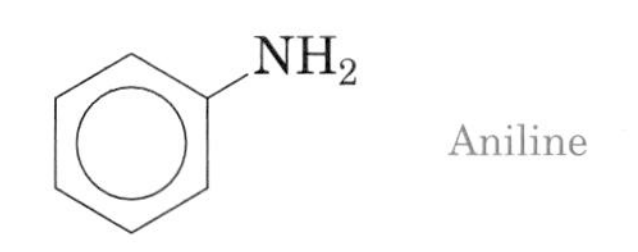

Simple secondary and tertiary amines are named the same way as the primary amines. We merely place all the R groups before the suffix "-amine" and use "di-" and "tri-" as necessary:

Secondary amines:

$CH_3-NH-CH_3$ Dimethylamine

$CH_3CH_2-NH-CH(CH_3)_2$ Ethylisopropylamine

$C_6H_5-NH-C_6H_5$ Diphenylamine

Tertiary amines:

$CH_3CH_2-N(CH_2CH_3)-CH_2CH_3$ Triethylamine

$CH_3-CH(CH_3)-N(CH_3)-CH_2CH_3$ Ethylisopropylmethylamine

Problem 15.2

Name:

(a) $CH_3CH_2NH_2$

(b) $CH_3-NH-CH_2CH_3$

(c) $CH_3CH_2-N(CH_3)-CH_2CH_3$

When one R is aromatic, the amine is generally named as a derivative of aniline:

$C_6H_5-NH-CH_3$ *N*-Methylaniline

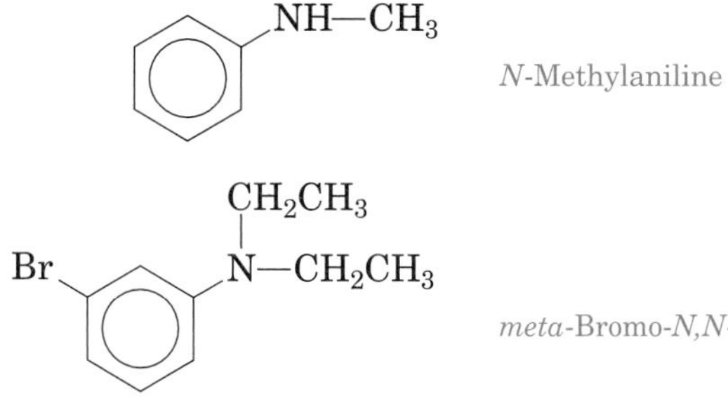

$m\text{-}Br-C_6H_4-N(CH_2CH_3)-CH_2CH_3$ *meta*-Bromo-*N*,*N*-diethylaniline

Do not confuse the capital letter *N* with the small *n*, which stands for *normal*.

The capital letter *N* is used to show that the group is attached to a nitrogen.

Problem 15.3

Name as a derivative of aniline:

$NH{-}CH_2CH_2CH_3$

Cl

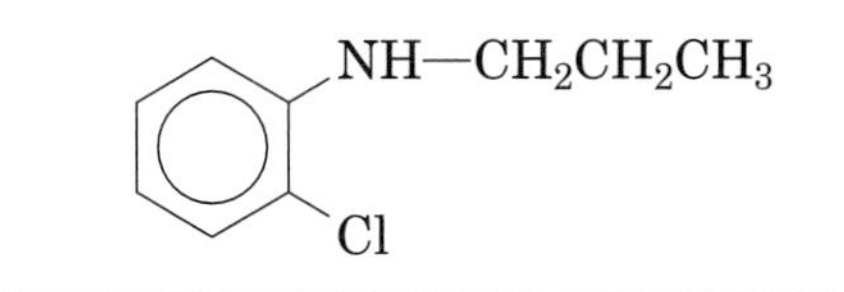

When it is necessary to name $-NH_2$ as a group, it is called the *amino group.*

$$\underset{\displaystyle NH_2}{CH_3{-}\underset{|}{C}H{-}CH_2{-}COOH}$$ 3-Aminobutanoic acid

Problem 15.4

Name:

NH_2

NH_2

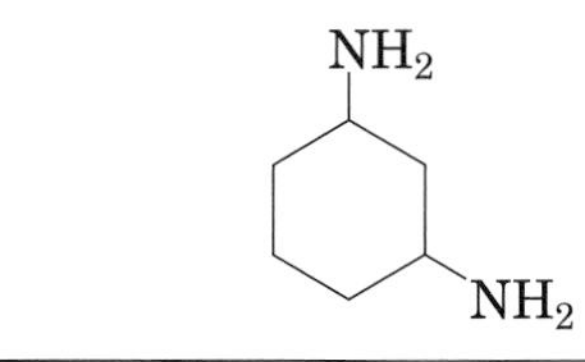

As with ethers, secondary and tertiary amines can be heterocyclic. As mentioned in Section 11.13, such compounds can be aromatic or nonaromatic. Five important ones are shown here:

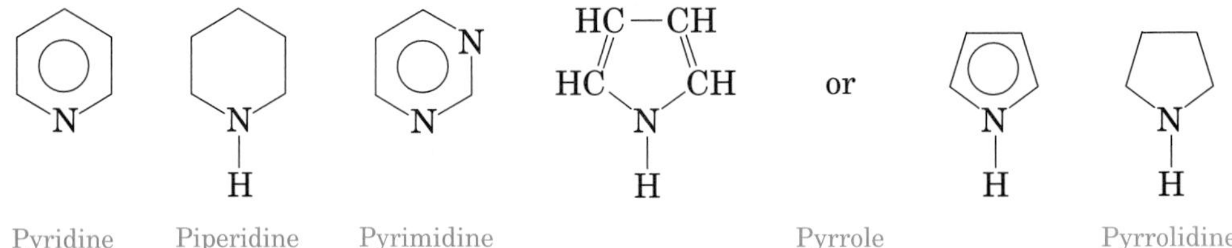

Pyridine Piperidine Pyrimidine Pyrrole Pyrrolidine

Pyrrole is an example of a five-membered heterocyclic aromatic compound. Even though there are only five atoms in the ring, there are six electrons (including the unshared pair on the nitrogen atom), so the ring is aromatic.

We shall meet other heterocyclic amines later in this chapter. Many of these compounds have important biological functions.

Problem 15.5

How many hydrogen atoms does pyridine have? How many does piperidine have?

Box 15A

Amphetamines (Pep Pills)

Amphetamine (benzedrine) and methamphetamine (methedrine) are synthetic amines that are powerful stimulants of the central nervous system. They reduce fatigue and diminish hunger by raising the glucose level of the blood. Because of these properties, they are widely used medicinally to counter mild cases of depression, to reduce hyperactivity in children, and as an appetite depressant for people who are trying to lose weight.

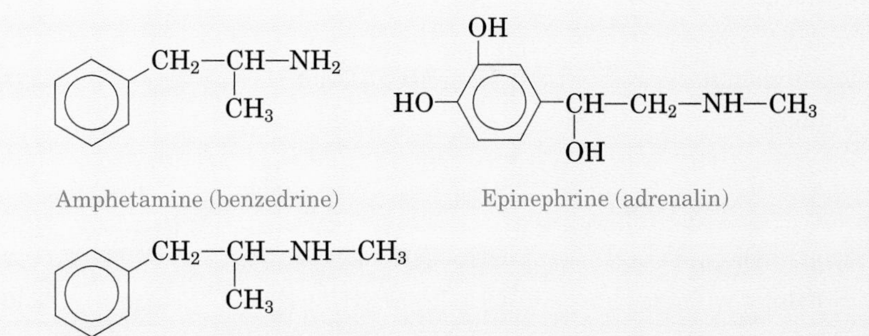

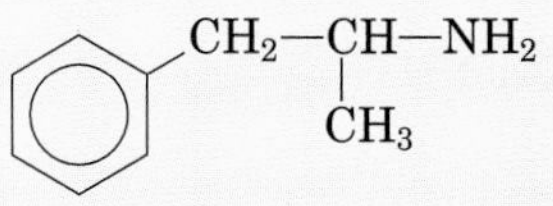

$C_6H_5-CH_2-CH(CH_3)-NH_2$

Amphetamine (benzedrine)

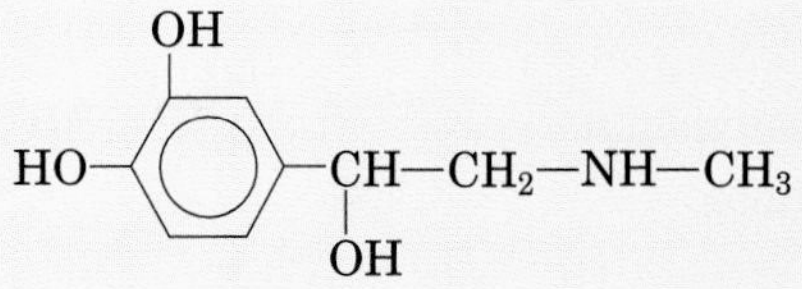

$HO-C_6H_3(OH)-CH(OH)-CH_2-NH-CH_3$

Epinephrine (adrenalin)

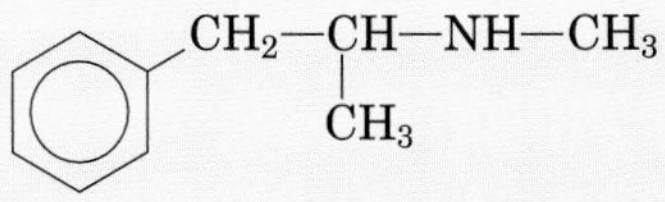

$C_6H_5-CH_2-CH(CH_3)-NH-CH_3$

Methamphetamine (methedrine)

The action of amphetamines is similar to that of epinephrine (adrenalin), which has a very similar structure. Epinephrine is both a neurotransmitter in the brain (Sec. 24.4) and a hormone in the blood stream (Sec. 24.6), secreted by the adrenal gland. When an individual feels excitement or fear, epinephrine helps to make glucose available to tissues that need it immediately, for example, when leg muscles must be used to run away from danger.

Amphetamines are used both legally (as prescribed by a physician) and illegally to elevate moods or reduce fatigue. They are called "uppers" or "pep pills." Methamphetamine is known as "speed," "ice," or "crank." Amphetamines are sometimes used by truck drivers on long-distance runs to stay awake all night and by athletes who want more strength in a game or a race. Abuse of these drugs can have severe effects on both the body and the mind. They are addictive, concentrate in the brain and nervous system, and can lead to long periods of sleeplessness, loss of weight, and paranoia.

15.3 Physical Properties of Amines

The boiling points of amines are between those of the alcohols on the one hand and the hydrocarbons and halides on the other. For example, the boiling point of *n*-propylamine, $CH_3CH_2CH_2NH_2$, is 48°C (Table 15.1). This is about halfway between the boiling point of butane (0°C) and that of 1-propanol (97°C), two compounds of about the same molecular weight. The reason is that a molecule of a primary or a secondary amine forms a hydrogen bond with another molecule of itself, but these hydrogen bonds are weaker than those formed by alcohols. As can be seen from the boiling points listed in Table 15.1, amines of up to two or three carbons are gases at room temperature. Amines are somewhat more soluble in water than alcohols of comparable molecular weight.

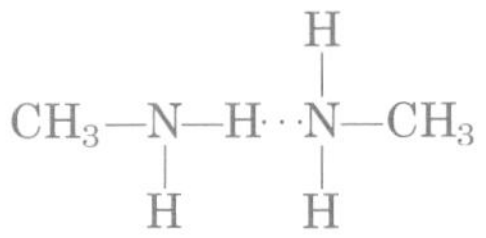

$CH_3-NH-H\cdots NH(H)-CH_3$

Table 15.1 Boiling Points and Solubilities of Ammonia and Some Primary, Secondary, and Tertiary Amines

Name	Formula	Boiling Point (°C)	Solubility in Water
Ammonia	NH_3	−33	Soluble
Methylamine	CH_3NH_2	−6	Soluble
Ethylamine	$CH_3CH_2NH_2$	17	Soluble
n-Propylamine	$CH_3CH_2CH_2NH_2$	48	Soluble
n-Butylamine	$CH_3CH_2CH_2CH_2NH_2$	78	Soluble
Aniline	[structure] NH_2	184	Soluble
Dimethylamine	$CH_3—NH—CH_3$	7	Soluble
Ethylmethylamine	$CH_3CH_2—NH—CH_3$	36	Soluble
Diethylamine	$CH_3CH_2—NH—CH_2CH_3$	56	Soluble
Piperidine	[structure] N—H	106	Soluble
Trimethylamine	$CH_3—N(CH_3)—CH_3$	3	Soluble
Triethylamine	$CH_3CH_2—N(CH_2CH_3)—CH_2CH_3$	89	Soluble
N,N-Dimethylaniline	[structure] $N(CH_3)—CH_3$	194	Insoluble
Pyridine	[structure] N	115	Soluble

The simple amines have strong odors resembling that of ammonia but not quite as sharp and pungent. The odor of amines also resembles that of raw fish, which is not surprising, since raw fish contains low-molecular-weight amines. Some diamines have especially bad odors. Two of these are

$$\underset{NH_2}{CH_2}—CH_2—CH_2—\underset{NH_2}{CH_2} \qquad \underset{NH_2}{CH_2}—CH_2—CH_2—CH_2—\underset{NH_2}{CH_2}$$

Putrescine (1,4-diaminobutane) Cadaverine (1,5-diaminopentane)

These compounds, which are among the end products in the decomposition of proteins, are found in decaying flesh.

The names give a hint of the odor.

15.4 The Basicity of Amines. Quaternary Salts

Amines are bases—this is their most important chemical property. They are the only important organic bases, and we can be fairly certain that any organic base we meet is an amine.

We saw in Chapter 8 that ammonia is a weak base. Since the amines are derivatives of ammonia, they share this property. When amines act as bases, they react with acids by accepting a proton (as do all Brønsted-Lowry bases). In Section 8.6 we saw that methylamine reacts with an acid such as HCl to give its conjugate acid, the methylammonium ion. Most other amines, whether primary, secondary, or tertiary, behave exactly the same way; they too are converted to positive ions:

Aliphatic amines are somewhat stronger bases than ammonia; aromatic amines are weaker.

$$CH_3NH_2 + HCl \longrightarrow CH_3\overset{+}{N}H_3\ Cl^-$$

Methylammonium chloride

$$CH_3NHCH_3 + HNO_3 \longrightarrow CH_3\overset{+}{N}H_2CH_3\ NO_3^-$$

Dimethylammonium nitrate

$$C_6H_5{-}N(CH_3){-}CH_3 + HF \longrightarrow C_6H_5{-}N^+H(CH_3){-}CH_3\ F^-$$

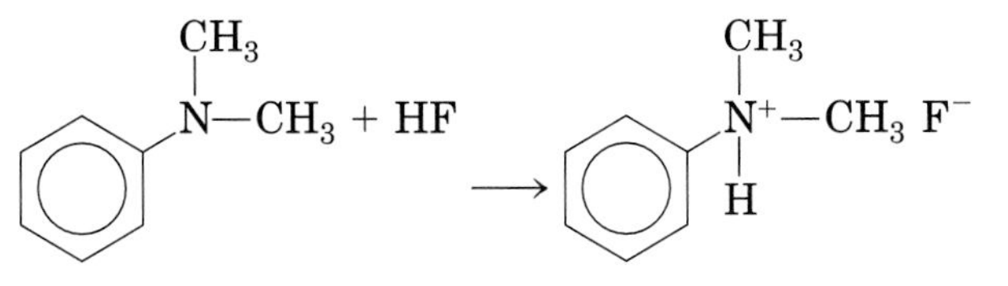

N,N-Dimethylanilinium fluoride

Carboxylic acids are strong enough to react with amines (primary, secondary, or tertiary) in the same way, for example,

$$\underset{\text{Ethylamine}}{CH_3CH_2{-}NH_2} + \underset{\text{Acetic acid}}{CH_3COOH} \longrightarrow \underset{\text{Ethylammonium acetate}}{CH_3CH_2{-}\overset{+}{N}H_3\ CH_3COO^-}$$

Problem 15.6

Draw the structural formulas for each salt formed:

(a) $CH_3CH_2CH_2NH_2 + HCOOH \longrightarrow$

(b) $C_6H_5{-}NHCH_3 + HBr \longrightarrow$

The ionic compounds produced when amines react with acids are called **amine salts.** Like all salts, they are solids at room temperature. As can be seen in the examples just given, they are named by changing "-amine" to "-ammonium" (or "-aniline" to "-anilinium") and adding the name of the negative ion.

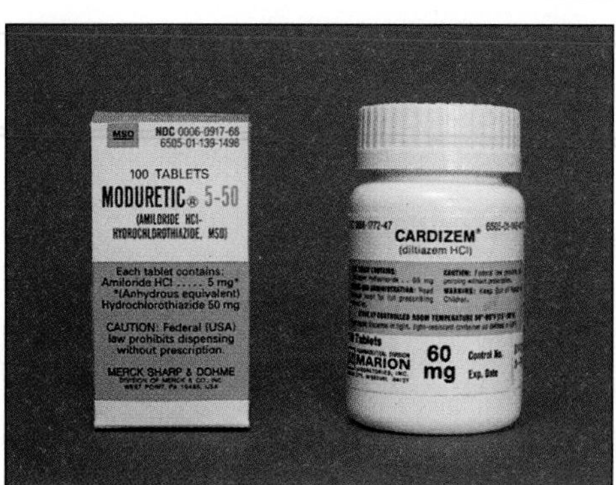

Figure **15.1**
Two drugs that are amine salts labeled as hydrochlorides. (Photograph by Beverly March.)

Drugs that are amine salts are commonly named this way (Fig. 15.1).

There is an older way to represent and name amine salts. Since these methods are still occasionally used, we will look at an example.

Modern way:

$$CH_3CH_2{-}\overset{\displaystyle CH_3\atop|}{N}H + HCl \longrightarrow CH_3CH_2{-}\overset{\displaystyle CH_3\atop|\,+}{N}H_2\ Cl^-$$

Ethylmethylammonium chloride

Older way:

The older method does not show the correct structural formula.

$$CH_3CH_2{-}\overset{\displaystyle CH_3\atop|}{N}H + HCl \longrightarrow CH_3CH_2{-}\overset{\displaystyle CH_3\atop|}{N}H \cdot HCl$$

Ethylmethylamine hydrochloride

In Section 14.6 we saw that the structure of a carboxylic acid depends on the pH. The same is true for amines, but when an amine (primary, secondary, or tertiary) is added to water, the pH is always above 7

Box 15B

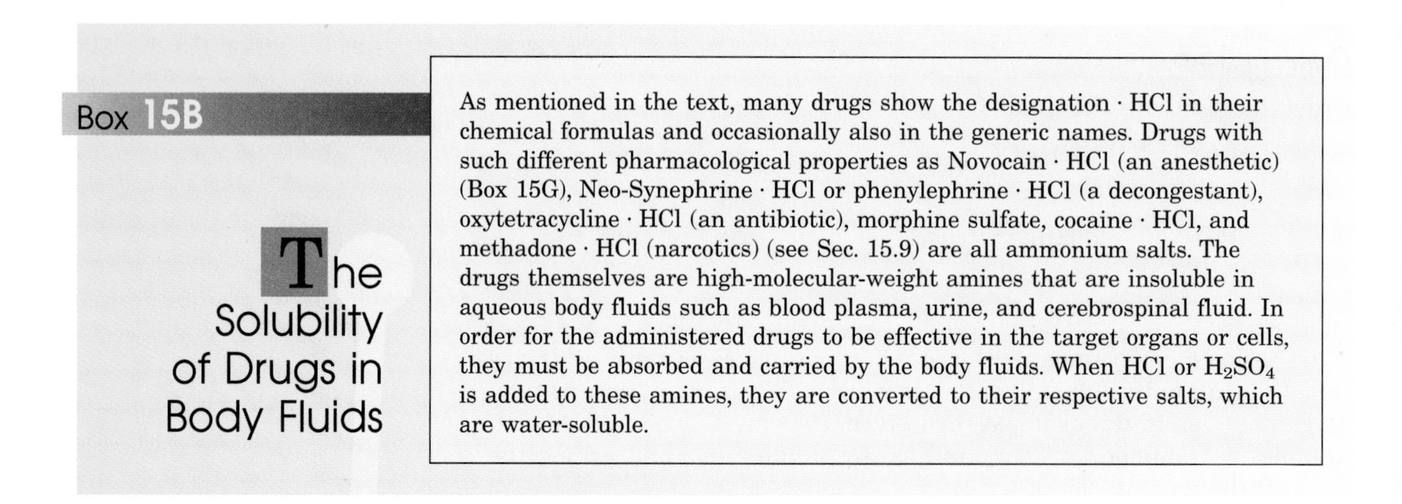

The Solubility of Drugs in Body Fluids

As mentioned in the text, many drugs show the designation · HCl in their chemical formulas and occasionally also in the generic names. Drugs with such different pharmacological properties as Novocain · HCl (an anesthetic) (Box 15G), Neo-Synephrine · HCl or phenylephrine · HCl (a decongestant), oxytetracycline · HCl (an antibiotic), morphine sulfate, cocaine · HCl, and methadone · HCl (narcotics) (see Sec. 15.9) are all ammonium salts. The drugs themselves are high-molecular-weight amines that are insoluble in aqueous body fluids such as blood plasma, urine, and cerebrospinal fluid. In order for the administered drugs to be effective in the target organs or cells, they must be absorbed and carried by the body fluids. When HCl or H_2SO_4 is added to these amines, they are converted to their respective salts, which are water-soluble.

because amines are basic. If we now add a strong acid like HCl, it neutralizes the amine (converts it to its salt) and lowers the pH. Thus,

an amine can exist in water only at a high pH; an amine salt can exist in water only at a low pH:

$$\underset{\text{High pH}}{RNH_2} \underset{OH^-}{\overset{H_3O^+}{\rightleftharpoons}} \underset{\text{Low pH}}{R\overset{+}{N}H_3}$$

Secondary and tertiary amines behave the same way.

As with carboxylic acids, we can easily go back and forth between amines and their salts. If we want to convert an amine to the salt, we just add a strong acid. If we want to convert an amine salt to the free amine, we just add a strong base.

This property of amines and their salts is quite useful in laboratories. Low-molecular-weight amines are soluble in water, but the higher ones are insoluble. However, virtually all amine *salts* are soluble in water. This provides a simple method for separating an insoluble amine from other organic compounds that are insoluble in water but are not basic. We simply add the mixture to a dilute aqueous solution of an acid such as HCl. The amine dissolves because it is converted to its (soluble) salt by the acid. The rest of the mixture is insoluble and thus easily removed.

Recall that ammonium salts are also soluble in water (Sec. 4.8).

Another important use of this property is in administering drugs (see Box 15B). Because drugs that are amines are often given in the form of their salts (Fig. 15.1), the amine itself is often called the **free base.**

Quaternary Ammonium Salts

Earlier in this section we saw examples of salts formed from primary, secondary, and tertiary amines that have three, two, and one hydrogen on the nitrogen, respectively:

$$R\overset{+}{N}H_3 \qquad R_2\overset{+}{N}H_2 \qquad R_3\overset{+}{N}H$$

It is also possible to have ions with *four* alkyl or aryl groups on the nitrogen, for example,

$$\left[CH_3CH_2-\overset{\overset{CH_3}{|}}{\underset{\underset{CH_3}{|}}{N^+}}-CH_2CH_3 \right] Cl^-$$

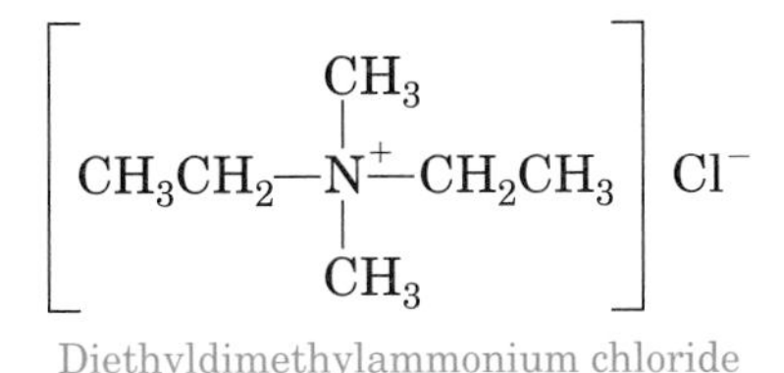

Diethyldimethylammonium chloride

Such salts are called **quaternary ammonium salts** and are named the same way as the other ammonium salts. A major difference between them and the other three types is that quaternary salts cannot be made by acidifying any amine, nor can a proton be removed from them by a base. Thus they have the same structure in water at all pH values. Otherwise, their properties are the same as those of the other three types of amine salts.

Certain quaternary salts are important biological compounds. In later chapters we will meet choline (Sec. 17.6), acetylcholine (Sec. 24.3), and other quaternary ammonium ions.

15.5 Amides: Nomenclature and Physical Properties

Amides are **compounds in which a carbonyl group is connected to a nitrogen atom.** The two remaining bonds of the nitrogen can be attached to hydrogens, alkyl groups, or aromatic rings:

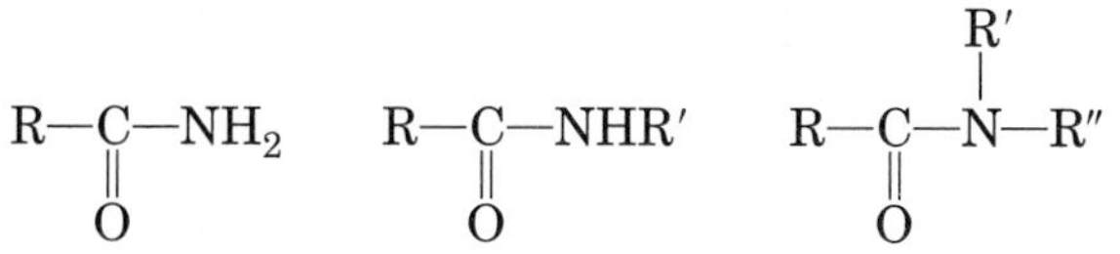

$$R{-}\overset{\overset{\displaystyle O}{\|}}{C}{-}NH_2 \qquad R{-}\overset{\overset{\displaystyle O}{\|}}{C}{-}NHR' \qquad R{-}\overset{\overset{\displaystyle O}{\|}}{C}{-}\underset{\underset{\displaystyle R'}{|}}{N}{-}R''$$

Amides

Amides are carboxylic acid derivatives. Just as a carboxylic ester can be thought of as being composed of a carboxylic acid and an alcohol, so can an amide be thought of as being composed of a carboxylic acid and ammonia or a primary or secondary amine (why not a tertiary amine?):

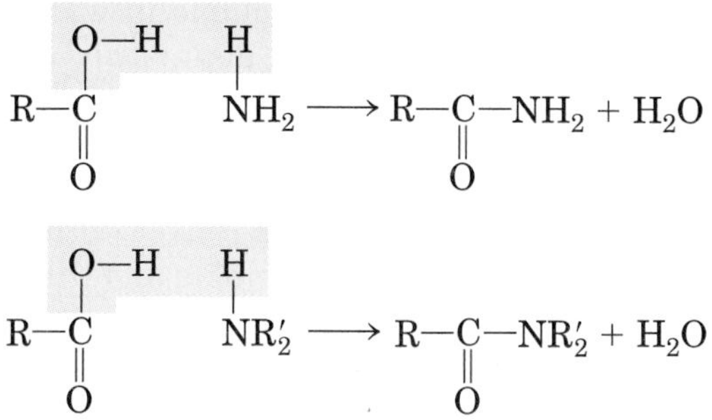

In Chapter 18 we will discuss proteins, which are polyamides.

$$R{-}\underset{\underset{\displaystyle O}{\|}}{\overset{\overset{\displaystyle O{-}H}{|}}{C}} \quad \overset{\overset{\displaystyle H}{|}}{NH_2} \longrightarrow R{-}\underset{\underset{\displaystyle O}{\|}}{C}{-}NH_2 + H_2O$$

$$R{-}\underset{\underset{\displaystyle O}{\|}}{\overset{\overset{\displaystyle O{-}H}{|}}{C}} \quad \overset{\overset{\displaystyle H}{|}}{NR'_2} \longrightarrow R{-}\underset{\underset{\displaystyle O}{\|}}{C}{-}NR'_2 + H_2O$$

Like other carboxylic acid derivatives, amides are named after the corresponding acids. The ending "-ic acid" or "-oic acid" is changed to "-amide."

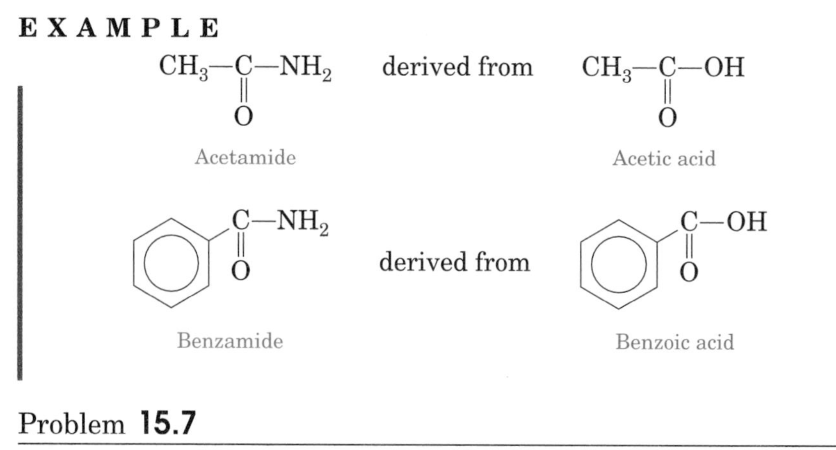

EXAMPLE

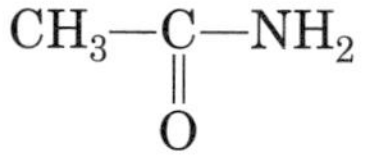

$$CH_3{-}\underset{\underset{\displaystyle O}{\|}}{C}{-}NH_2 \quad \text{derived from} \quad CH_3{-}\underset{\underset{\displaystyle O}{\|}}{C}{-}OH$$

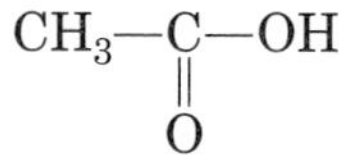

Acetamide — Acetic acid

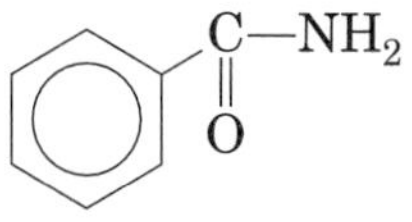

$$C_6H_5{-}\underset{\underset{\displaystyle O}{\|}}{C}{-}NH_2 \quad \text{derived from} \quad C_6H_5{-}\underset{\underset{\displaystyle O}{\|}}{C}{-}OH$$

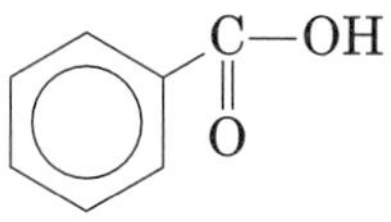

Benzamide — Benzoic acid

Problem 15.7

Name:

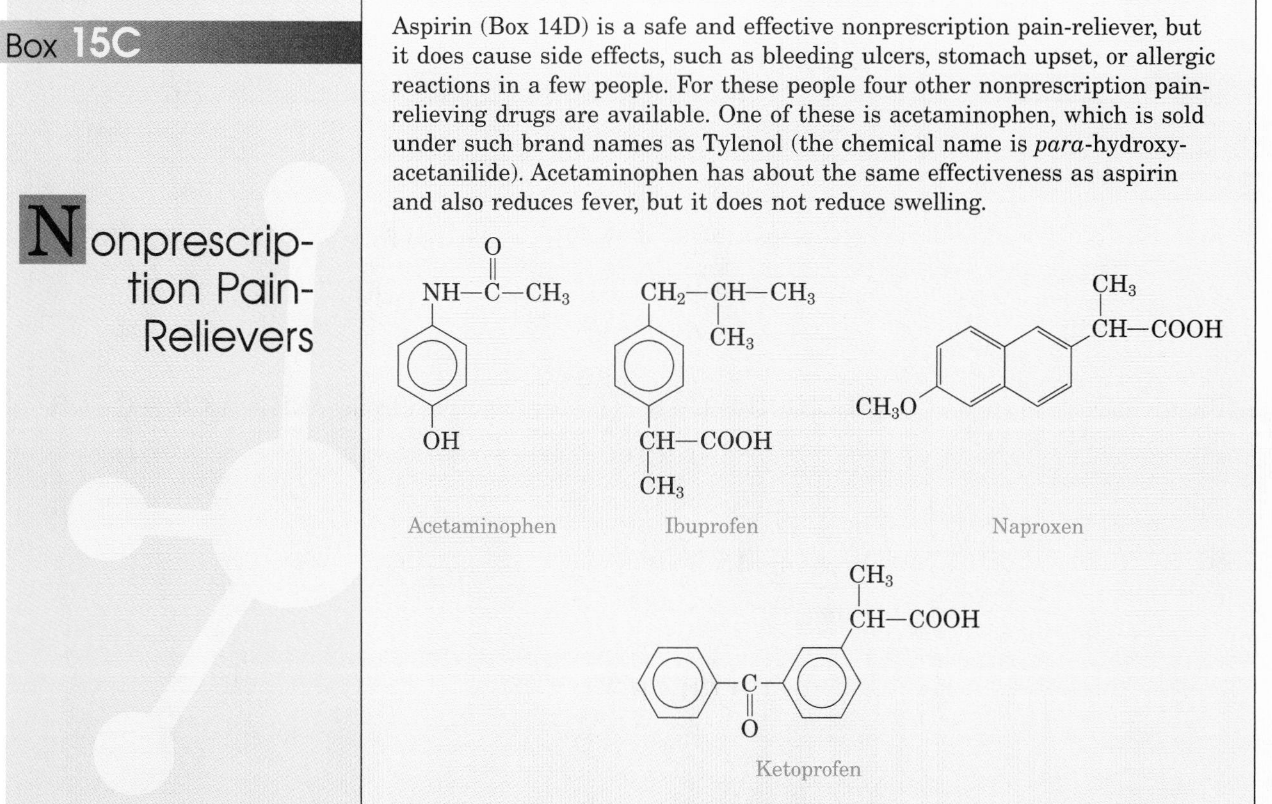

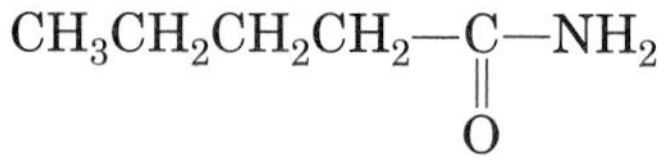

$$CH_3CH_2CH_2CH_2{-}\underset{\underset{\displaystyle O}{\|}}{C}{-}NH_2$$

Box 15C

Nonprescription Pain-Relievers

Aspirin (Box 14D) is a safe and effective nonprescription pain-reliever, but it does cause side effects, such as bleeding ulcers, stomach upset, or allergic reactions in a few people. For these people four other nonprescription pain-relieving drugs are available. One of these is acetaminophen, which is sold under such brand names as Tylenol (the chemical name is *para*-hydroxyacetanilide). Acetaminophen has about the same effectiveness as aspirin and also reduces fever, but it does not reduce swelling.

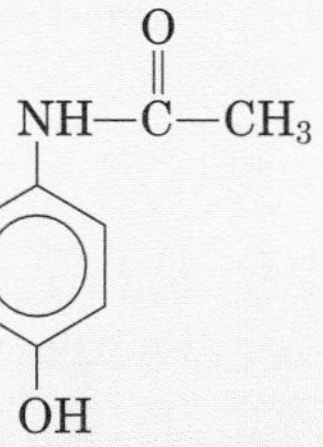

Acetaminophen

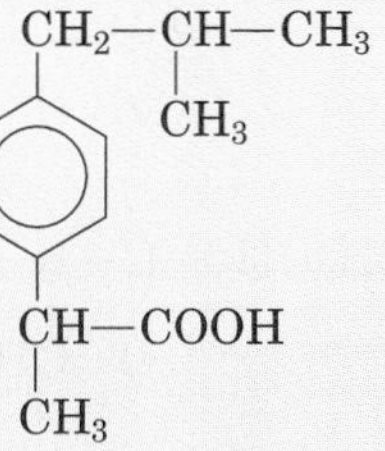

Ibuprofen

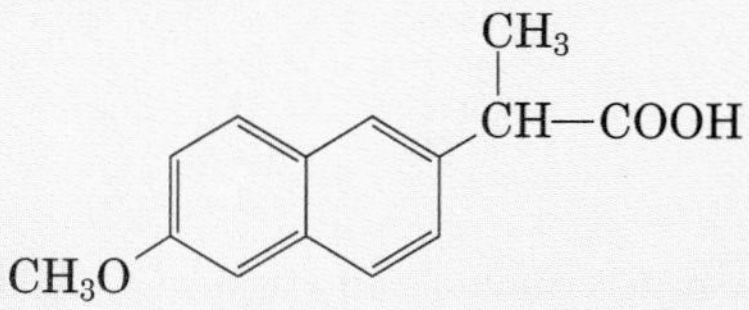

Naproxen

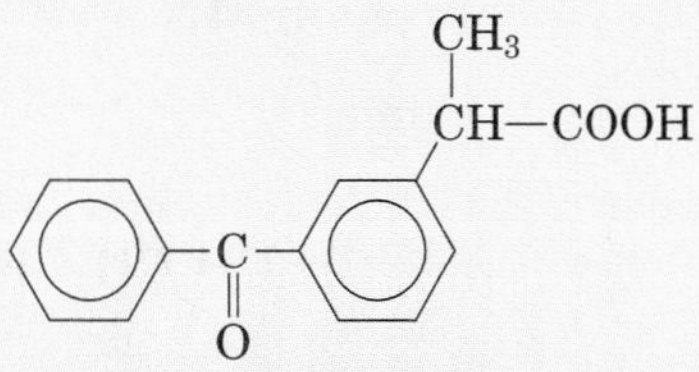

Ketoprofen

The other three nonprescription pain-relievers—ibuprofen, naproxen, and ketoprofen—are more recent. For some years they were available only by prescription, but the Food and Drug Administration now permits their sale without a prescription. Ibuprofen is available as Advil and Nuprin, naproxen as Aleve, and ketoprofen as Actron. Like aspirin, these three not only relieve pain, but also reduce joint inflammation associated with arthritis and athletic injuries, menstrual cramps, and fever. Both aspirin and ibuprofen act by inhibiting the synthesis of prostaglandins (Sec. 17.12). People with ulcers or certain other stomach and intestinal problems should not take aspirin or, to a lesser extent, the last three drugs. For them, the only safe choice is acetaminophen.

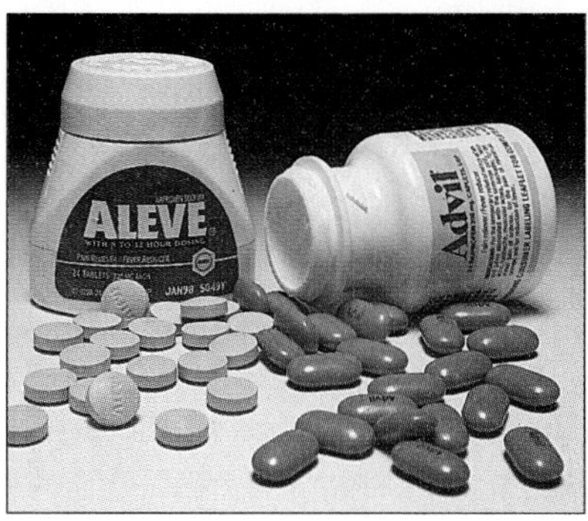

Two nonprescription pain-relievers. (Photograph by Charles D. Winters.)

When the nitrogen of an amide is connected to an alkyl or aryl group, the group is named as a prefix preceded by the letter *N*.

EXAMPLE

$H-C(=O)-NH-CH_3$ derived from $H-C(=O)-OH$

N-Methylformamide — Formic acid

$CH_3CH_2-C(=O)-N(CH_2CH_3)-CH_2CH_3$ derived from $CH_3CH_2-C(=O)-OH$

N,*N*-Diethylpropionamide — Propionic acid

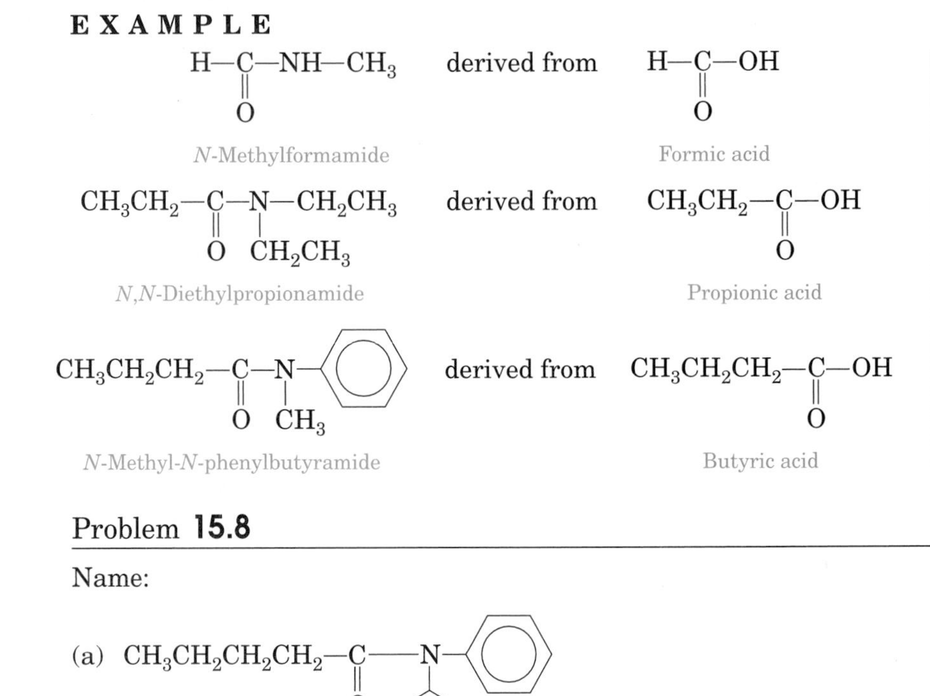

The IUPAC name is *N*-methyl-*N*-phenylbutanamide, derived from butanoic acid.

$CH_3CH_2CH_2-C(=O)-N(CH_3)-C_6H_5$ derived from $CH_3CH_2CH_2-C(=O)-OH$

N-Methyl-*N*-phenylbutyramide — Butyric acid

Problem **15.8**

Name:

(a) $CH_3CH_2CH_2CH_2-C(=O)-N(C_6H_5)-C_6H_5$

(b) $Br-C_6H_4-C(=O)-NH-CH_2CH_3$ (Br *para* to the amide group)

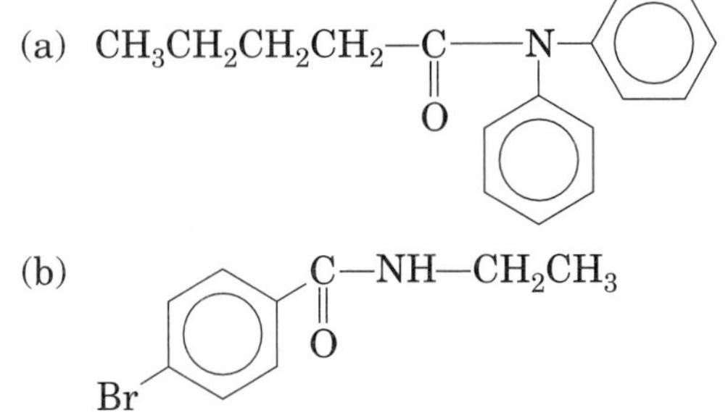

Amides (except for those with no hydrogen atom attached to the nitrogen) form strong hydrogen bonds. The smallest amide, formamide, and some of its *N*-substituted derivatives are liquids, but all other amides are solids at room temperature.

15.6 Acid–Base Properties of Organic Compounds. The Neutrality of Amides

Because basicity is the most important property of amines, it is only natural to wonder if amides are also basic. The answer is no. Amides are neither acidic nor basic; they are neutral.

At this point, it is well to review the acid–base properties of the organic compounds we have studied. Table 15.2 lists the oxygen-, nitrogen-, and sulfur-containing compounds we have learned about. We can assume that any organic compound that contains only carbon, hydrogen, and/or halogens is neither acidic nor basic.

Table 15.2 Acid–Base Properties of Organic Compounds

Acidic	Basic	Neutral
Sulfonic acids (strong)	Amines (weak)	Alcohols
Carboxylic acids (weak)	Carboxylate ions (very weak)	Ethers
Phenols (very weak)		Thioethers
Thiols (very weak)		Aldehydes
Amine salts (very weak)		Ketones
		Carboxylic esters
		Acyl chlorides
		Anhydrides
		Amides

These listings are according to the Arrhenius definitions of acid and base (Sec. 8.1) and not the Brønsted-Lowry definitions (Sec. 8.3).

15.7 Preparation of Amides

In Section 15.5 we saw that amides can be thought of as having a carboxylic acid portion and an ammonia or amine portion. However, amides cannot be made simply by adding ammonia or an amine to a carboxylic acid. Why not? Because, as we saw in Section 15.4, when we add ammonia or an amine to a carboxylic acid, we get a salt:

$$R{-}\underset{\underset{O}{\|}}{C}{-}OH + NH_3 \longrightarrow R{-}\underset{\underset{O}{\|}}{C}{-}O^- \, NH_4^+$$

We can get around this problem by starting with the acyl chloride or anhydride instead of the carboxylic acid (similar to the preparation of carboxylic esters, Sec. 14.10). Because acyl chlorides and anhydrides are neutral compounds, salts do not form.

General reactions:

$$\underset{\text{Acyl chloride}}{R{-}\underset{\underset{O}{\|}}{C}{-}Cl} + \underset{\text{Primary amine}}{R'NH_2} \longrightarrow R{-}\underset{\underset{O}{\|}}{C}{-}NH{-}R' + HCl$$

$$\underset{\text{Anhydride}}{R{-}\underset{\underset{O}{\|}}{C}{-}O{-}\underset{\underset{O}{\|}}{C}{-}R''} + \underset{\text{Secondary amine}}{R'_2NH} \longrightarrow R{-}\underset{\underset{O}{\|}}{C}{-}NR'_2 + R''COOH$$

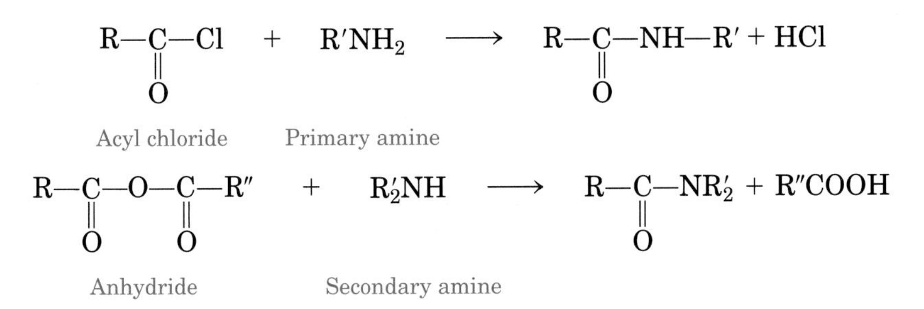

Specific examples:

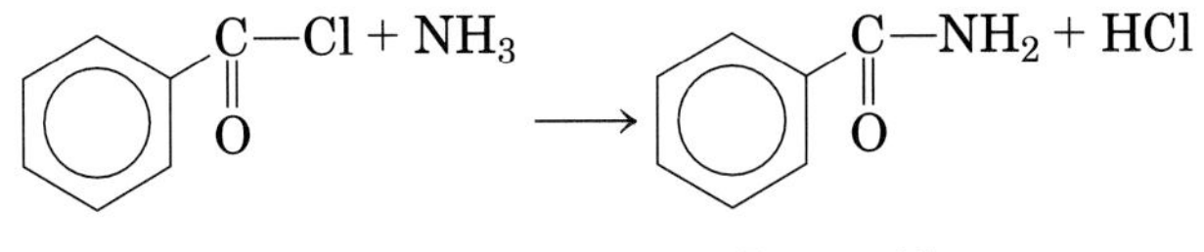

$$C_6H_5{-}\underset{\underset{O}{\|}}{C}{-}Cl + NH_3 \longrightarrow \underset{\text{Benzamide}}{C_6H_5{-}\underset{\underset{O}{\|}}{C}{-}NH_2} + HCl$$

Box 15D

Nylon

In Box 14E we met the polyesters, which are important synthetic fibers. Synthetic polyamides, especially nylon, are even more important.

Nylon is used to make shirts, dresses, stockings, underwear, and other kinds of clothing, as well as carpets, tire cord, rope, and parachutes. It also has nonfabric uses; for example, it is used in paint brushes, electrical parts, valves, gears, clips, and fasteners (Fig. 15D.1). It is very tough and strong, nontoxic, nonflammable, and resistant to chemicals.

Figure **15D.1** • Nylon is used to make carpeting and stockings, the latter shown at 300 times magnification. (Courtesy of E.I. Dupont de Nemours & Company, Inc.)

There are several types of nylon, but all are polyamides. The most important is made from hexamethylenediamine and adipoyl chloride (Fig. 15D.2). Like polyesters, polyamides are condensation polymers; each molecule has two ends that are "tied" together by amide linkages:

$$CH_3-\underset{\underset{O}{\|}}{C}-O-\underset{\underset{O}{\|}}{C}-CH_3 + C_6H_5NH_2 \longrightarrow CH_3-\underset{\underset{O}{\|}}{C}-NH-C_6H_5 + CH_3COOH$$

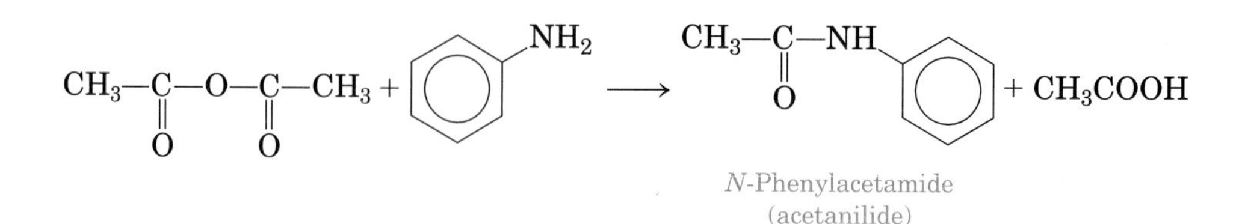

N-Phenylacetamide (acetanilide)

Tertiary amines do not undergo this reaction because they have no N—H bond.

Both acyl chlorides and anhydrides react readily with ammonia and with primary and secondary amines.

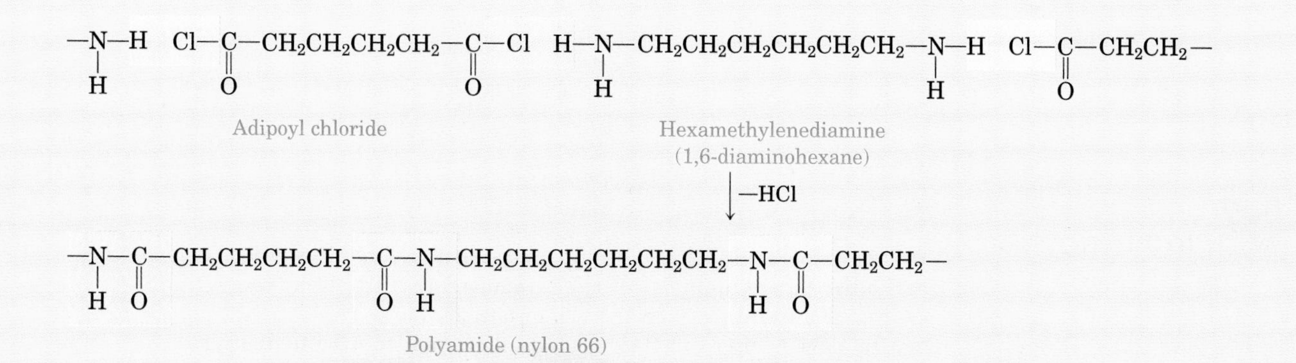

$$-\underset{\underset{\displaystyle H}{|}}{N}-H \quad Cl-\underset{\underset{\displaystyle O}{\|}}{C}-CH_2CH_2CH_2CH_2-\underset{\underset{\displaystyle O}{\|}}{C}-Cl \quad H-\underset{\underset{\displaystyle H}{|}}{N}-CH_2CH_2CH_2CH_2CH_2CH_2-\underset{\underset{\displaystyle H}{|}}{N}-H \quad Cl-\underset{\underset{\displaystyle O}{\|}}{C}-CH_2CH_2-$$

Adipoyl chloride — Hexamethylenediamine (1,6-diaminohexane)

$$\Big\downarrow -HCl$$

$$-\underset{\underset{\displaystyle H}{|}}{N}-\underset{\underset{\displaystyle O}{\|}}{C}-CH_2CH_2CH_2CH_2-\underset{\underset{\displaystyle O}{\|}}{C}-\underset{\underset{\displaystyle H}{|}}{N}-CH_2CH_2CH_2CH_2CH_2CH_2-\underset{\underset{\displaystyle H}{|}}{N}-\underset{\underset{\displaystyle O}{\|}}{C}-CH_2CH_2-$$

Polyamide (nylon 66)

Figure **15D.2** • The formation of nylon 66. Hexamethylenediamine is dissolved in water (bottom layer), and adipoyl chloride is dissolved in hexane (top layer). The two compounds meet at the interface between the two layers to form nylon, which is being wound onto a stirring rod. (Photograph by Charles D. Winters.)

An HCl molecule comes out every time an amide linkage is created. The nylon formed from adipoyl chloride and hexamethylenediamine is called nylon 66 because the acid and amine each have six carbons. Other nylons can be made from other dicarboxylic acid chlorides and diamines.

Problem **15.9**

Draw the structural formula for each product:

(a) $CH_3-\underset{\underset{\displaystyle O}{\|}}{C}-Cl + NH_3 \longrightarrow$

(b) $CH_3CH_2CH_2-\underset{\underset{\displaystyle O}{\|}}{C}-O-\underset{\underset{\displaystyle O}{\|}}{C}-CH_2CH_2CH_3 + CH_3-NH-CH_2CH_3 \longrightarrow$

Box 15E

Barbiturates

Urea is a *diamide* of carbonic acid:

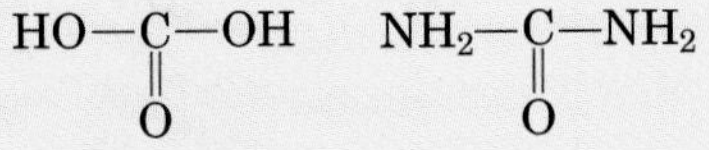

$$\underset{\text{Carbonic acid}}{HO{-}\overset{\displaystyle O}{\overset{\|}{C}}{-}OH} \qquad \underset{\text{Urea}}{NH_2{-}\overset{\displaystyle O}{\overset{\|}{C}}{-}NH_2}$$

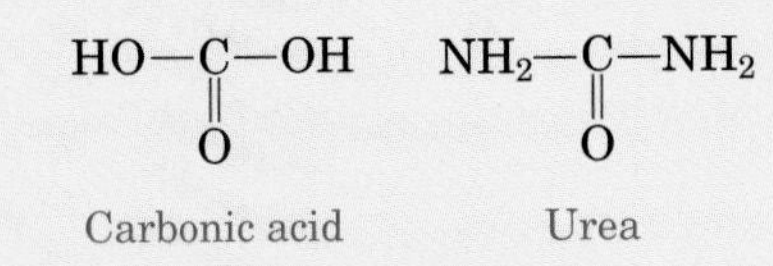

It is an important biological compound that is soluble in water. Humans and other higher animals get rid of their waste nitrogen by converting it to urea and excreting it in the urine (Sec. 21.8).

In 1864 Adolph von Baeyer (1835–1917) added urea to diethyl malonate (the diethyl ester of malonic acid) in the presence of ethoxide ion ($CH_3CH_2O^-$). The product was a cyclic compound that Baeyer called barbituric acid:

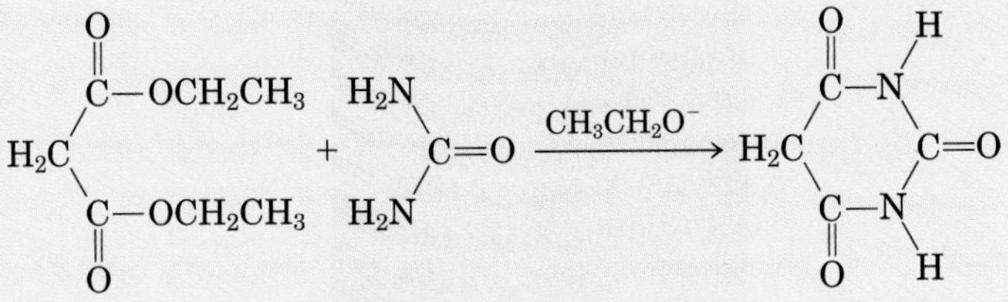

$$\underset{\text{Diethyl malonate}}{H_2C(COOCH_2CH_3)_2} + \underset{\text{Urea}}{(H_2N)_2C{=}O} \xrightarrow{CH_3CH_2O^-} \underset{\text{Barbituric acid}}{H_2C(C(=O){-}NH)_2C{=}O}$$

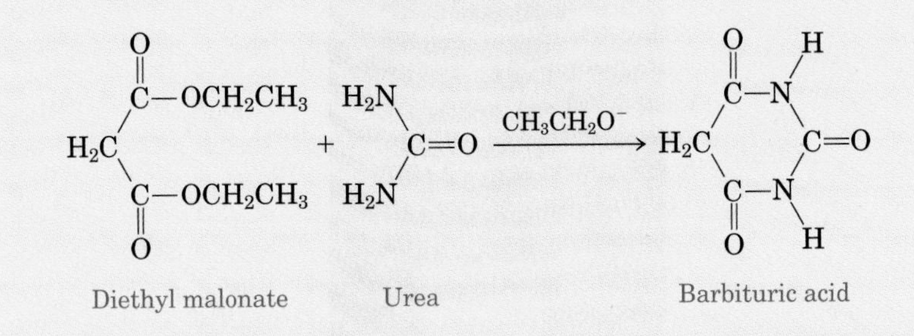

As you can see, this compound is a cyclic amide. It was later discovered that certain derivatives of barbituric acid had powerful physiological effects. They could put people to sleep. Barbituric acid itself is not used for this purpose, but a number of derivatives, substituted at the CH_2 group, are important drugs. Some of them are

15.8 Hydrolysis of Amides

The most important reaction of amides is hydrolysis. This reaction is analogous to the hydrolysis of carboxylic esters; that is, the amide is cleaved into two pieces: the carboxylic acid part and the amine or ammonia part:

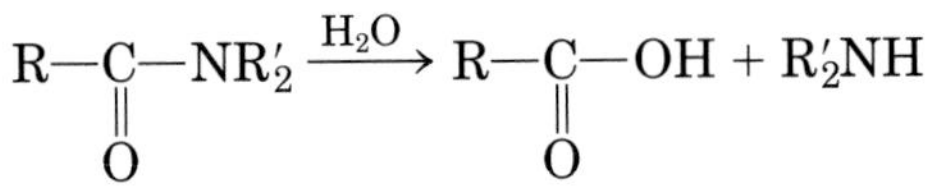

$$R{-}\overset{\displaystyle O}{\overset{\|}{C}}{-}NR'_2 \xrightarrow{H_2O} R{-}\overset{\displaystyle O}{\overset{\|}{C}}{-}OH + R'_2NH$$

As with carboxylic ester hydrolysis, amide hydrolysis requires the presence of a strong acid or base, but there is a difference. In Section 14.11 we saw that under basic conditions ester hydrolysis is not reversible

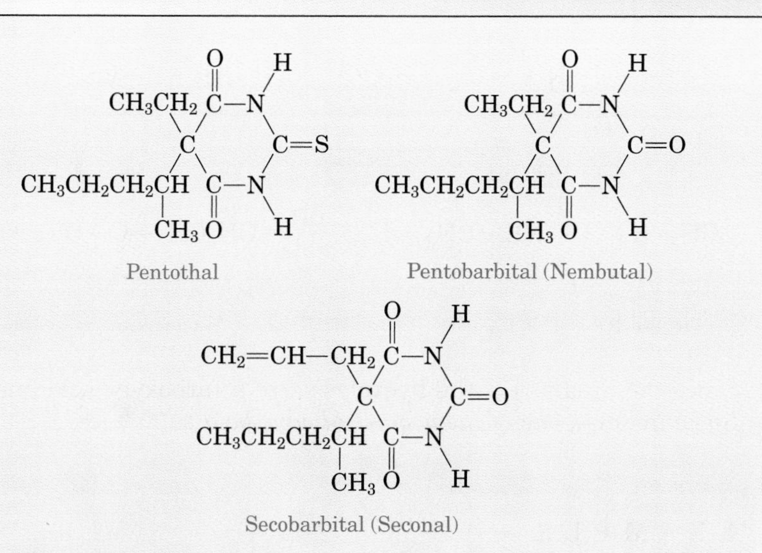

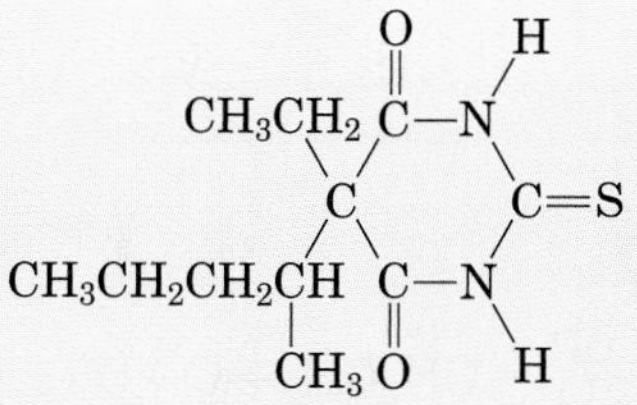

```
        O    H
        ‖   /
CH3CH2  C—N
     \ /     \
      C       C=S
     / \     /
CH3CH2CH2CH  C—N
     |   ‖     \
    CH3  O      H
```

Pentothal

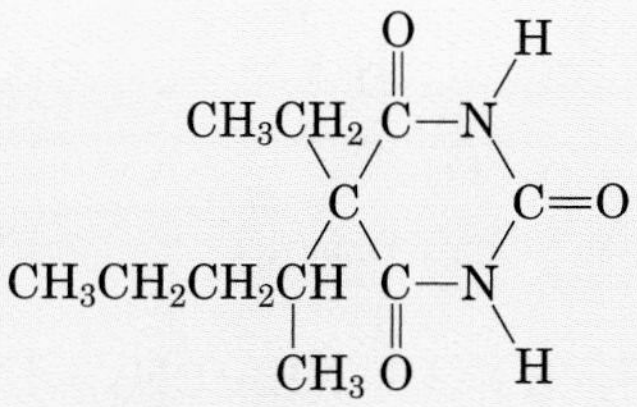

```
        O    H
        ‖   /
CH3CH2  C—N
     \ /     \
      C       C=O
     / \     /
CH3CH2CH2CH  C—N
     |   ‖     \
    CH3  O      H
```

Pentobarbital (Nembutal)

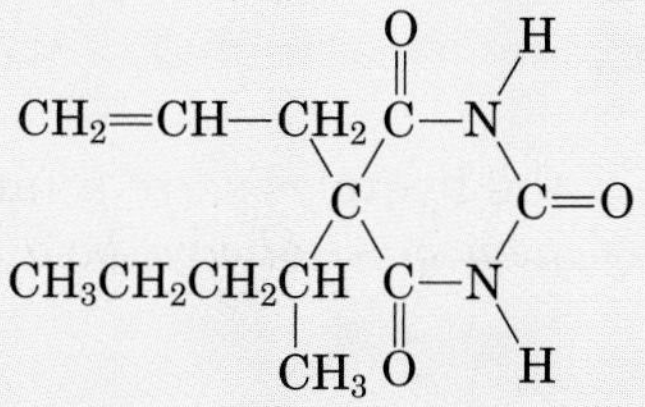

```
              O    H
              ‖   /
CH2=CH—CH2    C—N
           \ /     \
            C       C=O
           / \     /
 CH3CH2CH2CH  C—N
           |  ‖    \
          CH3 O     H
```

Secobarbital (Seconal)

To increase solubility in water, these drugs are often given in the form of their salts. Technically, only the salts should be called **barbiturates,** but in practice, all these compounds are given that name, whether in the free form or the salt form.

Barbiturates have two principal effects. In small doses they are sedatives (tranquilizers); in larger doses they induce sleep. Pentothal is used as a general anesthetic by dentists. Pentobarbital and secobarbital are often used to prepare patients for surgery.

Barbiturates are addictive. Because they act as tranquilizers, they are called "downers" (the opposite of amphetamines, Box 15A) or "goof balls." Barbiturates are responsible for many deaths. Barbiturates are especially dangerous when combined with alcohol, as they often are. The combined effect (called a **synergistic effect**) is usually greater than the sum of the effects from the two drugs taken separately.

(because the carboxylate *ion* is formed) but that acid-catalyzed ester hydrolysis is reversible because the carboxylic acid and alcohol that are produced can recombine to give the ester. With amides neither reaction is reversible because ions are formed in *both* cases. Let us look at the hydrolysis of a mono-*N*-substituted amide under both acidic and basic conditions:

General reactions:

$$\mathrm{R{-}\underset{\underset{\displaystyle O}{\|}}{C}{-}NHR' + H_2O \xrightarrow{H^+} R{-}\underset{\underset{\displaystyle O}{\|}}{C}{-}OH + R'\overset{+}{N}H_3}$$

$$\mathrm{R{-}\underset{\underset{\displaystyle O}{\|}}{C}{-}NHR' + H_2O \xrightarrow{OH^-} R{-}\underset{\underset{\displaystyle O}{\|}}{C}{-}O^- + R'NH_2}$$

The H^+, of course, is added in the form of an acid, such as HCl, and the $R'NH_3^+$ will actually be the salt $R'NH_3^+\ Cl^-$. Similarly, in the basic reaction, the OH^- may be in the form of NaOH, and the product will be the salt $RCOO^-\ Na^+$.

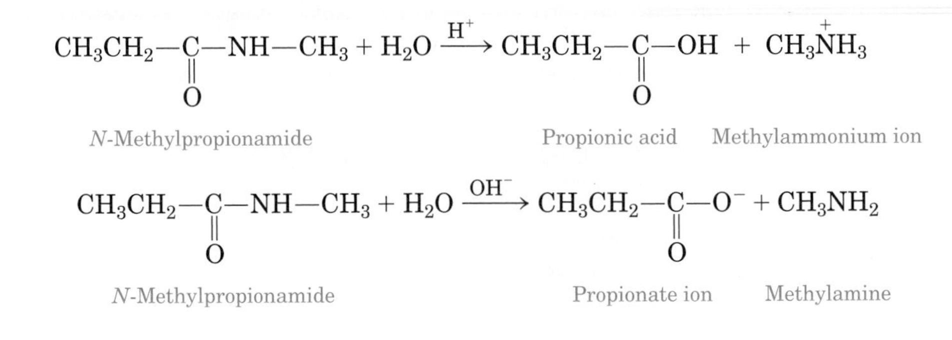

Specific examples:

$$CH_3CH_2-\underset{\underset{O}{\|}}{C}-NH-CH_3 + H_2O \xrightarrow{H^+} CH_3CH_2-\underset{\underset{O}{\|}}{C}-OH + CH_3\overset{+}{N}H_3$$

N-Methylpropionamide — Propionic acid — Methylammonium ion

$$CH_3CH_2-\underset{\underset{O}{\|}}{C}-NH-CH_3 + H_2O \xrightarrow{OH^-} CH_3CH_2-\underset{\underset{O}{\|}}{C}-O^- + CH_3NH_2$$

N-Methylpropionamide — Propionate ion — Methylamine

Since the products of the hydrolysis are a carboxylic acid and an amine (or ammonia), *one of them must always be a salt.*

EXAMPLE

Show the products, under acidic and basic conditions, of hydrolysis of

(a) $C_6H_5-\underset{\underset{O}{\|}}{C}-NH_2$ Benzamide

(b) $CH_3-\underset{\underset{O}{\|}}{C}-\underset{\underset{CH_3}{|}}{N}-CH_3$ *N,N*-Dimethylacetamide

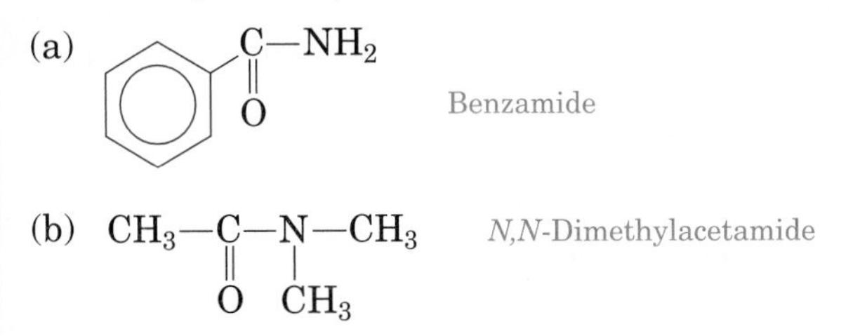

Answer • In both types of hydrolysis, cleavage takes place between the C=O carbon and the N:

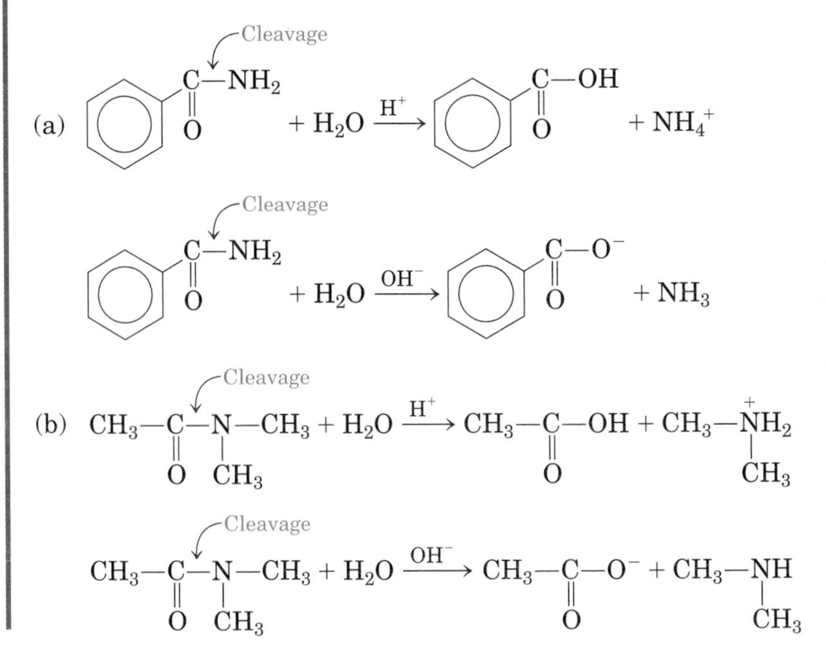

(a) (Cleavage between C and NH_2)

$$C_6H_5-\underset{\underset{O}{\|}}{C}-NH_2 + H_2O \xrightarrow{H^+} C_6H_5-\underset{\underset{O}{\|}}{C}-OH + NH_4^+$$

(Cleavage between C and NH_2)

$$C_6H_5-\underset{\underset{O}{\|}}{C}-NH_2 + H_2O \xrightarrow{OH^-} C_6H_5-\underset{\underset{O}{\|}}{C}-O^- + NH_3$$

(b) (Cleavage between C and N)

$$CH_3-\underset{\underset{O}{\|}}{C}-\underset{\underset{CH_3}{|}}{N}-CH_3 + H_2O \xrightarrow{H^+} CH_3-\underset{\underset{O}{\|}}{C}-OH + CH_3-\underset{\underset{CH_3}{|}}{\overset{+}{N}H_2}$$

(Cleavage between C and N)

$$CH_3-\underset{\underset{O}{\|}}{C}-\underset{\underset{CH_3}{|}}{N}-CH_3 + H_2O \xrightarrow{OH^-} CH_3-\underset{\underset{O}{\|}}{C}-O^- + CH_3-\underset{\underset{CH_3}{|}}{NH}$$

Problem 15.10

Show the products of hydrolysis, under acidic and basic conditions, of

$$CH_3CH_2-\underset{\underset{O}{\|}}{C}-NH-CH_2CH_3.$$

The hydrolysis of amides is a very important reaction in biochemistry. Most amide hydrolysis in the body is catalyzed by enzymes rather than by acids or bases. Enzymes are much better catalysts, and biological amide hydrolysis is a very rapid reaction.

15.9 Alkaloids

For thousands of years, primitive tribes in various parts of the world have known that physiological effects can be obtained by eating or chewing the leaves, roots, or bark of certain plants. The effects vary with the plant. Some cure disease; others, such as opium, are addictive drugs. Still others are deadly poisons, such as the leaves of the belladonna plant.

Chewing the bark of the cinchona tree, found in the Andes, is an effective treatment for malaria (it cured the Spanish viceroy to Peru in 1638).

The nature of the compounds that cause these effects was an interesting problem for chemists, and beginning around 1800, when analytical methods became advanced enough for the task, they began to extract the compounds responsible. They discovered that all these compounds are weak bases and called them alkaloids (meaning alkali-like). Today we define an **alkaloid** as **a basic compound obtained from a plant.** Many of them have physiological activity of one kind or another. They are very widespread in nature, and their structures vary widely. We have already seen that virtually all organic bases are amines, and so we are not surprised to learn that all alkaloids are amines.

We will look at a few of the more important alkaloids. Note that they generally have common names, usually taken from the name of the plant from which they are extracted, and, like other amines, their names end in "-ine."

Thousands of different alkaloids have been extracted from the leaves, bark, roots, flowers, and fruits of plants.

Coniine

One of the simplest alkaloids, coniine,

N–H (ring) with $CH_2CH_2CH_3$

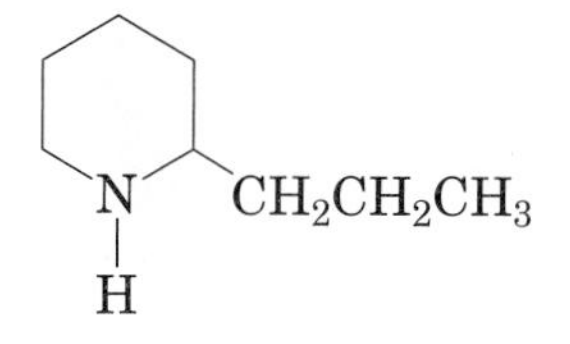

Coniine (2-propylpiperidine)

is very poisonous. It is found in the hemlock plant. The great philosopher Socrates (469–399 BC) committed suicide by drinking a cup of hemlock extract after he was found guilty of corrupting the youth of Athens (Fig. 15.2).

Nicotine

The active ingredient in tobacco is the alkaloid nicotine:

Nicotine has two heterocyclic rings—one aromatic (a pyridine ring) and one aliphatic (a pyrrolidine ring).

Nicotine

In large doses nicotine is a poison. Solutions of it are used as an insecticide.

Many people ingest it by smoking or chewing. In small doses nicotine is a stimulant. It is not especially harmful in itself, but it is habit-forming and thus exposes smokers to the harmful effects of the other components of cigarette smoke (tars, carbon monoxide, the carcinogens mentioned in Box 11G, and other poisons).

Caffeine

As you may already know, caffeine is the compound that is mainly responsible for the stimulating action of coffee and tea. It is also present in chocolate, cocoa, colas, and "stay-awake" pills (No-Doz). It is not only

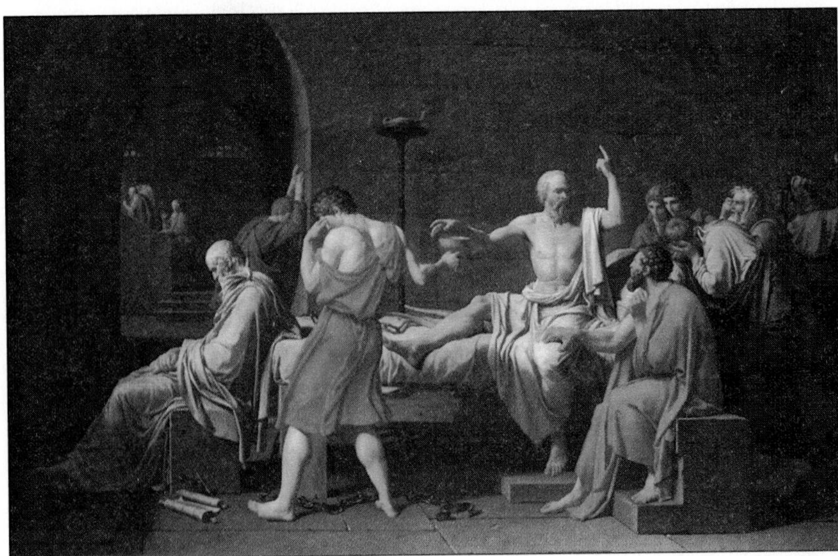

Figure **15.2** • *The Death of Socrates,* as painted by the French artist Jacques David (1748–1825) in 1787. Socrates was sentenced to death but could have escaped. He chose to drink the cup of hemlock. (Courtesy of the Metropolitan Museum of Art, New York.)

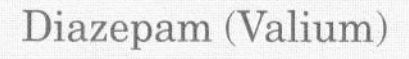

Box 15F

Tranquilizers

In our modern society the pressures and conflicts of daily life are much greater than they were in earlier times, when everything moved at a slower tempo. Drugs called **tranquilizers** are prescribed by physicians for the relief of anxiety disorders or for the short-term relief of the symptoms of anxiety or tension.

Most of these drugs belong to a family called the benzodiazepines, all of which have similar structures. The most popular is diazepam (Valium), but clorazepate and others are also frequently prescribed. These drugs all have a seven-membered ring containing two nitrogens:

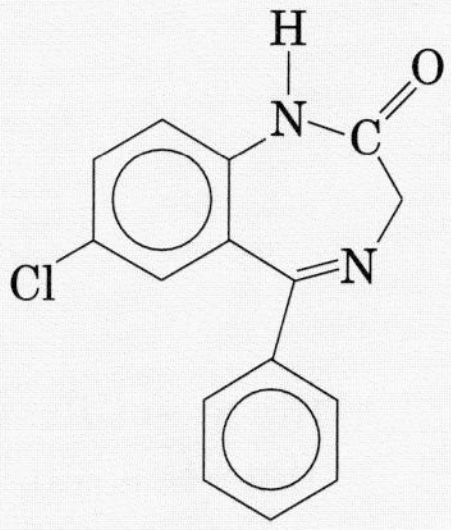

Diazepam (Valium)

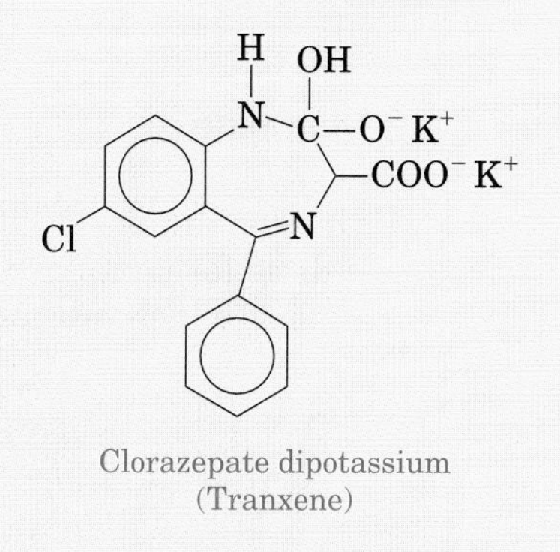

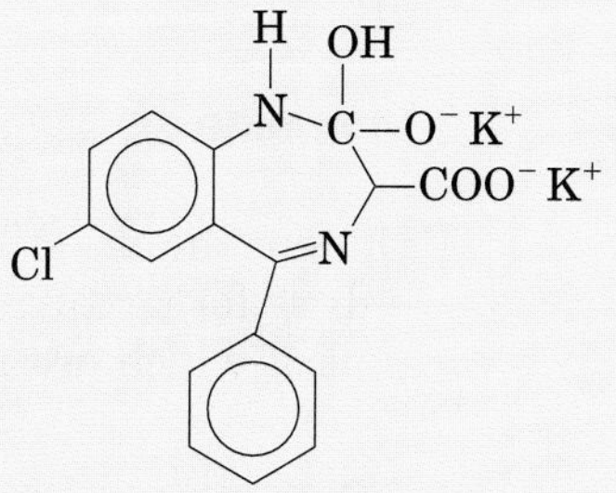

Clorazepate dipotassium (Tranxene)

These compounds are useful for relieving anxiety symptoms, but they are all habit-forming and must be used with care.

an alkaloid but also an amide (can you see why?) and a purine because it has the same ring system as the compound purine:

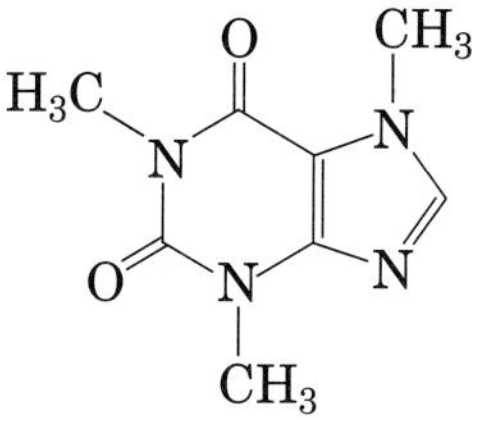

Caffeine

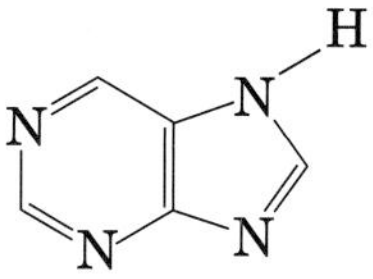

Purine

We shall meet other purines in Chapter 23.

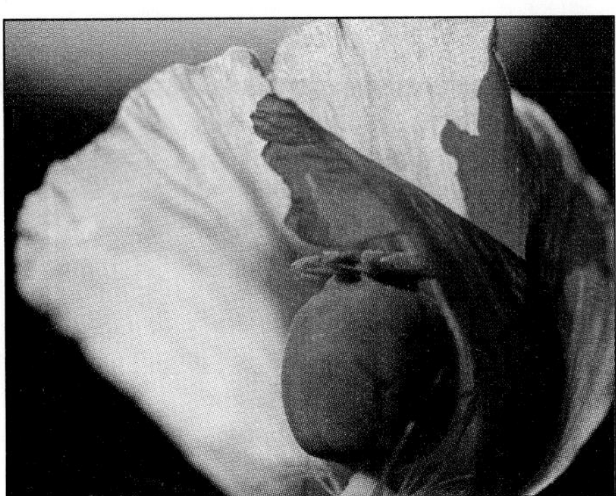

Figure **15.3**
A poppy plant. (FourByFive.)

Atropine

Several poisonous plants, including henbane and belladonna, contain atropine:

Atropine

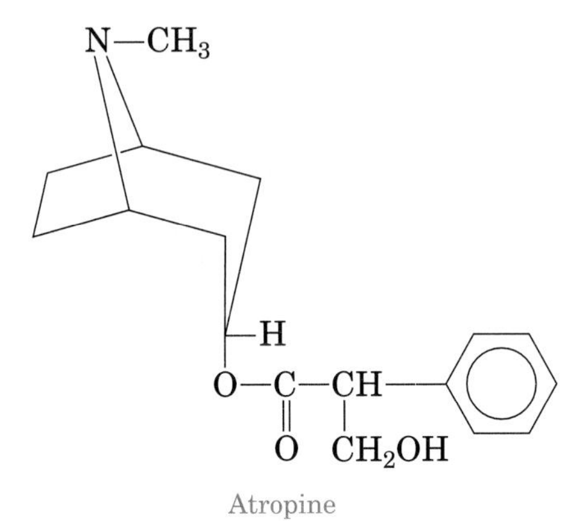

Dilute solutions of atropine have been used for thousands of years to dilate the pupils of the eyes (make them larger). The name "belladonna" (which means beautiful lady) probably comes from the fact that Roman women used extracts from this plant to make themselves more attractive.

The American writer Ambrose Bierce defined "belladonna" in his satirical *Devil's Dictionary* as follows: "In Italian a beautiful lady; in English a deadly poison. A striking example of the essential identity of the two tongues."

Morphine

Opium, the dried latex of the poppy plant (Fig. 15.3), has been used as a drug for centuries. It contains a number of alkaloids, the most important of which are morphine and its methyl ether, codeine.

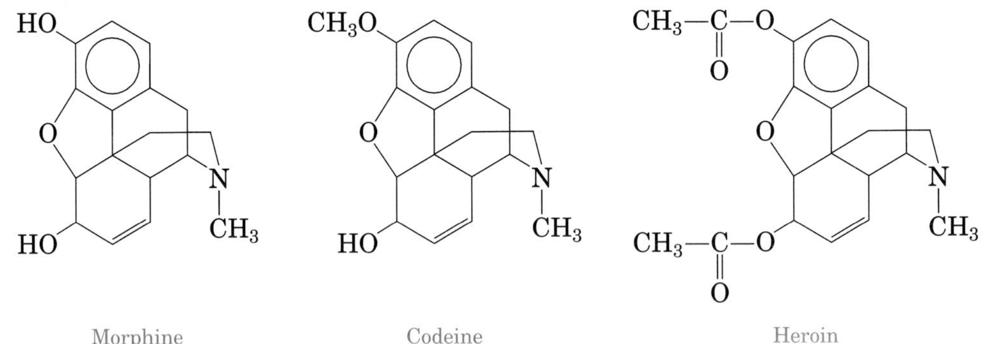

Morphine Codeine Heroin

Morphine is a valuable drug in medicine, being one of the most effective pain-killers known. Drugs of this type are called *narcotics.* The major drawback of these drugs is that they are addictive. Heroin is even more harmful than morphine. Heroin is not found in nature, but is made from morphine by esterification of the OH groups with acetic anhydride (Sec. 14.10). The ester groups make the heroin less polar, and as a consequence it enters the brain more rapidly than morphine. Once there, however, it is converted back to morphine by ester hydrolysis and is trapped in the brain cells. The use of heroin is illegal, and morphine is restricted to hospital use. Codeine is sometimes used in cough syrups to decrease throat discomfort.

In general, nonpolar molecules pass from blood vessels into the brain more easily than polar molecules, because they have to cross the nonpolar membranes of the endothelial cells surrounding the blood vessels. (See the blood-brain barrier, Sec. 25.1.)

Chemists today are still searching for safe, effective substitutes for these compounds. A purely synthetic substitute for the pain-killing effect of morphine is Demerol, and this drug is widely prescribed as a pain-killer. Methadone, another synthetic compound, is used as a substitute for heroin to aid in withdrawal from that drug. However, methadone is itself addictive and cannot represent a complete solution to the problem of heroin addiction.

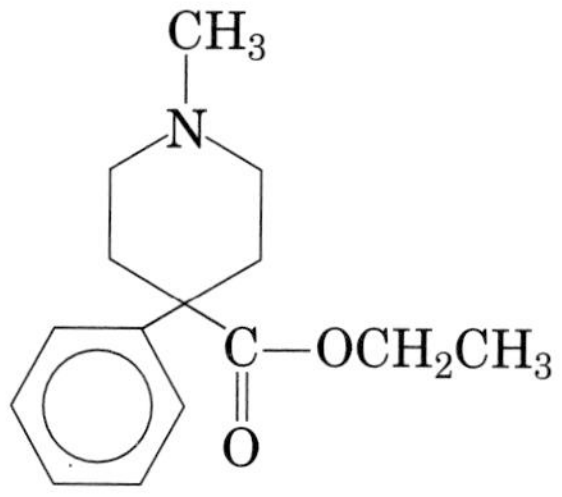

Demerol

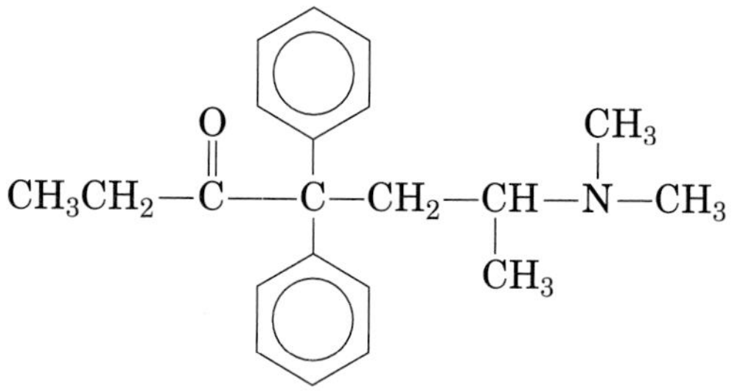

Methadone

Other Alkaloids

Other alkaloids worth mentioning are quinine, strychnine, and reserpine, all of which have complicated formulas with several rings:

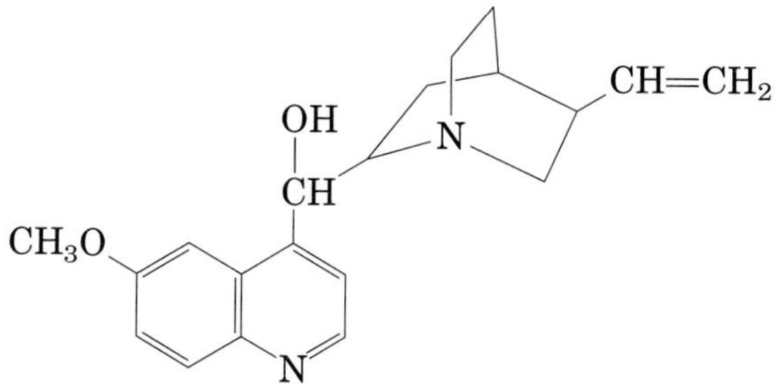

Quinine

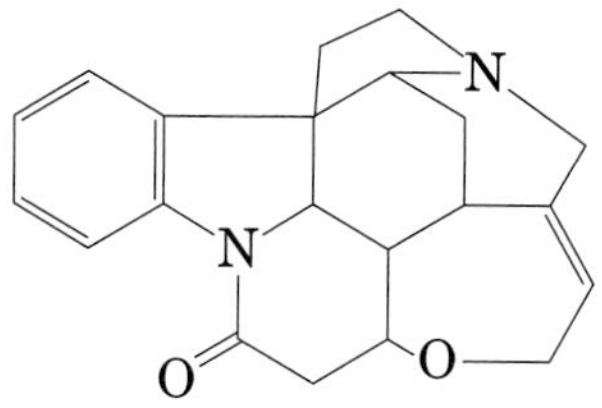

Strychnine

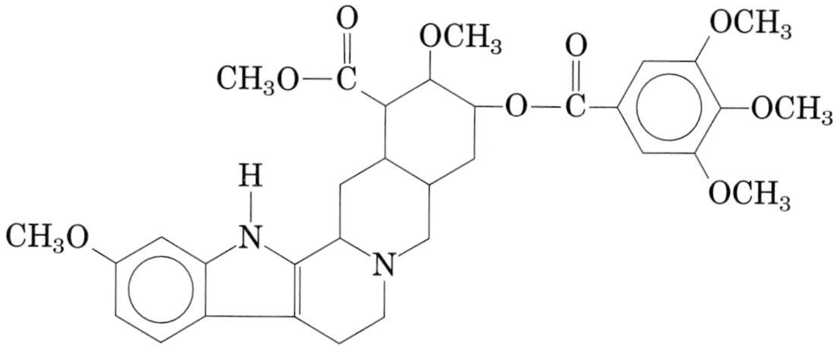

Reserpine

Box 15G

Cocaine and Local Anesthetics

Cocaine has a formula very similar to that of atropine:

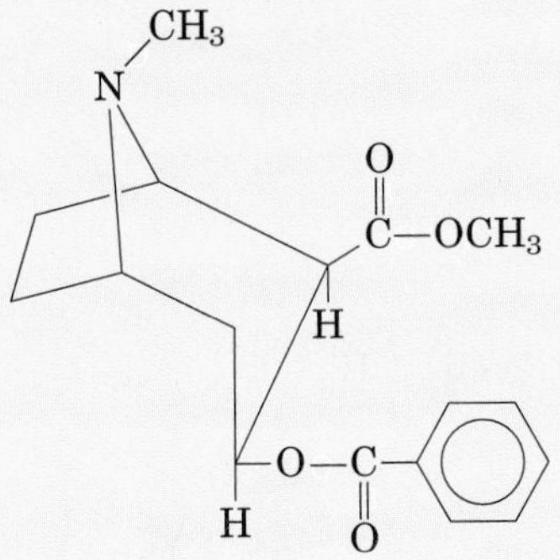

Cocaine

The leaves of the coca bush, found in Peru and Bolivia, are rich in cocaine, and the local inhabitants have chewed them for centuries to obtain its stimulating action. Cocaine reduces fatigue, permits greater physical endurance, and gives a feeling of tremendous confidence and power. In some of the Sherlock Holmes stories, the great detective injects himself with a 7 percent solution of cocaine to overcome boredom. When the addictive property of cocaine became apparent, it was made illegal but is still in widespread use. In recent years a cheap, smokable form of cocaine, known as "crack," has created many new addicts and caused enormous social and criminal problems. A particularly unfortunate example of this is that the addiction of pregnant women to crack is often transferred to their babies.

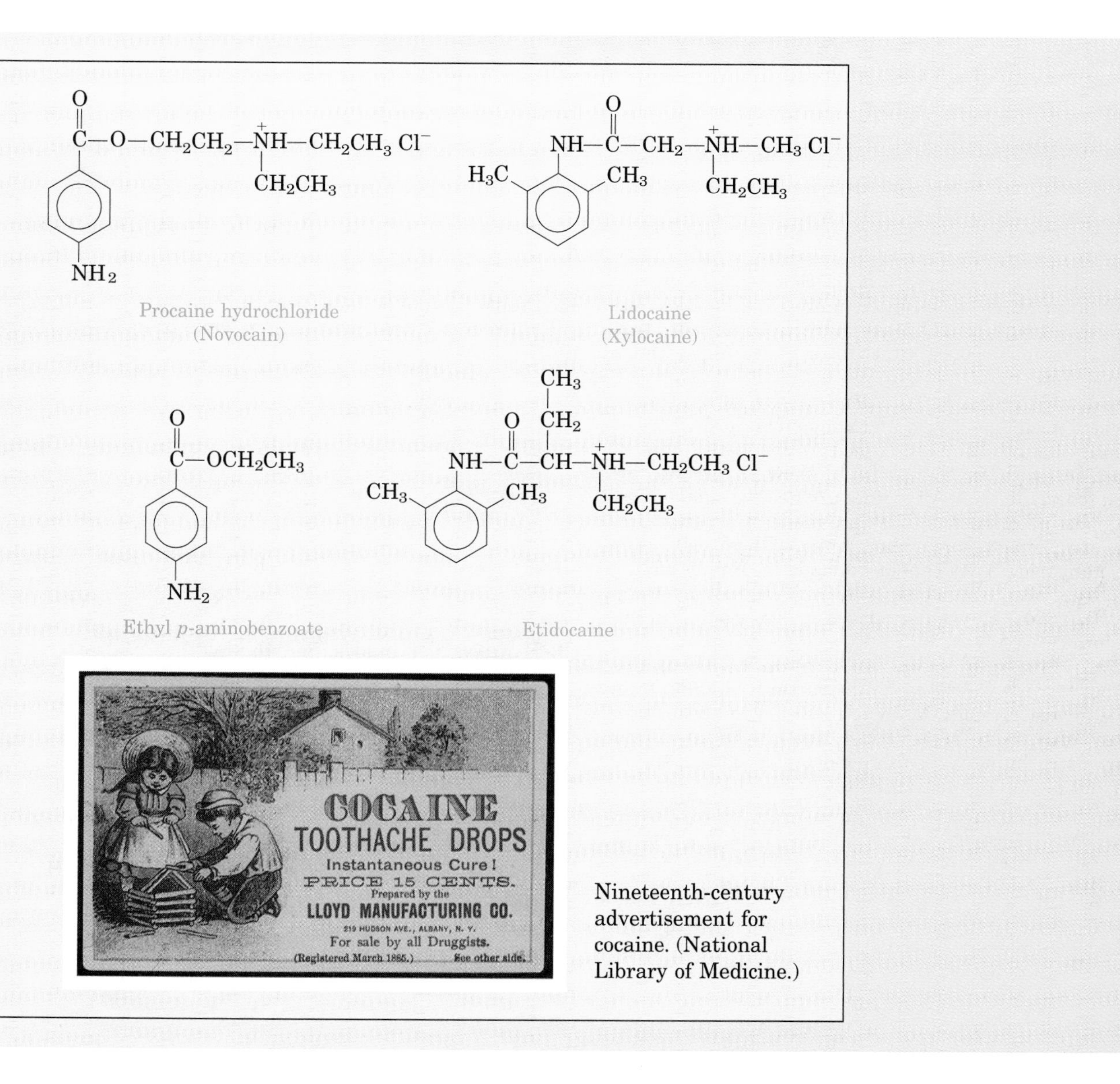

In the nineteenth century cocaine was used as a local anesthetic for such things as tooth extractions. Because of its toxicity, however, it was replaced by synthetic compounds. The first of these was procaine hydrochloride (Novocain), but this has now largely been replaced for dental use by lidocaine (Xylocaine). Ethyl *para*-aminobenzoate (Benzocaine) is present in sunburn lotions. Another local anesthetic is Etidocaine.

We mentioned at the beginning of this section that the bark of the cinchona tree is used to treat malaria. *Quinine,* which is extracted from this bark, is the actual compound responsible and is still used today for this purpose. *Strychnine,* which comes from the seeds of a plant called *Strychnos nux vomica,* has been used for many years as a poison for rats (sometimes for humans, too). It is also used medicinally, in low dosage, to counteract poisoning by central nervous system depressants. *Reserpine,* extracted from a shrub called Indian snake root, is used as a tranquilizer for patients in mental hospitals and as a drug that lowers blood pressure.

$$p\text{-}H_2N\text{-}C_6H_4\text{-}C(=O)\text{-}O\text{-}CH_2CH_2\text{-}\overset{+}{N}H(CH_2CH_3)\text{-}CH_2CH_3\ Cl^-$$

Procaine hydrochloride
(Novocain)

$$2,6\text{-}(CH_3)_2C_6H_3\text{-}NH\text{-}C(=O)\text{-}CH_2\text{-}\overset{+}{N}H(CH_2CH_3)\text{-}CH_3\ Cl^-$$

Lidocaine
(Xylocaine)

$$p\text{-}H_2N\text{-}C_6H_4\text{-}C(=O)\text{-}OCH_2CH_3$$

Ethyl *p*-aminobenzoate

$$2,6\text{-}(CH_3)_2C_6H_3\text{-}NH\text{-}C(=O)\text{-}CH(CH_2CH_3)\text{-}\overset{+}{N}H(CH_2CH_3)\text{-}CH_2CH_3\ Cl^-$$

Etidocaine

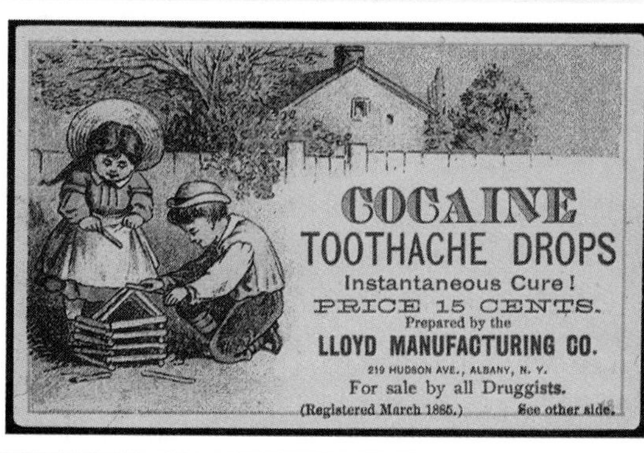

Nineteenth-century advertisement for cocaine. (National Library of Medicine.)

SUMMARY

Amines are derivatives of ammonia, NH_3. They are classified as **primary (RNH_2), secondary ($RNHR'$),** and **tertiary ($RR'NR''$),** depending on how many R groups are attached to the nitrogen. They are named by adding the suffix "-amine" to the name of the R groups. Their physical properties are similar to those of alcohols and ethers. The most important chemical property of amines is their basicity. They form salts with acids. Since these salts are soluble in water, even insoluble amines can be dissolved by treatment with acids. **Quaternary ammonium ions, R_4N^+,** have four groups attached to the nitrogen and a positive charge.

Amides are carboxylic acid derivatives that are combinations of a carboxylic acid and ammonia or a primary or secondary amine. They are named by changing "-ic acid" to "-amide," with an "*N*-" prefix to show any groups attached to the nitrogen. All amides are neutral compounds, and almost all are solids. They are made by treating ammonia or a primary or secondary amine with an acyl halide or an anhydride. They cannot be made by simply adding a carboxylic acid to ammonia or an amine (a salt is formed instead). Amides can be hydrolyzed in either acidic or basic solution.

Alkaloids are amines isolated from plants. Thousands are known, with widely varying structures ranging from simple to very complex. Many of them have physiological activity of various sorts. For example, quinine cures malaria, strychnine is a poison, morphine reduces severe pain, and nicotine is a stimulant.

Summary of Reactions

1. Reaction of amines with acids (Sec. 15.4):

Primary $RNH_2 + HCl \longrightarrow R\overset{+}{N}H_3\ Cl^-$

Secondary $R_2NH + R'COOH \longrightarrow R_2\overset{+}{N}H_2\ R'COO^-$

Tertiary $R_3N + HNO_3 \longrightarrow R_3\overset{+}{N}H\ NO_3^-$

2. Preparation of amides (Sec. 15.7):

a. From acyl chlorides and ammonia or amines:

$$R{-}\underset{\underset{O}{\|}}{C}{-}Cl + R_2'NH \longrightarrow R{-}\underset{\underset{O}{\|}}{C}{-}NR_2' + HCl$$

b. From anhydrides and ammonia or amines:

$$R{-}\underset{\underset{O}{\|}}{C}{-}O{-}\underset{\underset{O}{\|}}{C}{-}R + NH_3 \longrightarrow R{-}\underset{\underset{O}{\|}}{C}{-}NH_2 + R{-}\underset{\underset{O}{\|}}{C}{-}OH$$

3. Hydrolysis of amides (Sec. 15.8):

a. In acid solution:

$$R{-}\underset{\underset{O}{\|}}{C}{-}NH_2 + H_2O \xrightarrow{H^+} R{-}\underset{\underset{O}{\|}}{C}{-}OH + NH_4^+$$

b. In basic solution:

$$R{-}\underset{\underset{O}{\|}}{C}{-}NHR' + H_2O \xrightarrow{OH^-} R{-}\underset{\underset{O}{\|}}{C}{-}O^- + R'NH_2$$

KEY TERMS

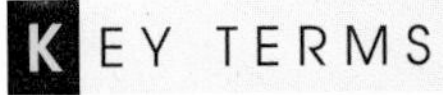

Alkaloid (Sec. 15.9)
Amide (Sec. 15.5)
Amine (Sec. 15.2)
Amine salt (Sec. 15.4)
Amphetamine (Box 15A)
Barbiturate (Box 15E)
Free base (Sec. 15.4)
Narcotic (Sec. 15.9)
Polyamide (Box 15D)
Primary amine (Sec. 15.2)
Quaternary ammonium salt (Sec. 15.4)
Secondary amine (Sec. 15.2)
Synergistic effect (Box 15E)
Tertiary amine (Sec. 15.2)
Tranquilizer (Box 15F)

PROBLEMS

Nomenclature of Amines

15.11 What is the difference between primary, secondary, and tertiary amines? Compare with the difference between primary, secondary, and tertiary alcohols.

15.12 In a secondary amine, how many hydrogens are attached to the nitrogen atom?

15.13 Classify each of the following as primary, secondary, or tertiary amines:

(a) 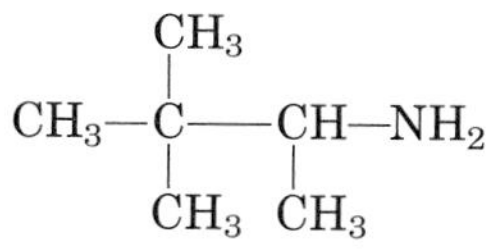 CH_3 above; CH_3—C——CH—NH_2; CH_3 CH_3 below

(b) N—CH_3 CH_3

(c) NH_2

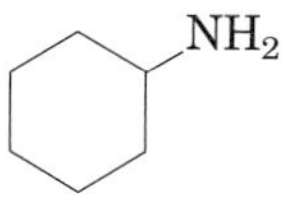

(d) 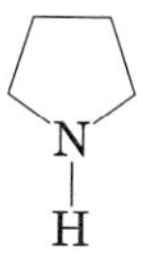 N H

(e) N CH_3

(f) CH_3 CH_3 above; CH_3—C—NH—C—CH_3; CH_3 CH_3 below

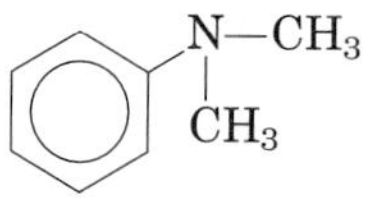

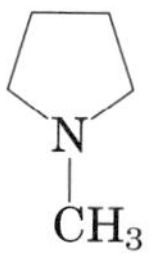

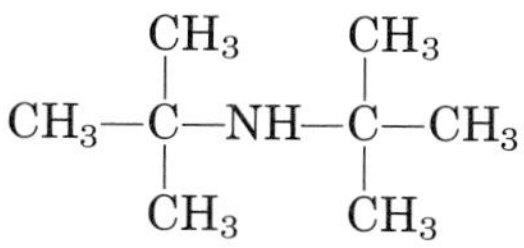

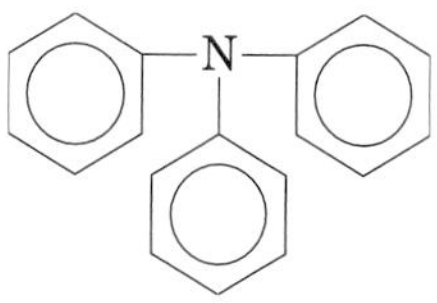

(g) N

(h) NH_2

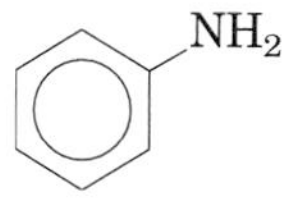

(i) 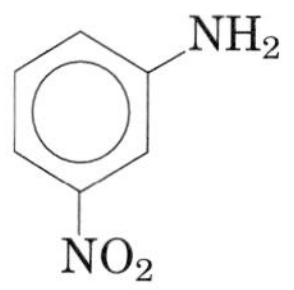 NH_2 NO_2

(j) 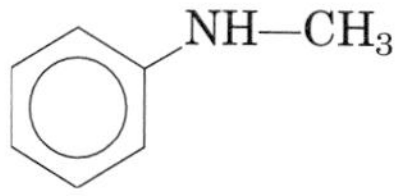 NH—CH_3

*(k) 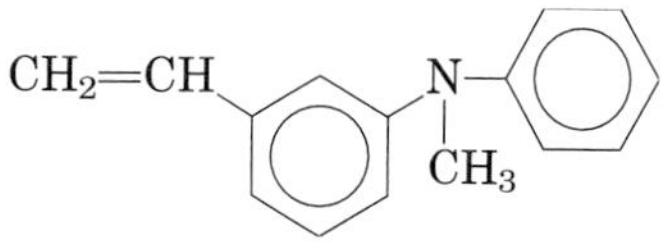 CH_2=CH, N, CH_3 (rings)

(l) 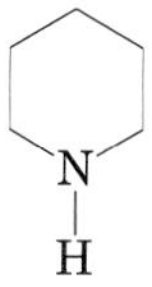 N H

(m) 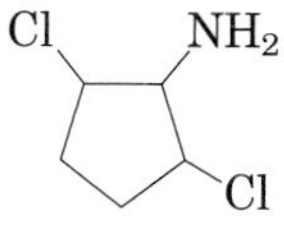 Cl NH_2 Cl

15.14 Name:

(a) $CH_3CH_2CH_2NH_2$

(b) CH_3CH_2—NH—CH_2CH_3

(c) $CH_3CH_2CH_2$—NH—CH_2CH_3

(d) 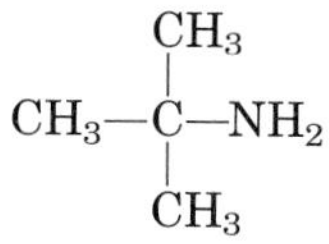 CH_3 above; CH_3—C—NH_2; CH_3 below

(e) 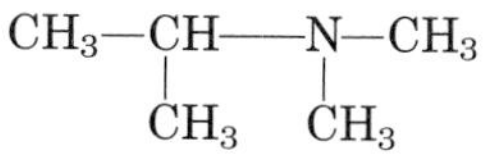 CH_3—CH——N—CH_3; CH_3 CH_3 below

(f) 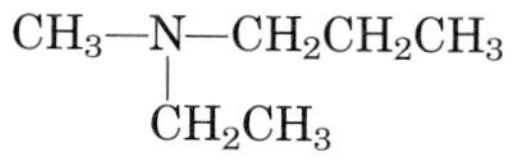 CH_3—N—$CH_2CH_2CH_3$; CH_2CH_3 below

15.15 Tell whether each amine in Problem 15.14 is primary, secondary, or tertiary.

15.16 Which of the following contain (1) heterocyclic aromatic rings, (2) heterocyclic nonaromatic rings, (3) no heterocyclic rings at all:

(a) atropine
(b) nicotine
(c) coniine
(d) morphine
(e) caffeine
(f) urea
(g) lidocaine
(h) Demerol
(i) pentobarbital
(j) quinine

15.17 Draw the structural formula for

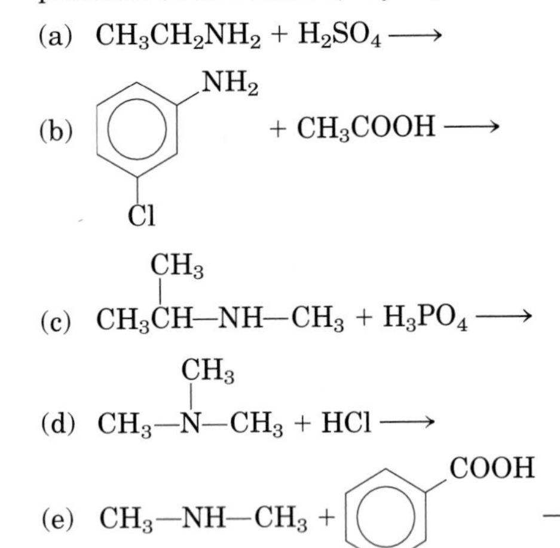

(a) methylamine
(b) di-*n*-propylamine
(c) *p*-methylaniline
(d) *N*-methylaniline
(e) triethylamine
(f) *sec*-butyl-*tert*-butylamine
(g) cadaverine

15.18 Unlike benzene, pyrrole has only five atoms in its ring. Why is it considered to be an aromatic compound?

Physical Properties of Amines

15.19 Without consulting any tables, arrange the following in order of increasing boiling point:

(a) $CH_3CH_2—O—CH_3$

(b) $CH_3—O—CH_3$

(c) $CH_3CH_2CH_2OH$

(d) $CH_3CH_2CH_2NH_2$

15.20 Explain why amines have about the same solubility in water as alcohols of the same molecular weights.

15.21 *n*-Propylamine ($CH_3CH_2CH_2NH_2$), ethylmethylamine ($CH_3CH_2NHCH_3$), and trimethylamine ($CH_3—N(CH_3)—CH_3$) all have the same molecular weight (they are isomers). Yet trimethylamine has a much lower boiling point than the other two (Table 15.1). Explain why.

Basicity of Amines. Quaternary Salts

15.22 Write the structural formulas for the principal products (if no reaction, say so).

(a) $CH_3CH_2NH_2 + H_2SO_4 \longrightarrow$

(b) 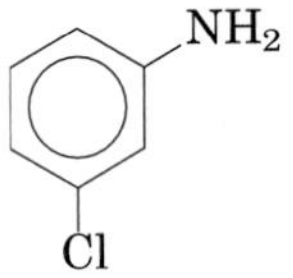 3-chloroaniline (benzene ring with NH_2 and Cl) $+ CH_3COOH \longrightarrow$

(c) $CH_3CH(CH_3)—NH—CH_3 + H_3PO_4 \longrightarrow$

(d) $CH_3—N(CH_3)—CH_3 + HCl \longrightarrow$

(e) $CH_3—NH—CH_3 +$ benzoic acid (benzene ring with COOH) $\longrightarrow$

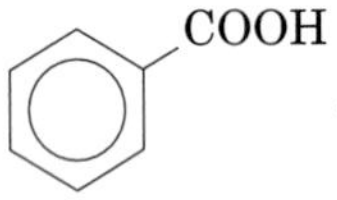

15.23 Draw the structural formulas for
(a) dimethylammonium iodide
(b) ethyltrimethylammonium hydroxide
(c) tetraethylammonium chloride
(d) anilinium nitrate

15.24 Name these salts:

(a) $CH_3CH_2—NH_3^+\ Cl^-$

(b) $CH_3CH_2—NH_2—CH_2CH_3^+\ Br^-$

(c) $CH_3CH_2—N(CH_2CH_3)(CH_2CH_3)—CH_2CH_3^+\ OH^-$

(d) 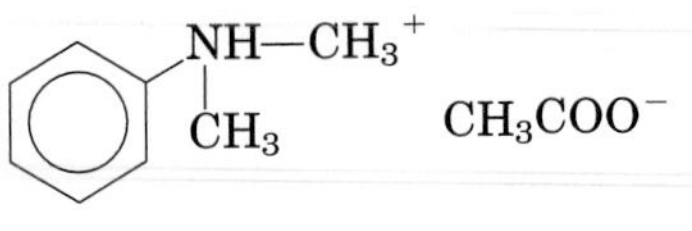 benzene ring with $NH(CH_3)—CH_3^+$ $\quad CH_3COO^-$

15.25 A student looking in an old chemistry book found the following name and formula:

$$CH_3—CH(CH_3)—NH_2 \cdot HBr$$

Isopropylamine hydrobromide

What are the modern name and formula for this compound?

15.26 If you dissolved $CH_3CH_2CH_2NH_2$ (*n*-propylamine) and $CH_3CH_2CH_2OH$ (*n*-propyl alcohol) in the same container of water and lowered the pH to 2 (by the addition of HCl), would anything happen to the structures of these compounds? If so, write the formulas for the species in solution at pH 2.

15.27 Explain why the drug morphine is often administered in the form of its salt, morphine sulfate.

15.28 How could you convert ethyl-*n*-propylamine, $CH_3CH_2CH_2—NH—CH_2CH_3$, to its salt?

15.29 Which of these are quaternary ammonium salts?

(a) $CH_3—N(CH_3)—CH_3$

(b) $CH_3CH_2—N^+(CH_3)(CH_3)—CH_3\ Cl^-$

(c) 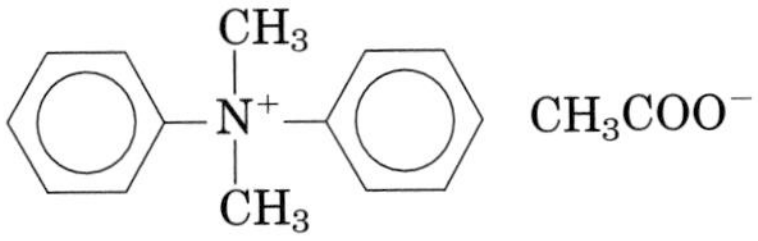 two benzene rings joined by $N^+(CH_3)(CH_3)$ $\quad CH_3COO^-$

(d) 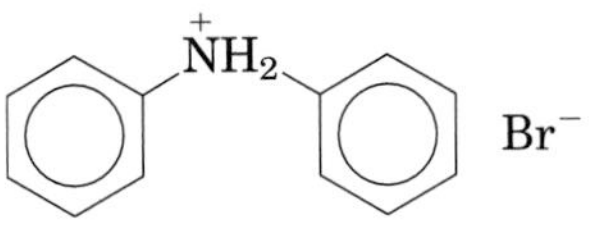 two benzene rings joined by $\overset{+}{N}H_2$ $\quad Br^-$

(e) 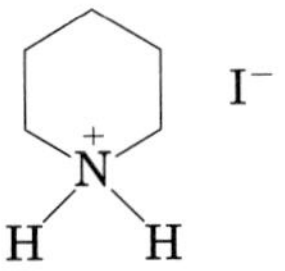 piperidine ring with $\overset{+}{N}$ bearing H and H $\quad I^-$

(f) 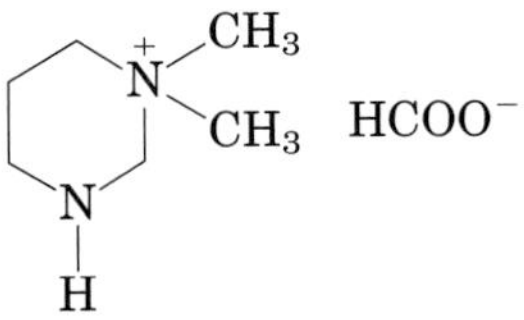 piperazine ring with one $\overset{+}{N}(CH_3)(CH_3)$ and one N—H $\quad HCOO^-$

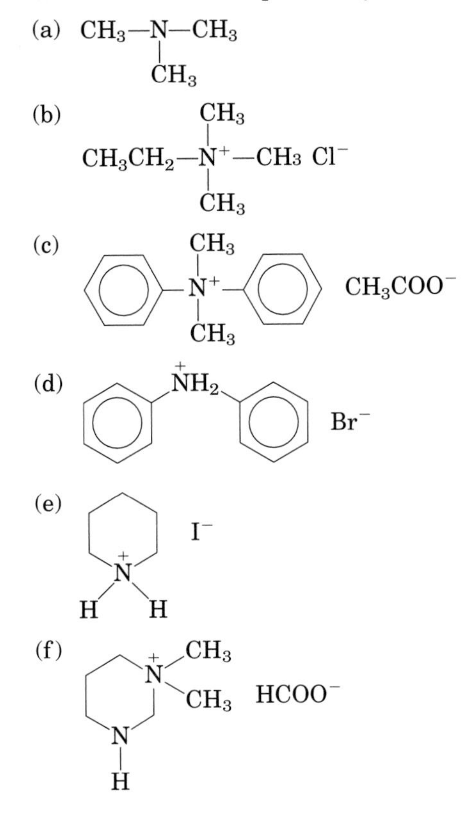

Structure and Nomenclature of Amides

15.30 How does the structure of an amide differ from that of an amine?

15.31 Tell whether each of the following is an amine, an amide, both, or neither.

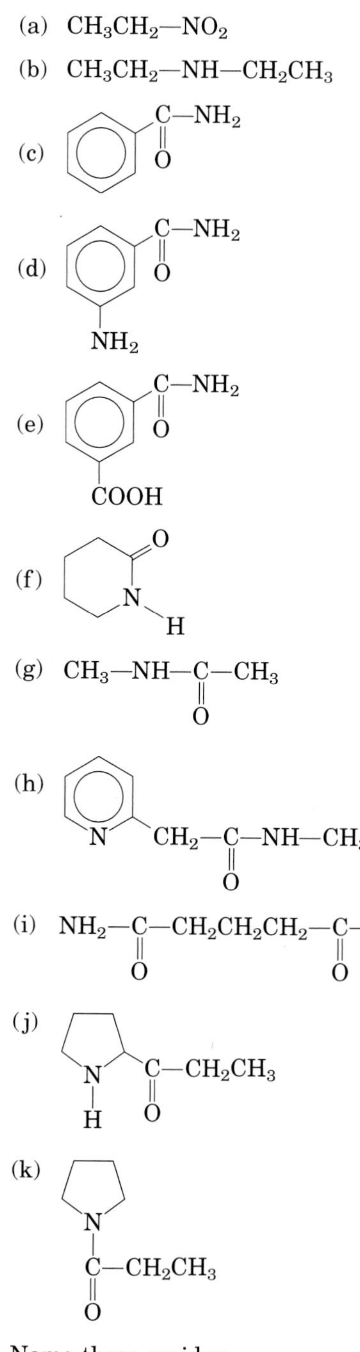

(a) $CH_3CH_2—NO_2$

(b) $CH_3CH_2—NH—CH_2CH_3$

(c) 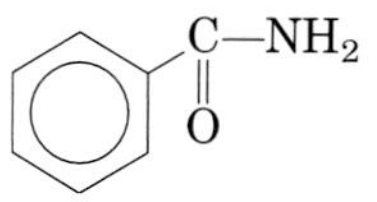 $C—NH_2$ ‖ O

(d) 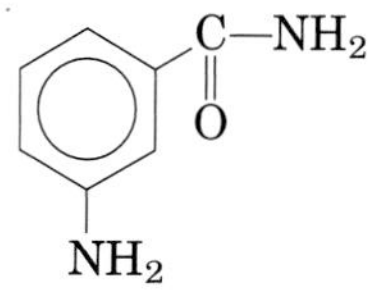 $C—NH_2$ ‖ O

NH_2

(e) 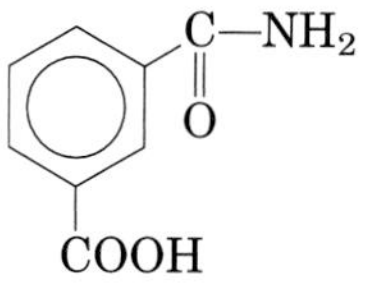 $C—NH_2$ ‖ O

COOH

(f) 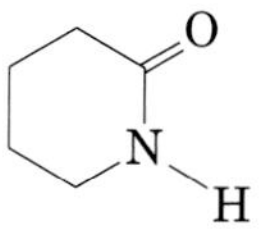 O

N

H

(g) $CH_3—NH—C—CH_3$ ‖ O

(h) 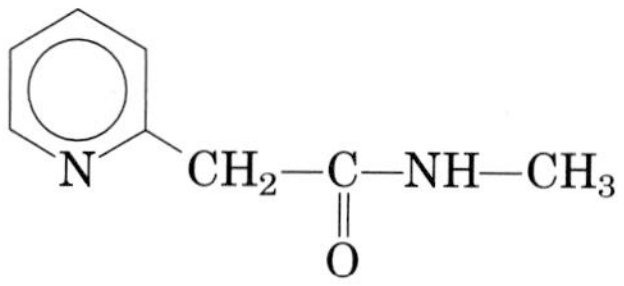 N $CH_2—C—NH—CH_3$ ‖ O

(i) $NH_2—C—CH_2CH_2CH_2—C—NH_2$ ‖ O ‖ O

(j) 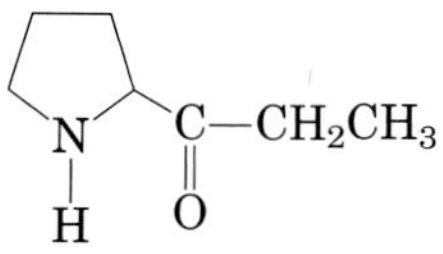 N $C—CH_2CH_3$ | H ‖ O

(k) 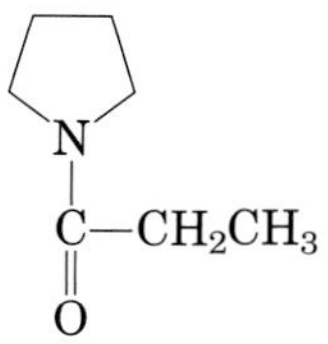 N | $C—CH_2CH_3$ ‖ O

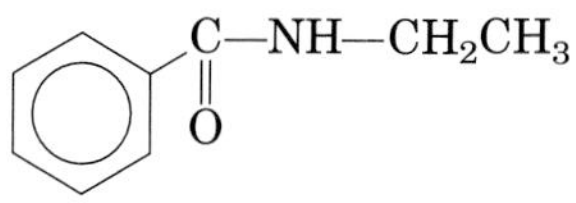

15.32 Name these amides:

(a) $H—C—NH_2$ ‖ O

(b) $C—NH—CH_2CH_3$ ‖ O

(c) $CH_3—C—N—CH_3$ ‖ O | CH_3

(d) 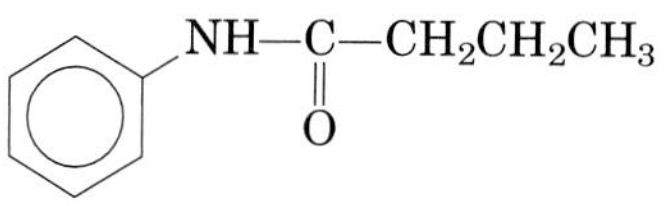 $NH—C—CH_2CH_2CH_3$ ‖ O

(e) $CH_3(CH_2)_{14}—C—NH—CH_3$ ‖ O

(f) 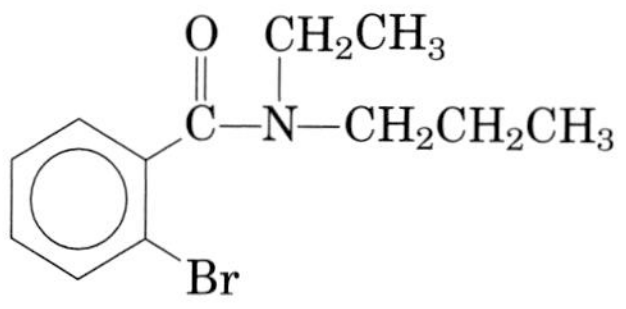 O CH_2CH_3 ‖ | $C—N—CH_2CH_2CH_3$

Br

15.33 Draw the structural formulas for
(a) propionamide
(b) *N*-ethylbutyramide
(c) *N,N*-dimethylacetamide
(d) *N*-phenylbenzamide
(e) *N*-ethyl-*N*-phenyl-*para*-bromobenzamide
(f) succindiamide

15.34 Which amides are liquids at room temperature?

Acid–Base Properties of Organic Compounds

15.35 Predict whether each of these compounds is acidic, basic, or neutral:

(a) CH_3OH

(b) CH_3NH_2

(c) CH_3SH

(d) 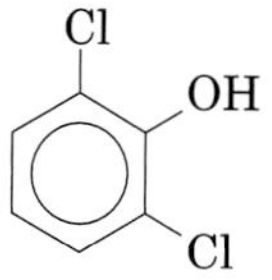 Cl OH Cl

(e) 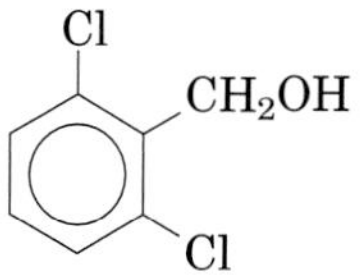 Cl CH_2OH Cl

(f) Cl CH_2NH_2 Cl

(g) Cl $CH_2—NH—C—CH_3$ ‖ O Cl

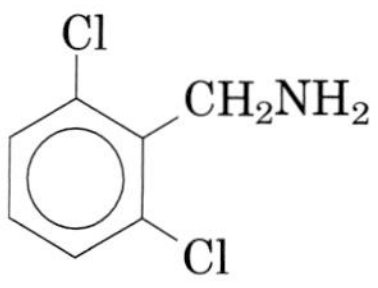

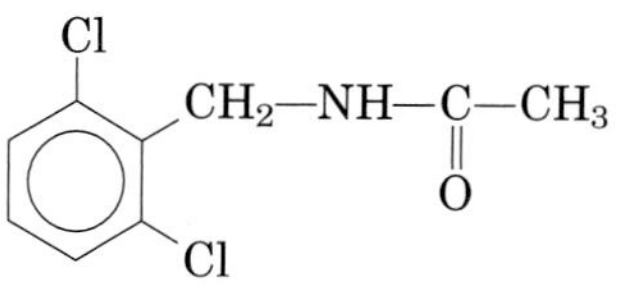

(h) $CH_3—C—H$ ‖ O

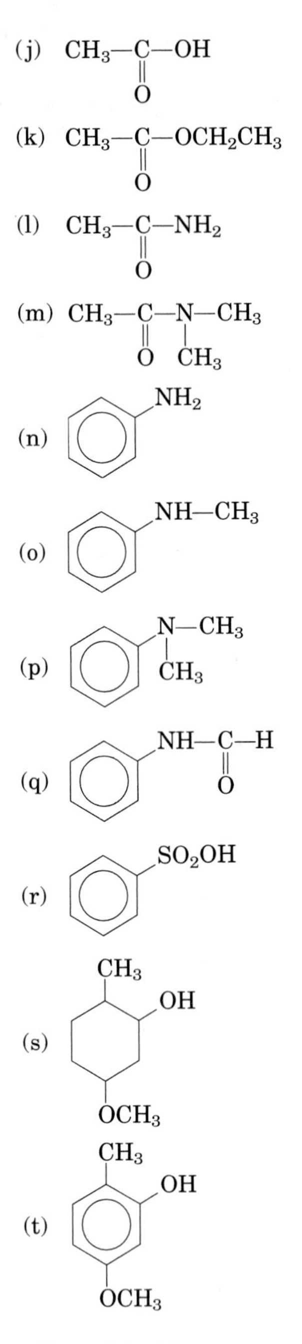

(i) $CH_3—C—CH_3$ ‖ O

(j) $CH_3-C(=O)-OH$

(k) $CH_3-C(=O)-OCH_2CH_3$

(l) $CH_3-C(=O)-NH_2$

(m) $CH_3-C(=O)-N(CH_3)-CH_3$

(n) 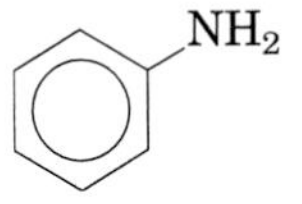 $C_6H_5-NH_2$

(o) 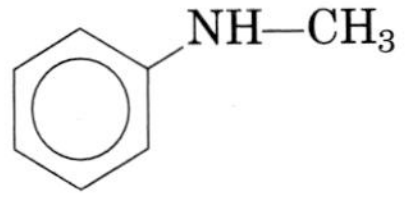 $C_6H_5-NH-CH_3$

(p) 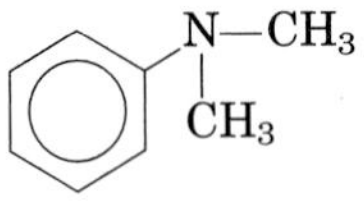 $C_6H_5-N(CH_3)-CH_3$

(q) 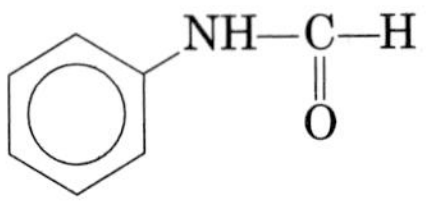 $C_6H_5-NH-C(=O)-H$

(r) 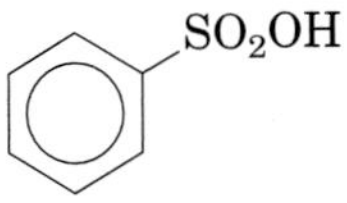 $C_6H_5-SO_2OH$

(s) 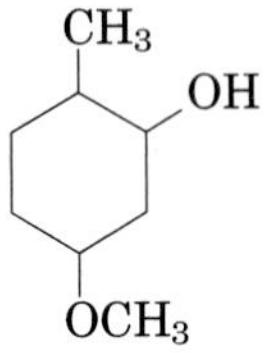 cyclohexane ring with CH_3, OH, and OCH_3 substituents

(t) 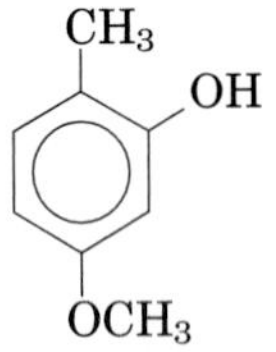 benzene ring with CH_3, OH, and OCH_3 substituents

Preparation of Amides

15.36 Is an amide formed when a carboxylic acid is added to an amine? If not, how can an amide be formed?

15.37 Write the structural formulas for the principal products (if no reaction, say so):

(a) $CH_3CH_2-C(=O)-OH + NH_3 \longrightarrow$

(b) 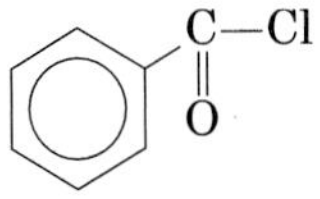 $C_6H_5-C(=O)-Cl + CH_3NH_2 \longrightarrow$

(c) $CH_3-C(=O)-O-C(=O)-CH_3 + CH_3CH_2-NH-CH_2CH_3 \longrightarrow$

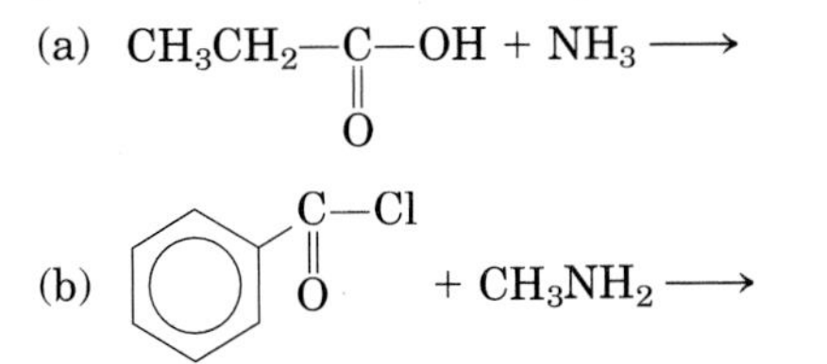

***15.38** Write the structural formulas of the principal organic products when putrescine (1,4-diaminobutane) reacts with (a) excess acetic acid (b) excess acetyl chloride.

15.39 Write the structural formulas for the principal organic products (if no reaction, say so):

(a) 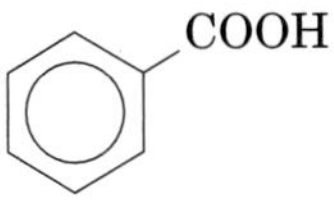 $C_6H_5-COOH + CH_3-NH-CH_3$

(b) $CH_3-C(=O)-Cl + CH_3CH_2-N(CH_2CH_3)-CH_3 \longrightarrow$

(c) $CH_3-C(=O)-O-C(=O)-CH_3 + CH_3NH_2 \longrightarrow$

(d) 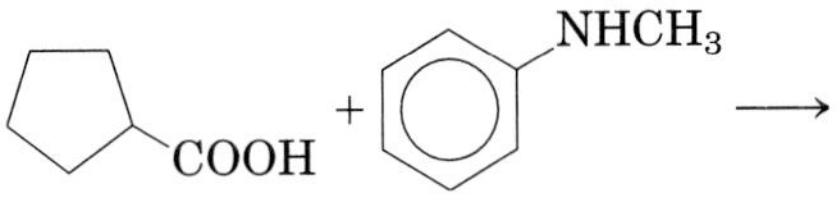 $C_5H_9-COOH + C_6H_5-NHCH_3 \longrightarrow$

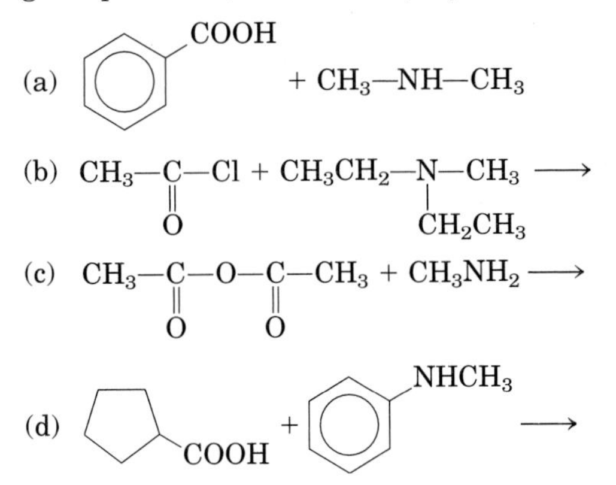

Hydrolysis of Amides

15.40 Tell which reagents can be used to bring about the following conversions:

(a) $CH_3CH_2-C(=O)-NH_2 \longrightarrow CH_3CH_2-C(=O)-O^-$

(b) $CH_3CH_2-C(=O)-NH_2 \longrightarrow CH_3CH_2-C(=O)-OH$

(c) 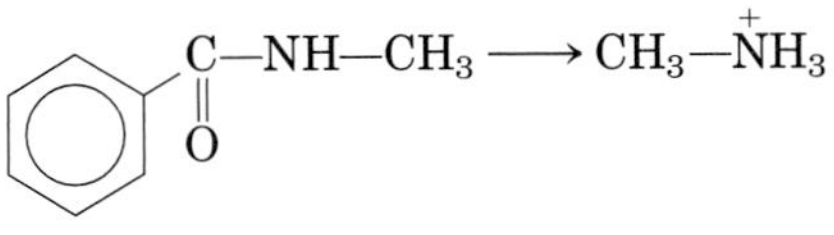 $C_6H_5-C(=O)-NH-CH_3 \longrightarrow CH_3-\overset{+}{N}H_3$

(d) 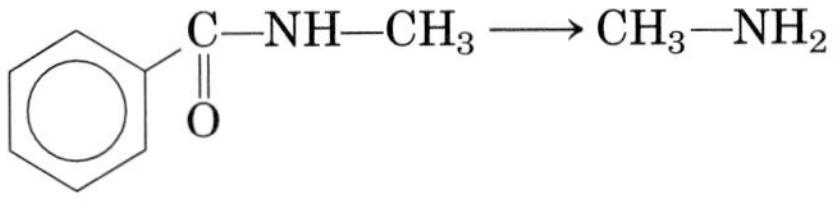 $C_6H_5-C(=O)-NH-CH_3 \longrightarrow CH_3-NH_2$

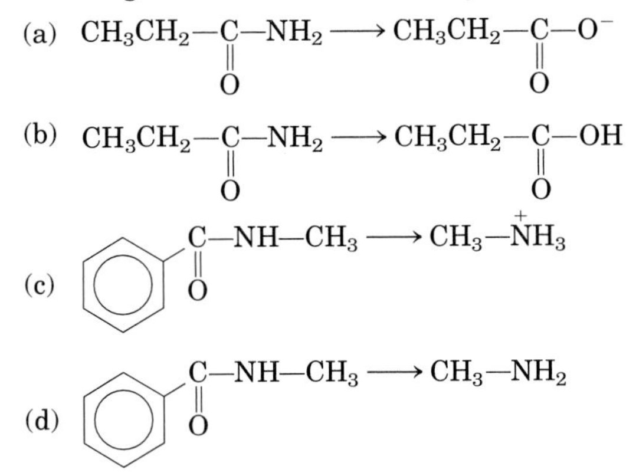

15.41 For the reactions shown in Problem 15.40, draw the structural formula for the other product formed in each case.

Alkaloids

15.42 Define alkaloid.

15.43 Which alkaloid

(a) is in coffee
(b) killed Socrates
(c) is in tobacco
(d) is used to dilate the pupils of eyes
(e) cures malaria
(f) is converted to heroin
(g) is medically used as an important pain-killer
(h) is a rat poison
(i) is an illegal drug that was taken by Sherlock Holmes

15.44 Identify all the functional groups in (a) seconal (b) coniine (c) morphine (d) Methadone. If alcohol or amine groups are present, tell whether they are primary, secondary, or tertiary.

***15.45** Which has more rings, strychnine or reserpine?

Boxes

15.46 (Box 15A) What are the possible negative side effects of illegal usage of "crank"?

15.47 (Box 15A) What is the basic structural difference between the natural hormone epinephrine and the synthetic pep pill methamphetamine?

15.48 (Box 15B) If you saw this label on a decongestant: phenylephrine · HCl, would you worry about being exposed to a strong acid such as HCl? Explain.

15.49 (Box 15C) What functional groups are present in (a) aspirin (b) acetaminophen (c) ibuprofen (d) naproxen (e) ketoprofen?

15.50 (Box 15C) Which pain-killer does not reduce swelling?

***15.51** (Box 15D) By analogy with the structure of nylon 66, draw the structure of nylon 46. If one of the starting compounds is adipoyl chloride, what is the name of the diamine used to make nylon 46?

***15.52** (Box 15D) As with polyesters, polyamides are also formed from acyl halides rather than from carboxylic acids. Draw the structure of the nylon that is formed from sebacyl chloride, Cl—C(=O)—$(CH_2)_8$—C(=O)—Cl, and hexamethylenediamine (1,6-diaminohexane).

15.53 (Box 15E) Urea is soluble in water. Show how water can form hydrogen bonds with urea.

15.54 (Box 15E) What is a synergistic effect? What is its importance when a patient takes barbiturates?

15.55 (Box 15E) Barbiturates are derivatives of urea. Identify the portion of the structure in secobarbital that contains the urea.

15.56 (Box 15F) Does diazepam contain a urea portion, as the barbiturates do?

15.57 (Box 15F) What structure is common to all the benzodiazepines?

15.58 (Box 15G) Cocaine has two carboxylic ester groups in its structure. Identify both.

15.59 (Box 15G) What is "crack"?

Additional Problems

15.60 Give IUPAC names for the following amphetamines:

(a) 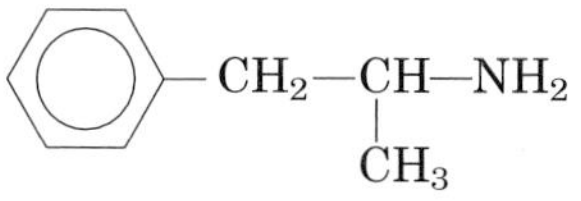 C_6H_5—CH_2—CH(CH_3)—NH_2

Amphetamine

(b) 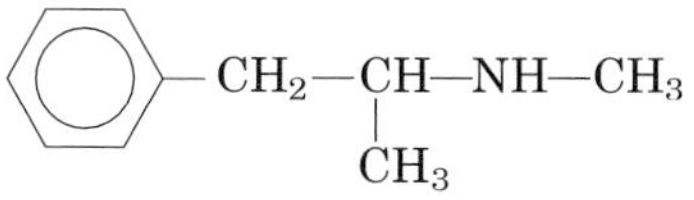 C_6H_5—CH_2—CH(CH_3)—NH—CH_3

Methamphetamine

15.61 Draw the structure of nicotinium chloride (nicotine hydrochloride).

15.62 Both *n*-decylamine and *n*-decyl alcohol are insoluble in water. If you were given a mixture of these two liquids, how could you separate them without distillation?

15.63 Draw the structural formula of any particular quaternary ammonium salt.

***15.64** Cetylpyridinium chloride has the structure

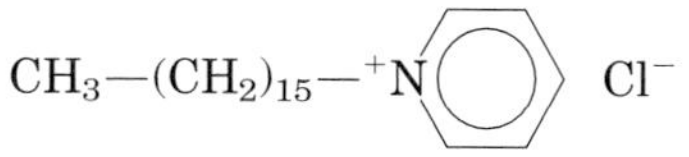

CH_3—$(CH_2)_{15}$—$^{+}$N(pyridinium ring) Cl^-

(a) What makes this compound, which bears a long-chain alkyl group, water-soluble?

(b) What class of compounds does it belong to?

15.65 There are four amines with the molecular formula C_3H_9N. Draw structural formulas for all of them and classify each as primary, secondary, or tertiary.

15.66 Why is each of the following not considered to be an alkaloid?

(a) Demerol
(b) lidocaine
(c) heroin
(d) marijuana (tetrahydrocannabinol)

15.67 Show the principal products:

(a) H—C(=O)—NH—CH_3 + H_2O $\xrightarrow{H^+}$

(b) CH_3CH_2—C(=O)—NH_2 + H_2O $\xrightarrow{OH^-}$

(c) $CH_3(CH_2)_8$—C(=O)—N(CH_3)—CH_3 + H_2O $\xrightarrow{OH^-}$

(d) 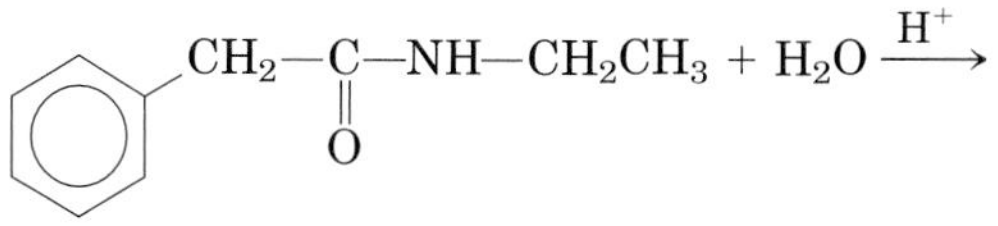 C_6H_5—CH_2—C(=O)—NH—CH_2CH_3 + H_2O $\xrightarrow{H^+}$

*(e) 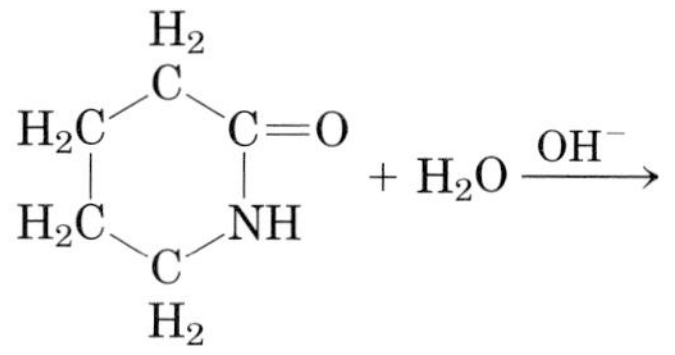 (six-membered ring: H_2C—CH_2—C(=O)—NH—CH_2—CH_2, i.e. ring of C=O, NH, and four CH_2 groups) + H_2O $\xrightarrow{OH^-}$

15.68 How could you convert butylammonium chloride, $CH_3CH_2CH_2CH_2\overset{+}{N}H_3\ Cl^-$, to the free amine, $CH_3CH_2CH_2CH_2NH_2$?

***15.69** When a certain amide, of molecular formula C_8H_9NO, was hydrolyzed with an acid catalyst, two organic products were isolated, one of which was benzoic acid. Draw structural formulas (a) for the other organic product and (b) for the original amide.

15.70 In each case, tell which compound has the higher boiling point:

(a) (1) $CH_3CH_2CH_2CH_2NH_2$ vs.
(2) $CH_3CH_2CH_2CH_2CH_3$

(b) (1) $CH_3CH_2CH_2CH_2NH_2$ vs.
(2) $CH_3CH_2CH_2CH_2OH$

(c) (1) $CH_3CH_2CH_2CH_2NH_2$ vs.
(2) $CH_3CH_2CH_2NH_2$

(d) (1) $CH_3CH_2CH_2CH_2NH_2$ vs.
(2) $CH_3CH_2—N(CH_3)—CH_3$

15.71 When an ophthalmologist uses atropine to dilate the pupils for an eye examination, he or she warns the patient not to go into bright sunlight without sunglasses for a few hours because too much light will temporarily blind the patient. What conclusion can you draw from this regarding the effect of a topical application of atropine?

***15.72** Name

(a) $NH_2—C(=O)—CH_2—C(=O)—NH_2$

(b) $NH—CH_3$

(c) CH_3, $N^+—CH_3$, CH_3 CH_3COO^-

(d) 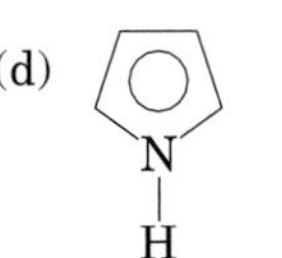

(e) $NH_2—(CH_2)_5—NH_2$

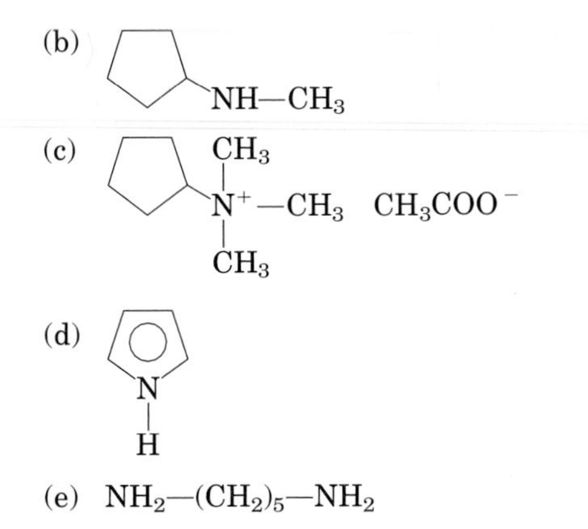

15.73 What kind of odors do low-molecular-weight amines have?

Part **three**

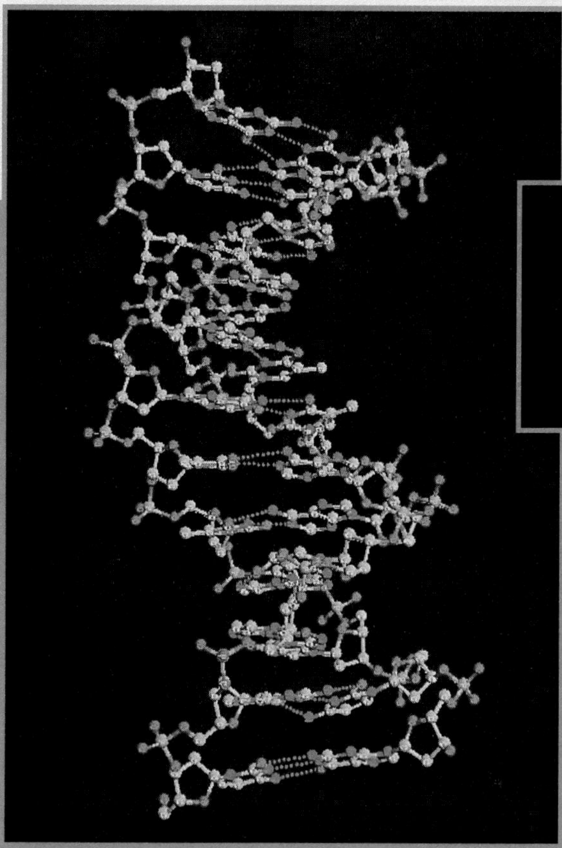

Photograph by Charles D. Winters.

Biochemistry

Jacqueline K. Barton

Dr. Jacqueline K. Barton is a native New Yorker and was educated in that city. She received her B.A. degree from Barnard College in 1974 and her Ph.D. from Columbia University in 1979. Following her Ph.D. work, Dr. Barton did further research at Yale University and Bell Laboratories and then joined the faculty of Hunter College. In 1983 she returned to Columbia University where she rose rapidly to the rank of full professor. In the fall of 1989 she assumed her present position as Professor of Chemistry at the California Institute of Technology.

Dr. Barton has done outstanding research in the field of biochemistry, particularly in the design of simple molecular probes to explore the variations in structure and conformation along the DNA helix (the subject of Chapter 23). In spite of having been a research scientist for a relatively short time, she has done important new work and has received many honors. In 1985 she received the Alan T. Waterman Award of the National Science Foundation as the outstanding young scientist in the United States. In 1987 she was the recipient of the American Chemical Society's Eli Lilly Award in Biological Chemistry, and the following year she received the Society's Award in Pure Chemistry. That same year, 1988, she also received the Mayor of New York's Award of Honor in Science and Technology.

We met in Dr. Barton's office at Cal Tech. As we find with so many chemists, she is interested in art and has a painting by the Spanish artist Miro on her office wall as well as a print by the French artist Vasarely. As we were to see in the interview, this interest in form and color in art carries over into her research.

Jacqueline K. Barton

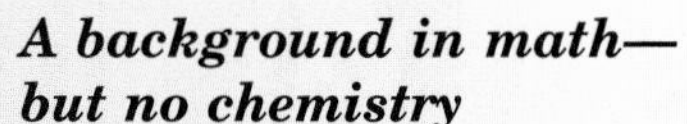

A background in math—but no chemistry

We have been interested in how the scientists we interviewed came to their careers in chemistry. In Dr. Barton's case she said, "I never took chemistry in high school. Maybe one shouldn't publicize that, but it's the truth. However, I was always very interested in mathematics, so I took a lot of calculus when I was in high school. I also took a course in geometry, and that interest in geometry has carried over into my research, since the sort of science I do now is very much governed by structures and shapes.

"When I went to college I thought that, in addition to taking math, I should take some science courses. I walked into the freshman chemistry class, and there were about 150 people there. However, there was also a small honors class with about 10 students. Even though I hadn't had chemistry before I thought I would try it—and I loved it. What chemistry allowed me to do was to combine the abstract and the real. I was very excited by it."

But, Dr. Barton says, it was really the experience of the laboratory that got her interested in chemistry. Like many of us in chemistry, she was fascinated by color changes in reactions and the significance of these observations. However, she was also interested in trying to predict what would happen in a reaction and, if her prediction was not correct, to try to explain this and then to do more experiments that would solve the puzzle. As she said, "That's really what got me started in science."

In addition, Dr. Barton also had an inspirational teacher and role model, Bernice Segal. "She was an absolute inspiration to me. She gave a magnificent course, and she was a very tough lady who asked a lot of you—and you did it!"

The platinum blues

Dr. Barton's Ph.D. thesis research was mostly on compounds known to chemists for years as the "platinum blues."

"Most platinum compounds are orange or red and yet there are these magnificent blue complexes. What are they? What are their structures, and why are they blue? We solved the structure of the molecule and found that it contains four platinum atoms in a line.

Research in bioinorganic chemistry

After talking about how Dr. Barton came into chemistry, we discussed her current research in bioinorganic chemistry, the role of metals in biological systems. She has received numerous awards for this work, reflecting the importance that the chemistry community places on the field and her contributions to it.

"The interest of my group is to exploit inorganic chemistry as a tool to ask questions of biological interest and to explore biological molecules. A lot of the work in bioinorganic chemistry thus far has been the exploration of metal centers in biology. Why is blood red? Why does the iron [in heme] do what it does? That's just one example, but there are hundreds of others. Many enzymes and proteins within the body in fact contain metals, and the reason we've looked at blood and then the heme center within it has been because it's colored. An obvious tool that transition metal chemistry provides is color, and so things change color when reactions occur. That is one of the things that fascinated me in the first place.

"Another wonderful thing about transition metal chemistry is that it allows us to build molecules that have interesting shapes and structures depending upon the coordination geometry. In fact, you can create a wealth of different shapes, several of which are chiral, and that's something we take advantage of in particular.[1] What we want to do is make a variety of molecules of different shapes, target these molecules to sites on a DNA strand, and then ask questions such as 'Does DNA vary in its shape as a function of sequence?' If we think about how proteins bind to DNA, do they also take advantage of shape recognition in binding to one site to activate one gene or turn off another gene? When scientists first wondered about these and other such problems, they would write down a one-dimensional sequence of DNA and would think about it in one-dimensional terms. How does the protein recognize a particular DNA sequence? DNA is clearly not one dimensional. It has a three-dimensional structure, and different sequences of bases will generate different shapes and different forms. Therefore, we think we can build transition metal complexes of particular shapes, target them to particular sequences of bases in DNA, and then use these complexes to plot out the topology of DNA. We can then ask how nature takes advantage of this topology. We want to develop a true molecular understanding, a three-dimensional understanding, of the structure and the shapes of biologically important molecules such as DNA and RNA."

A revolution in chemistry in the last ten years

One of the most important goals of these interviews has been to gain insight into the directions that science in general and chemistry in particular will move in this decade. Dr. Barton put her remarks in the context of recent developments in her field.

"I think our work may be an example of where chemistry is going in general. I think there has been a revolution in chemistry in the past 10 years. The revolution is at the interface between chemistry and biology where we can now ask chemical questions about biological molecules. First of all, we can make biological molecules that are pure. I can now go to a machine called a DNA synthesizer, and I can type in a sequence of DNA; from that sequence I can synthesize a pure material, with full knowledge of where all of the bonds are. Then I can run it through an HPLC and get it 100% pure.[2] Therefore, I can now talk about these biopolymers in chemical terms as molecules rather than as impure cellular extracts. I couldn't do that before."

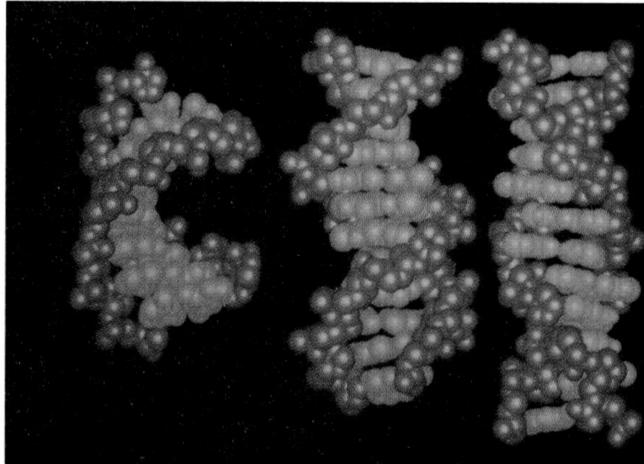

Some double helical conformations of DNA. Photo by Jacqueline Barton.

Not only do we have the ability to prepare biological molecules in pure form, but Dr. Barton stressed that we have the techniques to characterize them in ways that we chemists think about molecules. "The development of new techniques allows us to make a bridge between chemistry and biology and ask

[1]The chirality of molecules is discussed in Chapter 16.

[2]An HPLC is a high pressure liquid chromatograph, an instrument capable of separating one type of molecule from another.

chemical questions with molecular detail. It's an exciting time to be doing chemistry, and that is why I see it as a new frontier area."

New chemistry curricula

Dr. Barton stressed the point that it is chemists who are "making new materials and making and exploring biological systems. It's the chemist who looks at questions of molecular detail and asks about structure and its relationship to function." Since this involves so many areas of chemistry, she believes that "we are going to have to stop making divisions between inorganic, physical, analytical, and organic chemistry. We must all do a little bit of each." This is an attitude shared by many in chemical education today, and it means that we should perhaps rethink the curriculum in chemistry in particular and science in general.

No matter what the curricular structure, however, she believes what is important in the education of scientists is "to get across the excitement that now we can know what biologically important molecules look like. And, from knowing what they look like, we can manipulate them and change them a little. Then we ask how those changes affect the function, so we can relate the structure of the molecule and its macroscopic function."

"A protein molecule of average size is so small you could put more than a billion billion of them on the head of a pin. We now know we can manipulate molecules that are of those dimensions and can know exactly what they look like. I can't imagine that we can't get people interested in chemistry if we can get across the excitement that comes from the realization that we are looking at things so small and yet can do surgery on them."

Chemistry is fun

There was one basic theme in all of the interviews that we did and Dr. Barton expressed it well. "The bottom line is that chemistry is fun, it's addictive, and, if one has a sense of curiosity, it can be tremendously entertaining and appealing. And it is not so difficult. It's difficult when one thinks about it as rote memorization, which *is* difficult and boring. But that isn't what chemistry is. Chemistry is trying to understand the world around us in some detail. For example, we are interested in knowing such things as what makes skin soft, what makes things different in color, why sugar is sweet, or why a particular pharmaceutical agent makes us feel better."

Women in science

We were interested in Dr. Barton's perspective on the issue of women in science. Did she see special opportunities for women? Are there problems that women need to overcome or be aware of?

"Because I am a woman, and there are so few women currently in professional positions in chemistry, I'm asked those questions often. First of all, I am not an expert on the subject. What I like to think my best contribution to women in chemistry can be is to do the best science I can, and to be recognized for my science, not for being a woman in science. I think that it is generally important when women go into science that they should appreciate that there are no special opportunities; that is, you will be treated like any other person doing science. But just as there should be no special opportunities in that respect, happily—maybe this is naive of me—I think there are also no special detriments or obstacles that one need consider in this day and age. One shouldn't think that 'because I am a woman I can't do it.' That's patently false. In fact, everyone is extremely supportive of women who do science. However, I remember talking to Bernice Segal, my former teacher at Barnard College, and having her explain to me that when she was a graduate student she had to do things behind a curtain, because the women weren't supposed to be doing chemistry. Mildred Cohn, another one of my role models, took over 20 years to have her own independent position as a professor, as opposed to being a laboratory assistant working for someone else. The bottom line is that I don't have a story like that to tell. That's the good news. In my generation there are few such stories of blatant discrimination. Now the world is a much better place for a woman to do science."

Dr. Barton's enthusiasm for her work and for chemistry in general was obvious. It is evident that she will continue to do some of the most important work in science and that her infectious enthusiasm for chemistry will bring many more young people into our profession.

Carbohydrates

16.1 Introduction

Because H_2SO_4 added to carbohydrates removes —OH and —H groups and leaves black carbon, early chemists wrongly thought these compounds might be made of carbon and water and hence called them carbohydrates (*hydro* means water).

Carbohydrates are either **polyhydroxy aldehydes or ketones, or compounds that yield polyhydroxy aldehydes or ketones on hydrolysis.** This definition will become clearer as we proceed in this chapter.

Carbohydrates are of major importance to both plants and animals. It has been estimated that more than half of all the organic carbon atoms in the world are in carbohydrate molecules. Simple carbohydrate molecules are synthesized chiefly by chlorophyll-containing plants as long as the sun is shining. In this process, called **photosynthesis,** plants combine carbon dioxide from the air with water from the soil to give simple carbohydrates (mainly glucose, $C_6H_{12}O_6$):

It is the chlorophyll in leaves and grass that makes them green.

$$6CO_2 + 6H_2O \xrightarrow[\text{chlorophyll}]{\text{light}} C_6H_{12}O_6 + 6O_2$$

Plants have two main uses for the carbohydrates they make. They use them (1) as a means of storing energy (the energy comes, of course, from the sunlight) and (2) to provide supporting structures, such as the trunks of trees. Carbohydrates make up about three fourths of the dry weight of plants. Animals (including humans) get their carbohydrates by eating plants, but do not store much of what they consume. Less than 1 percent of the body weight of animals is made up of carbohydrates.

We also use carbohydrates for clothing (cotton, rayon, linen), and we use wood (which is chiefly carbohydrates) for building, burning, and making paper.

Humans use carbohydrates for many purposes. The most important is, of course, food. Typically, carbohydrates constitute about 65 percent of our diet. Our bodies use these compounds mostly for energy (Chap. 21) but also as a source of carbon atoms for the synthesis of many other compounds. They are found in nucleic acids and in connective tissue and are an essential part of the energy cycle of living things (Fig. 16.1).

Sulfuric acid reacting with sugar. Carbon can be seen forming as sulfuric acid is poured onto the sugar. (Photograph by Charles D. Winters.)

16.2 Enantiomers and Chirality

In order to understand the structure of carbohydrates and other biological molecules, we must consider a new form of stereoisomerism.

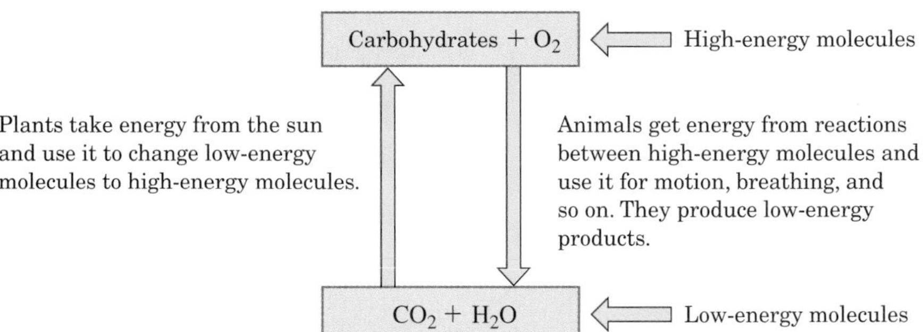

Figure **16.1** • The roles of plants and animals in the oxygen–carbon dioxide cycle.

Everything has a mirror image (except maybe a vampire). If we look at the world around us, we see that all objects can be divided into two classes: those that can be superimposed on their mirror image and those that cannot. **An object that cannot be superimposed on its mirror image** is called **chiral** (pronounced ky′ral). Chiral objects are nonsuperimposable on their mirror images.

A good example of a chiral object is your left hand. If you hold it up to a mirror, you see in the mirror an image of your right hand (Fig. 16.2), and you cannot superimpose the two. Some objects that are not chiral are golf balls, spoons, baseball bats, and wine glasses, all of which are superimposable on their mirror images (Fig. 16.3).

Putting the two hands palm to palm is not superimposing them. For one thing, the fingernails are on opposite sides.

The same is true for molecules: Some molecules are chiral; others are not. What kinds of molecules are chiral? Figure 16.4 shows a molecule that contains a carbon atom connected to four different groups. (Remember that carbon points its four bonds to the corners of a tetrahedron; Sec. 3.6.) The molecule on the left in Figure 16.4 is the mirror image of the molecule on the right, just as a left hand is the mirror image of a right hand, but the two molecules cannot be superimposed, any more than we can superimpose a left hand on a right hand (try putting a left-hand glove on your right hand). **A carbon atom connected to four different groups** is called a **chiral carbon** or an **asymmetric carbon.**

A chiral molecule is one that cannot be superimposed on its mirror image.

Because the presence of a single chiral carbon atom always makes a molecule chiral, it is important that you be able to recognize one when you see it. It is really not difficult. If a carbon atom is connected to four different groups, *no matter how slight the differences,* it is a chiral carbon. If any two groups are identical, it is not a chiral carbon.

EXAMPLE

Which of the carbon atoms shown in color are chiral?

(a) $CH_3—CH(OH)—CH_2—CH_3$

(b) $CH_3—CH_2—CH(OH)—CH_2—CH_3$

(c) $Br—CH_2—CH_2—CH_2—CH(CH_3)—CH_2—CH_2—CH_2—Cl$

(d) $CH_3—CH_2—C(=O)—CH_3$

(e)

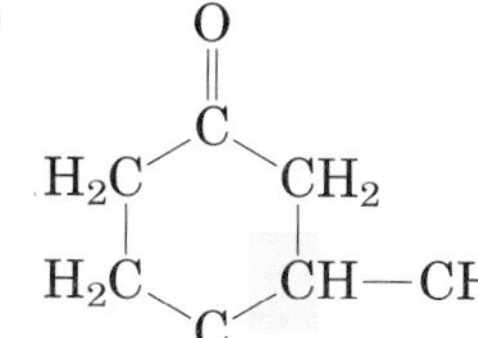

(f)

H2
C
H2C CH—Br
HC CH2
H3C C
H2

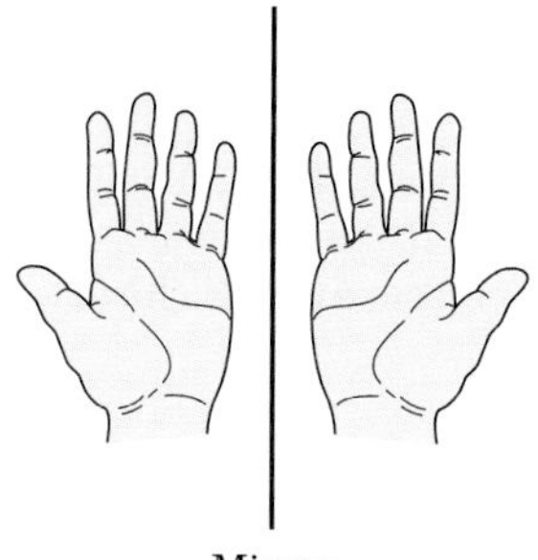

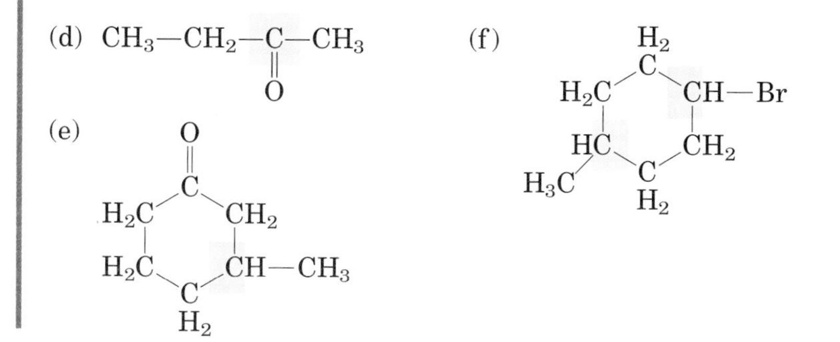

Figure **16.2**
Left and right hands are nonsuperimposable mirror images and thus are chiral objects.

Mirror

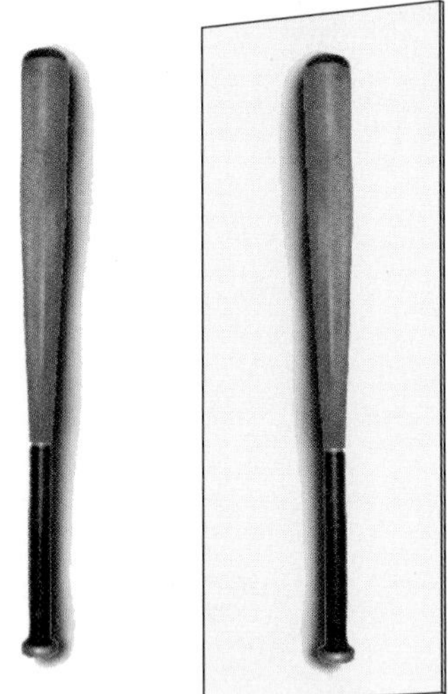

Figure **16.3**
Some objects that are superimposable on their mirror images. Such objects are not chiral.

Answer

(a) This is a chiral carbon. The four different groups are
1. $—CH_3$
2. $—CH_2CH_3$
3. $—H$
4. $—OH$

(b) This carbon is not chiral. The four groups are
1. $—H$
2. $—OH$
3. $—CH_2CH_3$
4. $—CH_2CH_3$

Groups 3 and 4 are identical.

(c) This is a chiral carbon. The four different groups are
1. $—H$
2. $—CH_3$
3. $—CH_2CH_2CH_2Cl$
4. $—CH_2CH_2CH_2Br$

Although the last two groups are similar, they are not identical. Even this small difference is enough to make the carbon chiral.

(d) This carbon is connected to only three groups:
1. $—CH_3$
2. $=O$
3. $—CH_2CH_3$

In order to be chiral, a carbon must be connected to *four* different groups. Thus, even though the three groups here are all different, they are not four, and this carbon is not chiral.

(e) The rules are no different for carbons in rings. This is a chiral carbon. The groups are
1. $—H$
2. $—CH_3$
3. $—CH_2—CO—$
4. $—CH_2—CH_2—$

The third group is different from the fourth.

(f) Neither of these is a chiral carbon. Whether we proceed clockwise around the ring or counterclockwise, we do not encounter the slightest difference. The two groups are identical:

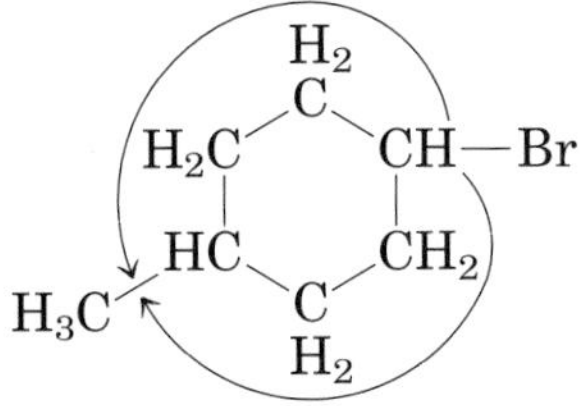

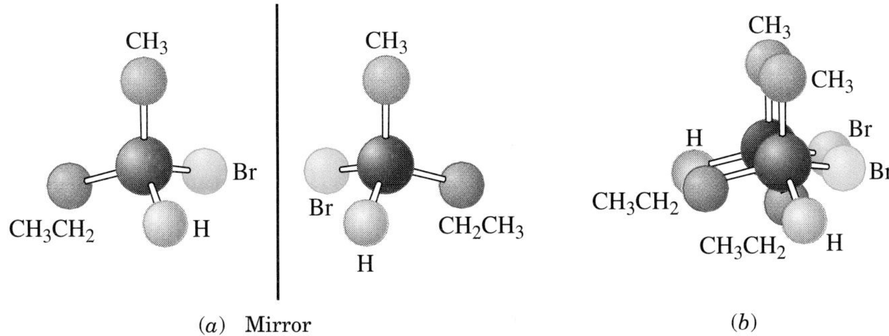

Figure **16.4** • (a) A molecule of 2-bromobutane and its mirror image. The two cannot be superimposed and thus are a pair of enantiomers. (b) If we try to superimpose them by, say, putting the CH_3 over the CH_3 and the Br over the Br, we find that the H goes over the CH_3CH_2 and the CH_3CH_2 goes over the H. We can superimpose any two groups, but the other two groups will not be superimposed.

Problem **16.1**

Which of the carbon atoms shown in color are chiral?

(a) $CH_3CH_2—CH—CH_2CH_3$ (with Br on the CH)

(b) H_3C–C$_6$H$_4$–CH(OH)–C$_6$H$_5$ (CH bearing OH, between a methyl-substituted benzene ring and a benzene ring)

(c) Ring: H_2C–$C(H_2)$–CH–CH_3; H_2C–$C(H_2)$–CH_2 (cyclohexane ring with CH–CH_3)

(d) $CH_3—CH—CH—CH_3$ (with CH_3 and Br on the two CH carbons)

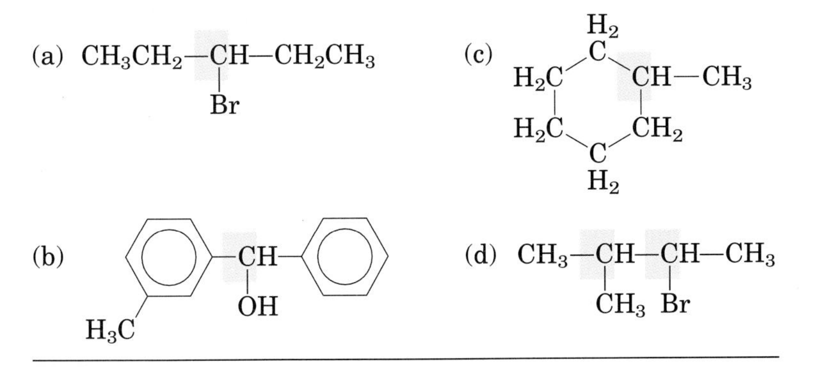

There are some chiral molecules that lack a chiral carbon, but these exceptions are not important enough for us to consider in this book. Also, some molecules that contain two or more chiral carbons are not chiral molecules, but these too are not very important. In Section 16.4 we will see what happens when molecules possess two or more chiral carbon atoms.

Now that we know how to recognize a chiral molecule, let us look once more at Figure 16.4. The two molecules of 2-bromobutane are not superimposable on each other. As we saw in Section 10.9, *superimposability is the same as identity.* If two objects are identical, we can superimpose one on the other (Fig. 16.5). Since the two molecules of 2-bromobutane are not superimposable, *they must be different.* They therefore meet our definition of stereoisomers (Sec. 10.9): two molecules with identical molecular and structural formulas but different three-dimensional shapes. Chiral molecules represent the most subtle kind of isomerism we have seen. Even the shapes of the molecules are identical in all respects except direction: One is "left-handed," the other "right-handed." **Isomers that are nonsuperimposable mirror images of each other** are called **enantiomers.**

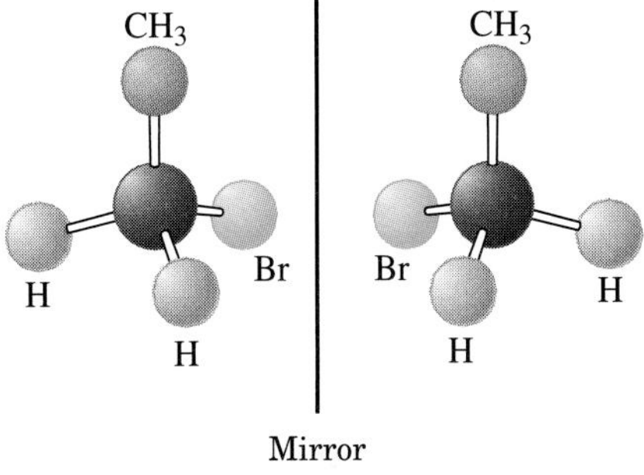

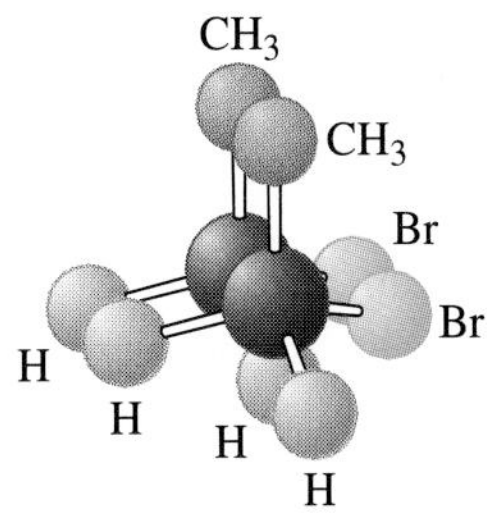

Figure **16.5**
(a) A molecule of bromoethane and its mirror image. This molecule is superimposable on its mirror image. All we have to do is (b) rotate it so that the Br is placed over the Br (the CH_3 is already over the CH_3). The two Hs are over the two Hs, and the molecules are superimposed and therefore identical. Note that bromoethane does not contain a chiral carbon.

16.3 Properties of Enantiomers. Optical Activity

In all cases of isomerism that we have previously seen, both structural isomerism and stereoisomerism, the isomers differ in almost all properties, but this is not the case for enantiomers. **All physical and chemical properties of enantiomers are identical, except two.**

1. The two members of a pair of enantiomers rotate the plane of polarized light in opposite directions. **Polarized light** is **light that vibrates in only one plane,** in contrast to ordinary light, which vibrates in all planes. When polarized light is passed through certain substances, the plane of its vibration is rotated through a certain angle, different for each substance (Fig. 16.6). An instrument that detects this rotation is called a **polarimeter.**

A substance that rotates the plane of polarized light is called **optically active.** Some substances rotate the plane to the right (clockwise). We call these **dextrorotatory** and use the symbol (+). Others rotate the plane to the left (counterclockwise). These are **levorotatory,** and their symbol is (−).

These words are from the Latin *dextro,* right, and *levo,* left.

There is a basic difference between chiral and nonchiral substances. **All chiral substances rotate the plane of polarized light** (they are optically active). Substances that are not chiral do not rotate the plane of polarized light. As we saw in the previous section, chiral substances always exist as two enantiomers. Both enantiomers rotate the plane of polarized light, in equal amounts **but in opposite directions.** In any pair of enantiomers, one always rotates the plane to the right and the other to the left. Only the direction is different; the amount of the rotation is the same.

This principle was first stated by Louis Pasteur in 1848.

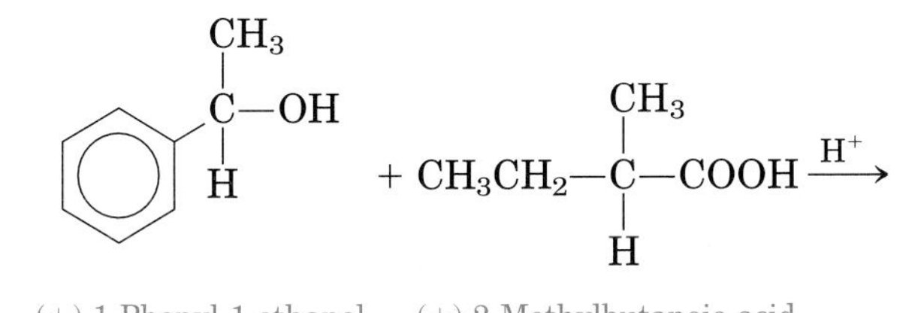

2. Enantiomers undergo chemical reactions, just as any other organic molecule would that has the same functional groups. When they react with another molecule *that is chiral,* the reaction rates are not the same for the two enantiomers. For example, (+)-1-phenyl-1-ethanol reacts with (+)-2-methylbutanoic acid to form a carboxylic ester:

Enantiomers react at the same rate with nonchiral molecules.

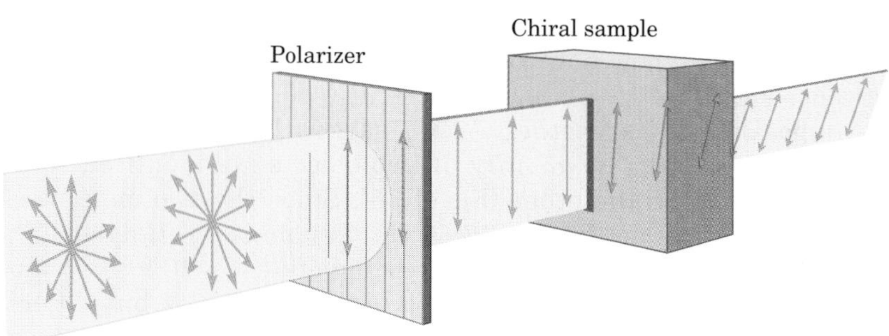

Figure 16.6 • Operation of a polarimeter. Ordinary light, which vibrates in all planes (indicated by arrows), becomes polarized when passed through a polarizer. When the polarized light passes through a chiral sample, the plane of the light is rotated.

$$\underset{\text{(+)-1-Phenyl-1-ethanol}}{C_6H_5\text{—}\overset{CH_3}{\underset{H}{C}}\text{—OH}} + \underset{\text{(+)-2-Methylbutanoic acid}}{CH_3CH_2\text{—}\overset{CH_3}{\underset{H}{C}}\text{—COOH}} \xrightarrow{H^+}$$

$$CH_3CH_2\text{—}\overset{CH_3}{\underset{H}{C}}\text{—}\underset{O}{\overset{\|}{C}}\text{—O—}\overset{CH_3}{\underset{H}{C}}\text{—}C_6H_5 + H_2O$$

The reaction between (+)-1-phenyl-1-ethanol and (−)-2-methylbutanoic acid takes place at a different rate. This is a very important difference in property, with major biological consequences (Sec. 16.13).

Emil Fischer. (Courtesy of E. F. Smith Memorial Collection, Center for History of Chemistry, University of Pennsylvania.)

In all properties except these two, enantiomers are identical. They have identical melting points, boiling points, densities, rates of reactions with molecules that are not chiral, and all other properties.

Although each enantiomer of a pair rotates the plane of polarized light, a mixture of equal amounts of the two does not, because the rotations cancel each other. Such a mixture is called a **racemic mixture** or a **racemate.**

In most cases, the only way to tell them apart is to pass polarized light through samples of the compounds and see if they rotate (+) or (−).

Fischer Projections

It is awkward to draw molecules in the three-dimensional shapes shown in Figures 16.4 and 16.5, but there is a way to represent these shapes correctly even if we are not very good artists (or don't want to take the time). This system was devised by the German organic chemist Emil Fischer (1852–1919). In a **Fischer projection,** you hold the molecule in such a way that the two bonds coming out at you are horizontal and the two bonds going behind the paper are vertical, for example:

Emil Fischer, who in 1902 became the second Nobel prize winner in chemistry, made many fundamental discoveries in the chemistry of carbohydrates, proteins, purines, and other areas of organic and biochemistry.

(a) [ball-and-stick model: CHO (top), HO (left), H (right), CH_3 (bottom)] = (b)

$$HO\text{—}\overset{CHO}{\underset{CH_3}{C}}\text{—}H$$

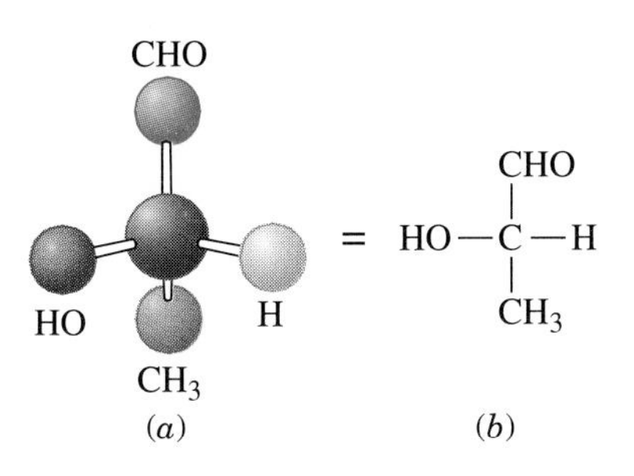

The —OH and —H are coming toward you, out of the page. The —CHO and —CH_3 are going away from you, into the page.

When you see the formula shown in (b), the Fischer projection, you must realize that the molecule is oriented in space as shown in (a).

We can use Fischer projections to determine (1) whether two molecules are identical or different and (2) whether or not two formulas are mirror images. But there are two rules to follow in using these projections:

(1) We must not take them out of the plane of the paper, and (2) we can rotate them 180° but not 90°.

EXAMPLE

Are these mirror images? Are they identical (superimposable)?

$$\text{(a)}\quad \begin{array}{ccc} & CH_3 & \\ HO & — & H \\ & CH_3 & \end{array} \quad \text{and} \quad \begin{array}{ccc} & CH_3 & \\ H & — & OH \\ & CH_3 & \end{array}$$

$$\text{(b)}\quad \begin{array}{ccc} & CH_3 & \\ HO & — & H \\ & | & \end{array} \quad \text{and} \quad \begin{array}{ccc} & CH_3 & \\ H & — & OH \\ & | & \end{array}$$

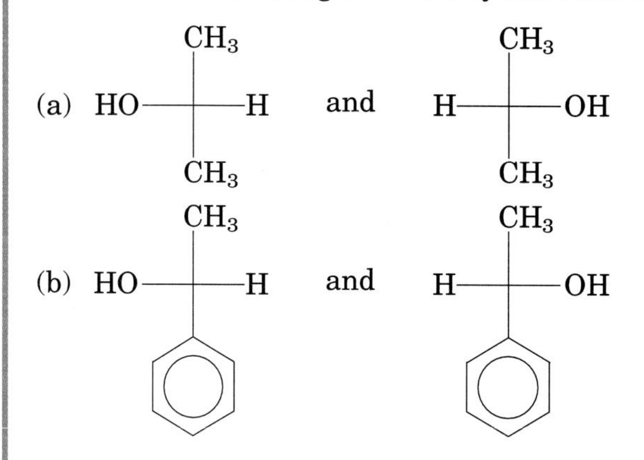

Answer • In both (a) and (b) the molecules are mirror images. If you held a mirror next to one you would see the other. In (a) they are superimposable. You cannot superimpose them by merely sliding one over onto the other, but you can if you rotate one 180° in the plane of the paper (which is allowed) and then slide it over. In (b) they are not superimposable. They would be if you could flip one over, but the rules say you must keep it in the plane of the paper.

Problem **16.2**

Are these mirror images? Are they identical?

$$\text{(a)}\quad \begin{array}{ccc} & CH_3 & \\ CH_3CH_2 & — & H \\ & Cl & \end{array} \quad \text{and} \quad \begin{array}{ccc} & CH_3 & \\ H & — & CH_2CH_3 \\ & Cl & \end{array}$$

$$\text{(b)}\quad \begin{array}{ccc} & CH_2CH_3 & \\ CH_3CH_2 & — & H \\ & Cl & \end{array} \quad \text{and} \quad \begin{array}{ccc} & CH_2CH_3 & \\ H & — & CH_2CH_3 \\ & Cl & \end{array}$$

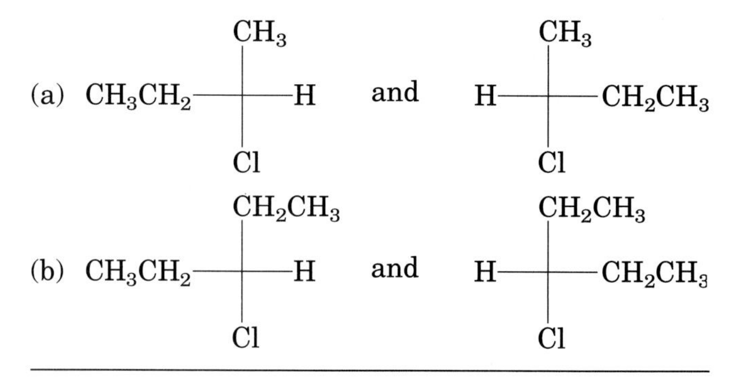

16.4 Compounds with More Than One Chiral Carbon

The **configuration** of a molecule is the three-dimensional shape of that molecule. Two stereoisomers are said to have different configurations.

Let us use Fischer projections to see what happens when a molecule possesses two chiral carbon atoms. Such a molecule is 2,3-dihydroxybutanal. (Three-dimensional configurations of these molecules are shown in Figure 16.7.) One stereoisomer is **A.**

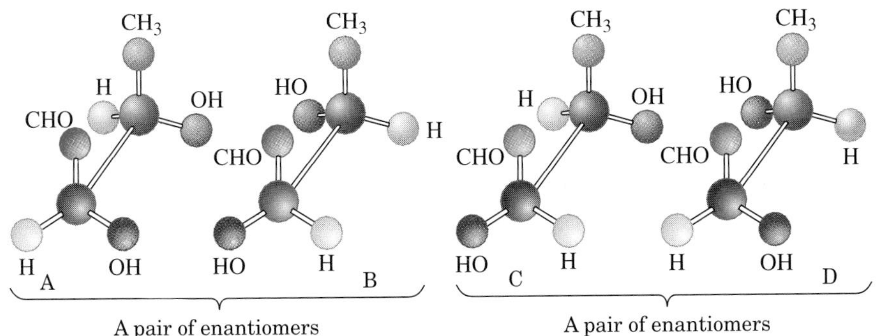

Figure **16.7** • The three-dimensional configurations of the four stereoisomers of 2,3-dihydroxybutanal.

A	B	C	D
CHO	CHO	CHO	CHO
H—┼—OH	HO—┼—H	HO—┼—H	H—┼—OH
H—┼—OH	HO—┼—H	H—┼—OH	HO—┼—H
CH_3	CH_3	CH_3	CH_3

A pair of enantiomers (**A**, **B**); Another pair of enantiomers (**C**, **D**)

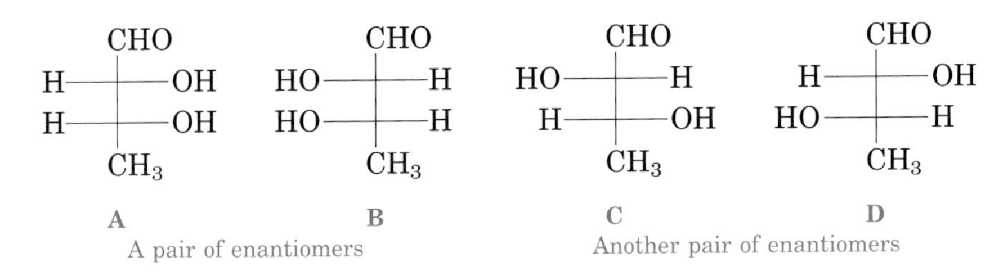

As can be seen from the formulas, **A** is not superimposable on its mirror image, **B.** Molecules **A** and **B** are therefore a pair of enantiomers and as such are identical in all properties except the two mentioned in the previous section.

Make sure that **B** is indeed the mirror image of **A** and that they are not superimposable.

But **C** is another stereoisomer of 2,3-dihydroxybutanal. If you inspect it carefully, you will see that it is not identical with (superimposable on) **A** or **B,** nor is it the mirror image of either of them. It is a third stereoisomer, different from **A** or **B.** Molecule **C** has its own mirror image, the compound **D.** Another inspection will show that **C** is not superimposable on its mirror image **D.** Molecules **C** and **D** are therefore a second pair of enantiomers. Their properties are also identical to each other except for the two properties mentioned earlier.

But are the properties of **A** and **B** identical to those of **C** and **D?** The answer is no. All physical and chemical properties of **A** and **B** are different from those of **C** and **D.** After all, **A** is not the mirror image of **C,** so there is no reason why their properties should be identical. The properties may be similar (they do have the same structural formula), but they are not identical.

Any molecule can have only one mirror image. Enantiomers always come in pairs; there are never more than two. **Stereoisomers that are not enantiomers** are called **diastereomers.** Molecule **C** is the enantiomer of **D** and a diastereomer of **A** and **B.**

What we saw for 2,3-dihydroxybutanal is also true for nearly all molecules that have two chiral carbon atoms: There are two pairs of enantiomers for a total of four stereoisomers. It does not matter whether the two chiral carbons are attached to each other or not.

In a similar manner, it can easily be shown (you may want to try it) that when three chiral carbons are present, there are eight stereoisomers (four pairs of enantiomers); four chiral carbons means 16 stereoisomers, and so forth. The general formula is

The number 2^n is actually a maximum. In some cases the actual number of stereoisomers is fewer.

$$\text{number of stereoisomers} = 2^n,$$

where n is the number of chiral carbon atoms.

Problem 16.3

How many stereoisomers are possible for

(a) 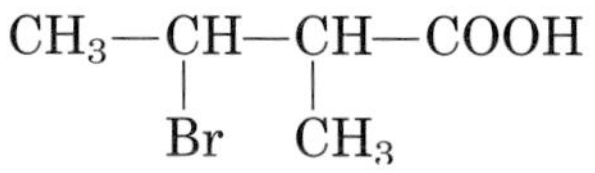 $CH_3-\underset{\displaystyle Br}{\underset{|}{CH}}-\underset{\displaystyle CH_3}{\underset{|}{CH}}-COOH$

(b) 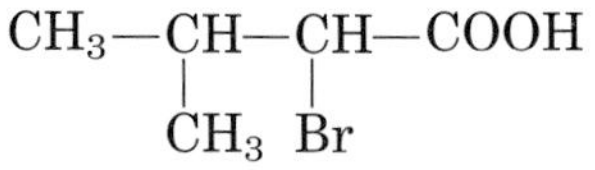 $CH_3-\underset{\displaystyle CH_3}{\underset{|}{CH}}-\underset{\displaystyle Br}{\underset{|}{CH}}-COOH$

(c) 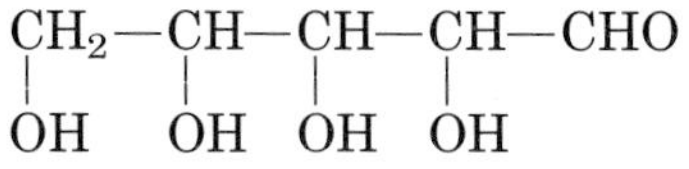 $\underset{\displaystyle OH}{\underset{|}{CH_2}}-\underset{\displaystyle OH}{\underset{|}{CH}}-\underset{\displaystyle OH}{\underset{|}{CH}}-\underset{\displaystyle OH}{\underset{|}{CH}}-CHO$

(d) 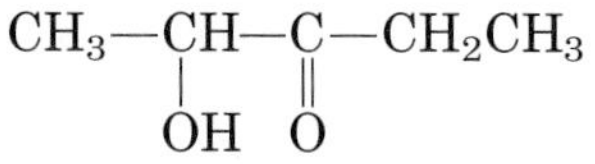 $CH_3-\underset{\displaystyle OH}{\underset{|}{CH}}-\underset{\displaystyle O}{\underset{\|}{C}}-CH_2CH_3$

16.5 Glucose

Now that we have been introduced to chirality, we are ready to return to carbohydrates. Carbohydrates can be divided into classes based on the size of the molecules. **Monosaccharides,** the basic units of carbohydrates, are simple sugars having the general formula $C_nH_{2n}O_n$, where n can vary from three to nine. **Disaccharides** consist of two monosaccharide units hooked together in a way that we shall discuss in Section 16.10. **Oligosaccharides** contain from three to ten units. **Polysaccharides** are also made of monosaccharides hooked together but may contain thousands of units (Fig. 16.8).

$C_nH_{2n}O_n$ means that any monosaccharide molecule contains equal numbers of O and C atoms and twice as many H atoms.

D-Glucose is also called **dextrose, grape sugar,** and **blood sugar.**

By far the most important monosaccharide is D-glucose. This sugar is not only the most important monosaccharide, but it is also the major constituent of disaccharides and polysaccharides.

D-Glucose is a white solid of molecular formula $C_6H_{12}O_6$. The structural formula of solid D-glucose is

A six-membered ring of this type is called a **pyranose ring.**

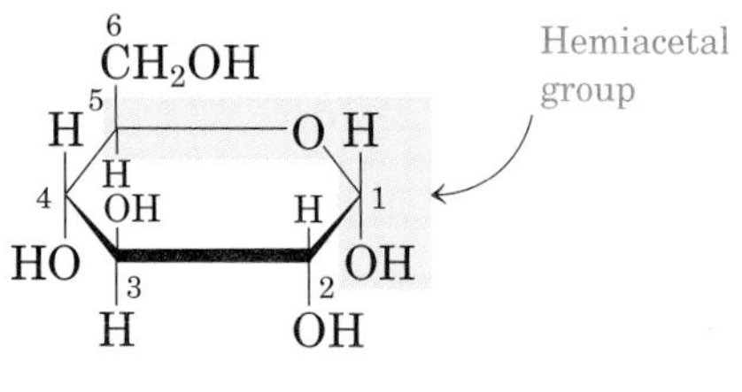

6
CH_2OH
5
H O H
H
4 OH H 1
HO OH
3 2
H OH

Hemiacetal group

This kind of drawing is called a **Haworth formula.** Note the numbering of the carbon atoms. If you look at C-1, you will see that it is connected to two oxygen atoms: one in an —OH group, the other in an —OR

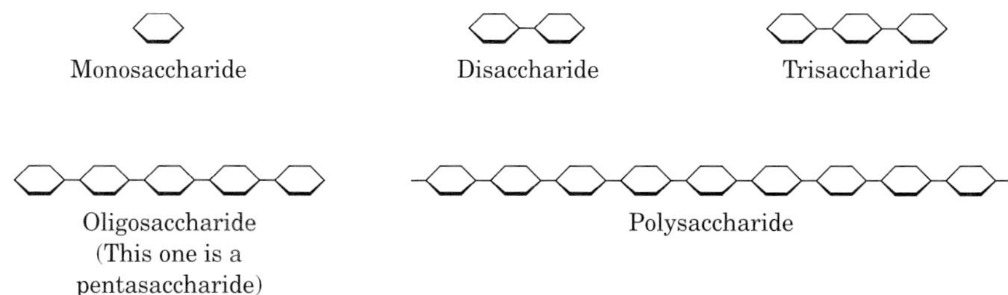

Figure **16.8** • Classification of carbohydrates.

group (the R begins with C-5). Such compounds are called **hemiacetals.** As we saw in Section 13.6, hemiacetals are not stable unless they are in a ring. Because of the ring, D-glucose is a perfectly stable molecule in the solid state. However, when it is dissolved in water, the instability of the hemiacetal shows itself, and the ring opens to turn C-1 into an aldehyde group and C-5 into an alcohol group:

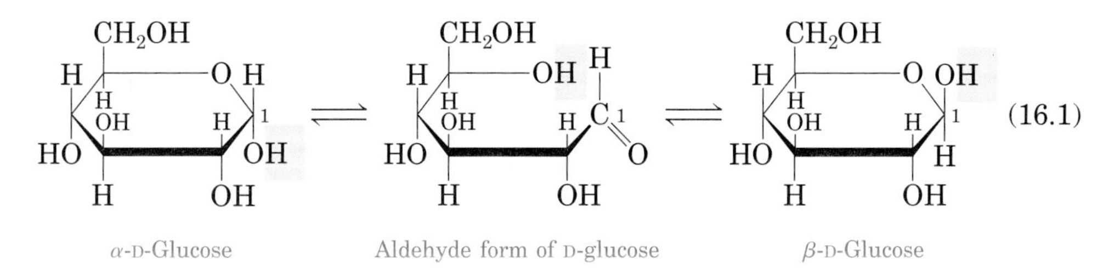

$$\alpha\text{-D-Glucose} \rightleftharpoons \text{Aldehyde form of D-glucose} \rightleftharpoons \beta\text{-D-Glucose} \qquad (16.1)$$

α-D-Glucose Aldehyde form of D-glucose β-D-Glucose

The hydrogen shown in color has traveled from the —OH oxygen to the —OR oxygen. But the aldehyde form of D-glucose is much less stable than the ring form. No sooner is it formed than it closes up again, but when it closes up there are two directions for the C-1 —OH to go: It can point down, which restores the original molecule, or it can point *up,* which gives a stereoisomer of the original molecule. The C-1 carbon is called an **anomeric carbon,** and the alpha and beta forms are called **anomers.** The α and β D-glucose molecules differ only at C-1. All other carbons have the same stereochemistry.

Anomers are two stereoisomers of a monosaccharide that differ only in the configuration at C-1 (aldoses) or C-2 (ketoses).

There are thus *three* forms of D-glucose. Two are ring forms; we call one alpha and the other beta. The third is the open-chain (aldehyde) form, which can also be drawn as a Fischer projection:

$$\begin{array}{rcl} & \text{H} & \\ & {}^{1}| & \\ & \text{C}{=}\text{O} & \\ & {}^{2}| & \\ \text{H}— & \text{C} & —\text{OH} \\ & {}^{3}| & \\ \text{HO}— & \text{C} & —\text{H} \\ & {}^{4}| & \\ \text{H}— & \text{C} & —\text{OH} \\ & {}^{5}| & \\ \text{H}— & \text{C} & —\text{OH} \\ & {}^{6}| & \\ & \text{CH}_2\text{OH} & \end{array}$$

Aldehyde form of D-glucose

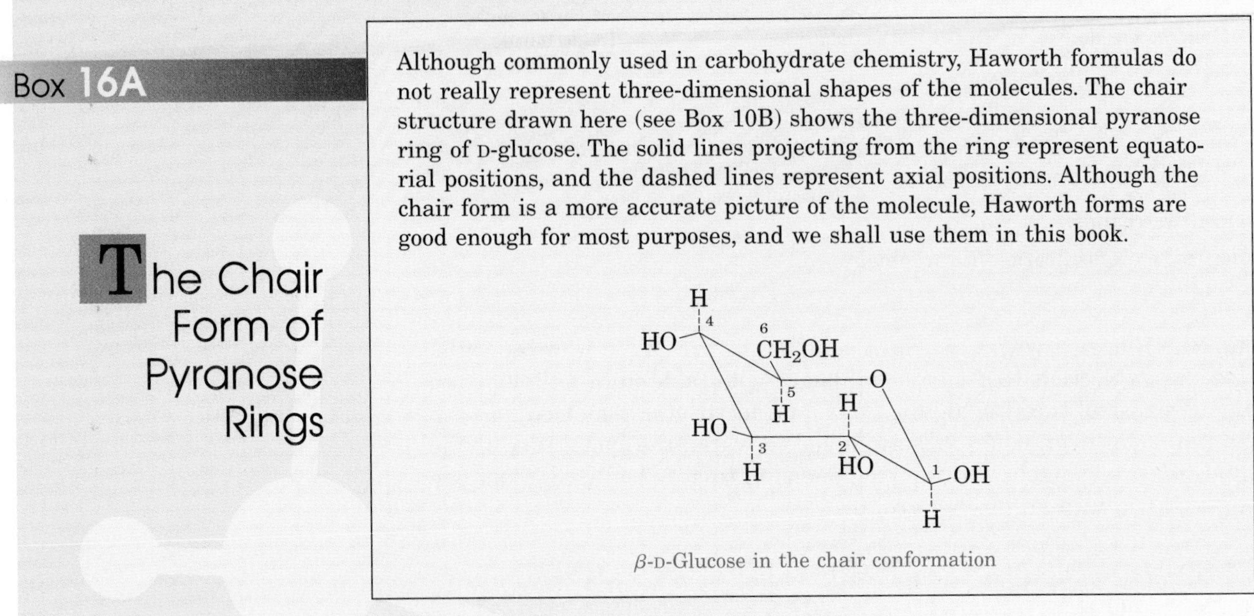

β-D-Glucose in the chair conformation

By evaporating the water from D-glucose solutions at different temperatures, we can obtain either the alpha or the beta form in the solid state. As long as either is a solid, its formula does not change. The aldehyde form cannot be obtained in the solid state. When we dissolve either the alpha or beta solid in water, we get the equilibrium shown in Equation 16.1: a mixture of all three forms. Because the aldehyde form is the least stable, it is present in very small amounts, much less than 1 percent. The main reason we know it is there at all is that the alpha and beta forms cannot change into one another except by going through the open-chain form. All three forms are optically active; each rotates the plane of polarized light by a different amount. When either the alpha or the beta form is dissolved in water, the amount of rotation gradually changes because a single form is changing to a mixture of forms. Because of this, the process is called **mutarotation** (meaning "change of rotation").

Box 16A
The Chair Form of Pyranose Rings
Although commonly used in carbohydrate chemistry, Haworth formulas do not really represent three-dimensional shapes of the molecules. The chair structure drawn here (see Box 10B) shows the three-dimensional pyranose ring of D-glucose. The solid lines projecting from the ring represent equatorial positions, and the dashed lines represent axial positions. Although the chair form is a more accurate picture of the molecule, Haworth forms are good enough for most purposes, and we shall use them in this book.
H
4
HO
6
CH_2OH
O
5
H
H
HO
3
2
H
HO
1
OH
H

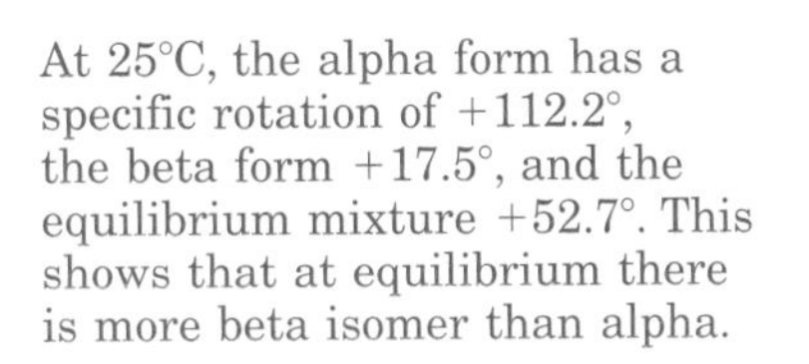

At 25°C, the alpha form has a specific rotation of +112.2°, the beta form +17.5°, and the equilibrium mixture +52.7°. This shows that at equilibrium there is more beta isomer than alpha.

Note that α and β D-glucose are not enantiomers. To be enantiomers, two molecules must be mirror images at every part of the molecule, but these two have opposite configurations only at C-1. Their configurations are the same everywhere else. Therefore they are diastereomers, not enantiomers.

16.6 Monosaccharides

Aldohexoses

Let us consider the open-chain form of D-glucose:

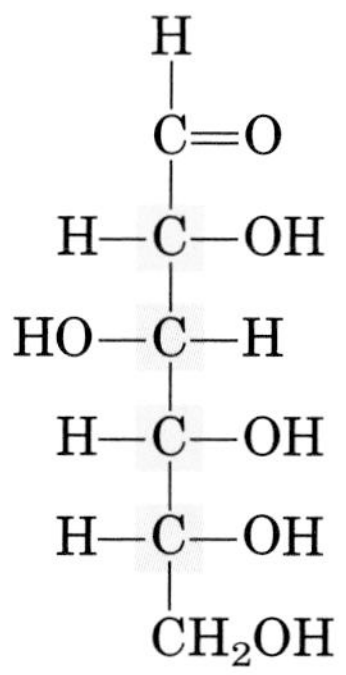

```
        H
        |
        C=O
        |
    H—C—OH
        |
   HO—C—H
        |
    H—C—OH
        |
    H—C—OH
        |
        CH2OH
```

It has four chiral (asymmetric) carbon atoms (shown in color). By the formula given in Section 16.4, this means that there are $2^4 = 16$ isomers, all of which have the same structure, except that they differ in configuration at the four chiral carbons. These 16 compounds are called **aldohexoses:**

aldo because they have an aldehyde group
hex because they have a chain of six carbons
ose because this is an ending used for carbohydrates

Aldohexoses differ in whether the —OH groups point left or right in the Fischer projection. In the cyclic Haworth forms this means up or down.

Two ways to show D-galactose:

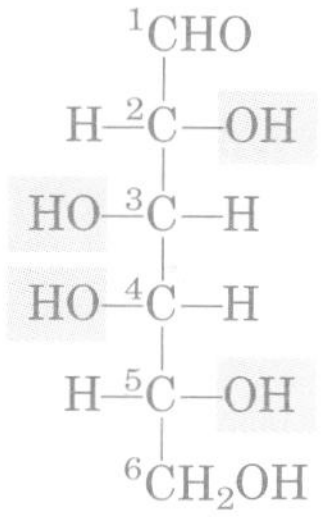

```
      1CHO
        |
   H—2C—OH
        |
  HO—3C—H
        |
  HO—4C—H
        |
   H—5C—OH
        |
      6CH2OH
```

—OH groups point left and right

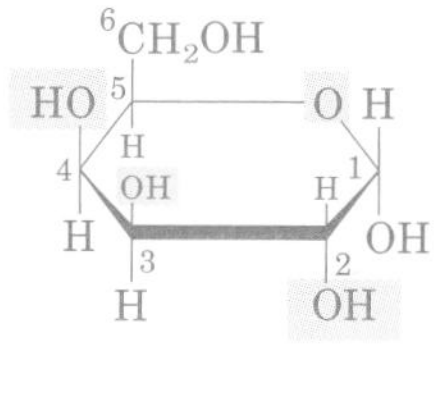

```
        6CH2OH
  HO  5 ———————— O  H
   4 H            H  1
     OH          OH
   H    3     2   OH
        H     OH
```

—OH groups point up and down

The 16 aldohexoses constitute 8 pairs of enantiomers. For example, the enantiomer of D-glucose is

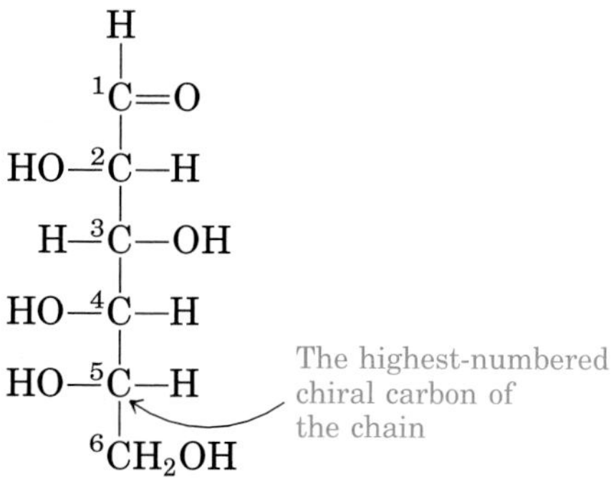

```
        H
        |
      1C=O
        |
  HO—2C—H
        |
   H—3C—OH
        |
  HO—4C—H
        |
  HO—5C—H
        |
      6CH2OH
```

The highest-numbered chiral carbon of the chain

L-Glucose

By convention, the configuration of the highest-numbered chiral carbon of the chain determines whether a monosaccharide is put into the D or the L series. If, in the Fischer projection, the OH points to the right, the whole molecule is placed in the D series. If it points to the left, the molecule is in the L series. Almost all natural monosaccharides belong to the D series. The eight D aldohexoses shown in Figure 16.9 all have different names. Their enantiomers are given the same names except that the prefix is L instead of D.

The three most important aldohexoses are D-glucose, D-mannose, and D-galactose.

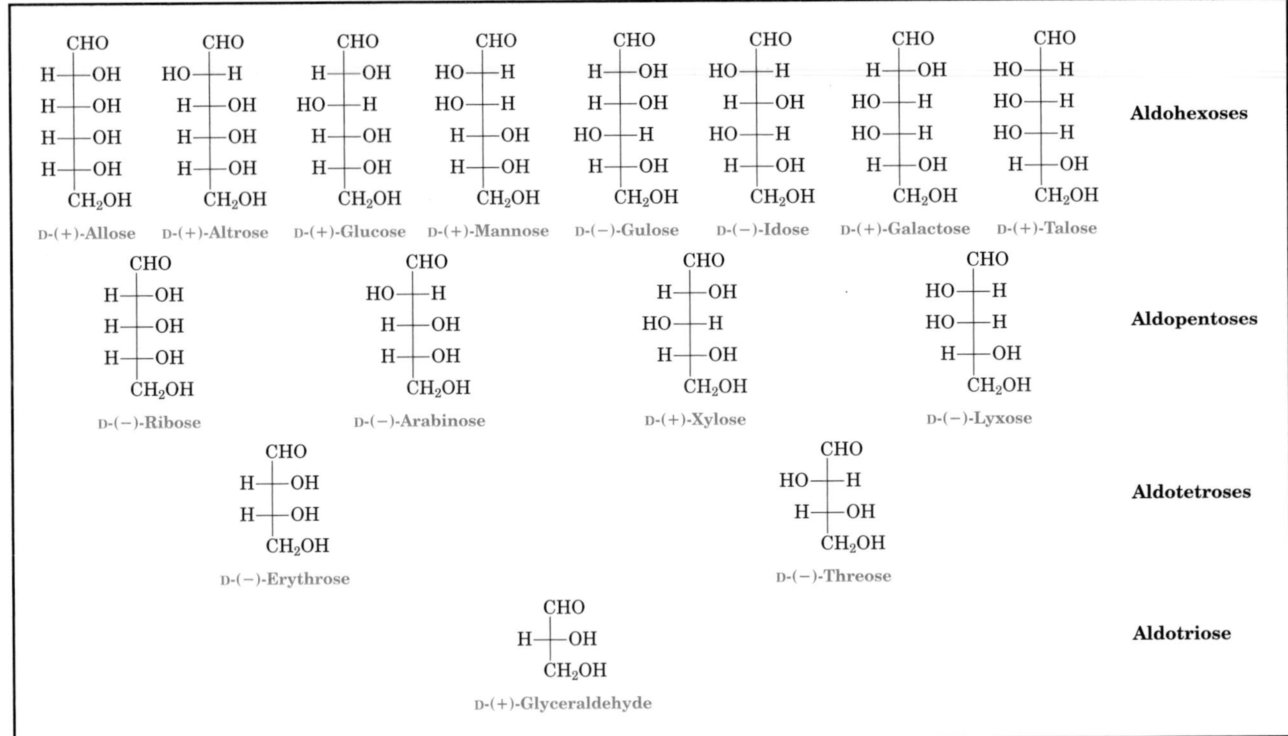

Figure **16.9**
The family of D aldoses, shown in Fischer projections. D-Glyceraldehyde, the smallest monosaccharide with a chiral carbon, is the standard on which the whole series is based. A (+) symbol means the compound shown rotates the plane of polarized light to the right; (−) indicates rotation to the left.

$2^3 = 8$
$2^5 = 32$

Other Aldoses

The aldohexoses have six-carbon chains. Similar compounds having chains of other lengths also exist. These are called aldotetroses, aldopentoses, aldoheptoses, and so on. The 2^n formula tells us there must be 8 aldopentoses, 32 aldoheptoses, and so forth. The D forms of some of these are shown in Figure 16.9. Besides glucose, only two aldoses are important enough to mention here. One of these is the aldohexose D-galactose, which differs from D-glucose only in the configuration at carbon 4 (see the formula in the margin on page 531). It is a constituent of plant polysaccharides and of lactose (milk sugar; Sec. 16.10). Galactose in the diet is converted by the body to glucose. The other is the aldopentose D-ribose:

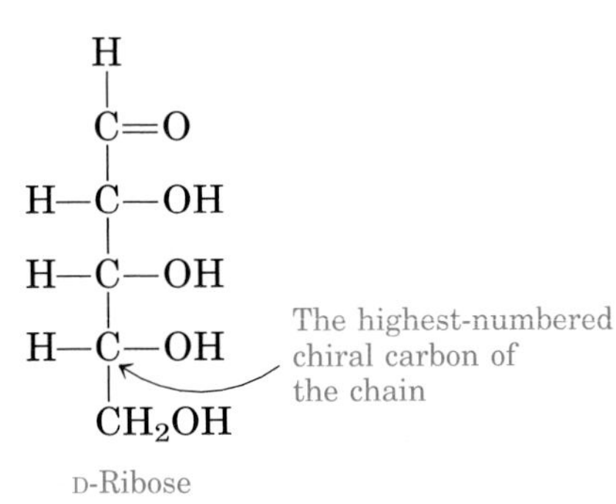

```
      H
      |
      C=O
      |
  H—C—OH
      |
  H—C—OH
      |
  H—C—OH      The highest-numbered
      |        chiral carbon of
    CH₂OH      the chain
```

D-Ribose

As we shall see in Section 23.2, D-ribose in its cyclic form is a component of ribonucleic acid (RNA).

Ketoses

Most monosaccharides are aldoses, but another type also exists, called **ketoses.** These compounds have a ketone group in the 2 position instead of an aldehyde group in the 1 position. Only one ketose is important; it has six carbons (so it is a **ketohexose**) and is called D-fructose:

$$\begin{array}{c} CH_2OH \\ | \\ C{=}O \\ | \\ HO{-}C{-}H \\ | \\ H{-}C{-}OH \\ | \\ H{-}C{-}OH \\ | \\ CH_2OH \end{array}$$

The highest-numbered chiral carbon of the chain

D-Fructose

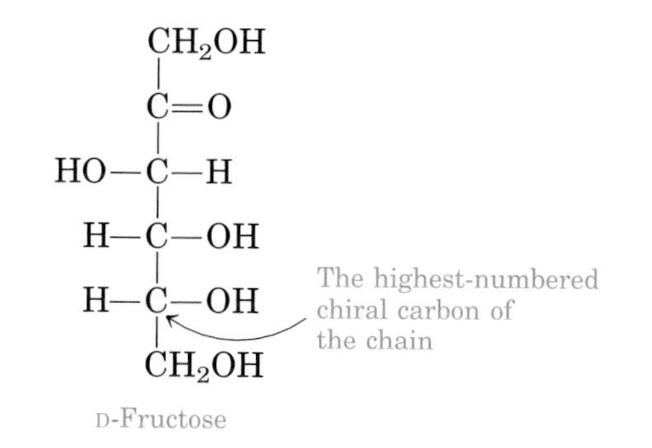

Note that D-fructose has the same configuration as D-glucose at C-3, C-4, and C-5. It differs from D-glucose only at C-1 and C-2. D-Fructose is found in honey and in many sweet fruits. It is also known as *fruit sugar.* D-Fructose is also the main sugar in semen. The sperm use it as a source of energy to propel themselves.

 16.7 Physical Properties of Monosaccharides

Sweetness

Most of the monosaccharides (and disaccharides) are sweet. For this reason they are called **sugars.** The degree of sweetness varies. As shown in Table 16.1, D-fructose is much sweeter than D-glucose. Ordinary table sugar, sucrose, is not a monosaccharide, but a disaccharide, the structure of which is given in Section 16.10.

We have no mechanical way to measure sweetness. It is done by having a group of people taste solutions of varying sweetness.

Artificial sweeteners are discussed in Box 18A.

Solubility

At room temperature carbohydrates are solids. Because of the many —OH groups, as well as the oxygen of the aldehyde or ketone group, monosaccharides are extremely soluble in water. These groups form numerous hydrogen bonds with the solvating water molecules. In the body this solubility allows a fast transport of energy supply in the form of sugar to different parts of the body through the circulatory system. Very concentrated solutions of carbohydrates in water are thick, viscous liquids. We are all familiar with such solutions as honey, molasses, and maple syrup.

Table 16.1 Comparative Sweetness of Some Sugars and Artificial Sweeteners

Sugar or Artificial Sweetener	Sweetness Relative to Sucrose	Type
Lactose	0.16	Disaccharide
Galactose	0.32	Monosaccharide
Maltose	0.33	Disaccharide
Glucose	0.74	Monosaccharide
Sucrose	1.00	Disaccharide (table sugar)
Invert sugar	1.25	Mixture of glucose and fructose
Fructose	1.74	Monosaccharide
Aspartame	100–150	Artificial sweetener
Acesulfame-K	200	Artificial sweetener
Saccharin	450	Artificial sweetener

Optical Activity

All three forms of D-glucose (alpha, beta, and open-chain) rotate the plane of polarized light to the right. For this reason it is often called D-(+)-glucose. Its enantiomer, L-glucose, must therefore rotate the plane to the left and is called L-(−)-glucose. However, this does not mean that all monosaccharides in the D series rotate the plane to the right. The letters D and L refer to the configuration of the molecules, and **there is no simple relationship between configuration and**

Collecting maple tree sap. (Courtesy of U. S. Department of Agriculture.)

rotation. D-Glucose just happens to rotate the plane to the right. Some of the other D monosaccharides also rotate to the right, but others rotate to the left (see Fig. 16.9). The ketose D-fructose is one of the D monosaccharides that rotate the plane to the left and hence is named D-(−)-fructose. Because of their rotation D-(+)-glucose is known as *dextrose* and D-(−)-fructose as *levulose.*

Of course, no matter which way any D monosaccharide rotates the plane of polarized light, its enantiomer always rotates it the opposite way.

16.8 Chemical Properties of Monosaccharides

Cyclic Forms

Like D-glucose, all monosaccharides with at least five carbons exist predominantly in the cyclic forms rather than in the open-chain aldehyde or ketone forms. Most of these cyclic forms have six-membered rings, called **pyranose** rings, similar to those of alpha- and beta-glucose. However, some of the monosaccharides form five-membered rings, called **furanose** rings, rather than the six-membered kind. Two examples are D-fructose and D-ribose:

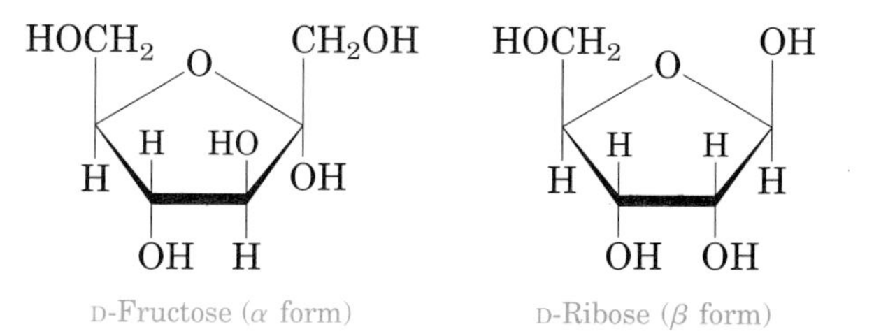

$HOCH_2$ O CH_2OH
H H HO OH
OH H

D-Fructose (α form)

$HOCH_2$ O OH
H H H H
OH OH

D-Ribose (β form)

Like D-glucose, all the monosaccharides that have cyclic structures show mutarotation when dissolved in water; they form equilibrium mixtures of the α, β, and open-chain forms.

EXAMPLE

The formula of D-xylose is shown in Figure 16.9. Draw the two furanose forms of this sugar.

Answer • Arrange the open-chain form so that C-1 lines up with the —OH group on C-4:

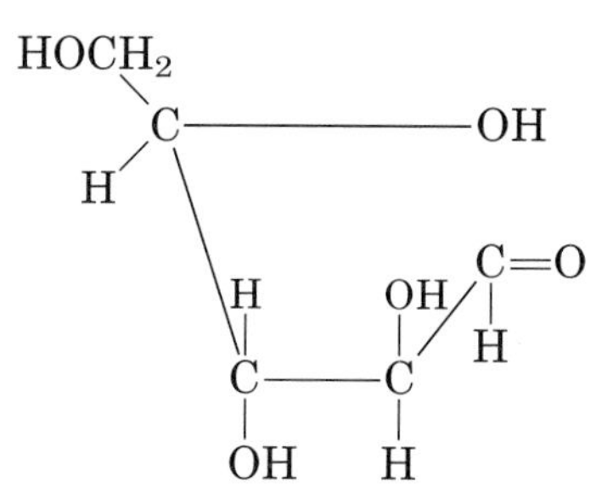

$HOCH_2$
C —OH
H
C=O
H OH H
C C
OH H

Form hemiacetals with both possible orientations of the —OH group on C-1:

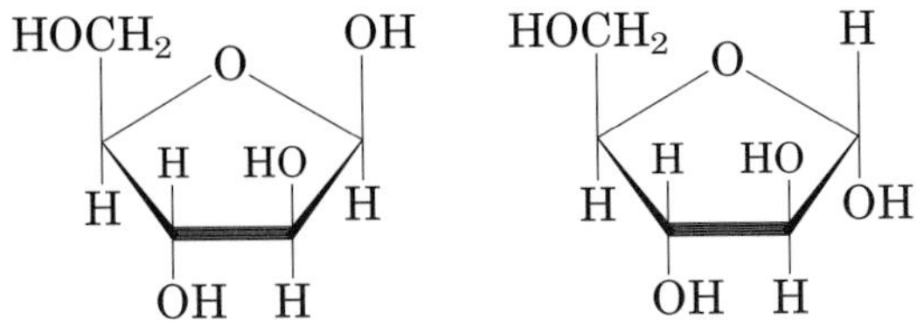

The two furanose forms of D-xylose

Problem 16.4

The formula of D-mannose is shown in Figure 16.9. Draw the two pyranose forms of this sugar.

Oxidation

Only sugars in which open-chain forms are in equilibrium with cyclic structures in solution undergo this kind of oxidation.

All aldoses have, in the open-chain form, an aldehyde group. Aldehydes are easily oxidized to carboxylic acids (Sec. 13.5), and so are aldoses. Because of this, aldoses are regarded as **reducing sugars,** meaning that they reduce oxidizing agents (oxidations of aldoses are further considered in Sec. 16.9). Although ordinary ketones are not easily oxidized, ketoses are because they have an —OH group on the carbon next to the C=O group (Sec. 13.5). Therefore, ketoses are also reducing sugars, which means that all monosaccharides are reducing sugars. Because of this property, the most common tests for monosaccharides (and other reducing sugars) consist of treating them with oxidizing agents, such as Benedict's solution (Sec. 13.5).

Glycoside Formation

We saw in Section 13.6 that aldehydes react with alcohols in the presence of H^+ to give acetals, with hemiacetals as intermediates. When D-glucose, a hemiacetal, is treated with methanol in the presence of HCl as a catalyst, it is converted to an acetal:

With glucose, the products are called **glucosides.**

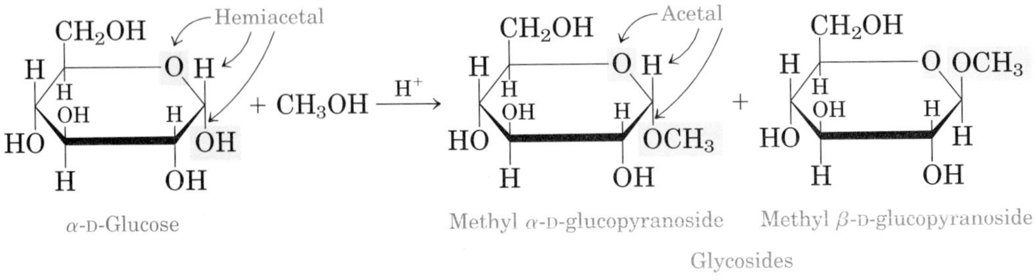

α-D-Glucose

Methyl α-D-glucopyranoside   Methyl β-D-glucopyranoside

Glycosides

Only the C-1 —OH is changed because all the other —OH groups are ordinary alcohol groups. The C-1 —OH is part of a hemiacetal group.

Other sugars also give this reaction, and the general name for all these products (including glucosides) is **glycosides.** Because glycosides are acetals rather than hemiacetals, they behave in some respects differently from monosaccharides. The main difference is that glycoside rings do not open when dissolved in water. (We say the ring is "locked.") This means that glycosides are *not* reducing sugars, because there is no aldehyde or keto group available, and do *not* exhibit mutarotation. It also means that α and β glycosides do not interconvert when dissolved in water.

16.9 Derivatives of Monosaccharides

Monosaccharides in which one or more functional groups have been altered through chemical reactions are *monosaccharide derivatives.* The carbonyl group of a monosaccharide can be reduced to give the corresponding sugar alcohol. For example, reduction of D-glucose gives D-sorbitol:

Aldose reductases are the enzymes that specifically catalyze this reaction in the cells.

H
|
C=O
|
H—C—OH
|
HO—C—H
|
H—C—OH
|
H—C—OH
|
CH_2OH

D-Glucose

—reduction→

CH_2OH
|
H—C—OH
|
HO—C—H
|
H—C—OH
|
H—C—OH
|
CH_2OH

D-Sorbitol

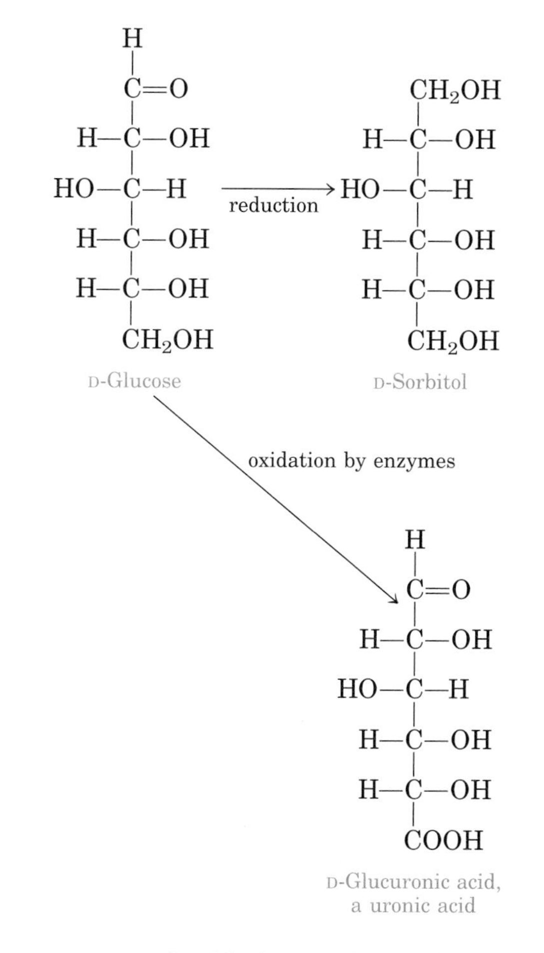

D-Sorbitol is also called D-glucitol. Analogously, the reduction product of D-galactose is D-galactitol, and so forth.

D-Glucose —oxidation by enzymes→

H
|
C=O
|
H—C—OH
|
HO—C—H
|
H—C—OH
|
H—C—OH
|
COOH

D-Glucuronic acid, a uronic acid

An important group of oxidation products of monosaccharides are the *uronic acids.* Oxidation by enzymes of the primary alcohol group at

C-6 yields these compounds, which are important building blocks of connective tissue acidic polysaccharides (Sec. 16.12).

Another group of acidic monosaccharides are phosphate esters of the sugars. Sugar phosphates are important intermediates in glycolysis (Sec. 21.2), for example:

Specific enzymes catalyze the esterification of the carbonyl group (C-1) or the primary alcohol group (C-6), to produce either glucose 1-phosphate or glucose 6-phosphate.

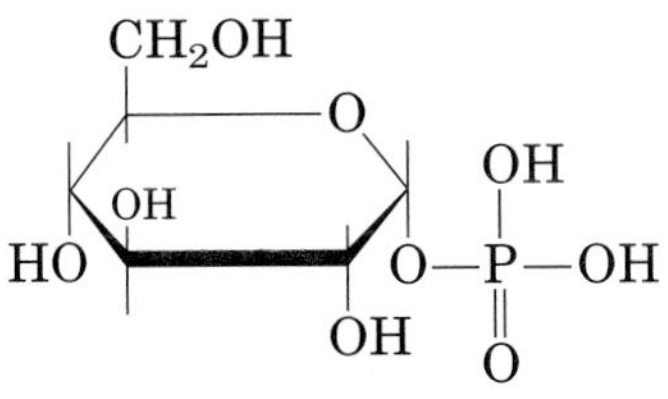

Glucose 1-phosphate

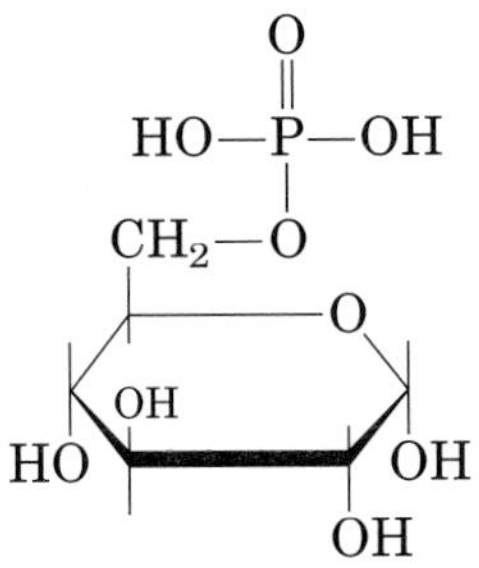

Glucose 6-phosphate

Amino sugars are produced by replacing the —OH group on C-2 of monosaccharides with an amino group. Amino sugars and their *N*-acetyl derivatives are important building blocks of the polysaccharides of connective tissue such as cartilage.

A typical *N*-acetyl derivative of an amino sugar.

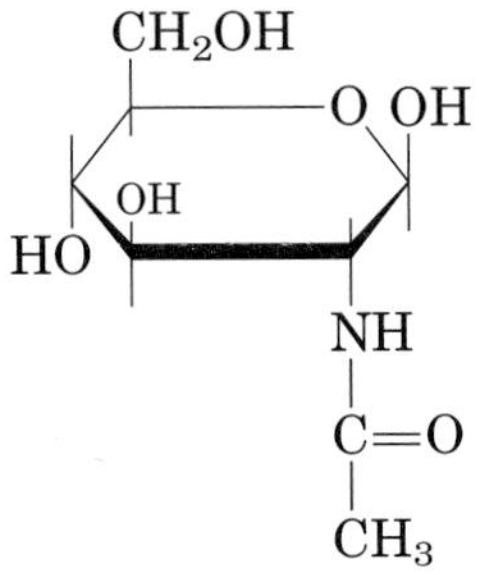

β-D-*N*-Acetylglucosamine

16.10 Disaccharides

We saw in Section 16.8 that monosaccharides can be converted to glycosides by treatment with an alcohol, ROH:

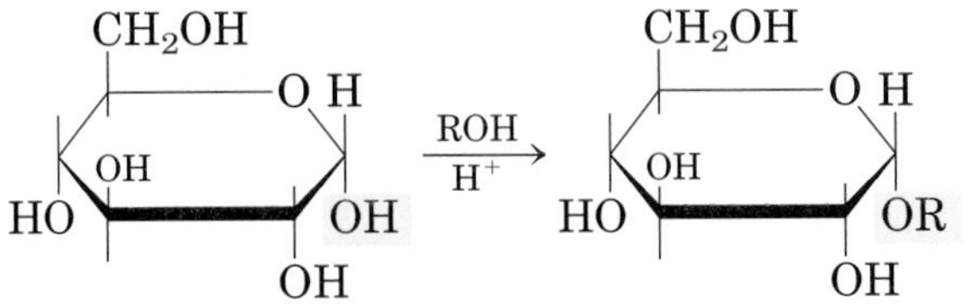

$$\xrightarrow[H^+]{ROH}$$

When the ROH is a second monosaccharide molecule, the product is a **disaccharide.**

Maltose

The disaccharide **maltose** contains two glucose rings connected in a $1 \rightarrow 4$ linkage:

The ether oxygen connects C-1 of one ring to C-4 of the other.

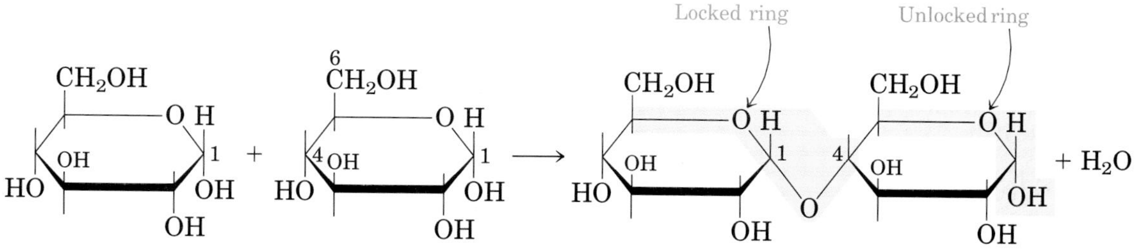

Locked ring

Unlocked ring

Maltose, a disaccharide

The ring on the left is locked. C-1 is an acetal function (a glycoside, shown in blue), and this ring will not open when maltose is dissolved in water. Furthermore, the C-1 oxygen in maltose points downward; as we have seen, this is an α linkage, and we use the symbol $\alpha(1 \rightarrow 4)$. The second ring, however, is unlocked. It has a hemiacetal at C-1, shown in yellow, so this ring opens in water to give an aldehyde. Because it has one unlocked ring, maltose shows mutarotation and is a reducing sugar.

Maltose is an ingredient in most syrups.

Lactose

Another disaccharide abundant in food is **lactose.** It occurs in mammalian milk in approximately 5 percent concentration. It consists of a β-D-galactose ring linked to a D-glucose ring through a $\beta(1 \rightarrow 4)$ glycosidic linkage:

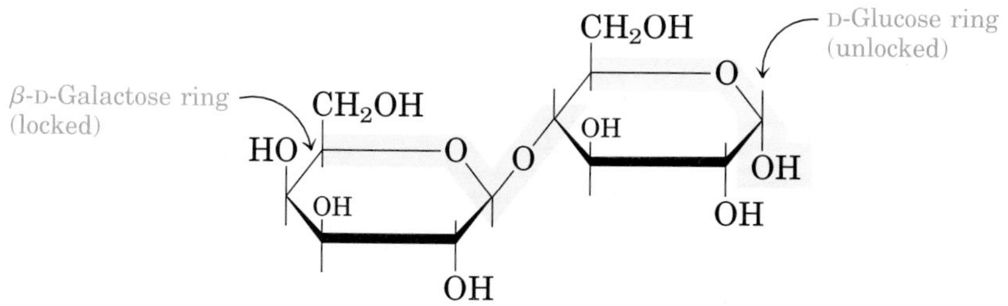

β-D-Galactose ring (locked)

D-Glucose ring (unlocked)

Lactose

Upon hydrolysis, lactose yields D-galactose and D-glucose.

Sucrose

In maltose and lactose, the rings are linked $1 \rightarrow 4$ so that one ring is locked and the other is unlocked. In **sucrose** the linkage is from C-1 of D-glucose (an aldose) to C-2 of D-fructose (a ketose):

Box 16B

Blood Groups

The ABO blood group system, discovered by Karl Landsteiner (1868–1943) in 1900, is of primary importance in safe blood transfusion. Whether an individual belongs to the A, B, AB, or O group is genetically determined and depends on the particular oligosaccharide assembly on the surface of the red blood cells. These oligosaccharides act as *antigens*. The four different blood types have the following antigen arrangements on the surface of their red blood cells:

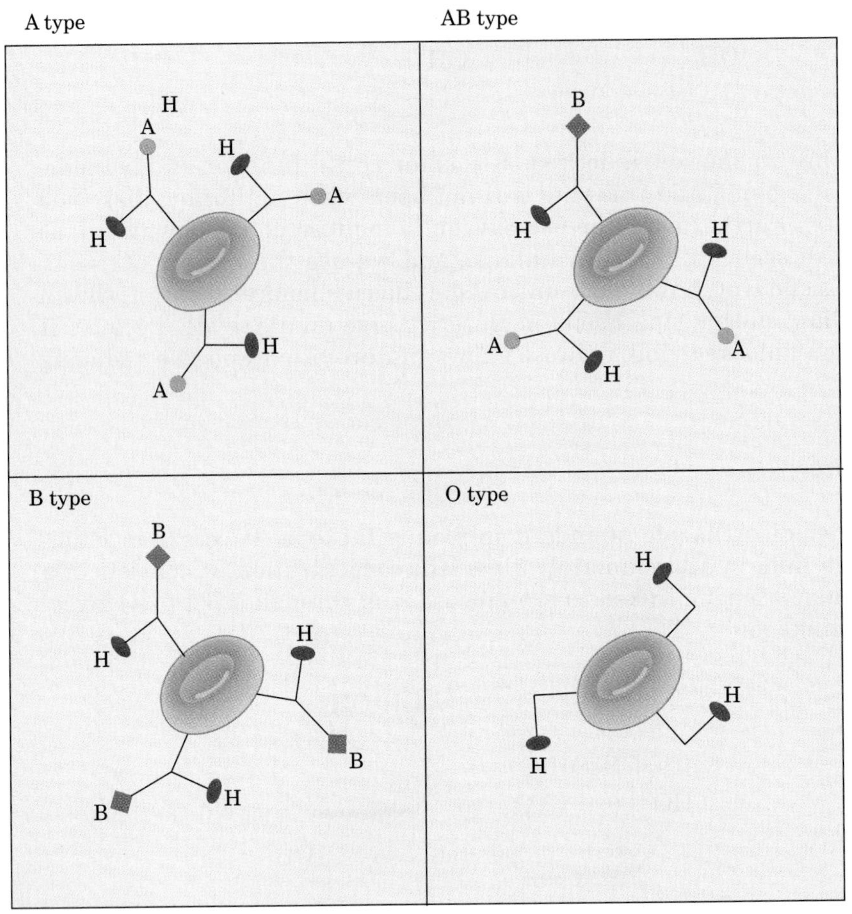

The A antigen is α-D-*N*-acetylgalactosamine:

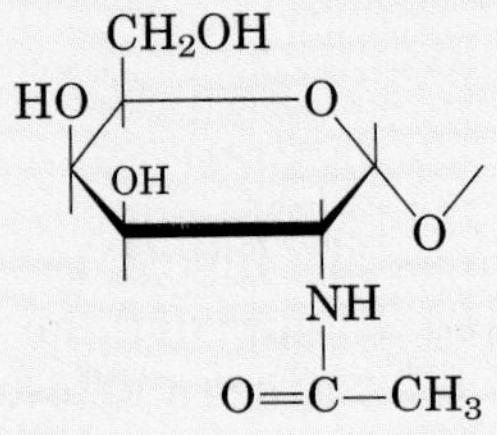

The B antigen is α-D-galactose:

The H antigen, common to all four types, is L-fucose:

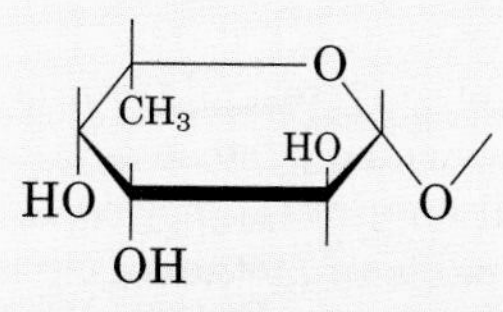

The blood carries antibodies against all foreign antigens. When a person receives a blood transfusion, the antibodies clump (aggregate) the foreign red blood cells. Thus, an individual having A-type blood has A antigens on the surface of the red blood cells and carries anti-B antibodies (against B antigens). A person of B-type blood carries B antigens on the surface of the red blood cells and has anti-A antibodies (against A antigens). Transfusion of A blood into a person with B blood would be fatal, and vice versa. O-type blood has neither A nor B antigens on the surface of the red blood cell but carries both anti-A and anti-B antibodies. The AB-type individual has both A and B antigens but no anti-A or anti-B antibodies. This gives rise to the following blood transfusion possibilities:

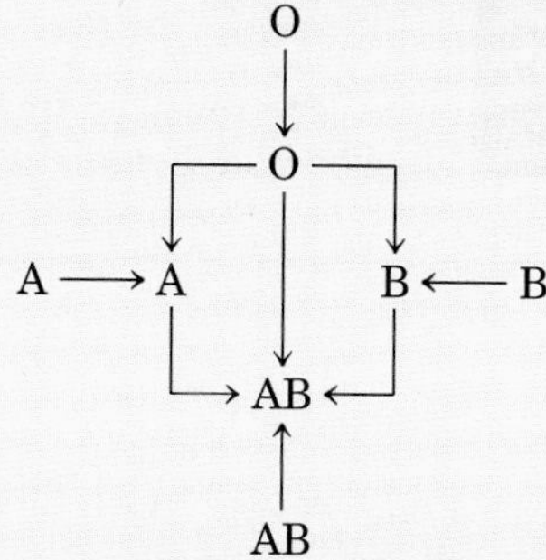

Thus, people with O-type blood are universal donors, and those with AB-type blood are universal acceptors. People with A-type blood can accept only from A- or O-type donors. B-type individuals can accept only B- or O-type blood. AB-type persons can accept blood from all groups. O-type individuals can accept blood only from O-type donors.

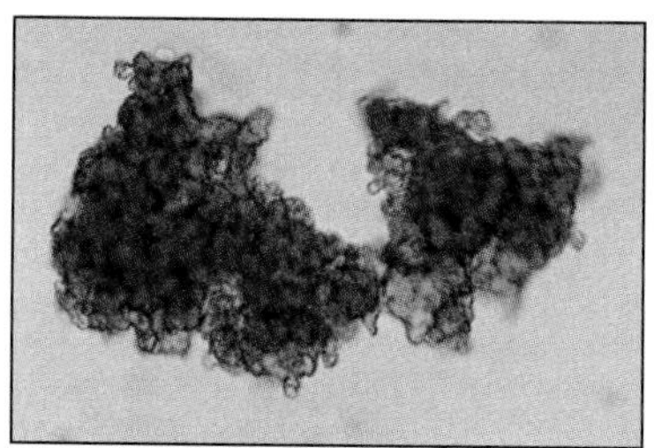

Blood clumping resulting from mixing of blood types A and B. (Phototake.)

Box 16C

Galactosemia

One child of every 18 000 is born with a genetic defect that makes the infant unable to utilize galactose. Galactose is part of milk sugar, lactose, and when the body cannot utilize it, it accumulates in the blood and in the urine (galactosuria). The accumulation is harmful because it leads to mental retardation, growth failure, cataract formation in the lens of the eye, and in some severe cases death due to liver damage. When galactose accumulation is due to a deficiency of the enzyme galactokinase, the disorder has only mild symptoms. When the enzyme galactose-1-phosphate uridinyltransferase is deficient, the disorder is called **galactosemia,** and its symptoms, as described above, are severe.

The deleterious effects of galactosemia can be avoided simply by giving the infant a milk formula in which sucrose is substituted for lactose. A galactose-free diet is critical only in infancy. With maturation the children develop another enzyme capable of metabolizing galactose. They are thus able to tolerate galactose as they mature.

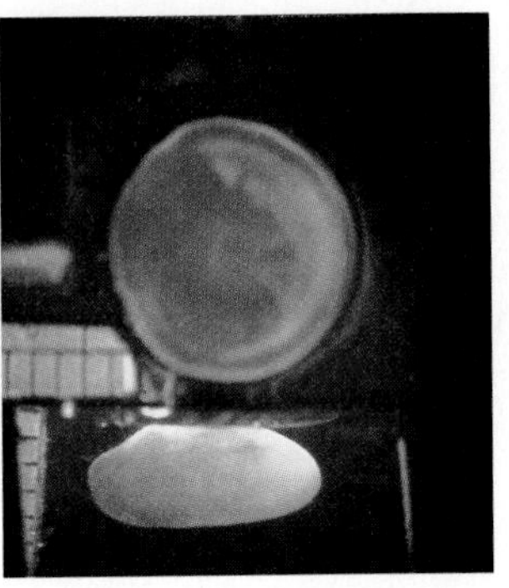

A lens removed from the eye of a patient who had a cataract.

α-D-Glucose ring (locked)

β-D-Fructose ring (locked)

Sucrose

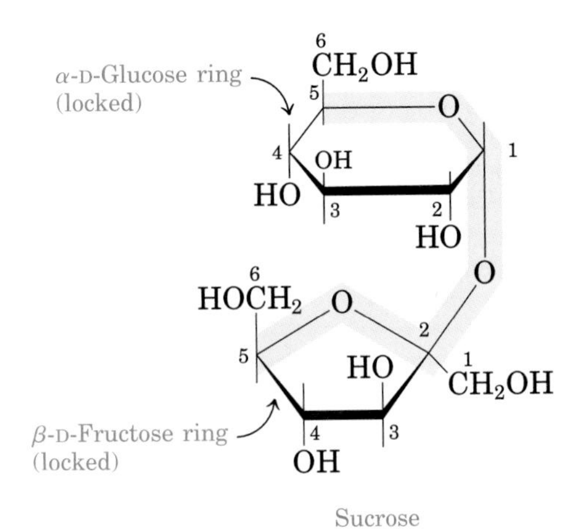

In this compound *both* rings are locked, so that sucrose is a nonreducing sugar and does not exhibit mutarotation. Sucrose, which is ordinary table sugar, is also called *cane sugar,* though it comes from sugar beets as well as from sugar cane and is also found in other plants. Sucrose is easily hydrolyzed by the enzyme invertase, producing a one-to-one mixture of D-glucose and D-fructose. This mixture, called *invert sugar,* is used as a food additive. It is sweeter than sucrose (Table 16.1).

16.11 Polysaccharides

The unlocked ring of maltose is still a hemiacetal, so a third monosaccharide could be added, giving a **trisaccharide.** If the connection is $1 \rightarrow 4$, there would still be a hemiacetal, and so we could add a fourth and continue the process until thousands of units have been added. Nature does just this in forming **polysaccharides.** The two most important polysaccharides, both made entirely of D-glucose units, are starch and cellulose.

Starch

When the $1 \rightarrow 4$ condensation is repeated many times with α-D-glucose molecules, a polyglucose named **amylose** is obtained. Amylose is one of the two components of **starch** and contains between 1000 and 2000 glucose units, all in the $\alpha(1 \rightarrow 4)$ glycosidic linkage (the same as in maltose). This linkage is flexible enough to allow the long-chain amylose molecules to curl up and form either random-coil or helical structures in solution (Fig. 16.10).

The second component of starch, called **amylopectin,** is also a polymer of α-D-glucose. Amylopectin is not entirely a straight-chain molecule but has random branches. The branching point is the $\alpha(1 \rightarrow 6)$ glycosidic linkage:

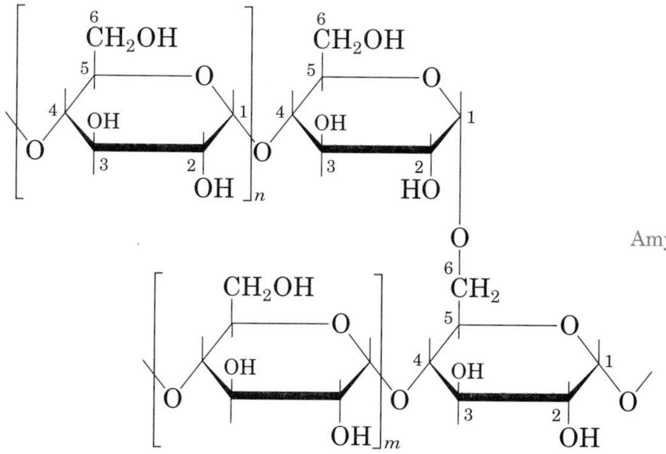

Amylopectin

The m and n refer to the number of glucose units in the linear portions of the chains.

There are, on the average, 20 to 25 glucose units in the straight-chain form, all connected by $\alpha(1 \rightarrow 4)$ linkages, between branching points (Fig. 16.11). Amylopectin contains as many as 10^5 to 10^6 glucose units in one gigantic macromolecule.

A **macromolecule** is a very large molecule.

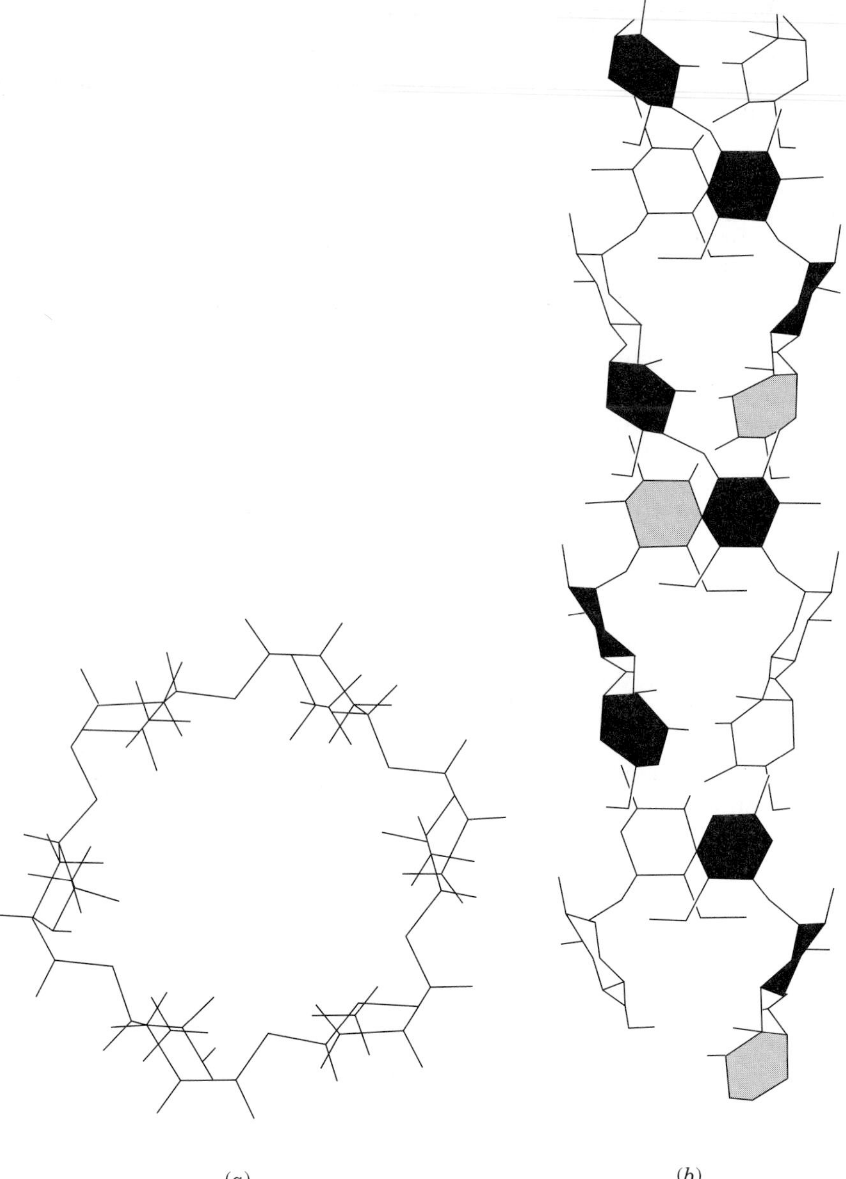

Figure **16.10**
(a) Cross section of a single helical form of amylose. There is a central cavity into which iodine and other small molecules can fit. (b) Double helix of amylose molecules as they exist in starch granules of plants. In both (a) and (b) the cyclic structures represent the pyranose ring of D-glucose.

An important test for the presence of starch is the addition of iodine, I_2, to give an intense, deep purple color (Fig. 16.12). In the presence of iodine, amylose and the straight-chain portion of amylopectin form helixes inside of which the iodine molecules assemble in long polyiodine chains. This assembly of iodine molecules gives the characteristic color.

Figure **16.12**
The starch in bread gives a positive iodine test. (Photograph by Beverly March.)

Glycogen

Amylose and amylopectin are storage polysaccharides in plants. **Glycogen,** also called **animal starch,** plays a similar role in animal tissue. Thus, when glucose is in abundance it is stored in the form of these polysaccharides in animal and plant tissues.

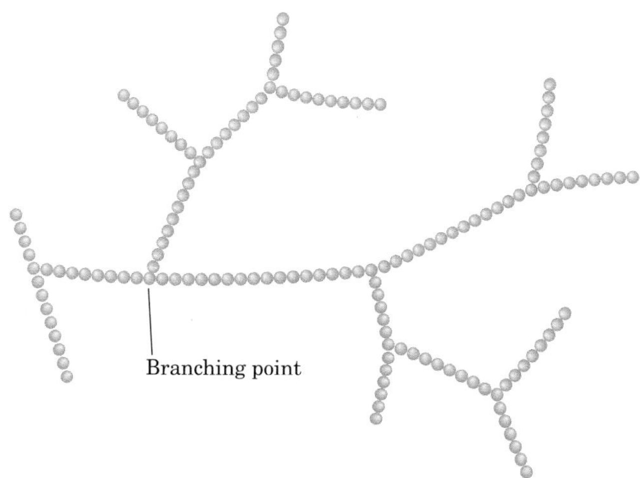

Figure **16.11** • Schematic representation of an amylopectin molecule.

Bread, grains, pasta, and rice are the main sources of starch for much of the world's population. (Photograph by Charles D. Winters.)

Glycogen is similar to amylopectin in that it has $\alpha(1 \rightarrow 6)$ branching points and $\alpha(1 \rightarrow 4)$ glycosidic linkages in the straight-chain portions. It is also a giant molecule, containing about 10^6 glucose units. The structural difference between glycogen and amylopectin is in the degree of branching. Glycogen contains smaller numbers of glucose units in the straight-chain part between branching points (on the average, 10 to 12 units) and consequently has more branching points per molecule than amylopectin.

Cellulose

Cellulose is a linear polymer containing only D-glucose units, all in the $1 \rightarrow 4$ linkage. Structurally, the difference between amylose and cellulose is that in cellulose all the $1 \rightarrow 4$ glycosidic linkages are β instead of α (the C-1 oxygen points up instead of down in the Haworth drawing):

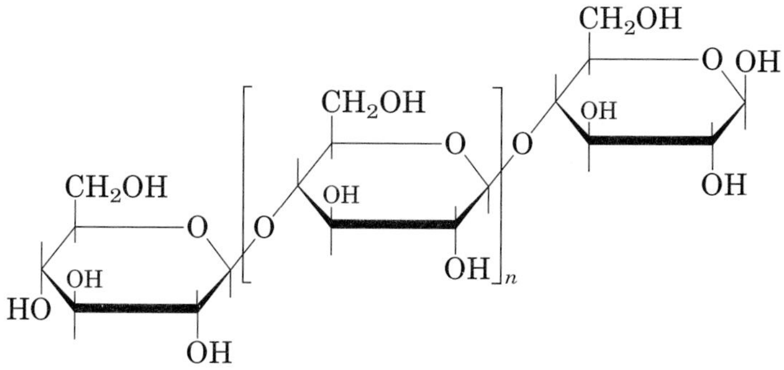

Cellulose

Cellulose is the most abundant molecule in living tissues. It makes up about 50 percent of the total organic carbon in the biosphere. Cotton is almost pure cellulose (95 percent), and wood is about 50 percent cellulose.

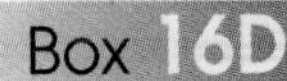

Box 16D

Polysaccharides as Fillers in Drug Tablets

Drugs that have the same generic name contain the same active ingredient (for example, aspirin). In spite of that, many physicians prefer one trade-name drug over another. In most cases there is some difference in the inactive ingredients, called **fillers.** Fillers are not completely inactive but may affect the rate of drug delivery.

Polysaccharides have been used as fillers because of their solubility in water and their thickening and gelling properties. For example, starch is used in pills, tablets, and capsules for oral delivery, as a filler and as a disintegrating or binding agent. Disintegration is mostly due to the rapid uptake of water. As a result the tablet swells and falls apart, allowing the drug to be delivered in the stomach and the intestines.

Cellulose is a structural polysaccharide that provides strength and rigidity to plants. It may contain between 300 and 3000 glucose units in one macromolecule.

Cellulose molecules act very much like extended stiff rods. This enables them to align themselves side by side into well-organized, water-insoluble crystalline lattices (fibers) in which the alcohol groups form numerous intermolecular hydrogen bonds with neighboring chains (Fig. 16.13). Thus, when a piece of cellulosic material is placed in water, there are not enough water molecules on the surface of the cellulose to pull one molecule away from the strongly hydrogen-bonded crystalline matrix. This is why cellulose is insoluble in water.

Cellulose is also a structural polysaccharide for humans since many of our houses, furniture, and other structures are made of wood. Large parts of the textile and plastic industries are based on cellulose (cotton) or cellulose derivatives (rayon) as well as on cellulose acetate:

Figure 16.13
Electron micrograph of cellulose fibers. (Biophoto Associates.)

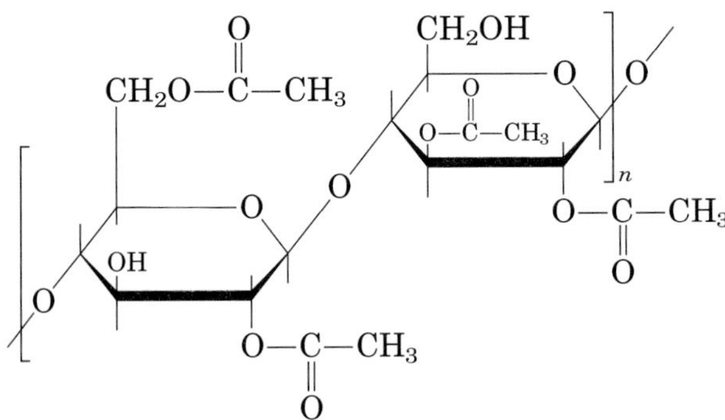

Cellulose acetate (—OH groups esterified at random)

Although the only difference between the basic structure of starch and cellulose is that starch has α linkages whereas cellulose has β linkages, this difference is enough to make starch one of the most important of our food sources, while we cannot digest cellulose at all. Although cellulose can be hydrolyzed to glucose by strong acids, human beings and other mammals do not have enzymes that are capable of hydrolyzing the $\beta(1 \rightarrow 4)$ linkages, so cellulose goes through the intestinal tract largely undigested. However, many bacteria and fungi possess cellulase, an enzyme that hydrolyzes cellulose to glucose. Ruminant mammals, such as the cow, can digest grass because microorganisms in their rumen produce the necessary cellulase. Termites eat wood, and in this case, too, it is not the termites that have the enzymes but microorganisms inside them.

16.12 Acidic Polysaccharides

Another group of polysaccharides, **acidic polysaccharides,** play an important role in the structure and function of connective tissues. These tissues are the matrix between organs and cells that provides mechanical strength and also filters the flow of molecular information between cells. Connective tissues are usually made up of collagen, a structural protein, and a variety of acidic polysaccharides that interact with the collagen to form tight or loose networks.

There is no generalized connective tissue, but there are a large number of highly specialized forms, such as cartilage, bone, synovial fluid, skin, tendons, blood vessel walls, intervertebral disks, and cornea.

Hyaluronic Acid

A typical and probably the simplest (structurally) acidic polysaccharide of connective tissue is **hyaluronic acid:**

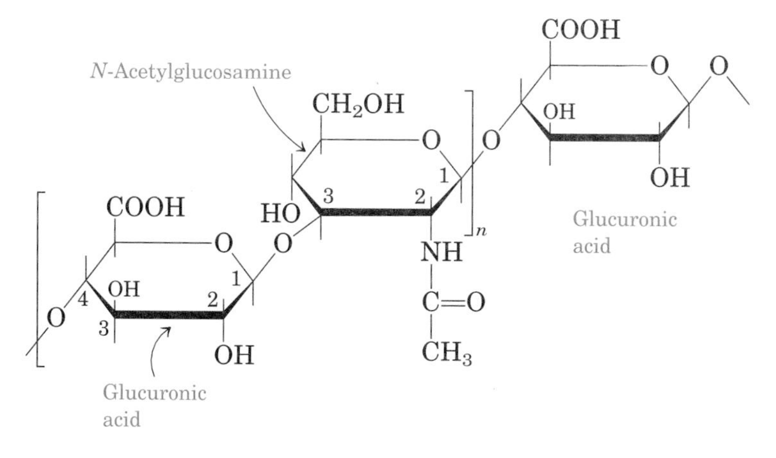

N-Acetylglucosamine

Glucuronic acid

Glucuronic acid

Hyaluronic acid

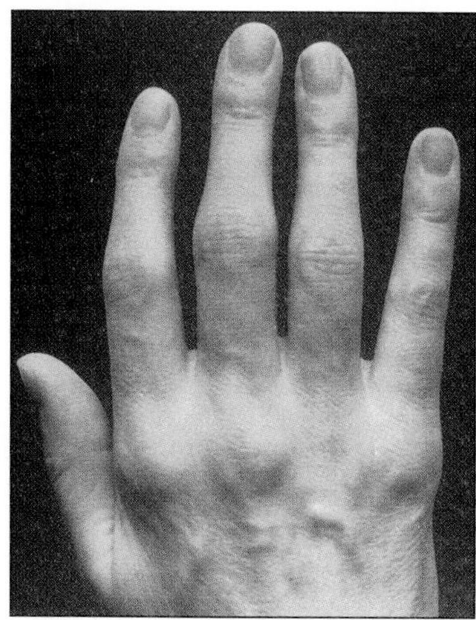

The joints of this hand are swollen by rheumatoid arthritis. (Courtesy of Drs. P. A. Dieppe, P. A. Bacon, A. N. Bamji, I. Watt, and the Gower Medical Publishing Ltd., London, England.)

In rheumatoid arthritis inflammation of the synovial tissue results in swelling of the joints.

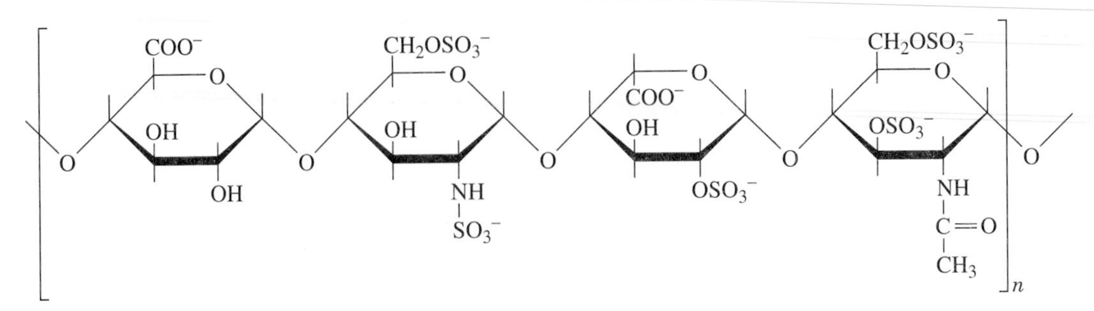

This giant molecule, with a molecular weight of between 10^5 and 10^7, may contain from 300 to 100 000 repeating units, depending on the organ in which it occurs. Hyaluronic acid is most abundant in embryonic tissues and in specialized connective tissues such as synovial fluid, where it acts as a lubricant, or the vitreous of the eye, where its function is to

Figure **16.14** • A tetrasaccharide repeating unit of heparin.

provide a clear, elastic gel that maintains the retina in its proper position (see Box 16E).

Hyaluronic acid is composed of D-glucuronic acid linked to *N*-acetylglucosamine by a $\beta(1 \rightarrow 3)$ linkage; the latter in turn is linked to the next glucuronic acid by a $\beta(1 \rightarrow 4)$ linkage.

Heparin

Heparin (Fig. 16.14) is one of the most common acidic polysaccharides. It occurs in such tissues as lung, liver, skin, and intestinal mucosa. It has many biological functions, the best known of which is its anticoagulant activity; that is, it prevents blood clotting. Although the details of this function are still unknown, the general features are as follows.

For blood to clot, an active enzyme called thrombin is necessary. Heparin helps to inactivate this enzyme by forming a complex (Fig. 16.15) consisting of itself, thrombin, and antithrombin (another blood constituent). Different heparin preparations have different anticoagulant activities. As is evident from its structure (Fig. 16.14), heparin is a heterogeneous polysaccharide in which the monosaccharide units are linked by $\alpha(1 \rightarrow 4)$ glycosidic bonds. The composition of heparin is not constant; it varies slightly from repeating unit to repeating unit. The structure in Figure 16.14 shows a tetrasaccharide as a repeating unit. A heparin preparation with good anticoagulant activity has a minimum size of at least eight to ten such repeating units. The larger the molecule, the better its anticoagulant activity.

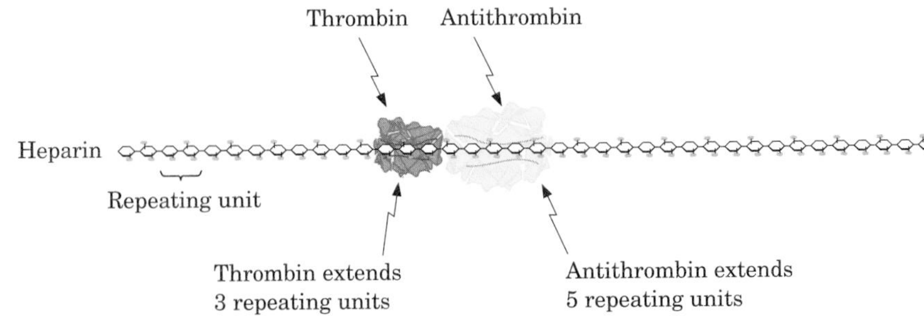

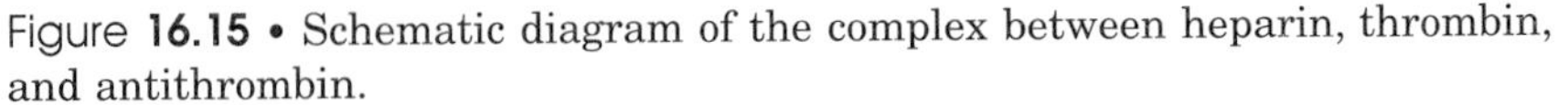

Figure **16.15** • Schematic diagram of the complex between heparin, thrombin, and antithrombin.

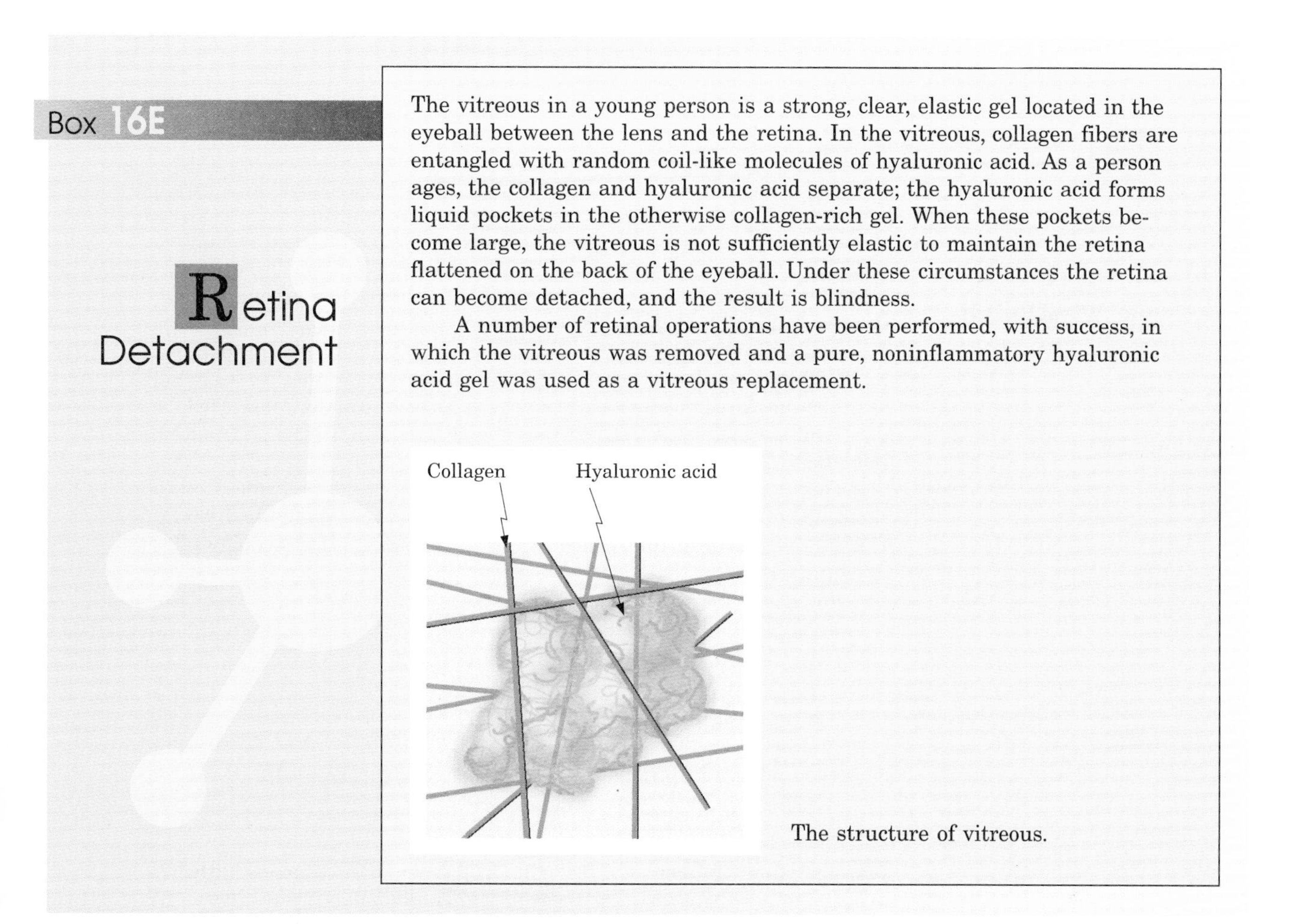

Box 16E Retina Detachment

The vitreous in a young person is a strong, clear, elastic gel located in the eyeball between the lens and the retina. In the vitreous, collagen fibers are entangled with random coil-like molecules of hyaluronic acid. As a person ages, the collagen and hyaluronic acid separate; the hyaluronic acid forms liquid pockets in the otherwise collagen-rich gel. When these pockets become large, the vitreous is not sufficiently elastic to maintain the retina flattened on the back of the eyeball. Under these circumstances the retina can become detached, and the result is blindness.

A number of retinal operations have been performed, with success, in which the vitreous was removed and a pure, noninflammatory hyaluronic acid gel was used as a vitreous replacement.

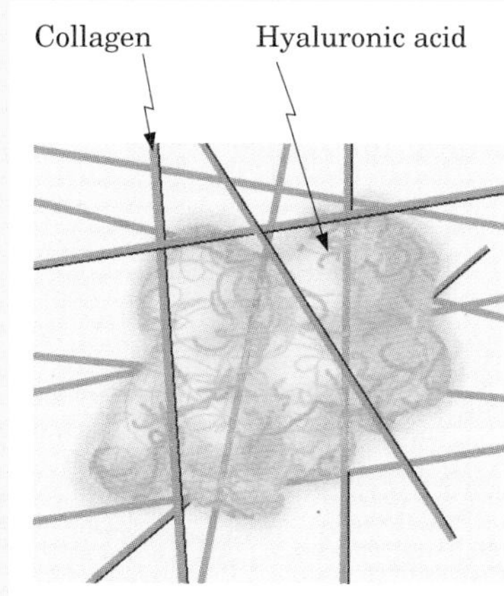

The structure of vitreous.

16.13 Chiral Compounds in Biological Organisms

Most organic compounds in the body contain one or more chiral carbons, which means that two or more stereoisomers are possible. The general rule is that, in most cases, **nature makes only one of the possible stereoisomers.** For example, the important steroid cholesterol (Sec. 17.9) contains eight chiral carbons. There are, therefore, $2^8 = 256$ stereoisomers of this compound. Of these 256, nature makes only one. No matter what the organism, the same isomer of cholesterol is found.

This is not an absolute rule, however. As we have seen, both D-glucose and D-galactose (diastereomers) are found in nature, and both the (+) and the (−) forms of lactic acid (enantiomers) are found in different organisms. There are even some cases where nature makes racemic mixtures (Sec. 16.3). But in most cases, where several isomers are possible, only one is made.

What is the reason for this? The answer is that biological molecules must often react with each other. We saw in Section 16.3 that enantiomers differ in the rates at which they react with other chiral molecules.

$$CH_3-\underset{\underset{OH}{|}}{CH}-COOH$$

Lactic acid

The (+) isomer of lactic acid is produced by muscle tissue in humans. The (−) isomer is found in sour milk.

Box 16F

Chiral Drugs

Most chiral compounds that are sold as drugs are racemic mixtures. It is well known that the members of any particular pair of enantiomers often have different biological effects. For example, ibuprofen (Box 15C), the most frequently used nonsteroidal anti-inflammatory drug, is sold as a racemic mixture. It is much cheaper to produce a racemic mixture than to synthesize one enantiomer or to separate one from a racemic mixture. However, when an enantiomer has a clear advantage over a racemic mixture (for example, it may have fewer side effects or act faster), then producing and marketing an enantiomer can be worthwhile. This is happening to ibuprofen. (+)-Ibuprofen reaches therapeutic concentrations in the human body in 12 minutes while the racemic mixture takes 30 minutes. This kind of situation, in which a previously approved racemic drug is replaced by one pure and more effective enantiomer, is now the trend in the pharmaceutical industry.

There is a caveat, however, as is demonstrated by the tragic story of the compound thalidomide. This racemic drug was marketed in 1960 in Europe (but never in the United States) as a combination sleeping pill and tranquilizer.

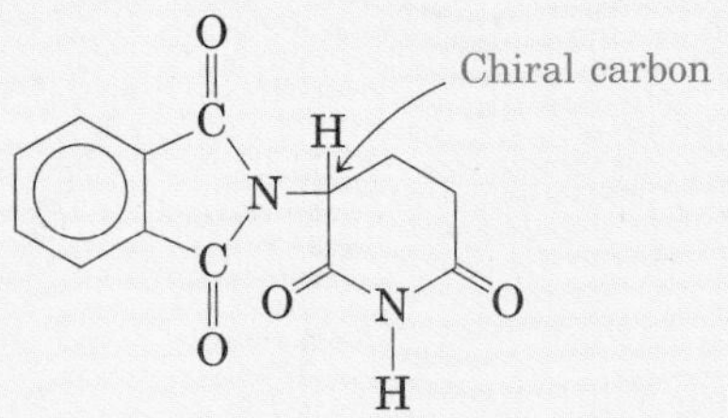

Chiral carbon

Many pregnant women who took the drug gave birth to babies without arms or fingers or with other deformities. Thalidomide is a *teratogen,* a substance that produces birth defects. Later it was found that only the (−) enantiomer is teratogenic. However, it was also found that the (+) enantiomer rapidly converts to the racemic mixture in rabbits, marmoset monkeys, and presumably in humans (although this was not directly tested). Thus, even if only the (+) enantiomer was given as a drug, there would be no advantage, since the body would convert it to the racemic mixture. In spite of its tragic past history, thalidomide is making a kind of a comeback but not as a sedative. It was found that inflammatory skin flare-ups in leprosy patients and canker sores in AIDS patients are eased and relieved by thalidomide. As long as the patients are not pregnant women, the teratogenic properties of thalidomide cannot be manifested.

Diastereomers differ even more. Thus it is likely that the other 255 isomers of cholesterol (including its enantiomer) would undergo cholesterol's biochemical reactions much more slowly than cholesterol itself.

It's a matter of *fit.* A chiral molecule fits another chiral molecule, where its enantiomer (or a diastereomer) does not fit. This behavior becomes extremely important when we consider that almost all biological reactions are catalyzed by enzymes. Enzymes, as we shall see in Chapter 19, are chiral molecules with chiral cavities. Figure 16.16 shows a

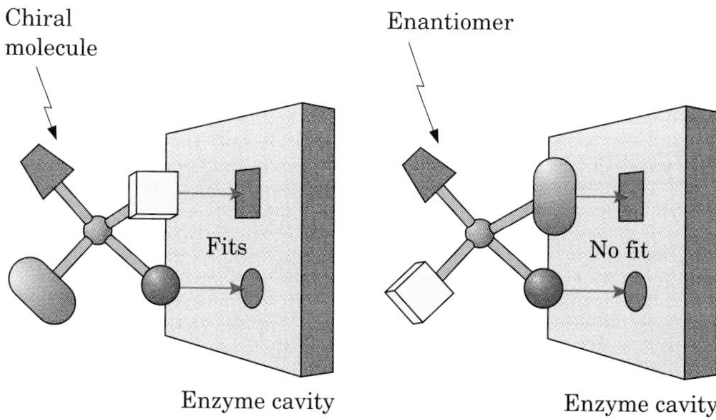

Figure **16.16** • Schematic diagram showing how one chiral molecule fits into an enzyme cavity but its enantiomer does not.

typical situation. One enantiomer fits into the cavity in a way that allows the reaction to proceed (like a right hand slipping into a right-handed glove). As seen in Figure 16.16, the enantiomer (the left hand) does not fit. This slows the reaction so much that, for all practical purposes, it does not take place at all.

When we know all this, we are not surprised to learn that the biological properties of enantiomers can differ greatly. D-Glucose is sweet and nutritious and is an important component of our diets. The L isomer, on the other hand, is tasteless, and the body cannot metabolize it. These substances have different tastes because taste arises from interactions between chiral molecules and chiral receptors on the taste buds of our tongues. Again, the enantiomers fit differently.

SUMMARY

Carbohydrates are polyhydroxy aldehydes or ketones or compounds that, on hydrolysis, yield polyhydroxy aldehydes or ketones. Their chief source in nature is photosynthesis, a process carried out by plants.

Molecules that are not superimposable on their mirror images are called **chiral.** Most chiral molecules contain a **chiral carbon** (also called an **asymmetric carbon**), which is one connected to four different groups. Since a chiral molecule is not superimposable on its mirror image, the two are different molecules. They are **a pair of enantiomers.** All chiral molecules rotate the plane of **polarized light** (light that vibrates in only one plane); molecules that are not chiral do not rotate the plane of polarized light.

Enantiomers have identical properties except for two: (1) they rotate the plane of polarized light in opposite directions, and (2) they react at different rates with other chiral compounds. This is very important in biological systems, where nature usually makes only one of a pair of enantiomers. A mixture of equal amounts of a (+) and (−) enantiomer is called a **racemic mixture** or a **racemate.** When a molecule has more than one chiral carbon, the total number of stereoisomers usually equals 2^n, where n is the number of chiral carbons. **Diastereomers,** which are stereoisomers that are not enantiomers, do not have identical properties.

Monosaccharides, the basic units of carbohydrates, are simple sugars of formula $C_nH_{2n}O_n$, where $n = 3$ to 9. The most important is D-glucose, which in the solid state has a six-membered (**pyranose**) ring containing a hemiacetal group. In solution, the ring opens to give a mixture of one open-chain and two cyclic forms. Glucose is an **aldose.** All aldoses, in the open-chain form, are straight-chain compounds containing an aldehyde group at C-1 and an —OH group on all other carbons. The 16 **aldohexoses** differ in their configuration at C-2 to C-5. **Ketoses** are similar but contain a keto group at C-2 instead of an aldehyde at C-1. Almost all monosaccharides are more stable in cyclic forms, either pyranose (six-membered) or **furanose** (five-membered) rings. Isomerism at the C-1 carbon produces α and β cyclic **anomers.** Monosaccharides are placed in the D or L series depending on whether the —OH group on the highest-numbered chiral carbon of the chain points left or right in the Fischer projection.

Most carbohydrates are soluble in water. Many mono- and disaccharides are sweet. All monosaccharides are **reducing sugars,** and almost all form **glycosides** when treated with alcohols and an acid catalyst. Derivatives of monosaccharides are formed by oxidation, reduction, and phosphorylation. When two or more monosaccharides are linked together by glycosidic bonds, we get **di-, tri-, oligo-,** and **polysaccharides.** Storage polysaccharides (starch and glycogen) are polymers of α-D-glucose linked by $1 \rightarrow 4$ and $1 \rightarrow 6$ glycosidic bonds. Cellulose, a structural polysaccharide, has $\beta(1 \rightarrow 4)$ glycosidic linkages. Acidic polysaccharides are important components of connective tissues and play roles in many biochemical processes.

KEY TERMS

Acidic polysaccharide (Sec. 16.12)
Aldohexose (Sec. 16.6)
Aldose (Sec. 16.6)
Anomers (Sec. 16.5)
Antibody (Box 16B)
Antigen (Box 16B)
Asymmetric carbon (Sec. 16.2)
Carbohydrate (Sec. 16.1)
Chiral carbon (Sec. 16.2)
Chirality (Sec. 16.2)
Configuration (Sec. 16.4)
Dextrorotatory (Sec. 16.3)
Diastereomer (Sec. 16.4)
Disaccharide (Sec. 16.10)
Enantiomers (Sec. 16.2)
Fischer projection (Sec. 16.3)
Furanose ring (Sec. 16.8)
Galactosemia (Box 16C)
Glucoside (Sec. 16.8)
Glycoside (Sec. 16.8)
Haworth formula (Sec. 16.5)
Ketose (Sec. 16.6)
Levorotatory (Sec. 16.3)
Locked ring (Sec. 16.8)
Macromolecule (Sec. 16.11)
Monosaccharide (Sec. 16.5)
Mutarotation (Sec. 16.5)
Oligosaccharide (Sec. 16.5)
Optical activity (Sec. 16.3)
Photosynthesis (Sec. 16.1)
Polarimeter (Sec. 16.3)
Polarized light (Sec. 16.3)
Polysaccharide (Sec. 16.11)
Pyranose ring (Sec. 16.8)
Racemate (Sec. 16.3)
Racemic mixture (Sec. 16.3)
Reducing sugar (Sec. 16.8)
Sugar (Sec. 16.7)
Teratogen (Box 16F)
Unlocked ring (Sec. 16.10)
Uronic acid (Sec. 16.9)

PROBLEMS

16.5 (a) What percent of the dry weight of plants is carbohydrates? (b) How does that compare with humans?

16.6 Which gas is produced (a) by plants during photosynthesis (b) by animals during metabolism?

Enantiomers and Chirality

16.7 Identify all the chiral carbons:

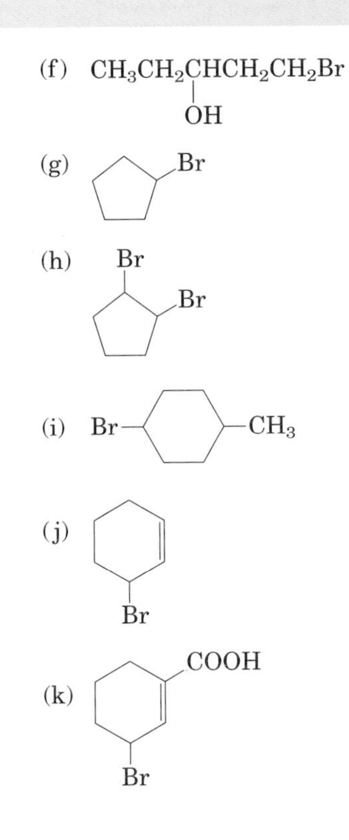

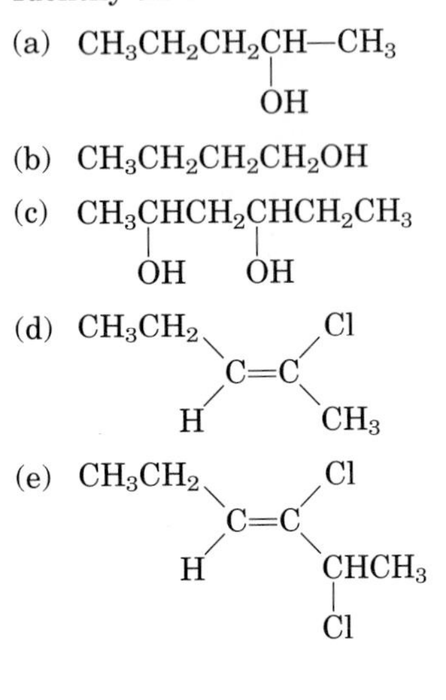

(a) $CH_3CH_2CH_2CH{-}CH_3$ (OH on the CH)

(b) $CH_3CH_2CH_2CH_2OH$

(c) $CH_3CHCH_2CHCH_2CH_3$ (OH on each CH)

(d) $CH_3CH_2(H)C{=}C(Cl)CH_3$

(e) $CH_3CH_2(H)C{=}C(Cl)CHCH_3$ (Cl on the CH)

(f) $CH_3CH_2CHCH_2CH_2Br$ (OH on the CH)

(g) Br

(h) Br Br

(i) Br— (cyclohexane ring) —CH_3

(j) (cyclohexene ring) Br

(k) COOH Br

(l) 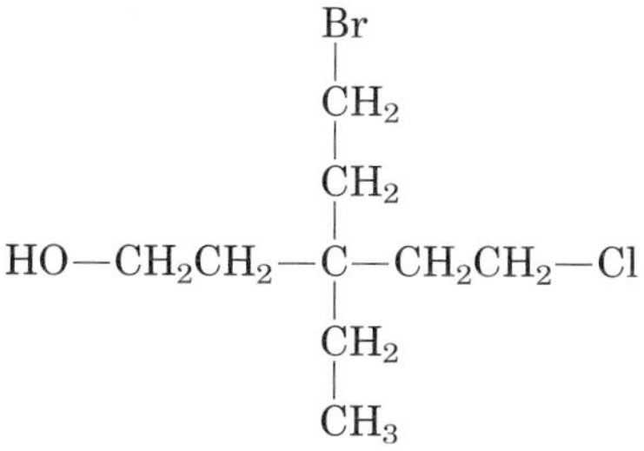

(m) 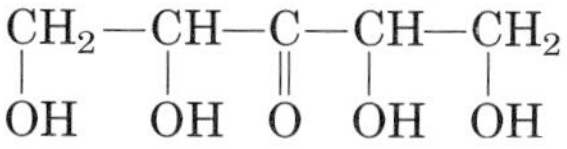

(n) 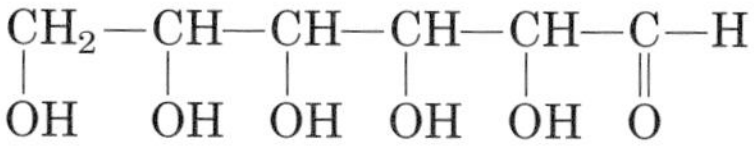

16.8 Which of these compounds can be optically active?

(a) CH_2BrCl

(b) 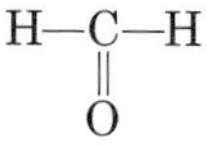

(c) 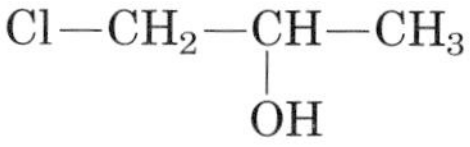

(d) 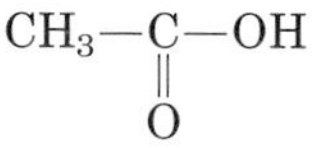

Properties of Enantiomers. Optical Activity

16.9 Here are some properties of D-isomenthol. Fill in the corresponding properties of its enantiomer, L-isomenthol.

Property	D-Isomenthol	L-Isomenthol
Melting point	82.5°C	
Boiling point	218.5°C	
Solubility in water	1.4 g/100 mL	
Density	0.9040 g/mL	
Rotation of the plane of polarized light	+24.1°	

16.10 In which two properties does the (+) form of the amino acid arginine differ from the (−) form?

*__16.11__ Pasteur noticed that a solution of a racemic mixture of tartaric acid in his laboratory was growing moldy. He filtered off the mold and found that the solution now rotated the plane of polarized light, though it had not done so before it became moldy. Explain how the mold caused the solution to become optically active.

Compounds with More Than One Chiral Carbon

16.12 How many stereoisomers are possible for each of these compounds?

(a) 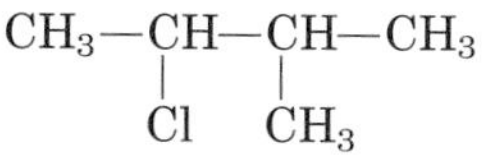

(b) 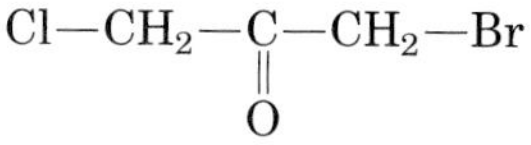

(c) 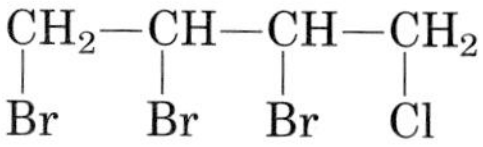

(d) 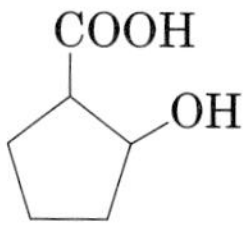

(e) 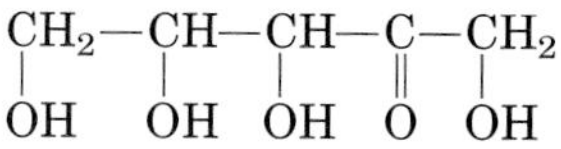

(f) 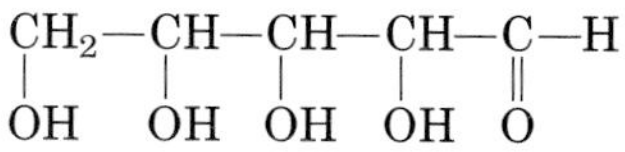

(g) 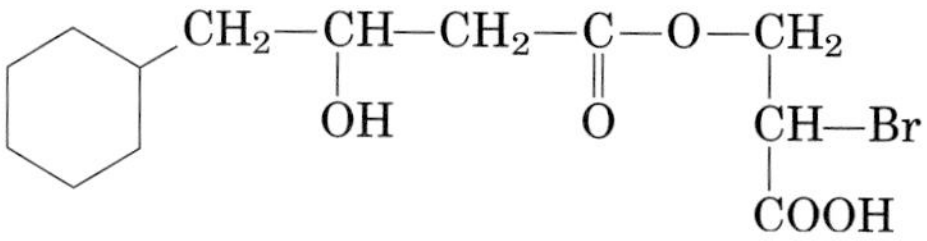

16.13 Define:

(a) enantiomer (c) polarized light
(b) diastereomer (d) racemic mixture

Glucose

16.14 What is the difference between oligo- and polysaccharides?

16.15 In glucose which carbon is the anomeric carbon?

16.16 Define: (a) carbohydrate (b) monosaccharide

16.17 What functional groups are present on each of these carbons of the glucose molecule?
(a) C-1 (b) C-2 (c) C-6

16.18 Are α-D-glucose and β-D-glucose enantiomers? Explain.

16.19 Explain why the optical rotation of D-glucose changes when it is dissolved in water.

16.20 (a) The specific rotation of α-D-glucose is +112.2°. What is the specific rotation of α-L-glucose? (b) When D-glucose is dissolved in water, the specific rotation changes to +52.7°. Does the specific rotation of L-glucose also change when it is dissolved in water? If so, what value does it change to?

Monosaccharides

16.21 Tell whether each of these monosaccharides is D or L:

(a) 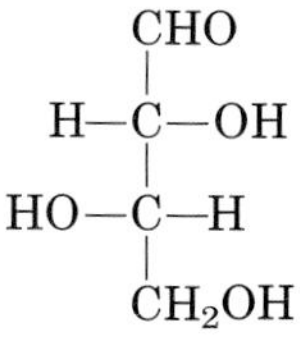

(b) 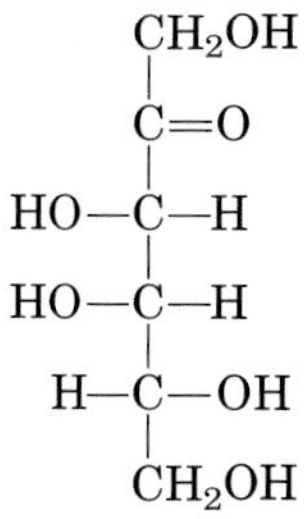

(c)

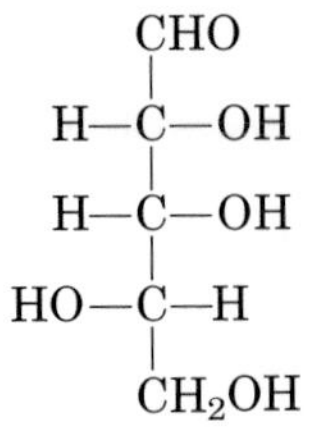

```
      CHO
      |
  H—C—OH
      |
  H—C—OH
      |
 HO—C—H
      |
      CH2OH
```

16.22 The structure of D-altrose is given in Figure 16.9. Draw its Haworth formula in both the alpha and beta configurations.

16.23 Consulting Figure 16.9, write the structure of L-arabinose in the Fischer projection and predict the direction of its optical rotation.

16.24 In an aldoheptose, configurational change about which carbon produces D and L isomerism?

***16.25** Draw the structure of L-(−)-mannose both in the straight-chain and cyclic forms (use Fig. 16.9).

Physical Properties of Monosaccharides

16.26 Explain why all monosaccharides and disaccharides are soluble in water.

16.27 The amino acid L-leucine rotates the plane of polarized light to the left. (a) In which direction does D-leucine rotate the light? (b) Can you predict in which direction L-valine (another amino acid) rotates the plane of the light?

Chemical Properties of Monosaccharides

16.28 Define: (a) aldose (b) ketose (c) pyranose (d) furanose

***16.29** The open-chain formula of D-arabinose is shown in Figure 16.9. Draw the structural formulas for the two furanose forms of this sugar.

16.30 (a) Why is D-fructose given the designation D-(−)? (b) How is this expressed in its common name?

16.31 Is this (a) an aldose or a ketose (b) a pyranose or a furanose (c) a pentose or a hexose:

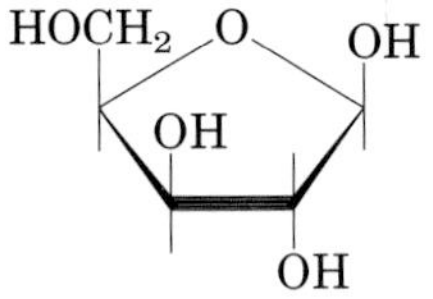

HOCH$_2$ O OH

OH

OH

Derivatives of Monosaccharides

16.32 What is the product when an aldose reductase catalyzes the reduction of D-glucose?

16.33 Draw the structural formula for each principal product:

(a) HOCH$_2$ O OH + $CH_3CH_2OH \xrightarrow{H^+}$

OH OH

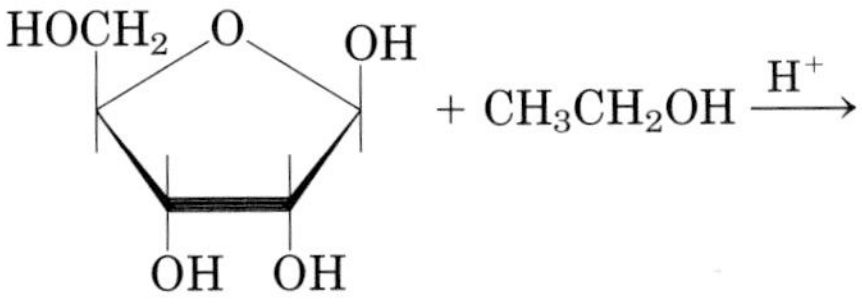

(b)

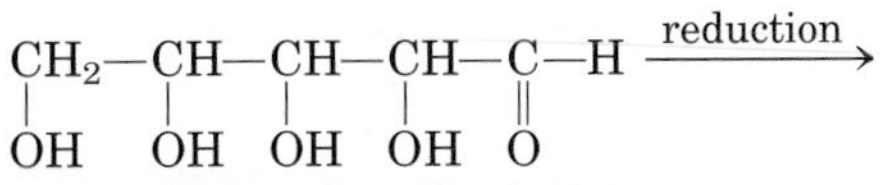

```
CH2—CH—CH—CH—C—H
|    |    |    |    ||
OH   OH   OH   OH   O
```

$\xrightarrow{\text{reduction}}$

(c) CH$_2$OH, HO, O OH, OH, OH + $CH_3OH \xrightarrow{H^+}$

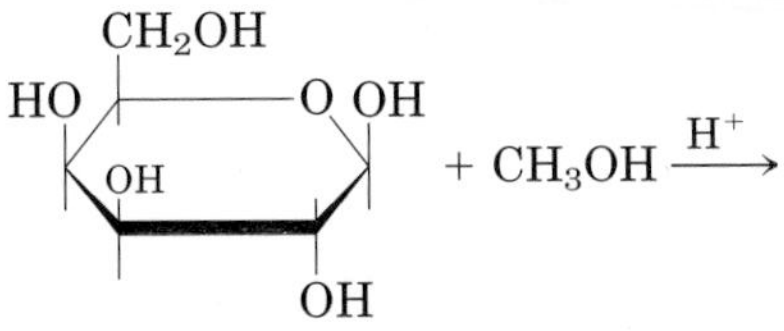

***16.34** Draw the structural formulas for (a) α-D-galacturonic acid and for (b) β-D-galacturonic acid.

16.35 In amino sugars the OH group on C-2 is replaced by an amino group. Write the structure of α-D-*N*-acetylgalactosamine.

Disaccharides

16.36 (a) Draw the structure of a disaccharide made of D-riboses linked together by a (1 → 3) glycosidic linkage. (b) Is this a reducing sugar?

***16.37** A disaccharide often found in mushrooms and other fungi is called trehalose. It is formed by the condensation of two α-D-glucose molecules through their anomeric carbons (1 → 1). Therefore, it is a nonreducing sugar. Draw the structure of trehalose.

16.38

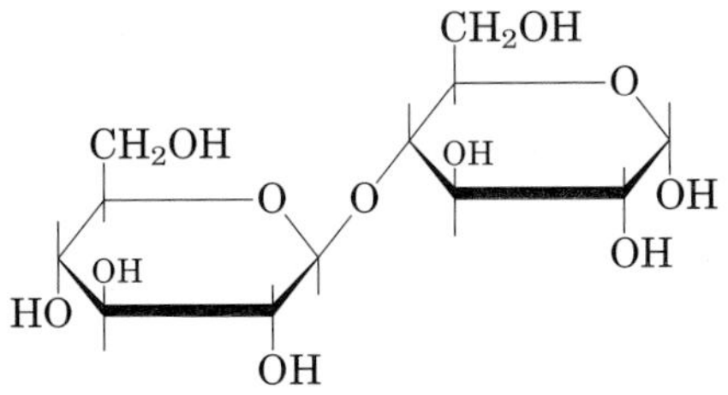

CH$_2$OH, O, OH, OH, OH (upper ring); CH$_2$OH, O, O, OH, HO, OH (lower ring)

(a) Does this compound have an acetal linkage? Find it. (b) Does it have a hemiacetal linkage? Find it. (c) Does it give a positive test with Benedict's solution? (d) How does it differ from maltose?

Polysaccharides

16.39 How is it possible that cows can digest grass but humans cannot?

16.40 How does iodine interact with amylose?

16.41 Draw the possible products of the reaction between H_2SO_4 and amylose:

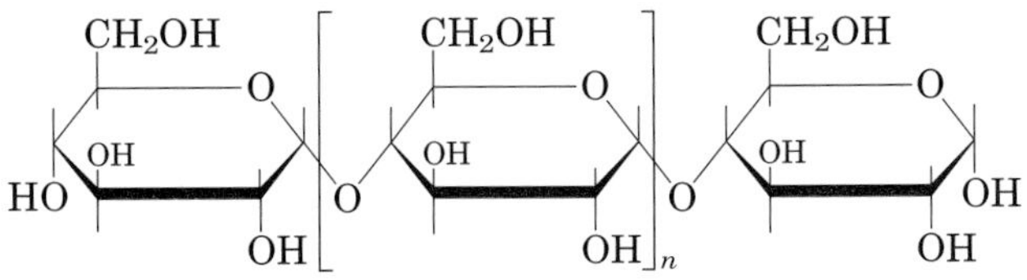

CH$_2$OH, O, OH, HO, OH — O — [CH$_2$OH, O, OH, OH]$_n$ — O — CH$_2$OH, O, OH, OH, OH

Acidic Polysaccharides

***16.42** Keratan sulfate is an important component of the cornea of the eye. The repeating unit of this acidic polysaccharide is

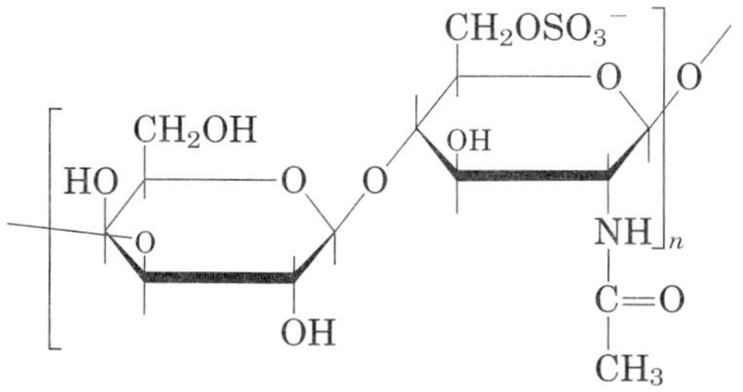

(a) What monosaccharides or derivatives of monosaccharides is keratan sulfate made of? (b) Describe the glycosidic linkages.

***16.43** Identify the products when hyalobiuronic acid is completely hydrolyzed by 6 M HCl:

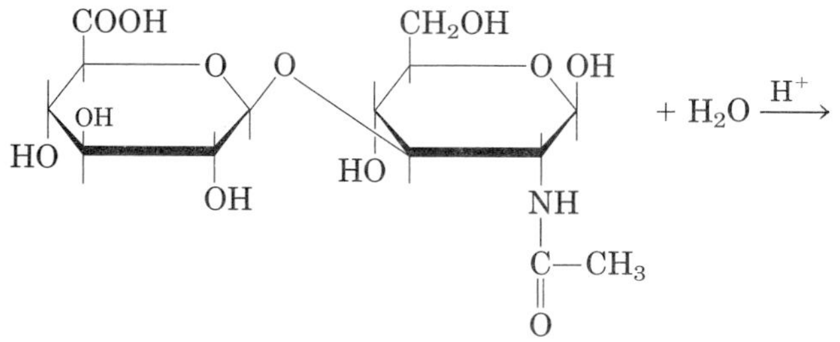

$$\text{hyalobiuronic acid} + H_2O \xrightarrow{H^+}$$

16.44 Hyaluronic acid acts as a lubricant in the synovial fluid of joints. In rheumatoid arthritis, inflammation breaks hyaluronic acid down to smaller molecules. What happens to the lubricating power of the synovial fluid?

***16.45** The anticlotting property of heparin is partly due to the negative charges it carries. (a) Identify the functional groups that provide the negative charges. (b) Which type of heparin is a better anticlotting agent, one with a high or a low degree of polymerization?

Chiral Compounds in Biological Organisms

16.46 Why is the study of chiral molecules important in biochemistry?

16.47 Explain why sucrose is sweet but its enantiomer is not.

Boxes

16.48 (Box 16A) Are all the alcohol groups in α-D-glucose in the equatorial position?

16.49 (Box 16A) Draw the formula of (a) α-L-glucose and (b) α-L-mannose (Fig. 16.9) in the chair forms.

16.50 (Box 16B) Does a person with O-type blood carry a terminal fucose molecule on the surface of her red blood cells?

16.51 (Box 16B) The H antigen that all blood types possess is L-fucose. (a) Is this a hexose? (b) Is this a pyranose or furanose ring? (c) What is unusual about this molecule?

16.52 (Box 16B) Why can't a person with B-type blood donate to a person with A-type blood?

16.53 (Box 16C) Why does congenital galactosemia appear only in infants?

16.54 (Box 16C) Why can galactosemia be relieved by feeding a baby a formula containing sucrose?

16.55 (Box 16D) How does a filler influence the effectiveness of a drug?

16.56 (Box 16E) What is the most important biological function of the vitreous and its component, hyaluronic acid?

16.57 (Box 16F) What is the advantage of taking (+)-ibuprofen rather than the corresponding racemic mixture?

16.58 (Box 16F) Thalidomide was a good sleeping pill. Why was it taken off the market?

Additional Problems

16.59 Why is cellulose insoluble in water?

16.60 What are the names of the compounds produced when mannose is (a) reduced at C-1 (b) oxidized at C-6?

***16.61** Dermatan sulfate is an important component of skin. It has a repeating unit in which β-L-iduronic acid is connected in a $\beta(1 \rightarrow 3)$ linkage to N-acetyl-α-D-glucosamine, which is sulfated on C-6. The N-acetyl group is on C-2. Draw the structure of this disaccharide.

16.62 Which is the most important monosaccharide in the human body?

16.63 Draw the structure of a disaccharide in which two D-glucose units are linked in a $\beta(1 \rightarrow 6)$ linkage.

16.64 A monosaccharide is dissolved in water. Its optical rotation is +17.5°. An hour later the rotation of the same solution is +15.1°, and 48 hours later it is −2.6°. What molecular process is going on?

16.65 You are given a sample of 2-butanol, a compound that contains a chiral carbon. Yet it fails to rotate the plane of polarized light. Explain how this is possible.

16.66 Which of these compounds are chiral?

(a) C_6H_5–CH_2CH_3

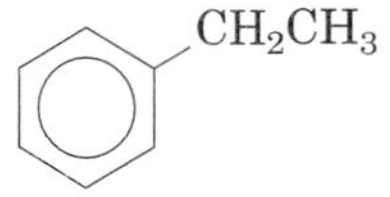

(b) C_6H_5–CH(Cl)–CH_3

(c) C_6H_5–CH(Cl)–C_6H_5

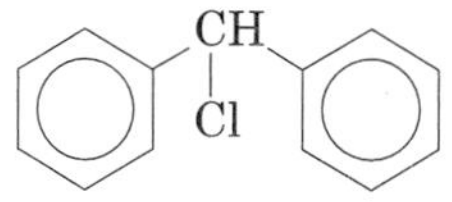

(d) H_3C–C_6H_4–CH(Cl)–C_6H_5

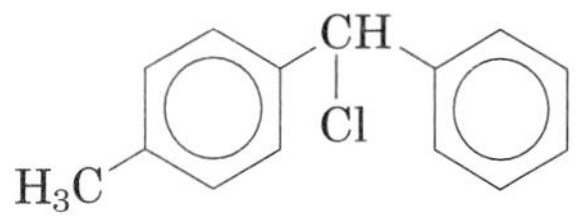

(e) H_3C–C_6H_4–CH(Cl)–C_6H_4–CH_3

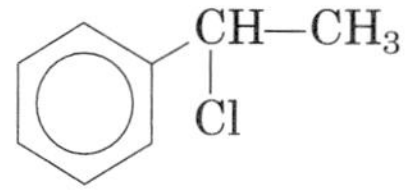

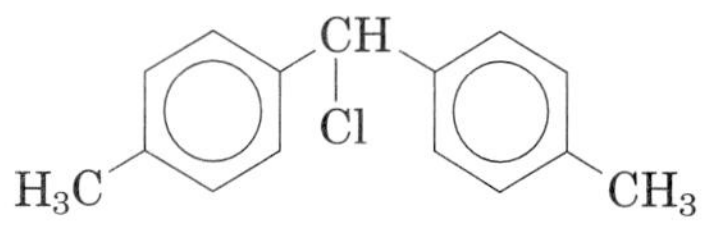

16.67 Ribitol and ribose-1-phosphate are derivatives of β-D-ribose. Draw the structures of these compounds.

16.68 Do glycosides exhibit mutarotation?

16.69 What role does chlorophyll play in carbohydrate synthesis by plants?

16.70 (a) Which disaccharide contains only α-D-glucose units? (b) Which polysaccharides?

16.71 How many different aldooctoses (8 carbons) can there be?

16.72 In the structure of heparin (Fig. 16.14) how many different positions do sulfate groups occupy? Name them.

***16.73** The structure of chondroitin-4-sulfate, an important component of cartilage, is similar to that of hyaluronic acid. The repeating unit consists of a D-glucuronic acid unit connected by a $\beta(1 \rightarrow 3)$ linkage to an *N*-acetylgalactosamine unit, the C-4 of which has a sulfate group. The *N*-acetyl group is on C-2. These disaccharide units are connected by a $\beta(1 \rightarrow 4)$ linkage as in hyaluronic acid. Write the structure of chondroitin-4-sulfate.

16.74 Why are monosaccharides and disaccharides called sugars?

16.75 How many chiral carbons are in (a) D-threose (b) D-xylose? (See Fig. 16.9).

16.76 Does ethyl α-D-galactopyranoside give a positive test with Benedict's solution?

16.77 Which is the anomeric carbon in a ketopentose?

Chapter 17

Lipids

17.1 Introduction

Note that, unlike the case of carbohydrates, we define lipids in terms of a property and not in terms of structure.

Lipids are substances found in living organisms that are **insoluble in water but soluble in nonpolar solvents and solvents of low polarity, such as diethyl ether.** This lack of solubility in water is an important property because our body chemistry is so firmly based on water. Most body constituents, including carbohydrates, are soluble in water. But the body also needs insoluble compounds for many purposes, including the separation of compartments containing aqueous solutions from each other, and that's where lipids come in.

"Hydrophobic" means water-hating.

The water-insolubility of lipids is due to the fact that the polar groups they contain are much smaller than their alkane-like (nonpolar) portions. These nonpolar portions provide the water-repellent, or *hydrophobic,* property.

An important use for lipids, especially in animals, is the storage of energy. As we saw in Section 16.1, plants store energy in the form of starch. Animals (including humans) find it more economical to use lipids (fats) instead. Although our bodies do store some carbohydrates in the form of glycogen for quick energy when we need it, energy stored in the form of fats is much more important. The reason is simply that the burning of fats produces more than twice as much energy (about 9 kcal/g) as the burning of an equal weight of carbohydrates (about 4 kcal/g). Lipids also serve as chemical messengers.

For purposes of study, we can divide lipids into four groups: (1) fats and waxes, (2) complex lipids, (3) steroids, and (4) prostaglandins and leukotrienes.

17.2 The Structure of Fats

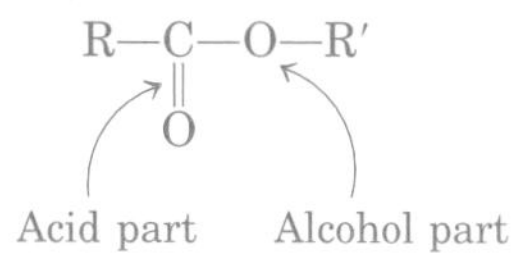

Fats are esters. We saw in Section 14.8 that esters are made up of an alcohol part and an acid part.

The Alcohol Part

In fats the alcohol part is always glycerol:

$$\begin{array}{l} CH_2\text{—OH} \\ | \\ CH\text{—OH} \\ | \\ CH_2\text{—OH} \end{array}$$

Glycerol

The Acid Part

In contrast to the alcohol part, the acid component of fats may be any number of acids, which do, however, have certain things in common:

1. They are practically all straight-chain carboxylic acids (virtually no branching).

2. They range in size from about 10 carbons to 20 carbons.

3. They have an even number of carbon atoms.

4. Apart from the —COOH group, they have no functional groups except that some do have double bonds.

The reason that only even-numbered acids are found in fats is that the body builds these acids entirely from acetic acid units and therefore puts the carbons in two at a time (Sec. 22.3). Table 17.1 shows the most important acids found in fats.

Since glycerol has three —OH groups, a single molecule of glycerol can be attached to three different acid molecules. Thus, a typical fat molecule might be

Remember (Sec. 14.2) that these are called fatty acids.

$$
\begin{array}{l}
CH_2{-}O{-}\overset{\overset{\displaystyle O}{\|}}{C}{-}(CH_2)_{12}{-}CH_3 \quad \leftarrow \text{Myristic acid, 14 carbons} \\
| \\
CH{-}O{-}\overset{\overset{\displaystyle O}{\|}}{C}{-}(CH_2)_7{-}CH{=}CH{-}(CH_2)_7{-}CH_3 \quad \leftarrow \text{Oleic acid, 18 carbons, 1 double bond} \\
| \\
CH_2{-}O{-}\underset{\underset{\displaystyle O}{\|}}{C}{-}(CH_2)_{16}{-}CH_3 \quad \leftarrow \text{Stearic acid, 18 carbons}
\end{array}
$$

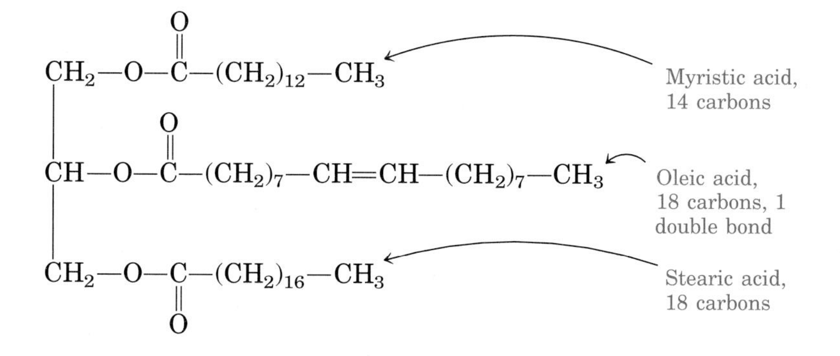

Such compounds are called **triglycerides** or **triacylglycerols:** All three —OH groups of glycerol are esterified. Triglycerides are the most common lipid materials, although **mono-** and **diglycerides** are not infrequent. In the latter two types, only one or two —OH groups of the glycerol are esterified by fatty acids.

Triglycerides are complex mixtures. Although some of the molecules have three identical fatty acids, in most cases two or three different acids are present.

The hydrophobic character of fats is caused by the long hydrocarbon chains. The ester linkages, though polar, are buried in a nonpolar environment, and this makes the fats insoluble in water.

The fatty acids can be divided into two groups: saturated and unsaturated. Saturated fatty acids have only single bonds in the hydrocarbon chain. Unsaturated fatty acids have at least one C=C double bond in the chains. All the unsaturated fatty acids listed in Table 17.1 are the cis isomers. This explains their physical properties, reflected in their melting points. Saturated fatty acids are solids at room temperature because the regular nature of their aliphatic chains allows the molecules to be packed in a close, parallel alignment:

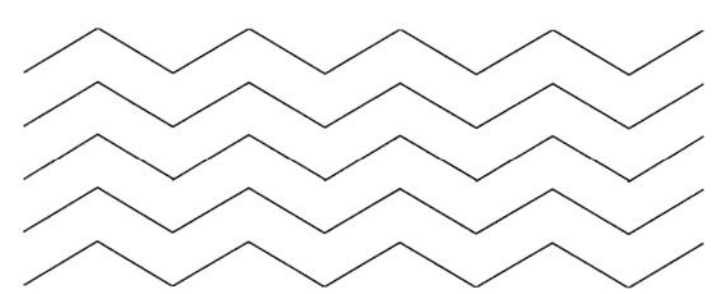

Table 17.1 The Most Important Acids in Fats

Acid	Number of Carbons	Number of Double Bonds	Formula	Schematic Structure	Melting Point (°C)
Lauric	12	0	$CH_3(CH_2)_{10}COOH$	COOH	44
Myristic	14	0	$CH_3(CH_2)_{12}COOH$	COOH	54
Palmitic	16	0	$CH_3(CH_2)_{14}COOH$	COOH	63
Stearic	18	0	$CH_3(CH_2)_{16}COOH$	COOH	70
Oleic	18	1	$CH_3(CH_2)_7CH{=}CH(CH_2)_7COOH$	COOH	4
Linoleic	18	2	$CH_3(CH_2)_4CH{=}CHCH_2CH{=}CH(CH_2)_7COOH$	COOH	−5
Linolenic	18	3	$CH_3CH_2CH{=}CHCH_2CH{=}CHCH_2CH{=}CH(CH_2)_7COOH$	COOH	−11
Arachidonic	20	4	$CH_3(CH_2)_4CH{=}CHCH_2CH{=}CHCH_2CH{=}CHCH_2CH{=}CH(CH_2)_3COOH$	COOH	−49.5

The interactions (attractive forces) between neighboring chains are weak, but the regular packing allows these forces to operate over a large portion of the chain so that a considerable amount of energy is needed in order to melt them.

The longer the aliphatic chain, the higher the melting point.

In contrast, unsaturated fatty acids are all liquids at room temperature because the cis double bonds interrupt the regular packing of the chains:

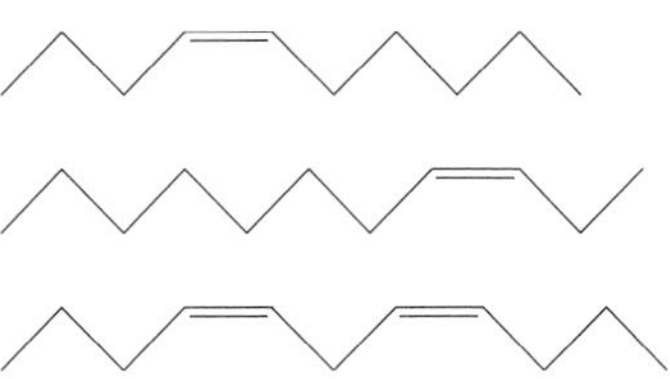

Thus much less energy is required to melt them. The greater the degree of unsaturation, the lower the melting point, because each double bond introduces more disorder into the packing of the molecules.

17.3 Properties of Fats

Physical State

With some exceptions, fats that come from animals are generally solids at room temperature, and those from plants or fish are usually liquids. Liquid fats are often called **oils,** though they are esters of glycerol just like solid fats and should not be confused with petroleum, which is mostly alkanes.

What is the structural difference between solid fats and liquid oils? In most cases it is the degree of unsaturation. The physical properties of the fatty acids are carried over to the physical properties of the triglycerides. Solid animal fats contain mainly *saturated fatty acids,* and vegetable oils contain high amounts of *unsaturated fatty acids.* Table 17.2 shows the average fatty acid content of some common fats and oils. Note that even solid fats contain some unsaturated acids and that liquid fats contain some saturated acids. Some unsaturated fatty acids (linoleic and linolenic acids) are called *essential fatty acids* because the body cannot synthesize them from precursors; they must therefore be included in the diet.

Though most vegetable oils have high amounts of unsaturated fatty acids, there are exceptions. Note that coconut oil has only a small amount of unsaturated acids. This oil is a liquid not because it contains many double bonds but because it is rich in low-molecular-weight fatty acids (chiefly lauric).

Oils with an average of more than one double bond per fatty acid chain are called *polyunsaturated.* For some years there has been a controversy about whether a diet rich in unsaturated and polyunsaturated fats helps to prevent heart attacks. (See Box 17G.)

Some vegetable oils. (Photograph by Charles D. Winters.)

Table 17.2 Average Percentage of Fatty Acids of Some Common Fats and Oils

	Lauric	Myristic	Palmitic	Stearic	Oleic	Linoleic	Linolenic	Other
Animal Fats								
Beef tallow	—	6.3	27.4	14.1	49.6	2.5	—	0.1
Butter	2.5	11.1	29.0	9.2	26.7	3.6	—	17.9
Human	—	2.7	24.0	8.4	46.9	10.2	—	7.8
Lard	—	1.3	28.3	11.9	47.5	6.0	—	5.0
Vegetable Oils								
Coconut	45.4	18.0	10.5	2.3	7.5	—	—	16.3
Corn	—	1.4	10.2	3.0	49.6	34.3	—	1.5
Cottonseed	—	1.4	23.4	1.1	22.9	47.8	—	3.4
Linseed	—	—	6.3	2.5	19.0	24.1	47.4	0.7
Olive	—	—	6.9	2.3	84.4	4.6	—	1.8
Palm	—	1.4	40.1	5.5	42.7	10.3	—	—
Peanut	—	—	8.3	3.1	56.0	26.0	—	6.6
Safflower	←——— 6.8 ———→				18.6	70.1	3.4	1.1
Soybean	0.2	0.1	9.8	2.4	28.9	52.3	3.6	2.7
Sunflower	—	—	5.6	2.2	25.1	66.2	—	—

Pure fats and oils are colorless, odorless, and tasteless. This statement may seem surprising, since we all know the tastes and colors of such fats and oils as butter and olive oil. The tastes, odors, and colors are caused by substances dissolved in the fat or oil.

Hydrogenation

In Section 11.6 we learned that double bonds can be reduced to single bonds by treatment with hydrogen (H_2) and a catalyst. It is therefore not difficult to convert unsaturated liquid oils to solids, for example,

$$\begin{array}{l} CH_2{-}O{-}\overset{O}{\overset{\|}{C}}{-}(CH_2)_7{-}CH{=}CH{-}(CH_2)_7{-}CH_3 \quad \text{(Oleic acid)} \\ | \\ CH{-}O{-}\overset{O}{\overset{\|}{C}}{-}(CH_2)_7{-}CH{=}CHCH_2CH{=}CH{-}(CH_2)_4{-}CH_3 \quad \text{(Linoleic acid)} \\ | \\ CH_2{-}O{-}\underset{\underset{O}{\|}}{C}{-}(CH_2)_7{-}CH{=}CHCH_2CH{=}CH{-}(CH_2)_4{-}CH_3 \quad \text{(Linoleic acid)} \end{array} + 5H_2 \xrightarrow{Pt} \begin{array}{l} CH_2{-}O{-}\overset{O}{\overset{\|}{C}}{-}(CH_2)_{16}{-}CH_3 \quad \text{(Stearic acid)} \\ | \\ CH{-}O{-}\overset{O}{\overset{\|}{C}}{-}(CH_2)_{16}{-}CH_3 \\ | \\ CH_2{-}O{-}\underset{\underset{O}{\|}}{C}{-}(CH_2)_{16}{-}CH_3 \end{array}$$

This hydrogenation is carried out on a large scale to produce the solid shortening sold in stores under such brand names as Crisco, Spry, and Dexo. In making such products, manufacturers must be careful not to hydrogenate all the double bonds because a fat with no double bonds at all would be too solid. Partial, but not complete, hydrogenation results in a product with the right consistency for cooking.

Margarine is also made by partial hydrogenation of vegetable oils. Because less hydrogen is used, margarine contains more unsaturation than hydrogenated shortenings.

Box 17A

Olestra

Obesity is a serious problem in the United States. Many people would like to lose weight but are unable to control their appetites enough to do so, which is why artificial sweeteners are popular (Box 18A). They have a sweet taste but do not add calories. But calories in the diet do not come only from sugar. Dietary fat is an even more important source (Sec. 21.6), and it has long been hoped that some kind of artificial fat, which would taste the same but would have no (or only a few) calories, would help in losing weight. Now one company, Procter & Gamble, has developed such a product, called olestra. Like the real fats, this molecule is a carboxylic ester, but instead of glycerol, the alcohol component is sucrose (Sec. 16.10). All eight of sucrose's OH groups are converted to ester groups; the carboxylic acids are the fatty acids containing 8–10 carbons. The one illustrated here is made with the C_{10} acid, capric.

$CH_2O{-}COC_9H_{19}$
O
$O{-}COC_9H_{19}$
$C_9H_{19}COO$
$C_9H_{19}COO$
O
$C_9H_{19}COOCH_2$ O
$C_9H_{19}COO$
$CH_2O{-}COC_9H_{19}$
$O{-}COC_9H_{19}$

Olestra

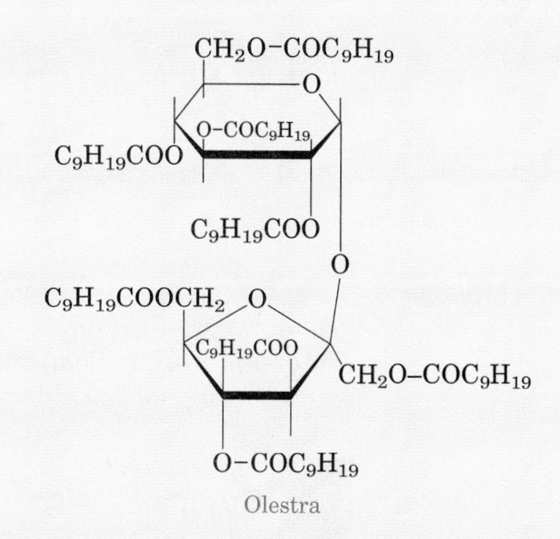

Although olestra has a chemical structure similar to a fat, the human body cannot digest it because the enzymes that digest ordinary fats are not designed for the particular size and shape of this molecule. Therefore it passes through the digestive system unchanged, and we derive no calories from it. Olestra can be used instead of an ordinary fat in cooking such items as cookies and potato chips. People who eat these will then be consuming fewer calories.

Early in 1996 the Food and Drug Administration approved the use of olestra (trade name Olean) for use in certain snack foods, including potato and tortilla chips. Later that year the chips were test marketed in several cities. Users stated that the products tasted essentially the same as chips made from real fat, although they contained 75 calories per ounce and no digestible fat, compared to 150 calories and 10 grams of fat for ordinary chips. However, olestra does cause diarrhea, cramps, and nausea in some people, effects which increase with the amount of olestra consumed, and those that are susceptible to such side effects will have to decide if the potential to lose weight is worth the discomfort involved. Furthermore, olestra dissolves and sweeps away some vitamins and nutrients in other foods being digested at the same time. To counter this, the manufacturers add some of these nutrients (vitamins A, D, E, and K) to the product. The Food and Drug Administration requires that all packages of olestra-containing foods carry a warning label about these side effects.

Many common products contain hydrogenated vegetable oils. (Photograph by Charles D. Winters.)

Saponification

Glycerides, being esters, are subject to hydrolysis, which can be carried out with either acids or bases. As we saw in Section 14.10, the use of bases is more practical. An example of the saponification of a typical fat is

$$\begin{array}{l} H_2C{-}O{-}\overset{\displaystyle O}{\overset{\|}{C}}{-}(CH_2)_{12}CH_3 \\ \;| \\ HC{-}O{-}\overset{\displaystyle O}{\overset{\|}{C}}{-}(CH_2)_{14}CH_3 \\ \;| \\ H_2C{-}O{-}\overset{\displaystyle O}{\overset{\|}{C}}{-}(CH_2)_7CH{=}CHCH_2CH{=}CH(CH_2)_4CH_3 \end{array} + 3NaOH \xrightarrow{H_2O}$$

$$\begin{array}{l} H_2C{-}OH \qquad CH_3(CH_2)_{12}{-}\overset{\displaystyle O}{\overset{\|}{C}}{-}O^-\ Na^+ \\ \;| \\ HC{-}OH + CH_3(CH_2)_{14}\overset{\displaystyle O}{\overset{\|}{C}}{-}O^-\ Na^+ \\ \;| \\ H_2C{-}OH \qquad CH_3(CH_2)_4CH{=}CHCH_2CH{=}CH(CH_2)_7\overset{\displaystyle O}{\overset{\|}{C}}{-}O^-\ Na^+ \end{array}$$

The mixture of sodium salts of fatty acids produced by saponification of fats or oils is **soap.** Soap has been used for thousands of years, and saponification is one of the oldest known chemical reactions (Box 17C).

17.4 Complex Lipids

The triglycerides discussed in the previous sections are significant components of fat storage cells. Other kinds of lipids, called **complex lipids,** are important in a different way. They constitute the main components

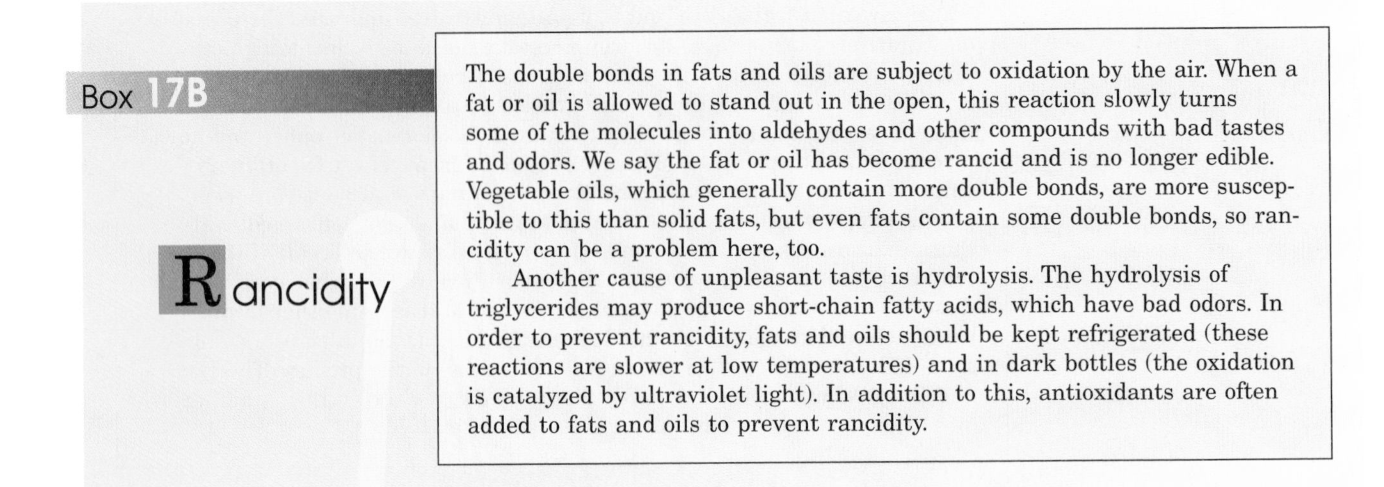

Box 17B

Rancidity

The double bonds in fats and oils are subject to oxidation by the air. When a fat or oil is allowed to stand out in the open, this reaction slowly turns some of the molecules into aldehydes and other compounds with bad tastes and odors. We say the fat or oil has become rancid and is no longer edible. Vegetable oils, which generally contain more double bonds, are more susceptible to this than solid fats, but even fats contain some double bonds, so rancidity can be a problem here, too.

Another cause of unpleasant taste is hydrolysis. The hydrolysis of triglycerides may produce short-chain fatty acids, which have bad odors. In order to prevent rancidity, fats and oils should be kept refrigerated (these reactions are slower at low temperatures) and in dark bottles (the oxidation is catalyzed by ultraviolet light). In addition to this, antioxidants are often added to fats and oils to prevent rancidity.

Box 17C

Soaps and Detergents

Soaps clean because each soap molecule has a hydrophilic head and a hydrophobic tail. The $—COO^-$ end of the molecule (the hydrophilic end), being ionic, is soluble in water. The other end (the hydrophobic end) is a long-chain alkane-like portion, which is insoluble in water but soluble in organic compounds. Dirt may contain water-soluble and water-insoluble portions. The water-soluble portion dissolves in water; no soap is needed for it. The soap is needed for the water-insoluble portion. The hydrophilic end of the soap molecule dissolves in the water; the hydrophobic end dissolves in the dirt. This causes an emulsifying action in which the soap molecules surround the dirt particles in an orderly fashion, the hydrophobic tail interacting with the hydrophobic dirt particle and the hydrophilic head providing the attraction for water molecules (Fig. 17C). The cluster of soap molecules in which the dirt is embedded is called a **micelle** (Fig. 17C(d)).

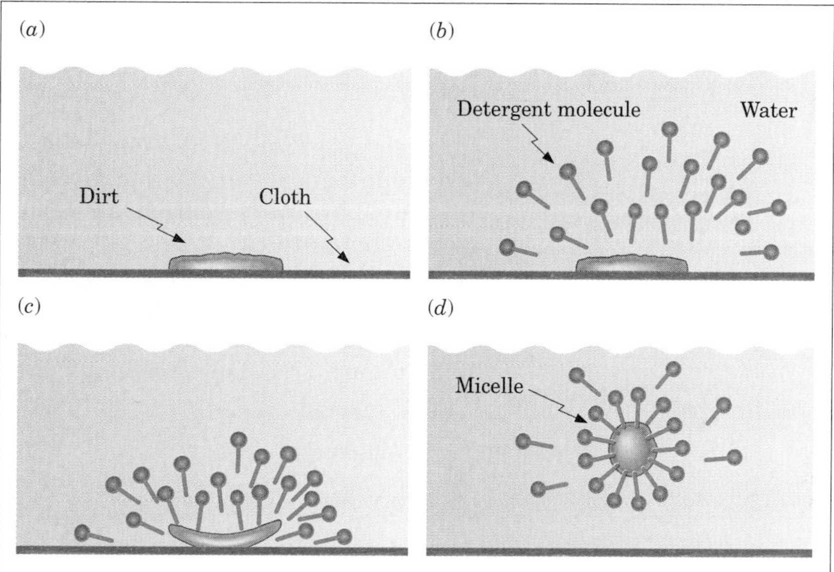

Figure **17C** • The cleaning action of soap. (a) Without soap, water molecules cannot penetrate through molecules of hydrophobic dirt. (b and c) Soap molecules line up at the interface between dirt and water. (d) Soap molecules carry dirt particles away.

The solid soaps and soap flakes with which we are all familiar are sodium salts of fatty acids. Liquid soaps are generally potassium salts of the same acids. Soap has the disadvantage of forming precipitates in *hard water,* which is water that contains relatively high amounts of Ca^{2+} and Mg^{2+} ions. Most community water supplies in the United States are hard water. Calcium and magnesium salts of fatty acids are insoluble in water. Therefore, when soaps are used in hard water, the Ca^{2+} and Mg^{2+} ions precipitate the fatty acid anions, with two unfortunate consequences: A certain amount of soap is wasted because the precipitated anions are not available for cleaning, and the precipitate itself is now a kind of dirt, which deposits on clothes, dishes, sinks, and bathtubs (it is the "ring" around the bathtub). Because of these difficulties, much of the soap used in the United States

Box 17C (continued)

has been replaced in the last 40 years by **detergents** for both household and industrial cleaning. Two typical detergent molecules are

$$CH_3(CH_2)_{11}—O—SO_2O^- \ Na^+$$

Sodium dodecyl sulfate
(sodium lauryl sulfate)

$$CH_3(CH_2)_9—CH(CH_3)—C_6H_4—SO_2O^- \ Na^+$$

Sodium *para*-(2-dodecyl) benzene sulfonate

It can be seen immediately that these molecules also have hydrophobic tails and hydrophilic heads and so can function in the same way as soap molecules. However, they do not precipitate with Ca^{2+} and Mg^{2+}, and so work very well in hard water.

of membranes (Sec. 17.5). Chemically, complex lipids can be divided into two groups: phospholipids and glycolipids.

Phospholipids contain an alcohol, fatty acids, and a phosphate group. There are two types: glycerophospholipids and sphingolipids. In **glycerophospholipids,** the alcohol is glycerol (see Sec. 17.6). In **sphingolipids,** the alcohol is sphingosine (see Sec. 17.7).

Glycolipids are complex lipids that contain carbohydrates (see Sec. 17.8). Figure 17.1 shows schematic structures for all of these.

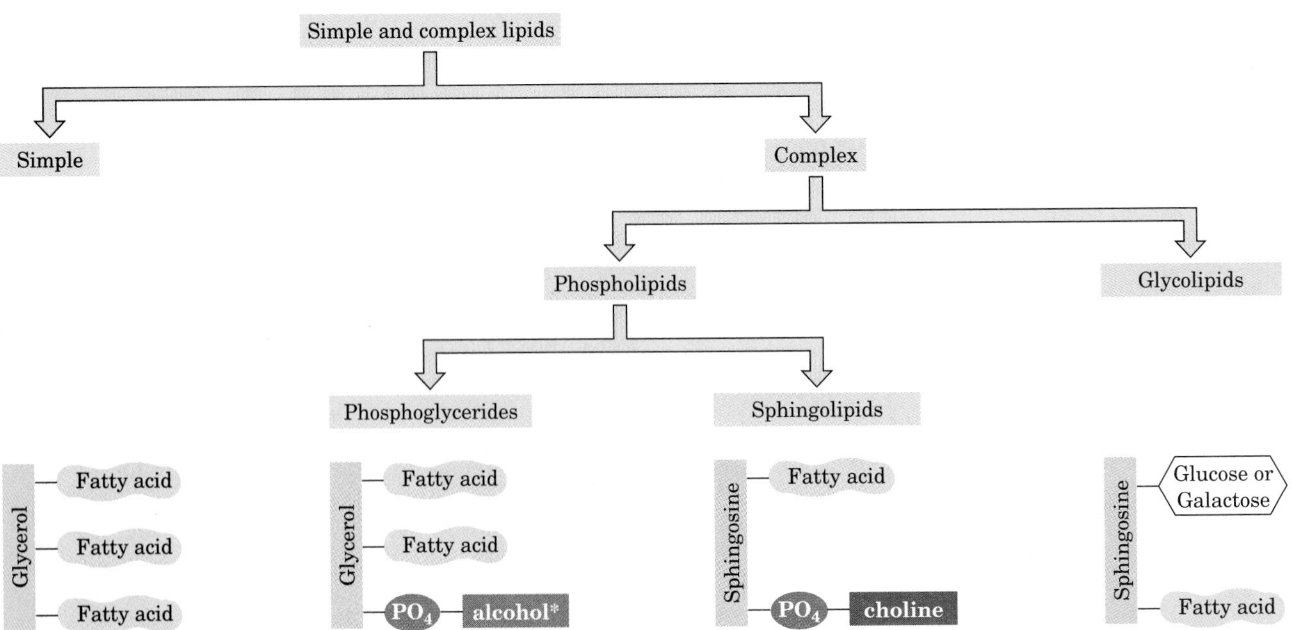

*The alcohol can be choline, serine, ethanolamine, or certain others.

Figure **17.1** • Schematic diagram of simple and complex lipids.

Box 17D

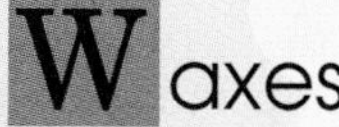

Waxes

Waxes are simple esters. They are solids because of their high molecular weights. As in fats, the acid portions of the esters consist of a mixture of fatty acids, but the alcohol portions are not glycerol but simple long-chain alcohols. For example, a major component of beeswax is 1-triacontyl palmitate:

$$CH_3(CH_2)_{14}-\overset{\overset{\displaystyle O}{\|}}{C}-O-(CH_2)_{29}-CH_3$$

Palmitic acid portion (arrow to $CH_3(CH_2)_{14}$–C(=O)) 1-Triacontanol portion (arrow to $(CH_2)_{29}-CH_3$)

Waxes generally have higher melting points than fats (60 to 100°C) and are harder. Animals and plants use them for protective coatings. The leaves of most plants are coated with wax, which helps to prevent microorganisms from attacking them and also allows them to conserve water. The feathers of birds and the fur of animals are also coated with wax. This is what allows ducks to swim.

Some important waxes are carnauba wax (from a Brazilian palm tree), lanolin (from lamb's wool), beeswax, and spermaceti (from whales). These are used to make cosmetics, polishes, candles, and ointments. Paraffin waxes are not esters. They are mixtures of high-molecular-weight alkanes. Neither is ear wax a simple ester. It is a gland secretion and contains a mixture of fats (triglycerides), phospholipids, and esters of cholesterol.

Figure **17D** • Bees making beeswax. (Photograph by Charles D. Winters.)

17.5 Membranes

It is the complex lipids mentioned in Section 17.4 that form the **membranes** around body cells and around small structures inside the cells. Unsaturated fatty acids are important components of these lipids. The purpose of these membranes is to separate cells from the external environment and to provide selective transport for nutrients and waste

These small structures inside the cell are called **organelles.**

products. That is, membranes allow the selective passage of substances into and out of cells.

These membranes are made of lipid bilayers (Fig. 17.2). In a **lipid bilayer** there are two rows (layers) of lipid molecules arranged tail to tail. The hydrophobic tails point toward each other because that enables them to get as far away from the water as possible. This leaves the hydrophilic heads projecting to the inner and outer surfaces of the membrane.

"Hydrophilic" means water-loving.

Most lipid molecules in the bilayer contain at least one unsaturated fatty acid.

The unsaturated fatty acids prevent the tight packing of the hydrophobic chains in the lipid bilayer and thereby provide a liquid-like character to the membranes. This is of extreme importance because many products of the body's biochemical processes must cross the membrane, and the liquid nature of the lipid bilayer allows such transport.

This effect is similar to that which causes unsaturated fatty acids to have lower melting points than saturated fatty acids.

The lipid part of the membrane serves as a barrier against any movement of ions or polar compounds into and out of the cells. In the lipid bilayer, protein molecules are either suspended on the surface or partly or fully embedded in the bilayer. These proteins stick out either on the inside or on the outside of the membrane; others are thoroughly embedded, going through the bilayer and projecting from both sides. The model shown in Figure 17.2, called the **fluid mosaic model** of membranes, allows the passage of nonpolar compounds by diffusion, since these compounds are soluble in the lipid membranes. Polar compounds are transported either through specific channels through the protein regions or by another mechanism called active transport (see Fig. 24.2). For any transport process, the membrane must behave like a nonrigid liquid so that the proteins can move sideways within the membrane.

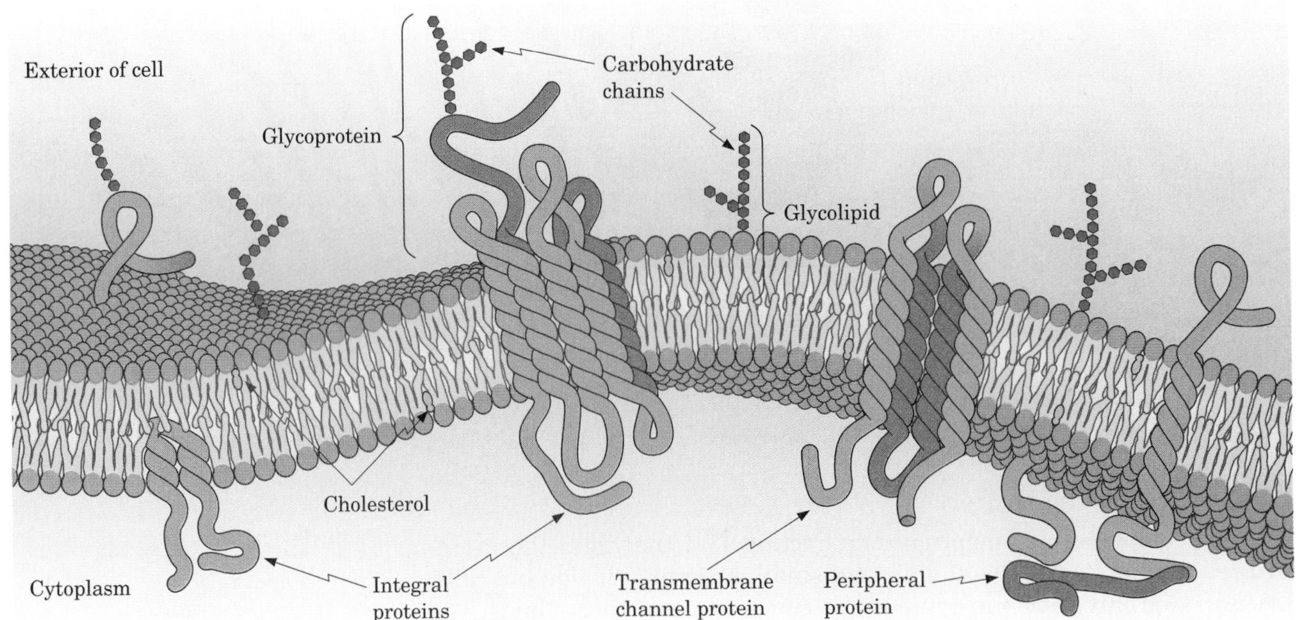

Figure **17.2** • The fluid mosaic model of membranes. Note that proteins are embedded in the lipid matrix.

17.6 Glycerophospholipids

The structure of glycerophospholipids (also called phosphoglycerides) is very similar to that of simple fats. The alcohol is glycerol. Two of the three —OH groups are esterified by fatty acids. As with the simple fats, these fatty acids may be any long-chain carboxylic acids with or without double bonds. The third —OH group is esterified not by a fatty acid but by a phosphate group, which is also connected to another alcohol. If the other alcohol is *choline,* a quaternary ammonium compound,

Glycerophospholipids are membrane components of cells throughout the body.

$$HO{-}CH_2CH_2{-}\overset{+}{N}(CH_3)_3$$

Choline

the glycerophospholipids are called **phosphatidylcholines** (common name **lecithin**):

This typical lecithin molecule has stearic acid on one end and linoleic acid in the middle. Other lecithin molecules contain other fatty acids, but the one on the end is always saturated and the one in the middle unsaturated.

Glycerol portion

$$CH_2{-}O{-}\overset{\overset{\displaystyle O}{\|}}{C}{-}(CH_2)_{16}{-}CH_3$$

Stearic acid portion

$$CH{-}O{-}\overset{\overset{\displaystyle O}{\|}}{C}{-}(CH_2)_7{-}CH{=}CH{-}CH_2{-}CH{=}CH{-}(CH_2)_4{-}CH_3$$

Linoleic acid portion

$$CH_2{-}O{-}\overset{\overset{\displaystyle O}{\|}}{P}{-}O^-$$
$$\qquad\qquad |$$
$$\qquad\qquad O{-}CH_2CH_2{-}\overset{+}{N}(CH_3)_3$$

Choline portion

Phosphate portion

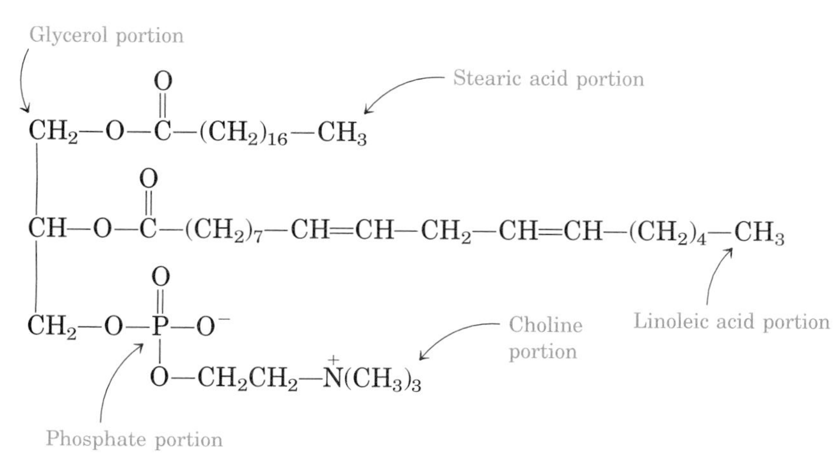

A typical lecithin

Note that lecithin has a negatively charged phosphate group and a positively charged quaternary nitrogen from the choline. These charged parts of the molecule provide a strongly hydrophilic head, while the rest of the molecule is hydrophobic. Thus, when a phospholipid such as lecithin is part of a lipid bilayer, the hydrophobic tail points toward the middle of the bilayer, and the hydrophilic heads line both the inner and outer surfaces of the membranes (Figs. 17.2 and 17.3).

Lecithins are just one example of glycerophospholipids. Another is the **cephalins,** which are similar to the lecithins in every way except that, instead of choline, they contain other alcohols, such as ethanolamine or serine:

$$HOCH_2CH_2NH_2$$

Ethanolamine

$$HOCH_2\underset{\underset{\displaystyle NH_2}{|}}{CH}{-}COOH$$

Serine

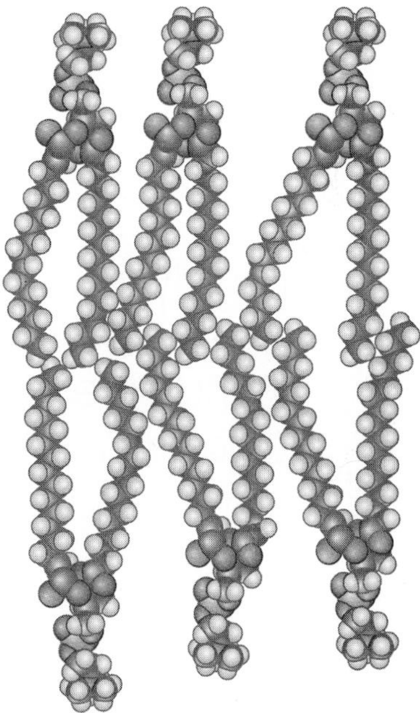

Figure **17.3**
Space-filling molecular models of complex lipids in a bilayer. (From L. Stryer, *Biochemistry,* 2nd ed. New York: W. H. Freeman, 1981. Copyright 1981 by W. H. Freeman and Co. All rights reserved.)

R = glycerol + fatty acid portions

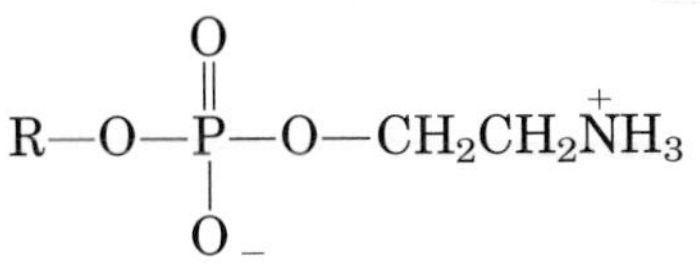

$$R-O-\underset{O^-}{\overset{O}{\overset{\|}{P}}}-O-CH_2CH_2\overset{+}{N}H_3$$

Phosphatidylethanolamine (cephalin)

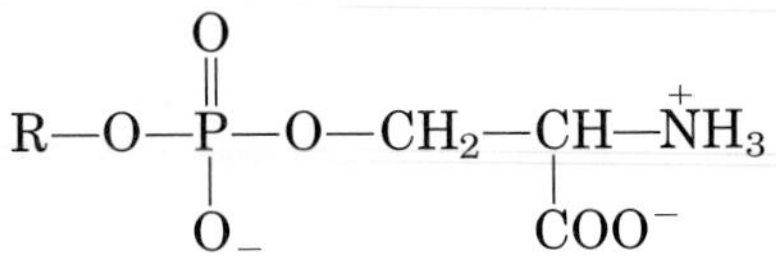

$$R-O-\underset{O^-}{\overset{O}{\overset{\|}{P}}}-O-CH_2-\underset{COO^-}{CH}-\overset{+}{N}H_3$$

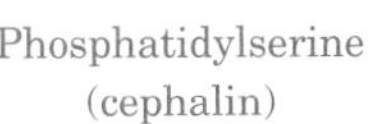

Phosphatidylserine (cephalin)

17.7 Sphingolipids

Johann Thudichum, who discovered sphingolipids in 1874, named these brain lipids after the monster of Greek mythology, the sphinx. Part woman and part winged lion, the sphinx devoured all who could not provide the correct answer to her riddles. Sphingolipids appeared to Thudichum as part of a dangerous riddle of the brain.

The coating of nerve axons (**myelin**) contains a different kind of complex lipid called **sphingolipids.** In sphingolipids the alcohol portion is sphingosine:

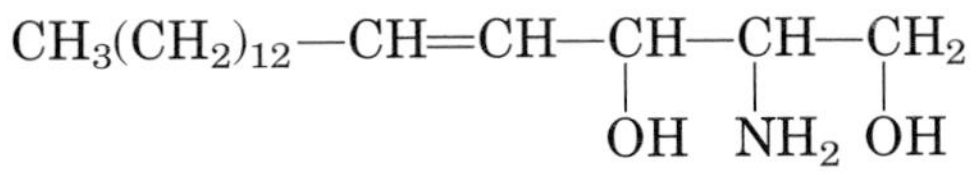

$$CH_3(CH_2)_{12}-CH=CH-\underset{OH}{CH}-\underset{NH_2}{CH}-\underset{OH}{CH_2}$$

Sphingosine

A long-chain fatty acid is connected to the $-NH_2$ group by an amide linkage, and the $-OH$ group at the end of the chain is esterified by phosphorylcholine:

Box 17E

The Myelin Sheath and Multiple Sclerosis

The human brain and spinal cord can be divided into gray and white regions. Forty percent of the human brain is white matter. Microscopic examination reveals that the white matter is made up of nerve axons wrapped in a white lipid coating, the **myelin sheath,** which provides insulation and thereby allows the rapid conduction of electrical signals. The myelin sheath consists of 70 percent lipids and 30 percent proteins in the usual lipid bilayer structure.

A specialized cell, the **Schwann cell,** wraps itself around the peripheral nerve axons to form numerous concentric layers (Fig. 17E). In the brain, other cells do the wrapping in a similar manner.

Multiple sclerosis affects 250 000 people in the United States. In this disease, a gradual degradation of the myelin sheath can be observed. The symptoms are muscle weariness, lack of coordination, and loss of vision. The symptoms may vanish for a time and return with greater severity. Autopsy of multiple sclerotic brains shows scar-like plaques of white matter, with bare axons not covered by myelin sheaths. The symptoms are produced because the demyelinated axons cannot conduct nerve impulses. A secondary effect of the demyelination is damage to the axon itself.

Similar demyelination occurs in the Guillain-Barré syndrome that follows certain viral infections. In 1976 fears of a "swine flu" epidemic prompted a vaccination program that precipitated a number of cases of Guillain-Barré syndrome. The main result is a paralysis that can cause death unless artificial breathing is supplied. The U.S. government assumed legal responsibility for the few bad vaccines and paid compensation to the victims and their families.

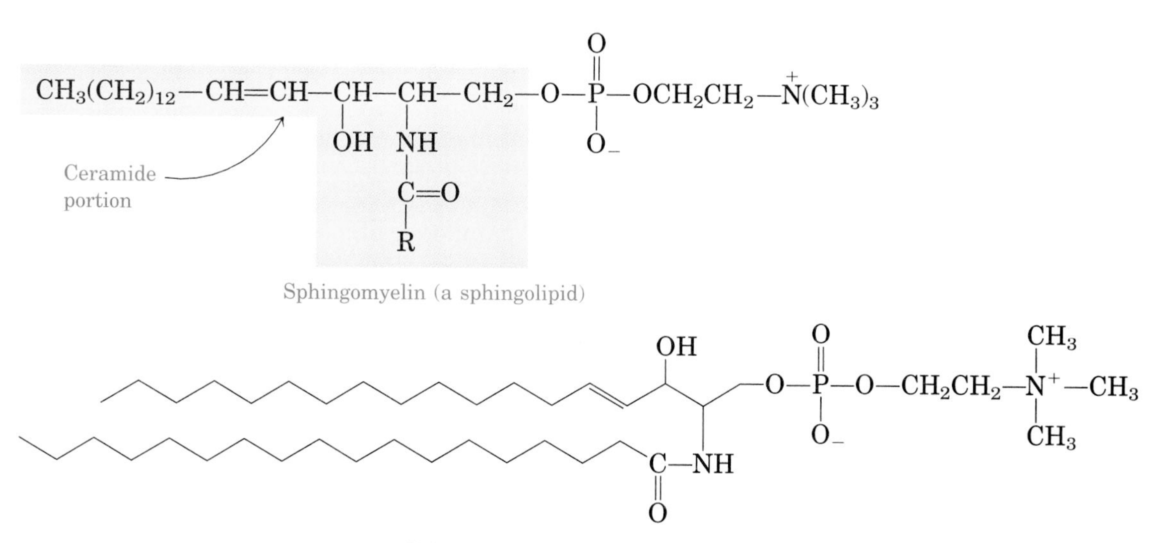

$$CH_3(CH_2)_{12}-CH{=}CH-\underset{\substack{|\\OH}}{CH}-\underset{\substack{|\\NH\\|\\C{=}O\\|\\R}}{CH}-CH_2-O-\overset{\substack{O\\||}}{\underset{\substack{|\\O^-}}{P}}-OCH_2CH_2-\overset{+}{N}(CH_3)_3$$

Ceramide portion

Sphingomyelin (a sphingolipid)

Sphingomyelin (schematic diagram)

Box 17E (continued)

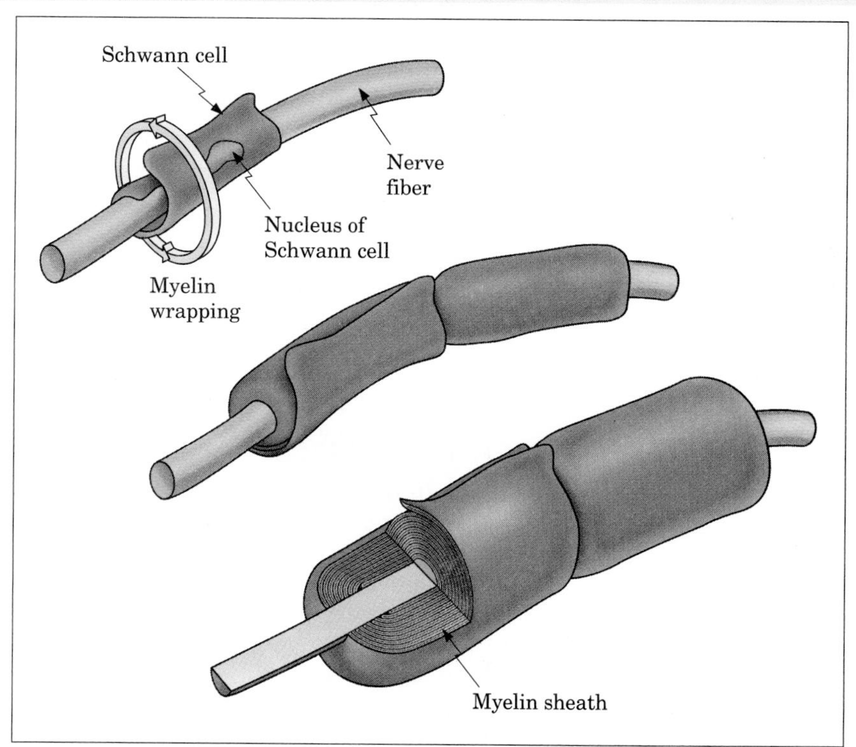

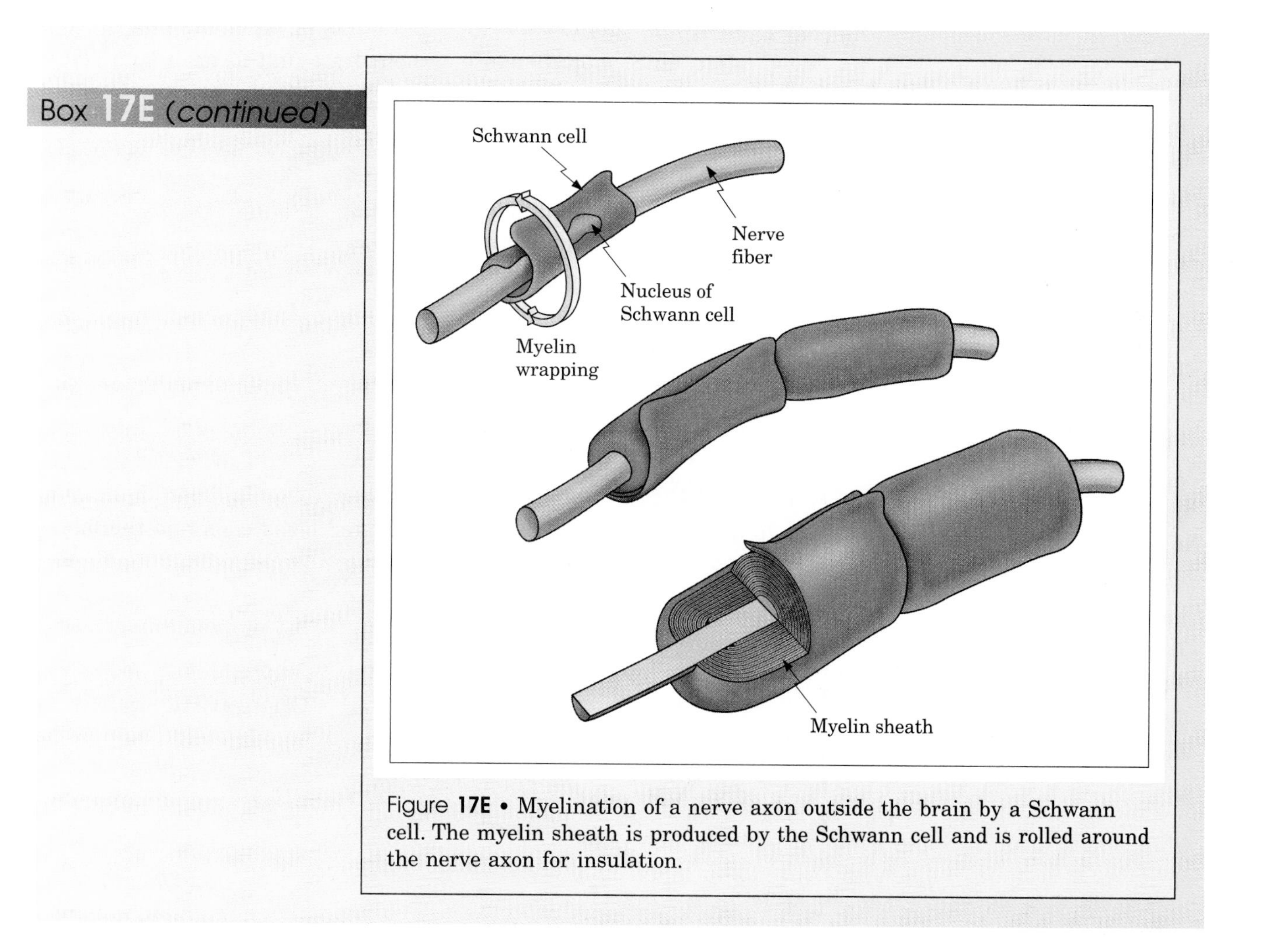

Figure **17E** • Myelination of a nerve axon outside the brain by a Schwann cell. The myelin sheath is produced by the Schwann cell and is rolled around the nerve axon for insulation.

The combination of a fatty acid and sphingosine (shown in color) is often referred to as the **ceramide** part of the molecule, since many of these compounds are also found in cerebrosides (see Sec. 17.8). The ceramide part of complex lipids may contain different fatty acids; stearic acid occurs mainly in sphingomyelin.

The phospholipids are not randomly distributed in membranes. In human red blood cells the sphingomyelin and phosphatidylcholine (lecithin) are on the outside of the membrane facing the blood plasma, while phosphatidylethanolamine and phosphatidylserine (cephalins) are on the inside of the membrane. In viral membranes most of the sphingomyelin is on the inside of the membrane.

17.8 Glycolipids

Glycolipids are complex lipids that contain carbohydrates. Among the glycolipids are the **cerebrosides,** which are ceramide mono- or oligosaccharides.

In cerebrosides the fatty acid of the ceramide part may contain either 18-carbon or 24-carbon chains, the latter found only in these complex lipids. A glucose or galactose carbohydrate unit forms a beta glycosidic bond with the ceramide portion of the molecule. The cerebrosides occur primarily in the brain (7 percent of the dry weight) and at nerve synapses.

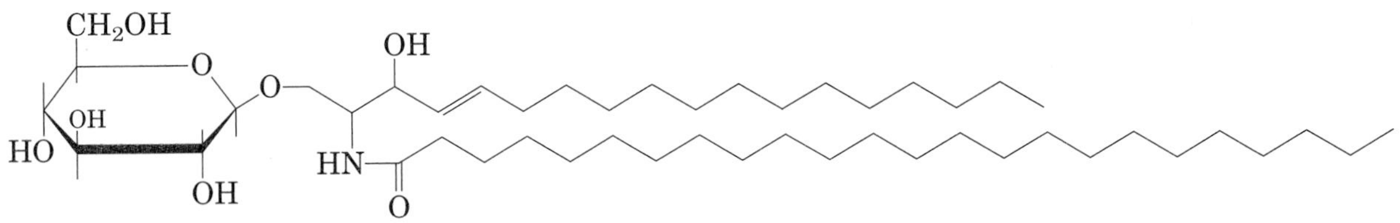

CH_2OH OH HO OH OH HN O O O

β-D-Glucose

Ceramide

Glucocerebroside

EXAMPLE

A lipid isolated from the membrane of red blood cells had the following structure:

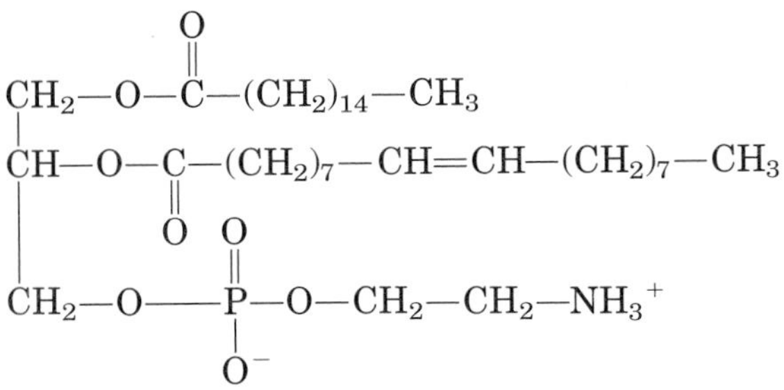

$$\begin{array}{l} CH_2-O-\overset{\overset{\displaystyle O}{\|}}{C}-(CH_2)_{14}-CH_3 \\ CH-O-\underset{\underset{\displaystyle O}{\|}}{C}-(CH_2)_7-CH=CH-(CH_2)_7-CH_3 \\ CH_2-O-\overset{\overset{\displaystyle O}{\|}}{\underset{\underset{\displaystyle O^-}{|}}{P}}-O-CH_2-CH_2-NH_3^+ \end{array}$$

(a) To what group of complex lipids does this compound belong?
(b) What are the components?

Answer • The molecule is an ester of glycerol and contains a phosphate group; therefore it is a glycerophospholipid. Besides glycerol and phosphate, it has a palmitic acid and an oleic acid component. The other alcohol is ethanolamine. Therefore, it belongs to the subgroup of cephalins.

Problem 17.1

A complex lipid had the following structure:

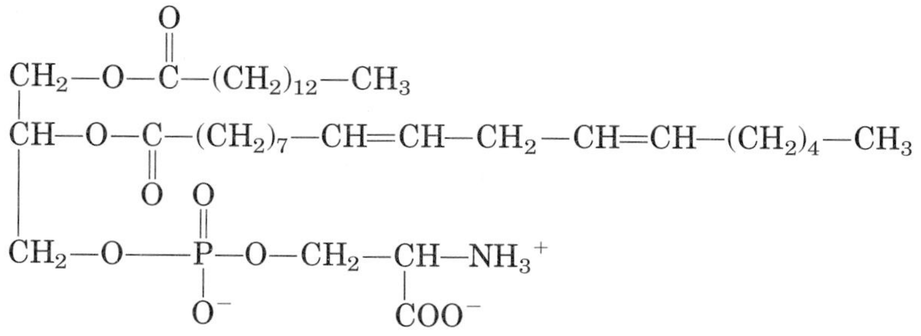

```
              O
              ‖
CH2—O—C—(CH2)12—CH3
|
CH—O—C—(CH2)7—CH=CH—CH2—CH=CH—(CH2)4—CH3
|           ‖
|           O   O
|               ‖
CH2—O———P—O—CH2—CH—NH3+
                |           |
                O−          COO−
```

(a) To what group of complex lipids does this compound belong?
(b) What are the components?

17.9 Steroids. Cholesterol

The third major class of lipids is the **steroids,** which are compounds containing this ring system:

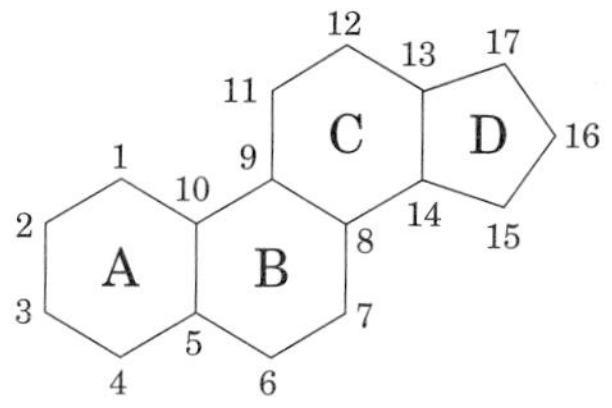

There are three cyclohexane rings (A, B, and C) connected in the same way as in phenanthrene (Sec. 11.12) and a fused cyclopentane ring (D). Steroids are thus completely different in structure from the lipids already discussed. Note that they are not necessarily esters, though some of them are.

Box 17F

Lipid Storage Diseases

Complex lipids are constantly being synthesized and decomposed in the body. Several genetic diseases are classified as lipid storage diseases. In these cases some of the enzymes needed to decompose the complex lipids are defective or altogether missing from the body. As a consequence, the complex lipids accumulate and cause enlarged liver and spleen, mental retardation, blindness, and, in certain cases, early death. Table 17F summarizes some of these diseases and indicates the missing enzyme and the accumulating complex lipid (Fig. 17F).

At present there is no treatment for these diseases. The best way to prevent them is by genetic counseling. Some of these diseases can be diagnosed during fetal development. For example, Tay–Sachs disease, which is carried by about 1 in every 30 Jewish Americans (versus 1 in 300 in the non-Jewish population), can be diagnosed from amniotic fluid obtained by amniocentesis.

However, there is now an experimental treatment available for patients with mild forms of certain lipid storage diseases such as Gaucher's disease. The human enzyme β-glucosidase can be extracted from placenta. It is encapsulated in a lipid bilayer and injected intravenously. The lipid bilayer capsule allows the drug to pass through the membranes of spleen cells and thus deliver the enzyme where it is needed. Unfortunately, because it is difficult to isolate the human enzyme from placenta, each injection costs thousands of dollars. If and when the human gene for the enzyme is isolated, the enzyme can be manufactured by the recombinant DNA technique (Sec. 23.14) and the cost drastically lowered.

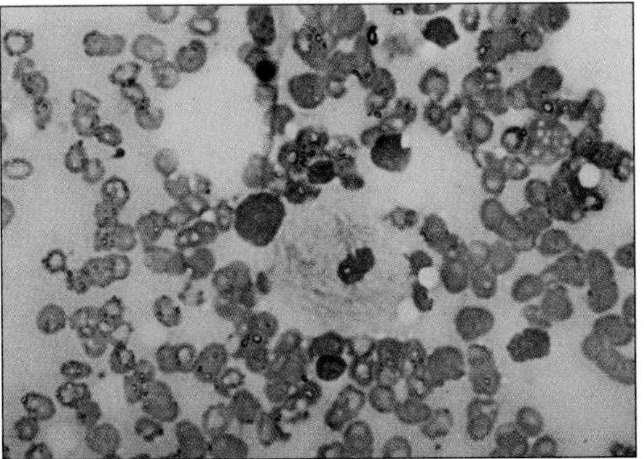

Figure **17F** • The accumulation of glucocerebrosides in the cell of a patient with Gaucher's disease. These cells (Gaucher cells) infiltrate the bone marrow. (Courtesy of Drs. P. G. Bullogh and V. J. Vigorita, and the Gower Medical Publishing Co., New York.)

Table 17F Lipid Storage Diseases

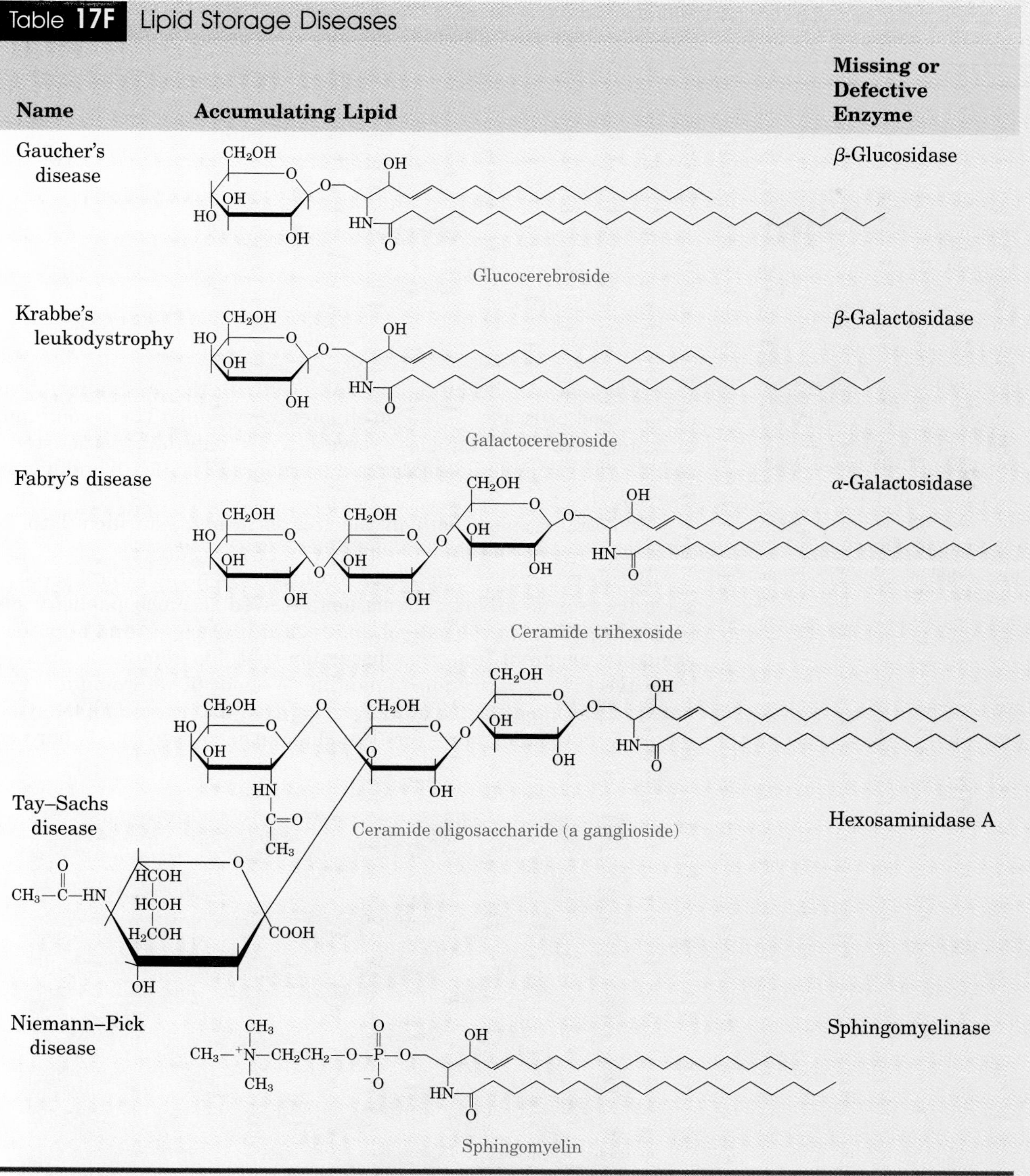

Name	Accumulating Lipid	Missing or Defective Enzyme
Gaucher's disease	Glucocerebroside	β-Glucosidase
Krabbe's leukodystrophy	Galactocerebroside	β-Galactosidase
Fabry's disease	Ceramide trihexoside	α-Galactosidase
Tay–Sachs disease	Ceramide oligosaccharide (a ganglioside)	Hexosaminidase A
Niemann–Pick disease	Sphingomyelin	Sphingomyelinase

Cholesterol

The most abundant steroid in the human body, and the most important, is cholesterol:

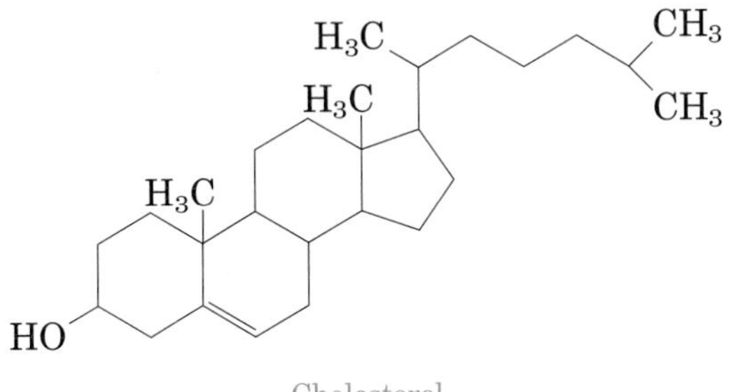

Cholesterol

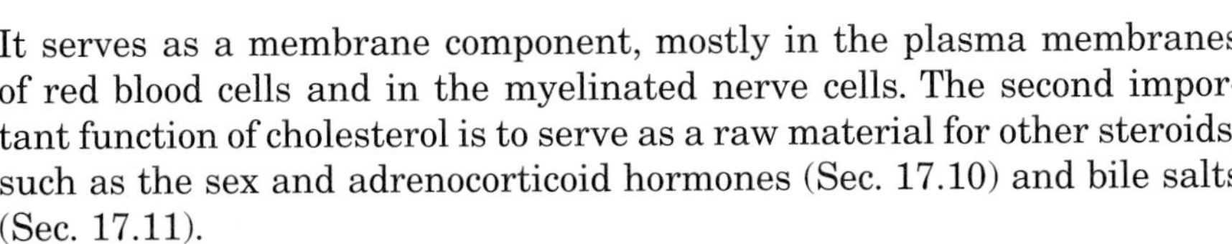

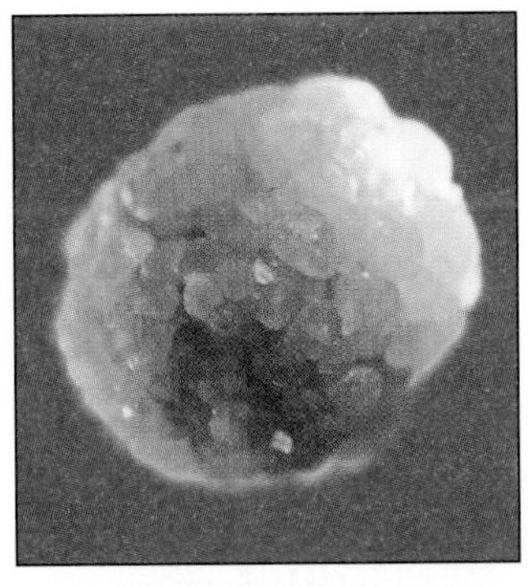

A human gallstone is almost pure cholesterol. This gallstone measures 5 mm in diameter. (© Carolina Biological Supply Company, Phototake, NYC)

It serves as a membrane component, mostly in the plasma membranes of red blood cells and in the myelinated nerve cells. The second important function of cholesterol is to serve as a raw material for other steroids, such as the sex and adrenocorticoid hormones (Sec. 17.10) and bile salts (Sec. 17.11).

Cholesterol exists both in the free form and esterified with fatty acids. Gallstones contain free cholesterol (Fig. 17.4).

Because the correlation between high serum cholesterol levels and such diseases as atherosclerosis has received so much publicity, many people are afraid of cholesterol and regard it as some kind of poison. It should be apparent from this discussion that, far from being poisonous, cholesterol is necessary for human life. Without it, we would die. Fortunately, there is no chance of that, since, even if it were completely eliminated from the diet, our livers would make enough to satisfy our needs.

Figure **17.4** • Cholesterol crystals, taken from fluid in the elbow of a patient suffering from bursitis, as seen in a polarizing light microscopic photograph. (Courtesy of Drs. P. A. Dieppe, P. A. Bacon, A. N. Bamji, I. Watt, and the Gower Medical Publishing Co. Ltd., London, England.)

Cholesterol in the body is in a dynamic state. Most of the cholesterol ingested and that manufactured by the liver is used by the body to make other molecules, such as bile salts. The serum cholesterol level controls the amount of cholesterol synthesized by the liver.

When the cholesterol level goes above 150 mg/100 mL, cholesterol synthesis in the liver is reduced to half the normal rate of production.

Cholesterol, along with fat, is transported from the liver to the peripheral tissues by lipoproteins. There are two important kinds: **high density lipoprotein (HDL),** which has a protein content of about 50 percent, and **low density lipoprotein (LDL),** in which the protein content is only about 25 percent. LDL contains about 45 percent cholesterol; HDL only 18 percent. High density lipoprotein transports cholesterol to the liver and also transfers cholesterol to LDL. Low density lipoprotein plays an important role in cholesterol metabolism. The core of LDL (Fig. 17.5) contains fats (triglycerides) and esters of cholesterol in which linoleic acid is an important constituent. On the surface of the LDL cluster are phospholipids, free cholesterol, and proteins. These molecules contain many polar groups so that water molecules in the blood plasma can solvate them by forming hydrogen bonds, making the LDL soluble in the blood plasma. The LDL carries cholesterol to the cells, where specific LDL-receptor molecules line the cell surface in certain concentrated areas called **coated pits.** One of the proteins on the surface of the LDL binds specifically to the LDL-receptor molecules in the coated pits. After such binding the LDL is taken inside the cell, where enzymes break it down, liberating the free cholesterol from cholesterol esters. In this manner cholesterol can be used by the cell, for example, as a component of a membrane. This is the normal fate of LDL and the normal course of cholesterol transport.

Lipoproteins are spherically shaped clusters containing both lipid molecules and protein molecules.

Michael Brown and Joseph Goldstein of the University of Texas shared the Nobel Prize in Medicine for 1986 for the discovery of the LDL-receptor-mediated pathway.

In certain cases, however, there are not enough LDL receptors. In the disease called familial hypercholesterolemia, the cholesterol level in the

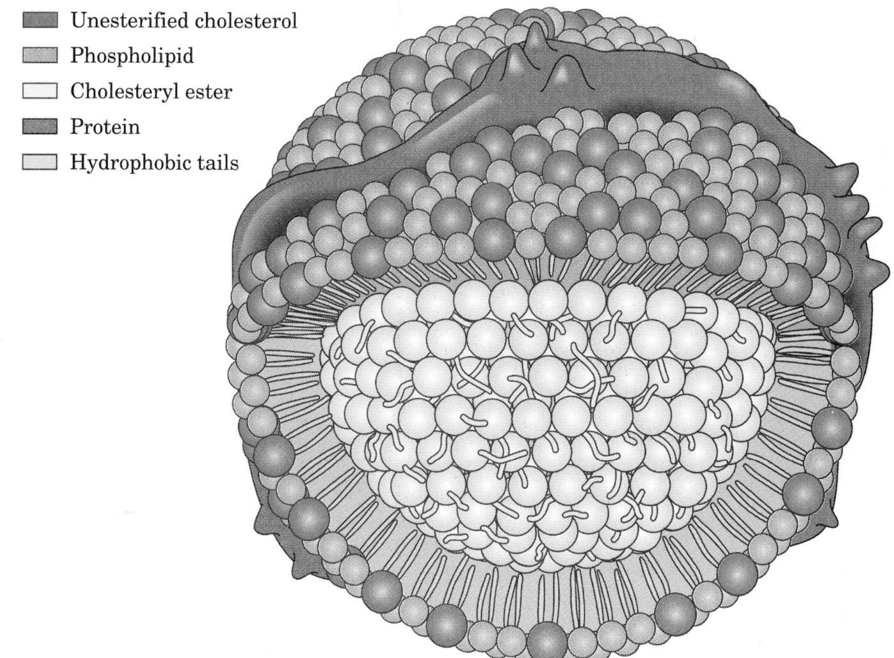

Figure **17.5** • Low density lipoprotein shown schematically.

Box 17G

Cholesterol and Heart Attacks

Like all lipids, cholesterol is insoluble in water, and if its level is elevated in the blood serum, plaque-like deposits may form on the inner surfaces of the arteries. This leads to a decrease in the diameter of the blood vessels, which may lead to a decrease in the flow of blood. *Atherosclerosis* (Fig. 17G) is the result, along with accompanying high blood pressure, which may lead to heart attack, stroke, or kidney dysfunction. Atherosclerosis enhances the possible complete blockage of some arteries by a clot at the point where the arteries are constricted by plaque. Furthermore, blockage may deprive cells of oxygen, and these may cease to function. The death of heart muscles due to lack of oxygen is called *myocardial infarction.*

The more general condition, *arteriosclerosis,* or hardening of the arteries with age, is also accompanied by increased levels of cholesterol in the blood serum. Whereas young adults have, on the average, 1.6 g of cholesterol per liter of blood, in people older than 55 this almost doubles to 2.5 g/L because the rate of metabolism slows with age. Diets low in cholesterol and saturated fatty acids usually reduce the serum cholesterol level, and a number of drugs are available that inhibit the synthesis of cholesterol in the liver. The commonly used drug, lovastatin, and related compounds act to inhibit one of the key enzymes in cholesterol synthesis, called HMG-CoA reductase. Thus, they block the synthesis of cholesterol inside the cells and stimulate the synthesis of LDL receptor proteins. In this way more LDL enters the cells, diminishing the amount of cholesterol that will be deposited on the inner walls of arteries. Although there is a good correlation between high serum cholesterol and various circulatory diseases, not everyone who suffers from hardening of the arteries has high serum cholesterol, nor do all patients with high serum cholesterol develop arteriosclerosis.

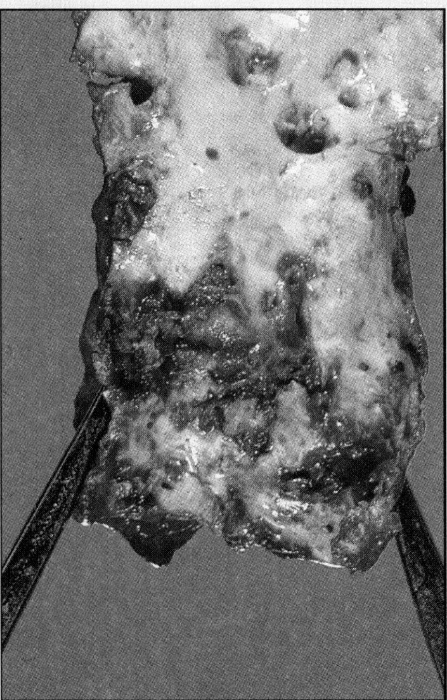

Figure **17G** • Effect of atherosclerosis in arteries. (© 1994 SIU, Photo Researchers, Inc.)

plasma may be as high as 680 mg/100 mL as opposed to 175 mg/100 mL in normal subjects. These high levels of cholesterol can cause atherosclerosis and heart attacks (Box 17G). The high plasma cholesterol levels in these patients are caused because there are not enough functional LDL receptors, or, if there are enough, they are not concentrated in the coated pits. Thus high LDL content means high cholesterol content in the plasma, since LDL cannot get into the cells and be metabolized. Therefore a high LDL level together with a low HDL level is a symptom of faulty cholesterol transport and a warning for possible atherosclerosis. Today it is generally considered desirable to have high levels of HDL and low levels of LDL in the bloodstream.

At the present time, our knowledge of the role played by serum cholesterol in atherosclerosis is incomplete. The best we can say is that it probably makes good sense to reduce the amount of cholesterol and saturated fatty acids in the diet.

17.10 Steroid Hormones

Cholesterol is the starting material for the synthesis of steroid hormones. In this process, the aliphatic side chain on the D ring is shortened by the removal of a six-carbon unit, and the secondary alcohol group on C-3 is oxidized to a ketone. The resulting molecule, *progesterone,* serves as the starting compound for both the sex hormones and the adrenocorticoid hormones (Fig. 17.6).

Adrenocorticoid Hormones

The adrenocorticoid hormones are products of the adrenal glands. We divide them into two groups according to function: *Mineralocorticoids* regulate the concentrations of ions (mainly Na^+ and K^+), and *glucocorticoids* control carbohydrate metabolism.

The term "adrenal" comes from *ad*jacent to the *renal* (which refers to the kidney). The name "corticoid" indicates that the site of the secretion is the cortex (outer part) of the gland.

Aldosterone is one of the most important mineralocorticoids. Increased secretion of aldosterone enhances the reabsorption of Na^+ and Cl^- ions in the kidney tubules and increases the loss of K^+. Since Na^+ concentration controls water retention in the tissues, aldosterone also controls tissue swelling.

Cortisol is the major glucocorticoid. Its function is to increase the glucose and glycogen concentrations in the body. This is done at the expense of other nutrients. Fatty acids from fat storage cells and amino acids from body proteins are transported to the liver, which, under the influence of cortisol, manufactures glucose and glycogen from these sources.

Cortisol and its ketone derivative, *cortisone,* have remarkable anti-inflammatory effects. These or similar synthetic derivatives, such as prednisolone, are used to treat inflammatory diseases of many organs, rheumatoid arthritis, and bronchial asthma.

The conversion of the male hormone to the female hormone results from the loss of a $—CH_3$ and a —H:

Sex Hormones

The most important male sex hormone is *testosterone* (Fig. 17.6). This hormone, which promotes the normal growth of the male genital organs, is synthesized in the testes from cholesterol. During puberty, increased testosterone production leads to such secondary male sexual characteristics as deep voice and facial and body hair.

Female sex hormones, the most important of which is *estradiol* (Fig. 17.6), are synthesized from the corresponding male hormone (testosterone) by aromatization of the A ring. Estradiol, together with its precursor progesterone, regulates the cyclic changes occurring in the uterus and ovaries known as the **menstrual cycle.** As the cycle begins, the level of estradiol in the body rises, and this causes the lining of the uterus to thicken. Then another hormone, called luteinizing hormone, triggers ovulation. If the ovum is fertilized, increased progesterone levels will inhibit

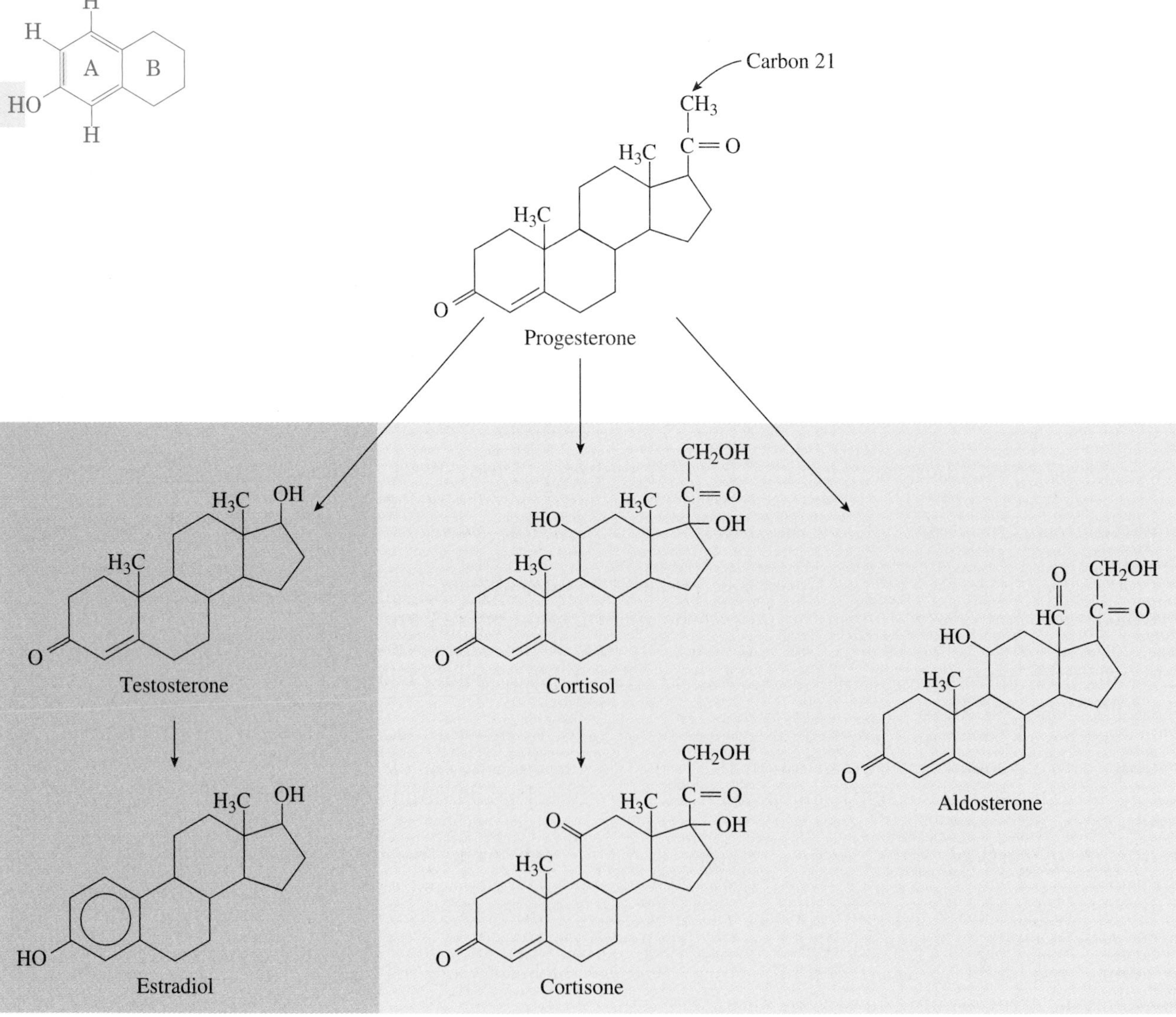

Figure 17.6 • The biosynthesis of hormones from progesterone.

Box 17H

Anabolic Steroids

Testosterone, the principal male hormone, is responsible for the buildup of muscles in men. Because of this, many athletes have taken this drug in an effort to increase muscular development. This is especially common among athletes in sports in which strength and muscle mass are important, including weight lifting, shot put, and hammer throw, but participants in other sports, such as running, swimming, and cycling, would also like larger and stronger muscles.

Although used by many athletes, testosterone has two disadvantages: (1) Besides its effect on muscles, it also affects secondary sexual characteristics, and too much of it can result in undesired side effects. (2) It is not very effective when taken orally, and must be injected for best results.

For these reasons, a large number of other anabolic steroids, all of them synthetic, have been developed. Some examples are the following:

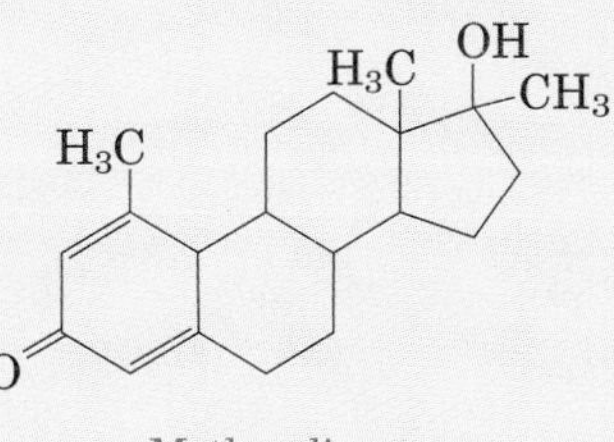

Methandienone

Methenolone

Nandrolone decanoate

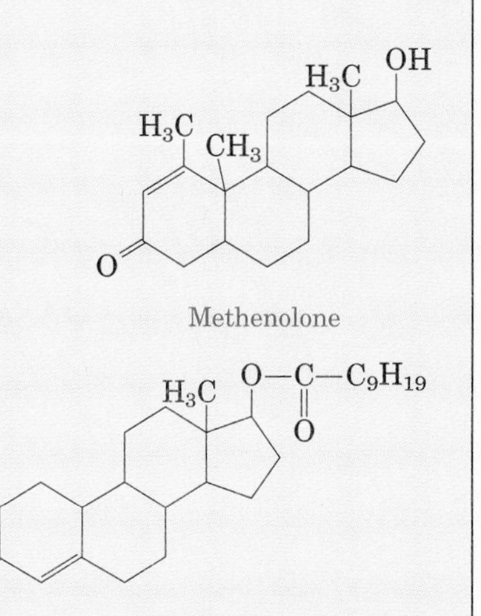

Of the three shown here, methandienone and methenolone can be taken orally, but nandrolone decanoate must be injected.

Anabolic steroids are also used by some women athletes. Since their bodies produce only small amounts of testosterone, women have much more to gain from anabolic steroids than men.

The use of anabolic steroids is forbidden in many sporting events, especially in international competition, largely for two reasons: (1) It gives some competitors an unfair advantage, and (2) these drugs can have many side effects, ranging from acne to liver tumors. Side effects can be especially disadvantageous for women, since they can include growth of facial hair, baldness, deepening of the voice, and menstrual irregularities. All athletes participating in the Olympic Games are required to pass a urine test for anabolic steroids. A number of winning athletes have had their victories taken away because they tested positive for steroid use. One tragic example is the world-class sprinter, the Canadian Ben Johnson, whose career was brought to an end at the 1988 Olympiad in Seoul, Korea. He had just won the 100-meter race in world record time when a urine analysis proved that he had taken steroid drugs. He was stripped of both his world record and his gold medal.

any further ovulation. Both estradiol and progesterone then promote further preparation of the uterine lining to receive the fertilized ovum. If no fertilization takes place, progesterone production stops altogether, and estradiol production decreases. This decreases the thickening of the uterine lining, and it is then sloughed off with accompanying bleeding. This is menstruation (Fig. 17.7).

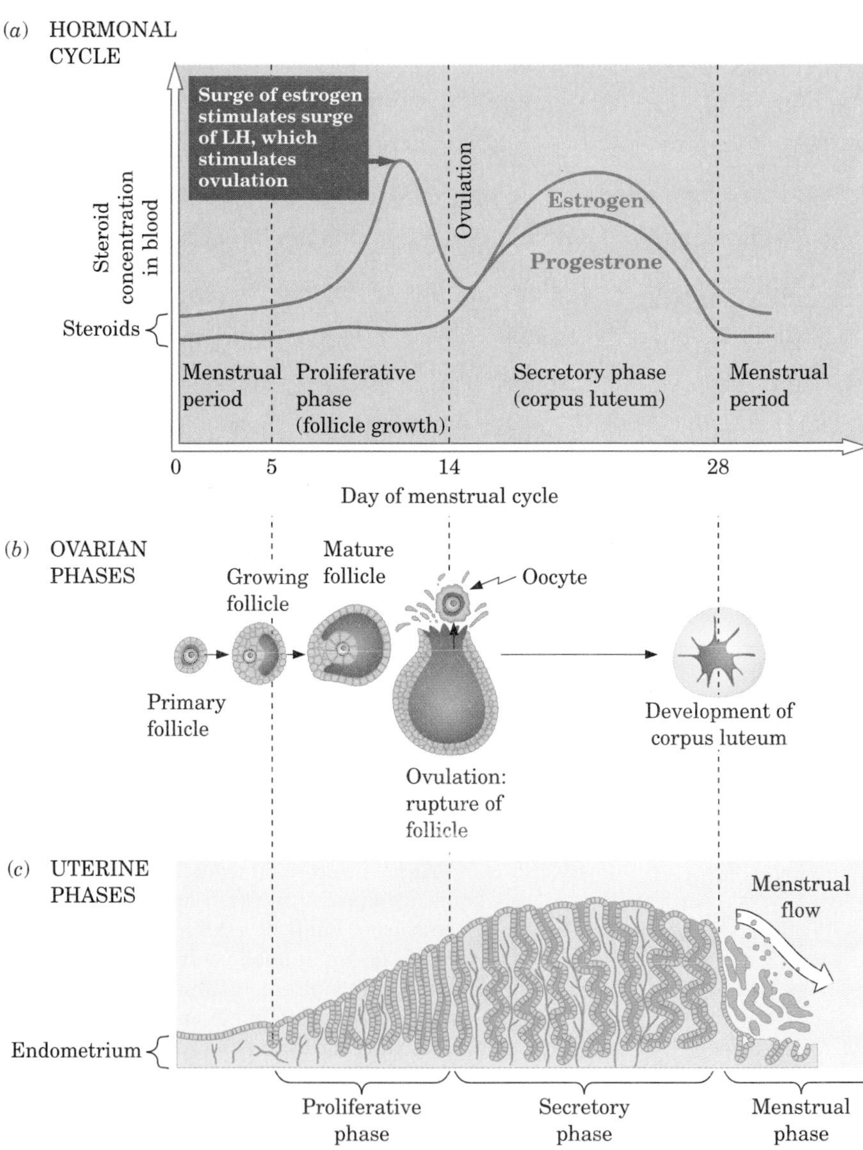

Figure **17.7** • Events of the menstrual cycle. (a) Levels of sex hormones in the blood stream during the phases of one menstrual cycle in which pregnancy does not occur. (b) Development of an ovarian follicle during the cycle. (c) Phases of development of the endometrium, the lining of the uterus. The endometrium thickens during the proliferative phase. In the secretory phase, which follows ovulation, the endometrium continues to thicken, and the glands secrete a glycogen-rich nutritive material in preparation to receive an embryo. When no embryo implants, the new outer layers of the endometrium disintegrate and the blood vessels rupture, producing the menstrual flow.

Because progesterone is essential for the implantation of the fertilized ovum, blocking its action leads to termination of pregnancy. Progesterone interacts with a receptor (a protein molecule) in the nucleus of cells. The receptor changes its shape when progesterone binds to it (see Sec. 25.3). A new drug, widely used in France and China, called mifepristone, or RU486,

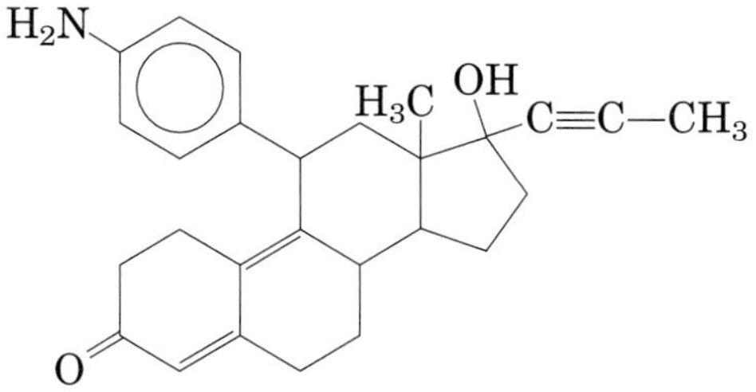

acts as a competitor to progesterone. It binds to the same receptor site. Because the progesterone molecule is prevented from reaching the receptor molecule, the uterus is not prepared for the implantation of the fertilized ovum, and the ovum is aborted. This is a "morning-after pill" (see also Box 17I). If taken orally within 72 hours after intercourse it

RU486 also binds to the receptors of glucocorticoid hormones. Its use as an antiglucocorticoid is also recommended to alleviate a disease known as Cushing syndrome, the overproduction of cortisone.

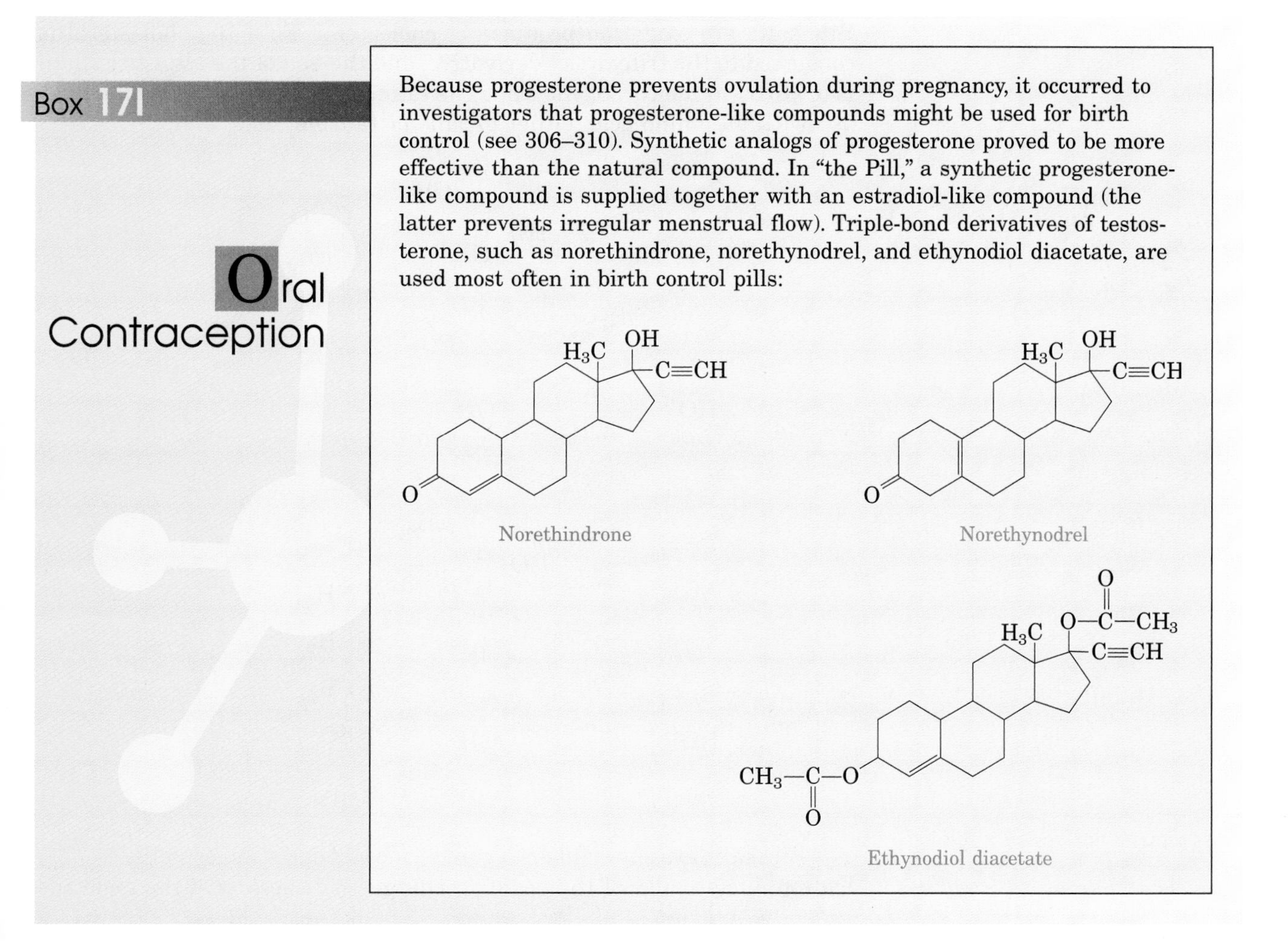

Box 17I

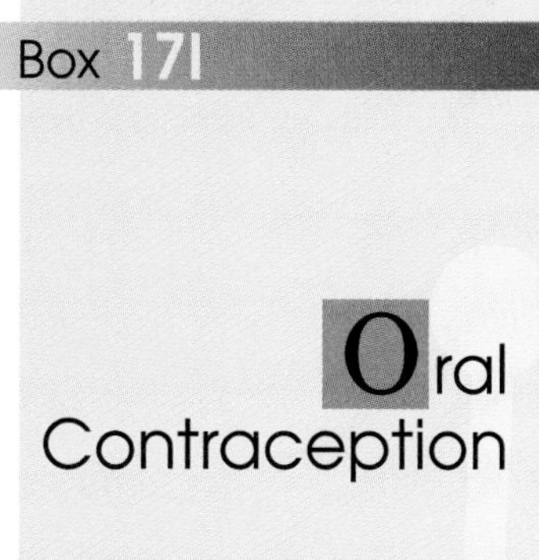

Because progesterone prevents ovulation during pregnancy, it occurred to investigators that progesterone-like compounds might be used for birth control (see 306–310). Synthetic analogs of progesterone proved to be more effective than the natural compound. In "the Pill," a synthetic progesterone-like compound is supplied together with an estradiol-like compound (the latter prevents irregular menstrual flow). Triple-bond derivatives of testosterone, such as norethindrone, norethynodrel, and ethynodiol diacetate, are used most often in birth control pills:

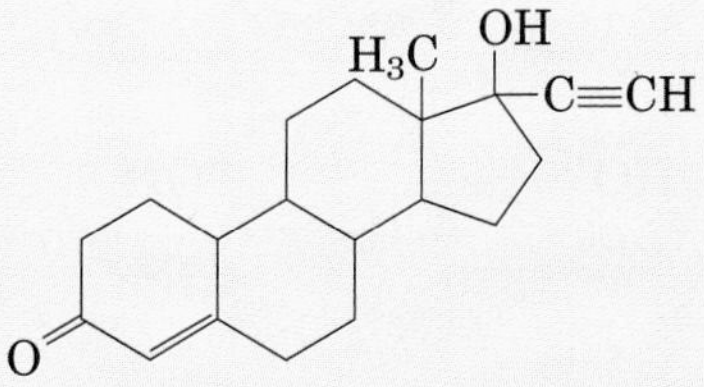

Norethindrone

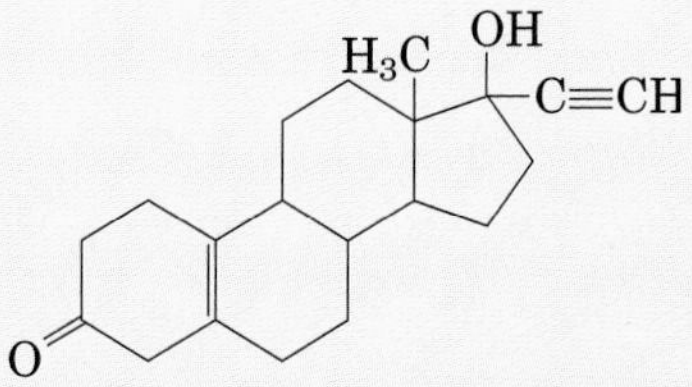

Norethynodrel

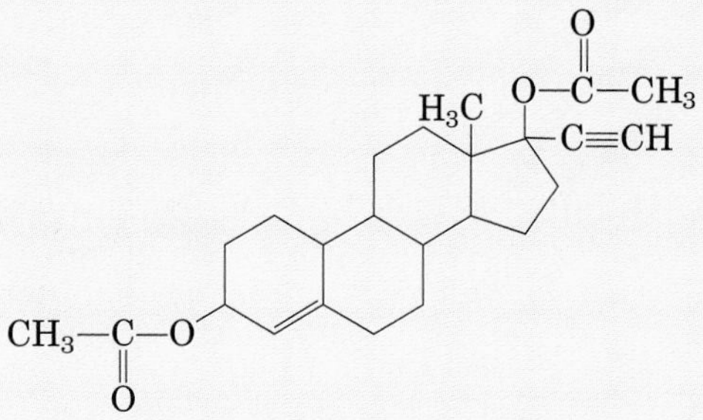

Ethynodiol diacetate

prevents pregnancy. However, the drug fails in 20 percent of cases unless followed 36 to 48 hours later by an injection, by an intravaginal suppository, or by oral administration of a prostaglandin (see Sec. 17.12). This chemical form of abortion has been in clinical trials in the United States as a supplement to surgical abortion for the last 3 years. The Food and Drug Administration has recently approved the use of this drug in the United States.

Estradiol and progesterone also regulate secondary female sex characteristics, such as the growth of breasts. Because of this, RU486, as an antiprogesterone, has also been reported to be effective against certain types of breast cancer.

Testosterone and estradiol are not exclusive to either males or females. A small amount of estradiol production occurs in males, and a small amount of testosterone production is normal in females. Only when the proportion of these two hormones (hormonal balance) is upset can one observe symptoms of abnormal sexual differentiation.

17.11 Bile Salts

Bile salts are oxidation products of cholesterol. First the cholesterol is converted to the trihydroxy derivative, and the end of the aliphatic chain is oxidized to the carboxylic acid. The latter in turn forms an amide linkage with an amino acid, either glycine or taurine:

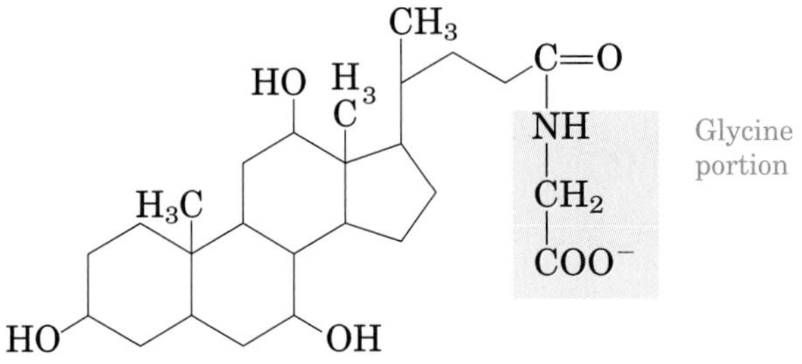

Glycine portion

Glycocholate

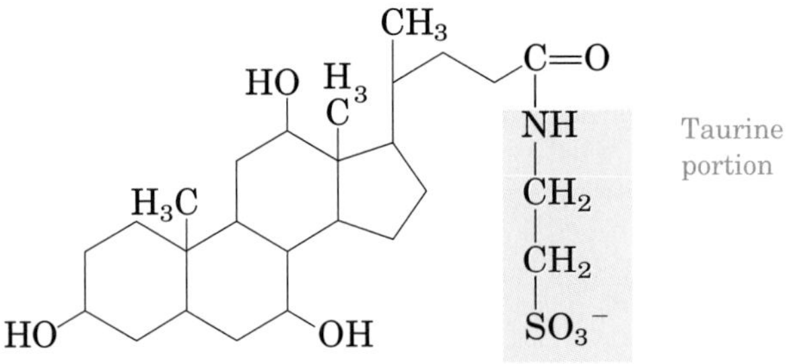

Taurine portion

Taurocholate

The dispersion of dietary lipids by bile salts is similar to the action of soap on dirt.

Bile salts are powerful detergents. One end of the molecule is strongly hydrophilic because of the negative charge, and the rest of the molecule is hydrophobic. Thus, bile salts can disperse dietary lipids in the small intestine into fine emulsions and thereby help digestion.

Since they are eliminated in the feces, bile salts also remove excess cholesterol in two ways: They themselves are breakdown products of cholesterol (thus cholesterol is eliminated via bile salts), and they solubilize deposited cholesterol in the form of bile salt–cholesterol particles.

17.12 Prostaglandins and Leukotrienes

This group of fatty acid–like substances was first discovered when it was demonstrated that seminal fluid caused a hysterectomized uterus to contract. The name implies that these substances are a product of the prostate gland, and in mature males the seminal gland does secrete 0.1 mg of prostaglandin per day. However, small amounts of prostaglandins are present throughout the body in both sexes.

Prostaglandins are synthesized in the body from arachidonic acid by a ring closure at C-8 and C-12:

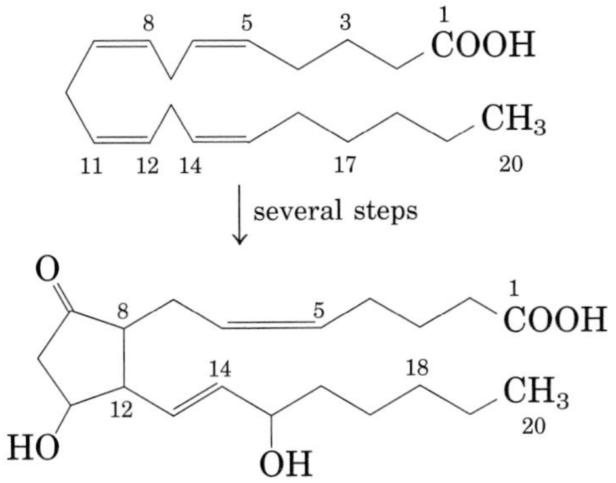

8 5 3 1 COOH
CH$_3$
11 12 14 17 20

Arachidonic acid

several steps

O 8 5 1 COOH
14 18 CH$_3$
12 20
HO OH

PGE_2

The prostaglandin E group (PGE) has a carbonyl group at C-9; the subscript indicates the number of double bonds. The prostaglandin F group (PGF) has two hydroxyl groups on the ring at positions C-9 and C-11.

Other prostaglandins (PGAs and PGBs) are derived from PGE.

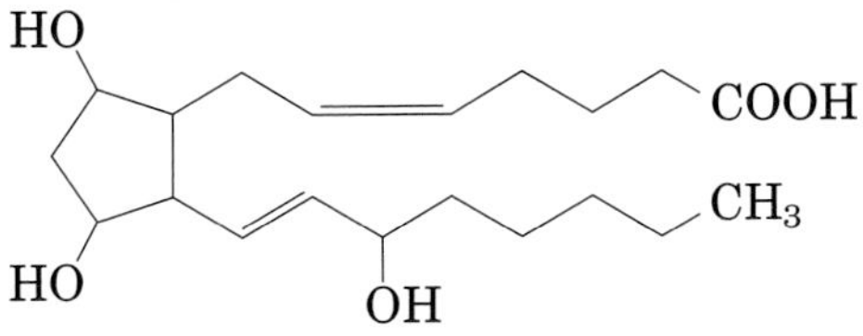

HO COOH
CH$_3$
HO OH

$PGF_{2\alpha}$

The prostaglandins as a group have a wide variety of effects on body chemistry. They seem to act as mediators of hormones. For example, PGE_2 and $PGF_{2\alpha}$ induce labor and are used for therapeutic abortion in early pregnancy. PGE_2 lowers blood pressure, but $PGF_{2\alpha}$ causes hypertension (increase of blood pressure). PGE_2 in aerosol form is used to treat asthma; it opens up the bronchial tubes by relaxing the surrounding muscles. PGE_1 is used as a decongestant; it opens up nasal passages by constricting blood vessels.

Many prostaglandins cause inflammation and fever. The anti-inflammatory effect of aspirin results from the inhibition of prostaglandin synthesis (Box 17J). PGA and PGE inhibit gastric secretions.

Another group of substances that act as mediators of hormonal responses are the leukotrienes. Like prostaglandins, leukotrienes are derived from arachidonic acid by an oxidative mechanism. However, in this case there is no ring closure.

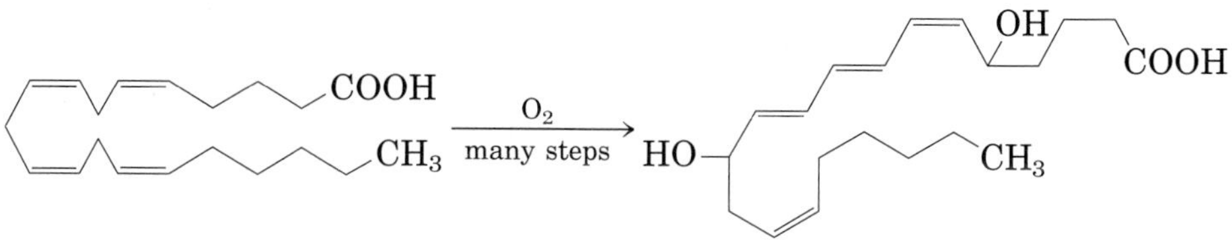

COOH
CH_3
O_2 → many steps
HO
OH
COOH
CH_3

Arachidonic acid

Leukotriene B4

Leukotrienes occur mainly in leukocytes but also in other tissues of the body. They produce long-lasting muscle contractions, especially in the lungs, and they cause asthma-like attacks. They are a hundred times more potent than histamines. Both prostaglandins and leukotrienes cause inflammation and fever. The inhibition of their production in the body is a major pharmacological concern. One way to counteract the effects of leukotrienes is to inhibit their uptake by leukotriene receptors (LTR) in the body. A new drug, zafirlukast, marketed under the name Accolate, an antagonist of LTRs, is used to treat and control chronic asthma.

Box 17J

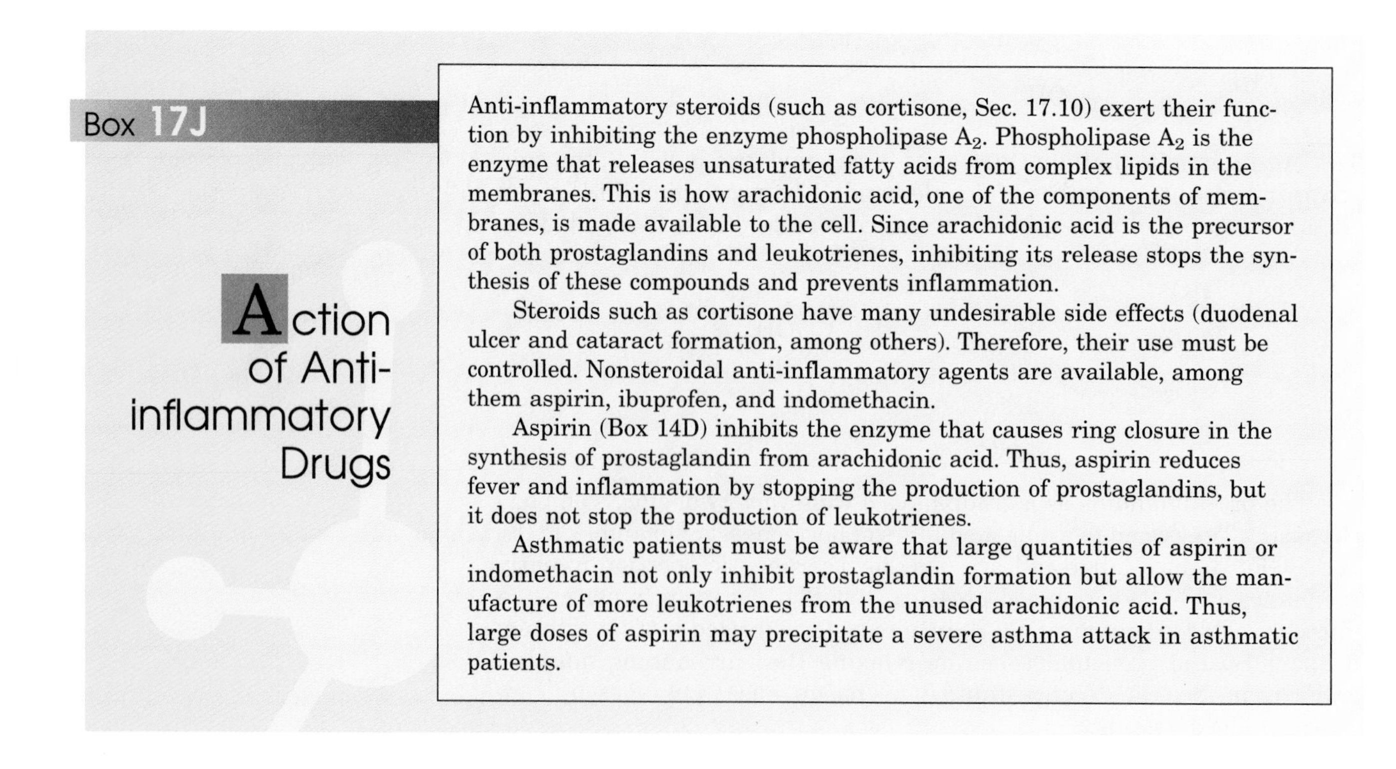

Action of Anti-inflammatory Drugs

Anti-inflammatory steroids (such as cortisone, Sec. 17.10) exert their function by inhibiting the enzyme phospholipase A_2. Phospholipase A_2 is the enzyme that releases unsaturated fatty acids from complex lipids in the membranes. This is how arachidonic acid, one of the components of membranes, is made available to the cell. Since arachidonic acid is the precursor of both prostaglandins and leukotrienes, inhibiting its release stops the synthesis of these compounds and prevents inflammation.

Steroids such as cortisone have many undesirable side effects (duodenal ulcer and cataract formation, among others). Therefore, their use must be controlled. Nonsteroidal anti-inflammatory agents are available, among them aspirin, ibuprofen, and indomethacin.

Aspirin (Box 14D) inhibits the enzyme that causes ring closure in the synthesis of prostaglandin from arachidonic acid. Thus, aspirin reduces fever and inflammation by stopping the production of prostaglandins, but it does not stop the production of leukotrienes.

Asthmatic patients must be aware that large quantities of aspirin or indomethacin not only inhibit prostaglandin formation but allow the manufacture of more leukotrienes from the unused arachidonic acid. Thus, large doses of aspirin may precipitate a severe asthma attack in asthmatic patients.

SUMMARY

Lipids are water-insoluble substances. They are divided into four groups: fats (glycerides), complex lipids, steroids, and prostaglandins and leukotrienes. **Fats** are made up of fatty acids and glycerol. In saturated fatty acids the hydrocarbon chains have only single bonds; unsaturated fatty acids have hydrocarbon chains with one or more double bonds, all in the cis configuration. Solid fats contain mostly saturated fatty acids, while **oils** contain substantial amounts of unsaturated fatty acids. The alkali salts of fatty acids are called **soaps.**

Complex lipids can be divided into two groups: phospholipids and glycolipids. **Phospholipids** are made of a central alcohol (glycerol or sphingosine), fatty acids, and a nitrogen-containing phosphate ester, such as phosphorylcholine. The **glycolipids** contain sphingosine and a fatty acid, which together are called the **ceramide** portion of the molecule, and a carbohydrate portion. Many phospholipids and glycolipids are important constituents of cell membranes.

Membranes are made of a **lipid bilayer** in which the hydrophobic parts of phospholipids (fatty acid residues) point to the middle of the bilayer, and the hydrophilic parts point toward the inner and outer surfaces of the membrane.

The third major group of lipids is the **steroids.** The basic feature of the steroid structure is a fused four-ring nucleus. The most common steroid is **cholesterol,** which also serves as a raw material for other steroids, such as bile salts and sex and other hormones. Cholesterol is also an integral part of membranes, occupying the hydrophobic region of the lipid bilayer. Because of its low solubility in water, cholesterol deposits are implicated in the formation of gallstones and the plaque-like deposits of atherosclerosis. Cholesterol is transported in the blood plasma by two kinds of lipoprotein: **HDL** and **LDL.** The LDL plays an important role in cholesterol metabolism. It is soluble in blood plasma because of polar groups on its surface. An oxidation product of cholesterol is **progesterone,** which is a sex hormone and also gives rise to the synthesis of other sex hormones, such as testosterone and estradiol, as well as to the **adrenocorticoid hormones.** Among the latter, the best known are cortisol and cortisone for their anti-inflammatory action. **Bile salts** are also oxidation products of cholesterol. They emulsify all kinds of lipids, including cholesterol, and are essential in the digestion of fats.

Prostaglandins and **leukotrienes** are derived from arachidonic acid. They have a wide variety of effects on body chemistry; among other things, they can lower or raise blood pressure, cause inflammation, and induce labor. They act generally as mediators of hormone action.

KEY TERMS

Adrenocorticoid hormone (Sec. 17.10)
Anabolic steroid (Box 17H)
Bilayer (Sec. 17.5)
Bile salt (Sec. 17.11)
Cephalin (Sec. 17.6)
Ceramide (Sec. 17.7)
Cerebroside (Sec. 17.8)
Cholesterol (Sec. 17.9)
Coated pits (Sec. 17.9)
Complex lipid (Sec. 17.4)
Detergent (Box 17C)
Diglyceride (Sec. 17.2)
Fat (Sec. 17.2)
Fluid mosaic model (Sec. 17.5)
Glycerophospholipid (Sec. 17.6)
Glycolipid (Sec. 17.8)
HDL (Sec. 17.9)
Hydrophilic (Sec. 17.5)
Hydrophobic (Sec. 17.1)
LDL (Sec. 17.9)
Lecithin (Sec. 17.6)
Leukotriene (Sec. 17.12)
Lipid (Sec. 17.1)
Lipid bilayer (Sec. 17.5)
Lipoprotein (Sec. 17.9)
Membrane (Sec. 17.5)
Micelle (Box 17C)
Monoglyceride (Sec. 17.2)
Myelin (Sec. 17.7)
Myelin sheath (Box 17E)
Oil (Sec. 17.3)
Organelle (Sec. 17.5)
Phospholipid (Sec. 17.4)
Polyunsaturated oils (Sec. 17.3)
Prostaglandin (Sec. 17.12)
Schwann cell (Box 17E)
Sex hormone (Sec. 17.10)
Soap (Sec. 17.3)
Sphingolipid (Sec. 17.7)
Steroid (Sec. 17.9)
Steroid hormone (Sec. 17.10)
Triacylglycerol (Sec. 17.2)
Triglyceride (Sec. 17.2)

PROBLEMS

17.2 Why are fats a good source of energy for storage in the body?

17.3 What is the meaning of the term hydrophobic? Why is the hydrophobic nature of lipids important?

The Structure of Fats

17.4 Draw the structural formula of a fat molecule (triglyceride) made of myristic acid, oleic acid, palmitic acid, and glycerol.

***17.5** How many glycerides (mono-, di-, and tri-) can be formed from glycerol and myristic acid? Draw their structures.

***17.6** (a) Draw schematically all possible diglycerides made up of glycerol, oleic acid, and/or stearic acid. (b) How many are there, all together? (c) Draw the detailed structure of one of the diglycerides.

Properties of Fats

17.7 For the diglycerides in Problem 17.6, predict which two will have the highest and which two will have the lowest melting points.

17.8 Predict which acid in each pair has the higher melting point, and explain why: (a) palmitic acid or stearic acid (b) arachidonic acid or arachidic acid (arachidic acid is the saturated 20-carbon acid)

17.9 Which has the higher melting point: (a) a triglyceride containing only lauric acid and glycerol or (b) a triglyceride containing only stearic acid and glycerol?

17.10 Explain why the melting points of the saturated fatty acids increase going from lauric to stearic acid (Table 17.1).

17.11 Predict the order of the melting points of triglycerides containing fatty acids, as follows:
(a) palmitic, palmitic, stearic
(b) oleic, stearic, palmitic
(c) oleic, linoleic, oleic

17.12 Coconut oil is a liquid fat (oil), yet it is saturated. Explain why.

***17.13** Rank, in order of increasing solubility in water (assuming that all are made with the same fatty acids) (a) triglycerides (b) diglycerides (c) monoglycerides. Explain.

17.14 How many moles of H_2 are used up in the catalytic hydrogenation of 1 mole of a triglyceride containing glycerol, palmitic acid, oleic acid, and linoleic acid?

17.15 Name the products of the saponification of this triglyceride:

```
CH2—O—C—(CH2)14CH3
|      ||
|      O
|
CH—O—C—(CH2)16CH3
|     ||
|     O                        CH3CH2—CH=CH
|                                         |
CH2—O—C—(CH2)7CH=CH—CH2—CH=CH—CH2
       ||
       O
```

17.16 Using the equation on page 564 as a guideline for stoichiometry, calculate the number of moles of NaOH it takes to saponify 5 moles of (a) triglycerides (b) diglycerides (c) monoglycerides.

Membranes

17.17 (a) Where in the body are membranes found? (b) What functions do they serve?

17.18 How do the unsaturated fatty acids of the complex lipids contribute to the fluidity of a membrane?

17.19 Which type of lipid molecules are most likely to be present in membranes?

17.20 What is the difference between an integral and a peripheral membrane protein?

Complex Lipids

17.21 Draw the structure of a cephalin that contains palmitic acid and oleic acid.

17.22 Complex lipids can act as emulsifying agents. The lecithin of egg yolk is used in making mayonnaise. Draw the structure of a lecithin. Identify the portion of the molecule that interacts with oil droplets and the portion that interacts with vinegar (acetic acid in water).

***17.23** Among the glycerophospholipids containing palmitic acid and linolenic acid, which will have the greatest solubility in water: (a) phosphatidylcholine (b) phosphatidylethanolamine (c) phosphatidylserine? Explain.

17.24 List the names of all the groups of complex lipids that contain ceramides.

17.25 Are the various phospholipids randomly distributed in membranes? Give an example.

Steroids

***17.26** Cholesterol has a fused four-ring core and is a part of body membranes. The —OH group on C-3 is the polar head, and the rest of the molecule provides the hydrophobic tail that does not fit into the zigzag packing of the hydrocarbon portion of the saturated fatty acids. Considering this structure, tell whether cholesterol contributes to the stiffening (rigidity) or to the fluidity of a membrane. Explain.

17.27 (a) Is cholesterol necessary for human life?
(b) Why do many people restrict cholesterol intake in their diet?

17.28 Where can pure cholesterol crystals be found in the body?

***17.29** (a) Find all the chiral carbons in cholesterol.
(b) How many total isomers are possible?
(c) How many of these do you think are found in nature?

17.30 Look at the structures of cholesterol and the hormones shown in Figure 17.6. Which ring of the steroid structure undergoes the most substitution?

17.31 Cholesterol linoleate is found in the core of LDL. Draw its structural formula.

17.32 How does LDL deliver its cholesterol to the cells?

Steroid Hormones and Bile Salts

17.33 What physiological functions are associated with cortisol?

17.34 Estradiol in the body is synthesized starting from progesterone. What chemical modifications occur when estradiol is synthesized?

17.35 Describe the chemical difference between the male hormone testosterone and the female hormone estradiol.

***17.36** Considering that RU486 can bind to the receptors of progesterone as well as those of cortisone and cortisol, what can you say regarding the importance of the functional group on C-11 of the steroid ring in drug and receptor binding?

17.37 (a) How does the structure of RU486 resemble that of progesterone? (b) How do the structures differ?

17.38 Explain how the constant elimination of bile salts through the feces can reduce the danger of plaque formation in atherosclerosis.

Prostaglandins and Leukotrienes

17.39 What is the basic chemical difference (a) between arachidonic acid and prostaglandin PGE_2? (b) Between PGE_2 and $PGF_{2\alpha}$?

17.40 Find and name all the functional groups in (a) glycocholate (b) cortisone (c) prostaglandin PGE_2 (d) leukotriene B4.

17.41 What is the major structural difference between prostaglandins and leukotrienes?

Boxes

17.42 (Box 17A) (a) Do you think olestra is soluble in water? Why or why not? (b) Why can't humans digest olestra, when we can digest fats, which are also esters?

17.43 (Box 17B) What causes rancidity? How can it be prevented?

17.44 (Box 17C) What is the major structural difference between soaps and detergents?

17.45 (Box 17C) Explain how soap acts as a cleaning agent.

17.46 (Box 17C) Why have detergents replaced much of the soap used in the United States in recent years?

17.47 (Box 17E) (a) What is the role of sphingomyelin in the conductance of nerve signals? (b) What happens to this process in multiple sclerosis?

***17.48** (Box 17F) Comparing the complex lipid structures listed for the lipid storage diseases (Table 17F) with the missing or defective enzymes, explain why in Fabry's disease the missing enzyme is α-galactosidase and not β-galactosidase.

17.49 (Box 17G) How does lovastatin prevent atherosclerosis?

17.50 (Box 17H) How does the oral anabolic steroid methenolone differ structurally from testosterone?

17.51 (Box 17I) What is the role of progesterone and similar compounds in contraceptive pills?

17.52 (Box 17J) How does cortisone prevent inflammation?

17.53 (Box 17J) How does indomethacin act in the body to reduce inflammation?

17.54 (Box 17J) How does aspirin reduce inflammation and fever?

Additional Problems

17.55 (a) Where in the body is cholesterol synthesized? (b) In what form is cholesterol secreted from the body?

17.56 What is the role of taurine in lipid digestion?

17.57 Draw a schematic diagram of a lipid bilayer. Show how the bilayer prevents the passage by diffusion of a polar molecule such as glucose. Show why nonpolar molecules, such as $CH_3CH_2—O—CH_2CH_3$, can diffuse through the membrane.

17.58 What is the role of progesterone in pregnancy?

***17.59** Suggest a reason why free cholesterol forms gallstones but the various esters of cholesterol do not.

17.60 What are the constituents of sphingomyelin?

17.61 What structural feature do detergents (Box 17C) and the bile salt, taurocholate, have in common?

17.62 What is the cholesterol level in normal blood plasma?

17.63 (a) Classify aldosterone into as many chemical families as appropriate. (b) In what functional groups does aldosterone differ from cortisone?

***17.64** What is the major difference between aldosterone and all the other hormones listed in Figure 17.6?

17.65 How is cholesterol converted to cortisone?

***17.66** How many grams of H_2 are needed to saturate 100.0 g of a triglyceride made of glycerol and one unit each of lauric, oleic, and linoleic acids?

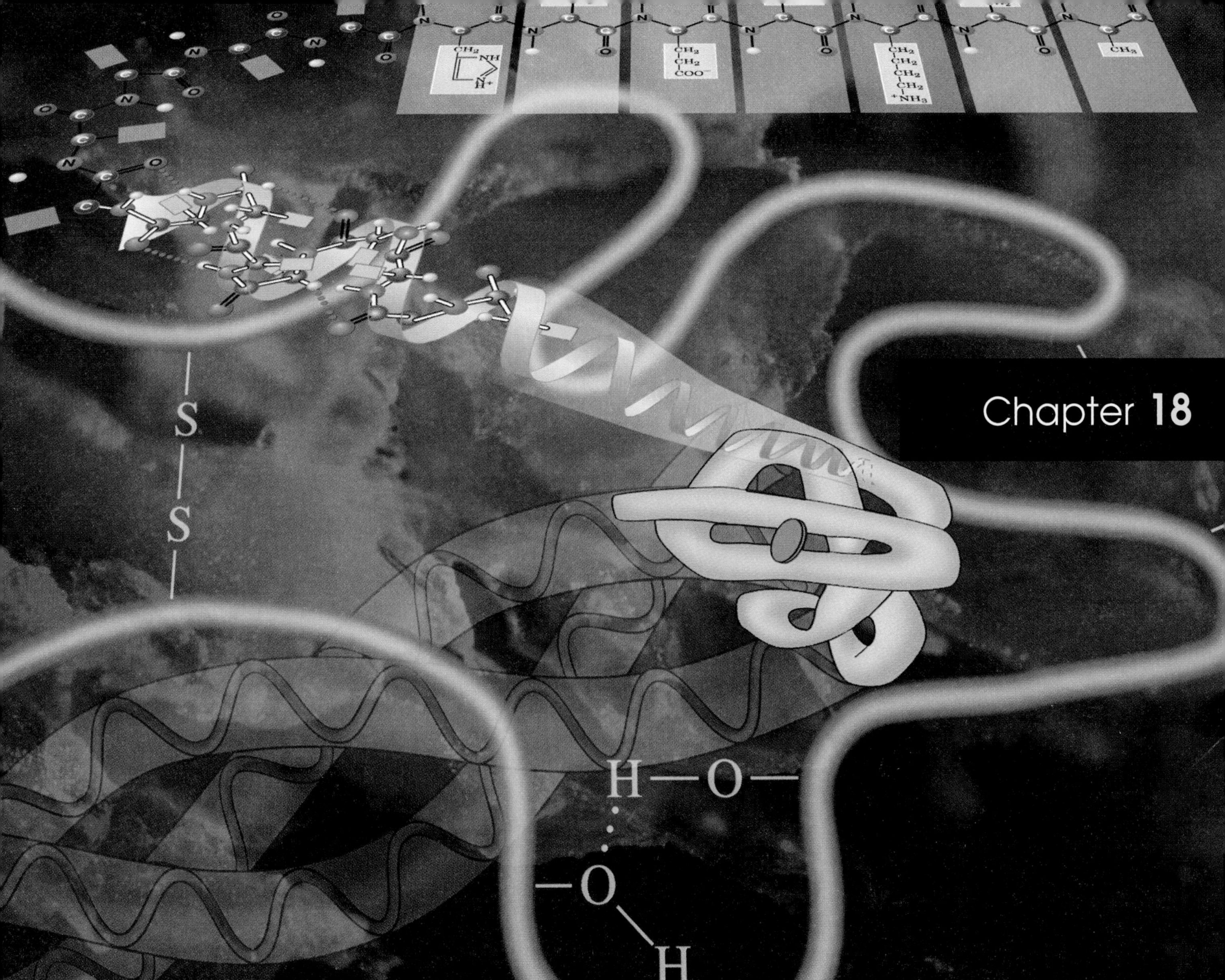

Chapter **18**

Proteins

18.1 Introduction

Proteins are by far the most important of all biological compounds. The very word "protein" is derived from the Greek *proteios,* meaning "of first importance," and the scientists who named these compounds more than 100 years ago chose an appropriate term. There are many types of proteins, and they perform many functions, some of which are as follows:

1. Structure We saw in Section 16.11 that the main structural material for plants is cellulose. For animals, it is structural proteins, which are the chief constituents of skin, bones, hair, and fingernails. Two important structural proteins are collagen and keratin.

2. Catalysis Virtually all the reactions that take place in living organisms are catalyzed by proteins called enzymes. Without enzymes, the reactions would take place so slowly as to be useless. We will discuss enzymes in Chapter 19.

The heart itself is a muscle, expanding and contracting about 70 to 80 times a minute.

3. Movement Every time we crook a finger, climb stairs, or blink an eye, we use our muscles. Muscle expansion and contraction are involved in every movement we make. Muscles are made up of protein molecules called myosin and actin.

4. Transport A large number of proteins fall into this category. Hemoglobin, a protein in the blood, carries oxygen from the lungs to the cells in which it is used and carbon dioxide from the cells to the lungs. Other proteins transport molecules across cell membranes.

5. Hormones Many hormones are proteins, among them insulin, oxytocin, and human growth hormone.

The antibodies are produced in the gamma globulin fraction of blood plasma.

6. Protection When a protein from an outside source or other foreign substance (called an antigen) enters the body, the body makes its own proteins (called antibodies) to counteract the foreign protein. This is the major mechanism the body uses to fight disease. Blood clotting is another protective device carried out by a protein, this one called fibrinogen. Without blood clotting, we would bleed to death from any small wound.

7. Storage Some proteins are used to store materials, in the way that starch and glycogen store energy. Examples are casein in milk and ovalbumin in eggs, which store nutrients for newborn mammals and birds. Ferritin, a protein in the liver, stores iron.

8. Regulation Some proteins control the expression of genes, and thereby regulate the kind of proteins manufactured in a particular cell, and control when such manufacture takes place.

These are not the only functions of proteins, but they are among the most important. It is very easy to see that any individual needs a great many proteins to carry out all these varied functions. A typical cell contains about 9000 different proteins; an individual human being has about 100 000 different proteins.

Collagen, actin, and keratin are some fibrous proteins. Albumin, hemoglobin, and immunoglobulins are some globular proteins.

We can divide proteins into two major types: **fibrous proteins,** which are insoluble in water and are used mainly for structural purposes, and **globular proteins,** which are more or less soluble in water and are used mainly for nonstructural purposes.

18.2 Amino Acids

Although there are so many different proteins, they all have basically the same structure: They are linear chains of amino acids. As its name implies, an **amino acid** is an **organic compound containing an amino group and an acid group.** Organic chemists can synthesize many thousands of amino acids, but nature is much more restrictive and uses only 20 different amino acids to make up proteins. Furthermore, all but one of the 20 fit the formula

$$\begin{array}{c} H \\ | \\ R{-}C{-}COOH \\ | \\ NH_2 \end{array}$$

$$\begin{array}{ll} H_2C{-}{-} & \overset{\displaystyle H}{\overset{|}{C}}{-}COOH \\ \;| & \;| \\ H_2C & NH \\ & \hspace{-2em}\underset{\displaystyle H_2}{C} \end{array}$$

Proline

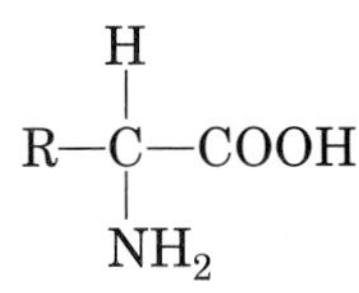

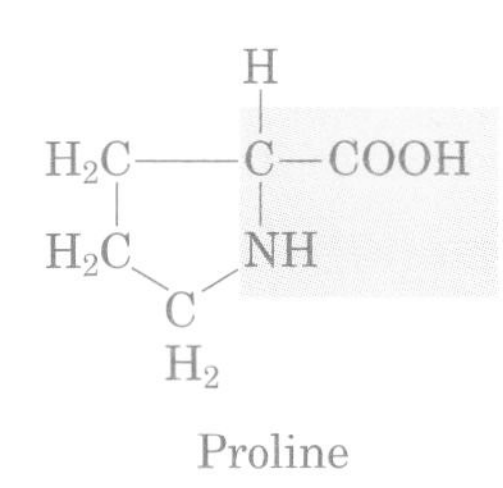

and even the one that doesn't fit the formula (proline) comes pretty close. Proline would fit except that it has a bond between the R and the N. The 20 amino acids found in proteins are called **alpha** amino acids. They are shown in Table 18.1, which also shows the one- and three-letter abbreviations that chemists use for them.

In these compounds, the $—NH_2$ is on the alpha carbon (the one next to the —COOH).

The one-letter abbreviations are more recent, but the three-letter abbreviations are still frequently used.

The most important aspect of the R groups is their polarity, and on this basis we can classify amino acids into the four groups shown in Table 18.1: nonpolar, polar but neutral, acidic, and basic. Note that the nonpolar side chains are *hydrophobic* (they repel water), whereas polar but neutral, acidic, and basic side chains are *hydrophilic* (attracted to water). This aspect of the R groups is very important in determining both the structure and the function of each protein molecule.

When we look at the general formula for the 20 amino acids

$$\begin{array}{c} H \\ | \\ R{-}C{-}COOH \\ | \\ NH_2 \end{array}$$

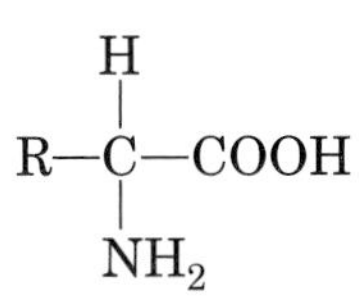

we see at once that all of them (except glycine, in which R = H) have a chiral carbon, since R, H, COOH, and NH_2 are four different groups. This means, as we saw in Section 16.2, that each of the amino acids (except glycine) exists as two enantiomers. As is the case for most examples of this kind, nature makes only one of the two possible forms for each amino acid, and it is virtually always the L form. Except for glycine, which exists in only one form, all the amino acids in all the proteins in your body are the L form. D amino acids are extremely rare in nature; some are found, for example, in the cell walls of a few types of bacteria.

Two of the amino acids in Table 18.1 have a second chiral carbon. Can you find them?

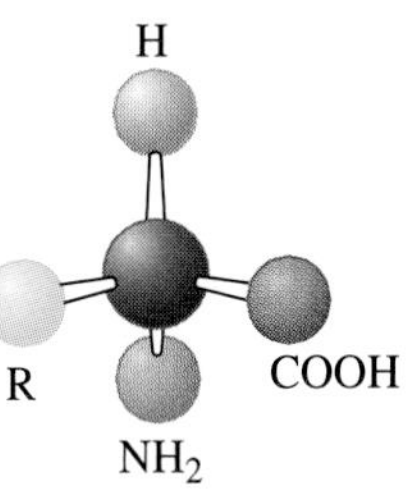

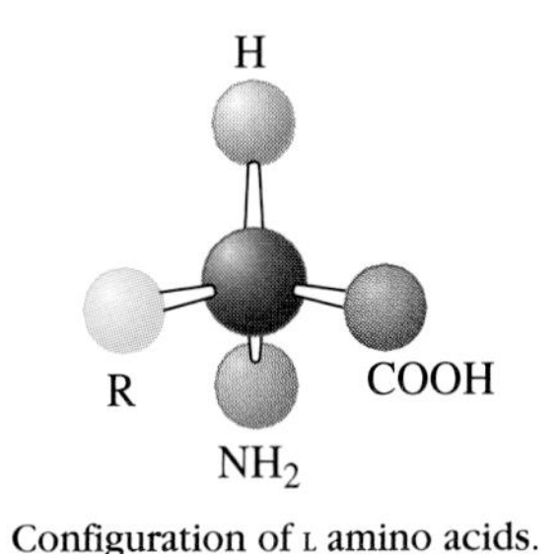

Configuration of L amino acids.

Table 18.1 The 20 Amino Acids Commonly Found in Proteins (both the three-letter and one-letter abbreviations are shown) and Their Isoelectric Points (pI)

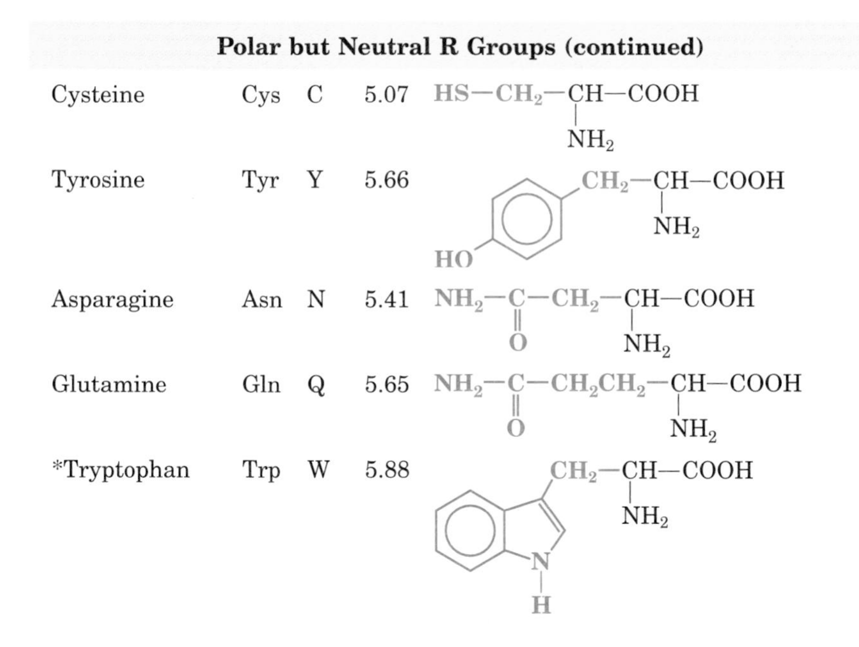

Nonpolar R Groups				
Glycine	Gly	G	5.97	$H-CH(NH_2)-COOH$
Alanine	Ala	A	6.01	$CH_3-CH(NH_2)-COOH$
*Valine[a]	Val	V	5.97	$CH_3-CH(CH_3)-CH(NH_2)-COOH$
*Leucine	Leu	L	5.98	$CH_3-CH(CH_3)-CH_2-CH(NH_2)-COOH$
*Isoleucine	Ile	I	6.02	$CH_3-CH_2-CH(CH_3)-CH(NH_2)-COOH$
Proline	Pro	P	6.48	—COOH, N—H
*Phenylalanine	Phe	F	5.48	$CH_2-CH(NH_2)-COOH$
*Methionine	Met	M	5.74	$CH_3-S-CH_2CH_2-CH(NH_2)-COOH$
Polar but Neutral R Groups				
Serine	Ser	S	5.68	$HO-CH_2-CH(NH_2)-COOH$
*Threonine	Thr	T	5.87	$CH_3-CH(OH)-CH(NH_2)-COOH$

18.3 Zwitterions

Up to now, we have shown the structural formula for amino acids as

$$R-\underset{NH_2}{\overset{H}{C}}-COOH$$

Polar but Neutral R Groups (continued)				
Cysteine	Cys	C	5.07	$HS-CH_2-CH(NH_2)-COOH$
Tyrosine	Tyr	Y	5.66	$HO-C_6H_4-CH_2-CH(NH_2)-COOH$
Asparagine	Asn	N	5.41	$NH_2-C(=O)-CH_2-CH(NH_2)-COOH$
Glutamine	Gln	Q	5.65	$NH_2-C(=O)-CH_2CH_2-CH(NH_2)-COOH$
*Tryptophan	Trp	W	5.88	(indol-3-yl)$-CH_2-CH(NH_2)-COOH$
Acidic R Groups				
Glutamic acid	Glu	E	3.22	$HO-C(=O)-CH_2CH_2-CH(NH_2)-COOH$
Aspartic acid	Asp	D	2.77	$HO-C(=O)-CH_2-CH(NH_2)-COOH$
Basic R Groups				
*Lysine	Lys	K	9.74	$NH_2-CH_2CH_2CH_2CH_2-CH(NH_2)-COOH$
*Arginine	Arg	R	10.76	$NH_2-C(=NH)-NH-CH_2CH_2CH_2-CH(NH_2)-COOH$
*Histidine	His	H	7.59	(imidazol-4-yl, N–H)$-CH_2-CH(NH_2)-COOH$

[a]Those marked with an asterisk are essential amino acids (see Sec. 26.4).

But in Section 14.7 we learned that carboxylic acids, RCOOH, cannot exist in the presence of a moderately weak base (such as NH_3). They donate a proton to become carboxylate ions, $RCOO^-$. Likewise, amines, RNH_2 (Sec. 15.4), cannot exist as such in the presence of a moderately weak acid (such as acetic acid). They gain a proton to become substituted ammonium ions, RNH_3^+.

An amino acid has —COOH and —NH_2 groups in the same molecule. Therefore, in water solution the —COOH donates a proton to the —NH_2, so that an amino acid actually has the structure

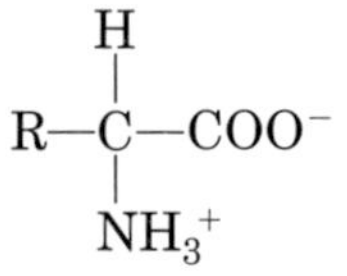

$$\mathrm{R{-}\overset{\displaystyle H}{\underset{\displaystyle NH_3^+}{C}}{-}COO^-}$$

"Zwitterion" comes from the German *zwitter*, meaning hybrid.

Compounds that have a positive charge on one atom and a negative charge on another are called **zwitterions.** Amino acids are zwitterions, not only in water solution but in the solid state as well. They are therefore ionic compounds, that is, internal salts. *Un-ionized RCH(NH$_2$)COOH molecules do not actually exist, in any form.*

The fact that amino acids are zwitterions explains their physical properties, which would otherwise be quite puzzling. All of them are solids with high melting points (for example, glycine melts at 262°C). This is just what we expect of ionic compounds (Table 10.1). The 20 amino acids are also fairly soluble in water, as ionic compounds generally are; if they had no charges, only the smaller ones would be expected to be soluble.

If we add an amino acid to water, it dissolves and then has the same zwitterionic structure that it has in the solid state. Let us see what happens if we change the pH of the solution, as we can easily do by adding a source of H_3O^+, such as HCl (to lower the pH), or a strong base such as NaOH (to raise the pH). Since H_3O^+ is a stronger acid than the typical carboxylic acid (Sec. 8.3), it donates a proton to the —COO^- group,

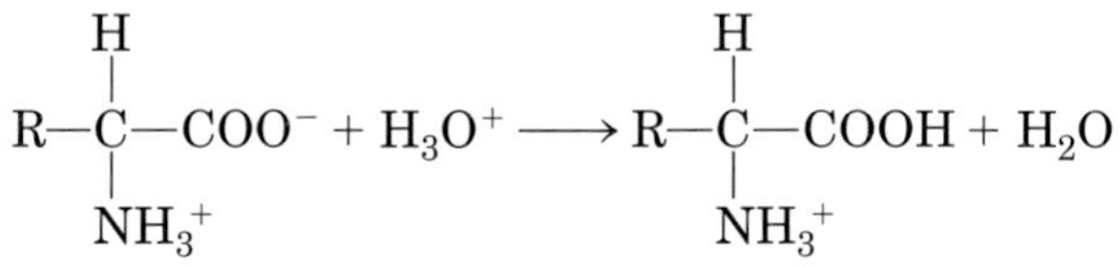

$$\mathrm{R{-}\overset{\displaystyle H}{\underset{\displaystyle NH_3^+}{C}}{-}COO^- + H_3O^+ \longrightarrow R{-}\overset{\displaystyle H}{\underset{\displaystyle NH_3^+}{C}}{-}COOH + H_2O}$$

turning the zwitterion into a positive ion. This happens to all amino acids if the pH is sufficiently lowered.

Addition of OH^- to the zwitterion causes the —NH_3^+ to donate its proton,

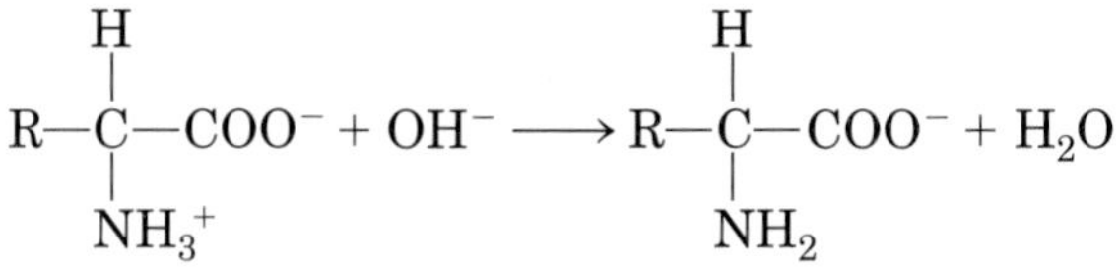

$$\mathrm{R{-}\overset{\displaystyle H}{\underset{\displaystyle NH_3^+}{C}}{-}COO^- + OH^- \longrightarrow R{-}\overset{\displaystyle H}{\underset{\displaystyle NH_2}{C}}{-}COO^- + H_2O}$$

turning the zwitterion into a negative ion. This happens to all amino acids if the pH is sufficiently raised.

Note that in both cases *the amino acid is still an ion,* so that it is still soluble in water. There is no pH at which an amino acid has no ionic character at all. If the amino acid is a positive ion at low pH and a negative ion at high pH, there must be some **pH at which all the molecules are in the zwitterionic form.** This pH is called the **isoelectric point** (symbol **pI**).

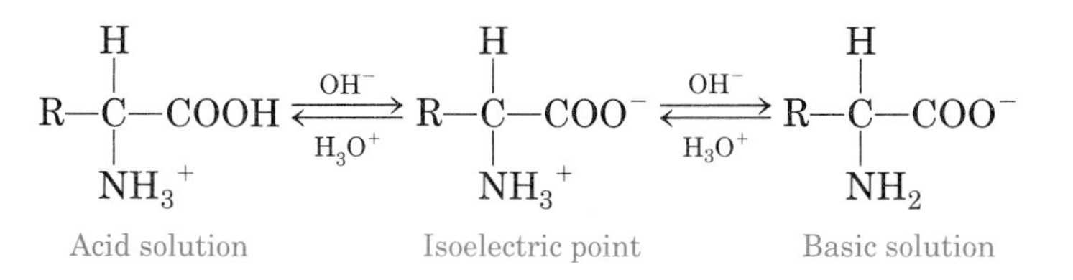

Every amino acid has a different isoelectric point, although most of them are not very far apart (see the values in Table 18.1). Fifteen of the

20 have isoelectric points near 6. However, the three basic amino acids have higher isoelectric points, and the two acidic amino acids have lower values.

At or near the isoelectric point, amino acids exist in aqueous solution largely or entirely as zwitterions. As we have seen, they react with either a strong acid, by taking a proton (the $—COO^-$ becomes $—COOH$), or a strong base, by giving a proton (the $—NH_3^+$ becomes $—NH_2$). To summarize:

$$\underset{\text{Acid solution}}{R{-}\overset{\displaystyle H}{\underset{\displaystyle NH_3^+}{C}}{-}COOH} \underset{H_3O^+}{\overset{OH^-}{\rightleftharpoons}} \underset{\text{Isoelectric point}}{R{-}\overset{\displaystyle H}{\underset{\displaystyle NH_3^+}{C}}{-}COO^-} \underset{H_3O^+}{\overset{OH^-}{\rightleftharpoons}} \underset{\text{Basic solution}}{R{-}\overset{\displaystyle H}{\underset{\displaystyle NH_2}{C}}{-}COO^-}$$

We learned in Section 8.3 that a compound that is both an acid and a base is called *amphoteric*. We also learned, in Section 8.10, that a solution that neutralizes both acid and base is a *buffer solution*. Amino acids are therefore amphoteric compounds, and aqueous solutions of them are buffers.

Proteins also have isoelectric points, and also act as buffers. This will be discussed in Section 18.6.

18.4 Cysteine: A Special Amino Acid

One of the 20 amino acids in Table 18.1 has a chemical property not shared by any of the others. This amino acid is cysteine. It can easily be dimerized by many mild oxidizing agents:

We met these reactions in Section 12.8.

$$\underset{\text{Cysteine}}{2HS{-}CH_2{-}\underset{\displaystyle NH_2}{CH}{-}COOH} \underset{[H]}{\overset{[O]}{\rightleftharpoons}} \underset{\text{Cystine}}{HOOC{-}\underset{\displaystyle NH_2}{CH}{-}CH_2{-}S{-}S{-}CH_2{-}\underset{\displaystyle NH_2}{CH}{-}COOH} + H_2O$$

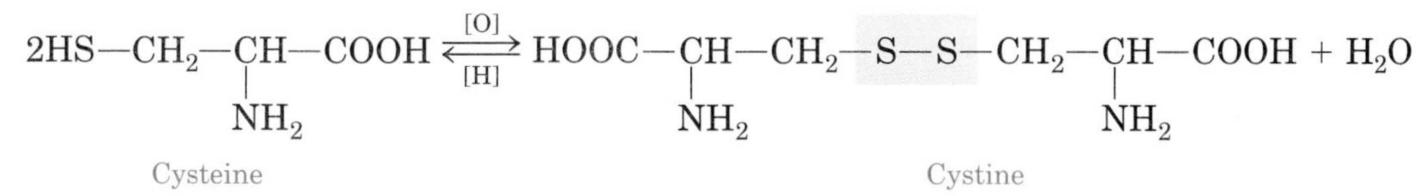

The dimer of cysteine, which is called **cystine,** can in turn be fairly easily reduced to give two molecules of cysteine. As we shall see, the presence of cystine has important consequences for the chemical structure and shape of protein molecules it is part of. **The S—S bond** (shown in color) is called a **disulfide linkage.**

A dimer is a molecule made up of two identical units.

18.5 Peptides and Proteins

Each amino acid has a carboxylic acid group and an amino group. In Section 15.5 we saw that a carboxylic acid and an amine could be combined to form an amide:

$$\underset{\text{Carboxylic acid}}{R{-}\underset{\displaystyle O}{\underset{\|}{C}}{-}OH} + \underset{\text{Amine}}{R'NH_2} \xrightarrow[-H_2O]{} \underset{\text{Amide}}{R{-}\underset{\displaystyle O}{\underset{\|}{C}}{-}NH{-}R'}$$

In the same way, it is possible for the COOH group of one amino acid molecule, say glycine, to combine with the amino group of a second molecule, say alanine:

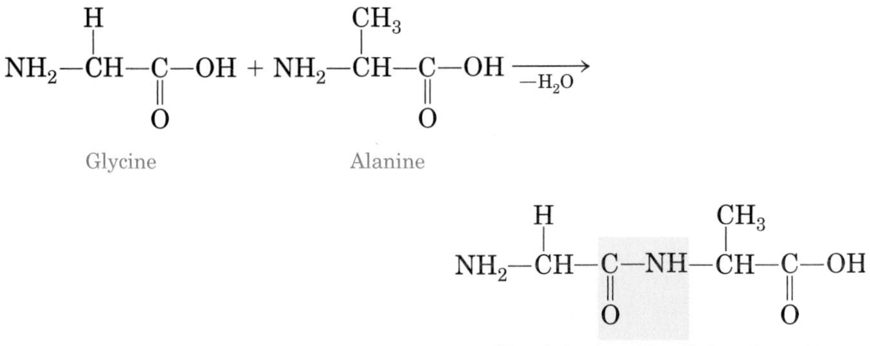

$$\underset{\text{Glycine}}{NH_2\text{—}\overset{\overset{\large H}{|}}{C}H\text{—}\underset{\underset{\large O}{\|}}{C}\text{—}OH} + \underset{\text{Alanine}}{NH_2\text{—}\overset{\overset{\large CH_3}{|}}{C}H\text{—}\underset{\underset{\large O}{\|}}{C}\text{—}OH} \xrightarrow[-H_2O]{}$$

$$NH_2\text{—}\overset{\overset{\large H}{|}}{C}H\text{—}\underset{\underset{\large O}{\|}}{C}\text{—}NH\text{—}\overset{\overset{\large CH_3}{|}}{C}H\text{—}\underset{\underset{\large O}{\|}}{C}\text{—}OH$$

Glycylalanine (Gly-Ala), a dipeptide

The $\text{—}\underset{\underset{O}{\|}}{C}\text{—}NH\text{—}$ is called the **peptide linkage.**

The synthesis of peptide bonds in cells is catalyzed by enzymes.

This reaction takes place in the cells by a mechanism that we shall examine in Section 23.10. The product is an amide made up of two amino acids joined together, called a **dipeptide.**

It is important to realize that glycine and alanine could also be linked the other way:

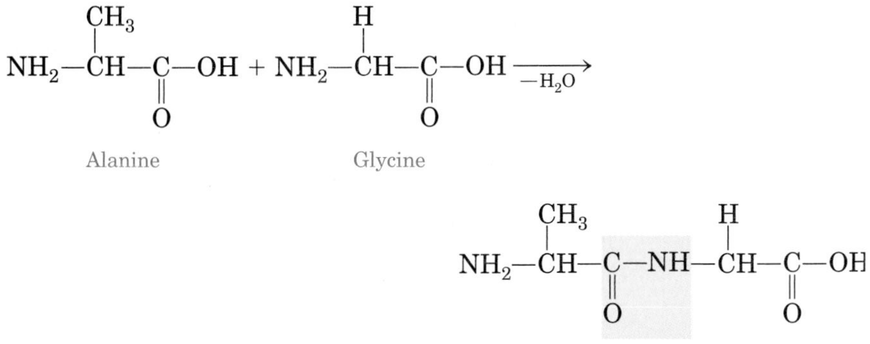

$$\underset{\text{Alanine}}{NH_2\text{—}\overset{\overset{\large CH_3}{|}}{C}H\text{—}\underset{\underset{\large O}{\|}}{C}\text{—}OH} + \underset{\text{Glycine}}{NH_2\text{—}\overset{\overset{\large H}{|}}{C}H\text{—}\underset{\underset{\large O}{\|}}{C}\text{—}OH} \xrightarrow[-H_2O]{}$$

$$NH_2\text{—}\overset{\overset{\large CH_3}{|}}{C}H\text{—}\underset{\underset{\large O}{\|}}{C}\text{—}NH\text{—}\overset{\overset{\large H}{|}}{C}H\text{—}\underset{\underset{\large O}{\|}}{C}\text{—}OH$$

Alanylglycine (Ala-Gly), a dipeptide

Note that in Ala-Gly the $\text{—}NH_2$ is connected to a $\text{—}CHCH_3$, while in Gly-Ala the $\text{—}NH_2$ is connected to a $\text{—}CH_2$.

In this case we get a *different* dipeptide. The two dipeptides are structural isomers, of course, but they are different compounds in all respects, with different properties.

EXAMPLE

Show how to form the dipeptide aspartylserine (Asp-Ser).

Answer • The name implies that this dipeptide is made of two amino acids, aspartic acid (Asp) and serine (Ser), with the amide being formed between the carboxyl group of aspartic acid and the amino group of serine. Therefore, we write the formula of aspartic acid with its amino group on the left side. Next we place the formula of serine to the right, with its amino group facing the carboxyl group of aspartic acid:

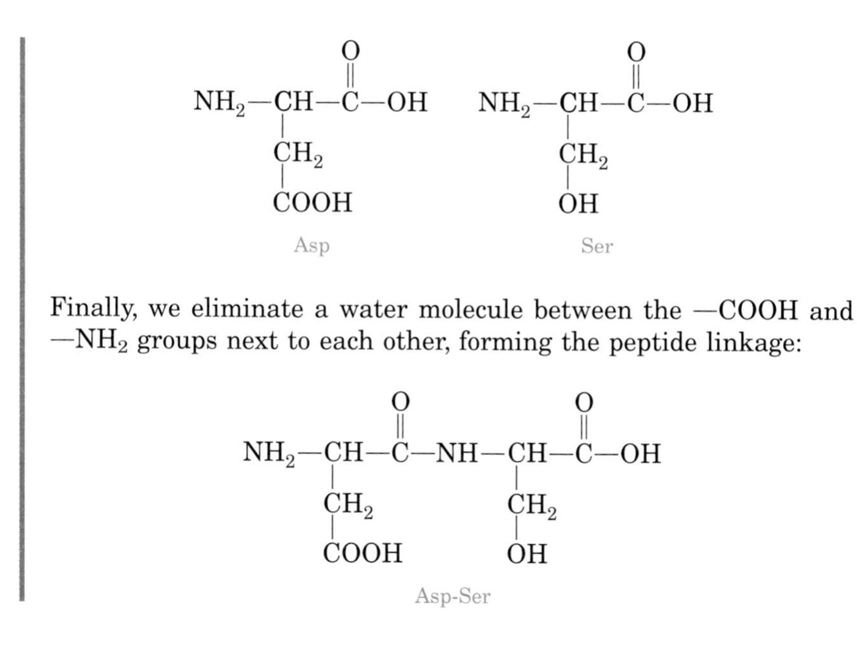

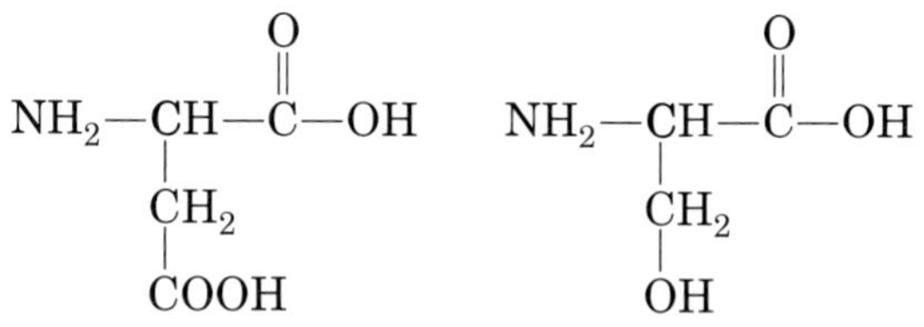

$$\underset{\text{Asp}}{NH_2\text{—}\underset{\underset{COOH}{|}}{\underset{|}{CH}}\text{—}\overset{O}{\overset{\|}{C}}\text{—}OH} \quad \text{(side chain: } CH_2\text{—}COOH) \qquad \underset{\text{Ser}}{NH_2\text{—}CH\text{—}\overset{O}{\overset{\|}{C}}\text{—}OH} \quad \text{(side chain: } CH_2\text{—}OH)$$

Asp Ser

Finally, we eliminate a water molecule between the —COOH and —NH_2 groups next to each other, forming the peptide linkage:

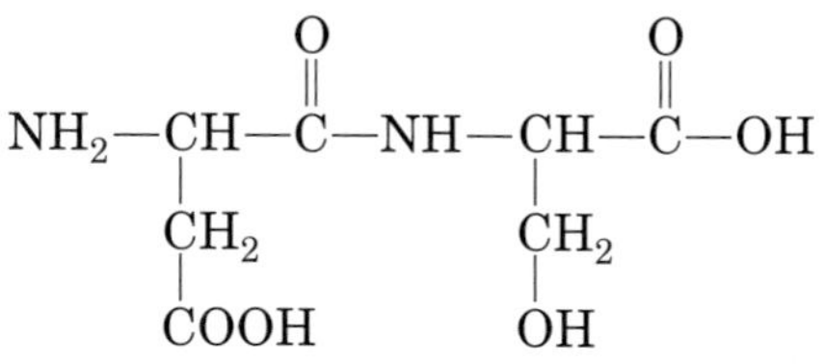

$$NH_2\text{—}\underset{\underset{COOH}{|}}{\underset{CH_2}{\underset{|}{CH}}}\text{—}\overset{O}{\overset{\|}{C}}\text{—}NH\text{—}\underset{\underset{OH}{|}}{\underset{CH_2}{\underset{|}{CH}}}\text{—}\overset{O}{\overset{\|}{C}}\text{—}OH$$

Asp-Ser

Problem 18.1

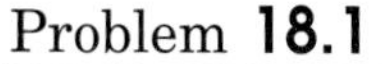

Show how to form the dipeptide valylphenylalanine (Val-Phe).

Any two amino acids, the same or different, can be linked together to form dipeptides in a similar manner. Nor does it end there. Each dipeptide still contains a COOH and an amino group. We can therefore add a third amino acid to alanylglycine, say lysine:

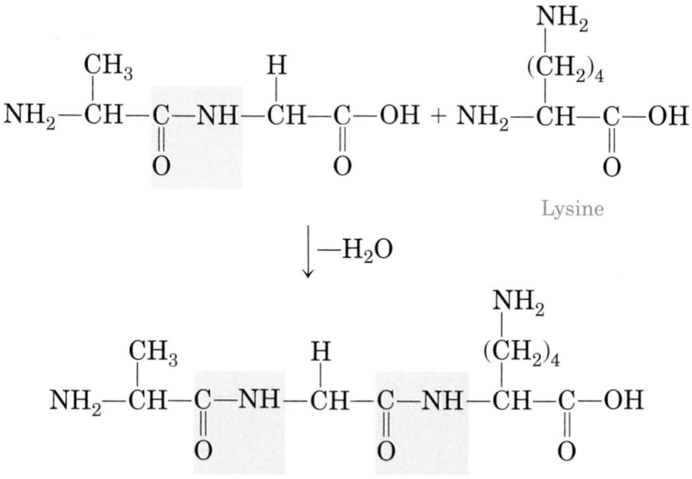

$$NH_2\text{—}\overset{CH_3}{\overset{|}{CH}}\text{—}\underset{O}{\underset{\|}{C}}\text{—}NH\text{—}\overset{H}{\overset{|}{CH}}\text{—}\underset{O}{\underset{\|}{C}}\text{—}OH + NH_2\text{—}\overset{\overset{NH_2}{|}}{\overset{(CH_2)_4}{\overset{|}{CH}}}\text{—}\underset{O}{\underset{\|}{C}}\text{—}OH$$

Lysine

$$\downarrow -H_2O$$

$$NH_2\text{—}\overset{CH_3}{\overset{|}{CH}}\text{—}\underset{O}{\underset{\|}{C}}\text{—}NH\text{—}\overset{H}{\overset{|}{CH}}\text{—}\underset{O}{\underset{\|}{C}}\text{—}NH\text{—}\overset{\overset{NH_2}{|}}{\overset{(CH_2)_4}{\overset{|}{CH}}}\text{—}\underset{O}{\underset{\|}{C}}\text{—}OH$$

Alanylglycyllysine (Ala-Gly-Lys), a tripeptide

The product is a **tripeptide.** Since it too contains a COOH and an NH_2 group, we can continue the process to get a tetrapeptide, a pentapeptide, and so on until we have a chain of hundreds or even thousands of amino acids. These **chains of amino acids** are the **proteins** that serve so many important functions in living organisms.

Box 18A

Artificial Sweeteners

Many people restrict their sugar intake. Some are forced to do so by diseases such as diabetes, others by the desire to lose weight. Since most of us like to eat sweet foods, artificial sweeteners are added to many foods and drinks for those who must (or want to) restrict their sugar intake. Three noncaloric artificial sweeteners are now approved by the Food and Drug Administration. The oldest of these, saccharin, is 450 times sweeter than sucrose (Table 16.1), yet it has very little caloric content. Saccharin has been in use for about a hundred years. Unfortunately, some tests have shown that saccharin, when fed in massive quantities to rats, caused some of these rats to develop cancer. Although other tests have given negative results, the question about the safety of saccharin is still not settled, though it remains on the market.

Saccharin   Aspartame   Acesulfame-K

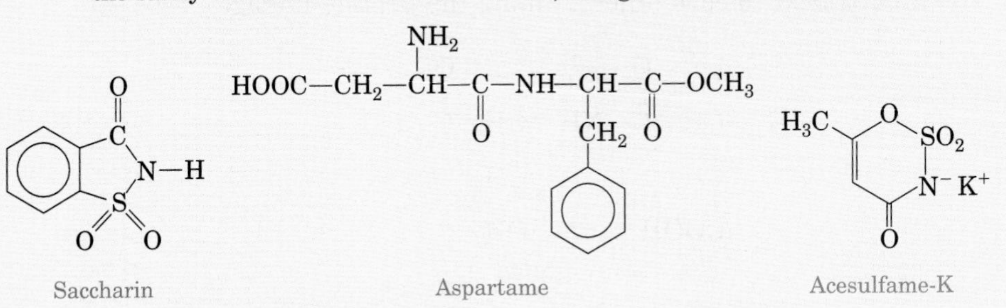

A newer artificial sweetener, aspartame, does not have the slight aftertaste that saccharin does. Aspartame is the methyl ester of a simple dipeptide, aspartylphenylalanine (Asp-Phe). Its sweetness was discovered in 1969, and after extensive biological testing it was approved by the U.S. Food and Drug Administration in 1981 for use in cold cereals, drink mixes, and gelatins and as tablets or powder to be used as a sugar substitute. Aspartame is 100 to 150 times sweeter than sucrose. It is made from natural amino acids, so that both the aspartic acid and the phenylalanine have the L configuration. The other possibilities have also been synthesized: the L-D, the D-L, and the D-D. They are all bitter rather than sweet. This is another example of the principle that, when a biological organism uses or makes a compound that has chiral carbons, in most cases only one stereoisomer is used or made. Aspartame is sold under such brand names as Equal and NutraSweet. The third artificial sweetener, acesulfame-K, is the most recent (approved in 1988). It is 200 times sweeter than sucrose and is used (under the name Sunette) in dry mixes. It is not metabolized in the body; that is, it passes through unchanged.

These are but a few of the many products that contain aspartame. (Photograph by Charles D. Winters.)

A very important tripeptide, present in high concentrations in all tissues, is **glutathione.** It contains L-glutamic acid, L-cysteine, and glycine. Its structure is unusual because the glutamic acid is linked to the cysteine by its γ-carboxyl group rather than by the α-carboxyl group as is usual in most peptides and proteins:

$$\underset{\text{Glu}}{NH_3^+{-}\underset{\displaystyle COO^-}{CH}{-}CH_2{-}CH_2{-}\underset{\displaystyle O}{\overset{\|}{C}}}{-}NH{-}\underset{\text{Cys}}{\underset{\displaystyle \underset{\displaystyle SH}{CH_2}}{CH}{-}\underset{\displaystyle O}{C}}{-}\underset{\text{Gly}}{NH{-}CH_2{-}COO^-}$$

Glutathione functions in the cells as a general protective agent. Oxidizing agents that would damage the cells, such as peroxides, will oxidize glutathione instead (the cysteine portion; see also Sec. 18.4), thus protecting the proteins and nucleic acids. Many foreign chemicals also get attached to glutathione, and in this sense it acts as a detoxifying agent.

A word must be said about the terms used to describe these compounds. The shortest chains are often simply called **peptides,** longer ones are **polypeptides,** and still longer ones are **proteins,** but chemists differ about where to draw the line. Many chemists use the terms "polypeptide" and "protein" almost interchangeably. We shall consider a protein to be a polypeptide chain that contains a minimum of 30 to 50 amino acids.

The amino acids in a chain are often called **residues.** It is customary to use either the one-letter or the three-letter abbreviations shown in Table 18.1 to represent peptides and proteins. For example, the tripeptide shown on page 599, alanylglycyllysine, is AGK or Ala-Gly-Lys. The amino acid residue with the free COOH group is called the **C-terminal** one (lysine in Ala-Gly-Lys), and the amino acid residue with the free amino group is the **N-terminal** one (alanine in Ala-Gly-Lys). It is the universal custom to write peptide and protein chains with the N-terminal residue on the left. No matter how long a protein chain gets—hundreds or thousands of units—it always has just two ends: one C-terminal and one N-terminal.

The naming of peptides also begins with the N-terminal amino acid.

18.6 Some Properties of Peptides and Proteins

The continuing pattern of

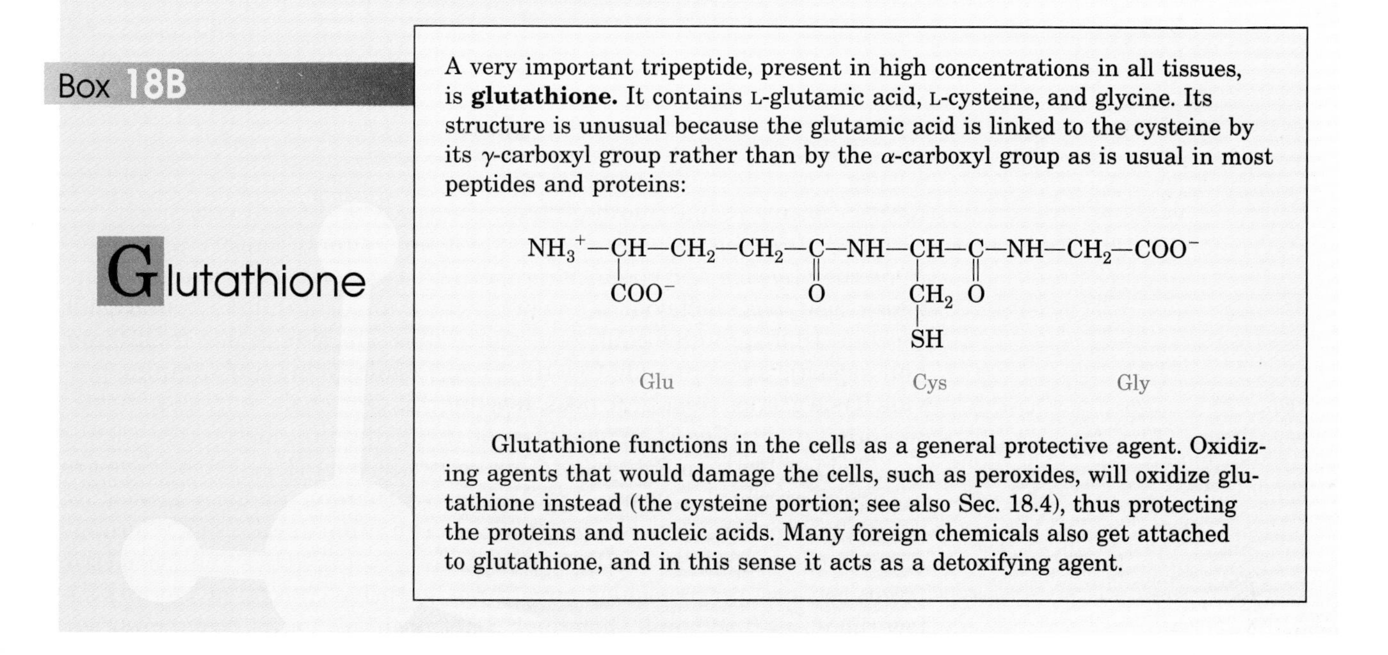

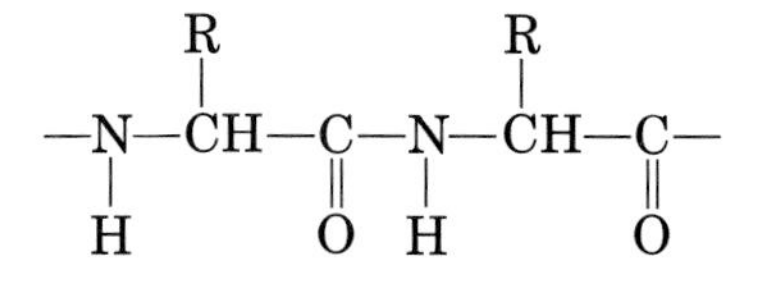

$$-\underset{\displaystyle H}{N}-\overset{\displaystyle R}{CH}-\underset{\displaystyle O}{\overset{\|}{C}}-\underset{\displaystyle H}{N}-\overset{\displaystyle R}{CH}-\underset{\displaystyle O}{\overset{\|}{C}}-$$

is called the **backbone** of the peptide or protein molecule; the R groups are called the **side chains.** The six atoms of the peptide linkage

$$\begin{array}{c} -\mathrm{C}-\underset{\underset{\mathrm{O}}{\|}}{\mathrm{C}}-\underset{\underset{\mathrm{H}}{|}}{\mathrm{N}}-\mathrm{C}- \end{array}$$

are rigid and lie in the same plane, and two adjacent peptide bonds can rotate relative to one another about the C—N and C—C bonds.

There is free rotation about most single bonds, but this is an exception because this resonance form:

$$-\mathrm{C}-\underset{\underset{^{-}\mathrm{O}}{|}}{\mathrm{C}}=\underset{\underset{\mathrm{H}}{|}}{\overset{+}{\mathrm{N}}}-\mathrm{C}-$$

contributes to the structure of the molecule (see Sec. 11.9), giving the C—N bond partial double bond character.

The 20 different amino acid side chains supply variety and determine the physical and chemical properties of proteins. Among these properties, acid–base behavior is one of the most important. Like amino acids (Section 18.3), proteins behave as zwitterions. The side chains of glutamic and aspartic acids provide COOH groups, while lysine and arginine provide basic groups (histidine does, too, but this side chain is less basic than the other two). (See the structures of these amino acids in Table 18.1.)

The terminal —COOH and $—NH_2$ groups also ionize, but these are only 2 out of 50 or more residues.

The isoelectric point of a protein occurs at the pH at which there are an equal number of positive and negative charges. At their isoelectric points, proteins behave like zwitterions: The number of $—COO^-$ groups equals the number of $—NH_3^+$ groups. At any pH above the isoelectric point, the protein molecules have a net negative charge; at any pH below the isoelectric point, they have a net positive charge. Some proteins, such as hemoglobin, have an almost equal number of acidic and basic groups; the isoelectric point of hemoglobin is at pH 6.8. Others, like serum albumin, have more acidic groups than basic groups; the isoelectric point of this protein is 4.9. In each case, however, because proteins behave like zwitterions, they act as buffers, for example, in the blood (Fig. 18.1).

Carbonates and phosphates are also blood buffers (Sec. 8.10).

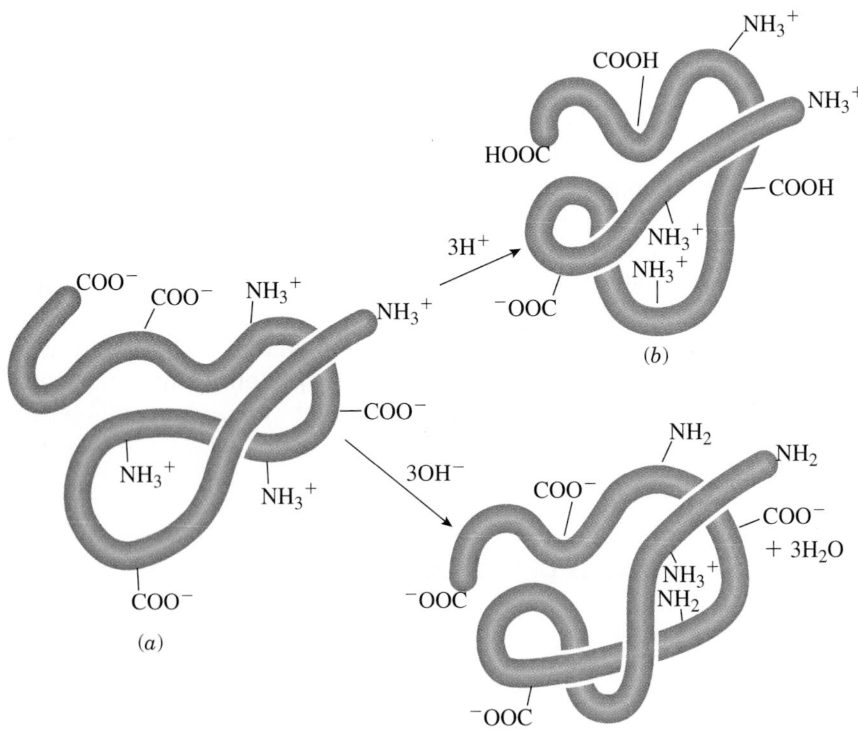

Figure **18.1**
Schematic diagram of a protein (a) at its isoelectric point and its buffering action when (b) H^+ or (c) OH^- ions are added.

The water-solubility of large molecules such as proteins often depends on the repulsive forces between like charges on their surfaces. When protein molecules are at a pH at which they have a net positive or negative charge, the presence of these like charges causes the protein molecules to repel each other. These repulsive forces are smallest at the isoelectric point, when the net charges are close to zero. When there are no repulsive forces, the protein molecules tend to clump together to form aggregates of two or more molecules, reducing their solubility. Therefore, *proteins are least soluble in water at their isoelectric points and can be precipitated from their solutions.*

We pointed out in Section 18.1 that proteins have many functions. In order to understand these functions, we must look at four levels of organization in their structures. The *primary structure* describes the linear sequence of amino acids in the polypeptide chain. *Secondary structure* refers to certain repeating patterns, such as the α-helix conformation or the pleated sheet (see Figs. 18.5 and 18.6). The *tertiary structure* describes the overall conformation of the polypeptide chain. A good analogy for all this is a coiled telephone cord (Fig. 18.2). The primary structure is the stretched-out cord. The secondary structure is the coil in the form of a helix. We can take the entire coil and twist it into various shapes. Any structure made by doing this is a tertiary structure. As we shall see, protein molecules twist and curl in a very similar manner.

Quaternary structure applies only to proteins with more than one chain and has to do with how the different chains are spatially related to each other.

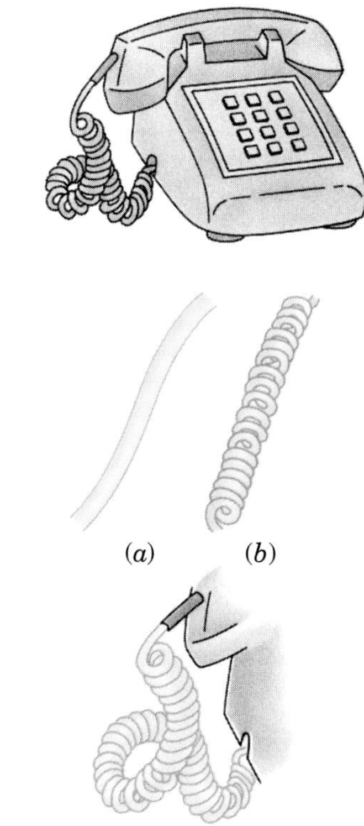

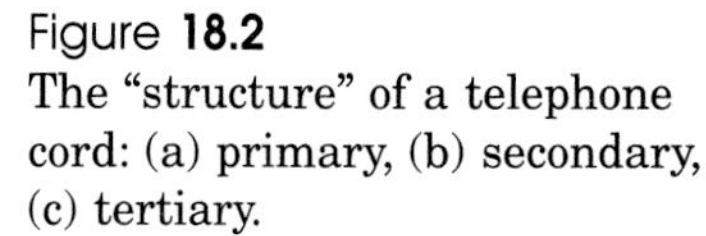

Figure **18.2**
The "structure" of a telephone cord: (a) primary, (b) secondary, (c) tertiary.

18.7 The Primary Structure of Proteins

Very simply, the **primary structure** of a protein consists of the sequence of amino acids that make up the chain. Each of the very large number of peptide and protein molecules in biological organisms has a different sequence of amino acids—and it is that sequence that allows the protein to carry out its function, whatever it may be.

Is it possible that so many different proteins can arise from different sequences of only 20 amino acids? Let us look at a little arithmetic, starting with a dipeptide. How many different dipeptides can be made from 20 amino acids? There are 20 possibilities for the N-terminal amino acid, and for each of these 20 there are 20 possibilities for the C-terminal amino acid. This means that there are

$$20 \times 20 = 400$$

different dipeptides possible from the 20 amino acids. What about tripeptides? We can form a tripeptide by taking any of the 400 dipeptides and adding any of the 20 amino acids. Thus there are

$$20 \times 20 \times 20 = 8000$$

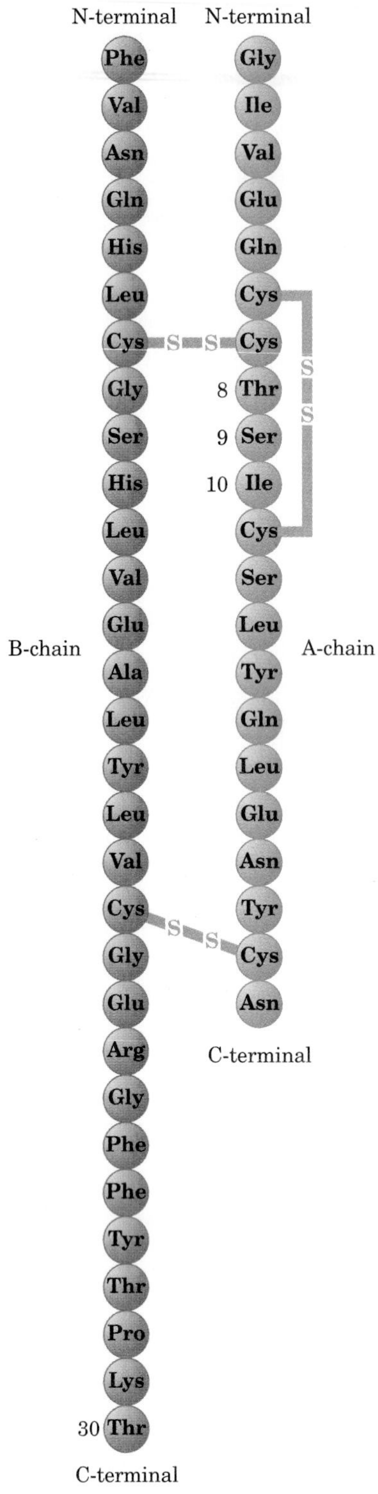

Figure **18.3**
The primary structure (amino acid sequence) of human insulin.

tripeptides, all different. It is easy to see that we can calculate the total number of possible peptides or proteins for a chain of n amino acids simply by raising 20 to the nth power (20^n).

Taking a typical small protein to be one with 60 amino acid residues, the number of proteins that can be made from the 20 amino acids is $20^{60} = 10^{78}$. This is an enormous number, possibly greater than the total number of atoms in the universe. It is clear that only a tiny fraction of all possible protein molecules has ever been made by biological organisms.

Each peptide or protein in the body has its own sequence of amino acids. We mentioned that proteins also have secondary, tertiary, and in some cases also quaternary structures. We will deal with these in Sections 18.8 and 18.9, but here we can say that **the primary structure of a protein determines to a large extent the native** (most frequently occurring) **secondary, tertiary, and quaternary structures.** That is, it is the particular sequence of amino acids on the chain that enables the whole chain to fold and curl in such a way as to assume its final shape. As we shall see in Section 18.10, without its particular three-dimensional shape, a protein cannot function.

Just how important is the exact amino acid sequence? Can a protein perform the same function if its sequence is a little different? The answer to this question is that a change in amino acid sequence may or may not matter, depending on what kind of a change it is. As an example, the enzyme ribonuclease is a protein chain consisting of 124 amino acid residues. It is possible, by means of another enzyme called carboxypeptidase, to remove the C-terminal amino acid residue of ribonuclease, leaving the rest of the chain intact. When this is done, the 124th amino acid, valine, is removed, leaving a chain of 123 units. This modified chain has the same biological activity as the original ribonuclease. In this case, removal of one amino acid makes no difference.

Another example is the hormone *insulin.* Human insulin consists of two chains having a total of 51 amino acids, connected by disulfide linkages. The sequence of amino acids is shown in Figure 18.3. Insulin is necessary for proper utilization of carbohydrates (Sec. 21.2), and people with severe diabetes (Box 24G) must take insulin injections. The amount of human insulin available is far too small to meet the need, so bovine insulin (from cattle) or insulin from hogs or sheep is used instead. Insulin from these sources is similar to human insulin, but not identical. The differences are entirely in the 8, 9, and 10 positions of the A chain and the C-terminal position (30) of the B chain:

	A Chain	**B Chain**
	8 9 10	***30***
Human	—Thr—Ser—Ile—	—Thr
Bovine	—Ala—Ser—Val—	—Ala
Hog	—Thr—Ser—Ile—	—Ala
Sheep	—Ala—Gly—Val—	—Ala

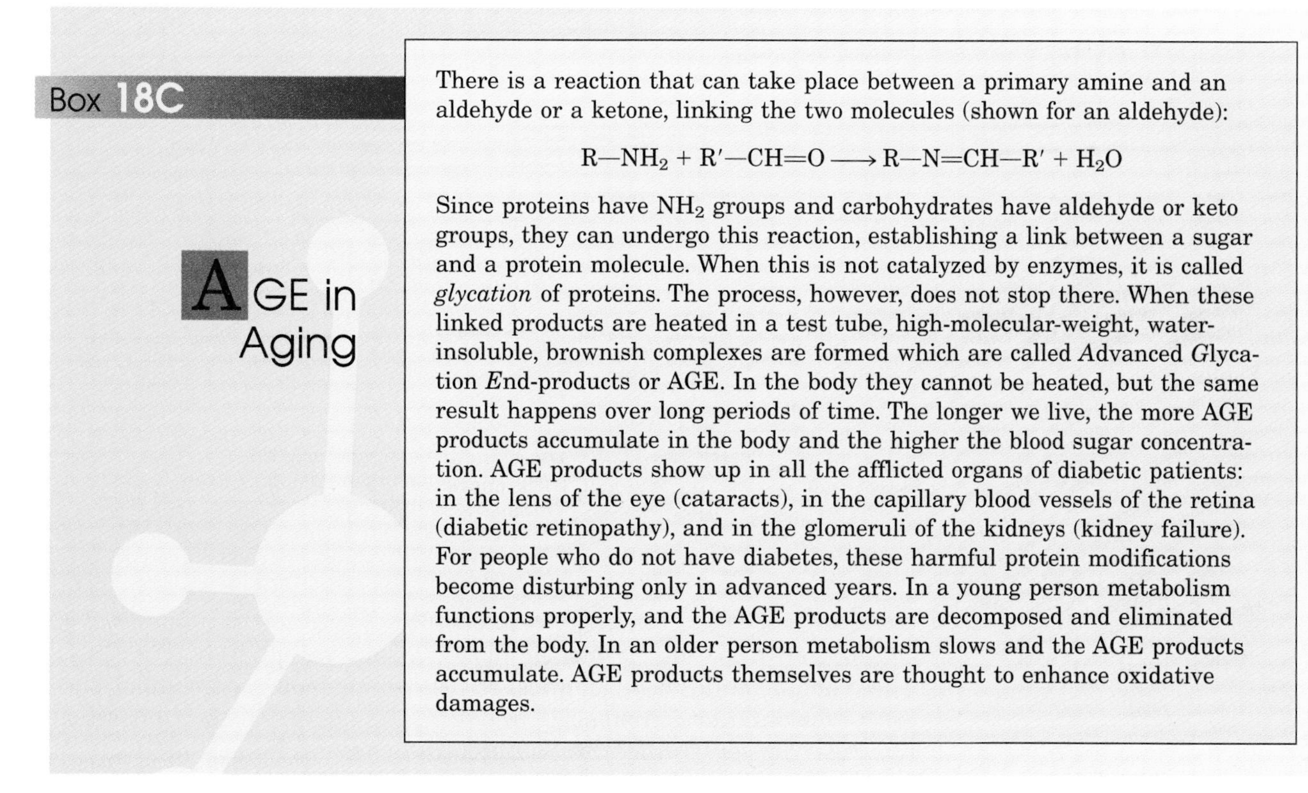

The remainder of the molecule is the same in all four varieties of insulin. Despite the slight differences in structure, all these insulins can be used

Box 18C

AGE in Aging

There is a reaction that can take place between a primary amine and an aldehyde or a ketone, linking the two molecules (shown for an aldehyde):

$$R{-}NH_2 + R'{-}CH{=}O \longrightarrow R{-}N{=}CH{-}R' + H_2O$$

Since proteins have NH_2 groups and carbohydrates have aldehyde or keto groups, they can undergo this reaction, establishing a link between a sugar and a protein molecule. When this is not catalyzed by enzymes, it is called *glycation* of proteins. The process, however, does not stop there. When these linked products are heated in a test tube, high-molecular-weight, water-insoluble, brownish complexes are formed which are called *A*dvanced *G*lycation *E*nd-products or AGE. In the body they cannot be heated, but the same result happens over long periods of time. The longer we live, the more AGE products accumulate in the body and the higher the blood sugar concentration. AGE products show up in all the afflicted organs of diabetic patients: in the lens of the eye (cataracts), in the capillary blood vessels of the retina (diabetic retinopathy), and in the glomeruli of the kidneys (kidney failure). For people who do not have diabetes, these harmful protein modifications become disturbing only in advanced years. In a young person metabolism functions properly, and the AGE products are decomposed and eliminated from the body. In an older person metabolism slows and the AGE products accumulate. AGE products themselves are thought to enhance oxidative damages.

by humans and perform the same function as human insulin. However, none of the other three is quite as effective as human insulin.

Another factor showing the effect of substituting one amino acid for another is that sometimes patients become allergic to, say, bovine insulin and can switch to hog or sheep insulin without causing allergies.

In contrast to the previous examples, there are small changes in amino acid sequence that make a great deal of difference. First we can

Box 18D

The Use of Human Insulin

Although human insulin, manufactured by recombinant DNA techniques (see Sec. 23.14), is on the market, many diabetic patients continue to use hog or sheep insulin because it is cheaper. Changing from animal to human insulin creates an occasional problem for diabetics. All diabetics experience an insulin reaction (hypoglycemia) when the insulin level in the blood is too high relative to blood sugar level. Hypoglycemia is preceded by symptoms of hunger, sweating, and poor coordination. These symptoms, called "hypoglycemic awareness," are a signal to the patient that hypoglycemia is coming and that it must be reversed, which the patient can do by eating sugar. Some diabetics who changed from animal to human insulin reported that the hypoglycemic awareness from recombinant DNA human insulin is not as strong as that from animal insulin. This can create some hazards. The effect is probably due to different rates of absorption in the body. The literature supplied with human insulin now incorporates a warning that hypoglycemic awareness may be altered.

Linus Pauling.

```
1                6      7      8       9
Cys— S — S —Cys—Pro —Arg—Gly—NH2
 |                |
2 Tyr            Asn 5
   \            /
    Phe — Gln
     3      4
```

Vasopressin

```
1                6      7      8       9
Cys— S — S —Cys—Pro —Leu—Gly—NH2
 |                |
2 Tyr            Asn 5
   \            /
    Ile — Gln
     3      4
```

Oxytocin

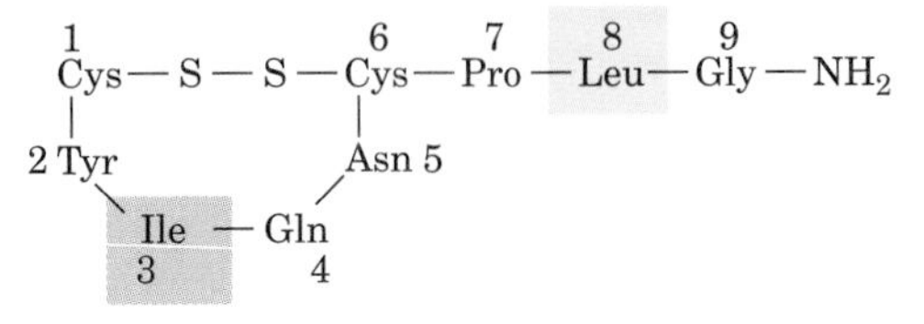

Figure **18.4** • The structures of vasopressin and oxytocin. Differences are shown in color.

Both oxytocin and vasopressin are secreted by the pituitary gland.

Both hormones are used as drugs, vasopressin to combat loss of blood pressure after surgery and oxytocin to induce labor.

consider two peptide hormones, oxytocin and vasopressin (Fig. 18.4). These nonapeptides have identical structures, including a disulfide bond, except for different amino acids in positions 3 and 8. Yet their biological functions are quite different. Vasopressin is an antidiuretic hormone. It increases the amount of water reabsorbed by the kidneys and raises blood pressure. Oxytocin has no effect on water in the kidneys and slightly *lowers* blood pressure. It affects contractions of the uterus in childbirth and the muscles in the breast that aid in the secretion of milk. Vasopressin also stimulates uterine contractions, but much less so than oxytocin.

Another instance where a minor change makes a major difference is in the blood protein hemoglobin. A change in only one amino acid in a chain of 146 is enough to cause a fatal disease—sickle cell anemia (Box 18E).

Although in some cases slight changes in amino acid sequence make little or no difference in the functioning of peptides and proteins, it is clear that the sequence is highly important in most cases. The sequences of a large number of protein and peptide molecules have now been determined. The methods for doing it are complicated and will not be discussed in this book.

The first sequence of an important protein, insulin, was obtained by Frederick Sanger (1918–) in England, for which he received the Nobel Prize in 1958.

18.8 The Secondary Structure of Proteins

Linus Pauling (1901–1994), who won the 1954 Nobel Prize in Chemistry, determined these structures. He also discovered or contributed much to our understanding of certain fundamental concepts, including electronegativity (Sec. 3.7), hybrid orbitals (Sec. 10.4), and resonance (Sec. 11.9).

Proteins can fold or align themselves in such a manner that certain patterns repeat themselves. These repeating patterns are referred to as **secondary structures.** The two most common secondary structures encountered in proteins are the α-helix and the β-pleated sheet (Fig. 18.5) originally proposed by Linus Pauling and Robert Corey. In contrast, those protein conformations that do not exhibit a repeated pattern are called **random coils** (Fig. 18.6).

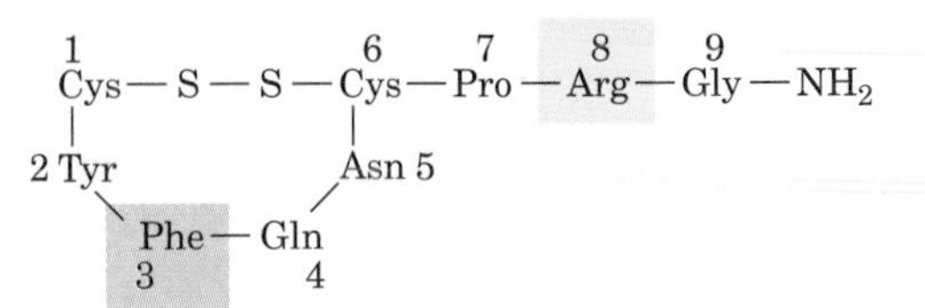

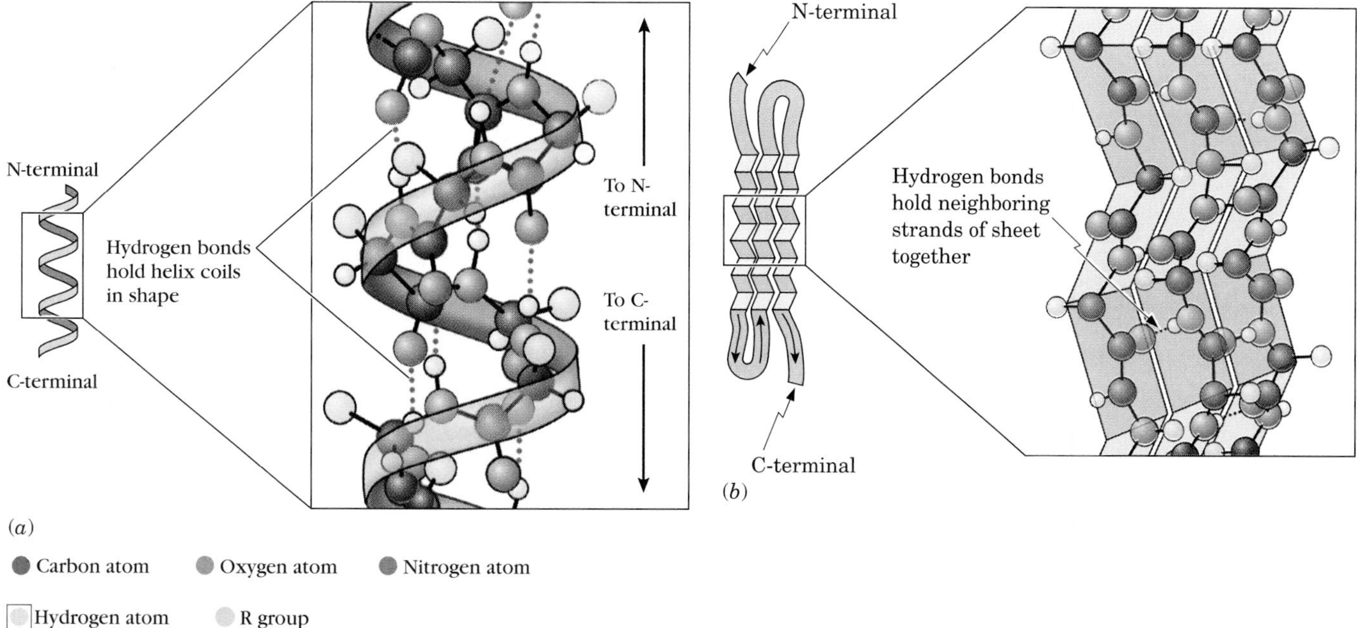

Figure **18.5** • (a) The α-helix. (b) The β-pleated sheet structure.

In the ***α*-helix form,** a single protein chain twists in such a manner that its shape resembles a coiled spring, that is, a helix. The shape of the helix is maintained by numerous intramolecular hydrogen bonds that exist between the *backbone* —C=O and H—N— groups. It can be seen from Figure 18.5 that there is a hydrogen bond between the —C=O oxygen atom of each peptide linkage and the —N—H hydrogen atom of another peptide linkage farther along the chain. These hydrogen bonds are in just the right position to cause the molecule (or a portion of it) to maintain a helical shape.

An *intra*molecular hydrogen bond goes from a hydrogen atom in a molecule to an O, N, or F atom in the same molecule.

The other important orderly structure in proteins is called the ***β*-pleated sheet.** In this case, the orderly alignment of protein chains is maintained by **intermolecular** hydrogen bonds. The β-pleated sheet structure can occur between molecules when polypeptide chains run *parallel* (all the N-terminal ends on one side) or *antiparallel* (neighboring N-terminal ends on opposite sides). β-pleated sheets can also occur intramolecularly, when the polypeptide chain makes a U-turn, forming a hairpin structure, and the pleated sheet is antiparallel (Fig. 18.5).

In all secondary structures the hydrogen bonding is between backbone —C=O and H—N— groups. This is the distinction between secondary and tertiary structure. In the latter, as we shall see, the hydrogen bonding is between R groups on the side chains.

Few proteins have predominantly α-helix or pleated sheet structures. Most proteins, especially globular ones, have only certain portions of their molecules in these conformations. The rest of the molecule is random coil. Many globular proteins contain all three kinds of secondary structure in different parts of their molecules: α-helix, β-pleated sheet, and random coil. A schematic representation of such a structure is shown in Figure 18.7.

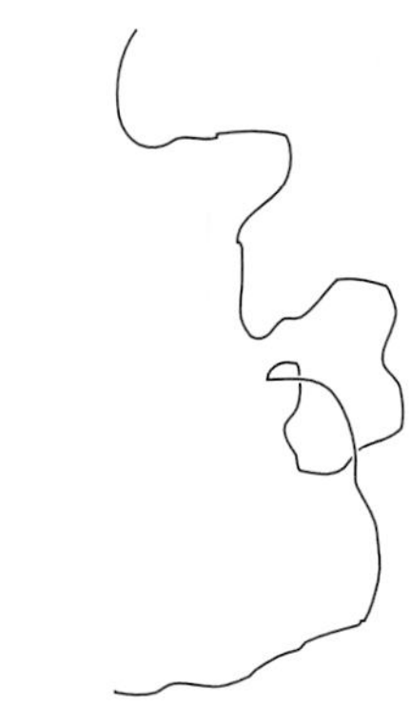

Figure **18.6**
A random coil.

Box 18E

Sickle Cell Anemia

Normal adult human hemoglobin has two alpha chains and two beta chains (Fig. 18.12). Some people, however, have a slightly different kind of hemoglobin in their blood. This hemoglobin (called HbS) differs from the normal type only in the beta chains and only in one position on these two chains: The glutamic acid in the sixth position of normal Hb is replaced by a valine residue in HbS:

	4	5	6	7	8	9
Normal Hb	—Thr—	Pro—	Glu—	Glu—	Lys—	Ala—
Sickle cell Hb	—Thr—	Pro—	Val —	Glu—	Lys—	Ala—

This change affects only two positions in a molecule containing 574 amino acid residues. Yet it is enough to result in a very serious disease, *sickle cell anemia.*

Red blood cells carrying HbS behave normally when there is an ample oxygen supply. When the oxygen pressure decreases, the red blood cells become sickle-shaped (Fig. 18E). This occurs in the capillaries. As a result of this change in shape, the cells may clog the capillaries. The body's defenses destroy the clogging cells, and the loss of the blood cells causes anemia.

This change at only a single position of a chain consisting of 146 amino acids is severe enough to cause a high death rate. A child who inherits two genes programmed to produce sickle cell hemoglobin (a *homozygote*) has an 80 percent smaller chance of surviving to adulthood than a child with only one such gene (a *heterozygote*) or a child with two normal genes. In spite of the high mortality of homozygotes, the genetic trait survives. In central Africa, 40 percent of the population in malaria-ridden areas carry the sickle cell gene, and 4 percent are homozygotes.

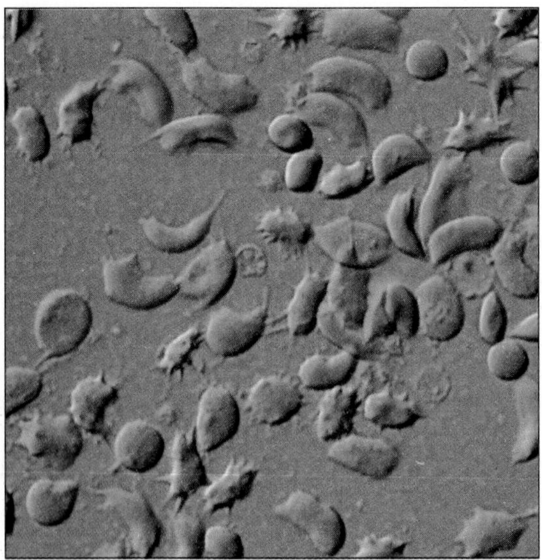

Figure **18E** • Blood cells from a patient with sickle cell anemia. Both normal cells (round) and sickle cells (shriveled) are visible. (G. W. Willis, M.D./Biological Photo Service.)

Box 18E (continued)

It seems that the sickle cell genes help to acquire immunity against malaria in early childhood, so that in malaria-ridden areas the transmission of these genes is advantageous.

In the black population of the United States today, the HbS gene frequency is about 10 percent, and 0.25 percent of the black population is homozygous. The sharp drop in the HbS gene frequency in the Afro-American population over a period of 200 to 300 years indicates the fast adaptation of human genetics to a new environment. People with only one HbS gene are said to carry the *sickle cell trait,* which is not generally harmful. However, when two people with sickle cell trait produce a child who inherits the HbS gene from both parents, that child is a homozygote and will have sickle cell anemia. Screening tests can determine whether prospective parents carry the sickle cell trait.

Keratin, a fibrous protein of hair, fingernails, horns, and wool, is one protein that does have a predominantly α-helix structure. Silk is made of fibroin, another fibrous protein, which exists mainly in the pleated sheet form. Silkworm silk and especially spider silk exhibit a combination of strength and toughness unmatched by high-performance synthetic fibers. In its primary structure silk contains sections that consist of only alanine (25%) and glycine (42%). The formation of pleated sheets, largely by the alanine sections, allows microcrystals to orient along the fiber axis. This accounts for the superior mechanical strength.

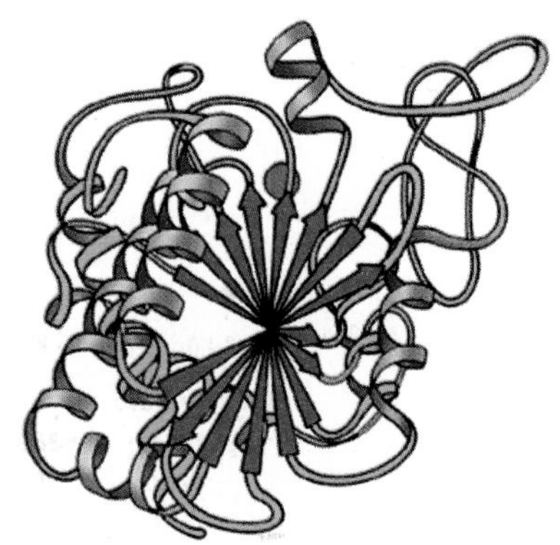

Figure **18.7**
Schematic structure of the enzyme carboxypeptidase. The β-pleated sheet portions are shown in blue, the green structures are the α-helix portions, and the orange strings are the random coil areas.

Another repeating pattern classified as a secondary structure is the **triple helix** of *collagen.* Collagen, which was mentioned in Section 16.12, is the structural protein of connective tissues (bone, cartilage, tendon, aorta, skin), where it provides strength and elasticity. It is the most abundant protein in the body, making up about 30 percent by weight of all the body's protein. The triple helix structure, which is unique to collagen, is made possible by the primary structure of collagen, which allows three polypeptide chains to come together. Each strand of collagen is made of repetitive units that can be symbolized as Gly-X-Y, that is, every third amino acid in the chain is glycine. Glycine, of course, has the shortest side chain (—H) of all amino acids, and this allows the three chains to come together. The X amino acid is frequently proline and the Y is

Box 18F

Collagen Abnormalities

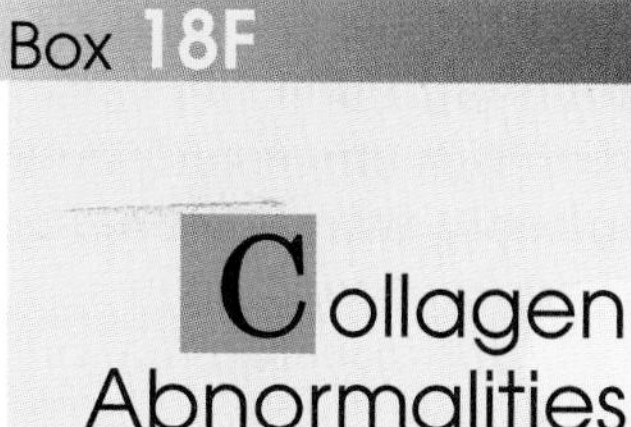

The amount of cross-linking in collagen controls its mechanical strength. The collagen in the Achilles tendon is highly cross-linked, but more flexible tendons are cross-linked to a lesser extent. **Marfan's syndrome** is a human hereditary disease in which collagen is insufficiently cross-linked. Some patients with Marfan's syndrome have excessive bone growth and unusual tallness, which encourages athletic activity. But the insufficiently cross-linked collagen in the blood vessels yields to stresses, especially during vigorous athletic activity, and leads to lethal rupture of the aorta.

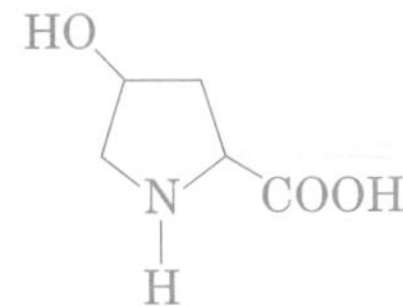

Hydroxyproline is not one of the 20 amino acids in Table 18.1, but the body makes it from proline and uses it in certain proteins, of which collagen is one example.

often hydroxyproline. The triple helix units, called *tropocollagen,* constitute the soluble form of collagen; they are stabilized by hydrogen bonding between the backbones of the three chains. Collagen is made of many tropocollagen units.

18.9 The Tertiary and Quaternary Structure of Proteins

Tertiary Structure

Tropocollagen is found only in fetal or young connective tissues. With aging, the triple helixes that organize themselves (Fig. 18.8) into fibrils cross-link and form insoluble collagen. This cross-linking of collagen is an example of the **tertiary structures** that stabilize the three-dimensional conformations of protein molecules. In collagen, the cross-linking consists of covalent bonds that link together two lysine residues on adjacent chains of the helix (Fig. 18.9).

A covalent bond that often stabilizes the tertiary structure of proteins is the **disulfide bridge.** In Section 18.4 we noted that the amino acid cysteine is easily converted to the dimer cystine. When a cysteine residue is in one chain and another cysteine residue is in another chain (or in another part of the same chain), formation of a disulfide bridge provides a covalent linkage that binds together the two chains or the two parts of the same chain:

$$\sim SH + HS \sim \xrightarrow{[O]} \sim S{-}S \sim + H_2O$$

Examples of both types are found in the structure of insulin (Fig. 18.3).

Besides covalent bonds, there are three other interactions that can stabilize tertiary structures: hydrogen bonding between side chains, salt bridges, and hydrophobic interactions (Fig. 18.10).

1. Hydrogen Bonding We saw in Section 18.8 that secondary structures are stabilized by hydrogen bonding between backbone —C=O and H—N— groups. **Tertiary structures are stabilized by hydrogen bonding between polar groups on side chains.**

2. Salt Bridges These occur only between two amino acids with ionized side chains, that is, between an acidic amino acid and a basic amino acid, each in its ionized form. The two are held together by simple ion-ion attraction.

3. Hydrophobic Interactions In aqueous solution, globular proteins usually turn their *polar* groups *outward,* toward the aqueous solvent, and their *nonpolar* groups *inward,* away from the water molecules. The nonpolar groups prefer to interact with each other, excluding water from these regions. This is called **hydrophobic interaction.** Although this type of interaction is weaker than hydrogen bonding or salt bridges, it usually acts over large surface areas so that cooperatively the interactions are strong enough to stabilize a loop or some other tertiary structure formation.

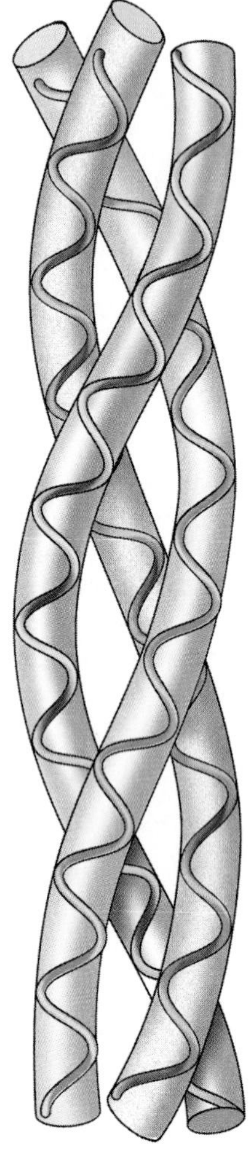

Figure **18.8**
The triple helix of collagen.

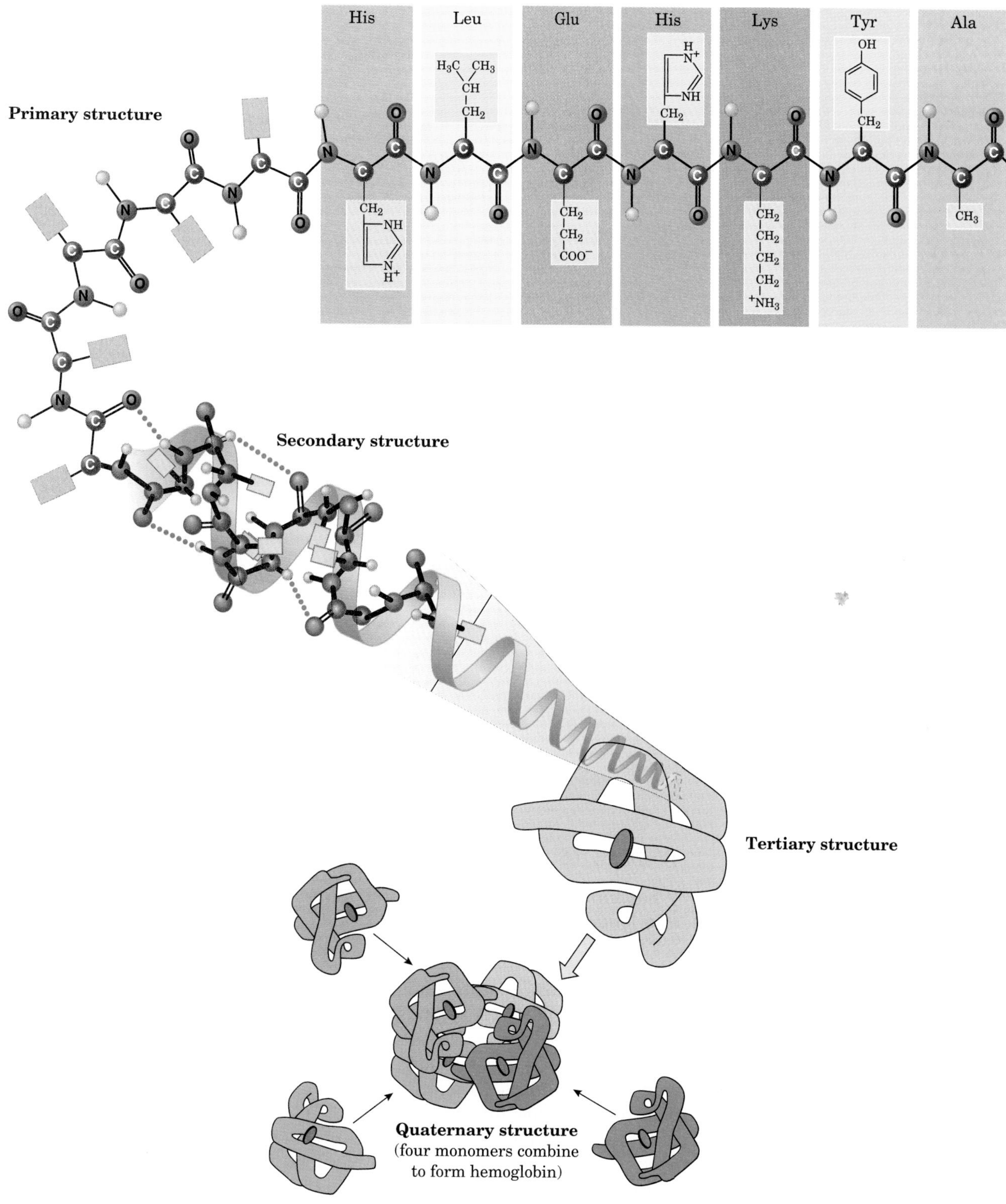

Primary, secondary, tertiary, and quaternary structures of protein.

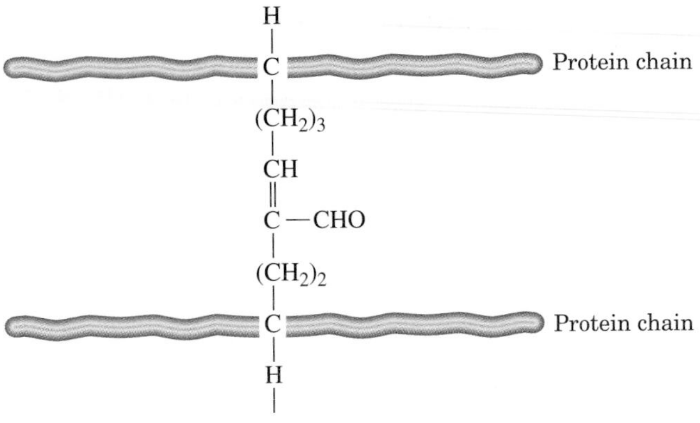

Figure **18.9**
Covalent cross-linking in collagen. The double bond and the aldehyde group are the result of oxidative processes, which include removal of the amino groups of lysine residues.

The four types of interaction that stabilize the tertiary structures of proteins are shown in Figure 18.11.

EXAMPLE

What kind of noncovalent interaction occurs between serine and lysine?

Answer • The side chain of serine ends in an —OH group, that of lysine in an $—NH_2$ group. The two groups can form a hydrogen bond.

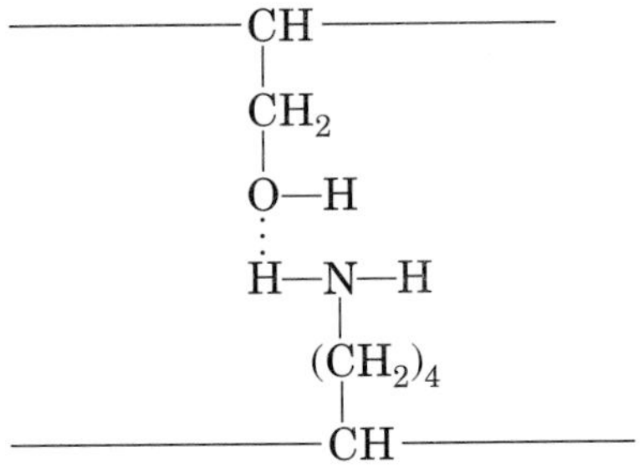

```
———————CH———————
       |
       CH2
       |
       O—H
       :
       H—N—H
         |
        (CH2)4
         |
———————CH———————
```

Problem **18.2**

What kind of noncovalent interaction occurs between phenylalanine and leucine?

Section 18.7 pointed out that the primary structure of a protein largely determines its secondary and tertiary structures. We can now see the reason for this. When the particular R groups are in the proper positions, all the hydrogen bonds, salt bridges, disulfide linkages, and hydrophobic interactions that stabilize the three-dimensional structure of that molecule are allowed to form.

The side chains of some protein molecules allow them to fold (form a tertiary structure) in only one possible way, but other proteins, especially those with long polypeptide chains, can fold in a number of possible ways. Certain proteins in living cells, called **chaperones,** help a newly synthesized polypeptide chain to assume the proper secondary and tertiary structures that are necessary for the functioning of that molecule, and prevent foldings that are not biologically active.

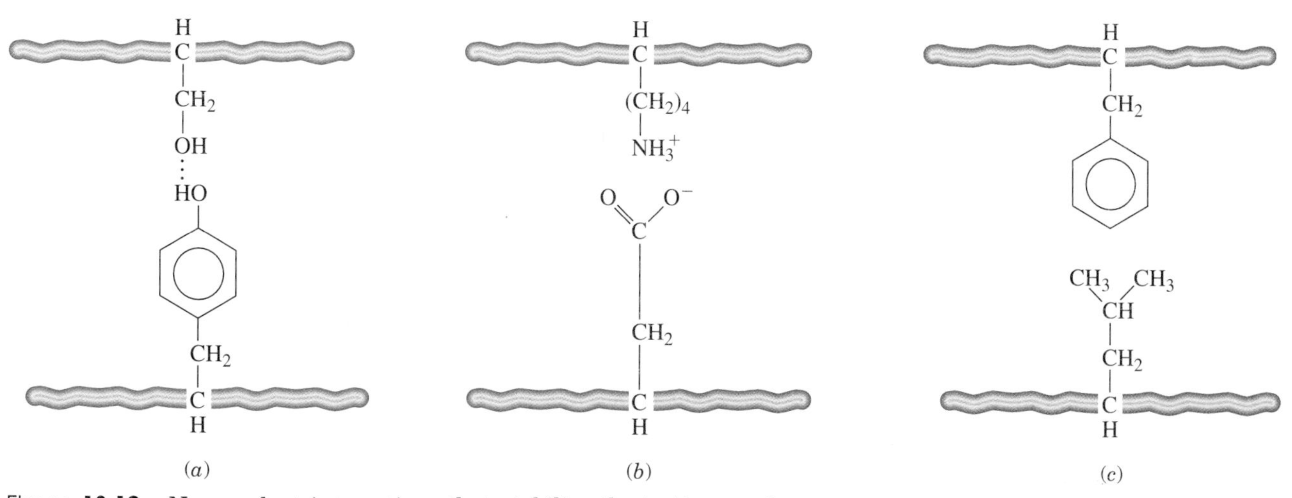

(a) (b) (c)

Figure **18.10** • Noncovalent interactions that stabilize the tertiary and quaternary structures of proteins: (a) hydrogen bonding, (b) salt bridge, (c) hydrophobic interaction.

Quaternary Structure

The highest level of protein organization is **quaternary structure,** which applies to proteins with more than one chain. Quaternary structure determines how the different subunits of the protein fit into an organized whole. The hemoglobin molecule provides an important example. Hemoglobin in adult humans is made of four chains (called **globins**): two identical chains (called alpha) of 141 amino acid residues each and two other identical chains (beta) of 146 residues each. Figure 18.12 shows how the four chains fit together.

When a protein consists of more than one polypeptide chain, each is called a **subunit.**

The hemoglobin molecule does not consist only of the four globin chains. Each globin chain surrounds a *heme* unit, the structure shown in Box 3C. Proteins that contain non-amino acid portions are called **conjugated proteins.** The non-amino acid portion of a conjugated protein

The O_2- and CO_2-carrying functions of hemoglobin are discussed in Sections 25.3 and 25.4.

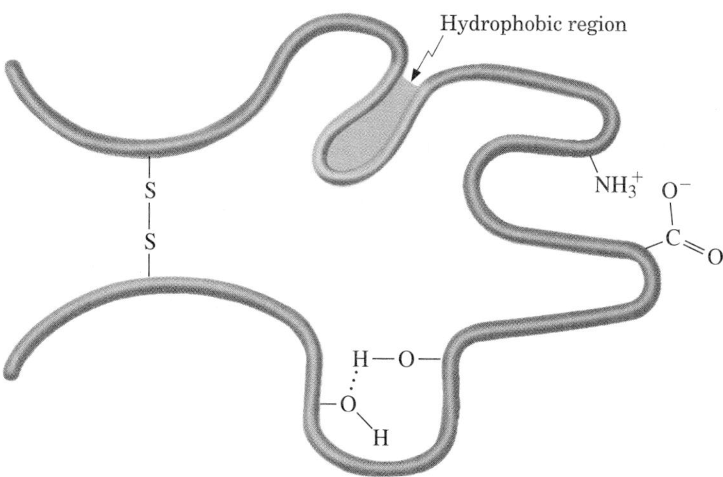

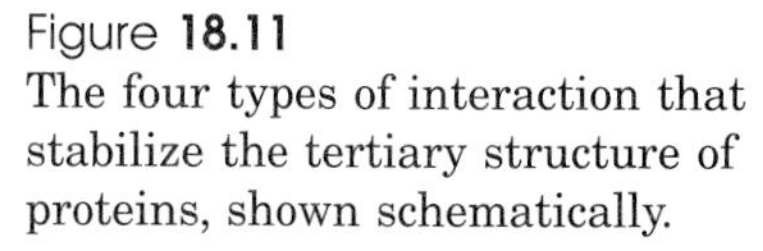

Figure **18.11**
The four types of interaction that stabilize the tertiary structure of proteins, shown schematically.

Figure **18.12**
Three-dimensional model of the hemoglobin molecule. The two alpha chains are in green, and the two beta chains are in blue. The red disks represent the four hemes. (From "The Hemoglobin Molecule" by M. F. Perutz. Copyright 1964 by Scientific American, Inc. All rights reserved.)

is called a **prosthetic group.** In hemoglobin, the globins are the amino acid portions and the heme units are the prosthetic groups.

The quarter-stagger alignment of tropocollagen in collagen fibrils and the twisting of these fibrils into a five-strand or six-strand helix in the collagen fibers also represent quaternary structures. (See Box 18G.)

18.10 Denaturation

Protein conformations are stabilized in their native states by secondary and tertiary structures and through the aggregation of subunits in quaternary structure. Any physical or chemical agent that destroys these stabilizing structures changes the conformation of the protein. We call this process **denaturation.** For example, heat cleaves hydrogen bonds, so boiling a protein solution destroys the α-helical structure (compare Figs. 18.5 and 18.6). In collagen the triple helixes disappear upon boiling, and the molecules have a largely random-coil conformation in the denatured state, which is *gelatin.* In other proteins, especially globular proteins, heat causes the unfolding of the polypeptide chains and, because of subsequent intermolecular protein-protein interactions, precipitation or coagulation takes place. That is what happens when we boil an egg.

Similar conformational changes can be brought about by the addition of denaturing chemicals. Solutions such as 6 M aqueous urea or guanidinium chloride break hydrogen bonds and cause the unfolding of globular proteins. Surface-active agents (detergents) change protein conformation by opening up the hydrophobic regions. Reducing agents, such as 2-mercaptoethanol ($HOCH_2CH_2SH$), can break the —S—S— disulfide bridges, while acids, bases, and salts affect the salt bridges as well as the hydrogen bonds. Ions of heavy metals (Hg^{2+}, Pb^{2+}) interact with the —SH groups.

Denaturation changes secondary, tertiary, and quaternary structure. It does not affect primary structure; that is, the sequence of amino acids

Box 18G

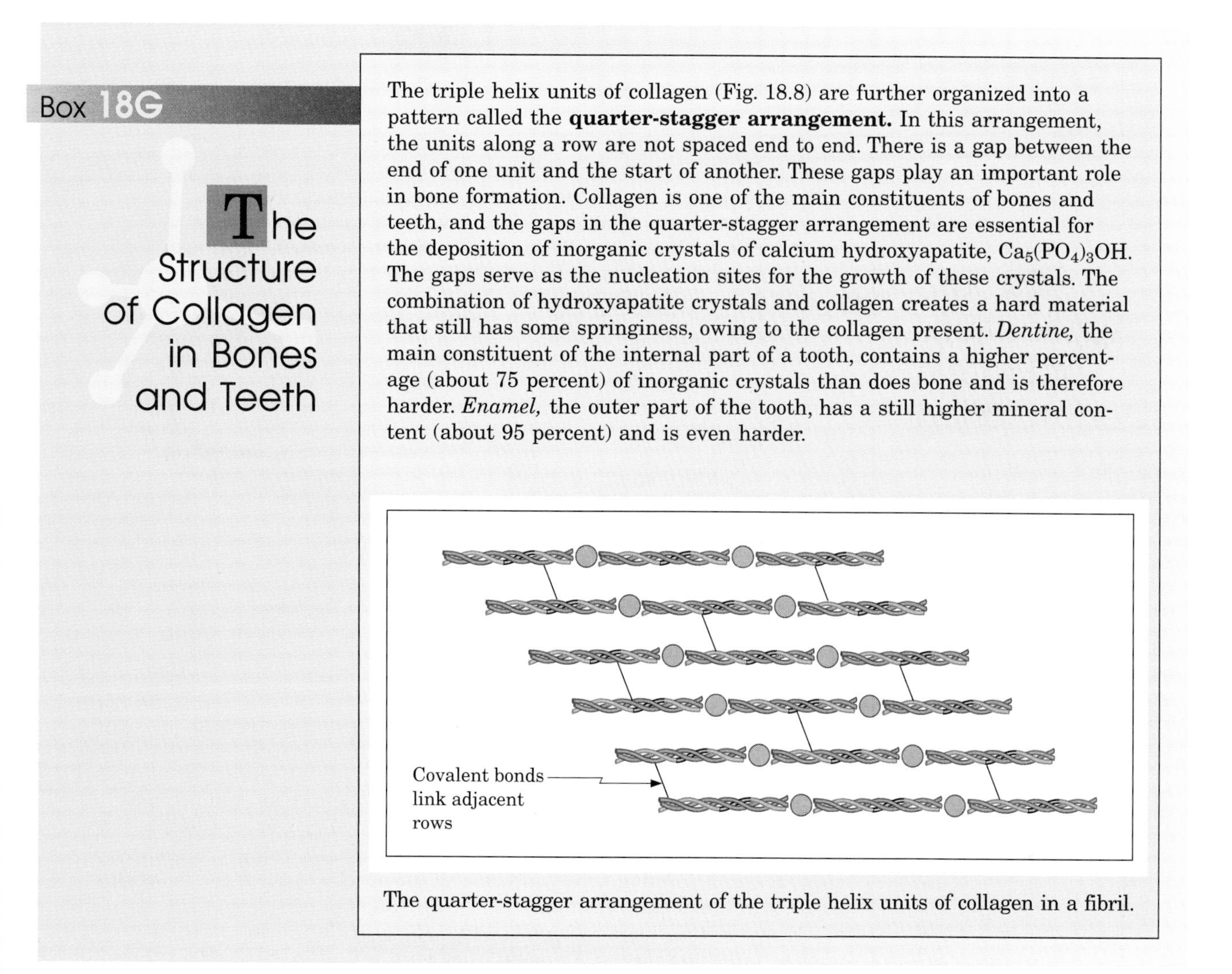

The Structure of Collagen in Bones and Teeth

The triple helix units of collagen (Fig. 18.8) are further organized into a pattern called the **quarter-stagger arrangement.** In this arrangement, the units along a row are not spaced end to end. There is a gap between the end of one unit and the start of another. These gaps play an important role in bone formation. Collagen is one of the main constituents of bones and teeth, and the gaps in the quarter-stagger arrangement are essential for the deposition of inorganic crystals of calcium hydroxyapatite, $Ca_5(PO_4)_3OH$. The gaps serve as the nucleation sites for the growth of these crystals. The combination of hydroxyapatite crystals and collagen creates a hard material that still has some springiness, owing to the collagen present. *Dentine,* the main constituent of the internal part of a tooth, contains a higher percentage (about 75 percent) of inorganic crystals than does bone and is therefore harder. *Enamel,* the outer part of the tooth, has a still higher mineral content (about 95 percent) and is even harder.

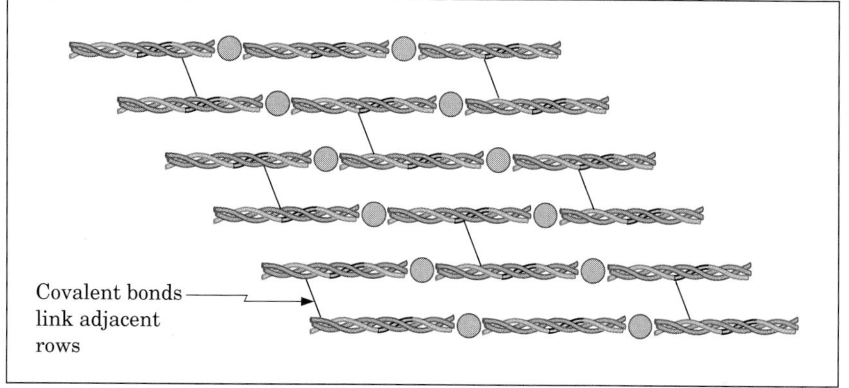

The quarter-stagger arrangement of the triple helix units of collagen in a fibril.

that make up the chain. If these changes occur to a small extent, denaturation can be reversed. For example, in many cases when we remove a denatured protein from a urea solution and put it back into water, it reassumes its secondary and tertiary structure. This is reversible denaturation. In living cells, some denaturation caused by heat can be reversed by chaperones (Sec. 18.9). These help a partially heat-denatured protein to regain its native secondary, tertiary, and quaternary structures. However, some denaturation is irreversible. We cannot unboil a hard-boiled egg.

18.11 Glycoproteins

Although many proteins such as serum albumin consist exclusively of amino acids, others also contain covalently linked carbohydrates and are

Box 18H

Some Applications of Protein Denaturation

As mentioned in the text, proteins can be denaturated by either chemical or physical means. Among the chemical means the most widely used process is attack at the disulfide bridges that influence the tertiary structure of proteins. The processes of permanent waving and straightening of curly hair are examples. The protein keratin, which makes up human hair, has a high percentage of disulfide bridges. These are primarily responsible for the shape of the hair, straight or curly. In either permanent waving or straightening, the hair is first treated with a reducing agent that cleaves some of the —S—S— bonds.

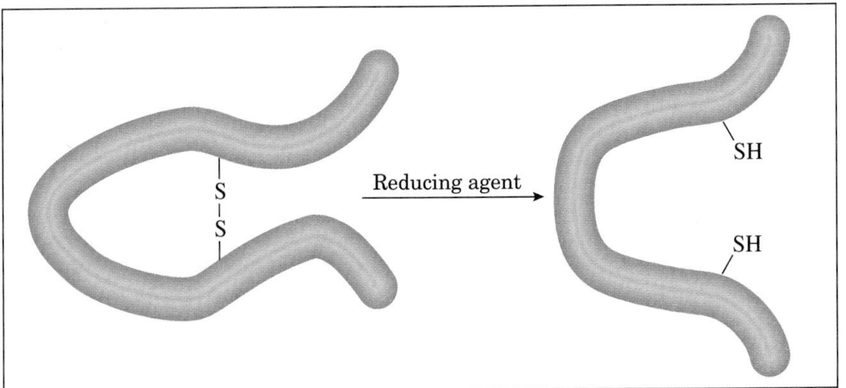

This allows the molecules to lose their rigid orientations and become more flexible. The hair is then set into the desired shape, using curlers or rollers, and an oxidizing agent is applied. The oxidizing agent reverses the above reaction, forming new disulfide bridges, which now hold the molecules together in the desired positions.

Heavy metal ions (for example, Pb^{2+}, Hg^{2+}, or Cd^{2+}) also denature proteins by attacking the —SH groups. They form —S—Hg—S— and similar bridges. This very feature is taken advantage of in the antidote: raw egg whites and milk. The egg and milk proteins are denatured by the metal ions, forming insoluble precipitates in the stomach. These must be pumped out or removed by inducing vomiting. In this way the poisonous metal ions

therefore classified as **glycoproteins.** These include most of the plasma proteins (for example, fibrinogen), enzymes such as ribonuclease, hormones such as thyroglobulin, storage proteins such as casein and ovalbumins, and protective proteins such as immunoglobulins and interferon. The carbohydrate content of these proteins may vary from a few percent (immunoglobulins) up to 85 percent (blood group substances, Box 16B). Most of the proteins in membranes (lipid bilayers, Sec. 17.5) are glycoproteins.

are removed from the body. If the antidote is not pumped out of the stomach, the digestive enzymes would degrade the proteins and release the poisonous heavy metal ions to be absorbed in the blood stream.

Other chemical agents such as alcohol also denature proteins, coagulating them. This process is used in sterilizing the skin before injections. At a concentration of 70 percent, ethanol penetrates bacteria and kills them by coagulating their proteins, whereas 95 percent alcohol denatures only surface proteins.

Proteins can also be denatured by physical means, most notably by heat. Bacteria are killed and surgical instruments are sterilized by heat. A special method of heat denaturation that is seeing increasing use in medicine is the use of lasers. A *laser beam* (a highly coherent light beam of a single wavelength) is absorbed by tissues, and its energy is converted to heat energy. This process can be used to cauterize incisions so that a minimum amount of blood is lost during the operation. Laser beams can be delivered by an instrument called a fiberscope. The laser beam is guided through tiny fibers, thousands of which are fitted into a tube only 1 mm in diameter. This way the energy for denaturation is delivered where it is needed, for example, to seal wounds or join blood vessels without the necessity of cutting through healthy tissues. Fiberscopes have been used successfully to diagnose and treat many bleeding ulcers in the stomach, intestines, and colon.

A novel use of the laser fiberscope is in treating tumors that cannot be reached for surgical removal. A drug called Photofrin, which is activated by light, is given to patients intravenously. The drug in this form is inactive and harmless. It is absorbed by all tissues but remains only in the cancerous tumors and is removed and excreted from healthy tissues. A laser fiberscope is then directed toward the tumor. The energy of the laser beam activates the Photofrin, which destroys the tumor. This is not a complete cure because the tumor may grow back, or it may have spread before the treatment.

Perhaps the most common laser surgery in the future will be photokeratectomy. This procedure, recently approved by the Food and Drug Administration, removes the outer layers of the cornea of the eye using the energy of a laser beam. In the computer-programmed process the curvature of the cornea is changed and the vision of a myopic eye is corrected. This procedure can in many cases make the use of prescription glasses and contact lenses unnecessary.

SUMMARY

Proteins are giant molecules made of amino acids linked together by **peptide bonds.** Proteins have many functions: structural (collagen), enzymatic (ribonuclease), carrier (hemoglobin), storage (casein), protective (immunoglobulin), and hormonal (insulin). **Amino acids** are organic compounds containing an amino (—NH_2) and a carboxylic acid (—COOH) group. The 20 amino acids found in proteins are classified by their side chains: nonpolar, polar but neutral, acidic, and basic. All amino acids in human tissues are L amino acids. Amino acids in the solid state, as well as in water, carry both positive and negative charges; they are called **zwitterions.** The pH at which the number of positive charges is the same as the number of negative charges is the **isoelectric point** of an amino acid or protein.

When the amino group of one amino acid condenses with the carboxyl group of another, an amide (**peptide**) linkage is formed, with the elimination of water. The two amino acids form a **dipeptide.** Three amino acids form a **tripeptide,** and so forth. Many amino acids form a **polypeptide** chain. Proteins are made of one or more polypeptide chains.

The linear sequence of amino acids is the **primary structure** of proteins. The repeating short-range conformations (**α-helix, β-pleated sheet, triple helix** of collagen, or **random coil**) are the **secondary structures.** The **tertiary structure** is the three-dimensional conformation of the protein molecule. Tertiary structures are maintained by covalent cross-links such as **disulfide bonds** and by **salt bridges, hydrogen bonds,** and **hydrophobic interactions.** The precise fit of polypeptide subunits into an aggregated whole is called the **quaternary structure.**

Secondary and tertiary structures stabilize the native conformation of proteins; physical and chemical agents, such as heat or urea, destroy these structures and **denature** the protein. Protein functions depend on native conformation; when a protein is denatured, it can no longer carry out its function. Some (but not all) denaturation is reversible; in some cases **chaperone** molecules reverse denaturation. Many proteins are classified as **glycoproteins** because they contain carbohydrate units.

KEY TERMS

AGE (Box 18C)
Alpha amino acid (Sec. 18.2)
Amino acid (Sec. 18.2)
Backbone (Sec. 18.6)
C-terminal amino acid (Sec. 18.5)
Chaperones (Sec. 18.9)
Conjugated protein (Sec. 18.9)
Cross-link (Sec. 18.9)
Denaturation (Sec. 18.10)
Disulfide linkage (Sec. 18.4)
Fiberscope (Box 18H)
Fibrous protein (Sec. 18.1)
Globular protein (Sec. 18.1)
Glycation (Box 18C)
Glycoprotein (Sec. 18.11)
α-Helix (Sec. 18.8)
Hydrophobic interaction (Sec. 18.9)
Hypoglycemic awareness (Box 18D)
Intramolecular hydrogen bond (Sec. 18.8)
Isoelectric point (Sec. 18.3)
Marfan's syndrome (Box 18F)
N-terminal amino acid (Sec. 18.5)
Peptide (Sec. 18.5)
Peptide linkage (Sec. 18.5)
β-Pleated sheet (Sec. 18.8)
Polypeptide (Sec. 18.5)
Primary structure (Sec. 18.7)
Prosthetic group (Sec. 18.9)
Protein (Sec. 18.1)
Quarter-stagger arrangement (Box 18G)
Quaternary structure (Sec. 18.9)
Random coil (Sec. 18.8)
Residue (Sec. 18.5)
Salt bridge (Sec. 18.9)
Secondary structure (Sec. 18.8)
Sickle cell anemia (Box 18E)
Side chain (Sec. 18.6)
Subunit (Sec. 18.9)
Tertiary structure (Sec. 18.9)
Triple helix (Sec. 18.8)
Zwitterion (Sec. 18.3)

PROBLEMS

18.3 The human body has about 100 000 different proteins. Why do we need so many?

18.4 The members of which class of proteins are insoluble in water and can serve as structural materials?

Amino Acids

18.5 Name two acidic and two basic amino acids.

18.6 Classify the following amino acids as nonpolar, polar but neutral, acidic, or basic:
(a) arginine
(b) leucine
(c) glutamic acid
(d) asparagine
(e) tyrosine
(f) phenylalanine
(g) glycine

***18.7** Which amino acid has the highest percent nitrogen (g N/100 g amino acid)?

18.8 Which amino acids have aromatic side chains?

***18.9** Draw the structure of proline. Which class of heterocyclic compounds does this molecule belong to?

18.10 Which amino acid is also a thiol?

18.11 Why is it necessary to have proteins in our diets?

18.12 Which amino acids in Table 18.1 have more than one chiral carbon?

18.13 What are the similarities and differences in the structures of alanine and phenylalanine?

***18.14** What special structural feature or property does each of the following amino acids have that makes it different from all the others:
(a) glycine (c) tyrosine
(b) cystine (d) proline

18.15 Draw the structures of L- and D-valine.

Zwitterions

18.16 Why are all amino acids solids at room temperature?

***18.17** Show how alanine, in solution at its isoelectric point, acts as a buffer (write equations to show why the pH does not change much if we add acid or base).

18.18 Explain why an amino acid cannot exist in an un-ionized form [$RCH(NH_2)COOH$] at any pH.

18.19 Draw the structure of valine at pH 1 and at pH 12.

Peptides and Proteins

18.20 Write by symbols all the tripeptides that can be formed from phenylalanine (Phe) and asparagine (Asn).

18.21 Draw the structure of a tripeptide made of threonine, arginine, and methionine.

18.22 (a) Use the three-letter abbreviations to write the representation of the following tetrapeptide. (b) Which amino acid is the C-terminal end and which the N-terminal end?

```
       CH3
       |
       CH—CH3
       |
NH2—CH—C—NH—CH—C—NH—CH—C—NH—CH—COOH
       ||    |   ||    |   ||    |
       O    CH2  O    CH2  O    CH2—C—NH2
             |          |            ||
         (phenyl)   (indolyl, N—H)    O
```

18.23 Draw the structure of the dipeptide leucylproline.

18.24 (a) What is a protein backbone? (b) What is an N-terminal end of a protein?

18.25 Show by chemical equations how alanine and glutamine can be combined to give two different dipeptides.

Properties of Peptides and Proteins

18.26 (a) How many atoms of the peptide bond lie in the same plane? (b) Which atoms are they?

18.27 (a) Draw the structural formula of the tripeptide Met-Ser-Cys. (b) Draw the different ionic structures of this tripeptide at pH 2.0, 7.0, and 10.0.

18.28 How can a protein act as a buffer?

18.29 Proteins are least soluble at their isoelectric points. What would happen to a protein precipitated at its isoelectric point if a few drops of dilute HCl were added?

Primary Structure of Proteins

18.30 How many different tripeptides can be made (a) using only leucine, threonine, and valine (b) using all 20 amino acids?

***18.31** How many different tetrapeptides can be made (a) if the peptides contain one residue each of asparagine, proline, serine, and methionine (b) if all 20 amino acids can be used

18.32 How many amino acid residues in the A chain of insulin are the same in insulin from humans, cattle (bovine), hogs, and sheep?

18.33 Based on your knowledge of the chemical properties of amino acid side chains, suggest a substitution for leucine in the primary structure of a protein that would probably not change the character of the protein very much.

Secondary Structure of Proteins

18.34 Is a random coil a (a) primary (b) secondary (c) tertiary or (d) quaternary structure?

18.35 Decide whether these types of structures that exist in collagen are primary, secondary, tertiary, or quaternary: (a) tropocollagen (b) collagen fibril (c) collagen fiber (d) the proline–hydroxyproline–glycine repeating sequence.

18.36 Proline is often called an α-helix terminator. That is, it is usually in the random-coil secondary structure following an α-helix portion of a protein chain. Why does proline not fit easily into an α-helix structure?

Tertiary and Quaternary Structure of Proteins

18.37 Polyglutamic acid (a polypeptide chain made only of glutamic acid residues) has an α-helix conformation below pH 6.0 and a random-coil conformation above pH 6.0. What is the reason for this conformational change?

18.38 Distinguish between inter- and intramolecular hydrogen bonding between backbone groups. Where in protein structures do you find one and where the other?

18.39 Identify the different primary, secondary, and tertiary structures in the numbered boxes:

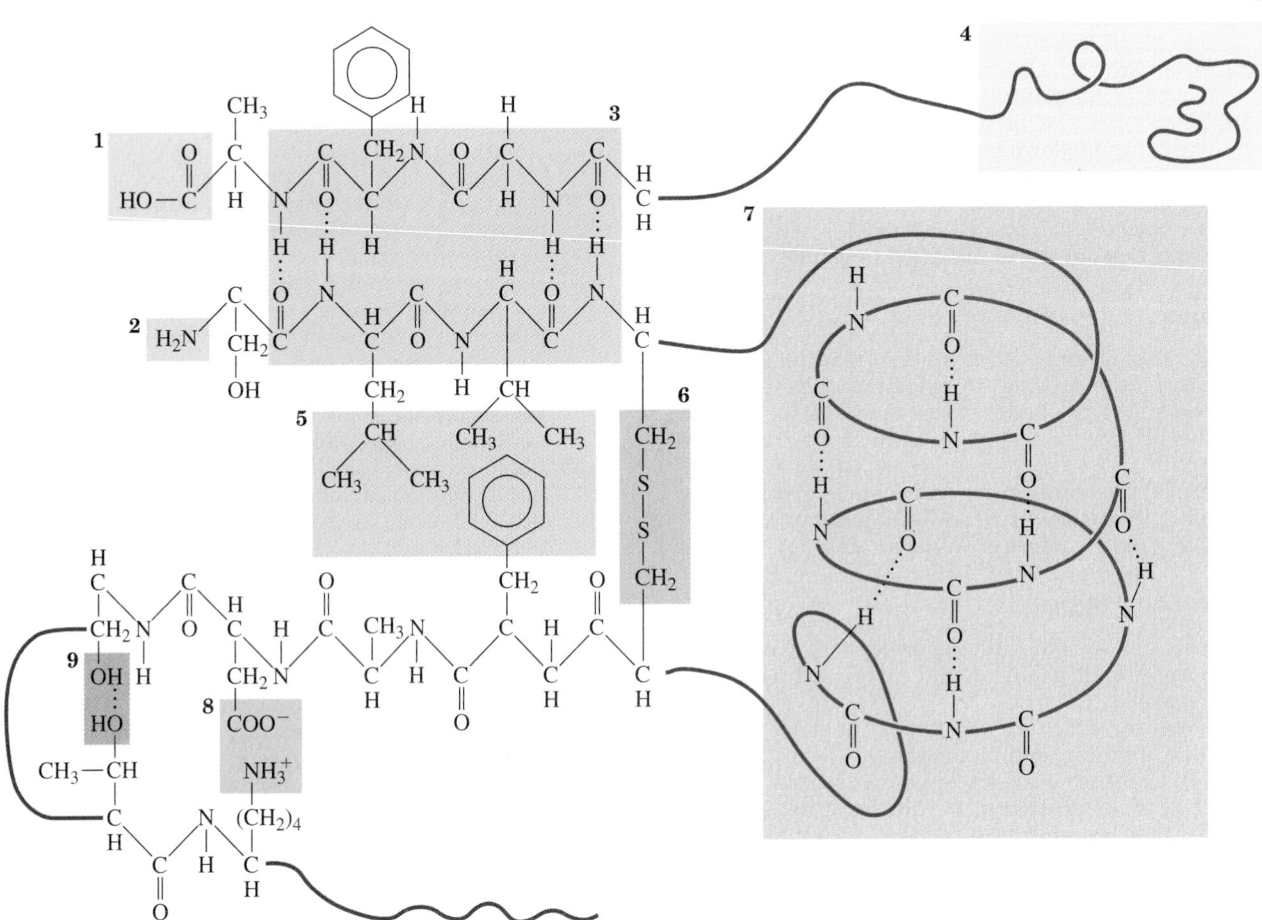

18.40 If both cysteine residues on the B-chain of insulin were changed to alanine residues, how would the tertiary structure of insulin be affected?

18.41 Which amino acid side chains can form hydrophobic bonds?

18.42 What kind of interaction operates when an aspartic acid residue is on one side chain and a lysine residue is on a neighboring one?

18.43 What is a conjugated protein?

Protein Denaturation

***18.44** Which amino acid side chain is most frequently involved in denaturation by reduction?

18.45 Does the primary structure of a protein always determine the secondary and tertiary structures?

18.46 What happens to the tertiary structure of a globular protein if 2-mercaptoethanol is added to it?

Boxes

18.47 (Box 18A) (a) Describe the difference between the structure of aspartame and the methyl ester of phenylalanylaspartic acid. (b) Do you expect this second compound to be as sweet as aspartame?

18.48 (Box 18B) What is unusual about the peptide bond of glutathione?

18.49 (Box 18C) AGE products become disturbing only in elderly people, even though they also form in younger persons. Why don't they harm these persons?

18.50 (Box 18D) Define hypoglycemic awareness.

18.51 (Box 18E) Why has the mutation in the beta chain of hemoglobin survived in many generations of heterozygotes, even though it causes sickle cell anemia?

18.52 (Box 18E) In sickle cell anemia, Val is substituted for Glu in the sixth position of the beta chain of hemoglobin. Some individuals have hemoglobin in which Asp is substituted for Glu. Would you expect this substitution also to be detrimental to health? Explain.

18.53 (Box 18F) Marfan's syndrome involves changes in the conformation of collagen. Which structure is abnormal in this syndrome—primary, secondary, tertiary, or quaternary? Explain.

18.54 (Box 18G) What constituent of bone determines its hardness?

18.55 (Box 18H) Silver nitrate, $AgNO_3$, is sometimes put into the eyes of newborn infants as a disinfectant against gonorrhea. Silver is a heavy metal. Explain how this may work against the bacteria.

18.56 (Box 18H) Why do nurses and physicians use 70 percent alcohol to wipe the skin before giving injections?

18.57 (Box 18H) What is the role of lasers in photokeratectomy?

18.58 (Box 18H) What makes natural hair either straight or curly?

Additional Problems

18.59 Hydrogen bonds stabilize both α-helix and β-pleated sheet structures. These bonds occur between the hydrogen of an N—H and the oxygen of a C═O. Can hydrogen bonding occur between two N—H groups?

18.60 Carbohydrates are integral parts of glycoproteins. Does the addition of carbohydrate to the protein core make the protein more or less soluble in water? Explain.

18.61 Can the hydrolysis of a peptide bond in a protein be considered a denaturation process?

18.62 How many different dipeptides can be made (a) using only alanine, tryptophan, glutamic acid, and arginine (b) using all 20 amino acids

***18.63** Denaturation is usually associated with transitions from helical structures to random coils. If an imaginary process were to transform the keratin in your hair from an α-helix to a β-pleated sheet structure, would you call the process denaturation?

18.64 In diabetes, insulin is administered intravenously. Explain why this hormone protein cannot be taken orally.

18.65 Draw the structure of lysine (a) above (b) below and (c) at its isoelectric point.

18.66 Cysteine plays an important role in forming the tertiary structures of proteins. What is it?

18.67 Considering the vast number of animal and plant species on earth (including those now extinct) and the large number of different protein molecules in each organism, have all possible protein molecules been used already by some species or other? Explain.

18.68 What kind of noncovalent interaction occurs between:
(a) valine and isoleucine
(b) glutamic acid and lysine
(c) tyrosine and threonine
(d) alanine and alanine

***18.69** How many different decapeptides (peptides containing 10 amino acids each) can be made from the 20 amino acids?

18.70 Which amino acid does not rotate the plane of polarized light?

18.71 Write the expected products of the acid hydrolysis of the following tripeptide:

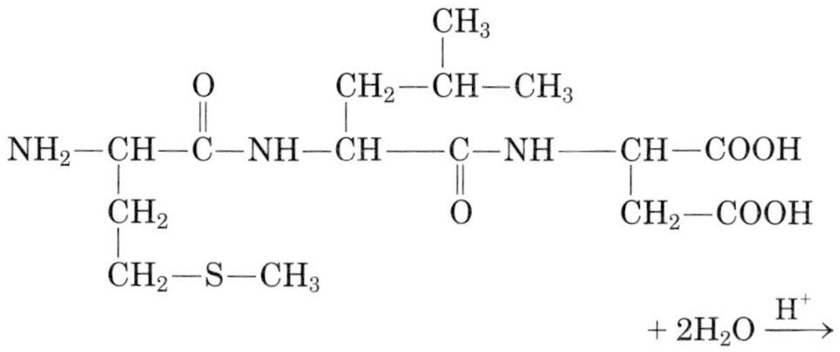

$$NH_2-\underset{\underset{\underset{CH_2-S-CH_3}{|}}{CH_2}}{\underset{|}{CH}}-\overset{\overset{O}{\|}}{C}-NH-\overset{\overset{\overset{\overset{CH_3}{|}}{CH_2-CH-CH_3}}{|}}{CH}-\underset{\underset{O}{\|}}{C}-NH-\underset{\underset{CH_2-COOH}{|}}{CH}-COOH + 2H_2O \xrightarrow{H^+}$$

18.72 What charges are there on aspartic acid at pH 2.0?

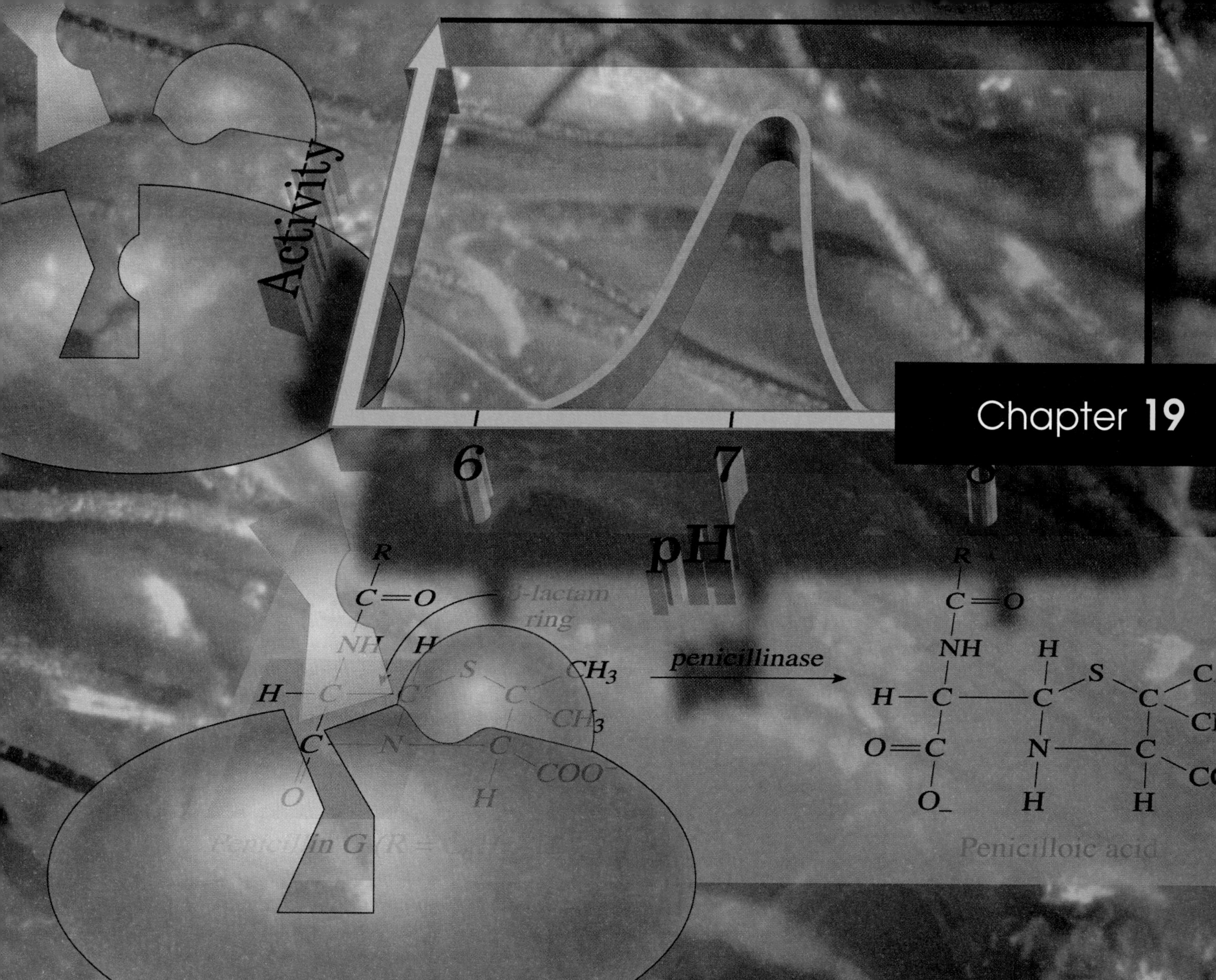

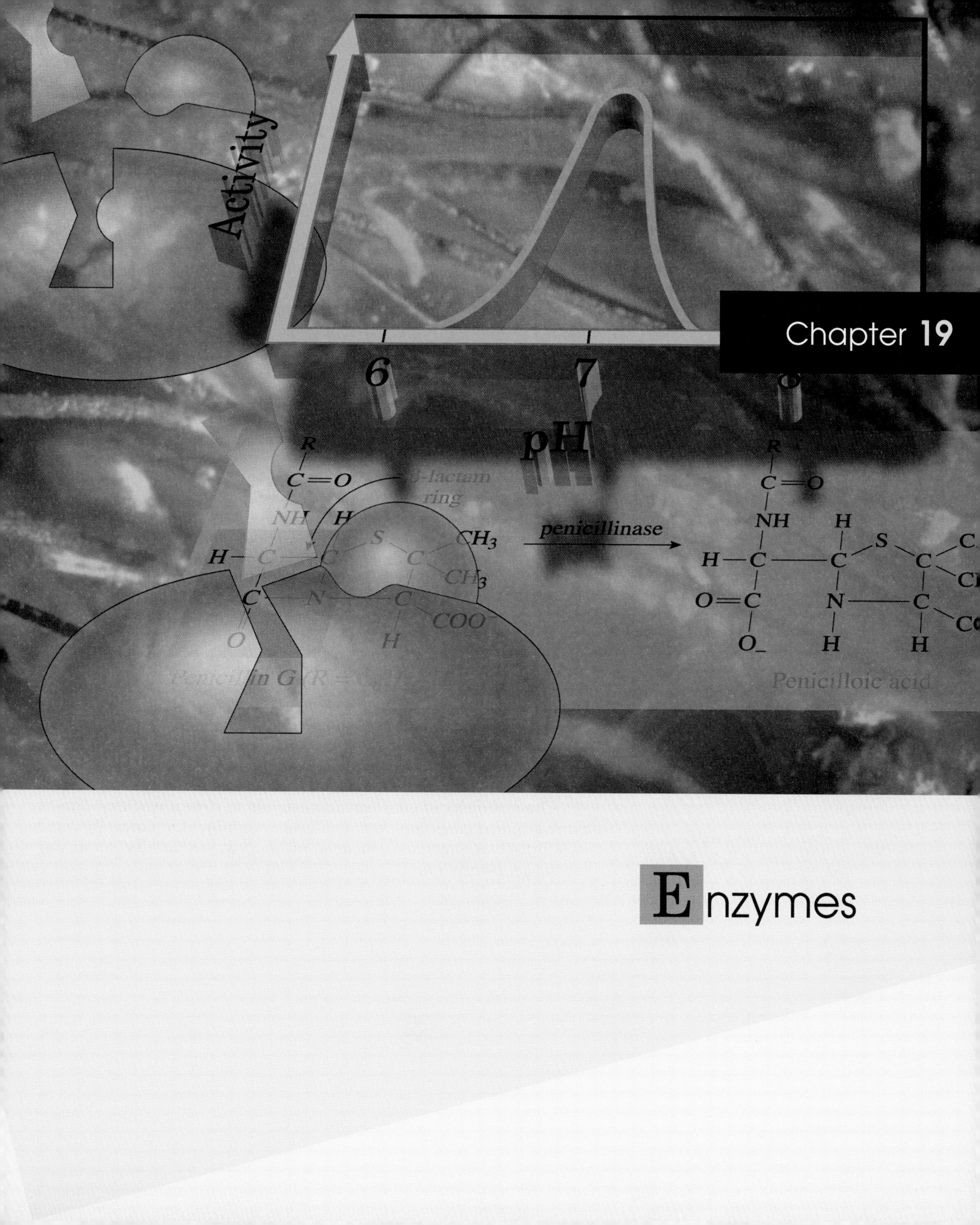

Chapter 19

Enzymes

19.1 Introduction

The cells in your body are chemical factories. Only a few of the thousands of compounds necessary for the operation of the human organism are obtained from the diet. Most of them are synthesized within the cells, which means that hundreds of chemical reactions are taking place in your cells every minute of your life.

Nearly all these reactions are catalyzed by **enzymes,** which are **protein molecules that increase the rates of chemical reactions without themselves undergoing any change.** Without enzymes, life as we know it would not be possible.

Proteins are not the only biological catalysts. Ribonucleic acids, especially introns (Sec. 23.11), can also act as enzymes in certain cases. Such biocatalysts are called **ribozymes.** This was discovered by Sidney Altman of Yale University and Thomas R. Cech of the University of Colorado, who were awarded the 1989 Nobel Prize in Chemistry for this work.

Like all catalysts, enzymes do not change the position of equilibrium. That is, enzymes cannot make a reaction take place that would not take place without them. What they do is increase the rate; they cause reactions to take place faster. As catalysts, enzymes are remarkable in two respects: (1) They are extremely effective, increasing reaction rates by anywhere from 10^9 to 10^{20} times, and (2) they are extremely specific.

As an example of their effectiveness, consider the oxidation of glucose. A lump of glucose or even a glucose solution exposed to oxygen under sterile conditions would show no appreciable change for months. In the human body the same glucose is oxidized within seconds.

Every organism has many enzymes—more than 3000 in a single cell. Presumably, each chemical reaction has one enzyme that catalyzes it. This means that enzymes are very specific, each of them speeding up only one particular reaction or class of reactions. For example, the enzyme urease catalyzes only the hydrolysis of urea and not that of other amides, even closely related ones. Another type of specificity can be seen with trypsin, an enzyme that cleaves the peptide linkages of protein molecules—but not every peptide linkage, only those on the carboxyl side of lysine and arginine residues (Fig. 19.1). The enzyme carboxypeptidase specifically cleaves only the last amino acid on a protein chain—the one at the C-terminal end. Lipases are less specific: They cleave any triglyceride, but they still don't affect carbohydrates or proteins.

The specificity of enzymes also extends to stereospecificity. The enzyme arginase converts the amino acid L-arginine (the naturally occurring form) to a compound called L-ornithine but has no effect on its mirror image, D-arginine.

The enzyme that oxidizes D-glucose does not work on L-glucose.

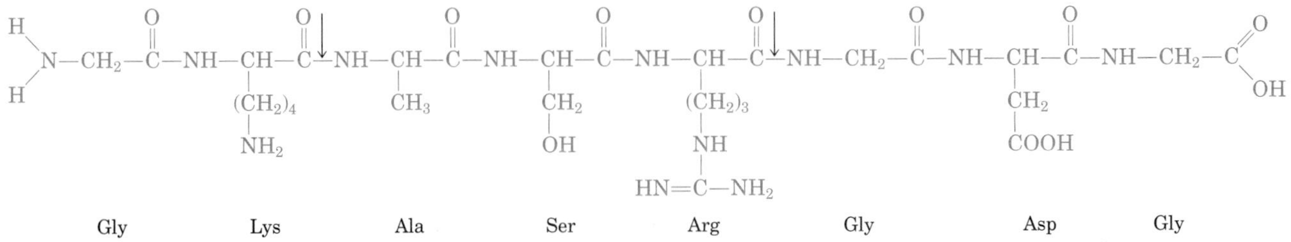

```
H         O           O ↓          O           O           O ↓            O           O
 \        ||          ||           ||          ||          ||             ||          ||          O
  N—CH2—C—NH—CH—C—NH—CH—C—NH—CH—C—NH—CH—C—NH—CH2—C—NH—CH—C—NH—CH2—C//
 /              |             |            |           |                        |             \
H            (CH2)4          CH3          CH2        (CH2)3                    CH2             OH
                |                          |           |                        |
               NH2                        OH          NH                       COOH
                                                       |
                                                    HN=C—NH2

      Gly       Lys          Ala          Ser         Arg         Gly          Asp          Gly
```

Figure **19.1** • A typical amino acid sequence. The enzyme trypsin catalyzes the hydrolysis of this chain only at the points marked with an arrow (the —COOH side of lysine and arginine).

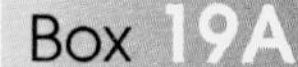

Box 19A

Muscle Relaxants and Enzyme Specificity

Acetylcholine is a neurotransmitter (Sec. 24.1) that operates between the nerve endings and muscles. It attaches itself to a specific receptor in the muscle end plate. This transmits a signal to the muscle to relax. A specific enzyme, acetylcholinesterase, then catalyzes the hydrolysis of the acetylcholine,

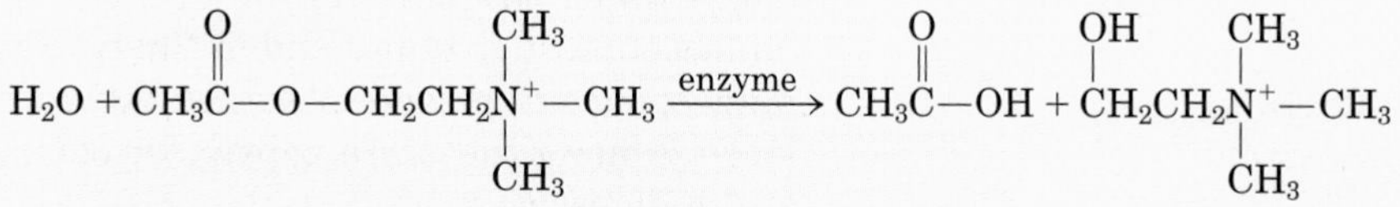

$$H_2O + \underset{\text{Acetylcholine}}{CH_3\overset{O}{\overset{\|}{C}}{-}O{-}CH_2CH_2\overset{CH_3}{\overset{|}{\underset{\underset{CH_3}{|}}{N^+}}}{-}CH_3} \xrightarrow{\text{enzyme}} \underset{\text{Acetic acid}}{CH_3\overset{O}{\overset{\|}{C}}{-}OH} + \underset{\text{Choline}}{\overset{OH}{\overset{|}{CH_2}}CH_2\overset{CH_3}{\overset{|}{\underset{\underset{CH_3}{|}}{N^+}}}{-}CH_3}$$

removing it from the receptor site. Succinylcholine,

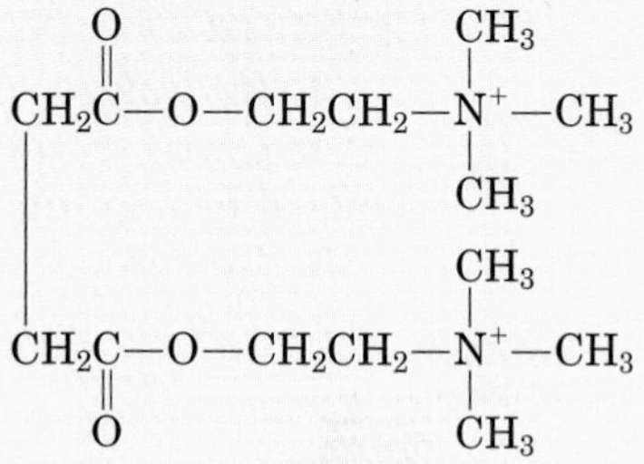

$$\begin{array}{l} CH_2\overset{O}{\overset{\|}{C}}{-}O{-}CH_2CH_2{-}\overset{CH_3}{\overset{|}{\underset{\underset{CH_3}{|}}{N^+}}}{-}CH_3 \\ | \\ CH_2\underset{\underset{O}{\|}}{C}{-}O{-}CH_2CH_2{-}\overset{CH_3}{\overset{|}{\underset{\underset{CH_3}{|}}{N^+}}}{-}CH_3 \end{array}$$

Succinylcholine

is sufficiently similar to acetylcholine so that it too attaches itself to the receptor of the muscle end plate and causes muscle relaxation. However, acetylcholinesterase can hydrolyze succinylcholine only very slowly, and thus the muscle stays relaxed for a long time.

This feature makes succinylcholine a good muscle relaxant during minor surgery, especially when a tube must be inserted into the bronchus (bronchoscopy). For example, after intravenous administration of 50 mg of succinylcholine, paralysis and respiratory arrest are observed within 30 seconds. While respiration is carried on artificially, the bronchoscopy can be performed within minutes.

Enzymes are distributed according to the body's need to catalyze specific reactions. A large number of protein-splitting enzymes are in the blood, ready to promote clotting. Digestive enzymes, which also split proteins, are located in the secretions of the stomach and pancreas. Even within the cells themselves, some enzymes are localized according to the need for specific reactions. The enzymes that help the oxidation of compounds that are part of the citric acid cycle (Sec. 20.4) are located in the mitochondria of the cells, and special organelles such as those called lysosomes contain an enzyme (lysozyme) that aids the dissolution of bacterial cell walls.

19.2 Naming and Classifying Enzymes

Enzymes are commonly given names derived from the reaction they catalyze and/or the compound or type of compound they act on. For example, lactate dehydrogenase is the enzyme that speeds up the removal of hydrogen from lactate. Acid phosphatase helps to cleave phosphate ester bonds under acidic conditions. As can be seen from these examples, the names of most enzymes end in "**-ase.**" Some enzymes, however, have older names, ones that were assigned before their actions were clearly understood. Among these are pepsin, trypsin, and chymotrypsin—all enzymes of the digestive tract.

Enzymes can be classified into six major groups according to the type of reaction they catalyze (see also Table 19.1):

1. *Oxidoreductases* catalyze oxidations and reductions.

2. *Transferases* catalyze the transfer of a group of atoms, such as CH_3, CH_3CO, or NH_2, from one molecule to another.

3. *Hydrolases* catalyze hydrolysis reactions.

4. *Lyases* catalyze the addition of a group to a double bond or the removal of a group to create a double bond.

5. *Isomerases* catalyze isomerization reactions.

6. *Ligases,* or *synthetases,* catalyze the joining of two molecules.

19.3 Common Terms in Enzyme Chemistry

Some enzymes, such as pepsin and trypsin, consist of polypeptide chains only. Other enzymes contain **nonprotein portions** called **cofactors.** The protein (polypeptide) portion of the enzyme is called an **apoenzyme.** The cofactors may be metallic ions, such as Zn^{2+} or Mg^{2+}, or they may be organic compounds. Organic cofactors are called **coenzymes.** An important group of coenzymes are the B vitamins, which are essential to the action of many enzymes (see Sec. 20.3). Another important coenzyme is heme (Box 3C), which is part of a number of oxidoreductases as well as of hemoglobin. In any case, an apoenzyme cannot catalyze a reaction without its cofactor, nor can the cofactor function without the apoenzyme. When a metal ion is a cofactor, it can be bound directly to the protein or to the coenzyme, if the enzyme contains one.

Some enzymes have two kinds of cofactor: a coenzyme and a metallic ion.

Box 19B Meat Tenderizers

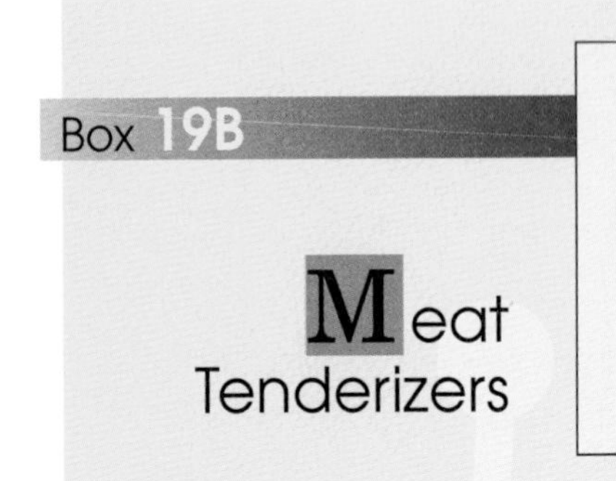

Pepsin and trypsin are proteases, a class of hydrolases that catalyze the hydrolysis of proteins. Some of the meat that we eat is tough and difficult to chew. Since meat contains a lot of protein, coating the meat with a protease before cooking hydrolyzes some of the long protein chains, breaking them into shorter chains and making the meat easier to chew. Meat tenderizers do just this. They contain proteases such as papain. Proteases are also used to increase the yield of meat from bones and meat scraps in the manufacture of processed meats, such as bologna and frankfurters.

Table 19.1 Classifications of Enzymes

Class	Typical Example	Reaction Catalyzed	Page Number in this Book
1. Oxidoreductases	Lactate dehydrogenase	$CH_3-C(=O)-COO^- \longrightarrow CH_3-CH(OH)-COO^-$ Pyruvate ⟶ L-(+)-Lactate	668
2. Transferases	Aspartate amino transferase or Aspartate transaminase	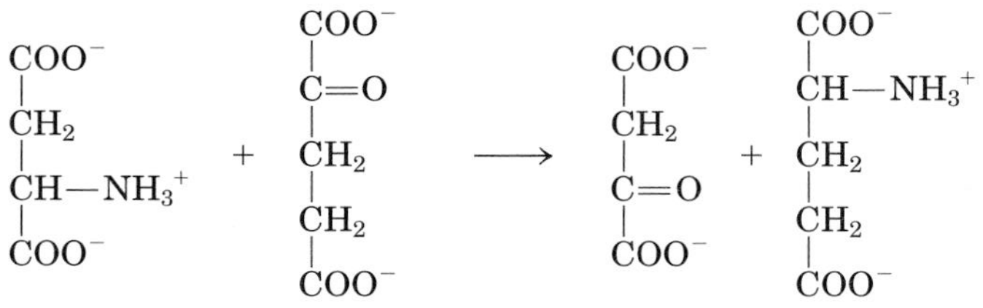 $^-OOC-CH_2-CH(NH_3^+)-COO^- + {}^-OOC-C(=O)-CH_2-CH_2-COO^- \longrightarrow {}^-OOC-CH_2-C(=O)-COO^- + {}^-OOC-CH(NH_3^+)-CH_2-CH_2-COO^-$ Aspartate + α-Ketoglutarate ⟶ Oxaloacetate + Glutamate	679
3. Hydrolases	Acetylcholinesterase	$CH_3-C(=O)-OCH_2CH_2\overset{+}{N}(CH_3)_3 + H_2O \longrightarrow CH_3COOH + HOCH_2CH_2\overset{+}{N}(CH_3)_3$ Acetylcholine ⟶ Acetic acid + Choline	625
4. Lyases	Aconitase	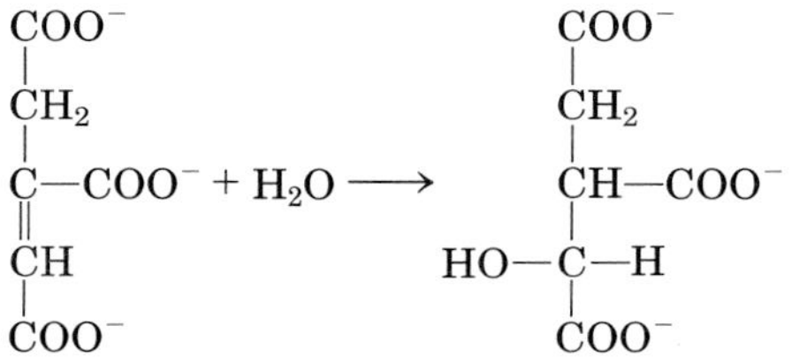 $^-OOC-CH_2-C(COO^-)=CH-COO^- + H_2O \longrightarrow {}^-OOC-CH_2-CH(COO^-)-CH(OH)-COO^-$ *cis*-Aconitate ⟶ Isocitrate	651
5. Isomerases	Phosphohexose isomerase	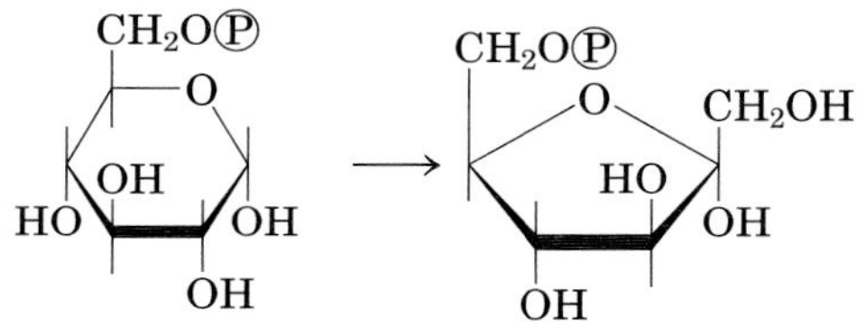 Glucose-6-phosphate ⟶ Fructose-6-phosphate	668
6. Ligases	Tyrosine-tRNA synthetase	ATP + L-tyrosine + tRNA ⟶ L-tyrosyltRNA + AMP + PP_i	724

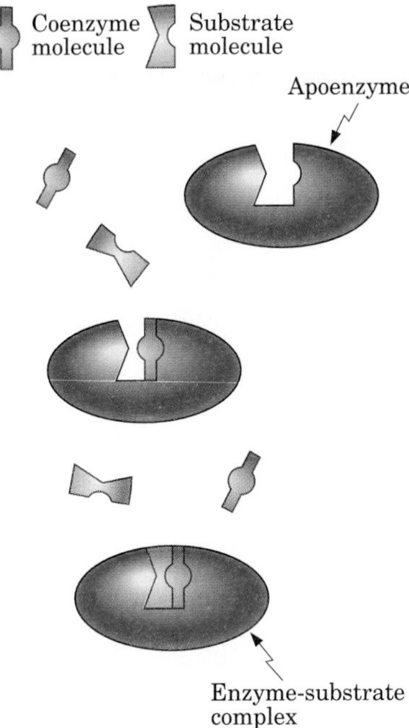

Figure 19.2
Schematic diagram of the active site of an enzyme and the participating components.

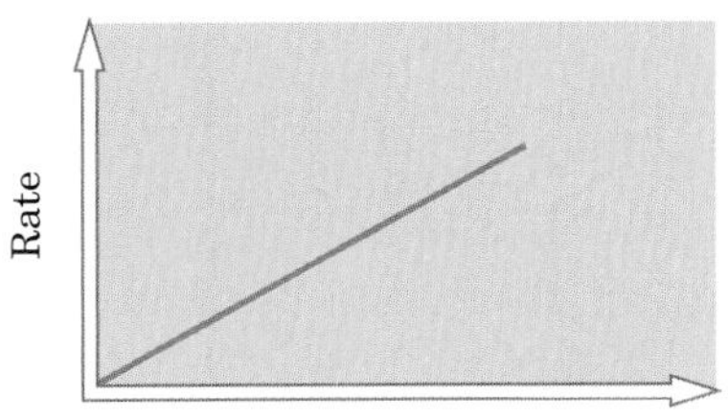

Figure 19.3
The effect of enzyme concentration on the rate of an enzyme-catalyzed reaction. Substrate concentration, temperature, and pH are constant.

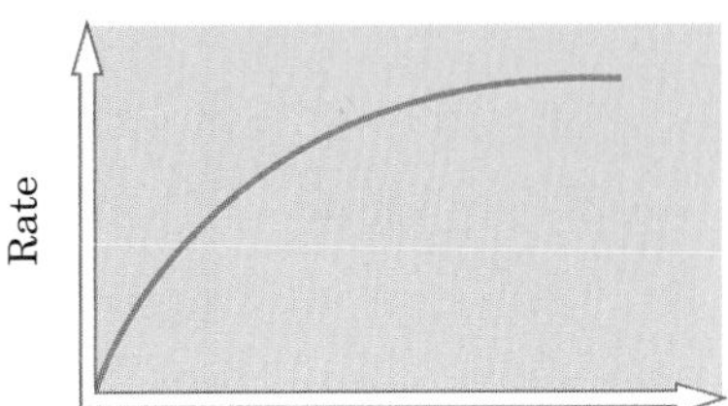

Figure 19.4
The effect of substrate concentration on the rate of an enzyme-catalyzed reaction. Enzyme concentration, temperature, and pH are constant.

The compound on which the enzyme works, the one whose reaction it speeds up, is called the **substrate.** The substrate usually adheres to the enzyme surface while it undergoes the reaction. There is a specific portion of the enzyme to which the substrate binds during the reaction. This part is called the **active site.** If the enzyme has coenzymes, they are located at the active site. Therefore the substrate is simultaneously surrounded by parts of the apoenzyme, coenzyme, and metal ion cofactor (if any), as shown in Figure 19.2.

Activation is **any process that makes an inactive enzyme active.** This can be the simple addition of a cofactor to an apoenzyme or the cleavage of a polypeptide chain of a proenzyme (Sec. 19.6). **Inhibition** is the opposite—**any process that makes an active enzyme less active or inactive** (Sec. 19.5). **Inhibitors** are compounds that accomplish this. Some inhibitors bind to the active site of the enzyme surface, thus preventing the binding of substrate. These are **competitive inhibitors.** Others, which bind to some other portion of the enzyme surface, may sufficiently alter the tertiary structure of the enzyme so that its catalytic effectiveness is slowed down. These are called **noncompetitive inhibitors.** Both competitive and noncompetitive inhibition are reversible, but there are some compounds that alter the structure of the enzyme *permanently* and thus make it *irreversibly* inactive.

19.4 Factors Affecting Enzyme Activity

Enzyme activity is a measure of how much reaction rates are increased. In this section we examine the effects of concentration, temperature, and pH on enzyme activity.

Enzyme and Substrate Concentration

If we keep the concentration of substrate constant and increase the concentration of enzyme, the rate increases linearly (Fig. 19.3). That is, if the enzyme concentration is doubled, the rate also doubles; if the enzyme concentration is tripled, the rate also triples. This is the case in practically all enzyme reactions because the molar concentration of enzyme is almost always much lower than that of substrate (that is, there are almost always many more molecules of substrate present than molecules of enzyme).

On the other hand, if we keep the concentration of enzyme constant and increase the concentration of substrate, we get an entirely different type of curve, called a saturation curve (Fig. 19.4). In this case the rate does not increase continuously. Instead, a point is reached after which the rate stays the same even if we increase the substrate concentration further. This happens because, at the saturation point, substrate molecules are bound to all the available active sites of the enzymes. Since the active sites are where the reactions take place, once they are all occupied the reaction is going at its maximum rate. Increasing the substrate concentration can no longer increase the rate because the excess substrate cannot find any active sites to attach to.

Temperature

Temperature affects enzyme activity because it changes the three-dimensional structure of the enzyme. In uncatalyzed reactions, the rate usually increases as the temperature increases. The effect of temperature on enzyme-catalyzed reactions is different. When we start at a low temperature (Fig. 19.5), an increase in temperature first causes an increase in rate. However, protein conformations are very sensitive to temperature changes. Once an optimum temperature is reached, any further increase in temperature causes changes in enzyme conformation. The substrate may then not fit properly onto the changed enzyme surface. Therefore the rate of reaction *decreases.*

After a *small* temperature increase above optimum, the (decreased) rate could still be increased again by lowering the temperature because, over a narrow temperature range, changes in conformation are reversible. However, at some higher temperature above optimum, we reach a point where the protein denatures (Sec. 18.10); the conformation is altered irreversibly, and the polypeptide chain cannot refold. At this point, the enzyme is completely inactivated. The inactivation of enzymes at low temperatures is used in the preservation of food by refrigeration.

The effect of temperature on reaction rates is discussed in Section 7.4.

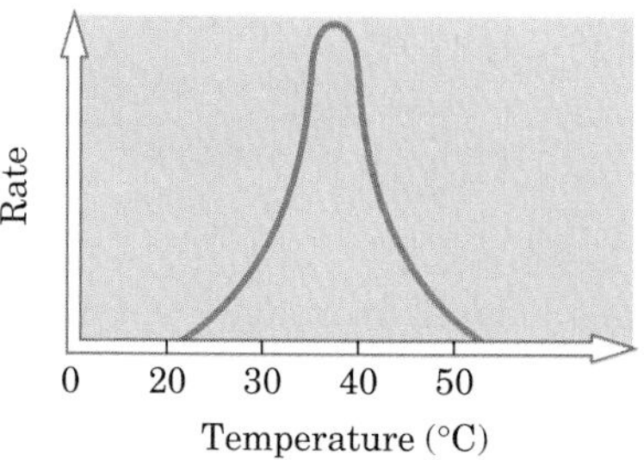

Figure **19.5**
The effect of temperature on the rate of an enzyme-catalyzed reaction. Substrate and enzyme concentrations and pH are constant.

pH

Since the pH of its environment changes the conformation of a protein (Sec. 18.10), we expect effects similar to those observed when the temperature is changed. Each enzyme operates best at a certain pH (Fig. 19.6). Once again, within a narrow pH range, changes in enzyme activity are reversible. However, at extreme pH values (either acidic or basic), enzymes are denatured irreversibly, and enzyme activity cannot be restored by changing back to the optimal pH.

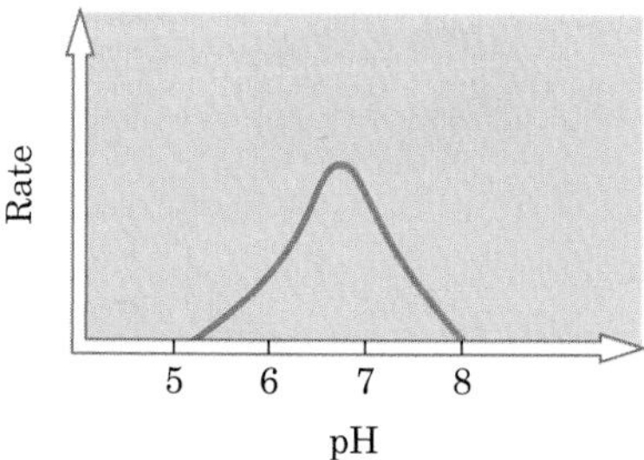

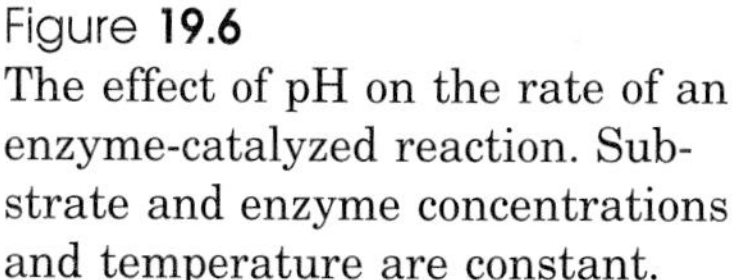

Figure **19.6**
The effect of pH on the rate of an enzyme-catalyzed reaction. Substrate and enzyme concentrations and temperature are constant.

19.5 Mechanism of Enzyme Action

We have seen that the action of enzymes is highly specific. What kind of mechanism can account for such specificity? It was suggested by Arrhenius about 100 years ago that catalysts speed up reactions by combining with the substrate to form some kind of intermediate compound. In an enzyme-catalyzed reaction, the intermediate is the **enzyme-substrate complex.**

To account for the high specificity of most enzyme-catalyzed reactions, a number of models have been proposed. The simplest and most frequently quoted is the **lock-and-key model** (Fig. 19.7). This model assumes that the enzyme is a rigid three-dimensional body. The surface that contains the active site has a restricted opening into which only one kind of substrate can fit, just as only the proper key can fit exactly into a lock.

According to the lock-and-key mechanism, an enzyme molecule has its own particular shape because that shape is necessary to maintain the active site in exactly the geometric alignment required for that particular

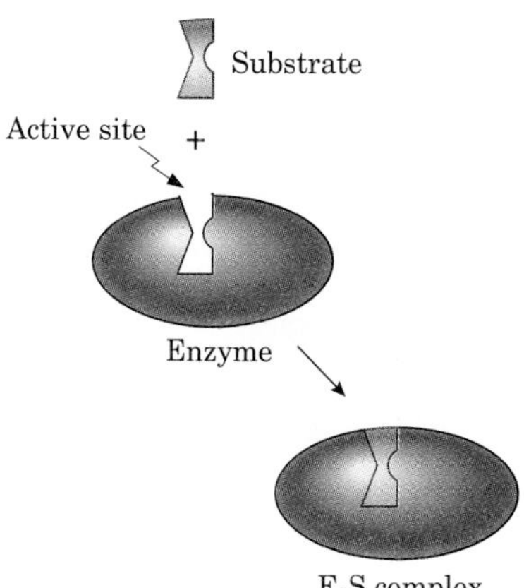

Figure **19.7**
The lock-and-key model of enzyme mechanism.

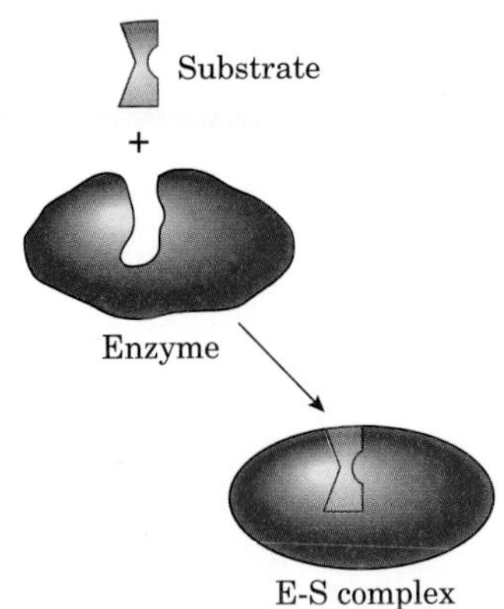

Figure 19.8
The induced-fit model of enzyme mechanism.

reaction. An enzyme molecule is very large (typically 100 to 200 amino acid residues), but the active site is usually composed of only two or a few amino acid residues, which may well be located at different places in the chain. The other amino acids—those not part of the active site—are located in the sequence in which we find them because that is the sequence that causes the whole molecule to fold up in exactly the required way.

The lock-and-key model explains the action of many enzymes. But for other enzymes, there is evidence that this model is too restrictive. Enzyme molecules are in a dynamic state, not a static one. There are constant motions within them, so that the active site has some flexibility.

From x-ray diffraction we know that the size and shape of the active site cavity change when the substrate enters. The American biochemist Daniel Koshland introduced the **induced-fit model** (Fig. 19.8), in which he compared the changes occurring in the shape of the cavity upon substrate binding to the changes in the shape of a glove when a hand is inserted. That is, the enzyme modifies the shape of the active site to accommodate the substrate.

Both the lock-and-key and the induced-fit models explain the phenomenon of competitive inhibition (Sec. 19.3). The inhibitor molecule fits into the active site cavity in the same way the substrate does (Fig. 19.9), preventing the substrate from entering. The result is that whatever reaction is supposed to take place on the substrate does not take place.

Many cases of noncompetitive inhibition can also be explained by either model. In this case, the inhibitor does not bind to the active site but to a different part of the enzyme. Nevertheless, the binding causes a change in the three-dimensional shape of the enzyme molecule, and this so alters the shape of the active site (the lock) that the substrate (the key) can no longer fit (Fig. 19.10).

If we compare enzyme activity in the presence and absence of an inhibitor, we can tell whether competitive or noncompetitive inhibition is taking place (Fig. 19.11). The maximum reaction rate is the same without an inhibitor and in the presence of a competitive inhibitor. The only difference is that this maximum rate is achieved at a low substrate con-

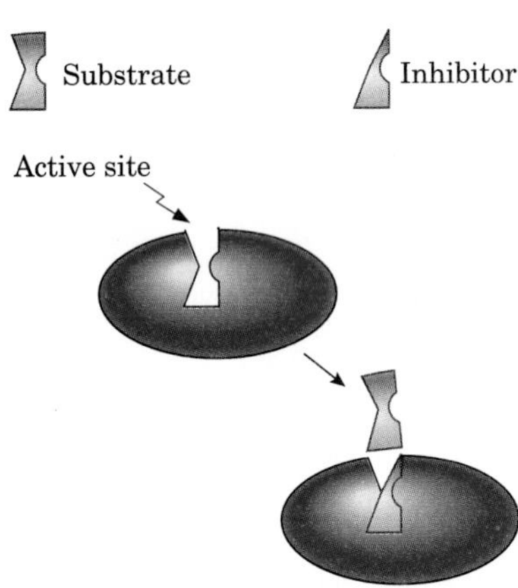

Figure 19.9
The mechanisn of competitive inhibition. When a competitive inhibitor enters the active site, the substrate cannot get in.

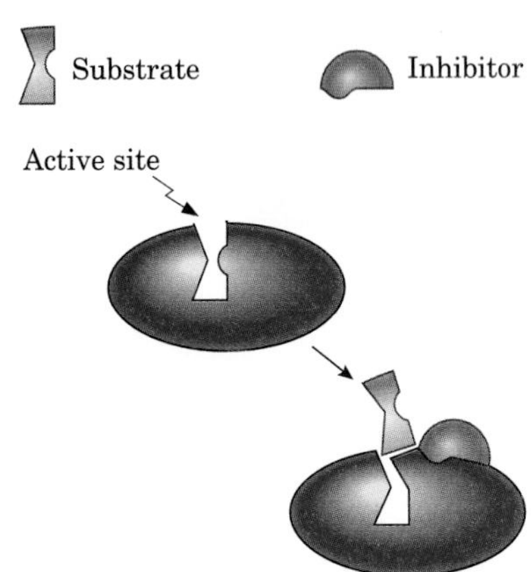

Figure 19.10
Mechanism of noncompetitive inhibition. The inhibitor attaches itself to a site other than the active site (allosterism) and thereby changes the conformation of the active site.

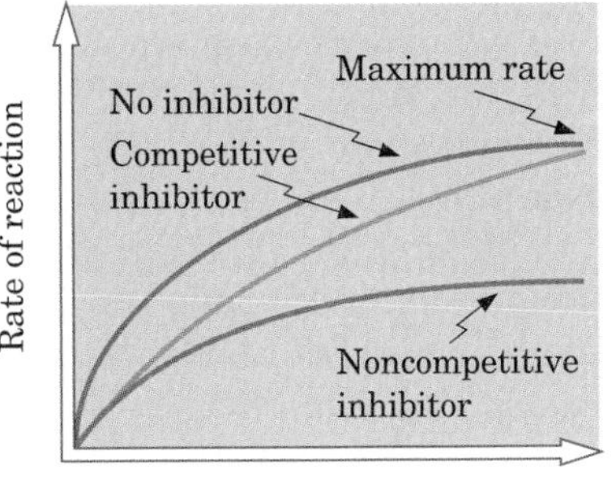

Figure 19.11
Enzyme kinetics in the presence and absence of inhibitors.

Box 19C

Active Sites

The perception of the active site as either a rigid (lock-and-key model) or a partly flexible template (induced-fit model) is an oversimplification. Not only is the geometry of the active site important, but so are the specific interactions that take place between enzyme surface and substrate. To illustrate, we take a closer look at the active site of the enzyme pyruvate kinase. This enzyme catalyzes the transfer of the phosphate group from phosphoenol pyruvate (PEP) to ADP, an important step in glycolysis (Sec. 21.2):

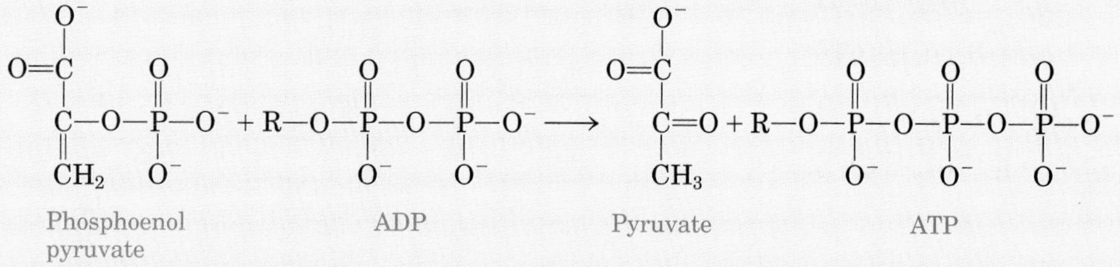

$$\text{Phosphoenol pyruvate} + \text{ADP} \longrightarrow \text{Pyruvate} + \text{ATP}$$

Phosphoenol pyruvate ADP Pyruvate ATP

The active site of the enzyme binds both substrates, PEP and ADP. The enzyme has two cofactors, K^+ and Mg^{2+}. The K^+ binds the carboxyl group of the PEP, and the Mg^{2+} anchors two phosphate groups, one from the PEP and one from the ADP. Other side chains of the apoenzyme bind the rest of the ADP into the active site. All these acids are in the form of their ions. There is also a hydrophobic area on the enzyme that binds the nonpolar $=CH_2$ unit.

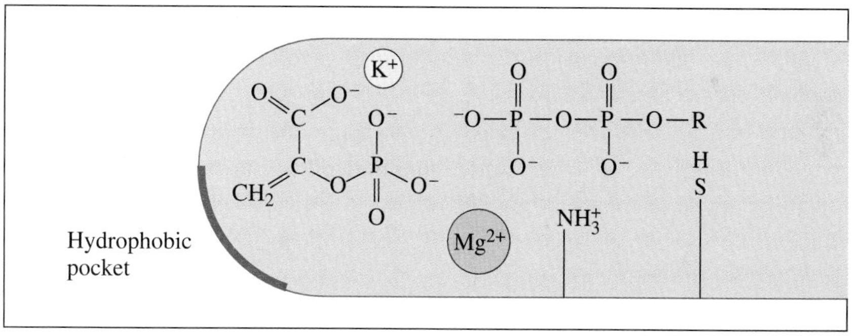

Hydrophobic pocket

The active site and the substrates of pyruvate kinase.

centration with no inhibitor but at a high substrate concentration when an inhibitor is present. This is the true sign of competitive inhibition because here the substrate and the inhibitor are competing for the same active site. If the substrate concentration is sufficiently increased, the inhibitor will be displaced from the active site by Le Chatelier's principle.

If, on the other hand, the inhibitor is noncompetitive, it cannot be displaced by addition of excess substrate, since it is bound to a different site. In this case the enzyme cannot be restored to its maximum activity, and the maximum rate of the reaction is lower than it would be in the absence of the inhibitor.

Box 19D

Sulfa Drugs as Competitive Inhibitors

The vitamin folic acid is a coenzyme in a number of biosynthetic processes, such as the synthesis of amino acids and of nucleotides. Humans obtain folic acid from the diet or from microorganisms in the intestinal tract. These microorganisms can synthesize folic acid if *para*-aminobenzoic acid is available to them:

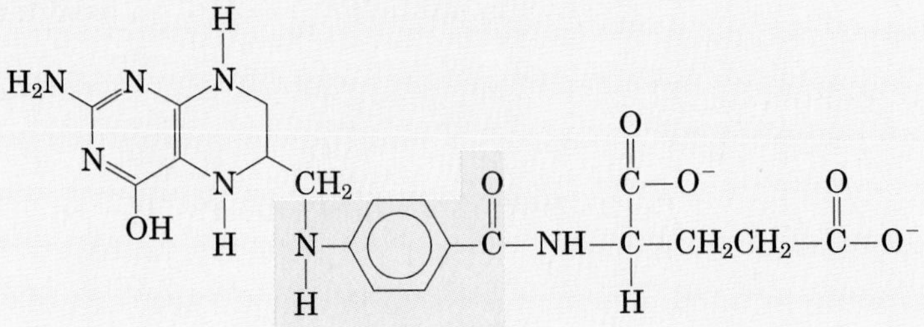

Folic acid

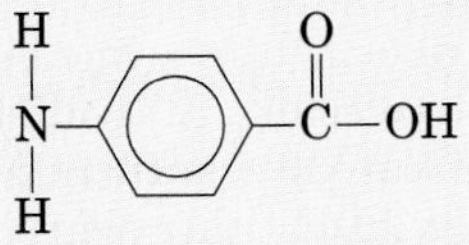

p-Aminobenzoic acid

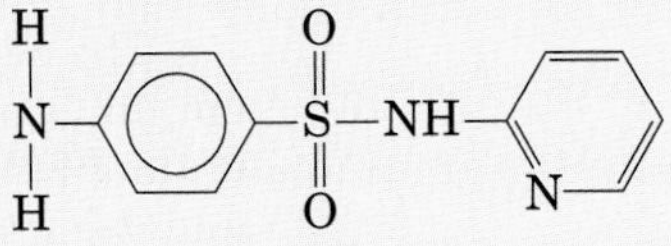

Sulfapyridine

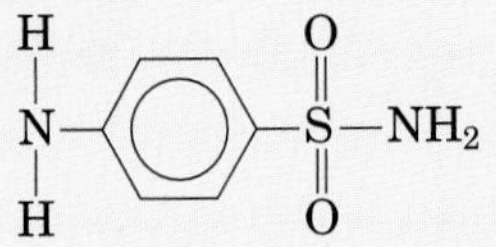

Sulfanilamide

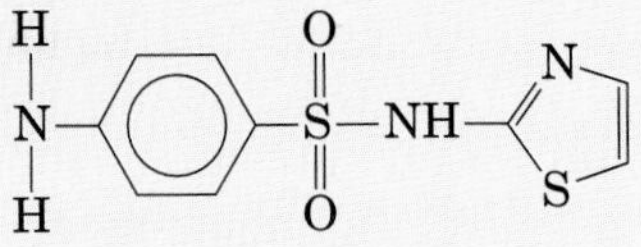

Sulfathiazole

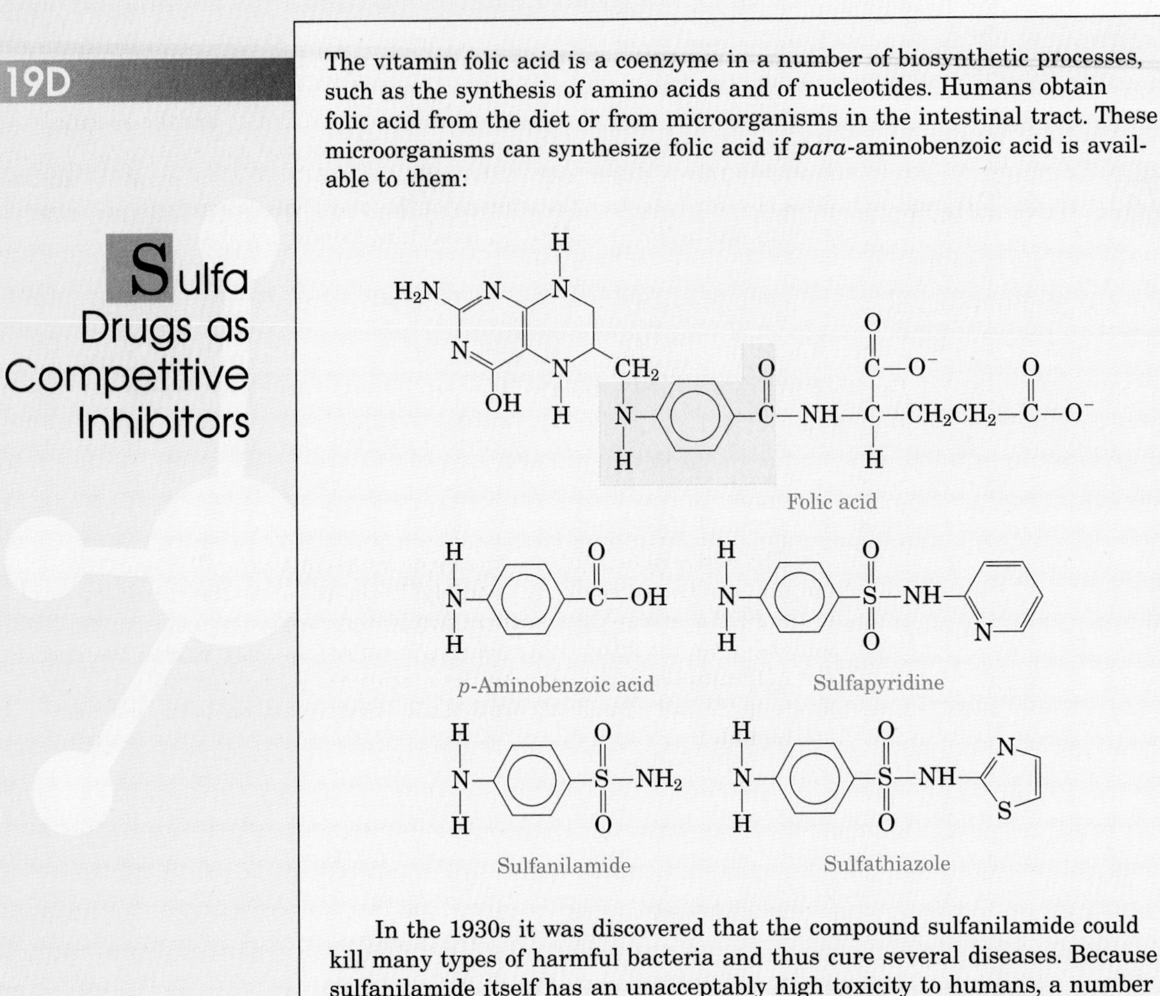

In the 1930s it was discovered that the compound sulfanilamide could kill many types of harmful bacteria and thus cure several diseases. Because sulfanilamide itself has an unacceptably high toxicity to humans, a number of derivatives of this compound, such as sulfapyridine and sulfathiazole, which act in a similar way, are used instead. These drugs work by "tricking" the bacteria, which normally use *para*-aminobenzoic acid as a raw material in the synthesis of folic acid. When the bacteria get a sulfa drug instead, they cannot tell the difference and use it to make a molecule that also has a folic acid type of structure but is not exactly the same. When they try to use this fake folic acid as a coenzyme, not only doesn't it work, but it is now a competitive inhibitor of the enzyme's action, so that many of the bacteria's amino acids and nucleotides cannot be made, and the bacteria die.

19.6 Enzyme Regulation

Feedback Control

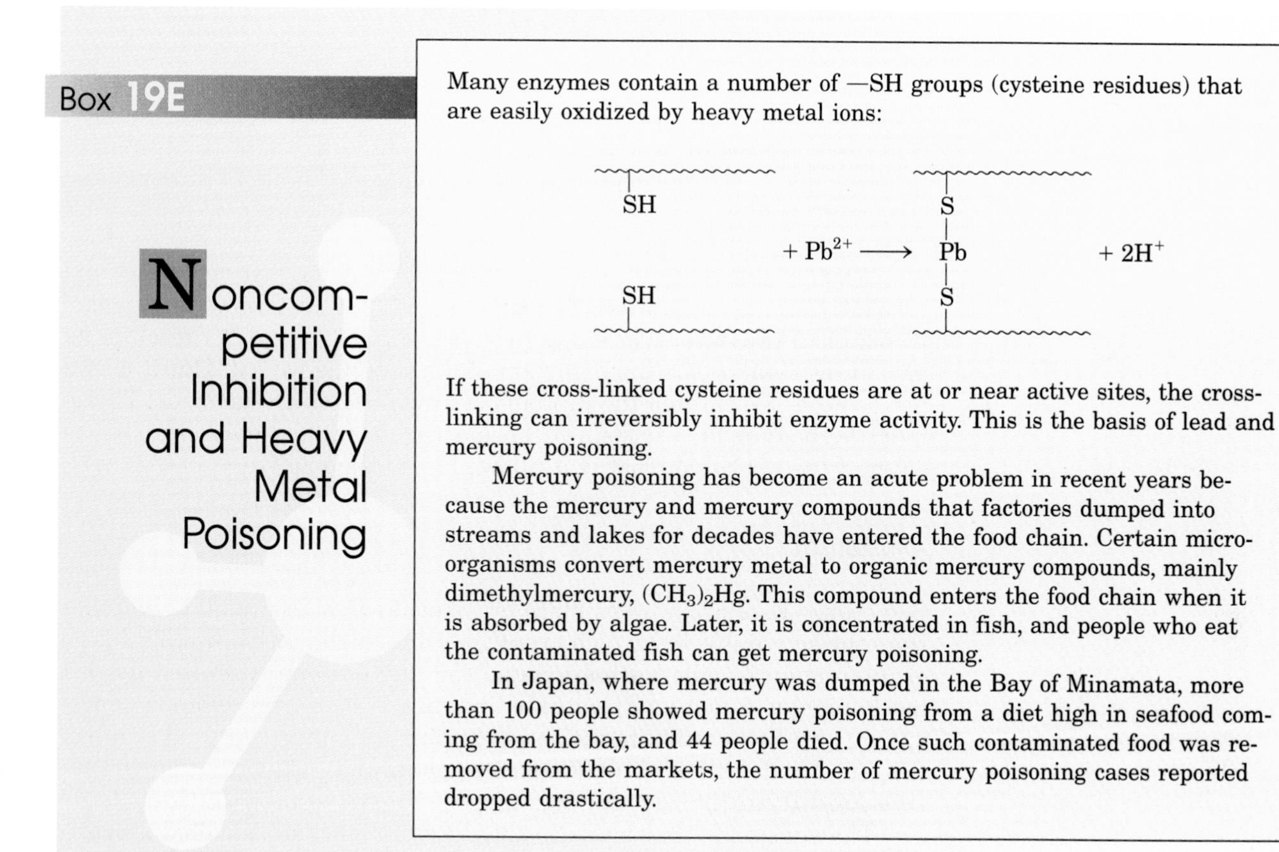

Enzymes are often regulated by environmental conditions. The reaction product of one enzyme may control the activity of another, especially in

Box 19E Noncompetitive Inhibition and Heavy Metal Poisoning

Many enzymes contain a number of —SH groups (cysteine residues) that are easily oxidized by heavy metal ions:

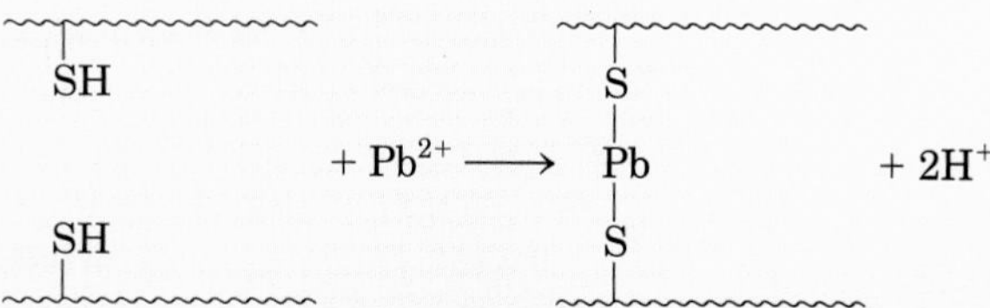

If these cross-linked cysteine residues are at or near active sites, the cross-linking can irreversibly inhibit enzyme activity. This is the basis of lead and mercury poisoning.

Mercury poisoning has become an acute problem in recent years because the mercury and mercury compounds that factories dumped into streams and lakes for decades have entered the food chain. Certain microorganisms convert mercury metal to organic mercury compounds, mainly dimethylmercury, $(CH_3)_2Hg$. This compound enters the food chain when it is absorbed by algae. Later, it is concentrated in fish, and people who eat the contaminated fish can get mercury poisoning.

In Japan, where mercury was dumped in the Bay of Minamata, more than 100 people showed mercury poisoning from a diet high in seafood coming from the bay, and 44 people died. Once such contaminated food was removed from the markets, the number of mercury poisoning cases reported dropped drastically.

a complex system in which enzymes work cooperatively. For example, in a system

$$A \xrightarrow{E_1} B \xrightarrow{E_2} C \xrightarrow{E_3} D$$

each step is catalyzed by a different enzyme. The last product in the chain, D, may inhibit the activity of enzyme E_1 (by competitive or noncompetitive inhibition). When the concentration of D is low, the three reactions proceed rapidly, but as the concentration of D increases, the action of E_1 becomes inhibited and eventually stops. In this manner, the accumulation of D is a *message* telling enzyme E_1 to shut down because the cell has enough D for its present needs. Shutting down E_1 stops the whole process. Enzyme regulation by a process in which formation of a product inhibits an earlier reaction in the sequence is called **feedback control.**

Proenzymes

Some enzymes are manufactured by the body in an inactive form. In order to make them active, a small part of their polypeptide chain must be removed. These inactive forms of enzymes are called **proenzymes** or **zymogens.** Once the excess polypeptide chain is removed, the enzyme becomes active. For example, trypsin is manufactured as the inactive molecule trypsinogen (a zymogen). When a fragment containing six amino acid residues is removed from the N-terminal end, the molecule becomes a fully active trypsin molecule. Removal of the fragment not only shortens the chain but also changes the three-dimensional structure (the tertiary structure), allowing the molecule to achieve its active form.

The removal of the six-amino-acid fragment is of course also catalyzed by an enzyme.

Why does the body go to all this trouble? Why not just make the fully active trypsin to begin with? The reason is very simple. As we have seen, trypsin is a protease—it cleaves proteins (Fig. 19.1) and is therefore an important catalyst for the digestion of the proteins we eat. But it would not be good if it cleaved the proteins our own bodies are made of! Therefore, the body makes trypsin in an inactive form, and only after it has entered the digestive tract is it allowed to become active.

Allosterism

Sometimes regulation takes place by an event that occurs at a site other than the active site, but which eventually affects the active site. This type of interaction is called **allosterism,** and any enzyme regulated by this mechanism is called an **allosteric enzyme.** If a substance binds noncovalently and reversibly to a site *other than the active site,* it may affect the enzyme in either of two ways: It may inhibit enzyme action (**negative modulation**) or it may stimulate it (**positive modulation**).

The substance that binds to the allosteric enzyme is called a **regulator,** and the site it attaches to is called the **regulatory site.** In most cases, allosteric enzymes contain more than one polypeptide chain (subunits); the regulatory site is on one polypeptide chain and the active site on another.

Specific regulators can bind reversibly to the regulatory sites. For example, the enzyme protein kinase is an allosteric enzyme. In this case the enzyme has only one polypeptide chain, so it carries both the active site and the regulatory site at different parts of this chain (Fig. 19.12). The regulator is another protein molecule, one that binds reversibly to the regulatory site. As long as the regulator is bound to the regulatory site, the total enzyme-regulator complex is inactive. When the regulator is removed from the regulatory site the protein kinase becomes active. Thus **the allosteric enzyme action is controlled by the regulator.**

Allosteric regulation may occur with proteins other than enzymes. Section 25.3 explains how the oxygen-carrying ability of hemoglobin, an allosteric protein, is affected by modulators.

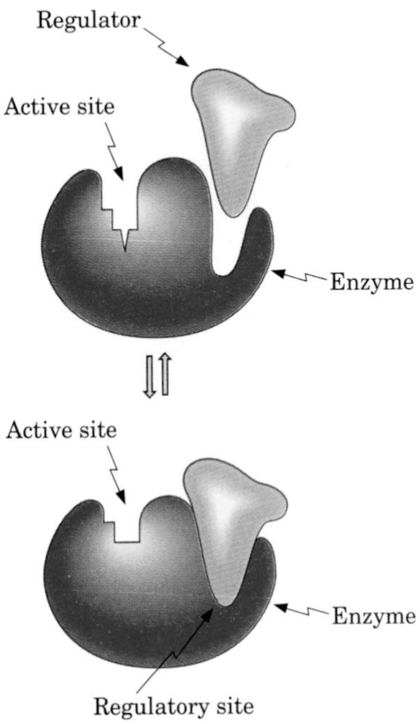

Figure **19.12**
The allosteric effect. Binding of a regulator to a site other than the active site changes the shape of the active site.

Isoenzymes

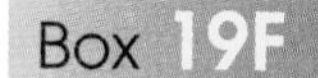

Another form of regulation of enzyme activity occurs when the same enzyme appears in different forms in different tissues. Lactate dehydrogenase catalyzes the conversion of lactate to pyruvate, and vice versa (Fig. 21.3, step ⑪). The enzyme has four subunits. Two kinds of subunits exist, called H and M. The enzyme that dominates in the heart is an H_4

Box 19F

Penicillin: War of Enzyme Against Enzyme

Since World War II, the most widely used antibacterial agent has been penicillin, which is produced by a mold and prevents the growth of bacteria. Penicillin was discovered accidentally by Alexander Fleming (1881–1955) in 1929 and is now produced in large quantities. It inhibits an enzyme called transpeptidase. Bacteria need this enzyme to make their cell walls rigid and cross-linked. In the presence of penicillin, bacterial cell walls are not cross-linked, so that the contents of the bacterial cells cannot be maintained. The cytoplasm spills out, and the bacteria die.

Penicillin caused a revolution in medicine and paved the way for a host of antibiotics, all products of microorganisms directed against other microorganisms. In the last 40 years penicillin has been overused. In many countries it is sold without prescription and has also been used in animal feed. This happens in spite of the fact that some people are severely allergic to penicillin, so that administering the drug may cause shock and fatal coma.

New bacterial strains that can resist penicillin have appeared in the last 25 to 30 years. These new bacterial strains contain the enzyme penicillinase. Penicillin contains a *beta*-lactam ring. The penicillinase opens this ring and makes penicillin ineffective:

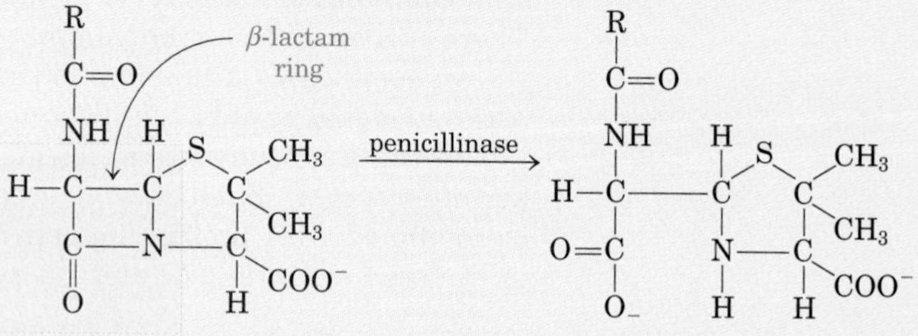

β-lactam ring

penicillinase →

Penicillin G (R = $C_6H_5CH_2$)

Penicilloic acid

The rapid evolution of the new strains of bacteria threatened the effectiveness of this antibiotic, but research is always a step ahead in this continuous war. New synthetic penicillins have been developed with R groups that prevent the attack of penicillinase.

Even more effective is a drug called clavulinic acid. This compound has no antibacterial property; that is, it does not kill bacteria. However, it acts as an irreversible inhibitor of penicillinase. It can do so because, like penicillin, it has its own *beta*-lactam ring that irreversibly binds to the active site of penicillinase. Therefore, when given in combination with penicillin, it makes the penicillin effective against all strains of bacteria.

enzyme, meaning that all four subunits are of the H type, although some M type subunits are also present. In the liver and skeletal muscles the M type dominates. Other types of tetramer (four units) combinations exist in different tissues: H_3M, H_2M_2, and HM_3. These different forms of the same enzyme are called **isozymes** or **isoenzymes.** The H_4 enzyme is allosterically inhibited by high levels of pyruvate, while the M_4 is not. In diagnosing the severity of heart attacks (Sec. 19.7), the release of H_4 isoenzyme is monitored in the serum.

19.7 Enzymes in Medical Diagnosis and Treatment

Most enzymes are confined within the cells of the body. However, small amounts of enzymes can also be found in body fluids such as blood, urine, and cerebrospinal fluid. The level of enzyme activity in these fluids can easily be monitored. It has been found that abnormal activity (either high or low) of particular enzymes in various body fluids signals either the onset of certain diseases or their progression. Table 19.2 lists some enzymes used in medical diagnosis and their activities in normal body fluids.

A number of enzymes are assayed (measured) during myocardial infarction in order to diagnose the severity of the heart attack. Dead heart muscle cells spill their enzyme contents into the serum. Thus, the level of aspartate aminotransferase (AST) (formerly called glutamate-oxaloacetate transaminase, or GOT) in the serum rises rapidly after a heart attack. In addition to AST, lactate dehydrogenase (LD-P) and creatine kinase (CK) levels are also monitored. In infectious hepatitis, the alanine aminotransferase (ALT) (formerly called glutamate-pyruvate transaminase, or GPT) level in the serum can rise to ten times normal. There is also a concurrent increase in AST activity in the serum.

In some cases, the administration of an enzyme is part of therapy. After duodenal or stomach ulcer operations, patients are advised to take tablets containing digestive enzymes that are in short supply in the stomach after surgery. Such enzyme preparations contain lipases, either alone or combined with proteolytic enzymes, and are sold under such names as Pancreatin, Acro-lase, and Ku-zyme.

Table 19.2 Enzyme Assays Useful in Medical Diagnosis

Enzyme	Normal Activity	Body Fluid	Disease Diagnosed
Alanine aminotransferase (ALT)	3–17 U/L[a]	Serum	Hepatitis
Acid phosphatase	2.5–12 U/L	Serum	Prostate cancer
Alkaline phosphatase (ALP)	13–38 U/L	Serum	Liver or bone disease
Amylase	19–80 U/L	Serum	Pancreatic disease or mumps
Aspartate aminotransferase (AST)	7–19 U/L	Serum	Heart attack or hepatitis
	7–49 U/L	Cerebrospinal fluid	Heart attack or hepatitis
Lactate dehydrogenase (LD-P)	100–350 WU/mL	Serum	Heart attack
Creatine kinase (CK)	7–60 U/L	Serum	Heart attack
Phosphohexose isomerase (PHI)	15–75 U/L	Serum	Heart attack

[a] U/L = International units per liter; WU/mL = Wrobleski units per milliliter.

SUMMARY

Enzymes are proteins that catalyze chemical reactions in the body. Most enzymes are very specific—they catalyze only one particular reaction. The compound whose reaction is catalyzed by an enzyme is called the **substrate.** Enzymes are classified into six major groups according to the type of reaction they catalyze. Most enzymes are named after the substrate and the type of reaction they catalyze, adding the ending "**-ase.**"

Some enzymes are made of polypeptide chains only. Others have, besides the polypeptide chain (the **apoenzyme**), nonprotein **cofactors,** either organic compounds (**coenzymes**) or inorganic ions. Only a small part of the enzyme surface participates in the actual catalysis of chemical reactions. This part is called the **active site.** Cofactors, if any, are part of the active site.

Compounds that slow enzyme action are called **inhibitors.** A **competitive inhibitor** attaches itself to the active site. A **noncompetitive inhibitor** binds to other parts of the enzyme surface. The higher the enzyme and substrate concentrations, the higher the enzyme activity, except that, at sufficiently high substrate concentrations, a saturation point is reached. After this, increasing substrate concentration no longer increases the rate. Each enzyme has an optimum temperature and pH at which it has its greatest activity.

Two closely related mechanisms by which enzyme activity and specificity are explained are the **lock-and-key model** and the **induced-fit model.**

Enzyme activity is regulated by four mechanisms. In **feedback control,** the concentration of products influences the rate of the reaction. In **allosterism,** an interaction takes place at a position other than the active site but affects the active site, either positively or negatively. Some enzymes, called **proenzymes** or **zymogens,** must be activated by removing a small portion of the polypeptide chain. Finally, enzyme activity is also regulated by **isozymes,** which are different forms of the same enzyme.

Abnormal enzyme activity can be used to diagnose certain diseases.

KEY TERMS

Activation (Sec. 19.3)
Active site (Sec. 19.3)
Allosterism (Sec. 19.6)
Apoenzyme (Sec. 19.3)
Coenzyme (Sec. 19.3)
Cofactor (Sec. 19.3)
Competitive inhibition (Sec. 19.3)
Enzyme (Sec. 19.1)
Enzyme activity (Sec. 19.4)
Enzyme-substrate complex (Sec. 19.5)
Feedback control (Sec. 19.6)
Induced-fit model (Sec. 19.5)
Inhibition (Sec. 19.3)
Isoenzyme (Sec. 19.6)
Isozyme (Sec. 19.6)
Lock-and-key model (Sec. 19.5)
Noncompetitive inhibition (Sec. 19.3)
Proenzyme (Sec. 19.6)
Regulator (Sec. 19.6)
Ribozyme (Sec. 19.1)
Substrate (Sec. 19.3)
Zymogen (Sec. 19.6)

PROBLEMS

Enzymes

19.1 What is the difference between the terms catalyst and enzyme?

19.2 Are all enzymes proteins?

19.3 How many enzymes are in one single cell?

19.4 What does an enzyme do to the energy of activation of a reaction?

19.5 Why does the body need so many different enzymes?

19.6 Trypsin splits polypeptide chains at the carboxyl side of a lysine or arginine residue (Fig. 19.1). Chymotrypsin splits polypeptide chains on the carboxyl side of an aromatic residue or any other nonpolar, bulky residue. Which enzyme is more specific?

Naming and Classifying Enzymes

***19.7** Lipases are less specific than trypsin. Explain why this is so.

19.8 Monoamine oxidases are important enzymes in brain chemistry. Judging from the name which of these would be a suitable substrate for this class of enzymes:

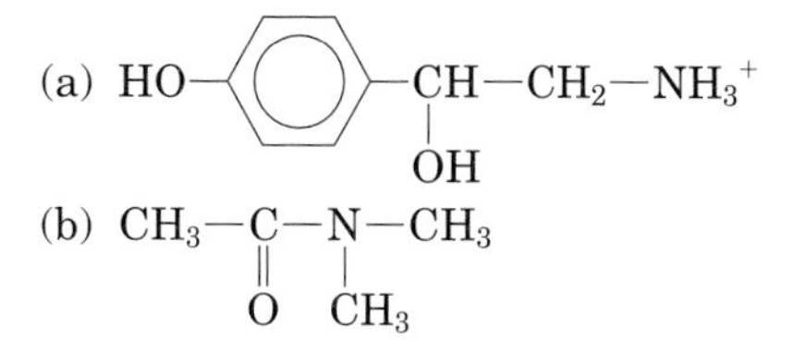

(a) $HO-C_6H_4-CH(OH)-CH_2-NH_3^+$

(b) $CH_3-C(=O)-N(CH_3)-CH_3$

(c) $C_6H_5-NO_2$

***19.9** On the basis of the classification given in Section 19.2, decide to which group each of the following enzymes belongs:

Enzyme	Reaction
(a) phosphoglyceromutase	
	$^{-}OOC-CH(OH)-CH_2-OPO_3^{2-} \rightleftharpoons$
	3-Phosphoglycerate
	$^{-}OOC-CH(OPO_3^{2-})-CH_2OH$
	2-Phosphoglycerate
(b) urease	
	$H_2N-C(=O)-NH_2 + H_2O \rightleftharpoons 2NH_3 + CO_2$
	Urea
(c) succinate dehydrogenase	
	$^{-}OOC-CH_2-CH_2-COO^{-}$ + FAD $\rightleftharpoons$
	Succinate, Coenzyme
	$^{-}OOC-CH{=}CH-COO^{-} + FADH_2$
	Fumarate, Reduced coenzyme
(d) aspartase	
	$^{-}OOC-CH{=}CH-COO^{-} + NH_4^{+} \rightleftharpoons$
	Fumarate
	$^{-}OOC-CH_2-CH(NH_2)-COO^{-} + H^{+}$
	L-Aspartate

19.10 What kind of reaction does each of the following enzymes catalyze?
(a) deaminases (c) dehydrogenases
(b) hydrolases (d) isomerases

19.11 What is the difference between a coenzyme and a cofactor?

19.12 In the citric acid cycle an enzyme converts succinate to fumarate. The enzyme consists of a protein portion and an organic molecule portion called FAD. What term do we use to refer to (a) the protein portion and (b) the organic molecule portion?

19.13 What is the difference between reversible and irreversible noncompetitive inhibition?

19.14 Which kind of inhibitor enters the active site: (a) a competitive inhibitor or (b) a noncompetitive inhibitor?

Factors Affecting Enzyme Activity

***19.15** At a very low concentration of a certain substrate we find that when the substrate concentration is doubled the rate of the enzyme-catalyzed reaction is also doubled. Would you expect to find the same at a very high substrate concentration?

19.16 If we wish to double the rate of an enzyme-catalyzed reaction, can we do this by increasing the temperature 10°C?

19.17 A bacterial enzyme has a temperature-dependent activity as follows:

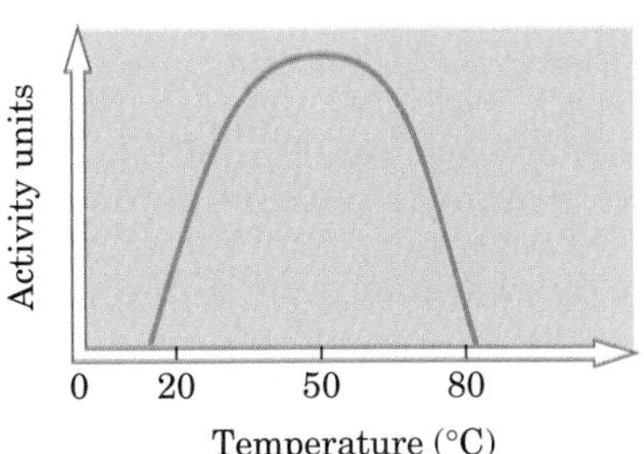

(a) Is this enzyme more or less active at normal body temperature than when a person has a fever? (b) What happens to the enzyme activity if the patient's temperature is lowered to 35°C?

19.18 The optimum temperature for the action of lactate dehydrogenase is 36°C. It is irreversibly inactivated at 85°C, but a yeast containing this enzyme can survive for months at −10°C. Explain how this can happen.

Mechanism of Enzyme Action

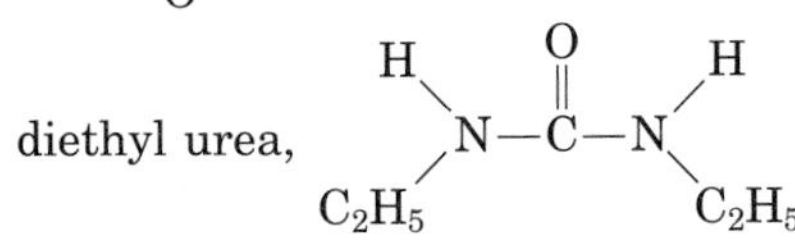

19.19 Urease can catalyze the hydrolysis of urea, $H_2N-C(=O)-NH_2$, but not the hydrolysis of diethyl urea, $(C_2H_5)(H)N-C(=O)-N(H)(C_2H_5)$. Explain why diethyl urea is not hydrolyzed.

19.20 Which is a correct statement describing the induced-fit model of enzyme action? Substrates fit into the active site (a) because both are exactly the same size and shape (b) by changing their size and shape to match those of the active site (c) by a change in the size and shape of the active site upon binding.

***19.21** What is the maximum rate that can be achieved in competitive inhibition compared with noncompetitive inhibition?

19.22 Enzymes are long protein chains, usually containing more than 100 amino acid residues. Yet the active site contains only a few amino acids.

Explain why all the other amino acids of the chain are present and what would happen to the enzyme activity if significant changes were made in the structure.

Enzyme Regulation

19.23 The decomposition of glycogen to yield glucose is catalyzed by the enzyme phosphorylase. Epinephrine, a hormone, increases the activity of the phosphorylase by converting the inactive phosphorylase *b* to the active *a* form, without binding to the active site of phosphorylase. What kind of regulatory mechanism is at work?

19.24 Can the product of a reaction in a sequence act as an inhibitor for another reaction of the sequence? Explain.

19.25 What is the difference between a zymogen and a proenzyme?

19.26 The enzyme trypsin is synthesized by the body in the form of a long chain (trypsinogen) from which a piece must be cut before the trypsin can be active. Why does the body not synthesize the trypsin directly?

19.27 Discuss four ways by which the body regulates enzyme action.

19.28 What is an isozyme?

Enzymes in Medical Diagnosis and Treatment

***19.29** The enzyme formerly known as GPT (glutamate-pyruvate transaminase) has a new name: ALT (alanine aminotransferase). Looking at the equation at the bottom of page 682, which is catalyzed by this enzyme, what prompted this change of name?

19.30 If an examination of a patient indicated elevated levels of AST but normal levels of ALT, what would be your tentative diagnosis?

19.31 Which LD-P isozyme is monitored in the case of a heart attack?

19.32 Chemists who have been exposed for years to organic vapors usually show higher-than-normal activity when given the alkaline phosphatase test. What organ in the body is affected by organic vapors?

19.33 What enzyme preparation is given to patients after duodenal ulcer surgery?

19.34 Chymotrypsin is secreted by the pancreas and passed into the intestine. The optimum pH for this enzyme is 7.8. If a patient's pancreas cannot manufacture chymotrypsin, would it be possible to supply it orally? What happens to the chymotrypsin activity during its passage through the gastrointestinal tract?

Boxes

19.35 (Box 19A) Acetylcholine causes muscles to contract. Succinylcholine, a close relative, is a muscle relaxant. Explain the different effects of these related compounds.

19.36 (Box 19A) An operating team usually administers succinylcholine before bronchoscopy. What is achieved by this procedure?

***19.37** (Box 19B) Which enzyme resembles papain in its action: trypsin or carboxypeptidase? Explain.

19.38 (Box 19C) What is the function of the hydrophobic pocket in the active site of pyruvate kinase?

***19.39** (Box 19C) Oxalate ion, $^{-}O-\underset{\underset{O}{\|}}{C}-\underset{\underset{O}{\|}}{C}-O^{-}$, acts as a competitive inhibitor of pyruvate kinase. Identify the potential sites where oxalate ion can bind to the active site of the enzyme.

19.40 (Box 19D) What part of folic acid is the key structure that sulfa drugs mimic?

***19.41** (Box 19D) Sulfa drugs kill certain bacteria by preventing the synthesis of folic acid, a vitamin. Why don't these drugs kill people?

19.42 (Box 19E) Which functional group is affected in mercury poisoning?

19.43 (Box 19F) How does clavulinic acid accomplish the killing of penicillin-resistant bacteria?

19.44 (Box 19F) (a) What enzyme does penicillin inhibit? (b) What is the function of this enzyme?

Additional Problems

***19.45** Food can be preserved by inactivation of enzymes that would cause spoilage, for example, by refrigeration. Give an example of food preservation in which the enzymes are inactivated by (a) heat and (b) lowering the pH.

19.46 Why is enzyme activity during myocardial infarction measured in the serum of patients rather than in the urine?

***19.47** The activity of pepsin was measured at various pH values. The temperature and the concentrations of pepsin and substrate were held constant. The following activities were obtained:

pH	Activity
1	0.5
1.5	2.6
2	4.8
3	2.0
4	0.4
5	0.0

(a) Plot the pH dependence of pepsin activity. (b) What is the optimum pH? (c) Predict the activity of pepsin in the blood at pH 7.4.

***19.48** What is the common characteristic of the amino acids of which the carboxyl groups of the peptide linkages can be hydrolysed by trypsin?

19.49 Many enzymes are active only in the presence of Zn^{2+}. What common term is used for ions like this when discussing enzyme activity?

19.50 An enzyme has the following pH dependence:

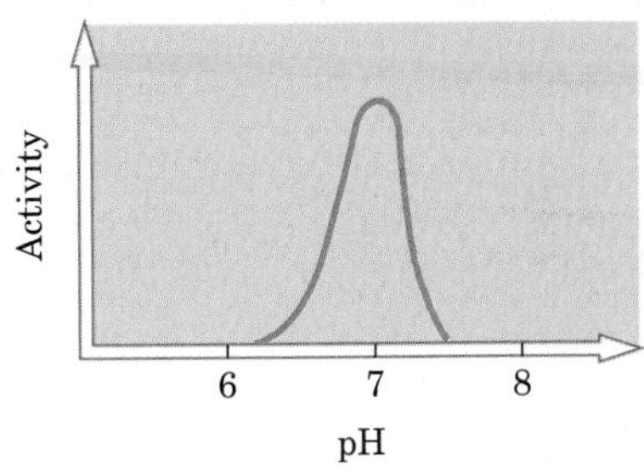

Where do you think this enzyme works best?

19.51 What enzyme is monitored in the diagnosis of infectious hepatitis?

19.52 The enzyme chymotrypsin catalyzes the following type of reaction:

$$\text{R—CH(CH}_2\text{—C}_6\text{H}_5\text{)—}\overset{\text{O}}{\overset{\|}{\text{C}}}\text{—NH—CH(CH}_3\text{)—R}' + \text{H}_2\text{O}$$

$$\downarrow$$

$$\text{R—CH(CH}_2\text{—C}_6\text{H}_5\text{)—}\overset{\text{O}}{\overset{\|}{\text{C}}}\text{—OH} + \text{H}_2\text{N—CH(CH}_3\text{)—R}'$$

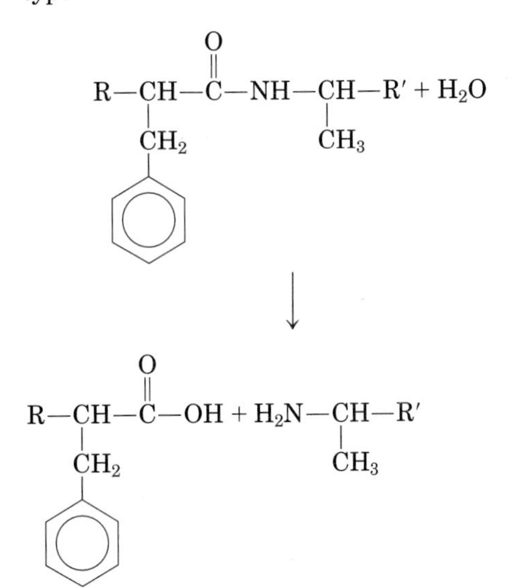

On the basis of the classification given in Section 19.2, which group of enzymes does chymotrypsin belong to?

***19.53** Nerve gases operate by forming covalent bonds at the active site of cholinesterase. Is this an example of competitive inhibition? Can the nerve gas molecules be removed by simply adding more substrate (acetylcholine) to the enzyme?

***19.54** What would be the appropriate name for an enzyme that catalyzes each of the following reactions.

(a) $\text{CH}_3\text{CH}_2\text{OH} \longrightarrow \text{CH}_3\underset{\text{O}}{\underset{\|}{\text{C}}}\text{—H} + \text{H}_2$

(b) $\text{CH}_3\underset{\text{O}}{\underset{\|}{\text{C}}}\text{—O—CH}_2\text{CH}_3 + \text{H}_2\text{O} \longrightarrow$

$$\text{CH}_3\underset{\text{O}}{\underset{\|}{\text{C}}}\text{—OH} + \text{CH}_3\text{CH}_2\text{OH}$$

19.55 On page 701 a reaction between pyruvate and glutamate, to form alanine and α-ketoglutarate, is given. How would you classify the enzyme that catalyzes this reaction?

19.56 Most enzymes operate best at one particular pH. Enzyme activity usually decreases at both higher and lower pH values. Give a possible explanation for this.

19.57 To what extent can enzymes increase the rate of reactions?

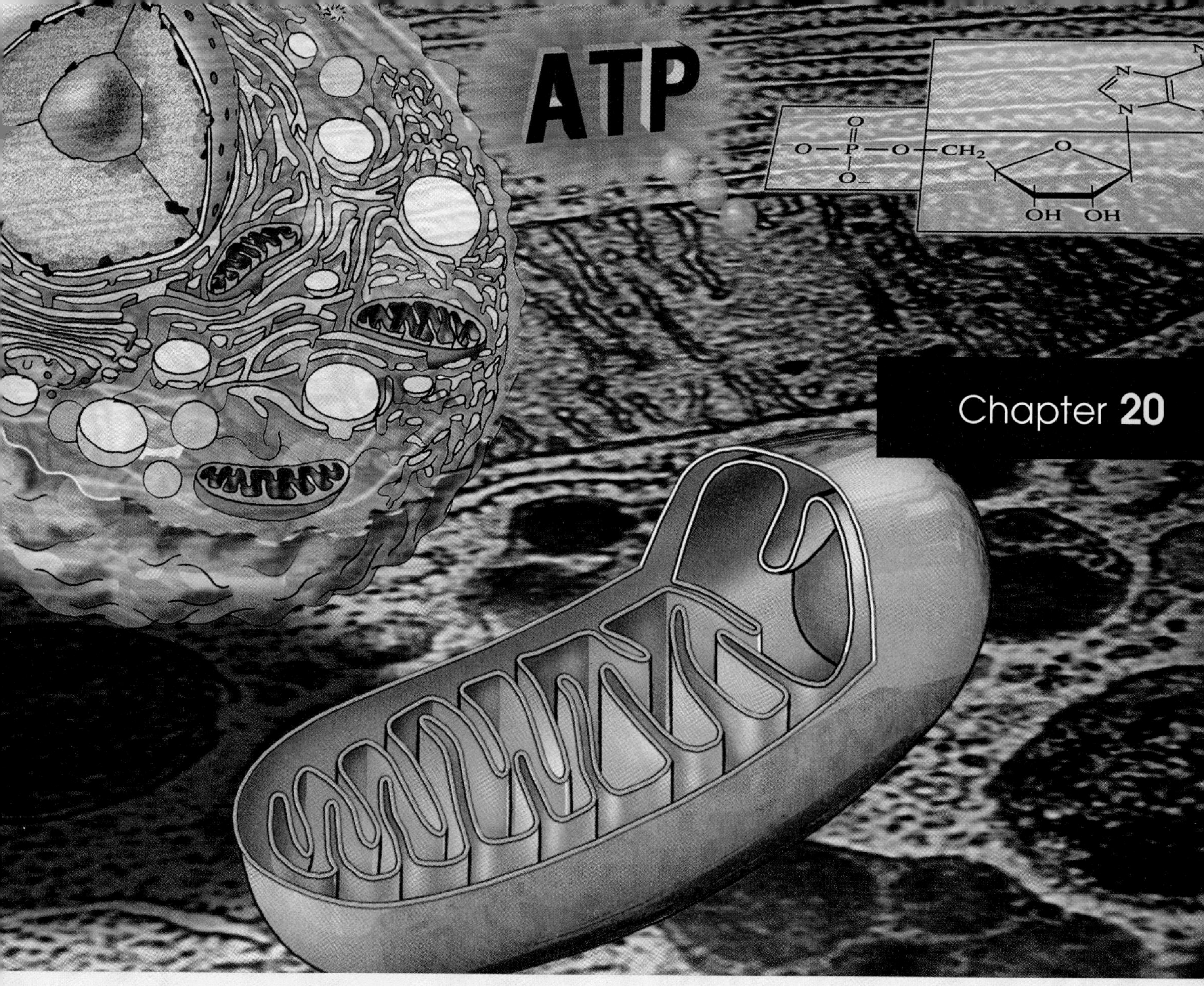

Chapter 20

Bioenergetics. How the body converts food to energy

20.1 Introduction

The same compounds may be synthesized in one part of a cell and broken down in a different part of the cell.

Living cells are in a dynamic state, which means that compounds are constantly being synthesized and then broken down into smaller fragments. Thousands of different reactions are taking place at the same time.

The sum total of all the chemical reactions involved in maintaining the dynamic state of the cell is called **metabolism.**

In general, we can divide metabolic reactions into two broad groups: (1) those in which molecules are broken down to provide the energy needed by the cell and (2) those that synthesize the compounds needed by the cell—both simple and complex.

The process of breaking down molecules to supply energy is **catabolism.**
The process of building up molecules (synthesis) is **anabolism.**

In spite of the large number of chemical reactions, there are only a few that dominate cell metabolism. In this chapter and the next, we focus our attention on the catabolic pathways that yield energy. A **biochemical pathway** is **a series of consecutive biochemical reactions.** We will see the actual reactions by means of which the chemical energy stored in our food is converted to the energy we use every minute of our lives—to think, to breathe, and to use our muscles to walk, write, eat, and everything else. In Chapter 22 we will look at some synthetic (anabolic) pathways.

The common catabolic pathway consists of two sequences:

1. the citric acid cycle (Sec. 20.4)
2. oxidative phosphorylation (Secs. 20.5, 20.6)

The food we eat consists of many types of compounds, largely the ones we discussed: carbohydrates, lipids, and proteins. All of them can serve as fuel, and we derive our energy from them. To convert those compounds to energy, the body uses a different pathway for each type of compound. *However, all these diverse pathways converge to one* **common catabolic pathway,** which is illustrated in Figure 20.1. The diverse pathways are shown as different food streams. The small C_2 and C_4 molecules produced from the original large molecules in food drop into an imaginary collecting funnel that represents the common catabolic pathway. At the end of the funnel appears the energy carrier molecule adenosine triphosphate (ATP).

The whole purpose of catabolic pathways is to convert the chemical energy in foods to molecules of ATP. In this chapter we deal with the common catabolic pathway only. In Chapter 21 we will discuss the ways in which the different types of food (carbohydrates, lipids, and proteins) feed molecules into the common catabolic pathway.

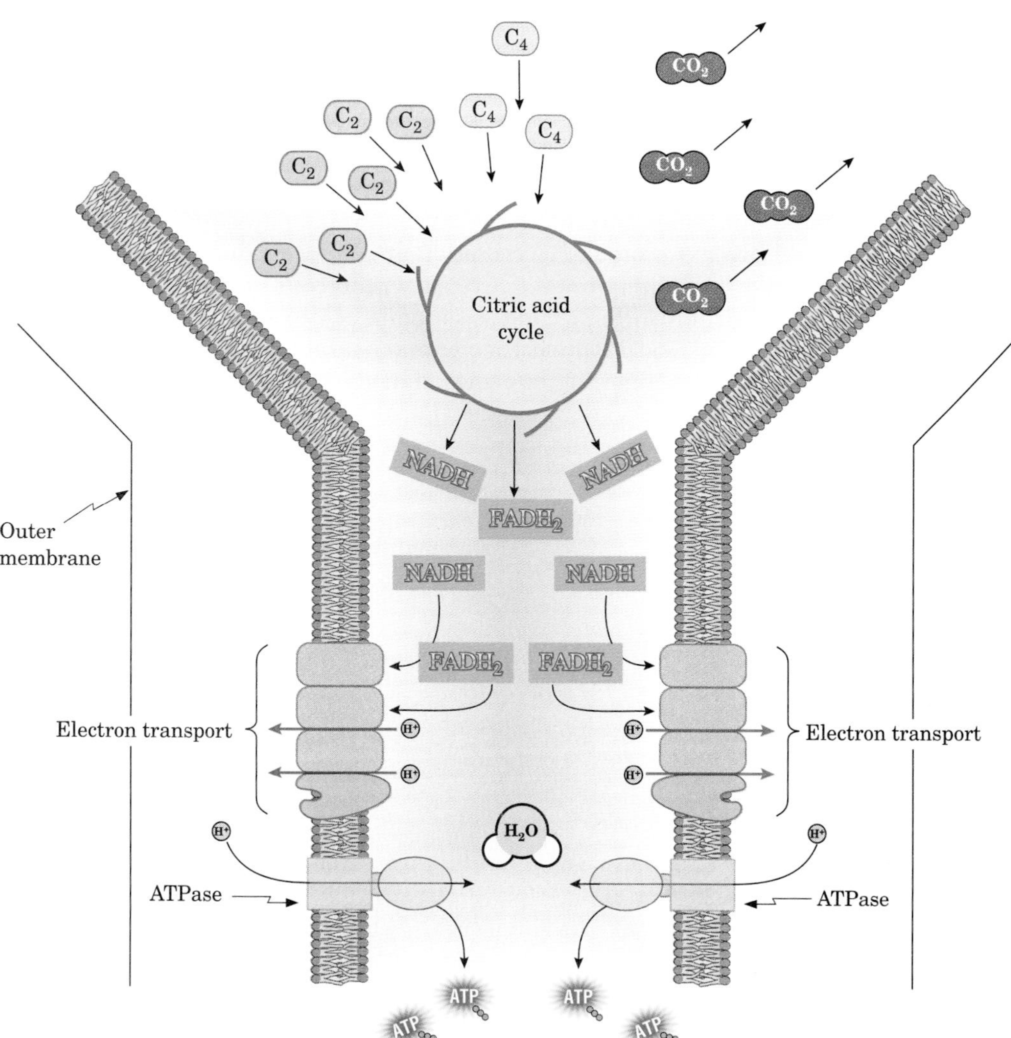

Figure **20.1** • In this schematic diagram of the common catabolic pathway, an imaginary funnel represents what happens in the cell. The diverse catabolic pathways drop their products into the funnel of the common catabolic pathway, mostly in the form of C_2 fragments. (Sec. 20.4. The source of the C_4 fragments will be shown in Sec. 21.9.) The spinning wheel of the citric acid cycle breaks these molecules down further. The carbon atoms are released in the form of CO_2, and the hydrogen atoms and electrons are picked up by special compounds such as NAD^+ and FAD. Then the reduced NADH and $FADH_2$ cascade down into the stem of the funnel, where the electrons are transported inside the walls of the stem and the H^+ ions (represented by dots) are expelled to the outside. In their drive to get back, the H^+ ions form the energy carrier ATP. Once back inside, they combine with the oxygen that picked up the electrons and produce water.

20.2 Cells and Mitochondria

A typical animal cell has many components, as shown in Figure 20.2. Each serves a different function. For example, the replication of DNA (Sec. 23.4) takes place in the **nucleus; lysosomes** remove damaged cellular components and some unwanted foreign materials; and **Golgi bodies** package and process proteins for secretion and delivery to other cellular compartments. The specialized structures within cells are called **organelles.**

Mitochondria is the plural form, and mitochondrion is the singular.

The **mitochondria,** which are made of two membranes (Fig. 20.3), are the organelles in which the common catabolic pathway takes place in higher organisms. The enzymes that catalyze the common pathway are all located in these organelles. Because the enzymes are located inside the mitochondria, the starting material of the reactions in the com-

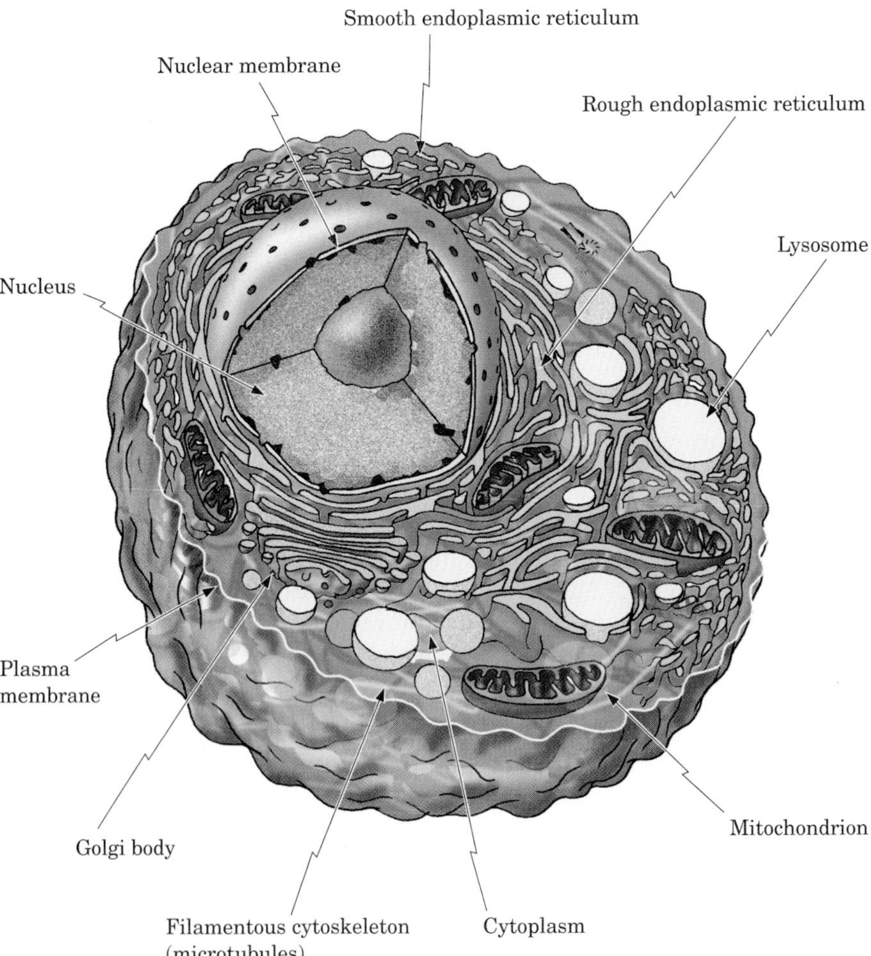

Figure **20.2** • Diagram of a rat liver cell, a typical higher animal cell. (Adapted from R. H. Garrett and C. M. Grisham, *Biochemistry,* Philadelphia: Saunders College Publishing, 1995.)

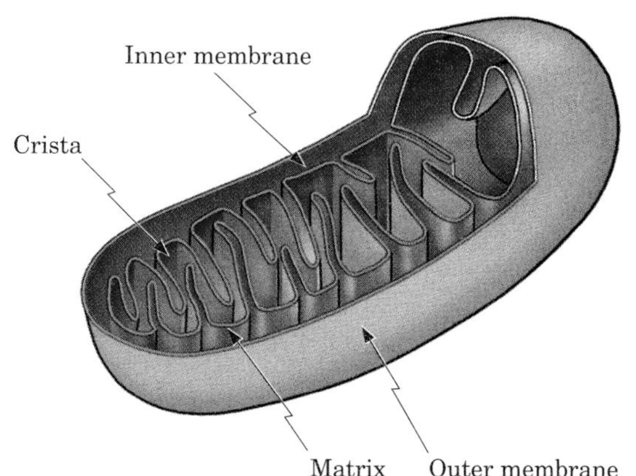

Figure **20.3**
A schematic drawing of a mitochondrion, cut to reveal the internal organization.

mon pathway must get through the two membranes to enter the mitochondria, and products must leave the same way.

The inner membrane of a mitochondrion is quite resistant to the penetration of any ions and of most uncharged molecules. However, ions and molecules can still get through the membrane—they are transported across it by the numerous protein molecules embedded in it (Fig. 17.2, p. 568). The outer membrane, on the other hand, is quite permeable to small molecules and ions and does not need many different kinds of transporting membrane proteins.

The inner membrane is highly corrugated and folded, and the folds are called **cristae.** One can compare the organization of a mitochondrion to that of the galley of an ancient ship. The mitochondrion as a whole is the ship. The cristae are the benches to which the enzymes of the oxidative phosphorylation cycle are chained like ancient slaves, who provide the driving power. The space between the inner and outer membranes is like the space within the double hull of a ship.

The enzymes of the citric acid cycle are located in the matrix, which is the inner nonmembranous portion of a mitochondrion (Fig. 20.3). We shall soon see in detail how the specific sequence of the enzymes causes the chain of events in the common catabolic pathway. Beyond that, we must also discuss the ways in which nutrients and reaction products move into and out of the mitochondria.

 20.3 The Principal Compounds of the Common Catabolic Pathway

There are two parts to the common catabolic pathway. The first is the **citric acid cycle** (also called the **Krebs cycle**), and the second is the **oxidative phosphorylation** pathway, also called the **electron transport chain** or the **respiratory chain.**

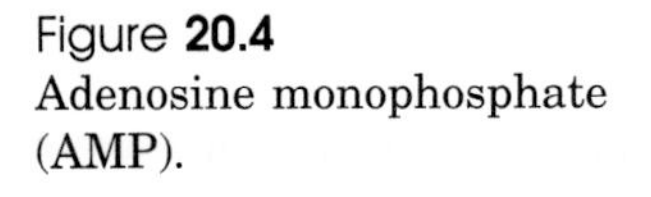

In order to understand what is actually happening in these reactions, we must first introduce the principal compounds participating in the common catabolic pathway. The most important of these are three rather complex compounds: **adenosine monophosphate (AMP),**

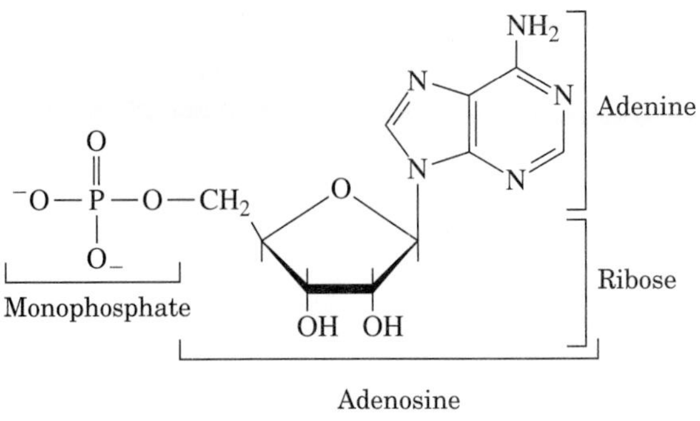

Figure **20.4**
Adenosine monophosphate (AMP).

adenosine diphosphate (ADP), and **adenosine triphosphate (ATP)** (Figs. 20.4 and 20.5). All three of these molecules contain a heterocyclic amine portion called **adenine** and a sugar (furanose) portion called **ribose** (Sec. 16.8). Taken together, this portion of the molecules is called **adenosine.**

Adenine is a nucleotide base that frequently appears in biochemical compounds, especially nucleic acids.

AMP, ADP, and ATP all contain adenosine connected to phosphate groups. The only difference between AMP, ADP, and ATP is the number of phosphate groups. As you can see from Figure 20.5, each phosphate is attached to the next by an anhydride linkage (Sec. 14.11). ATP contains three phosphates and two anhydride linkages. In all three molecules, the first phosphate is attached to the ribose by an inorganic ester linkage (Sec. 14.11).

A phosphoric anhydride linkage, P—O—P, contains more chemical energy than a phosphate-ester linkage, C—O—P. This means that when ATP and ADP are hydrolyzed to yield inorganic phosphate (Fig. 20.5), they release more energy per phosphate group than does AMP. Conversely, when inorganic phosphate bonds to AMP or ADP, greater amounts of energy are added to the chemical bond than when it bonds to adenosine. ADP and ATP contain *high-energy* phosphate bonds.

The PO_4^{3-} ion is generally called **inorganic phosphate.**

Of the three phosphate groups, the one closest to the adenosine is the least energetic; the other two possess a good deal more energy. Consequently, ATP releases the most energy and AMP the least when each gives up one phosphate group. This makes ATP a very useful compound for energy storage and release. The energy gained in the oxidation of food is stored in the form of ATP. However, this is only a short-term storage. ATP molecules in the cells normally do not last longer than about a minute. They are hydrolyzed to ADP and inorganic phosphate to yield

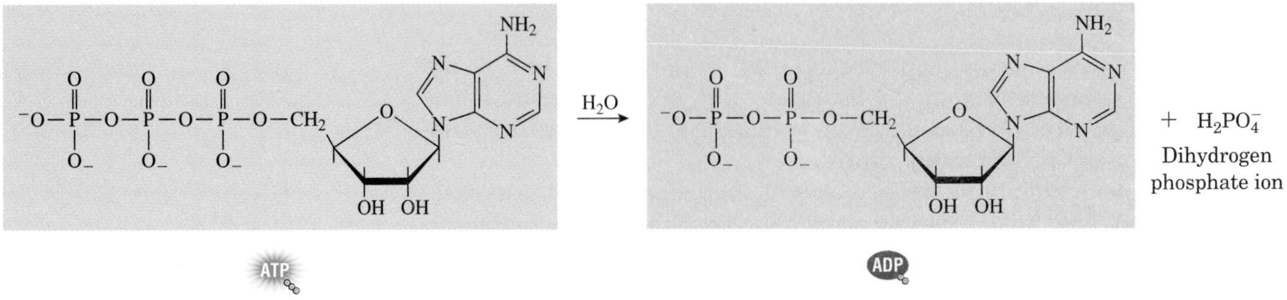

When one phosphate group is hydrolyzed from each, the following energy yields are obtained: ATP = 7.3 kcal/mol; ADP = 7.3 kcal/mol; AMP = 3.4 kcal/mol.

Figure **20.5** • Hydrolysis of ATP produces ADP plus inorganic phosphate.

energy for other processes, such as muscle contraction and nerve signal conduction. This means that ATP is constantly being formed and decomposed. Its turnover rate is very high. It has been estimated that during strenuous exercise the human body manufactures and degrades as much as 1 kg (more than 2 lb) of ATP every 2 minutes.

In summary, when the body takes in food, some of it goes to produce energy and some is used for building molecules and for other purposes. All the energy that is extracted from the food is converted to ATP. This is the form in which the body stores its energy. In order to release this energy, the body hydrolyzes the ATP to ADP (sometimes to AMP). Exactly how these things happen is what we will be discussing in the rest of this chapter and in the next.

The body is able to extract only 40 to 60 percent of the total caloric content of food.

Two other actors in this drama are the coenzymes (Sec. 19.3) NAD^+ and FAD (Fig. 20.6), both of which contain an ADP core. In NAD^+ the operative part of the coenzyme is the nicotinamide part. In FAD the operative part is the flavin. In both molecules, the ADP is the *handle* by which the apoenzyme holds onto the coenzyme, and it is the other end of the molecule that carries out the actual chemical reaction. For example, when NAD^+ is reduced, it is the nicotinamide part of the molecule that gets reduced:

The + in NAD^+ refers to the positive charge on the nitrogen.

Nicotinamide and riboflavin are both members of the vitamin B group.

$$NAD^+ + H^+ + 2e^- \rightleftharpoons NADH$$

NAD⁺ NADH

The "R" stands for the rest of the NAD^+ molecule; "R′" stands for the rest of the FAD molecule.

The reduced form of NAD^+ is called NADH. The same reduction happens on two nitrogens of the flavin portion of FAD:

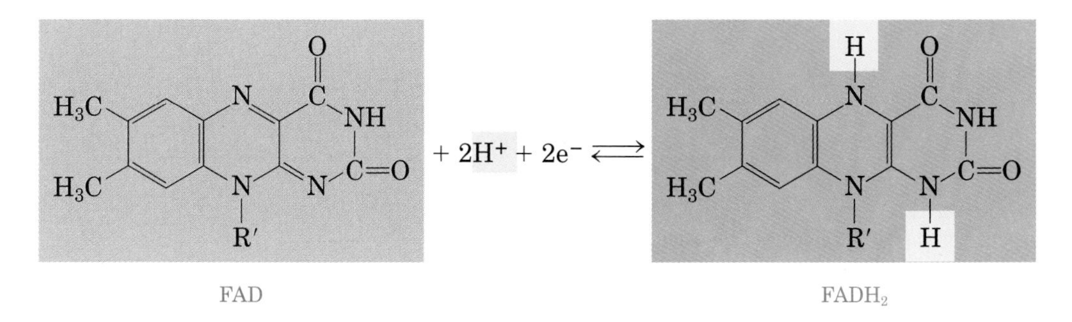

$$FAD + 2H^+ + 2e^- \rightleftharpoons FADH_2$$

FAD $FADH_2$

The reduced form of FAD is called $FADH_2$.

We view **NAD^+** and **FAD coenzymes** as the **hydrogen ion** and **electron transporting molecules.**

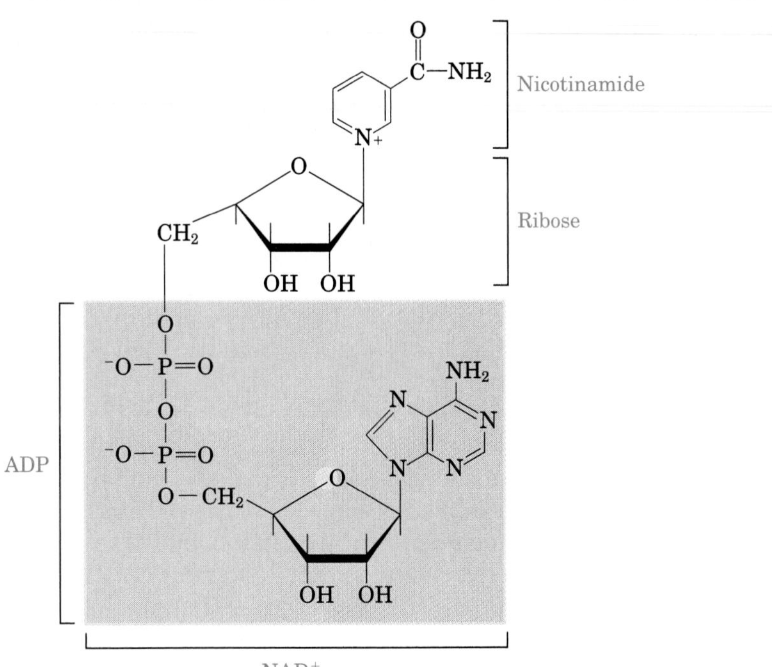

The final principal compound in the common catabolic pathway is **coenzyme A** (CoA, Fig. 20.7), which is the **acetyl carrying group.** Coenzyme A also contains ADP, but here the next structural unit is pantothenic acid, another B vitamin. Just as ATP can be looked upon as an ADP molecule to which a $—PO_3$ is attached by a high-energy bond, so

Nicotinamide

Ribose

ADP

NAD^+
*N*icotinamide *a*denine *d*inucleotide

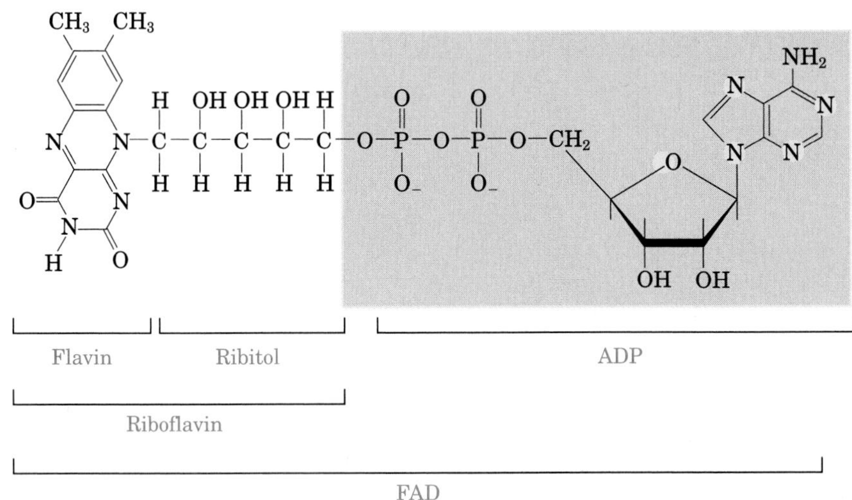

FAD
*F*lavin *a*denine *d*inucleotide

Figure **20.6** • The structures of NAD^+ and FAD.

can **acetyl coenzyme A** be considered a CoA molecule linked to an acetyl group by another high-energy bond. The active part of coenzyme A is the mercaptoethylamine. The acetyl group of acetyl coenzyme A is attached to the SH group:

$$\text{CoA—S—}\underset{\underset{\text{O}}{\|}}{\text{C}}\text{—CH}_3$$

Acetyl coenzyme A

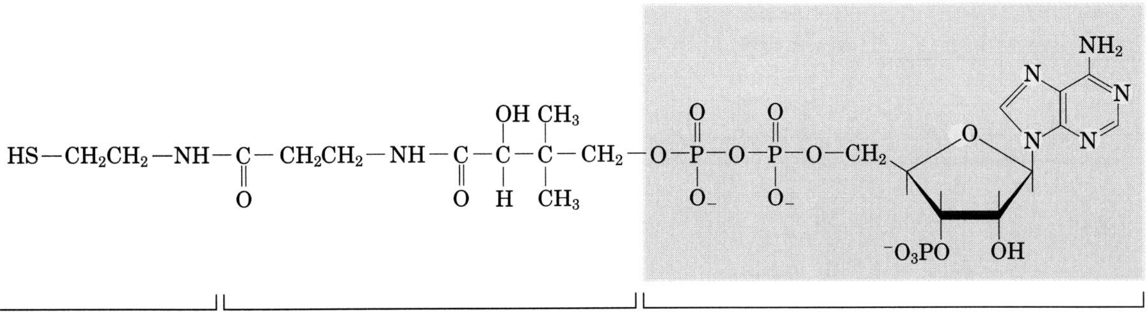

$$HS{-}CH_2CH_2{-}NH{-}\underset{\underset{O}{\|}}{C}{-}CH_2CH_2{-}NH{-}\underset{\underset{O}{\|}}{C}{-}\underset{\underset{H}{|}}{\overset{\overset{OH}{|}}{C}}{-}\underset{\underset{CH_3}{|}}{\overset{\overset{CH_3}{|}}{C}}{-}CH_2{-}O{-}\underset{\underset{O^-}{|}}{\overset{\overset{O}{\|}}{P}}{-}O{-}\underset{\underset{O^-}{|}}{\overset{\overset{O}{\|}}{P}}{-}O{-}CH_2{-}\text{(ribose-3'-}OPO_3^{-}\text{, 2'-}OH\text{, adenine)}$$

Mercaptoethylamine | Pantothenic acid | Phosphorylated ADP

Figure **20.7** • The structure of coenzyme A.

20.4 The Citric Acid Cycle

The process of catabolism begins when carbohydrates and lipids have been broken down into two-carbon pieces. The two-carbon fragments are the acetyl ($CH_3CO—$) portions of acetyl coenzyme A. The acetyl is now fragmented further in the **citric acid cycle** (named after the main com-

Box 20A Nicotine and Nicotinic Acid

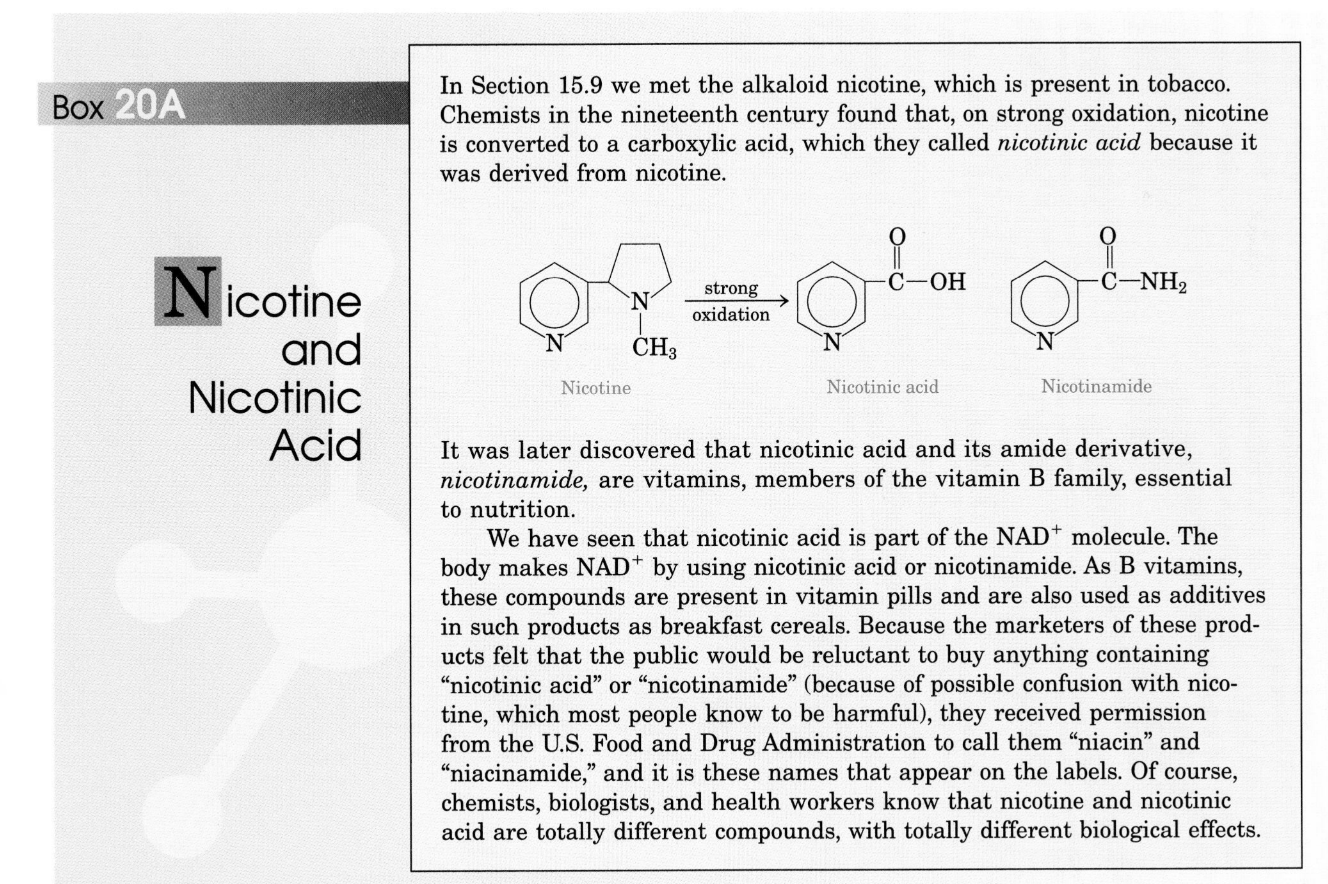

In Section 15.9 we met the alkaloid nicotine, which is present in tobacco. Chemists in the nineteenth century found that, on strong oxidation, nicotine is converted to a carboxylic acid, which they called *nicotinic acid* because it was derived from nicotine.

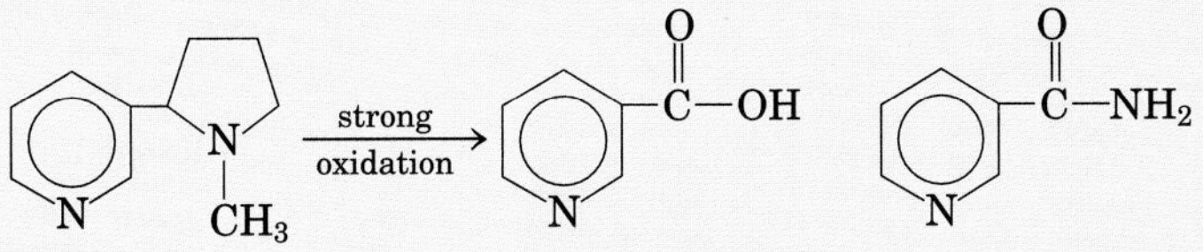

Nicotine $\xrightarrow[\text{oxidation}]{\text{strong}}$ Nicotinic acid (pyridine–C(=O)–OH); Nicotinamide (pyridine–C(=O)–NH_2)

It was later discovered that nicotinic acid and its amide derivative, *nicotinamide,* are vitamins, members of the vitamin B family, essential to nutrition.

We have seen that nicotinic acid is part of the NAD^+ molecule. The body makes NAD^+ by using nicotinic acid or nicotinamide. As B vitamins, these compounds are present in vitamin pills and are also used as additives in such products as breakfast cereals. Because the marketers of these products felt that the public would be reluctant to buy anything containing "nicotinic acid" or "nicotinamide" (because of possible confusion with nicotine, which most people know to be harmful), they received permission from the U.S. Food and Drug Administration to call them "niacin" and "niacinamide," and it is these names that appear on the labels. Of course, chemists, biologists, and health workers know that nicotine and nicotinic acid are totally different compounds, with totally different biological effects.

Hans Krebs (1900–1981). Nobel laureate in 1953, established the relationships among the different components of the cycle.

ponent of the cycle), which is also called the **Krebs cycle** and, sometimes, the **tricarboxylic acid cycle.**

The details of the citric acid cycle are given in Figure 20.8. A good way to gain an insight is to use Figure 20.8 in connection with the simplified schematic diagram shown in Figure 20.9 which shows only the carbon balance.

We will now follow the two carbons of the acetate (C_2) through each step in the citric acid cycle. The circled numbers correspond to those in Figure 20.8.

In step ⑧ we will see where the oxaloacetate comes from.

① Acetyl coenzyme A enters the cycle by combining with a C_4 compound called oxaloacetate:

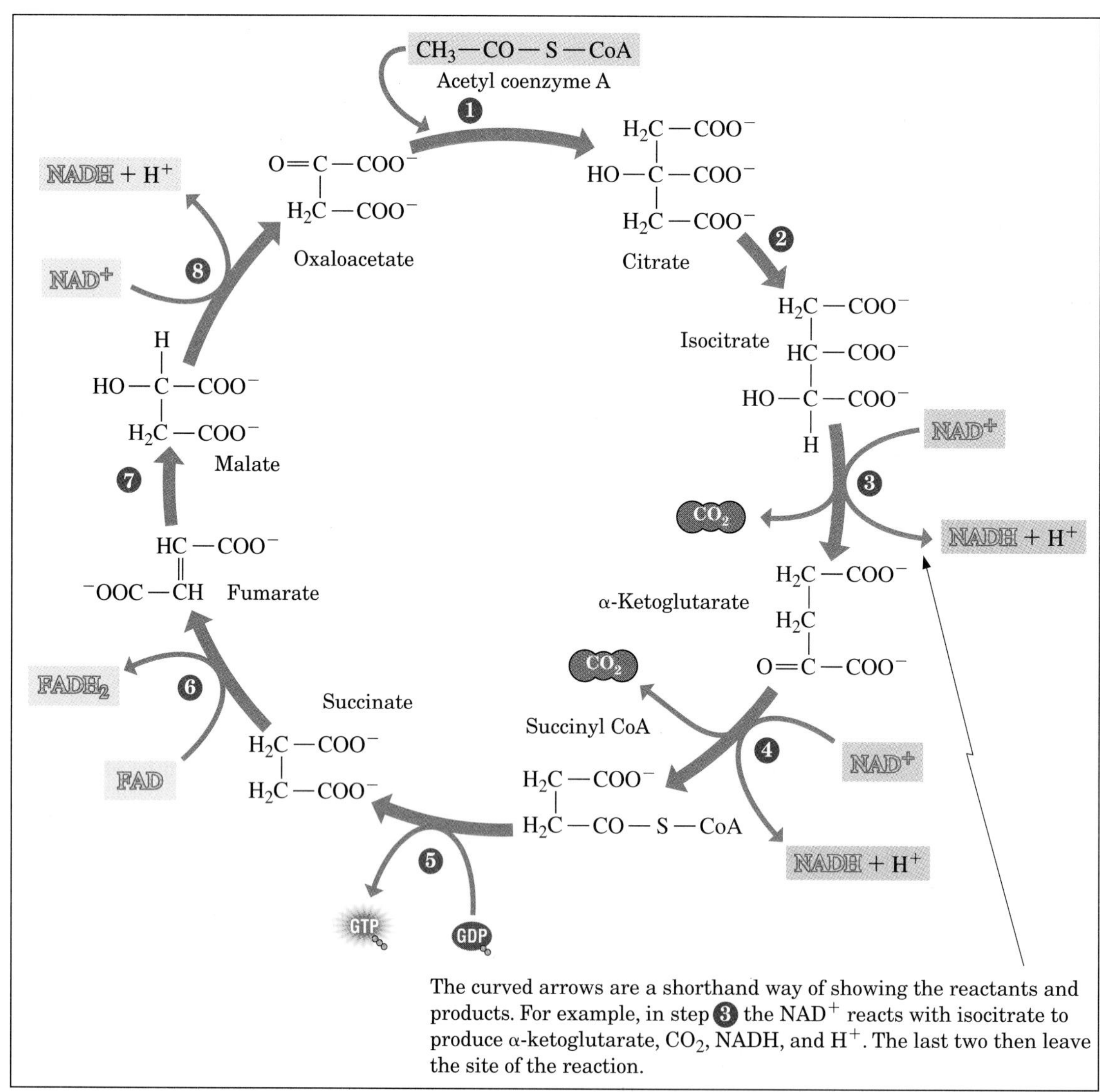

Figure **20.8** • The citric acid (Krebs) cycle. The numbered steps are explained in detail in the text.

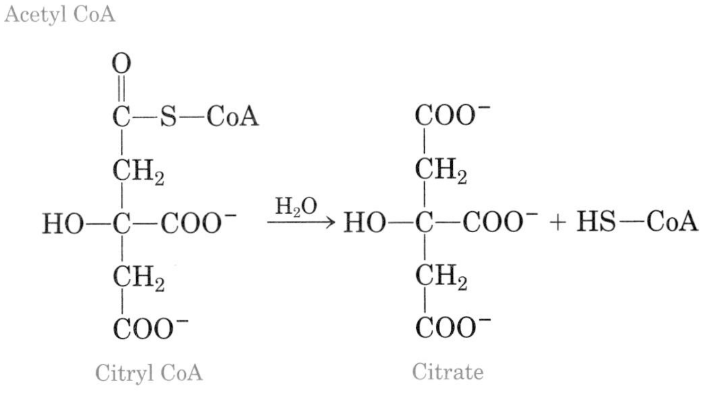

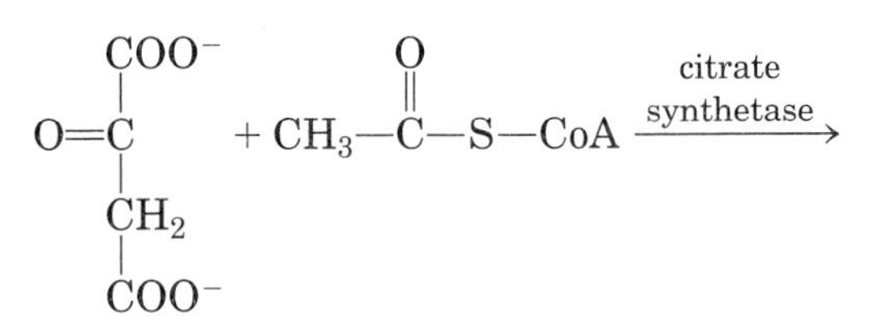

$$\begin{array}{c}\mathrm{COO^-}\\ |\\ \mathrm{O{=}C}\\ |\\ \mathrm{CH_2}\\ |\\ \mathrm{COO^-}\\ \text{Oxaloacetate}\end{array} + \begin{array}{c}\mathrm{O}\\ \|\\ \mathrm{CH_3{-}C{-}S{-}CoA}\\ \\ \\ \\ \\ \text{Acetyl CoA}\end{array} \xrightarrow[\text{synthetase}]{\text{citrate}}$$

$$\begin{array}{c}\mathrm{O}\\ \|\\ \mathrm{C{-}S{-}CoA}\\ |\\ \mathrm{CH_2}\\ |\\ \mathrm{HO{-}C{-}COO^-}\\ |\\ \mathrm{CH_2}\\ |\\ \mathrm{COO^-}\\ \text{Citryl CoA}\end{array} \xrightarrow{H_2O} \begin{array}{c}\mathrm{COO^-}\\ |\\ \mathrm{CH_2}\\ |\\ \mathrm{HO{-}C{-}COO^-}\\ |\\ \mathrm{CH_2}\\ |\\ \mathrm{COO^-}\\ \text{Citrate}\end{array} + \mathrm{HS{-}CoA}$$

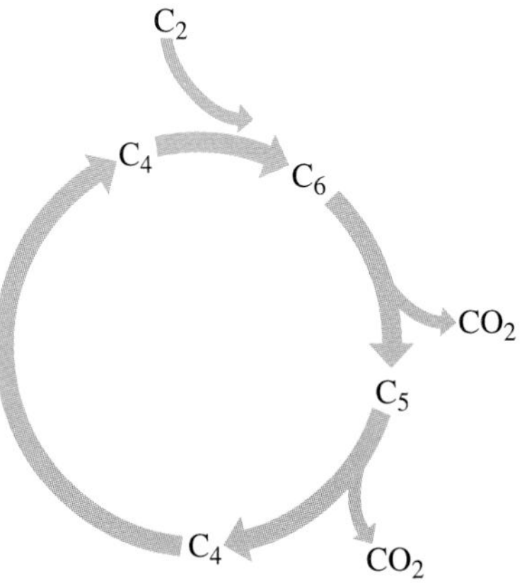

Figure **20.9**
A simplified view of the citric acid cycle, showing only the carbon balance.

The first thing that happens is the addition of the CH_3 group of the acetyl CoA to the C=O bond of the oxaloacetate, catalyzed by the enzyme citrate synthetase. This is followed by hydrolysis to produce the C_6 compound citrate ion and CoA. Therefore, step ① is a building up rather than a degradation.

② The citrate ion is dehydrated to *cis*-aconitate:

$$\begin{array}{c}\mathrm{COO^-}\\ |\\ \mathrm{CH_2}\\ |\\ \mathrm{HO{-}C{-}COO^-}\\ |\\ \mathrm{CH_2}\\ |\\ \mathrm{COO^-}\\ \text{Citrate}\end{array} \xrightarrow{\text{aconitase}} \begin{array}{c}\mathrm{COO^-}\\ |\\ \mathrm{CH_2}\\ |\\ \mathrm{C{-}COO^-}\\ \|\\ \mathrm{C{-}H}\\ |\\ \mathrm{COO^-}\\ \textit{cis}\text{-Aconitate}\end{array} + \mathrm{H_2O}$$

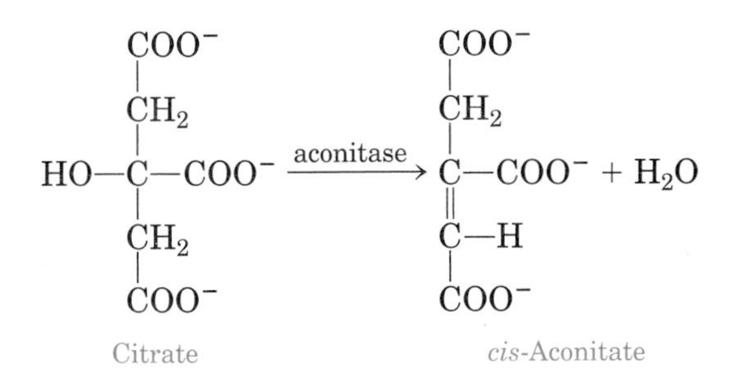

after which the *cis*-aconitate is hydrated again, but this time to isocitrate instead of citrate:

$$\begin{array}{c}\mathrm{COO^-}\\ |\\ \mathrm{CH_2}\\ |\\ \mathrm{C{-}COO^-}\\ \|\\ \mathrm{C{-}H}\\ |\\ \mathrm{COO^-}\\ \textit{cis}\text{-Aconitate}\end{array} \xrightarrow[\text{aconitase}]{H_2O} \begin{array}{c}\mathrm{COO^-}\\ |\\ \mathrm{CH_2}\\ |\\ \mathrm{CH{-}COO^-}\\ |\\ \mathrm{HO{-}CH}\\ |\\ \mathrm{COO^-}\\ \text{Isocitrate}\end{array}$$

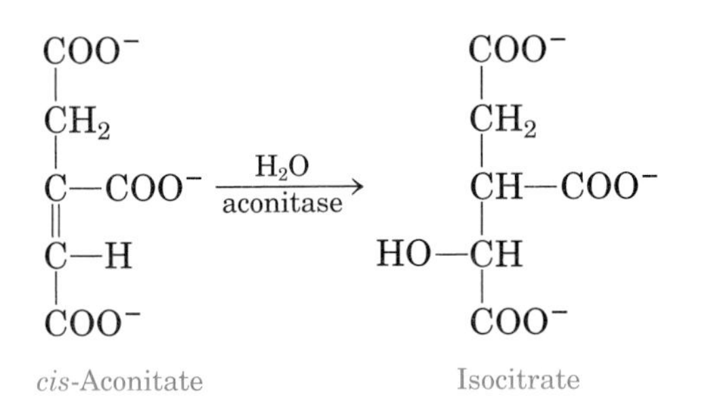

In citric acid the alcohol is a tertiary alcohol. We learned in Section 12.3 that tertiary alcohols cannot be oxidized. The alcohol in the isocitrate is a secondary alcohol, which upon oxidation yields a ketone.

Decarboxylation means loss of CO_2.

③ The isocitrate is now oxidized and decarboxylated at the same time:

$$\begin{array}{l} COO^- \\ | \\ CH_2 \\ | \\ CH{-}COO^- \\ | \\ HO{-}CH \\ | \\ COO^- \end{array} + \text{NAD}^+ \xrightarrow[\text{dehydrogenase}]{\text{isocitrate}} \begin{array}{l} COO^- \\ | \\ CH_2 \\ | \\ CH_2 \\ | \\ C{=}O \\ | \\ COO^- \end{array} + \text{NADH} + CO_2$$

Isocitrate α-Ketoglutarate

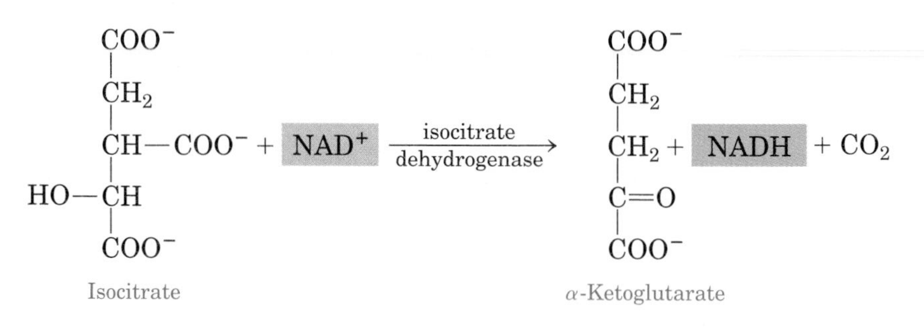

The oxidizing agent NAD^+ has removed two hydrogens. One of them has been used to reduce NAD^+ to NADH. The other has replaced the COO^- that goes into making CO_2. Note that the CO_2 given off comes from the original oxaloacetate and not from the two carbons of the acetyl CoA. Both of these carbons are still present in the α-ketoglutarate. Also note that we are now down to a C_5 compound, α-ketoglutarate.

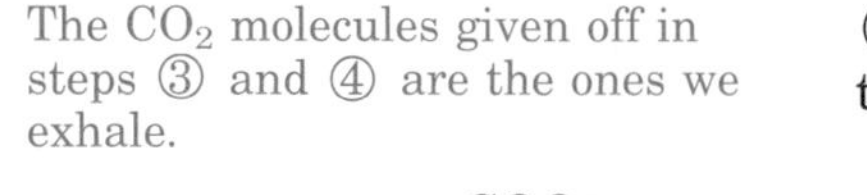

The CO_2 molecules given off in steps ③ and ④ are the ones we exhale.

④ and ⑤ Next a complex system removes another CO_2, once again from the original oxaloacetate portion rather than from the acetyl CoA portion:

The "P_i" is inorganic phosphate.

$$\begin{array}{l} COO^- \\ | \\ CH_2 \\ | \\ CH_2 \\ | \\ C{=}O \\ | \\ COO^- \end{array} + \text{NAD}^+ + \text{GDP} + P_i + H_2O \xrightarrow[\text{system}]{\text{complex enzyme}} \begin{array}{l} COO^- \\ | \\ CH_2 \\ | \\ CH_2 \\ | \\ COO^- \end{array} + CO_2 + \text{NADH} + H^+ + \text{GTP}$$

α-Ketoglutarate Succinate

We are now down to a C_4 compound, succinate. This oxidative decarboxylation is more complex than the first. It occurs in many steps and requires a number of cofactors. For our purpose it is sufficient to know that, during this second oxidative decarboxylation, a high-energy compound called **guanosine triphosphate** (GTP) is also formed.

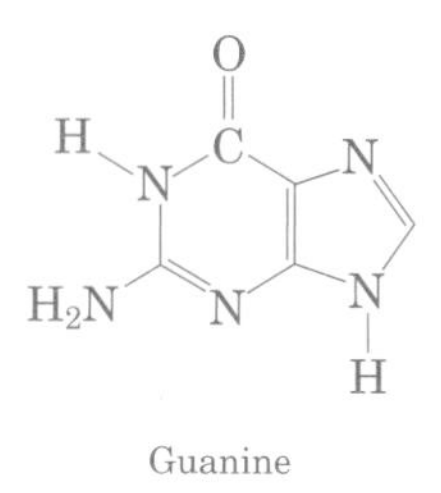

GTP is similar to ATP except that guanine replaces adenine. Otherwise, the linkages of the base to ribose and the phosphates are exactly the same as in ATP. The function of GTP is also similar to that of ATP, namely, to store energy in the form of high-energy phosphate bonds (chemical energy).

Guanine

⑥ In this step, the succinate is oxidized by FAD, which removes two hydrogens to give fumarate (the double bond in this molecule is trans):

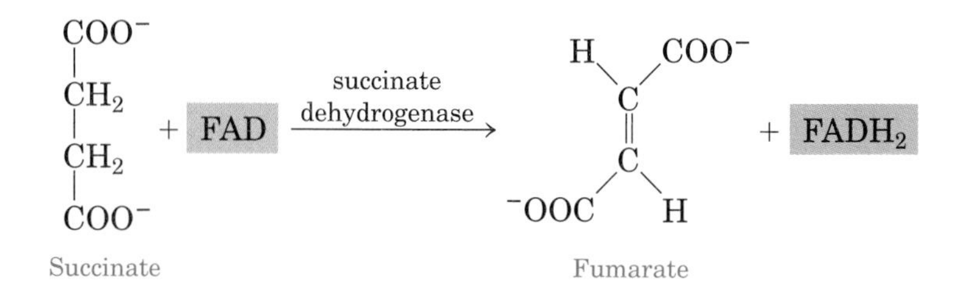

$$\begin{array}{l} COO^- \\ | \\ CH_2 \\ | \\ CH_2 \\ | \\ COO^- \end{array} + \text{FAD} \xrightarrow[\text{dehydrogenase}]{\text{succinate}} \begin{array}{c} H \quad COO^- \\ \diagdown \ \diagup \\ C \\ \| \\ C \\ \diagup \ \diagdown \\ {}^-OOC \quad H \end{array} + \text{FADH}_2$$

Succinate Fumarate

This reaction cannot be carried out in the laboratory, but with the aid of an enzyme catalyst, the body does it easily.

⑦ The fumarate is now hydrated to give the malate ion:

$$\underset{\text{Fumarate}}{\begin{matrix} H & & COO^- \\ & C & \\ & \| & \\ & C & \\ ^-OOC & & H \end{matrix}} + H_2O \xrightarrow{\text{fumarase}} \underset{\text{Malate}}{\begin{matrix} COO^- \\ | \\ CH{-}OH \\ | \\ CH_2 \\ | \\ COO^- \end{matrix}}$$

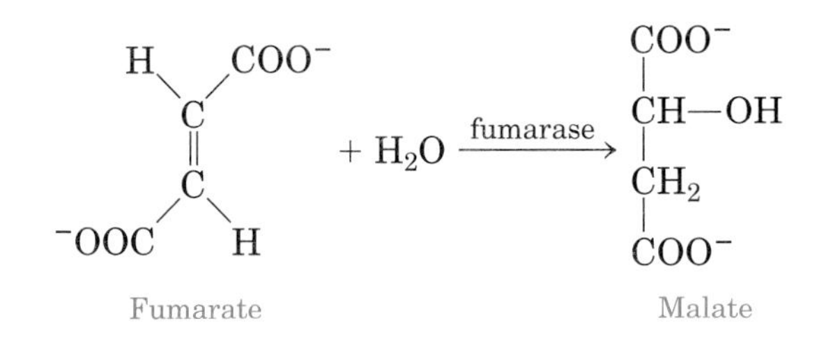

⑧ In the final step of the cycle, malate is oxidized by NAD^+ to give oxaloacetate:

$$\underset{\text{Malate}}{\begin{matrix} COO^- \\ | \\ CH{-}OH \\ | \\ CH_2 \\ | \\ COO^- \end{matrix}} + NAD^+ \xrightarrow[\text{dehydrogenase}]{\text{malate}} \underset{\text{Oxaloacetate}}{\begin{matrix} COO^- \\ | \\ C{=}O \\ | \\ CH_2 \\ | \\ COO^- \end{matrix}} + NADH + H^+$$

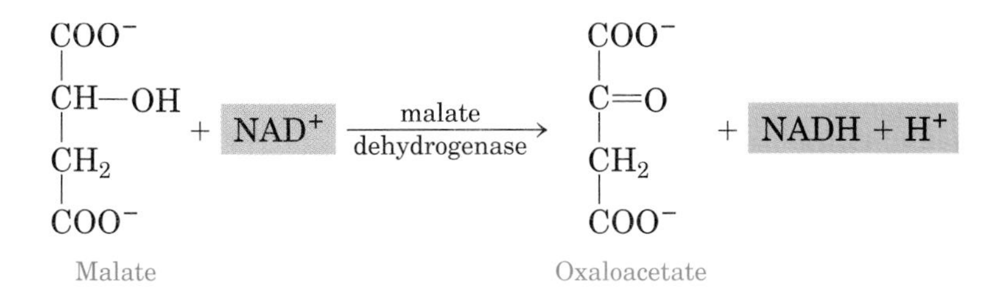

Thus, the final product of the Krebs cycle is oxaloacetate, which is the compound that we started with in step ①.

What has happened in the entire process is that the original two acetate carbons of acetyl CoA were added to the C_4 oxaloacetate to produce a C_6 unit, which then lost two carbons in the form of CO_2, to produce, at the end of the process, the C_4 unit oxaloacetate. The *net* effect is the conversion of the two acetate carbons of acetyl CoA to two molecules of carbon dioxide.

How does this produce energy? We have already learned that one step in the process produces a high-energy molecule of GTP. But other steps contribute also. In several of the steps, the citric acid cycle converts NAD^+ to NADH and FAD to $FADH_2$. These reduced coenzymes carry the H^+ and electrons that eventually will provide the energy for the synthesis of ATP (discussed in detail in Secs. 20.5 and 20.6).

This stepwise degradation and oxidation of acetate in the citric acid cycle results in the most efficient extraction of energy. Rather than in one burst, the energy is released in small packets carried away step by step in the form of NADH and $FADH_2$.

But the cyclic nature of this acetate degradation has other advantages besides maximizing energy yield: (1) the citric acid cycle components also provide raw materials for amino acid synthesis as the need arises (Chap. 22), and (2) the many-component cycle provides an excellent method for regulating the speed of catabolic reactions. The regulation can occur at many different parts of the cycle, so that feedback information can be used at many points to speed up or slow down the process as necessary.

For example, α-ketoglutaric acid is used to synthesize glutamic acid.

The following equation represents the overall reactions in the citric acid cycle.

$$CH_3COOH + 2H_2O + 3NAD^+ + FAD \longrightarrow 2CO_2 + 3NADH + FADH_2 + 3H^+ \quad (20.1)$$

The citric acid cycle is controlled by a feedback mechanism. When the essential products of the cycle, such as ATP and NADH + H^+, accumulate, they inhibit some of the enzymes in the cycle. Citrate synthase (step ①), isocitrate dehydrogenase (step ③), and α-ketoglutarate dehydrogenase (step ④) are inhibited by ATP and/or by NADH + H^+. This slows down or shuts off the cycle. On the other hand, when the feed material, acetyl CoA, is in abundance, the cycle is speeded up. The enzyme isocitrate dehydrogenase (step ③) is stimulated by ADP and NAD^+.

20.5 Electron and H^+ Transport (Oxidative Phosphorylation)

The reduced coenzymes NADH and $FADH_2$ are end products of the citric acid cycle. They carry hydrogen ions and electrons and thus the potential to yield energy when these combine with oxygen to form water:

The oxygen in this reaction is the oxygen we breathe.

$$4H^+ + 4e^- + O_2 \longrightarrow 2H_2O$$

This simple exothermic reaction is carried out in many steps.

A number of enzymes are involved, all embedded in the inner membrane of the mitochondria. These enzymes are situated in a particular *sequence* in the membrane so that the product from one enzyme can be passed on to the next enzyme, in a kind of assembly line. The enzymes are arranged in order of increasing affinity for electrons, so electrons flow through the enzyme system. The sequence of electron-carrying enzymes in the mitochondrial membrane is

The letters used to designate the cytochromes were given in order of their discovery.

$$\text{flavoprotein} \longrightarrow \text{FeS protein} \longrightarrow \text{Q (quinone) enzyme} \longrightarrow \text{cytochrome b} \longrightarrow \text{cytochrome } c_1 \longrightarrow \text{cytochrome c} \longrightarrow \text{cytochrome a} \longrightarrow \text{cytochrome } a_3$$

The fact that these enzymes are so close together allows the electrons to pass from one to another.

In Section 20.4 we saw that NADH + H^+ is a product of the citric acid cycle. The **oxidative phosphorylation** process (Fig. 20.10) starts with the NADH that arrives from the interior of the mitochondrion. This carries two electrons and one H^+ ion. The two electrons are passed on to the flavoprotein, which has a riboflavin coenzyme. The H^+ from the NADH and also another H^+ from the interior of the mitochondrion are expelled by the flavoprotein into the intermembrane space. The two electrons pass through the FeS protein and Q enzyme. At this point, two more H^+ ions are pumped out to the intermembrane space. The cy-

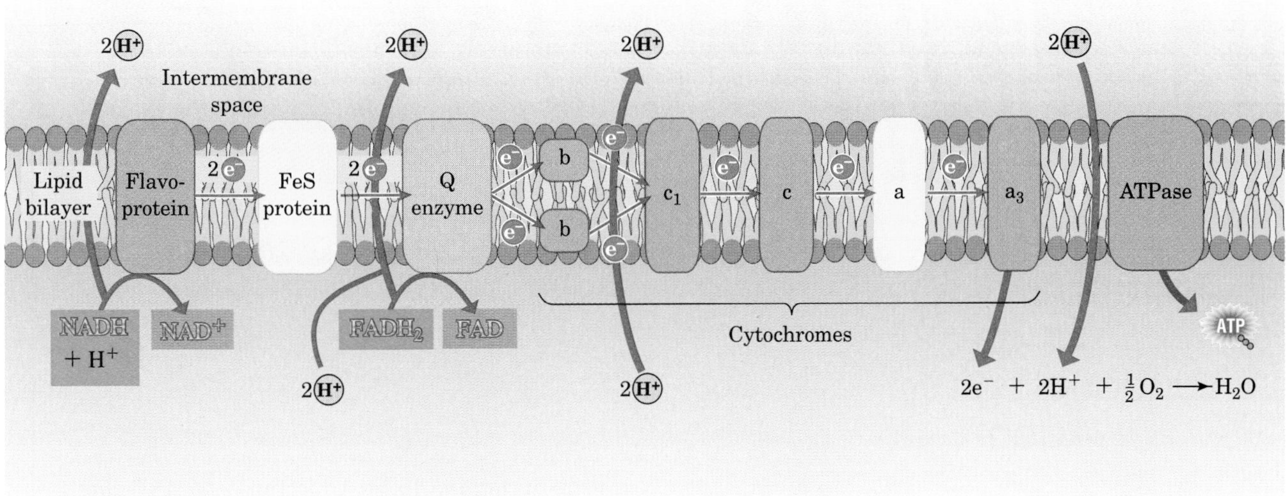

Figure **20.10** • Schematic diagram of oxidative phosphorylation, also called the electron transport chain.

tochromes in turn pick up only one electron at a time. Therefore, the two electrons from the Q enzyme are passed on to two molecules of cytochrome b. When the electrons are transferred to cytochrome c_1, another two H^+ ions are pumped out from the matrix of the mitochondrion. The electrons finally end up, via cytochrome c $\rightarrow$ cytochrome a $\rightarrow$ cytochrome a_3, on the inside of the membrane, where they combine with oxygen, O_2, and H^+ ions to form water. During this process of transporting the two electrons from one molecule of NADH, a total of six H^+ ions are pumped out of the mitochondrion.

When the $FADH_2$ carries the electrons, it transfers them directly to the Q enzyme rather than to the flavoprotein. Otherwise, the electron flow follows the same route as with NADH. However, since the flavoprotein stage is skipped during the electron transport, only four H^+ ions are expelled into the intermembrane space for each $FADH_2$ molecule.

 20.6 Phosphorylation and the Chemiosmotic Pump

In 1961 Peter Mitchell (1920–1992), an English chemist, proposed the **chemiosmotic hypothesis:** The energy in the electron transfer chain creates a proton gradient. A **proton gradient** is a continuous variation in the H^+ concentration along a given region. In this case there is a higher concentration of H^+ in the intermembrane space than inside the mitochondrion. The protons pumped out of the mitochondrion provide a driving force. This driving force, which is the result of the spontaneous flow of ions from a region of high concentration to a region of low concentration, propels the protons back to the mitochondrion through a complex that is given the name of **proton translocating ATPase.** This compound is located on the inner membrane of the mitochondrion and is the

Mitchell received the Nobel Prize in Chemistry in 1978.

Box 20B

2,4-Dinitrophenol as an Uncoupling Agent

Nitrated aromatic compounds are highly explosive. Trinitrotoluene (TNT) is the best known. During World War I, many ammunition workers were exposed to 2,4-dinitrophenol (DNP), a compound used to prepare the explosive picric acid.

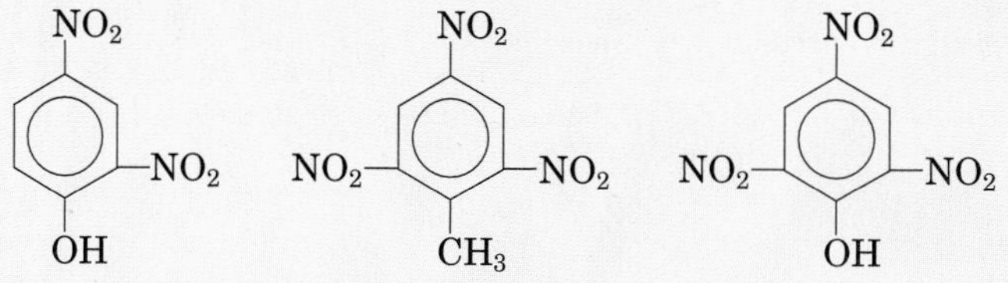

DNP — TNT — Picric acid

It was observed that these workers lost weight. As a consequence, DNP was used as a weight-reducing drug during the 1920s. Unfortunately, DNP eliminated not only the fat but sometimes also the patient, and its use as a diet pill was discontinued after 1929.

Today we know why DNP works as a weight-reducing drug. It is an effective *protonophore,* which is a compound that transports H^+ ions through a membrane passively, without the expenditure of energy. We have seen that H^+ ions accumulate in the intermembrane space of mitochondria and, under normal conditions, drive the synthesis of ATP while they are going back to the inside. This is Mitchell's chemiosmotic principle. When DNP is ingested, it transfers the H^+ back to the mitochondrion easily, and no ATP is manufactured. The energy of the electron separation is dissipated as heat and is not built in as chemical energy in ATP. The loss of this energy-storing compound makes the utilization of food much less efficient, resulting in weight loss.

Because of this property of DNP, it has been used in biochemical research on oxidative phosphorylation. It is an *uncoupling agent*—it uncouples the oxidative process from the phosphorylation and allows us to study the steps in the electron transport chain without the simultaneous phosphorylation. There exists in the medical literature a case history of a woman whose muscles contained mitochondria in which electron transport was not coupled to oxidative phosphorylation. This unfortunate woman could not utilize electron transport to generate ATP. As a consequence she was severely incapacitated and bedridden. When the uncoupling worsened, the heat generated by the uncontrolled oxidation (no ATP production) was so severe that the patient required continuous cooling.

active enzyme that catalyzes the conversion of ADP and inorganic phosphate to ATP (the reverse of the reaction shown in Fig. 20.5):

$$\text{ADP} + P_i \xrightleftharpoons{\text{ATPase}} \text{ATP} + H_2O$$

The proton translocating ATPase can catalyze the reaction in both directions. When protons that have accumulated on the outer surface of the mitochondrion stream inward, the enzyme manufactures ATP and

Box 20C

Protection Against Oxidative Damage

As we have seen, most of the oxygen we breathe in is used in the final step of the common pathway, namely, in the conversion of our food supply to ATP. This reaction occurs in the mitochondria of cells. However, some 10% of the oxygen we consume is used elsewhere, in specialized oxidative reactions. One such reaction is the hydroxylation of compounds, for example steroids. This is carried out by the enzyme cytochrome P-450, which usually resides outside the mitochondria in the endoplasmic reticulum of cells. The reaction catalyzed by P-450 is

$$RH + O_2 + NADPH + H^+ \longrightarrow ROH + H_2O + NADP^+$$

where RH represents a steroid or other molecule. This reaction is also used by our body to detoxify foreign substances, for example barbiturates. But P-450 is not always beneficial. Some of the most powerful carcinogens are converted to their active forms by P-450.

Another use of oxygen for detoxification is in the fight against infection by bacteria. Specialized white blood cells called leukocytes (Sec. 25.2) that engulf bacteria, kill them by generating superoxide, O_2^-, a highly reactive form of oxygen. But superoxide and other highly reactive forms of oxygen, such as ·OH radicals and hydrogen peroxide (H_2O_2), not only destroy foreign cells but damage our own cells as well. They especially attack unsaturated fatty acids and thereby damage cell membranes (Sec. 17.5). One theory that has been advanced to explain why humans and animals show symptoms of old age as they grow older is that these oxidizing agents damage healthy cells over the years. Our bodies have their own defenses against these very reactive compounds. One example is glutathione (Box 18B), which protects us against oxidative damages. Furthermore, we have in our cells two powerful enzymes that decompose these highly reactive oxidizing agents:

$$2O_2^- + 2H^+ \xrightarrow{\text{superoxide dismutase}} H_2O_2 + O_2$$

$$2H_2O_2 \xrightarrow{\text{catalase}} 2H_2O + O_2$$

All the enzymes involved in reactions with highly reactive forms of oxygen contain a heavy metal: iron in the case of P-450 and catalase and copper and/or zinc in the case of superoxide dismutase. If the aging theory mentioned above is correct, then these protecting agents keep us alive for many years, but eventually the oxidizing processes do so much damage that we exhibit symptoms of aging and eventually die.

stores the electrical energy in the form of chemical energy. On the other hand, the enzyme can also hydrolyze ATP and as a consequence pump out H^+ from the mitochondrion. Each pair of protons that is translocated gives rise to the formation of one ATP molecule.

The protons that enter a mitochondrion combine with the electrons transported through the **electron transport chain** and with oxygen to form water. The net result of the two processes (electron/H^+ transport and ATP formation) is that the oxygen we breathe in combines with four H^+ ions and four electrons to give two water molecules. The four H^+ ions and four electrons come from the NADH and $FADH_2$ molecules produced in the citric acid cycle. The functions of the oxygen, therefore, are (1) to oxidize NADH to NAD^+ and $FADH_2$ to FAD so that all these molecules can go back and participate in the citric acid cycle and (2) to provide energy for the conversion of ADP to ATP.

The latter function is accomplished indirectly. The entrance of the H^+ ions into the mitochondrion drives the ATP formation, but the H^+ ions enter the mitochondrion because the O_2 depleted the H^+ ion concentration when water was formed. It is a rather complex process involving the transport of electrons along a whole series of enzyme molecules (which catalyze all these reactions), but without it the cell cannot utilize the O_2 molecules and eventually dies. The following equations represent the overall reactions in oxidative phosphorylation:

$$\text{NADH} + 3\text{ADP} + \tfrac{1}{2}\text{O}_2 + 3\text{P}_\text{i} + \text{H}^+ \longrightarrow \text{NAD}^+ + 3\text{ATP} + \text{H}_2\text{O} \quad (20.2)$$

$$\text{FADH}_2 + 2\text{ADP} + \tfrac{1}{2}\text{O}_2 + 2\text{P}_\text{i} \longrightarrow \text{FAD} + 2\text{ATP} + \text{H}_2\text{O} \quad (20.3)$$

20.7 The Energy Yield

The energy released during electron transport is now finally built into the ATP molecule. Therefore, it is instructive to look at the energy yield in the universal biochemical currency: the number of ATP molecules.

Each pair of protons entering a mitochondrion results in the production of one ATP molecule. For each NADH molecule, three pairs of protons are pumped into the intermembrane space in the electron transport process. Therefore, for each NADH molecule, we get three ATP molecules, as can be seen in Equation 20.2. For each $FADH_2$ molecule, we have seen that only four protons are pumped out of the mitochondrion. Therefore, only two ATP molecules are produced for each $FADH_2$, as seen in Equation 20.3.

Now we can produce the energy balance for the whole common catabolic pathway (citric acid cycle and oxidative phosphorylation combined). For each C_2 fragment entering the citric acid cycle, we obtain three NADH and one $FADH_2$ (Eq. 20.1) plus one GTP, which is equivalent in energy to one ATP. Thus, the total number of ATP molecules produced per C_2 fragment is

$$\begin{aligned} 3\text{NADH} \times 3\text{ATP/NADH} &= 9\text{ATP} \\ 1\text{FADH}_2 \times 2\text{ATP/FADH}_2 &= 2\text{ATP} \\ 1\text{GTP} &= \underline{1\text{ATP}} \\ & 12\text{ATP} \end{aligned}$$

Each C_2 fragment that enters the cycle produces 12 ATP molecules and uses up two O_2 molecules. The total effect of the energy-production chain of reactions we have discussed in this chapter (the common catabolic pathway) is to oxidize one C_2 fragment with two molecules of O_2 to produce two molecules of CO_2 and 12 molecules of ATP:

$$C_2 + 2O_2 + 12ADP + 12P_i \longrightarrow 12ATP + 2CO_2$$

This is the *net* reaction for the whole common pathway.

The important thing is not the waste product, CO_2, but the 12 ATP molecules, since these will now release their energy when they are converted to ADP.

20.8 Conversion of Chemical Energy to Other Forms of Energy

As mentioned in Section 20.3, the storage of chemical energy in the form of ATP lasts only a short time. Usually within a minute the ATP is hydrolyzed and thus releases its chemical energy. How does the body utilize this chemical energy? To answer this question, let us look at the different forms in which energy is needed in the body.

Conversion to Other Forms of Chemical Energy

The activity of many enzymes is controlled and regulated by phosphorylation. For example, the enzyme phosphorylase, which catalyzes the breakdown of glycogen (Box 21B), occurs in an inactive form, phosphorylase b. The enzyme has a single serine residue at its active site. When ATP transfers a phosphate group to this serine, the enzyme becomes active. Thus the chemical energy of ATP is used in the form of chemical energy to activate phosphorylase b so that glycogen can be utilized. We shall see several other examples of this in Chapters 21 and 22.

Electrical Energy

The body maintains a high concentration of K^+ ions inside the cells despite the fact that outside the cells the K^+ concentration is low. The reverse is true for Na^+. In order that K^+ should not diffuse out of the cells and Na^+ should not penetrate in, there are special transport proteins in the cell membranes that constantly pump K^+ into the cells and Na^+ out. This pumping requires energy, which is supplied by the hydrolysis of ATP to ADP. With this pumping, the charges inside and outside the cell are unequal, and this generates an electric potential. Thus the chemical energy of ATP is transformed into electrical energy.

Chemical energy is converted to electrical energy in nerve transmission.

Mechanical Energy

ATP is the immediate source of energy in muscle contraction. In essence, muscle contraction takes place when thick and thin filaments slide past each other (Fig. 20.11). The thick filament is *myosin,* an ATPase enzyme—that is, one that hydrolyzes ATP. The thin filament, *actin,* binds strongly to myosin in the contracted state. However, when ATP binds to myosin, the actin–myosin dissociates, and the muscle relaxes. When myosin hydrolyzes ATP, it interacts with actin once more, and a new contraction occurs. Therefore, the hydrolysis of ATP drives the alternating association and dissociation of actin and myosin and consequently the contraction and relaxation of the muscle.

Heat Energy

One molecule of ATP upon hydrolysis to ADP yields 7.3 kcal/mole. Some of this energy is released as heat and is used by the body to maintain body temperature. If we estimate that the specific heat of the body is about the same as that of water, a person weighing 60 kg would need to hydrolyze approximately 99 moles of ATP to raise the temperature of the body from room temperature, 25°C, to 37°C. Not all body heat is derived from ATP hydrolysis; some exothermic reactions in the body also contribute.

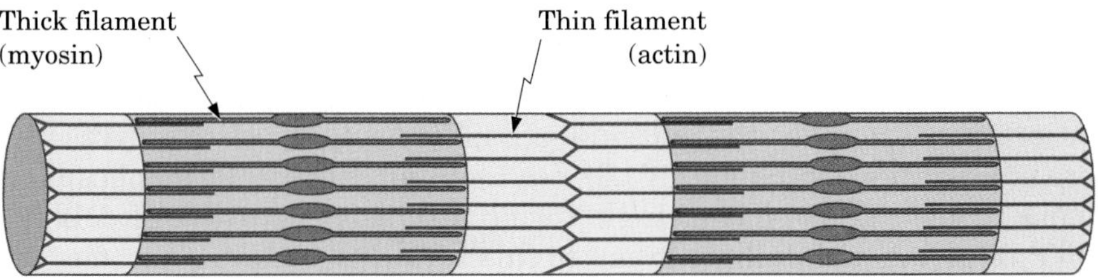

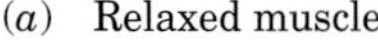
(*a*) Relaxed muscle

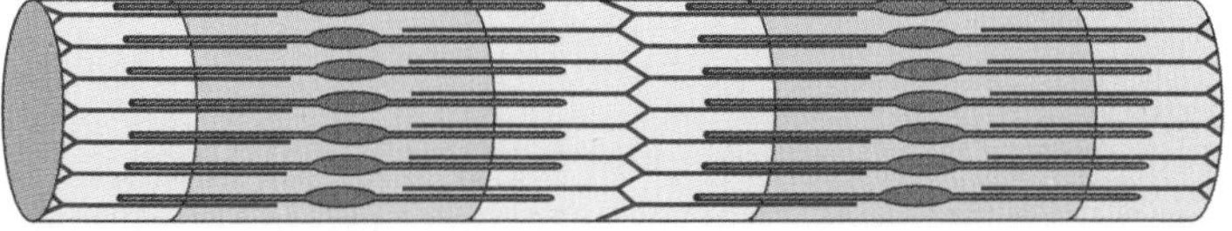
(*b*) Contracted muscle

Figure **20.11** • Schematic diagram of muscle contraction.

SUMMARY

The sum total of all the chemical reactions involved in maintaining the dynamic state of cells is called **metabolism.** The breaking down of molecules is **catabolism;** the building up of molecules is **anabolism.** The **common metabolic pathway** uses a two-carbon C_2 fragment (acetyl) from different foods. Through the **citric acid cycle** and **oxidative phosphorylation** (**electron transport chain**), the C_2 fragment is oxidized. The products formed are water and carbon dioxide. The energy from oxidation is built into the high-chemical-energy-storing molecule ATP.

Both the citric acid cycle and oxidative phosphorylation take place in the **mitochondria.** The enzymes of the citric acid cycle are located in the matrix, while the enzymes of the oxidative phosphorylation chain are on the inner mitochondrial membrane. Some of them project into the intermembrane space.

The principal carriers in the common pathway are as follows: **ATP** is the phosphate carrier, **CoA** is the C_2 fragment carrier, and **NAD$^+$** and **FAD** carry the hydrogen ions and electrons. The unit common to all these carriers is **ADP.** This is the nonactive end of the carriers, which acts as a handle that fits into the active sites of the enzymes.

In the citric acid cycle the C_2 fragment first combines with a C_4 fragment (oxaloacetate) to yield a C_6 fragment (citrate). An oxidative decarboxylation yields a C_5 fragment. One CO_2 is released, and one NADH + H^+ is passed on to the electron transport chain. Another oxidative decarboxylation provides a C_4 fragment. Once again a CO_2 is released and another NADH + H^+ is passed on to the electron transport chain. Subsequently, two dehydrogenation (oxidation) steps yield one $FADH_2$ and one additional NADH + H^+, along with an analog of ATP called GTP. The cycle is controlled by a feedback mechanism.

The NADH enters the electron transport chain at the flavoprotein dehydrogenase stage. Two H^+ ions are expelled into the intermembrane space of the mitochondrion, and the electrons are passed along the enzymes of the electron transport chain. Each enzyme transfers electrons to the next in the series through redox reactions. As the electrons are transported, H^+ ions are expelled from the interior of the mitochondrion to the intermembrane space. For each NADH, six H^+ ions are expelled. Finally, the electrons inside the mitochondrion combine with oxygen and H^+ to form water. When the expelled H^+ ions stream back into the mitochondrion, they drive a complex enzyme called **proton translocating ATPase,** which makes one ATP molecule for each two H^+ ions that enter the mitochondrion. Therefore, for each NADH + H^+ coming from the citric acid cycle, three ATP molecules are formed. The $FADH_2$ enters the respiratory chain later. While the electrons of the $FADH_2$ are transported, only four H^+ ions are expelled into the intermembrane space. As a consequence, only two ATP molecules are formed for each $FADH_2$. The overall result is that for each C_2 fragment entering the citric acid cycle, 12 ATP molecules are produced.

The chemical energy is stored in ATP only for a short time. ATP is hydrolyzed, usually within a minute. This chemical energy is used to do chemical, mechanical, and electrical work in the body and to maintain body temperature.

KEY TERMS

Acetyl coenzyme A (Sec. 20.3)
Anabolism (Sec. 20.1)
ATP (Sec. 20.3)
Biochemical pathway (Sec. 20.1)
Catabolism (Sec. 20.1)
Chemiosmotic hypothesis (Sec. 20.6)
Citric acid cycle (Sec. 20.4)
Coenzyme A (Sec. 20.3)
Common catabolic pathway (Sec. 20.1)
Cristae (Sec. 20.2)
Electron transport chain (Sec. 20.6)
Energy yield (Sec. 20.7)
FAD (Sec. 20.3)
Inorganic phosphate (Sec. 20.3)
Krebs cycle (Sec. 20.4)
Metabolism (Sec. 20.1)
Mitochondria (Sec. 20.2)
NAD$^+$ (Sec. 20.3)
Oxidative phosphorylation (Sec. 20.5)
Pathway (Sec. 20.1)
Proton gradient (Sec. 20.6)
Respiratory chain (Sec. 20.3)
Tricarboxylic acid cycle (Sec. 20.4)

PROBLEMS

Metabolism

20.1 Distinguish between metabolism and catabolism.

20.2 (a) How many sequences are there in the common catabolic pathway? (b) Name them.

Cells and Mitochondria

20.3 What name is given to the organelles in which the common catabolic pathway takes place?

20.4 How do ions and molecules enter the mitochondria?

20.5 What is the name of the nonmembranous portion of a mitochondrion?

20.6 (a) Where are the enzymes of the citric acid cycle located? (b) Where are the enzymes of oxidative phosphorylation located?

Principal Compounds of the Common Catabolic Pathway

20.7 Which is more energetic, the phosphate-ester linkage or the phosphoric anhydride linkage?

20.8 What are the products of the reaction

$$AMP + H_2O \xrightarrow{H^+} ?$$

20.9 Which yields more energy, the hydrolysis (a) of ATP to ADP or (b) of ADP to AMP?

20.10 ATP is often called the molecule for storing chemical energy. Does "storage" mean a long-term preservation of energy? Explain.

***20.11** Does the amount of ATP produced by the metabolism of a given amount of food correspond to all the caloric energy that is in the food?

***20.12** When NAD^+ is reduced, two electrons enter the molecule, together with one H^+. Exactly where in the product will the two electrons be located?

20.13 Which atoms in the flavin portion of FAD are reduced to yield $FADH_2$?

20.14 Which part of the FAD molecule is (a) the operative part and (b) the handle?

20.15 In the common catabolic pathway, there are a number of important molecules that act as carriers (transfer agents). (a) What is the carrier of phosphate groups? (b) Which are the coenzymes transferring hydrogen ions and electrons? (c) What kind of groups does coenzyme A carry?

***20.16** The sugar units in FAD and NAD^+ are different. Explain the difference.

20.17 Name all the functional groups in the structure of CoA.

20.18 Name the vitamin B molecules that are a part of the structure of (a) NAD^+ (b) FAD (c) coenzyme A

***20.19** In both NAD^+ and FAD the vitamin B portion of the molecule is the active part. Is this also true for CoA?

20.20 What type of compound is formed when coenzyme A reacts with acetate?

20.21 The fats and carbohydrates metabolized by our bodies are eventually converted to a single compound. What is it?

The Citric Acid Cycle

20.22 The first step in the citric acid cycle is abbreviated as

$$C_2 + C_4 = C_6$$

(a) What do these symbols stand for? (b) What are the common names of the three compounds involved in this reaction?

20.23 What is the only C_5 compound in the citric acid cycle?

20.24 Identify by number those steps of the citric acid cycle that are not redox reactions.

20.25 What kind of reaction occurs in the citric acid cycle when a C_6 compound is converted to a C_5 compound?

20.26 How many CO_2 molecules are produced in one citric acid cycle?

20.27 What is the role of succinate dehydrogenase in the citric acid cycle?

20.28 (a) In the citric acid cycle how many steps can be classified as decarboxylation reactions? (b) In each case what is the concurrent oxidizing agent?

20.29 Is ATP directly produced during any step of the citric acid cycle? Explain.

20.30 There are four dicarboxylic acid compounds, each containing four carbons, in the citric acid cycle. Which is (a) the least and (b) the most oxidized?

20.31 Why is a many-step cyclic process more efficient in utilizing energy from food than a single-step combustion?

20.32 Did the two CO_2 molecules given off in one citric acid cycle originate from the entering acetyl group?

20.33 Which intermediates of the citric acid cycle contain C═C double bonds?

20.34 Most of the reactions of the citric acid cycle can also be carried out in the laboratory, without the appropriate enzymes, albeit at a much slower rate. Which reaction in the citric acid cycle cannot be carried out without enzyme catalysis?

***20.35** Oxidation is defined as loss of electrons. When oxidative decarboxylation occurs, as in step ④, where do the electrons of the α-ketoglutarate go?

Oxidative Phosphorylation

20.36 What is the main function of oxidative phosphorylation (the electron transport chain)?

20.37 How many *pairs* of H^+ ions are expelled for each (a) NADH molecule (b) $FADH_2$ molecule?

20.38 We breathe in O_2 and breathe out CO_2. Explain how the body uses the O_2 to convert the foods we eat to CO_2, H_2O, and energy.

20.39 In oxidative phosphorylation water is formed from H^+, e^-, and O_2. Where does this take place?

20.40 At what points in oxidative phosphorylation are the H^+ ions and the electrons separated from each other?

20.41 How many ATP molecules are generated (a) for each H^+ translocated through the ATPase complex (b) for each C_2 fragment that goes through the complete common catabolic pathway?

20.42 Since H^+ is pumped out into the intermembrane space, what is the pH there compared with that in the matrix?

The Chemiosmotic Pump

20.43 What is the channel through which H^+ ions reenter the matrix of mitochondria?

***20.44** The proton gradient accumulated in the intermembrane area of a mitochondrion drives the ATP-manufacturing enzyme ATPase. Why do you think Mitchell called this the "chemiosmotic hypothesis"?

The Energy Yield

20.45 If each mole of ATP yields 7.3 kcal of energy upon hydrolysis, how many kcal of energy would you get from 1 g of CH_3COO^- (C_2) entering the citric acid cycle?

20.46 If a C_{18} compound is metabolized and enters the citric acid cycle as nine C_2 fragments (acetyl groups), how many molecules of ATP are produced in the common metabolic pathway from the C_{18} compound?

Conversion of Chemical Energy to Other Forms

20.47 (a) How do muscles contract? (b) Where does the energy in muscle contraction come from?

20.48 Give an example of the conversion of the chemical energy of ATP to electrical energy.

20.49 How is the enzyme phosphorylase activated?

Boxes

20.50 (Box 20A) (a) Why was nicotinic acid given its name? (b) Why is it often called niacin?

***20.51** (Box 20B) Oligomycin is an antibiotic that allows electron transport to continue but stops phosphorylation, in bacteria as well as in humans. Would you use this as an antibacterial drug for people? Explain.

20.52 (Box 20B) 2,4-Dinitrophenol is a protonophore, or proton carrier. What is special about its mode of transportation?

20.53 (Box 20C) (a) What kind of reaction is catalyzed by cytochrome P-450? (b) Where does the oxygen reactant come from?

Additional Problems

20.54 (a) What is the difference between ATP and GTP? (b) Compared with ATP, would you expect GTP to carry more, less, or about the same amount of energy?

***20.55** How many grams of CH_3COOH molecules must be metabolized in the common metabolic pathway to yield 87.6 kcal of energy?

20.56 What happens to the end products of the common pathway?

20.57 What structural characteristics do citric acid and malic acid have in common?

***20.58** Two keto acids are important in the citric acid cycle. Identify the two keto acids and tell how they are manufactured.

20.59 Which filament of muscles is an enzyme, catalyzing the reaction that converts ATP to ADP?

20.60 One of the end products of food metabolism is water. How many molecules of H_2O are formed from the entry of each molecule of (a) NADH + H^+ (b) $FADH_2$? (Use Fig. 20.10.)

20.61 What is the chief energy-carrying molecule in the body?

***20.62** Acetyl CoA is labeled with radioactive carbon as shown: $CH_3{}^*CO$—S—CoA. This enters the citric acid cycle. If the cycle is allowed to progress only to the α-ketoglutarate level, will the CO_2 expelled by the cell be radioactive?

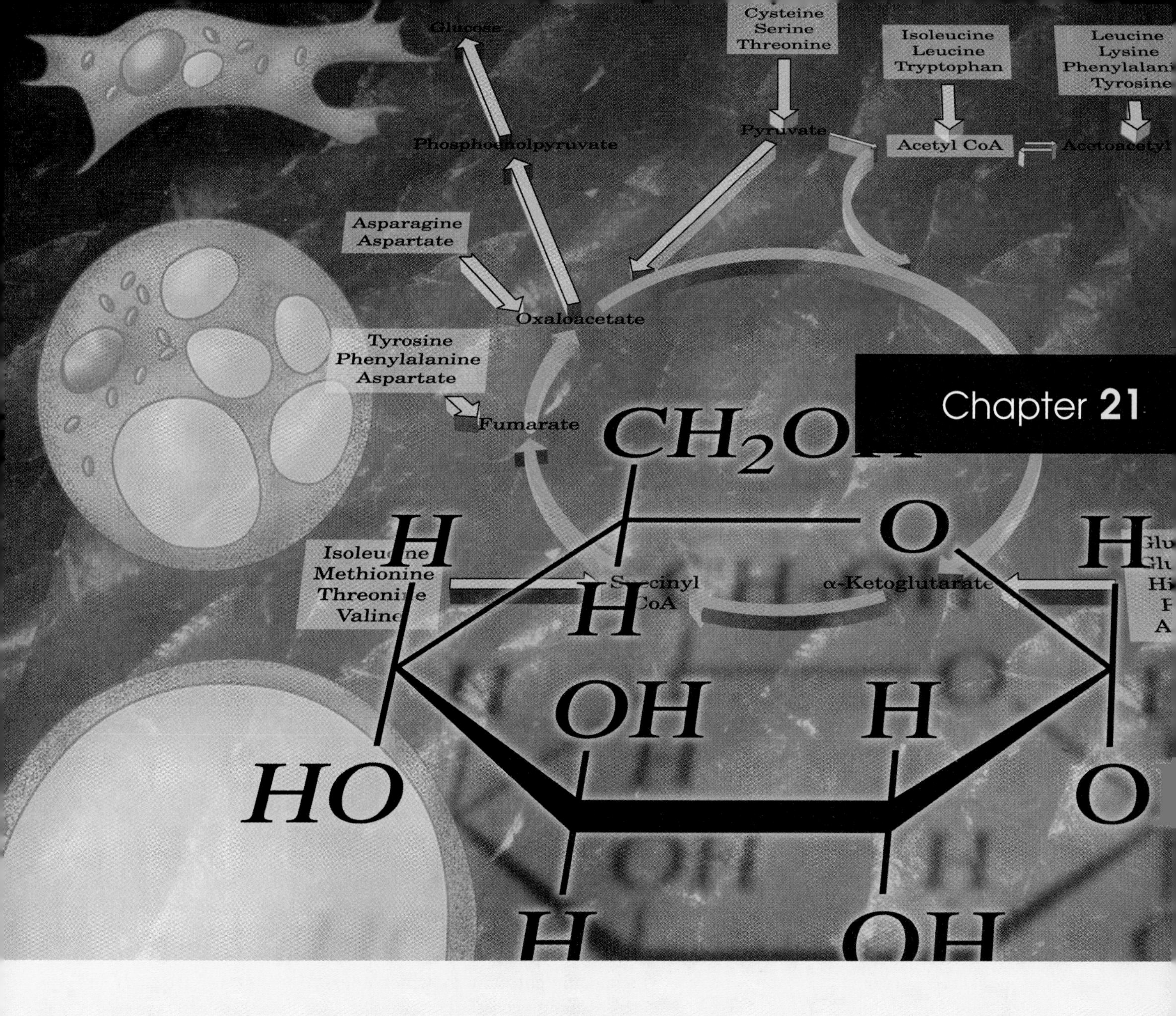

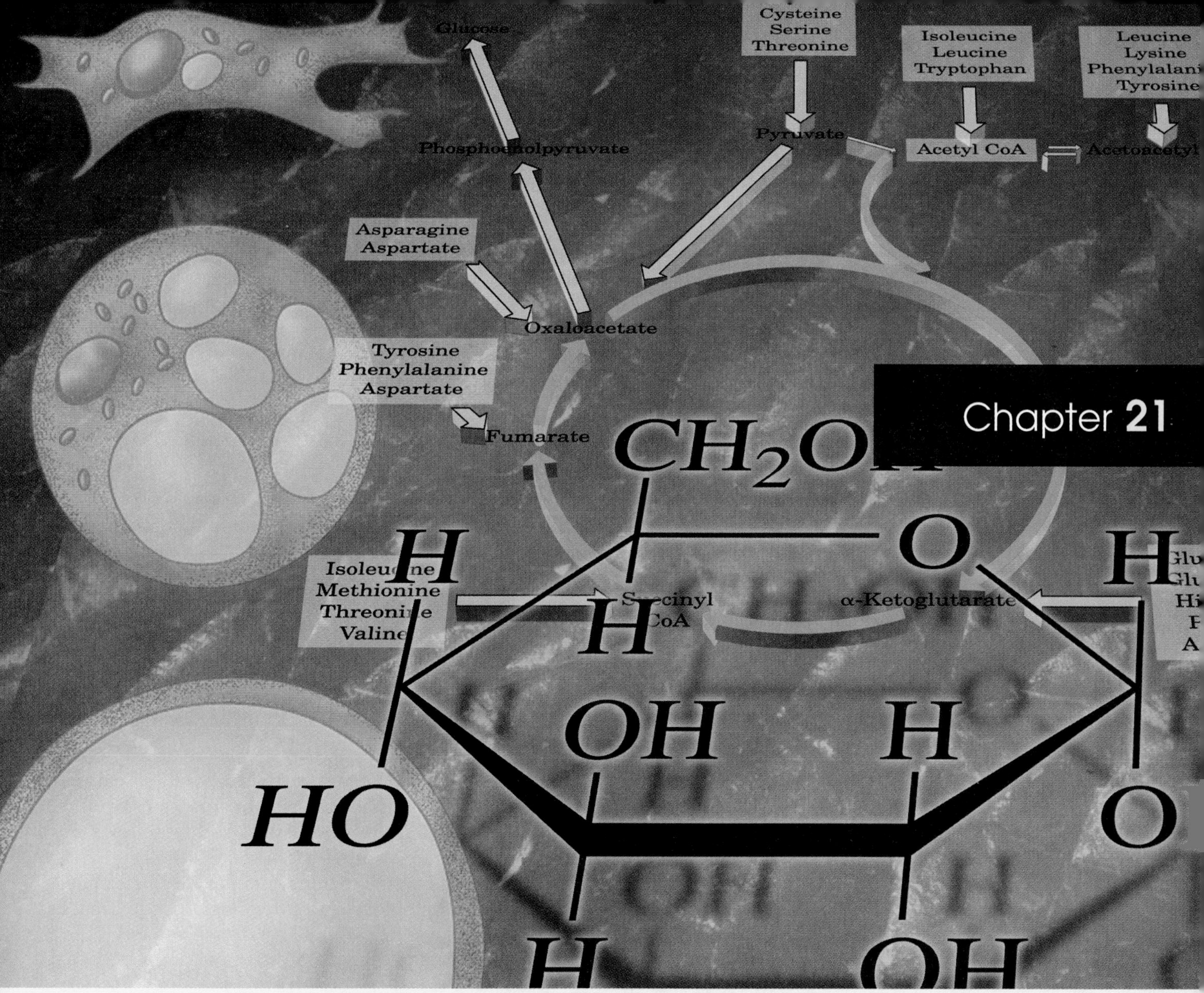

Chapter 21

Specific catabolic pathways: Carbohydrate, lipid, and protein metabolism

21.1 Specific Pathways and Their Convergence to the Common Pathway

The food we eat serves two main purposes: (1) It fulfills our energy needs, and (2) it provides the raw materials to build the compounds our bodies need. Before either of these processes can take place, the food—carbohydrates, fats, and proteins—must be broken down to small molecules that can be absorbed through the intestinal walls.

Carbohydrates

Complex carbohydrates (di- and polysaccharides) are broken down by stomach acid and enzymes to produce monosaccharides (Sec. 26.8). Monosaccharides, the most important of which is glucose, may also come from the enzymatic breakdown of glycogen. As you may recall, this highly branched polymer stores carbohydrates in the liver and muscles until needed. Once monosaccharides are produced, they can be used either to build new oligo- and polysaccharides or to provide energy. The specific pathway by which energy is extracted from monosaccharides is called **glycolysis** (Secs. 21.2 and 21.3).

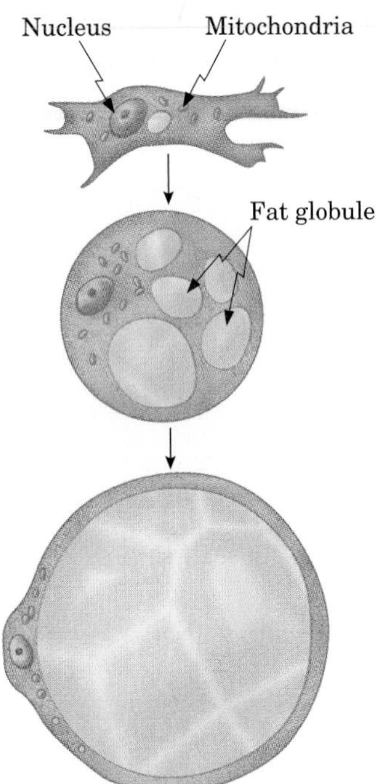

Figure **21.1**
Storage of fat in a fat cell. As more and more fat droplets accumulate in the cytoplasm, they coalesce to form a very large globule of fat. Such a fat globule may occupy most of the cell, pushing the cytoplasm and the organelles to the periphery. (Modified from C. A. Villee, E. P. Solomon, and P. W. Davis, *Biology,* Philadelphia: Saunders College Publishing, 1985.)

Lipids

Ingested fats are broken down by lipases to glycerol and fatty acids or to monoglycerides, which are absorbed through the intestine (Sec. 26.9). In a similar way, complex lipids are also hydrolyzed to smaller units before absorption. As with carbohydrates, these smaller molecules (fatty acids, glycerol, and so on) can be used to build complex molecules needed in membranes, or they can be oxidized to provide energy, or they can be stored in fat storage depots (Fig. 21.1). The stored fats can then be broken down later to glycerol and fatty acids whenever they are needed as fuel.

Specialized cells that store fats are called **fat depots.**

The specific pathway by which energy is extracted from glycerol involves the same glycolysis pathway as that used for carbohydrates (Sec. 21.4). The specific pathway used by the cells to obtain energy from fatty acids is called **β-oxidation** (Sec. 21.5).

Proteins

The amino acid pool is the name used for the free amino acids found both inside and outside cells throughout the body.

As you might expect from a knowledge of their structures, proteins are broken down by HCl in the stomach and by digestive enzymes in the stomach and intestines (pepsin, trypsin, chymotrypsin, and carboxypeptidases) to produce their constituent amino acids. The amino acids absorbed through the intestinal wall enter the **amino acid pool.** They serve as building blocks for proteins as needed and, to a smaller extent (especially during starvation), as a fuel for energy. In the latter case, the nitrogen of the amino acids is catabolized through **oxidative deamination** and the **urea cycle** and is expelled from the body as urea in the

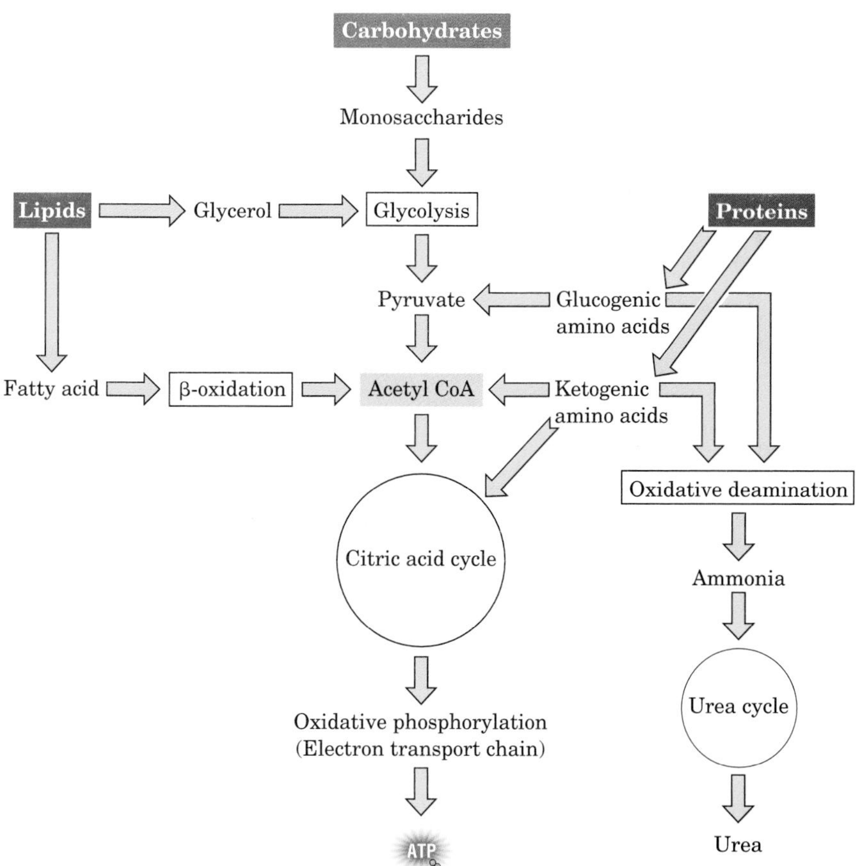

Figure **21.2**
The convergence of the specific pathways of carbohydrate, fat, and protein catabolism into the common catabolic pathway, which is made up of the citric acid cycle and oxidative phosphorylation.

urine (Sec. 21.8). The carbon skeletons of the amino acids enter the common catabolic pathway (Chap. 20) either as α-keto acids (pyruvic, oxaloacetic, and α-ketoglutaric acids) or as acetyl coenzyme A (Sec. 21.9). In all cases, the **specific pathways of carbohydrate, fat, and protein degradation converge to the common catabolic pathway** (Fig. 21.2). This way, the body needs fewer enzymes to get energy from diverse food materials. Efficiency is thus achieved because a minimal number of chemical steps are required and also because the energy-producing factories of the body are localized in the mitochondria.

21.2 Glycolysis

Glycolysis is **the specific pathway by which the body gets energy from monosaccharides.** The detailed steps are shown in Figure 21.3, and the most important features are shown schematically in Figure 21.4.

In the first steps of glucose metabolism, energy is consumed rather than released. At the expense of two molecules of ATP (which are converted to ADP), glucose (C_6) is phosphorylated; first glucose 6-phosphate

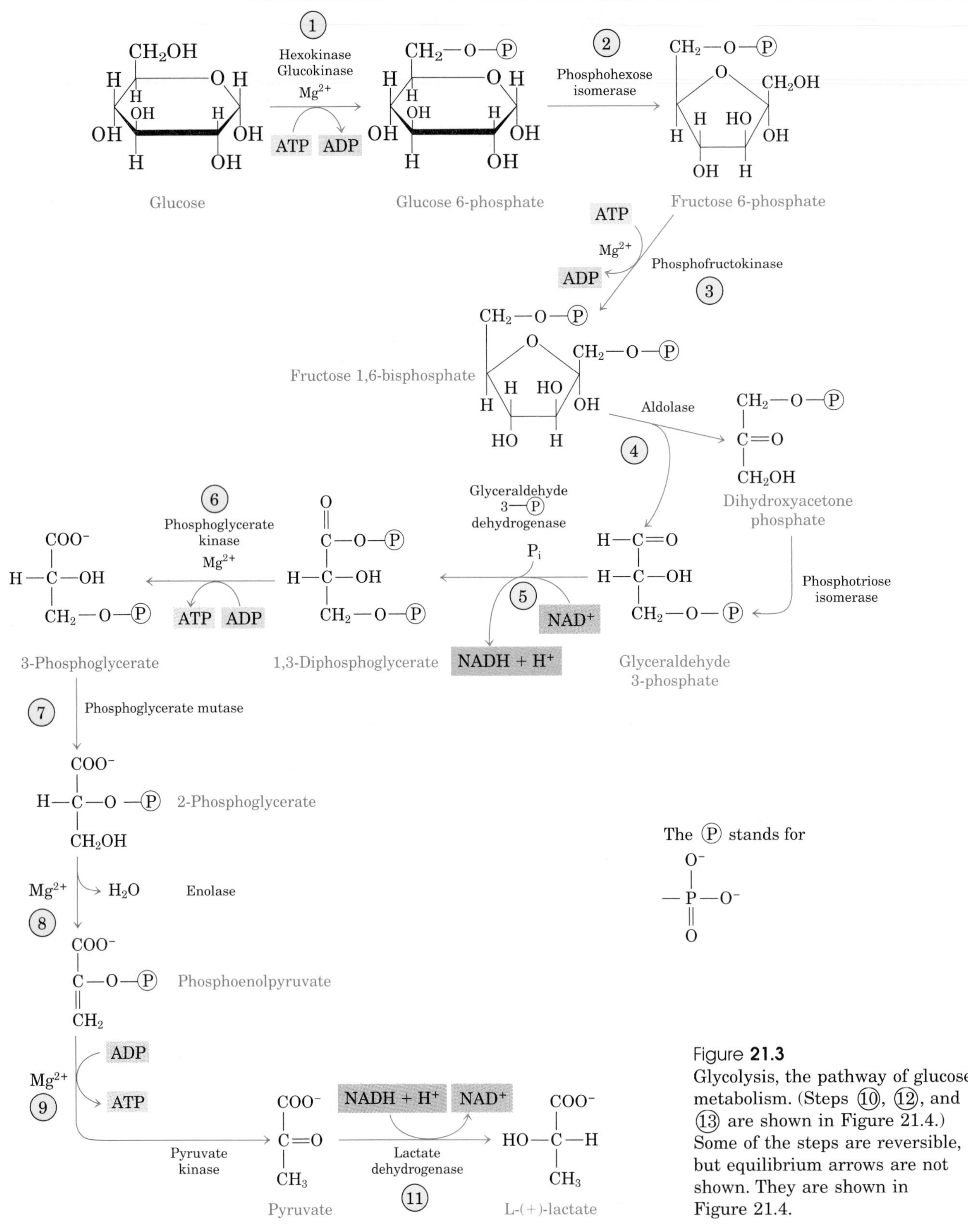

Figure **21.3**
Glycolysis, the pathway of glucose metabolism. (Steps (10), (12), and (13) are shown in Figure 21.4.) Some of the steps are reversible, but equilibrium arrows are not shown. They are shown in Figure 21.4.

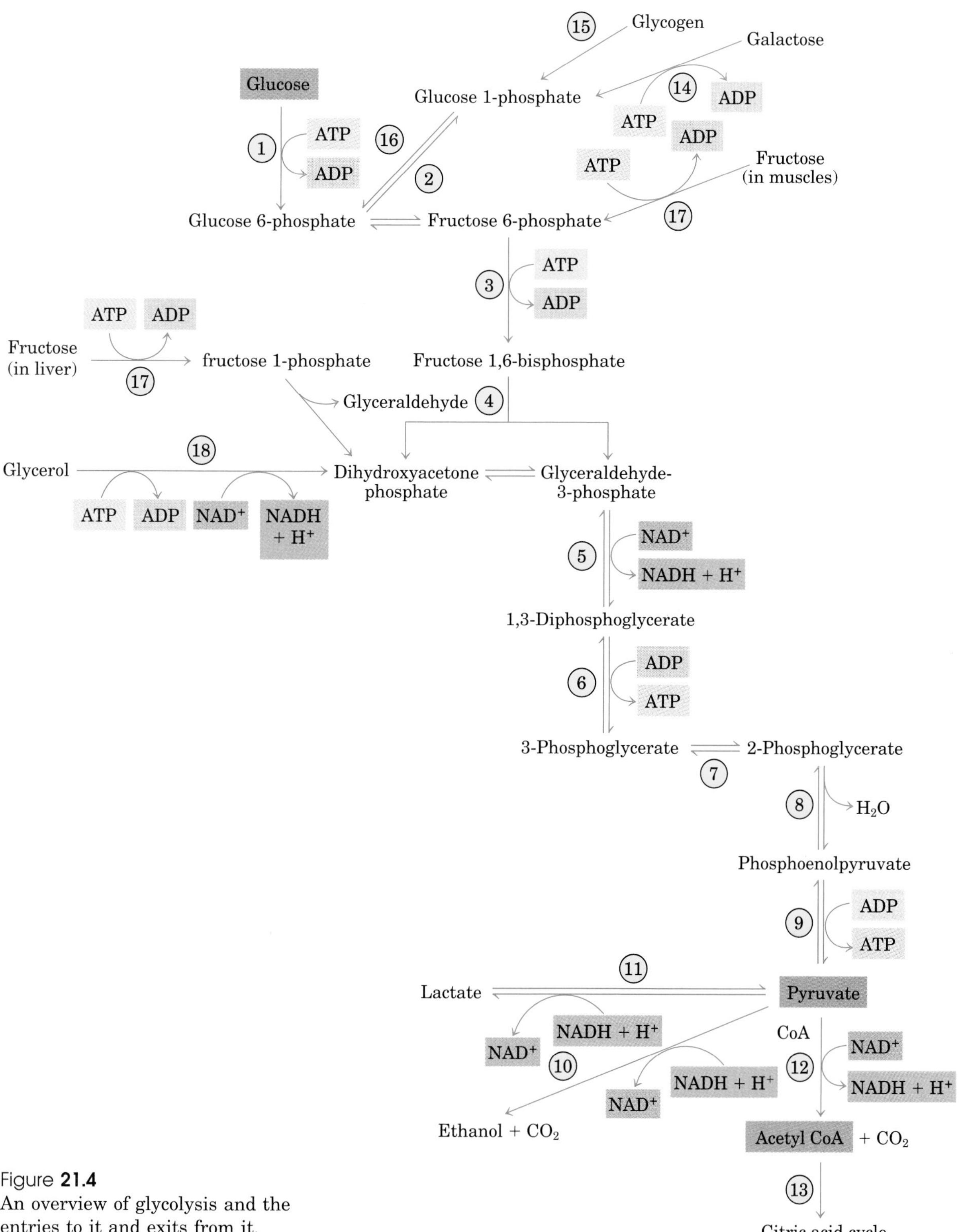

Figure **21.4**
An overview of glycolysis and the entries to it and exits from it.

is formed ①, and then, after isomerization to fructose 6-phosphate ②, a second phosphate group is attached to yield fructose 1,6-bisphosphate ③. We can consider these steps as the activation process.

In the second stage the C_6 compound fructose 1,6-bisphosphate is broken ④ into two C_3 fragments. The two C_3 fragments, glyceraldehyde 3-phosphate and dihydroxyacetone phosphate, are in equilibrium (they can be converted to each other). Only the glyceraldehyde 3-phosphate is oxidized in glycolysis, but as this species is removed from the equilibrium mixture, Le Chatelier's principle ensures that the dihydroxyacetone phosphate is converted to glyceraldehyde 3-phosphate.

In the third stage, the glyceraldehyde 3-phosphate is oxidized to 1,3-diphosphoglycerate ⑤. The hydrogen of the aldehyde group is removed by the NAD^+ coenzyme. In the next step ⑥, the phosphate from the carboxyl group is transferred to ADP, yielding ATP and 3-phosphoglycerate. The latter, after isomerization ⑦ and dehydration ⑧, is converted to phosphoenolpyruvate, which loses its remaining phosphate ⑨ and yields pyruvate ion and another ATP molecule.

In step ⑨, after hydrolysis of the phosphate, the resulting enol of pyruvic acid tautomerizes to the more stable keto form (Sec. 13.6).

The cytosol is the fluid inside the cell that surrounds the nucleus and organelles such as mitochondria.

All these glycolysis reactions occur in the cytosol outside the mitochondria. They occur in the absence of O_2 and are therefore also called reactions of the **anaerobic pathway.** As indicated in Figure 21.4, the end product of glycolysis, pyruvate, does not accumulate in the body. In certain bacteria and yeast, pyruvate undergoes reductive decarboxylation ⑩ to produce ethyl alcohol (Sec. 12.4). In some bacteria, and also in mammals in the absence of oxygen, pyruvate is reduced to lactate ⑪. But most importantly pyruvate goes through an oxidative decarboxylation in the presence of coenzyme A ⑫ to produce acetyl CoA:

$$\underset{\text{Pyruvate}}{CH_3-C(=O)-C(=O)-O^-} + CoASH \longrightarrow \underset{\text{Acetyl coenzyme A}}{CH_3-C(=O)-S-CoA} + CO_2$$

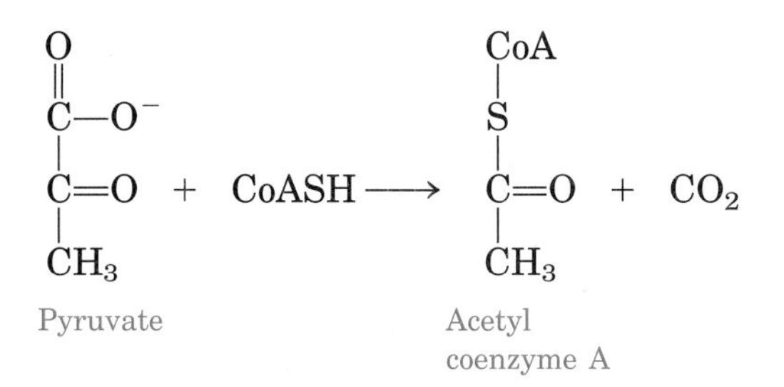

This reaction is catalyzed by a complex enzyme system, pyruvate dehydrogenase, that sits on the inner membrane of the mitochondrion. The reaction produces acetyl CoA, CO_2, and NADH + H^+. The acetyl CoA then enters the citric acid cycle ⑬ and goes through the common pathway.

In summary, after converting complex carbohydrates to glucose, the body gets energy from glucose by converting it to acetyl CoA (by way of pyruvate) and then using the acetyl CoA as a raw material for the common pathway.

21.3 The Energy Yield from Glucose

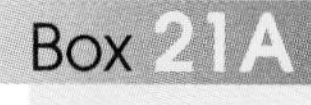

In conjunction with Figure 21.4, we can sum up the energy derived from glucose catabolism in terms of ATP production. However, before we be-

Box 21A

Lactate Accumulation

Many athletes suffer muscle cramps when they undergo strenuous exercise. This is the result of a shift from normal glucose catabolism (glycolysis → citric acid cycle → oxidative phosphorylation) to that of lactate production ⑪ (Fig. 21.4). During exercise, oxygen is used up rapidly, and this slows down the rate of the common pathway. The demand for energy makes *anaerobic* glycolysis proceed at a high rate, but because the *aerobic* (oxygen-demanding) pathways are slowed down, not all the pyruvate produced in glycolysis can enter the citric acid cycle. The excess pyruvate ends up as lactate, which causes painful muscle contractions.

The same shift in catabolism also occurs in heart muscle when coronary thrombosis leads to cardiac arrest. The oxygen supply is cut off by the blockage of the artery to the heart muscles. The common pathway and its ATP production are shut off. Glycolysis proceeds at an accelerated rate, accumulating lactate. The heart muscle contracts, producing a cramp. Just as in skeletal muscle, massage of heart muscles can relieve the cramp and start the heart beating. Even if heartbeat is restored within 3 minutes (the amount of time the brain can survive without being damaged), acidosis develops as a result of the cardiac arrest. Therefore, at the same time that efforts are under way to start the heart beating by chemical, physical, or electrical means, an intravenous infusion of 8.4 percent bicarbonate solution is given to combat acidosis.

(Visuals, Unlimited)

gin we must take into account that glycolysis takes place in the cytosol, while oxidative phosphorylation occurs in the mitochondria. Therefore, the NADH + H^+ produced in glycolysis must penetrate the mitochondrial membrane in order to be utilized in oxidative phosphorylation.

There are two routes available to get the NADH + H^+ into the mitochondria; these have different efficiencies. In one, which operates in muscle and nerve cells, only two ATP molecules are produced for each NADH + H^+. In the other, which operates in the heart and the liver, three ATP molecules are produced for each NADH + H^+ produced in the cytosol, just as is the case in the mitochondria (Sec. 20.7). Since most energy production takes place in skeletal muscle cells, when we construct the energy balance sheet we use two ATP molecules for each NADH + H^+ produced in the cytosol.

Muscles attached to bones are called **skeletal muscles.**

With this knowledge we are ready to calculate the energy yield of glucose in terms of ATP molecules produced. This is shown in Table 21.1. In the first stage of glycolysis ① ② ③, two ATP molecules are used up, but this is more than compensated for by the production of 14 ATP molecules in ⑤, ⑥, ⑨, and ⑫ and in the conversion of pyruvate to acetyl CoA. The *net* yield of these steps is 12 ATP molecules. As we saw in Section 20.7, the oxidation of one acetyl CoA produces 12 ATP molecules, and one glucose molecule provides two acetyl CoA molecules. Therefore, the total net yield from metabolism of one glucose molecule in skeletal muscle,

$$C_6H_{12}O_6 + 6O_2 \longrightarrow 6CO_2 + 6H_2O$$

is 36 molecules of ATP.

This calculation applies to glucose metabolism in the skeletal muscle cells, which is what happens most frequently. The two NADH produced in the glycolysis yield only a total of four ATP because of the efficiency of the glycerol 3-phosphate transport. Thus, a total of 36 ATP

Table 21.1 ATP Yield from Complete Glucose Metabolism

Step Number in Fig. 21.4	Chemical Steps	Number of ATP Molecules Produced
① ② ③	Activation (glucose → 1,6-fructose bisphosphate)	−2
⑤	Oxidative phosphorylation 2(glyceraldehyde 3-phosphate → 1,3-diphosphoglycerate), producing 2 NADH + H^+ in cytosol	4
⑥ ⑨	Dephosphorylation 2(1,3-diphosphoglycerate → pyruvate)	4
⑫	Oxidative decarboxylation 2(pyruvate → acetyl CoA), producing 2 NADH + H^+ in mitochondrion	6
⑬	Oxidation of two C_2 fragments in citric acid and oxidative phosphorylation common pathway, producing 12 ATP for each C_2 fragment	24
	Total	36

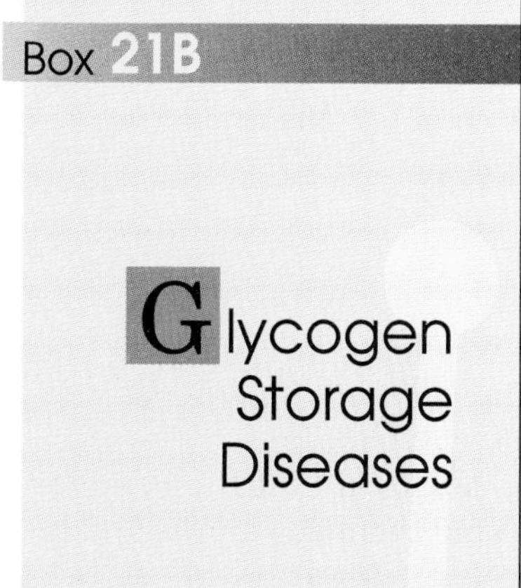

Missing enzymes that metabolize glycogen give rise to a number of genetically inherited glycogen storage diseases. One is McArdle's disease. Patients with this disease have a limited capacity for exercise, suffering from painful muscle cramps. The genetic defect is the absence of phosphorylase in the muscles. Thus, glycogen cannot be utilized.

Another glycogen storage disease is Cori's disease. Patients suffering from this have no debranching enzyme in their liver and muscles. Thus, only the outer branches of glycogen are utilized and the rest accumulates, resulting in an enlarged liver. Hypoglycemia (Box 26C) and ketosis (high amounts of ketone bodies as a result of dominant fat metabolism) are other symptoms of Cori's disease.

Still another example of glycogen accumulation in the muscles is caused by a deficiency of the enzyme maltase. Maltase cuts glycogen into disaccharide units (maltose), and if the enzyme is absent, glycogen will not be hydrolyzed.

molecules are produced from one glucose molecule. However, if the same glucose is metabolized in the heart or liver, the two NADH produced in the glycolysis are transported into the mitochondrion. Therefore, they yield a total of six ATP molecules, so that in this case, there are 38 ATP molecules produced for each glucose molecule.

Glucose is not the only monosaccharide that can be used as an energy source. Other hexoses, such as galactose ⑭ or fructose ⑰, enter the glycolysis pathway at the stages indicated in Figure 21.4. They also yield 36 molecules of ATP per hexose molecule. Furthermore, the glycogen stored in the liver and muscle cells and elsewhere can also be converted by enzymatic breakdown and phosphorylation to glucose 1-phosphate ⑮. This in turn isomerizes to glucose 6-phosphate, providing an entry to the glycolytic pathway ⑯. The pathway in which glycogen breaks down to glucose is called **glycogenolysis.**

21.4 Glycerol Catabolism

The glycerol hydrolyzed from neutral fats or complex lipids (Chap. 17) can also be a rich energy source. The first step in glycerol utilization is an activation step. The body uses one ATP molecule to form glycerol 1-phosphate:

Glycerol 1-phosphate is the same as glycerol 3-phosphate.

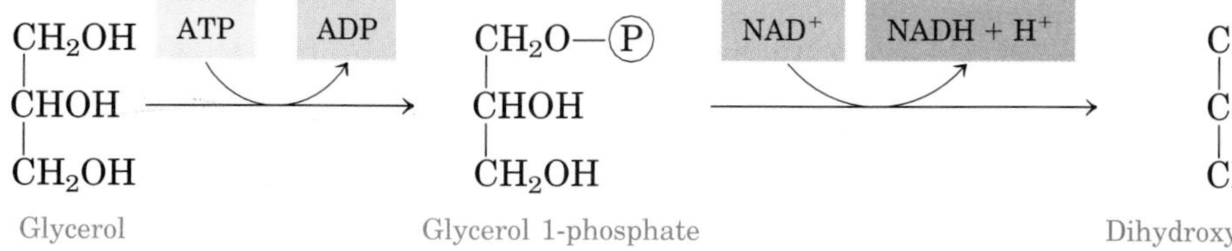

$$\begin{array}{c} CH_2OH \\ | \\ CHOH \\ | \\ CH_2OH \end{array} \xrightarrow{ATP \ \to \ ADP} \begin{array}{c} CH_2O{-}Ⓟ \\ | \\ CHOH \\ | \\ CH_2OH \end{array} \xrightarrow{NAD^+ \ \to \ NADH + H^+} \begin{array}{c} CH_2O{-}Ⓟ \\ | \\ C{=}O \\ | \\ CH_2OH \end{array}$$

Glycerol — Glycerol 1-phosphate — Dihydroxyacetone phosphate

The glycerol phosphate is oxidized by NAD^+ to dihydroxyacetone phosphate, yielding NADH + H^+ in the process. Dihydroxyacetone phosphate then enters the glycolysis pathway ⑱ and is isomerized to glyceraldehyde phosphate, as shown in Figure 21.4. A net yield of 20 ATP molecules is produced from each glycerol molecule, which is 6.7 ATP molecules per carbon atom.

21.5 β-Oxidation of Fatty Acids

As early as 1904, Franz Knoop in Germany proposed that the body utilizes fatty acids as an energy source by breaking them down into C_2 fragments. Prior to fragmentation, the β carbon is oxidized:

The β carbon is the second carbon from the COOH group.

$$-\mathrm{C}-\mathrm{C}-\mathrm{C}-\overset{\beta}{\mathrm{C}}-\overset{\alpha}{\mathrm{C}}-\mathrm{COOH}$$

Thus, the name **β-oxidation** has its origin in Knoop's prediction. It took about 50 years to establish the mechanism by which fatty acids are utilized as an energy source.

The overall process of fatty acid metabolism is shown in Figure 21.5. As is the case with the other foods we have seen, the first step is an activation step. This occurs in the cytosol, where the fat was previously hydrolyzed to glycerol and fatty acids. The activation ① converts ATP to AMP and inorganic phosphate. This is equivalent to the cleavage of two high-energy phosphate bonds. The chemical energy derived from the splitting of ATP is built into the compound acyl CoA, which is formed when the fatty acid combines with coenzyme A (Fig. 21.5). The fatty acid oxidation occurs inside the mitochondrion, so the acyl CoA must pass through the mitochondrial membrane.

Transfer into the mitochondrion is accomplished by an enzyme system called carnitine acyltransferase.

Once the fatty acid in the form of acyl CoA is inside the mitochondrion, the β-oxidation starts. In the first oxidation (dehydrogenation) ②, two hydrogens are removed, creating a double bond between the alpha and beta carbons of the acyl chain. The hydrogens and electrons are picked up by FAD.

In the next step ③, the double bond is hydrated. An enzyme specifically places the hydroxy group on C-3. The second oxidation (dehydrogenation) ④ requires NAD^+ as a coenzyme. The two hydrogens and electrons removed are transferred to the NAD^+ to form NADH + H^+. In the process, a secondary alcohol group is oxidized to a keto group at the beta carbon. In the final step ⑤, the enzyme thiolase cleaves the terminal C_2 fragment from the chain, and the rest of the molecule is attached to a new molecule of coenzyme A.

The cycle now starts again with the remaining acyl CoA, which is now two carbon atoms shorter. At each turn of the cycle, one acetyl CoA is produced. Most fatty acids contain an even number of carbon atoms. The cyclic spiral is continued until we reach the last four carbon atoms. When this fragment enters the cycle at the end, two acetyl CoA molecules are produced in the fragmentation step.

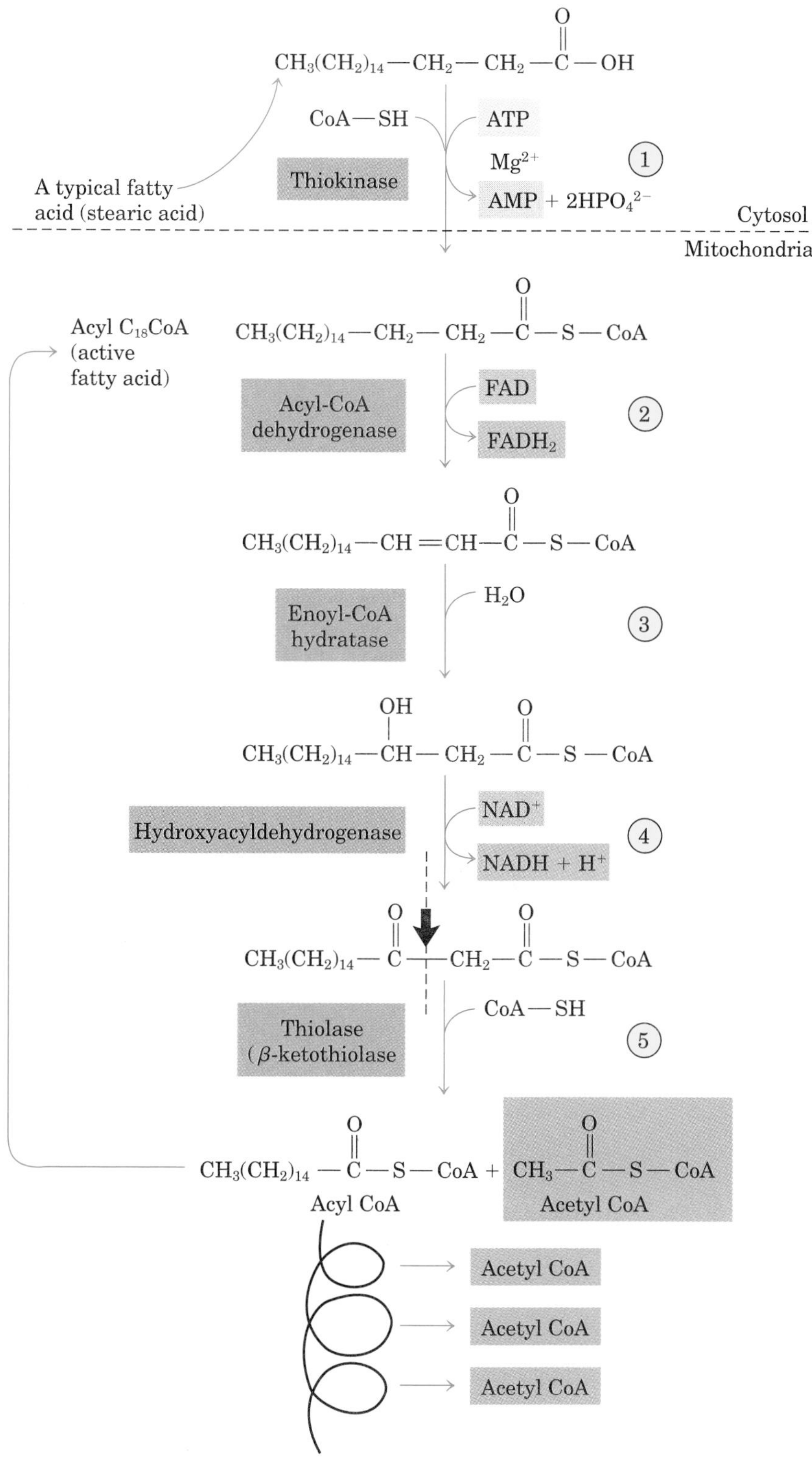

Figure **21.5** • The β-oxidation spiral of fatty acids. Each loop in the spiral contains two dehydrogenations, one hydration, and one fragmentation. At the end of each loop, one acetyl CoA is released.

The β-oxidation of unsaturated fatty acids proceeds in the same way. There is an extra step involved, in which the cis double bond is isomerized to a trans bond, but otherwise the spiral is the same.

21.6 The Energy Yield from Stearic Acid

In order to compare the energy yield from fatty acids with that of other foods, let us select a typical and quite abundant fatty acid: stearic acid, the C_{18} saturated fatty acid.

We start with the initial step, in which energy is used up rather than produced. The reaction

$$\text{ATP} \longrightarrow \text{AMP} + 2\text{P}_i$$

breaks two high-energy phosphate bonds. This is equivalent to hydrolyzing two molecules of ATP to ADP. In each cycle of the spiral we obtain one $FADH_2$, one NADH + H^+, and one acetyl CoA. Stearic acid (C_{18}) goes through *seven cycles* in the spiral before it reaches the final C_4 stage. In the last (eighth) cycle one $FADH_2$, one NADH + H^+, and two acetyl CoA molecules are produced. Now we can add up the energy. Table 21.2 shows that for a C_{18} compound we obtain a total of 146 ATP molecules.

It is instructive to compare the energy yield from fats with that from carbohydrates, since both are important constituents of the diet. In Section 21.3 we saw that glucose, $C_6H_{12}O_6$, produces 36 ATP molecules, that is, 6 for each carbon atom. For stearic acid, there are 146 ATP molecules

Table 21.2 ATP Yield from Complete Stearic Acid Metabolism

Step Number in Fig. 21.5	Chemical Steps	Happens	Number of ATP Molecules Produced
①	Activation (stearic acid → stearyl CoA)	Once	−2
②	Dehydrogenation (acyl CoA → transenoyl CoA), producing $FADH_2$	8 times	16
④	Dehydrogenation (hydroxyacyl CoA → keto acyl CoA), producing NADH + H^+	8 times	24
	C_2 fragment (acetyl CoA → common catabolic pathway), producing 12 ATP for each C_2 fragment	9 times	108
		Total	146

and 18 carbons, or 146/18 = 8.1 ATP molecules per carbon atom. Since additional ATP molecules are produced from the glycerol portion of fats (Sec. 21.4), fats have a higher caloric value than carbohydrates.

21.7 Ketone Bodies

In spite of the high caloric value of fats, the body preferentially uses glucose as an energy supply. When an animal is well fed (plenty of sugar intake), fatty acid oxidation is inhibited, and fatty acids are stored in the form of neutral fat in fat depots. Only when the glucose supply dwindles, as in fasting or starvation, or when glucose cannot be utilized, as in the case of diabetes, is the β-oxidation pathway of fatty acid metabolism mobilized.

In some pathological conditions, it is possible for glucose not to be available at all.

Unfortunately, low glucose supply also slows down the citric acid cycle. This happens because oxaloacetate is produced from the carboxylation of pyruvate. This oxaloacetate normally enters the citric acid cycle (Fig. 20.8) where it is essential for the continuous operation of the cycle. But if there is no glucose, there will be no glycolysis, no pyruvate formation, and therefore no oxaloacetate production.

Thus, even though the fatty acids are oxidized, not all the resulting C_2 fragments (acetyl CoA) can enter the citric acid cycle because there is not enough oxaloacetate. Therefore, acetyl CoA builds up in the body, with the following consequences.

The liver is able to condense two acetyl CoA molecules to produce acetoacetyl CoA:

$$\underset{\text{Acetyl CoA}}{2CH_3{-}\overset{\overset{\large O}{\|}}{C}{-}SCoA} \longrightarrow \underset{\text{Acetoacetyl CoA}}{CH_3{-}\overset{\overset{\large O}{\|}}{C}{-}CH_2{-}\overset{\overset{\large O}{\|}}{C}{-}SCoA} + CoASH$$

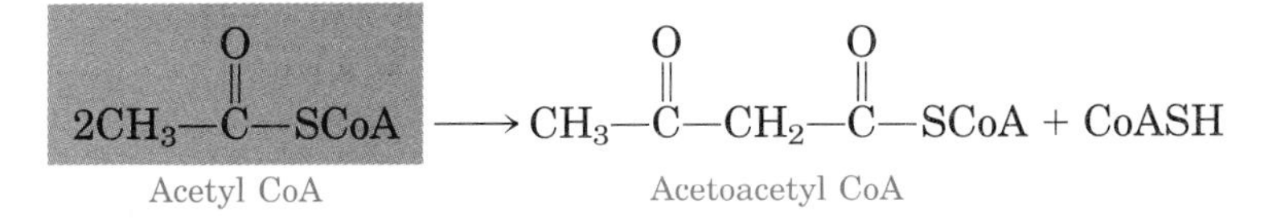

When the acetoacetyl CoA is hydrolyzed, it yields acetoacetate and its reduced form, β-hydroxybutyrate:

$$\underset{\text{Acetoacetyl CoA}}{CH_3{-}\overset{\overset{\large O}{\|}}{C}{-}CH_2{-}\overset{\overset{\large O}{\|}}{C}{-}SCoA} \xrightarrow{H_2O} \underset{\text{Acetoacetate}}{CH_3{-}\overset{\overset{\large O}{\|}}{C}{-}CH_2{-}\overset{\overset{\large O}{\|}}{C}{-}O^-} + CoASH + H^+$$

Acetoacetate → (NADH + H⁺ → NAD⁺) → β-Hydroxybutyrate; Acetoacetate → (H⁺ → CO₂) → Acetone

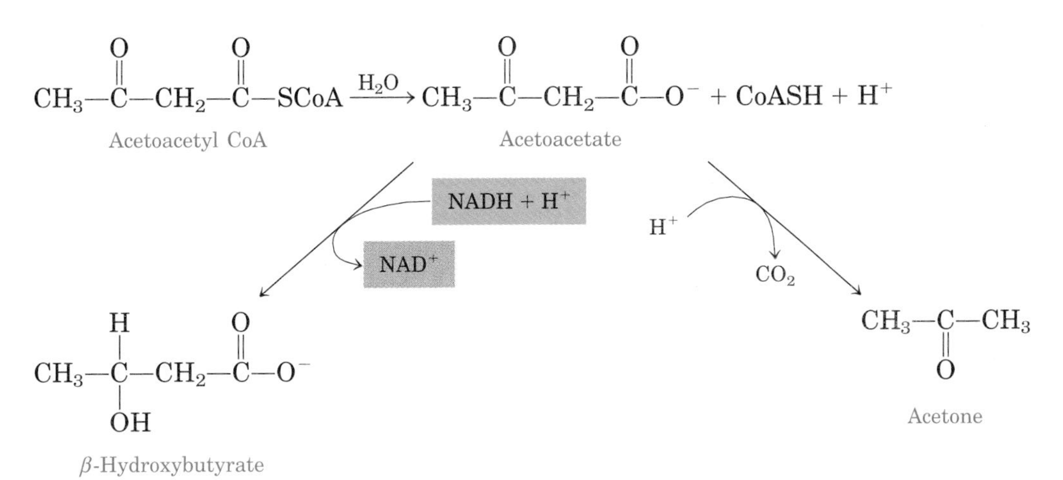

$$\underset{\beta\text{-Hydroxybutyrate}}{CH_3{-}\overset{\overset{\large H}{|}}{\underset{\underset{\large OH}{|}}{C}}{-}CH_2{-}\overset{\overset{\large O}{\|}}{C}{-}O^-} \qquad \underset{\text{Acetone}}{CH_3{-}\underset{\underset{\large O}{\|}}{C}{-}CH_3}$$

Box 21C

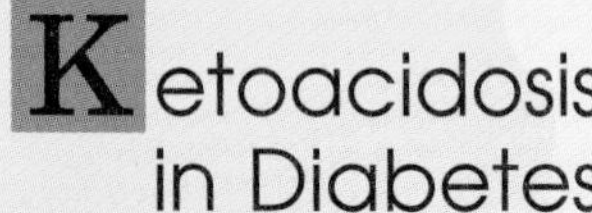

Ketoacidosis in Diabetes

In untreated diabetes the glucose concentration in the blood is high because the lack of insulin prevents utilization of glucose by the cells. Regular injections of insulin remedy this situation. However, in some stressful conditions ketoacidosis can still develop. A typical case was a diabetic patient admitted to the hospital in semicoma. He showed signs of dehydration, his skin was inelastic and wrinkled, his urine showed high concentrations of glucose and ketone bodies, and his blood contained excess glucose and had a pH of 7.0, a drop of 0.4 pH units from normal, which is an indication of severe acidosis. The urine also contained the bacteria *E. coli.* This indication of urinary tract infection explained why the normal doses of insulin were insufficient to prevent ketoacidosis.

The stress of infection can upset the normal control of diabetes by upsetting the balance between administered insulin and other hormones produced in the body. This happened during the infection mentioned above, and his body started to produce ketone bodies (Sec. 21.7) in large quantities. Both glucose and ketone bodies appear in the blood before they show up in the urine.

The acidic nature of ketone bodies (acetoacetic acid and β-hydroxybutyric acid) lowers the blood pH. A large drop in pH is prevented by the bicarbonate/carbonic acid buffer (Sec. 8.10), but even a drop of 0.3 to 0.5 pH units is sufficient to decrease the Na^+ concentration. The decrease of Na^+ in the interstitial tissues draws out K^+ ions from the cells. This in turn impairs brain function and leads to coma. During the secretion of ketone bodies and glucose in the urine, a lot of water is lost, and the body becomes dehydrated. This means the blood volume shrinks. Thus the blood pressure drops, and the pulse rate increases to compensate. Smaller quantities of nutrients reach the brain cells, and this too can cause coma.

The patient mentioned above was infused with physiological saline solution to remedy the dehydration. Extra doses of insulin restored his glucose level to normal, and antibiotics cured the urinary infection.

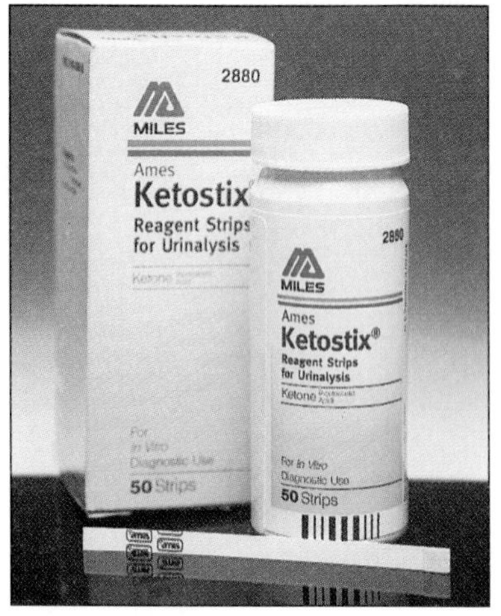

Test kit for the presence of ketone bodies in the urine. (Photograph by Charles D. Winters.)

These two compounds, together with smaller amounts of acetone, are collectively called **ketone bodies.** Under normal conditions, the liver sends these compounds into the blood stream to be carried to the tissues and utilized there via the common catabolic pathway. Normally the concentration of ketone bodies in the blood is low. But during starvation and in untreated diabetes mellitus, ketone bodies accumulate in the blood and can reach high concentrations. When this occurs, the excess is secreted in the urine. A check of urine for ketone bodies is used in the diagnosis of diabetes.

21.8 Catabolism of the Nitrogen of Amino Acids

The proteins of our foods are hydrolyzed to amino acids in digestion. These amino acids are primarily used to synthesize new proteins. However, unlike carbohydrates and fats, they cannot be stored, so excess amino acids are catabolized for energy production. What happens to the carbon skeleton of the amino acids is dealt with in the next section. Here we discuss the catabolic fate of the nitrogen

An overview of the whole process of protein catabolism is shown in Figure 21.6.

In the tissues, amino ($—NH_2$) groups freely move from one amino acid to another. The enzymes that catalyze these reactions are the **transaminases.**

In essence there are three stages in nitrogen catabolism in the liver. The final product of the three stages is urea, which is excreted in the urine of mammals. The first stage is a **transamination.** Amino acids transfer their amino groups to α-ketoglutaric acid:

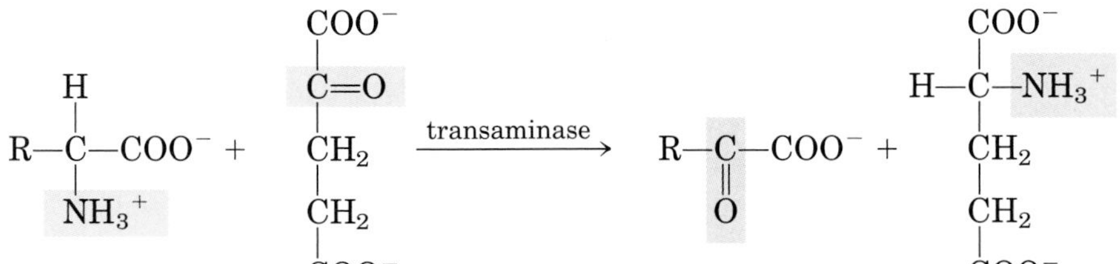

$$\underset{\text{Amino acid (zwitterion form)}}{R—\overset{H}{\underset{NH_3^+}{C}}—COO^-} + \underset{\alpha\text{-Ketoglutarate}}{\begin{matrix} COO^- \\ | \\ C{=}O \\ | \\ CH_2 \\ | \\ CH_2 \\ | \\ COO^- \end{matrix}} \xrightarrow{\text{transaminase}} \underset{\alpha\text{-Keto acid}}{R—\overset{}{\underset{\underset{O}{\|}}{C}}—COO^-} + \underset{\text{Glutamate}}{\begin{matrix} COO^- \\ | \\ H—C—NH_3^+ \\ | \\ CH_2 \\ | \\ CH_2 \\ | \\ COO^- \end{matrix}}$$

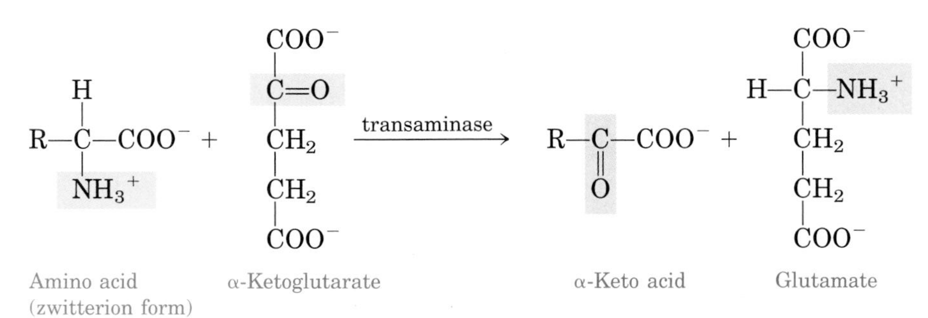

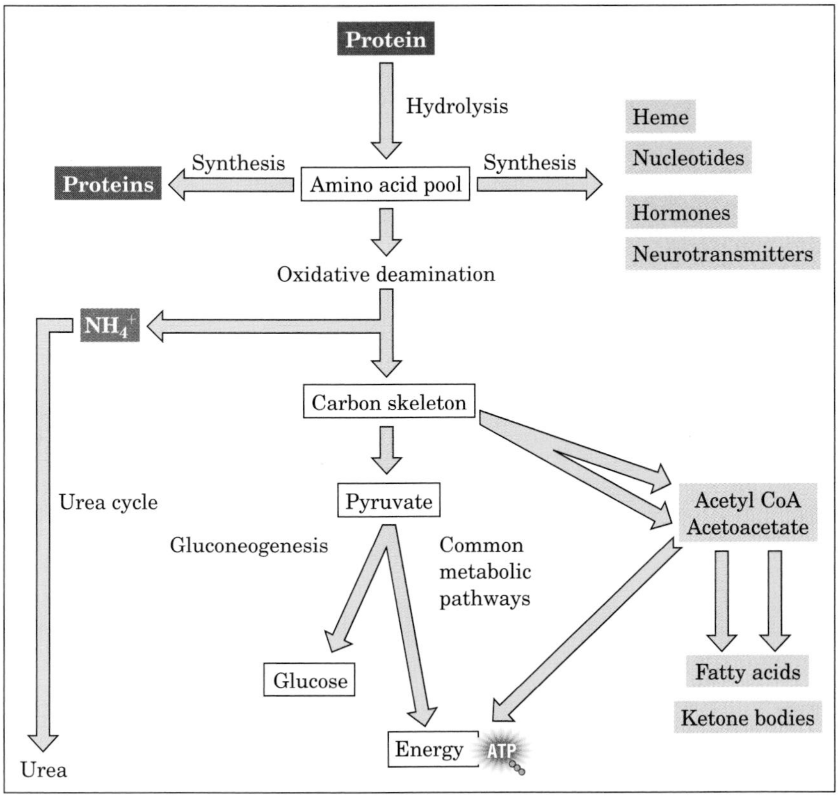

Figure **21.6** • An overview of pathways in protein catabolism.

The catabolism of the α-keto acid is discussed in the next section.

The carbon skeleton of the amino acid remains behind as an α-keto acid. The second stage of the nitrogen catabolism is the **oxidative deamination** of glutamate, which occurs in the mitochondrion:

$$\underset{\text{Glutamate}}{{}^-OOC-CH_2-CH_2-CH(NH_3^+)-COO^-} + NAD^+ + H_2O \rightleftharpoons NH_4^+ + \underset{\alpha\text{-Ketoglutarate}}{{}^-OOC-CH_2-CH_2-C(=O)-COO^-} + NADH + H^+$$

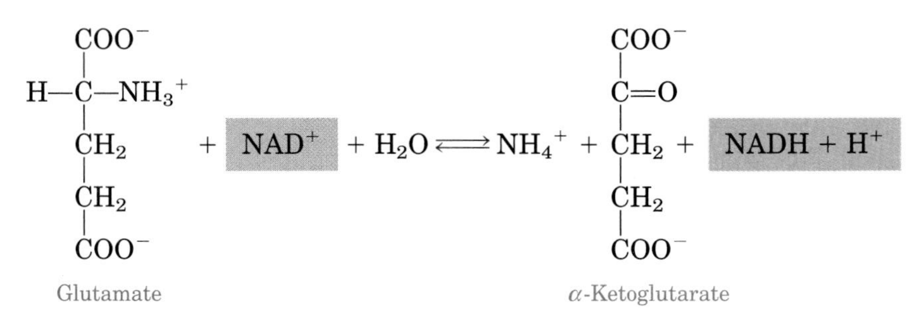

The body must get rid of NH_4^+ because both it and NH_3 are toxic.

The oxidative deamination yields NH_4^+ and regenerates α-ketoglutarate, which can again participate in the first stage (transamination). The NADH + H^+ produced in the second stage enters the oxidative phosphorylation pathway and eventually produces three ATP molecules.

In the third stage the NH_4^+ is converted to urea through the **urea cycle** (Fig. 21.7). First the NH_4^+ is condensed with CO_2 in the mitochondrion to form an unstable compound, carbamoyl phosphate ①. This

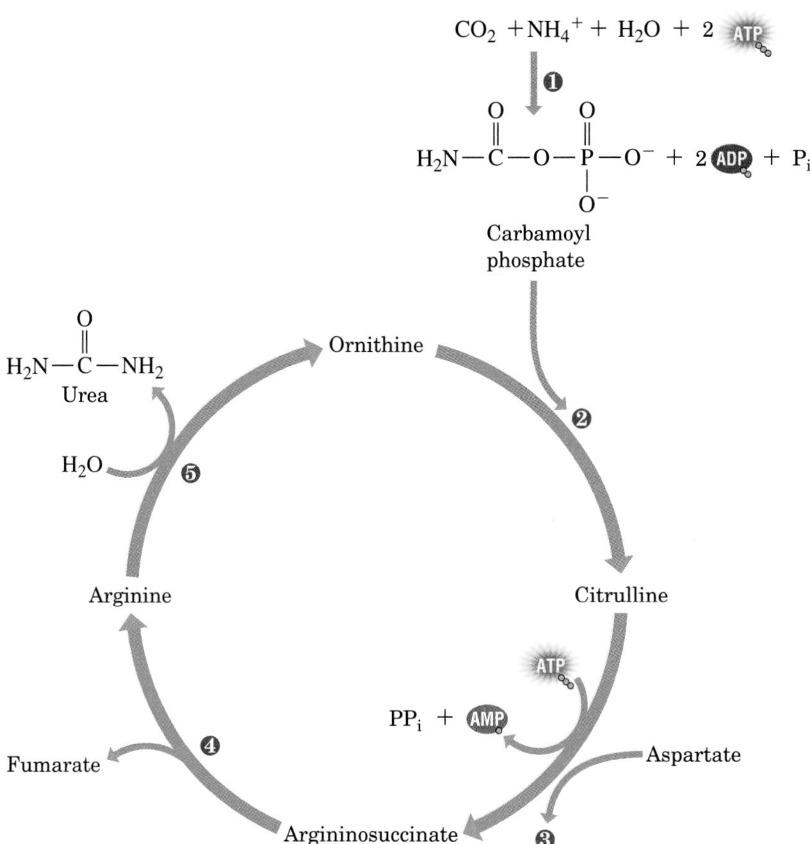

Figure **21.7** • The urea cycle.

condensation occurs at the expense of two ATP molecules. In the next step ②, carbamoyl phosphate is condensed with ornithine, a basic amino acid similar in structure to lysine:

Ornithine does not occur in proteins.

$$\underset{\text{Ornithine}}{\overset{+}{H_3N}-CH_2-CH_2-CH_2-CH(NH_3^+)-COO^-} + \underset{\text{Carbamoyl phosphate}}{H_2N-C(=O)-O-P(=O)(O^-)-O^-} \longrightarrow \underset{\text{Citrulline}}{H_2N-C(=O)-NH-CH_2-CH_2-CH_2-CH(NH_3^+)-COO^-} + PO_4^{3-}$$

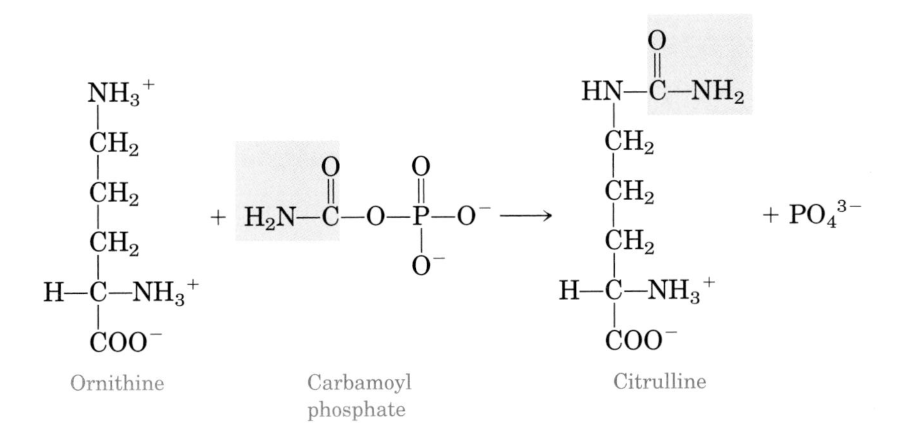

The result is citrulline, which diffuses out of the mitochondrion into the cytosol.

A second condensation reaction in the cytosol takes place between citrulline and aspartate, forming argininosuccinate ③:

$$ATP + \underset{\text{Citrulline}}{H_2N-C(=O)-NH-(CH_2)_3-CH(NH_3^+)-COO^-} + \underset{\text{Aspartate}}{\overset{+}{H_3N}-CH(COO^-)-CH_2-COO^-} \longrightarrow \underset{\text{Argininosuccinate}}{\overset{+}{N}H_2=C(-NH-CH(COO^-)-CH_2-COO^-)-NH-(CH_2)_3-CH(NH_3^+)-COO^-} + AMP + PP_i$$

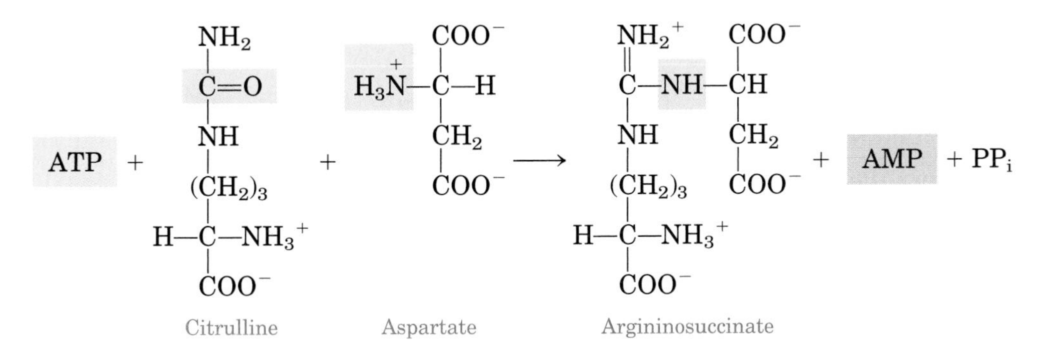

The energy for this reaction comes from the hydrolysis of ATP to AMP and pyrophosphate (PP_i).

The PP_i stands for pyrophosphate:

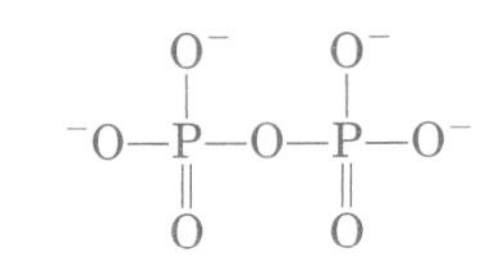

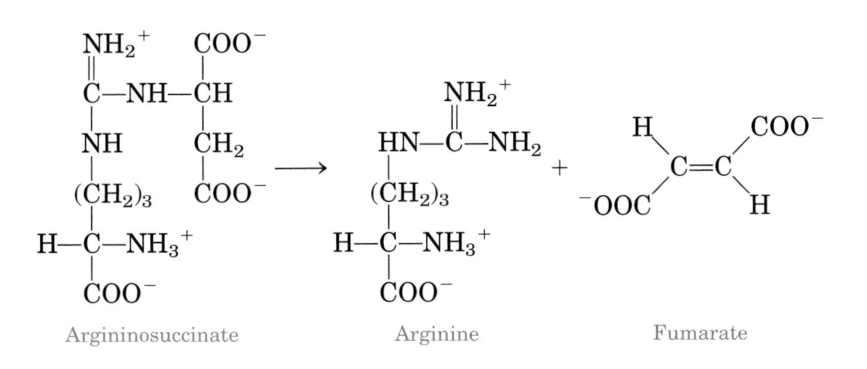

$$^-O-P(=O)(O^-)-O-P(=O)(O^-)-O^-$$

In the fourth step, the argininosuccinate is split into arginine and fumarate ④:

$$\underset{\text{Argininosuccinate}}{\overset{+}{N}H_2=C(-NH-CH(COO^-)-CH_2-COO^-)-NH-(CH_2)_3-CH(NH_3^+)-COO^-} \longrightarrow \underset{\text{Arginine}}{\overset{+}{N}H_2=C(NH_2)-NH-(CH_2)_3-CH(NH_3^+)-COO^-} + \underset{\text{Fumarate}}{{}^-OOC(H)C=C(H)COO^-}$$

In the final step ⑤, arginine is hydrolyzed to urea and ornithine:

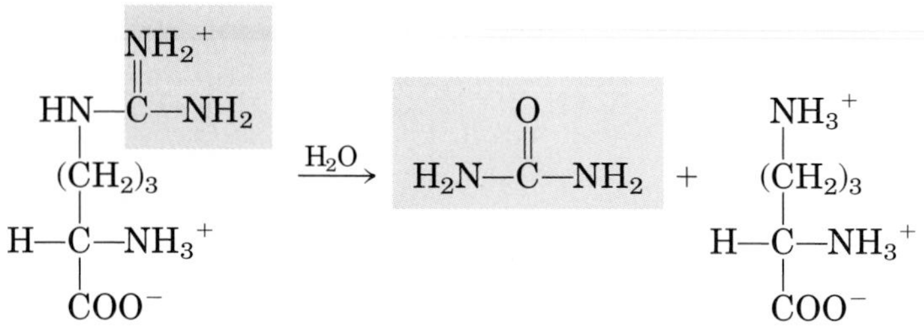

$$\underset{\text{Arginine}}{\mathrm{HN{-}C({=}NH_2^+){-}NH_2;\ (CH_2)_3;\ H{-}C{-}NH_3^+;\ COO^-}} \xrightarrow{H_2O} \underset{\text{Urea}}{\mathrm{H_2N{-}C({=}O){-}NH_2}} + \underset{\text{Ornithine}}{\mathrm{NH_3^+{-}(CH_2)_3{-}CH(NH_3^+){-}COO^-}}$$

Hans Krebs, who elucidated the citric acid cycle, was also instrumental in establishing the urea cycle.

The urea is excreted in the urine. The ornithine re-enters the mitochondrion and thus completes the cycle. It is now ready to pick up another carbamoyl phosphate ②.

An important aspect of carbamoyl phosphate as an intermediate is that it can be used for synthesis of nucleotide bases (Chap. 23). Furthermore, the urea cycle is linked to the citric acid cycle because both involve fumarate.

21.9 Catabolism of the Carbon Skeleton of Amino Acids

Once the alpha amino group is removed from an amino acid by oxidative deamination (Sec. 21.8), the remaining carbon skeleton is used as an energy source (Fig. 21.8). We are not going to study the pathways involved except to point out the eventual fate of the skeleton. Not all the carbon skeletons of amino acids are used as fuel. Some of them may be degraded up to a certain point, and the resulting intermediate then used as a building block to construct another needed molecule. For example, if the carbon skeleton of an amino acid is catabolized to pyruvate, there are two possible choices for the body: (1) to use the pyruvate as an energy supply via the common pathway or (2) to use it as a building block to synthesize glucose (Sec. 22.2). **Those amino acids that yield a carbon skeleton that is degraded to pyruvate or another intermediate capable of conversion to glucose** (such as oxaloacetate) are called **glucogenic.** One example is alanine. When alanine reacts with α-ketoglutaric acid, the transamination produces pyruvate directly, as shown in Figure 21.8:

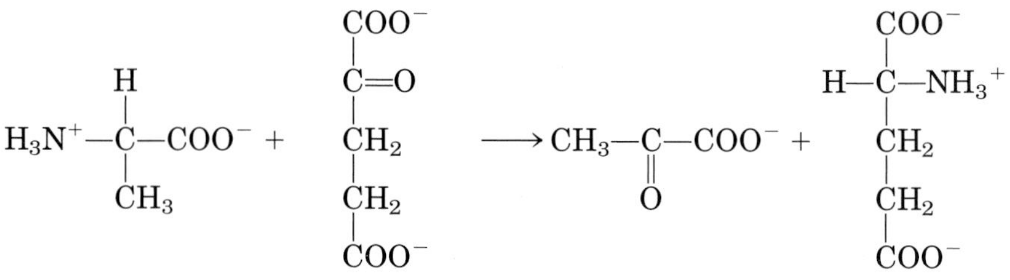

$$\underset{\text{Alanine}}{\mathrm{H_3N^+{-}CH(CH_3){-}COO^-}} + \underset{\alpha\text{-Ketoglutarate}}{\mathrm{^-OOC{-}C({=}O){-}CH_2{-}CH_2{-}COO^-}} \longrightarrow \underset{\text{Pyruvate}}{\mathrm{CH_3{-}C({=}O){-}COO^-}} + \underset{\text{Glutamate}}{\mathrm{^-OOC{-}CH(NH_3^+){-}CH_2{-}CH_2{-}COO^-}}$$

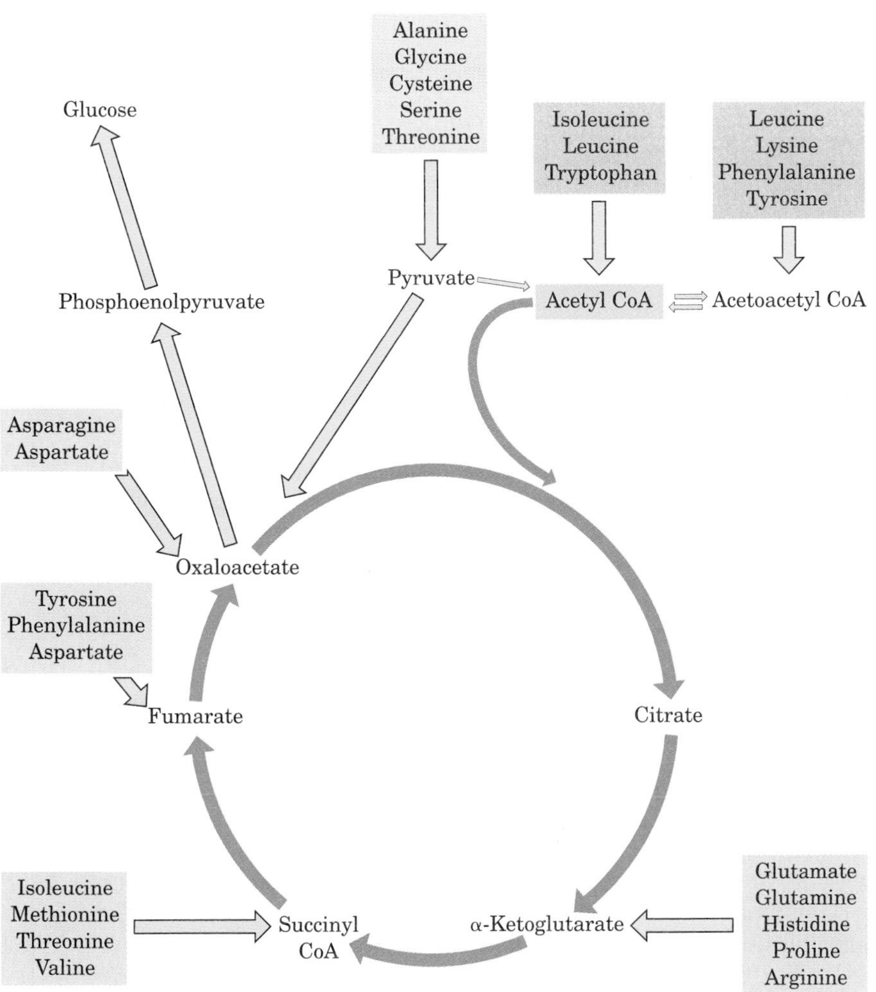

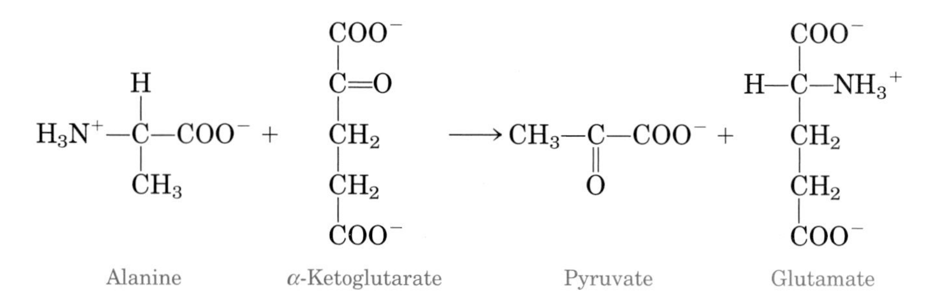

Figure **21.8**
Catabolism of the carbon skeletons of amino acids. The glucogenic amino acids are in the blue boxes; the ketogenic ones in the brown boxes.

Box 21D

MSG

Glutamic acid is one of the 20 amino acids found in proteins (Sec. 18.2). Its monosodium salt, **monosodium glutamate (MSG),** enhances the flavor of many foods without itself contributing any significant taste. It makes such foods as meats, vegetables, soups, and stews taste better. Because of this, MSG is added to many canned and frozen foods. In pure form it is sold in supermarkets under such brand names as Accent. The spice racks of many kitchens contain a bottle labeled MSG, next to the nutmeg, thyme, and basil. In particular, Chinese restaurants in this country use copious quantities of MSG.

Although most people can consume MSG with no problems, there have been claims that some are allergic to it and get symptoms that include dizziness, numbness spreading from the jaw to the back of the neck, hot flashes, and sweating. Collectively, these symptoms are known as Chinese restaurant syndrome. However, recent studies cast doubt on the theory that these symptoms are caused by MSG.

On the other hand, many amino acids are degraded to acetyl CoA and acetoacetic acid. These compounds cannot form glucose but are capable of yielding ketone bodies; they are called **ketogenic.** Leucine is an example of a ketogenic amino acid.

Both glucogenic and ketogenic amino acids, when used as an energy supply, enter the citric acid cycle at some point (Fig. 21.8) and are eventually oxidized to CO_2 and H_2O. The oxaloacetate (a C_4 compound) produced in this manner enters the citric acid cycle, adding to the oxaloacetate produced in the cycle itself.

Box 21E

Ubiquitin and Protein Targeting

In previous sections we dealt with the catabolism of dietary proteins, especially with their use as an energy source. The body's own cellular proteins are also broken down and degraded. This occurs sometimes as a response to a stress such as starvation, but more often it is done to maintain a steady-state level of regulatory proteins or to eliminate damaged proteins. But how does the cell know which proteins to degrade and which to leave alone? One pathway that targets proteins for destruction depends on an ancient protein, called ubiquitin. (As the name implies, it is present in all cells belonging to higher organisms.) Its sequence of 76 amino acids is essentially identical in yeast and in humans. The C-terminal amino acid of ubiquitin is glycine, the carboxyl group of which forms an amide linkage with the side chain of a lysine residue in the protein that is targeted for destruction.

This process, known as *ubiquitinylation,* requires three enzymes and costs energy by using up one ATP molecule per ubiquitin molecule. In many cases, a polymer of ubiquitin molecules (polyubiquitin) is attached to the target protein molecule. Once a protein molecule becomes "flagged" by several ubiquitin molecules, it is delivered to an organelle containing the machinery for proteolysis. Part of the proteolytic system is *proteasome,* a large water-soluble complex assembly of proteases and regulatory proteins. In this assembly the polyubiquitin is removed, and the protein is fully or partially degraded.

Cleavage of a protein is called *proteolysis.*

One role of the ubiquitin targeting is that of "garbage collector." Damaged proteins that the chaperones (Sec. 18.9) cannot salvage by proper refolding must be removed. For example, oxidation of hemoglobin causes unfolding and rapid degradation by ubiquitin targeting follows. A second purpose is to control the concentration of some regulatory proteins. For example, the protein products of some oncogenes (Box 23G) are also removed by ubiquitin targeting.

Both of these cases result in complete degradation of the targeted proteins. A third case involves only a partial degradation of the target protein. For example, a virus invades a cell. The infected cell recognizes a foreign protein from the virus and targets it for destruction through the ubiquitin system. The targeted protein is partially degraded. A peptide segment of the viral protein is incorporated into a system that rises to the surface of the infected cell and displays the viral peptide as an antigen. The T cells (Sec. 25.10) can now recognize the infected cell as a foreign cell; they attack and kill it.

An antigen is a foreign body recognized by the immune system (see Sec. 25.10).

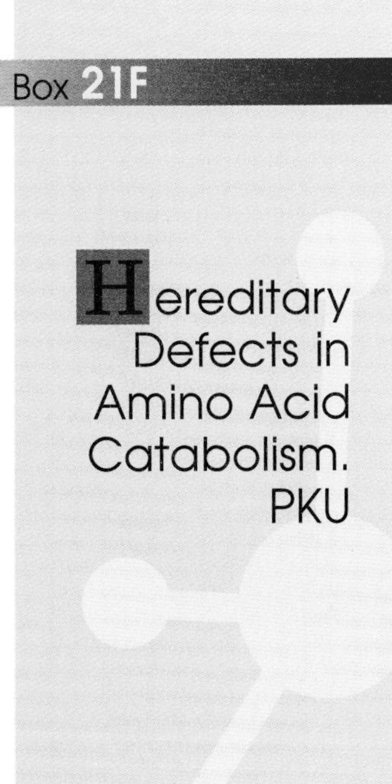

Box 21F
Hereditary Defects in Amino Acid Catabolism. PKU

Many hereditary diseases involve missing or malfunctioning enzymes that catalyze the breakdown of amino acids. The oldest-known of such diseases is cystinuria, which was described as early as 1810. In this disease, cystine shows up as flat hexagonal crystals in the urine. Stones form because of the low solubility of cystine in water. This leads to blockage in the kidneys or the ureters and requires surgery. One way to reduce the amount of cystine secreted is to remove as much methionine as possible from the diet. Beyond that, an increased fluid intake increases the volume of the urine, reducing the solubility problem. It has been found that penicillamine can also prevent cystinuria.

An even more important genetic defect is the absence of the enzyme phenylalanine hydroxylase, causing a disease called phenylketonuria (PKU). In normal catabolism, this enzyme helps degrade phenylalanine by converting it to tyrosine. If the enzyme is defective, phenylalanine is converted to phenylpyruvate (see the conversion of alanine to pyruvate in Section 21.9). Phenylpyruvate (a ketoacid) accumulates in the body and inhibits the conversion of pyruvate to acetyl CoA, thus depriving the cells of energy via the common pathway. This is most important in the brain, which gets its energy from the utilization of glucose. The result is mental retardation. This genetic defect can be detected early because phenylpyruvic acid appears in the urine. A federal regulation requires that all infants be tested for this disease. Once PKU is detected, mental retardation can be prevented by restricting the intake of phenylalanine in the diet. Patients with PKU should avoid the artificial sweetener aspartame (Box 18A) because it yields phenylalanine when hydrolyzed in the stomach.

21.10 Catabolism of Heme

Red blood cells are continuously being manufactured in the bone marrow. Their life span is relatively short—about four months. Aged red blood cells are destroyed in the phagocytic cells. When a red blood cell is destroyed, its hemoglobin is metabolized: The globin (Sec. 18.9) is hydrolyzed to amino acids, the heme is converted to *bilirubin,* and the iron is preserved in ferritin, an iron-carrying protein, and is reused. The bilirubin enters the liver via the blood and is then transferred to the gallbladder, where it is stored in the bile and finally excreted via the small intestine. The color of feces is provided by *urobilin,* an oxidation product of bilirubin.

Phagocytes are specialized blood cells that destroy foreign bodies.

Box 21G

Jaundice

In normal individuals there is a balance between the destruction of heme in the spleen and the removal of bilirubin from the blood by the liver. When this balance is upset, the bilirubin concentration in the blood increases. If this condition continues, the skin and whites of the eyes both become yellow, in a condition known as **jaundice.** Therefore jaundice can signal a malfunctioning of the liver, the spleen, or the gallbladder, where bilirubin is stored before excretion. Hemolytic jaundice is the acceleration of the destruction of red blood cells in the spleen to such a level that the liver cannot cope with the bilirubin production. When gallstones obstruct the bile ducts and the bilirubin cannot be excreted in the feces, the backup of bilirubin causes jaundice. In infectious hepatitis the liver is incapacitated, and bilirubin is not removed from the blood, causing jaundice.

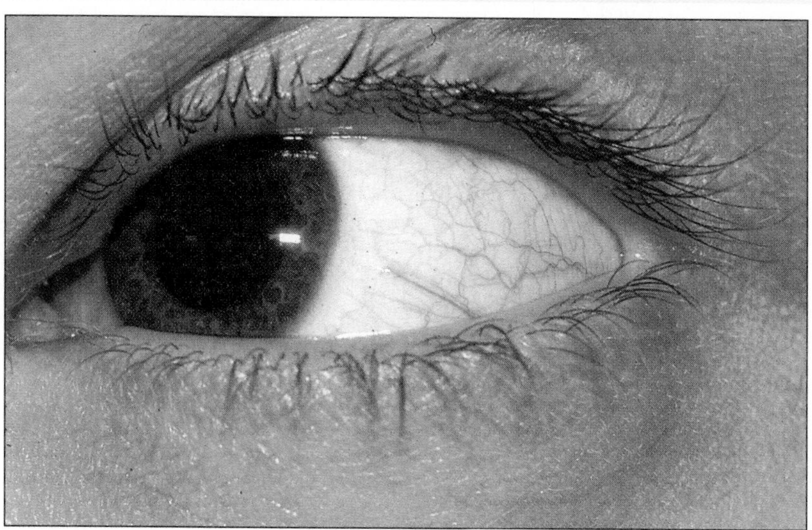

Yellowing of the sclera (white outer coat of eyeball) in jaundice. The yellowing is caused by an excess of bilirubin, a bile pigment, in the blood. (Science Photo Library/Photo Researchers, Inc.)

SUMMARY

The foods we eat are broken down into small molecules in the stomach and intestines before being absorbed. These small molecules serve two purposes: They can be the building blocks of new materials the body needs to synthesize (anabolism), or they can be used for energy supply (catabolism). Each group of compounds—carbohydrates, fats, and proteins—has its own catabolic pathway. All the different catabolic pathways converge to the common pathway.

The specific pathway of carbohydrate catabolism is **glycolysis.** In this process, hexose monosaccharides are activated by ATP and eventually converted to two C_3 fragments, dihydroxyacetone phosphate and glyceraldehyde phosphate. The glyceraldehyde phosphate is further oxidized and eventually ends up as pyruvate. All these reactions occur in the cytosol. Pyruvate is converted to acetyl CoA, which is further catabolized in the common pathway. When completely metabolized, a hexose yields the energy of 36 ATP molecules.

Fats are broken down to glycerol and fatty acids. Glycerol is catabolized in the glycolysis pathway and yields 20 ATP molecules.

Fatty acids are broken down into C_2 fragments in the **β-oxidation spiral.** At each turn of the spiral one acetyl CoA is released together with one $FADH_2$ and one $NADH + H^+$. These products go through the common pathway. Stearic acid, a C_{18} compound, yields 146 molecules of ATP. In starvation and under certain pathological conditions, not all the acetyl CoA produced in the β-oxidation of fatty acids enters the common pathway. Some of it forms acetoacetate, β-hydroxybutyrate, and acetone, commonly called **ketone bodies.** Excess ketone bodies in the blood are secreted in the urine.

Proteins are broken down to amino acids. The nitrogen of the amino acids is first transferred to glutamate. This in turn is **oxidatively deaminated** to yield ammonia. Mammals get rid of the toxic ammonia by converting it to urea in the **urea cycle.** Urea is secreted in the urine. The carbon skeletons of amino acids are catabolized via the citric acid cycle. Some of these enter as pyruvate or other intermediates of the citric acid cycle; these are **glucogenic amino acids.** Others are incorporated into acetyl CoA or ketone bodies and are called **ketogenic amino acids.** Heme is catabolized to bilirubin, which is excreted in the feces.

KEY TERMS

Amino acid pool (Sec. 21.1)
Anaerobic pathway (Sec. 21.2)
Cytosol (Sec. 21.2)
Fat depot (Sec. 21.1)
Glucogenic amino acid (Sec. 21.9)
Glycogenolysis (Sec. 21.3)
Glycolysis (Sec. 21.2)
Ketoacidosis (Box 21C)
Ketogenic amino acid (Sec. 21.9)
Ketone bodies (Sec. 21.7)
β-Oxidation (Sec. 21.5)
Oxidative deamination (Sec. 21.8)
Transamination (Sec. 21.8)
Ubiquitinylation (Box 21E)
Urea cycle (Sec. 21.8)

PROBLEMS

Specific Pathways

21.1 What are the units into which large carbohydrate molecules are broken down in the digestive process?

21.2 What are the products of the lipase-catalyzed hydrolysis of fats?

21.3 What is the main use of amino acids in the body?

Glycolysis

21.4 Although catabolism of a glucose molecule eventually produces a lot of energy, the first step uses up energy. Explain why this step is necessary.

21.5 In one step of the glycolysis pathway, a C_6 chain is broken into two C_3 fragments, only one of which can be further degraded in the glycolysis pathway. What happens to the other C_3 fragment?

21.6 Why is glycolysis called an anaerobic pathway?

21.7 (a) Which steps in glycolysis of glucose need ATP? (b) Which steps in glycolysis yield ATP directly?

21.8 What two C_3 fragments are obtained from splitting fructose 1,6-bisphosphate?

21.9 Find all the oxidation steps in the glycolysis pathway shown in Figure 21.3.

21.10 The end product of glycolysis, pyruvate, cannot enter as such into the citric acid cycle. What is the name of the process that converts this C_3 compound to a C_2 compound?

21.11 Where do the anaerobic reactions of glycolysis occur?

***21.12** Which of these steps yields energy and which consumes energy?
(a) pyruvate → lactate
(b) pyruvate → acetyl CoA + CO_2

21.13 How many moles of lactate are produced from three moles of glucose?

***21.14** How many moles of net $NADH + H^+$ are produced from one mole of glucose going to (a) acetyl CoA? (b) lactate?

Energy Yield from Glucose

21.15 Of the 36 molecules of ATP produced by the complete metabolism of glucose, how many are produced in glycolysis alone, that is, before the common pathway?

21.16 How many *net* ATP molecules are produced in the skeletal muscles for each glucose molecule
(a) in glycolysis alone (up to pyruvate)?
(b) in converting pyruvate to acetyl CoA?
(c) in the total oxidation of glucose to CO_2 and H_2O?

***21.17** (a) If fructose is metabolized in the liver, how many moles of net ATP are produced from each mole during glycolysis? (b) How many moles are produced if the same thing occurs in a muscle cell?

21.18 What is the difference between glycolysis and glycogenolysis?

Glycerol Catabolism

21.19 Explain why dihydroxyacetone phosphate is called an oxidation product of glycerol 1-phosphate.

***21.20** Which yields more energy upon hydrolysis, ATP or glycerol 1-phosphate? Why?

β-Oxidation of Fatty Acids

21.21 Why is the breakdown of fatty acids called β-oxidation?

21.22 (a) Which part of the cells contains the enzymes needed for the β-oxidation of fatty acids? (b) How does the activated fatty acid get there?

21.23 Assume that lauric acid is metabolized through β-oxidation. What are the products of the reaction after three turns of the spiral?

21.24 How many turns of the spiral are there in the β-oxidation of (a) lauric acid? (b) palmitic acid?

Energy Yield from Fatty Acids

21.25 Calculate the number of ATP molecules obtained in the β-oxidation of myristic acid, $CH_3(CH_2)_{12}COOH$.

21.26 Two of the enzymes in the metabolism of fatty acids have *thio* in their names: thiolase and thiokinase. What processes do these enzymes catalyze?

21.27 Assuming that both fats and carbohydrates are available, which does the body preferentially use as an energy source?

21.28 If equal weights of fats and carbohydrates are eaten, which will give more calories? Explain.

Ketone Bodies

21.29 Why do starving people have ketone bodies in their urine?

21.30 Do ketone bodies have nutritional value?

21.31 What happens to the oxaloacetate produced from carboxylation of pyruvate?

Catabolism of Amino Acids

21.32 What kind of reaction is this and what is its function in the body?

$$(H_3C)_2CH-CH(NH_3^+)-COO^- + {}^-OOC-C(=O)-CH_2-CH_2-COO^- \longrightarrow$$

$$(H_3C)_2CH-C(=O)-COO^- + {}^-OOC-CH(NH_3^+)-CH_2-CH_2-COO^-$$

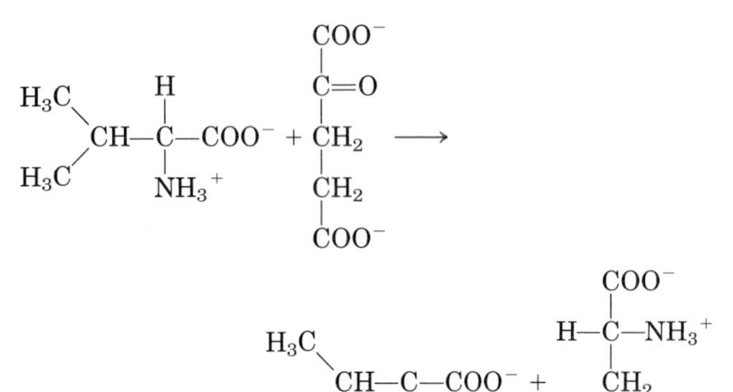

21.33 Write an equation for the oxidative deamination of alanine.

21.34 Ammonia, NH_3, and ammonium ion, NH_4^+, are both soluble in water and could easily be excreted in the urine. Why does the body convert them to urea rather than excreting them directly?

21.35 Write a balanced equation for the transamination reaction of tyrosine by α-ketoglutarate.

***21.36** (a) How many moles of ATP are used up in the conversion of one mole of ammonia to urea? (b) How many moles of high-energy bonds are hydrolyzed in the same conversion?

21.37 What compound is common to both the urea and citric acid cycles?

21.38 (a) What is the toxic product of the oxidative deamination of glutamate? (b) How does the body get rid of it?

***21.39** If leucine is the only carbon source in the diet of an experimental animal, what would you expect to find in the urine of such an animal?

Catabolism of Heme

***21.40** A high bilirubin content in the blood may indicate liver disease. Explain why.

21.41 When hemoglobin is fully metabolized, what happens to the iron in it?

21.42 Which component of the metabolized hemoglobin molecule ends up in the amino acid pool?

Boxes

21.43 (Box 21A) How does coronary thrombosis precipitate cardiac arrest?

21.44 (Box 21B) (a) What are the symptoms of Cori's disease? (b) What causes it? (c) Answer the same questions for McArdle's disease.

21.45 (Box 21C) What system counteracts the acidic effect of ketone bodies in the blood?

21.46 (Box 21D) What causes the accumulation of γ-aminobutyric acid in the Chinese restaurant syndrome?

21.47 (Box 21D) Draw the structure of monosodium glutamate.

21.48 (Box 21E) What is meant by the expression: "the amino acid sequence of ubiquitin is highly conserved"?

21.49 (Box 21E) How does ubiquitin attach itself to a target protein?

21.50 (Box 21F) How can you prevent mental retardation in infants once PKU has been diagnosed?

***21.51** (Box 21F) Draw structural formulas for each reaction component, and complete the equation.

$$\text{phenylalanine} \longrightarrow \text{phenylpyruvate} + ?$$

21.52 (Box 21G) What compound causes the yellow coloration of jaundice?

Additional Problems

21.53 How is sucrose metabolized through glycolysis?

21.54 Ornithine is a basic amino acid similar to lysine. (a) What is the structural difference between the two? (b) Is ornithine a constituent of proteins?

***21.55** (a) At which step of the glycolysis pathway does NAD^+ participate (Figs. 21.3 and 21.4)? (b) At which step does $NADH + H^+$ participate? (c) As a result of the overall pathway, is there a net increase of NAD^+, of $NADH + H^+$, or of neither?

***21.56** What is the net energy yield in moles of ATP produced when yeast converts one mole of glucose to ethanol?

***21.57** Can the intake of alanine, glycine, and serine relieve hypoglycemia caused by starvation? Explain.

21.58 If you received a laboratory report showing the presence of a high concentration of ketone bodies in the urine of a patient, what disease would you suspect?

21.59 Write the products of the transamination reaction between alanine and oxaloacetate:

$$CH_3—\underset{\displaystyle NH_3^+}{\underset{|}{CH}}—COO^- + \underset{\displaystyle \underset{\displaystyle COO^-}{\underset{|}{CH_2}}}{\underset{|}{\overset{\displaystyle COO^-}{\overset{|}{C}}=O}} \longrightarrow$$

***21.60** Phosphoenolpyruvate (PEP) has a high-energy phosphate bond that has more energy than the anhydride bonds in ATP. What step in glycolysis suggests that this is so?

***21.61** Suppose that a fatty acid labeled with radioactive carbon-14 is fed to an experimental animal. Where would you look for the radioactivity?

***21.62** What functional groups are in carbamoyl phosphate (p. 681)?

21.63 Are ketone bodies produced in normal metabolism?

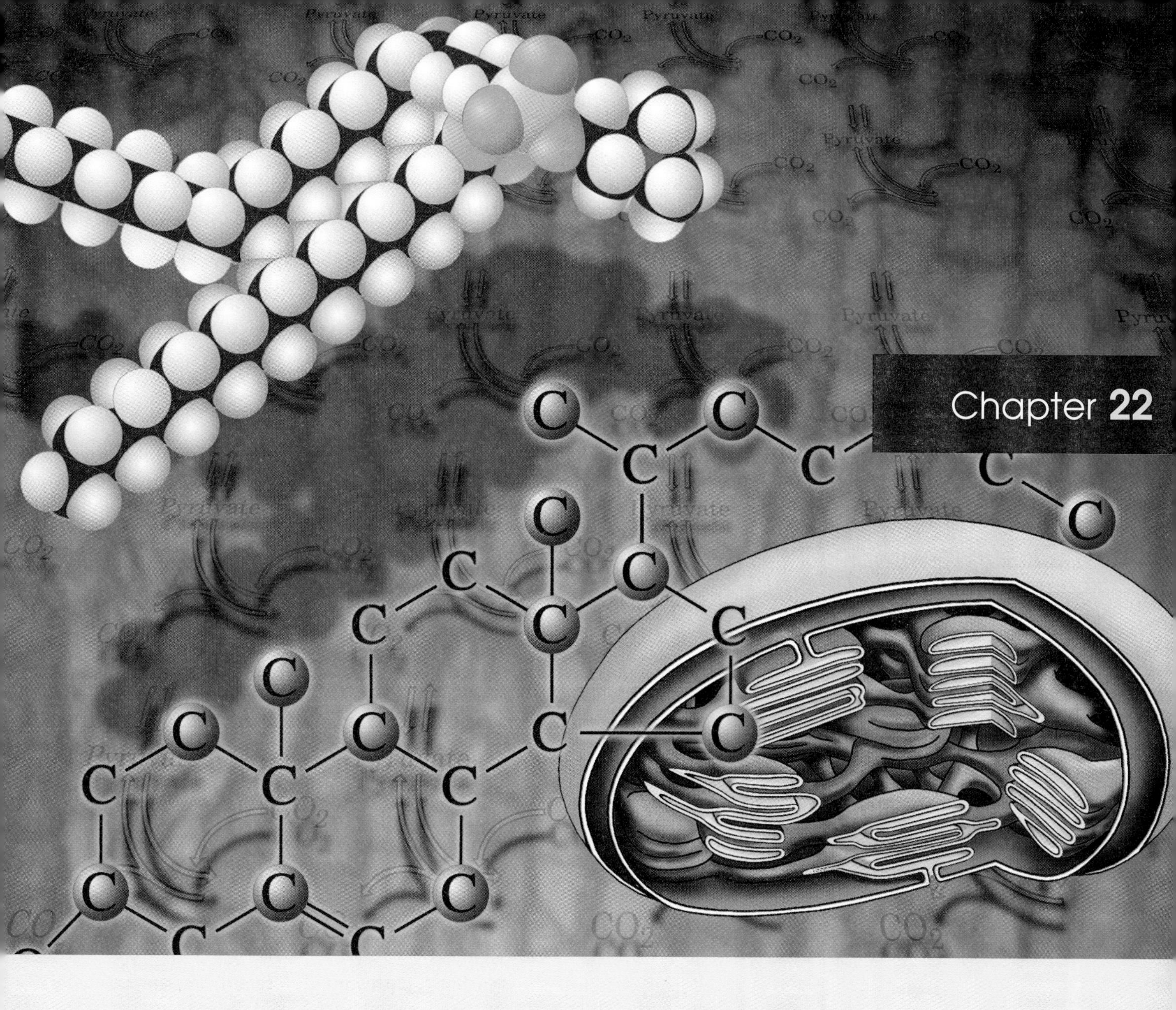

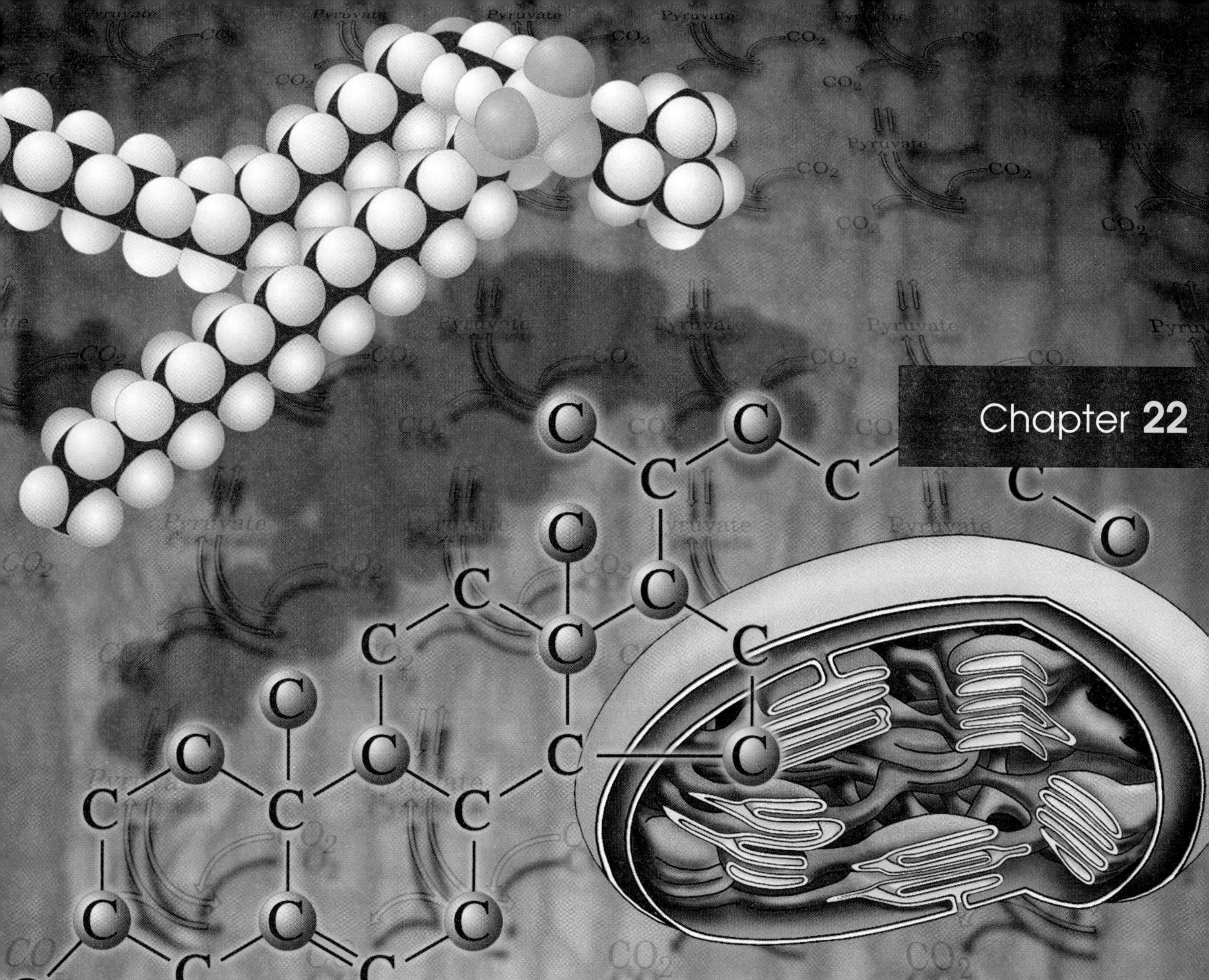

Chapter 22

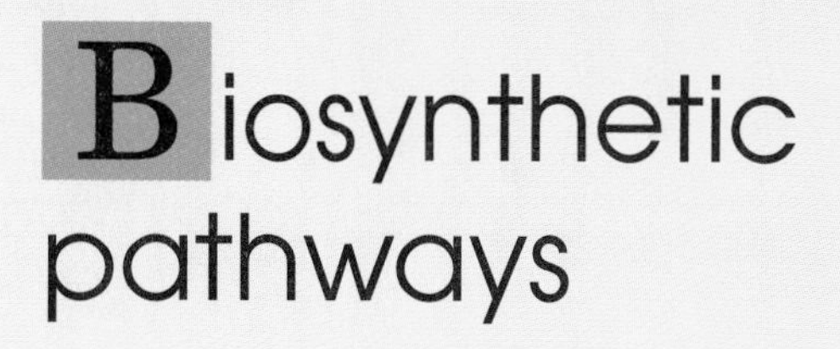

Biosynthetic pathways

22.1 Introduction

Anabolic pathways are also called **biosynthetic pathways.**

In the human body, and in most other living tissues, the pathways by which a compound is synthesized (anabolism) are usually different from the pathways by which it is degraded (catabolism). There are several reasons why it is biologically advantageous for anabolic and catabolic pathways to be different. We will give two of them:

1. Flexibility If the normal biosynthetic pathway is blocked, the body can often use the reverse of the degradation pathway instead (remember that most steps in degradation are reversible), thus providing another way to make the necessary compounds.

2. Overcoming the effect of Le Chatelier's principle This can be illustrated by the cleavage of a glucose unit from a glycogen molecule, an equilibrium process:

$$\underset{\text{Glycogen}}{(\text{Glucose})_n} + P_i \xrightleftharpoons{\text{phosphorylase}} \underset{\substack{\text{Glycogen} \\ \text{(one unit smaller)}}}{(\text{glucose})_{n-1}} + \text{glucose 1-phosphate} \quad (22.1)$$

Phosphorylase catalyzes not only glycogen degradation (the forward reaction) but also glycogen synthesis (the reverse reaction). However, in the body there is a large excess of inorganic phosphate, P_i. This would drive the reaction, on the basis of Le Chatelier's principle, to the right, which represents glycogen degradation. In order to provide a method for the *synthesis* of glycogen even in the presence of excess inorganic phosphate, a different pathway is needed in which P_i is not a reactant. Thus the body uses the following synthetic pathway:

The structure of UDP glucose is shown in Section 22.2.

$$\underset{\text{Glycogen}}{(\text{Glucose})_{n-1}} + \text{UDP glucose} \longrightarrow \text{UDP} + \underset{\substack{\text{Glycogen} \\ \text{(one unit larger)}}}{(\text{glucose})_n} \quad (22.2)$$

Not only are the synthetic pathways different from the catabolic pathways, but the energy requirements are also different, and so is the location. Most catabolic reactions occur in the mitochondria, whereas anabolic reactions generally take place in the cytosol. We shall not go into the energy balances of the biosynthetic processes as we did for catabolism. However, it must be kept in mind that, while energy (in the form of ATP) is *obtained* in the degradative processes, biosynthetic processes *consume* energy.

22.2 Biosynthesis of Carbohydrates

We discuss the biosynthesis of carbohydrates under three headings: (a) conversion of atmospheric CO_2 to glucose in plants, (b) synthesis of glucose in animals and humans, and (c) conversion of glucose to other carbohydrate molecules in animals and humans.

Conversion of Atmospheric CO_2

The most important biosynthesis of carbohydrates takes place in plants. This is **photosynthesis.** In this process the energy of the sun is built into chemical bonds; the overall reaction is

$$n\mathrm{H_2O} + n\mathrm{CO_2} \xrightarrow[\text{chlorophyll}]{\substack{\text{energy in}\\ \text{the form}\\ \text{of sunlight}}} (\mathrm{CH_2O})_n + n\mathrm{O_2} \qquad (22.3)$$

where $(CH_2O)_n$ is a general formula for carbohydrates. This is a very complicated process and takes place only in plants, not in animals (Box 22A). We shall not discuss it further here except to note that the carbohydrates of plants—starch, cellulose, and other mono- and polysaccharides—serve as the basic carbohydrate supply of all animals, including humans.

Synthesis of Glucose

We saw in Chapter 21 that when the body needs energy, carbohydrates are broken down via glycolysis. When energy is not needed, glucose can be synthesized from the intermediates of the glycolytic and citric acid pathways. This process is called **gluconeogenesis.** As shown in Figure 22.1, a large number of intermediates—pyruvate, lactate, oxaloacetate, malate, and several amino acids (the glucogenic amino acids we met in Sec. 21.9)—can serve as starting compounds. Gluconeogenesis proceeds in reverse order from glycolysis, and many of the enzymes of glycolysis also catalyze gluconeogenesis. However, at four points there are unique enzymes (marked in Fig. 22.1) that catalyze only gluconeogenesis and not the breakdown reactions. These four enzymes make gluconeogenesis a pathway that is distinct from glycolysis.

Gluconeogenesis means synthesis of "new" glucose.

ATP is used up in gluconeogenesis and produced in glycolysis.

Conversion of Glucose to Other Carbohydrates

The third important biosynthetic pathway for carbohydrates is the conversion of glucose to other hexoses and to hexose derivatives and the synthesis of di-, oligo-, and polysaccharides. The common step in all of these is the activation of glucose by uridine diphosphate (UDP) to form UDP glucose:

Box 22A

Photosynthesis

Photosynthesis requires sunlight, water, CO_2, and pigments found in plants, mainly chlorophyll. The overall reaction shown in Equation 22.3 in the text actually occurs in two distinct steps. First, the light interacts with the pigments that are located in highly membranous organelles of plants, called **chloroplasts.** Chloroplasts resemble mitochondria (Sec. 20.2) in many respects: They contain a whole chain of oxidation–reduction enzymes similar to the cytochrome and iron–sulfur complexes of mitochondrial membranes, and they also contain a proton translocating ATPase. In a manner similar to mitochondria, in chloroplasts too the proton gradient accumulated in the intermembrane region drives the synthesis of ATP (see the chemiosmotic pump, Sec. 20.6). The chlorophylls themselves, buried in a complex protein that traverses the chloroplast membranes, are molecules similar to the heme we have already encountered in hemoglobin (Box 3C). In contrast to heme, the chlorophylls contain Mg^{2+} instead of Fe^{2+}:

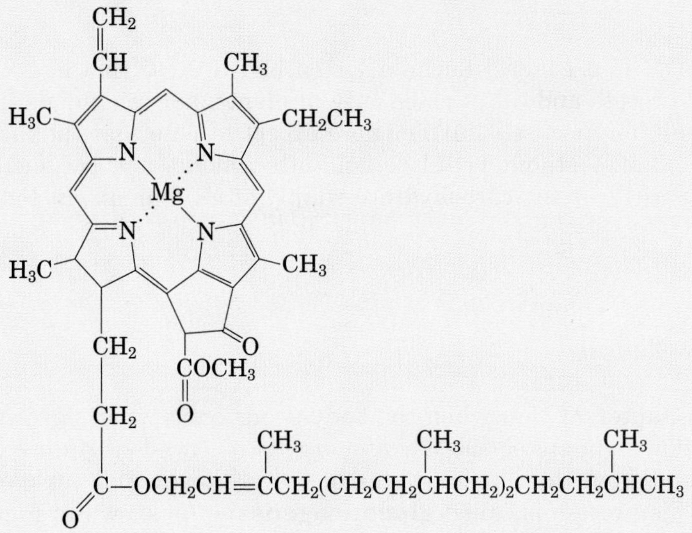

Chlorophyll a

The first reaction that takes place in photosynthesis is called the light reaction because chlorophyll captures the energy of sunlight and with its aid strips the electrons and protons from water to form oxygen, ATP, and NADPH + H^+ (See Fig. 22.3):

$$H_2O + ADP + P_i + NADP^+ + \text{sunlight} \longrightarrow O_2 + ATP + NADPH + H^+$$

The second reaction, called the dark reaction because it does not need light, in essence converts CO_2 to carbohydrates:

$$CO_2 + ATP + NADPH + H^+ \longrightarrow \text{carbohydrates} + ADP + P_i + NADP^+$$

In this step the energy, now in the form of ATP, is used to help NADPH + H^+ reduce carbon dioxide to carbohydrates. Thus, the protons and electrons stripped in the light reaction are added to the carbon dioxide in the dark reaction. This reduction takes place in a multistep cyclic process called the **Calvin cycle,** named after its discoverer, Melvin Calvin (1911–1997), who was awarded the 1961 Nobel Prize in Chemistry for this work. In this cycle the CO_2 is first attached to a C_5 fragment, making a C_6 fragment, which breaks down to two C_3 fragments (triose phosphates) that are eventually, through a complex series of steps, converted to a C_6 compound (glucose).

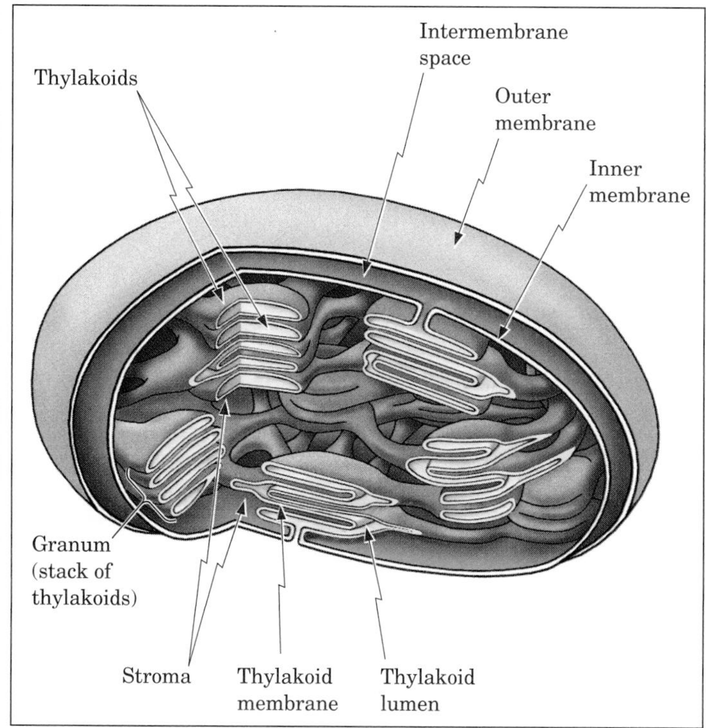

Inside of chloroplast.

Photosynthesis is the process responsible for the growth of all green plants. (Photograph by Beverly March.)

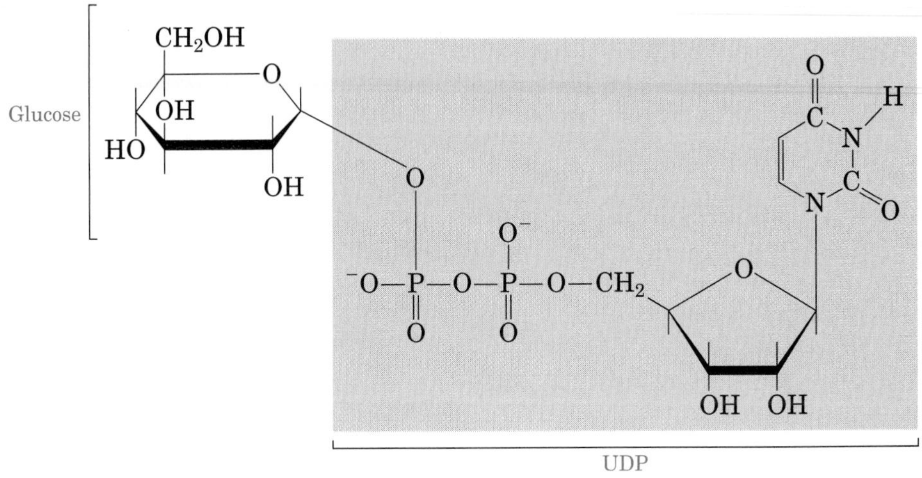

Glucose
UDP
UDP glucose

UDP is similar to ADP except that the base is uracil instead of adenine. UTP, an analog of ATP, contains two high-energy phosphate bonds. For example, when the body has excess glucose and wants to store it as glycogen (this process is called **glycogenesis**), the glucose is first converted to glucose 1-phosphate, but then a special enzyme catalyzes the reaction:

$$\text{glucose 1-phosphate} + \text{UTP} \longrightarrow \text{UDP glucose} + {}^{-}O{-}\overset{O}{\overset{\|}{\underset{O^{-}}{\underset{|}{P}}}}{-}O{-}\overset{O}{\overset{\|}{\underset{O^{-}}{\underset{|}{P}}}}{-}O^{-}$$

$$\text{UDP glucose} + \underset{\text{Glycogen}}{(\text{glucose})_n} \longrightarrow \text{UDP} + \underset{\substack{\text{Glycogen}\\ \text{(one unit larger)}}}{(\text{glucose})_{n+1}}$$

The biosynthesis of many other di- and polysaccharides and their derivatives also uses the common activation step: forming the appropriate UDP compound.

22.3 Biosynthesis of Fatty Acids

The body can synthesize all the fatty acids it needs except for linoleic and linolenic acids (essential fatty acids, Sec. 17.3). The source of carbon in this synthesis is acetyl CoA. Since acetyl CoA is also a degradation product of the β-oxidation spiral of fatty acids (Sec. 21.5), we might expect that the synthesis is the reverse of the degradation. This is not the case. For one thing, fatty acid synthesis occurs in the cytosol, while degradation takes place in the mitochondria. Fatty acid synthesis is catalyzed by a multienzyme system.

However, there is one aspect of fatty acid synthesis that is the same as in fatty acid degradation: Both processes involve acetyl CoA, and therefore both proceed in units of two carbons. Fatty acids are built up two carbons at a time, just as they are broken down two carbons at a time (Sec. 21.5).

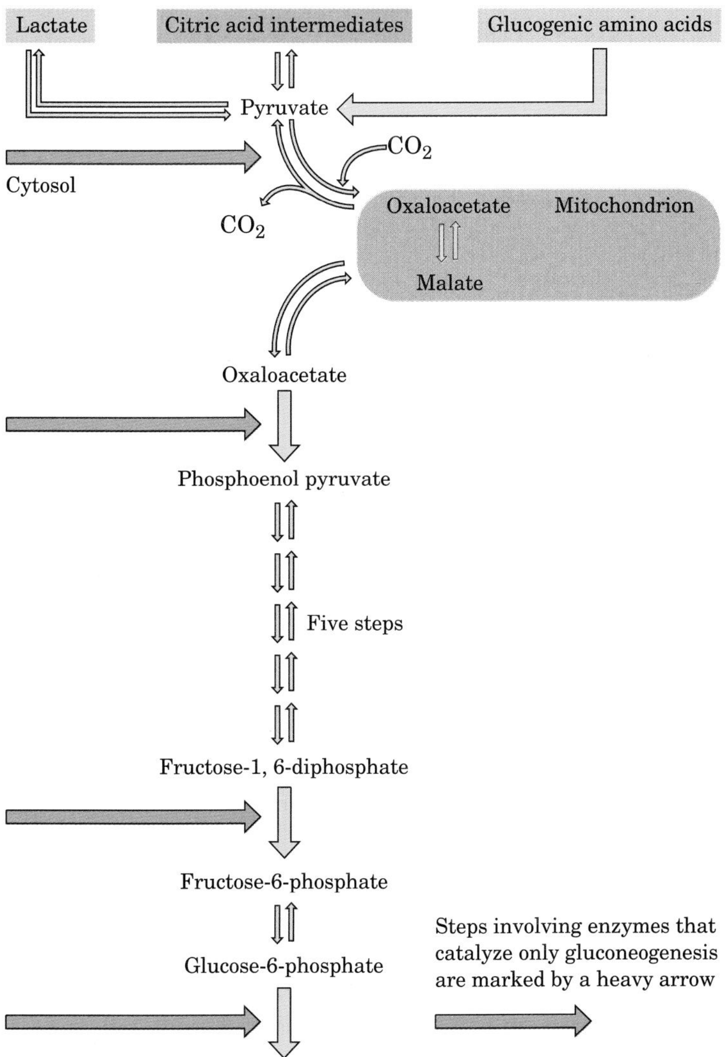

Figure **22.1**
Gluconeogenesis. All reactions take place in the cytosol, except for those shown in the mitochondria.

Most of the time, fatty acids are synthesized when excess food is available. That is, when we eat more food than we need for energy, our bodies turn the excess acetyl CoA (produced by catabolism of carbohydrates—Sec. 21.2) into fatty acids and then to fats. The fats are stored in the fat depots, which are specialized fat-carrying cells (see Fig. 21.1).

The key to fatty acid synthesis is an **a**cyl **c**arrier **p**rotein called **ACP.** This can be looked upon as a merry-go-round—a rotating protein molecule to which the growing chain of fatty acids is attached. As the growing chain rotates with the ACP, it sweeps over the multienzyme complex, and at each enzyme one reaction of the chain is catalyzed (Fig. 22.2).

At the beginning of this cycle, the ACP picks up an acetyl group from acetyl CoA and delivers it to the first enzyme, called synthase:

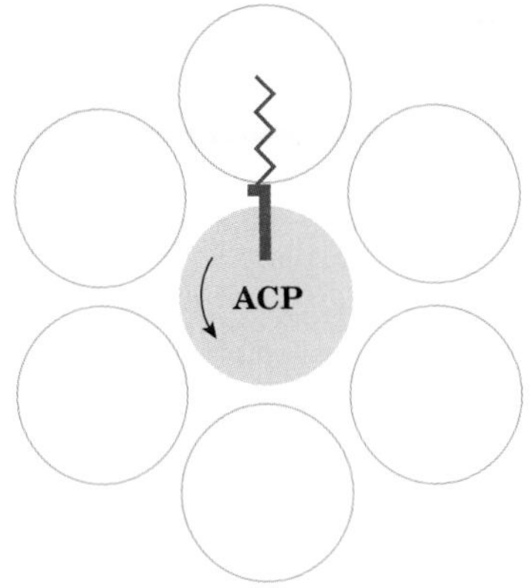

Figure **22.2**
The biosynthesis of fatty acids. The ACP (central dark sphere) has a long side chain (▬) that carries the growing fatty acid (〰). The ACP rotates counterclockwise, and its side chain sweeps over a multienzyme system (empty spheres). As each cycle is completed, a C_2 fragment is added to the growing fatty acid chain.

$$\mathrm{CH_3\overset{O}{\overset{\|}{C}}{-}S{-}CoA + HS{-}ACP \longrightarrow HS{-}CoA + CH_3\overset{O}{\overset{\|}{C}}{-}S{-}ACP}$$

$$\mathrm{CH_3\overset{O}{\overset{\|}{C}}{-}S{-}ACP + synthase{-}SH \longrightarrow CH_3\overset{O}{\overset{\|}{C}}{-}S{-}synthase + HS{-}ACP}$$

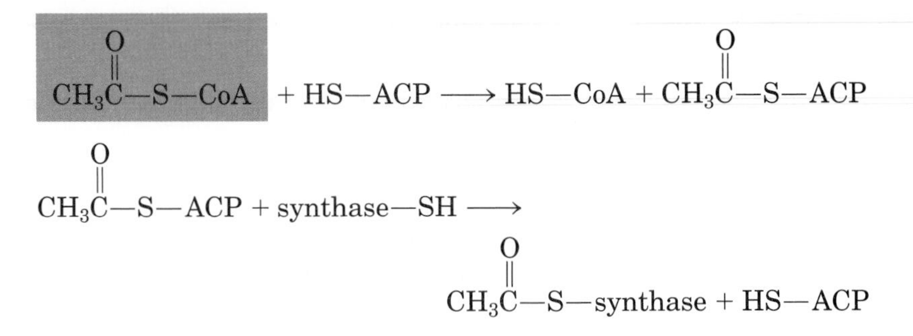

The C_2 fragment on the synthase is condensed with a C_3 fragment attached to the ACP in a process in which CO_2 is given off:

$$\mathrm{synthase{-}S{-}\overset{O}{\overset{\|}{C}}{-}CH_3 + \underset{\text{Malonyl ACP}}{ACP{-}S{-}\overset{O}{\overset{\|}{C}}{-}CH_2{-}COOH} \rightleftharpoons \underset{\text{Acetoacetyl ACP}}{ACP{-}S{-}\overset{O}{\overset{\|}{C}}{-}CH_2{-}\overset{O}{\overset{\|}{C}}{-}CH_3} + CO_2 + synthase{-}SH}$$

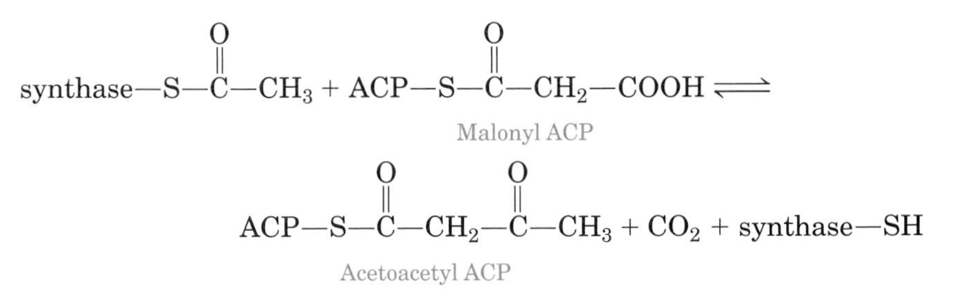

The result is a C_4 fragment, which is reduced twice and dehydrated before it becomes a fully saturated C_4 group. This is the end of one cycle of the merry-go-round.

In the next cycle, the C_4 fragment is transferred to the synthase, and another malonyl ACP (C_3 fragment) is added. When CO_2 splits out, a C_6 fragment is obtained. The merry-go-round continues to turn. At each turn, another C_2 fragment is added to the growing chain. Chains up to C_{16} (palmitic acid) can be obtained in this process. If the body needs longer fatty acids—for example, stearic (C_{18})—another C_2 fragment is added to palmitic acid by a different enzyme system.

Unsaturated fatty acids are obtained from saturated fatty acids by an oxidation step in which hydrogen is removed and combined with O_2 to form water:

$$\mathrm{R{-}\underset{H}{\overset{H}{C}}{-}\underset{H}{\overset{H}{C}}{-}(CH_2)_nCOOH + O_2 + NADPH + H^+ \xrightarrow{\text{enzyme}} NADP^+ + 2H_2O + R{-}\overset{H}{C}{=}\overset{H}{C}{-}(CH_2)_n{-}COOH}$$

The structure of $NADP^+$ is the same as that of NAD^+, except that an additional phosphate group is attached to one of the ribose units (Fig. 22.3).

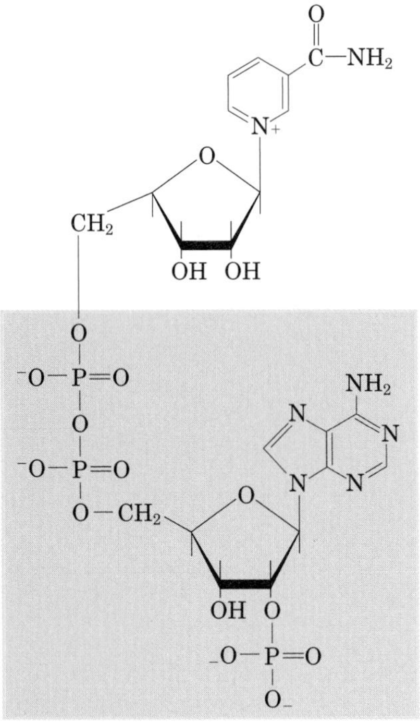

Figure **22.3**
The structure of $NADP^+$.

22.4 Biosynthesis of Membrane Lipids

The various membrane lipids (Secs. 17.6 to 17.8) are assembled from their constituents. We just saw how fatty acids are synthesized in the body.

These fatty acids are activated by CoA, forming acyl CoA. The compound glycerol 3-phosphate, which is obtained from the reduction of dihydroxyacetone (a C_3 fragment of glycolysis, Fig. 21.4), is the second building block of glycerophospholipids. This compound combines with two acyl CoA molecules, which can be the same or different:

We encountered glycerol 3-phosphate as a vehicle for transporting electrons in and out of mitochondria (Sec. 21.3).

$$\begin{array}{l} H_2C{-}OH \\ \;\;| \\ HC{-}OH \\ \;\;| \\ H_2C{-}O{-}\overset{\overset{\displaystyle O}{\|}}{\underset{\underset{\displaystyle O^-}{|}}{P}}{-}O^- \end{array} + \begin{array}{l} R{-}\overset{\overset{\displaystyle O}{\|}}{C}{-}S{-}CoA \\ R'{-}\overset{\overset{\displaystyle O}{\|}}{C}{-}S{-}CoA \end{array} \longrightarrow \begin{array}{l} H_2C{-}O{-}\overset{\overset{\displaystyle O}{\|}}{C}{-}R \\ \;\;| \\ HC{-}O{-}\overset{\overset{\displaystyle O}{\|}}{C}{-}R' \\ \;\;| \\ H_2C{-}O{-}\overset{\overset{\displaystyle O}{\|}}{\underset{\underset{\displaystyle O^-}{|}}{P}}{-}O^- \end{array} + 2CoA{-}SH$$

Glycerol 3-phosphate Acyl CoA Phosphatidate

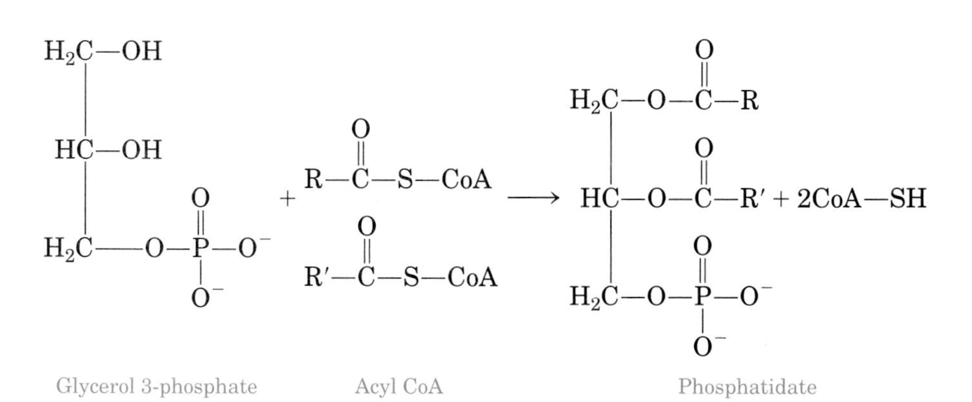

To complete the molecule, an activated serine or an activated choline is added to the $-OPO_3^{2-}$ group (see the structures in Sec. 17.6; a model of phosphatidylcholine is shown in Fig. 22.4).

Sphingolipids (Sec. 17.7) are similarly built up from smaller molecules. An activated phosphocholine is added to sphingosine to make sphingomyelin.

Each constituent is added in its activated form (that is, combined with CoA).

The glycolipids are made in a similar fashion. Ceramide is assembled as above, and the carbohydrate is added one unit at a time in the form of activated monosaccharides (UDP glucose and so on).

Cholesterol, the molecule that controls the fluidity of membranes and is a precursor of all steroid hormones, is also synthesized by the human

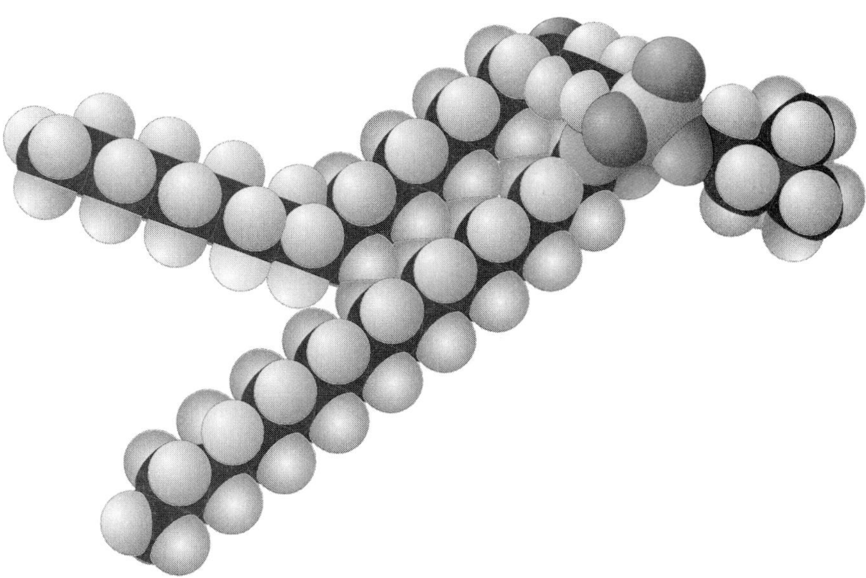

Figure **22.4** • A model of phosphatidylcholine, commonly called lecithin.

Box 22B

Phospholipid Synthesis and Hyaline Membrane Disease

Hyaline membrane disease, which affects the breathing of infants and has caused numerous sudden infant deaths, is not completely understood at this time. Analysis has shown that there is a great difference in the phospholipid composition of the lung between normal and affected individuals. The phospholipid of the lung is mainly phosphatidylcholine, which in normal lungs contains large quantities of unsaturated fatty acids. In contrast, the phosphatidylcholine in the lungs of patients with hyaline membrane disease is largely saturated. Furthermore, these patients have less phosphatidylcholine in their lung tissue than normal individuals. This indicates a faulty biosynthetic apparatus, but the exact cause has not yet been determined.

body. It is assembled in the liver from C_2 fragments in the form of acetyl CoA. All the carbon atoms of cholesterol come from the carbons of acetyl CoA molecules (Fig. 22.5). The drug lovastatin inhibits the biosynthesis of cholesterol and is frequently prescribed to control the cholesterol level in the blood (see Box 17G).

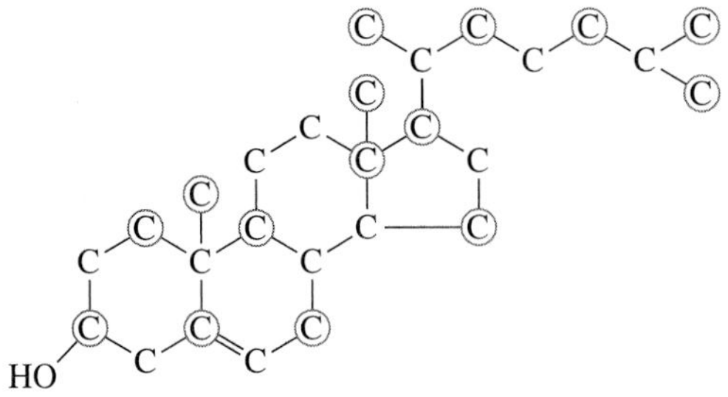

Figure **22.5**
Biosynthesis of cholesterol. The circled carbon atoms come from the $—CH_3$ group, and the others come from the —CO— group of the acetate.

Table 18.1 shows the essential and nonessential amino acids.

22.5 Biosynthesis of Amino Acids

The human body needs 20 different amino acids to make its protein chains—all 20 are found in a normal diet. Some of the amino acids can be synthesized from other compounds: These are the nonessential amino acids. Others cannot be synthesized by the human body and must be supplied in the diet. These are the **essential amino acids** (see Sec. 26.4). Most nonessential amino acids are synthesized from some intermediate of either glycolysis (Sec. 21.2) or the citric acid cycle (Sec. 20.4). Glutamic acid plays a central role in the synthesis of five nonessential amino acids. Glutamic acid itself is synthesized from α-ketoglutaric acid, one of the intermediates in the citric acid cycle:

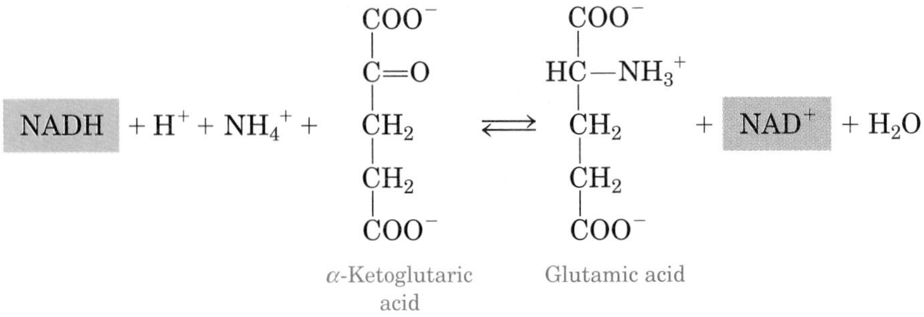

$$\text{NADH} + H^+ + NH_4^+ + \begin{array}{c} COO^- \\ | \\ C{=}O \\ | \\ CH_2 \\ | \\ CH_2 \\ | \\ COO^- \end{array} \rightleftharpoons \begin{array}{c} COO^- \\ | \\ HC{-}NH_3^+ \\ | \\ CH_2 \\ | \\ CH_2 \\ | \\ COO^- \end{array} + \text{NAD}^+ + H_2O$$

α-Ketoglutaric acid — Glutamic acid

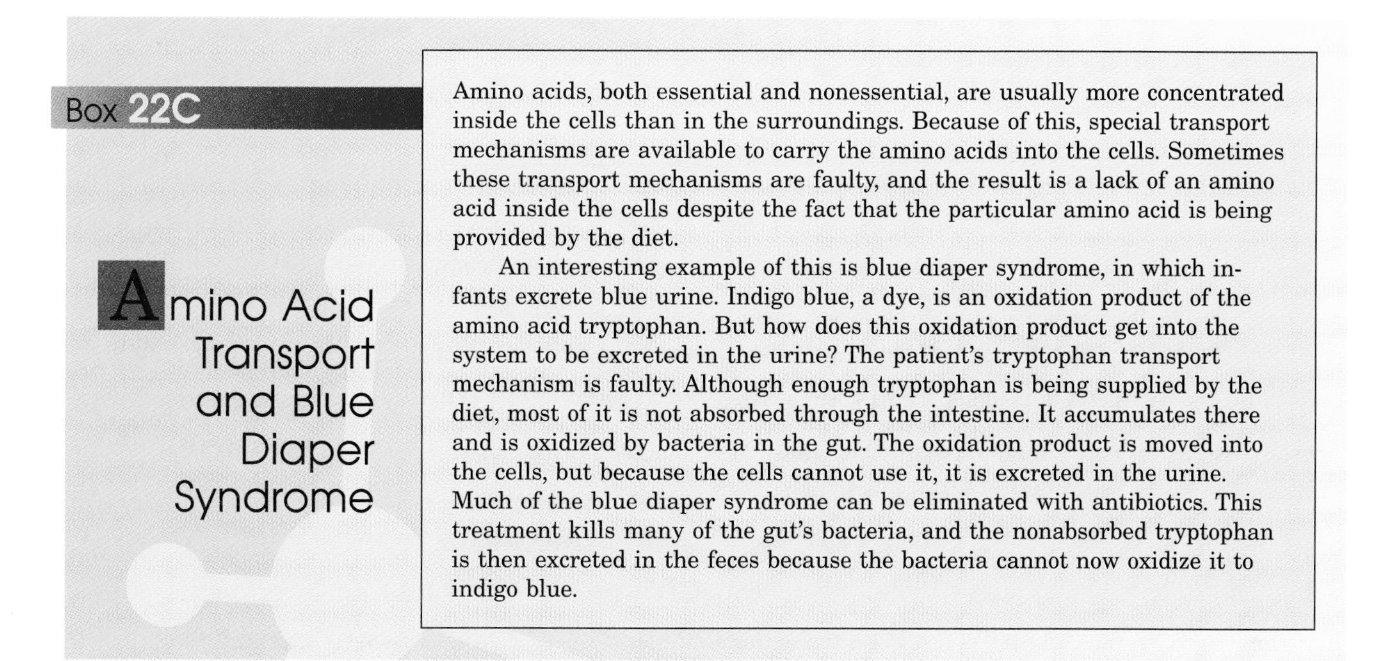

The forward reaction is the synthesis, and the reverse reaction is the oxidative deamination (degradation) reaction we encountered in the ca-

Box 22C

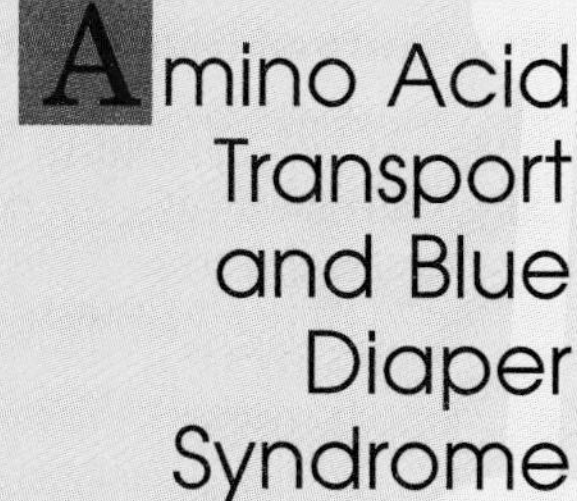

Amino Acid Transport and Blue Diaper Syndrome

Amino acids, both essential and nonessential, are usually more concentrated inside the cells than in the surroundings. Because of this, special transport mechanisms are available to carry the amino acids into the cells. Sometimes these transport mechanisms are faulty, and the result is a lack of an amino acid inside the cells despite the fact that the particular amino acid is being provided by the diet.

An interesting example of this is blue diaper syndrome, in which infants excrete blue urine. Indigo blue, a dye, is an oxidation product of the amino acid tryptophan. But how does this oxidation product get into the system to be excreted in the urine? The patient's tryptophan transport mechanism is faulty. Although enough tryptophan is being supplied by the diet, most of it is not absorbed through the intestine. It accumulates there and is oxidized by bacteria in the gut. The oxidation product is moved into the cells, but because the cells cannot use it, it is excreted in the urine. Much of the blue diaper syndrome can be eliminated with antibiotics. This treatment kills many of the gut's bacteria, and the nonabsorbed tryptophan is then excreted in the feces because the bacteria cannot now oxidize it to indigo blue.

tabolism of amino acids (Sec. 21.8). This is one case in which the synthetic and degradative pathways are exactly the reverse of each other.

Glutamic acid can serve as an intermediate in the synthesis of alanine, serine, aspartic acid, asparagine, and glutamine. For example, the transamination reaction we saw in Section 21.8 leads to alanine formation:

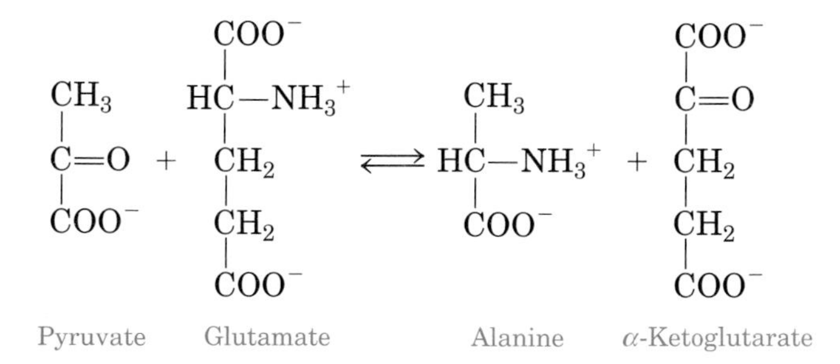

$$\begin{array}{c} CH_3 \\ | \\ C{=}O \\ | \\ COO^- \end{array} + \begin{array}{c} COO^- \\ | \\ HC{-}NH_3^+ \\ | \\ CH_2 \\ | \\ CH_2 \\ | \\ COO^- \end{array} \rightleftharpoons \begin{array}{c} CH_3 \\ | \\ HC{-}NH_3^+ \\ | \\ COO^- \end{array} + \begin{array}{c} COO^- \\ | \\ C{=}O \\ | \\ CH_2 \\ | \\ CH_2 \\ | \\ COO^- \end{array}$$

Pyruvate Glutamate Alanine α-Ketoglutarate

Besides being the building blocks of proteins, amino acids also serve as intermediates for a large number of biological molecules. We have already seen that serine is needed in the synthesis of membrane lipids (Sec. 22.4). Certain amino acids are also intermediates in the synthesis of heme and of the purines and pyrimidines that are the raw materials for DNA and RNA (Chap. 23).

SUMMARY

For most biochemical compounds, the biosynthetic pathways are different from the degradation pathways. Carbohydrates are synthesized in plants from CO_2 and H_2O, using sunlight as an energy source (**photosynthesis**). Glucose can be synthesized by animals from the intermediates of glycolysis, from those of the citric acid cycle, and from glucogenic amino acids. This process is called **gluconeogenesis.** When glucose or other monosaccharides are built into di-, oligo-, and polysaccharides, each monosaccharide unit in its activated form is added to a growing chain.

Fatty acid biosynthesis is accomplished by a multienzyme system. The key to this process is the **acyl carrier protein, ACP,** which acts as a merry-go-round transport system; it carries the growing fatty acid chain over a number of enzymes, each of which catalyzes a specific reaction. With each complete turn of the merry-go-round, a C_2 fragment is added to the growing fatty acid chain. The source of the C_2 fragment is malonyl ACP, a C_3 compound attached to the ACP. This becomes C_2 by loss of CO_2. Membrane lipids are synthesized in the body by assembling the constituent parts. The fatty acids are activated by conversion to acyl CoA.

Many nonessential amino acids are synthesized in the body from the intermediates of glycolysis or of the citric acid cycle. In half of these, glutamic acid is the donor of the amino group in transamination. Amino acids serve as building blocks for proteins.

KEY TERMS

Acyl carrier protein (Sec. 22.3)
Biosynthetic pathway (Sec. 22.1)
Chloroplasts (Box 22A)
Essential amino acid (Sec. 22.5)
Gluconeogenesis (Sec. 22.2)
Glycogenesis (Sec. 22.2)
Photosynthesis (Sec. 22.2)

PROBLEMS

Biosynthesis

22.1 Why are the pathways the body uses for anabolism and catabolism mostly different?

22.2 What compound provides the energy for biosynthesis?

22.3 Glycogen can be synthesized in the body by the same enzymes that degrade it. Why is this process utilized in glycogen synthesis only to a small extent, while most glycogen biosynthesis occurs via a different synthetic pathway?

22.4 Do most anabolic and catabolic reactions take place in the same location?

Carbohydrate Biosynthesis

22.5 Photosynthesis is the reverse of what process?

22.6 In photosynthesis, what is the source of (a) carbon (b) hydrogen (c) energy

22.7 Name a compound that can serve as a raw material for gluconeogenesis and is (a) from the glycolytic pathway (b) from the citric acid cycle (c) an amino acid

22.8 How does the structure of UDP differ from that of ADP?

22.9 Which steps in gluconeogenesis are not reversible?

22.10 Are the enzymes that combine two C_3 compounds to a C_6 compound in gluconeogenesis the same as or different from those that cut the C_6 compound into two C_3 compounds in glycolysis?

22.11 Which part of gluconeogenesis occurs in the mitochondria?

22.12 Glycogen is written as $(\text{glucose})_n$. (a) What does the n stand for? (b) What is the value of n?

22.13 What are the constituents of UTP?

Fatty Acid Biosynthesis

22.14 What is the source of carbon in fatty acid synthesis?

22.15 (a) Where in the body does fatty acid synthesis occur?
(b) Does fatty acid degradation occur in the same location?

22.16 When you eat more food than your body needs, excess acetyl CoA is synthesized by the pathways outlined in Chapter 21. What does the body do with this excess acetyl CoA?

22.17 In fatty acid biosynthesis which compound is added repeatedly to the synthase?

***22.18** (a) What is the name of the first enzyme in fatty acid synthesis? (b) What does it do?

22.19 From what compound is the CO_2 released in fatty acid synthesis?

22.20 What is the functional group of the synthase to which the C_2 fragment is attached?

***22.21** In the synthesis of unsaturated fatty acids, $NADPH + H^+$ is converted to $NADP^+$. Yet this is an oxidation step and not a reduction step. Explain.

22.22 Which of these fatty acids can be synthesized by the multienzyme fatty acid synthesis complex alone?
(a) oleic
(b) stearic
(c) myristic
(d) arachidonic
(e) lauric

22.23 What is the difference between the structures of NAD^+ and $NADP^+$?

22.24 Under what conditions does the body synthesize fatty acids?

***22.25** Linoleic and linolenic acids cannot be synthesized in the human body. Does this mean that the human body cannot make an unsaturated fatty acid from a saturated one?

Biosynthesis of Membrane Lipids

22.26 When the body synthesizes this membrane lipid, from what ingredients does it assemble it?

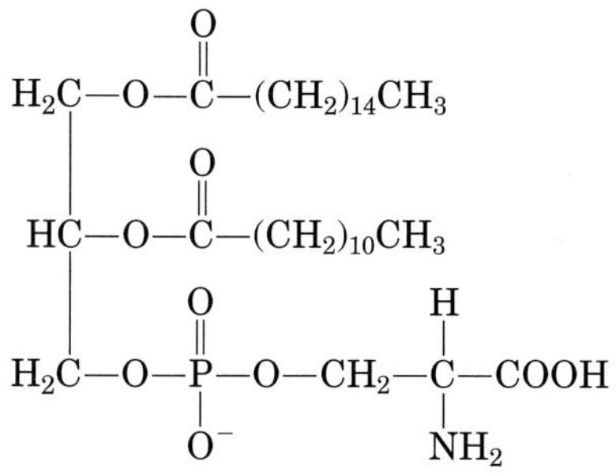

$$\begin{array}{l} H_2C-O-\overset{O}{\overset{\|}{C}}-(CH_2)_{14}CH_3 \\ | \\ HC-O-\overset{O}{\overset{\|}{C}}-(CH_2)_{10}CH_3 \\ | \\ H_2C-O-\underset{O^-}{\underset{|}{\overset{O}{\overset{\|}{P}}}}-O-CH_2-\underset{NH_2}{\underset{|}{\overset{H}{\overset{|}{C}}}}-COOH \end{array}$$

***22.27** Name the activated constituents necessary to form the glycolipid glucoceramide.

22.28 Would the cell membranes of a person on a diet containing no cholesterol stiffen?

Biosynthesis of Amino Acids

***22.29** What compound reacts with glutamate in a transamination process to yield serine?

22.30 What reaction is the reverse of the synthesis of glutamate from α-ketoglutarate, ammonia, and $NADH + H^+$?

***22.31** Which amino acid will be synthesized by this process?

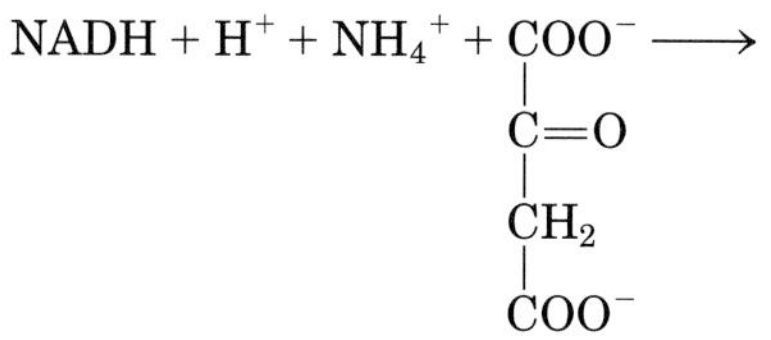

$$NADH + H^+ + NH_4^+ + \begin{array}{c} COO^- \\ | \\ C=O \\ | \\ CH_2 \\ | \\ COO^- \end{array} \longrightarrow$$

***22.32** Draw the structure of the compound needed to synthesize asparagine from glutamate by transamination.

22.33 Name the products of the transamination reaction

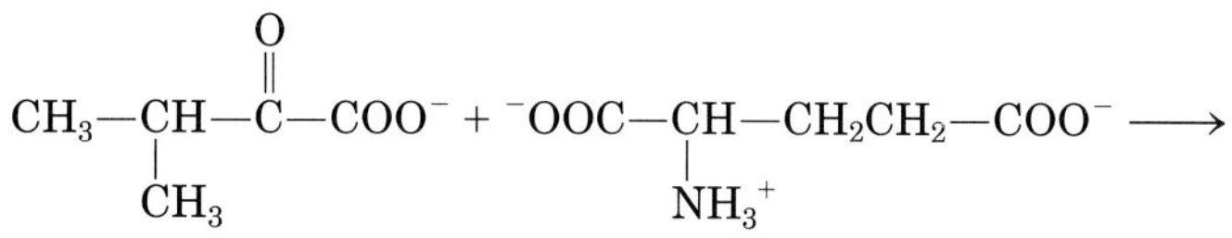

$$CH_3-\underset{CH_3}{\underset{|}{CH}}-\overset{O}{\overset{\|}{C}}-COO^- + {}^-OOC-\underset{NH_3^+}{\underset{|}{CH}}-CH_2CH_2-COO^- \longrightarrow$$

Boxes

22.34 (Box 22A) What is the major difference in structure between the chlorophylls and heme?

22.35 (Box 22A) What happens in the light reaction of photosynthesis?

22.36 (Box 22C) (a) What compound produces the colored urine of blue diaper syndrome? (b) Where does this compound come from?

Additional Problems

22.37 How are unsaturated fatty acids synthesized in the body?

22.38 What is the activated building block of cholesterol?

***22.39** What C_3 fragment carried by ACP is used in fatty acid synthesis?

22.40 When glutamate transaminates phenylpyruvate, what amino acid is produced?

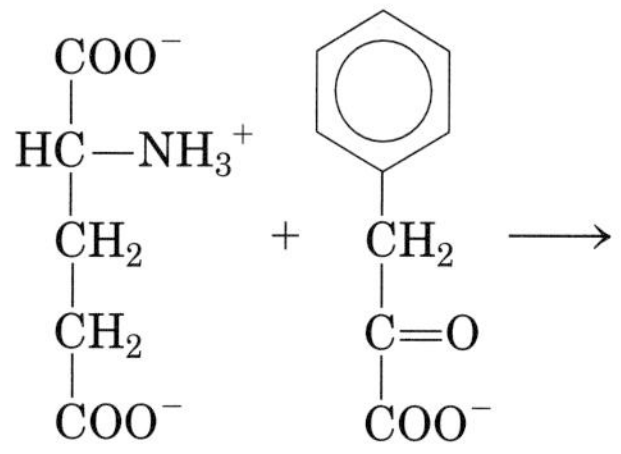

$$\begin{array}{c} COO^- \\ | \\ HC-NH_3^+ \\ | \\ CH_2 \\ | \\ CH_2 \\ | \\ COO^- \end{array} + \begin{array}{c} C_6H_5 \\ | \\ CH_2 \\ | \\ C=O \\ | \\ COO^- \end{array} \longrightarrow$$

22.41 Can the complex enzyme system participating in every fatty acid synthesis manufacture fatty acids of any length?

***22.42** Each activation step in the synthesis of complex lipids occurs at the expense of one ATP molecule. How many ATP molecules are used up in the synthesis of one molecule of lecithin?

22.43 Would a person on a completely cholesterol-free diet totally lack the cholesterol necessary to synthesize steroid hormones?

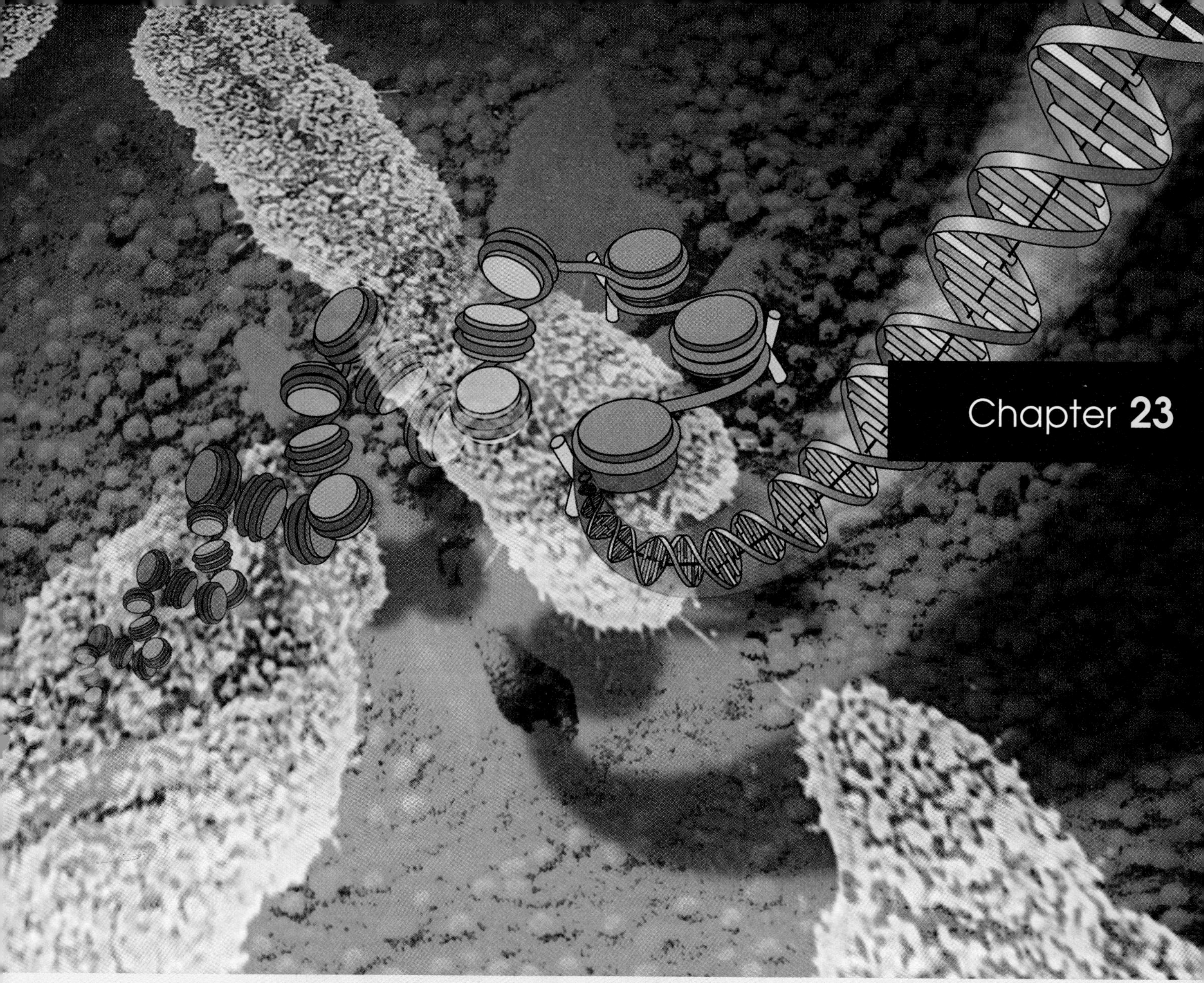

Chapter 23

Nucleic acids and protein synthesis

23.1 Introduction

Each cell of our bodies contains thousands of different protein molecules. We recall from Chapter 18 that all these molecules are made up of the same 20 amino acids, but in different sequences. A lobster has a different set of protein molecules than a mushroom or a zebra. Though the lobster and the zebra have some proteins in common, each has many protein molecules not found in the other. Within a species also, different individuals may have some differences in their proteins, though the differences are much less than those between species. This shows up graphically in cases where people have such conditions as hemophilia, albinism, or color-blindness, because they lack certain proteins that normal people have.

There are two parents if reproduction is sexual; one parent if it is asexual.

Once scientists understood this, the next question was, how do the cells know which proteins to synthesize out of the extremely large number of possible amino acid sequences? The answer to this question is that an individual gets the information from its parents; this is called *heredity.* We all know that a pig gives birth to a pig and a mouse to a mouse. A pig does not give birth to a mouse. When an egg cell joins a sperm cell, a new cell called a *zygote* is produced. A pig zygote looks very much like a mouse zygote. Yet the pig zygote will produce the enzymes necessary to cause the zygote to grow into a pig and not a mouse.

It was easy to determine that the information is obtained from the parent or parents, but what form does this information take? During the last 50 years, revolutionary developments have led to an answer to this question—the transmission of heredity on the molecular level.

Some simple organisms, such as bacteria, do not have a nucleus. Their chromosomes are condensed in a central region.

From about the end of the nineteenth century, biologists suspected that the transmission of hereditary information from one generation to another takes place in the *nucleus* of the cell. More precisely, it was felt that structures within the nucleus, called **chromosomes,** have something to do with heredity. Each species has a different number of chromosomes in the nucleus. The information that determines external characteristics (red hair, blue eyes) and internal characteristics (blood group, hereditary diseases) was thought to reside in **genes** located inside the chromosomes.

Studies on the nuclei of fruit flies in the early 1930s revealed that the genes that carry the different traits lie in sequences along the chromosomes. Chemical analysis of nuclei showed that they are largely made up of special basic proteins called *histones* and a type of compound called **nucleic acids.** By 1940 it became clear through the work of Oswald Avery (1877–1955) that of all the material in the nucleus, only a nucleic acid called deoxyribonucleic acid (DNA) carries the hereditary information. That is, the genes are located in the DNA. Other work in the 1940s by George Beadle (1903–1989) and Edward Tatum (1909–1975) demonstrated that each gene controls the manufacture of one enzyme, and through this the external and internal characteristics are expressed. Thus, the expression of the gene (DNA) in terms of an enzyme (protein) led to the study of protein synthesis and its control. **The information**

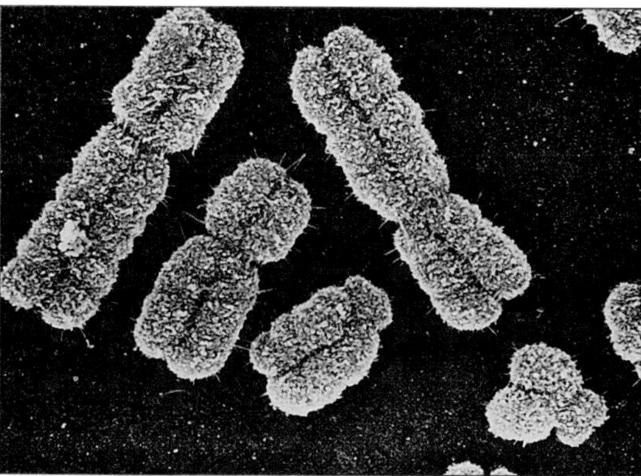

Human chromosomes magnified about 8000 times. (Biophoto Associates/Photo Researchers.)

that tells the cell which proteins to manufacture is carried in the molecules of DNA.

In the following sections we provide some of the highlights of this rapidly developing field.

23.2 Components of Nucleic Acids

There are two kinds of nucleic acids found in cells—each has its own role in the transmission of hereditary information. The two types are **ribonucleic acid (RNA)** and **deoxyribonucleic acid (DNA).** As we just saw, DNA is present in the chromosomes of the nucleus. RNA is not found in the chromosomes. It is located elsewhere in the nucleus and even outside the nucleus, in the cytoplasm. As we will see in Section 23.5, there are three types of RNA, all with similar structures.

Both DNA and RNA are polymers. Just as proteins consist of chains of amino acids and polysaccharides of chains of monosaccharides, nucleic acids are also chains. The building blocks (monomers) of nucleic acid chains are **nucleotides.** Nucleotides themselves, however, are composed of three simpler units: a *base,* a *sugar,* and *phosphate.* We will look at each of these in turn.

Bases

The bases found in DNA and RNA are chiefly those shown in Figure 23.1. All of them are basic because they are heterocyclic amines (Sec. 15.4). Two of these bases, adenine and guanine, are purines (they have the same ring system as the purines shown in Sec. 15.9), and the other three—cytosine, thymine, and uracil—are pyrimidines (Sec. 11.13). The two purines (A and G) and one of the pyrimidines (C) are found in both DNA and RNA, but uracil (U) is found *only* in RNA, while thymine (T) is found *only* in DNA. Note that thymine differs from uracil only in the methyl group in the 5 position. Thus, both DNA and RNA contain four bases:

The three pyrimidines and guanine are in their keto rather than their enol forms (see Sec. 13.6).

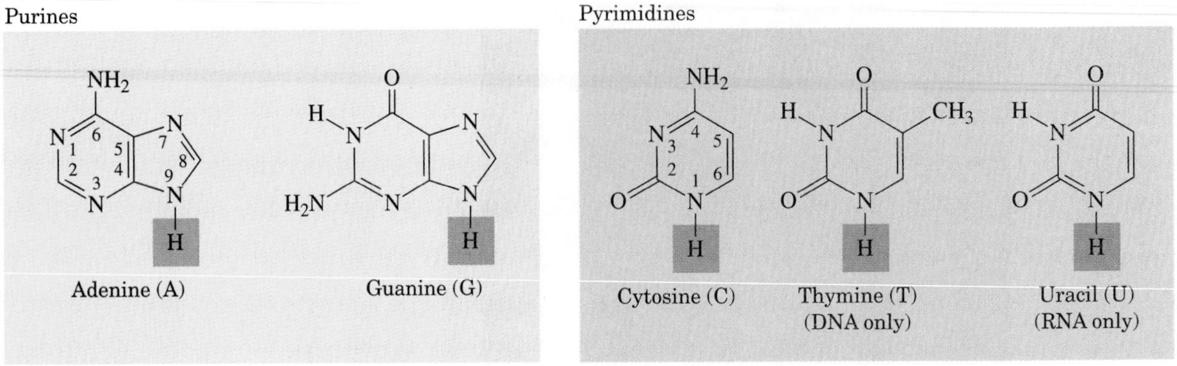

Purines

Pyrimidines

Adenine (A)

Guanine (G)

Cytosine (C)

Thymine (T) (DNA only)

Uracil (U) (RNA only)

Figure **23.1** • The five principal bases of DNA and RNA. Note how the rings are numbered. The hydrogens shown in blue are lost when the bases are connected to the sugars.

The initial letter of each base is used as an abbreviation for that base.

two pyrimidines and two purines. For DNA the bases are A, G, C, and T; for RNA the bases are A, G, C, and U.

Sugars

The sugar component of RNA is D-ribose (Sec. 16.6). In DNA it is D-deoxyribose (hence the name deoxyribonucleic acid):

The only difference between these molecules is that ribose has an OH group in the 2 position not found in deoxyribose.

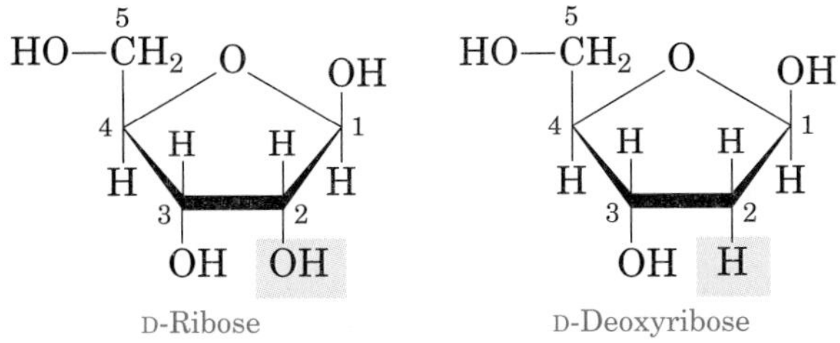

D-Ribose

D-Deoxyribose

The combination of sugar and base is known as a **nucleoside.** The purine bases are linked to the C-1 of the sugar through one of the nitrogens of the five-membered ring (the N connected to an H in color in Fig. 23.1):

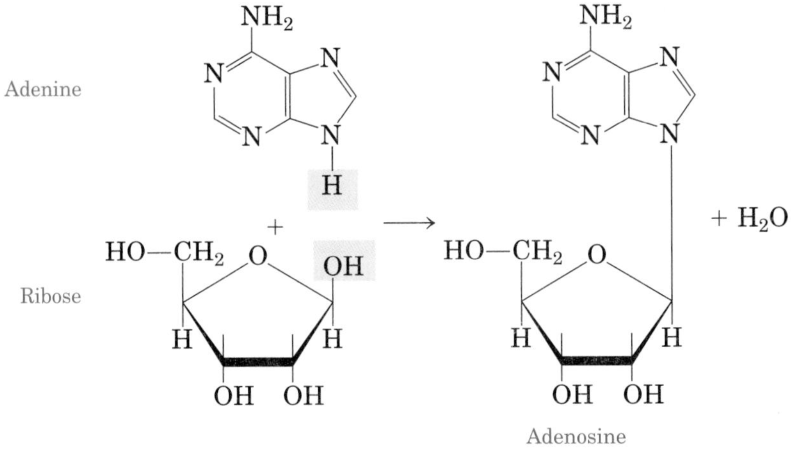

Adenine

Ribose

$+ H_2O$

Adenosine

Table 23.1 The Names of the Eight Nucleosides and Eight Nucleotides in DNA and RNA

Base	Nucleoside	Nucleotide
		DNA
Adenine (A)	Deoxyadenosine	Deoxyadenosine 5′-monophosphate (dAMP)[a]
Guanine (G)	Deoxyguanosine	Deoxyguanosine 5′-monophosphate (dGMP)
Thymine (T)	Deoxythymidine	Deoxythymidine 5′-monophosphate (dTMP)
Cytosine (C)	Deoxycytidine	Deoxycytidine 5′-monophosphate (dCMP)
		RNA
Adenine (A)	Adenosine	Adenosine 5′-monophosphate (AMP)
Guanine (G)	Guanosine	Guanosine 5′-monophosphate (GMP)
Uracil (U)	Uridine	Uridine 5′-monophosphate (UMP)
Cytosine (C)	Cytidine	Cytidine 5′-monophosphate (CMP)

[a]The d indicates that the sugar is *d*eoxyribose.

The nucleoside made of guanine and ribose is called guanosine. The names of the other nucleosides are given in Table 23.1.

The pyrimidine bases are linked to the C-1 of the sugar through their N-1 nitrogen:

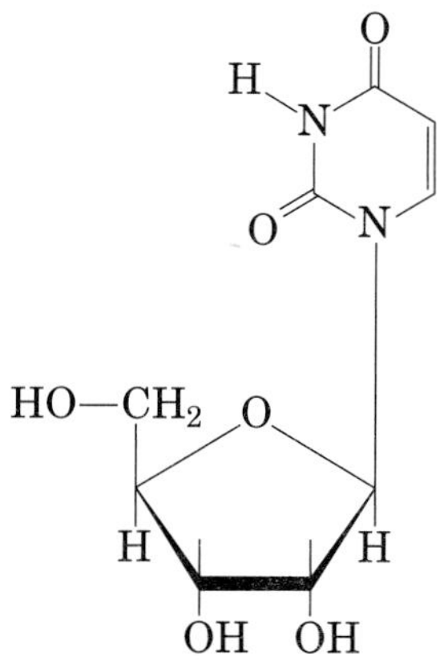

Uridine

Phosphate

The third component of nucleic acids is phosphoric acid. When this group is linked to the —CH_2OH group of a nucleoside, the result is a compound known as a **nucleotide.** For example, adenosine combines with phosphate to form the nucleotide adenosine monophosphate, AMP:

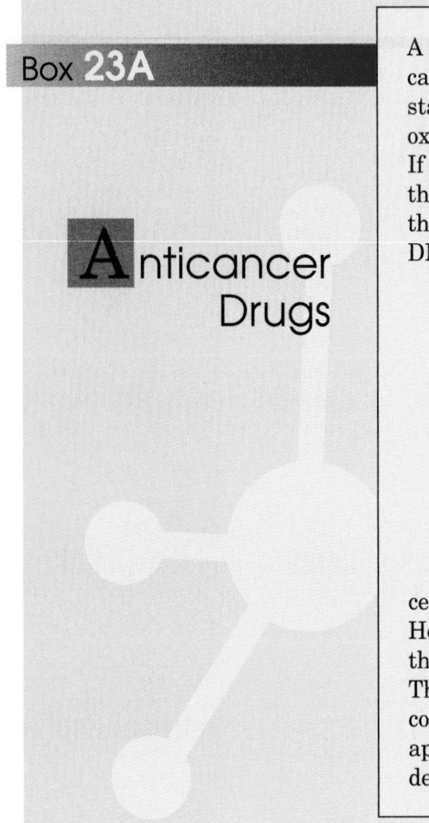

Box 23A

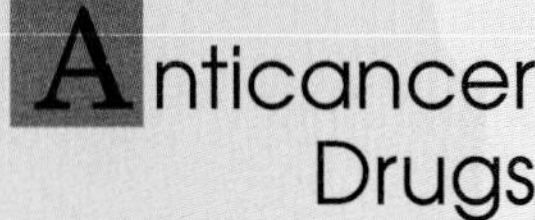

A major difference between cancer cells and most normal cells is that the cancer cells divide much more rapidly. Rapidly dividing cells require a constant new supply of DNA. One component of DNA is the nucleoside deoxythymidine, which is synthesized in the cell by the methylation of uridine. If fluorouracil is administered to a cancer patient as part of chemotherapy, the body converts it to fluorouridine, a compound that irreversibly inhibits the enzyme that manufactures thymidine from uridine, greatly decreasing DNA synthesis.

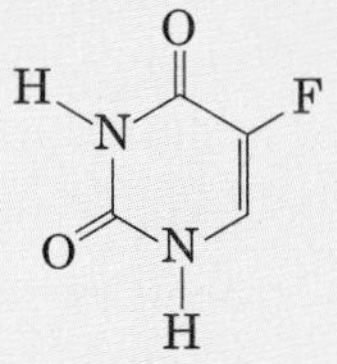

Fluorouracil

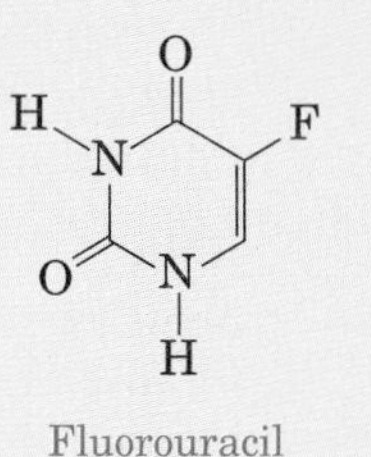

Since this affects the fast-dividing cancer cells more than the healthy cells, the growth of the tumor and the spread of the cancer are arrested. However, chemotherapy with fluorouracil or other anticancer drugs weakens the body because it also interferes with DNA synthesis in normal cells. Therefore, chemotherapy is used intermittently to give the body time to recover from the side effects of the drug. During the period after chemotherapy, special precautions must be taken so that bacterial infections do not debilitate the already weakened body.

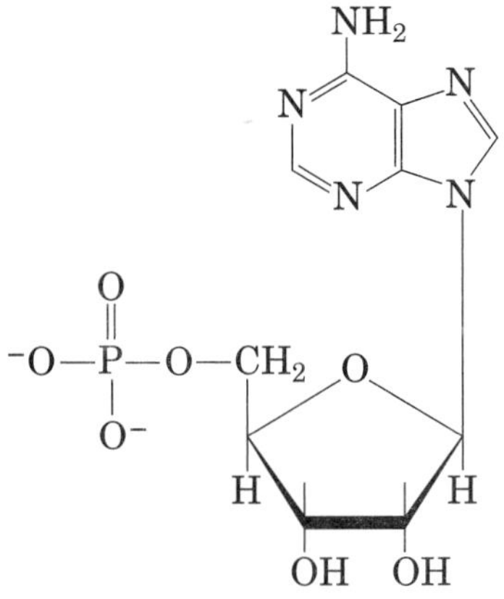

AMP

The names of the other nucleotides are given in Table 23.1. We are already familiar with some of these nucleotides. We have seen how they are part of the structure of key coenzymes, cofactors, and activators (Secs. 20.3 and 22.2). In Section 23.3 we will see how DNA and RNA are chains of nucleotides. In summary:

A nucleoside = base + sugar
A nucleotide = base + sugar + phosphoric acid
A nucleic acid = a chain of nucleotides

 23.3 Structure of DNA and RNA

In Chapter 18 we saw that proteins have primary, secondary, and higher structures. Nucleic acids, which are also chains of monomers, also have primary and secondary structures.

Primary Structure

Primed numbers are used for the ribose and deoxyribose portions of nucleosides, nucleotides, and nucleic acids. Unprimed numbers are used for the bases.

Nucleic acids are chains of nucleotides, as shown schematically in Figure 23.2. As the figure shows, the structure can be divided into two parts: (1) the backbone of the molecule and (2) the bases that are the side-chain groups. The backbone in DNA consists of alternating deoxyribose and phosphate groups. Each phosphate group is linked to the 3′ carbon of one deoxyribose unit and simultaneously to the 5′ carbon of the next deoxyribose unit (Fig. 23.3). Similarly, each sugar unit is linked to one phosphate at the 3′ position and to another at the 5′ position. The primary structure of RNA is the same except that each sugar is ribose (so there is an —OH group in the 2′ position) rather than deoxyribose.

Thus, the backbone of the DNA and RNA chains has two ends: a 3′ —OH end and a 5′ —OH end. These two ends have roles similar to those of the C-terminal and N-terminal residues in proteins. This backbone provides the structural stability of the DNA and RNA molecules.

The bases that are linked, one to each sugar unit, are the side chains and carry all the information necessary for protein synthesis. Analysis of the base composition of DNA molecules from many different species was done by Erwin Chargaff (1905–   ), who showed that in DNA taken from many different species, the quantity of adenine (in moles) is always approximately equal to that of thymine, and the quantity of guanine is always approximately equal to that of cytosine, though the adenine/guanine ratio varies widely from species to species. This important information helped to establish the secondary structure of DNA, as we shall see below.

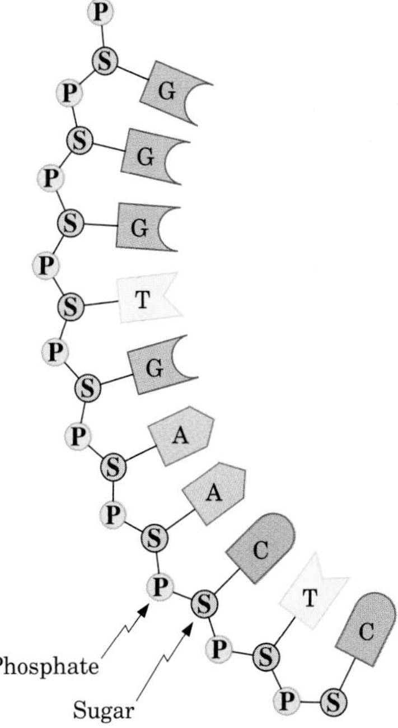

Figure **23.2**
Schematic diagram of a nucleic acid molecule (DNA or RNA). The four bases of each nucleic acid are arranged in various specific sequences.

Just as the order of the amino acid residues of protein side chains determines the primary structure of the protein (for example, —Ala—Gly—Glu—Met—), **the order of the bases** (for example, —ATTGAC—) **provides the primary structure of DNA.** As with proteins, we need a convention to tell us which end to start with when we are writing the sequence of bases. For nucleic acids the convention is to begin the sequence with the nucleotide that has the free 5′ —OH terminal. Thus, the sequence AGT means that adenine is the base at the 5′ terminal and thymine at the 3′ terminal.

The sequence TGA is not the same as AGT, but its opposite.

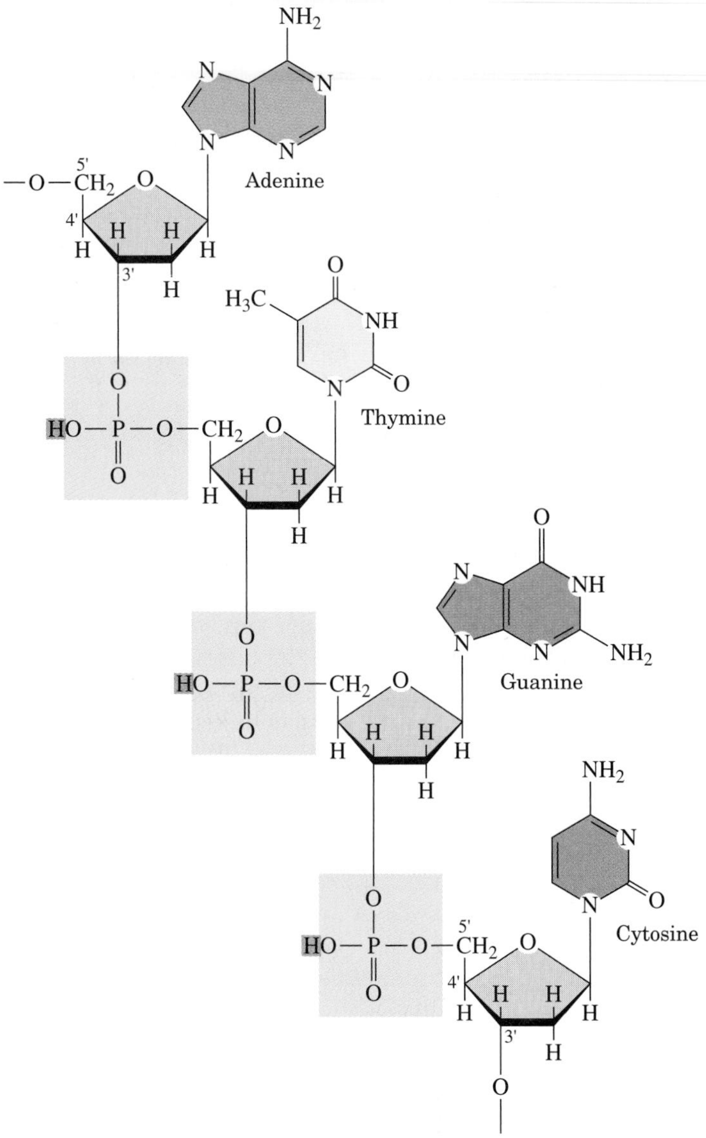

Figure **23.3** • Primary structure of the DNA backbone. The hydrogens shown in blue cause the acidity of nucleic acids. In the body, at neutral pH, the Hs are replaced by Na^+ and K^+.

Secondary Structure of DNA

Watson, Crick, and Wilkins were awarded the 1962 Nobel Prize in Medicine for their discovery. Franklin had died in 1958.

In 1953 James Watson (1928–) and Francis Crick (1916–) established the three-dimensional structure of DNA. Their work is a cornerstone in the history of biochemistry. The model of DNA established by Watson and Crick was based on two important pieces of information obtained by other workers: (1) the Chargaff rule that (A and T) and (G and C) are present in equimolar quantities and (2) x-ray diffraction photographs obtained by Rosalind Franklin (1920–1958) and Maurice

Watson and Crick with their model of the DNA molecule.

Vilkins (1916–). By the clever use of these facts, Watson and Crick oncluded that DNA is composed of *two* strands entwined around each ther in a double helix, as shown in Figure 23.4.

In the DNA double helix the two polynucleotide chains run in oppo-ite directions. This means that at each end of the double helix there is ne 5′ —OH and one 3′ —OH terminal. The sugar–phosphate backbone s on the outside, and the bases point inward. The bases are paired ac-ording to Chargaff's rule: For each adenine on one chain a thymine is ligned opposite it on the other chain; each guanine on one chain has a

Recall from Section 18.8 that a helix has a shape like a coiled spring or a spiral staircase. It was Pauling's discovery that human hair protein is helical that led Watson and Crick to look for helixes in DNA.

Rosalind Franklin (1920–1958). (Chemical Heritage Foundation.)

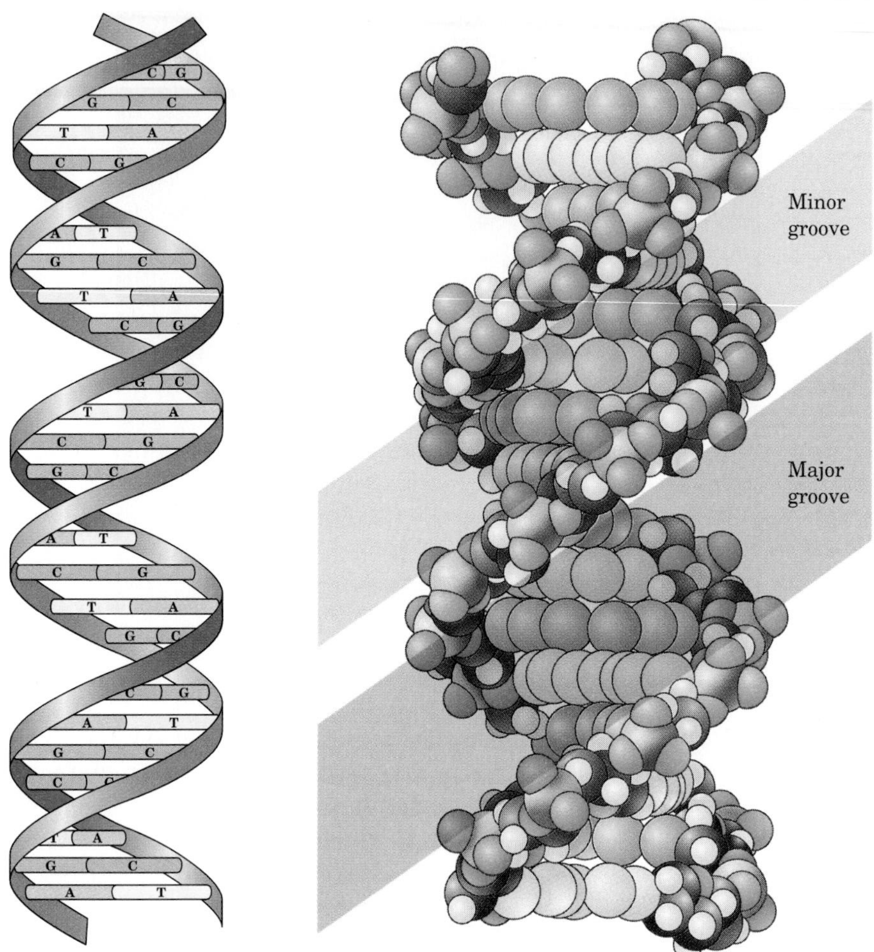

Figure **23.4** • Three-dimensional structure of the DNA double helix.

cytosine aligned with it on the other chain. **The bases so paired form hydrogen bonds with each other, thereby stabilizing the double helix** (Fig. 23.5). They are called **complementary base pairs.**

The important thing here, as Watson and Crick realized, is that *only* adenine could fit with thymine and *only* guanine with cytosine. Let us consider the other possibilities. Can two purines (AA, GG, or AG) fit opposite each other? Figure 23.6 shows that they would overlap. How about two pyrimidines (TT, CC, or CT)? As shown in Figure 23.6, they would be too far apart. *There must be a pyrimidine opposite a purine.* But could A fit opposite C or G opposite T? Figure 23.7 shows that the hydrogen bonding would be much weaker.

The entire action of DNA—and of the heredity mechanism—depends on the fact that, **wherever there is an adenine on one strand of the helix, there must be a thymine on the other strand because that is the only base that fits and forms strong hydrogen bonds, and similarly for G and C.** The entire heredity mechanism rests on these slender hydrogen bonds (Fig. 23.7), as we shall see in Section 23.4.

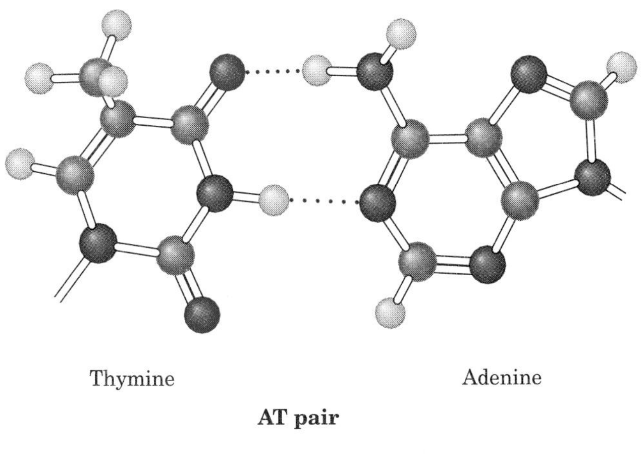

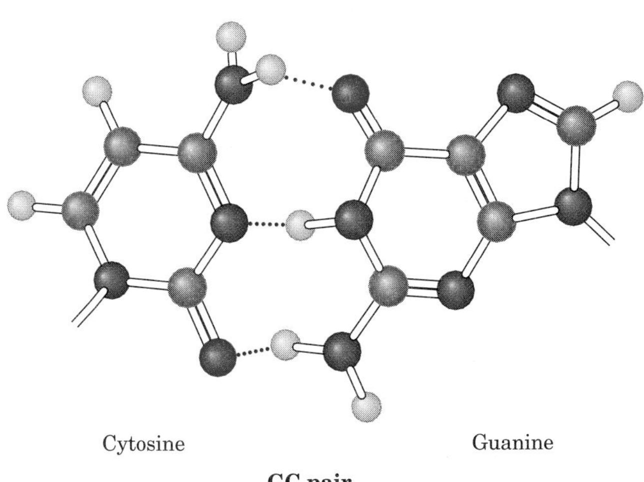

Figure **23.5** • A and T pair up by forming two hydrogen bonds; G and C, by forming three hydrogen bonds.

The beauty of establishing the three-dimensional structure of the DNA molecule was that the knowledge of this structure immediately led to the explanation for the transmission of heredity: how the genes transmit traits from one generation to another. Before we look at the mechanism of DNA replication (in the next section), let us summarize the three differences in structure between DNA and RNA:

1. DNA has the four bases A, G, C, and T. RNA has three of these—A, G, and C—but the fourth base is uracil, not thymine.

2. In DNA the sugar is deoxyribose. In RNA it is ribose.

3. DNA is almost always double-stranded, with the helical structure shown in Figure 23.4. There are several kinds of RNA (as we shall see in Sec. 23.5); all of them are single-stranded, though base-pairing can occur within a chain (see, for example, Fig. 23.9). When it does, adenine pairs with uracil, since thymine is not present.

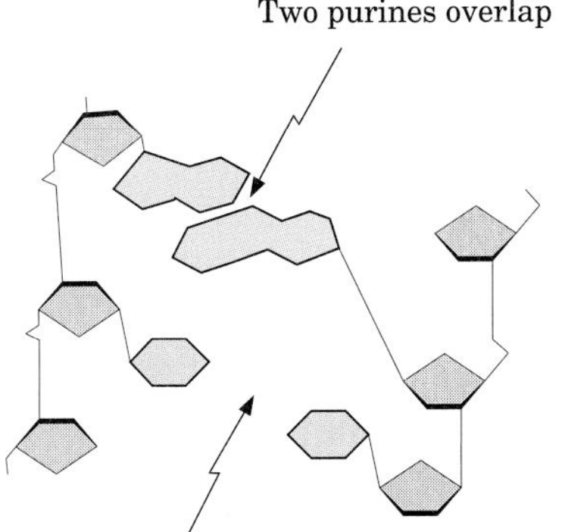

Figure **23.6**
The bases of DNA cannot stack properly in the double helix if a purine is opposite a purine or a pyrimidine is opposite a pyrimidine.

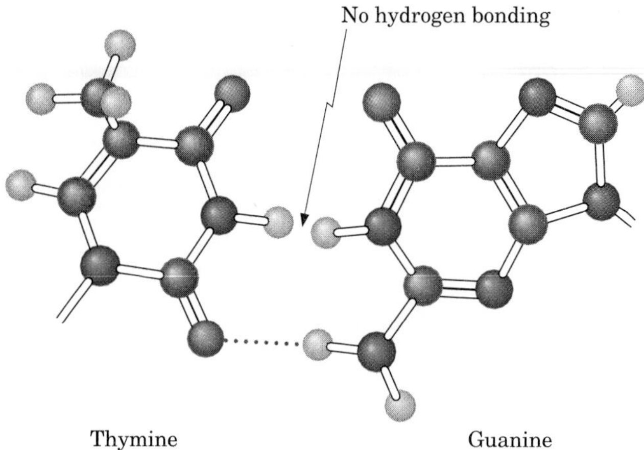

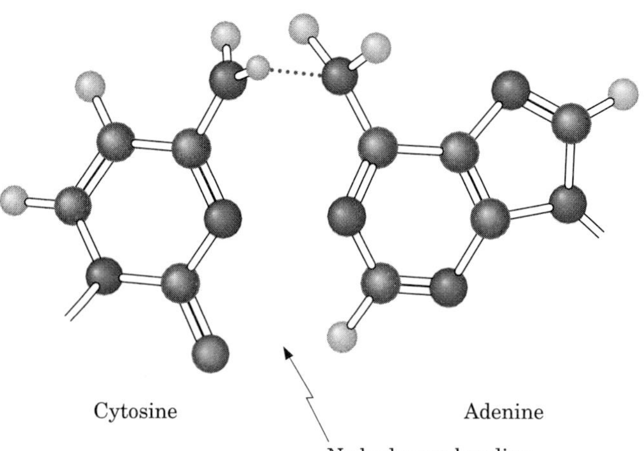

Figure **23.7** • Only one hydrogen bond is possible for GT or CA. These combinations are not found in DNA. Compare this with Figure 23.5.

23.4 DNA Replication

The DNA in the chromosomes carries out two functions: (1) It reproduces itself, and (2) it supplies the information necessary to make all the proteins in the body, including enzymes. The second function is covered in Sections 23.6 to 23.10. Here we are concerned with the first, **replication.**

Each gene is a section of a DNA molecule that contains a specific sequence of the four bases A, G, T, and C, typically containing about 1000 to 2000 nucleotides. The base sequence of the gene carries the information necessary to produce one protein molecule. If the sequence is changed (for example, if one A is replaced by a G, or if an extra T is inserted), a different protein is produced instead, which might mean that the individual would have brown eyes instead of blue or perhaps would not have some vital metabolic protein such as insulin.

The pigments responsible for the blue or brown color of eyes are synthesized with the help of specific enzymes. If one of these enzymes is lacking, the eye color may be different.

But consider the task that must be accomplished by the organism. When an individual is conceived, the egg and sperm cells unite to form the zygote. This cell, which is very tiny in most mammals, contains a small amount of DNA, but this DNA contains *all* the genetic information the individual will ever have. A fully grown large mammal, such as a human being or a horse, may contain more than a trillion cells. Each cell (except the egg and sperm cells) contains the same amount of DNA as the original single cell. Furthermore, cells are constantly dying and being replaced. Thus, there must be a mechanism by which DNA molecules can be copied (just as we can copy a letter on a photocopying machine) over and over again, millions of times, *without error.* In Section 23.13 we shall see that such errors sometimes do happen and can have serious consequences, but here we want to examine this remarkable mechanism that takes place every day in billions of organisms, from microbes to whales, and has been taking place for billions of years—with only a tiny percentage of errors.

The DNA double helix contains millions of bases. One DNA strand may carry many inheritable genes, each of which is a stretch of DNA a few hundred or thousand bases long. Genetic information is transmitted from one cell to the next when cell division occurs. The two new cells carry all the information that the original cell possessed. Where originally there was one set of DNA molecules, there will now be two sets of DNA molecules, one set in each new cell.

The replication of DNA molecules starts with the unwinding of the double helix. This can occur at either end or in the middle. Special molecules called **unwinding proteins** attach themselves to one DNA strand (Fig. 23.8) and cause the separation of the double helix. All four kinds of free DNA nucleotide molecules are present in the vicinity. These nu-

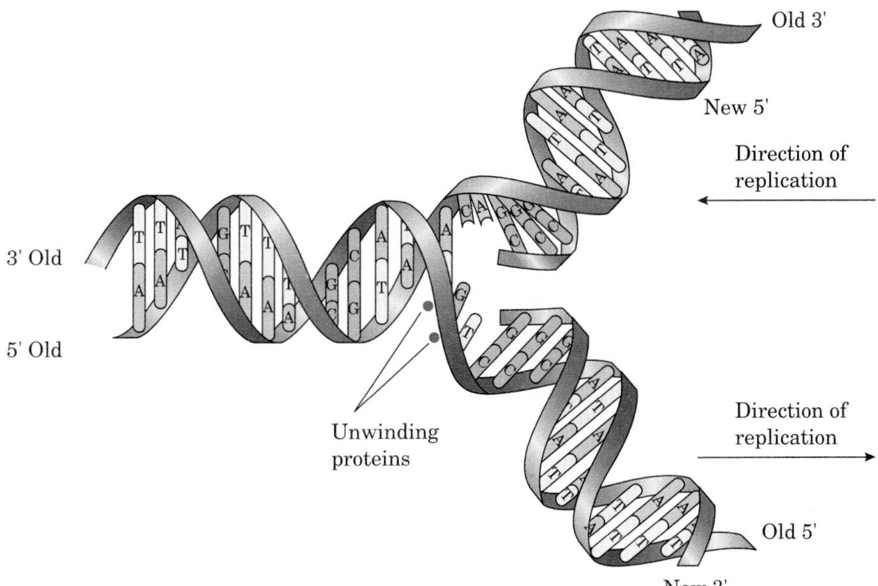

Figure **23.8**
The replication of DNA. The two strands of the DNA double helix are shown separating. (Modified from P. W. Davis and E. P. Solomon, *The World of Biology.* Philadelphia: Saunders College Publishing, 1986.)

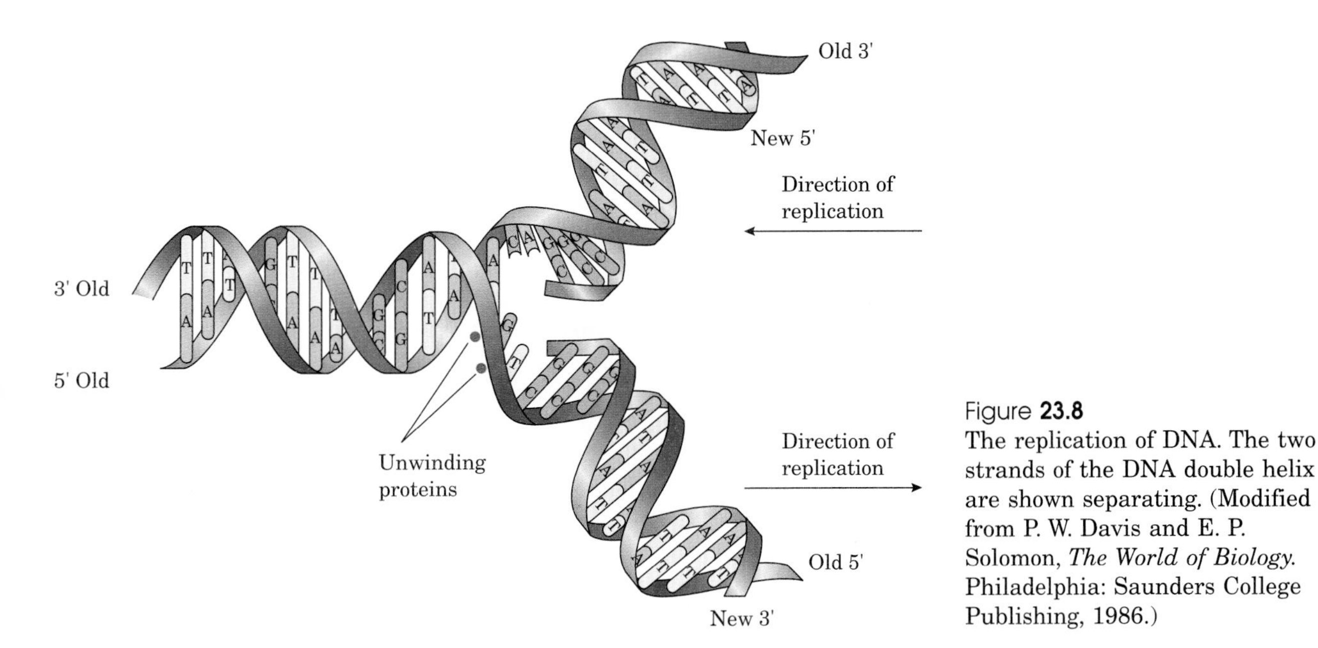

cleotides constantly move into the area and try to fit themselves into new chains. The key to the process is that, as we saw in Section 23.3, **only thymine can fit opposite adenine and only cytosine opposite guanine.** Wherever a cytosine, for example, is present on one of the strands of an unwound portion of the helix, all four nucleotides may approach, but three of them are turned away because they do not fit. Only the nucleotide of guanine fits.

While the bases of the newly arrived nucleotides are being hydrogen bonded to their partners, enzymes called polymerases join the nucleotide backbones. At the end of the process, there are two double-stranded DNA molecules, each exactly the same as the original one because only T fits opposite A and only G against C. The process is called **semiconservative** because only two of the four strands are new; the other two were present in the original molecule.

The base sequence of each newly synthesized DNA chain is **complementary** to the chain already there.

If the unwinding begins in the middle, the synthesis of new DNA molecules on the old templates continues in both directions until the whole molecule is duplicated. This is the more common pathway. The unwinding can also start at one end and proceed in one direction until the whole double helix is unwound.

An interesting detail of DNA replication is that the two daughter strands are synthesized in different ways. One of the syntheses is continuous in the $5' \rightarrow 3'$ direction, and the enzymes involved are capable of linking millions of phosphate ester linkages continuously in this direction. However, since the strands in the double helix run in opposite directions, only one strand of the double helix runs $5' \rightarrow 3'$; the other runs $3' \rightarrow 5'$. Along this second strand there is no possibility of continuous synthesis. What happens is that along the $3' \rightarrow 5'$ strand, the enzymes can synthesize only short fragments because the only way they can work is from $5'$ to $3'$. These short fragments consist of about 1000 nucleotides each and are called **Okazaki fragments,** after their discoverer. They are eventually joined together by the enzyme DNA ligase. The two newly formed strands run in opposite directions. The new strand synthesized continuously is called the **leading strand;** the one assembled from Okazaki fragments is called the **lagging strand.**

23.5 RNA

We previously noted that there are three types of RNA.

1. Messenger RNA (mRNA) carries the genetic information from the DNA in the nucleus directly to the cytoplasm, where the protein is synthesized. It consists of a chain of nucleotides whose sequence is exactly complementary to that of one of the strands of the DNA. This type of RNA is not very stable. It is synthesized as needed and then degraded. Thus its concentration at any time is rather low.

In DNA–RNA interactions, the complementary bases are

DNA	RNA
A—	U
G—	C
C—	G
T—	A

2. Transfer RNA (tRNA) are relatively small molecules, containing from 73 to 93 nucleotides per chain. There is at least one different tRNA molecule for each of the 20 amino acids from which the body makes its proteins. The tRNA molecules are L-shaped, but they can be repre-

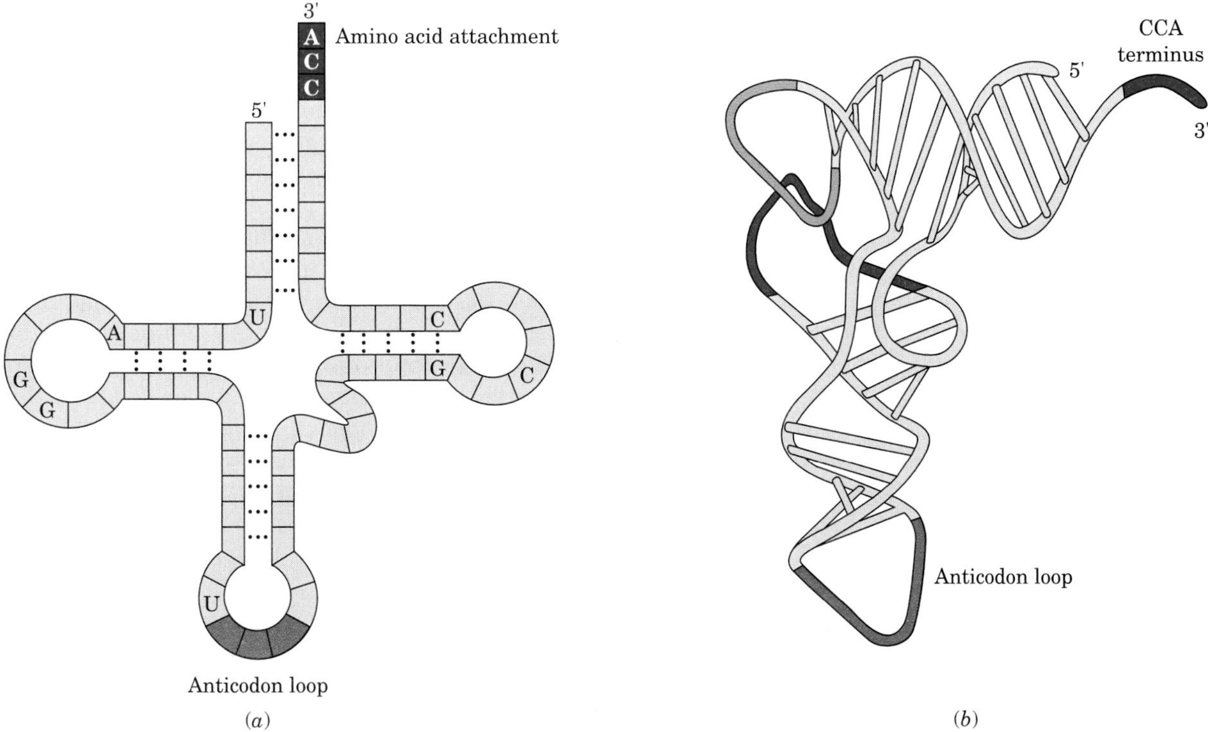

Figure **23.9** • Structure of tRNA. (a) Two-dimensional simplified cloverleaf structure. (b) Three-dimensional structure. (From *Biochemistry,* 2nd ed., by Lubert Stryer. Copyright 1981 by W. H. Freeman and Co. All rights reserved. Part (b) also courtesy of Dr. Sung-Hou Kim.)

sented as a cloverleaf in two dimensions. A typical one is shown in Figure 23.9. Transfer RNA molecules contain not only cytosine, guanine, adenine, and uracil but also several other modified nucleotides.

3. Ribosomal RNA (rRNA) is found in the **ribosomes,** which are small spherical bodies located in the cells but outside the nuclei. They consist of about 35 percent protein and 65 percent of a type of RNA called ribosomal RNA. These are large molecules with molecular weights up to 1 million. As we shall see in Section 23.10, protein synthesis takes place on the ribosomes.

23.6 Transmission of Information

We have seen that the DNA molecule is a storehouse of information. We can compare it to a loose-leaf cookbook, each page of which contains one recipe. The pages are the genes. In order to prepare a meal, we use a number of recipes. Similarly, to provide a certain inheritable trait, a number of genes, segments of DNA, are needed.

However, the recipe itself is not the meal. The information in the recipe must be expressed in the proper combination of food ingredients. Similarly, the information stored in DNA must be expressed in the proper combination of amino acids representing a particular protein. The way this works is now so well established that it is called the **central dogma** of molecular biology. The dogma states that the **information contained in DNA molecules is transferred to RNA molecules, and then from the RNA molecules the information is expressed in the structure of proteins.** Two steps are involved.

Although this statement is correct in the vast majority of cases, in certain viruses the flow of information goes from RNA to DNA.

1. Transcription Since the information (that is, the DNA) is in the nucleus of the cell and the amino acids are assembled outside the nucleus, the information must first be carried out of the nucleus. This is analogous to copying the recipe from the cookbook. All the necessary information is copied, though in a slightly different format, as if we were converting the printed page into handwriting. On the molecular level this is accomplished by transcribing the information from the DNA molecule onto a molecule of messenger RNA, so named because it carries the message from the nucleus to the site of protein synthesis. The transcribed information on the mRNA molecule is then carried out of the nucleus.

2. Translation The mRNA serves as a template on which the amino acids are assembled in the proper sequence. In order for this to happen, the information that is written in the language of nucleotides must be translated into the language of amino acids. The translation is done by the second type of RNA, transfer RNA. There is an exact word-to-word translation. Each amino acid in the protein language has a corresponding word in the RNA language. Each word in the RNA language is a sequence of three bases. This correspondence between three bases and one amino acid is called the **genetic code** (we will discuss the code in Sec. 23.9).

A summary of the process is

$$\text{DNA} \xrightarrow{\text{replication}} \text{DNA} \xrightarrow{\text{transcription}} \text{mRNA} \xrightarrow{\text{translation}} \text{protein}$$

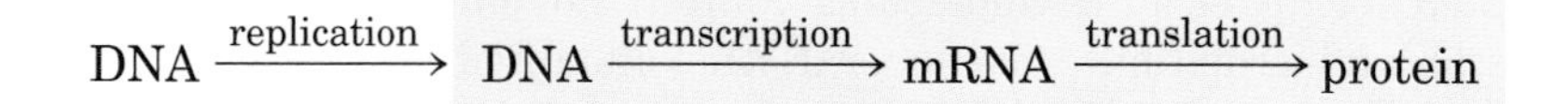

23.7 Transcription

The copying of the information (the recipe from the cookbook) is done with the help of an enzyme called RNA polymerase, which catalyzes the synthesis of mRNA. First the DNA double helix begins to unwind at a point near the gene that is to be transcribed. Only one strand of the DNA molecule is transcribed. Ribonucleotides assemble along the unwound DNA strand in the complementary sequence. Opposite each C on the DNA there is a G on the growing mRNA, and the other complementary bases follow the patterns G $\rightarrow$ C, A $\rightarrow$ U, and T $\rightarrow$ A.

These are not the deoxyribonucleotides used in DNA replication.

Note again that RNA contains no thymine but has uracil instead.

On the DNA strand, there is always a sequence of bases that the RNA polymerase recognizes as an **initiation signal,** saying, in essence, "Start here." At the end of the gene, there is a **termination sequence** that tells the enzyme, "Stop the synthesis." Between these two signals, the enzyme zips up the complementary bases by combining each ribose to the next phosphate. The enzyme synthesizes the mRNA molecule from

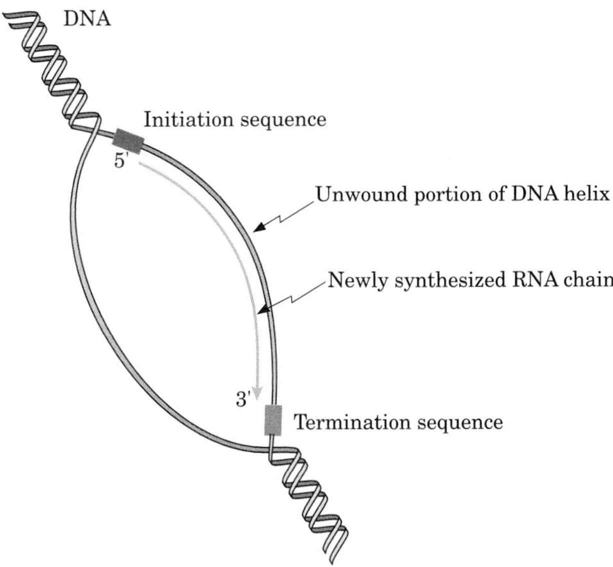

Figure **23.10** • Transcription of a gene. The information in one DNA strand is transcribed to a strand of RNA.

the 5′ to the 3′ end (the zipper can move only in one direction). But because the complementary chains (RNA and DNA) run in opposite directions, the enzyme must move along the DNA template in the $3' \rightarrow 5'$ direction of the DNA (Fig. 23.10). Once the mRNA molecule has been synthesized, it moves away from the DNA template, which then rewinds to the original double-helix form.

After RNA molecules are synthesized, they move out of the nucleus and into the cytoplasm.

Transfer RNA and ribosomal RNA are also synthesized in this manner, on DNA templates.

23.8 The Role of RNA in Translation

Translation is the process by which the genetic information preserved in the DNA and transcribed into the mRNA is converted to the language of proteins, that is, the amino acid sequence. All three types of RNA participate in the process.

The synthesis of proteins takes place on the ribosomes (Sec. 23.5). These spheres dissociate into two parts—a larger and a smaller body. Each of these bodies contains rRNA and some polypeptide chains that act as enzymes, speeding up the synthesis. In higher organisms, including humans, the larger ribosomal fragment is called the 60S ribosome and the smaller one the 40S ribosome. As the mRNA is being made on the DNA template (Sec. 23.7), the 5′ end of the mRNA, coming off this assembly, is first attached to the smaller ribosomal body and later joined by the larger body. Together they form a unit on which the mRNA is stretched out. Once the mRNA is attached to the ribosome in this way, the 20 amino acids are brought to the site, each carried by its own particular tRNA molecule.

The S, or Svedberg unit, is a measure of the size of these bodies.

The 20 amino acids are always available in the cytoplasm, near the site of protein synthesis.

The most important segments of the tRNA molecule are (1) the site to which enzymes attach the amino acids and (2) the recognition site. Figure 23.9 shows that the 3′ terminal of the tRNA molecule carries the amino acid. As we have said, each tRNA is specific for one amino acid only. How does the body make sure that, say, alanine attaches only to the one tRNA molecule that is specific for alanine? The answer is that each cell carries at least 20 specific enzymes for this purpose. Each of these enzymes recognizes only one amino acid and only one tRNA. The enzyme attaches the activated amino acid to the 3′ terminal of the tRNA.

The second important segment of the tRNA molecule carries the *recognition site,* which is a sequence of three bases called an **anticodon** (Fig. 23.9), which is located at the opposite end of the molecule in the three-dimensional structure of tRNA. This triplet of bases can align itself in a complementary fashion to another triplet on mRNA. The triplets of bases on the mRNA are called **codons.**

23.9 The Genetic Code

Marshall Nirenberg. (Courtesy of National Institutes of Health.)

By 1961 it was apparent that the order of bases in a DNA molecule corresponds to the order of amino acids in a particular protein. But the code was unknown. Obviously, it could not be a one-to-one code. There are only four bases, so if, say, A coded for glycine, G for alanine, C for valine, and T for serine, there would be 16 amino acids that could not be coded.

In 1961 Marshall Nirenberg (1927–   ) and his coworkers attempted to break the code in a very ingenious way. They made a synthetic molecule of mRNA consisting of uracil bases only. They put this into a cellular system that synthesized proteins and then supplied the system with all 20 amino acids. The only polypeptide produced was a chain consisting solely of the amino acid phenylalanine. This showed that the code for phenylalanine must be UUU or some other multiple of U.

A series of similar experiments by Nirenberg and other workers followed, and by 1967 the entire genetic code had been broken. **Each amino acid is coded for by a sequence of three bases,** called a **codon.** The complete code is shown in Table 23.2.

A very few exceptions to the genetic code in Table 23.2 occur in mitochondrial RNA.

The first important aspect of **the genetic code** is that it is almost **universal.** In virtually every organism, from a bacterium to an elephant to a human, the same sequence of three bases codes for the same amino acid. The universality of the genetic code implies that all living matter on earth arose from the same primordial organisms. This is perhaps the strongest evidence for Darwin's theory of evolution.

Because of this multiple coding, the genetic code is called a **multiple** code or a **degenerate** code.

There are 20 amino acids in proteins, but there are 64 possible combinations of four bases into triplets. All 64 codons (triplets) have been deciphered. Three of them, UAA, UAG, and UGA, are stop signs. They terminate protein synthesis. The remaining 61 codons all code for amino acids. Since there are only 20 amino acids, there must be more than one codon for each amino acid. Indeed, some amino acids are coded for by as

Table 23.2 The Genetic Code

UUU	Phe	UCU	Ser	UAU	Tyr	UGU	Cys
UUC	Phe	UCC	Ser	UAC	Tyr	UGC	Cys
UUA	Leu	UCA	Ser	UAA	END[a]	UGA	END[a]
UUG	Leu	UCG	Ser	UAG	END[a]	UGG	Trp
CUU	Leu	CCU	Pro	CAU	His	CGU	Arg
CUC	Leu	CCC	Pro	CAC	His	CGC	Arg
CUA	Leu	CCA	Pro	CAA	Gln	CGA	Arg
CUG	Leu	CCG	Pro	CAG	Gln	CGG	Arg
AUU	Ile	ACU	Thr	AAU	Asn	AGU	Ser
AUC	Ile	ACC	Thr	AAC	Asn	AGC	Ser
AUA	Ile	ACA	Thr	AAA	Lys	AGA	Arg
AUG[b]	Met	ACG	Thr	AAG	Lys	AGG	Arg
GUU	Val	GCU	Ala	GAU	Asp	GGU	Gly
GUC	Val	GCC	Ala	GAC	Asp	GGC	Gly
GUA	Val	GCA	Ala	GAA	Glu	GGA	Gly
GUG	Val	GCG	Ala	GAG	Glu	GGG	Gly

[a] END refers to signals indicating chain endings.
[b] This codon also signals the beginning of the chain.

many as six codons. Alanine, for example, is coded for by four: GCU, GCC, GCA, and GCG.

As there are three stop signs in the code, there is also an initiation sign. The initiation sign is AUG, which is also the codon for the amino acid methionine. This means that, in all protein synthesis, the first amino acid is always methionine. Methionine can also be put into the middle of the chain because there are two kinds of tRNA for it.

Although all protein synthesis starts with methionine, most proteins in the body do not have a methionine residue at the beginning of the chain. In most cases the initial methionine is removed by an enzyme before the polypeptide chain is completed. The code on the mRNA is always read in the $5' \rightarrow 3'$ direction, and the first amino acid to be linked to the initial methionine is the N-terminal end of the translated polypeptide chain.

23.10 Translation and Protein Synthesis

So far we have met the molecules that participate in protein synthesis (Sec. 23.8) and the dictionary of the translation, the genetic code. Now let us look at the actual mechanism by which the polypeptide chain is assembled.

There are four major stages in protein synthesis: activation, initiation, elongation, and termination.

All protein synthesis takes place outside the nucleus, in the cytosol.

In the activation step energy is used up: Two high-energy phosphate bonds are broken for each amino acid added to the chain: ATP $\rightarrow$ AMP + PP_i and $PP_i \rightarrow 2P_i$.

Activation

Each amino acid is first activated by reacting with a molecule of ATP:

$$\underset{\text{ATP}}{\text{Adenosine}{-}O{-}\overset{\overset{\displaystyle O}{\|}}{\underset{\underset{\displaystyle O^-}{|}}{P}}{-}O{-}\overset{\overset{\displaystyle O}{\|}}{\underset{\underset{\displaystyle O^-}{|}}{P}}{-}O{-}\overset{\overset{\displaystyle O}{\|}}{\underset{\underset{\displaystyle O^-}{|}}{P}}{-}O^-} + \underset{\text{Amino acid}}{H\overset{\overset{\displaystyle R}{|}}{\underset{\underset{\displaystyle NH_2}{|}}{C}}{-}COOH} \longrightarrow$$

$$\underset{\text{Amino acid-AMP}}{\text{Adenosine}{-}O{-}\overset{\overset{\displaystyle O}{\|}}{\underset{\underset{\displaystyle O^-}{|}}{P}}{-}O{-}\overset{\overset{\displaystyle O}{\|}}{C}{-}\overset{\overset{\displaystyle R}{|}}{\underset{\underset{\displaystyle NH_2}{|}}{C}}H} + \underset{\text{Pyrophosphate}}{{}^-O{-}\overset{\overset{\displaystyle O}{\|}}{\underset{\underset{\displaystyle O^-}{|}}{P}}{-}O{-}\overset{\overset{\displaystyle O}{\|}}{\underset{\underset{\displaystyle O^-}{|}}{P}}{-}O^-}$$

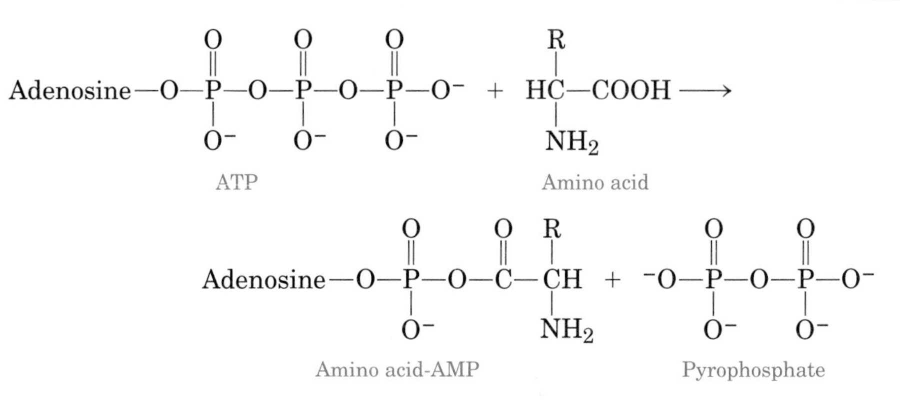

The activated amino acid is then attached to its own particular tRNA molecule with the aid of an enzyme (a synthetase) that is specific for that particular amino acid and that particular tRNA molecule:

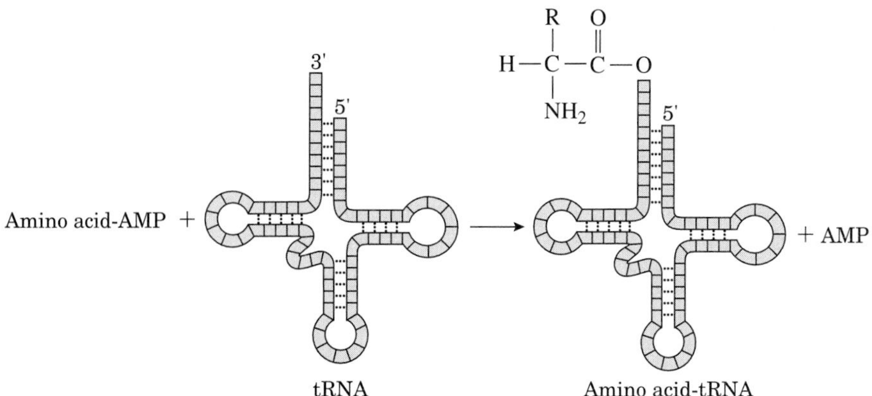

Initiation

This stage consists of three steps: (a) The mRNA molecule, which carries the information necessary to synthesize one protein molecule, attaches itself to the 40S ribosome (the smaller body; see Sec. 23.8). This is shown in Figure 23.11(a). (b) The anticodon of the first tRNA molecule (which is always a methionine tRNA) binds to the codon of the mRNA that represents the initiation signal. (c) The 60S ribosome (the larger portion) now combines with the 40S body, as shown in Figure 23.11(c). The 60S body carries two binding sites. The one shown on the left in Figure 23.11(c) is called the **P site** because that is where the growing peptide chain will bind. The one right next to it is called the **A site** because that is where the incoming tRNA will bring the next amino acid. When the 60S ribosome attaches itself to the 40S one, it does so in such a way that the P site is right where the methionine tRNA already is.

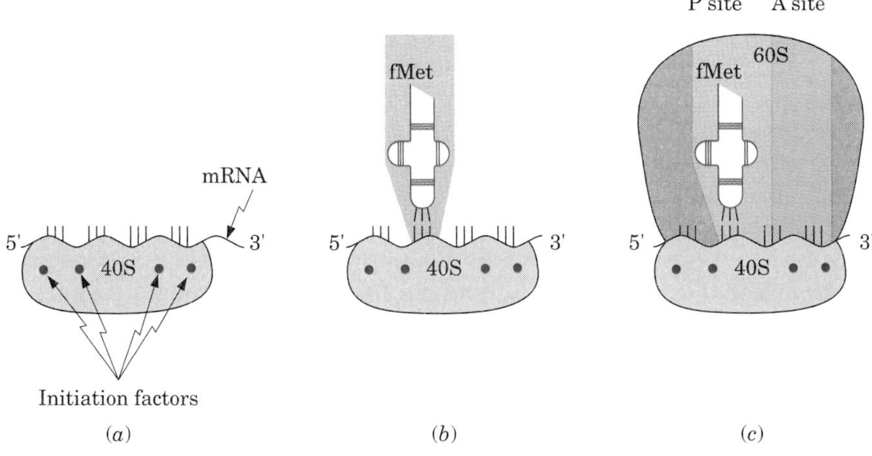

Figure **23.11** • The initiation of protein synthesis. (a) The mRNA attaches to the 40S ribosomal body. (b) The first tRNA anticodon binds to the initiation mRNA codon. (c) The 60S ribosomal body joins the unit.

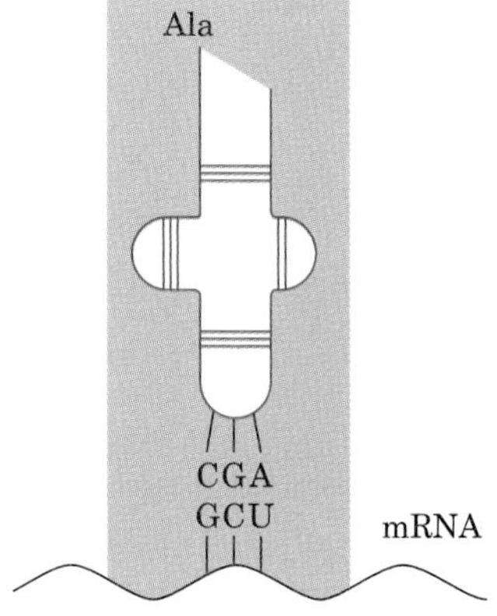

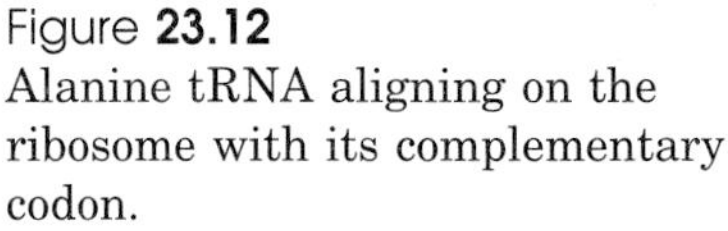

Figure **23.12**
Alanine tRNA aligning on the ribosome with its complementary codon.

Elongation

At this point the A site is vacant, and each of the 20 tRNA molecules can come in and try to fit itself in. *But only one of the 20 carries exactly the right anticodon that corresponds to the next codon on the mRNA.* (In Figures 23.12 and 23.13 we have made this an alanine tRNA.) The binding of this tRNA to the A site takes place with the aid of proteins called **elongation factors.** Energy for the process is obtained by hydrolyzing a guanosine triphosphate (GTP) to GDP and inorganic phosphate. Once at the A site, the new amino acid (Ala) is linked to the Met in a peptide bond by the enzyme **transferase.** The empty tRNA remains on the P site.

This enzyme, which is part of the 60S ribosome unit, is not a protein but a ribozyme.

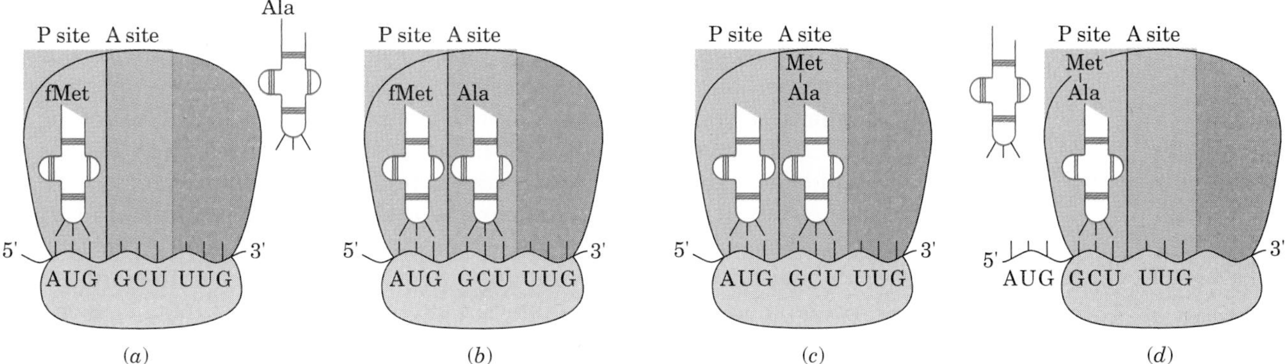

Figure **23.13** • Phases of elongation in protein synthesis. (a) After initiation, the tRNA of the second amino acid (alanine in this case) approaches the ribosome. (b) The tRNA binds to the A site. (c) The enzyme transferase connects methionine to alanine, forming a peptide bond. (d) The peptide tRNA is moved (translocated) from the A to the P site while the ribosome moves to the right and simultaneously releases the empty tRNA. In this diagram, the alanine tRNA is shown in blue.

In the next phase of elongation, the whole ribosome moves one codon along the mRNA. Simultaneously with this move, the dipeptide is **translocated** from the A site to the P site, as shown in Figure 23.13(d), while the empty tRNA dissociates and goes back to the tRNA pool to pick up another amino acid. After the translocation, the A site is associated with the next codon on the mRNA, which is UUG in Figure 23.13(d). Once again, each tRNA can try to fit itself in, but only the one whose anticodon is AAC can align itself with UUG. This one, the tRNA that carries Leu, now comes in. The transferase now establishes a new peptide linkage between Leu and Ala, moving the dipeptide from the P site to the A site and forming a tripeptide. These elongation steps are repeated until the last amino acid is attached.

Termination

After the last translocation, the next codon reads "STOP" (UAA, UGA, or UAG). No more amino acids can be added. Releasing factors then cleave the polypeptide chain from the last tRNA in a mechanism not yet fully understood. The tRNA itself is released from the P site. While the mRNA is attached to the ribosomes, many polypeptide chains are synthesized on it simultaneously. At the end, the whole mRNA is released from the ribosome.

Their work done, the two parts of the ribosome separate.

23.11 Genes, Exons, and Introns

A **gene** is a stretch of DNA that carries one particular message; for example, "make a globin molecule." In bacteria this message is continuous. The series of codons (triplets of bases) spell out which amino acids must be assembled to make the globin molecule, and in what sequence. This series of codons lies between an initiation signal and a termination signal.

In higher organisms the message is not continuous. Stretches of DNA that spell out the amino acid sequence to be assembled are interrupted by long stretches that seemingly do not code for anything. The coding sequences are called **exons** and the noncoding sequences, **introns.** For example, the globin gene has three exons broken up by two introns. Since DNA contains exons and introns, the mRNA transcribed from it also contains exons and introns. The introns are cut out by enzymes (and the exons spliced together) before the mRNA is actually used to synthesize a protein (Fig. 23.14). In other words, the introns function as spacers. In-

Philip A. Sharp (1944–   ) and Richard J. Roberts (1943–   ) were awarded the 1993 Nobel Prize in Medicine for the discovery of the noncoding nature of introns.

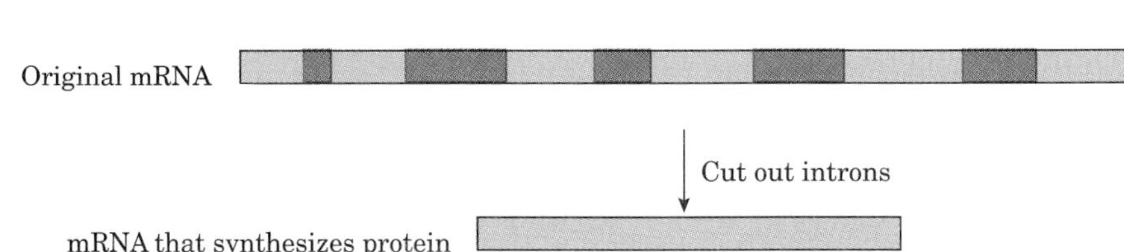

Figure **23.14** • Introns are cut out of mRNA before the protein is synthesized.

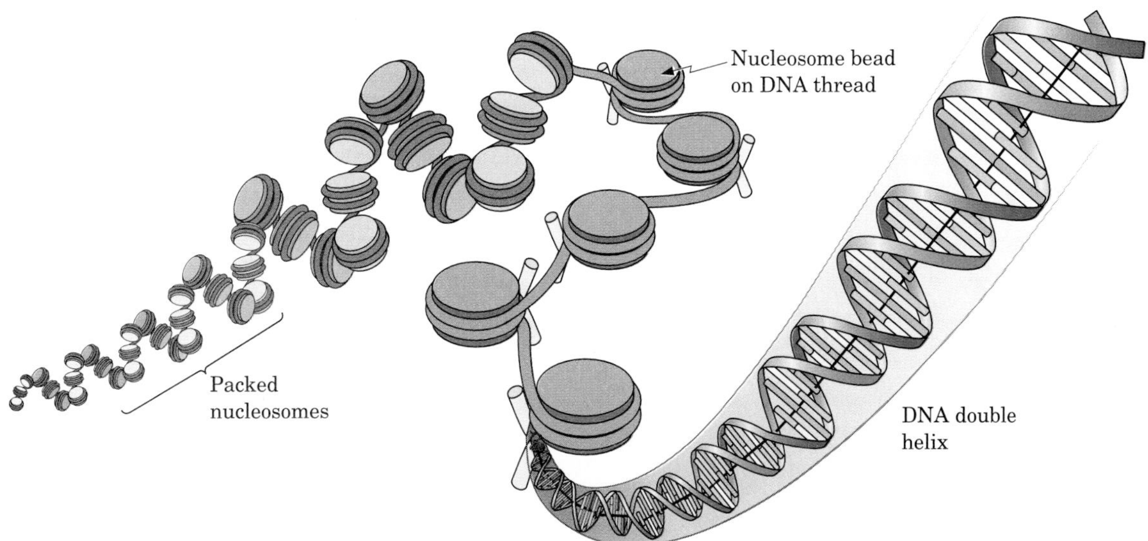

Figure **23.15** • Schematic diagram of a nucleosome. The band-like DNA double helix winds around cores consisting of eight histones.

trons can also function as enzymes catalyzing the splicing of exons into "mature mRNA" (that is, with no intervening sequences). They can also catalyze the splicing of tRNAs from larger precursor molecules.

In humans only 3% of the DNA codes for proteins or RNA with clear functions. Introns are not the only noncoding DNA sequences. Other DNAs, in which short nucleotide sequences are repeated hundreds or thousands of times, are called **satellites.** Large satellite stretches appear at the ends and centers of chromosomes and are necessary for the stability of the chromosomes. Smaller repetitive sequences, **mini-** or **micro-satellites,** when they mutate, are associated with cancer.

One DNA molecule may have between 1 million and 100 million bases. Therefore, there are many genes in one DNA molecule. If a DNA molecule were fully stretched out, its length would be perhaps 1 cm. However, the DNA molecules in the nuclei are not stretched out. They are coiled around basic protein molecules called **histones.** The acidic DNA and the basic histones attract each other by electrostatic (ionic) forces. The DNA and histones combine to form units called **nucleosomes.** A nucleosome is a core of eight histone molecules around which the DNA double helix is wrapped (Fig. 23.15).

Cells of an organism die in large clusters when exposed to toxins or deprived of oxygen. Such a typical necrosis occurs in heart attack and stroke. **Apoptosis** or programmed death of cells is clearly distinguishable from necrosis. Only scattered cells die at a time and no inflammation or scar results. In apoptosis, the cell cleaves into apoptotic bodies containing intact organelles and large nuclear fragments. These bodies are swallowed up by adjacent cells. The distinguishing fingerprint of death by apoptosis is the cleavage of the nucleosomes and the appearance of DNA fragments of multiples of 180 base pairs: 180; 360; or 540 base pairs.

23.12 Gene Regulation

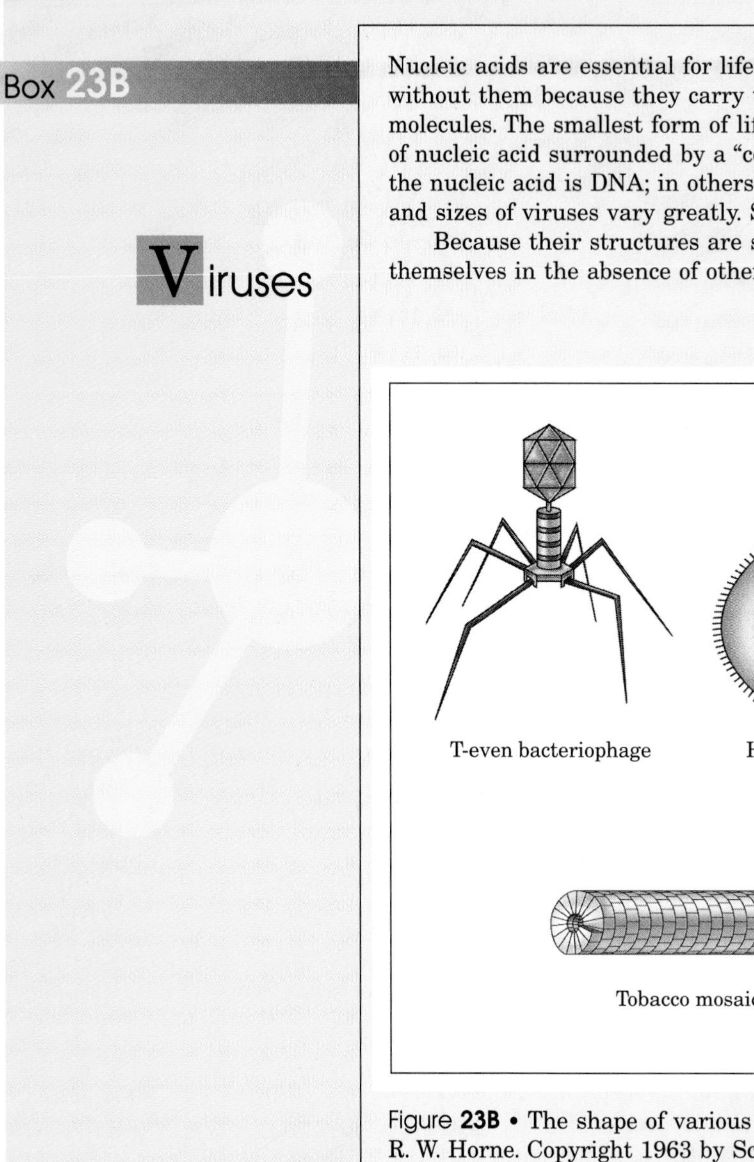

Every embryo that is formed by sexual reproduction inherits its genes from the parent sperm and egg cells. But the genes in its chromosomal DNA are not active all the time. They are switched on and off during development and growth of the organism. Soon after formation of the embryo, the cells begin to differentiate. Some cells become neurons,

Box 23B

Viruses

Nucleic acids are essential for life as we know it. No living thing can exist without them because they carry the information necessary to make protein molecules. The smallest form of life, the viruses, consist only of a molecule of nucleic acid surrounded by a "coat" of protein molecules. In some viruses the nucleic acid is DNA; in others it is RNA. No virus has both. The shapes and sizes of viruses vary greatly. Some of them are shown in Figure 23B.

Because their structures are so simple, viruses are unable to reproduce themselves in the absence of other organisms. They carry DNA or RNA but

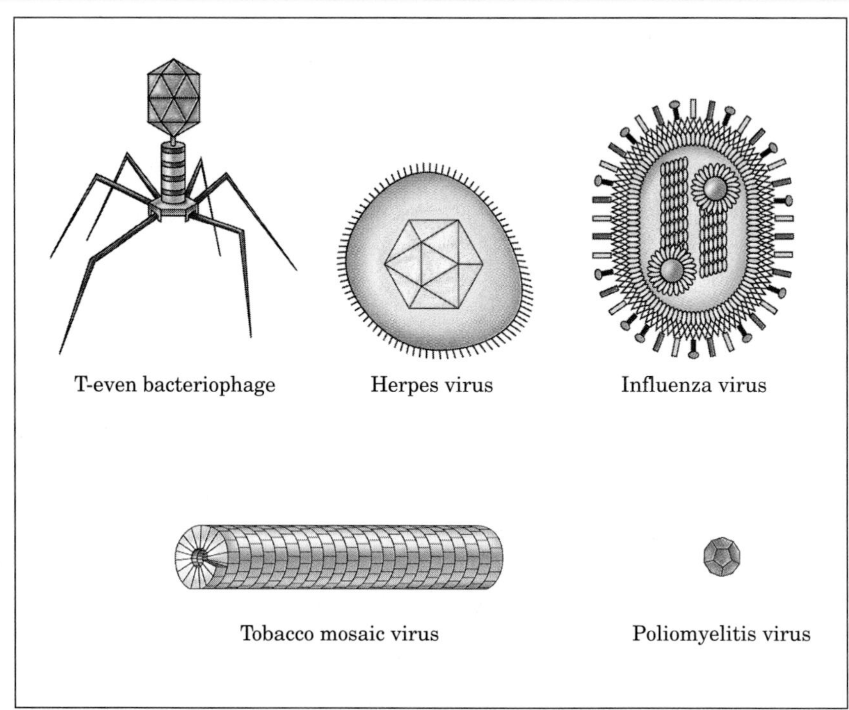

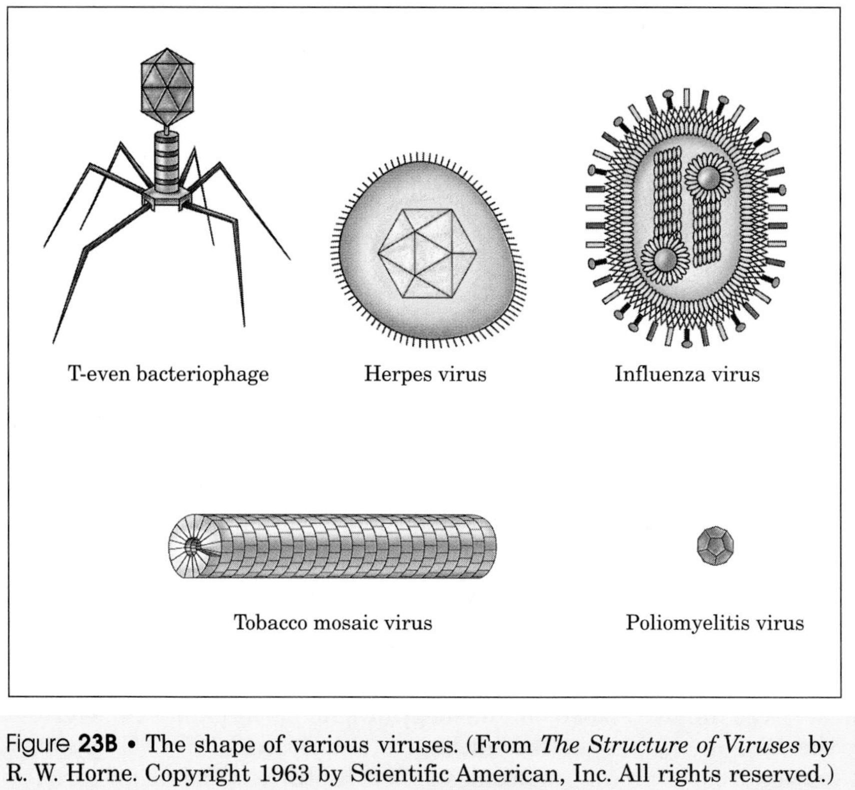

Figure **23B** • The shape of various viruses. (From *The Structure of Viruses* by R. W. Horne. Copyright 1963 by Scientific American, Inc. All rights reserved.)

The turning on, or activation, of a gene is called **gene expression.**

Prokaryotes are simple unicellular organisms with no true nucleus and no organelles surrounded by membranes (example: bacteria). **Eukaryotes** are organisms with cells that have visibly evident nuclei, each surrounded by a membrane, and also have organelles.

some cells become muscle cells, some liver cells, and so on. Each cell is a specialized unit that uses only some of the many genes it carries in its DNA. This means that each cell must switch some of its genes on and off—either permanently or temporarily. How this is done is the subject of **gene regulation.** We know less about gene regulation in eukaryotes than in the simpler prokaryotes. Even with our limited knowledge, however, we can state that organisms do not have a single, unique way of controlling genes. Many gene regulations occur at the transcriptional level (DNA → RNA). Some operate at the translational level (mRNA → protein). Among these ways are the following:

do not have the nucleotides, enzymes, amino acids, and other molecules necessary to replicate their nucleic acid (Sec. 23.4) or to synthesize proteins (Sec. 23.10). Instead, viruses invade the cells of other organisms and cause those cells (the hosts) to do these tasks for them. Typically, the protein coat of a virus remains outside the host cell, attached to the cell wall, while the DNA or RNA is pushed inside. Once the viral nucleic acid is inside the cell, the cell stops replicating its own DNA and making its own proteins and now replicates the viral nucleic acid and synthesizes the viral protein, according to the instructions on the viral nucleic acid. One host cell can make many copies of the virus.

In many cases, the cell bursts when a large number of new viruses have been synthesized, sending the new viruses out into the intercellular material, where they can infect other cells. This kind of process causes the host organisms to get sick, perhaps to die. Among the many human diseases caused by viruses are measles, hepatitis, mumps, influenza, the common cold, rabies, and smallpox. There is no cure for most viral diseases. Antibiotics, which can kill bacteria, have no effect on viruses. So far, the best defense against these diseases has been immunization (Box 25F), which under the proper circumstances can work spectacularly well. Smallpox, once one of the most dreaded diseases, has been totally eradicated from this planet by many years of vaccination, and comprehensive programs of vaccination against such diseases as polio and measles have greatly reduced the incidence of these diseases.

Lately, a number of antiviral agents have been developed that completely stop the reproduction of viral nucleic acids (DNA or RNA) inside infected cells without preventing the DNA of normal cells from replicating. One such drug is called vidarabine, or Ara-A, and is sold under the trade name Vira-A. Antiviral agents often act like the anticancer drugs (Box 23A), in that they have structures similar to one of the nucleotides necessary for the synthesis of nucleic acids. Vidarabine is the same as adenosine, except that the sugar is arabinose instead of ribose. Vidarabine is used to fight a life-threatening viral illness, herpes encephalitis. It is also effective in neonatal herpes infection and chicken pox. But as with many other anticancer and antiviral drugs, vidarabine is toxic, causing nausea and diarrhea. In some cases it has caused chromosomal damage.

Many other antiviral drugs are now in clinical trials. It is hoped that ways will be found to reduce their toxicity and improve their specificity so that they will act against one particular virus only.

1. Operon, a case of negative regulation This process operates only in prokaryotes. In the common species of bacteria *Escherichia coli,* the genes for a number of enzymes are organized in a unit. This unit contains DNA sequences that code for proteins and also sequences that play a part in gene regulation. Sequences that code for proteins are called **structural genes.** Preceding the structural genes is a DNA sequence called the **control sites,** made up of two parts called **operator site** and **promoter site.** Preceding the control sites is a DNA sequence called the

Since RNA synthesis proceeds in one direction ($5' \rightarrow 3'$) (see Fig. 23.11), the gene (DNA) to be transcribed runs from $3' \rightarrow 5'$. Thus the control sites are in front of, or upstream of, the $3'$ end of the structural gene.

Box 23C

AIDS

In recent years a deadly virus, commonly called the AIDS virus (for Acquired Immune Deficiency Syndrome), or HIV (for Human Immunodeficiency Virus), has spread alarmingly all over the world. The AIDS virus invades the human immune system, and especially enters the T (lymphocyte) cells and kills them. The AIDS virus, an RNA-containing or **retrovirus,** decreases the population of T cells (blood cells that fight invading foreign cells; see Sec. 25.10), and thus allows other opportunistic invaders, such as the protozoan *Pneumocystis carinii,* to proliferate, causing pneumonia and eventual death.

The progressive paralysis of the immune system makes the AIDS virus a lethal threat. Those who are ill with AIDS or have been exposed to the virus carry antibodies against the virus in their blood. One can detect these antibodies through patented tests, and all blood banks are now testing donated blood for AIDS infection. In the United States currently two to three out of 10 000 blood samples are found to carry antibodies against AIDS. With such screening, infections through blood transfusions are minimized.

At the time of writing, there is no known cure for AIDS. At least two approaches have been tried. One involves an antiviral agent, azidothymidine (AZT). AIDS patients with pneumonia caused by *Pneumocystis carinii* show some improvement after taking this drug. However, AZT is toxic, and most patients cannot tolerate therapeutic doses. Also, AZT provides only transient improvement—after about two years the virus develops resistance to the drug. Other antiviral agents, such as dideoxyinosine (DDI), dideoxycytidine (DDC), and (−)-2-deoxy-3′-thiacytidine (3TC), have been found beneficial in patients who cannot tolerate AZT or whose HIV becomes resistant to AZT. None of these drugs provides a cure. They only slow the development of the disease. All four are nucleoside-analog inhibitors, which inhibit a key viral enzyme, reverse transcriptase.

A new group of drugs approved to treat AIDS patients provides a second approach acting against a different viral enzyme, protease. These new drugs, indinavir, ritonavir, and saquinavir, are protease inhibitors, entering into the active site of the enzyme. By slowing down the action of protease they suppress viral replication. These drugs have proved to be effective even in AIDS patients who are very ill, with few of the side effects of AZT and the other drugs of its type. However, to suppress the viral infection of lymphocytes, large doses are required. Furthermore, as with the first class of drugs, the HIV virus quickly mutates and after two to three months of treatment, protease-inhibitor-drug-resistant strains of HIV appear. To slow down the mutation of the virus, even larger doses of protease inhibitors are required. The present strategy is to treat AIDS patient with a battery of different drugs, for example AZT and indinavir, or ritonavir, DDC and 3TC, to minimize the viral mutations. It has been shown that employment of such two- or three-drug cocktails greatly reduces the presence of HIV virus in the patients' blood for short periods, up to eight weeks. Whether the virus can be eradicated with long-term treatment is a question yet to be answered, though at least temporarily, this treatment reduces the death rate.

A vaccine would be the ideal remedy. But because of the deadly nature of the virus, one cannot use attenuated or even "killed" virus for immunization as was done with polio vaccine. Even if only one virus out of a million survived, it could kill the vaccinated person.

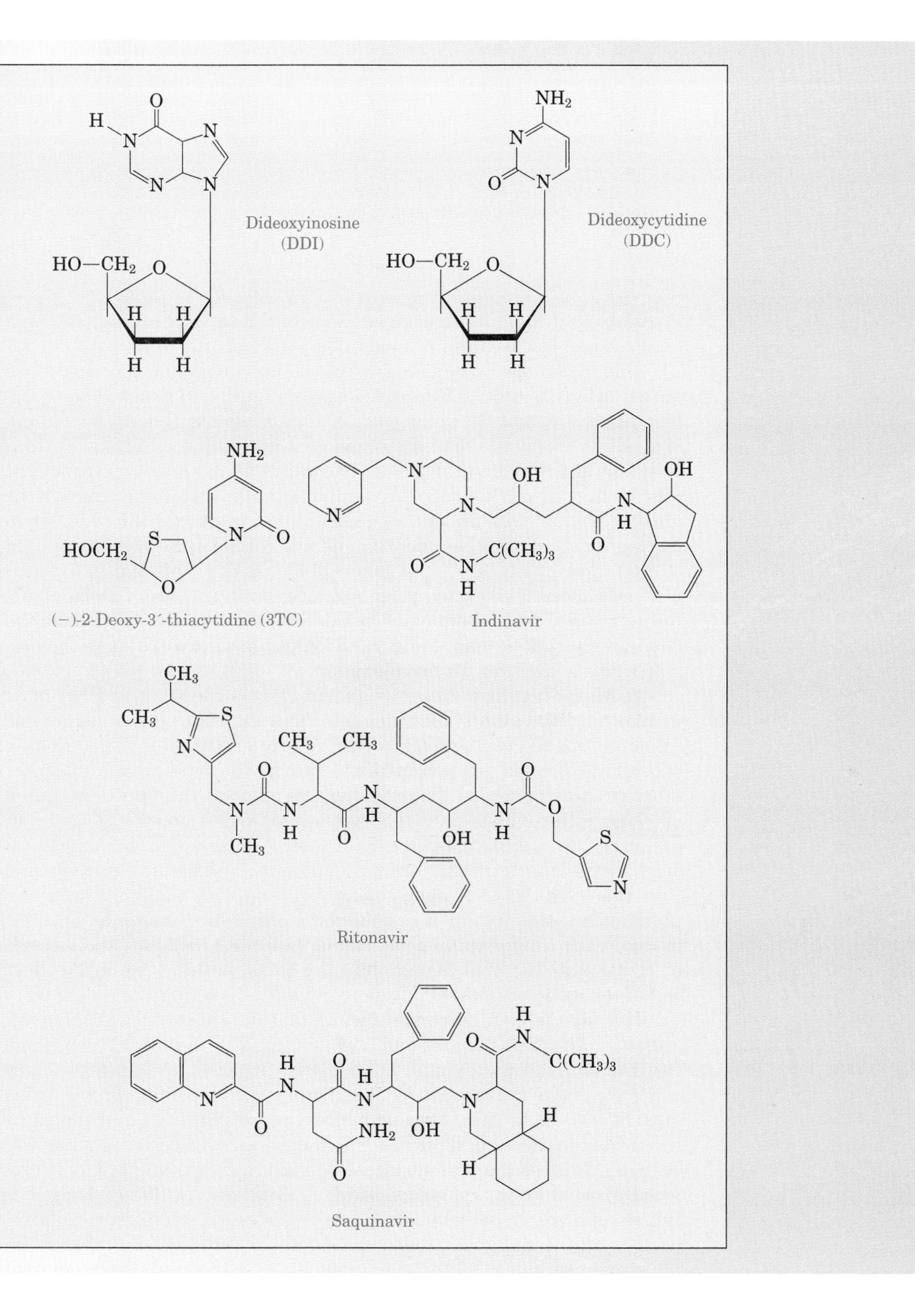

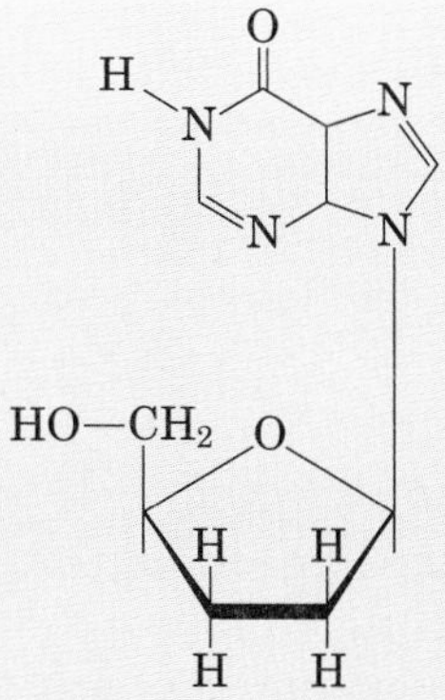

Dideoxyinosine (DDI)

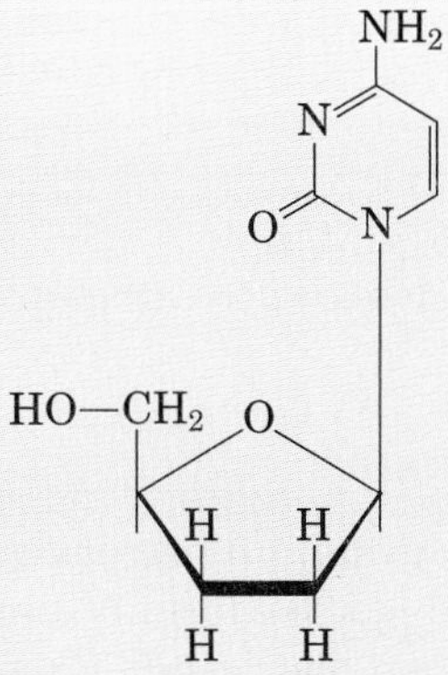

Dideoxycytidine (DDC)

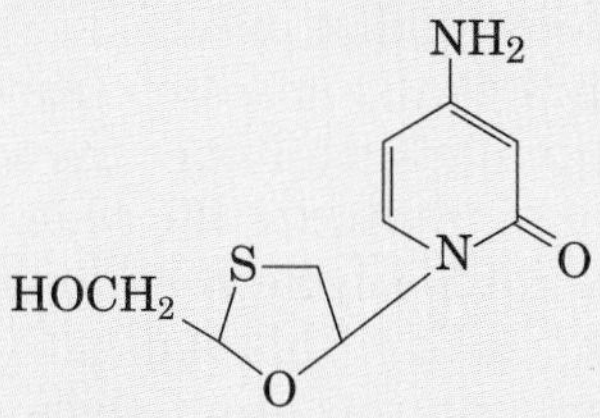

(−)-2-Deoxy-3′-thiacytidine (3TC)

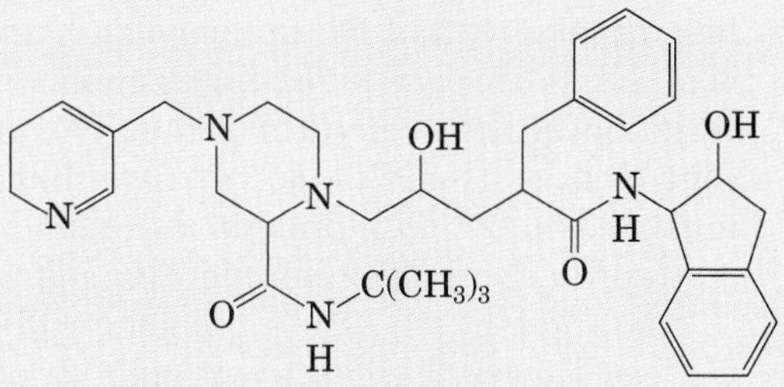

Indinavir

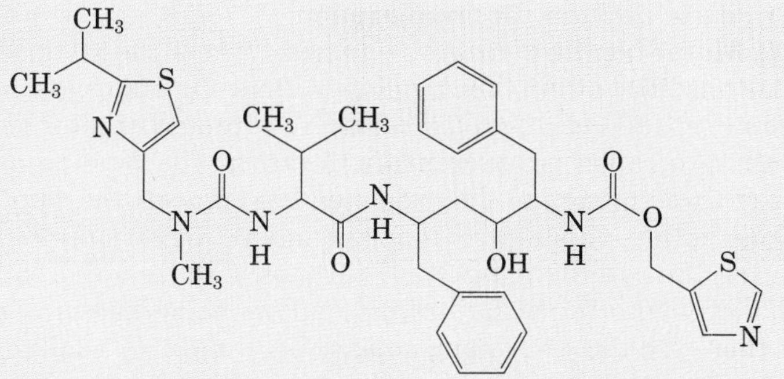

Ritonavir

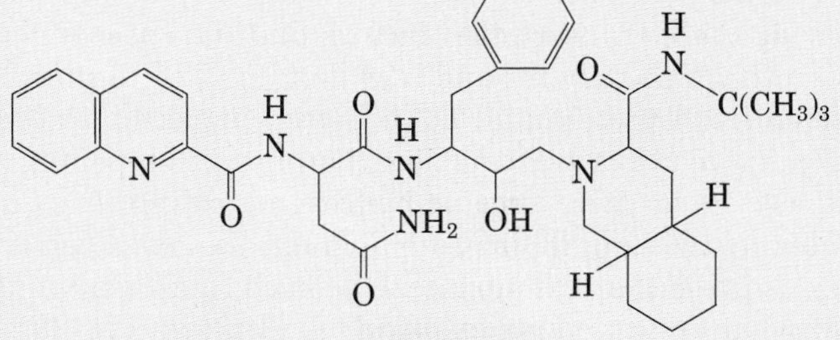

Saquinavir

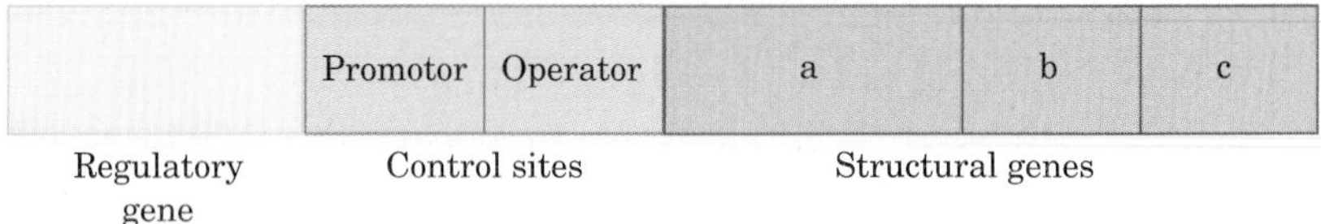

Figure **23.16** • A schematic view of the *lac* operon in *E. coli.* The a, b, and c correspond, respectively, to the genes that code for the proteins β-galactosidase, lactose permease, and thiogalactosidase transacetylase.

regulatory gene. The structural genes, the control sites, and the regulatory gene together form a unit called the **operon.**

One of the operons of *E. coli,* called the *lac* operon, contains three structural genes that code for three enzymes, called β-galactosidase, lactose permease, and thiogalactosidase transacetylase (Fig. 23.16). There is an allosteric protein, called a repressor protein, that prevents the transcription of the structural genes. This binds to the operator site of the control sites. The RNA polymerase that actually synthesizes the mRNA must be bound to the promoter site in order to transcribe the gene. When the repressor occupies the operator site, the RNA polymerase cannot bind to the promoter site. Thus, the gene will be inactive. This is negative regulation. However, when the proper inducer is present—in this case, lactose—it will form a complex with the repressor. As a consequence, the operator site will be empty, and the RNA polymerase can bind to the promoter site to perform the transcription.

2. Metal-binding fingers In eukaryotes the enzyme RNA-polymerase has little affinity for binding to DNA. In eukaryotic cells proteins called *selective-binding-proteins* bind to a promoter site. These binding proteins are called **transcription factors.** The binding one after another of these factors to the promoter site controls the rate of initiation of transcription. This control is very selective. While a prokaryotic operon can vary the rate by a factor of perhaps 1 thousand, a eukaryotic assembly of transcription factors may allow the synthesis of mRNA (and from there the target protein) to vary by a factor of 1 million. An example of such a wide variation in eukaryotic cells is that a specific gene, for example, the αA-crystallin gene, can be expressed in the lens of the eye, at a rate a million-fold higher than the same protein gene in the liver cell of the same organism.

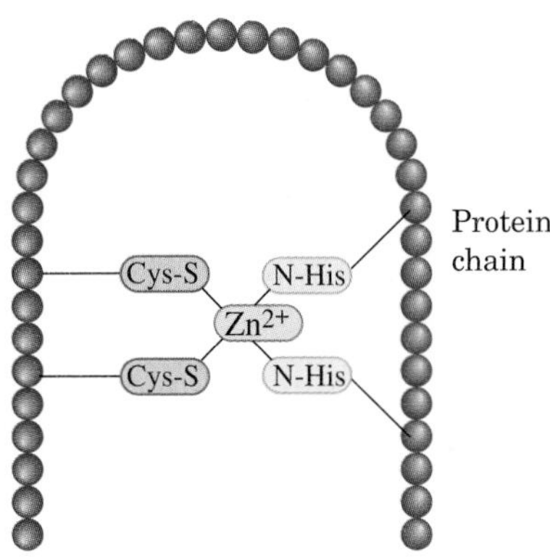

Figure **23.17**
Schematic view of a metal-binding finger (a region of a protein called a transcription factor). The zinc ion forms coordinate-covalent bonds to two cysteines and two histidines.

How do these transcription factors find the specific gene control sequences into which they fit, and how do they bind to them? The interaction between the protein and DNA is by nonspecific electrostatic interactions (positive ions attracting negative ones and repelling other positive ions) as well as by more specific hydrogen bonding. They find their targeted sites by twisting their protein chains so that a certain amino acid sequence is present at the surface. One such conformational twist is provided by what are called **metal-binding fingers** (Fig. 23.17). These finger shapes are created by Zn^{2+} ions, which form coordinate covalent bonds with the amino acid side chains of the protein. The zinc fingers interact with specific DNA (or sometimes RNA) sequences. The recognition comes by hydrogen bonding between a nucleotide (for example, guanine)

Box 23D

The Promoter. A Case of Targeted Expression

The process of transferring a gene from one organism to another is called *transfection,* and the animal or plant carrying the transfected gene is called *transgenic.*

Although the major part of a gene contains the region that codes for a particular protein, a portion of the gene does not code for any part of the protein. Instead it acts as a switch that turns the transcription process on or off. This regulatory part of the gene is called the **promoter.** By using the gene insertion technique discussed in Sec. 23.14, one can combine the coding sequence of one gene with the promoter of another gene, thereby targeting the expression of a protein to a certain organ.

A recent development in AIDS research (Box 23C) involved an enzyme called HIV-1 protease. This enzyme is essential if the virus that causes AIDS is to multiply inside the host cell. It has been argued that if suitable drugs that can inhibit HIV-1 protease could be found, the development of AIDS could be arrested, and perhaps the disease cured. But how can we test the hundreds of organic compounds that give promise of being HIV-1 protease inhibitors?

This is where the concept of the promoter comes in. For example, if the coding region of the HIV-1 protease is linked to the promoter of αA-crystallin, a protein molecule that appears only in the lens of the eye, the expression of the HIV-1 protease could be targeted to the eye only. True to the design, a DNA sequence containing the gene for HIV-1 protease and the promoter of αA-crystallin were transfected into a mouse. The result was transgenic mice that became blind 24 days after birth. Otherwise, the mice were not affected at all—they ate, ran, and bred as normal mice do. The blindness was the result of the action of HIV-1 protease in the lens, causing cataract formation (Fig. 23D).

These newly created transgenic mice could now be used to test drugs to find out if they would be good inhibitors of HIV-1 protease. If injection of a potential drug delayed cataract formation, that would be proof that it truly acts as an inhibitor of HIV-1 protease in a living organism and not just in test tubes. Many of the new drugs now used in treatment of AIDS patients (see Box 23C), for example, indinavir, have first been tested in this manner. This is clearly more efficient than testing all such potential inhibitors directly in humans.

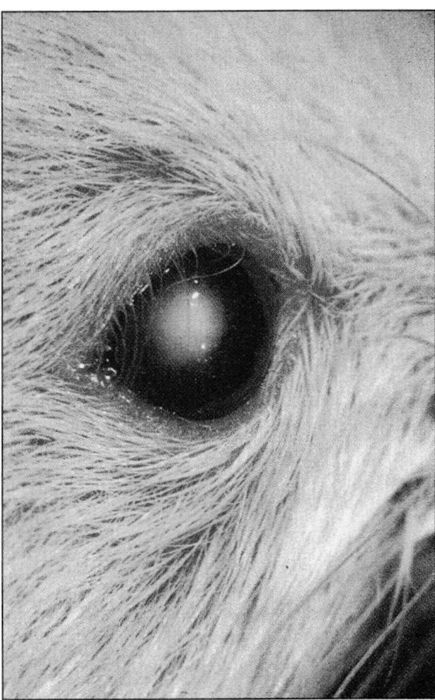

Figure **23D** • Eye with cataractous lens of a transgenic mouse containing the HIV-1 protease linked to αA-crystallin promoter, 27 days after birth. (Photo courtesy of Dr. Paul Russell, National Eye Institute, NIH.)

and the side chain of a specific amino acid (for example, arginine). Besides metal-binding fingers, at least two other prominent transcription factors exist, called helix-turn-helix and leucine zipper. Transcription factors can also be repressors, reducing the rate of transcription. Some can serve as both activators and repressors.

23.13 Mutations, Mutagens, and Genetic Diseases

In Section 23.4 we saw that the base-pairing mechanism provides an almost perfect way to copy a DNA molecule during replication. The key word here is "almost." No machine, not even the copying mechanism of DNA replication, is totally without error. It has been estimated that, on average, there is one error for every 10^{10} bases (that is, one in 10 billion). An error in the copying of a sequence of bases is called a **mutation.** Mutations can occur during replication. Base errors can also occur during transcription in protein synthesis (a nonheritable error). These errors may have widely varying consequences. For example, the codon for valine in mRNA can be GUA, GUG, GUC, or GUU. In DNA these correspond to CAT, CAC, CAG, and CAA, respectively. Let us assume that the original codon in the DNA is CAT. If during replication a mistake is made and the CAT was spelled as CAG in the copy, there will be no harmful mutation because when a protein is synthesized, the CAG will be transcribed onto the mRNA as GUC, which also codes for valine. Therefore, although a mutation occurred, the same protein is manufactured.

Sickle cell anemia (Box 18E) is caused by a single amino acid change: valine for glutamic acid. In this case CTT is mutated to CAT.

On the other hand, assume that the original sequence in the DNA is CTT, which transcribes onto mRNA as GAA and which codes for glutamic acid. If during replication a mutation occurs and CTT becomes ATT, the new cells will probably die. The reason is that ATT transcribes to UAA, which does not code for any amino acid but is a stop signal. Thus, instead of continuing to build a protein chain with glutamic acid, the synthesis stops altogether. An important protein is not manufactured, and the organism may die. In this way, *very* harmful mutations are not carried over from one generation to the next.

This is discussed in Section 9.5.

Examples of mutagens are benzene, carbon tetrachloride, and vinyl chloride.

Ionizing radiation (x-rays, ultraviolet light, gamma rays) can cause mutations. Furthermore, a large number of chemicals can induce mutation by reacting with DNA. Such chemicals are called **mutagens.** Many changes caused by radiation and chemicals do not become mutations because the cell has its own repair mechanism, which can prevent mutations by cutting out damaged areas and resynthesizing them. In spite of this defense mechanism, certain errors in copying that result in mutations do slip by.

Not all mutations are harmful. Certain ones are beneficial because they enhance the survival rate of the species. For example, mutation is used to develop new strains of plants that can withstand pests.

A **recessive gene** is one that expresses its message only if both parents transmit it to the offspring.

Specific diseases that result from defective or missing genes are discussed in Boxes 17F, 18E, 18F, 21B, and 21F.

If a mutation is harmful, it results in an inborn genetic disease. This may be carried as a recessive gene from generation to generation with no individual demonstrating the symptoms of the disease. Only when both parents carry recessive genes does an offspring have a 25 percent chance of inheriting the disease. If the defective gene is dominant, on the other hand, every carrier will develop symptoms.

Box 23E

Mutations and Biochemical Evolution

We can trace the genetic relationship of different species through the variability of their amino acid sequences in different proteins. For example, the blood of all mammals contains hemoglobin, but the amino acid sequences of the hemoglobins are not identical. In Table 23E we see that the first ten amino acids in the β-globin of humans and gorillas are exactly the same. As a matter of fact, there is only one amino acid difference, at position 104, between us and apes. The β-globin of the pig differs from ours at ten positions, of which two are in the N-terminal decapeptide. That of the horse differs from ours in 26 positions, of which four are in this decapeptide.

β-Globin seems to have gone through many mutations during the evolutionary process, since only 26 of the 146 sites are invariant, that is, exactly the same in all species studied so far.

The relationship between different species can also be established by similarities in mRNA primary structures. The proteins of closely related species, such as humans and apes, have very similar primary structures, presumably because these two species diverged on the evolutionary tree only recently. On the other hand, species far removed from each other diverged long ago and have undergone more mutations, which show up in differences in primary structures of their proteins.

Not only the *number* of amino acid substitutions is significant in the evolutionary process caused by mutation, but the *kind* of substitution is even more important. If the substitution is by an amino acid with physicochemical properties similar to those of the amino acid in the ancestor protein, the mutation is most probably viable. For example, whereas in human and gorilla β-globin position 4 is occupied by threonine, in the pig and horse it is occupied by serine. Both amino acids provide an —OH-carrying side chain.

Table 23E Amino Acid Sequence of the N-Terminal Decapeptides of β-Globin in Different Species

	Position									
Species	***1***	***2***	***3***	***4***	***5***	***6***	***7***	***8***	***9***	***10***
Human	Val	His	Leu	Thr	Pro	Glu	Glu	Lys	Ser	Ala
Gorilla	Val	His	Leu	Thr	Pro	Glu	Glu	Lys	Ser	Ala
Pig	Val	His	Leu	Ser	Ala	Glu	Glu	Lys	Ser	Ala
Horse	Val	Glu	Leu	Ser	Gly	Glu	Glu	Lys	Ala	Ala

23.14 Recombinant DNA

There are no cures for the inborn genetic diseases that we discussed in the preceding section. The best we can do is detect the carriers and, through genetic counseling of prospective parents, try not to perpetuate the defective genes. However, a technique called the **recombinant DNA technique** gives some hope for the future. At this time recombinant DNA

Box 23F

Mutagens and Carcinogens

There are many chemicals (both synthetic and natural) that cause cancer when introduced into the body. These are called carcinogens (Box 11G). One of the main tasks of the U.S. Food and Drug Administration and the Environmental Protection Agency is to identify these chemicals and eliminate them from our food, drugs, and environment. To prove that a chemical is a carcinogen requires long and costly laboratory testing with animals. Because we know that most carcinogens are also mutagens (that is, they induce mutations in bacteria), we now have available a much faster and cheaper screening test (the Ames test). In this test, one takes a bacterial strain, for example, *Salmonella,* that needs the amino acid histidine in its diet. When the bacteria are grown on a medium that does not contain histidine, very few survive. If a mutagen is added to the medium, however, some of the *Salmonella* may undergo mutations that can live without a supply of histidine. These will multiply and show up as a heavy growth of bacterial colonies. With such a simple test, many chemicals can be tested for mutagenic activity. Those that are found to be mutagens then can be further tested in animals to see if they are also carcinogens.

techniques are used mostly in bacteria, but it is possible that they may someday be extended to human cells as well.

As used today, the recombinant DNA technique begins with certain circular DNA molecules found in the cells of the bacteria *E. coli.* These molecules, called **plasmids** (Fig. 23.18), consist of double-stranded DNA arranged in a ring. Certain highly specific enzymes called **restriction endonucleases** cleave DNA molecules at specific locations (a different

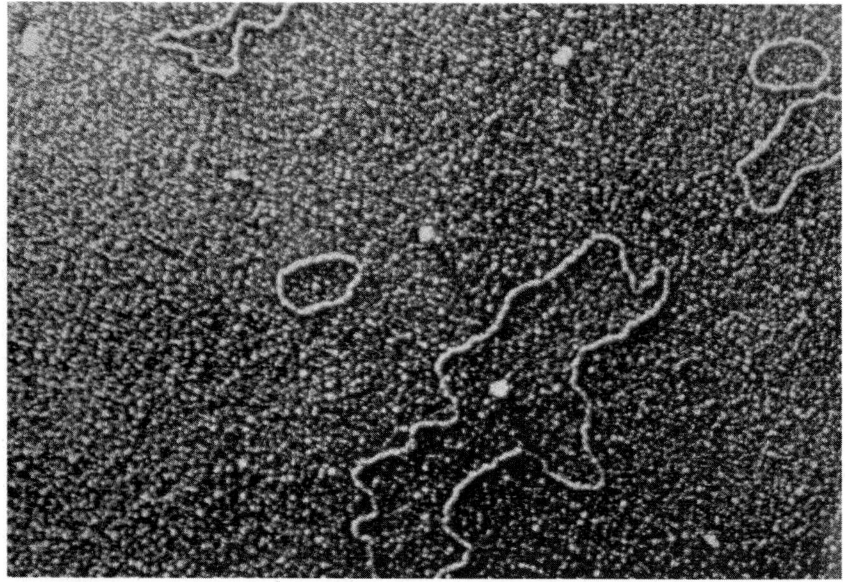

Figure **23.18** • Plasmids from a bacterium used in the recombinant DNA technique. (Thomas Broker/Cold Spring Harbor Laboratory.)

Box 23G

Oncogenes

An **oncogene** is a gene that in some way or other participates in the development of cancer. Cancer cells differ from normal cells in a variety of structural and metabolic ways; however, the most important difference is their uncontrolled proliferation. Most ordinary cells are quiescent. When they lose their controls they give rise to tumors, benign or malignant. The uncontrolled proliferation allows these cells to spread, invade other tissues and colonize them, a process called **metastasis.** Malignant tumors can be caused by a variety of agents: chemical carcinogens such as benzo(e)pyrene (Box 11G), ionizing radiation such as x-rays, and viruses. However, for a normal cell to become a cancer cell certain transformations must occur. Many of these transformations take place on the level of the genes. Certain genes in normal cells have counterparts in viruses, especially in retroviruses (which carry their genes in the form of RNA rather than DNA). This means that a normal cell in a human being and in a virus both have genes that code for similar proteins. Usually these genes code for proteins that control cell growth. Such a protein is epidermal growth factor (EGF). When a retrovirus gets into the cell it takes control of the cell's own EGF gene. At that point the EGF gene becomes an oncogene, a cancer-causing gene, because it produces large amounts of the EGF protein, which in turn allows the epidermal cells to grow in an uncontrolled manner.

Because the normal EGF gene in the normal cell was turned into an oncogene, we call it a **proto-oncogene.** Such conversions of a proto-oncogene to an oncogene can occur not just by viral invasion but also by a mutation of the gene. For example, a form of cancer called human bladder carcinoma is caused by a single mutation of a gene; a change from guanine to thymine. The resulting protein has a valine in place of a glycine. This change is sufficient to transform the cell to a cancerous state, because the protein for which the gene is coded is a protein that amplifies signals (Sec. 24.4). Thus, because of the mutation that caused the proto-oncogene to become an oncogene, the cell stays "on"—its metabolic process is not shut off, and it is now a cancer cell that proliferates without control. There are a number of other mechanisms by which a proto-oncogene is transformed to an oncogene, but in each case the gene product has something to do with growth, hormonal action, or some other kind of cell regulation. About 50 viral and cellular oncogenes have been identified.

location for each enzyme). For example, one of these enzymes may split a double-stranded DNA as follows:

```
~~~~ GAATTC ~~~~   enzyme    ~~~~ G          AATTC ~~~~
                   ------>                +
~~~~ CTTAAG ~~~~             ~~~~ CTTAA          G ~~~~
```

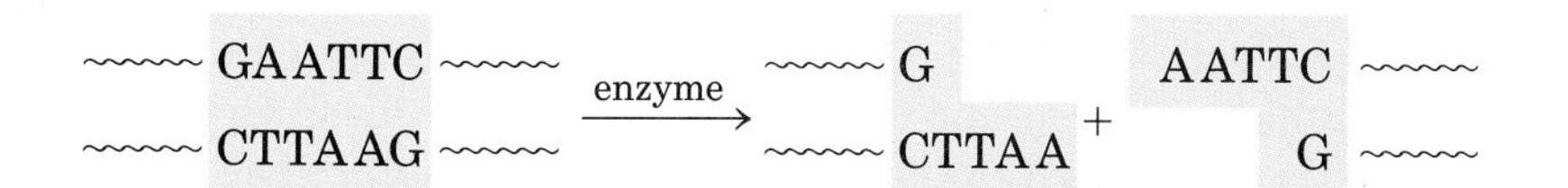

The enzyme is so programmed that whenever it finds this specific sequence of bases in a DNA molecule, it cleaves it as shown. Since a plasmid is circular, cleaving it this way produces a double-stranded chain with two ends (Fig. 23.19). These are called "sticky ends" because each has on one strand several free bases that are ready to pair up with a complementary section if they can find one.

The next step is to give them one. This is done by adding a gene from some other species. The gene is a strip of double-stranded DNA that has the necessary base sequence. For example, we can put in the human gene that manufactures insulin, which we can get in two ways: (1) It can be made in a laboratory by chemical synthesis; that is, chemists can combine the nucleotides in the proper sequence to make the gene, or (2) we can cut a human chromosome with the same restriction enzyme. Since it is the same enzyme, it cuts the human gene so as to leave the same sticky ends:

We use "H" to indicate a human gene.

```
H—GAATTC—H         H—G              AATTC—H
            ——→                 +
H—CTTAAG—H         H—CTTAA              G—H
```

The human gene must be cut at two places, so that a piece of DNA that carries two sticky ends is freed. To splice the human gene into the plasmid, the two are mixed in the presence of DNA ligase, and the sticky ends come together:

We use "B" to indicate a bacterial plasmid.

```
H—G              AATTC—B         H—GAATTC—B
            +             ——→
H—CTTAA              G—B         H—CTTAAG—B
```

This reaction takes place at both ends of the human gene, and the plasmid is a circle once again (Fig. 23.19).

Millions of copies of selected DNA fragments can be made within a few hours with high precision by a technique called **polymerase chain reaction,** discovered by Kary B. Mullis (1945–), who shared the 1993 Nobel Prize in Chemistry for this discovery.

Human insulin is now marketed by the Lilly Corporation in two forms, called Humulin R and Humulin N (see Box 18D). The R form has a faster onset action. Both of them are manufactured by the recombinant DNA technique.

The modified plasmid is then put back into a bacterial cell, where it replicates naturally every time the cell divides. Bacteria multiply quickly, and soon we have a large number of bacteria, all containing the modified plasmid. All these cells now manufacture human insulin by transcription and translation. We can thus use bacteria as a factory to manufacture specific proteins. This new industry has tremendous potential for lowering the price of drugs that are now manufactured by isolation from human or animal tissues (for example, human interferon, a molecule that fights infection). Not only bacteria but also plant cells can be used (Fig. 23.20). Ultimately, if recombinant DNA techniques can be applied to humans and not just to bacteria, it is possible that genetic diseases might someday be cured by this powerful technique. An infant or fetus missing a gene might be given that gene. Once in the cells, the gene would reproduce itself for an entire lifetime.

At the time of writing (late 1996), such gene therapy has been sanctioned by the National Institutes of Health for, among other things, some patients suffering from cystic fibrosis, a hereditary disease. Seventy percent of cystic fibrosis patients have one identified faulty gene. The plan is to spray a solution of virus into which the healthy gene has been cloned into the nose or lungs of the patient. The virus, which is otherwise harmless, enters the cells of the patients, carrying with it the healthy gene. The hope is that the product of the healthy gene will compete with the product of the faulty gene, alleviating the symptoms of the disease or curing it altogether.

By far the greatest potential of **genetic engineering,** that is, inserting new genes into cells, is in the field of agriculture. Genetic engineering has produced tomatoes that can be picked in their naturally ripened states and yet do not spoil quickly on the shelves of supermarkets. Other genetically engineered plants, corn and cotton, for example, can resist damages by insects or fungi or herbicides. In addition to herbicide and insect-resistant plants, genetic engineering offers opportunities to improve crop yields and resist freezing temperatures. All these will contribute to the increase of the food supply necessary to feed our ever-growing world population.

Some people feel that genetically engineered organisms may create havoc with the ecology of the planet by introducing new, dominant species. But plant breeding has been going on for centuries trying to achieve exactly the same goals. Which is better: a genetically engineered ripe tomato or one picked green and ripened in the warehouse by ethylene gas? Obviously, care must be taken and controls exercised so as not to release harmful newly engineered life-forms into the environment.

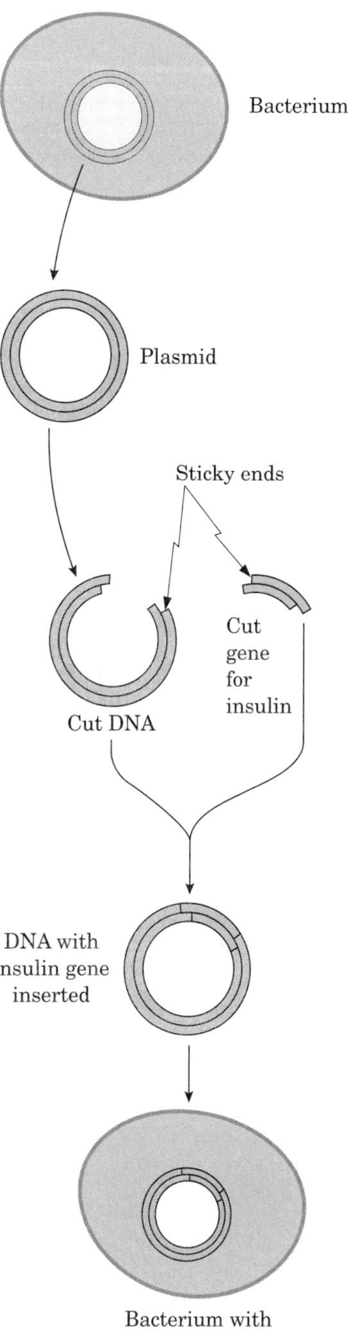

Figure **23.19**
The recombinant DNA technique can be used to turn a bacterium into an insulin "factory." (From P. B. Berlow, et al., *Introduction to the Chemistry of Life.* Philadelphia: Saunders College Publishing, 1982.)

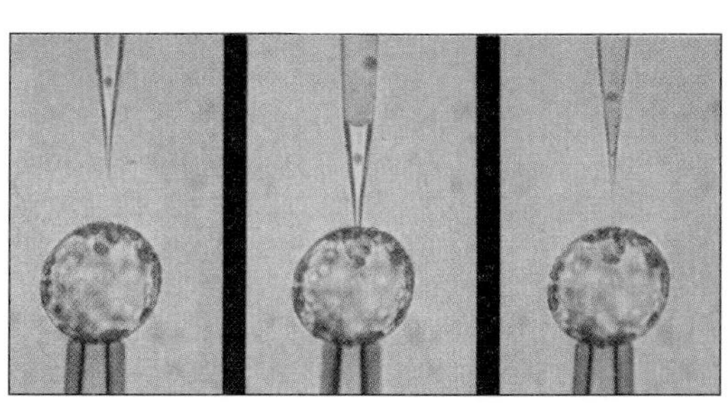
(*a*)

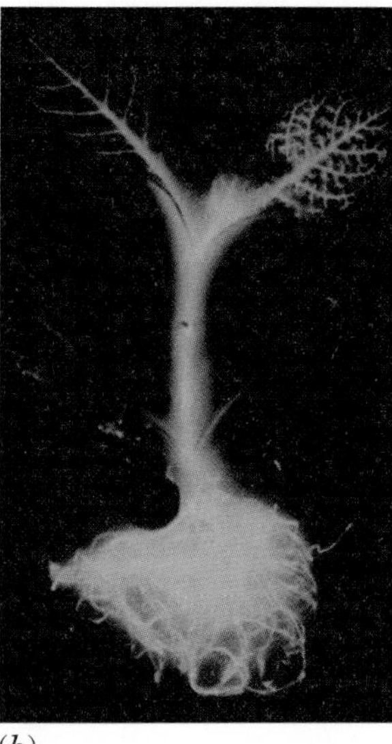
(*b*)

Figure **23.20** • (a) Injection of an aqueous DNA solution into the nucleus of a protoplast. (Courtesy of Dr. Anne Crossway, Calgene, Inc., Davis, California, and *Biotechniques,* Eaton Publishing, Natick, Massachusetts.) (b) A luminescent tobacco plant. The gene of the enzyme luciferase (from a firefly) has been incorporated into the genetic material of the tobacco plant. (Courtesy of Dr. Marlene DeLuca, University of California at San Diego.)

Box 23H

DNA Fingerprinting

Every person has a genetic makeup consisting of about 3 billion pairs of nucleotides, distributed over 46 chromosomes. The base sequence in the nucleus of every one of our billions of cells is identical. However, except for people who have an identical twin, the base sequence in the total DNA of one person is different from that of every other person. This makes it possible to identify suspects in criminal cases from a bit of skin or a trace of blood left at the scene of the crime and to prove the identity of a child's father in paternity cases. The nuclei of these cells are extracted, and restriction enzymes are used to cut the DNA molecules at specific points. The resulting DNA fragments are put on a gel and undergo a process called electrophoresis. In this process the DNA fragments move with different velocities; the smaller fragments move faster and the larger fragments

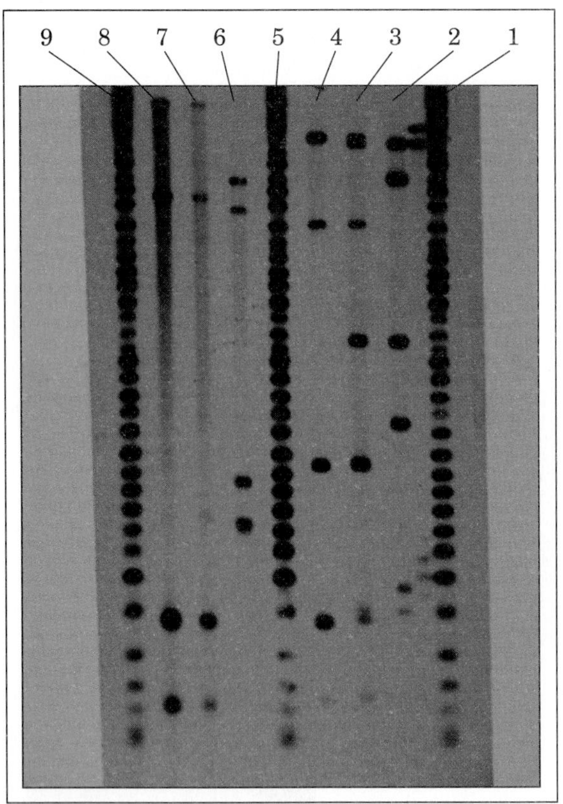

Figure **23H** • DNA fingerprint. (Courtesy of Dr. Lawrence Kobilinsky.)

Box 23H (continued)

slower. After a sufficient amount of time the fragments separate, and when they are made visible in the form of an autoradiogram, one can see bands in a lane. This is called a **DNA fingerprint.**

When the DNA fingerprint made from a sample taken from a suspect matches that from a sample obtained at the scene of the crime, the police have a positive identification. Figure 23H shows DNA fingerprints derived by using one particular restriction enzyme. A total of nine lanes can be seen. Three (numbers 1, 5, and 9) are control lanes. They contain the DNA fingerprint of a virus, using one particular restriction enzyme. Three other lanes (2, 3, and 4) were used in a paternity suit: These contain the DNA fingerprints of the mother, the child, and the alleged father. This is a positive identification. The child's DNA fingerprint contains six bands. The alleged father's DNA fingerprint also contains six bands, of which three match those of the child. The mother's DNA fingerprint has five bands, all of which match those of the child. In such cases one cannot expect a perfect match even if the man is the actual father because the child has inherited only half of its genes from the father. Thus, of six bands only three are expected to match. In the case just described the paternity suit was won on the basis of the DNA fingerprint matching.

In the left area of the radiogram are three more lanes (6, 7, and 8). These DNA fingerprints were used in an attempt to identify a rapist. In lanes 7 and 8 are the DNA fingerprints of semen obtained from the rape victim. In lane 6 is the DNA fingerprint of the suspect. The DNA fingerprints of the semen do not match that of the suspect. This is a negative identification and excluded the suspect from the case. When positive identification occurs, the probability that a positive match is due to chance is 1 in 100 billion.

However, a certain caution has developed lately in court cases involving DNA fingerprinting. Scientific committees and expert witnesses now demand that in addition to matching or not matching autoradiograms, some internal controls should appear on the gel. This is demanded in order to prove that the two lanes to be compared were run and processed under identical conditions. Furthermore, identifications in courts do not rely solely on DNA fingerprinting. Additional evidence is provided by analysis of blood group substances (see Box 16B), either directly from blood or from secretions such as saliva or cervical mucus samples, which also contain them.

Box 23I

The Human Genome Project

The complete DNA sequence of any organism is called a **genome.** The genome of an average human being contains approximately 3 billion base pairs. These are distributed among 22 pairs of chromosomes plus two sex chromosomes. Each chromosome consists of a single DNA molecule. Among the 3 billion pairs that constitute the human genome are some 300 million base pairs, which represent 100 000 genes. It is the task of the Human Genome Project to determine the total sequences of the genome and in the process to identify the sequence and location of the genes.

A subsidiary but also an essential goal is to determine the total DNA sequence of model systems of simpler organisms: The bacterium *E. coli* (3 million base pairs); the yeast *S. cerevisiae* (14 million); the nematode *C. elegans* (80 million); the fruit fly, *Drosophila melanogaster* (165 million); and the mouse, *Mus musculus* (3 billion). The first organism of which the complete genome was established was the bacterium, *Hemophilus influenzae,* in 1995. This bacterium has only 1.8 million base pairs, which code for 1743 genes. Shortly thereafter the even smaller genome of *Mycobacterium genitalium* (0.6 million base pairs) was obtained, and work is underway or has been completed on all the other targeted model systems. It is hoped that the whole Human Genome Project will be finished by 2001. Knowing the genomes of the model systems will help the workers studying the human genome to elucidate how the different genes are linked together. Many laboratories around the world are participating in this project.

In the course of the project, it is hoped that many of the genes that govern human diseases, such as cystic fibrosis and breast cancer, just to name a few, will be identified. With such identification it is expected that strategies can be developed for the diagnosis, prevention, and perhaps therapy of these diseases.

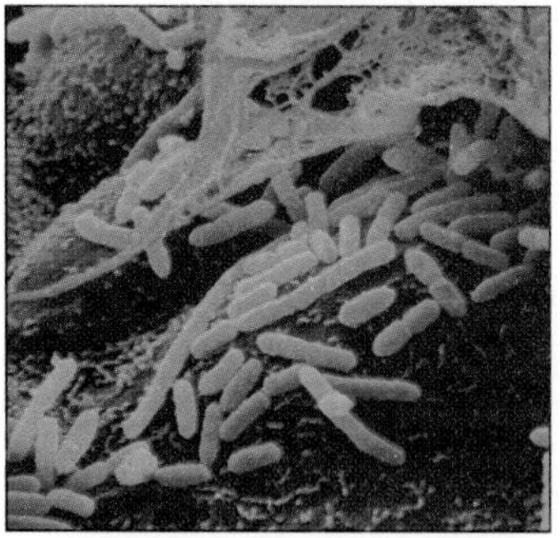

Hemophilus influenzae bacteria resting on nasal tissue. (© Dr. Tony Brain/SPL/Photo Researchers.)

SUMMARY

Nucleic acids are composed of sugars, phosphates, and organic bases. There are two kinds: **ribonucleic acid (RNA)** and **deoxyribonucleic acid (DNA).** In DNA the sugar is deoxyribose; in RNA it is ribose. In DNA the bases are adenine (A), guanine (G), cytosine (C), and thymine (T). In RNA they are A, G, C, and uracil (U). Nucleic acids are giant molecules with backbones made of alternating units of sugar and phosphate. The bases are side chains linked to the sugar units.

DNA is made of two strands that form a double helix. The sugar–phosphate backbone runs on the outside of the double helix, and the bases point inward. There is **complementary pairing** of the bases in the double helix. Each A on one strand is hydrogen-bonded to a T on the other, and each G is hydrogen-bonded to a C. No other pairs fit. The DNA molecule carries, in the sequence of its bases, all the information necessary to maintain life. When cell division occurs and this information is passed from parent cell to daughter cells, the sequence of the parent DNA is copied. The copying involves Okazaki fragments.

A **gene** is a segment of a DNA molecule that carries the sequence of bases that directs the synthesis of one particular protein. The information stored in the DNA is transcribed onto RNA and then expressed in the synthesis of a protein molecule. This is done in two steps: **transcription** and **translation.** There are three kinds of RNA: **messenger RNA (mRNA), transfer RNA (tRNA),** and **ribosomal RNA (rRNA).** In transcription, the information is copied from DNA onto mRNA by complementary base-pairing ($A \rightarrow U$, $T \rightarrow A$, $G \rightarrow C$, $C \rightarrow G$). There are also start and stop signals. The mRNA is strung out along the ribosomes. Transfer RNA carries the individual amino acids. Each tRNA goes to a specific site on the mRNA. A sequence of three bases (a triplet) on mRNA constitutes a **codon.** It spells out the particular amino acid that the tRNA brings to this site. Each tRNA has a recognition site, the **anticodon,** that pairs up with the codon. When two tRNA molecules are aligned at adjacent sites, the amino acids that they carry are linked by an enzyme. The process continues until the whole protein is synthesized. The genetic code is multiple. In most cases there is more than one codon for each amino acid. DNA in higher organisms contains sequences, called **introns,** that do not code for proteins. The sequences that do code for proteins are called **exons.**

There are a number of mechanisms for **gene regulation,** the switching on and off of the action of genes. In *E. coli,* portions of DNA called **operons** contain not only structural genes but also control sites and a regulatory gene. **Transcription factors** regulate gene expression by positive enhancement. Most of human DNA (96 to 98 percent) does not code for proteins.

A change in the sequence of bases is called a **mutation.** Mutations can be caused by an internal mistake or can be induced by chemicals or radiation. A change in just one base can cause a mutation. This may be harmful or beneficial or may cause no change whatsoever in the amino acid sequence. If a mutation is very harmful, the organism may die. Chemicals that cause mutations are called **mutagens.**

With the discovery of **restriction enzymes** that can cut DNA molecules at specific points, scientists have found ways to splice DNA segments together. In this manner a human gene—for example, the one that codes for insulin—can be spliced into a bacterial **plasmid.** Then the bacteria, when multiplied, transmit this new information to the daughter cells. Therefore, the ensuing generations of bacteria are able to manufacture human insulin. This powerful method is called the **recombinant DNA technique. Genetic engineering** is the process by which genes are inserted into cells.

KEY TERMS

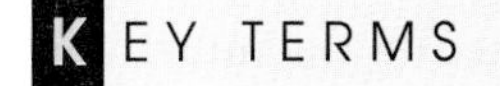

A site (Sec. 23.10)
Activation (Sec. 23.10)
Anticodon (Sec. 23.8)
Apoptosis (Sec. 23.11)
Base (Sec. 23.2)
Central dogma (Sec. 23.6)
Chromosome (Sec. 23.1)
Codon (Secs. 23.8, 23.9)
Complementary sequence (Sec. 23.4)
Control sites (Sec. 23.12)
Degenerate code (Sec. 23.9)
Deoxyribonucleic acid (Sec. 23.2)
DNA (Sec. 23.2)
DNA fingerprinting (Box 23H)
Double helix (Sec. 23.3)
Elongation (Sec. 23.10)
Elongation factor (Sec. 23.10)
Eukaryote (Sec. 23.12)
Exon (Sec. 23.11)
Gene (Secs. 23.1, 23.11)
Gene expression (Sec. 23.12)
Gene regulation (Sec. 23.12)
Genetic code (Sec. 23.9)
Genetic engineering (Sec. 23.14)
Genome (Box 23I)
Histone (Secs. 23.1, 23.11)
Initiation (Sec. 23.10)
Initiation signal (Sec. 23.7)
Intron (Sec. 23.11)
Lagging strand (Sec. 23.4)
Leading strand (Sec. 23.4)
Messenger RNA (mRNA) (Sec. 23.5)
Metal-binding fingers (Sec. 23.12)
Metastasis (Box 23G)
Multiple code (Sec. 23.9)
Mutagen (Sec. 23.13)
Mutation (Sec. 23.13)
Nucleic acid (Sec. 23.1)
Nucleoside (Sec. 23.2)
Nucleosome (Sec. 23.11)
Nucleotide (Sec. 23.2)
Okazaki fragment (Sec. 23.4)

Oncogene (Box 23G)
Operator site (Sec. 23.12)
Operon (Sec. 23.12)
P site (Sec. 23.10)
Plasmid (Sec. 23.14)
Polymerase chain reaction (Sec. 23.14)
Prokaryote (Sec. 23.12)
Promoter (Box 23D)
Promoter site (Sec. 23.12)
Recognition site (Sec. 23.8)
Recombinant DNA (Sec. 23.14)
Regulatory gene (Sec. 23.12)
Replication (Sec. 23.4)
Restriction endonuclease (Sec. 23.14)
Retrovirus (Box 23C)
Ribonucleic acid (Sec. 23.2)
Ribosomal RNA (rRNA) (Sec. 23.5)
Ribosome (Sec. 23.5)
RNA (Sec. 23.2)
Satellites (Sec. 23.11)
Structural genes (Sec. 23.12)
Termination (Sec. 23.10)
Termination sequence (Sec. 23.7)
Transcription (Sec. 23.6)
Transfection (Box 23D)
Transfer RNA (tRNA) (Sec. 23.5)
Transgenic (Box 23D)
Translation (Sec. 23.6)
Translocation (Sec. 23.10)
Unwinding protein (Sec. 23.4)

PROBLEMS

Nucleic Acids and Heredity

23.1 Explain, in terms of DNA, why a pig cannot give birth to a mouse.

23.2 What structures of the cell, visible in a microscope, contain hereditary information?

23.3 Name one hereditary disease.

Components of Nucleic Acids

23.4 (a) Where in a cell is the DNA located? (b) Where in a cell is the RNA located?

23.5 What are the components of
(a) a nucleotide
(b) a nucleoside

23.6 Draw the structures of (a) guanine (b) adenine

23.7 What is the difference in structure between thymine and uracil?

23.8 Which DNA and RNA bases contain a carbonyl group?

23.9 Draw the structures of (a) cytidine (b) deoxycytidine

23.10 Which DNA and RNA bases are primary amines?

23.11 What is the difference in structure between ribose and deoxyribose?

23.12 What is the difference between a nucleoside and a nucleotide?

Structure of DNA and RNA

23.13 In RNA which carbons of the ribose are linked to the phosphate group and which to the base?

23.14 What constitutes the backbone of DNA?

23.15 Draw the structures of (a) UMP (b) dAMP

23.16 In DNA, which carbon atoms of the sugar are linked to the phosphate groups?

23.17 The sequence of a short DNA segment is ATGGCAATAC. (a) What name do we give to the two ends (terminals) of a DNA molecule? (b) In this segment, which end is which?

***23.18** Chargaff showed that in samples of DNA from many different species, the molar quantity of A was always approximately equal to the molar quantity of T, and the same for C and G. How did this information help to establish the structure of DNA?

23.19 What makes nucleic acids acidic?

23.20 What interactions stabilize the three-dimensional structure of DNA?

23.21 Which nucleic acid is single-stranded?

DNA Replication

23.22 A DNA molecule normally replicates itself millions of times, with almost no errors. What single fact about the structure is most responsible for this?

23.23 What functional groups on the bases form hydrogen bonds in the DNA double helix?

23.24 Draw the structures of adenine and thymine, and show with a diagram the two hydrogen bonds that stabilize A—T pairing in DNA.

23.25 Draw the structures of cytosine and guanine, and show with a diagram the three hydrogen bonds that stabilize C—G pairing in nucleic acids.

23.26 How many bases are there in a DNA double helix?

23.27 Where does the unwinding of DNA occur?

23.28 What is the function of the unwinding proteins?

23.29 What do we call the enzymes that join nucleotides into a DNA strand?

23.30 In which direction is the DNA molecule synthesized continuously?

23.31 Why does the body synthesize the leading strand continuously and the lagging strand discontinuously (that is, in Okazaki fragments)?

RNA, Transmission, Transcription, and Translation

23.32 How many nucleotides are there in a tRNA chain?

23.33 Which has the longest chains: tRNA, mRNA, or rRNA?

23.34 What is the central dogma of molecular biology?

23.35 Which kind of RNA has a sequence complementary to that of DNA?

23.36 Which kind of RNA is associated with proteins?

23.37 (a) What is the shape of a tRNA molecule in a two-dimensional diagram? (b) In a three-dimensional diagram?

23.38 Which end of the DNA contains the initiation signal?

23.39 In what part of the cell does transcription occur?

23.40 (a) Which ribosome portion has specific A and P sites? (b) What happens at each site during translation?

23.41 What are the two most important sites on tRNA molecules?

23.42 What is the function of the ribosomes?

The Genetic Code

23.43 (a) If a codon is GCU, what is the anticodon? (b) What amino acid does this codon code for?

***23.44** If a segment of DNA is 981 units long, how many amino acids appear in the protein this DNA segment codes for? (Assume that the entire segment is used to code for the protein and that there is no methionine at the N-terminal end of the protein.)

23.45 Which codons are stop signals?

Translation and Protein Synthesis

23.46 (a) What is the main role of the 40S ribosome? (b) of the 60S ribosome?

23.47 What is the function of elongation proteins?

Genes, Exons, and Introns

23.48 What is the nature of the interaction between histones and DNA in nucleosomes?

23.49 Define (a) intron (b) exon

23.50 Does mRNA also have introns and exons? Explain.

23.51 (a) What percentage of human DNA codes for proteins? (b) What is the function of the rest of the DNA?

23.52 What is the distinguishing fingerprint of death by apoptosis?

23.53 What are satellites in DNA?

Gene Regulation

23.54 When gene regulation occurs at the transcriptional level, what molecules are involved?

23.55 Name the two kinds of control sites.

23.56 What is an operon?

23.57 Compare the interactions of RNA polymerase with the DNA chain in prokaryotes as opposed to eukaryotes.

23.58 What kind of interactions exist between metal-binding fingers and DNA?

Mutations

23.59 Using Table 23.2, give an example of a mutation that (a) does not change anything in a protein molecule (b) might cause fatal changes in a protein

Recombinant DNA

23.60 How do restriction endonucleases operate?

23.61 What are sticky ends?

23.62 A new genetically engineered corn has been approved by the Food and Drug Administration. This new corn shows increased resistance to a destructive insect called a corn borer. What is the difference, in principle, between this genetically engineered corn and one that developed insect resistance by mutation (natural selection)?

Boxes

23.63 (Box 23A) Draw the structure of the fluorouridine nucleoside that inhibits DNA synthesis.

23.64 (Box 23A) Give an example of how anticancer drugs work in chemotherapy.

23.65 (Box 23B) Why are viruses considered to be parasites?

***23.66** (Box 23B) What is a viral "coat"? Where do the ingredients—amino acids, enzymes, and so forth—necessary to synthesize the "coat" come from?

23.67 (Box 23C) What are the target cells for the AIDS virus?

23.68 (Box 23C) What HIV-1 viral enzymes are inhibited by (a) AZT (b) indinavir?

23.69 (Box 23D) What was the role of the αA-crystallin promoter in the development of protease-inhibitor drugs as a treatment for AIDS?

23.70 (Box 23F) If 1-amino-2-cyanoethane were found to be mutagenic by the Ames test, how could you prove that it is not carcinogenic?

23.71 (Box 23G) What kind of mutation transforms the EGF proto-oncogene to an oncogene in human bladder carcinoma?

23.72 (Box 23H) After having been cut by restriction enzymes, how are the DNA fragments separated from each other?

23.73 (Box 23H) How is DNA fingerprinting used in paternity suits?

23.74 (Box 23I) What was the first organism to have its genome completely sequenced?

***23.75** (Box 23I) Can you give a logical explanation why the genome of a bacterium is so much smaller than that of a mammal? (Hint: Look at the difference in introns between prokaryotes and eukaryotes, Sec. 23.11.)

Additional Problems

23.76 Why is it important that a DNA molecule be able to replicate itself millions of times without error?

23.77 Why is DNA replication called semiconservative?

23.78 How does an inducer such as lactose enhance the production of the enzyme β-galactosidase?

23.79 In the tRNA structure there are stretches where complementary base pairing is necessary and other areas where it is absent. Describe two functionally critical areas (a) where base pairing is mandatory and (b) where it is absent.

23.80 Is there any way to prevent a hereditary disease? Explain.

***23.81** Which nuclear superstructure is stabilized by an acid–base interaction?

23.82 How does the cell make sure that a specific amino acid (say valine) attaches itself only to the one tRNA molecule that is specific for valine?

23.83 Draw the structures of (a) uracil (b) uridine

23.84 (a) What is a plasmid? (b) How does it differ from a gene?

23.85 Why do we call the genetic code degenerate?

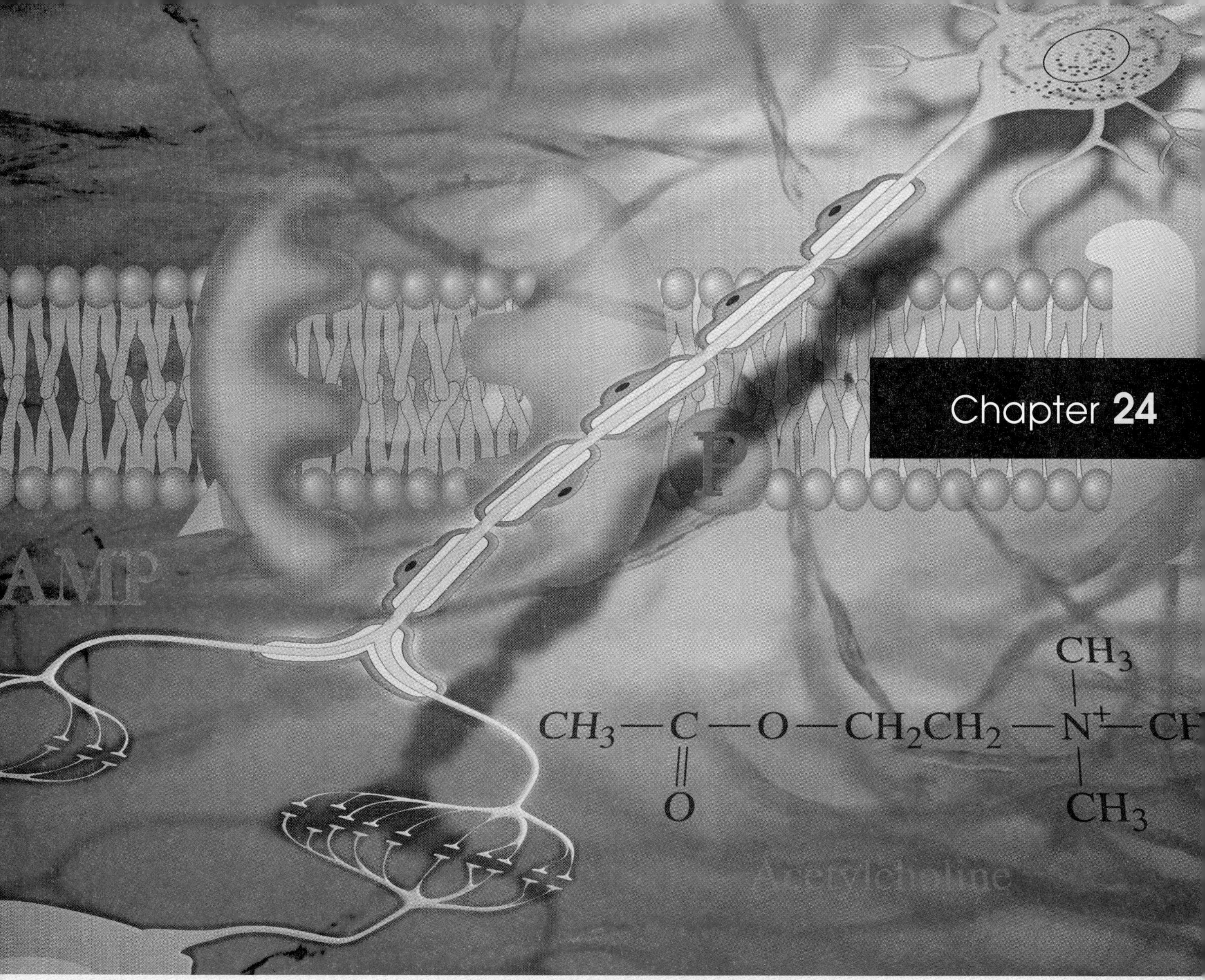

Chemical communication: Neurotransmitters and hormones

24.1 Introduction

Each cell in the body is an isolated entity enclosed in its own membrane. The thousands of reactions in each cell would be uncoordinated unless cells could communicate with each other. Such communication allows the activity of a cell in one part of the body to be coordinated with the activity of cells in a different part of the body. The body uses two principal types of molecules for these communications: (1) relatively small molecules called **chemical messengers** and (2) protein molecules on the surface of cell membranes called **receptors.**

When your house is on fire and the fire is threatening your life, external signals—light, smoke, and heat—register alarm at specific receptors in your eyes, nose, and skin. From there the signals are transmitted by specific compounds to nerve cells, or **neurons.** In the neurons the signals travel as electric impulses along the axons (Fig. 24.1). When they reach the end of the neuron, the signals are transmitted to adjacent neurons by specific compounds called **neurotransmitters.** Communication between the eyes and the brain, for example, is by neural transmission.

Nerve cells are present throughout the body and, together with the brain, constitute the nervous system.

Once the danger signals are processed in the brain, other neurons carry messages to the muscles and to the endocrine glands. The message to the muscles is to run away or to take some other action in response to the fire (save the baby or run to the fire extinguisher, for example). In order to do one of these things, the muscles must be activated. Again, neurotransmitters carry the messages from the neurons to the muscle cells and to endocrine glands. The endocrine glands are stimulated, and a different chemical signal, called a **hormone,** is secreted into your blood stream. "The adrenalin begins to flow." The danger signal carried by adrenalin (Box 15A) makes quick energy available so that the muscles can contract and relax rapidly, allowing your body to take quick action to avoid the danger.

If during the process you get cut and foreign bodies enter your blood stream, special cells in your blood, the **lymphocytes,** recognize the foreign bodies and manufacture still other communication compounds, called **immunoglobulins,** to fight and eliminate the foreign bodies.

Without these chemical communicators, the whole organism—you—would not survive because there is a constant need for coordinated efforts to face a complex outside world.

In this chapter we will investigate the chemistry and the mode of communication achieved by two groups of communication chemicals: neurotransmitters and hormones. The action of immunoglobulins will be discussed in Chapter 25. (The classification is made on the basis of how each works rather than on chemical structure.)

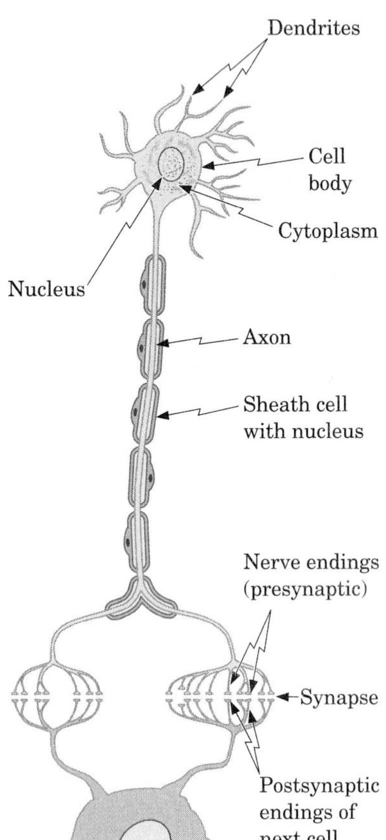

Figure **24.1**
Neuron and synapse.

24.2 Neurotransmitters

Neurotransmitters are compounds that communicate between two nerve cells or between a nerve cell and another cell (such as a muscle cell). If we look at a nerve cell (Fig. 24.1), we see that it consists of a main cell body from which projects a long, fiber-like part called an **axon.** Coming off the other side of the main body are hair-like structures called **dendrites.**

Neurons do not typically touch each other. Between the axon end of one neuron and the cell body or dendrite end of the next, there is a space filled with an aqueous fluid. This fluid-filled space is called a **synapse.** If the chemical signal travels, say, from top to bottom, we call the nerve ends on the top **presynaptic** and those on the bottom **postsynaptic.**

The neurotransmitters are stored at the presynaptic site in **vesicles,** which are small, membrane-enclosed packages. Events begin when a message is transmitted from one neuron to the next by neurotransmitters. The message is carried by calcium ions. When the Ca^{2+} concentration in a neuron reaches a certain level (more than 10^{-4} *M*), the vesicles fuse with the synaptic membrane of the nerve cells. Then they empty the neurotransmitters into the synapse. They travel across the synapse and are adsorbed onto specific receptor sites. There are, broadly speaking, three classes of neurotransmitters: *cholinergic, adrenergic,* and *peptidergic neurotransmitters.* This classification is based on the chemical nature of the most important neurotransmitters in each group.

The neurotransmitters fit into the receptor sites in a manner reminiscent of the lock-and-key model mentioned in Section 19.5.

24.3 Acetylcholine

The main **cholinergic neurotransmitter** is acetylcholine:

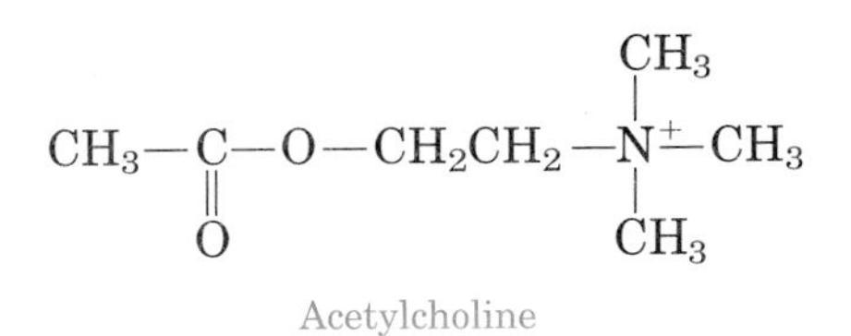

$$CH_3-\underset{\underset{O}{\|}}{C}-O-CH_2CH_2-\overset{\overset{CH_3}{|}}{\underset{\underset{CH_3}{|}}{N^+}}-CH_3$$

Acetylcholine

When an electric nerve impulse moves along the neuron and reaches the vesicles, it causes acetylcholine molecules to be released into the synapse. These molecules travel across the short synapse to the next neuron, where they are adsorbed onto specific receptor sites on the postsynaptic membrane. The presence of the acetylcholine molecules at the postsynaptic receptor site then triggers chemical reactions in which ions can freely cross membranes. Because it involves ions, which carry electric charges, this process is translated into an electric signal that now travels along this neuron until it reaches the other end, where the process is repeated.

By this means, the message moves from neuron to neuron until finally it gets transmitted, again by acetylcholine molecules, to the muscles or endocrine glands that are the ultimate target of the message.

Meanwhile, what happens to the postsynaptic receptor sites of the neurons after the message gets transmitted? If the acetylcholine molecules remained there, no signal would be transmitted even though a message was being received from the previous neuron. Therefore, the acetyl-

Box 24A

Alzheimer's Disease and Acetylcholine Transferase

Alzheimer's disease is the name given to the symptoms of senile behavior (second childhood) that afflict about 1.5 million older people in the United States. One of the postmortem indications for this disease is the presence of a neurofibrillary tangle. The tangle disrupts nerve cell functions creating neurons that are distrophied. These neurons congregate and form plaques that are filled with a protein called β-amyloid. This water-soluble protein is present in normal patients but becomes water insoluble in Alzheimer's patients. The result is that these nerve cells in the cerebral cortex die, the brain becomes smaller, and part of the cortex atrophies. People with Alzheimer's disease are forgetful, especially about recent events. As the disease advances they become confused and in severe cases lose their ability to speak; then they need total care. There is as yet no cure for this disease.

It has been found that patients with Alzheimer's disease have significantly diminished acetylcholine transferase activity in their brains. This enzyme synthesizes acetylcholine by transferring the acetyl group from acetyl CoA to choline:

$$CH_3\overset{\overset{O}{\|}}{C}-S-CoA + HO-CH_2CH_2-\underset{\underset{CH_3}{|}}{\overset{\overset{CH_3}{|}}{N^{\pm}}}-CH_3 \longrightarrow CH_3\overset{\overset{O}{\|}}{C}-O-CH_2CH_2-\underset{\underset{CH_3}{|}}{\overset{\overset{CH_3}{|}}{N^{\pm}}}-CH_3 + CoA-SH$$

The diminished concentration of acetylcholine can be partially compensated for by inhibiting the enzyme acetylcholinesterase, which decomposes acetylcholine. Certain drugs, which act as acetylcholinesterase inhibitors, have been shown to improve memory and other cognitive functions in some people with the disease. Two such drugs, Cognex (tacrine) and Aricept, are already on the market. The alkaloid huperzine A, an active ingredient of Chinese herb tea, used for centuries to improve memory, is also a potent inhibitor of acetylcholinesterase.

choline must be removed so that the neuron is reactivated. The acetylcholine is removed rapidly from the receptor site by the enzyme acetylcholinesterase, which hydrolyzes it.

$$H_2O + \underbrace{CH_3\overset{\overset{O}{\|}}{C}-O-CH_2CH_2\underset{\underset{CH_3}{|}}{\overset{\overset{CH_3}{|}}{N^{\pm}}}-CH_3}_{\text{Acetylcholine}} \xrightarrow{\text{acetylcholin-esterase}} \underbrace{CH_3\overset{\overset{O}{\|}}{C}-OH}_{\text{Acetic acid}} + \underbrace{HO-CH_2CH_2\underset{\underset{CH_3}{|}}{\overset{\overset{CH_3}{|}}{N^{\pm}}}-CH_3}_{\text{Choline}}$$

The removal of acetylcholine from the receptor site opens the site once again and gets it ready for the next message. This rapid removal enables the nerves to transmit more than 100 signals per second.

The action of the acetylcholinesterase enzyme is obviously essential to the whole process. When this enzyme is inhibited, the removal of

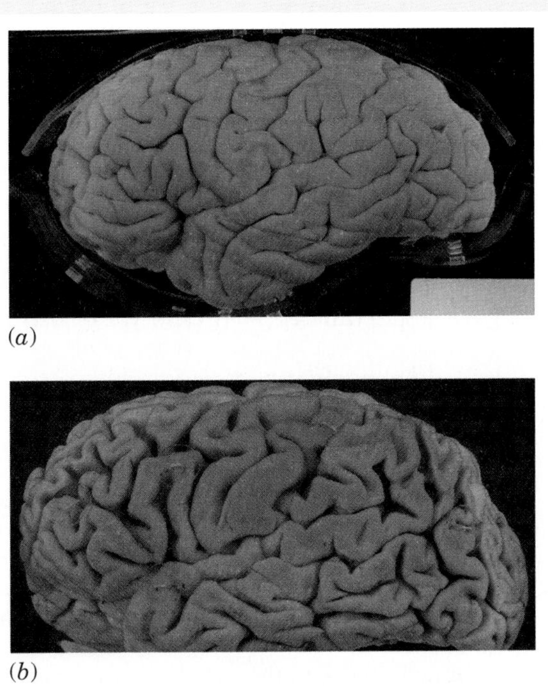

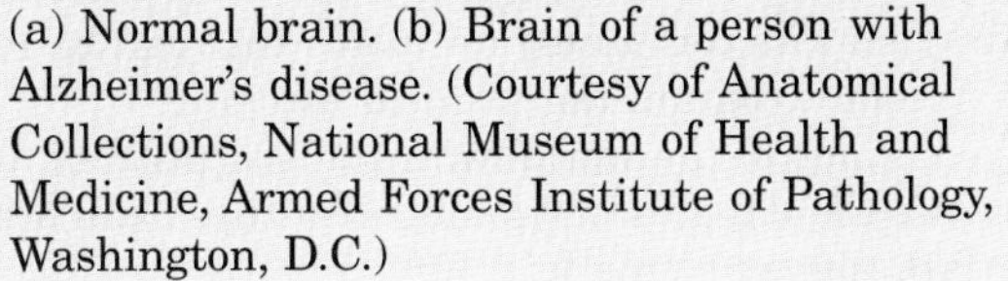

(a) Normal brain. (b) Brain of a person with Alzheimer's disease. (Courtesy of Anatomical Collections, National Museum of Health and Medicine, Armed Forces Institute of Pathology, Washington, D.C.)

acetylcholine is incomplete, and nerve transmission ceases. For example, the plant extract *curare* inhibits acetylcholinesterase and in large doses can cause death by paralysis. This is how the poisoned arrows of the Amazon Indians work. In small doses, curare is used as a muscle relaxant. A less dangerous muscle relaxant is decamethionium:

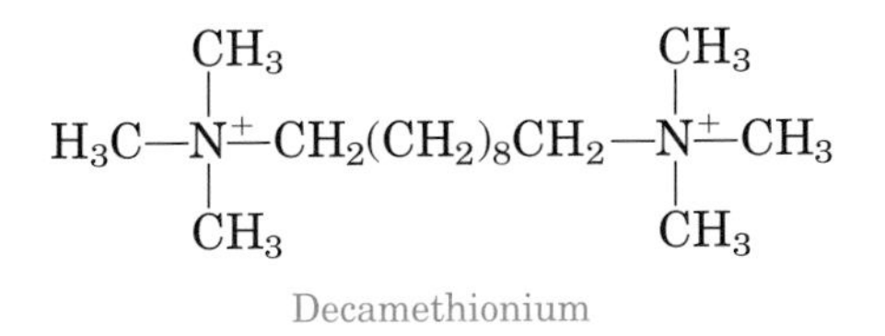

$$H_3C-\overset{\displaystyle CH_3}{\underset{\displaystyle CH_3}{\overset{|}{\underset{|}{N^+}}}}-CH_2(CH_2)_8CH_2-\overset{\displaystyle CH_3}{\underset{\displaystyle CH_3}{\overset{|}{\underset{|}{N^+}}}}-CH_3$$

Decamethionium

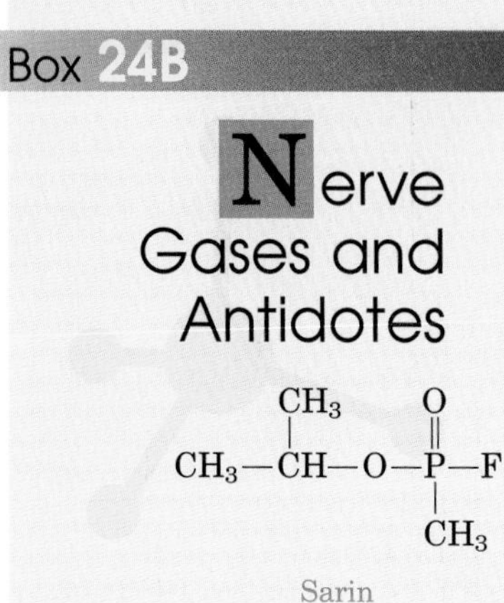

This molecule resembles the choline end of acetylcholine and therefore acts as a competitive inhibitor of acetylcholinesterase. Succinylcholine,

Box 24B

Nerve Gases and Antidotes

Most nerve gases in the military arsenal exert their lethal effect by binding to acetylcholinesterase. Under normal conditions, this enzyme decomposes the synaptic neurotransmitter acetylcholine within a few milliseconds after it is released at the nerve endings. Nerve gases such as Sarin (agent GB, also called Tabun), Soman (agent GD), and agent VX are organic phosphonates related to such pesticides as parathion (the latter being much less lethal, of course):

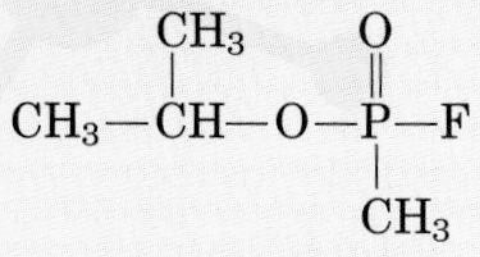

$$CH_3-CH(CH_3)-O-P(=O)(CH_3)-F$$

Sarin (agent GB)

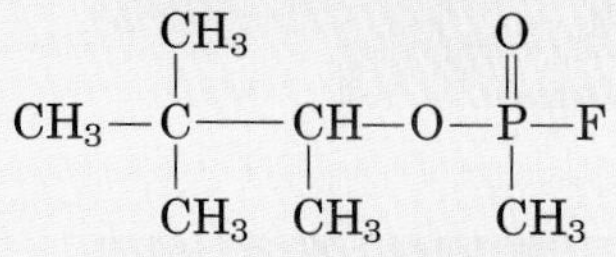

$$CH_3-C(CH_3)_2-CH(CH_3)-O-P(=O)(CH_3)-F$$

Soman (agent GD)

$$(CH_3)_2CH-N(CH(CH_3)_2)-CH_2CH_2-S-P(=O)(CH_3)-O-CH_2CH_3$$

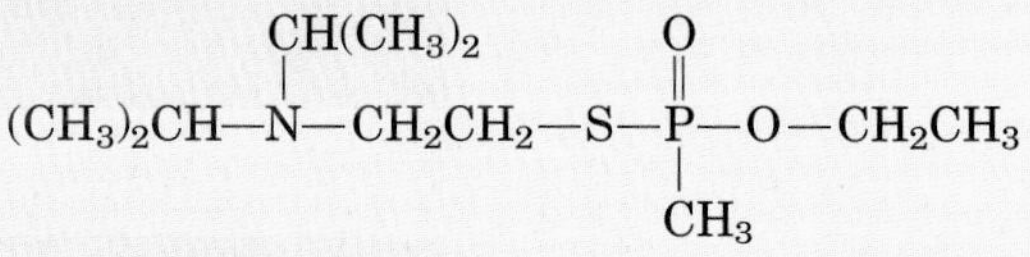

Agent VX

If any of these phosphonates bind to acetylcholinesterase, the enzyme is completely inactivated and the transmission of nerve signals stops. The result is a cascade of symptoms: sweating, bronchial constriction due to mucus buildup, dimming of vision, vomiting, choking, convulsions, paralysis, and respiratory failure. Direct inhalation of as little as 0.5 mg can cause death within a few minutes. If the dosage is less or if the nerve gas is absorbed through the skin, the lethal effect may take several hours. In warfare, protective clothing and gas masks are effective countermeasures. Also, first aid kits containing antidotes that can be injected are available. These antidotes contain the alkaloid atropine.

The nations of the world, with the exception of a few Arab countries, recognizing the dangers of proliferating these and other chemicals used for warfare, signed a treaty in 1993. The treaty obliges all the signatory governments to destroy their stockpiles of chemical weapons and to dismantle the plants that manufacture them within 10 years.

another competitive inhibitor (Box 19A), is also an excellent muscle relaxant.

24.4 Monoamines and Amino Acids: Adrenergic Neurotransmitters

The second class of neurotransmitters, the adrenergic neurotransmitters, includes such monoamines as epinephrine (Box 15A), serotonin, and dopamine. These monoamines transmit nerve signals by a mechanism whose beginning is similar to the action of acetylcholine. But once the monoamine neurotransmitter (for example, norepinephrine) is adsorbed onto the receptor site, it activates a secondary messenger inside the cell, called **cyclic AMP (cAMP).** The manufacture of cAMP activates processes that result in the transmission of an electrical signal. The cAMP is manufactured by adenylate cyclase from ATP:

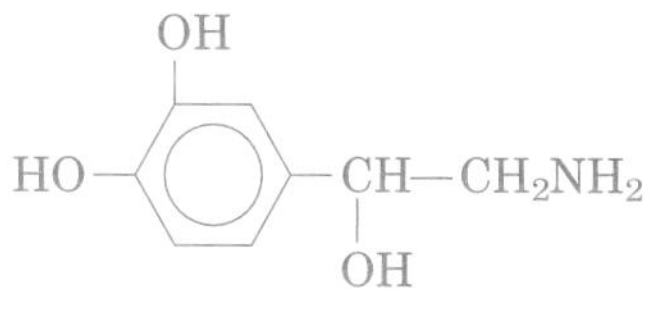

$$HO-C_6H_3(OH)-CH(OH)-CH_2NH_2$$

Norepinephrine

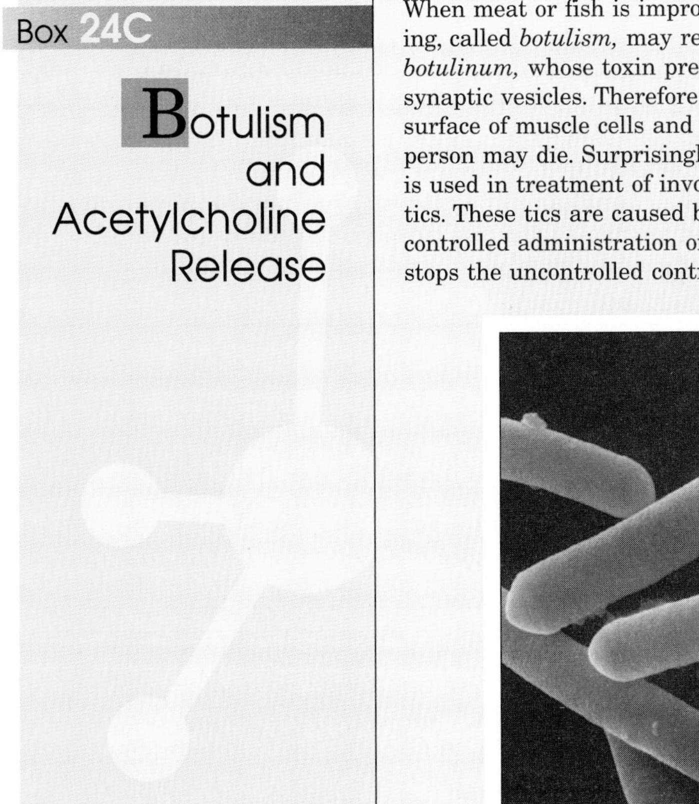

Box 24C Botulism and Acetylcholine Release

When meat or fish is improperly cooked or preserved a deadly food poisoning, called *botulism,* may result. The culprit is the bacterium, *Clostridium botulinum,* whose toxin prevents the release of acetylcholine from the presynaptic vesicles. Therefore no neurotransmitter reaches the receptors on the surface of muscle cells and the muscles no longer contract. If untreated the person may die. Surprisingly, the botulin toxin actually has a medical use. It is used in treatment of involuntary muscle spasms, for example, in facial tics. These tics are caused by uncontrolled release of acetylcholine. Therefore, controlled administration of the toxin, applied locally to the facial muscles, stops the uncontrolled contractions and relieves the facial distortions.

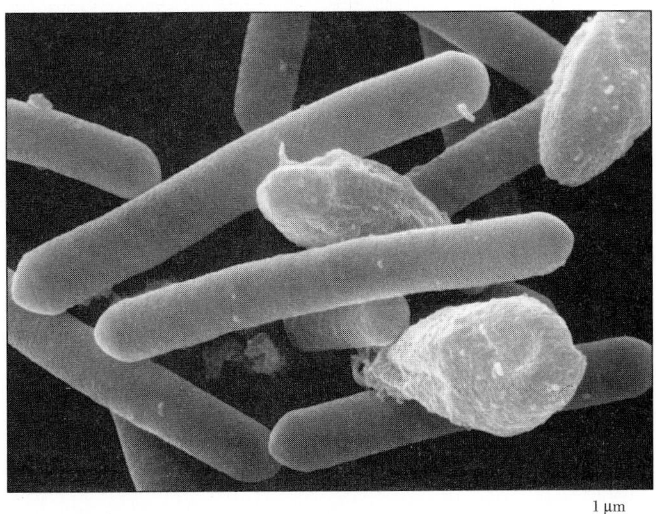

Clostridium botulinum is a food-poisoning bacteria. (U.S. Department of Agriculture.)

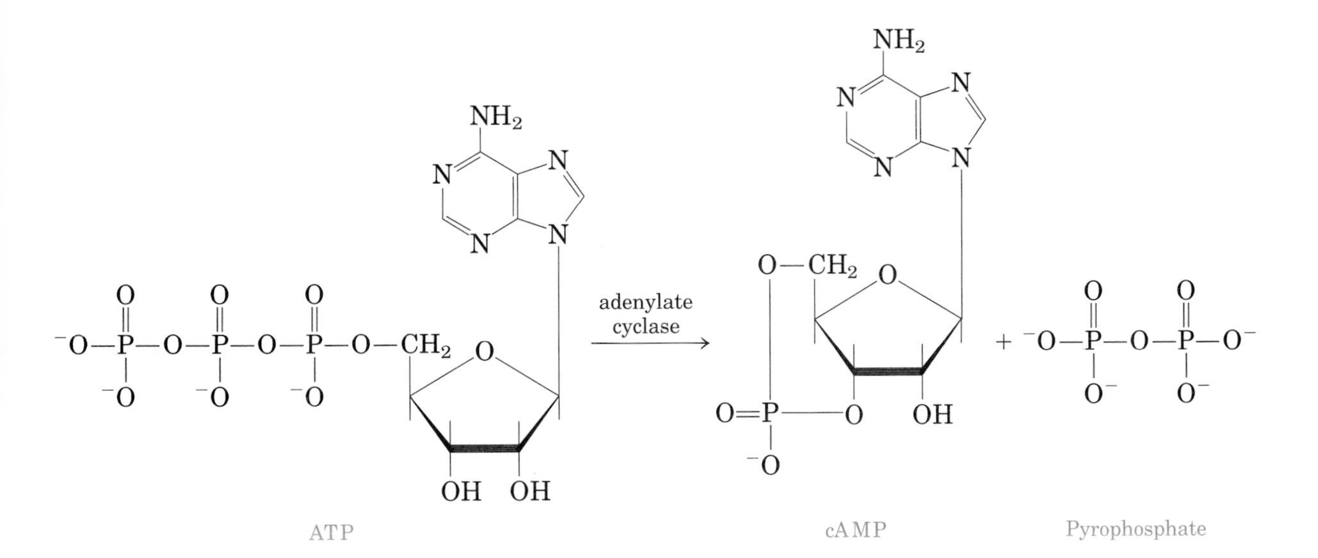

$$\text{ATP} \xrightarrow[\text{cyclase}]{\text{adenylate}} \text{cAMP} + {}^{-}O{-}\overset{O}{\overset{\|}{P}}(O^{-}){-}O{-}\overset{O}{\overset{\|}{P}}(O^{-}){-}O^{-}$$

ATP cAMP Pyrophosphate

The action of cAMP is depicted in Figure 24.2. The activation of adenylate cyclase accomplishes two important goals: (1) It converts an event occurring at the outer surface of the target cell (adsorption onto receptor site) to a change inside the target cell (release of cAMP). Thus the primary messenger (neurotransmitter or hormone) does not have to cross the membrane. (2) It *amplifies* the signal. One molecule adsorbed on the receptor triggers the adenylate cyclase to make many cAMP molecules. Thus the signal is amplified many thousands of times.

How does this signal amplification stop? When the neurotransmitter or hormone dissociates from the receptor, the adenylate cyclase stops

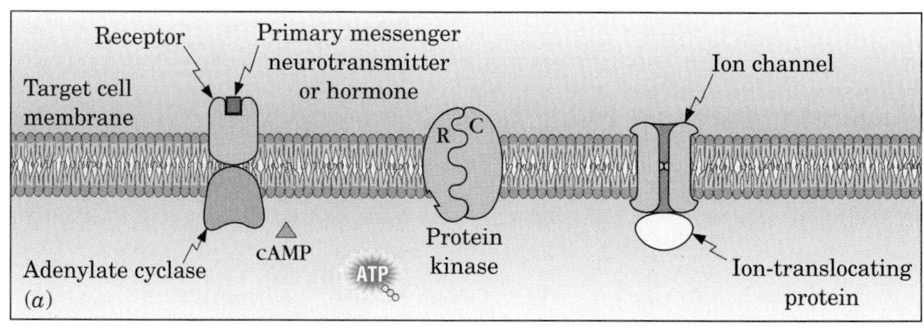

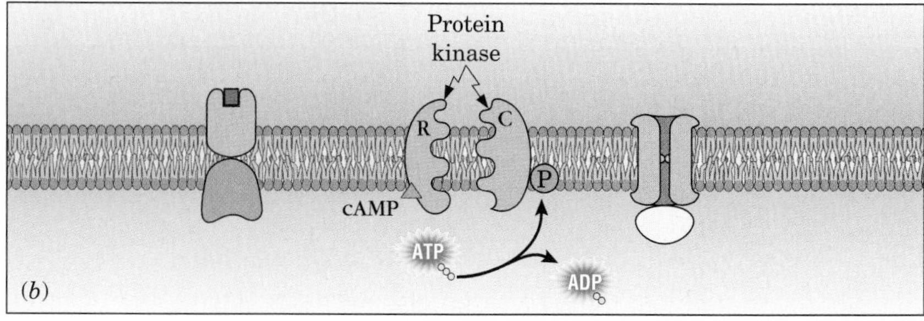

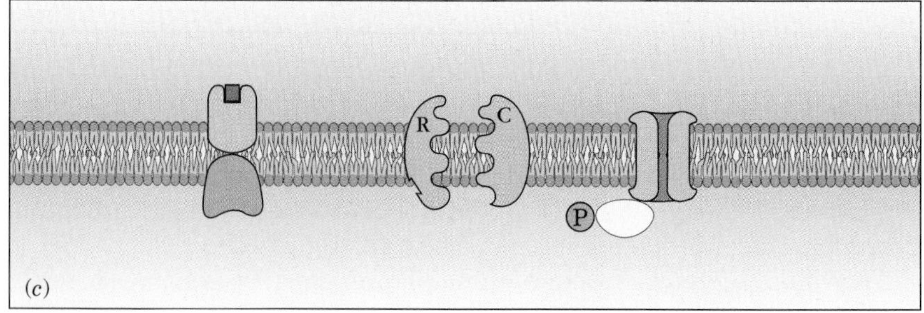

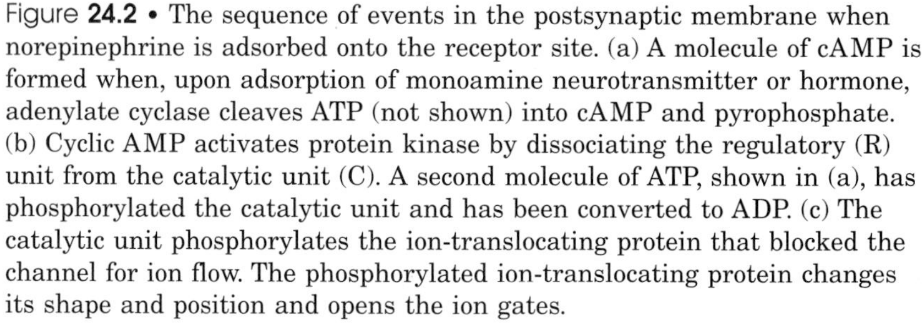

Figure **24.2** • The sequence of events in the postsynaptic membrane when norepinephrine is adsorbed onto the receptor site. (a) A molecule of cAMP is formed when, upon adsorption of monoamine neurotransmitter or hormone, adenylate cyclase cleaves ATP (not shown) into cAMP and pyrophosphate. (b) Cyclic AMP activates protein kinase by dissociating the regulatory (R) unit from the catalytic unit (C). A second molecule of ATP, shown in (a), has phosphorylated the catalytic unit and has been converted to ADP. (c) The catalytic unit phosphorylates the ion-translocating protein that blocked the channel for ion flow. The phosphorylated ion-translocating protein changes its shape and position and opens the ion gates.

the manufacture of cAMP. The cAMP already produced is destroyed by the enzyme phosphodiesterase.

The amplification through the secondary messenger (cAMP) is a relatively slow process. It may take from 0.1 second to a few minutes. Therefore, in cases where the transmission of signals must be fast, in milli- or microseconds, a neurotransmitter such as acetylcholine acts on membrane permeability directly, without the mediation of a secondary messenger.

The inactivation of the adrenergic neurotransmitters is somewhat different from that of the cholinergic transmitters. While acetylcholine is decomposed by acetylcholinesterase, most of the adrenergic neurotransmitters are inactivated in a different way. The body inactivates monoamines by oxidizing them to aldehydes. Enzymes that catalyze these reactions are very common in the body. They are called *monoamine oxidases* (MAOs). For example, there is an MAO that converts both epinephrine and norepinephrine to the corresponding aldehyde:

OH, HO—(benzene ring)—CH(OH)—CH_2—$\overset{+}{N}H_2$—CH_3

Epinephrine (salt form)

OH, HO—(benzene ring)—CH(OH)—CH_2—NH_3^+

Norepinephrine (salt form)

$\xrightarrow{\text{MAO}}$

OH, HO—(benzene ring)—CH(OH)—C(=O)—H

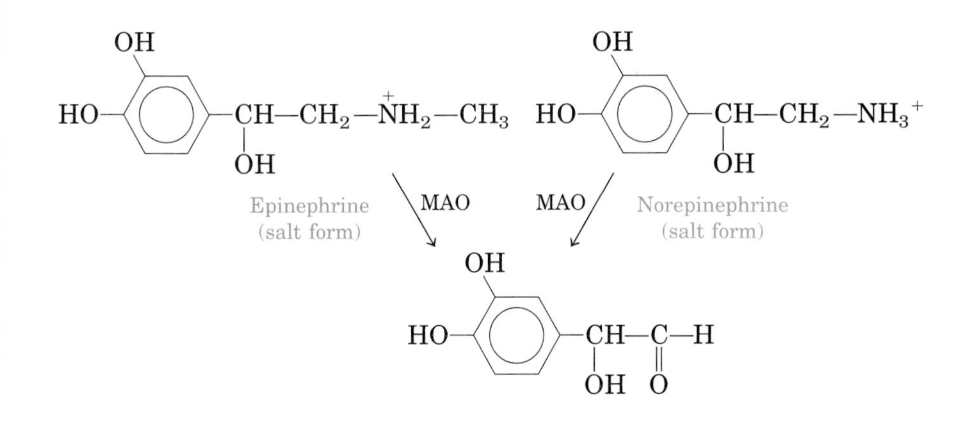

Many drugs that are used as antidepressants or antihypertensive agents are inhibitors of monoamine oxidases, for example, Tofranil and Elavil. These inhibitors prevent MAOs from converting monoamines to aldehydes, thus increasing the concentration of the active adrenergic neurotransmitters.

Shortly after adsorption onto the postsynaptic membrane, the neurotransmitter comes off the receptor site and is reabsorbed through the presynaptic membrane and stored again in the vesicles.

The neurotransmitter histamine is present in mammalian brains:

COO^-—HC(NH_3^+)—CH_2—(imidazolium ring: $\overset{+}{H}N$, NH) $\xrightarrow{H^+}$ H—HC(NH_3^+)—CH_2—(imidazolium ring: $\overset{+}{H}N$, NH) + CO_2

Histidine Histamine

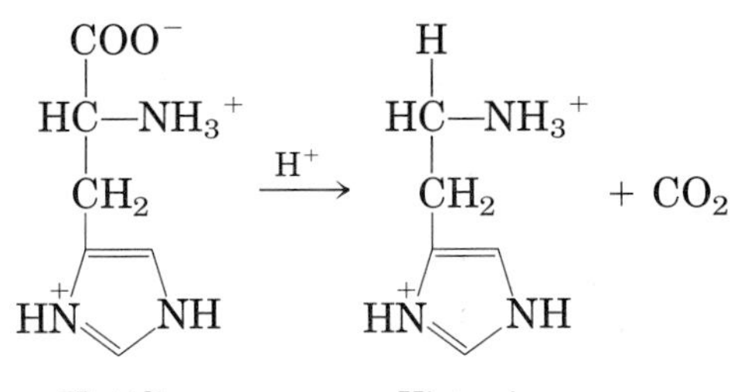

Histamine cannot readily pass the blood-brain barrier (Sec. 25.1) and must be synthesized in the brain neurons by a one-step decarboxylation of the amino acid histidine.

Box 24D

Parkinson's Disease—Depletion of Dopamine

Parkinson's disease is characterized by spastic motion of the eyelids as well as rhythmic tremors of the hands and other parts of the body, often when the patient is at rest. The posture of the patient changes to a forward, bent-over position; walking becomes slow, with shuffling footsteps. This is a degenerative nerve disease. The neurons affected contain, under normal conditions, mostly dopamine as a neurotransmitter. People with Parkinson's disease have depleted amounts of dopamine in their brains. However, the dopamine receptors are not affected. Thus, when L-dopa is administered, many of these patients are able to synthesize dopamine and resume normal nerve transmission. In these patients, L-dopa reverses the symptoms of Parkinson's disease, although the respite is only temporary. In other patients the L-dopa regimen provides little benefit.

Dopamine cannot be administered directly because it cannot penetrate the blood-brain barrier and therefore does not reach the tissue where its action is needed. L-Dopa, on the other hand, is transported through the arterial wall and is converted to dopamine in the brain:

$$\text{HO(OH)}C_6H_3\text{—}CH_2\text{—}\underset{COO^-}{\overset{NH_3^+}{C}}\text{—H} \longrightarrow \text{HO(OH)}C_6H_3\text{—}CH_2\text{—}\underset{H}{\overset{NH_2}{C}}\text{—H} + CO_2$$

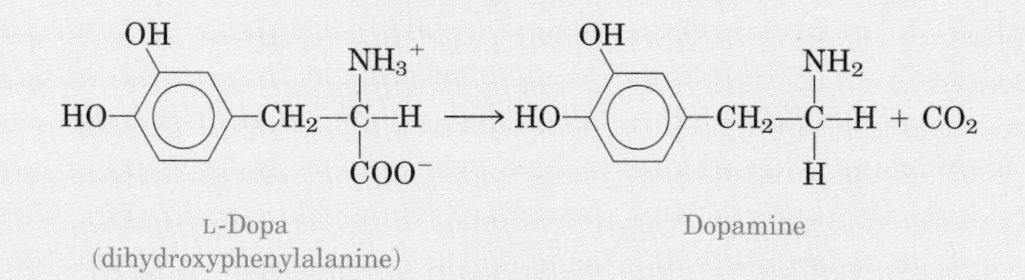

L-Dopa (dihydroxyphenylalanine) — Dopamine

Recently, a new drug treatment has shown promise as a possible route for slowing neuron degeneration. Deprenyl, a monoamine oxidase (MAO) inhibitor, given together with L-dopa, reduced the symptoms of Parkinson's disease and even increased the life span of patients. It appears that Deprenyl not only increases the level of dopamine by preventing its oxidation by MAOs but also reduces the degeneration of neurons.

Synthesis and degradation of dopamine is not the only way the brain keeps its concentration at a steady state. The concentration is also controlled by specific proteins, called transporters, that ferry the used dopamine from the receptor back across the synapse into the original neuron from whence it came. Cocaine addiction (Box 15G) works through such a transporter. Cocaine binds to the dopamine transporter, like a reversible inhibitor, preventing the uptake of dopamine. Thus, the dopamine is not transported back to the original neuron and stays in the synapse, increasing the continuous firing of signals, which is the psychostimulatory effect associated with a cocaine "high."

Sir James W. Black of England received the 1988 Nobel Prize in Medicine for the invention of cimetidine and such other drugs as propranolol (Table 24.1).

The action of histamine as a neurotransmitter is very similar to that of other monoamines. There are two kinds of receptors for histamine. One receptor, H_1, can be blocked by classic antihistamines such as dimenhydrinate (Dramamine) or diphenhydramine (Benadryl). The other receptor, H_2, can be blocked by ranitidine (Zantac) and cimetidine (Tagamet). H_1 receptors are found in the respiratory tract. They affect the vascular, muscular, and secretory changes associated with hay fever and

Box 24E

Chemically Induced Headaches

There are a number of compounds that cause headaches because they dilate the arteries in the head. This dilation occurs as a response to signals generated either at certain receptor sites in the wall of the arteries or in neurons in the brain stem. In any case, after the compounds bind to these receptors, muscle relaxation occurs in the arterial wall, which causes the dilation. Some of these compounds are nitroglycerine; nitrites in smoked and cured meats; monosodium glutamate, a seasoning used especially in Chinese food (Box 21D); tyramine in cheeses; and phenylethylamine in red wines (frequently the cause of hangover headaches).

asthma. Therefore, the classic antihistamines relieve these symptoms. The H_2 receptors are mainly in the stomach and affect the secretion of HCl. Cimetidine and ranitidine, both H_2 blockers, reduce acid secretion and thus are effective drugs for ulcer patients. However, once the patients are off the medication, the ulcer frequently recurs. This and other developments prompted a new strategy. It was found that a bacterium called *Helicobacter pylori* was almost always present in patients with certain types of ulcers or gastric cancers. These bacteria live under the mucus coating of the stomach and release a toxin that causes the ulceration. The new treatment for ulcers and gastric disorders uses antibiotics such as amoxicillin or tetracycline, or a combination of antibiotics and H_2 blockers. Once the *Helicobacter pylori* bacteria are killed by the antibiotics, recurrence of the ulcer is prevented.

Amino acids distributed throughout the neurons also act as neurotransmitters. Some of them, such as glutamic acid, aspartic acid, and cysteine, act as excitatory neurotransmitters similar to acetylcholine and norepinephrine. Others, such as glycine, β-alanine, taurine, and, mainly, γ-aminobutyric acid (GABA)

$H_2NCH_2CH_2SO_2OH$	$H_2NCH_2CH_2COOH$	$H_2NCH_2CH_2CH_2COOH$
Taurine	β-Alanine	γ-Aminobutyric acid (GABA)

reduce neurotransmission. Note that some of these neurotransmitter amino acids are not found in proteins.

24.5 Peptidergic Neurotransmitters

In the last few years, scientists have isolated a number of brain peptides that have affinity for certain receptors and therefore act as if they were neurotransmitters in the classic sense. Some 25 or 30 such peptides are now known, and not all of them behave as true neurotransmitters.

The first brain peptides isolated were the enkephalins. These pentapeptides are present in certain nerve cell terminals. They bind to spe-

Box 24F

Nitric Oxide and Long-Term Memory

It has been known for some time that a primitive part of the brain, the *hippocampus,* has something to do with long-term memory. In the 1950s a physician found that a patient suffering from frequent epileptic seizures was cured by removal of the hippocampus. However, the patient also permanently lost his ability to remember new things. Faces or facts learned before the operation could be recalled with ease, but not those encountered after the operation. Since then the hippocampus has been studied not so much as a suspected storehouse of memories but for its "long-term potentiation" ability, or LTP. LTP is turned on by a brief experience and persists for a long time.

The molecular mechanism proposed for LTP is a process in which the primary messenger (the neurotransmitter glutamic acid) that shuttles between two nerve cells across the synapse also elicits the production of a secondary messenger. This secondary messenger goes back to the first nerve cell and prompts it to release more glutamic acid. Such a process would strengthen the connections between the two nerve cells, thereby inducing LTP.

The most likely candidate for this secondary messenger is nitric oxide, NO. This simple molecule is synthesized in the cells when arginine is converted to citrulline. (These two compounds appear in the urea cycle, Sec. 21.8.) Nitric oxide is a relatively nonpolar molecule, and once it has been produced in the second nerve cell, it quickly diffuses across the lipid bilayer membrane. Within its short half-life, 4 to 6 seconds, it can reach the neighboring first nerve cell and deliver its message. Four seconds are sufficient to induce LTP. It has been found that NO can act as an intercellular messenger in other parts of the body as well. It appears in the intercellular communication between blood vessels and the smooth muscles surrounding them. Thus it controls the constriction and relaxation of blood vessels.

Furthermore, it appears that nitric oxide is used by the immune system (Sec. 25.10) to fight infections by viruses. Studies have shown that NO is a potent agent against certain viruses, including the herpes simplex type 1 virus, which causes cold sores. When the viral infection begins the immune system manufactures the protein interferon, which in turn activates a gene that produces the enzyme nitric oxide synthase.

While the toxic effect of NO against viral invaders is beneficial to the host, in other situations the toxicity of NO can be detrimental. In strokes and neurodegenerative diseases large amounts of NO are formed, and these contribute to the death of neurons. For this reason pharmaceutical companies are conducting research to develop nitric oxide synthase inhibitors to be used as antistroke drugs, and as a potential remedy for Parkinson's disease.

cific pain receptors and seem to control pain perception. Since they bind to the receptor site that also binds the pain-killing alkaloid morphine (Sec. 15.9), it is assumed that it is the N-terminal end of the pentapeptide that fits the receptor (Fig. 24.3).

Table 24.1 lists some typical examples of the three main classes of neurotransmitters and drugs that affect their action.

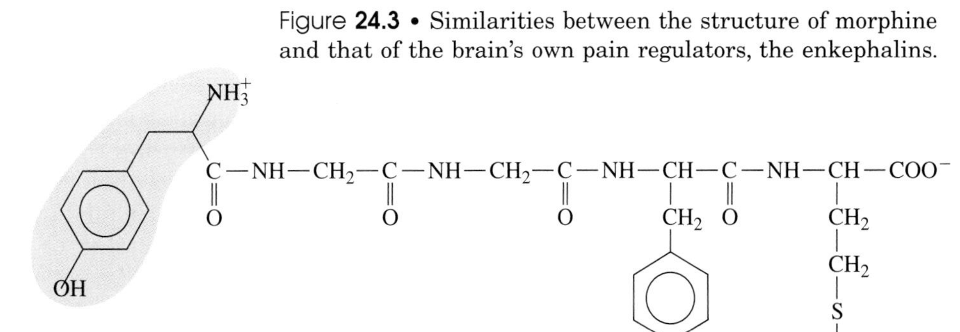

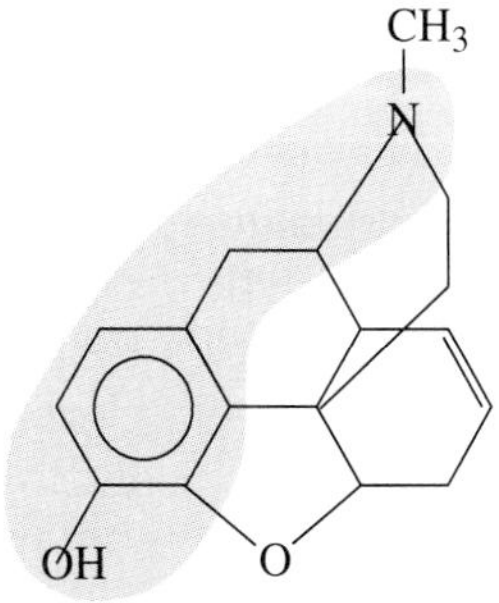

Morphine

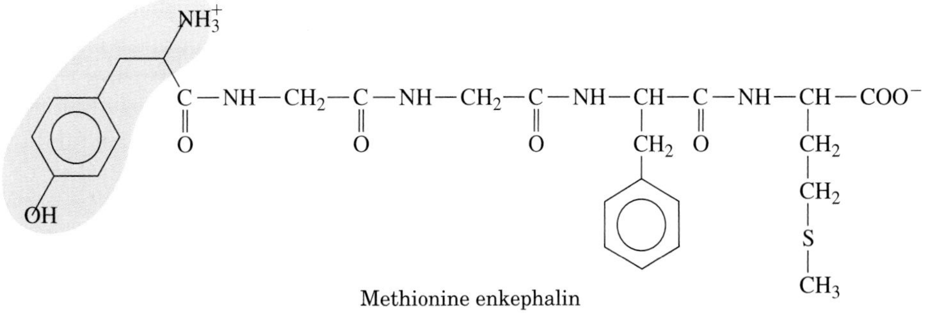

Methionine enkephalin

Figure **24.3** • Similarities between the structure of morphine and that of the brain's own pain regulators, the enkephalins.

Table 24.1 Drugs That Affect Nerve Transmission

	Drugs That Affect Receptor Sites		Drugs That Affect Available Concentration of Neurotransmitters or Their Removal from Receptors	
Neurotransmitter	***Agonists (Activate Receptor Sites)***	***Antagonists (Block Receptor Sites)***	***Increase Concentration***	***Decrease Concentration***
Acetylcholine (cholinergic)	Nicotine Pilocarpine Carbachol	Curare Succinylcholine Atropine Propantheline (Pro-Banthine)	Malathion Nerve gases	*Clostridium botulinum* toxin
Norepinephrine (adrenergic)	Phenylephrine (Neo-Synephrine) Epinephrine (adrenalin)	Methyldopa (Aldomet) Propranolol (Inderal, Lopresor)	Amphetamines Iproniazide Antidepressants (Tofranil, Elavil)	Reserpine
Dopamine (adrenergic)		Clozapine (Clorazil)	Deprenyl	
Serotonin (adrenergic)			Antidepressant Fluoxetin (Prozac)	
Histamine (adrenergic)	2-Methylhistamine Betazole Pentagastin	Fexofenadine (Allegra) Promethazine (Phenergan) Diphenylhydramine (Benadryl) Ranitidine (Zantac) Cimetidine (Tagamet)	Histidine	Hydrazino histidine
Enkephalin (peptidergic)	Opiate	Morphine Heroin Demerol		Naloxone (Narcan)

24.6 Hormones

Glands that secrete hormones are called **endocrine glands.**

Hormones are diverse compounds secreted by specific tissues (glands), released into the blood stream, and then adsorbed onto specific receptor sites, usually relatively far from their source. This is the classic definition of a hormone. Some of the principal hormones are tabulated in Table 24.2. Figure 24.4 shows the location of the major hormone-secreting glands, and Figure 24.5 shows the target organs of hormones secreted by the pituitary gland.

This classic definition is too rigid, however, for what modern science knows about hormones. For example, in Section 24.4 we saw that epinephrine and norepinephrine are neurotransmitters. But these compounds also fit the classic definition of a hormone. This shows that the distinction between hormones and neurotransmitters is physiological (hormones are secreted into the blood stream) and not chemical.

Table 24.2 The Principal Hormones and Their Action

Gland	Hormone	Action	Structures Shown on Page
Parathyroid	Parathyroid hormone	Increases blood calcium	
		Excretion of phosphate by kidney	
Thyroid	Thyroxine (T_4)	Growth, maturation, and metabolic rate	417
	Triiodothyronine (T_3)	Metamorphosis	
Pancreatic islets			
Beta cells	Insulin	Hypoglycemic factor	604
		Regulation of carbohydrates, fats, and proteins	
Alpha cells	Glucagon	Liver glycogenolysis	
Adrenal medulla	Epinephrine	Liver and muscle glycogenolysis	755
	Norepinephrine		
Adrenal cortex	Cortisol	Carbohydrate metabolism	580
	Aldosterone	Mineral metabolism	580
	Adrenal androgens	Androgenic activity (esp. females)	
Kidney	Renin	Hydrolysis of blood precursor protein to yield angiotensin	
Anterior pituitary	Luteinizing hormone	Causes ovulation	
	Interstitial cell-stimulating hormone	Formation of testosterone and progesterone in interstitial cells	
	Prolactin	Growth of mammary gland	
	Mammotropin	Lactation	
		Corpus luteum function	
Posterior pituitary	Vasopressin	Contraction of blood vessels	606
		Kidney reabsorption of water	
	Oxytocin	Stimulates uterine contraction and milk ejection	606
Ovaries	Estradiol	Estrous cycle	580
	Progesterone	Female sex characteristics	580
Testes	Testosterone	Male sex characteristics	580
	Androgens	Spermatogenesis	

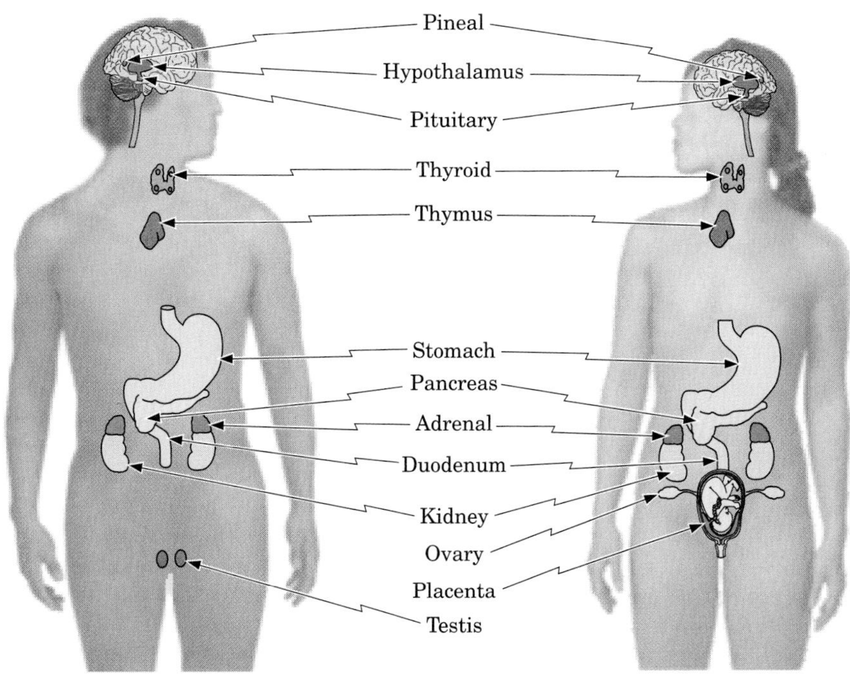

Figure **24.4** • Location of the major hormone-secreting (endocrine) organs.

Whether a certain compound is considered to be a neurotransmitter or a hormone depends on whether it acts over a short distance across a synapse (2×10^{-6} cm), in which case it is a neurotransmitter, or over a long distance (20 cm) from secretory gland through the blood stream to target cell, in which case it is a hormone.

Chemically, hormones can be classified into three groups: (1) *small molecules* derived from amino acids, for example, epinephrine and thyroxine; (2) *peptides* and *proteins,* for example, insulin, glucagon, and vasopressin; (3) *steroids* derived from cholesterol, such as testosterone, cortisol, and aldosterone (Sec. 17.10).

Hormones can also be classified according to how they work. Some of them—epinephrine, for example—activate enzymes. Others *affect the synthesis of enzymes and proteins* by working on the transcription of genes (Sec. 23.7); steroid hormones work in this manner. Finally, some hormones affect the *permeability of membranes;* insulin and glucagon belong to this class.

Still another way of classifying hormones is according to their potential to act directly or through a secondary messenger. The steroid hormones can penetrate the cell membrane and also pass through the membrane of the nucleus. For example, estradiol stimulates uterine growth.

Other hormones act through secondary messengers, as mentioned in Section 24.4. For example, epinephrine, glucagon, luteinizing hormone, norepinephrine, and vasopressin use cAMP as a secondary messenger.

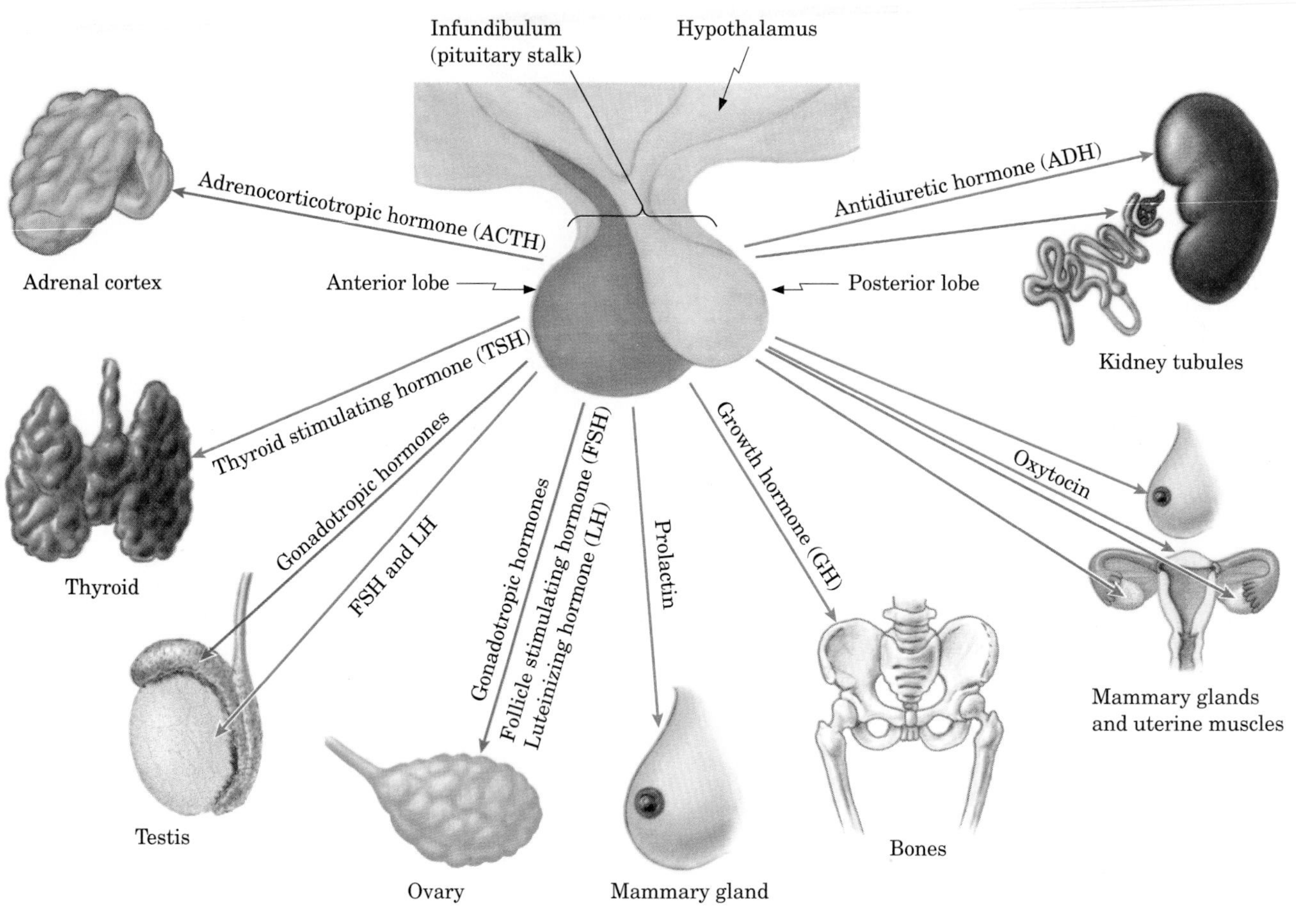

Figure **24.5** • The pituitary gland is suspended from the hypothalamus by a stalk of neural tissue. The hormones secreted by the anterior and posterior lobes of the pituitary gland and the target tissues they act upon are shown. (Modified from P. W. Davis and E. P. Solomon, *The World of Biology.* Philadelphia: Saunders College Publishing, 1986.)

Box 24G

Diabetes

The disease *diabetes mellitus* affects about 10 million people in the United States. In a normal person the pancreas, a large gland behind the stomach, secretes the hormone insulin, as well as other hormones. Diabetes usually results from low insulin secretion. Insulin is necessary for glucose molecules to penetrate such cells as brain, muscle, and fat cells, where they can be used. Insulin accomplishes this task by being adsorbed onto the receptors in the target cells. This adsorption triggers the manufacture of cyclic GMP (not cAMP), and this secondary messenger increases the transport of glucose molecules into the target cells.

In diabetic patients the glucose level rises to 600 mg/100 mL of blood or higher (normal is 80 to 100 mg/100 mL). There are two kinds of diabetes. In insulin-dependent diabetes, patients do not manufacture enough of this hormone in the pancreas. This disease develops early, before the age of 20, and must be treated with daily injections of insulin. Even with daily injection of insulin the blood sugar level fluctuates, and the fluctuation may cause other disorders, such as cataracts, blindness, kidney disease, heart attack, and nervous disorders.

In non-insulin-dependent diabetes the patient has enough insulin in the blood but cannot utilize it properly because there is an insufficient number of receptors in the target cells. These patients usually develop the disease after age 40 and are likely to be obese. Overweight people usually have a lower-than-normal number of insulin receptors in their adipose (fat) cells.

There are several oral drugs that help this second type of diabetic patient. These are mostly sulfonyl urea compounds, such as

$$CH_3-C_6H_4-S(=O)_2-NH-C(=O)-NH-CH_2CH_2CH_2CH_3$$

Tolbutamide (Orinase)

$$Cl-C_6H_4-S(=O)_2-NH-C(=O)-NH-CH_2CH_2CH_3$$

Chlorpropamide (Diabinese)

These drugs seem to control the symptoms of diabetes, but it is not known exactly how.

SUMMARY

Neurotransmitters send **chemical messengers** across a short distance—the **synapse** between two **neurons** or between a neuron and a muscle or endocrine gland cell. This communication occurs in milliseconds. Hormones transmit their signals more slowly and over a longer distance, from the source of their secretion (**endocrine gland**) through the blood stream into target cells. Chemical communicators interact with specific molecules called **receptors.**

There are three kinds of **neurotransmitters: cholinergic, adrenergic,** and **peptidergic.** Acetylcholine belongs to the first class, epinephrine (adrenalin) and norepinephrine to the second class, and enkephalins to the third class. Nerve transmission starts with the neurotransmitters packaged in **vesicles** in the **presynaptic end** of neurons. When these neurotransmitters are released, they cross the membrane and the synapse and are adsorbed onto receptor sites on the **postsynaptic membranes.** This adsorption triggers an electrical response. Some neurotransmitters act directly, while others act through a secondary messenger, **cyclic AMP.** After the electrical signal is triggered, the neurotransmitter molecules must be removed from the postsynaptic end. In the case of acetylcholine, this is done by an enzyme called acetylcholinesterase; in the case of monoamines, by enzymes (MAOs) that oxidize them to aldehydes.

Hormones can be classified into three groups: small molecules, peptides and proteins, and steroids. The first two classes bind to receptors on the target cell membrane and use secondary messengers to exert their influence. The third class penetrates the cell membrane, and their receptors are in the cytoplasm. Together with their receptors, they penetrate the cell nucleus. Hormones can act in three ways: (1) They activate enzymes, (2) they affect the gene transcription of an enzyme or protein, and (3) they change membrane permeability.

KEY TERMS

Adrenergic neurotransmitter (Sec. 24.4)
Axon (Sec. 24.2)
Chemical messenger (Sec. 24.1)
Cholinergic neurotransmitter (Sec. 24.3)
Dendrite (Sec. 24.2)
Endocrine gland (Sec. 24.6)
Hormone (Sec. 24.6)
Neuron (Sec. 24.1)
Neurotransmitter (Sec. 24.2)
Peptidergic neurotransmitter (Sec. 24.5)
Postsynaptic (Sec. 24.2)
Presynaptic (Sec. 24.2)
Receptor (Sec. 24.1)
Synapse (Sec. 24.2)
Vesicle (Sec. 24.2)

PROBLEMS

Chemical Communication

24.1 What are the names given to the two principal types of molecules used in cell-to-cell communication?

24.2 What kind of signal travels along the axon of a neuron?

Neurotransmitters

24.3 What is the difference between the presynaptic and postsynaptic ends of a neuron?

24.4 Define: (a) synapse (b) receptor (c) presynaptic (d) postsynaptic (e) mediator (f) vesicle

24.5 What is the role of Ca^{2+} in releasing neurotransmitters into the synapse?

Cholinergic Neurotransmitters

24.6 List all functional groups in the products of the reaction catalyzed by acetylcholinesterase.

24.7 How does acetylcholine transmit an electric signal from neuron to neuron?

***24.8** Which end of the acetylcholine molecule fits into the receptor site?

24.9 Explain how succinylcholine acts as a muscle relaxant (see Box 19A).

24.10 Which carbons of cAMP contain the phosphate linkages?

Adrenergic Neurotransmitters

24.11 (a) Find two monoamine neurotransmitters in Table 24.1. (b) Explain how they act. (c) What medication controls the particular diseases caused by the lack of monoamine neurotransmitters?

***24.12** How much energy is used up in manufacturing one mole of cAMP?

***24.13** How is the catalytic unit of protein kinase activated in adrenergic neurotransmission?

24.14 The formation of cyclic AMP is described in Section 24.4. Show by analogy how cyclic GMP is formed from GTP.

***24.15** By analogy to the action of MAO on epinephrine, write the structural formula of the product of the corresponding oxidation of dopamine.

24.16 What happens to the cAMP in the cell once the neurotransmitter is removed from the receptor?

24.17 Explain how adrenergic neurotransmission is affected by (a) amphetamines (b) reserpine (see Table 24.1)

***24.18** Which step in the events depicted in Figure 24.2 provides an electrical signal?

24.19 What kind of product is epinephrine oxidized to by MAO?

24.20 How is histamine removed from the receptor site?

24.21 Cyclic AMP affects the permeability of membranes for ion flow. (a) What blocks the ion channel? (b) How is this blockage removed? (c) What is the direct role of cAMP in this process?

24.22 Dramamine and cimetidine are both antihistamines. Would you expect Dramamine to cure ulcers and cimetidine to relieve the symptoms of asthma? Explain.

24.23 How do we treat ulcers caused by *Helicobacter pylori?*

24.24 Is there any relation between the structure of bile salts (Sec. 17.11) and that of one of the amino acid neurotransmitters?

24.25 What is unique in the structure of GABA that distinguishes it from all the amino acids that are present in proteins?

Peptidergic Neurotransmitters

24.26 What is the chemical nature of enkephalins?

24.27 What is the mode of action of Demerol as a pain killer? (See Table 24.1.)

Hormones

***24.28** The structures of vasopressin and oxytocin are given in Figure 18.4. They have completely different functions and different receptors. At which amino acid position, 3 or 8, is the difference greater? Explain.

24.29 What gland controls lactation?

24.30 To which of the three chemical groups do these hormones belong? (a) norepinephrine (b) thyroxine (c) oxytocin (d) progesterone

Boxes

24.31 (Box 24A) Alzheimer's disease causes loss of memory. What kind of drugs may provide some relief for, if not cure, this disease? How do they act?

24.32 (Box 24A) What is the difference in the β-amyloid protein of a brain from a normal patient compared with that from a patient with Alzheimer's disease?

24.33 (Box 24B) (a) What is the effect of nerve gases? (b) What molecular substance do they affect?

24.34 (Box 24C) What is the mode of action of botulinum toxin?

24.35 (Box 24C) How can a deadly botulinum toxin cure facial tics?

24.36 (Box 24D) Why would a dopamine pill be ineffective in treating Parkinson's disease?

24.37 (Box 24D) What is the mechanism by which cocaine stimulates the continuous firing of signals between neurons?

24.38 (Box 24E) How can a common food such as cheese cause headaches?

24.39 (Box 24F) How is nitric oxide synthesized in the cells?

24.40 (Box 24F) How is the toxicity of NO detrimental in strokes?

24.41 (Box 24G) What is the common chemical feature of oral antidiabetic drugs?

24.42 (Box 24G) What is the difference between insulin-dependent and non-insulin-dependent diabetes?

Additional Problems

24.43 (a) In terms of their action, what do the hormone vasopressin and the neurotransmitter dopamine have in common? (b) What is the difference in their mode of action?

***24.44** Considering its chemical nature, how does aldosterone (Sec. 17.10) affect mineral metabolism (Table 24.2)?

24.45 What is the function of the ion-translocating protein in adrenergic neurotransmission?

***24.46** Decamethionium acts as a muscle relaxant. If an overdose of decamethionium occurs, can paralysis be prevented by administering large doses of acetylcholine? Explain.

24.47 Endorphin, a potent pain-killer, is a peptide containing 22 amino acids, among them the same five N-terminal amino acids found in the enkephalins. Does this explain its pain-killing action?

24.48 How do alpha and beta alanine differ in structure?

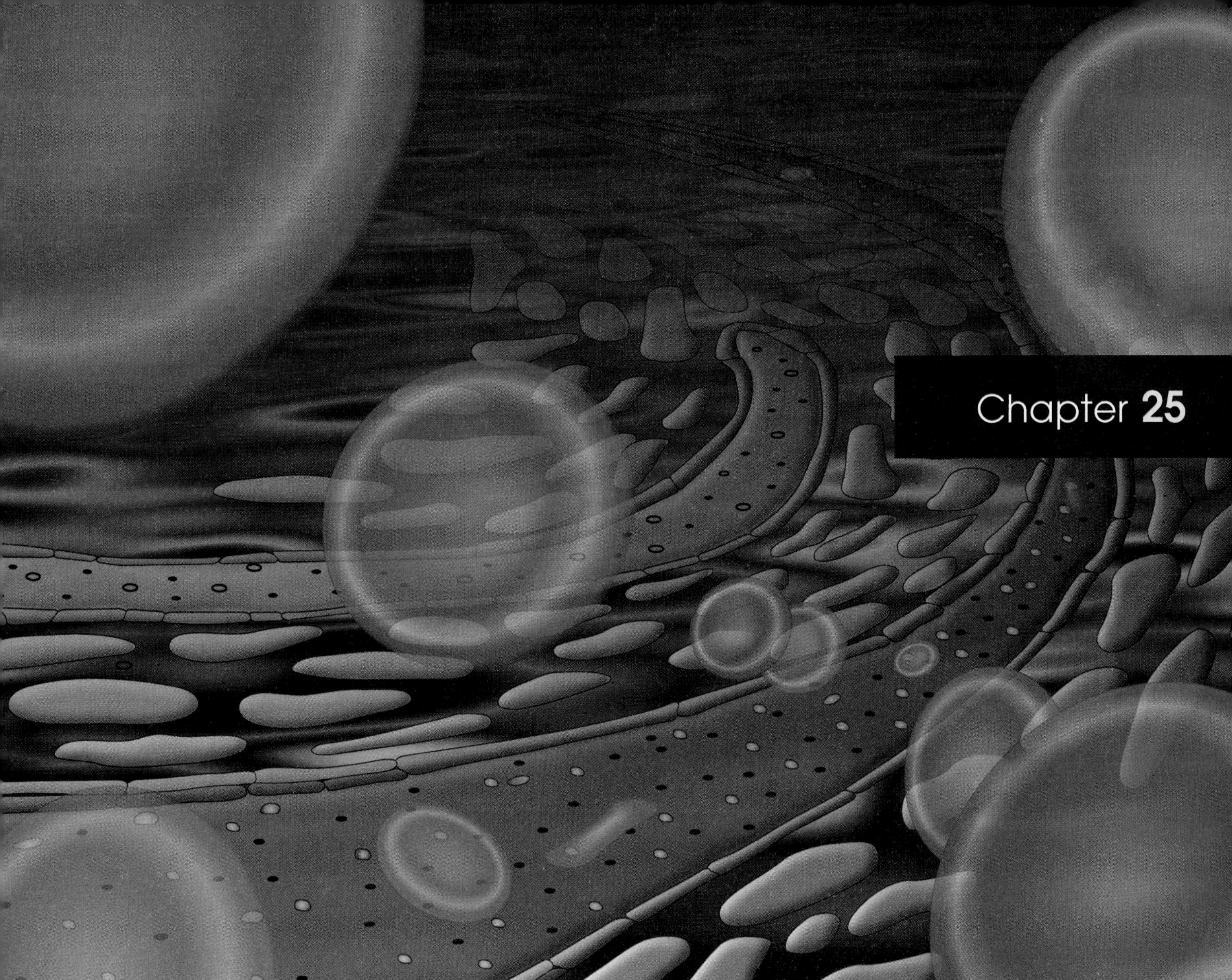

Chapter 25

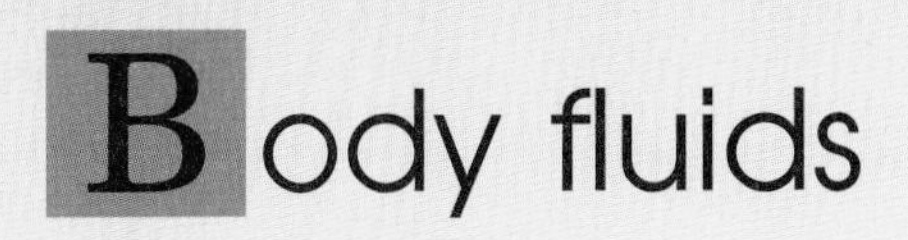

Body fluids

25.1 Introduction

Single-cell organisms receive their nutrients directly from the environment and discard waste products directly into it. In multicellular organisms, the situation is not so simple. There, too, each cell needs nutrients and produces wastes, but most of the cells are not directly in contact with the environment.

Body fluids serve as a medium for carrying nutrients to and waste products from the cells and also for carrying the chemical communicators (Chapter 24) that coordinate activities among cells.

All the body fluids that are not inside the cells are collectively known as **extracellular fluids.** These fluids make up about one quarter of a person's body weight. The most abundant is **interstitial fluid,** which directly surrounds most cells and fills the spaces between them. It makes up about 17 percent of body weight. Another body fluid is **blood plasma,** which flows in the arteries and veins. This makes up about 5 percent of body weight. Other body fluids that occur in lesser amounts are urine, lymph, cerebrospinal fluid, aqueous humor, and synovial fluid. All body fluids are aqueous solutions. Water is the only solvent in the body.

The fluid inside the cells is called **intracellular fluid.**

The blood plasma circulates in the body and is in contact with the other body fluids through the semipermeable membranes of the blood vessels. Therefore, blood can exchange chemical compounds with other body fluids (Fig. 25.1) and, through them, with the cells and organs of the body.

The lymphatic capillary vessels that drain the interstitial fluids (Fig. 25.1) enter certain organs, called *lymphoid organs,* such as the thymus, the spleen, and the lymph nodes. These are the organs where many of the **lymphocytes** that play an important role in the immune system are located. The lymph nodes essentially filter out all foreign material from the lymph. The smaller lymphatic vessels eventually come together in a large vessel, the *thoracic duct,* which drains into major blood vessels, completing the circuit.

There is only a limited exchange between blood and cerebrospinal fluid, on the one hand, and blood and the interstitial fluid of the brain,

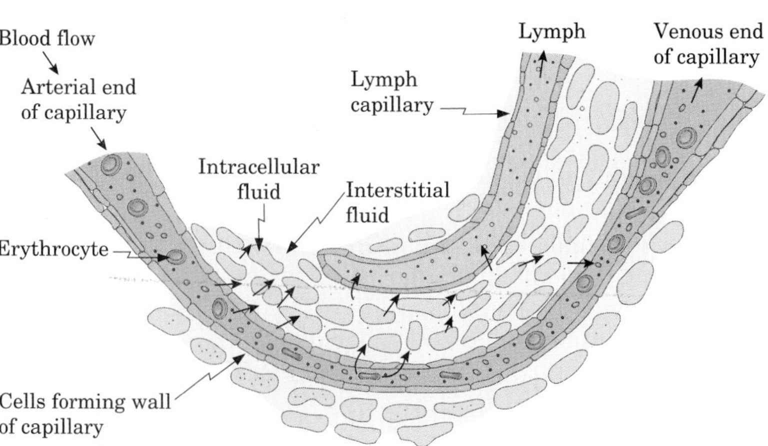

Figure **25.1**
Exchange of compounds among three body fluids: blood, interstitial fluid, and lymph. (After J. R. Holum, *Fundamentals of General, Organic and Biological Chemistry.* New York: John Wiley & Sons, 1978, p. 569.)

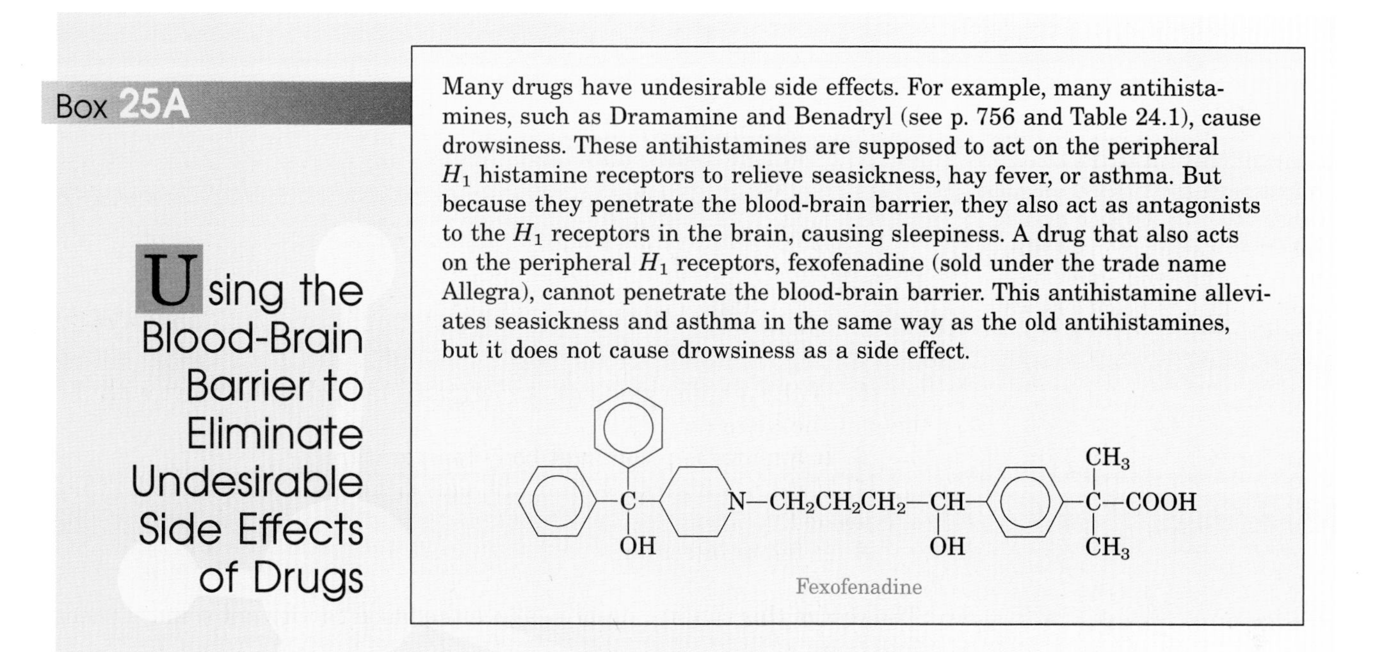

Box 25A Using the Blood-Brain Barrier to Eliminate Undesirable Side Effects of Drugs

Many drugs have undesirable side effects. For example, many antihistamines, such as Dramamine and Benadryl (see p. 756 and Table 24.1), cause drowsiness. These antihistamines are supposed to act on the peripheral H_1 histamine receptors to relieve seasickness, hay fever, or asthma. But because they penetrate the blood-brain barrier, they also act as antagonists to the H_1 receptors in the brain, causing sleepiness. A drug that also acts on the peripheral H_1 receptors, fexofenadine (sold under the trade name Allegra), cannot penetrate the blood-brain barrier. This antihistamine alleviates seasickness and asthma in the same way as the old antihistamines, but it does not cause drowsiness as a side effect.

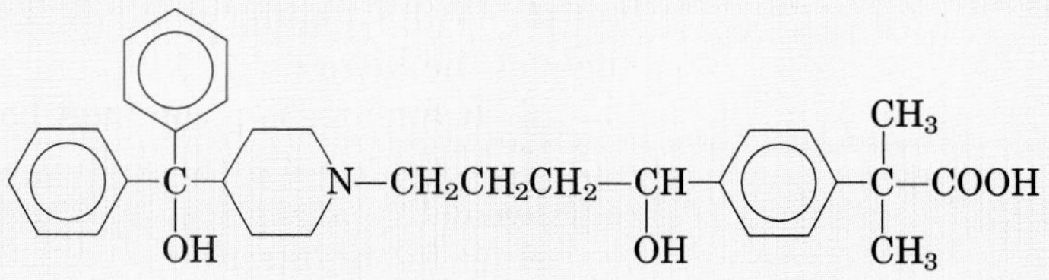

Fexofenadine

on the other. The limited exchange between blood and interstitial fluid is referred to as the **blood-brain barrier.** This is permeable to water, oxygen, carbon dioxide, glucose (Fig. 9E.4), alcohols, and most anesthetics, but only slightly permeable to electrolytes such as Na^+, K^+, and Cl^- ions. Many higher-molecular-weight compounds are also excluded.

The blood-brain barrier protects the cerebral tissue from detrimental substances in the blood and allows it to maintain low K^+ concentration, which is needed to generate the high electrical potential essential for neurotransmission.

It is vital that the body maintain in the blood a proper balance of levels of salts, proteins, and all other components. The process of maintaining these levels as well as the temperature is called **homeostasis.**

Body fluids have special importance for the health care professions. Samples of body fluid can be taken with relative ease. The chemical analysis of blood plasma, blood serum, urine, and occasionally cerebrospinal fluid is of major importance in diagnosing disease.

25.2 Functions and Composition of Blood

It has been known for centuries that "life's blood" is essential to human life. The blood has many functions, including the following:

Figure 25.2 shows some of these functions.

1. It carries O_2 from the lungs to the tissues.
2. It carries CO_2 from the tissues to the lungs.
3. It carries nutrients from the digestive system to the tissues.
4. It carries waste products from the tissues to the excretory organs.

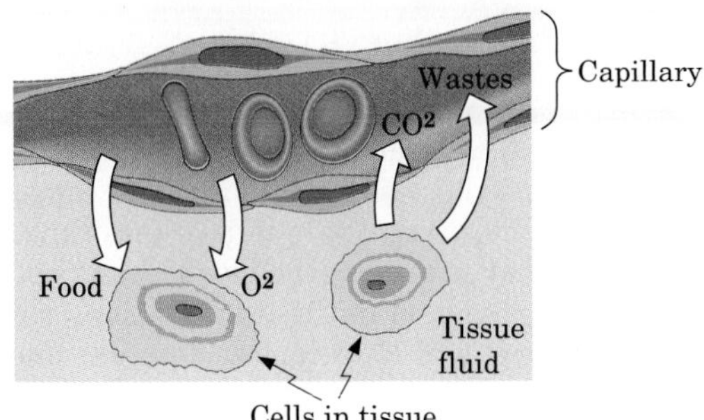

Figure **25.2**
Nutrients, oxygen, and other materials diffuse out of the blood and through the tissue fluid that bathes the cells. Carbon dioxide and other waste products diffuse out of the cells and enter the blood through the capillary wall.

5. With its buffer systems, it maintains the pH of the body (with the help of the kidneys).

6. It maintains a constant body temperature.

7. It carries hormones from the endocrine glands to wherever they are needed.

8. It fights infection.

The rest of this chapter describes how the blood carries out some of these functions.

A 150-pound (68-kg) man has about 6 L of whole blood, 50 to 60 percent of which is plasma.

Whole blood is a complicated mixture. It contains several types of cells (Fig. 25.3) and a liquid, noncellular portion called plasma, in which many substances are dissolved (Table 25.1). There are three main types of **cellular elements** of blood: erythrocytes, leukocytes, and platelets.

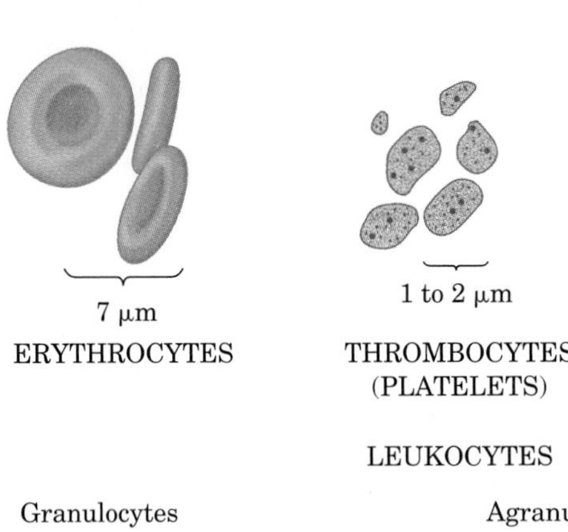

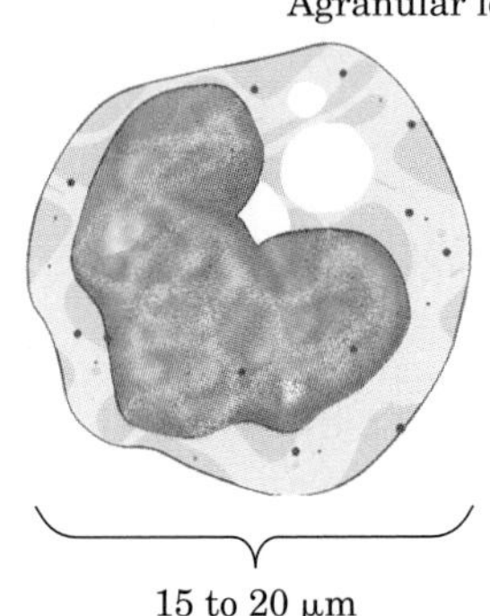

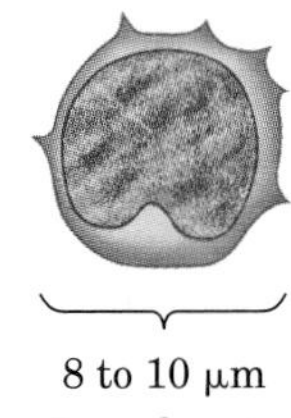

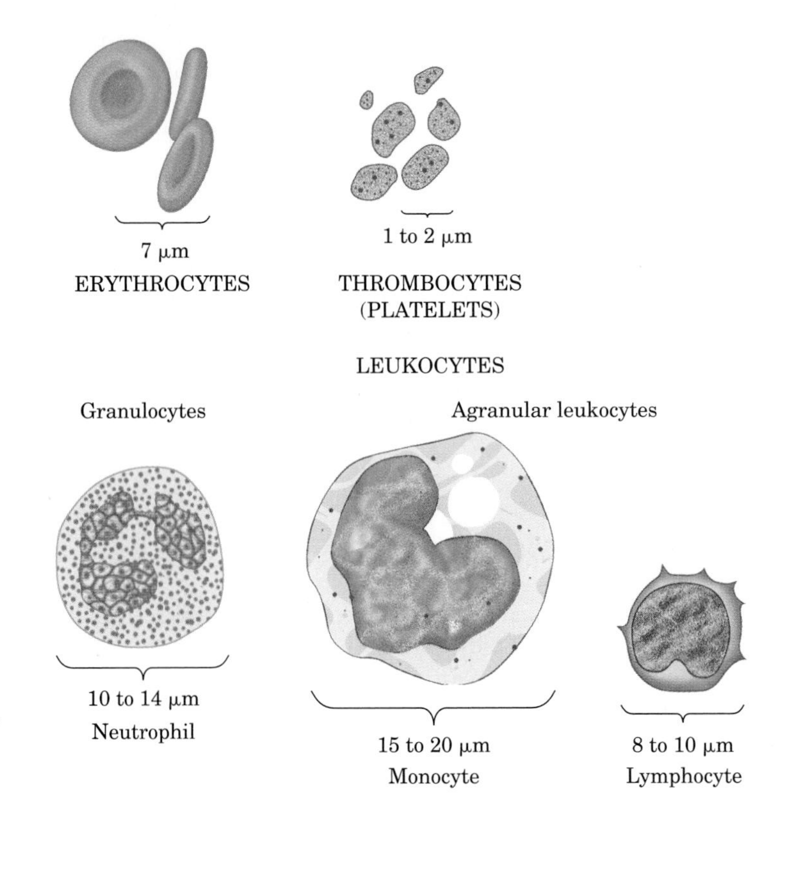

Figure **25.3**
Some cellular elements of blood. The dimensions are shown in microns (μm; also called micrometers).

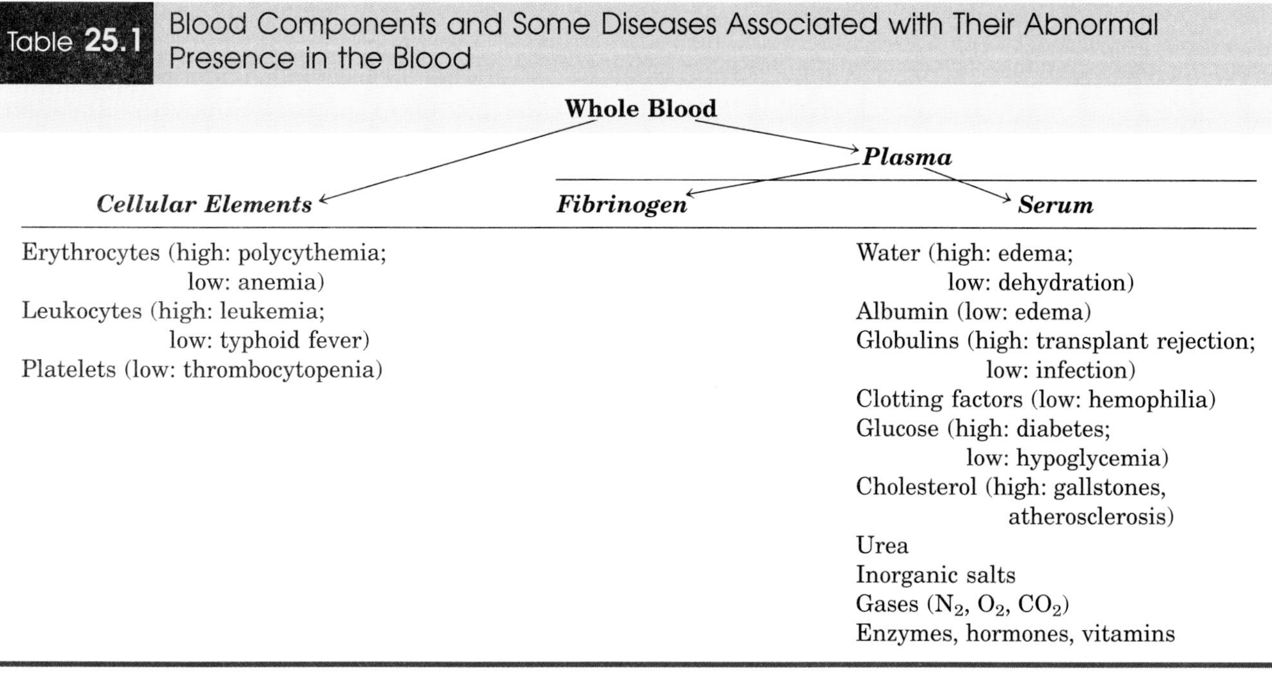

Table 25.1 Blood Components and Some Diseases Associated with Their Abnormal Presence in the Blood

Whole Blood → *Cellular Elements*; *Plasma*

Plasma → *Fibrinogen*; *Serum*

Cellular Elements	*Fibrinogen*	*Serum*
Erythrocytes (high: polycythemia; low: anemia)		Water (high: edema; low: dehydration)
Leukocytes (high: leukemia; low: typhoid fever)		Albumin (low: edema)
Platelets (low: thrombocytopenia)		Globulins (high: transplant rejection; low: infection)
		Clotting factors (low: hemophilia)
		Glucose (high: diabetes; low: hypoglycemia)
		Cholesterol (high: gallstones, atherosclerosis)
		Urea
		Inorganic salts
		Gases (N_2, O_2, CO_2)
		Enzymes, hormones, vitamins

Erythrocytes

The most prevalent cells of blood are the red blood cells, which are called **erythrocytes.** There are about 5 million red blood cells in every cubic millimeter of blood, or roughly 100 million in every drop. Erythrocytes are very specialized cells. They have no nuclei and hence no DNA. Their main function is to carry oxygen to the cells and carbon dioxide away from the cells. Erythrocytes are formed in the bone marrow. They stay in the blood stream for about 120 days. Old erythrocytes are removed by the liver and spleen and destroyed. The constant formation and destruction of red blood cells maintain a steady number of erythrocytes.

In an adult male, there are approximately 30 trillion erythrocytes.

Leukocytes

Leukocytes (white blood cells) are relatively minor cellular components (minor in numbers, not in function). For each 1000 red blood cells, there are only one or two white blood cells. Most of the different leukocytes destroy invading bacteria or other foreign substances by devouring them **(phagocytosis).** Like erythrocytes, leukocytes are manufactured in the bone marrow. Specialized white blood cells formed in the lymph nodes and in the spleen are called lymphocytes. They synthesize immunoglobulins (antibodies, Sec. 25.10) and store them.

Platelets

When a blood vessel is cut or injured, the bleeding is controlled by a third type of cellular element: the **platelets,** or **thrombocytes.** They are formed in the bone marrow and spleen and are more numerous than leukocytes but less numerous than erythrocytes.

There are about 300 000 platelets in each cubic millimeter of blood, or one for every 10 or 20 red blood cells.

Plasma

If all the cellular elements from whole blood are removed by centrifugation, the resulting liquid is the **plasma.** The cellular elements—mainly red blood cells, which settle at the bottom of the centrifuge tube—occupy between 40 and 50 percent of the blood volume.

Blood plasma is 92 percent water. The dissolved solids in the plasma are mainly proteins (7 percent). The remaining 1 percent contains glucose, lipids, enzymes, vitamins, hormones, and such waste products as urea and CO_2. Of the plasma proteins, 55 percent is albumin, 38.5 percent is globulin, and 6.5 percent is fibrinogen. If plasma is allowed to stand, it forms a clot, a gel-like substance. We can squeeze out a clear liquid from the clot. This fluid is the **serum.** It contains all the components of the plasma except the fibrinogen. This protein is involved in the complicated process of clot formation (Box 25B).

A dried clot becomes a scab.

As for the other plasma proteins, most of the globulins take part in the immune reactions (Sec. 25.10), and the albumin provides proper osmotic pressure. If the albumin concentration drops (from malnutrition or from kidney disease, for example), the water from the blood oozes into the interstitial fluid and creates the swelling of tissues called **edema.**

See the discussion of osmotic pressure in Section 6.9.

25.3 Blood as a Carrier of Oxygen

One of the most important functions of blood is to carry oxygen from the lungs to the tissues. This is done by hemoglobin molecules located inside the erythrocytes. As we saw in Section 18.9, hemoglobin is made up of two alpha and two beta protein chains, each attached to a molecule of heme (Fig. 18.12).

The active sites are the hemes, and at the center of each heme is an iron(II) ion. The heme with its central Fe^{2+} ion forms a plane. Because each hemoglobin molecule has four hemes, it can hold a total of four O_2 molecules. However, the ability of the hemoglobin molecule to hold O_2 depends on how much oxygen is in the environment. To see this, look at Figure 25.4, which shows how the oxygen-carrying ability of hemoglobin depends on oxygen pressure. When oxygen enters the lungs, the pressure is high (100 mm Hg). At this pressure, all the Fe^{2+} ions of the hemes bind oxygen molecules; they are fully saturated. By the time the blood reaches the muscles through the capillaries, the oxygen pressure in the muscle is only 20 mm Hg. At this pressure only 30 percent of the binding sites carry oxygen. Thus, 70 percent of the oxygen carried will be released to the tissues. The S shape of the binding (dissociation) curve (Fig. 25.4) implies that

The O_2 picked up by the hemoglobin binds to the Fe^{2+}.

$$HbO_2 \rightleftharpoons Hb + O_2$$

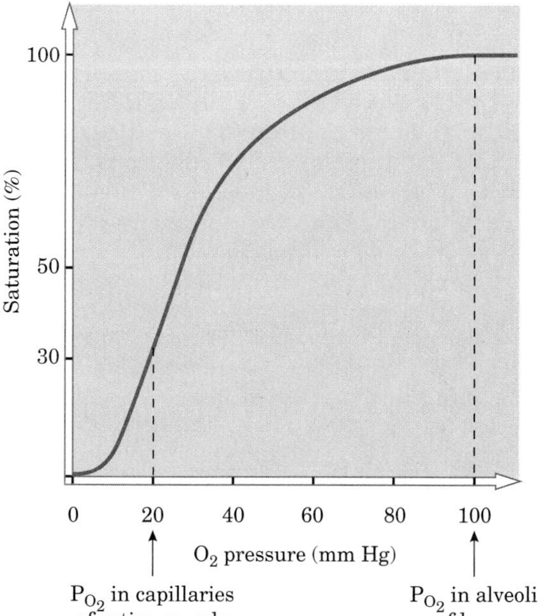

Figure **25.4**
An oxygen dissociation curve. Saturation (%) means the percentage of Fe^{2+} ions that carry O_2 molecules.

is not a simple equilibrium reaction. Each heme has a cooperative effect on the other hemes. This cooperative action allows hemoglobin to deliver twice as much oxygen to the tissues as it would if each heme acted independently. The reason for this is as follows: When a hemoglobin molecule is carrying no oxygen, the four globin units coil into a certain shape (Fig. 18.12). When the first oxygen molecule attaches to one of the heme subunits, it changes the shape not only of that subunit but also of a second one, making it easier for the second one to bind an oxygen. When the second heme binds, the shape changes once again, and the oxygen-binding ability of the two remaining subunits is increased still more.

This is an allosteric effect (Sec. 19.6) and explains the S-shaped curve shown in Figure 25.4.

The oxygen-delivering capacity of hemoglobin is also affected by its environment. A slight change in the pH of the environment changes the oxygen-binding capacity. This is called the **Bohr effect.** An increase in CO_2 pressure also decreases the oxygen-binding capacity of hemoglobin. When a muscle contracts, both H^+ ions and CO_2 are produced. The H^+ ions, of course, lower the pH of the muscle. Thus, at the same pressure (20 mm Hg), more oxygen is released for an active muscle than for a muscle at rest. Similarly, an active contracting muscle also produces CO_2, which then accumulates and further enhances the release of oxygen.

Lowering the pH decreases the oxygen-binding capacity of hemoglobin.

This is how the body delivers more oxygen to those tissues that need it.

25.4 Transport of Carbon Dioxide in the Blood

At the end of the previous section we saw that the waste products of tissue cells, H^+ ions and CO_2, facilitate the release of oxygen from hemoglobin. This enables the cells to receive the oxygen they need. What happens to the CO_2 and H^+? They bind to the hemoglobin. There is an

Box 25B

Blood Clotting

When body tissues are damaged, the blood flow must be stopped or else enough of it will pour out to cause death. The mechanism used by the body to stem leaks in the blood vessels is **clotting.** It is a complicated process involving many factors. Here we mention only a few important steps.

When a blood vessel is injured, the first line of defense is the platelets, which are constantly circulating in the blood. These rush to the site of injury, adhere to the collagen molecules in the capillary wall that are exposed by the cut, and form a gel-like plug. This plug is porous, however, and in order to seal the site a firmer gel (a **clot**) is needed. The clot is a three-dimensional network of fibrin molecules that also contains the platelets. The fibrin network is formed from the blood fibrinogen by the enzyme thrombin. Together with the embedded platelets this constitutes the blood clot.

The question arises: Why doesn't the blood clot in the blood vessels under normal conditions (with no injury or disease)? The reason is that the enzyme that starts clot formation, thrombin, exists in the blood only in its inactive form, called prothrombin. Prothrombin itself is manufactured in the liver, and vitamin K is needed for its production. Even when prothrombin is in sufficient supply, a number of proteins are needed to change it to thrombin. These proteins are given the collective name *thromboplastin.* Any thromboplastic substance can activate prothrombin in the presence of Ca^{2+} ions. Thromboplastic substances exist in the platelets, in the plasma, and in the injured tissue itself.

Clotting is nature's way of protecting us from loss of blood. However, we don't want the blood to clot during blood transfusions because this would stop the flow. To prevent this, we add sodium citrate. Since sodium citrate interacts with Ca^{2+} ions and removes them from the solution, the thromboplastic substances cannot activate prothrombin in the absence of Ca^{2+} ions, and no clot forms. After surgery, anticoagulant drugs are occasionally administered to prevent clot formation. A clot is not dangerous if it stays near the injury, because once the body repairs the tissue, the clot is digested

equilibrium in which, upon the release of oxygen, **carbaminohemoglobin** is formed:

$$HbO_2 + H^+ + CO_2 \rightleftharpoons Hb\begin{matrix} / H^+ \\ \backslash CO_2 \end{matrix} + O_2$$

Carbaminohemoglobin

The CO_2 is bound to the terminal $—NH_2$ groups of the four polypeptide chains so that each hemoglobin can carry a maximum of four CO_2 molecules, one for each chain. How much CO_2 each hemoglobin actually carries depends on the pressure of CO_2. The higher the CO_2 pressure, the more carbaminohemoglobin is formed.

The remaining 5 percent of the CO_2 is transported as CO_2 gas dissolved in the plasma.

But only 25 percent of the total CO_2 produced by the cells is transported to the lungs in the form of carbaminohemoglobin. Another 70 per-

and removed. However, a clot formed in one part of the body may break loose and travel to other parts, where it may lodge in an artery. This is **thrombosis.** If the clot then blocks oxygen and nutrient supply to the heart and brain, it can result in paralysis and death. The most common anticoagulants are bishydroxycoumarin (Dicumarol) and heparin. Heparin enhances the inhibition of thrombin by antithrombin (Fig. 16.15), and bishydroxycoumarin blocks the transport of vitamin K to the liver, preventing prothrombin formation.

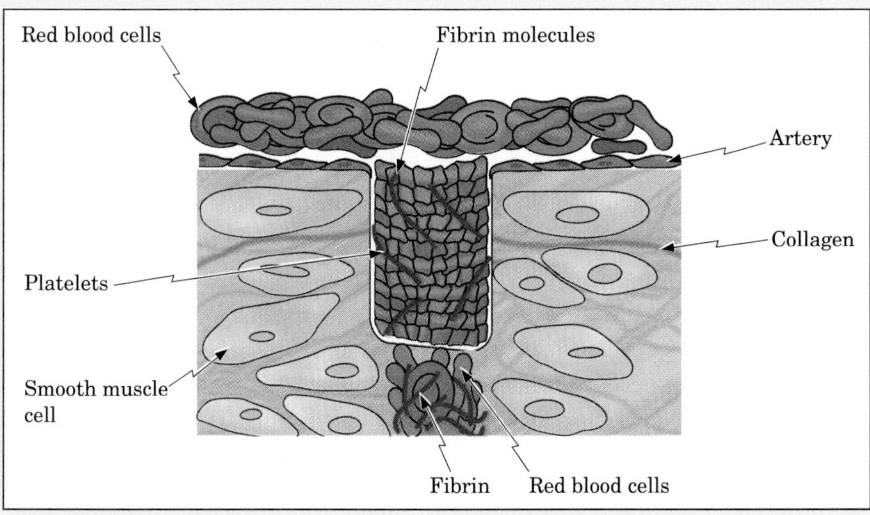

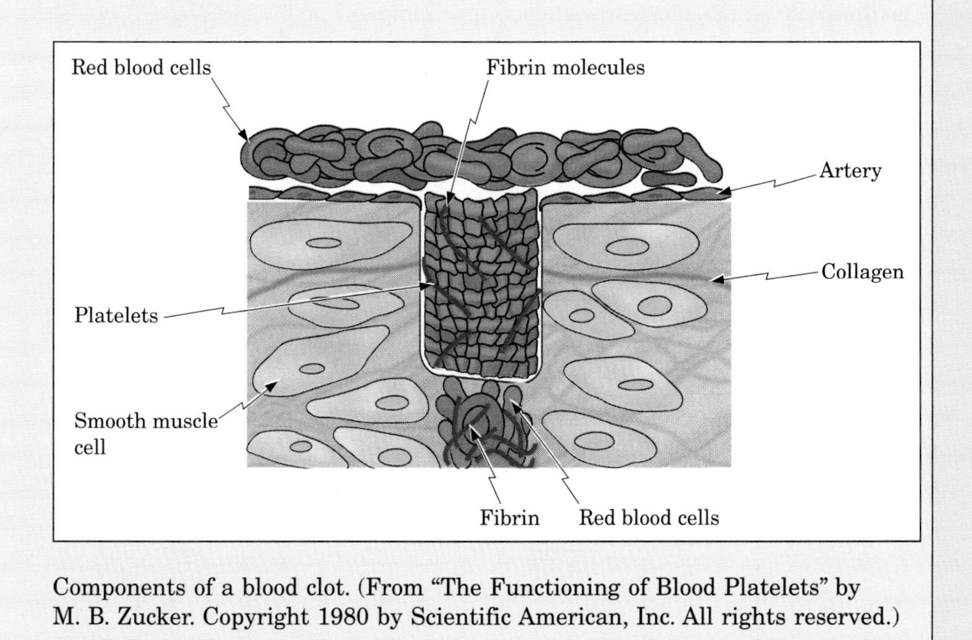

Components of a blood clot. (From "The Functioning of Blood Platelets" by M. B. Zucker. Copyright 1980 by Scientific American, Inc. All rights reserved.)

cent is converted in the red blood cells to carbonic acid by the enzyme carbonic anhydrase:

$$CO_2 + H_2O \underset{\text{anhydrase}}{\overset{\text{carbonic}}{\rightleftharpoons}} H_2CO_3$$

The reaction proceeds to the left because loss of CO_2 from the lungs causes this equilibrium to shift to the left (Le Chatelier's principle).

A large part of this H_2CO_3 is carried as such to the lungs, where it is converted back to CO_2 by carbonic anhydrase and released.

25.5 Blood Buffers

The average pH of human blood is 7.4. Any change larger than 0.1 pH unit in either direction is pathological, and if the pH goes below 6.8 or above 7.8, death can result. When the pH drops below normal, the con-

Box 25C

Composition of Body Fluids

In Section 25.1 it was pointed out that there are three main body fluids: intracellular fluid, interstitial fluid, and blood plasma. Figure 25C shows the average concentrations of the substances dissolved in these three fluids. Blood plasma and interstitial fluid have similar compositions (the main difference is that the blood has a higher protein concentration), but intracellular fluid is quite different. On the cation side, the intracellular fluid has a high concentration of K^+ and a low concentration of Na^+; in the other two fluids, these concentrations are reversed. Intracellular fluid also has a higher Mg^{2+} concentration. On the anion side, there is virtually no Cl^- inside cells, while this is the most abundant anion outside the cells. On the other hand, the concentrations of phosphate (mostly HPO_4^{2-}) and, to a lesser extent, of sulfate are much greater inside the cells. As might be expected, the protein concentration is much higher inside the cells.

Figure **25C** • The composition of body fluids.

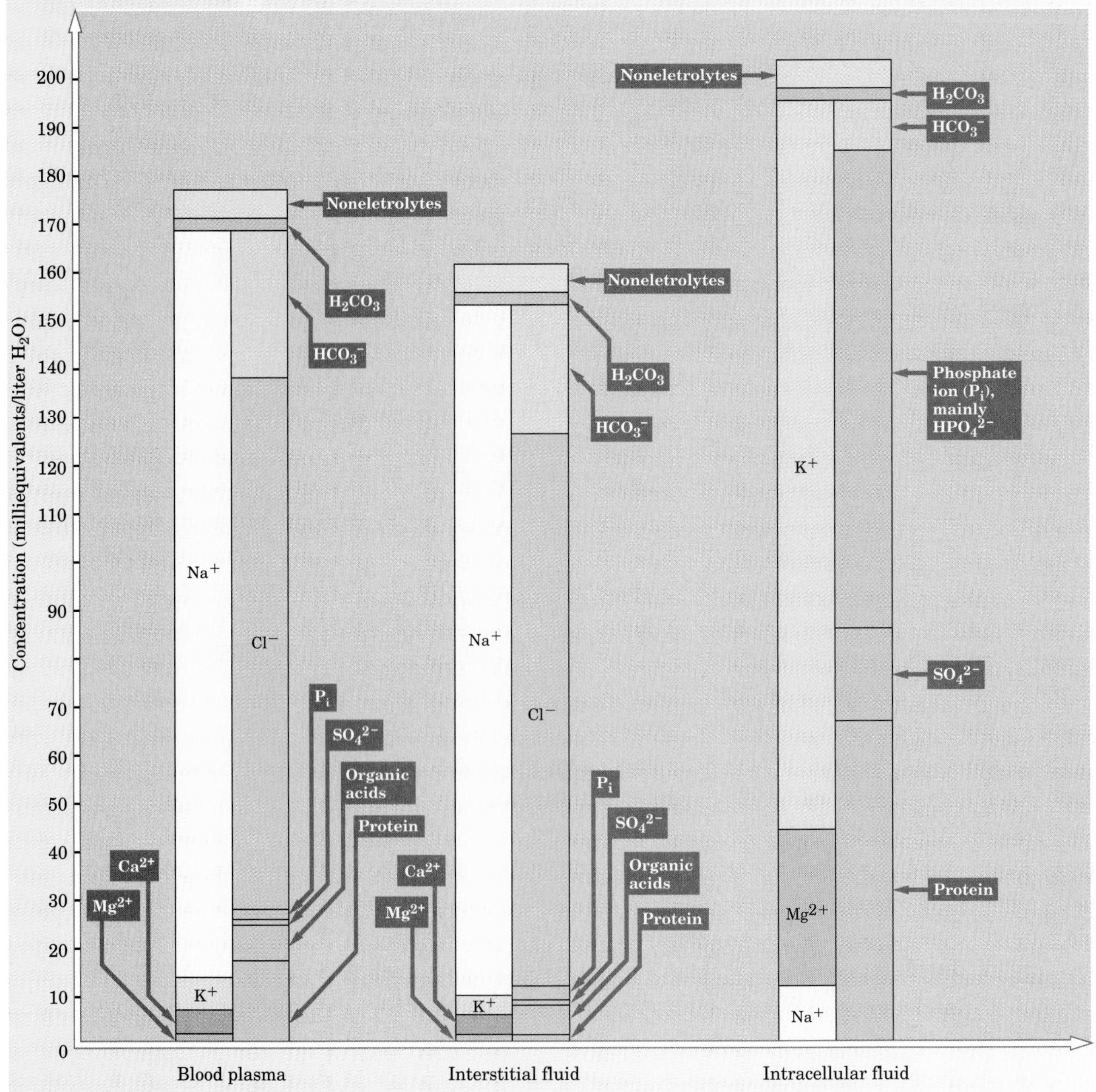

dition is called **acidosis;** when it rises above 7.45, it is called **alkalosis** (Box 8E). There are three buffer systems that maintain an average blood pH of 7.4. The most important is the HCO_3^-/H_2CO_3 buffer (Sec. 8.10). The other two are the $HPO_4^{2-}/H_2PO_4^-$ buffer (Sec. 8.10) and the protein molecules dissolved in the plasma (Sec. 18.6).

25.6 Blood Cleansing—A Kidney Function

We have seen that one of the functions of the blood is to carry waste products away from the cells. CO_2, one of the principal waste products of respiration, is carried by the blood to the lungs and exhaled. The other waste products are filtered out by the kidneys and eliminated in the urine. The kidney is a superfiltration machine. In the event of kidney failure, filtration can be done by hemodialysis (Fig. 25.5). About 100 L of blood pass through a normal human kidney daily. Of this, only about 1.5 L are excreted as urine. Obviously, we do not have here just a simple filtration system in which small molecules are lost and large ones are retained. The kidneys also reabsorb from the urine those small molecules that are not waste products.

Hemodialysis is discussed in Box 6F.

The balance between filtration and reabsorption is controlled by a number of hormones.

The biological units inside the kidneys that perform these functions are called **nephrons,** and each kidney contains about a million of them. A nephron is made up of a filtration head called a **Bowman's capsule** connected to a tiny, U-shaped tube called a **tubule.** The part of the tubule close to the Bowman's capsule is the **proximal tubule,** the U-shaped twist is called **Henle's loop,** and the part of the tubule farthest from the capsule is the **distal tubule** (Fig. 25.6).

Blood vessels penetrate the kidney throughout. The arteries branch into capillaries, and one tiny capillary enters each Bowman's capsule. Inside the capsule the capillary first branches into even smaller vessels, called **glomeruli,** and then leaves the capsule. The blood enters the

Arteries are blood vessels that carry oxygenated blood away from the heart.

The singular is glomerulus.

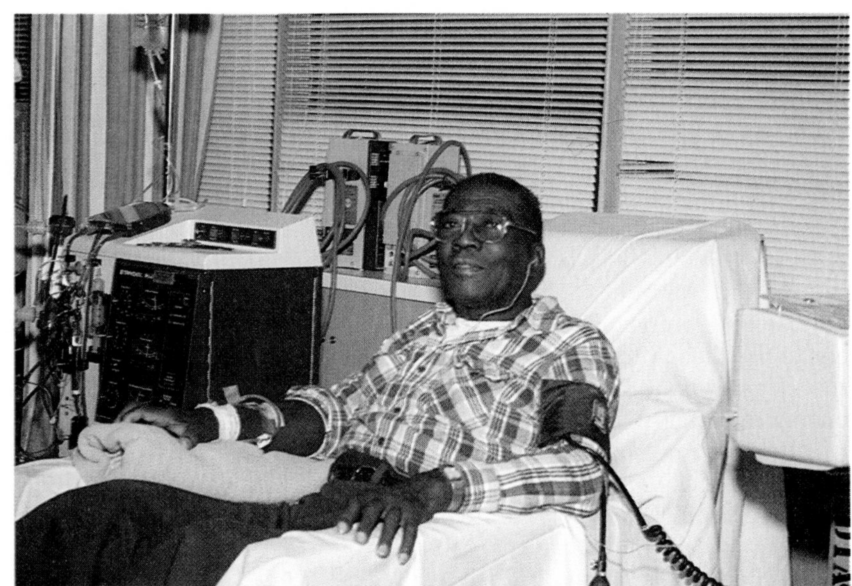

Figure **25.5**
A patient undergoing hemodialysis for kidney disease. (Photograph by Beverly March; courtesy of Long Island Jewish Hospital.)

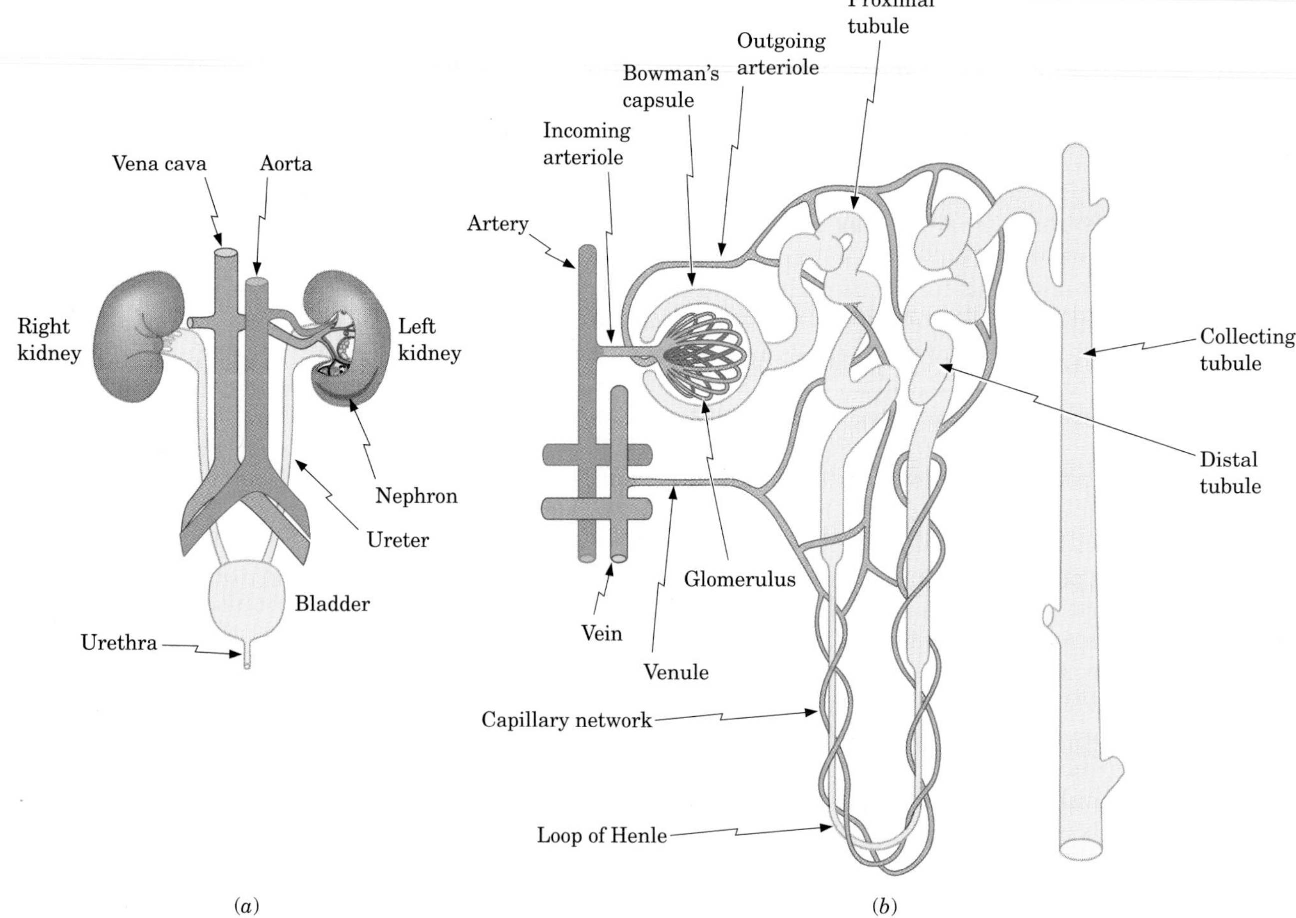

Figure **25.6** • Excretion through the kidneys. (a) The human urinary system. (b) A kidney nephron and its components, with the surrounding circulatory system.

glomeruli with every heartbeat, and the pressure forces the water, ions, and all small molecules (urea, sugars, salts, amino acids) through the walls of the glomeruli and the Bowman's capsule. These molecules and ions enter the proximal tubule. Blood cells and large molecules (proteins) are retained in the capillary and exit with the blood.

As shown in Figure 25.6, the tubules and Henle's loop are surrounded by blood vessels; these blood vessels reabsorb vital nutrients. Eighty percent of the water is reabsorbed in the proximal tubule. Almost all the glucose and amino acids are reabsorbed here. When excess sugar is present in the blood (diabetes, Box 24G), some of it passes into the urine, and the measurement of glucose concentration in the urine is used in diagnosing diabetes.

By the time the glomerular filtrate reaches Henle's loop, the solids and most of the water have been reabsorbed, and only wastes (such as urea, creatinine, uric acid, ammonia, and some salts) pass into the collecting tubules that bring the urine to the ureter, from which it passes to the bladder, as shown in Figure 25.6(a).

Urine

Normal urine contains about 4 percent dissolved waste products. The rest is water. The daily amount of urine varies greatly but averages about 1.5 L per day. The pH varies from 5.5 to 7.5. The main solute is urea, the end product of protein metabolism (Sec. 21.8). Other nitrogenous waste products, such as creatine, creatinine, ammonia, and hippuric acid,

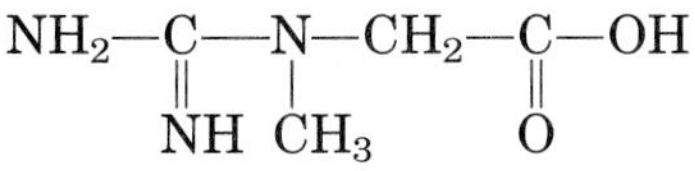

$NH_2-C(=NH)-N(CH_3)-CH_2-C(=O)-OH$

Creatine

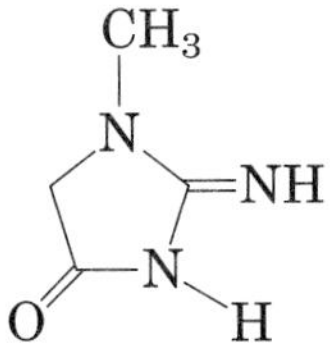

Creatinine

$C_6H_5-C(=O)-NH-CH_2-COOH$

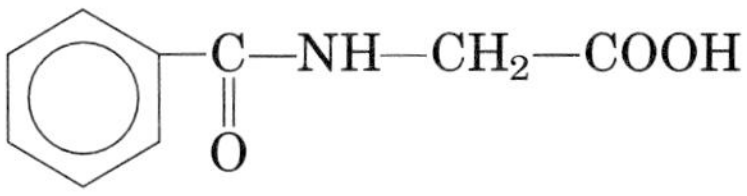

Hippuric acid

Box 25D

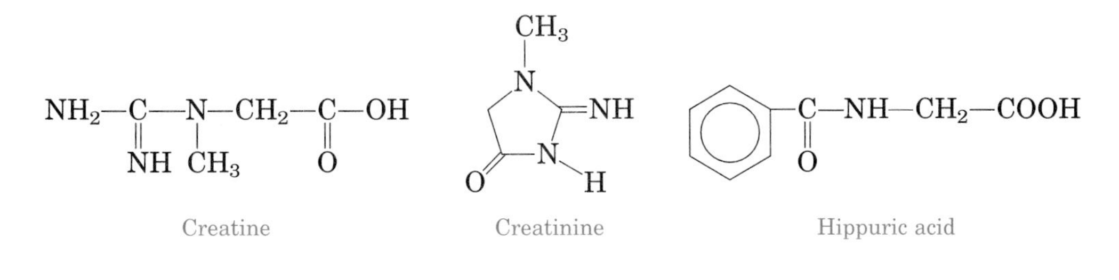

Sex Hormones and Old Age

The same sex hormones that provide a healthy reproductive life in the young can and do cause problems in the aging body. For male fertility the prostate gland, among others, provides the semen. The prostate gland is situated at the exit of the bladder surrounding the ureter. Its growth is promoted by the conversion of the male sex hormone, testosterone, to 5-alpha-dihydrotestosterone by the enzyme 5-alpha-reductase. The majority of men over the age of 50 have enlarged prostate glands, a condition known as benign prostatic hyperplasia. This condition affects more than 90 percent of men over the age of 70. The enlarged prostate gland constricts the ureter and causes a restricted flow of urine.

This symptom can be relieved by oral intake of a drug called terazosin (trade name Hytrin). This drug was originally approved for high blood pressure, where its action results from blocking the alpha-1-adenoreceptors, reducing the constriction around peripheral blood vessels. In a similar way, it reduces constriction around the ureter (the duct that conveys urine from the kidney to the bladder), allowing an unrestricted flow of urine.

Estradiol and progesterone control sexual characteristics in the female body (Sec. 17.10). With the onset of the menopause, smaller amounts of these and other sex hormones are secreted. This not only stops fertility but occasionally creates medical problems. The most serious is *osteoporosis,* which is the loss of bone tissue through absorption. This results in brittle and occasionally deformed bone structure. Estrogens (a general name for female sex hormones other than progesterone) in the proper amount help to absorb calcium from the diet and thus prevent osteoporosis. The lack of enough female sex hormones can also result in atrophic vaginitis and atrophic urethritis. A large number of postmenopausal women take pills that contain a synthetic progesterone analog and natural estrogens. However, continuous medication with these pills, for example, Premarin (mixed estrogens) and Provera (a progesterone analog), is not without danger. A slight but statistically significant increase in cancer of the uterus has been reported in postmenopausal women who use such medication. A physician's advice is needed in weighing the benefits against the risks.

are also present, though in much smaller amounts. In addition, normal urine contains inorganic ions such as Na^+, Ca^{2+}, Mg^{2+}, Cl^-, PO_4^{3-}, SO_4^{2-}, and HCO_3^-.

Under certain pathological conditions, other substances can appear in the urine. It is because of this that urine analysis is such an important part of diagnostic medicine. We have already noted that the presence of glucose and ketone bodies is an indication of diabetes. Among other abnormal constituents may be proteins, which can indicate such kidney diseases as nephritis.

25.7 Buffer Production—Another Kidney Function

Among the waste products sent into the blood by the tissues are H^+ ions. These are neutralized by the HCO_3^- ions that are a part of the blood's buffer system:

$$H^+ + HCO_3^- \rightleftharpoons H_2CO_3$$

When the blood reaches the lungs, the H_2CO_3 is decomposed by carbonic anhydrase, and the CO_2 is exhaled. If the body had no mechanism for replacing HCO_3^-, we would lose most of the bicarbonate buffer from the blood, and the blood pH would eventually drop (acidosis). The lost HCO_3^- ions are continuously replaced by the kidneys—this is another principal kidney function. The replacement takes place in the distal tubules. The cells lining the walls of the distal tubules reabsorb the CO_2 that was lost in the glomeruli. With the aid of carbonic anhydrase, carbonic acid forms quickly and then dissociates to HCO_3^- and H^+ ions:

$$CO_2 + H_2O \rightleftharpoons H_2CO_3 \rightleftharpoons HCO_3^- + H^+$$

The H^+ ions move from the cells into the urine in the tubule, where they are partially neutralized by a phosphate buffer. To compensate for the lost positive ions, Na^+ ions from the tubule enter the cells. When this happens, Na^+ and HCO_3^- ions move from the cells into the capillary. Thus, the H^+ ions picked up at the tissues and temporarily neutralized in the blood by HCO_3^- are finally pumped out into the urine. At the same time, the HCO_3^- ions lost in the lungs are regained by the blood in the distal tubules.

25.8 Water and Salt Balance in Blood and Kidneys

We mentioned in Section 25.6 that the balance in the kidneys between filtration and reabsorption is under hormonal control. The reabsorption of water is promoted by vasopressin, a small peptide hormone manufactured in the hypophysis (Sec. 18.7, Fig. 18.4). In the absence of this hormone, only the proximal tubules—not the distal tubules or the collecting tubules—reabsorb water. As a consequence, without vasopressin too much water passes into the urine. In the presence of vasopressin, water is also reabsorbed in these parts of the nephrons; thus, vasopressin causes the blood to retain more water and produces a more concentrated urine. The production of urine is called **diuresis.** Any agent that reduces the volume of urine is called an **antidiuretic.**

Patients with diabetes insipidus suffer from this condition.

For this reason vasopressin is also called ADH, for *anti*d*i*uretic *h*ormone.

Usually, the vasopressin level in the body is sufficient to maintain the proper amount of water in tissues under various levels of water intake. However, when severe dehydration occurs (as a result of diarrhea, excessive sweating, or insufficient water intake), another hormone also helps to maintain proper fluid level. This hormone is aldosterone (Sec. 17.10); it controls the Na^+ ion concentration in the blood. In the presence of aldosterone, the reabsorption of Na^+ ions is increased. When more Na^+ ions enter the blood, more Cl^- ions follow (to maintain electroneutrality), as well as more water to solvate these ions. Thus, increased aldosterone production enables the body to retain more water. When the Na^+ ion and water levels in the blood return to normal, aldosterone production stops.

25.9 Blood Pressure

Blood pressure is produced by the pumping action of the heart. The blood pressure at the arterial capillary end is about 32 mm Hg. This pressure is higher than the osmotic pressure of the blood (18 mm Hg). The osmotic pressure of the blood is caused by the fact that more solutes are dissolved in the blood than in the interstitial fluid. At the capillary end, therefore, nutrient solutes flow from the capillary into the interstitial fluid and from there into the cell (Fig. 25.1). On the other hand, in the venous capillary the blood pressure is about 12 mm Hg. This is less than the osmotic pressure, so solutes (waste products) from the interstitial fluid flow into the capillaries.

The blood pressure is maintained by the total volume of blood, by the pumping of the heart, and by the muscles that surround the blood vessels and provide the proper resistance to blood flow. Blood pressure is controlled by several very complex systems, some of them acting within seconds and some that take days to respond after a change in blood pressure occurs. For example, if a patient hemorrhages, three different nervous control systems begin to function within seconds. The first is the **baroreceptors** in the neck, which detect the consequent drop in pressure and send appropriate signals to the heart to pump harder and to the muscles surrounding the blood vessels to contract and thus restore pressure. Chemical receptors on the cells that detect less O_2 delivery or CO_2 removal also send nerve signals. Finally, the central nervous system also reacts to oxygen deficiency with a feedback mechanism.

Hormonal controls act somewhat more slowly and may take minutes or even days. The kidneys secrete an enzyme called **renin,** which acts on an inactive blood protein called **angiotensinogen,** converting it to **angiotensin,** a potent vasoconstrictor. The action of this peptide increases blood pressure. Aldosterone (Sec. 25.8) also increases blood pressure by increasing Na^+ ion and water reabsorption in the kidneys.

Box 25E

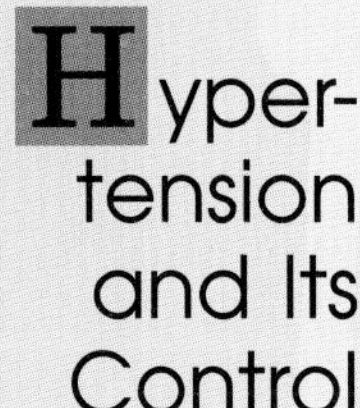

Hypertension and Its Control

Nearly 60 million people in the United States have high blood pressure **(hypertension).** Epidemiologic studies have shown that even mild hypertension (diastolic pressure between 90 and 104 mm Hg) brings about the risk of cardiovascular disease. This can lead to heart attack, stroke, or kidney failure. Hypertension can be managed effectively with diet and drugs. As noted in the text, blood pressure is under a complex control system, so hypertension must be managed on more than one level. Recommended dietary practices are low Na^+ ion intake and abstention from caffeine and alcohol.

The most common drugs for lowering blood pressure are the **diuretics.** There are a number of synthetic organic compounds that increase urine excretion. By doing so, they decrease the blood volume as well as the Na^+ ion concentration and thus lower blood pressure. Other drugs affect the nerve control of blood pressure. Propranolol blocks the adrenoreceptor sites of adrenergic neurotransmitters (Sec. 24.4). By doing so, it reduces the flow of nerve signals and thereby reduces the blood output of the heart; it affects the responses of baroreceptors and chemoreceptors in the central nervous system and the adrenergic receptors in the smooth muscles surrounding the blood vessels. Propranolol and other blockers not only lower blood pressure but also reduce the risk of myocardial infarction. Similar lowering of blood pressure is achieved by drugs that block calcium channels (one example is the drug Cardizem). For proper muscle contraction, calcium ions must pass through muscle cell membranes. This is facilitated by calcium channels. If these channels are blocked by the drug, the heart muscles will contract less frequently, pumping less blood and reducing blood pressure.

Other drugs that reduce hypertension are the vasodilators that relax the vascular smooth muscles. However, many of these have side effects causing fast heartbeat (tachycardia) and also promote Na^+ and water retention. Some drugs on the market act on one specific site rather than on many sites as the blockers do. This obviously reduces the possible side effects. An example is an enzyme inhibitor that prevents the production of angiotensin. Captopril and similar specific enzyme inhibitors lower blood pressure only if the sole cause of hypertension is angiotensin production. Prostaglandins PGE and PGA (Sec. 17.12) lower the blood pressure and, taken orally, have prolonged duration of action. Since hypertension is such a common problem, and it is manageable by drugs, there is great activity in drug research to come up with even more effective and safer antihypertension drugs.

Finally, there is a long-term renal blood control for volume and pressure. When blood pressure falls, the kidneys retain more water and salt, thus increasing blood volume and pressure.

25.10 Immunoglobulins

A very important function of blood is to fight invading substances that enter the blood from outside the body. These may be bacteria, viruses, molds, pollen grains, or anything else. A body with no defense against such invaders could not survive. There is a rare genetic disease in which a person is born without the immune system. Attempts have been made to bring such children up in an enclosure totally sealed from the environment. While in this environment, they can survive, but when the environment is removed, such people always die within a short time. Furthermore, the disease AIDS (Box 23D) slowly destroys the immune system, leaving its victims to die from some invader that a person without this disease would easily be able to fight.

As we shall see, the beauty of the body's immune system is that it is so flexible. The system is capable of making millions of potential defenders, so that it can almost always find just the right one to counter the invader, even when it has never seen that particular invader before. The cells that the body uses for these fights are white cells (lymphocytes) of two types, called B cells and T cells.

The names come from the fact that B cells mature in the bone marrow, while the T cells mature in the thymus gland.

The molecules produced by the B cells are the **immunoglobulins.** Foreign substances that invade the body are called **antigens.** Molecules that counteract these substances are called **antibodies.** Immunoglobulins are antibodies.

The process of finding and making just the right immunoglobulin to fight a particular invader is relatively slow, compared to the chemical messengers discussed in Chapter 24. While neurotransmitters act within a millisecond and hormones within seconds, minutes, or hours, immunoglobulins respond to an antigen over a longer span of time: weeks and months.

The classes of immunoglobulins are shown in Table 25.2.

Immunoglobulins are glycoproteins, that is, carbohydrate-carrying protein molecules. Not only do the different classes of immunoglobulins vary in molecular weight and carbohydrate content, but there is considerable variation in their concentration in the blood. The IgG and IgM antibodies are the most important antibodies in the blood. They interact with antigens and trigger the swallowing up (phagocytosis) of these cells by such specialized cells as the microphages and the leukocytes (white blood cells).

The IgA molecules are found mostly in secretions: tears, milk, and mucus. Therefore these immunoglobulins attack the invading material before it gets into the blood stream. IgE immunoglobulins play a part in allergic reactions: asthma and hay fever.

In immunoglobulin IgG, the light chains have a molecular weight of 25 000 and the heavy chains of 50 000.

Each immunoglobulin molecule is made of four polypeptide chains. There are two identical light chains and two identical heavy chains in

Table 25.2 Immunoglobulin Classes

Class	MW	Carbohydrate Content (%)	Concentration in Serum (mg/100 mL)
IgA	200 000–700 000	7–12	90–420
IgD	160 000	<1	1–40
IgE	190 000	10–12	0.01–0.1
IgG	150 000	2–3	600–1800
IgM	950 000	10–12	50–190

There are several diseases for which vaccines are available (polio, measles, smallpox, and others). A vaccine may be made up of dead or weakened viruses or bacteria. Small doses of the vaccine eventually confer immunity. For example, the Salk polio vaccine is a polio virus that has been made harmless by treatment with formaldehyde; it is given by intramuscular injection. In contrast, the Sabin polio vaccine is a mutated form of the wild virus; the mutation makes it sensitive to temperature. Thus attenuated, the live virus is taken orally. The treatment, the body temperature, and the gastric juices render it harmless before it penetrates the blood stream.

Vaccines change lymphocytes into plasma cells that produce large quantities of antibodies to fight any invading antigens. This is, however, only the immediate, short-term response. Some lymphocytes become *memory cells* rather than *plasma cells.* These memory cells do not secrete the antibody; they store it to serve as a detecting device for future invasion of the same foreign cells. In this way, long-term immunity is conferred. If a second invasion occurs, these memory cells divide directly into antibody-secreting plasma cells as well as into more memory cells. This time the response is faster, because it does not have to go through the process of activation and differentiation into plasma cells, which usually takes two weeks.

Smallpox, which was once one of humanity's worst scourges, has been totally wiped out, and smallpox vaccination is no longer required.

each molecule. The four polypeptide chains are arranged symmetrically, forming a Y shape (Fig. 25.7). Three disulfide bridges link the four chains into a unit. Both light and heavy chains have constant and variable regions. The constant regions have the same amino acid sequences in different antibodies, and the variable regions have different amino acid sequences in each antibody.

Studies have shown that many other molecules that function as communication signals between cells also have the same basic structural design as the immunoglobulins. They belong to what is known as the **immunoglobulin superfamily.** Glycoprotein molecules called cell-adhesion molecules, from nerve or liver cells, operate between cells and cause loose cells to adhere to each other to form layers and then tissues such as epithelium. They also play a significant part in the development of embryos.

Any macromolecule, whether protein, carbohydrate, or nucleic acid, that is foreign to the body will cause an immune response. If the foreign material is a bacterium or virus, then macromolecules on the surfaces of these microorganisms initiate the immune response (Box 16B).

The variable regions of the antibody recognize the foreign substance (the antigen) and bind to it (Fig. 25.8). Since each antibody contains two variable regions, it can bind two antigens, and this results in a large aggregate (Fig. 25.7).

These forces are discussed in Section 5.7.

The binding of the antigen to the variable region of the antibody is not by covalent bonds but by much weaker forces such as London forces, dipole–dipole interactions, and hydrogen bonds. It is similar to the binding of neurotransmitters and hormones to a receptor site. The antigen

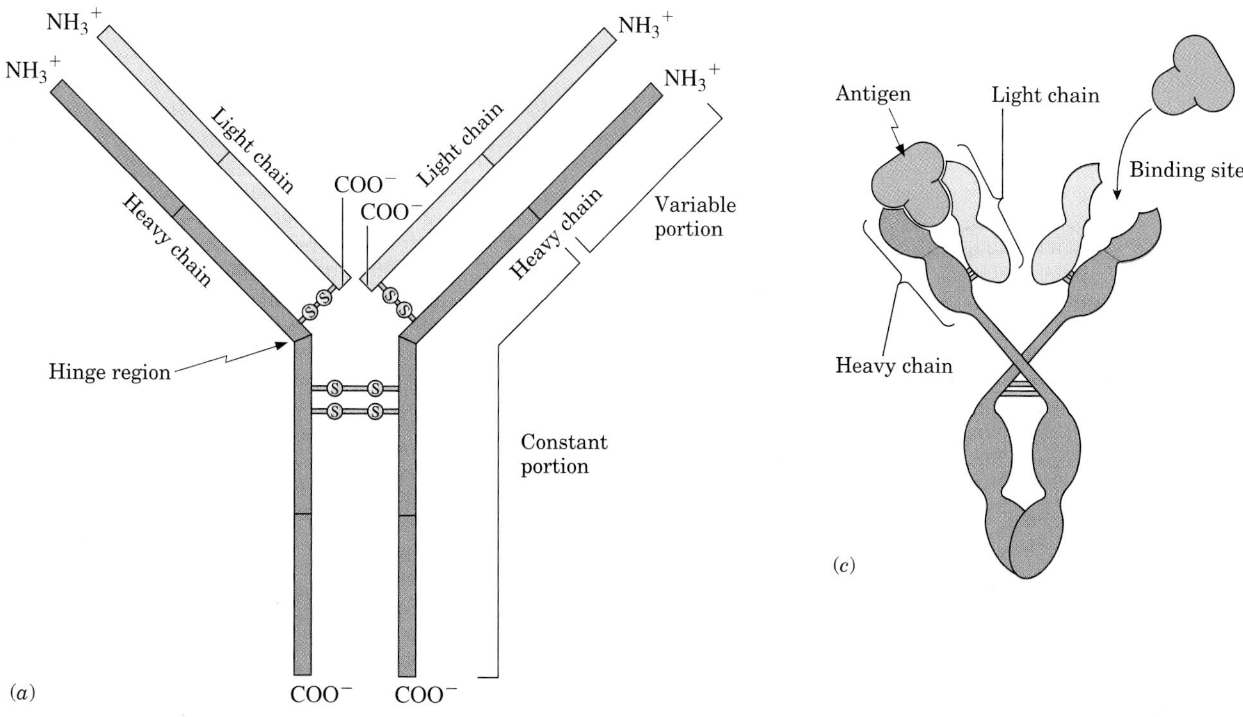

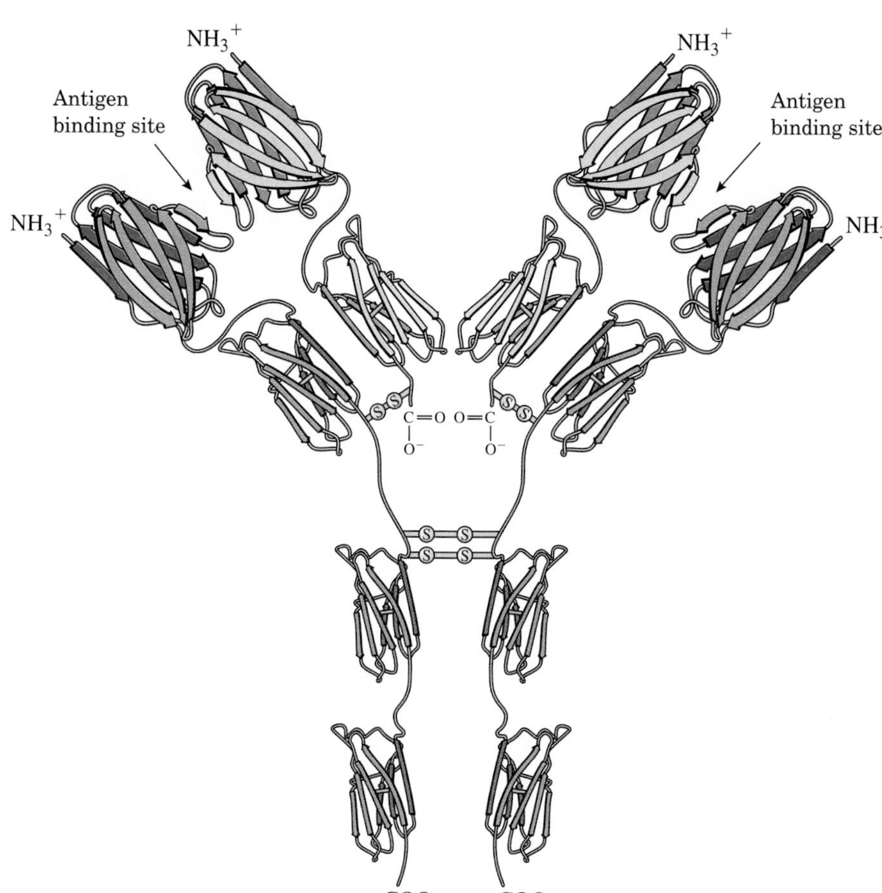

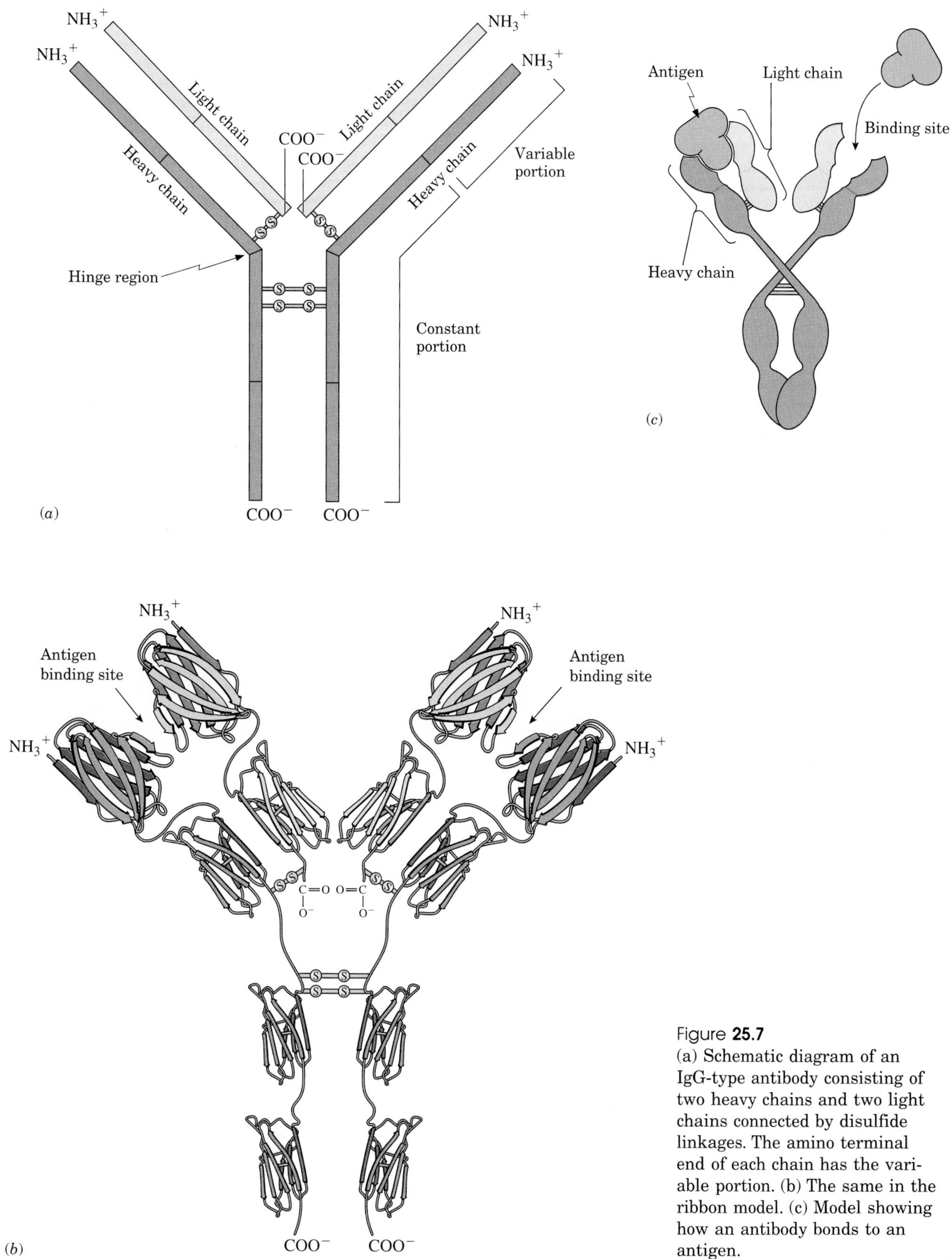

Figure **25.7**
(a) Schematic diagram of an IgG-type antibody consisting of two heavy chains and two light chains connected by disulfide linkages. The amino terminal end of each chain has the variable portion. (b) The same in the ribbon model. (c) Model showing how an antibody bonds to an antigen.

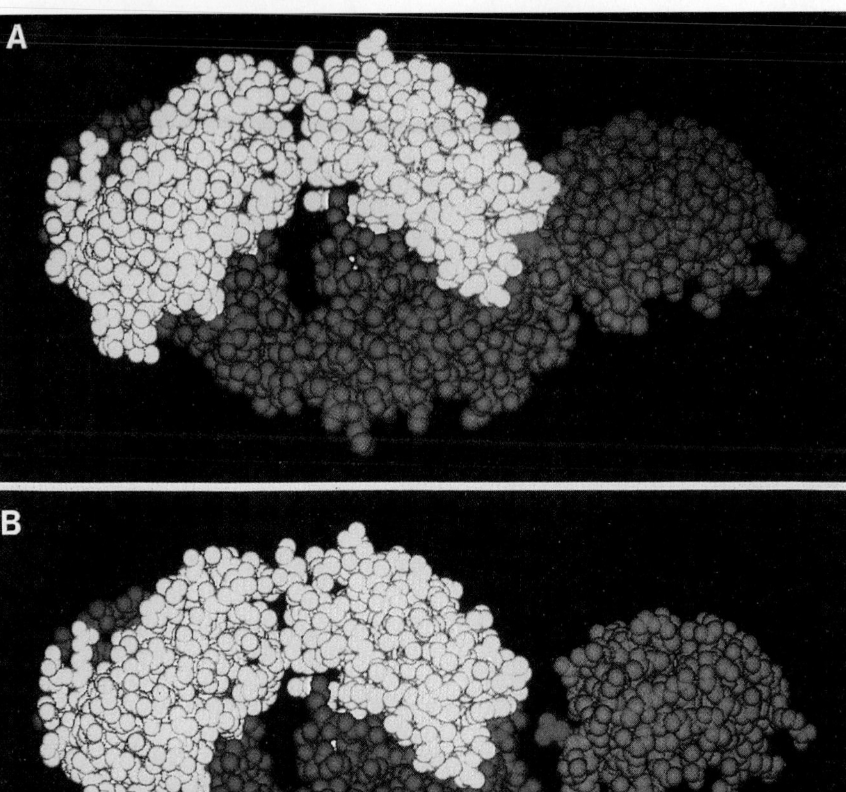

Figure **25.8** • (a) An antigen-antibody complex. The antigen (shown in green) is lysozyme. The heavy chain of the antibody is shown in blue; the light chain in yellow. The most important amino acid residue (glutamine in the 121 position) on the antigen is the one that fits into the antibody groove (shown in red). (b) The antigen-antibody complex has been pulled apart. Note how they fit each other. (Courtesy of Dr. R. J. Poijak, Pasteur Institute, Paris, France.)

has to fit into the antibody surface. Since there are a large number of different antigens against which the human body must fight, there are more than 10 000 different antibodies in our systems.

All of the above describes the action of antibodies that are secreted by the B cells. The action of the T cells is somewhat different. They come into action when they encounter an antigen. The initially formed T cells differentiate. Some of them become effective T cells that kill the invading foreign cell at short range or by cell-to-cell contact. Other T cells become **memory T cells** (Box 25F). These will remain in the blood stream so that if the same antigens enter again, even years later, the body does not need time to build up a defense, but is ready to kill them immediately.

This is why, for example, if we have measles once, we will not get this disease a second time.

SUMMARY

The most important body fluid is **blood plasma,** which is whole blood from which cellular elements have been removed. Other important body fluids are **urine** and the **interstitial fluids** directly surrounding the cells of the tissues. The **cellular elements** of the blood are the **red blood cells (erythrocytes),** which carry O_2 to the tissues and CO_2 away from the tissues; the **white blood cells (leukocytes),** which fight infection; and the **platelets,** which control bleeding. The plasma contains **fibrinogen,** which is necessary for blood clot formation. When fibrinogen is removed from the plasma, what remains is the **serum.** The serum contains albumins, globulins, nutrients, waste products, inorganic salts, enzymes, hormones, and vitamins dissolved in water. The lymphoid organs contain the **lymphocytes** that play an important role in the immune system. The body maintains **homeostasis.**

The red blood cells carry the O_2. The Fe(II) in the heme portion of hemoglobin binds the oxygen, which is then released at the tissues by a combination of factors: low O_2 pressure in the tissue cells, high H^+ ion concentration **(Bohr effect),** and high CO_2 concentration. Of the CO_2 in the venous blood, 25 percent binds to the terminal $—NH_2$ of the polypeptide chains of hemoglobin to form carbaminohemoglobin, 70 percent is carried in the plasma as H_2CO_3, and 5 percent is carried as dissolved gas.

Waste products are removed from the blood in the **kidneys.** The filtration units—**nephrons**—contain an entering blood vessel that branches out into fine vessels called **glomeruli.** Water and all the small molecules in the blood diffuse out of the glomeruli, enter the **Bowman's capsules** of the nephrons, and from there go into the **tubules.** The blood cells and large molecules remain in the blood vessels. The water, inorganic salts, and organic nutrients are then reabsorbed into the blood vessels. The waste products plus water form the urine, which goes from the tubule into the ureter and finally into the bladder.

The kidneys not only filter out waste products but also produce HCO_3^- ions for buffering, replacing those lost through the lungs. The water balance and salt balance of the blood and urine are under hormonal control. Vasopressin helps to retain water, and aldosterone increases the blood Na^+ ion concentration by reabsorption. Blood pressure is controlled by a large number of factors.

Immunoglobulins (antibodies) are protein molecules containing carbohydrates. They are made of two heavy chains and two light chains. All four are linked together by disulfide bridges. Immunoglobulins contain variable regions in which the amino acid composition of each antibody is different. These regions interact with **antigens.** Immunoglobulins interact with antigens to form insoluble large aggregates.

KEY TERMS

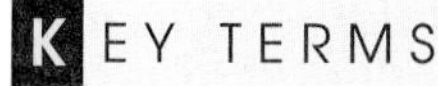

Antibody (Sec. 25.10)
Antidiuretic (Sec. 25.8)
Antigen (Sec. 25.10)
Baroreceptor (Sec. 25.9)
Blood-brain barrier (Sec. 25.1)
Blood plasma (Sec. 25.2)
Bohr effect (Sec. 25.3)
Cellular elements (Sec. 25.2)
Diuresis (Sec. 25.8)
Diuretic (Box 25E)
Edema (Sec. 25.2)
Erythrocyte (Sec. 25.2)
Extracellular fluid (Sec. 25.1)
Glomeruli (Sec. 25.6)
Homeostasis (Sec. 25.1)
Hypertension (Box 25E)
Immunoglobulin (Sec. 25.10)
Immunoglobulin superfamily (Sec. 25.10)
Interstitial fluid (Sec. 25.1)
Intracellular fluid (Sec. 25.1)
Leukocyte (Sec. 25.2)
Lymphocyte (Sec. 25.2)
Nephron (Sec. 25.6)
Phagocytosis (Sec. 25.2)
Plasma (Sec. 25.2)
Platelet (Sec. 25.2)
Serum (Sec. 25.2)
Thrombocyte (Sec. 25.2)
Thrombosis (Box 25B)

PROBLEMS

Body Fluids

25.1 What is the common solvent in all body fluids?
25.2 Which organs are included among the lymphoid organs?
25.3 (a) Is the blood plasma in contact with the other body fluids?
(b) If so, in what manner?

Functions and Composition of Blood

25.4 Name the three principal types of blood cells.
25.5 What is the function of erythrocytes?
25.6 Which blood cells are the largest?
25.7 Define:
(a) interstitial fluid
(b) plasma
(c) serum
(d) cellular elements

25.8 Where are the following manufactured?
(a) erythrocytes (c) lymphocytes
(b) leukocytes (d) platelets

25.9 How does a low albumin content of the serum cause edema?

25.10 State the function of
(a) albumin (c) fibrinogen
(b) globulin

25.11 Which blood plasma component protects against infections?

25.12 How is serum prepared from blood plasma?

Blood as a Carrier of Oxygen

25.13 How many molecules of O_2 are bound to a hemoglobin molecule at full saturation?

***25.14** A chemist breathing air contaminated in an industrial accident analyzes it and finds that the partial pressure of oxygen is only 38 mm Hg. How many O_2 molecules, on the average, will be transported by each hemoglobin molecule? (Consult Figure 25.4.)

25.15 What is the oxygen pressure in the muscles?

***25.16** From Figure 25.4, predict the O_2 pressure at which the hemoglobin molecule binds 4, 3, 2, 1, and zero O_2 molecules to its heme.

25.17 Explain how the absorption of O_2 on one binding site in the hemoglobin molecule increases the absorption capacity at other sites.

25.18 (a) What happens to the oxygen-carrying capacity of hemoglobin when the pH is lowered? (b) What is the name of this effect?

Transport of CO_2 in the Blood

25.19 (a) Where does CO_2 bind to hemoglobin?
(b) What is the name of the complex formed?

25.20 How many CO_2 molecules can bind to a hemoglobin molecule at full saturation?

***25.21** If the plasma carries 70 percent of its carbon dioxide in the form of H_2CO_3 and 5 percent in the form of dissolved CO_2 gas, predict the equilibrium constant of the reaction catalyzed by carbonic anhydrase:

$$H_2O + CO_2 \rightleftharpoons H_2CO_3$$

(Assume that the H_2O concentration is included in the equilibrium constant.)

Blood Buffers

25.22 After reviewing Box 8E, explain why emphysema may lower the pH of the blood, causing acidosis.

25.23 Which is the most important blood buffer?

Kidney Functions and Urine

25.24 What nitrogenous waste products are eliminated in the urine?

25.25 Besides nitrogenous waste products, what other components are normal constituents of urine?

25.26 What happens to the H^+ ions sent into the blood as waste products by the cells of the tissues?

Water and Salt Balance in the Blood and Kidneys

25.27 In which part of the kidneys does vasopressin change the permeability of membranes?

25.28 How does aldosterone production counteract the excessive sweating that usually accompanies a high fever?

Blood Pressure

25.29 What would happen to the glucose in the arterial capillary if the blood pressure at the end of the capillary were 12 mm Hg rather than the normal 32 mm Hg?

***25.30** Some drugs used as antihypertensive agents, for example, Captopril (Box 25E), inhibit the enzyme that activates angiotensinogen. What is the mode of action of such drugs in lowering blood pressure?

Immunoglobulins, Antibodies, and Antigens

25.31 Distinguish among the roles of IgA, IgE, and IgG immunoglobulins.

25.32 (a) Which immunoglobulin has the highest carbohydrate content and the lowest concentration in the serum? (b) What is its main function?

***25.33** Box 16B states that the antigen in the red blood cells of a person with B-type blood is a galactose unit. Show schematically how the antibody of a person with A-type blood would aggregate the red blood cells of a B-type person if such a transfusion were made by mistake.

25.34 In the immunoglobulin structure there is a region called the "hinge region" that joins the stem of the Y to the arms. The hinge region can be cleaved by a specific enzyme to yield one Fc fragment (the stem of the Y) and two Fab fragments (the two arms). Which of these two kinds of fragments can interact with an antigen? Explain.

25.35 How are the light and heavy chains of an antibody held together?

25.36 What do we mean by the term immunoglobulin superfamily?

25.37 What is the function of cell adhesion molecules?

25.38 What kind of interaction takes place between an antigen and an antibody?

25.39 How does the action of B cells and T cells differ?

Boxes

25.40 (Box 25A) What side effect of antihistamines is avoided by use of the drug fexofenadine?

25.41 (Box 25B) What is the first line of defense when a blood vessel is cut?

25.42 (Box 25B) What are the functions of (a) thrombin, (b) vitamin K, and (c) thromboplastin in clot formation?

25.43 (Box 25C) (a) Which body fluid(s) is (are) much richer in K^+ ions than in Na^+ ions? (b) Which is (are) much richer in Na^+ ions than in K^+ ions?

25.44 (Box 25C) (a) In which body fluid(s) is (are) the negative ions mostly Cl^-? (b) In which is Cl^- essentially absent?

25.45 (Box 25D) What are the symptoms of enlarged prostate gland (benign prostatic hyperplasia)?

25.46 (Box 25D) What causes osteoporosis in postmenopausal women?

25.47 (Box 25E) How do diuretics act in lowering blood pressure?

25.48 (Box 25E) People with high blood pressure are advised to have a low-sodium diet. How does such a diet lower blood pressure?

25.49 (Box 25F) What is the difference between memory cells and plasma cells?

Additional Problems

25.50 Albumin makes up 55 percent of the protein content of plasma. Is the albumin concentration of the serum higher, lower, or the same as that of the plasma? Explain.

25.51 Which nitrogenous waste product of the body can be classified as an amino acid?

25.52 Which immunoglobulins form the first line of defense against invading bacteria?

25.53 Where do all the lymphatic vessels come together?

25.54 What happens to the oxygen-carrying ability of hemoglobin when acidosis occurs in the blood?

25.55 What feature of the immunoglobulins enables them to interact with thousands of different antigens?

25.56 Does the heme of hemoglobin participate in CO_2 transport as well as in O_2 transport?

***25.57** The variable regions of immunoglobulins bind the antigens. How many polypeptide chains carry variable regions in one immunoglobulin molecule?

25.58 What is a diuretic drug?

***25.59** When kidneys fail the body swells, a condition called edema. Why and where does the water accumulate?

25.60 Which hormone controls water retention in the body?

25.61 Describe the action of renin.

***25.62** The oxygen dissociation curve of fetal hemoglobin is lower than that of adult hemoglobin. How does the fetus obtain its oxygen supply from the mother's blood when the two circulations meet?

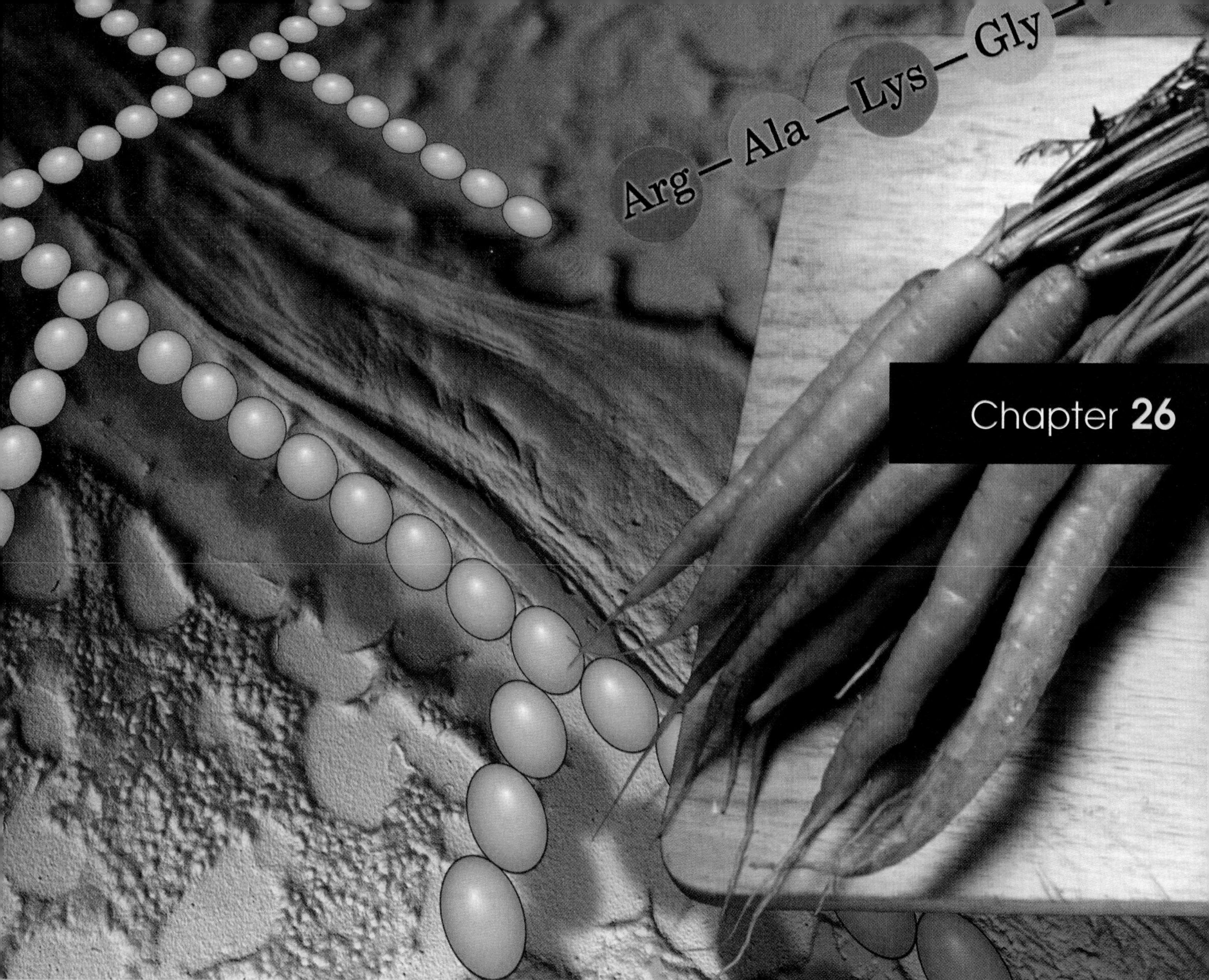

Chapter **26**

Nutrition and digestion

26.1 Introduction

In Chapters 20 and 21 we saw what happens to the food that we eat in its final stages—after the proteins, lipids, and carbohydrates have been broken down into their components. In this chapter we discuss the earlier stages—nutrition and diet—and then the digestive processes that break down the large molecules to the small ones that undergo metabolism. We have seen that the purpose of food is to provide energy and new molecules to replace those that the body uses. The synthesis of new molecules is especially important for the period during which a child becomes an adult.

26.2 Nutrition

A late nineteenth-century advertisement for a cure-all tonic. (Courtesy of the National Library of Medicine.)

Discriminatory curtailment diets are those that avoid certain food ingredients that are considered harmful to the health of an individual; for example, low-sodium diets for people with high blood pressure.

The components of food and drink that provide growth, replacement, and energy are called **nutrients.** Not all components of food are nutrients. Some components of food and drinks, such as those that provide flavor, color, or aroma, enhance our pleasure in the food but are not themselves nutrients.

Nutritionists classify nutrients into six groups:

1. carbohydrates
2. lipids
3. proteins
4. vitamins
5. minerals
6. water

A healthy body needs the proper intake of all nutrients. However, nutrient requirements vary from one person to another. For example, it requires more energy to maintain the body temperature of an adult than that of a child. For this reason, nutritional requirements are usually given per kilogram of body weight. Furthermore, the energy requirements of a physically active body are greater than those of people in sedentary occupations. Therefore, when average values are given, as in recommended daily allowances (RDA), one should be aware of the wide range that these average values represent.

The public interest in nutrition and diet changes with time and geography. Seventy or eighty years ago the main nutritional interest of most Americans was getting enough food to eat and avoiding diseases caused by vitamin deficiency, such as scurvy or beriberi. This is still the main concern of the large majority of the world's population. Today, in affluent societies such as ours, the nutritional message is no longer "eat more" but rather "eat less and discriminate more in your selection of food." For example, a sizable percentage of the American population avoids foods containing substantial amounts of cholesterol (Box 17G) and saturated fatty acids to reduce the risk of heart attacks.

Along with discriminatory curtailment diets came many different faddish diets. Diet faddism is an exaggerated belief in the effects of nutrition upon health and disease. This is not new; it has been prevalent for many years. A recommended food is rarely as good and a condemned food is never as bad as faddists claim. Scientific studies prove, for instance, that food grown with chemical fertilizers is just as healthy as food grown with organic (natural) fertilizers.

Each food contains a large variety of nutrients. For example, a typical breakfast cereal lists as its ingredients: milled corn, sugar, salt, malt flavoring, vitamins A, B, C, and D, plus flavorings and preservatives. Consumer laws require that most packaged food be labeled, in a uniform manner, to show the nutritional values of the food. Figure 26.1 shows a typical label of the type found on almost every can, bottle, or box of food that we buy.

These labels must list the percentages of daily values of four key vitamins and minerals: vitamins A and C, calcium, and iron. If other vitamins or minerals have been added, or if the product makes a nutritional claim about other nutrients, their values must be shown as well. The percent daily values on the labels are based on a daily intake of 2000 Calories. For anyone who eats more than that, the actual percentage figures would be lower (and higher for those who eat less). Note that each label specifies the serving size; the percentages are based on that, not on the entire contents of the package. The section at the bottom of the label is exactly the same on all labels, no matter what the food, and shows the daily amounts of nutrients recommended by the government, based on a consumption of either 2000 or 2500 Cal. Some food packages are allowed to carry shorter labels, either because they have only a few nutrients or because the package has limited label space. The labels make it much easier for the consumer to know exactly what he or she is eating.

The U. S. Department of Agriculture issues occasional guidelines as to what constitutes a healthy diet. The latest, issued in 1992, depicts this in the form of a pyramid (Fig. 26.2). As shown, they consider the basis of a healthy diet to be foods richest in starch (bread, rice, etc.), plus lots of fruits and vegetables (which are rich in vitamins and minerals). Protein-rich foods (meat, fish, dairy products) are to be consumed more sparingly, and fats, oils, and sweets are not considered necessary at all.

An important nonnutrient in some foods is **fiber.** This generally consists of the indigestible portion of vegetables such as lettuce, cabbage, and celery. Chemically, it is made of cellulose, which, we saw in Section 16.11, cannot be digested by humans. But though we cannot digest it, fiber is necessary for proper operation of the digestive system; without it constipation may result. More seriously, a diet lacking sufficient fiber may also lead to colon cancer. Among other foods high in fiber are whole wheat, brown rice, peas, and beans.

Nutrition Facts

Serving size 1 Bar (28g)
Servings per Container 6

Amount Per Serving	
Calories 120	Calories from Fat 35
	% Daily Value*
Total Fat 4g	6%
Saturated Fat 2g	10%
Cholesterol 0mg	0%
Sodium 45mg	2%
Potassium 100mg	3%
Total Carbohydrate 19g	6%
Dietary fiber 2g	8%
Sugars 13g	
Protein 2g	

Vitamin A 15%	Vitamin C 15%
Calcium 15%	Iron 15%
Vitamin D 15%	Vitamin E 15%
Thiamin 15%	Riboflavin 15%
Niacin 15%	Vitamin B_6 15%
Folate 15%	Vitamin B_{12} 15%
Biotin 10%	Pantothenic Acid 10%
Phosphorus 15%	Iodine 2%
Magnesium 4%	Zinc 4%

*Percent Daily Values are based on a 2,000 calorie diet. Your daily values may be higher or lower depending on your calorie needs.

	Calories:	2,000	2,500
Total Fat	Less than	65g	80g
Sat Fat	Less than	20g	25g
Cholesterol	Less than	300mg	300mg
Sodium	Less than	2,400mg	2,400mg
Potassium		3,500mg	3,500mg
Total Carbohydrate		300g	375g
Dietary Fiber		25g	30g

Figure **26.1**
A food label for a peanut butter crunch bar. The portion at the bottom (following the asterisk) is the same on all labels that carry it.

26.3 Calories

The largest part of our food supply goes to provide energy for our bodies. As we saw in Chapters 20 and 21, the energy comes from the oxidation of carbohydrates, fats, and proteins. The energy derived from food is usually measured in calories. One nutritional calorie (Cal) equals 1000 cal or 1 kcal. Thus, when we say that the average daily nutritional requirement for a young adult male is 2900 Cal, we mean the same amount of energy needed to raise the temperature of 2900 kg of water by 1°C or 29 kg (64 lb) of water by 100°C. A young adult female needs 2100 Cal per day. These are peak requirements. Children and older people, on the average, require less energy. Keep in mind that these energy requirements are for active people. For bodies completely at rest, the corre-

One calorie is the amount of heat necessary to raise the temperature of 1 g of water by 1°C (Sec. 1.9).

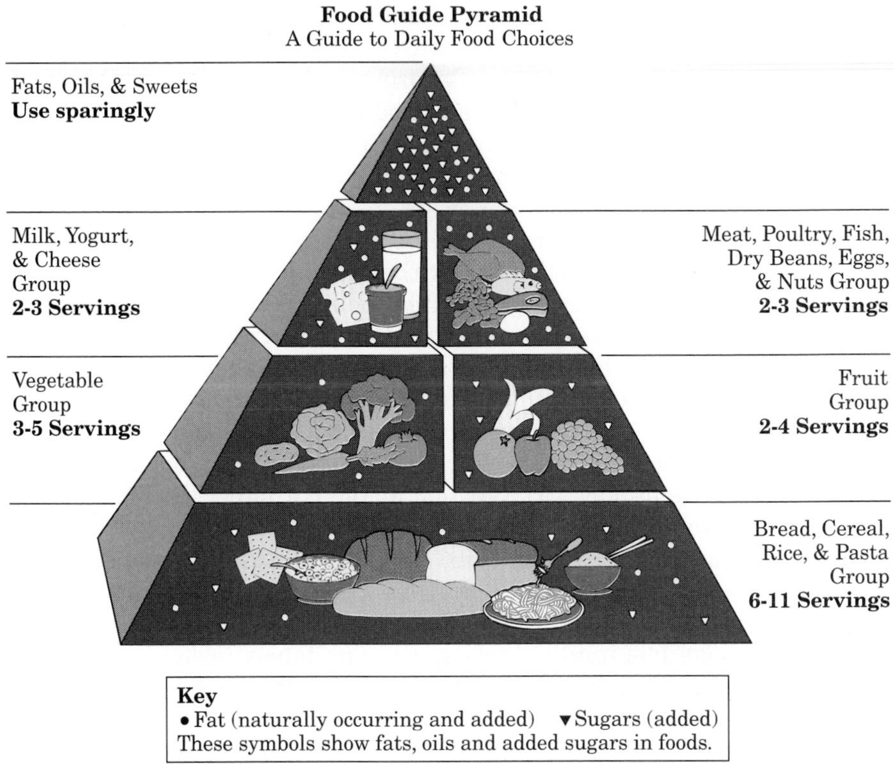

Figure **26.2**
The Food Guide Pyramid developed by the U. S. Department of Agriculture as a general guide to a healthful diet.

FOOD GROUP SERVING SIZES

Bread, Cereal, Rice, and Pasta
- 1/2 cup cooked cereal, rice, pasta
- 1 ounce dry cereal
- 1 slice bread
- 2 cookies
- 1/2 medium doughnut

Vegetables
- 1/2 cup cooked or raw chopped vegetables
- 1 cup raw leafy vegetables
- 3/4 cup vegetable juice
- 10 french fries

Fruit
- 1 medium apple, banana, or orange
- 1/2 cup chopped, cooked, or canned fruit
- 3/4 cup fruit juice
- 1/4 cup dried fruit

Milk, Yogurt, and Cheese
- 1 cup milk or yogurt
- $1\frac{1}{2}$ ounces natural cheese
- 2 ounces process cheese
- 2 cups cottage cheese
- $1\frac{1}{2}$ cups ice cream
- 1 cup frozen yogurt

Meat, Poultry, Fish, Dry Beans, Eggs, and Nuts
- 2–3 ounces cooked lean meat, fish, or poultry
- 2–3 eggs
- 4–6 tablespoons peanut butter
- $1\frac{1}{2}$ cups cooked dry beans
- 1 cup nuts

Fats, Oils, and Sweets
- Butter, mayonnaise, salad dressing, cream cheese, sour cream, jam, jelly

sponding energy requirement for young adult males is 1800 Cal/day, and that for females is 1300 Cal/day. The requirement for a resting body is called the **basal caloric requirement.**

An imbalance between the caloric requirement of the body and the caloric intake creates health problems. Chronic caloric starvation exists in many parts of the world, where people simply do not have enough food to eat because of prolonged drought, devastation by war, natural disasters, or simply overpopulation. Famine particularly affects infants and children. Chronic starvation, called **marasmus,** increases infant mortality up to 50 percent. It results in arrested growth, muscle wasting, anemia, and general weakness. Even if starvation is later alleviated, it leaves permanent damage, insufficient body growth, and lowered resistance to disease.

At the other end of the caloric spectrum is excessive caloric intake. This results in obesity, or the accumulation of body fat. Obesity increases the risk of hypertension, cardiovascular disease, and diabetes. Reducing diets aim at decreased caloric intake without sacrificing any essential nutrients. A combination of exercise and lower caloric intake can eliminate obesity, but usually these diets must achieve their goal over an extended period. Crash diets give the illusion of quick weight loss, but most of this is due to loss of water, which can be regained very quickly. To reduce obesity we must lose body fat and not water; this takes a lot of effort, because fats contain so much energy. A pound of body fat is equivalent to 3500 Cal. Thus, to lose 10 pounds it is necessary either to consume 35 000 fewer Calories, which can be achieved if one reduces caloric intake by 350 Cal every day for 100 days (or by 700 Cal daily for 50 days), or to use up, through exercise, the same numbers.

Excessive caloric intake results from all sources of food: carbohydrates and proteins as well as fats.

26.4 Carbohydrates, Fats, and Proteins in the Diet

Carbohydrates are the major source of energy in the diet. They also furnish important compounds for the synthesis of cell components (Chapter 22). The main dietary carbohydrates are the polysaccharide starch, the disaccharides lactose and sucrose, and the monosaccharides glucose and fructose. Nutritionists recommend a minimum carbohydrate intake equivalent to 500 Cal per day. In addition to these digestible carbohydrates, cellulose in plant materials provides dietary fiber. Artificial sweeteners (Box 18A) can be used to reduce mono- and disaccharide intake.

Fats are the most concentrated source of energy. About 98 percent of dietary fats are triglycerides; the remaining 2 percent consists of complex lipids and cholesterol. Only two fatty acids are essential in higher animals, including humans: linolenic and linoleic acids (Sec. 17.3). These fatty acids are needed in the diet because our bodies cannot synthesize them. Nutritionists occasionally list arachidonic acid as an essential fatty acid. However, our bodies can synthesize arachidonic acid from linoleic acid.

Although the proteins in our diet can be used for energy (Sec. 21.9), their main use is to furnish amino acids from which the body synthesizes its own proteins (Sec. 23.10). The human body is incapable of synthesizing ten of the amino acids needed to make proteins. These ten (the **essential amino acids**) must be obtained from our food; they are shown in Table 18.1. The body breaks down food proteins into their amino acid constituents and then puts the amino acids together again to make body proteins. For proper nutrition, the foods we eat should contain about 20 percent protein.

A dietary protein that contains all the essential amino acids is called a **complete protein.** Casein, the protein of milk, is a complete protein, as are most other animal proteins—those found in meat, fish, and eggs. People who eat adequate quantities of meat, fish, eggs, and dairy products get all the amino acids they need to keep healthy. About 50 g per day of complete proteins constitutes an adequate quantity.

Different activity levels require different caloric needs. (Photographs by Charles D. Winters.)

In clinical practice it occasionally happens that a patient cannot be fed through the gastrointestinal tract. Under these conditions intravenous feeding or **parenteral nutrition** is required. In the past, patients could be maintained on parenteral nutrition only for short periods. Calories were provided in the form of glucose, and amino acids were supplied in the form of protein hydrolysates. Over long periods, however, the use of parenteral nutrition resulted in deterioration and progressive emaciation of the patient. More recently, the solutions used in parenteral nutrition are being manufactured to contain both essential and nonessential amino acids in fixed proportions. Calories now do not come only from glucose. Glucose supplies only about 20 percent of the caloric needs. Fat emulsions provide the rest of the caloric requirements and the essential fatty acids. Finally, the full vitamin and mineral requirements are also supplied in the parenteral solution.

An important animal protein that is *not* complete is gelatin, which is made by denaturing collagen (Sec. 18.10). Gelatin lacks tryptophan and is low in several other amino acids, including isoleucine and methionine. Many people on quick reducing diets take "liquid protein." This is nothing but denatured and partially hydrolyzed collagen (gelatin). Therefore, if this is the only protein source in the diet, some essential amino acids will be lacking.

Most plant proteins are incomplete. For example, corn protein lacks lysine and tryptophan; rice protein lacks lysine and threonine; wheat protein lacks lysine; and even soy protein, one of the best plant proteins, is very low in methionine. Adequate amino acid nutrition is possible with a vegetarian diet, but only if a range of different vegetables is eaten.

In many underdeveloped countries, protein deficiency diseases are widespread because the people get their protein mostly from plants. Among these is a disease called kwashiorkor, the symptoms of which are a swollen stomach, skin discoloration, and retarded growth.

26.5 Vitamins

Vitamins and minerals are essential for good nutrition. Animals maintained on diets that contain sufficient carbohydrates, fats, and proteins and provided with an ample water supply cannot survive on these alone. They also need the essential organic components called **vitamins** and inorganic ions called **minerals** (Sec. 26.6). Deficiencies in vitamins and minerals lead to many nutritionally controllable diseases (one example is shown in Fig. 26.3). The Food and Drug Administration lists a recommended daily allowance (RDA) for each vitamin and mineral. These are given in Table 26.1.

The name "vitamin" comes from a mistaken generalization. Casimir Funk (1884–1967) discovered that certain diseases such as beriberi, scurvy, and pellagra are caused by lack of certain nutrients. He found that the "antiberiberi factor" compound is an amine, and mistakenly thought that all such nutrients are amines. Hence he coined the name "Vitamine."

Vitamins can conveniently be divided into two groups: water-soluble and fat-soluble. Water-soluble vitamins cannot be stored by the body; if they are not metabolized, they are washed out by the fluids we consume

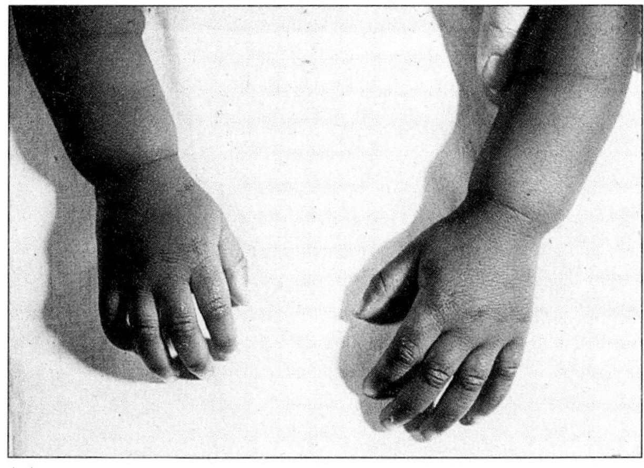

(a)

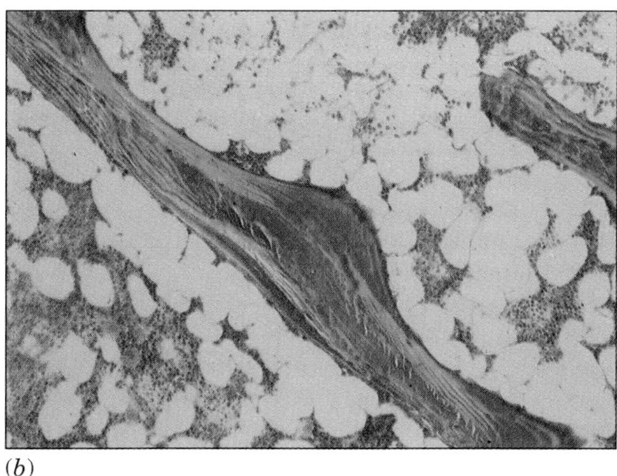

(b)

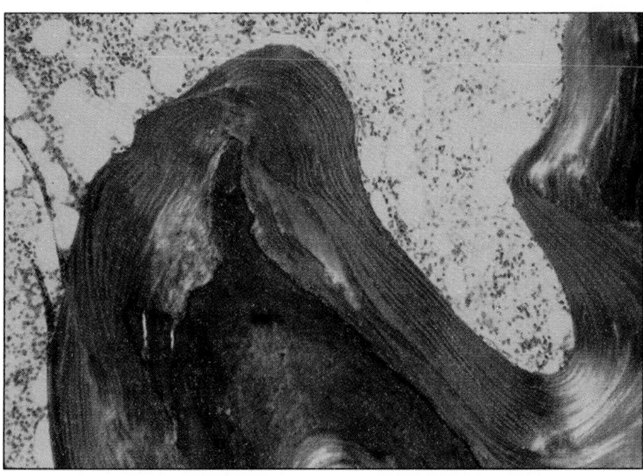

(c)

Figure **26.3**
(a) Symptoms of rickets, a vitamin D deficiency in children. The nonmineralization of the bones of the radius and the ulna results in prominence of the wrist. (Courtesy of Drs. P. G. Bullogh and V. J. Vigorita and the Gower Medical Publishing Co., New York.) Histology of (b) normal and (c) osteomalacic bone. The latter shows the accumulation of osteoids (red stain) due to vitamin D deficiency. (Courtesy of Drs. P. A. Dieppe, P. A. Bacon, A. N. Bamji, and I. Watt and Gower Medical Publishing Ltd., London, England.)

each day. In contrast, unused fat-soluble vitamins can accumulate in body tissues.

Fat-Soluble Vitamins

The fat-soluble vitamins are A, D, E, and K. Being fat-soluble, these vitamins are stored in body fat, especially in the liver. Because they cannot be washed out by water, they can accumulate to the point where they become harmful (Box 26B).

Vitamin A is important in vision. As can be seen from its structure, it is the source of retinal, which is used in the retina of the eye (Box 11C).

H_3C CH_3 CH_3 CH_3 CH_2OH CH_3

Vitamin A

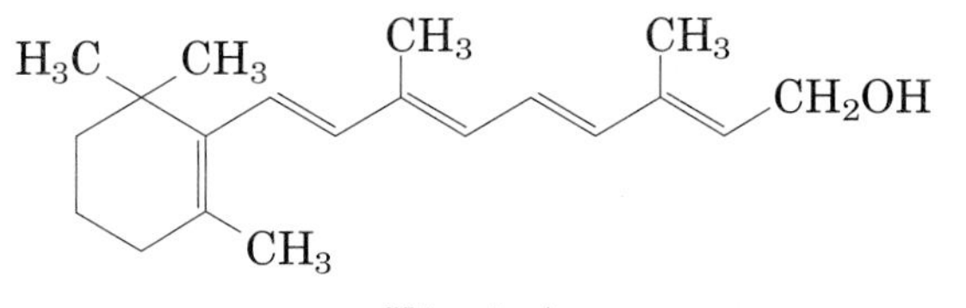

Liver and carrots have a high vitamin A content. (Photograph by Charles D. Winters.)

Table 26.1 Recommended Daily Allowances for Vitamins and Minerals

Vitamin or Mineral	Recommended Daily Allowance[a]
Fat-soluble vitamins	
A	750 μg (1500 μg)[b]
D	10 μg; exposure to sunlight
E	4–9 mg
K	70–140 μg
Water-soluble vitamins	
C	60 mg
B_1 (thiamine)	1 mg
B_2 (riboflavin)	1.4 mg
Nicotinic acid (niacin)	18 mg
B_6 (pyridoxal)	2 mg
Folic acid	400 μg
Pantothenic acid	7 mg
Biotin	300 μg
B_{12}	3 μg
Major minerals	
Sodium	1100–3300 mg
Chloride	1700–5100 mg
Potassium	2000–5600 mg
Phosphorus	800–1500 mg
Calcium	800 mg
Magnesium	400 mg
Trace elements	
Iron	12 mg; in menstruating women, 17–23 mg
Zinc	15 mg
Copper	2–3 mg
Manganese	2.5–5.0 mg
Chromium	0.05–0.2 mg
Cobalt	0.05–1.8 mg (20–30 mg)[b]
Iodide	150 μg (1000 μg)[b]
Fluoride	1.5–4.0 mg (8–20 mg)[b]
Selenide	0.05–0.2 mg (2.4–3.0 mg)[b]

[a]The U. S. RDAs are set by the Food and Nutrition Board of the National Research Council. The numbers given here are based on the 1985 recommendations. The RDA varies with age, sex, and level of activity; the numbers given are average values for both sexes between the ages of 18 and 54.
[b]Toxic if doses above the level shown in parentheses are taken.

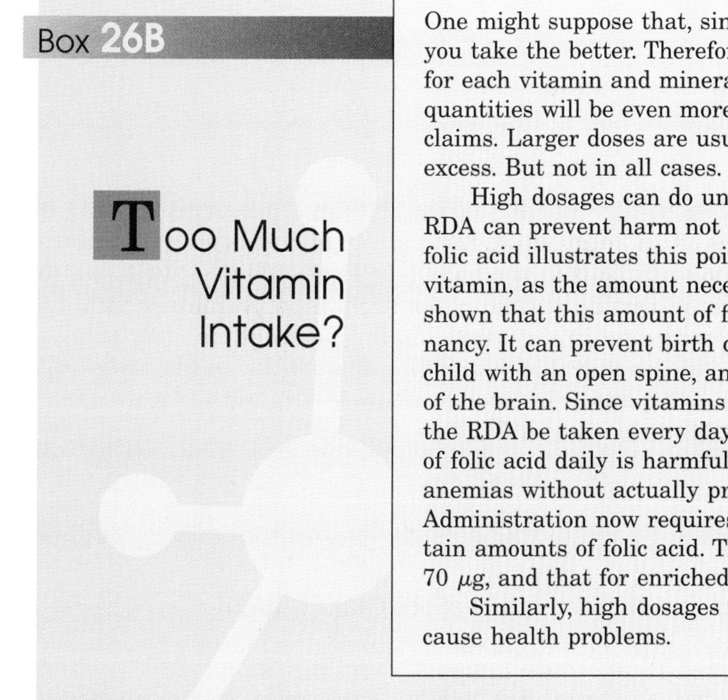

Lack of vitamin A causes night blindness, a condition in which vision is impaired under low light levels. Vitamin A also helps to heal eye and skin injuries and prevents keratinization, a condition wherein mucous membranes of the eye, digestive tract, and other areas become dry and hard. We obtain this vitamin by eating such foods as liver, butter, and egg yolks. Certain vegetables, such as carrots, spinach, and sweet potatoes are also sources of this vitamin, but indirectly. These

Box 26B

Too Much Vitamin Intake?

One might suppose that, since vitamins are essential nutrients, the more you take the better. Therefore, some people consider that the RDAs listed for each vitamin and mineral in Table 26.1 are only a minimum, and larger quantities will be even more beneficial. No scientific proof exists for such claims. Larger doses are usually harmless, because the body excretes the excess. But not in all cases.

High dosages can do unexpected harm in certain instances, while the RDA can prevent harm not usually associated with the vitamin. The case of folic acid illustrates this point. Table 26.1 shows an RDA of 400 μg for this vitamin, as the amount necessary to prevent anemia. Lately it has been shown that this amount of folic acid is especially important during pregnancy. It can prevent birth defects that result in *spina bifida,* the birth of a child with an open spine, and *anencephaly,* the birth of a child without most of the brain. Since vitamins are not stored in the body, it is important that the RDA be taken every day. *But not more.* An intake of 1000 μg or more of folic acid daily is harmful because it can hide the symptoms of certain anemias without actually preventing these diseases. The Food and Drug Administration now requires that food labeled as enriched must carry certain amounts of folic acid. The amount for a slice of enriched bread is 40 to 70 μg, and that for enriched cereals is 100 μg per serving.

Similarly, high dosages for niacin and of vitamin A have been shown to cause health problems.

vegetables contain β-carotene, which the body converts to vitamin A (Box 11A).

Vitamin D is a mixture of several similar compounds, one of which is cholecalciferol (also called vitamin D_3).

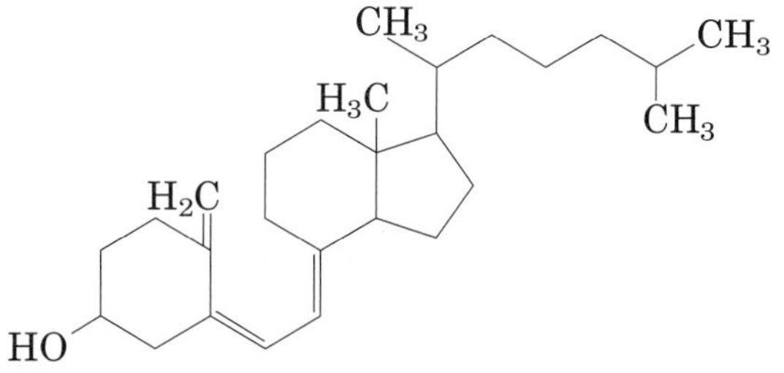

CH_3 CH_3 H_3C CH_3 H_2C HO

Vitamin D_3

This vitamin is important in bone formation. Lack of it in children causes rickets [Fig. 26.3(a)], a disease in which the bones are misformed, but adults also need this vitamin to maintain healthy bones. Good sources of vitamin D are salmon, sardines, butter, eggs, and milk (milk in the United States is usually fortified with this vitamin). Furthermore, human skin contains certain steroids that can be converted to vitamin D by sunlight, so that moderate sunlight is a source of this vitamin (excessive sunlight is harmful to the skin).

Vitamin E Once again, there are several forms of this vitamin, the most important of which is α-tocopherol.

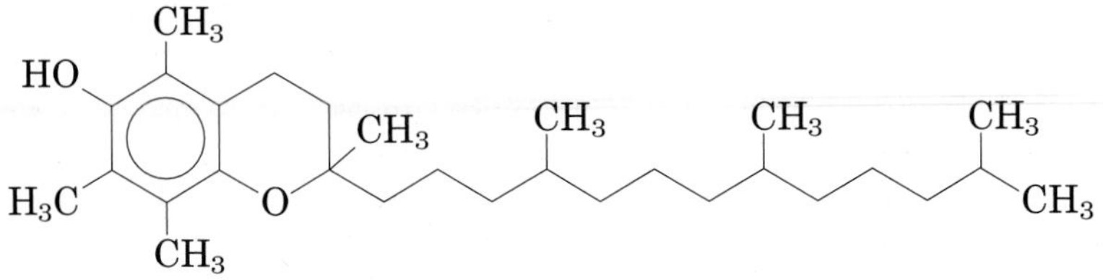

α-Tocopherol (vitamin E)

Less is known about the function of this vitamin, but it seems that it mostly functions as an antioxidant, removing oxidizing agents that cause damage to the various cells in the body (see Boxes 9D and 20C). It also prevents anemia in certain special cases such as in premature infants. Vitamin E is so prevalent in vegetable and fish oils that virtually nobody suffers from a deficiency of this vitamin.

Vitamin K is important in blood clotting: It is essential for the manufacture of prothrombin (Box 25B).

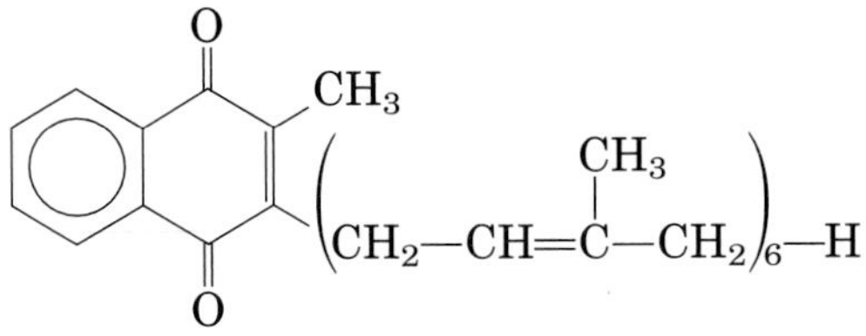

Vitamin K

Like vitamin E, this vitamin is present in many foods, especially leafy vegetables like spinach and cauliflower, as well as potatoes and beef liver. Deficiencies are very rare.

Water-Soluble Vitamins

Apart from vitamin C, all the other vitamins in this class belong to the B group, whether they are called by a name such as B_1 or not.

Vitamin C (ascorbic acid) is essential to the hydroxylation of collagen, the structural protein of connective tissue (skin, bone, cartilage, etc.).

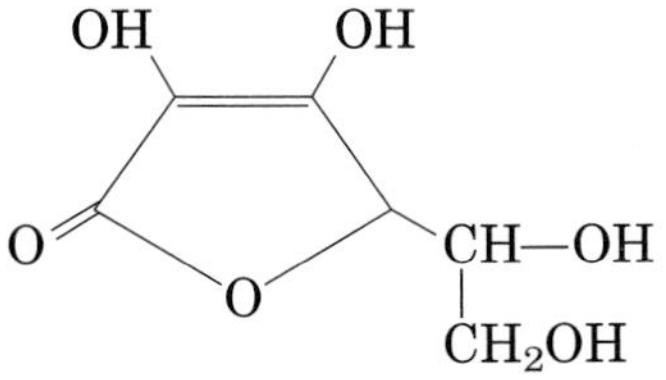

Vitamin C (ascorbic acid)

In the days of sailing ships, many seamen got scurvy because they had no source of fresh fruits or vegetables on voyages that could last many weeks away from land. In response to this, British ships began carrying limes, which could be stored for long periods. British sailors acquired the nickname "limey" from this practice, a name that was often extended to any Englishman.

This was one of the first vitamins to be discovered, when it was realized that a lack of it caused scurvy, a disease prevalent among sailors who did not have a source of fresh fruits on voyages that lasted many months. The best source of vitamin C is citrus fruits (oranges, lemons, limes, grapefruit, etc.), but it is also present in other fruits and vegetables, including berries, broccoli, cabbage, peppers, and tomatoes.

Whether vitamin C has other effects than hydroxylation of collagen is a source of much controversy. Claims have been made that it prevents and cures the common cold as well as other diseases, but the verdict is still not in on these claims, even though much research has been done. Vitamin C does appear to be an antioxidant, similar to vitamin E.

Vitamin B_1 (thiamine), present in beans, soybeans, cereals, ham, and liver, is a coenzyme in oxidative decarboxylation and in the pentose phosphate shunt. It prevents a dietary disease called beriberi, a nervous system disorder. It is often added to white bread.

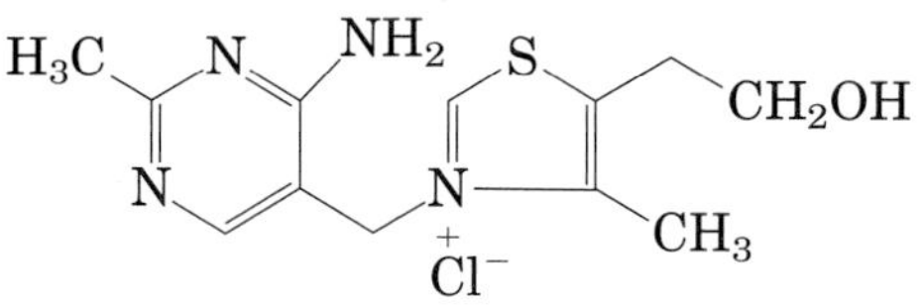

Vitamin B_1 (thiamine)

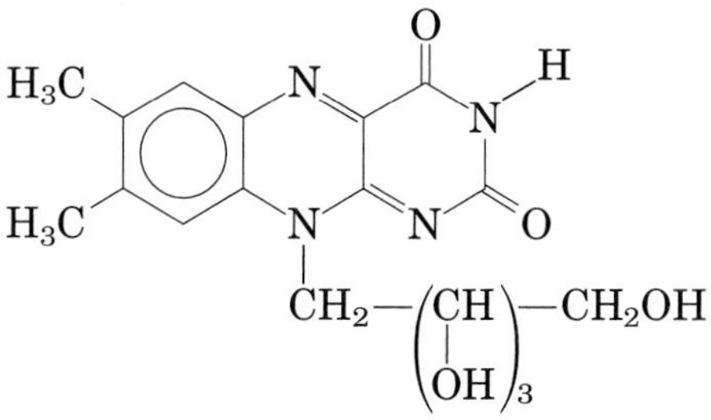

Vitamin B_2 (riboflavin)

Vitamin B_2 (riboflavin) is used by the body to make FAD (Sec. 20.3), which is essential to many metabolic processes. Good food sources are liver, yeast, almonds, mushrooms, beans, milk, and cheese. Deficiencies are rare, but result in eye and skin diseases.

Nicotinic acid (niacin), as well as the amide form nicotinamide, are similarly used by the body to make NAD^+ (Sec. 20.3; see also Box 20A). A deficiency results in a disease called pellagra, characterized by skin changes and severe nerve dysfunction. Nicotinic acid is found in many foods, including chickpeas, lentils, peaches, figs, fish, meat, bread, beans, rice, and berries.

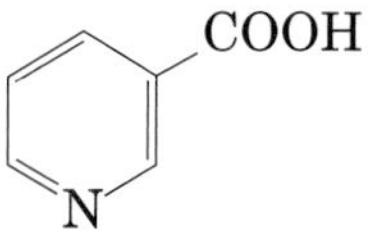

Foods with a high content of riboflavin and other B vitamins. (Photograph by Charles D. Winters.)

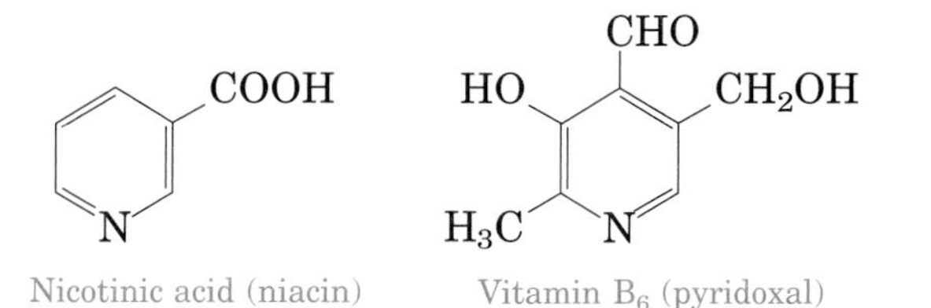

Nicotinic acid (niacin)

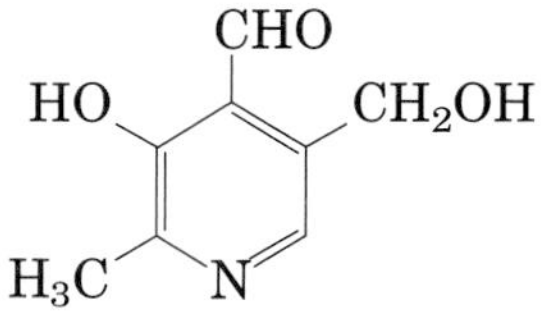

Vitamin B_6 (pyridoxal)

Vitamin B_6 (pyridoxal) is also a coenzyme, used by the body in transamination and heme synthesis. Deficiency symptoms include convulsions, chronic anemia, and peripheral neuropathy. Good sources are meat, fish, nuts, oats, and wheat germ.

Folic acid is a coenzyme in methylation and DNA synthesis. Anemia is the chief deficiency disease. It is found in liver, kidney, eggs, spinach, beets, orange juice, avocados, and cantaloupe. It has recently been found that pregnant women need more folic acid than other people, in order to prevent birth defects.

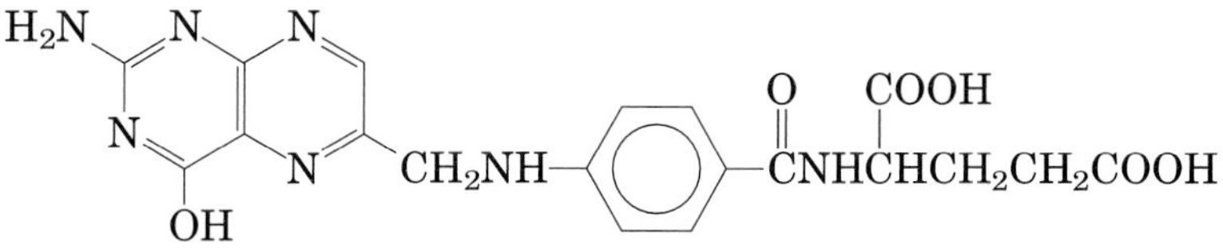

Folic acid

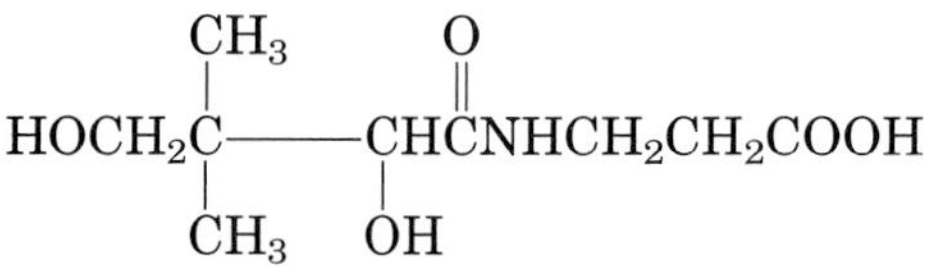

Pantothenic acid

Pantothenic acid, still another coenzyme (it is part of coenzyme A, Sec. 20.3), is found in so many foods (for example, soybeans, peanuts, wheat, vegetables, liver, eggs, and milk) that a deficiency disease has not been definitely established in humans, though it has been linked to intestinal disturbances and depression.

Biotin is involved in the synthesis of fatty acids, and deficiency can result in dermatitis, nausea, and depression. Foods rich in biotin are yeast, liver, kidney, nuts, tomatoes, and egg yolks.

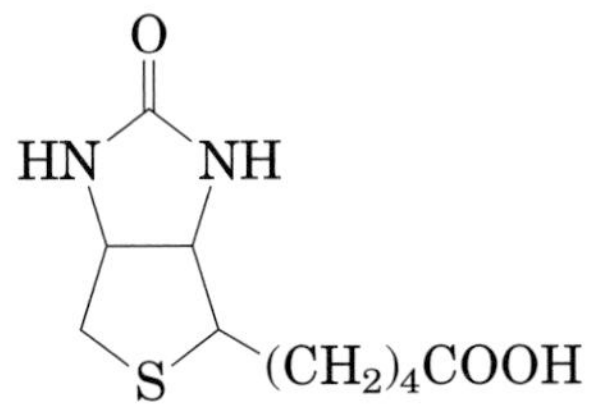

O
HN NH
S $(CH_2)_4COOH$

Biotin

Vitamin B_{12} has the most formidable structural formula of all the vitamins, and is the only one that contains cobalt.

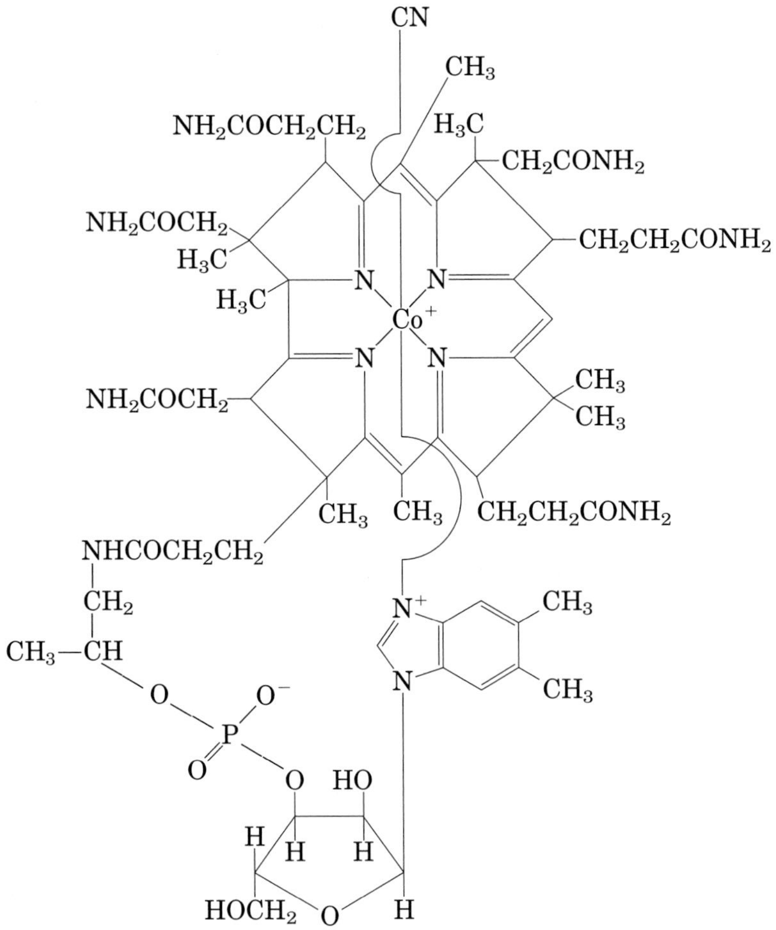

CN
CH_3
$NH_2COCH_2CH_2$ H_3C
CH_2CONH_2
NH_2COCH_2
H_3C $CH_2CH_2CONH_2$
H_3C N N
Co^+
N N
NH_2COCH_2 CH_3
CH_3
CH_3 CH_3 $CH_2CH_2CONH_2$
$NHCOCH_2CH_2$
CH_2
CH_3—CH N^+ CH_3
O O^- N CH_3
P
O O HO
H H H
$HOCH_2$ O H

Vitamin B_{12}

Compared to the other vitamins, the body needs very little of this one (only 3 micrograms per day), but a deficiency results in a disease called pernicious anemia, and degradation of nerves, spinal cord, and brain. This vitamin is present in many animal-derived foods, such as liver, kidney, salmon, oysters, milk products, and eggs, and humans need so little that getting enough is no problem for most people. However, strict vegetarians may need to take vitamin B_{12} supplements. The vitamin is used by the body as part of the methyl-removing enzyme in folate metabolism. Without it, neither red blood cells nor DNA can be synthesized.

26.6 Water and Minerals

Water

Water makes up 60 percent of our body weight. Most of the compounds in our body are dissolved in water, which also serves as a transporting medium to carry nutrients and waste materials. These functions are discussed in Chapter 25. We must maintain a proper balance between water intake and water excretion via urine, feces, sweat, and exhalation of breath. A normal diet requires about 1200 to 1500 mL of water per day. This is in addition to the water content of our foods.

Major Minerals

In this class are six elements, all of which should be present in the diet every day, in their ionic forms.

- Sodium ion Na^+
- Chloride ion Cl^-
- Potassium ion K^+

These ions are used by the body to carry electrical charges, and to maintain osmotic pressure in the cells. They are present in many foods.

- Phosphorus, in the form of phosphate ion PO_4^{3-}

As we have seen, both metabolism and the synthesis and functions of nucleic acids rely on phosphate and its esters and anhydrides. It is also an essential component of bones. It can be obtained from lentils, nuts, oats, egg yolk, cheese, and meat.

- Calcium ion Ca^{2+}

This ion performs many functions in the body, being involved in bone and teeth formation, blood coagulation, muscle contraction, and hormonal functions. About 90% of body calcium is in the bones and teeth. Milk and milk products are the best source of calcium, but eggs and nuts also contain it.

- Magnesium ion Mg^{2+}

This ion is also present in bones and teeth, but it is also an important cofactor for enzymes. It can be obtained from cheese, cocoa, chocolate, nuts, soybeans, and beans.

Trace Elements

The body also uses other elements, in smaller amounts, though by no means does it use all the elements in the periodic table. The following are recognized as important trace elements that must be obtained in the diet. Some of them are poisonous in larger quantities.

- Iron ions Fe^{2+} and Fe^{3+}

The body needs Fe^{2+} (dietary Fe^{3+} is reduced by the body) for its hemoglobin molecules, and for some of the enzymes that take part in oxidative phosphorylation (Sec. 20.5). A deficiency results in anemia. Meat, raisins, beans, chickpeas, smoked fish, liver, and clams are among the foods plentiful in iron. Women generally need more dietary iron than men because they lose hemoglobin each month in menstruation.

- Zinc ion Zn^{2+}
- Copper ion Cu^{2+}

These ions are cofactors necessary for the operation of certain enzymes. Lack of zinc causes retarded growth and enlarged liver; copper deficiency results in anemia and loss of hair color. If an infant gets insufficient zinc, it may grow up as a dwarf. Zinc may be obtained from yeast, soybeans, nuts, corn, cheese, and meat, while copper-containing foods include oysters, sardines, lamb, and liver.

- Manganese ion Mn^{2+}

Manganese is necessary for bone formation and proper nerve action. Lack of it may result in low levels of serum cholesterol and retarded growth of hair and nails. Foods rich in manganese include nuts, fruits, vegetables, and whole-grain cereals.

- Chromium ion Cr^{3+}

This ion is essential to normal glucose metabolism, and lack of it impairs growth and reduces life span. It is found in meat and whole grains, such as wheat and rye flour.

- Cobalt ion Co^{2+}

This ion is a part of vitamin B_{12}, which does not function without it. It has no other known biochemical use in humans and is obtained in the diet along with vitamin B_{12}.

- Iodide ion I^-

Iodine is needed for the thyroxine hormone which controls thyroid function (Box 12I). Lack of it causes the disease called goiter. The best dietary source is seafood, but many people buy iodized salt, which contains a small amount of sodium iodide.

- Fluoride ion F^-

This ion is essential for sound teeth, especially in children. Lack of it enables bacteria to cause tooth decay. It is not found in foods but can be obtained from certain municipally fluoridated water supplies. In the absence of this source, fluoridated tooth paste will supply some fluoride.

- Selenium ion Se^{2-}

Selenium, in very small amounts, is involved in fat metabolism; a deficiency results in muscle disorders. In areas where the selenium content of the soil is relatively high, the residents have a relatively low level of heart attacks and stroke. The element is found in meat and seafood.

26.7 Digestion

In order for food to be used in our bodies, it must be absorbed through the intestinal walls into the blood stream or lymph system. Some nutrients, such as vitamins, minerals, glucose, and amino acids, can be absorbed directly. Others, such as starch, fats, and proteins, must first be broken down into smaller components before they can be absorbed. **This breakdown process** is called **digestion.**

26.8 Digestion of Carbohydrates

Before the body can absorb carbohydrates, it must break down di-, oligo-, and polysaccharides into monosaccharides because only monosaccharides can pass into the blood stream.

The monosaccharide units are connected to each other by glycosidic (acetal) linkages. As we saw in Section 13.6, the cleavage of acetals by water is called *hydrolysis.* In the body this hydrolysis is catalyzed by acids and by enzymes. When the metabolic need arises, storage polysaccharides are hydrolyzed to yield glucose and maltose.

The three storage polysaccharides are amylose, amylopectin, and glycogen.

The hydrolysis is aided by a number of enzymes: α-amylase attacks all three storage polysaccharides at random, hydrolyzing the $\alpha(1 \rightarrow 4)$ glycosidic linkages, and β-amylase also hydrolyzes the $\alpha(1 \rightarrow 4)$ glycosidic linkages but in an orderly fashion, cutting disaccharidic maltose units one by one from the nonreducing (locked) end of a chain. A third enzyme, called the **debranching enzyme,** attacks and hydrolyzes the $\alpha(1 \rightarrow 6)$ glycosidic linkages (Fig. 26.4). In acid-catalyzed hydrolysis, storage polysaccharides are attacked at random points, although acid catalysis is slower than enzyme hydrolysis at body temperature.

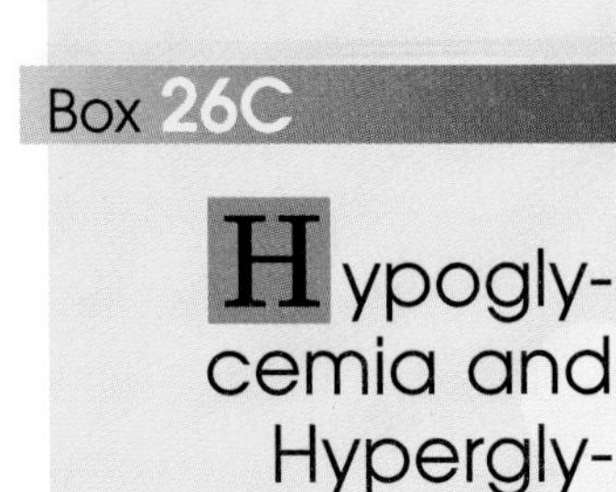

Box 26C

Hypoglycemia and Hyperglycemia

The glucose content of normal blood varies between 65 and 105 mg/100 mL. Usually this is measured after 8 to 12 hours of fasting, and therefore this range is known as the normal fasting level. **Hypoglycemia** develops when the glucose concentration falls below this level. Since the brain cells normally use glucose as their only energy source, hypoglycemic patients experience dizziness and fainting spells. In more severe cases convulsion and shock may occur.

In **hyperglycemia,** when the blood glucose level reaches 140 to 160 mg/100 mL, the kidneys begin to filter glucose out of the blood stream, and it appears in the urine.

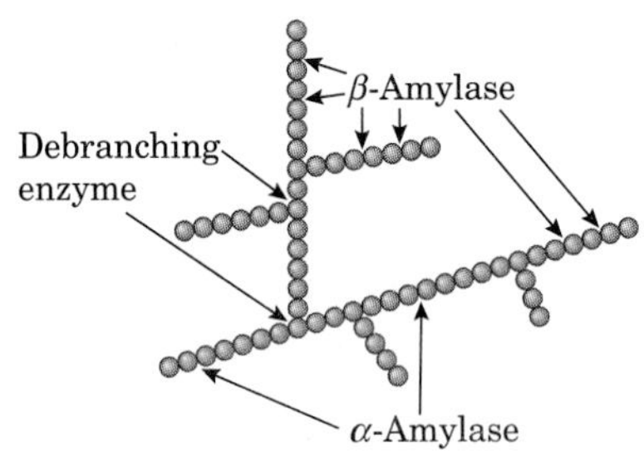

Figure **26.4**
The action of different enzymes on glycogen and starch.

The digestion (hydrolysis) of starch and glycogen in our food supply starts in the mouth, where α-amylase is one of the main components of saliva. The hydrochloric acid in the stomach and the other hydrolytic enzymes in the intestinal tract decompose starch and glycogen to produce mono- and disaccharides (D-glucose and maltose).

The D-glucose produced by hydrolysis of the di-, oligo-, and polysaccharides enters the blood stream and is carried to the cells to be utilized (Sec. 21.2). For this reason, D-glucose is often called blood sugar (Box 26C). In healthy people little or none of this sugar ends up in the urine except for short periods of time. In the condition known as **diabetes mellitus,** however, glucose is not completely metabolized and does appear in the urine (Box 24G). Because of this, it is necessary to test the urine of diabetic patients for the presence of glucose.

26.9 Digestion of Lipids

The lipids in the food we eat must be broken down into smaller components before they can be absorbed into the blood or lymph system through the intestinal walls. The enzymes that promote this breakdown are located in the small intestine and are called **lipases.** However, since lipids are insoluble in the aqueous environment of the gastrointestinal tract, they must be dispersed into fine colloidal particles before the enzymes can act on them.

The bile salts (Sec. 17.11) perform this important function. Bile salts are manufactured in the liver from cholesterol and stored in the gallbladder. From there they are secreted through the bile ducts into the intestine. The emulsion produced by the bile salts and dietary fats is acted upon by the lipases. These break fats down into glycerol and fatty acids and complex lipids into fatty acids, alcohols (glycerol, choline, ethanolamine, sphingosine), and carbohydrates. All these breakdown products are absorbed through the intestinal walls.

26.10 Digestion of Proteins

The digestion of dietary proteins begins with cooking, which denatures proteins. (Denatured proteins are broken down more easily by the hydrochloric acid in the stomach and by digestive enzymes than are native proteins.) Stomach acid contains about 0.5 percent HCl. The HCl both denatures the proteins and cleaves the peptide linkages randomly. Pepsin, the proteolytic enzyme of stomach juice, breaks peptide bonds on the C=O side of three amino acids only: tryptophan, phenylalanine, and tyrosine (Fig. 26.5).

Most protein digestion occurs in the small intestine. There, the enzyme chymotrypsin breaks internal peptide bonds at the same positions as does pepsin, while another enzyme, trypsin, breaks them only on the C=O side of arginine and lysine. Other enzymes, such as carboxypeptidase, cut amino acids one by one from the C-terminal end of the protein. The amino acids and small peptides are then absorbed through the intestinal walls.

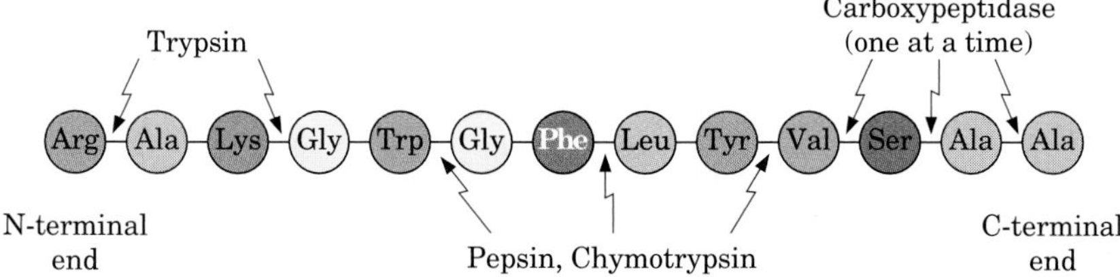

Figure **26.5** • Different enzymes hydrolyze peptide chains in different but specific ways.

SUMMARY

Nutrients are components of foods that provide growth, replacement, and energy. Nutrients are classified into six groups: carbohydrates, lipids, proteins, vitamins, minerals, and water. Each food contains a variety of nutrients. The largest part of our food intake is used to provide energy for our bodies. A typical young adult needs 2900 Cal (male) or 2100 Cal (female) as an average daily caloric intake. **Basal caloric requirements** are the energy need when the body is completely at rest. These are less than the normal requirements. Imbalance between need and caloric intake may create health problems; chronic starvation increases infant mortality, while obesity leads to hypertension, cardiovascular diseases, and diabetes.

Carbohydrates are the major source of energy in our diet. Fats are the most concentrated source of energy. Essential fatty and amino acids are needed as building blocks because our bodies cannot synthesize them. Vitamins and minerals are essential constituents of diets that are needed in small quantities. The fat-soluble vitamins are A, D, E, and K. Vitamin C and the B group are water-soluble vitamins. Most of the B vitamins are essential coenzymes. The most important dietary minerals are Na^+, Cl^-, K^+, PO_4^{3-}, Ca^{2+}, and Mg^{2+}, but trace minerals are also necessary. Water makes up 60 percent of body weight.

The digestion of carbohydrates, fats, and proteins is aided by the stomach acid, HCl, and by enzymes that reside in the mouth, stomach, and intestines.

KEY TERMS

Basal caloric requirement (Sec. 26.3)
Complete protein (Sec. 26.4)
Debranching enzyme (Sec. 26.8)
Fiber (Sec. 26.2)
Hyperglycemia (Box 26C)
Hypoglycemia (Box 26C)
Kwashiorkor (Sec. 26.4)
Lipase (Sec. 26.9)
Marasmus (Sec. 26.3)
Mineral (Secs. 26.5, 26.6)
Nutrient (Sec. 26.2)
Parenteral nutrition (Box 26A)
RDA (Sec. 26.2)
Trace element (Sec. 26.6)
Vitamin (Sec. 26.5)

PROBLEMS

Nutrition

26.1 Are nutrient requirements uniform for everyone?

26.2 (a) Is raspberry flavoring a nutrient? (b) What is the chemical nature of this flavoring? (See Box 14C.)

***26.3** If sodium benzoate, a food preservative, is excreted as such, and if calcium propionate, another food preservative (see Box 14B), is hydrolyzed in the body and metabolized to CO_2 and H_2O, would you consider either of these preservatives as nutrients? If so, why?

26.4 Is corn grown solely with organic fertilizers more nutritious than that grown with chemical fertilizers?

26.5 Which part of the Nutrition Facts label found on food packages is the same for all labels carrying it?

26.6 Which kinds of food does the government recommend that we have the most servings of each day?

26.7 What is the importance of fiber in the diet?

Calories

26.8 Define basal caloric requirement.

26.9 A young adult female needs a caloric intake of 2100 Cal/day. Her basal caloric requirement is only 1300 Cal/day. Why is the extra 800 Cal needed?

26.10 What ill effects may obesity bring?

26.11 Assume that you want to lose 20 pounds of body fat in 60 days. Your present dietary intake is 3000 Cal/day. What should your caloric intake be, in Cal/day, in order to achieve this goal, assuming no change in exercise habits?

Carbohydrates, Fats, and Proteins in the Diet

26.12 Which nutrient provides energy in its most concentrated form?

26.13 Is it possible to get a sufficient supply of nutritionally adequate proteins by eating only vegetables?

26.14 How many (a) essential fatty acids and (b) essential amino acids do humans need in their diets?

26.15 What is the precursor of arachidonic acid in the body?

26.16 Suggest a way to cure kwashiorkor.

Vitamins, Minerals, and Water

26.17 In a prison camp during a war, the prisoners are fed plenty of rice and water but nothing else. What would the result of such a diet be in the long run?

26.18 (a) How many mL of water per day does a normal diet require? (b) How many calories does this contribute?

26.19 Why did British sailing ships carry a supply of limes?

26.20 What are the symptoms of vitamin A deficiency?

26.21 What is the function of vitamin K?

26.22 (a) Which vitamin contains cobalt? (b) What is the function of this vitamin?

26.23 Vitamin C is recommended in megadoses by some people for all kinds of symptoms from colds to cancer. What disease has been scientifically proven to be prevented when sufficient daily doses of vitamin C are in the diet?

26.24 What are the best dietary sources of calcium, phosphorus, and cobalt?

26.25 Which vitamin contains a sulfur atom?

26.26 What are the symptoms of vitamin B_{12} deficiency?

Digestion

26.27 What chemical processes take place during digestion?

26.28 What is the product of the reaction when β-amylase acts on amylose?

***26.29** Do lipases degrade (a) cholesterol (b) fatty acids?

26.30 Does the debranching enzyme help in digesting amylose?

26.31 Does HCl in the stomach hydrolyze both the $(1 \rightarrow 4)$ and $(1 \rightarrow 6)$ glycosidic linkages?

26.32 What is the difference between protein digestion by trypsin and by HCl?

Boxes

26.33 (Box 26A) Why cannot glucose supply all the caloric needs in parenteral nutrition over a long period?

26.34 (Box 26B) The RDA for niacin is 18 mg. Would it be still better to take a 100-mg tablet of this vitamin every day?

26.35 (Box 26C) A patient complains of dizziness. Blood analysis shows that he has 30 mg of glucose in 100 mL of blood. What would you recommend to alleviate the dizziness?

Additional Problems

26.36 Which vitamin is part of coenzyme A (CoA)?

26.37 (a) Name the essential fatty acids. (b) Are they saturated or unsaturated?

26.38 Why is it necessary to have proteins in our diets?

***26.39** Which vitamin is part of the coenzyme NAD^+?

26.40 According to the government's food pyramid, are there any foods that we can completely omit from our diets and still be healthy?

26.41 What kind of carbohydrate provides dietary fiber?

***26.42** As an employee of a company that markets walnuts, you are asked to provide information for an ad that would stress the nutritional value of walnuts. What information would you provide?

***26.43** In diabetes, insulin is administered intravenously. Explain why this hormone protein cannot be taken orally.

***26.44** What kind of supplemental enzyme would you recommend for a patient after a peptic ulcer operation?

26.45 Name three essential amino acids.

26.46 Why is zinc needed in the body?

26.47 In a trial a woman was accused of poisoning her husband by adding arsenic to his meals. Her attorney stated that this was done to promote her husband's health, arsenic being an essential nutrient. Would you accept this argument?

Significant figures

Appendix

If you measure the volume of a liquid in a graduated cylinder (Fig. 1.6), you might find that it is 36 mL, to the nearest milliliter, but you cannot tell if it is 36.2, or 35.6, or 36.0 mL because this measuring instrument does not give the last digit with any certainty. A buret (Fig. 1.6) gives more digits, and if you use one you should be able to say, for instance, that the volume is 36.3 mL and not 36.4 mL. But even with a buret, you could not say whether the volume is 36.32 or 36.33 mL. For that, you would need an instrument that gives still more digits. This example should show you that *no measured number can ever be known exactly.* No matter how good the measuring instrument, there is always a limit to the number of digits it can measure with certainty.

Scientists have found it useful to establish a method for telling to what degree of certainty any measured number is known. The method is very simple; it consists merely of writing down all the digits that are certain and not writing down any that are not certain. We define the number of **significant figures** in a measured number as **the number of digits that are known with certainty.**

What do we mean by this? Assume that you are weighing a small object on a laboratory balance that can weigh to the nearest 0.1 g, and you find that the object weighs 16 g. Because the balance weighs to the nearest 0.1 g, you can be sure that the object does not weigh 16.1 g or 15.9 g. In this case, you would write the weight as 16.0 g. To a scientist, there is a difference between 16 g and 16.0 g. Writing 16 g says that you don't know the digit after the 6. Writing 16.0 g says that you do know it: It is 0. However, you don't know the digit after that.

There are several rules governing the use of significant figures in reporting measured numbers.

Determining the Number of Significant Figures

All digits written down are significant except two types:

1. Zeros that come before the first nonzero digit are *not* significant.

EXAMPLE

How many significant figures are there in each of these numbers?

Number	Number of Significant Figures
23.742	5
332	3
0.023	2
0.230	3
0.000023	2
3.004	4
0.050008	5

2. Zeros that come after the last nonzero digit are significant if the number is a decimal, but if it is a whole number (no decimal point) they may or may not be significant. We cannot tell without knowing something about the number. This is the ambiguous case.

EXAMPLE

How many significant figures are there in each of these numbers?

Number	Number of Significant Figures
32.0400	6
0.0002300	4
1.02000	6
32500	3, 4, or 5

The ambiguous case is the only flaw in the system. If you know that a certain small business made a profit of $36,000 last year, you can be sure that the 3 and 6 are significant, but what about the rest? It might have been $36,126, or $35,786.53, or maybe even exactly $36,000. We just don't know because it is customary to round such numbers off. On the other hand, if the profit were reported as $36,000.00, then all seven digits are significant.

In science, we often get around the ambiguous case by using exponential notation (Sec. 1.3). Suppose a measurement comes out to be 2500 g. If we made the measurement, we of course know whether the two zeros are significant, but we need to tell others. If these digits are *not* significant, we write our number as 2.5×10^3. If one zero is significant, we write 2.50×10^3. If both zeros are significant, we write 2.500×10^3. Since we now have a decimal point, all the digits shown are significant.

Multiplying and Dividing

The rule in multiplication and division is that the final answer should have the *same* number of significant figures as there are in the number with the *fewest* significant figures.

This means that most of those beautiful digits on your calculator display are usually meaningless (insignificant).

EXAMPLE

Do the following multiplications and divisions:

(a) 3.6×4.27

(b) 0.004×217.38

(c) $\dfrac{42.1}{3.695}$

(d) $\dfrac{0.30652 \times 138}{2.1}$

Answer

(a) 15 (3.6 has two significant figures)
(b) 0.9 (0.004 has one significant figure)
(c) 11.4 (42.1 has three significant figures)
(d) 2.0×10^1 (2.1 has two significant figures)

Adding and Subtracting

In addition and subtraction, the rule is completely different. The number of significant figures in each number doesn't matter. The answer is given to the same number of *decimal places* as the term with the fewest decimal places.

EXAMPLE

Add or subtract:

(a)		(b)		(c)	
	320.084		61.4532		
	80.47		13.7		
	200.23		22		14.26
	20.0		0.003		− 1.05041
	620.8		97		13.21

Answer • In each case, we add or subtract in the normal way but then round off so that the only digits that appear in the answer are those in the columns in which every digit is significant.

Rounding Off

When we have too many significant figures in our answer, it is necessary to round off. In this book we have used the rule that *if the first digit dropped* is 5, 6, 7, 8, or 9, we raise *the last digit kept* to the next higher number; otherwise, we do not.

EXAMPLE

In each case, drop the last two digits:

33.679 2.4715 1.1145
0.001309 3.52

Answer

33.679 = 33.7
2.4715 = 2.47
1.1145 = 1.11
0.001309 = 0.0013
3.52 = 4

Counted or Defined Numbers

All the preceding rules apply to *measured* numbers and **not** to any numbers that are *counted* or *defined.* Counted and defined numbers are known exactly. For example, a triangle has 3 sides, not 3.1 or 2.9. Here, we treat the number 3 as if it has an infinite number of zeros following the decimal point.

EXAMPLE

Multiply 53.692 (a measured number) $\times$ 6 (a counted number).

Answer

322.152

Because 6 is a counted number, we know it exactly, and 53.692 is the number with the fewest significant figures. All we really are doing is adding 53.692 six times.

For problems relating to significant figures, see Chapter 1, Problems 1.62 to 1.68.

Answers*

CHAPTER 1 Matter, Energy, and Measurement

1.1 (a) 47 100 (b) 0.00000793
1.2 1.17×10^6
1.3 1.45×10^5
1.4 (a) 2.9×10^8 (b) 2.3×10^{-1}
1.5 (a) 1.94×10^6 (b) 1.51×10^{-6}
1.6 (a) 147°F (b) 8.3°C
1.7 109 kg
1.8 13.8 km
1.9 743 mph
1.10 78.5 g
1.11 2.43 g/mL
1.12 1.016 g/mL
1.13 48×10^3 cal = 48 kcal
1.14 540 cal
1.15 0.0430 cal/g · deg
1.17 See Sec. 1.1.
1.19 A fact is a statement based on experience. A hypothesis is a statement offered to explain the facts.
1.21 Chemical changes: (a) (e) (f) (g) Physical changes: (b) (c) (d)
1.23 (a) 403 000 (b) 3200 (c) 0.0000713 (d) 0.000000000555
1.25 (a) 2.00×10^{18} (b) 1.37×10^5 (c) 2.36×10^{-10} (d) 8.8×10^1 (e) 1.56×10^{-8}
1.27 (a) 2.80×10^5 (b) 5.03×10^3 (c) 1.572×10^{-7}
1.29 3.25×10^{-5}
1.31 (a) 1000 g (b) 0.001 g
1.33 (a) 100 cm (b) 230 mL (c) 75 kg (d) 15 mL (e) 50 mg (f) 100 mm (g) 40 g
1.35 (a) 160°C, 433 K (b) 100°C, 373 K (c) −18°C, 255 K (d) −157°C, 116 K
1.37 (a) 93.9 lb (b) 735 g (c) 86 cm (d) 23.1 miles (e) 10.3 L (f) 2.2 oz (avdp) (g) 31.80 L (h) 11.5 gal (i) 1.8 km (j) 1.18 fl oz
1.39 A student weighing 127 lb weighs 57.6 kg. A student 5 feet 5 inches tall (65.0 in.) is 1.65 m tall.
1.41 2.99×10^{10} cm/s
1.43 bottom: manganese; top: sodium acetate; middle: calcium chloride
1.45 1.023 g/cc
1.47 water
1.49 They are equally good buys. Butter A costs \$5.00 for 1135 g, or 0.44 cents/g; butter B costs \$4.00 for 908 g, or 0.44 cents/g.
1.51 1200 cal
1.53 0.34 cal/g · deg
1.55 They did three tests. See Box 1A.
1.57 334 g
1.59 The body shivers. Further temperature lowering results in unconsciousness and then death.
1.61 methyl alcohol, because its higher specific heat allows it to retain the heat longer
1.63 (a) 3 (b) 2 (c) 1 (d) 4 (e) 5
1.65 (a) 25 000 (b) 4.1 (c) 15.5
1.67 (a) 10963.1 (b) 244 (c) 172.34
1.69 0.75 cal/g · deg
1.71 0.732
1.73 No. mL is a unit of volume, g a unit of mass.
1.75 (a) 33°C (b) 38°C (c) 93°C (d) 222°C
1.77 (a) 0.225 km (b) 0.416 qt (c) 93.8 oz (avdp) (d) 0.0397 lb
1.79 The largest is 41 g. The smallest is 4.1×10^{-8} kg.
1.81 10.9 h
1.83 5.9 lb/gal
1.85 (a) 1.57 g/mL (b) 1.25 g/mL

*Answers to in-text and odd-numbered end-of-chapter problems.

CHAPTER 2 Atoms

2.1 (a) 109 amu (b) 230 amu
2.2 (a) silver (b) protactinium
2.3 (a) Zr: 40, Hg: 80 (b) Zr: 40, Hg: 80
2.4 (a) 46 (b) 48 (c) 50
2.5 lithium-7
2.6 (a) V: $1s^22s^22p^63s^23p^64s^23d^3$
(b) As: $1s^22s^22p^63s^23p^64s^23d^{10}4p^3$
2.7 They were not based on evidence.
2.9 See Sec. 2.2.
2.11 The mixture is heated; the liquid with the lower boiling point evaporates first and is condensed. The other liquid remains in the flask.
2.13 the law of conservation of mass
2.15 (a) nucleus (b) outside nucleus (c) nucleus
2.17 (a) chromium (b) fluorine (c) selenium (d) bismuth (e) radium
2.19 the same, since the number of protons would not change
2.21 All contain 86 protons. They have 124, 132, and 136 neutrons, respectively.
2.23 (a) 10p, 12n (b) 46p, 58n (c) 17p, 18n (d) 52p, 76n (e) 3p, 4n (f) 92p, 146p
2.25 tin-120, tin-121, tin-124
2.27 (a) F^- (b) K^+ (c) Ar (d) H^+ (e) Na^+ (f) S^{2-} (g) Br^- (h) Mn^{2+}
2.29 63.62 amu
2.31 62.5% ^{121}Sb and 37.5% ^{123}Sb
2.33 (a) (c) (d) (f)
2.35 metals: (c) (e) (i); nonmetals: (a) (g) (j) (k); metalloids: (b) (d) (f) (h)
2.37 (a) $1s^22s^1$ (b) $1s^22s^22p^6$ (c) $1s^22s^2$ (d) $1s^22s^22p^2$ (e) $1s^2$ (f) $1s^22s^22p^63s^23p^64s^23d^5$ (g) $1s^22s^22p^63s^23p^5$ (h) $1s^22s^22p^63s^23p^3$
2.39 (a) Mg: $1s^22s^22p^63s^2$; Mg^{2+}: $1s^22s^22p^6$
(b) Na: $1s^22s^22p^63s^1$; Na^+: same as Mg^{2+}
(c) Al: $1s^22s^22p^63s^23p^1$; Al^{3+}: same as Mg^{2+}
2.41 (a) $1s^22s^22p^63s^23p^63d^{10}4s^24p^64d^15s^2$
(b) $1s^22s^22p^63s^23p^63d^{10}4s^24p^64d^{10}5s^25p^5$
(c) $1s^22s^22p^63s^23p^63d^{10}4s^24p^64d^{10}5s^25p^66s^2$
2.43 *6s:* 2; *6p:* 6; *6d:* 10; *6f:* 14; 32 altogether
2.45 The properties are similar because all of them have the same outer-shell configurations. They are not identical because each has a different number of filled inner shells.
2.47 C, H, O, N
2.49 Some chemicals are; some are not; and for many it depends on the dose. Since everything is made of chemicals, many chemicals must be nonpoisonous.
2.51 copper and tin
2.53 (a) O, Si (b) O, C
2.55 metals: (a) (c) (e); nonmetals: (b) (d)
2.57 (a) 17p, 17e, 18n (b) 42p, 42e, 56n (c) 20p, 18e, 24n (d) 53p, 54e, 74n (e) 64p, 64e, 94n (f) 83p, 80e, 129n
2.59 (a) 126 (b) 141 (c) 122 (d) 151
2.61 (d)
2.63 47.9 amu
2.65 (a) 4 (b) 7 (c) 6 (d) 1 (e) 8 (f) 2 (g) 8 (h) 4 (i) 6 (j) 6
2.67 (a) 18 (b) fifth
2.69 (a) electron: −1; proton: +1; neutron: no charge
(b) electron 1/1835, proton 1, neutron 1
2.71 76.5% rubidium-85 and 23.5% rubidium-87

CHAPTER 3 Chemical Bonds

3.1 (a) KCl (b) CaF_2 (c) Fe_2O_3

3.2
```
 ..   ..   ..
:Br—P—Br:
 ..   |    ..
     :Br:
      ..
```

3.3
```
     H
     |      ..
H—C—C—O—H
     |   ||   ..
     H  O:
         ..
```

3.4
```
 [  ..   ..   ..   ]2−
 [ :O—S—O:  ]
 [  ..   |    ..   ]
 [     :O:       ]
 [      ..        ]
```

3.5
```
 ..  .    ..
:O=N—O:
          ..
```

3.6 (a) F (b) Br
3.7 (a) nonpolar (b) polar covalent (c) ionic
3.8 magnesium oxide, barium iodide, sodium bromide
3.9 SrS, NaF
3.10 copper(I) bromide or cuprous bromide; mercury(II) bromide or mercuric bromide; iron(III) oxide or ferric oxide
3.11 potassium hydrogen phosphate, aluminum sulfate, iron(II) sulfite or ferrous sulfite
3.12 nitrogen dioxide, phosphorus tribromide, sulfur dichloride
3.13 The patient gets calcium ions, Ca^{2+}.

3.15 (a) Mg^{2+} (b) F^- (c) Al^{3+} (d) S^{2-} (e) K^+ (f) Br^-

3.17 Li^- has two electrons in its outer shell, so it does not have a complete octet.

3.19 C^{4+}, C^{4-}, Si^{4+}, and Si^{4-} all have charges that are too concentrated for small ions.

3.21 No. Solid KF consists of potassium ions, K^+, and fluoride ions, F^-, held together in a three-dimensional crystal similar to that shown in Figure 3.2.

3.23 The 2 means that in the crystal there are two Na^+ ions for every SO_4^{2-} ion.

3.25 (a) NaBr (b) K_2O (c) $AlCl_3$ (d) $BaCl_2$ (e) CaO

3.27 (a) Ca^{2+} S^{2-} (b) Mg^{2+} F^- (c) Cs^+ O^{2-} (d) Sc^{3+} Cl^- (e) Al^{3+} S^{2-}

3.29 (a) 3 (b) 1 (c) 4 (d) 1 (e) 2

3.31 (a) H—C—H with H above and H below C (b) H—C≡C—H (c) H(H)C=C(H)H (d) F—B—F with F below B (e) H—C—H with =O below C (f) Cl—C—C—C—Cl with Cl above and Cl below each C

3.33 (a) Br—C=C—Br with H below each C (b) CH_3—C—H with =O below C (c) H—C=NH with H below C

3.35 An argon atom has a complete octet, so (a) it does not lose or gain any electrons, and (b) it has no need to share any.

3.37 (a) K· (b) :S̈e· (c) ·N̈· (d) :Ï· (e) :Är: (f) Be (g) :C̈l·

3.39 (a) :B̈r—B̈r: (b) H—S̈: with H below S (c) H—N̈=N̈—H (d) $[:C\equiv N:]^-$ (e) $[NH_4]^+$: H—N—H with H above and H below N (f) $[BF_4]^-$: :F̈—B—F̈: with :F̈: above and below B

3.41 These are completely different species, whose Lewis structures are

bromine atom :B̈r·

bromine molecule :B̈r—B̈r:

bromide ion $[:\ddot{Br}:]^-$

3.43 (a) 12 (b) 8 (c) 12 (d) 8 (e) 8

3.45 H—N̈(H)—H + H^+ ⟶ $[$H—N(H)(H)—H$]^+$

3.47 (a) tetrahedral (b) pyramidal (c) planar triangular (d) linear (e) angular (f) planar triangular (g) each carbon is planar triangular (h) angular (i) pyramidal (j) planar triangular (k) angular (l) pyramidal

3.49 most polar: (c); least polar: (b)

3.51 (a), (c), and (d) are dipoles.

3.53 (a) yes, covalent (b) no (c) yes, ionic (d) yes, covalent (e) no (f) no (g) no (h) no

3.55 (a) $[ClO_4]^-$: :Ö—Cl—Ö: with :Ö: above and :Ö: below Cl (b) $Ca(ClO_4)_2$

3.57 (a) sodium fluoride (b) magnesium sulfide (c) aluminum oxide (d) barium chloride (e) calcium bisulfite (f) potassium iodide (g) strontium phosphate (h) iron(II) hydroxide or ferrous hydroxide (i) sodium dihydrogen phosphate (j) lead acetate (k) barium hydride (l) ammonium hydrogen phosphate

3.59 (a) dihydrogen sulfide (b) iodine fluoride (c) iodine heptafluoride (d) nitrogen oxide (e) dinitrogen trioxide (f) sulfur dioxide (g) selenium trioxide

3.61 K^+, HPO_4^{2-}

3.63 It is needed for the thyroid gland.

3.65 an external anti-infective

3.67 When inhaled, it decreases the ability of the hemoglobin in the blood to carry oxygen.

3.69 BF_3 + :PH_3 ⟶ F_3B—PH_3 (:F̈—B(F)(F) + :P(H)(H)—H ⟶ :F̈—B(F)(F)—P(H)(H)—H)

3.71 tetrahedral

3.73 (a) This ion has one more electron than is needed for a complete octet.
(b) The charge density is too high.
(c) This ion does not have a complete octet.

3.75 No. Ionic bonds point in every direction in space (they are spherically symmetrical).

3.77 (·) + (·) ⟶ (· ·)

3.79 (a) The carbon on the right has only six electrons in its outer shell. (b) There are too many electrons. The compound C_2H_4 has a total of 12 valence electrons, but this formula shows 14.

3.81 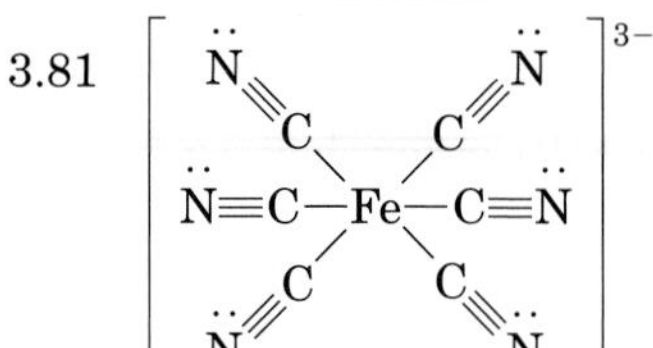

CHAPTER 4 Chemical Reactions

4.1 (a) 180.0 amu (b) 601.9 amu
4.2 5.545 moles
4.3 222 g
4.4 15 moles of C atoms, 30 moles of H atoms, 15 moles of O atoms
4.5 4.9×10^{-4} moles of Cu^+ ions
4.6 7.86×10^{24} molecules of water
4.7 $As_2O_5 + 3H_2O \longrightarrow 2H_3AsO_4$
4.8 $2C_6H_{14} + 19O_2 \longrightarrow 14H_2O + 12CO_2$
4.9 $3K_2C_2O_4 + Ca_3(AsO_4)_2 \longrightarrow 2K_3AsO_4 + 3CaC_2O_4$
4.10 8.3 moles of CH_4
4.11 1.25×10^3 g of $C_2H_2Br_4$
4.12 17.1 g of H_2
4.13 80.7%
4.14 $Cu^{2+} + S^{2-} \longrightarrow CuS$
4.15 (a) Cr is oxidized; Ni^{2+} is reduced. Ni^{2+} is the oxidizing agent; Cr the reducing agent.
(b) H_2 is oxidized; H_2CO is reduced. H_2CO is the oxidizing agent; H_2 the reducing agent.
4.17 (a) 74.6 amu (b) 164.0 amu (c) 89.8 amu (d) 342.0 amu (e) 210.2 amu (f) 77.0 amu (g) 342.3 amu (h) 96.0 amu (i) 354.5 amu
4.19 (a) 81.4 g NO_2 (b) 50 g 2-propanol (c) 1.33×10^3 g WCl_5 (d) 62.6 g galactose (e) 8.6 g vitamin C
4.21 (a) 3.71 moles Ag^+ ions (b) 2.72 moles Na^+ ions (c) 0.74 mole SO_4^{2-} ions
4.23 (a) 70.9 amu (b) 39.9 amu (c) 123.9 amu (d) 28.0 amu (e) 4.0 amu
4.25 (a) 1.7×10^{24} molecules of TNT
(b) 1.67×10^{21} molecules of water
4.27 1.1×10^{-19} g
4.29 (a) $HI + NaOH \longrightarrow NaI + H_2O$
(b) $Ba(NO_3)_2 + H_2S \longrightarrow BaS + 2HNO_3$
(c) $CH_4 + 2O_2 \longrightarrow CO_2 + 2H_2O$
(d) $2C_4H_{10} + 13O_2 \longrightarrow 8CO_2 + 10H_2O$
(e) $2Fe + 3CO_2 \longrightarrow Fe_2O_3 + 3CO$
(f) $2Al + 6HBr \longrightarrow 2AlBr_3 + 3H_2$
(g) $P_4 + 5O_2 \longrightarrow P_4O_{10}$
4.31 (a) 2/3 mole N_2 (b) 2/3 mole N_2O_3 (c) 12 moles O_2
4.33 719 g Br_2
4.35 879 g O_2
4.37 4660 g CO_2
4.39 111 g of aspirin
4.41 (a) H_2 (b) 14.3 g N_2 (c) 18.8 g NH_3
4.43 94.0%
4.45 (a) Na^+, Cl^-
(b) $Sr^{2+}(aq) + CO_3^{2-}(aq) \longrightarrow SrCO_3(s)$
4.47 $Pb^{2+}(aq) + 2Cl^-(aq) \longrightarrow PbCl_2(s)$
4.49 $2H^+(aq) + SO_3^{2-}(aq) \longrightarrow SO_2(g) + H_2O(l)$
4.51 Soluble: (a) (b) (d) (e)
Insoluble: (c) (f)
4.53 (a) $Ca_3(PO_4)_2$ (c) $BaCO_3$ (d) $Fe(OH)_2$ (f) Sb_2S_3 (g) $PbSO_4$
No precipitate will form in (b) and (e).
4.55 No, one gains electrons and the other loses electrons. Electrons are not destroyed, but transferred.
4.57 (a) C_7H_{12} gets oxidized; O_2 gets reduced.
(b) O_2 is the oxidizing agent; C_7H_{12} the reducing agent.
4.59 exothermic: (b) (c) (e); endothermic: (a) (d)
4.61 160 000 cal
4.63

Fe electrode — e^- → — e^- → Pt electrode
Water
Aqueous Zn^{2+} solution
Salt bridge

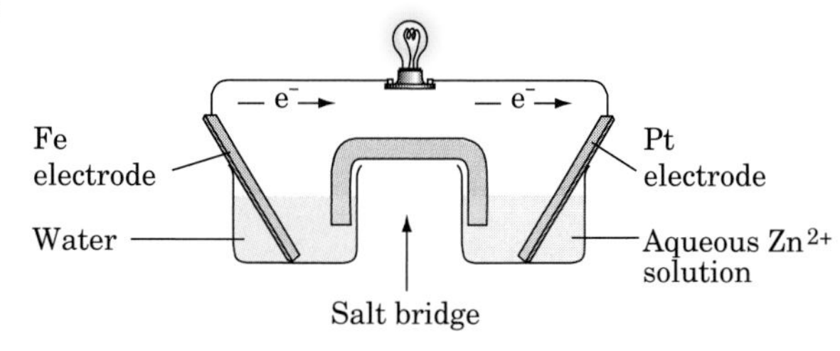

4.65 In those parts of the film where the light hits, AgBr is reduced to silver. These parts turn black. In the parts where the light does not hit, no silver forms. This creates the image.
4.67 2.7 oz
4.69 It was oxidized.
4.71 More than 90 percent of the energy needed to operate our cars, trucks, airplanes, ships, and farm and factory machinery and to light, heat, and cool our buildings comes from the combustion of oil, coal, and natural gas.
4.73 0.00271 mole aspirin; 1.63×10^{21} molecules aspirin

4.75 (a) 5.68 moles CO_2 (b) 7.00 moles CO_2
4.77 (a) $AlCl_3$ is in excess (b) 82.5%
4.79 9.961 g
4.81 4×10^{10} molecules

CHAPTER 5 Gases, Liquids, and Solids

5.1 0.4 atm
5.2 16.4 atm
5.3 13 atm
5.4 4.84 atm
5.5 0.422 mole
5.6 MW = 82
5.7 0.107 atm
5.8 (a) yes (b) no
5.9 83.3 g
5.10 229 cal
5.11 strong forces in the nucleus, electrostatic attractions between nucleus and electrons, covalent and ionic bonds within molecules, and intermolecular forces between molecules
5.13

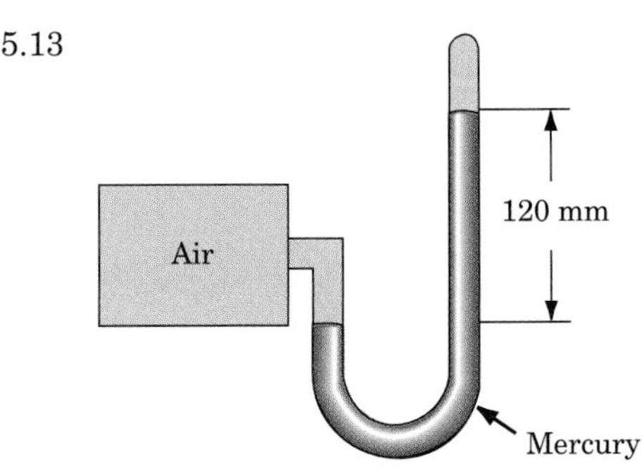

5.15 The barometric pressure drops because the air column above us weighs less than before.
5.17 0.50 atm
5.19 1.87 L
5.21 475°C
5.23 836 mL
5.25 61.0 mL
5.27 1.87 atm
5.29 1.17 atm
5.31 2.4 L
5.33 8040 mm Hg
5.35 93.2 mL
5.37 12 L
5.39 MW = 52
5.41 the same
5.43 324 mm Hg
5.45 (c)
5.47 London dispersion forces provide attraction that at low temperatures overcomes the kinetic energy of helium gas.
5.49 Yes, a small dipole, as in NO, may provide weaker interaction forces than London dispersion forces acting over large surfaces.
5.51 See Sec. 5.9.
5.53 Molecules of water escape the surface and go into the vapor phase, but many fewer come back. Eventually all of them are in the vapor.
5.55 When the temperature decreases, the kinetic energy decreases; it is not lost but converted to potential energy or energy of interaction. More interaction means more attraction and hence condensation.
5.57 In diamond each carbon atom is bonded to four other carbon atoms, forming a large network crystal (10^{22} to 10^{23} atoms); a molecule of buckminsterfullerene contains only 60 atoms.
5.59

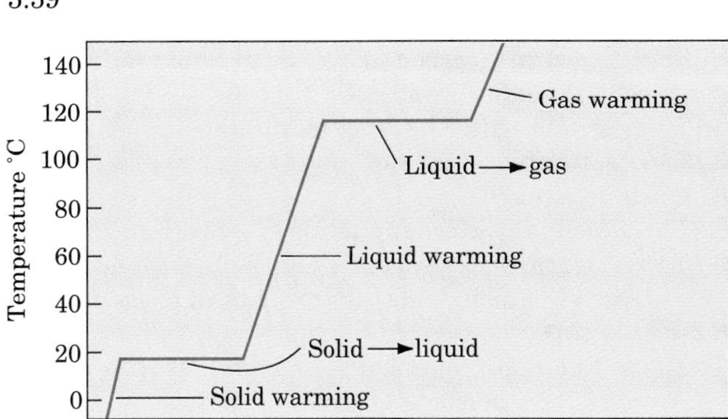

5.61 The gas at 100°C. At the lower temperature the molecules move slower, so there is more order, and hence a lower entropy, than at the higher temperature.
5.63 Because mercury vapors are poisonous. If spilled and not cleaned up, some of the mercury will go into the vapor phase, and people will inhale it.
5.65 The volume of the chest cavity increases, lowering the pressure. Air flows in from outside.
5.67 highest in venous blood, lowest in arterial blood
5.69 The volume of water increases when it freezes. This expansion will crack the bottle.
5.71 112 mL
5.73 CO
5.75 0.34 mole
5.77 MW = 91.9
5.79 pentane
5.81 0.633 g/L
5.83 MW = 28

CHAPTER 6 Solutions and Colloids

6.1 To 11 g KBr add enough water to make 250 mL of solution.
6.2 1.9% w/v
6.3 To 158.2 g KCl (1.06 moles × 2) add enough water to make 2.0 L of solution.
6.4 0.055 *M*
6.5 185 mL
6.6 15 g
6.7 To 4.00 mL of stock solution add enough water to make 200.0 mL of solution.
6.8 To 0.13 mL of the concentrated solution add enough water to make 20.0 mL solution.
6.9 −12.5°C
6.10 −18.6°C
6.11 0.80 osmol
6.12 (a)
6.13 A homogeneous mixture appears uniform throughout; a heterogeneous mixture shows boundaries.
6.15 (a) solid in solid (b) solid in liquid (c) gas in gas (d) gas (carbon dioxide) and liquid (ethyl alcohol) in liquid (water)
6.17 True solutions. They mix in all proportions.
6.19 unsaturated
6.21 (b), (a), (c), (d)
6.23 by cooling it without stirring
6.25 The warm water contains less oxygen than cold water. The fish die because of lack of oxygen.
6.27 (a)
6.29 (a) 76 mL ethyl alcohol (b) 7.8 mL ethyl acetate (c) 132 mL benzene
6.31 (a) 1.85% w/v (b) 5.2% w/v (c) 0.566% w/v
6.33 In each case add enough water to the given weight of solute to make the total volume specified: (a) 19.5 g of NH_4Br—175 mL of solution (b) 167 g NaI—1.35 L of solution (c) 2.4 g C_2H_5OH—330 mL of solution
6.35 0.17 *M*
6.37 0.405 *M*
6.39 0.568 *M*
6.41 25.0 mL
6.43 To 2.08 mL of the stock solution add enough water to make 250 mL of solution.
6.45 In each case add enough water to the given volume of concentrated solution to make the total volume specified: (a) 0.12 L—550 mL (b) 0.36 L—1.75 L (c) 1.8 mL—385 mL
6.47 No, the concentration is 0.01 ppb.
6.49 See Sec. 6.6.
6.51 (b)
6.53 soluble: (b), (c), (d)
6.55 homogeneous: (a); heterogeneous: (d); colloidal: (b), (c), (e), (f)
6.57 Water molecules in the solvation layer of a colloidal particle move together with the particle; water molecules in the bulk move randomly.
6.59 (a) −3.72°C (b) −5.58°C (c) −5.58°C (d) −7.44°C
6.61 344 g methanol
6.63 5.4 moles
6.65 (a) B (b) B (c) A (d) B (e) A (f) no rise. In each case, the side with the higher osmolarity rises.
6.67 (a) 12 (b) 2.5 (c) 8.4 (d) 0.036 osmol
6.69 (b)
6.71 HNO_2, HNO_3
6.73 1.768 g CO_2 lost
6.75 $CaCO_3$
6.77 Because of perspiration they lose potassium ions, which are needed for proper muscle action.
6.79 egg yolk
6.81 (a) 26.3% w/v (b) 25.0% w/w
6.83 4 ppm
6.85 Methanol is more efficient. It has a lower molecular weight; hence there are more moles in the same mass.
6.87 $CO_2(g) + H_2O(l) \longrightarrow H_2CO_3(aq)$, $SO_2(g) + H_2O(l) \longrightarrow H_2SO_3(aq)$
6.89 To 25.0 mL of the stock solution add enough water to make the volume of the solution 250 mL.
6.91 0.006 g
6.93 At least one of the solids must be melted. The other solid is then dissolved in the newly formed liquid; the mixture is stirred and allowed to solidify.

CHAPTER 7 Reaction Rates and Equilibrium

7.1 $K = \dfrac{[H_2SO_3]}{[SO_2][H_2O]}$
7.2 $K = \dfrac{[N_2][H_2]^3}{[NH_3]^2}$
7.3 $K = 0.602$
7.4 The equilibrium shifts to the left.
7.5 The equilibrium shifts to the right, because SO_2 is removed by the water.

7.6 Endothermic. The increase in temperature drives an endothermic reaction toward the products.
7.7 The change in concentration of a reactant or product per unit time. Moles per liter per minute.
7.9 6.14×10^{-3} moles/L per min
7.11 Two are broken and two are formed.
7.13 The activation energy is low because the breaking of covalent bonds is not required.
7.15 The spoiling is caused by chemical reactions initiated by bacteria. These reactions are much faster at room temperature than at refrigerator temperature.
7.17 50°C
7.19 (1) increase the concentrations (2) increase the temperature (3) add a catalyst
7.21 It provides an alternate pathway, of lower activation energy.
7.23 digesting a piece of candy; explosion of TNT; rusting of iron; addition of Na or K metal to water
7.25 (a) $K = \frac{[CO][Cl_2]}{[COCl_2]}$

(b) $K = \frac{[H_2S]^2[O_2]^3}{[H_2O]^2[SO_2]^2}$

(c) $K = \frac{[CO_2]^4[H_2O]^6}{[C_2H_6]^2[O_2]^7}$

7.27 $K = 1.5$
7.29 $K = 9.9 \times 10^{-2}$
7.31 products: (a), (b), (c); reactants: (d), (e)
7.33 (a) No, the speed of reaction depends on the energy of activation, not on the exothermic character.
(b) Yes, endothermic reactions always have high energies of activation (Fig. 7.5) and so must be very slow.
7.35 (a) right (b) left (c) left (d) left (e) no shift
7.37 It shifts to the right.
7.39 (a) no change (b) no change (c) K smaller
7.41 The higher temperature causes chemical reactions in the body to go faster. If the reactions go too fast, it is dangerous to the health of the patient. At low temperature, the reactions are too slow.
7.43 See Box 7C.
7.45 because the reaction is too slow at low temperature
7.47

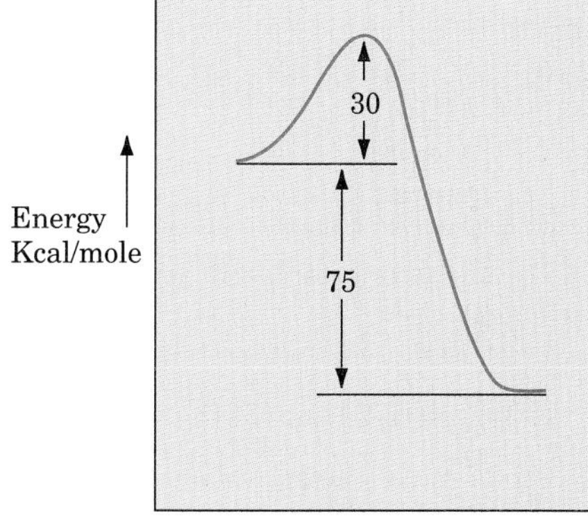

7.49 The heat from the match provides the necessary energy of activation. Once it starts, the combustion is exothermic; the heat liberated provides the energy of activation for the other molecules.
7.51 2.0×10^{-3} mole of N_2O_4 per L per sec
7.53 0.4 M
7.55 0.8 M
7.57 A, because the orientation does not matter, but rod-like molecules need special orientations in order for effective collisions to occur.
7.59 0.030 mole/L per min

CHAPTER 8 Acids and Bases

8.1 (b)
8.2 1×10^{-2} M
8.3 3
8.4 1×10^{-4} M
8.5 3
8.6 8.2
8.7 4×10^{-8} M
8.8 9.3
8.9 (a) 9.14 (b) 10.25
8.10 Use $CH_3CH(OH)COO^-/CH_3CH(OH)COOH$ in a ratio of 2 to 1.
8.11 4.27
8.12 (a) 60.0 (b) 85.7 (c) 32.7
8.13 (a) 1.55 equiv (b) 0.0430 equiv
8.14 0.093 N
8.15 0.0836 N
8.17 $LiOH(s) \xrightarrow{H_2O} Li^+(aq) + OH^-(aq)$
$(CH_3)_2NH + H_2O \rightleftharpoons (CH_3)_2NH_2^+ + OH^-$
8.19 (a) H_2O (b) NH_3 (c) NH_4^+ (d) C_6H_5OH (e) HCO_3^- (f) H_2CO_3 (g) H_3O^+ (h) $H_2PO_4^-$ (i) $CH_3{-}NH_3^+$ (j) HPO_4^{2-}
8.21 No. The Brønsted-Lowry definitions are simply an extension of the Arrhenius definitions.
8.23 (a) HCOOH (b) HF (c) CH_3COOH
8.25 (a) $CO_3^{2-} + 2H_3O^+ \rightleftharpoons H_2CO_3 + 2H_2O$ or
$Na_2CO_3 + 2HCl \rightleftharpoons H_2CO_3 + 2NaCl$
(b) $Mg + 2H_3O^+ \rightleftharpoons Mg^{2+} + H_2 + 2H_2O$ or
$Mg + 2HCl \rightleftharpoons MgCl_2 + H_2$

(c) $OH^- + H_3O^+ \rightleftharpoons 2H_2O$ or
$NaOH + HCl \rightleftharpoons H_2O + NaCl$
(d) $NH_3 + H_3O^+ \rightleftharpoons NH_4^+ + H_2O$ or
$NH_3 + HCl \rightleftharpoons NH_4^+ + Cl^-$
(e) $CH_3NH_2 + H_3O^+ \rightleftharpoons CH_3NH_3^+ + H_2O$ or
$CH_3NH_2 + HCl \rightleftharpoons CH_3NH_3^+ + Cl^-$
(f) $HCO_3^- + H_3O^+ \rightleftharpoons H_2CO_3 + H_2O$ or
$NaHCO_3 + HCl \rightleftharpoons H_2CO_3 + NaCl$

8.27 (a) 10^{-4} *M* (b) 10^{-11} *M* (c) 10^{-7} *M* (d) 10^{-14} *M*

8.29 (a) 9 (b) 5 (c) 1 (d) 7 (e) 12
acidic: (b) (c); basic: (a) (e); neutral (d)

8.31 (a) 1×10^{-12} *M* (b) 1×10^{-7} *M* (c) 1×10^{-2} *M* (d) 1×10^{-13} *M* (e) 4×10^{-9} *M* (f) 8×10^{-13} *M* (g) 3×10^{-4} *M*; acidic: (c) (g); basic: (a) (d) (e) (f); neutral: (b)

8.33 (a) butyric 1.5×10^{-5}; barbituric 1.0×10^{-5}; lactic 1.4×10^{-4} (b) strongest: lactic acid; weakest: barbituric acid

8.35 Na^+ does not react with water, but CO_3^{2-} does:

$$CO_3^{2-} + H_2O \rightleftharpoons HCO_3^- + OH^-$$

The OH^- ions make the solution basic.

8.37 Buffer action is the result of Le Chatelier's principle. Adding acid or base disturbs an equilibrium, which shifts in the direction that minimizes the effect of the addition.

8.39 (a) $H_3O^+ + HPO_4^{2-} \rightleftharpoons H_2PO_4^- + H_2O$
(b) $OH^- + H_2PO_4^- \rightleftharpoons HPO_4^{2-} + H_2O$

8.41 The ability of the buffer to neutralize added amounts of H_3O^+ or OH^-. The greater the ability, the greater the capacity.

8.43 carbonate and phosphate; see Sec. 8.10

8.45 (a) 80.9 (b) 31.0 (c) 23.9 (d) 37.0 (e) 26.0 (f) 17.0 (g) 47.3

8.47 0.348 *N*

8.49 5.60 *N*

8.51 (a) 3.0 *N* (b) 9.0 *N* (c) 0.34 *N* (d) 5.4 *N*

8.53 11.3 mL

8.55 0.00330 equiv

8.57 To 2.0 g of $NaHCO_3$ add enough water to make a total volume of 1 L.

8.59 immediate washing with a steady stream of cold water

8.61 Alka-Seltzer, Brioschi, Chooz, Riopan, Tums

8.63 $NaHCO_3$ and an acid

8.65 (a) 3.6 (b) 8.6 (c) 4.0

8.67 Solving the Henderson-Hasselbalch equation,

$$4.55 = 4.75 + \log(CH_3COO^-/CH_3COOH)$$

we obtain a CH_3COOH:CH_3COO^- ratio of 1.6:1. We therefore make up a solution that contains 1.6 times as many moles of CH_3COOH as of CH_3COO^- ions.

8.69 The HCl solution is more acidic because its $[H_3O^+]$ is higher.

8.71 (a) An acid is a substance that produces H_3O^+ ions in aqueous solution. A base is a substance that produces OH^- ions in aqueous solution.
(b) An acid is a proton donor. A base is a proton acceptor.

8.73 0.00740 *N*

8.75 The concentrations, 1×10^{-7} *M* for H_3O^+ and OH^- are too low.

8.77 No. HF, being a stronger acid, reacts with water to a greater extent, producing a higher $[H_3O^+]$, so the pH of its solution is lower.

8.79 EW = 172

8.81 Yes. The ethoxide ion will not react with NH_3, but it will react with water, since the latter is a stronger acid than ethyl alcohol (Table 8.2):

$$C_2H_5O^- + H_2O \rightleftharpoons C_2H_5OH + OH^-$$

CHAPTER 9 Nuclear Chemistry

9.1 xenon-139

9.2 radium-219

9.3 oxygen-15

9.4 0.31 g

9.5 3.3×10^{-3} mCi

9.6 1.5 mL

9.7 one photon

9.9 $5.2 \times 10^9\ s^{-1}$

9.11 because they are not emitted from atomic nuclei

9.13 (a) $^{19}_{9}F$ (b) $^{32}_{15}P$ (c) $^{87}_{37}Rb$

9.15 none

9.17 (a) $^{159}_{64}Gd$ (b) $^{141}_{57}La$ (c) $^{242}_{96}Cm$

9.19 (a) $^{206}_{81}Tl$ (b) $^{234}_{92}U$ (c) $^{170}_{70}Yb$

9.21 $^{29}_{15}P \longrightarrow {}^{0}_{+1}e + {}^{29}_{14}Si$

9.23 $^{238}_{92}U \longrightarrow {}^{234}_{90}Th + {}^{4}_{2}He$; $^{234}_{90}Th \longrightarrow {}^{234}_{91}Pa + {}^{0}_{-1}e$

9.25 (a) $^{28}_{14}Si$ (b) $^{143}_{54}Xe$ (c) $^{1}_{0}n$ (d) $^{210}_{84}Po$ (e) $^{13}_{7}N$

9.27 140 years

9.29 No, the conversion $Ra \longrightarrow Ra^{2+}$ involves electrons from the electron cloud. Radioactivity is a nuclear phenomenon.

9.31 15 min

9.33 We cannot see or feel it nor can our senses detect it in any other way.

9.35 (a)

9.37 (a), (e) the amount of radiation absorbed by tissues

(b), (g) the effective dose absorbed by humans
(c) the effective dose delivered
(d), (f) intensity of radiation

9.39 The penetrating power of alpha particles is so low that most of them are stopped by the outer skin. But when they get into the lung, the membranes offer little resistance so the particles can damage the cells of the lung.

9.41 alpha particles

9.43 iodine-131, because any iodine taken in by the body goes to the thyroid gland

9.45 The half-life is too short. It would not last long enough to reach a hospital.

9.47 ${}^{1}_{1}H$

9.49 curium-248

9.51 neutrons. They are controlled by inserting or removing boron control rods.

9.53 The three rays have different charges: positive, negative, and neutral.

9.55 one-tenth of a half-life

9.57 It gives off an alpha particle.

9.59 See Box 9E.

9.61 because they are destroyed by collisions with electrons, and electrons are in all matter

9.63 If ingested, iodine concentrates in the thyroid gland, where radioactive iodine can cause cancer.

9.65 fluorine-19 and neon-20

9.67 the same

9.69 (a) 82% (b) 11% (c) 0.1%

9.71 x-rays

9.73 two neutrons

9.75 from the small amount of mass converted to energy during the breakup of heavy atomic nuclei

9.77 (a) because a radioactive sample decays, and the decay products are always mixed with the starting isotope (b) because a neutron in the nucleus is converted to a proton

9.79 Oxygen-16 is stable because it has an equal number of protons and neutrons. The others are unstable because the numbers are unequal; in this case the less equal, the faster the isotope decays.

CHAPTER 10 Organic Chemistry. Alkanes

10.1 (a) The carbon uses one sp^3 orbital to overlap with a p orbital of the fluorine. Each of the other three sp^3 orbitals of the carbon overlaps with an s orbital of a hydrogen.
(b) Two sp^3 orbitals, one from each carbon, overlap to form the C—C sigma (σ) bond. Each of the remaining sp^3 orbitals of the left carbon overlaps with one s orbital of a hydrogen. One of the three remaining sp^3 orbitals of the right carbon overlaps with a p orbital of a fluorine; each of the other two overlaps with an s orbital of a hydrogen.

10.2 (a) and (b)

10.3 (a) butane (b) 2-methylbutane
(c) 3-ethylheptane

10.4 (a) $CH_3—CH(CH_3)—CH_2—CH_3$

(b) $CH_3—CH_2—CH(CH_3)—CH_2—CH_2—CH_3$

10.5 (a) 2,3-dimethylpentane
(b) 2,2,4-trimethylpentane

10.6 (a) $CH_3—C(CH_3)_2—CH_3$

(b) $CH_3—CH_2—C(CH_2CH_2CH_3)(CH_2CH_3)—CH(CH_2CH_3)—CH_2—CH_2—CH_3$

10.7 4-ethyl-3,4-dimethylheptane

10.8 $CH_3—CH_2—CH(CH_3)—C(CH_3)(CH_2CH_3)—CH_2—CH_2—CH_3$

10.9 (a) 2,3-dichloropentane (b) 3-bromo-2-chloro-2,3-diiodohexane

10.10 (a) 2-chloropropane; isopropyl chloride
(b) 1-iodo-2,2-dimethylpropane; neopentyl iodide

10.11 (a) chlorocyclopentane
(b) 1,1-diethylcyclobutane
(c) 1,2,4-trimethylcyclohexane
(d) 1,3-dichloro-5-ethylcyclopentane

10.13 No, they are nonelectrolytes.

10.15 organic

10.17 no difference

10.19 (1): (a) (c); (2): (d) (e) (f) (g) (h); (3): (b) (i)

10.21 (1) $CH_3CH_2CH_2—OH$ (2) $CH_3—CH(OH)—CH_3$

(3) $CH_3—O—CH_2—CH_3$

10.23 (a) $CH_3—CH{=}CH—CH(CH_3)—CH_3$

(b) $NH_2—CH_2—CH_2—O—CH_2—CH(CH_3)—Br$

(c) $Cl—CH{=}CH—C{\equiv}C—N(CH_3)—CH_3$

10.25 (a) spherical (b) dumbbell shape (c) a mixture of the spherical and dumbbell shapes

10.27 (a) Of the four sp^3 orbitals of the carbon one overlaps with a p orbital of one chlorine, the second with a p orbital of the other chlorine; each of the remaining two overlaps with the 1s orbital of a hydrogen.
(b) Each carbon uses four sp^3 orbitals. One sp^3 orbital of one carbon overlaps with one sp^3 orbital of the other carbon. One sp^3 orbital of one carbon overlaps with the p orbital of the bromine. Each of the other five sp^3 orbitals overlaps with the s orbital of a different hydrogen.

10.29 (a) $CH_3—CH(CH_3)—CH_2CH_3$ (b) $CH_3—C(CH_3)_2—CH_3$

(c) CH_3—(cyclopentane ring)

10.31 (1) $CH_3CH_2CH_2CH_2CH_2CH_3$ hexane
(2) $CH_3—CH(CH_3)—CH_2CH_2CH_3$ 2-methylpentane
(3) $CH_3CH_2—CH(CH_3)—CH_2CH_3$ 3-methylpentane
(4) $CH_3—C(CH_3)_2—CH_2CH_3$ 2,2-dimethylbutane
(5) $CH_3—CH(CH_3)—CH(CH_3)—CH_3$ 2,3-dimethylbutane

10.33 (a) heptane (b) 2-methylpentane (c) 2,4-dimethylhexane (d) 2,5-dimethylheptane (e) 2,3,3-trimethylpentane (f) 3,5-dimethylheptane (g) 4,6-diethyl-2,7-dimethyl-6-propylnonane

10.35 (a) 1-bromopropane
(b) 1,4-difluoro-2-methylpentane
(c) 1,2,4-tribromobutane
(d) 2,5-dichloro-3-iodo-4-methylhexane
(e) 1,2,3-triiodo-2-methylbutane
(f) 1,1,3-trichlorobutane

10.37 (a) methylpropane (b) 2-chloropropane (c) 1-iodo-2,2-dimethylpropane (d) 2-chloro-2-methylpropane (e) 2-bromobutane

10.39 (a) bromocyclopentane
(b) 1-chloro-4-methylcyclohexane
(c) 1,3-dibromo-1-ethyl-3-methylcyclopentane
(d) 1,1,2,4-tetrachlorocyclohexane
(e) 1-fluoro-2,2,3,3-tetramethylcyclobutane

10.41 (a) (cyclopentane) (b) (cyclopropane ring with substituents CH_3, Br, CH_3)

(c) (cyclohexane ring with substituents CH_3CH_2, H_3C, H_3C, CH_3)

10.43 (a) (cyclobutane ring with H_3C, H, H, CH_3) (b) (cyclohexane ring with H, H, Cl, Cl)

(c) (cyclopentane ring with CH_3, CH_3, CH_2CH_3, CH_2CH_3)

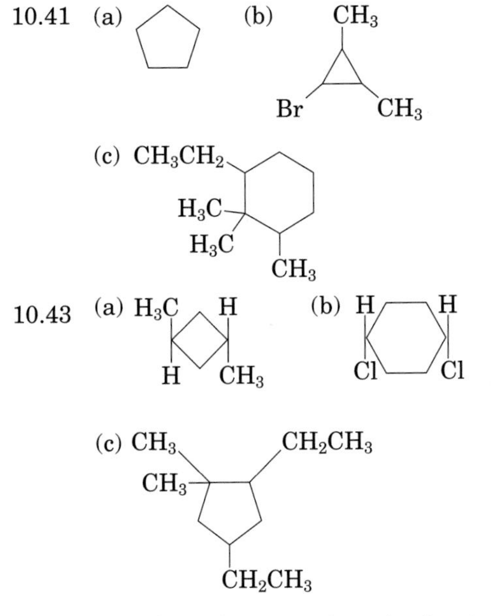

10.45 They burn but are otherwise inert.

10.47 (a) alcohol (b) carboxylic acid (c) ketone (d) alkane (e) amine (f) halide (g) alkyne

10.49 $CH_3CH{=}CH_2$ (b) $CH_3—C({=}O)—OH$
(c) $CH_3CH_2—NH_2$ (d) $H—C({=}O)—NH_2$
(e) CH_3CH_2OH (f) $CH_3—O—CH_3$
(g) $CH_3—SH$ (h) $CH_3CH_2—C({=}O)—H$
(i) $CH_3—C({=}O)—O—CH_3$ (j) $CH_3CH_2—C({=}O)—CH_3$

10.51 (a) carboxylic acid group, carboxylic ester group
(b) halide, ether, carboxylic acid group
(c) (d) carboxylic acid group, amine

10.53 Each molecule rapidly and spontaneously converts to the other.

10.55 Petroleum is a mixture of compounds with a wide range of boiling points. Each use requires a much narrower boiling point range.

10.57 They absorb infrared light from the earth and reemit it back to the earth, preventing energy from escaping.

10.59 one at a tertiary, two at secondary carbons

10.61 $CH_3{-}NH{-}CH_3$ $CH_3CH_2NH_2$

10.63 (a) Two sp^3 orbitals, one from each carbon, overlap to form the C—C σ bond. Each of the three remaining orbitals of the carbon on the right overlaps with a *p* orbital of the fluorine, as does one of the three remaining sp^3 orbitals of the carbon on the left. Each of the last two remaining sp^3 orbitals of the left carbon overlaps with the *s* orbital of a hydrogen.
(b) The two C—C σ bonds are formed by overlap of one sp^3 orbital from each of two carbons. The central carbon has two sp^3 orbitals left: Each of these overlaps with an *s* orbital of a hydrogen. Each of the end carbons has three sp^3 orbitals left. In both cases, one of these overlaps with a *p* orbital of a chlorine; each of the other two, with the *s* orbital of a hydrogen.

10.65 (a) Pentane has five carbons. There cannot be a group only in the 4 position, since a lower number would be obtained counting from the other end.
(b) There cannot be a methyl group at the end of a parent chain, since that makes the parent chain longer.
(c) The prefix di requires two numbers. Only one is given.
(d) If two numbers are given, the prefix di must be used.
(e) If there is only one group on a ring, it gets no number (the number 1 is understood).
(f) There cannot be an ethyl group in the 2 position of an alkane, since that only makes the parent chain longer.
(g) same answer as (f)
(h) There cannot be a two-membered ring.

10.67 They look the same. Neither has any color. You could probably tell them apart by measuring a physical property, such as boiling point.

10.69 (a) $CH_3{-}CH(Cl){-}CH(Cl){-}CH_2CH_2CH_3$

(b) Cl Cl Cl

(c) cyclobutane ring with two CH_3 groups on the same carbon

(d) cycloheptane ring bearing $CH(CH_3){-}CH_3$; $CH(CH_3){-}CH_2CH_3$; and $CH_2{-}C(CH_3)_2{-}CH_3$ on adjacent carbons

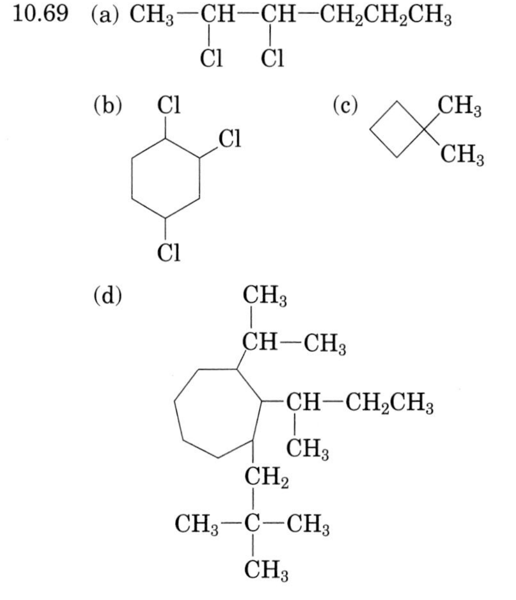

10.71 (a) alcohol, ketone (b) carboxylic acid, carboxylic ester (c) aldehyde, alcohol

CHAPTER 11 Alkenes, Alkynes, and Aromatic Compounds

11.1 2,5-dichloro-2-hexene

11.2 3-ethyl-2-hexene

11.3 (a) 1,4-dimethylcyclopentene (b) 5-chlorocycloheptene

11.4 (a) yes (b) no (c) no

11.5 $CH_3CH_2CH_2CH_2CH_3$

11.6 cyclopentane with Cl, Cl / CH_3, H + cyclopentane with Cl, H / CH_3, Cl

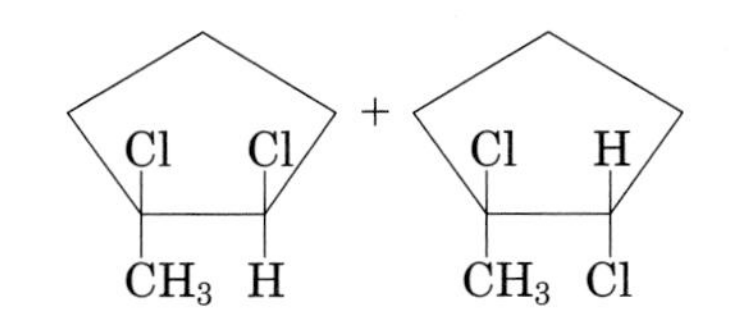

11.7 $CH_3CH_2CH_2{-}C(Br)(CH_3){-}CH_3$

11.8 $CH_3{-}CH(OH){-}CH_3$

11.9 $-CH_2{-}CH(CN){-}CH_2{-}CH(CN){-}CH_2{-}CH(CN){-}$

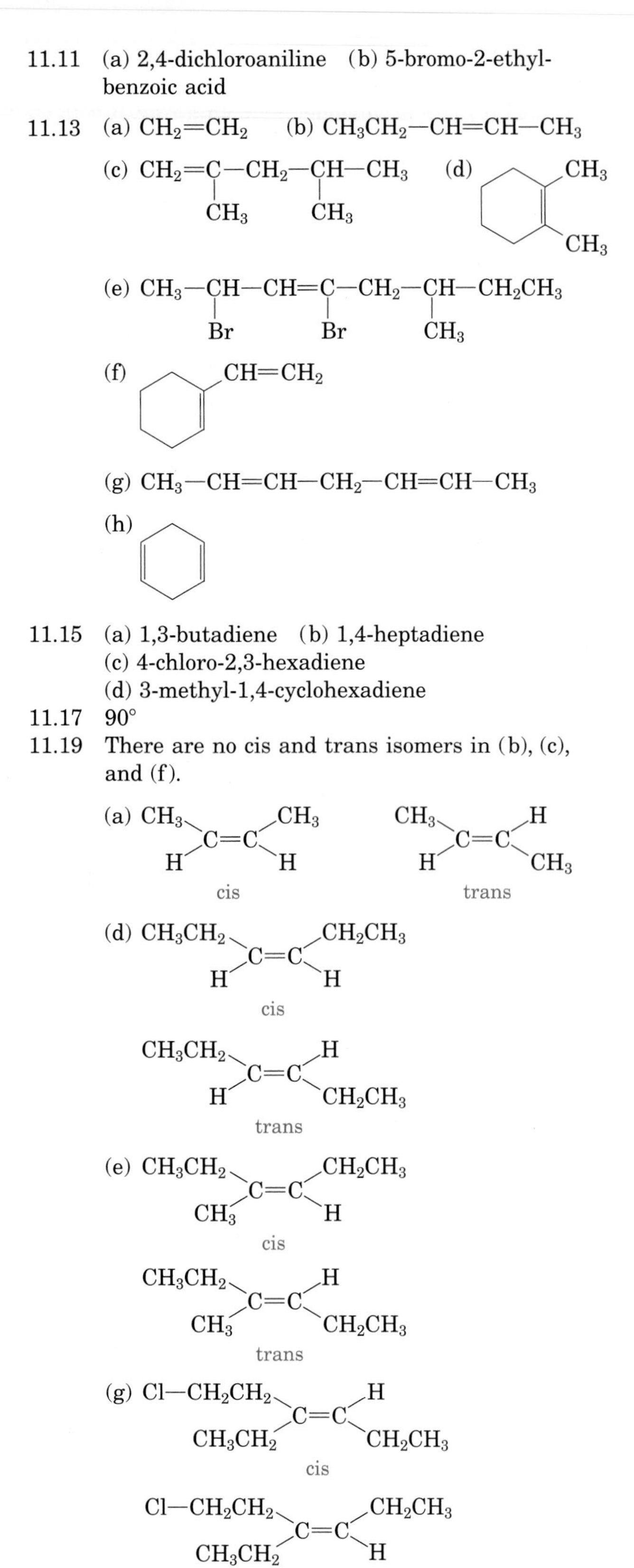

11.10 (a) *meta*-bromochlorobenzene (b) *para*-bromotoluene (c) *ortho*-nitrophenol

11.11 (a) 2,4-dichloroaniline (b) 5-bromo-2-ethylbenzoic acid

11.13 (a) $CH_2{=}CH_2$ (b) $CH_3CH_2{-}CH{=}CH{-}CH_3$

(c) 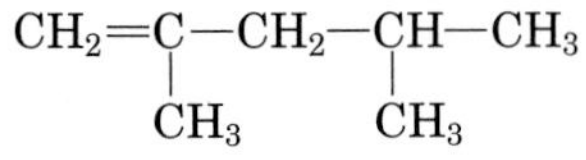 $CH_2{=}\underset{\displaystyle CH_3}{\underset{|}{C}}{-}CH_2{-}\underset{\displaystyle CH_3}{\underset{|}{CH}}{-}CH_3$ (d) 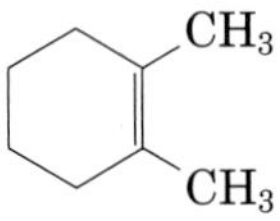 CH_3 CH_3

(e) 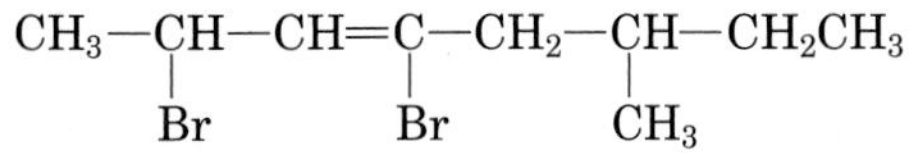 $CH_3{-}\underset{\displaystyle Br}{\underset{|}{CH}}{-}CH{=}\underset{\displaystyle Br}{\underset{|}{C}}{-}CH_2{-}\underset{\displaystyle CH_3}{\underset{|}{CH}}{-}CH_2CH_3$

(f) 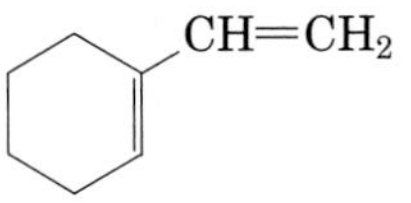 $CH{=}CH_2$

(g) $CH_3{-}CH{=}CH{-}CH_2{-}CH{=}CH{-}CH_3$

(h)

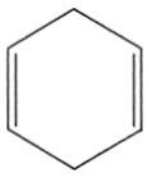

11.15 (a) 1,3-butadiene (b) 1,4-heptadiene
(c) 4-chloro-2,3-hexadiene
(d) 3-methyl-1,4-cyclohexadiene

11.17 90°

11.19 There are no cis and trans isomers in (b), (c), and (f).

(a) 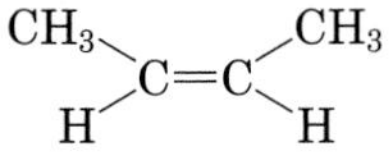 CH_3, H / C=C / CH_3, H — cis; CH_3, H / C=C / H, CH_3 — trans

(d) 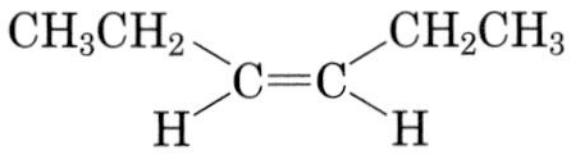 CH_3CH_2, H / C=C / CH_2CH_3, H — cis

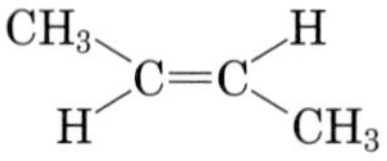

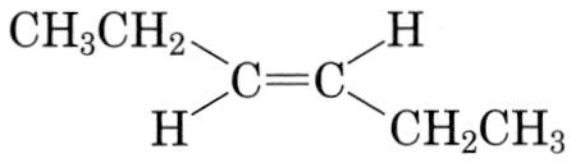

CH_3CH_2, H / C=C / H, CH_2CH_3 — trans

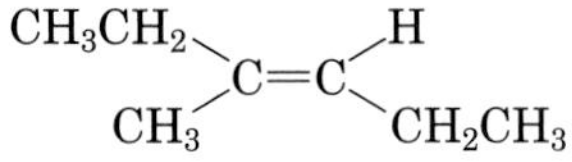

(e) 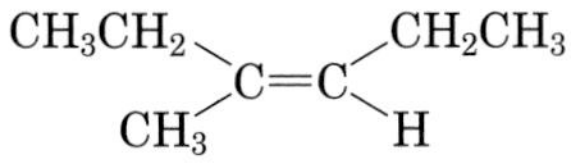 CH_3CH_2, CH_3 / C=C / CH_2CH_3, H — cis

CH_3CH_2, CH_3 / C=C / H, CH_2CH_3 — trans

(g) 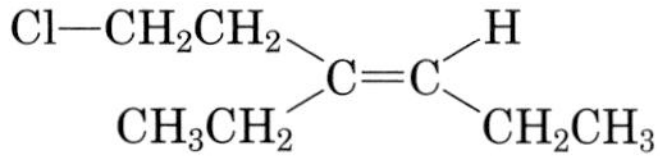 $Cl{-}CH_2CH_2$, CH_3CH_2 / C=C / H, CH_2CH_3 — cis

$Cl{-}CH_2CH_2$, CH_3CH_2 / C=C / CH_2CH_3, H — trans

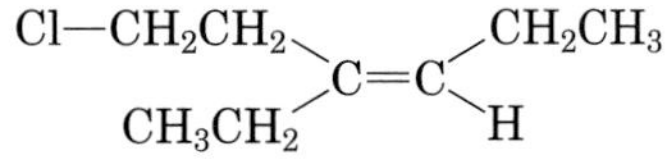

11.21 reactions in which something is added to a double or triple bond (for examples, see Sec. 11.6)

11.23 (a) $CH_3CH_2CH_2CH_3$

(b) 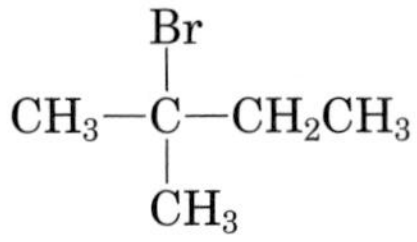 $CH_3{-}\overset{\displaystyle Br}{\overset{|}{\underset{\displaystyle CH_3}{\underset{|}{C}}}}{-}CH_2CH_3$

(c) CH_3, Cl, Cl

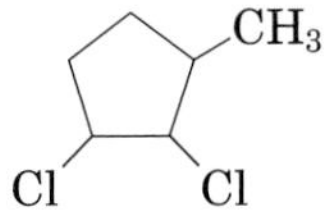

(d) 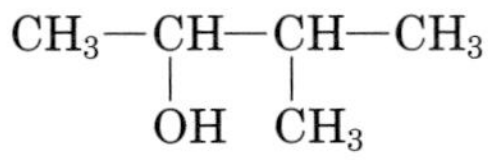 $CH_3{-}\underset{\displaystyle OH}{\underset{|}{CH}}{-}\underset{\displaystyle CH_3}{\underset{|}{CH}}{-}CH_3$

(e) 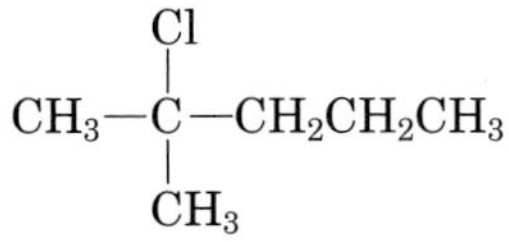 $CH_3{-}\overset{\displaystyle Cl}{\overset{|}{\underset{\displaystyle CH_3}{\underset{|}{C}}}}{-}CH_2CH_2CH_3$

(f) 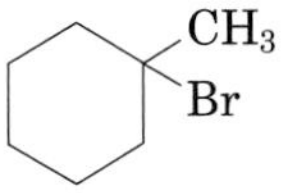 CH_3, Br

11.25 (a) 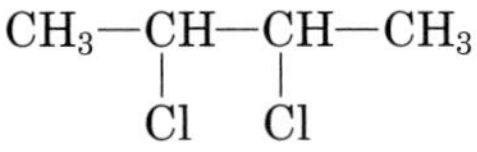 $CH_3{-}\underset{\displaystyle Cl}{\underset{|}{CH}}{-}\underset{\displaystyle Cl}{\underset{|}{CH}}{-}CH_3$

(b), (c) no reaction

(d) 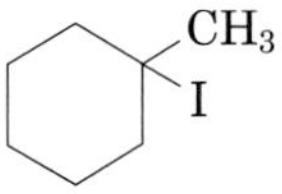 CH_3, I

11.27 (a) HBr (b) H_2O, H_2SO_4 (c) HI (d) H_2, Pt (e) Cl_2

11.29 Add a drop or two of Br_2 dissolved in CCl_4. If the brown color of Br_2 disappears, the compound is cyclohexene. If it does not disappear, the compound is cyclohexane.

11.31 (a) $CH_3CH{=}CHCl$ (b) $CH_2{=}CH{-}COOCH_3$

11.33 (a) 2-pentyne (b) 1,3-dichloro-1-butyne
(c) 1-bromo-4-methyl-2-pentyne

11.35 because the two triple-bonded carbons and the two atoms connected to them form a straight line

11.37 one that contains an aromatic loop (sextet) of electrons

11.39 (a) 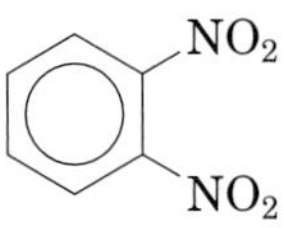 NO_2, NO_2 (b) 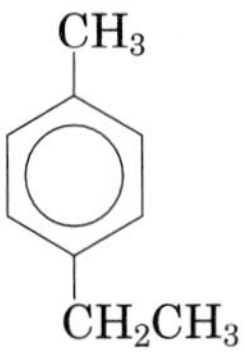 CH_3, CH_2CH_3

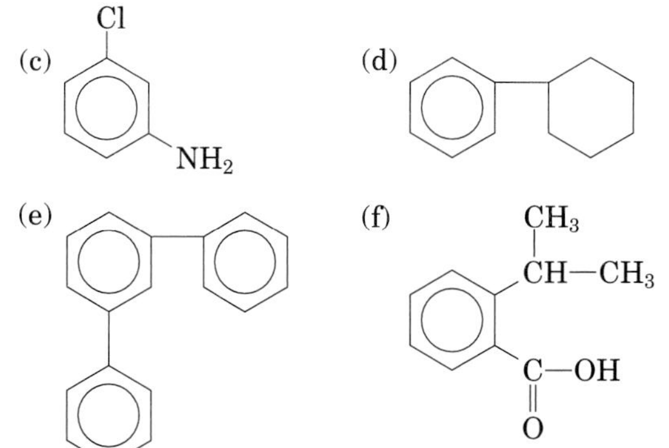

11.41 (a) iodobenzene (b) phenol (c) *meta*-dibromobenzene (d) *meta*-ethylphenol (e) *para*-bromonitrobenzene (f) *meta*-*tert*-butylbenzoic acid (g) *ortho*-isopropylaniline

11.43 If they underwent addition, the very stable aromatic sextet would be destroyed.

11.45 (a) 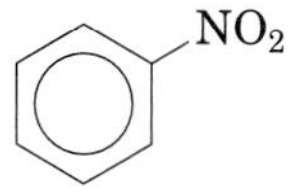 NO_2 (b) 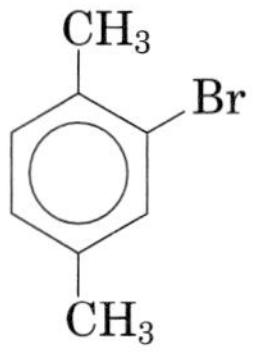 CH_3, Br, CH_3

(c) 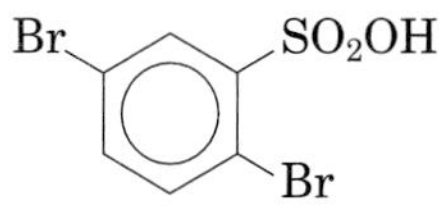 Br, SO_2OH, Br

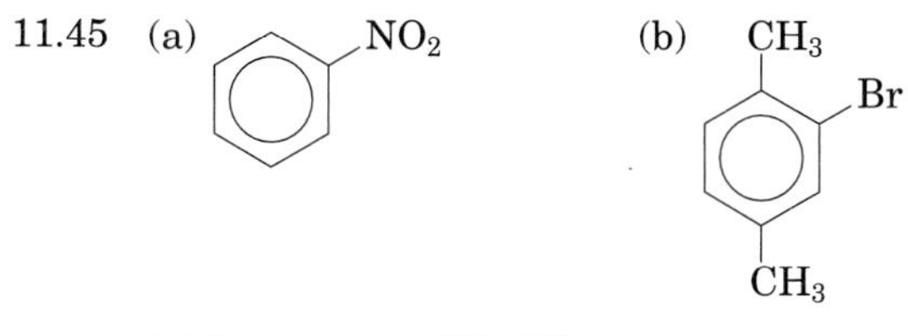

11.47 (a) no reaction (b) 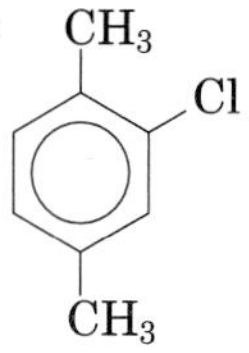 CH_3, Cl, CH_3

11.49

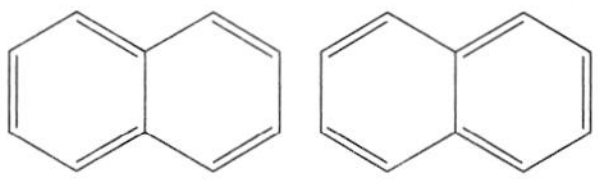

11.51 N, CH_3 N, CH_3 N, CH_3

11.53 CH_3, CH, CH_2, CH, CH_2, CH, CH, C, H_3C, CH_2

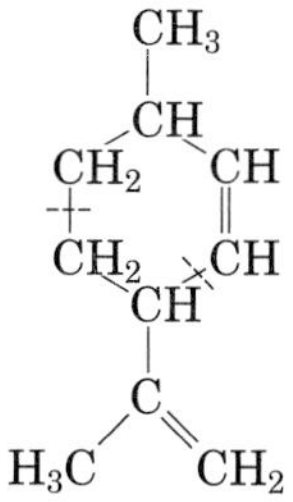

11.55 1.9×10^9 molecules, or 1.9 billion molecules

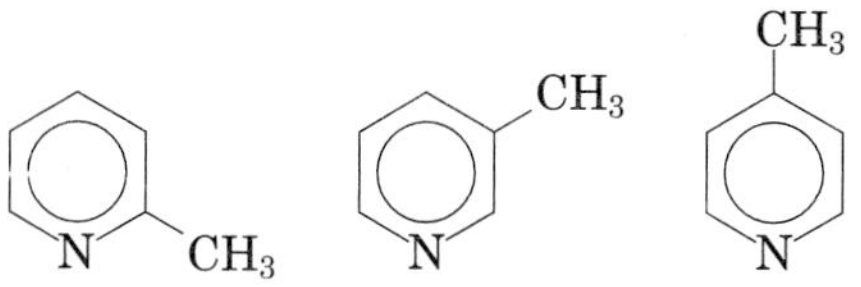

11.57 (a) the *all-trans* isomer (b) the *all-trans* isomer

11.59 (1) $CH_3—C(CH_3)=CH_2 + H^+ \longrightarrow CH_3—\overset{+}{C}(CH_3)—CH_3$

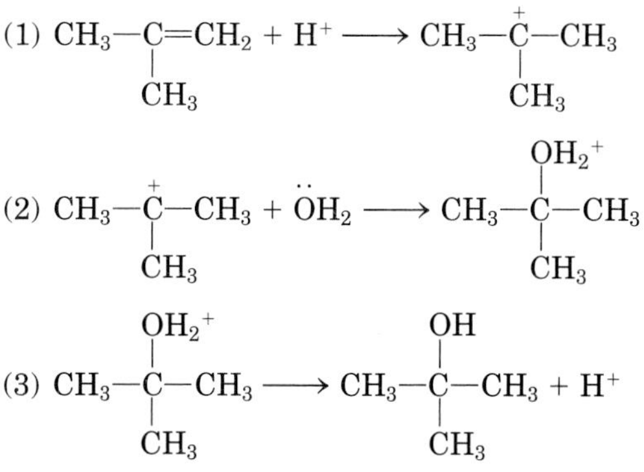

(2) $CH_3—\overset{+}{C}(CH_3)—CH_3 + \ddot{O}H_2 \longrightarrow CH_3—C(OH_2^+)(CH_3)—CH_3$

(3) $CH_3—C(OH_2^+)(CH_3)—CH_3 \longrightarrow CH_3—C(OH)(CH_3)—CH_3 + H^+$

11.61 $—CH_2—C(Cl)=CH—CH_2—CH_2—C(Cl)=CH—CH_2—$

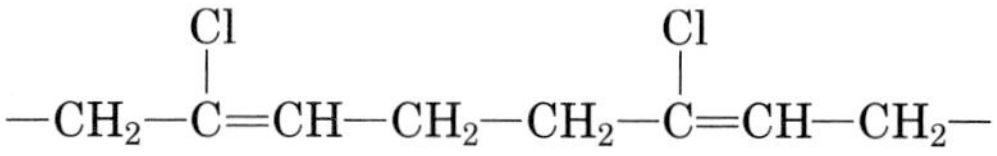

11.63 aromatic hydrocarbons with at least four rings and at least one angular junction

11.65 Aromatic rings do not have double bonds nor are they unsaturated. They have a closed loop of electrons.

11.67 Br Br

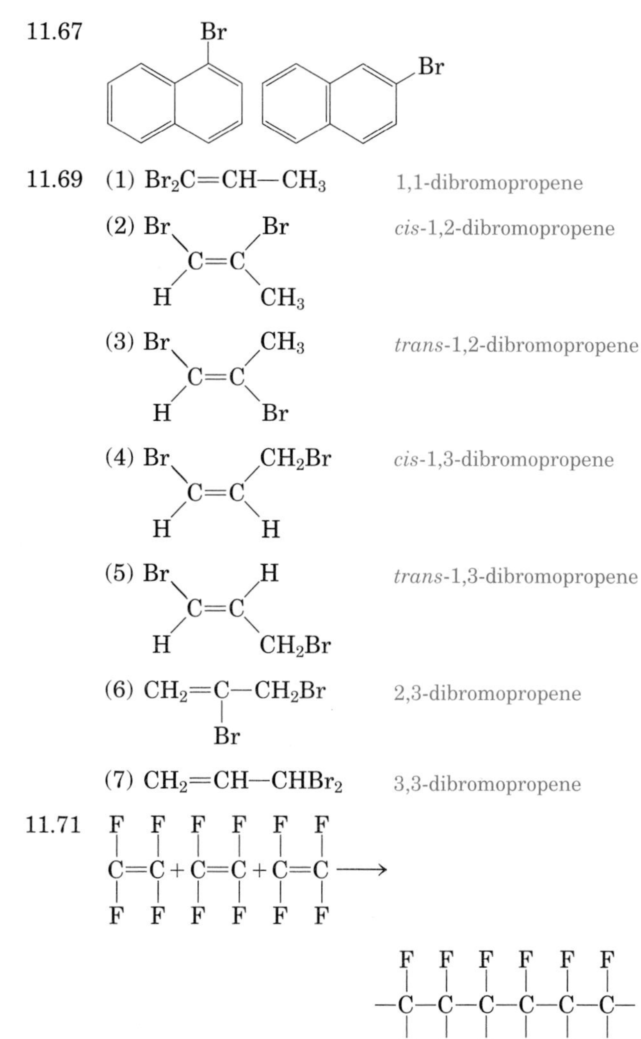

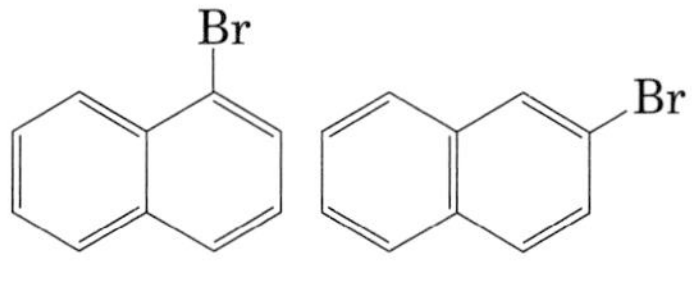

11.69 (1) $Br_2C=CH—CH_3$ 1,1-dibromopropene

(2) 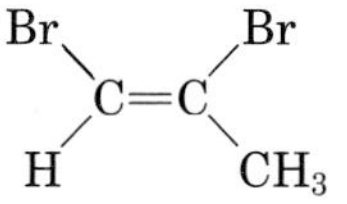 Br, Br / C=C / H, CH_3 *cis*-1,2-dibromopropene

(3) 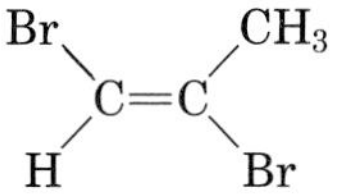 Br, CH_3 / C=C / H, Br *trans*-1,2-dibromopropene

(4) 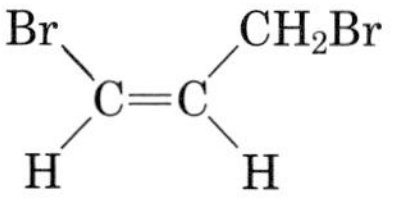 Br, CH_2Br / C=C / H, H *cis*-1,3-dibromopropene

(5) 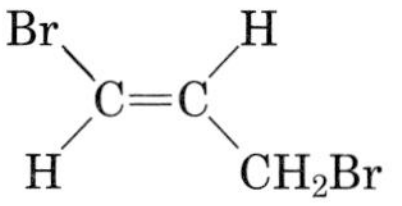 Br, H / C=C / H, CH_2Br *trans*-1,3-dibromopropene

(6) 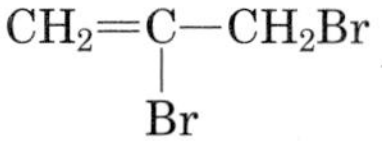 $CH_2=C(Br)—CH_2Br$ 2,3-dibromopropene

(7) $CH_2=CH—CHBr_2$ 3,3-dibromopropene

11.71 $F_2C=CF_2 + F_2C=CF_2 + F_2C=CF_2 \longrightarrow$

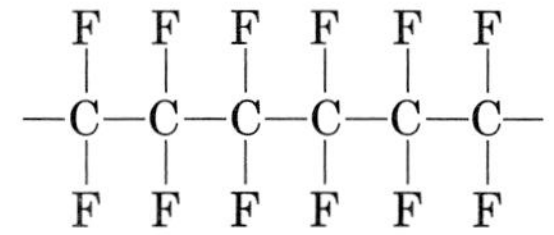

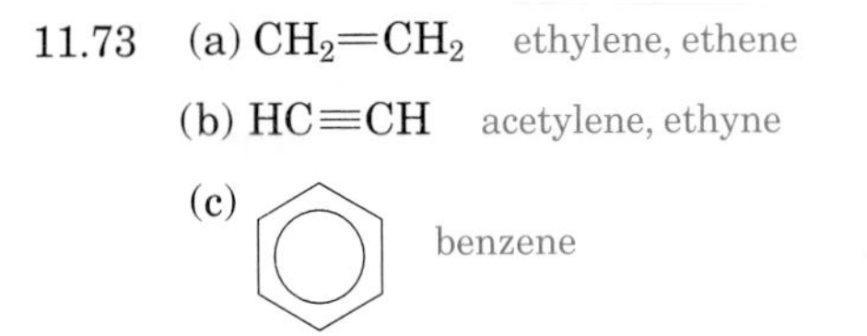

11.73 (a) $CH_2{=}CH_2$ ethylene, ethene

(b) $HC{\equiv}CH$ acetylene, ethyne

(c)  benzene

11.75 $C_7H_8 + 9O_2 \longrightarrow 7CO_2 + 4H_2O$

11.77 (1) 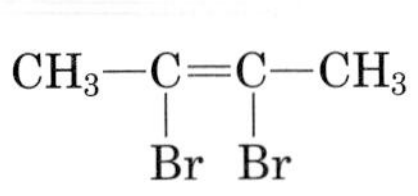 $CH_3{-}C(Br){=}C(Br){-}CH_3$ (2) 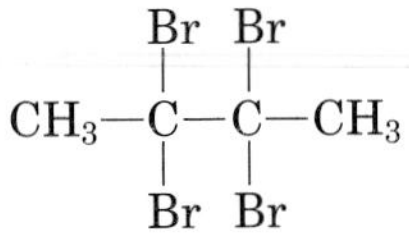 $CH_3{-}CBr_2{-}CBr_2{-}CH_3$

CHAPTER 12 Alcohols, Phenols, Ethers, and Halides

12.1 (a) secondary (b) tertiary (c) primary

12.2 (a) 2,3-dimethyl-3-pentanol (b) 5-ethyl-3-octanol

12.3 $CH_3CH_2{-}CH{=}CH{-}CH_3$

12.4 $CH_3{-}CH{=}CH{-}CH_2{-}CH_3$

12.5 $CH_3{-}CH(CH_3){-}C({=}O){-}H$ $CH_3{-}CH(CH_3){-}C({=}O){-}OH$

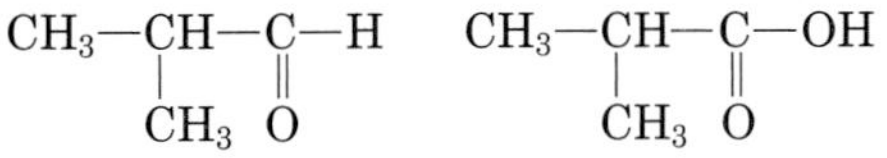

12.6 $C_6H_5{-}CH_2{-}C({=}O){-}CH_3$

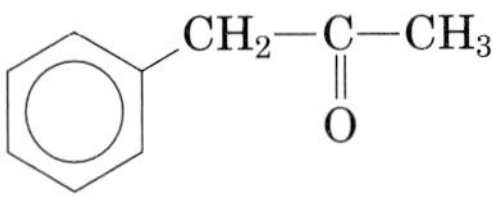

12.7 (a) diphenyl ether (b) ethyl isopropyl ether

12.9 (a) l-propanol, *n*-propyl alcohol

(b) 2-propanol, isopropyl alcohol

(c) 1,2,3-propanetriol, glycerol

(d) 2-methyl-2-propanol, *tert*-butyl alcohol

(e) 3-bromo-l-butanol

(f) 2,4-dimethyl-3-hexanol

(g) 5-bromo-4-methyl-2-pentanol

(h) 4-chloro-2-propyl-l-butanol

(i) 3-methyl-3-pentanol

(j) 3-bromo-2,6-dimethylcyclohexanol

(k) 1,3-dimethylcyclopentanol

12.11 See Sec. 12.2.

12.13 (a) $CH_3CH_2{-}OH$ (b) 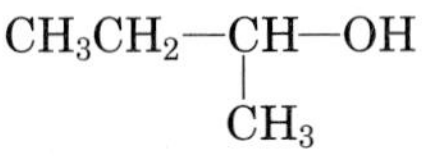 $CH_3CH_2{-}CH(CH_3){-}OH$

(c) (d)

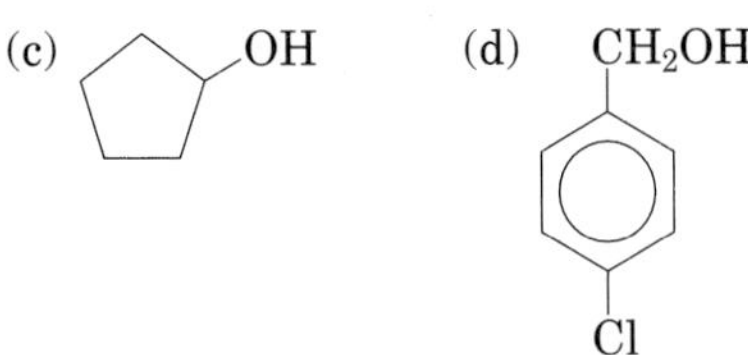

(e) 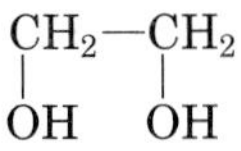 $CH_2(OH){-}CH_2(OH)$ (f) CH_3OH

12.15 (a) $CH_2{=}CH_2$ (b) CH_3COOH

12.17 The product actually formed in each case is written first.

(a) $CH_3{-}CH{=}CH{-}CH_3$

$CH_2{=}CH{-}CH_2{-}CH_3$

(b) 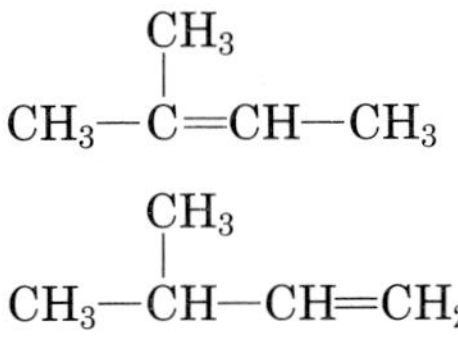 $CH_3{-}C(CH_3){=}CH{-}CH_3$

$CH_3{-}CH(CH_3){-}CH{=}CH_2$

(c) $CH_3{-}CH_2{-}C(CH_3){=}C(CH_3){-}CH_3$

$CH_3{-}CH{=}C(CH_3){-}CH(CH_3){-}CH_3$

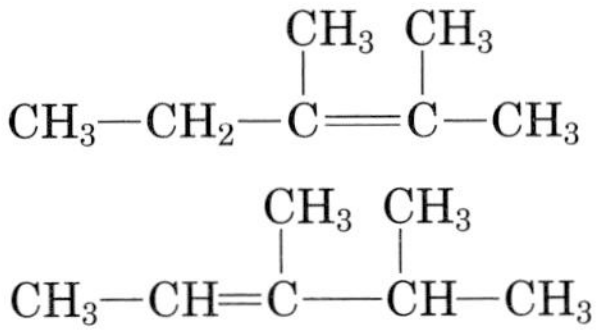

(d) 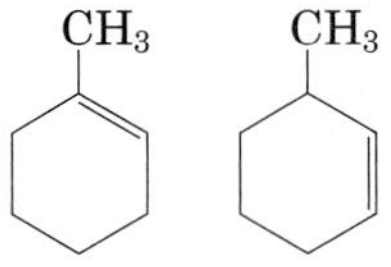 CH_3 CH_3

12.19 (a) ethylene glycol (b) glycerol (c) ethanol, isopropyl alcohol (d) methanol (e) ethanol (f) methanol (g) glycerol

12.21 (a) 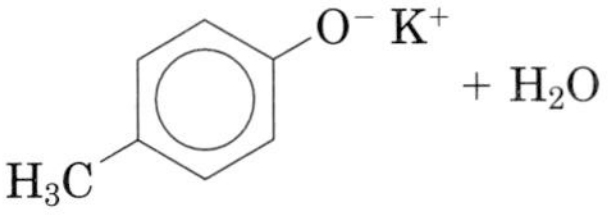 $H_3C{-}C_6H_4{-}O^-\,K^+ + H_2O$

(b) 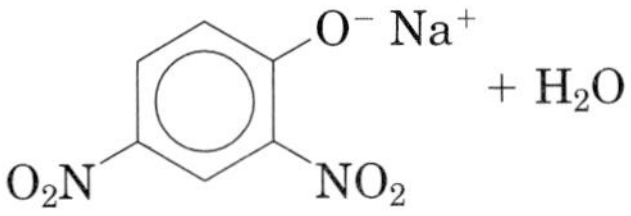 $O_2N{-}C_6H_3(NO_2){-}O^-\,Na^+ + H_2O$

12.23 (a) $CH_3{-}O{-}CH_3$

(b) $CH_3CH_2{-}O{-}CH_2CH_2CH_3$

(c) 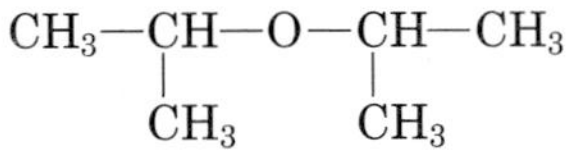 $CH_3{-}CH(CH_3){-}O{-}CH(CH_3){-}CH_3$

(d) 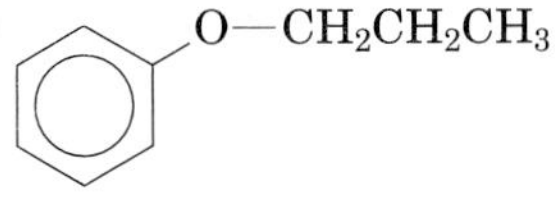 $C_6H_5{-}O{-}CH_2CH_2CH_3$

(e) $CH_3CH_2{-}O{-}CH{=}CH_2$

12.25 They are inert.

12.27 (a) (1)(2) (b) (1)(5) (c) (4)(5) (d) (2)(5) (e) (3)(5) (f) (1)(2) (g) (1)(4) (h) (1)(2)(5)

12.29 (1) $CH_3CH_2CH_2CH_2—OH$

(2) $CH_3—CH(OH)—CH_2CH_3$

(3) $CH_3—CH(CH_3)—CH_2—OH$ (4) $CH_3—C(CH_3)_2—OH$

(5) $CH_3—O—CH_2CH_3CH_3$

(6) $CH_3—O—CH(CH_3)—CH_3$

(7) $CH_3CH_2—O—CH_2CH_3$

12.31 (a) CH_3CH_2OH (b) $CH_3CH_2CH_2OH$ (c) $CH_3CH_2CH_2OH$

12.33 They are soluble. Two and three —OH groups, respectively, form extensive hydrogen bonds with water.

12.35 (a) $CH_3CH_2—S—S—CH_2CH_3$ (b) $CH_3—SH$

12.37 $HOOC—CH(NH_2)—CH_2—S—S—CH_2—CH(NH_2)—COOH$

12.39 (a) perchloroethylene (b) 1,1,1,2-tetrafluoroethane

12.41 (a) 24 proof (b) 38 proof (c) 96 proof

12.43 by distillation

12.45 Ethanol can cause brain damage in a developing fetus.

12.47 No. One is $—O—NO_2$; the other —O—NO.

12.49 2-chloro-1,1,2-trifluoroethyl difluoromethyl ether

12.51 the action of certain skin bacteria on natural body secretions

12.53 [structure: OH ← phenol; O ← ether]

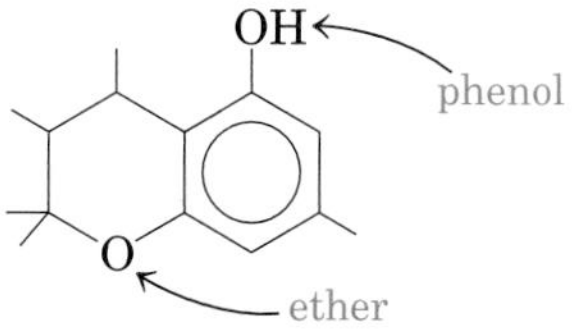

12.55 A biodegradable substance can be decomposed by natural processes in the environment.

12.57 No. (1) Zaitsev's rule does not apply to enzyme-catalyzed reactions. (2) Even if it did, in this case the same product is obtained whichever way the double bond forms.

12.59 (a)  [tetrahydrofuran ring, O] (b) 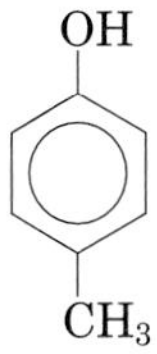 [4-methylphenol: OH, CH_3] (c)  [epoxide ring, O]

12.61 Molecules of 1-butanol form hydrogen bonds with each other. Molecules of methyl propyl ether cannot do this.

12.63 A Cl group increases the acidity of a phenol; a CH_3 group decreases it.

12.65 $C_6H_{12}O_6 \xrightarrow{\text{zymase}} 2CH_3CH_2OH + 2CO_2$

12.67 (a) $CH_3CH_2—S^- \ Na^+$

(b) 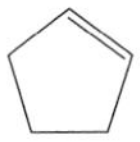 [cyclopentene]

(c) NR

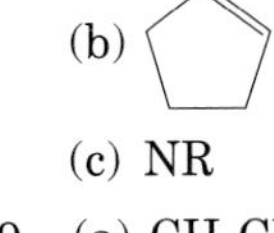

12.69 (a) $CH_3CH_2CH_2—C(=O)—H$ and

$CH_3CH_2CH_2—C(=O)—OH$

(b) the second one (the carboxylic acid)

(c) by distilling the reaction mixture before the aldehyde is further oxidized

12.71 (a) thiols (b) ethers (c) phenols (d) alcohols

CHAPTER 13 Aldehydes and Ketones

13.1 (a) 2-chlorobutanal (b) 4-chloro-2-butanone

13.2 (a) $CH_3CH_2—C(=O)—OH$ (b) NR

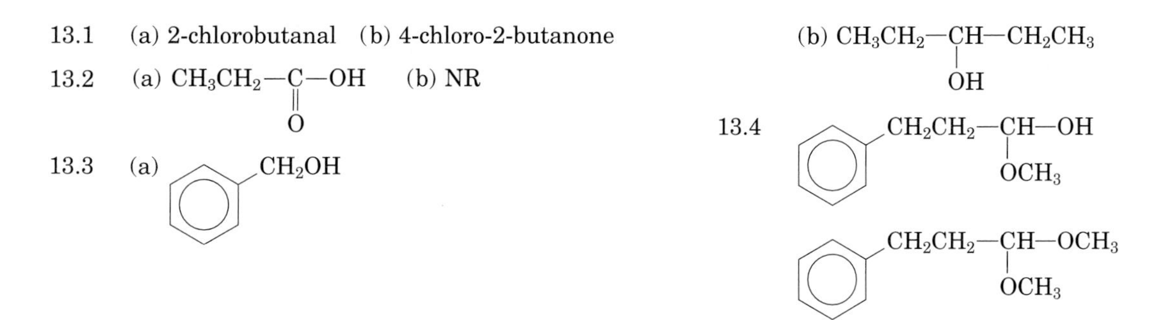

13.3 (a) 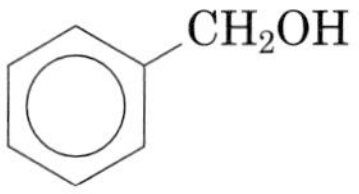 [benzene ring]—CH_2OH

(b) $CH_3CH_2—CH(OH)—CH_2CH_3$

13.4 [benzene ring]—$CH_2CH_2—CH(OCH_3)—OH$

[benzene ring]—$CH_2CH_2—CH(OCH_3)—OCH_3$

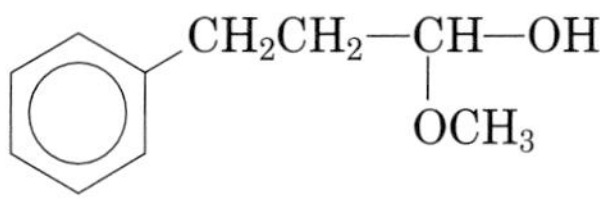

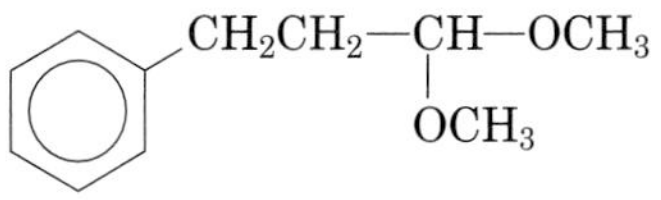

13.5 (a) acetal from an aldehyde
(b) hemiacetal from an aldehyde
(c) hemiacetal from a ketone
(d) acetal from a ketone

13.6 (a) (b) $CH_3CH_2-\underset{\underset{O}{\|}}{C}-H + CH_3OH$

(c) (cyclobutane ring)$=O + CH_3CH_2OH$

(d) CH_3 (cyclohexane ring)$=O + CH_3OH$

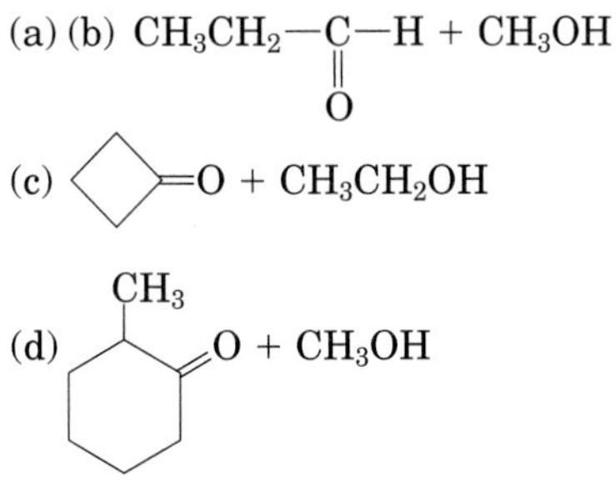

13.7 (b) (c) (d) (f)

13.9 aldehyde: (a) (e); ketone: (b); both: (d); neither: (c)

13.11 (a) 4-chlorohexanal (b) 4-hydroxy-3-hexanone
(c) 2,5-dimethylcyclohexanone (d) 3-butenal

13.13 (a) $H-\underset{\underset{O}{\|}}{C}-H$

(b) $CH_3CH_2-\underset{\underset{O}{\|}}{C}-H$

(c) $CH_3CH_2CH_2-\underset{\underset{CH_3}{|}}{CH}-\underset{\underset{CH_3}{|}}{CH}-CH_2-\underset{\underset{O}{\|}}{C}-H$

(d) $CH_3CH_2CH_2-\underset{\underset{O}{\|}}{C}-H$

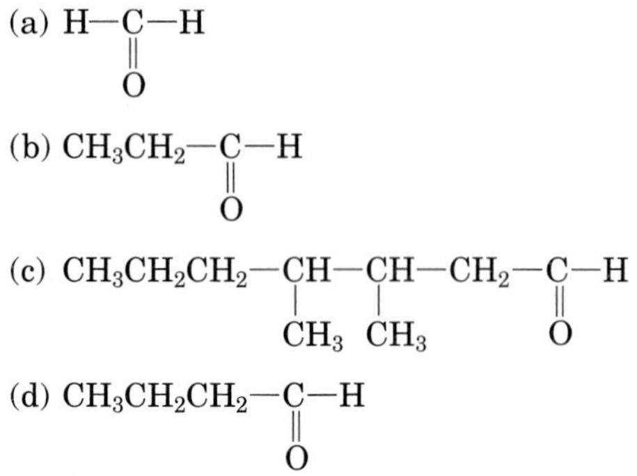

13.15 (a) 2 (b) 2 (c) 1 (d) 2 (e) 1

13.17 Acetone is more polar than methyl ethyl ether.

13.19 $CH_3-\underset{\underset{O\cdots H-O-H}{\|}}{C}-H$ $\quad CH_3-\underset{\underset{O\cdots H-O-H}{\|}}{C}-CH_3$

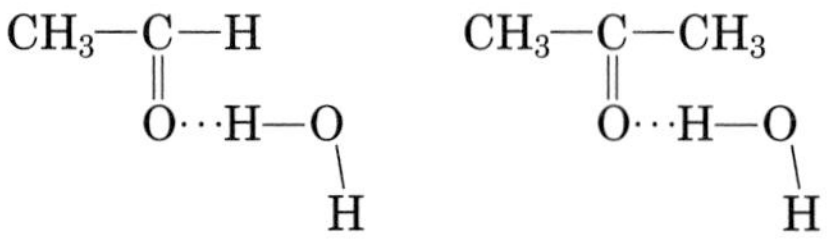

13.21 Benedict's or Tollens' reagents will give positive tests with the aldehyde, but not with the ketone.

13.23 $R-\underset{\underset{O}{\|}}{C}-H + 2Cu^{2+} + 5OH^- \longrightarrow$

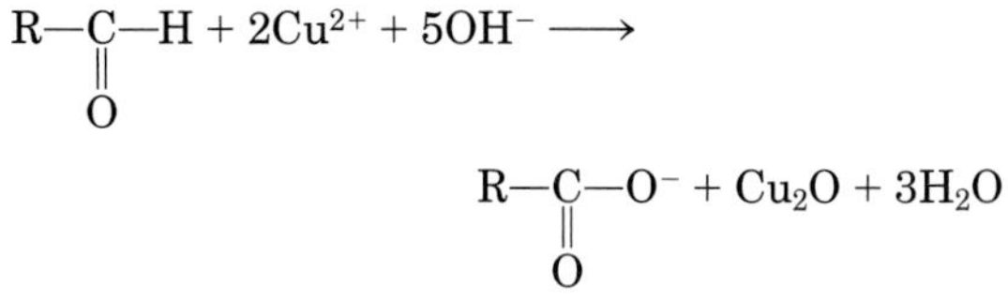

$R-\underset{\underset{O}{\|}}{C}-O^- + Cu_2O + 3H_2O$

13.25 (a) $CH_3-\underset{\underset{OH}{|}}{CH}-CH_3$ (b) $CH_3CH_2CH_2OH$

(c) (cyclopentane ring)$-OH$

(d) $-\underset{\underset{OH}{|}}{CH}-$

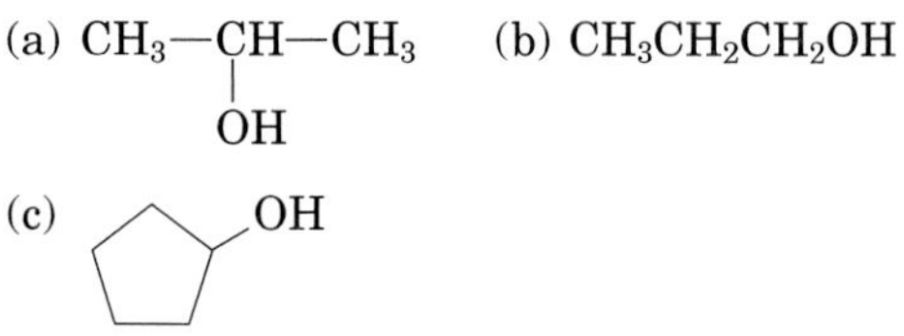

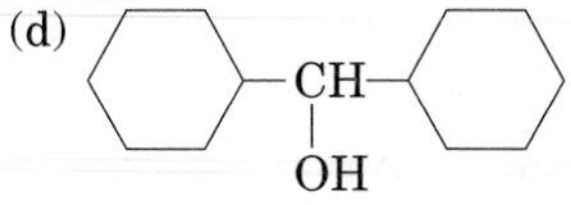

(e) $CH_3CH_2CH_2CH_2-OH$

13.27 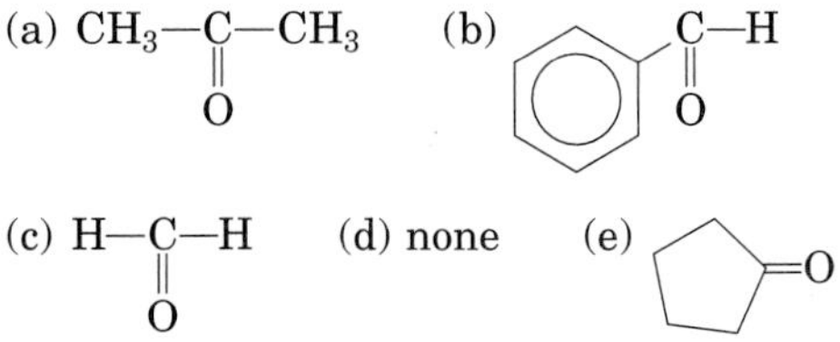 (a) $CH_3-\underset{\underset{O}{\|}}{C}-CH_3$ (b) (benzene ring)$-\underset{\underset{O}{\|}}{C}-H$

(c) $H-\underset{\underset{O}{\|}}{C}-H$ (d) none (e) (cyclopentane ring)$=O$

13.29 $CH_3-\overset{\overset{CN}{|}}{\underset{\underset{OH}{|}}{C}}-H$

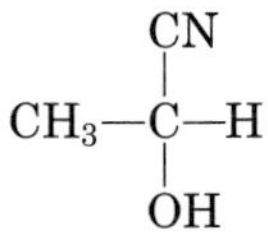

13.31 Acetals: (b)(c)(e) Hemiacetals: (a)(f)
Neither: (d)(g)

13.33 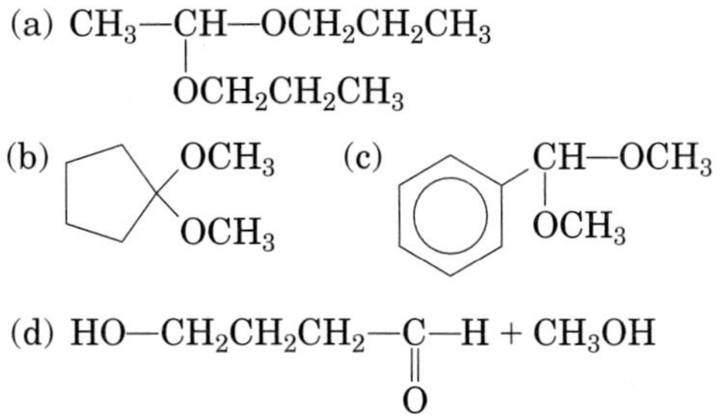 (a) $CH_3-\underset{\underset{OCH_2CH_2CH_3}{|}}{CH}-OCH_2CH_2CH_3$

(b) (cyclopentane ring)$<^{OCH_3}_{OCH_3}$ (c) (benzene ring)$-\underset{\underset{OCH_3}{|}}{CH}-OCH_3$

(d) $HO-CH_2CH_2CH_2-\underset{\underset{O}{\|}}{C}-H + CH_3OH$

13.35 (a) $CH_3CH_2-\underset{\underset{OH}{|}}{CH}-OCH_2CH_3$

$CH_3CH_2-\underset{\underset{OCH_2CH_3}{|}}{CH}-OCH_2CH_3$

(b) (benzene ring)$-\underset{\underset{OH}{|}}{CH}-OCH_3$

(benzene ring)$-\underset{\underset{OCH_3}{|}}{CH}-OCH_3$

(c) CH_3-(cyclopentane ring)$<^{OCH_3}_{OCH_3}$ $\quad CH_3-$(cyclopentane ring)$<^{OH}_{OCH_3}$

(d) (tetrahydropyran ring, O)$-OH$

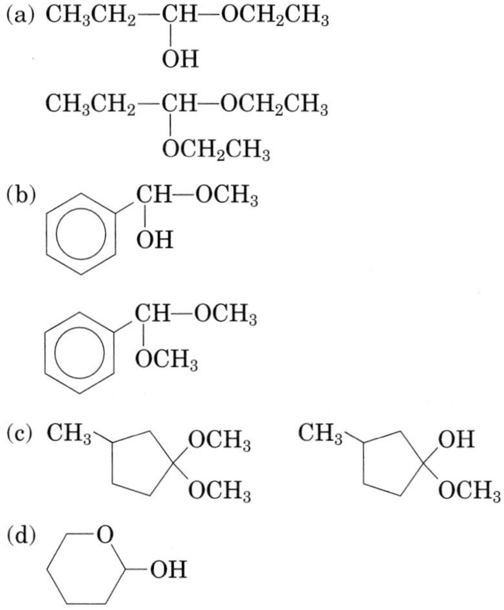

13.37 (a)(b)(d)(f)

13.39 (a) $CH_3-\underset{\underset{O}{\|}}{C}-H$ (b) $CH_3CH_2-\underset{\underset{O}{\|}}{C}-CH_3$

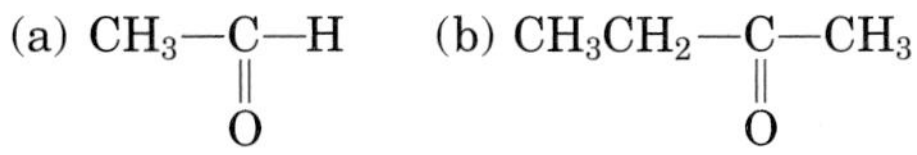

(c) $H{-}\overset{\underset{\|}{O}}{C}{-}CH_2CH_2{-}\overset{\underset{\|}{O}}{C}{-}H$ (d) cyclopentanone (five-membered ring, =O)

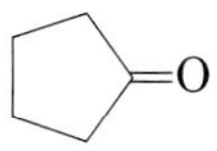

13.41 H, H_3C, $C{=}O$ … $O{=}C$, H, CH_3 … $C{=}O$, H, CH_3 (three acetaldehyde molecules, arrows showing ring formation) → H_3C–CH–O–CH–CH_3, O–CH–O, CH–CH_3 (six-membered ring: three CH groups alternating with three O atoms, each CH bearing CH_3)

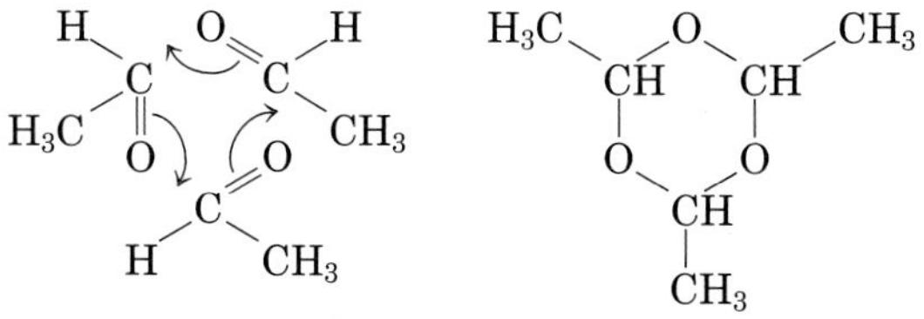

13.43 α-chloroacetophenone

13.45 $Cl{-}C(Cl)(Cl){-}CH(OCH_2CH_3){-}OH$

13.47 a three-dimensional polymer network that is formed at high temperatures and becomes rigid once cooled so that it cannot be melted again

13.49 17 mg

13.51 muscone, β-ionone

13.53 Acetone, $CH_3{-}CO{-}CH_3$, does not have an H connected to an O, N, or F.

13.55 Some of the butanal had become oxidized on standing, to give butanoic acid.

13.57 A 37 percent solution of formaldehyde in water. It is used as a disinfectant and to embalm cadavers.

13.59 A molecule of propanol forms hydrogen bonds with other molecules of propanol. Molecules of propanal cannot do this.

13.61 for example:

(a) 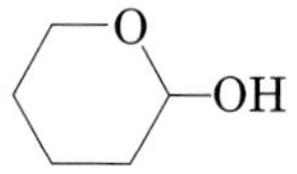 six-membered ring containing O, with –OH on the carbon next to O

(b) 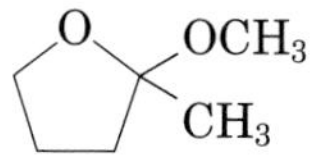 five-membered ring containing O, with OCH_3 and CH_3 on the carbon next to O

(c) $CH_3{-}C(OH)(OCH_3){-}CH_3$

13.63 $CH_2(OH){-}CH(OH){-}CH_2(OH)$ glycerol

13.65 (a) $CH_3{-}\overset{\underset{\|}{O}}{C}{-}CH_2CH_2CH_3 + CH_3OH$

(b) NR

(c) $CH_3{-}CH(OCH_2CH_3){-}OCH_2CH_3$

(d) $CH_3CH_2CH_2CH_2{-}\overset{\underset{\|}{O}}{C}{-}CH_3$

13.67 (a) 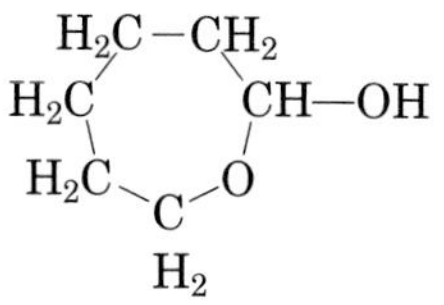 $H_2C{-}CH_2$, H_2C, $CH{-}OH$, H_2C, O, C H_2 (seven-membered ring: H_2C–CH_2–CH(OH)–O–CH_2–H_2C–H_2C)

(b) $H_2C{-}CH_2$, HC, $CH{-}OH$, O; $CH_3CH_2{-}CH(I)$– attached to HC (five-membered ring)

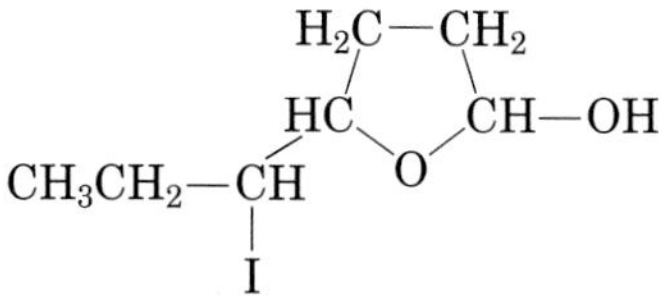

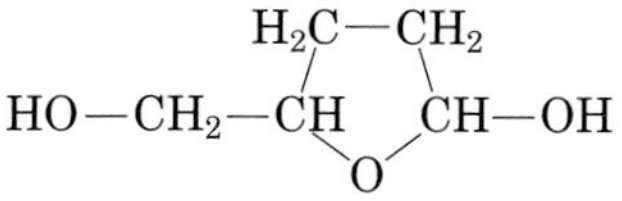

(c) $H_2C{-}CH_2$; $HO{-}CH_2{-}CH$, $CH{-}OH$, O (five-membered ring)

and $H_2C{-}CH_2$; $HO{-}CH$, $CH{-}OH$; $H_2C{-}O$ (six-membered ring)

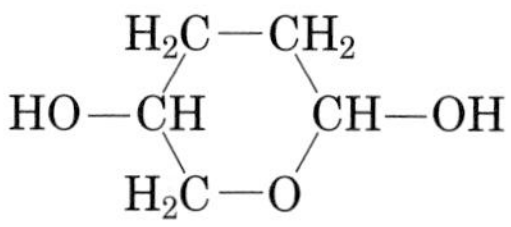

CHAPTER 14 Carboxylic Acids and Esters

14.1 (a) decanoic acid (b) 2-bromo-4-hydroxypentanoic acid

14.2 $CH_3CH_2CH_2{-}\overset{\underset{\|}{O}}{C}{-}O^- + H_2O$

14.3 (a) potassium laurate (b) sodium salicylate

14.4 (a) ethyl propionate (b) *tert*-butyl benzoate

14.5 (a) $H{-}\overset{\underset{\|}{O}}{C}{-}OCH_3$

(b) $CH_3CH_2CH_2{-}\overset{\underset{\|}{O}}{C}{-}O{-}CH(CH_3){-}CH_3$

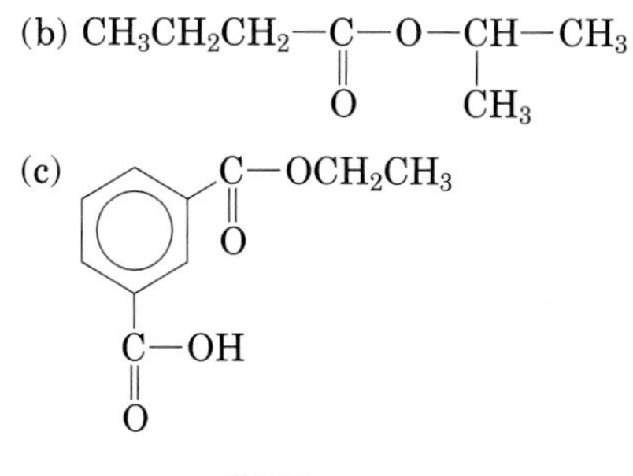

(c) 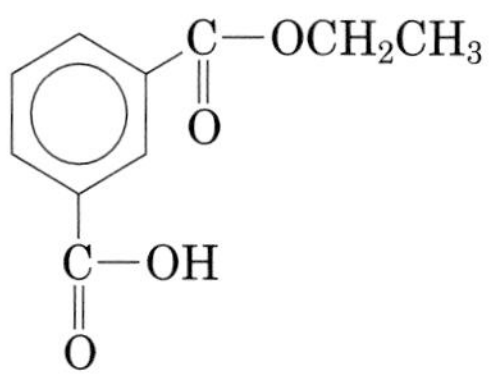 benzene ring bearing $\overset{\underset{\|}{O}}{C}{-}OCH_2CH_3$ and $\overset{\underset{\|}{O}}{C}{-}OH$ (meta)

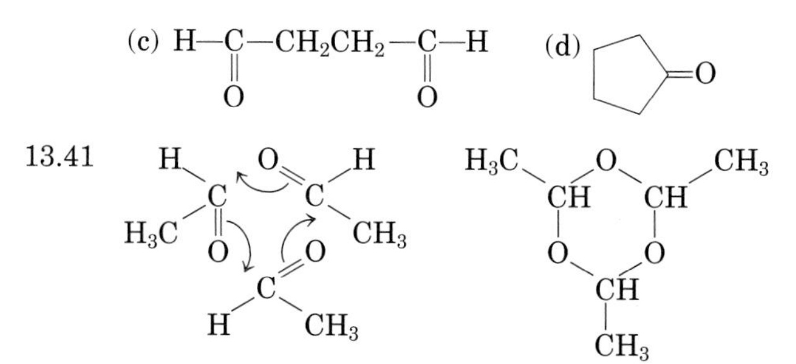

mono

[benzene ring with two $-\overset{O}{\overset{\|}{C}}-OCH_2CH_3$ groups]

di

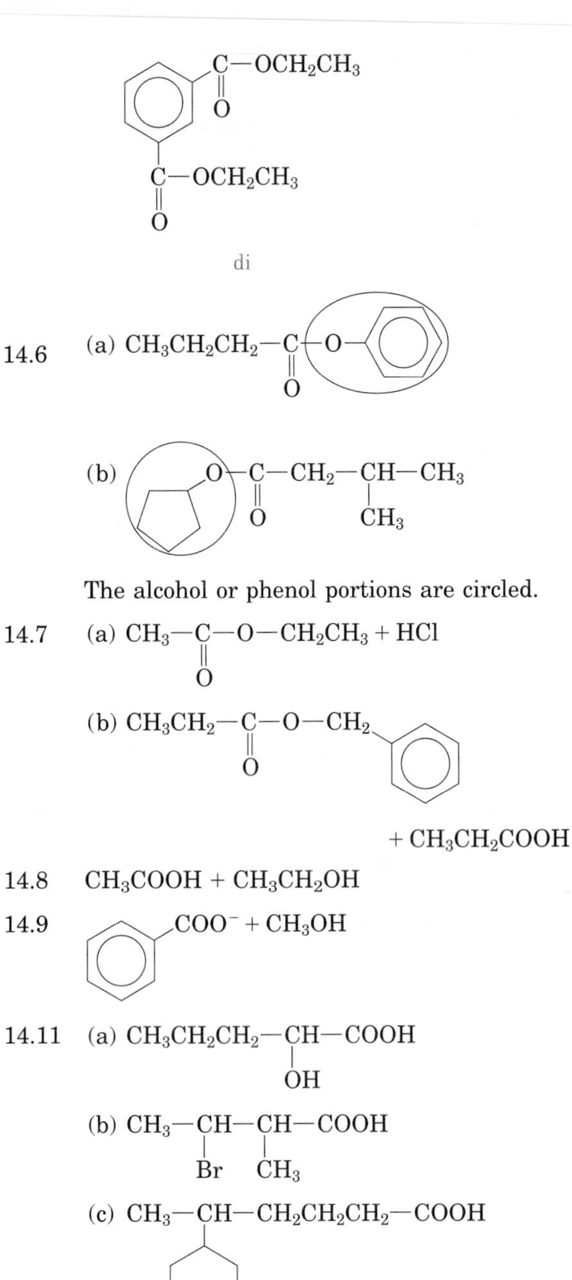

14.6 (a) $CH_3CH_2CH_2-\underset{\underset{O}{\|}}{C}-O-C_6H_5$ [phenyl portion circled]

(b) $C_5H_9-O-\underset{\underset{O}{\|}}{C}-CH_2-\underset{\underset{CH_3}{|}}{CH}-CH_3$ [cyclopentyl portion circled]

The alcohol or phenol portions are circled.

14.7 (a) $CH_3-\underset{\underset{O}{\|}}{C}-O-CH_2CH_3 + HCl$

(b) $CH_3CH_2-\underset{\underset{O}{\|}}{C}-O-CH_2-C_6H_5$

$+ CH_3CH_2COOH$

14.8 $CH_3COOH + CH_3CH_2OH$

14.9 $C_6H_5COO^- + CH_3OH$

14.11 (a) $CH_3CH_2CH_2-\underset{\underset{OH}{|}}{CH}-COOH$

(b) $CH_3-\underset{\underset{Br}{|}}{CH}-\underset{\underset{CH_3}{|}}{CH}-COOH$

(c) $CH_3-\underset{\underset{C_6H_{11}}{|}}{CH}-CH_2CH_2CH_2-COOH$ [cyclohexyl ring]

(d) $CH_3CH_2CH_2-CH=\underset{\underset{CH_3}{|}}{C}-CH_2COOH$

14.13 (a) CH_3-COOH (b) CH_3CH_2-COOH

(c) $HOOC-COOH$ (d) $HOOCCH_2COOH$

(e) [benzene ring with COOH and ortho OH] (f) $CH_3(CH_2)_4COOH$

14.15 It was named after the Latin word for ant (*formica*), because it was first isolated from red ants.

14.17 An acetic acid molecule forms two hydrogen bonds with another acetic acid molecule (Sec. 14.3). Two propanol molecules form only one hydrogen bond.

14.19 [six-membered ring: $HO-\underset{\underset{O}{\|}}{C}-\overset{H_2}{C}-C(=O)-O-H\cdots O$ — intramolecular hydrogen bond between the O–H and the C=O oxygen]

14.21 Bad odors: thiols, carboxylic acids
Pleasant odors: ketones

14.23 (a) $CH_3-\underset{\underset{OH}{|}}{CH}-COOH$

(b) $HOOCCH_2-\overset{\overset{OH}{|}}{\underset{\underset{COOH}{|}}{C}}-CH_2COOH$

14.25 to cauterize gums and canker sores

14.27 $CH_3CH_2OH + O_2 \longrightarrow CH_3COOH + H_2O$

14.29 (a) $HCOO^-\ K^+$

(b) $CH_3(CH_2)_3COO^-\ NH_4^+$

(c) $HOOC-COO^-\ Na^+$

(d) $(C_6H_5COO^-)_3\ Fe^{3+}$

(e) $[CH_3(CH_2)_4COO^-]_2\ Mg^{2+}$

(f) $\left(\begin{matrix} H & & H \\ & C=C & \\ {}^-OOC & & COO^- \end{matrix}\right)\ 2Li^+$

14.31 (a) $CH_3COO^- + H_2O$

(b) $HCOO^-\ NH_4^+$

(c) ${}^-OOC-CH_2-COO^- + 2H_2O$

(d) $CH_3-\underset{\underset{OH}{|}}{CH}-COO^- + H_2CO_3$

(e) $CH_3COOH + Cl^-$

(f) $C_6H_5COOH + HSO_4^-$

14.33 (a) $CH_3CH_2CH_2-\underset{\underset{O}{\|}}{C}-OH$

(b) It is converted to its salt:
$CH_3CH_2CH_2-\underset{\substack{\|\\ O}}{C}-O^-$

(c) same as (a)

14.35 a compound in which the H of an O—H acid is replaced by an R

14.37 (a) $H-\underset{\substack{\|\\ O}}{C}-OCH_2CH_3$

(b) $CH_3(CH_2)_3-\underset{\substack{\|\\ O}}{C}-OCH_3$

(c) $CH_3CH_2CH_2-\underset{\substack{\|\\ O}}{C}-O-\underset{\substack{|\\ CH_3}}{CH}-CH_3$

(d) $CH_3(CH_2)_{16}-\underset{\substack{\|\\ O}}{C}-O-C_6H_5$

(e) $CH_3O-\underset{\substack{\|\\ O}}{C}-CH=CH-\underset{\substack{\|\\ O}}{C}-OCH_3$

(f) $C_6H_5-\underset{\substack{\|\\ O}}{C}-O-CH=CH_2$

(g) $CH_3CH_2CH_2CH_2-\underset{\substack{|\\ CH_3}}{CH}-CH_2-\underset{\substack{\|\\ O}}{C}-O-\overset{\substack{CH_3\\ |}}{\underset{\substack{|\\ CH_3}}{C}}-CH_3$

14.39 (b) (d) (e) (a) (c)

14.41 (a) $CH_3-\underset{\substack{\|\\ O}}{C}-O-CH_3$

(b) $H-\underset{\substack{\|\\ O}}{C}-O-\underset{\substack{|\\ CH_3}}{CH}-CH_3$

(c) $CH_3-\underset{\substack{|\\ CH_3}}{CH}-\underset{\substack{\|\\ O}}{C}-O-CH_2CH_3$

(d) $CH_3-\underset{\substack{|\\ OH}}{CH}-\underset{\substack{\|\\ O}}{C}-O-CH_2-C_6H_5$

(e) $CH_3-O-\underset{\substack{\|\\ O}}{C}-(CH_2)_4-\underset{\substack{\|\\ O}}{C}-O-CH_3$

(f) $Cl-CH_2CH_2-\underset{\substack{\|\\ O}}{C}-O-C_6H_4-CH_3$

14.43 (a) $C_6H_5-\underset{\substack{\|\\ O}}{C}-OCH_2CH_3 + H_2O$

(b) $CH_3(CH_2)_6-\underset{\substack{\|\\ O}}{C}-OCH_3 + HCl$

(c) $CH_3-\underset{\substack{\|\\ O}}{C}-O-\underset{\substack{|\\ CH_3}}{CH}-CH_3 + CH_3COOH$

(d) $C_5H_9-\underset{\substack{\|\\ O}}{C}-O-C_6H_4Br + HCl$

(e) NR

(f) $CH_3-\underset{\substack{\|\\ O}}{C}-O-CH_2CH_2CH_2-O-\underset{\substack{\|\\ O}}{C}-CH_3$

$+ 2CH_3COOH$

14.45 the base-catalyzed hydrolysis of carboxylic esters

14.47 (a) $HO-CH_2CH_2CH_2CH_2COOH$

(b) $HO-CH_2CH_2-\underset{\substack{|\\ CH_3}}{CH}-COO^-$

14.49 $HO-CH_2-\underset{\substack{\|\\ O}}{C}-CH_2-OH + H_3PO_4$

$\longrightarrow HO-CH_2-\underset{\substack{\|\\ O}}{C}-CH_2-O-\overset{\substack{O\\ \|}}{\underset{\substack{|\\ OH}}{P}}-OH + H_2O$

14.51 (a) $HO-\overset{\substack{O\\ \|}}{\underset{\substack{|\\ OH}}{P}}-O-\overset{\substack{O\\ \|}}{\underset{\substack{|\\ OH}}{P}}-OH$

(b) $HO-\overset{\substack{O\\ \|}}{\underset{\substack{|\\ OH}}{P}}-O-\overset{\substack{O\\ \|}}{\underset{\substack{|\\ OH}}{P}}-O-\overset{\substack{O\\ \|}}{\underset{\substack{|\\ OH}}{P}}-OH$

14.53 $HO-P(=O)(OH)-O-P(=O)(OH)-O-P(=O)(OH)-OCH_2CH_3$

$^{-}O-P(=O)(O^{-})-O-P(=O)(O^{-})-O-P(=O)(O^{-})-OCH_2CH_3$

14.55 $\left(CH_3CH_2-C(=O)-O^{-}\right)_2 Ca^{2+}$

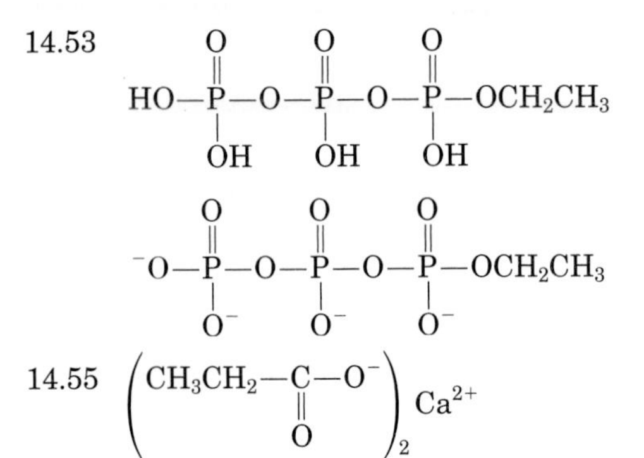

14.57 Salicylic acid has a sour taste and irritates the mouth and stomach lining. Aspirin does not have these effects.

14.59 $-C(=O)-CH=CH-C(=O)-O-(CH_2)_6-O-C(=O)-CH=CH-C(=O)-O-(CH_2)_6-O-$ (each C=C bearing H atoms on opposite sides)

14.61 (a) yes

(b) $CH_3-CH-C=O$ (CH and C=O joined through a ring O) $\quad CH_2-C=O$ (joined through a ring O)

(c) The polymers would not dissolve in water; the lactones would.

14.63 (a) $CH_3(CH_2)_6-C(=O)-O-C(=O)-(CH_2)_6CH_3 +$ cyclopentyl$-OH \longrightarrow$ cyclopentyl$-O-C(=O)-(CH_2)_6CH_3 + CH_3(CH_2)_6COOH$

(b) $(CH_3)_2C_6H_3-C(=O)-Cl + CH_3-CH(CH_3)-CH(CH_3)-CH_2-OH \longrightarrow (CH_3)_2C_6H_3-C(=O)-O-CH_2-CH(CH_3)-CH(CH_3)-CH_3 + HCl$

14.65 $CH_3O-S(=O)_2-OCH_3$

14.67 thiols, phenols, carboxylic acids, sulfonic acids
14.69 only (a)
14.71 for example:

$HO-P(=O)(OH)-O-P(=O)(OH)-OCH_3$

14.73 No. When one molecule of anhydride reacts with one molecule of alcohol to produce one molecule of carboxylic ester, the other product is a molecule of carboxylic acid, which is much less reactive toward the alcohol.

CHAPTER 15 Amines and Amides

15.1 (a) secondary (b) primary (c) tertiary

15.2 (a) ethylamine (b) ethylmethylamine (c) diethylmethylamine

15.3 *ortho*-chloro-*N*-propylaniline

15.4 1,3-diaminocyclohexane

15.5 pyridine: 5 piperidine: 11

15.6 (a) $CH_3CH_2CH_2NH_3^+\ HCOO^-$

(b) $\overset{+}{N}H_2CH_3\ Br^-$

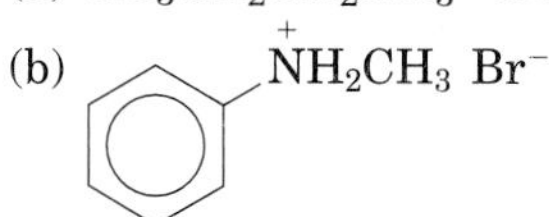

15.7 valeramide

15.8 (a) *N,N*-diphenylvaleramide (b) *para*-bromo-*N*-ethylbenzamide

15.9 (a) $CH_3{-}\underset{\underset{O}{\|}}{C}{-}NH_2 + HCl$

(b) $CH_3CH_2CH_2{-}\underset{\underset{O}{\|}}{C}{-}\underset{\underset{CH_3}{|}}{N}{-}CH_2CH_3 +$ $CH_3CH_2CH_2COOH$

15.10 acid: $CH_3CH_2COOH + CH_3CH_2NH_3^+$
base: $CH_3CH_2COO^- + CH_3CH_2NH_2$

15.11 See Sec. 15.2.

15.13 Primary: (a) (c) Secondary: (d) (f) Tertiary: (b) (e)

15.15 Primary: (a) (d) (h) (i) (m)
Secondary: (b) (c) (j) (l)
Tertiary: (e) (f) (g) (k)

15.17 (a) CH_3NH_2

(b) $CH_3CH_2CH_2{-}\underset{\underset{H}{|}}{N}{-}CH_2CH_2CH_3$

(c) NH_2 / CH_3 (d) $NH{-}CH_3$

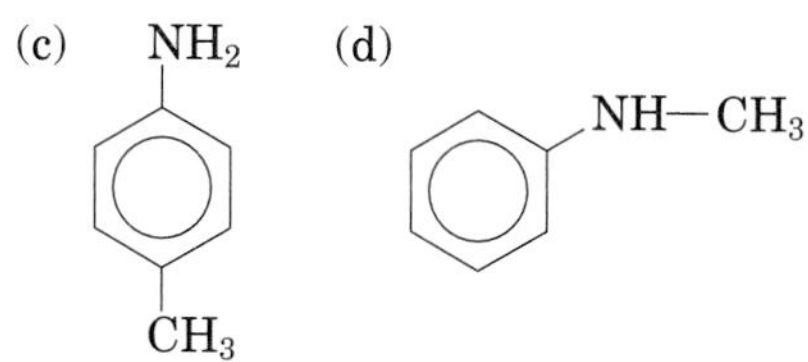

(e) $CH_3CH_2{-}\underset{\underset{CH_2CH_3}{|}}{N}{-}CH_2CH_3$

(f) $CH_3{-}\overset{\overset{CH_3}{|}}{\underset{\underset{CH_3}{|}}{C}}{-}NH{-}\underset{\underset{CH_3}{|}}{CH}{-}CH_2CH_3$

(g) $NH_2{-}(CH_2)_5{-}NH_2$

15.19 (b), (a), (d), (c)

15.21 Trimethylamine is the only one of the three that has no N—H bond, and so cannot form a hydrogen bond with another molecule of itself.

15.23 (a) $CH_3{-}\overset{+}{N}H_2{-}CH_3\ I^-$

(b) $\left[CH_3CH_2{-}\overset{\overset{CH_3}{|}}{\underset{\underset{CH_3}{|}}{N}}{-}CH_3\right]^+ OH^-$

(c) $\left[CH_3CH_2{-}\overset{\overset{CH_2CH_3}{|}}{\underset{\underset{CH_2CH_3}{|}}{N}}{-}CH_2CH_3\right]^+ Cl^-$

(d) $C_6H_5{-}NH_3^+\ NO_3^-$

15.25 isopropylammonium bromide
$CH_3{-}\underset{\underset{CH_3}{|}}{CH}{-}NH_3^+\ Br^-$

15.27 The salt is more soluble in body fluids.

15.29 (b) (c) (f)

15.31 Amine: (b) (j) Amide: (c) (e) (f) (g) (i) (k)
Both: (d) (h) Neither: (a)

15.33 (a) $CH_3CH_2{-}\underset{\underset{O}{\|}}{C}{-}NH_2$

(b) $CH_3CH_2CH_2{-}\underset{\underset{O}{\|}}{C}{-}NH{-}CH_2CH_3$

(c) $CH_3{-}\underset{\underset{O}{\|}}{C}{-}\underset{\underset{CH_3}{|}}{N}{-}CH_3$

(d) $C_6H_5{-}\underset{\underset{O}{\|}}{C}{-}NH{-}C_6H_5$

(e) $Br{-}C_6H_4{-}\underset{\underset{O}{\|}}{C}{-}\overset{\overset{CH_2CH_3}{|}}{N}{-}C_6H_5$

(f) $NH_2{-}\underset{\underset{O}{\|}}{C}{-}CH_2CH_2{-}\underset{\underset{O}{\|}}{C}{-}NH_2$

15.35 acidic: (c) (d) (j) (r) (t)
basic: (b) (f) (n) (o) (p)
neutral: (a) (e) (g) (h) (i) (k) (l) (m) (q) (s)

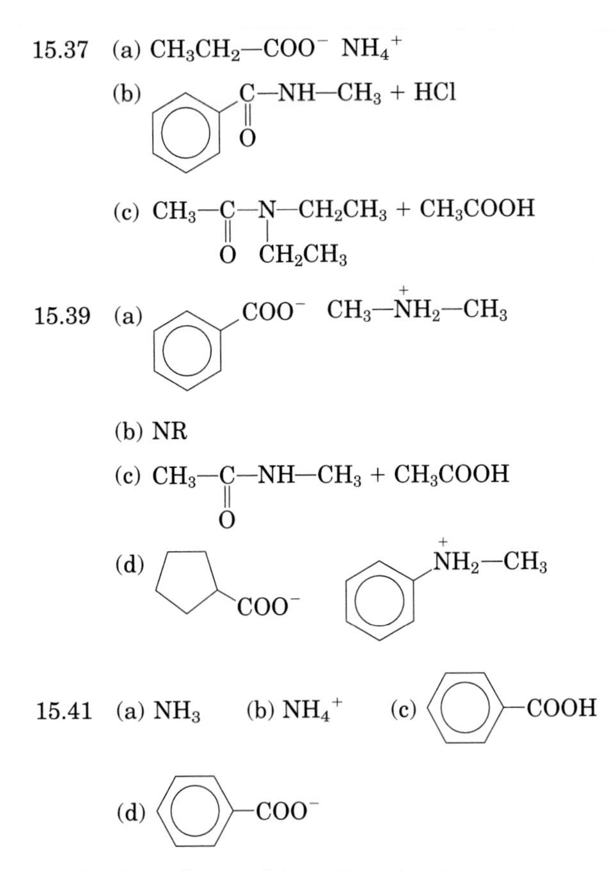

15.37 (a) $CH_3CH_2-COO^-\ NH_4^+$

(b) 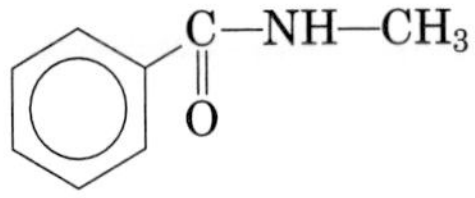 $C_6H_5-C(=O)-NH-CH_3 + HCl$

(c) $CH_3-C(=O)-N(CH_2CH_3)-CH_2CH_3 + CH_3COOH$

15.39 (a) 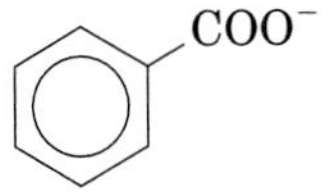 $C_6H_5-COO^-\ CH_3-\overset{+}{N}H_2-CH_3$

(b) NR

(c) $CH_3-C(=O)-NH-CH_3 + CH_3COOH$

(d) 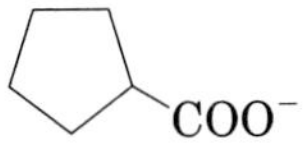 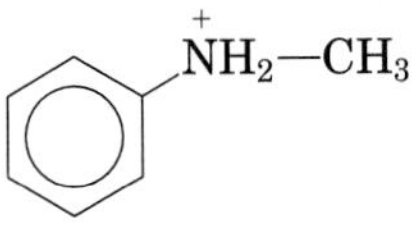 $C_5H_9-COO^-$ (cyclopentyl) $\quad C_6H_5-\overset{+}{N}H_2-CH_3$

15.41 (a) NH_3 (b) NH_4^+ (c) 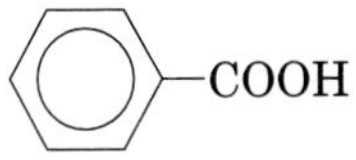 C_6H_5-COOH

(d) 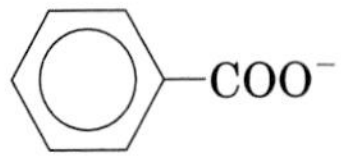 $C_6H_5-COO^-$

15.43 (a) caffeine (b) coniine (c) nicotine (d) atropine (e) quinine (f) morphine (g) morphine (h) strychnine (i) cocaine

15.45 Strychnine has seven rings; reserpine has six.

15.47 Epinephrine has three —OH groups not present in methamphetamine.

15.49 (a) carboxylic acid, carboxylic ester (b) amide, phenol (c) carboxylic acid (d) carboxylic acid, ether (e) carboxylic acid, ketone

15.51 $-NH-C(=O)-(CH_2)_4-C(=O)-NH-(CH_2)_4-NH-C(=O)-$

1,4-diaminobutane

15.53 $NH_2-C(=O)-N(H)-H\cdots O(H)-H$, with the carbonyl $O\cdots H-O-H$

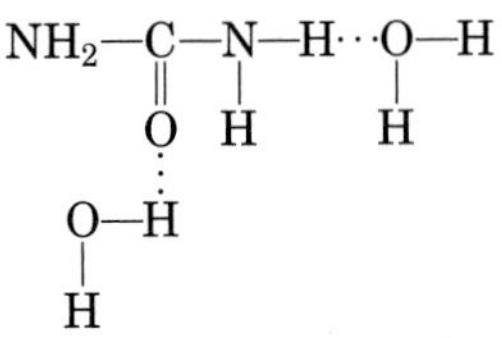

15.55 Ring structure: $C(=O)-NH-C(=O)-NH-C(=O)$ joined through a central C; the $N-C(=O)-N$ part is circled and labelled "urea portion"

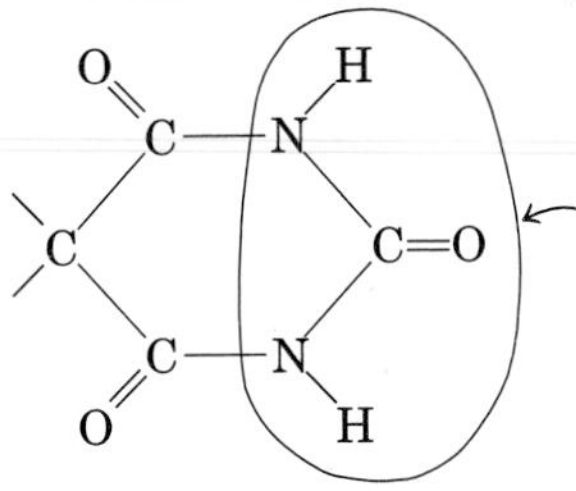

15.57 a seven-membered ring containing two nitrogens

15.59 a cheap, smokable form of cocaine

15.61 Pyridine ring bonded to a pyrrolidine ring whose nitrogen carries H_3C and H, $\overset{+}{N}$, Cl^-

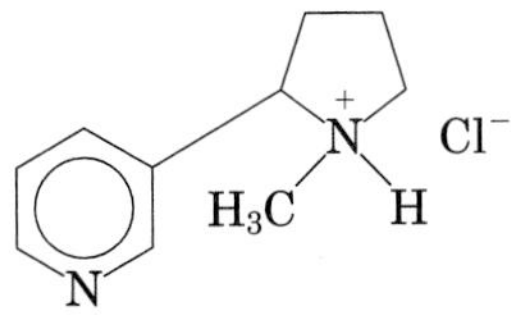

15.63 See the second part of Sec. 15.4.

15.65 primary: $CH_3CH_2CH_2NH_2$, $CH_3-CH(CH_3)-NH_2$

secondary: $CH_3CH_2-NH-CH_3$

tertiary: $CH_3-N(CH_3)-CH_3$

15.67 (a) $HCOOH + CH_3NH_3^+$

(b) $CH_3CH_2COO^- + NH_3$

(c) $CH_3(CH_2)_8COO^- + CH_3-NH-CH_3$

(d) 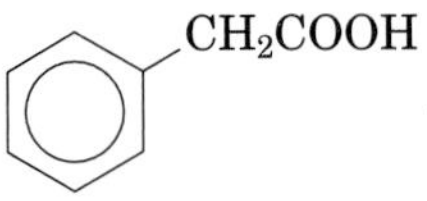 $C_6H_5-CH_2COOH + CH_3CH_2NH_3^+$

(e) $NH_2CH_2CH_2CH_2CH_2COO^-$

15.69 (a) $CH_3NH_3^+$ (b) 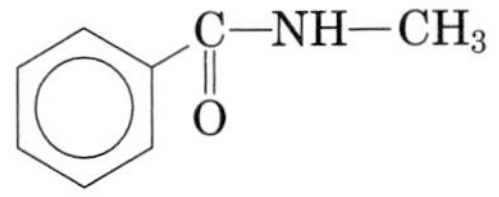 $C_6H_5-C(=O)-NH-CH_3$

15.71 The effect (enlarged pupils) lasts for hours.

15.73 The odors resemble that of ammonia (but not as sharp or pungent), and that of raw fish.

CHAPTER 16 Carbohydrates

16.1 (b); the right carbon in (d)

16.2 (a) They are mirror images. They are not identical.
(b) They are mirror images. They are identical.

16.3 (a) 4 (b) 2 (c) 8 (d) 2

16.4

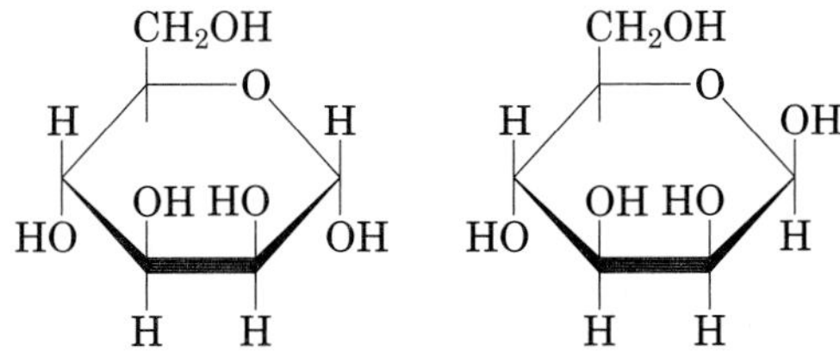

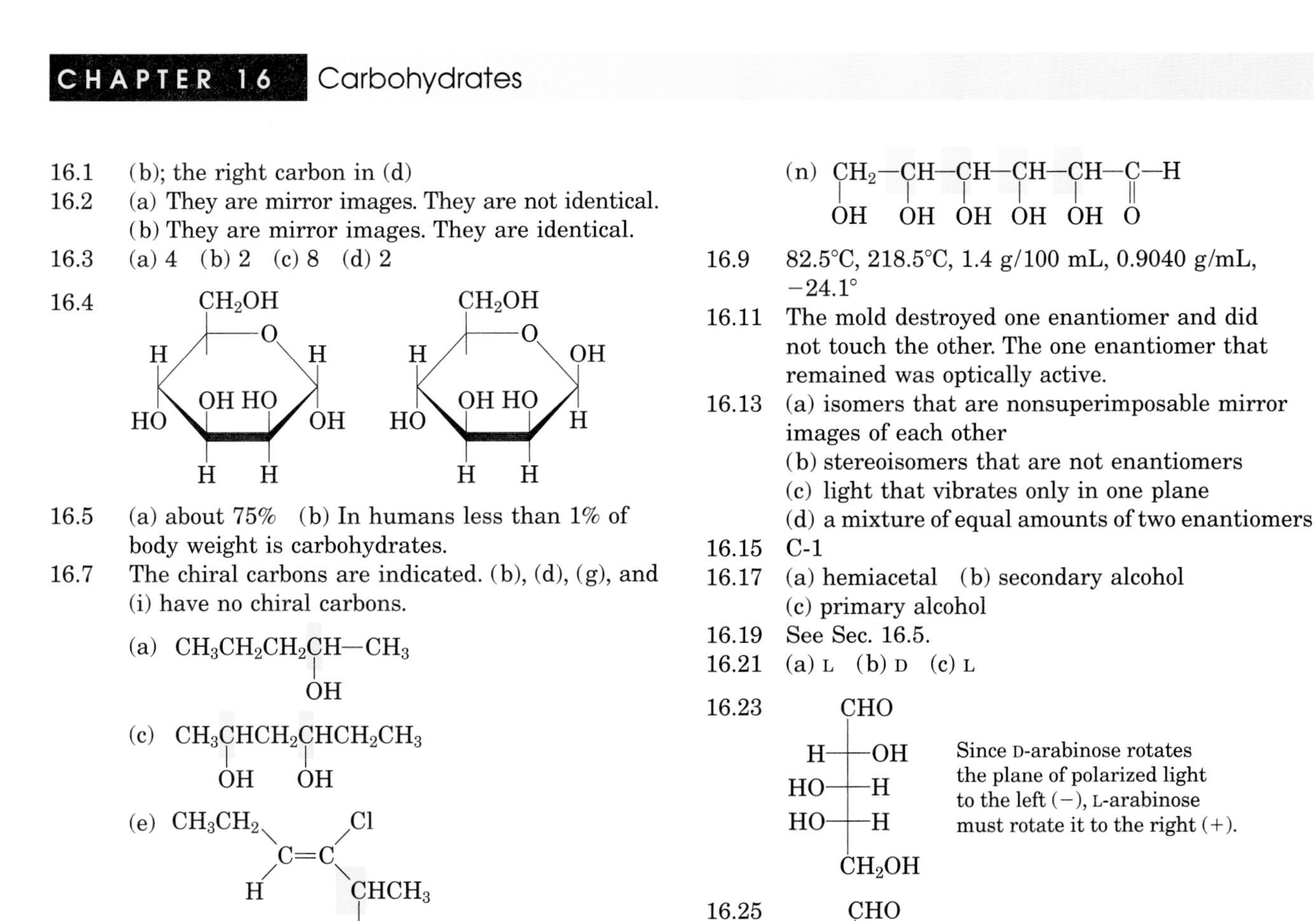

16.5 (a) about 75% (b) In humans less than 1% of body weight is carbohydrates.

16.7 The chiral carbons are indicated. (b), (d), (g), and (i) have no chiral carbons.

(a) 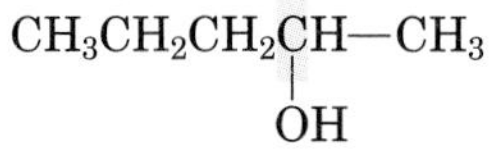 $CH_3CH_2CH_2CH(OH)—CH_3$

(c) 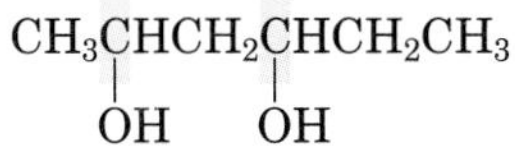 $CH_3CH(OH)CH_2CH(OH)CH_2CH_3$

(e) 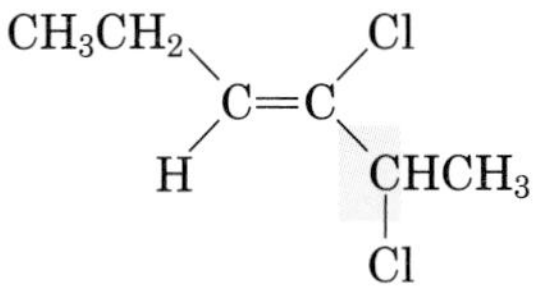 $CH_3CH_2(H)C{=}C(Cl)CH(Cl)CH_3$

(f) $CH_3CH_2CH(OH)CH_2CH_2Br$

(h) 1,2-dibromocyclopentane (Br, Br)

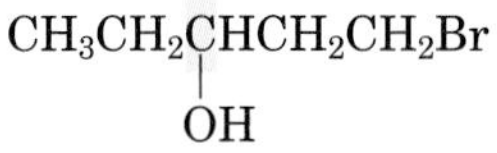

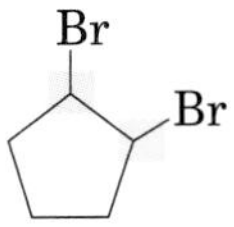

(j) 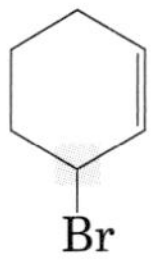 cyclohexene ring with Br

(k) 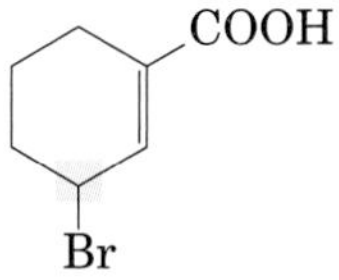 cyclohexene ring with COOH and Br

(l) 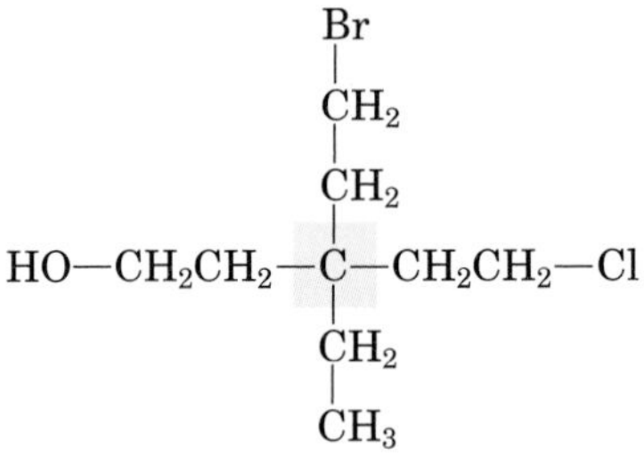 $HO—CH_2CH_2—C(CH_2CH_2Br)(CH_2CH_3)—CH_2CH_2—Cl$

(m) 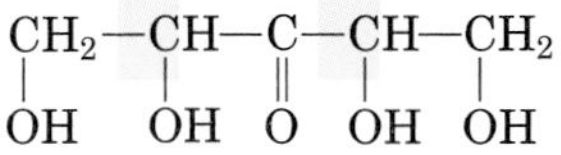 $CH_2(OH)—CH(OH)—C({=}O)—CH(OH)—CH_2(OH)$

(n) 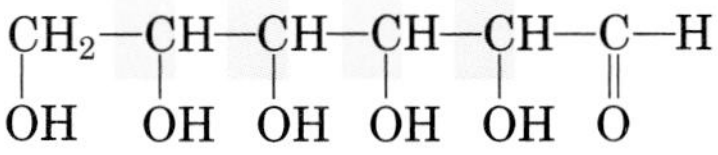 $CH_2(OH)—CH(OH)—CH(OH)—CH(OH)—CH(OH)—C({=}O)—H$

16.9 82.5°C, 218.5°C, 1.4 g/100 mL, 0.9040 g/mL, −24.1°

16.11 The mold destroyed one enantiomer and did not touch the other. The one enantiomer that remained was optically active.

16.13 (a) isomers that are nonsuperimposable mirror images of each other
(b) stereoisomers that are not enantiomers
(c) light that vibrates only in one plane
(d) a mixture of equal amounts of two enantiomers

16.15 C-1

16.17 (a) hemiacetal (b) secondary alcohol
(c) primary alcohol

16.19 See Sec. 16.5.

16.21 (a) L (b) D (c) L

16.23

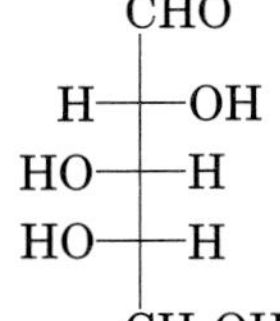

Since D-arabinose rotates the plane of polarized light to the left (−), L-arabinose must rotate it to the right (+).

16.25

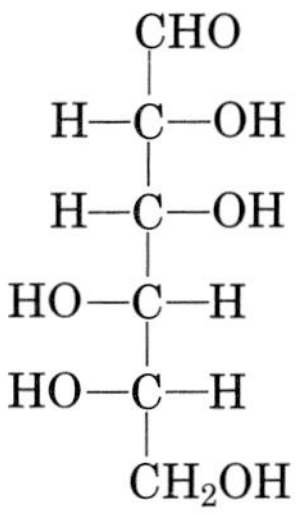

CHO
H—C—OH
H—C—OH
HO—C—H
HO—C—H
CH_2OH

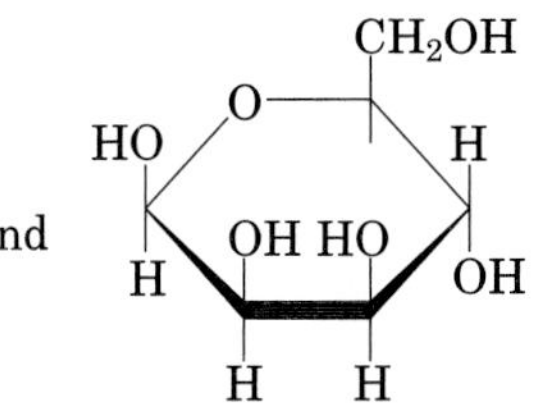

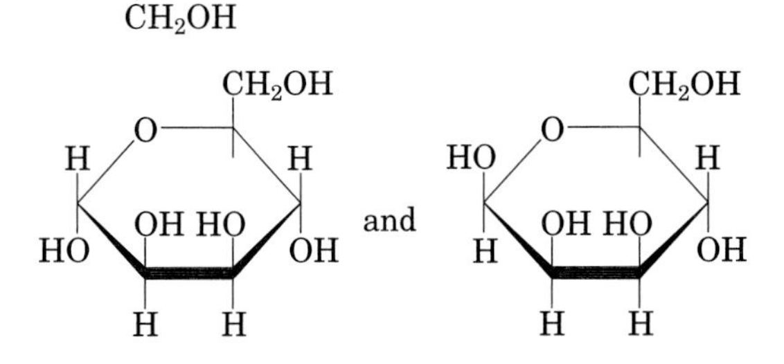

16.27 (a) to the right (b) No, it must be measured.

16.29

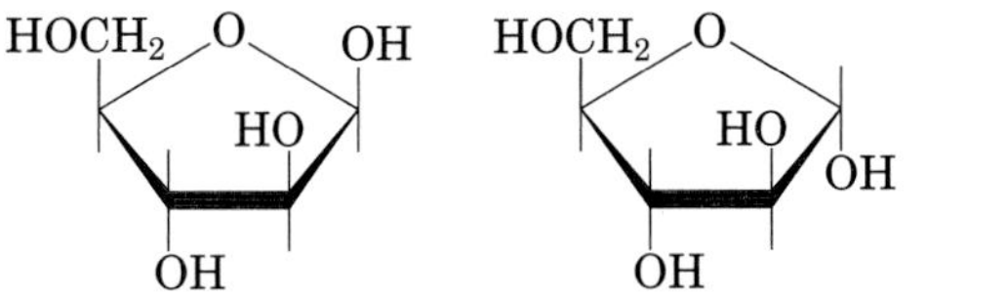

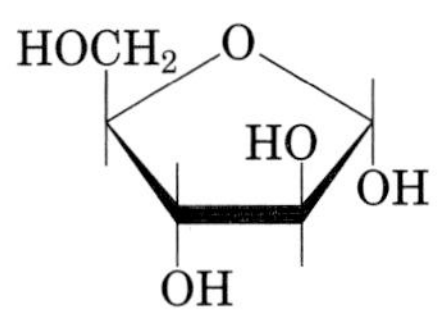

16.31 (a) an aldose (b) a furanose (c) a pentose

16.33 (a) $HOCH_2$ O OCH_2CH_3; OH OH and $HOCH_2$ O OCH_2CH_3; OH OH

(b) $\underset{OH}{CH_2}-\underset{OH}{CH}-\underset{OH}{CH}-\underset{OH}{CH}-\underset{OH}{CH_2}$

(c) CH_2OH, HO, O, OCH_3, OH, OH and CH_2OH, HO, O, OH, OCH_3, OH

16.35 CH_2OH, HO, O, OH, OH, NH, C=O, CH_3

16.37 CH_2OH, O, OH, HO, OH, O, OH, OH, $HOCH_2$, O, OH

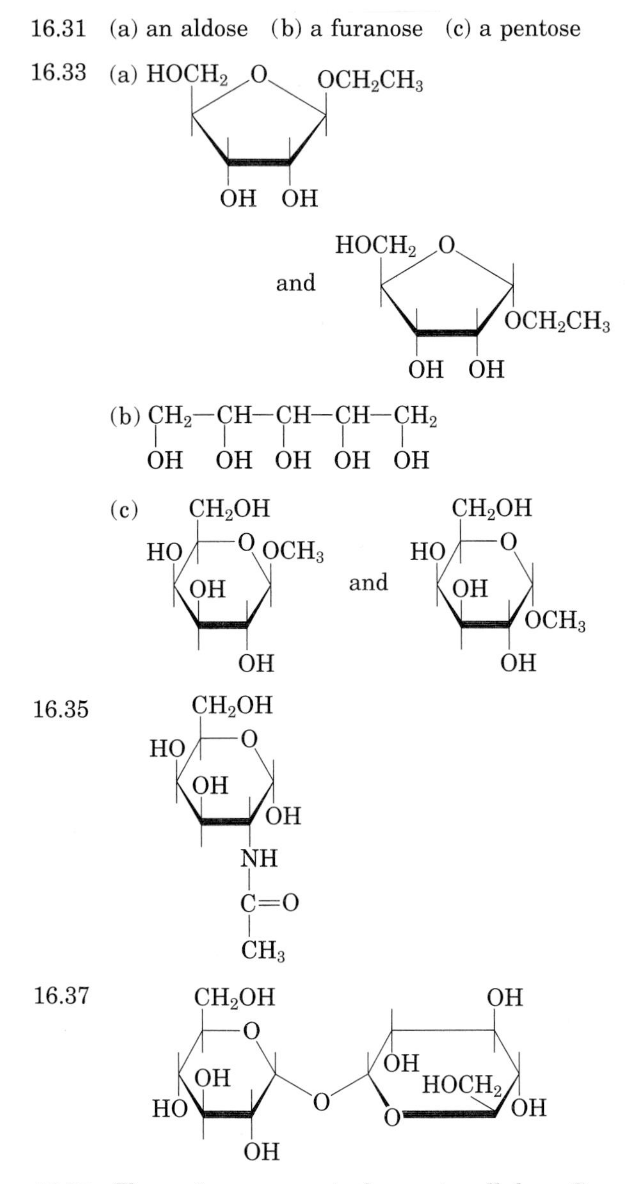

16.39 The main component of grass is cellulose. Cows have microorganisms in their stomachs that possess the enzyme cellulase, which hydrolyzes cellulose. Humans do not have these microorganisms.

16.41 CH_2OSO_3H, O, OH, HO, OSO_3H, O [CH_2OSO_3H, O, OH, OH]$_n$ O, CH_2OH, O, OH, OSO_3H, OSO_3H

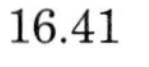

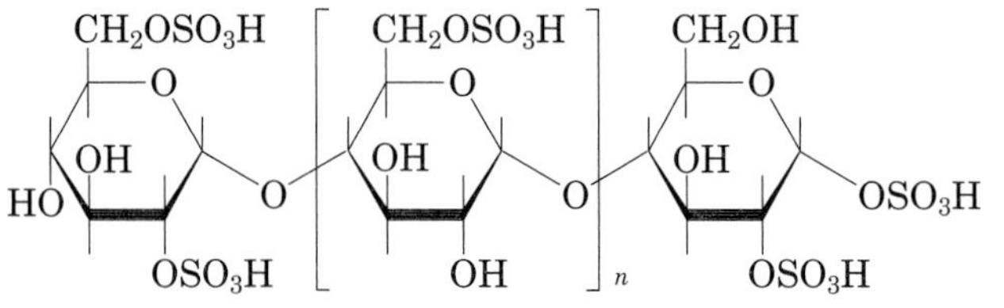

16.43 (1) glucuronic acid (2) D-glucosamine (3) acetic acid

16.45 (a) COO^-, OSO_3^-, $NHSO_3^-$ (b) high degree

16.47 Sweetness depends on the proper fitting of molecules into receptors on the taste buds. Sucrose fits the receptors. Its enantiomer does not.

16.49 (a) OH, CH_2, H, OH, O, H, OH, H, H, OH, OH, H

(b) HO, CH_2, H, HO, O, OH, H, H, HO, H, H, OH

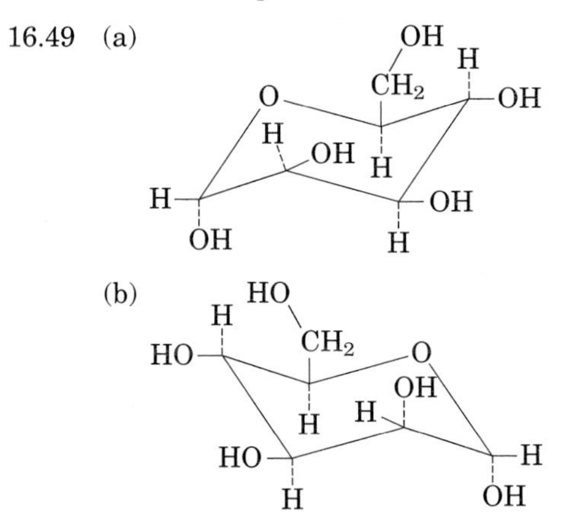

16.51 (a) yes (b) pyranose (c) it has no OH group on C-6.

16.53 With maturation children develop another enzyme that metabolizes galactose.

16.55 By influencing the rate of delivery. If the filler disintegrates rapidly in the stomach, it can deliver the drug rapidly.

16.57 The (+) form is absorbed faster than the racemic mixture, thus reaching a therapeutic concentration faster.

16.59 See Sec. 16.11.

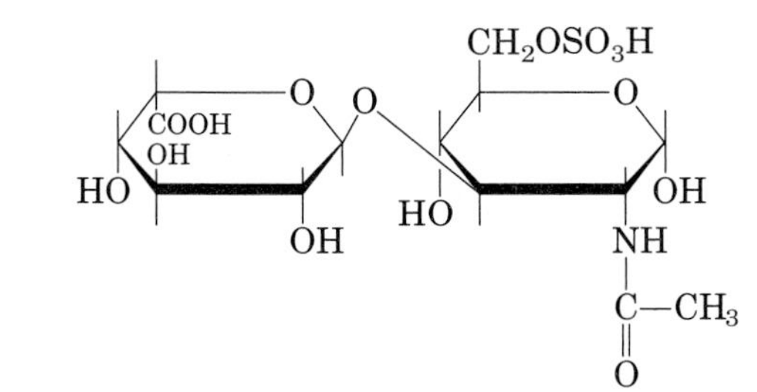

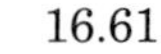

16.61 COOH, O, OH, HO, OH, O, CH_2OSO_3H, O, HO, OH, NH, $\overset{O}{\overset{\|}{C}}-CH_3$

16.63 CH_2OH, O, OH, HO, OH, $O-CH_2$, O, OH, HO, OH, OH

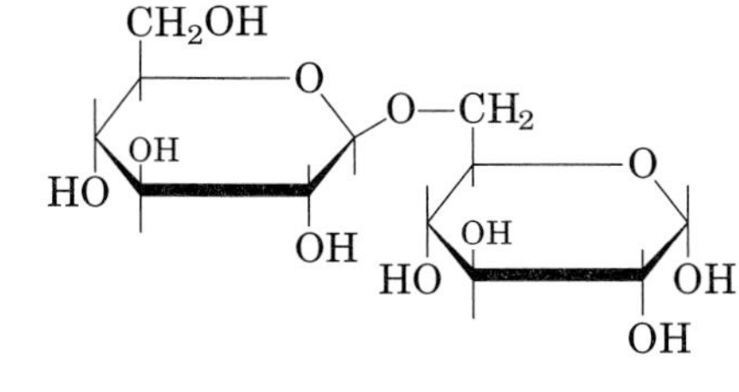

16.65 It is a racemic mixture.

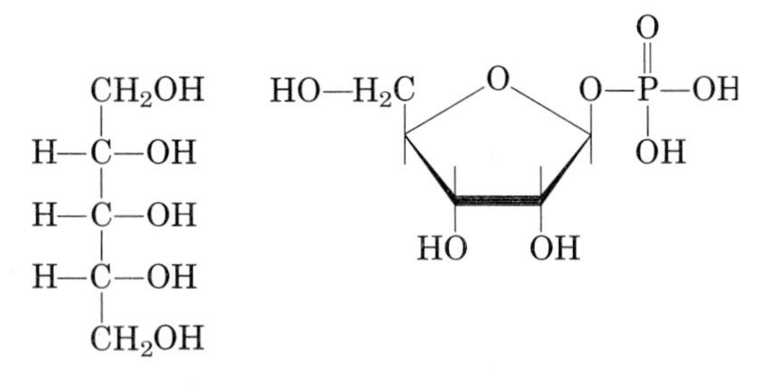

16.67 CH_2OH, $H-C-OH$, $H-C-OH$, $H-C-OH$, CH_2OH

$HO-H_2C$, O, $O-\overset{O}{\overset{\|}{P}}-OH$, OH, HO, OH

16.69 It is a catalyst.

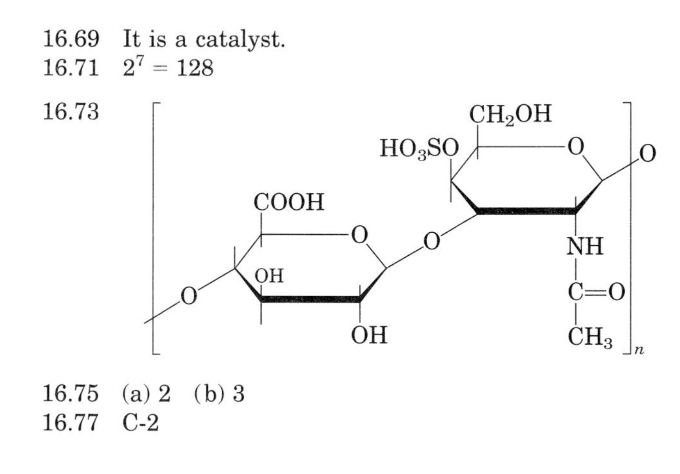

16.71 $2^7 = 128$

16.73

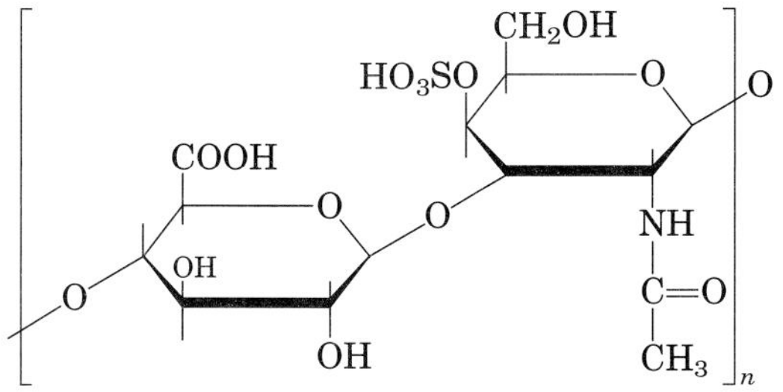

CH_2OH, HO_3SO, O, O, COOH, O, O, OH, O, NH, C=O, CH_3, OH, $]_n$

16.75 (a) 2 (b) 3

16.77 C-2

CHAPTER 17 Lipids

17.1 It is an ester of glycerol and contains a phosphate group; therefore it is a glycerophospholipid. Besides glycerol and phosphate it has a myristic acid and a linoleic acid component. The other alcohol is serine. Therefore, it belongs to the subgroup of cephalins.

17.3 Hydrophobic means water hating. It is important because if the body did not have such molecules there could be no structure since the water would dissolve everything.

17.5 There are five:

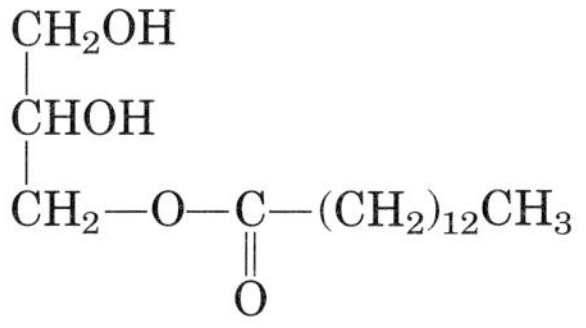

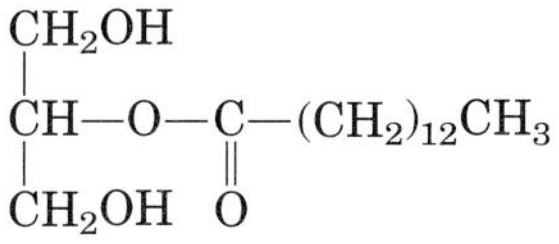

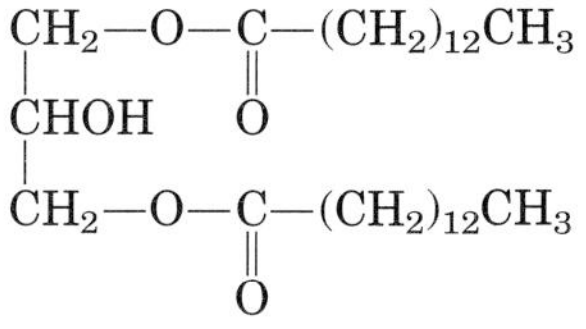

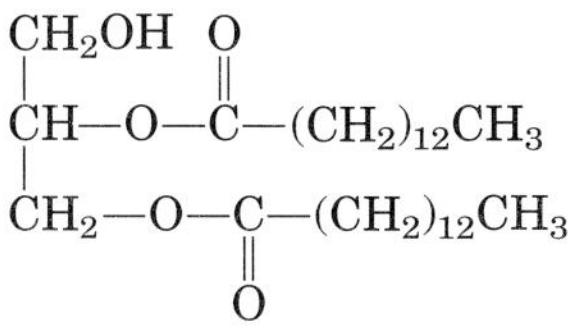

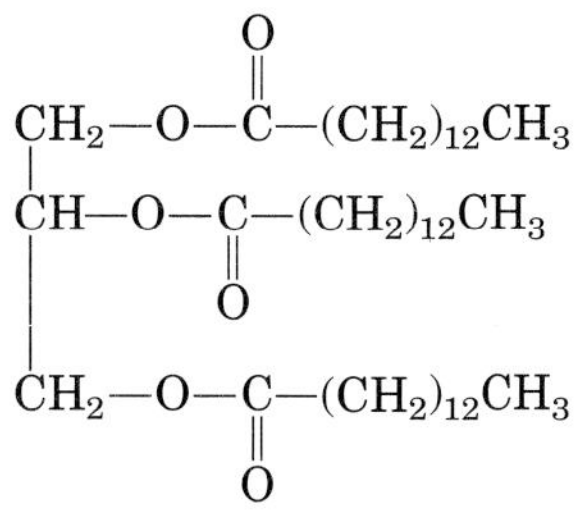

CH_2OH
$CHOH$
CH_2—O—C(=O)—$(CH_2)_{12}CH_3$

CH_2OH
CH—O—C(=O)—$(CH_2)_{12}CH_3$
CH_2OH

CH_2—O—C(=O)—$(CH_2)_{12}CH_3$
CHOH
CH_2—O—C(=O)—$(CH_2)_{12}CH_3$

CH_2OH
CH—O—C(=O)—$(CH_2)_{12}CH_3$
CH_2—O—C(=O)—$(CH_2)_{12}CH_3$

CH_2—O—C(=O)—$(CH_2)_{12}CH_3$
CH—O—C(=O)—$(CH_2)_{12}CH_3$
CH_2—O—C(=O)—$(CH_2)_{12}CH_3$

17.7 highest: A and B (no double bonds); lowest C and D (two double bonds in each diglyceride)

17.9 (b) because its molecular weight is higher

17.11 lowest (c); then (b); highest (a)

17.13 the more long chain groups the lower the solubility; lowest (a); then (b); highest (c)

17.15 glycerol, sodium palmitate, sodium stearate, and sodium linolenate

17.17 (a) They are found around cells and around small structures inside cells. (b) They separate cells from the external environment and allow selective passage of nutrients and waste products into and out of cells.

17.19 complex lipids and cholesterol

17.21

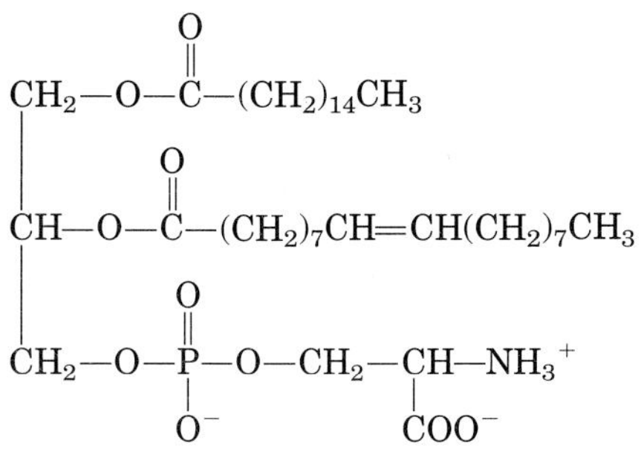

CH_2—O—C(=O)—$(CH_2)_{14}CH_3$
CH—O—C(=O)—$(CH_2)_7CH{=}CH(CH_2)_7CH_3$
CH_2—O—P(=O)(O^-)—O—CH_2—CH(COO^-)—NH_3^+

17.23 (c) because it has three charges, while the others have only two

17.25 No. For example, in red blood cells lecithins are on the outside, facing the plasma; cephalins are on the inside.

17.27 (a) yes (b) because high serum cholesterol has been correlated with such diseases as atherosclerosis

17.29 (a) Chiral carbons are circled.

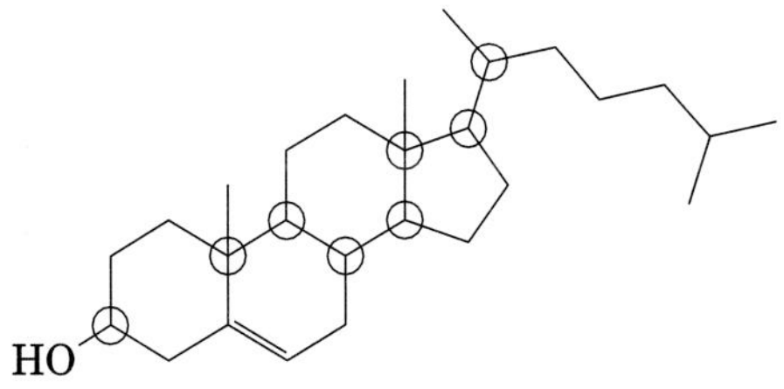

HO

(b) $2^8 = 256$ (c) only 1

17.31 $CH_3(CH_2)_4CH{=}CHCH_2CH{=}CH(CH_2)_7{-}C({=}O){-}O{-}$

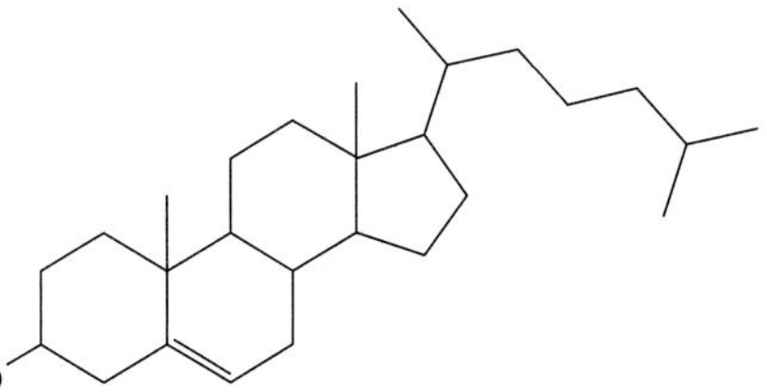

17.33 It increases the glucose and glycogen concentrations in the body. It is also an anti-inflammatory agent.

17.35 Loss of a methyl group at the junction of rings A and B, together with a hydrogen at position 1, produces a double bond. This causes tautomerization of the keto group at C-3, resulting in an aromatic ring A, with a phenolic group at the C-3 position.

17.37 (a) They both have a steroid ring structure. (b) RU486 has a *para*-aminophenyl group on ring C and a triple-bond group on ring D.

17.39 (a) PGE_2 has a ring, a ketone group, and two OH groups. (b) The C=O group on PGE_2 is an OH group on $PGF_{2\alpha}$.

17.41 Prostaglandins contain a 5-membered ring; leukotrienes have no ring.

17.43 (a) oxidation of the double bond to aldehydes and other compounds (b) exclude oxygen and sunlight; keep refrigerated

17.45 The hydrophobic tail of the soap penetrates the fat globules, and the hydrophilic head stays on the surface to interact with the water and solubilize the grease particles.

17.47 (a) Sphingomyelin acts as an insulator. (b) The insulator is degraded, impairing nerve conduction.

17.49 It blocks cholesterol synthesis, thus helping to remove cholesterol from the blood.

17.51 They prevent ovulation.

17.53 It inhibits prostaglandin formation by preventing ring closure.

17.55 (a) in the liver (b) as bile salts

17.57 (See Fig. 17.2.) Polar molecules cannot penetrate the bilayer. They are insoluble in lipids. Nonpolar molecules can interact with the interior of the bilayer (like dissolves like).

17.59 The various esters of cholesterol have different fatty acids (saturated, unsaturated, etc.). This prevents them from packing into a crystal lattice. Cholesterol, being uniform, does form a crystal lattice.

17.61 They are both salts of sulfonic acids.

17.63 (a) aldehyde, ketone, alcohol, C=C double bond (b) Cortisone has all these groups except the aldehyde.

17.65 See Sec. 17.10.

CHAPTER 18 Proteins

18.1

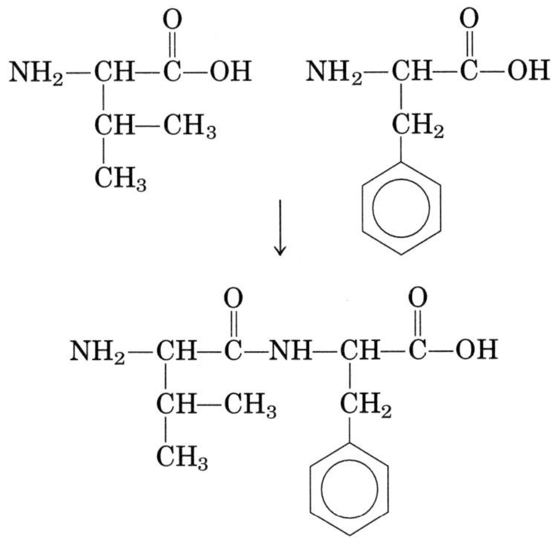

$$NH_2-CH(CH(CH_3)-CH_3)-C(=O)-OH \quad NH_2-CH(CH_2-C_6H_5)-C(=O)-OH$$

$$\downarrow$$

$$NH_2-CH(CH(CH_3)-CH_3)-C(=O)-NH-CH(CH_2-C_6H_5)-C(=O)-OH$$

18.2 hydrophobic interactions

18.3 Most proteins are highly specific in their actions, so a great many are needed to do all the tasks required.

18.5 acidic: glutamic acid, aspartic acid; basic: lysine, arginine, histidine

18.7 arginine

18.9

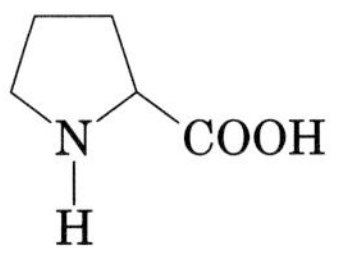

(pyrrolidine ring, N–H, with COOH on the carbon adjacent to N)

pyrrolidines (saturated pyrroles)

18.11 They supply most of the amino acids we need in our bodies.

18.13 Their structures are the same, except that a hydrogen of alanine is replaced by a phenyl group in phenylalanine.

18.15

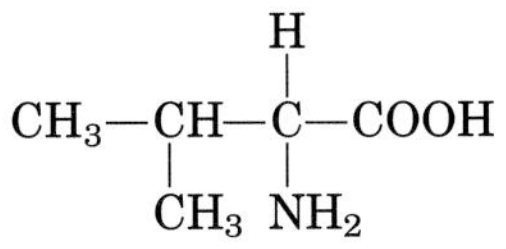

$$CH_3-CH(CH_3)-C(H)(NH_2)-COOH$$

L-Valine

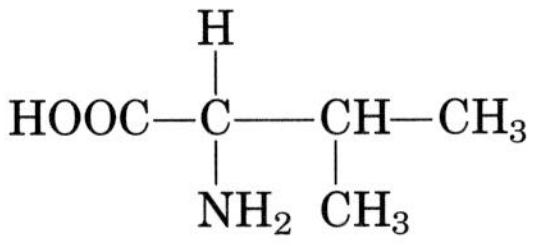

$$HOOC-C(H)(NH_2)-CH(CH_3)-CH_3$$

D-Valine

18.17 $CH_3-CH(NH_3^+)-COO^- + H_3O^+ \longrightarrow CH_3-CH(NH_3^+)-COOH + H_2O$

$CH_3-CH(NH_3^+)-COO^- + OH^- \longrightarrow CH_3-CH(NH_2)-COO^- + H_2O$

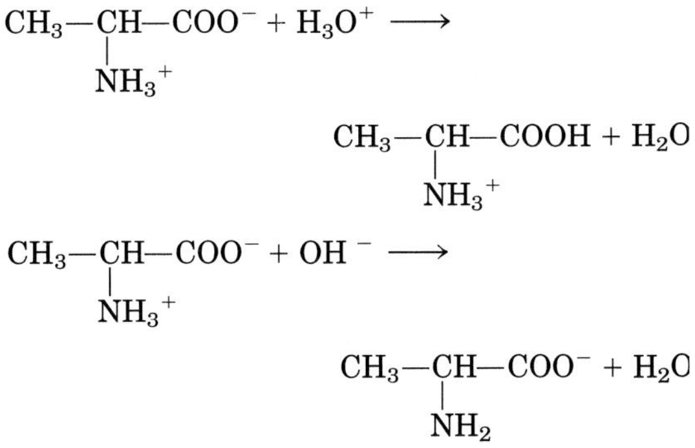

18.19 at pH 1 $CH_3-CH(CH_3)-CH(NH_3^+)-COOH$

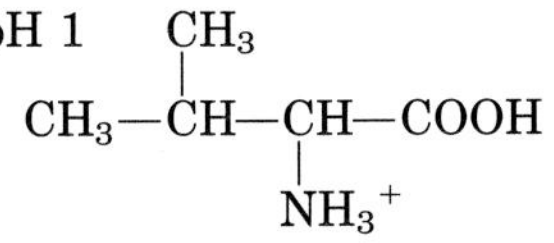

at pH 12 $CH_3-CH(CH_3)-CH(NH_2)-COO^-$

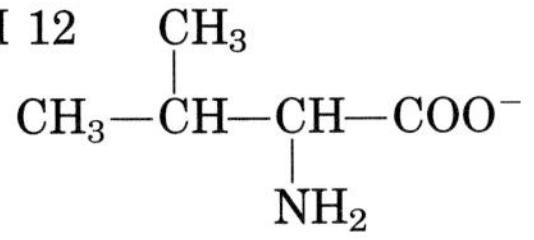

18.21 There are six. One of them is

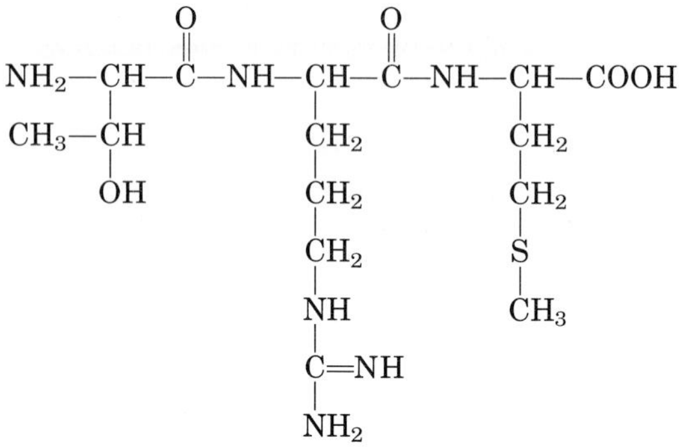

```
             O                O
             ‖                ‖
NH2—CH—C—NH—CH—C—NH—CH—COOH
     |          |          |
CH3—CH         CH2        CH2
     |          |          |
     OH        CH2        CH2
                |          |
               CH2         S
                |          |
               NH         CH3
                |
               C=NH
                |
               NH2
```

18.23

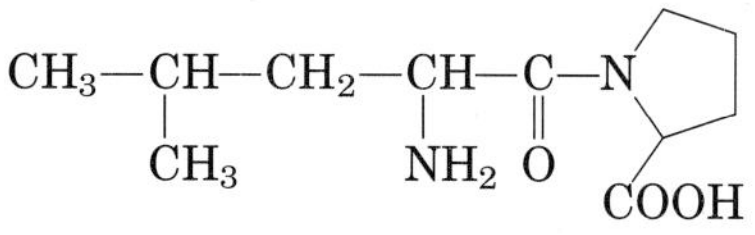

```
CH3—CH—CH2—CH—C—N⟨pyrrolidine ring⟩
     |         |   ‖      \
    CH3       NH2  O      COOH
```

18.25

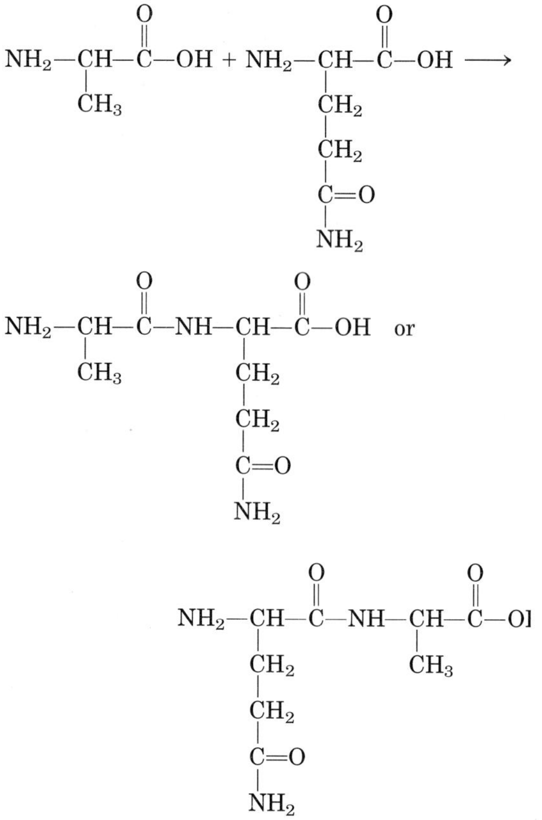

```
           O                      O
           ‖                      ‖
NH2—CH—C—OH + NH2—CH—C—OH ⟶
     |                 |
    CH3               CH2
                       |
                      CH2
                       |
                      C=O
                       |
                      NH2

           O           O
           ‖           ‖
NH2—CH—C—NH—CH—C—OH   or
     |          |
    CH3        CH2
                |
               CH2
                |
               C=O
                |
               NH2

           O           O
           ‖           ‖
NH2—CH—C—NH—CH—C—OH
     |          |
    CH2        CH3
     |
    CH2
     |
    C=O
     |
    NH2
```

18.27 at pH 2.0

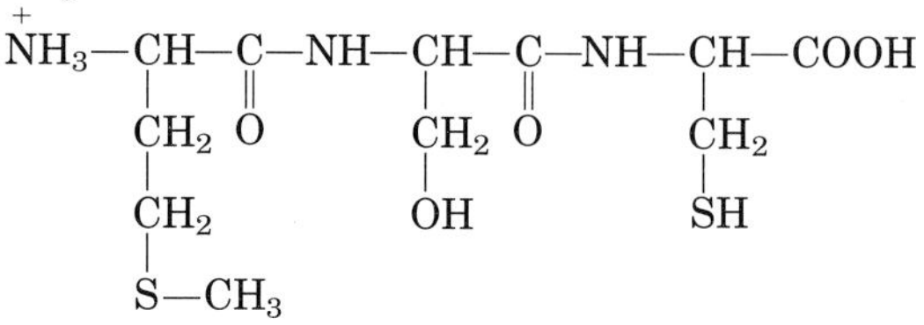

```
+
NH3—CH—C—NH—CH—C—NH—CH—COOH
     |  ‖       |  ‖       |
    CH2 O      CH2 O      CH2
     |          |          |
    CH2        OH         SH
     |
    S—CH3
```

at pH 7.0

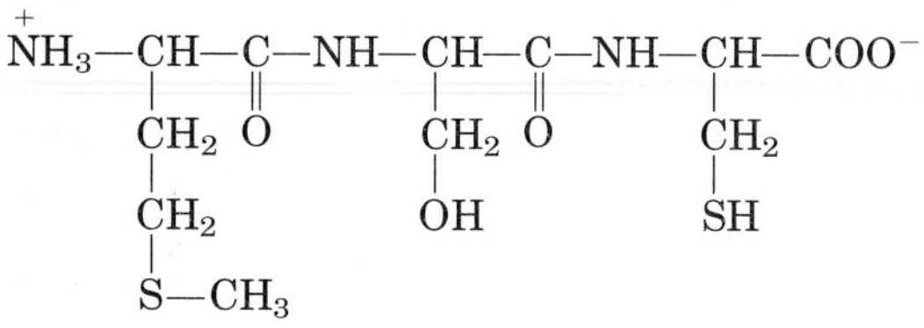

```
+
NH3—CH—C—NH—CH—C—NH—CH—COO⁻
     |  ‖       |  ‖       |
    CH2 O      CH2 O      CH2
     |          |          |
    CH2        OH         SH
     |
    S—CH3
```

at pH 10.0

```
NH2—CH—C—NH—CH—C—NH—CH—COO⁻
     |  ‖       |  ‖       |
    CH2 O      CH2 O      CH2
     |          |          |
    CH2        OH         SH
     |
    S—CH3
```

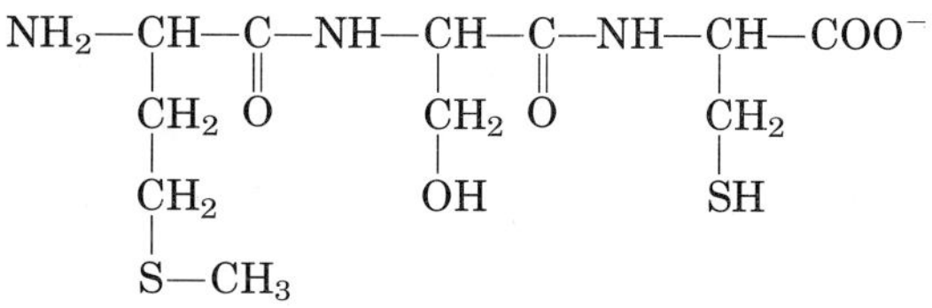

18.29 It would acquire a net positive charge and become more water-soluble.

18.31 (a) 24 (b) $20^4 = 160\ 000$

18.33 valine or isoleucine

18.35 (a) secondary (b) tertiary and quaternary (c) quaternary (d) primary

18.37 Above pH 6.0 the COOH groups are converted to COO^- groups. The negative charges repel each other, disrupting the compact α-helix and converting it to a random coil.

18.39 (1) C-terminal end (2) N-terminal end (3) pleated sheet (4) random coil (5) hydrophobic interaction (6) disulfide bridge (7) α-helix (8) salt bridge (9) hydrogen bonds

18.41 valine, phenylalanine, alanine, leucine, isoleucine

18.43 a protein that contains a non–amino acid portion called a prosthetic group

18.45 No. Sometimes a helper molecule, called a chaperone, is needed to attain the biologically active species.

18.47 (a) The sequence of amino acids in the two compounds is reversed. (b) no

18.49 See Box 18C.

18.51 because this mutation provided resistance to malaria in heterozygotes

18.53 Tertiary. The cross-linking stabilizes the tertiary structure between adjacent chains.

18.55 Ag^+ denatures the proteins of bacteria and kills them.

18.57 See Box 18H.

18.59 yes

18.61 no

18.63 Yes. Any change that destroys native protein structure without hydrolyzing the chain is a denaturation.

18.65

```
(a) NH2—(CH2)4—CH—COO⁻
                |
               NH2

     +
(b) NH3—(CH2)4—CH—COOH
                |
               NH3⁺
```

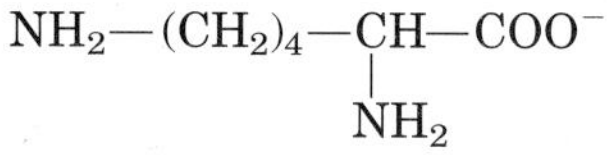

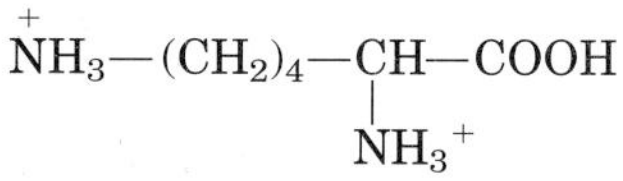

(c) $NH_2-(CH_2)_4-\underset{\displaystyle NH_3^+}{\underset{|}{CH}}-COO^-$

18.67 No; that number is $20^{60} = 10^{78}$, even if we limit it to 60 amino acids. This number is possibly greater than the total number of atoms in the universe.

18.69 $20^{10} = 1.0 \times 10^{13}$

18.71 $\overset{+}{N}H_3-\underset{\displaystyle CH_2CH_2-S-CH_3}{\underset{|}{CH}}-COOH$ + $\overset{+}{N}H_3-CH(CH_2-CH(CH_3)-CH_3)-COOH$ + $\overset{+}{N}H_3-\underset{\displaystyle CH_2-COOH}{\underset{|}{CH}}-COOH$

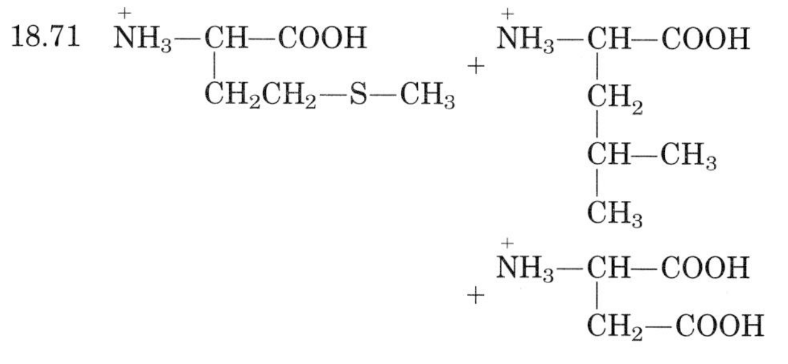

CHAPTER 19 Enzymes

19.1 A catalyst is a substance that speeds up a reaction without being used up itself. An enzyme is a catalyst that is a protein or a nucleic acid.

19.3 more than 3000

19.5 because enzymes are highly specific, and there are a great many different reactions that must be catalyzed

19.7 Trypsin catalyzes the hydrolysis of peptides only at specific points (for example, at a lysine), while lipases catalyze the hydrolysis of all kinds of triglycerides.

19.9 (a) isomerase (b) hydrolase (c) oxidoreductase (d) lyase

19.11 A cofactor is a nonprotein portion of an enzyme. A coenzyme is an organic cofactor.

19.13 See Sec. 19.3.

19.15 No, at high substrate concentration the enzyme surface is saturated, and doubling of the substrate concentration will produce only a slight increase in the rate of the reaction or no increase at all.

19.17 (a) less active at normal body temperature (b) The activity decreases.

19.19 It is too bulky to fit into the cavity where the active site is located.

19.21 In competitive inhibition the same maximum rate can be obtained as in the reaction without inhibitor, although at a higher substrate concentration. In noncompetitive inhibition the maximum rate is always lower than that without inhibitor.

19.23 an allosteric mechanism

19.25 There is no difference. They are the same.

19.27 1. feedback control 2. allosterism 3. zymogens 4. isoenzymes (See Sec. 19.6.)

19.29 The enzyme is now named for the compound (alanine) that donates the amino group. The former name refers to the products.

19.31 H_4

19.33 digestive enzymes such as Pancreatin or Acro-lase

19.35 Succinylcholine is a competitive inhibitor of acetylcholinesterase and prevents acetylcholine from triggering muscle contractions.

19.37 Trypsin. Both trypsin and papain convert proteins to peptides. Carboxypeptidase yields amino acids by cleaving peptide bonds starting at the C-terminal end.

19.39 It binds to the K^+ and to the Mg^{2+} cofactors.

19.41 Humans do not synthesize their folic acid; they get it in their diets.

19.43 It inhibits penicillinase, which bacteria synthesize to deactivate penicillin.

19.45 (a) pasteurized milk, canned tomatoes (b) yogurt, pickles

19.47 (a)

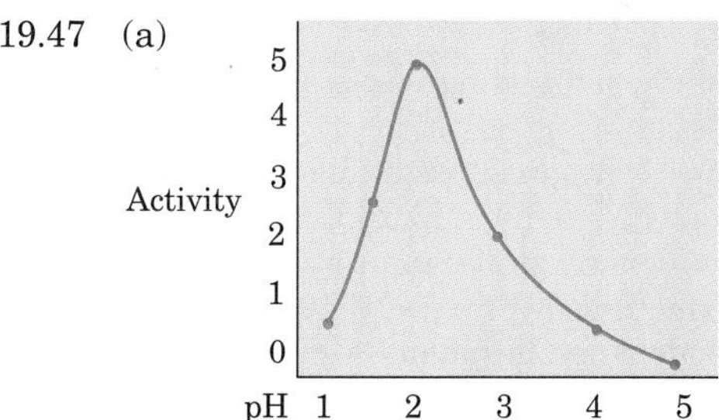

(b) 2 (c) zero activity

19.49 cofactors

19.51 ALT

19.53 No, the nerve gas binds irreversibly by a covalent bond, so it can't be removed simply by adding substrate.

19.55 a transferase

19.57 from 10^9 to 10^{20} times

CHAPTER 20 Bioenergetics. How the Body Converts Food to Energy

20.1 Metabolism is the sum total of all the chemical reactions involved in maintaining the dynamic state of the cell. Catabolism is the breaking down of molecules to supply energy.
20.3 mitochondria
20.5 the matrix
20.7 the phosphoric anhydride linkage
20.9 Neither; they yield the same energy.
20.11 No. The body is able to extract only from 40 to 60 percent of the caloric value in the form of ATP.
20.13 the 2 N atoms that are part of C=N bonds
20.15 (a) adenosine (b) NAD^+ and FAD (c) acetyl groups
20.17 thiol, amide, hydroxy, phosphoric anhydride, phosphoric ester, ether, amino
20.19 No. The pantothenic acid portion is not the active part.
20.21 acetyl coenzyme A
20.23 α-ketoglutarate
20.25 oxidative decarboxylation
20.27 It removes two hydrogens from succinate to produce fumarate.
20.29 No, but GTP is produced in step ⑤.
20.31 It allows the energy to be released in small packets.
20.33 *cis*-aconitate and fumarate
20.35 to NAD^+, which becomes $NADH + H^+$
20.37 (a) 3 (b) 2
20.39 in the inner membranes of the mitochondria
20.41 (a) 0.5 (b) 12
20.43 the proton translocating ATPase
20.45 1.5 kcal
20.47 (a) by sliding the thick filaments (myosin) and the thin filaments (actin) past each other (b) from the hydrolysis of ATP
20.49 ATP transfers a phosphate group to the serine residue at the active site of the enzyme.
20.51 No. It would harm humans because they would not synthesize enough ATP molecules.
20.53 (a) hydroxylation (b) the air we breathe in
20.55 60 g or 1 mole of CH_3COOH
20.57 They are both hydroxy acids.
20.59 myosin
20.61 ATP

CHAPTER 21 Specific Catabolic Pathways: Carbohydrate, Lipid, and Protein Metabolism

21.1 monosaccharides
21.3 They serve as building blocks for the synthesis of proteins.
21.5 The two C_3 fragments are in equilibrium, and as the glyceraldehyde phosphate is used up, the equilibrium shifts and converts the other C_3 fragment (dihydroxyacetone phosphate) to glyceraldehyde phosphate.
21.7 (a) steps ① and ③ (b) steps ⑥ and ⑨
21.9 step ⑤
21.11 in the cytosol outside the nucleus
21.13 6 moles
21.15 12
21.17 (a) 3 (b) 2
21.19 The secondary OH group has been oxidized to a carbonyl group.
21.21 because the third carbon of the chain (the β carbon) is oxidized prior to fragmentation
21.23 $CH_3(CH_2)_4CO$—CoA, three acetyl CoA, three $FADH_2$, three $NADH + H^+$
21.25 112
21.27 carbohydrates
21.29 See Sec. 21.7.
21.31 It enters the citric acid cycle.
21.33

$$CH_3-\underset{\underset{NH_3^+}{|}}{CH}-COO^- + NAD^+ + H_2O \longrightarrow$$

$$CH_3-\underset{\underset{O}{\|}}{C}-COO^- + NADH + H^+ + NH_4^+$$

21.35

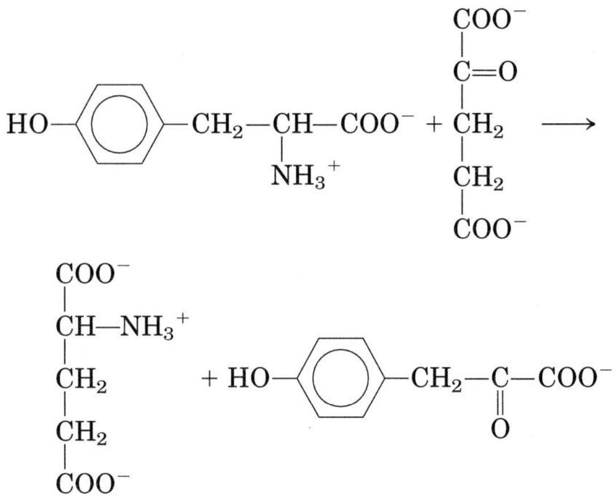

$$HO-C_6H_4-CH_2-\underset{\underset{NH_3^+}{|}}{CH}-COO^- + \begin{array}{c} COO^- \\ | \\ C{=}O \\ | \\ CH_2 \\ | \\ CH_2 \\ | \\ COO^- \end{array} \longrightarrow$$

$$\begin{array}{c} COO^- \\ | \\ CH-NH_3^+ \\ | \\ CH_2 \\ | \\ CH_2 \\ | \\ COO^- \end{array} + HO-C_6H_4-CH_2-\underset{\underset{O}{\|}}{C}-COO^-$$

21.37 fumarate
21.39 ketone bodies
21.41 It is stored in ferritin and reused.
21.43 It cuts off the oxygen supply, enhancing the glycolysis that produces lactate, which contracts the heart muscle.
21.45 the bicarbonate/carbonic acid buffer
21.47 $HOOC—CH_2CH_2—CH(NH_3^+)—COO^-\ Na^+$ or

$Na^+\ ^-OOC—CH_2CH_2—CH(NH_3^+)—COOH$

21.49 Glycine, the C-terminal amino acid of ubiquitin, forms an amide linkage with a lysine residue in the target protein.

21.51 $C_6H_5—CH_2—CH(NH_3^+)—COO^- + {}^-OOC—C(=O)—CH_2—CH_2—COO^- \longrightarrow C_6H_5—CH_2—C(=O)—COO^- + {}^-OOC—CH(NH_3^+)—CH_2—CH_2—COO^-$

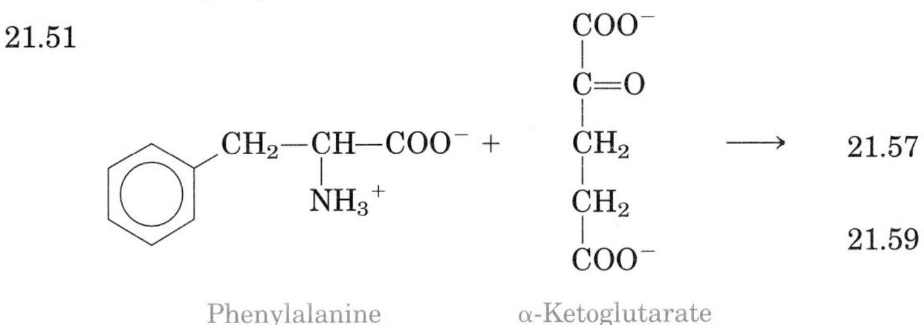

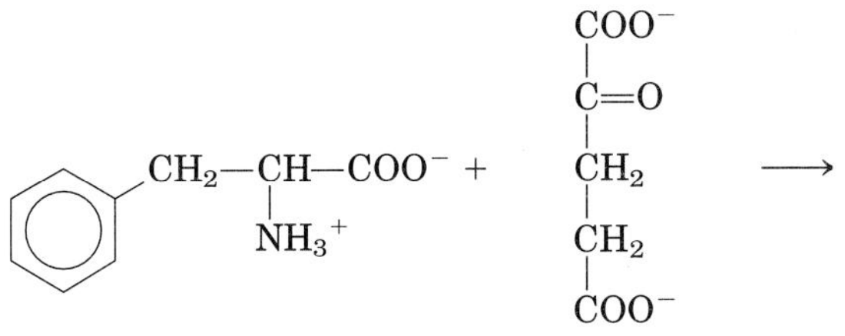

Phenylalanine α-Ketoglutarate 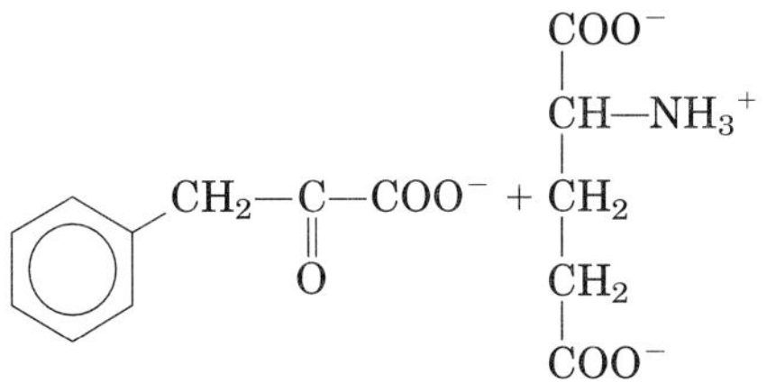 Phenylpyruvate Glutamate

21.53 It is first hydrolyzed to glucose and fructose. Metabolism of these compounds is shown in Figure 21.4.

21.55 (a) in steps ⑤ and ⑫

(b) in steps ⑪ and ⑩

(c) If enough O_2 is present in muscles, a net increase of NADH + H^+ occurs. If not enough O_2 is present (as in yeast), there is no net increase of either NAD^+ or NADH + H^+.

21.57 Yes, they are glucogenic and can be converted to glucose.

21.59 $CH_3—C(=O)—COO^- + {}^-OOC—CH(NH_3^+)—CH_2—COO^-$

21.61 in the carbon dioxide exhaled by the animal
21.63 yes, in small quantities

CHAPTER 22 Biosynthetic Pathways

22.1 for flexibility and to overcome unfavorable equilibria
22.3 because the presence of a large inorganic phosphate pool would shift the reaction to the degradation process so that no substantial amount of glycogen would be synthesized
22.5 the aerobic metabolism of carbohydrates
22.7 (a) pyruvate (b) oxaloacetate (c) alanine
22.9 conversions of (1) pyruvate to oxaloacetate (2) oxaloacetate to phosphoenol pyruvate (3) fructose 1,6-bisphosphate to fructose 6-phosphate (4) glucose 6-phosphate to glucose
22.11 the conversion of oxaloacetate to malate
22.13 uracil, ribose, and three phosphates
22.15 (a) the cytosol (b) no
22.17 malonyl ACP
22.19 malonyl ACP
22.21 It is an oxidation step because hydrogen is removed from the substrate. The oxidizing agent is O_2. NADPH is also oxidized during this step.
22.23 $NADP^+$ has an extra phosphate.
22.25 No, the body makes other unsaturated fatty acids such as oleic and arachidonic.
22.27 palmitoyl CoA, serine, acyl CoA, UDP-glucose
22.29 $CH_2(OH)—C(=O)—COO^-$
22.31 aspartic acid
22.33 valine and α-ketoglutarate
22.35 The chlorophyll captures the energy of light and with its aid strips the electrons and protons from water.
22.37 by oxidation of a saturated fatty acid by certain enzymes and NADPH + H^+
22.39 malonyl ACP
22.41 No, it can add C_2 fragments to a growing chain up to C_{18}. After that another enzyme system takes over.
22.43 No, cholesterol is constantly synthesized in the liver.

CHAPTER 23 Nucleic Acids and Protein Synthesis

23.1 because it can transmit only its own hereditary information (DNA molecules) to the offspring

23.3 hemophilia, sickle-cell anemia, etc.

23.5 (a) base, sugar, phosphate (b) base, sugar

23.7 Thymine has a methyl group in the 5 position; uracil has a hydrogen there.

23.9 (a)

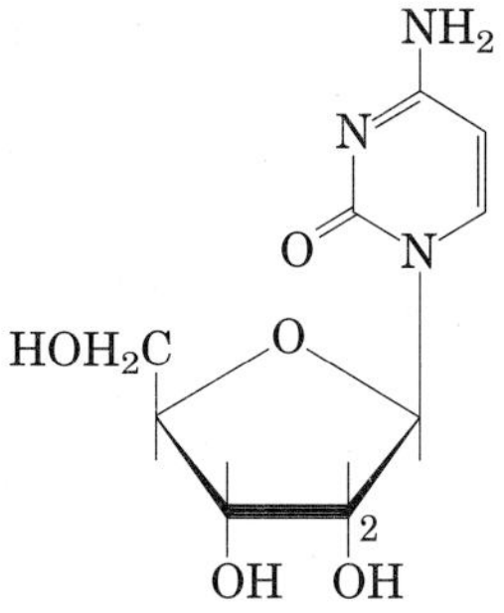

(b) the same, but without the —OH in the 2 position.

23.11 Ribose has an —OH in the 2 position; deoxyribose has an H there.

23.13 C-3 and C-5 to the phosphate group; C-1 to the base

23.15 (a)

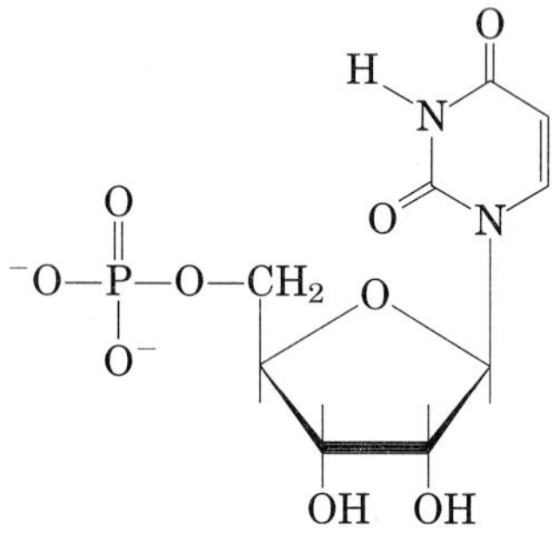

(b)

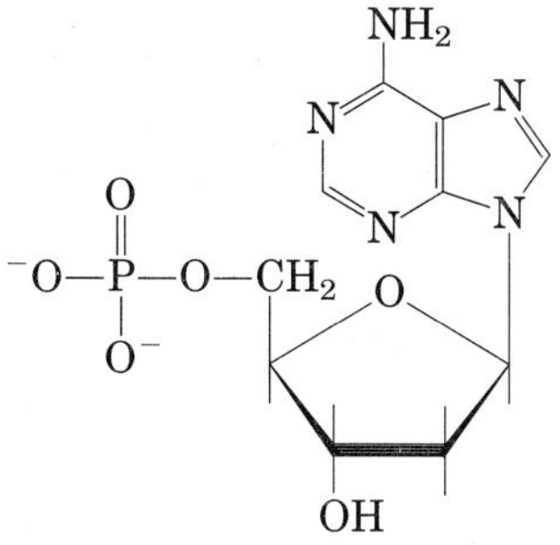

23.17 (a) the 3′ —OH end and the 5′ —OH end (b) A is the 5′ —OH end. C is the 3′ —OH end.

23.19 the phosphoric acid groups in the backbone

23.21 RNA

23.23 amino and carbonyl groups

23.25 See Figure 23.5.

23.27 at either end of the chain or in the middle

23.29 polymerases

23.31 The enzymes can synthesize DNA only in the $5' \longrightarrow 3'$ direction. The leading strand runs this way. The lagging strand runs $3' \longrightarrow 5'$ and must be synthesized in short fragments.

23.33 rRNA

23.35 mRNA

23.37 (a) cloverleaf (b) L-shaped

23.39 in the nucleus

23.41 the recognition site (anticodon) and the 3′ terminal to which the specific amino acid is attached

23.43 (a) CGA (b) alanine

23.45 UAA, UAG, UGA

23.47 They enable the tRNA to bind to the A site of the rRNA.

23.49 An exon is a portion of a DNA molecule that codes for proteins; an intron is a portion that does not.

23.51 (a) about 3 percent (b) Some of it codes for RNA, and some are regulatory sequences, but the function of most of it is unknown.

23.53 DNAs in which short nucleotide sequences are repeated hundreds or thousands of times

23.55 operator and promoter

23.57 In prokaryotes the RNA polymerase binds directly to the promoter. In eukaryotes the RNA polymerase has little affinity for the DNA; it needs transcription factors for binding.

23.59 (a) UAU and UAC. Both code for tyrosine. (b) If UAU is changed to UAA, then instead of tyrosine, the chain will be terminated. This might well be fatal.

23.61 the ends of a strand of DNA that has been cut by endonucleases to leave several free bases that can pair up with a complementary section

23.63

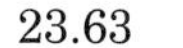

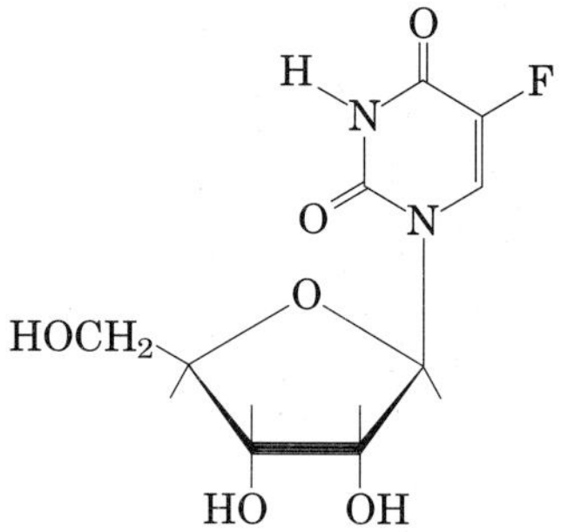

23.65 They do not have the molecules necessary to reproduce themselves and must rely on their hosts.

23.67 T (lymphocyte) cells

23.69 See Box 23D.

23.71 A change from guanine to thymine in the gene results in a valine in place of a glycine.

23.73 The DNA fragments of the child, mother, and alleged father are run on the same gel and their patterns are compared.

23.75 Human DNA contains introns and other noncoding sequences. Bacterial DNA, like that of all prokaryotes, does not have such sequences, and so the total genome is much smaller.

23.77 because only two of the four strands of the products are newly formed

23.79 (a) The 5′ end and the section just before the 3′ end must be base paired. (b) There must be no base pairing in the anticodon loop.

23.81 the nucleosomes

23.83 See Sec. 23.2.

23.85 because for most amino acids there is more than one codon

CHAPTER 24 Chemical Communication: Neurotransmitters and Hormones

24.1 (1) chemical messengers (2) receptors

24.3 The presynaptic end stores vesicles containing neurotransmitters; the postsynaptic end contains receptors.

24.5 The concentration of Ca^{2+} in neurons controls the process. When it reaches 10^{-4} *M* the vesicles release the neurotransmitters into the synapse.

24.7 The acetylcholine molecules are released into the synapse and travel to the next neuron. Their presence triggers reactions in which ions can cross membranes.

24.9 See Box 19A.

24.11 (a) norepinephrine and histamine (b) They activate a secondary messenger, cAMP, inside the cell. (c) amphetamines and histidine

24.13 It is phosphorylated by an ATP molecule.

24.15 (HO)$_2C_6H_3$—CH_2—C(=O)—H (ring with two HO groups)

24.17 (a) Amphetamines increase and (b) reserpine decreases the concentration of the adrenergic neurotransmitter.

24.19 the corresponding aldehyde

24.21 (a) the ion-translocating protein (b) It gets phosphorylated and changes its shape. (c) It activates the protein kinase that does the phosphorylation of the ion-translocating protein.

24.23 with antibiotics or a combination of antibiotics and H_2 blockers

24.25 The amino group in GABA is in the gamma position; proteins contain only alpha amino acids.

24.27 It blocks the pain receptors.

24.29 the anterior pituitary

24.31 Acetylcholinesterase inhibitors, such as Cognex, inhibit the enzyme that decomposes the neurotransmitter.

24.33 (a) They stop nerve transmission. (b) They bind to acetylcholinesterase.

24.35 Facial tics are caused by uncontrolled release of acetylcholine, which when absorbed by muscle cells, causes uncontrolled contractions. The botulinum prevents release of the acetylcholine, thus preventing the contractions.

24.37 See Box 24D.

24.39 in the conversion of arginine to citrulline

24.41 C_6H_5—S(=O)$_2$—NH—C(=O)—NH—$CH_2CH_2CH_2$—

24.43 (a) Both deliver a message to a receptor. (b) Neurotransmitters act rapidly over short distances. Hormones act more slowly, and over longer distances. They are carried by the blood from the source to the target cell.

24.45 It opens the ion gates.

24.47 Yes. It fits into the same receptor site as enkephalins.

CHAPTER 25 Body Fluids

25.1 water

25.3 (a) yes (b) through semipermeable membranes

25.5 They carry oxygen to the cells and carbon dioxide away from the cells.

25.7 See Secs. 25.1 and 25.2.
25.9 Proper osmotic pressure is not provided, and water from the blood oozes into the interstitial fluid and causes swelling of tissues (edema).
25.11 globulins
25.13 four
25.15 20 mm Hg
25.17 Each heme has a cooperative effect on the other hemes.
25.19 (a) at the terminal NH_2 groups of the four polypeptide chains
(b) carbaminohemoglobin
25.21 $K = 10$
25.23 bicarbonate/carbonic acid
25.25 inorganic ions
25.27 the distal tubule and the collecting tubule
25.29 No glucose reaches the tissues or the cells. Because 12 mm Hg is lower than the osmotic pressure, nutrients flow from the interstitial fluid into the blood.
25.31 See Sec. 25.10.
25.33 See Figure 25.7.
25.35 by disulfide bridges
25.37 They allow cells to adhere to each other to form layers and eventually tissues.
25.39 See Sec. 25.10.
25.41 platelets
25.43 (a) intracellular fluid (b) blood plasma, interstitial fluid
25.45 restricted flow of urine
25.47 by lowering the volume of blood
25.49 See Box 25F.
25.51 creatine
25.53 in the thoracic duct
25.55 They have variable regions that react with antigens.
25.57 four
25.59 When kidneys fail, no filtration of blood occurs, and urine excretion diminishes. Water accumulates in the interstitial fluid, hence the swelling.
25.61 It converts angiotensinogen to angiotensin, which increases blood pressure.

CHAPTER 26 Nutrition and Digestion

26.1 No, they differ according to body weight, age, occupation, and sex.
26.3 Only the calcium propionate. When this is metabolized it yields energy.
26.5 the section at the bottom that gives daily caloric needs
26.7 It is necessary for proper operation of the digestive system, and a lack of it may lead to colon cancer.
26.9 for daily activities involving motion
26.11 1830 Cal/day
26.13 Yes, but one must eat a wide range of vegetables and cereals.
26.15 linoleic acid
26.17 dietary deficiency diseases, since rice cannot supply the essential amino acids lysine or threonine, or essential fatty acids
26.19 to prevent scurvy
26.21 It assists blood clotting.
26.23 scurvy
26.25 biotin, thiamine
26.27 hydrolysis of acetal, ester, and amide linkages
26.29 neither
26.31 yes, randomly
26.33 If glucose supplied all caloric needs, it would cause deterioration of the patient.
26.35 Eat some candy or other sugar-containing food.
26.37 (a) linoleic acid, linolenic acid
(b) unsaturated
26.39 nicotinic acid (niacin)
26.41 cellulose
26.43 The body would treat it as a food protein and digest it (hydrolyze it) before it could get into the blood serum.
26.45 See Table 18.1.
26.47 No. Arsenic is not an essential nutrient.

Glossary–Index

C

D

E

I

J

K

Q

R

T

U